ANNUAL REVIEW OF PLANT PHYSIOLOGY AND PLANT MOLECULAR BIOLOGY

ANNUAL REVIEW OF PLANT PHYSIOLOGY AND PLANT MOLECULAR BIOLOGY

VOLUME 45, 1994

RUSSELL L. JONES, *Editor*
University of California, Berkeley

CHRISTOPHER R. SOMERVILLE
Carnegie Institution of Washington, Stanford, California

VIRGINIA WALBOT, *Associate Editor*
Stanford University

ANNUAL REVIEWS INC. 4139 EL CAMINO WAY P.O. BOX 10139 PALO ALTO CALIFORNIA 94303-0897 USA

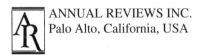

ANNUAL REVIEWS INC.
Palo Alto, California, USA

International Standard Serial Number: 1040-2519
International Standard Book Number: 0-8243-0645-7
Library of Congress Catalog Card Number:A-51-1660

Annual Reviews Inc. and the Editors of its publications assume no responsibility for the statements expressed by the contributors to this *Review*.

Typesetting by Ruth McCue-Saavedra and the Annual Reviews Inc. Editorial Staff

PRINTED AND BOUND IN THE UNITED STATES OF AMERICA

PREFACE

This volume presages several editorial changes in the *Annual Review of Plant Physiology and Plant Molecular Biology* (*ARPP&PMB*). Winslow Briggs retired as Editor after publication of Volume 44. Win Briggs has been associated with this series for more than 30 years. He was appointed Associate Editor of the *Annual Review of Plant Physiology* in 1960 and after 13 years in this position he succeeded Leonard Machlis as Editor in 1973. The series evolved considerably under Briggs' Editorship. Changes in the organization and format of the volume were followed in 1988 by a change in focus that was formalized with a new name in recognition of the influence of molecular biology on plant science. Win Briggs worked tirelessly for this series and his leadership will be sorely missed. Briggs' influence at Annual Reviews Inc. has extended beyond his service as Editor. He was appointed Chair of the Editorial Affairs Committee of Annual Reviews Inc. in 1978 and was made a member of the Board of Directors in 1982. He will continue to serve in these capacities. His influence on what scientists read and his impact on the field of plant physiology have been enormous. He deserves our heartfelt thanks.

Christopher Somerville will join Virginia Walbot and me in editing this series along with the five outstanding members of the Editorial Committee. Several other changes in editorial policy accompany this changing of the guard at *ARPP&PMB*. Members of the Editorial Committee are appointed to five-year terms, and beginning in 1993, Editors are appointed for fixed terms. I look forward to serving as Editor of *ARPP&PMB* for the next five years.

There have also been changes in the position of Production Editor at *ARPP&PMB*. Ike Burke, who had been Production Editor since 1988, was appointed Technology Applications Manager at Annual Reviews Inc. in 1992. Amy Marks took over as Production Editor as was responsible for producing Volumes 44 and 45. Amy joins an enthusiastic team at *ARPP&PMB* and we know her outstanding technical skills will lead to the production of reviews of the highest quality.

The Editorial Committee meets annually, generally in the fall, to plan the volume that will be published two years hence and we welcome your input into this process. Please send suggestions for topics and authors to either Amy Marks or me. On behalf of the Editorial Committee and the Production Editor, I thank you, our readers and contributors, for your continued support.

Russell L. Jones
Editor

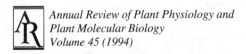

Annual Review of Plant Physiology and
Plant Molecular Biology
Volume 45 (1994)

CONTENTS

PREFATORY CHAPTER

Chapters From My Life, *James Bonner*   1

GENETICS AND MOLECULAR BIOLOGY

The Genetic and Molecular Basis of Root Development, *Roger A.*
Aeschbacher, John W. Schiefelbein, and Philip N. Benfey   25

Messenger RNA 3′ End Formation in Plants, *Arthur G. Hunt*   47

The Plant Mitochondrial Genome: Physical Structure, Information
Content, RNA Editing, and Gene Migration to the Nucleus,
Wolfgang Schuster and Axel Brennicke   61

Geminiviruses and Their Uses as Extrachromosomal Replicons,
Marja C. P. Timmermans, O. Prem Das, and Joachim Messing   79

Gene Expression Regulated by Abscisic Acid and its Relation to
Stress Tolerance, *Peter M. Chandler and Masumi Robertson*   113

TISSUE, ORGAN, AND WHOLE PLANT EVENTS

The Transduction of Blue Light Signals in Higher Plants,
Timothy W. Short and Winslow R. Briggs   143

Cytokinin Accumulation and Action: Biochemical, Genetic, and
Molecular Approaches, *Andrew N. Binns*   173

PPFMs and other Covert Contaminants: Is There More to Plant
Physiology than Just Plant?, *Mark A. Holland and Joseph C.*
Polacco   197

BIOCHEMISTRY AND BIOSYNTHESIS

Molecular Analysis of Proteins in the Plant Plasma Membrane,
M. R. Sussman   211

Diurnal Regulation of Photosynthetic Carbon Metabolism in
C_3 Plants, *Donald R. Geiger and Jerome C. Servaites*   235

Alkaloid Biosynthesis: Molecular Aspects, *T. Hashimoto and*
    Y. Yamada   257
Molecular Biology of Carotenoid Biosynthesis in Plants, *Glenn E.*
    Bartley, Pablo A. Scolnik, and Giovanni Giuliano   287
Storage Proteins of Vegetative Plant Tissues, *P. E. Staswick*   303
The Glycine Decarboxylase Complex from Plant Mitochondria,
    David J. Oliver   323
Inhibitors of Photosynthetic Enzymes/Carriers and Metabolism,
    Leszek A. Kleczkowski   339
The Role of Carbonic Anhydrase in Photosynthesis, *Murray R.*
    Badger and G. Dean Price   369
Auxin-Binding Proteins, *Alan M. Jones*   393
The Ribonucleases of Higher Plants, *Pamela J. Green*   421
Malate Compartmentalization—Responses to a Complex
    Metabolism, *Enrico Martinoia and Doris Rentsch*   447
Structural and Functional Aspects of Chaperonin-Mediated
    Protein Folding, *Anthony A. Gatenby and Paul V. Viitanen*   469
The Use and Characteristics of the Photoacoustic Method in the
    Study of Photosynthesis, *Shmuel Malkin and Ora Canaani*   493

CELL DIFFERENTIATION

Plant Hormone-Induced Changes in the Orientation of Cortical
    Microtubules: Alterations in the Cross-linking Between
    Microtubules and the Plasma Membrane, *Hiroh Shibaoka*   527
Protein Import into Plant Mitochondria, *Anthony L. Moore,*
    Carlton K. Wood, and Felicity Z. Watts   545
Integration of Carbon and Nitrogen Metabolism in Plant and
    Algal Cells, *H. C. Huppe and D. H. Turpin*   577
Endocytosis in Plants, *Philip S. Low and Sreeganga Chandra*   609

ACCLIMATION AND ADAPTATION

Photoinhibition of Photosynthesis in Nature, *S. P. Long,*
    S. Humphries, and P. G. Falkowski   633
Taxol, *P. F. Heinstein and C.-j. Chang*   663

INDEXES

Author Index   675
Subject Index   704
Cumulative Index of Contributing Authors, Volumes 36–45   721
Cumulative Index of Titles, Volumes 36–45   724

SOME RELATED ARTICLES IN OTHER *ANNUAL REVIEWS*

From the *Annual Review of Biochemistry,* Volume 63 (1994)

Quinoenzymes in Biology, J. P. Klinman and D. Mu

GTPases: Multifunctional Molecular Switches Regulating Vesicular Traffic, C. Nuoffer and W. E. Balch

Regulation of Eukaryotic DNA Replication, D. Coverley and R. A. Laskey

5-Lipoxygenase, A. W. Ford-Hutchinson, M. Gresser, and R. N. Young

Nitrogenase: A Nucleotide-Dependent Molecular Switch, J. B. Howard and D. C. Rees

Structure, Function, Regulation, and Assembly of D-Ribulose-1,5-Bisphosphate Carboxylase/Oxygenase, F. C. Hartman and M. R. Harpel

Genetic and Biochemical Studies of Protein N-Myristoylation, D. R. Johnson, R. S. Bhatnagar, L. J. Knoll, and J. I. Gordon

Intermediate Filaments: Structure, Dynamics, Function, and Disease, E. Fuchs and K. Weber

Structure and Function of G Protein-Coupled Receptors, C. D. Strader, T. M. Fong, M. R. Tota, D. Underwood, and R. A. F. Dixon

Role of Chromatin Structure in the Regulation of Transcription by RNA Polymerase II, S. M. Paranjape, R. T. Kamakaka, and J. T. Kadonaga

From the *Annual Review of Cell Biology,* Volume 9 (1993)

The Spectrin-Based Membrane Structure and Micron-Scale Organization of the Plasma Membrane, V. Bennett and D. M. Gilligan

Signal-Dependent Membrane Protein Trafficking in the Endocytic Pathway, I. S. Trowbridge, J. F. Collan, and C. R. Hopkins

Macromolecular Domains Within the Cell Nucleus, D. L. Spector

Protein Import into Peroxisomes and Biogenesis of the Organelle, S. Subramani

The Role of GTP-Binding Proteins in Transport Along the Exocytic Pathway, S. Ferro-Novick and P. Novick

Role of the Major Heat Shock Proteins as Molecular Chaperones, C. Georgopoulos and W. J. Welch

From the *Annual Review of Entomology,* Volume 39 (1994)

Evolution of Resistance to Bacillus thuringiensis, B. E. Tabashnik

From the *Annual Review of Genetics,* Volume 27 (1993)

The Molecular Genetics of Nitrate Assimilation in Fungi and Plants, N. M. Crawford and H. N. Arst, Jr.

Dosage Analysis of Maize Endosperm Development, J. A. Birchler

Mapping Polygenes, S. D. Tanksley

Molecular Aspects of Self-Incompatibility in Flowering Plants, A. E. Clarke and E. Newbigin

The Function of Heat-Shock Proteins in Stress Tolerance Degradation and Reactivation of Damaged Proteins, D. A. Parsell and S. Lindquist

From the *Annual Review of Microbiology,* Volume 47 (1993)

Transport of Nucleic Acids Through Membrane Channels: Snaking Through Small Holes, V. Citovsky and P. Zambryski
Genetically Engineered Protection Against Viruses in Transgenic Plants, J. H. Fitchen and R. N. Beachy
Release of Recombinant Microorganisms, M. Wilson and S. E. Lindow

From the *Annual Review of Phytopathology,* Volume 32 (1994)

The Virulence System of Agrobacterium Tumefaciens, P. J. J. Hooykas and A. G. M. Beijersbergen
Plasmodesmata in Relation to Viral Movement Within Leaf Tissues, W. J. Lucas and R. L. Gilbertson
Early Events in the Activation of Plant Defense Responses, R. A. Dixon, M. J. Harrison, and C. J. Lamb

James Bonner

Annu. Rev. Plant Physiol. Plant Mol. Biol. 1994. 45:1–23

CHAPTERS FROM MY LIFE

James Bonner

Division of Biology, California Institute of Technology, Pasadena, California 91125

CONTENTS

IN THE BEGINNING 1
UNDERGRADUATE SCHOOL.. 3
    We Have Physical Chemistry 4
    Theodosius Dobzhansky 5
GRADUATE SCHOOL.. 7
POSTDOCTORAL WORK.. 11
PROFESSIONAL WORK.. 13
    Photoperiodism Work at the University of Chicago 14
    The War Over Rubber.. 16
    Fresh Beginnings—Cell Biology 18
A NEW PARADIGM 19
FROM KATMANDU TO TIMBUKTU TO KOTA KINABALU 22

IN THE BEGINNING

I was born in Ansley, Nebraska on September 1, 1910. When I was six weeks old my father, mother, and I returned by train to Kingston, Ontario, where my father was an Assistant Professor of Chemistry at Queen's University. My father had obtained his PhD in physical chemistry from the University of Toronto in 1910. My paternal grandfather was a very conservative Presbyterian minister, as well as a Nebraska homesteader, and my father had run away from home to go to Nebraska Wesleyan University in Lincoln. There he met my mother, who had come to Nebraska in a covered wagon when she was six years old and who had grown up on a homesteaded farm near Ansley. After graduating from high school, she had become a school teacher to earn money for college. She attended Nebraska Wesleyan for two years, taught school again for a year, and then returned to complete her bachelors degree. At Nebraska Wesleyan my parents met a new member of the faculty, Dr. Frederick Alway, who had just returned from Germany, where he had received his

PhD in chemistry. He was a good teacher and he persuaded many of the undergraduates to major in chemistry. He later went on to the University of Minnesota to perform similar marvels. Dr. Alway persuaded my father and others, like R. A. Gortner, the biochemist, to apply for graduate school fellowships. My father received a fellowship to go to Princeton and become an organic chemist. While at Princeton, he discovered that he really wanted to be a physical chemist. It was easier to leave Princeton, go to another university, and find a physical chemist who would accept a new graduate student, than it was to simply change fields at Princeton. So my father transferred to the University of Toronto and became W. Lash Miller's student. Both of my parents were the first in their respective families to go to and graduate from university. That they were both students of Dr. Alway and that they both had a missionary zeal for producing more chemists explains at least in part the events I will subsequently reveal.

My only sister, Priscilla, was born in the fall of 1914. World War I also started that year. We left Kingston in 1915—my father had been invited to become head of the Chemistry Department of the University of Utah. After we arrived in Utah, my parents bought a house on the edge of the city, overlooking the Wasatch Mountains. Behind the house was a beautiful prairie, bounded by a gigantic gully. A cowboy brought his cows to graze on our prairie every day and at night the coyotes sang to us.

My mother, who deemed her skill as a teacher to be greater than that of any elementary school teacher in the Salt Lake City school system, taught me at home until I was eight years old. I learned to read, but not to write. I didn't learn any mathematics and I didn't learn to read music. That all came later. When I was eight years old, my mother placed me in the fourth grade despite considerable resistance from the school. There were disadvantages to starting school two years younger than others in the class: I was a social misfit. I also didn't know how to play baseball, I couldn't read music, and my handwriting (which I learned in fourth grade) was then, and still is, terrible.

Of all the things I learned in elementary school, I remember geography the most. In ninth grade, I learned about the geography of Africa, Australia, and Asia, and decided I wanted to visit all those places I'd never heard of before. That same year my parents gave me a copy of *My Life as an Explorer* by Sven Hedin, who was a professional explorer. He knew when he was a little boy that he wanted to be an explorer, so he trained himself to be used to hardships and to be able to find his way without road signs or asking other people. Hedin explored a lot of Central Asia and after reading of his adventures, I decided that I wanted to become an explorer too. My parents coldwatered this idea, however, by saying that most of the world had already been explored and it was very hard to find places to try to find out something new about. Of course, I later found that my parents were wrong in this regard and that there were a lot of places that still needed to be explored, some of them right there in Utah and lots of them still in Central Asia.

I received another kind of education while living in Utah. My parents had both grown up on farms and left their farms to attend college. They both felt strongly that (*a*) capable people, such as their children must be, should go to college and (*b*) farms were the best place for children to be brought up because there was physical work to do. My father wanted us to understand "the dignity of labor," so when I was eight years old we moved from the home with the prairie, the cowboys, and the coyotes to a three-acre farmlet, which we would now call a family orchard. My responsibility, as the eldest, was to change the water from ditch to ditch to irrigate the entire farm over a period of several days. My other responsibilities included pruning, spraying for insects, suppressing fires, and training my younger siblings to take over the irrigation in case I was shunted off into some other activity. The chores changed with the seasons. They never ended.

UNDERGRADUATE SCHOOL

It was taken for granted in our family that everyone wanted to become a chemist, and it turned out to be true. All seven of my parent's children received bachelors degrees in chemistry from the University of Utah. Four of them became biological chemists, two became physical chemists, and one became an applied mathematician, having given up on chemistry, but only after receiving his bachelors degree in chemistry. As the oldest child, it was my duty to start the procession off to the university to get an AB degree in chemistry. I graduated from high school in the summer of 1927 and entered the University of Utah in the fall of that year. I spent two years at the University of Utah as a chemistry major with a minor in mathematics and played flute in the university orchestra, sitting next to my brother Lyman who played the first oboe, all under the auspices of the concert master Simon Ramo who later became the "R" of TRW. The calculus and differential equations were all fascinating and easy. The chemistry was all fascinating and easy. English was more difficult. Each week we had to turn in two essays to Professor Crabtree, who would read them and mark them up before handing them back. After a year of having my essays scribbled on by an exacting person like him, I learned a lot about good writing. According to Professor Crabtree, the only way you learned about writing was by reading other peoples' writing and by having your own work criticized by people like him.

The year after my sophomore year in college, my father had another sabbatical leave (his first had been spent at the University of California at Berkeley in 1922–1923). This time he was going to the new, but rapidly becoming well known, California Institute of Technology (Caltech). His former student, Don

Yost, had received his PhD from Caltech and was now an Assistant Professor there. My father planned to work with Don Yost on an interesting physical chemical problem that both of them considered potentially fruitful for them. My parents rented out their house and farmlet to my uncle Truesdale, who was now a chemical engineer employed by a company headquartered in Salt Lake City.

With the entire family packed up into our 1921 Essex, we drove to Pasadena in about four days. My parents found a real estate agent who assured them that "Pasadena is the city of churches and schools." They rented a house in northwest Pasadena, where they were indeed surrounded by schools. My brother Lyman and I had taken the required Caltech admission exams and had been admitted and awarded tuition scholarships. Registration at Caltech was not as difficult as you might imagine. There are a limited number of courses and the curriculum and requirements dictate completely which ones you must take. Thus I registered for Physical Chemistry, the great bug bear of the Caltech undergraduate chemist.

We Have Physical Chemistry

Physical chemistry was taught using Noyes & Sherrill's textbook, *Chemical Principles.* The book contained a lot of problems, and each day we were assigned a few to work out on paper and hand in at the next class. You couldn't solve subsequent problems unless you'd solved previous ones and totally understood the principles that you were dealing with. In this course I learned how to study and how to be absolutely sure I wasn't fooling myself into thinking I understood something that I didn't.

The fall term had its signs and portents for the future. On October 29, 1929, the great stock market crash occurred. Rich people became poor, etc. None of this affected Caltech, my family, or me right away, but such effects did come. Even so, the first term was perfectly satisfactory. I got *A*s in everything, including Physical Chemistry. It was a lot of hard work, but I felt then and I still feel today that this first term at Caltech really turned me on. The lectures were outstanding. E. C. Watson's weekly lectures in physics were always wonderful. We learned everything by solving problems. This was true in Physical Chemistry, Physics, and even in Quantitative Analysis. We were taught a whole new way—no rote. Or as my colleague of later years, Herman Kalckar, used to say, "No rosary."

For the winter term, everything was the same, except for Biology. The biology class was being taught for the second time. We had three lectures a week and laboratory class twice a week. The lectures were given by the chairman of the Division of Biology, Professor Thomas Hunt Morgan. Professor Morgan was famous as the father of modern genetics. He was also famous because he was a good geneticist, a good embryologist, a good biologist, and he was good at choosing his colleagues. Alfred H. Sturtevant and Calvin B. Bridges, two of the three Columbia University undergraduates who had, with him, developed genetics, became faculty members at Caltech. Dr. Morgan gave absolutely fascinating lectures concentrating on genetics with a bit of embryology tossed in. Slightly more than halfway through the term, Dr. Morgan turned the class over to Henry Borsook, from the University of Toronto, who lectured on biochemistry and metabolism. Dr. Borsook was interested in the mechanism of protein synthesis and tried to prove that proteins could be synthesized by proteolytic enzymes working in the reverse direction, but he didn't succeed. Biology was fascinating and I loved the class, but it didn't seem to require any thinking and I remember musing at the time that this wasn't right. There ought to be some problems!

Theodosius Dobzhansky

Once again, I received As in all my classes, including Physical Chemistry, and at the beginning of the third term, the only decision I had to make was whether to take Introduction to Astronomy or a new biology class. I chose the latter. The new class had a laboratory, which was presided over by a brand new Assistant Professor, Theodosius Dobzhansky. He had been discovered, I believe, at the University of Leningrad, by a roving scout of The Rockefeller Foundation. He was awarded a Rockefeller Foundation Fellowship to come to the United States to study with Dr. Morgan at Columbia, and had arrived in the summer of 1927. He was amazed to find that at the end of the 1927–1928 school year, the whole laboratory was moving to Pasadena, California. His fellowship was for only one year, so it would run out sometime soon after he got to Pasadena. Dr. Morgan invited him to come along, saying, with typical Morgan offhandedness, "When your fellowship ends I'll make you an Assistant Professor." It could be done that way in those days.

Theodosius Dobzhansky and his wife, Natasha, arrived in Pasadena in the fall of 1928 and moved into the new Kerckhoff Laboratories of Biology (the west half of what we now know as the Kerckhoff Laboratories). Theodosius not only ran the laboratory, but also invented interesting things for us to do. We went on field trips. We went to the Marine Station at Corona del Mar, where we trapped *Drosophila* using half-pint milk bottles with fermenting yeast solutions soaked on a piece of paper towel. We brought the bottles back to the station and counted how many of the flies were *Drosophila melanogaster* and how many were other species of *Drosophila*. We learned a little bit about the phyla of animals (but not of plants). We talked about the natural history of animals in the wild. Dobzhansky frequently told stories about his two trips to Central Asia, where he had been sent with some colleagues to study the fauna of the region, particularly the Altai and the Tien Shan, and to see whether there were any of the wild horses, Prezhevalsky's horse, that were known to inhabit these remote areas.

Toward the end of the term, I learned that because of the financial stringency of the Institute after the beginning of the Great Depression, they would be unable to fund my scholarship for the coming year. A few days later and just in time to really cheer me up a lot, Dobzhansky asked me if I would accept a position as a research assistant to him for the summer of 1930. Dr. Morgan had authorized him to offer me the position for something like $250 for four months, and I accepted. The academic procession during commencement was the most impressive I had ever seen, and it included my father, wearing the colorful regalia of the PhD from the University of Toronto. At the end of the academic year, my parents and siblings packed up all their belongings, shipped what couldn't be carried, and drove back to Salt Lake City. I remained in Pasadena.

On the first day of the summer term, I went to Dobzhansky's laboratory to find out what I was to do for work. He was working on *Drosophila,* of course (that's what he had come to the United States to learn about), and he was determining the breakage points of translocations. Dobzhansky was trying to find out exactly where the break was in each case, and that meant finding out where it was with respect to mutant marker genes along each of the chromosomes. I had to be able to distinguish between mutant flies and wild type flies, and between the different kinds of mutant, and I had to keep good records, all of which I could do easily.

For the first week or so I was immersed in this project but interesting as it was, it was not as amazing as the next development. On Thursday, I think, of the first week, Theodosius said out of the clear blue sky, "We have worked very hard this week. I think tomorrow we go to the beach." We spent the next day at the Corona del Mar beach with Natasha, taking a final shower at the end of the day in fresh water at the Caltech Marine Station. After another week in the lab Theodosius suddenly said, "We have been working very hard now for two weeks. I think it is time we take a vacation for a couple of days." So he packed up his camping gear, I brought a blanket and a sweater, and we went up into the San Jacinto Mountains and camped there. Dobzhansky always carried an insect-collecting net with him wherever he went out-of-doors, so he collected insects and told me about how many different kinds of insects there were and how he was a specialist on a particular kind of beetle, the Coccinnellidae. He said that he was going to catch all of the wild beetles of this group to be found in California and then he would write a taxonomic description of each and write a key on how to distinguish them from one another (which he did). This was a seductive plot. Biologists got to take trips outdoors. They had an excuse to go and see nature in the wild. That was something to ponder.

Theodosius also gave me a second task. He was a prolific writer. He often stressed that "a month without a paper sent to press is a wasted month," and he fulfilled this objective for all of his scientific life. After he passed away and we counted the number of papers and how many years he'd been working on them, it was almost exactly one paper a month for his entire 55 years or so of scientific productivity. But it was hard work even for him, since he was trying to write one paper a month while he was learning English. He and Natasha had arrived in New York City not knowing a word of English. They learned by the total immersion method. He recruited me to read the things he wrote and change them into proper English. I was pretty good at this and I'd had a lot of writing experience, so Dobzhansky learned to be an excellent writer. He wrote grammatically proper English, with an enormous vocabulary, but he always kept the real honest-to-God Dobzhansky accent, as long as he lived.

At the end of the summer I had to return to the University of Utah to complete my bachelors degree, but I planned to return to Caltech afterward. So I packed up, paid my rent, and asked my landlady, Maria Planas, to save me a room for next year. I hitchhiked home. This was early in the fall of 1930 and I passed through Las Vegas. President Hoover had announced that he was about to begin construction of Boulder Dam and people from all over the United States had congregated in Las Vegas looking for jobs working on the dam. The Depression had really started to hit. I walked out of town a little way to North Las Vegas and was lucky to get a ride that took me all the way to Salt Lake City.

GRADUATE SCHOOL

In my senior year I had applied for admission to graduate school in biology at Caltech and just to keep things happy in the family, I applied for admission in the Division of Chemistry at Caltech as well. I was admitted to both departments. I immediately accepted admission in the Division of Biology and was awarded a teaching fellowship for $750 per year. It was 1931 and the Depression was deepening. The $750 looked pretty good so I accepted. I graduated from the University of Utah with high honors (with a major in chemistry and a minor in mathematics).

Immediately after graduation, I did a few assays of ore samples for a man who was eager to have their precious metals content in a hurry. This gave me a little money to start me on my way. Toward the end of the summer, I boarded a bus, went back to Pasadena, and reported to the Planas' house, where I got my old room back. I walked to Kerckhoff but there was not a soul in the building. There was no one at work. I then rode my bicycle out to the Caltech farm in Arcadia. E. G. Anderson was there, hoeing the weeds out of his corn, and someone else, whom I didn't know, was there with him. He introduced me to the new person, George Beadle, then a postdoc from Cornell. That was the first time I saw George Wells Beadle, my friend and confidant for many years.

After talking with George Beadle for a while, I rode back to Caltech and noticed that there was somebody working in a new building on the corner of Michigan and San Pasqual. I walked up to the building, knocked on the door, and was let in. The inhabitants introduced themselves as Herman E. Dolk, Assistant Professor of Biology, a plant physiologist; and Kenneth V. Thimann, an Instructor of Biochemistry. They were working on a problem that sounded quite interesting. They told me that Frits Went in Holland had discovered a plant hormone that caused cells to grow longer in plants and that this hormone had a lot to do with all kinds of plant growth responses. They were studying the responses of plants to the hormone, using a fungus, *Rhizopus suinus,* as a source of the hormone. After the *Rhizopus* had grown on a culture medium for a few days and the culture medium was extracted with ether, the ether extract contained active plant growth hormone. They told me that the structure of the hormone was being worked on in Holland where the whole matter had been discovered, but the synthetic stuff was not yet available. We talked for quite a while and the next day I went back and we talked some more. There was still no one else in the Division of Biology, so finally Professor Dolk said, "Of course I know you've come here to become a geneticist, but since there isn't any geneticist around to consult with right now, why don't you pitch in and help us produce some more of this plant growth hormone? We really need somebody more to help us in our work and get it going a little faster." So I agreed and started working on the production of plant growth substance by the fungus *Rhizopus.*

Dr. Dolk thought I should try to grow *Rhizopus* on different media and see if I could find a medium that would cause the *Rhizopus* to make a lot more growth substance. This would be an enormous help in providing enough growth substance for physiological experiments. I started doing what Dolk had suggested. One day I discovered something: Instead of feeding the *Rhizopus* an ordinary synthetic medium like I'd been doing, and on which the *Rhizopus* grew very well, I put various sources of nitrogen into the medium, including, for no particular reason, the bactopeptone that one uses to make bacterial media for fastidious bacteria that need a complicated source of nitrogen. On the bactopeptone medium the *Rhizopus* not only grew like mad, but it produced over 50 times as much growth substance per unit volume of growth medium as it did on any of the media that I'd used before. In fact, if it was grown on bactopeptone and also aerated so that the fungus had a lot of oxygen available to it, it produced 100 or 200 times as much growth substance per unit volume of medium as I'd ever seen before. This was spectacular! I had found a medium that would produce growth substance until it would run out of Dolk's and Thimann's ears. I wrote a paper on my discovery and after Dr. Thimann edited the paper, we sent it off to be published. I was hooked. I'd seen my name in print and decided that I wanted to write a lot of publications just like Theodosius Dobzhansky did.

In retrospect, I had discovered something really important. My discovery, if followed up, would lead to the identification of the chemical in the bactopeptone that is responsible for helping *Rhizopus* make growth substance out of it. We now know that the growth substance is indole-3-acetic acid, so my best bet would have been that it was tryptophan in the bactopeptone that was oxidized and decarboxylated and made into indole-3-acetic acid by *Rhizopus,* especially in the presence of a lot of oxygen. However, nobody suggested that it would be good to use this as a springboard from which to determine the nature of the plant growth substance. It was hailed as a great achievement in making it easier to produce an active substance to use in experiments, but nobody pointed out to me that I had discovered an important clue and I didn't see it for myself. I had an opportunity to become the discoverer of the nature of the plant growth substance, but I let it slide and went over and worked with Dobzhansky for a couple of months on *Drosophila* genetics. My unfinished work on plant growth substance kept gnawing at my conscience, and ultimately, I did return to the study of plant growth substance and its chemistry.

By this time Dr. Morgan had assigned me to Room 307 Kerckhoff and since there weren't many graduate students, I had the whole room to myself. I played the flute an hour a day, most often in my office. I opened the windows, of course, because it was hot, so my flute music went wafting out over the southern part of Pasadena, but nobody complained. I was also invited to play in the Pasadena Civic Symphony and I did do this for three years until I got my PhD and went to Europe. I worked very hard, but didn't accomplish any major goals. I tried to invent ways to find out how the growth substance made plants grow. As it turned out, of course, the science of biology was not advanced enough to permit the study of this subject. I did only two things worthy of remembrance. First, I found that oat coleoptile sections cut out of the oat coleoptile grow wonderfully when floated on a solution of plant growth substance. If they aren't given any plant growth substance (the sections just floated on plain buffer solution or water), they grow just a tiny bit. If they are put on a solution containing a lot of plant growth substance, they elongate rapidly, and this growth can be measured by looking at 5 mm long sections under a microscope with a graduated scale. This is a very quick, easy, and quantitative way to measure amounts of growth substance in a medium. It's not as inventive as the method invented by Frits Went, but it is simple and has been used a lot by many people. Second, I discovered that if you put oat coleoptile sections in a solution of, say, pH 4.5 they grow quite rapidly for a while, and if they're put in a similar solution, but at, say, pH 7.5, they don't grow. By putting them in solutions of different pHs one can plot a sort of titration curve and find the pK of what it is that is being influenced by hydrogen ions that makes oat coleoptiles grow. This discovery, independent of everything else known about growth substances in plants, has started a whole new chain of thought in the minds of several people in recent years, and it has engendered a secondary fall-out of papers on the subject.

Professor Dolk was killed in an auto accident in March 1932 and Frits Went came to Caltech early in 1933 as an Assistant Professor of Biology. We had many interesting seminars at Caltech. As J. B. S. Haldane talked on evolution, his cigar got too short, so he took out his pipe, stuffed the cigar stump in it, and smoked it all the way. A seminar by Von Wettstein on maternal inheritance of chloroplasts was bitterly attacked as unlikely, if not impossible.

My $750 per year turned out to be a lot more than expected. It was Depression time, remember, and on principle, I would never buy any food that cost more than ten cents for three pounds, so feeding myself didn't cost much. My brother Lyman had come to Caltech to become a graduate student in Chemistry, starting one year after I did, and we lived together in a two-room-and-bath apartment on the lower floor of the Planas' house. I bought a used Pasadena Police motorcycle—a Henderson four cylinder. I rode it a lot, exploring every canyon and riding up to Santa Barbara to visit my aunt. I soon found a friend who had a Henderson exactly like mine and we rode together. One day when I came home from a Sunday trip I had a message from Mrs. Planas that he had called and asked me to visit him at Huntington hospital. I found my friend in a very sad condition: he had been hit by a car when his motorcycle was standing still. He never recovered fully from this accident. I could see a whole tragic story coming, so I put a for sale sign on the motorcycle. I never rode it again. In a few days a man offered to trade me a 1924 Chevrolet Superior roadster for my Henderson. I took it. The Chevrolet had a four-cylinder, overhead valve engine and would go forever without boiling. It was reliable enough to drive to Salt Lake City and back.

Toward the end of 1933 (at the beginning of my third year of graduate study), Dr. Morgan suggested that I go to Europe for the next year. I said I would like to go to Utrecht where they knew all about plant growth substance, and maybe other places as well. I received a letter from Professor William J. Robbins, the Chairman of the National Research Council Selection Committee, saying that I had been awarded a one-year NRC postdoctoral fellowship with a stipend of $1625/year and all travel expenses. I accepted!

I wrote my PhD thesis and my final PhD exam was scheduled. Professor Sturtevant was the chairman of my thesis committee in the absence of Dr. Morgan, who had acted as the chairman of all previous PhD examinations, but who was away collecting his Nobel Prize. Dr. Sturtevant, I remember very clearly, put his feet on the table and said, "James, when you think of a gene, tell us what you envisage the gene looks like and is." Luckily I've forgotten what I said in reply. We had a commencement, of course, to which my father came. I think that one of the joys of his mature years was seeing his children get PhDs one by one. Dr. Millikan hung the PhD hood around my neck and it was all over.

POSTDOCTORAL WORK

We drove back to Salt Lake City in my 1924 Chevrolet roadster, and after a few days I took the Greyhound bus to New York, where I boarded the *Pennland*, the steamer that was to take me to Europe. It was a wonderful trip. You would hardly know there was a Depression going. The ship was filled with college students going to Europe for the summer. Upon arrival at Antwerp, I found that the National Research Council had thoughtfully arranged for my train ticket to Utrecht. When I got to Utrecht, I rented a room, put my few belongings in it, and found, as I had expected, that the universities are empty in the summer. From June to September, professors and students take their summer vacations—certainly very non-Caltechian behavior—but I had expected this, so I was prepared. I bought a bicycle and made an outline of a trip to see Europe by train and bicycle. I spent about two and a half months on this European tour, having a wonderful time. I took the train to Berlin and found, to my amazement, that I could understand people and they could understand my German. I knew a few people in Berlin and indeed, this was true everywhere I went. I'd written ahead to several people, and I went to the famous laboratories in Berlin, Jena, Leipzig, Dresden, Prague, Münich, Heidelberg, Innsbruck, Zürich, Bern, Basel, Köln, and finally back to Utrecht.

Now it was time to go to work. I had arranged to go to the laboratory of Professor Kruyt, the most notable polymer chemist of the time, and I was under the impression that I would learn something about colloid chemistry. At that time it was fashionable to think that colloid chemistry would give us an understanding of how protoplasm was made and what its properties were. This was an incorrect view of the matter, as it turned out, and I certainly didn't learn much by working in that lab. I did learn Dutch quite well, however. I also got to know all of the notable plant physiologists in Utrecht. I worked in the laboratory that had just been given by Frits Went's father to the new Professor Konigsberger. A. N. J. Heyn was the chief assistant in the botany department. Professor Kögl was the head of the organic chemistry department. A. J. Haagen-Smit was the first assistant. Kögl, Haagen-Smit, and Hanni Erxleben jointly had isolated and were determining the structure of the growth substance that Frits Went had found. They had isolated and were in the midst of determining the structure of auxin a. At the end of 1934, while I was still in Utrecht, they published the isolation of indole-3-acetic acid from urine and showed that it was active as an auxin. Since its structure was obviously very different from the putative structure of the putative auxin a, the new auxin was called heteroauxin. Even though this was published in 1934, I heard nothing about it in Holland. There was no seminar, no celebration of this great new finding. As it turned out, of course, Kögl and Haagen-Smit had found the correct auxin. The substance is really the auxin of plants. Auxin a and auxin b were fantasies apparently invented by Hanni Erxleben.

I also had an opportunity to work for a couple of months at the University of Leiden Medical School in the Department of Biochemistry, which was presided over by Professor Bungenberg de Jong. Bungenberg de Jong was the great expert on coacervates. A coacervate is a colloidal system in which one colloid is dispersed in a second. I learned little from the two colloid chemistry laboratories. In the beginning of 1935, I moved to Zürich to A. Frey-Wyssling's laboratory at the Swiss Institute of Technology (ETH). Professor Frey-Wyssling was an expert on the use of the polarizing microscope for the study of cell wall properties. The manuscript that resulted from this work, and which I'm happy to point out was written and published in German, showed quantitatively that under the influence of auxin, the microfibrils of cell walls become much more readily separated into independent fibrils. That is, the interaction of the fibrils is greatly diminished. This is what makes it easier to stretch the cell wall as a result of auxin action. I think this paper was a considerable contribution to science.

During the fall term, Dr. Morgan invited me to come back to Caltech as a research fellow, which in time would turn into an instructorship. Of course I accepted. Where else would I want to go? And anyway, that was the only job offer I received. I looked no further.

Later in the fall, I attended an international congress of botany in Amsterdam. It was exciting to see the bigwigs of plant biology—whose names I knew, but whose faces I did not. I think I met Dennis Hoagland of the University of California at Berkeley for the first time. I also met and made a life-long friend of Hiroshi Tamiya. Hiroshi was professor of botany at the University of Tokyo and a world's authority on many aspects of photosynthesis. Although at this time, before World War II, he published papers in German (and we talked in German) as most Japanese did, he transformed himself into an English writer by the end of the war, and it was in his English speaking and writing incarnation that he became well known in America.

At the end of the year, I made a reservation for my return trip on the North German Lloyd Line. We crossed the Atlantic in less than five days on the *Bremen,* in contrast to the almost two weeks it took on the *Pennland.* I landed in New York, went to Detroit by train, bought a 1934 Ford roadster, and drove across the country, visiting my relatives in Nebraska, and finally home to Salt Lake.

PROFESSIONAL WORK

By the time I returned to Caltech, and by the time Dr. Morgan had given me back Room 307 Kerckhoff and had confirmed that I was an Instructor in Biology, it was 1936. I was 25 years old and I had to decide what to investigate. The study of root growth was one of my initial areas of investigation. Phillip White, of The Rockefeller Institute, had successfully grown tomato roots in culture through repeated transfers. A 1 cm long tip would grow into a big, long root; he'd cut the tip off of that, put it in fresh medium, and it would grow again. The medium consisted of sucrose, inorganic salts, and yeast extract, and I thought the thing to do was to find out what was in the yeast extract.

I decided to start a root organ culture program. I tried growing isolated pea roots, leaving the tomato roots to White temporarily. I germinated pea seeds aseptically, and after the roots had grown for a week and were several cm long, I cut the apical 1 cm off of each root and transferred the tips to fresh petri dishes containing liquid medium. The medium contained an inorganic salt concoction that I had devised with sucrose as a carbon source. I found that the roots would grow very well for a week, but when transferred to fresh medium, they would give out. So I tried yeast extract, like Phillip White had used, and found that yeast extract would make pea roots grow quite well. There had to be something in yeast extract that plants needed. I knew that yeast extract was good because it contained the B vitamins, including vitamin B1, which had just recently been made available synthetically by Merck. So I wrote to Dr. Randolph Major, Director of Research at Merck, and asked him if I could get a bit of crystalline vitamin B1, which he kindly sent. I put the crystalline vitamin B1 in the medium together with sucrose as a carbon source and the inorganic salts, and I found that pea roots grew wonderfully well with synthetic vitamin B1 instead of yeast extract as a supplement. They would go through maybe six or eight transfers and then the rate of growth would decrease slowly. The pea roots needed something else besides vitamin B1, but they certainly needed vitamin B1. So I wrote a letter to Phillip White telling him the joyous news that I had discovered that synthetic vitamin B1 could replace yeast extract to a considerable extent in the growing of roots, but he didn't answer my letter. Instead, he repeated my experiment and published the results in the *Proceedings of the National Academy of Sciences,* which only took a few weeks from submission to publication in those days. His paper on the subject appeared in 1936, whereas mine, which was written earlier, but submitted to *Science,* didn't appear until 1937. My long paper on the subject also didn't appear until 1937, in the *American Journal of Botany.* So I learned something: Be careful how you spread the joyous news.

I had a good time with roots. My first graduate student, Fred Addicott, came from Stanford to join in the fun. We grew isolated roots of many different dicotyledonous species. Roots of monocotyledons wouldn't grow in our medium. Most successfully cultured roots of dicotyledons require thiamine. Many plant roots require both vitamin B1 and the B vitamin, niacin. The tomato requires thiamine and pyridoxine (vitamin B6). Thiamine is synthesized in leaves and is transported downward to the roots where it is essential for their growth. Even though the B vitamins are so clearly root growth hormones for those species that require them for continued growth in culture, it is interesting that thiamine, for example, has not been generally classified as a plant hormone.

My other initial topic of investigation was the wound hormone. For the study of wound hormones, after much consideration and advice from Frits Went, I decided to study the wound hormone of the string bean, which Wehnelt had studied previously. The pod of the string bean may be cut in half longitudinally along the suture, and the seeds removed. In each little cup from which the seed has been removed there is undamaged epidermis, subtended by undamaged parenchymal cells. If a drop of juice of ground bean pods is placed on the surface of that little cup, it will cause cell division and growth of the uninjured bean tissue under it—the ground pea pod juice contains a wound hormone that can induce undamaged cells to grow. My first postdoctoral fellow, James English, Jr., a chemist fresh from Yale University, joined me in this study. We found that a good source of wound hormone is the ground up bean pods themselves. The ground up bean pods were extracted with water, the water extract concentrated, and after six fractionation steps, a crystalline active material was obtained. This crystalline active material was characterized in a variety of ways, and was shown to possess the structure of 1-decene-1,10-dicarboxylic acid. We called this previously unknown compound traumatic acid. Traumatic acid has biological activity as a wound hormone both on bean pods and on potato tuber slices. Today, it appears that traumatic acid is probably produced as a non-enzymatic oxidative product of 12-oxo-trans-10-dodecenoic acid, the first compound in the jasmonic acid pathway. This revisionist suggestion was made by Zimmerman & Coudron 40 years after our group made the initial observations. In any case, studies of the B vitamins as plant hormones and the isolation and structure determination of traumatic acid consumed almost but not quite all of my time up to the beginning of World War II.

Photoperiodism Work at the University of Chicago

After Christmas of the 1937–1938 academic year, I received a letter from E. J. Kraus, inviting me to work on photoperiodism with Karl Hamner in the University of Chicago's wonderfully outfitted laboratories and greenhouses.

Several people had suggested that flowering, which is induced in many plants by the appropriate length of day and night, is mediated by a hormone made in leaves that goes to the buds and induces them to transform from vegetative to flower buds. Kraus asked me to spend a summer and then perhaps consider staying at Chicago after that. I consulted with Dr. R. A. Millikan, the de facto president of Caltech, about this proposition. Dr. Millikan had been a Professor of Physics at the University of Chicago and had come to Caltech at the invitation of Dr. Noyes and of the Trustees to become the chairman of its executive committee. Concerning my invitation to go to Chicago, Dr. Millikan said, "Yes, go ahead and do it. You won't like the University of Chicago and you won't want to stay there so don't let them give you any money. We'll pay you all the time and you come back when you're all done and want to. But, as I said, you won't like it. I didn't. I didn't stay and I don't think you will either." So I prepared my thinking about Chicago, along the lines suggested by Dr. Millikan, and I also began to think about photoperiodism. I told my graduate student, Fred, that I was going to be gone for a little while. After the end of school in 1938, I got in my car (a 1937 Ford sedan), put the camping gear and a suitcase full of clothes in it, and drove to Chicago, where I was greeted by Hamner and Kraus. They had arranged a room for me in the Quadrangle Club, the faculty club of the University of Chicago. I was also introduced to the botany greenhouses, which were not only extensive, but also had good facilities for working on photoperiodism.

Garner & Allard first recognized in 1920 that the relative lengths of day and night are important for plants. They worked on Maryland Mammoth tobacco, a short-day plant (i.e. it flowers only when the days are short and the nights are long). Further studies showed that the effect of day length on plants is perceived by the leaves. The leaves then send a message to the bud to turn into a flower bud. Hamner also had worked on this phenomenon with the cocklebur (*Xanthium pennsylvanicum,*) another short-day plant. If the day is 15.5 hr or less, the cocklebur will flower. In addition, Hamner's previous work had showed that to induce flowering, it is sufficient to expose a cocklebur plant to one short day. If we then transferred the plants to long days, the cocklebur would flower.

Hamner and I spent the entire summer studying the photoperiodic response of the cocklebur. We showed that it is the length of the night, and not the day, that determines the flowering response. There is a critical night length that has to be longer than 8.5 hr to induce the plant to flower. Short-day plants are really long-night plants.

Let us consider a plant that is left in a long night for 16 hr. It will, of course, produce the floral stimulus and will subsequently flower. Suppose, however, that we interrupt the dark period by a short pulse of light given in the middle of the night, thus dividing the 16-hour dark period into 2 slightly shorter than

8-hour dark periods. Such a light break, as it is often called, even though it is less than 1 min long, is sufficient to cause the long, dark night to be ineffective in causing flowering. The light pulse has effectively replaced continuous light in making the cocklebur think that it has seen a long day. This middle-of-the-night light-break can be of quite low intensity light, just a few foot candles at the leaf surface for a minute, let's say. The discovery during the summer of 1938 of the effectiveness of the light break in controlling flowering of cocklebur was shown subsequently by others to be true of other short-day plants as well. This finding has had considerable subsequent application. Several years after the publication of the 1938 work, the light-break principle was used by Sterling Hendricks and his colleague, Harry Borthwick, at the US Department of Agriculture in Beltsville to make an action spectrum (i.e. to determine the effectiveness of different wavelengths of light in causing the suppression of flowering of cocklebur) and with this knowledge, Hendricks & Borthwick discovered phytochrome.

By the end of the summer of 1938, Hamner and I wrote up the work we had done and sent it in immediately for publication in *Botanical Gazette,* a University of Chicago publication noted for its rapid publication. Our contribution appeared in the December issue. I have no hesitation in describing this paper as a minor classic.

Since this summer-long exposure to the study of photoperiodism, I have maintained my interest in the subject. I did not get seduced into the study of photomorphogenesis and the properties of the pigment, phytochrome. I tried to discover the nature of the hormone that is produced in leaves and sent to buds to make them turn into flower buds. I tried many approaches. All have been fruitless. To this day, I do not even have a good idea about the chemical nature of the hormone.

The War Over Rubber

Germany invaded Poland on September 1, 1939. This set in motion the train of events of World War II. Japan, which was now counted among the Axis Powers, pronounced the desirability of the Southeast Asian Co-Prosperity Sphere. This included all of the lands that normally supplied the entire world with natural rubber for making automobile and truck tires and all of the other products made of rubber that contributed to our culture. All natural rubber then and now comes from tropical plants that grow principally in Southeast Asia. The question was, what to do? Organic chemists had not yet produced a suitable polymer for making automobile tires. Rubber trees (*Hevea brasiliensis*), the only source of natural rubber, can be grown commercially on a large scale only in Southeast Asia, even though the trees are native to Brazil.

Guayule (*Parthenium argentatum*) is the one plant in the western world that has been a serious rubber producer. Guayule grows in Northern Mexico and a bit of Texas. After caucusing on the subject of rubber, Frits Went and I decided to prepare ourselves for a possible war by making ourselves masters of all guayule knowledge, learning about how it was grown, and what one could do to improve it. We would become the world's authority on the subject. In the fall of 1939, Frits Went arranged for us to meet with Mr. Carnahan, the president of the Intercontinental Rubber Company (IRC), which was the only large company involved in the production of rubber from guayule.

On the appointed day, Frits and I met with Mr. Carnahan in Salinas, California, where the IRC had large guayule plantations. Carnahan was a mining engineer and an enthusiast about guayule. The IRC, which had been in operation at that time for over 50 years, had hired a professional plant physiologist and plant breeder, Dr. William MacCallum, who had harvested seeds from strongly growing plants of high rubber content, grown them in the nursery, and multiplied their seeds. The strain that he commended most highly was strain 593. It was said to be the best grower and the highest yielder of rubber under field conditions. The discouraging thing we found out about guayule was that the IRC always planted their guayule out in the field on a ten-year rotation basis. In this way they achieved a claimed rubber content of 8 to 10% of the dry weight of the plant.

We learned about growing guayule too. Irrigation is said to decrease rubber content of the plant and so they are dry farmed. That's why they take so long to grow. We were informed that drought stress is required for rubber production.

As a result of our meeting, we formalized an agreement to do research on the guayule in cooperation with the IRC. We would, in return for supplies of the plants and seeds of guayule, return to them what we found out in the form of reports. When we returned to Pasadena we brought with us as a gift, a small bale of rubber seedlings, as well as a supply of seeds of strain 593. We set up a special laboratory for the rapid analyses of large numbers of samples per day for rubber content. We practiced growing guayule plants. We studied the nutrient requirements of the guayule, as well as how to kill the various pests that attack it.

When the Japanese started their penetration of Southeast Asia after December 7, 1941, the United States government, with all deliberate speed, developed the "Emergency Rubber Program," which consisted of a multi-pronged attack on what to do about getting rubber. One such prong was to grow guayule and grow it as quickly as possible on as large a scale as possible. This prong was called the Emergency Rubber Project (ERP).

I was immediately commandeered by the ERP. This project was allotted to the United States Forest Service (USFS) for administration. Congress directed the Department of Agriculture and the USFS to immediately purchase the buildings and holdings of the IRC and to get going on guayule. In February 1942 the guayule project of the ERP was set up in Salinas. Major Evan Kelly was the director of the ERP, succeeded later by Paul Roberts of the USFS. Dr. William McGinnies of the USFS became the director of research for the guayule project. My laboratories and greenhouse facilities were declared a special laboratory of the ERP, and I was appointed a non–civil servant special agent of the USFS assigned to the ERP. On January 2, 1942 I began an experiment that was described in a report to Bill McGinnies concerning the nature of the factors that control the rubber content of guayule. This was published in the *Botanical Gazzette* in 1943 and I believe it was the most important finding we made. We showed that the formation of rubber by guayule is controlled almost entirely by night temperature. The temperature during the day has to be high enough for the guayule to photosynthesize. It can be 60° or 80°F—it doesn't matter. The temperature at night has to be below 50°F, preferably at the optimum of 45°F. The plant only grows in weight—it does not grow in height and it does not flower. It concentrates on making rubber. A high proportion of the photosynthate made during the day is converted to rubber under these low night temperature conditions. It has since been shown by Chauncy R. Benedict of College Station, Texas, that low night temperature causes the production in guayule of a polyprenyl transferase, which catalyzes the polymerization of isopentenyl pyrophosphate into polyisoprene. The low temperature induction of the gene for the production of polyprenyl transferase is the reason that cold makes guayule make rubber. These findings did not, of course, help the ERP. They came too late. My own finding that low night temperature increases rubber formation in guayule was simply disregarded; even though true, it didn't jibe with the then current central dogma.

In late 1945 it became quite clear the war was going to end in the not-too-distant future, so it was time to think of other things to do. Early in 1946, the ERP was officially declared ended.

Fresh Beginnings—Cell Biology

The end of the war, gasoline rationing, and guayule husbandry all encouraged me to start a new program of research. With the encouragement of my new associate, Samuel G. Wildman, this new start included what we today call cell biology. We would isolate chloroplasts, mitochondria, cytoplasm, and lots of enzymes!

We first wished to isolate chloroplasts. These consisted of membranes, chlorophyll contained in grana, and the soluble stroma that bathes the grana. Spinach leaves were ground in a colloid mill. This device grinds up plant tissue wonderfully well and leaves intact grana but no intact chloroplasts. The grindate is then centrifuged at 20,000 xg. This completely pellets the grana, as well as mitochondria, etc. The supernatant from this centrifugation comprises the soluble leaf proteins that in turn make up about 16% of the dry weight of the leaf. An interesting discovery was that upon analysis, whether by centrifugation, ultracentrifugation, or electrophoresis in the Tiselius electrophoresis apparatus, over half of the total soluble protein consists of Fraction I, a single component of molecular weight approximately 500,000. Sam found similar results with leaf proteins of several other kinds of plants and in 1947, he published the first of a series of papers on the proteins of green leaves.

Fraction I was subsequently shown by our former postdoctoral fellow, John Littleton of Palmerston North, New Zealand, to be the main constituent of the stroma protein of chloroplasts. Littleton and Paul Ts'o, as well as others, then went on to show that Fraction I is, in fact, ribulose-1,5-bisphosphate carboxylase, the enzyme of the first step in the photosynthetic path of carbon. There is so much of it in leaves because it is not a very good enzyme (i.e. its turnover number is low). However, Fraction I is the most abundant protein in the world and it is central not only to the life of plants, but to all life: "All flesh is grass." Sam Wildman has continued to work on Fraction I protein from the day he discovered it to the present time—it's that important a protein!

I moved next to the mitochondria. Understanding of plant respiration lagged until great forward strides were made in a rush in 1948–1951. Adele Millerd and I had first met at the University of Sydney in 1949. She then knew that succinate was oxidized in vitro by cytoplasmic particles of a suspiciously mitochondria-like nature. While at Caltech from 1950 to 1953, she established that these particles were indeed mitochondria, very much like those of animal tissues. Millerd further showed that these plant mitochondria were capable not only of using the Krebs cycle to oxidize pyruvate, but they also carried out coupled oxidative phosphorylation with production of ATP. The paper describing these developments in 1951 brought our previously fractional knowledge of plant respiration up to the level of that known for other organisms.

A NEW PARADIGM

Even plant biologists can sometimes take a hint from what is going on in the world around them. Early in 1960, I said to myself, "It's rather late now to get to the basic problems of biology. Many other people have started already. But there is one basic problem remaining, namely, how does RNA get made? I will work on that." A new postdoctoral fellow, Ru-chih C. Huang, joined me in the middle of 1960. She had been a graduate student of Joe Varner's and came highly recommended by Joe. I suggested to Ru-chih that she try isolating native chromatin (i.e. DNA strands with proteins attached) of the nucleus of pea epicotyls, and that she find out if preparations of this kind were capable of making RNA from the four riboside triphosphates. She found very rapidly that isolated crude nuclear extract was capable of catalyzing the incorporation of C^{14}-labeled nucleoside triphosphate into TCA-insoluble material. The next thing was to separate and purify the enzyme responsible for this activity and to determine whether or not it depended on the presence of DNA. Ru-chih purified the enzymatic activity, and found that it caused the incorporation of all four riboside triphosphates into something degradable by RNase and therefore, RNA. The activity also depended entirely on the presence of DNA in the reaction mixture. We published a short paper on this subject at the end of 1960. Demonstrating the importance of this subject, three other groups published papers at the end of 1960 noting the existence of an enzyme that we now call RNA polymerase or DNA-dependent RNA polymerase.

In further study of the enzyme preparation, it soon turned out that completely deproteinized DNA is a much better template for DNA-dependent RNA polymerase than is either crude or purified chromatin made from nuclear extracts. This is because DNA in the nucleus is complexed with a class of proteins called histones. These proteins are complexed with DNA by ionic bonds and can be dissociated from DNA by high concentrations of salt such as 2 M sodium chloride. It is therefore possible to remove histones from the DNA of nuclear extracts and to recover the DNA histone-free. Histone-free DNA is a 10- or 20-fold more effective template for RNA synthesis than the original histone-covered DNA.

Study of the literature on histones made it clear that not much was known about how many kinds of histones there were or whether there were different histones in different creatures or in different specialized cells. Nobody knew what histones were for and no one had studied histones in plants. My colleague, Paul Ts'o, encouraged me to arrange a conference on histone biology and chemistry to see if we could make any sense out of the histones. We obtained money from the Office of Naval Research, the National Science Foundation, and private donors, and organized a conference for everyone in the world who knew something about histones. We planned our conference at the Rancho Santa Fe, which at that time was a pretty secluded resort hotel, but has since become a small city. We consumed the whole hotel by our conference of 56 participants. It became clear that there was complete confusion about the number of molecular species of histones in nature (estimates varied from a dozen to thousands), as well as complete confusion about the similarity or non-similarity of the histones of different species and of the histones of different cells of a single species. No pure histone had been prepared up to the time of our conference.

I returned from that meeting, one that included almost every one of the world's known experts on histones, certain that the future of histone chemistry would depend on a new generation of histone chemists. Ru-chih Huang and I agreed. We picked as a candidate a new graduate student, Douglas Fambrough. Doug was first sent to Stanford University for a month to work with Kenneth Murray, who instructed him on the use of amberlite CG-50 chromatography for the separation of lysine-rich histones from arginine-rich histones. He then came back and got to work. Doug used for the first time polyacrylamide gel electrophoresis to monitor the purity of individual histone fractions. He ultimately used it as a final purification step for histones destined for amino acid sequencing.

In all of this work, we compared the histones of the pea plant (bud) and of the calf thymus. Interestingly enough, cysteine, previously thought to be absent from histones, is found in both pea and calf in one fraction, now known as Histone III. Pea Histone III contains a single cysteine per molecule and is capable of forming dimers in solution. The Histone III of calf thymus contains two cysteines per molecule and is able to form a great variety of multimers upon oxidation of solutions just by sitting in the refrigerator. The cysteine of Histone III undoubtedly has been one of the main sources of confusion concerning the heterogeneity of the histones.

The histones of pea and of calf thymus are similar with respect to N-terminal amino acids, molecular size (as estimated by polyacrylamide gel electrophoresis in denaturing medium), and amino acid composition. Further, the histones of different tissues of the same organism appear to be identical with a few exceptions such as the disappearance of Histone I from the erythrocytes of birds. As a postdoctoral fellow, Keije Marushige published a much read and studied paper in the *Journal of Molecular Biology,* entitled simply "Properties of rat liver chromatin." The moral of this article was that what you can do with pea plants you can do with liver. The article found a great many readers.

Now we knew how to prepare pure histones, and now that we saw that the histones of peas and cows looked so similar in size, terminal amino acids, and so forth, the time had come to do sequence analysis and comparison between the different histones. This was started in collaboration with my long-time friend Emil Smith, who at that time was Chairman of the Department of Biological Chemistry of the UCLA Medical School. We decided to start with Histone IV, which is the smallest of the four species of histone molecules and therefore the most easily separated from all the others. Emil Smith declared that he needed to have two grams, an amount that one would now consider obscenely large, but two grams he got. It's no trouble preparing two grams of pure Histone IV from calf thymus glands, but it's quite a chore to prepare two grams of pure Histone IV from the apical buds of pea seedlings. We mechanized the procedures for growing pea seedlings in barrels under a shower of rain and for subsequently separating the shoots from the two cotyledons, which were discarded. In the year we spent preparing pure Histone IV from pea seedlings, we were the largest user in the western world of the Alaska cultivar of peas—about 25 tons of (dry) pea seeds.

The Histone IV of peas and of cows were eventually sequenced and shown to be essentially identical. There are two conservative amino acid replacements between the two species. The conservation of primary structure in histones and the probability that similar observations would be made with histones other than Histone IV caused biologists to reevaluate the role of histones. A more specific and important role in some aspect of chromatin structure or function might be forthcoming.

The sequencing of histones continued in our laboratory for a while and we contributed substantially to the knowledge concerning Histone II and Histone III. However, in the world as a whole, histone sequencing became a growth industry and became one of those things that's best to turn over to others.

Bill Garrard and Bill Pearson's program for taking the output from a gel scanner and plotting the acquired information as a series of Gaussians makes it easy to determine the relative amount of each individual histone species. By this method and with our knowledge of the molecular weights of the individual histones, it is possible to determine their stoichiometry. There are five species of histone molecules, with one molecule of Histone I (HI) for each two molecules of Histone IIA, Histone IIB, Histone III, and Histone IV, and therefore, one molecule of Histone I for eight molecules of other histones. Douglas Brutlag discovered that if chromatin treated with a low concentration (.02%) of formaldehyde for 24 hr at ice bucket temperature is then banded in CsCl, it loses all histones except HI. When the disposition of the remaining HI molecules along the DNA chain is investigated by electron microscopy, it turns out that they are spaced at approximately 200 base pair intervals.

The main hero of histone chemistry is Douglas Fambrough, who made order out of chaos. Unluckily, it also made a neuroscientist of our hero.

From this point on, I have followed the now well-traveled trail to gene isolation and study, and to the production of transgenic creatures. Stay tuned.

FROM KATMANDU TO TIMBUKTU TO KOTA KINABALU

Some will no doubt complain that it is more profitable for the serious scientist to stick to his problem and flog it to death. To them I say, for myself, browsing in far-flung pastures is more fun: Dark CO_2 fixation by succulents, chemical plant ecology, the path of carbon from CO_2 to rubber, plant taxonomy, and treatment of plant-chemical interaction by enzyme kinetics are all matters that I have also touched, and they have all been fun.

Probably, though, my most important contribution to science so far lies in the 108 graduate students to whom I have acted as advisor and mentor during their work toward the PhD degree and thereafter. The first, Frederick Addicott, received his PhD in 1939. The last, so far, is Carlotta Glackin, 1988. I am proud of my graduate students. A further contribution I have made has been through the some 200 postdoctoral fellows, visiting professors, and others who have worked in our laboratory and learned new skills and gained new insights and who have returned to the great outer world the better for it. I am proud of them as well.

Finally, I spoke earlier about the world that awaits exploration. I have studied it pretty thoroughly. It's all wonderful. From Katmandu to Timbuktu to Kota Kinabalu and beyond. Do not miss it!!

ACKNOWLEDGMENTS

I thank my wife, Ingelore, for her close and ever helpful collaboration and for her wise counsel as well as for her continuing inspiration. The preparation of this manuscript could not have been accomplished without the skilled and cheerily helpful work of Stephanie Canada. I thank her both for her help and her enthusiasm.

Annu. Rev. Plant Physiol. Plant Mol. Biol. 1994. 45:25–45

THE GENETIC AND MOLECULAR BASIS OF ROOT DEVELOPMENT

Roger A. Aeschbacher

Department of Biology, New York University, New York, New York 10003

John. W. Schiefelbein

Department of Biology, University of Michigan, Ann Arbor, Michigan 48109-1048

Philip N. Benfey[1]

Department of Biology, New York University, New York, New York 10003

KEY WORDS: *Arabidopsis,* differentiation, cell expansion, embryo, meristem, morphogenesis, nematode, nodule, signal transduction

CONTENTS

INTRODUCTION 26
AXIS FORMATION AND ROOT MERISTEM INITIATION 28
MERISTEM STRUCTURE IN SEED AND PRIMARY ROOT 30
LATERAL AND ADVENTITIOUS ROOT MERISTEMS 34
ELABORATION OF THE MERISTEMATIC PROGRAM—CELL DIVISION AND
    EXPANSION 35
DIFFERENTIATION OF ROOT HAIR CELLS 36
ROOT CELL–SPECIFIC GENE EXPRESSION 38
ROOT DEVELOPMENT IN RESPONSE TO ENVIRONMENTAL STIMULI 39
FUTURE PROSPECTS 41

1
   To whom correspondence should be addressed.

0066-4294/94/0601-0025$05.00

INTRODUCTION

The study of plant development has made striking progress in the last few years with the isolation of several genes that were first identified as mutations causing abnormal development, (reviewed in 1). Particularly rapid progress has been made in characterizing the genes that specify organ identity in flowers (13); however, little is known about the genes that regulate organ formation.

Recently, attention has turned to the root as a useful system for understanding organ development (reviewed in 57). Roots are relatively simple organs. Their growth pattern is uniform and continuous, they have a small number of differentiated cell types, and they have radial symmetry. These features facilitate the identification and characterization of mutant plants that undergo abnormal root development.

This review focuses on the current knowledge of the stages of root development and the genetic lesions that disrupt the processes required for passage from one stage to the next. We also review recent work on the developmental changes that occur in the root in response to environmental stimuli.

Much of the recent work on the molecular and genetic basis of root development has used *Arabidopsis thaliana*. This is partly because of the advantages of using *Arabidopsis* for genetic and molecular analyses (45). Another major advantage of using *Arabidopsis* is its small size, which allows many plants to be grown in petri dishes where the roots are visible. This has permitted facile genetic screens for plants that exhibit abnormal root development. Table I gives an overview of the *A. thaliana* root mutants described in this review. Very few mutants affecting root development had been identified previously (for review see 57), probably because of the difficulty of examining an organ that normally grows underground.

Historically, most descriptive and experimental analyses of root development were performed in species other than *A. thaliana* (for review see 21). To characterize mutants in this species, an accurate description of the developmental stages of the wild-type root was required. What has emerged is a picture of a plant organ with a remarkably simple organization and growth pattern. There is already a reasonably detailed description of the steps that lead from axis formation in the embryo to the growth pattern of the mature root. A complete description of the typical clonal history of every cell in the *A. thaliana* root should soon be available. This full description, combined with the small size and transparency of the root, and the ease of genetic manipulation allows us the claim that the *Arabidopsis* root is the *Caenorhabditis elegans* of the plant world. We review the progress that has been made in understanding the development of the root of *A. thaliana* and point out the areas that await characterization.

Table 1 Root development mutants of *Arabidopsis*

Mutant	Developmental Stage Affected	Mutagen	Complements	Map Position (Chromosome)	Ref
gnom (*gn*)	zygotic cell division	EMS	*mp*	1	42
monopteros (*mp*)	embryonic basal region formation	EMS	*gn*	1	10
hobbit	embryonic root meristem formation	EMS		n.a.	55
short-root (*shr*)	radial pattern formation and meristem maintenance	T-DNA	*cob, lit, sab*	n.a.	6
cobra (*cob*)	post-embryonic cell expansion	EMS	*shr, lit, sab, pom*	5	6
lion's tail (*lit*)	post-embryonic cell expansion	EMS	*shr, cob, sab, pom*	n.a.	6
pom-pom (*pom*)	post-embryonic cell expansion	EMS	*lit, sab, cob*	n.a.	29
sabre (*sab*)	post-embryonic cell expansion	T-DNA	*shr, lit, cob, pom*	1	6
rsw1	post-embryonic cell expansion	EMS	*rsw2, rsw3, reb1*	n.a.	4
rsw2	post-embryonic cell expansion	EMS	*rsw1, rsw3, reb1*	n.a.	4
rsw3	post-embryonic cell expansion	EMS	*rsw1, rsw2, reb1*	n.a.	4
reb1	post-embryonic cell expansion	EMS	*rsw1, rsw2, rsw3*	n.a.	4
rhd1	root hair cell differentiation	EMS	*rhd2, rhd3, rhd4, tip1*	1	60
rhd2	root hair cell differentiation	EMS	*rhd1, rhd3, rhd4, tip1*	5	60
rhd3	root hair cell differentiation	EMS	*rhd1, rhd2, rhd4, tip1*	3	60
dwf	stimulus response— gravity; Dominant	EMS	*aux1, agr1, axr1, axr2*	3	46
axr1	stimulus response— gravity	EMS	*aux1, agr1, dwf, axr2*	1	19
axr2	stimulus response— gravity; Dominant	EMS	*aux1, agr1, dwf, axr1*	3	70
rpt1, wav1	stimulus response— light, obstacle avoidance	EMS	*rpt2, wav2, wav3, wav4, wav5, wav6*	n.a.	47, 48, 49
rpt2	stimulus response— light	EMS	*rpt1, wav2, wav3, wav4, wav5, wav6*	n.a.	48

Mutant	Developmental Stage Affected	Mutagen	Complements	Map Position (Chromosome)	Ref
wav3	stimulus response—obstacle avoidance	EMS	rpt1, rpt2, wav2, wav4, wav5, wav6	n.a.	47
wav4	stimulus response—obstacle avoidance	EMS	rpt1, rpt2, wav2, wav3, wav5, wav6	n.a.	47

Notes: Mutations are recessive unless otherwise noted. Mutants listed under "Complements" are those that have been tested. Abbreviations: n.a.; not available.

AXIS FORMATION AND ROOT MERISTEM INITIATION

In classical descriptions of root morphology one of the most striking features is that differentiated cells are found in files that can be traced back to a small number of cells termed *initials* (18, 22). Similar simplicity was not found in morphological analyses of shoots. This feature of root morphology led to the *histogen* concept, which stated that the cells at the beginning of the files were actively dividing progenitors of the more differentiated cells (18). This idea was challenged when it was discovered that some of the cells at the beginning of the files divided infrequently. This discovery led to the concept of the *quiescent center* in which a number of cells at the apex did not actively participate in forming the differentiated cells that emanate from the apex (22). The role of the infrequently dividing cells in the quiescent center is still an issue. Surgical dissection experiments indicated that these cells are required to organize a newly proliferating root, which suggests that they may be a source of replacement cells for the actively dividing cells that are normally located adjacent to the quiescent center cells (2).

In this review we use the term *initials* to refer to the cells that are actively dividing and are the earliest detectable progenitors of the more differentiated cells. These cells have many of the characteristics of pluripotent animal stem cells. They are undifferentiated cells capable of proliferation, self-maintenance, and the production of several differentiated progeny (51). The less frequently dividing cells of the quiescent center have the characteristics of potential stem cells (51) if they are truly the source of replacement for the initials. In this review we include both the initials and the quiescent center cells when we refer to the root meristem.

The first fundamental question in root development is how and when the root originates. An answer to this question is beginning to emerge from analysis of genetic *pattern mutants*. In a large-scale screen of plants grown from ethyl methanesulfonate (EMS)-treated seeds, nine genes, represented by an average of eight mutants each, were identified in which the phenotype could be traced back to a morphological change in the embryo (43). Four of the nine genes appear to cause deletions of regions along the apical-basal axis. Based on the portions deleted, the apical-basal axis could be partitioned into three regions: 1. the apical portion, which comprises the cotyledons and the shoot apical meristem, 2. the central region, which encompasses the hypocotyl, and 3. the basal portion, which includes the root as well as the hypocotyl. There is also a *terminal* deletion represented by a mutant that appears to delete both the shoot and root apical meristems. This classification of mutants led to the suggestion that combinations of two of the genes defined above are necessary to specify each of the major parts of the embryo (34).

In this model, determination of the lower portion of the apical-basal axis, including formation of the embryonic root and root meristem, depends on the *MONOPTEROS* and the *GNOM* genes. Detailed characterization of mutations in both of these genes indicates a more complex situation. The *gnom* mutant was initially described as a terminal deletion because it appeared to lack both shoot and root apical structures (34). Analysis of 24 alleles indicated that all give rise to a heterogeneous phenotype (42). The phenotype ranged from ball-shaped, in which no apical-basal axis was detectable, to cone-shaped, in which relatively normal cotyledons were formed on a truncated hypocotyl with no root. Most combinations in trans of two alleles gave rise to a similar range of phenotypes as single alleles. One class of three alleles in combination with any of 19 other alleles yielded a new weak phenotype, in which a rudimentary root was formed. These results suggested that instead of forming an allelic series, all the alleles represented the inactive state of the *GNOM* gene. The weak phenotype that was observed in some trans-heterozygotes could be attributed to recovery of partial activity of a dimeric complex made up of two identical subunits that are mutated in different domains.

To better understand the origin of the phenotypic variation observed among the *gnom* alleles, a detailed comparison of mutant and wild-type embryogenesis was performed (42). In *A. thaliana,* the zygotic embryo divides asymmetrically to give a smaller apical and a larger basal cell. The apical cell gives rise to most of the embryo while the basal cell goes through a series of transverse divisions to form the suspensor. Partitioning of the apical cell by a series of precise cell divisions results in the octant-stage embryo. The *gnom* mutant appeared to be defective in the first asymmetric division. In mutant embryos the apical and basal cells appeared to be of similar size. Subsequent abnormal and variable placement of cell division planes was also observed. The phenotypic heterogeneity may be correlated with the variability in the placement of these subsequent division planes. These observations suggested that *GNOM* gene activity is required for the asymmetric division of the zygote and that in its absence, a symmetrical division occurs which leads to other embryonic defects. One observation suggested a more direct role for the *GNOM* gene in root development. Bisection of *gnom* mutant seedlings never gave rise to roots at the cut edges as is normally observed in wild-type. One interpretation of this result is that *GNOM* gene activity is needed to effect the asymmetric cell divisions required to organize root formation from a wound site.

In a detailed analysis of *monopteros* mutants, 10 strong alleles that lack both the hypocotyl and the root, and 3 weaker alleles that have a short hypocotyl were identified (10). All 13 alleles have relatively normal apical development, although the number and placement of cotyledons and leaves is variable. Abnormal embryonic development was evident from the octant stage onward. *Monopteros* mutant embryos had four tiers of cells instead of the normal two tiers. In the normal embryo, the upper tier of the octant stage gives rise to the cotyledons and shoot apical meristem, while the lower tier develops into the hypocotyl, the embryonic root and part of the root meristem. The remainder of the root meristem is contributed by the derivatives of the hypophysis, which is the suspensor cell closest to the embryo (see below). In the *monopteros* mutant embryos, the cells in the lower tier failed to elongate and instead retained the characteristics of upper tier cells. In addition the hypophysis of mutant embryos failed to undergo the series of stereotyped cell divisions that results in the formation of a portion of the root meristem. This analysis indicated that the *MONOPTEROS* gene may play a role in coordinating the formation of the embryonic root by facilitating communication between cells derived from the lower tier and those derived from the hypophysis. In bisection experiments, apparently normal roots developed from the wound sites of mutant seedlings. This indicated that the gene's activity is required primarily in organizing the basal region of the embryo and is not required for ectopic root meristem formation in post-germination stage roots.

The finding that *GNOM* and *MONOPTEROS* appear to act at different stages of development argues against a combinatorial role in regulating root development. A genetic analysis revealed that *gnom* is epistatic to *monopteros* as would be predicted from disruption of very early embryonic development by the *gnom* mutation and disruption of later embryonic development by the *monopteros* mutation (10, 42).

MERISTEM STRUCTURE IN SEED AND PRIMARY ROOT

The structure of the root apical meristem in the mature *A. thaliana* embryo has been analyzed by sectioning desiccated seeds (16). A simple, highly conserved pattern of cells was found. The interpretation of the pattern was based on analysis of cell lineages that could be traced back to putative initials. The initials appeared to be arranged in three tiers indicating that *A. thaliana* has a closed meristem (versus an open meristem in which the initials are not separated into tiers). At the tip of the embryonic root meristem there are 12 columella cells (four cells surrounded by eight) that are the putative initials of the columella root cap cells located below them. Immediately above these columella initials are four central cells. No mitotic figures were ever observed in these cells. In addition, division of these cells would not be required to maintain the integrity of the root meristem. Radioactive labeling experiments with DNA precursors indicated that the central cells had a very low incidence of labeling while the cells adjacent to the central cells all had an incidence of labeling consistent with their role as putative initials. The conclusion drawn from these experiments was that the central cells function as the quiescent center in the *A. thaliana* root. Analysis of earlier embryonic stages indicated that the columella and central cells are the derivatives of the hypophysis, while the other cells that make up the embryonic meristem are derived from the apical cell. The central cells are flanked by eight cells that appear to be the initials of the cortex and endodermis. These are surrounded by the putative initials of the epidermis and lateral root cap. The putative stele initials are above the central cells in the uppermost tier (Figure 1).

These anatomical studies indicated that different regions of the root cap are descended from different initials. There is a set of initials for the columella or central region of the root cap, and a second set of initials that appears to give rise to both the lateral portions of the root cap and to the epidermis. The number of cell layers that are derived from these two sets of initials is always the same. This indicates that the structural integrity of the root is maintained by synchronized cell division of these initials (16). Division of the cortex/endodermis initials must also be synchronized with those for the epidermis,

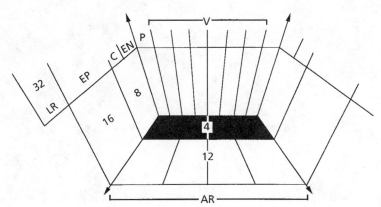

Figure 1 Schematic drawing of a median longitudinal section of the root apical meristem of the *A. thaliana* seed prior to germination. Four central cells (black) are surrounded by initials for the epidermis (EP), lateral root cap (LR), apical root cap or columella (AR), cortex (C) and endodermis (EN), pericycle (P), and the various vascular cell types (V). The number of cells that constitute the different initials in the meristem is indicated as well as the number of lateral root cap (LR) initials. (courtesy of B. Scheres)

because root cells have limited freedom of movement in relation to each other (16).

This organization of the embryonic root meristem is maintained in the primary root. Dramatic changes in morphology accompany differentiation of the cells and cell files appear to mature at different rates (16). The radial organization of the primary root is also remarkably simple. There is only a single layer of each of the outer four cell types. The differentiated cortex and endodermal layers have an invariant eight cells each (6, 16). External to the cortex layer, the epidermis varies in cell number but root hair cells are invariably found at the junction of two underlying cortex cells. Internal to the endodermal layer is a single layer of pericycle cells. Within the pericycle is the central cylinder, which is always diarch with two phloem elements at 90° to the xylem elements (16) (See Figure 2).

In older roots a process known as *secondary thickening* occurs. In *A. thaliana* this is accompanied by loss of the outer three layers of cells and formation of a vascular cambium and secondary xylem. These new cells are derived from cells that are found within the stele of the primary root (16).

In some plants, root development originates in and is maintained by a single apical cell. A complete anatomical description of the program of cell division

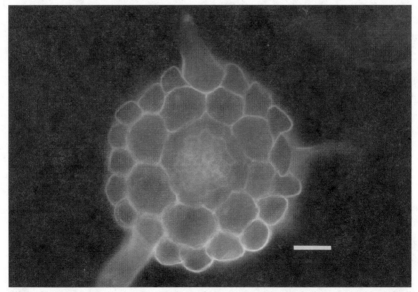

Figure 2   A transverse section through the *A. thaliana* root, stained to indicate the position of cell walls. The outer layer of cells is the epidermis. Internal to the epidermis is the cortex. Note that the three root hair cells in this section are each located external to the junction of two adjacent cortex cells. The cortex and the endodermis, which is the next internal cell layer, both have an invariant eight cells in the primary root. Internal to the endodermis is the stele, which consists of the pericycle and the vascular tissue. Bar = 20 μm. (Courtesy of Moira Galway)

of the root apical cell is available for the fern, *Azolla pinnata* (27). From embryogenesis through cessation of root growth, every cell division and cell expansion step contributing to root morphogenesis has been described. The apical cell of *A. pinnata* acts as a pluripotent stem cell giving rise to all the differentiated derivatives of the root. The remarkably reproducible pattern of cell division and cell expansion in *A. pinnata* raised the question of how this program is regulated. Three possibilities were proposed: 1. long-range gradients of positional information, 2. short-range cell-cell interactions that provide the positional information, or 3. cell-autonomous genetic programming (27). Clearly, these three possibilities could apply to multicellular organ development in any context. Analysis of drug-induced disruptions of cell wall placement that did not cause changes in subsequent cell development led to the conclusion that in *A. pinnata* cell-autonomous genetic programming is the most probable major form of regulation (27). In *A. thaliana,* different portions of the meristem are derived from different embryonic tissues, and this creates a more complex situation with an increased likelihood that cell-cell interactions coordinate the formation and activity of the meristem. Mutations that uncouple this coordination should help to resolve the relative contribution of cell-cell interactions versus cell-autonomous processes. *Monopteros* may be one such mutation. Another may be the *hobbit* mutation that appears to affect specifically the formation of the root meristem while leaving the rest of the embryonic root intact (55).

Another indication of potential cell-cell interactions that communicate positional information comes from experiments in which carrot embryos were surgically transected. After transection of the hypocotyl, regeneration of the lower portion of the embryo occurred. The final size and structure of the regenerated embryo were similar to that of the original embryo (56). Based on these experiments, it was suggested that the positional information necessary to organize an embryo was retained upon removal of part of the embryo. It is possible that cell-cell communication plays a role in this regeneration process.

One of the major challenges in understanding root meristem formation is to describe the events that lead to the formation of the simple organization of the root apex and to identify mutations that disrupt this process. The *monopteros* and *gnom* mutations appear to disrupt events that are necessary to obtain a root meristem but they do not contribute directly to its organizational pattern. Genes that disrupt later processes may have more subtle effects (e.g. changing the number of cells in one of the cell layers). The simplicity and predictability of the root apical pattern make identification and analysis of such mutants practical.

LATERAL AND ADVENTITIOUS ROOT MERISTEMS

The formation of new roots from existing roots or other organs provides a highly accessible opportunity to examine the steps required to originate a root meristem. Unlike most aerial organs that form from the shoot apical meristem, lateral and adventitious roots develop from differentiated cells that must re-differentiate to initiate a new root meristem (65). In the case of lateral roots, the first steps in the formation of the new meristem involve changes in the pericycle cells that lie opposite xylem elements (65). Although there currently is no way to predict precisely where lateral roots will form, they are usually not initiated near the root tip of higher plants. It has been suggested that an inhibitor of lateral root formation may diffuse from the primary root apex (65).

Many questions concerning lateral root initiation remain unanswered. Are all pericycle cells competent to initiate lateral roots? Does a single pericycle cell first receive a signal to form a meristem and then recruit neighboring cells [in a situation somewhat analogous to vulval development in *Caenorhabditis elegans* (31)] or is more than one cell involved from the earliest stages of initiation? Are the steps required to initiate a lateral root meristem similar to those involved in forming the embryonic root meristem? What processes are required for the lateral root to push through the cortex and epidermis and emerge into the external environment?

An anatomical examination of lateral root structure in *A. thaliana* indicated that there was considerably more variability in cell number in each layer of the lateral root than in the primary root (16). For example, instead of the invariant eight cortex cells, there were seven to eleven cells (16). This variability was interpreted as an indication of greater plasticity in lateral root organization as compared to primary roots (16). The organization of the lateral root could be controlled by a different set of factors than is the primary root. The greater variability in lateral roots compared to the primary root may reflect the rigor-ously controlled patterns of cell division during embryonic development that may be essential for initiating the stereotyped structure of the primary root (16).

One approach to determining whether lateral root initiation uses a different set of regulatory controls than does embryonic root initiation is to look for mutations that affect only lateral roots. Few such mutants have been found to date, and those that have been analyzed appear to affect plant hormone produc-tion or perception such as the super root (*Sur*) mutant (11). More extensive screens need to be undertaken. Many, if not most, of the regulatory steps in lateral root formation may also occur in embryonic root formation. If true, then it is of interest to know, for example, whether the regulation of the loci involved is identical in primary and lateral roots.

An important feature of lateral root growth is that the root primordium has to force its way through the cortex and epidermis to emerge into the external environment. A hydroxyproline-rich glycoprotein, HRGPnt3, is expressed transiently in a subset of cells in the pericycle and endodermis during initiation of the lateral root of tobacco (36). HRGPnt3 is a cell wall protein, and it may modify the structure of the cell wall of those cells that penetrate the cortex and epidermis upon lateral root induction (36).

ELABORATION OF THE MERISTEMATIC PROGRAM—CELL DIVISION AND EXPANSION

The division of an initial cell gives rise to a new initial and to a cell that enters the differentiation pathway. Subsequent cell divisions are principally anticlinal, giving rise to the files of cells that are characteristic of root morphology. The differentiation of cells can be followed by analysis of cells along each file. The younger, less differentiated cells are present near the root apex and progressively larger and more specialized cells are found as one moves up the file. This has led to the definition of three zones of development along the root: 1. the meristematic zone, which includes the initials and the quiescent center; 2. the elongation zone, in which cells divide, expand, and differentiate; and 3. the specialization zone, in which the final differentiated characteristics of the cells are apparent (57, 65). The processes that determine the shape of the root must take place primarily in the meristematic and elongation zones. Because there appear to be no morphogenetic cell movements, the primary parameters that determine shape are the timing of cell division, the orientation of the cell division plane, and cell expansion. These are interrelated processes because the degree and direction of cell expansion depends, in part, on when and in what plane cell division occurs. By identifying mutants that affect root shape, it should be possible to analyze the genetic program that controls morphogenesis.

Microtubules have long been thought to play a major role in controlling the orientation of cell expansion because depolymerization of microtubules leads to loss of control over cell expansion (26). The internal cytoskeleton must communicate in some way with the external cell wall to coordinate and regulate cell expansion. Molecular analysis of mutations affecting root cell expansion may shed light on the regulation of this process. Mutants with abnormal cell expansion have been identified among *A. thaliana* plants that have roots with an increased diameter. Three mutants with altered cell expansion appear to be affected primarily in a single cell layer. Abnormal expansion is most pronounced in the epidermal cells of *cobra,* in the cortex cells of *sabre,* and in the stele cells of *lion's tail* (6). These mutations are nonallelic (6). A fourth mutant, *pom-pom* has the greatest expansion in both the epidermis and cortex

(29). Quantification of cross-sectional areas confirmed a significant difference in the degree of expansion of various cell types in each of the mutants (6, 29). This suggests that expansion can be regulated differentially in individual tissues. The expansion of cells in *cobra, lion's-tail,* and *pom-pom* appears to depend on the rate of growth of the root. When this rate is suboptimal, the expanded cell phenotype is not expressed. Expansion of the cortex cell layer in the *sabre* mutant is principally in the radial direction and is not growth-rate dependent (6). The *sabre, lion's tail,* and *pom-pom* mutations are recessive while the *cobra* mutation is semi-dominant (6, 29). Double mutant combinations of *sabre* with either *cobra* or *lion's tail* have an additive phenotype with expanded cortex and expanded epidermal or stele tissues (6).

In an independent screen of seedlings, three nonallelic mutants that exhibited radial swelling of the root were isolated: *rsw1, rsw2,* and *rsw3.* These mutants appear wild-type at 18° C, but exhibit radial swelling at the root apex when transferred to 31° C. Which cell layers expand in these mutants has not been reported, although *rsw3* is the only one in which epidermal cells appear expanded (4). It is not known whether these three mutations are allelic to any of the expansion mutations mentioned above.

The roots of many plants have no predetermined stage at which growth ceases. This is the case with *A. thaliana.* An interesting exception is the fern, *A. pinnata,* whose apical cell stops dividing after a maximum of 50–55 cell divisions (27). Several experiments have investigated how the growth potential of the meristem is maintained. Regrowth after surgical excision of large parts of the maize meristem suggests that new stem cells could be formed or recruited from existing tissue (65). One source for such recruitment may be the quiescent center (22). One hypothesis is that a regulatory function maintains the correct balance between cells that are recruited into the meristem and those that are lost to terminal differentiation. A mutant in *A. thaliana* that apparently affects the ability of the root meristem to maintain growth has been isolated. The root tips of the mutant *short-root* appear completely differentiated and lack elongation and meristematic zones (6). There is also no endodermal cell layer in the root of this mutant. Other growth processes seem relatively unaffected by the mutation (6). Analysis of mutants that have abnormal root growth may lead to a better understanding of the processes that maintain the division potential of the meristem.

DIFFERENTIATION OF ROOT HAIR CELLS

Root hairs are tubular-shaped extensions of epidermal cells that develop in a defined region near the root apex. The formation of root hair cells involves controlled changes in cell shape that provide an opportunity to study the regulation of cell morphogenesis. In addition, because not all epidermal cells normally form root hairs, the study of root hair development can be used as a paradigm of cell fate specification in plants.

Genetic analysis of root hair development is facilitated by the fact that root hairs are dispensable, and plants lacking root hairs are viable and able to reproduce normally (60). The loci that have been identified by mutation may be subdivided into two major groups: those that affect the initiation of root hairs and those that are required for root hair elongation (4, 60).

The root epidermis of most plants consists of two basic types of cells: ones that form root hairs and ones that are hairless. Depending on the particular plant species, the mechanism involved in specifying cell fate in the root epidermis is either associated with an asymmetric cell division or the position of the cell relative to cells in the underlying cortex (3). The former mechanism is used by many grasses (64), whereas the latter operates in crucifers. In *A. thaliana,* root hairs form only from epidermal cells that lie over a transverse wall separating two adjacent cortical cells (12, 16; see Figure 2). In addition to this apparent positional control, the number of hair-bearing and hairless cells in the epidermal layer of *A. thaliana* is relatively predictable, and loci involved in establishing or maintaining the normal epidermal cell pattern have been identified recently (M Galway & J Schiefelbein, unpublished information).

The first outward sign of root hair formation is the localized swelling of the epidermal cell at the site of root hair emergence. One *A. thaliana* gene that plays a role in this process is *RHD1* (60). The *rhd1* mutants form an abnormally large swelling or bulge on the epidermal cell surface during the initial phase of hair formation (60). The *reb1* mutants also display an abnormal epidermal cell-swelling phenotype and may define a second gene regulating the swelling process (4). The *COBRA* gene (discussed earlier) appears to represent yet another gene affecting epidermal cell expansion although its role in root hair development has not been characterized (6).

Root hair elongation normally proceeds by tip growth [a type of polarized cell expansion that also occurs during the growth of pollen tubes, fungal hyphae, and many other tubular-shaped cells (61, 62)]. Several loci involved in root hair tip growth, including *RHD2, RHD3, RHD4,* and *TIP1* have been described. These genes may encode products that affect known tip-growth factors, such as components of the actin cytoskeleton or Ca^{2+} fluxes (30, 59). The *rhd2* mutants possess "stubby" hairs that are apparently caused by an inability of the cells to expand beyond the initial swelling stage (60). The phenotypes of the *rhd3* ("wavy" hairs) and *rhd4* ("bulging" hairs) mutants are interpreted as defects in the control of polarized cell expansion at the root hair tip (60). Double mutant analyses indicate that the *RHD2* gene product is required before the *RHD3* or *RHD4* products, and the *RHD3* and *RHD4* products probably act in separate pathways. Although the *RHD2* and *RHD4* genes appear to specifically affect root hair cells, the *RHD3* gene is required for normal cell expansion in many plant tissues. For example, the roots of *rhd3* mutants are approximately 70% as long as the wild-type, because of a reduction in cell enlargement (60). The *TIP1* gene is interesting because it appears to encode a product required for tip growth in both root hairs and pollen tubes (58). The *tip1* mutant plants produce shortened, branched root hairs, and the *tip1* pollen tubes grow more slowly than wild-type through the transmitting tissue of wild-type flowers (58). Because pollen function is not impaired in the four *RHD* mutants, there appears to be some degree of redundancy in the genetic control of tip growth in different plant cell types.

Several additional *A. thaliana* loci are required for normal root hair differentiation, although these were originally identified because the mutants display other phenotypes. Some auxin-resistant mutants fall into this category. For example, the *dwf* and *axr2* mutants display defects in root hair formation (41, 70). Also, the cytokinin-resistant (*ckr1*) mutants produce shorter root hairs than normal (66), and the *hy3* mutants, which have mutations in the phytochrome B gene, produce longer root hairs than normal when grown in the light (52). Although the abnormal root hairs in these mutants need to be characterized in greater detail, these studies illustrate the utility of root-hair cells for analyzing the control of plant cell differentiation by diverse regulatory pathways.

ROOT CELL–SPECIFIC GENE EXPRESSION

Molecular methods complement genetic approaches for identifying and characterizing root-expressed genes that may play a role in root development. The screening of random clones from a pea root cDNA library led to the identification of genes that are highly, but not exclusively, expressed in roots (20). Differential hybridization was used to isolate four root-specific cDNA clones from tobacco (14). One of these clones, TobRB7, is particularly interesting because it possesses sequence similarity to proteins proposed to function as membrane channels, and its promoter specifically directed GUS reporter gene expression in the root meristem and central cylinder region (71, 72). Several root-specific genes expressed during *Rhizobium* infection of legume roots have also been characterized (25, 54).

The goal of many molecular studies has been to identify root-expressed genes and the location of specific gene products within roots. Developmental regulation of distinct α- and β-tubulin genes has been detected in the maize seedling root apex, with differences identified between the isotypes present in the distal root tip and the differentiated root tissue proximal to the tip (33). A β-tubulin isoform that is expressed preferentially in roots has been identified in *A. thaliana* (50), and actin isoforms have been detected in a characteristic pattern in the roots of soybean (44). Three members of the proline-rich (PRP) cell wall protein gene family of soybean are expressed in roots or root nodules: SbPRP1, expressed in apical and elongating regions of primary and lateral roots (67); SbPRP2, present in the intercellular regions of the root cortex (73); and ENOD2, expressed in nodule parenchyma cells (69). One of the tonoplast intrinsic protein (γ-TIP) genes of A. thaliana is expressed in root cells of the elongating region, which may indicate a role for this gene product in vacuole formation during cell enlargement (40). A plasma membrane H^+-ATPase isoform in A. thaliana is expressed preferentially in roots, and has been proposed to play a role in ion uptake in the root cells (28). A root-specific lectin identified in barley is expressed specifically in the outer layers of cells (38). In each of these instances, a complete understanding of the gene's role in root development will probably require that appropriate mutations be generated and characterized.

The analysis of root-specific gene expression has also included the characterization of promoter subdomains. For instance, the cauliflower mosaic virus 35S promoter has a subdomain that confers root-specific expression and binds a transcription factor (ASF1) that is preferentially expressed in root tissue (7, 8, 35). The petunia EPSP synthase promoter also has a subdomain that confers root-specific expression (9). Sequences 3' to the coding region of an oilseed rape gene are required for expression in the developing cortex of embryos and root apices (15).

The properties of some of these root-expressed genes and sequences indicate that they may regulate or participate in developmentally-important processes. Even if these genes are not directly involved in root development, they may still be useful as markers for examining the formation of roots. For example, transgenic plants have been generated with root cell- or region-specific promoters directing GUS gene expression in A. thaliana, including a root meristem/root cap–specific marker line (6). This line has been used to analyze the specific root cell types affected by a root morphogenesis mutation (6). Root-expressed gene products can also be used to generate antibodies that may act as cell-specific markers for characterizing root development in normal and mutant plants (37).

ROOT DEVELOPMENT IN RESPONSE TO ENVIRONMENTAL STIMULI

Plant roots can modify their development in response to environmental stimuli including gravity, light, touch, and invading pathogens. The alterations in development frequently take the form of changes in the direction of growth, a response that requires the perception of an external signal; transduction of the signal; alteration in gene regulation and/or protein activity; and modification of programs of cell division, expansion, and differentiation.

To dissect the pathways that lead from stimulus perception to developmental modification, genetic screens have been devised to identify mutants that respond abnormally to external stimuli. Most plant roots respond positively to gravity. Mutants in gravitropic response have been isolated in several species (53). In *A. thaliana,* such mutants have been identified by growing plants vertically on nutrient agar plates and then turning the plates 90°. The roots of gravitropic response mutants do not change their direction of growth when the plates are rotated. Five independent loci that respond abnormally to gravity have been identified: *aux1* and *dwf* (46), *agr1* (5), *axr1* (19), and *axr2* (70). Four of these mutations (*dwf, aux1, axr1,* and *axr2*) also show resistance to the inhibition of root growth by exogenously applied auxin. Recent experiments have shown that application of auxin transport inhibitors to wild-type plants can induce phenocopies of some of these mutations (48). The gravitropic signal appears to be perceived in the root cap, but the change in the direction of growth appears to occur in the elongation zone. These observations indicate that the auxin polar transport system may be involved in the transport or transduction of the signal from the root cap to the elongation zone (49).

Roots respond negatively to light by growing away from the light source. *A. thaliana* plants that are mutated in their phototropic response have been isolated by growing seedlings on vertical nutrient agar plates that receive illumination only from one side. Two phototropic response mutants (*rpt1* and *rpt2*) that did not change the direction of root growth when illuminated from the side have been isolated (48). The gravitropic response of these mutants was normal as was the phototropic response in the aerial part of the plant. In addition, auxin transport inhibitors had no effect on the normal phototropic response (48).

As a root grows through soil it must respond to obstacles in its path. In *A. thaliana* this appears to be accomplished by rotation of the root tip. To screen for mutants in touch stimulus response, seedlings were grown on hard agar in plates that were inclined at 45°. The root cannot enter the agar and thus rotates to avoid it. The gravitropic response then brings the root back into contact with the agar. The periodic movement away from and toward the surface of the plate causes a wavy growth pattern of the root. Several mutants that had abnormal response to this touch stimulus situation have been isolated (47). The mutant *wav1* is unable to induce wavy growth at all, whereas *wav2, wav3,* and *wav4* have abnormal growth patterns with waves of shorter pitch or irregular form. Two other mutants, *wav5* and *wav6,* were determined to be alleles of *aux1* and *agr1,* respectively. The mutants *wav2, wav3,* and *wav4* show normal gravitropic and phototropic responses; however, *wav1* is defective in phototropism and may be allelic to *rpt1* (49). Therefore, the initial steps in root-tip rotation and the phototropic response may be regulated by the same gene (49).

Plant roots change their development in response to invading bacteria, fungi, and parasites. A particularly striking example is the formation of nodules on legume roots in response to contact with *Rhizobium* (reviewed in 23). The factor that elicits at least some of the early developmental changes in the root has been isolated and purified (39). This mitogenic lipo-oligosaccharide factor is able to induce cell polarization and a structure called the pre-infection thread in the outer cortical cells (68). Identification of the plant receptor system that recognizes the *Rhizobium* morphogen factor is a challenge for the future. In addition, it may be possible to use this mitogenic factor to isolate mutants (in an appropriate legume) in the nodulation process.

Another system that holds promise for a genetic approach to understanding pathogen initiated developmental changes is the response of roots to nematode invasion. Cyst nematodes form a syncytium within the vascular cylinder by digestion of cell walls and cell hypertrophy (63). Upon infection by root-knot nematodes, parenchymatic and cortical cells enlarge and form root galls. Stele cells surrounding the nematode head undergo several nuclear divisions and enlarge to become specialized feeding cells, called giant cells (63). Infection by root-knot and cyst-forming nematodes damage roots partially by mechanical rupture of the root and partially by the development of the feeding structures that affect water transport (17).

Recently, infection of *A. thaliana* by cyst-forming as well as root-knot nematodes was established (17). The initial infection process was greatly affected by the culture conditions of the plant (17). The relatively small size and transparency of *A. thaliana* roots made it possible to follow infection and development of the nematodes in vivo using a light microscope. In the case of the root-knot nematode, *Meloidogyne incognita,* infection close to the root tip was achieved by separation of the epidermal layer and the cortex by the nematode stylet. The nematode then oriented itself toward the meristem and migrated between cells without apparent damage to them. Near the root apex, the nematode turned and invaded the vascular system. Cells in the differentiating vascular cylinder next to the nematode head then began to undergo nuclear divisions and become highly metabolically active nurse cells from which the nematode could feed (63). The ability to observe microscopically these morphogenetic changes (formation of syncytia, giant cells, and galls) in *A. thaliana* roots raises the possibility that mutants defective for such changes may be identified in a genetic screen.

FUTURE PROSPECTS

A promising beginning has been made toward the ultimate goal of understanding the molecular processes that lead from embryonic axis formation to mature root organization. Progress has been made in defining stages of root development that are amenable to genetic analysis. The root appears to be particularly suited to studies of morphogenesis and the response to environmental changes. To complete the description of the development of the *A. thaliana* root, fate mapping needs to be performed as has recently been done for the shoot apical meristem of this plant (24, 32). In the near future, screens for mutations that produce gross developmental abnormalities should achieve saturation, thus completing the list of loci involved. One challenge for morphogenesis studies will then be to devise more sensitive screens that will detect subtle but significant mutations. Enhancer trap screens for interesting patterns in root development should reveal other genes that may be redundant or pleiotropic in function and thus missed in the mutational screens. The primary challenge will be to define the genetic pathways and then comprehend the function of the genes thus identified. It is difficult, at this stage, to predict the nature of the gene products that regulate processes such as the orientation of the cell division plane or unidirectional cell expansion. Work in this area should lead to new insights into fundamental cell biological processes.

There is great potential for progress in understanding the genetic and molecular basis of developmental changes that occur in roots in response to environmental interactions. Imaginative screens have been devised for stimulus-response mutants. Other areas under investigation are responses to drought and water stress; response to invasion by bacteria, fungi, and parasites; and the response to gradients of inorganic compounds.

The remarkable simplicity of organization and growth pattern of the root (particularly in *A. thaliana*), is one of the principal reasons for us to expect that rapid progress can be made toward understanding the molecular and genetic basis of root development. Many of the fundamental developmental processes must be analogous (if not identical) to those that regulate formation of other plant organs. Therefore information gained from the study of root development should be applicable to other aspects of plant development.

ACKNOWLEDGMENTS

We would like to thank S. Poethig, K. Roberts, B. Scheres, L. Dolan, I. Sussex, C. Schultz, G. Coruzzi, M.-T. Hauser and L. Di Laurenzio for helpful discussions and careful reading of the manuscript. R. A. Aeschbacher was supported by a fellowship from the Swiss National Science Foundation. The work in J. W. Schiefelbein's laboratory was supported by a grant (PCM-9004568) from the NSF. The work in P. N. Benfey's laboratory was supported by a grant (GM43778) from the NIH.

Literature Cited

1. Aeschbacher RA, Benfey PN. 1992. Genes that regulate plant development. *Plant Sci.* 83:115–26
2. Barlow PW. 1976. Towards an understanding of the behavior of root meristems. *J. Theor. Biol.* 57:433–51
3. Barlow PW. 1984. Positional controls in root development. In *Positional Controls in Plant Development,* ed. PW Barlow, DJ Carr, 1:281–318. Cambridge: Cambridge Univ. Press
4. Baskin TI, Betzner AS, Hoggart R, Cork A, Williamson RE. 1992. Root morphology mutants in *Arabidopsis thaliana. Aust. J. Plant Physiol.* 19:427–38
5. Bell JC, Maher EP. 1990. Mutants of *Arabidopsis thaliana* with abnormal gravitropic responses. *Mol. Gen. Genet.* 220: 289–93
6. Benfey PN, Linstead PJ, Roberts K, Schiefelbein JW, Hauser M-T, Aeschbacher RA. 1993. Root development in *Arabidopsis*: four mutants with dramatically altered root morphogenesis. *Development* 119:57–70
7. Benfey PN, Ren L, Chua N-H. 1989. The CaMV 35S enhancer contains at least two domains which can confer different developmental and tissue-specific expression patterns. *EMBO J.* 8:2195–202
8. Benfey PN, Ren L, Chua N-H. 1990. Tissue-specific expression from CaMV 35S enhancer subdomains in early stages of plant development. *EMBO J.* 9:1677–84
9. Benfey PN, Takatsuji H, Ren L, Shah DM, Chua N-H. 1990. Sequence requirements of the 5-enolpyruvylshikimate 3-phosphate synthase promoter for tissue-specific expression in petals and seedlings. *Plant Cell* 2:849–56
10. Berleth T, Jurgens G. 1993. The role of the monopteros gene in organizing the basal body region of the *Arabidopsis* embryo. *Development* 118:575–87
11. Boerjan W, den Boer B, Van Montagu M. 1992. Molecular genetic approaches to plant development. *Int. J. Dev. Biol.* 36:59–66
12. Bunning E. 1951. Über die Differenzierungsvorgänge in der Kruziferenwurzel. *Planta* 39:126–53
13. Coen ES, Meyerowitz EM. 1991. The war of the whorls: genetic interactions controlling flower development. *Nature* 353:31–37
14. Conkling MA, Cheng C-L, Yamamoto YT, Goodman HM. 1990. Isolation of transcriptionally regulated root-specific genes from tobacco. *Plant Physiol.* 93:1203–11
15. Dietrich RA, Radke SE, Harada JJ. 1992. Downstream DNA sequences are required

16. Dolan L, Janmaat K, Willemsen V, Linstead P, Poethig S, et al. 1993. Cellular organization of the *Arabidopsis thaliana* root. *Development* 119:71–84
17. Dorhout R, Gommers FJ, Kolloffel C. 1991. Water transport through tomato roots infected with *Meloidogyne incognita. Phytopathology* 81:379–85
18. Esau K. 1977. *Anatomy of Seed Plants.* New York: Wiley
19. Estelle MA, Somerville C. 1987. Auxin-resistant mutants of *Arabidopsis thaliana* with an altered morphology. *Mol. Gen. Genet.* 206:200–6
20. Evans IM, Swinhoe R, Gatehouse LN, Gatehouse JA, Boulter D. 1988. Distribution of root messenger RNA species in other vegetative organs of pea *Pisum sativum* L.. *Mol. Gen. Genet.* 214:153–57
21. Feldman LJ. 1984. Regulation of root development. *Annu. Rev. Plant Physiol.* 35: 223–42
22. Feldman LJ. 1984. The development and dynamics of the root apical meristem. *Am. J. Bot.* 71:1308–14
23. Fisher RF, Long SR. 1992. Rhizobium-plant signal exchange. *Nature* 357:655–60
24. Furner IJ, Pumfrey JE. 1992. Cell fate in the shoot apical meristem of *Arabidopsis thaliana. Development* 115:755–64
25. Gloudemans T, Bisseling T. 1989. Plant gene expression in early stages of Rhizobium-legume symbiosis. *Plant Sci.* 65:1–14
26. Green PB. 1962. Mechanism for plant cellular morphogenesis. *Science* 138:1404–5
27. Gunning BES. 1982. The root of the water fern *Azolla*: cellular basis of development and multiple roles for cortical microtubules. In *Developmental Order: Its Origin and Regulation,* ed. B Subtelny, PB Green, pp. 379–421. New York: Liss
28. Harper JF, Manney L, DeWitt ND, Yoo MH, Sussman MR. 1990. The *Arabidopsis thaliana* plasma membrane H^+-ATPase multigene family. *J. Biol. Chem.* 265: 13601–8
29. Hauser M-T, Benfey PN. 1993. Genetic regulation of root expansion in *Arabidopsis thaliana. NATO-ASI Plant Mol. Biol. Ser.* In press
30. Heath IB. 1990. *Tip Growth in Plant and Fungal Cells.* San Diego: Academic
31. Horvitz HR, Sternberg PW. 1991. Multiple intercellular signalling systems control the development of the *Caenorhabditis elegans* vulva. *Nature* 351:535–41
32. Irish VF, Sussex IM. 1992. A fate map of

the *Arabidopsis* embryonic shoot apical meristem. *Development* 115:745–53

33. Joyce CM, Villemur R, Snustad DP, Silflow CD. 1992. Tubulin gene expression in maize (*Zea mays* L.): changes in isotype expression along the developmental axis of the seedling root. *J. Mol. Biol.* 227:97–107

34. Jurgens G, Mayer U, Torres Ruiz RA, Berleth T, Misera S. 1991. Genetic analysis of pattern formation in the *Arabidopsis* embryo. *Development* Suppl. 1:27–38

35. Katagiri F, Lam E, Chua N-H. 1989. Two tobacco DNA-binding proteins with homology to the nuclear factor CREB. *Nature* 340:727–30

36. Keller B, Lamb CJ. 1989. Specific expression of a novel cell wall hydroxyproline-rich glycoprotein gene in lateral root initiation. *Genes Dev.* 3:1639–46

37. Knox JP, Linstead PJ, Peart J, Cooper C, Roberts K. 1991. Developmentally regulated epitopes of cell surface arabinogalactan proteins and their relation to root tissue pattern formation. *Plant J.* 1:317–26

38. Lerner DR, Raikhel NV. 1989. Cloning and characterization of root-specific barley lectin. *Plant Physiol.* 91:124–29

39. Lerouge P, Roche P, Faucher C, Maillet F, Truchet G, et al. 1990. Symbiotic host-specificity of *Rhizobium meliloti* is determined by a sulphated and acylated glucosamine oligosaccharide signal. *Nature* 344:781–84

40. Ludevid D, Hofte H, Himelblau E, Chrispeels MJ. 1992. The expression pattern of the tonoplast intrinsic protein gamma-TIP in *Arabidopsis thaliana* is correlated with cell enlargement. *Plant Physiol.* 100:1633–39

41. Maher EP, Martindale SJB. 1980. Mutants of *Arabidopsis thaliana* with altered responses to auxins and gravity. *Biochem. Genet.* 18:1041–53

42. Mayer U, Buttner G, Jurgens G. 1993. Apical-basal pattern formation in the *Arabidopsis* embryo: studies on the role of the *gnom* gene. *Development* 117:149–62

43. Mayer U, Torres Ruiz RA, Berleth T, Misera S, Jurgens G. 1991. Mutations affecting body organisation in the *Arabidopsis* embryo. *Nature* 353:402–7

44. McLean BG, Eubanks S, Meagher RB. 1990. Tissue-specific expression of divergent actins in soybean root. *Plant Cell* 2:335–44

45. Meyerowitz EM. 1987. *Arabidopsis thaliana. Annu. Rev. Genet.* 21:93–111

46. Mirza JI, Olsen GM, Iversen T-H, Maher EP. 1984. The growth and gravitropic responses of wild-type and auxin-resistant mutants of *Arabidopsis thaliana. Physiol. Plantarum* 60:516–22

47. Okada K, Shimura Y. 1990. Reversible root tip rotation in *Arabidopsis* seedlings induced by obstacle-touching stimulus. *Science* 250:274–76

48. Okada K, Shimura Y. 1992. Mutational analysis of root gravitropism and phototropism of *Arabidopsis thaliana* seedlings. *Aust. J. Plant Physiol.* 19:439–48

49. Okada K, Shimura Y. 1992. Aspects of recent developments in mutational studies of plant signaling pathways. *Cell* 70:369–72

50. Oppenheimer DG, Haas N, Silflow CD, Snustad DP. 1988. The beta-tubulin gene family of *Arabidopsis thaliana*: preferential accumulation of the beta-1 transcript in roots. *Gene* 63:87–102

51. Potten CS, Loeffler M. 1990. Stem cells: attributes, cycles, spirals, pitfalls and uncertainties. Lessons for and from the crypt. *Development* 110:1001–20

52. Reed JW, Nagpal P, Poole DS, Furuya M, Chory J. 1993. Mutations in the gene for the red/far-red light receptor phytochrome B alter cell elongation and physiological responses throughout *Arabidopsis* development. *Plant Cell* 5:147–57

53. Roberts JA. 1987. *Developmental Mutants in Higher Plants.* Cambridge: Cambridge Univ. Press

54. Scheres B, van de Wiel C, Zalensky A, Horvath B, Spaink H, et al. 1990. The ENOD12 gene product is involved in the infection process during the pea-Rhizobium interaction. *Cell* 60:281–94

55. Scheres B, Willemsen V, Janmaat K, Wolkenfelt H, Dolan L, Weisbeek P. 1993. Analysis of root development in *Arabidopsis thaliana. NATO-ASI Plant Mol. Biol. Ser.* In press

56. Schiavone FM, Racusen RH. 1991. Regeneration of the root pole in surgically bisected carrot embryos occurs by position-dependent, proximodistal replacement of missing tissues. *Development* 113:1305–13

57. Schiefelbein JW, Benfey PN. 1991. The development of plant roots: new approaches to underground problems. *Plant Cell* 3:1147–54

58. Schiefelbein JW, Galway M, Masucci J, Ford S. 1993. Pollen tube and root hair tip growth is disrupted in a mutant of *Arabidopsis thaliana. Plant Physiol.* In press

59. Schiefelbein JW, Shipley A, Rowse P. 1992. Calcium influx at the tip of growing root hair cells of *Arabidopsis thaliana. Planta* 187:455–59

60. Schiefelbein JW, Somerville C. 1990. Genetic control of root hair development in *Arabidopsis thaliana. Plant Cell* 2:235–43

61. Schnepf E. 1986. Cellular polarity. *Annu. Rev. Plant Physiol.* 37:23–47

62. Sievers A, Schnepf E. 1981. *Cell Biology Monographs: Cytomorphogenesis in Plants.* New York: Springer-Verlag

63. Sijmons PC, Grundler FMW, von Mende N, Burrows PR, Wyss U. 1991. *Arabidopsis thaliana* as a new model host for plant-parasitic nematodes. *Plant J.* 1:245–54

64. Sinnot EW, Bloch R. 1939. Cell polarity and the differentiation of root hairs. *Proc. Natl. Acad. Sci. USA* 25:248–52

65. Steeves TA, Sussex IM. 1989. *Patterns in Plant Development.* Cambridge: Cambridge Univ. Press

66. Su W, Howell SH. 1992. A single genetic locus, *Ckr1,* defines *Arabidopsis* mutants in which root growth is resistant to low concentrations of cytokinin. *Plant Physiol.* 99:1569–74

67. Suzuki H, Fowler TJ, Tierney ML. 1993. Deletion analysis and localization of SbPRP1, a soybean cell wall protein gene, in roots of transgenic tobacco and cowpea. *Plant Mol. Biol.* 21:109–19

68. van Brussel AAN, Bakhuizen R, van Spronsen PC, Spaink HP, Tak T, et al. 1992. Induction of pre-infection thread structures in the leguminous host plant by mitogenic lipo-oligosaccharides of Rhizobium. *Science* 257:70–72

69. van de Wiel C, Scheres B, Franssen H, van Lierop M-J, van Lammeren A, et al. 1990. The early nodulin transcript ENOD2 is located in the nodule parenchyma (inner cortex) of pea and soybean root nodules. *EMBO J.* 9:1–7

70. Wilson AK, Pickett FB, Turner JC, Estelle MA. 1990. A dominant mutation in *Arabidopsis* confers resistance to auxin, ethylene and abscissic acid. *Mol. Gen. Genet.* 222: 377–83

71. Yamamoto YT, Cheng C-L, Conkling MA. 1990. Root specific genes from tobacco and *Arabidopsis* homologous to an evolutionarily conserved family of membrane channel proteins. *Nucleic Acids Res.* 18:7449

72. Yamamoto YT, Taylor CG, Acedo GN, Cheng C-L, Conkling MA. 1991. Characterization of cis-acting sequences regulating root-specific gene expression in tobacco. *Plant Cell* 3:371–82

73. Ye Z-H, Song Y-R, Marcus A, Varner JE. 1991. Comparative localization of 3 classes of cell wall proteins. *Plant J.* 1:175–83

Annu. Rev. Plant Physiol. Plant Mol. Biol. 1994. 45:47–60

MESSENGER RNA 3' END FORMATION IN PLANTS

Arthur G. Hunt

Department of Agronomy, University of Kentucky, Lexington, Kentucky 40546-0091

KEY WORDS: polyadenylation signals, RNA processing, gene expression

CONTENTS

INTRODUCTION 47
THE STRUCTURE OF PLANT POLYADENYLATION SIGNALS 48
   Far-upstream Elements 50
   Near-upstream Elements 52
   The Cleavage/Polyadenylation Site.. 53
   Downstream Sequences 54
DIFFERENTIAL UTILIZATION OF POLYADENYLATION SIGNALS IN PLANTS 54
DIRECTIONS FOR FUTURE RESEARCH 56

INTRODUCTION

Messenger RNA 3' end formation is an important process in gene expression in eukaryotic cells. This event initiates (at the very least) the process of transcription termination, thus completing the transcription cycle and freeing RNA polymerase II for further initiation events. In addition, a tract of poly-adenylate, which is required for subsequent posttranscriptional events, is added to most mRNAs, enabling expression of the mature mRNA in the cytoplasm.

The process of mRNA 3' end formation in mammals has been well charac-terized. The precursor mRNA is processed by endonucleolytic cleavage and subsequent addition of a polyadenylate tract to the 3' OH generated by this cleavage. These events are catalyzed by a group of at least four complexes, two of which recognize specific RNA sequences in the precursor and two of

0066-4294/94/0601-0047$05.00

which presumably possess the requisite enzymatic activities for cleavage and polyadenylation. This process has been reviewed recently (40, 51), and the reader is referred to these reviews for a detailed synopsis.

Messenger RNA polyadenylation is a common feature of gene expression, having been observed in eukaryotic nuclei (40), human mitochondria (35), and even with some bacterial genes (4 and references therein). Although our understanding of the process of mRNA 3′ end formation in other organisms does not approach that of the process in mammals, nature apparently has adopted some different strategies to generate a 3′-polyadenylated mRNA competent for translation. This is best reflected in the variations in sequence signals that direct RNA 3′ end formation in different organisms (38, 40). Each of these strategies probably reflects aspects of gene expression that are unique to the organism of interest, including plants. This review focuses mostly on the distinctive structure of plant polyadenylation signals and the implications of this understanding for aspects of gene expression in plants.

THE STRUCTURE OF PLANT POLYADENYLATION SIGNALS

Although early studies concluded that, as in mammals, AAUAAA is a polyadenylation [poly(A)] signal in plants (8), it became apparent that this sequence was not a universal feature near many polyadenylation sites in plant genes. This observation spurred comparative studies of the regions surrounding plant poly(A) sites (20, 29). From these studies, an exact match for the sequence AAUAAA could be found near poly(A) sites in only about 1/3 of plant genes, and 4–5 of 6 base matches for this sequence could be found in an additional 50% of plant gene 3′ regions. A significant proportion (15%) of plant poly(A) sites had no AAUAAA-like motif. Another distinctive feature of plant gene transcripts was the presence of multiple poly(A) sites in most instances (6, 14, 32).

These observations suggested that plant poly(A) signals may be quite different from their counterparts in other organisms. Studies showing that mammalian poly(A) signals were poorly functional in plants supported this conjecture (3, 16, 28). Moreover, although the cauliflower mosaic virus (CaMV) poly(A) signal was found to be utilized in yeast, this was because of the fortuitous occurrence of yeast poly(A) signals that were distinct from the sequences needed for functioning of this signal in plants (19). Two other regions containing plant poly(A) signals, from the octopine synthase (*ocs*) gene and from a pea gene encoding the small subunit for ribulose bisphosphate carboxylase (the *rbcS*-E9 gene), did not function in yeast (BD Mogen & AG Hunt, unpublished observations). Thus, plant poly(A) signals seem to differ from yeast and mammalian signals.

Our present understanding of plant poly(A) signals comes mostly from functional analyses of four genes: the CaMV 19S/35S transcription unit (11, 30, 45, 46), the *rbcS*-E9 gene (15, 30, 31), the *ocs* gene (27), and a zein 27-kDa gene (53). Nuclease protection or reverse transcription/polymerase chain reaction (RT-PCR) have been used to evaluate the effects of deletions and site-directed mutations on the function of particular poly(A) sites. Such approaches are necessary because of the redundancy of plant poly(A) signals, and of the frequent occurrence of weak, cryptic poly(A) sites in most commonly used plant transformation vectors. Reporter gene activity or RNA analysis using northern blotting fails to provide sufficient information to interpret properly the effects of mutations on poly(A) site function.

The genes listed above represent three rather different sources: viral and bacterial pathogens and two cellular genes. Nevertheless, these studies all lead to similar models, suggesting that the results can be generalized to most, if not

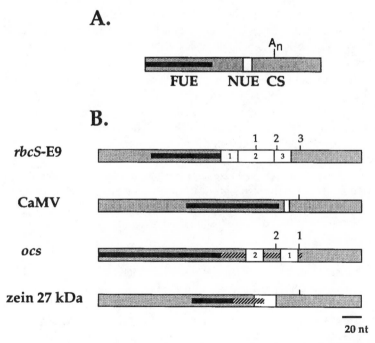

Figure 1 Organization of plant poly(A) signals. A. Structure of a typical poly(A) signal, showing the relative position of the far-upstream element (FUE; thick black line), near-upstream element (NUE; white boxes), and cleavage/poly(A) site (CS; tick mark above bar). B. Organization of the *rbcS*-E9, CaMV, *ocs,* and zein 27-kDa gene poly(A) signals, showing the relative positions of elements defined by functional analyses. For the *ocs* and zein gene signals, the 3′ limit of the FUE has not been defined accurately and could extend into the region defined by the thick dashed line. For the *rbcS*-E9 and *ocs* genes, the different poly(A) sites and their corresponding NUEs are noted with numbers as defined in Mogen et al (31) and MacDonald et al (27), respectively.

all, plant poly(A) signals. These studies indicate that plant poly(A) signals are complex, being composed of many distinct *cis* elements (Figure 1). These elements can be grouped into three classes: *far-upstream elements* (FUEs), which are relatively distant from their associated sites; *near-upstream elements* (NUEs), which are located within 40 nucleotides (nts) of their associated sites; and the actual cleavage/poly(A) sites (CSs) themselves. The known properties of each type of element are described below.

Far-upstream Elements

The most distinctive feature of plant poly(A) signals is the requirement of sequences upstream from possible AAUAAA motifs. Such distal elements (FUEs) have been identified in each of the four poly(A) signals mentioned above, including two from cellular genes, and are thus probably a general feature of poly(A) signals in plants. FUEs are novel in many respects. They can control more than one site in a gene, though not necessarily equally (27, 31), but they are not responsible for the actual 3' end profile in a given gene. FUEs from different genes are interchangeable (31). As noted below, there does not seem to be a strict consensus sequence that correlates with FUE function.

The FUEs from two signals have been characterized in some detail. In both cases, they seem to consist of an extended domain between 40 and 150 nts upstream from the sites they control. In the CaMV signal, the FUE seemed to consist, in part, of the sequence motif UAUUUGUA (46). Deletions spanning this motif had a marked negative effect on the functioning of the signal, and introduction of this sequence near a cryptic AAUAAA motif increased its ability to function as a poly(A) signal. However, deletions that did not include the UAUUUGUA motif also significantly impaired efficient 3'end processing (30, 46). Taken together, these studies indicate that the FUE from the CaMV poly(A) signal consists of several subelements, one of which has the sequence UAUUUGUA.

The FUE in the pea *rbcS*-E9 gene poly(A) signal lies between 40 and 160 nts upstream from its associated sites (15, 31). Interestingly, whereas deletion of this upstream region completely abolished functioning of three of the *rbcS*-E9 poly(A) sites (15), linker scanning mutations through this region showed that no 20 nt substitution could mimic the effect of a complete deletion (31). Progressively larger substitutions in the relevant region approached the deletion phenotype. These results imply a degree of functional redundancy analogous to that suggested for the FUE in the CaMV poly(A) signal. Although the sequence UAUUUGUA was not present in the *rbcS*-E9 FUE, two copies of a part of this sequence (UUGUA) did occur (Figure 2). The *rbcS*-E9 FUE was able to replace the CaMV FUE (31), indicating a functional equivalence between these two elements.

FUEs have been defined in two other plant poly(A) signals, those from the ocs and zein 27-kDa genes (27, 53). The FUEs have only been localized to relatively large regions, and these regions do not contain the sequence UAUUUGUA. The *ocs* FUE did have one UUGUA motif (Figure 2). Indeed, beyond a characteristic UG-richness, no common motif can be discerned in the four FUEs identified to date (Figure 2). It is possible that many different subelements contribute to FUE function, and that the particular combinations of subelements determine the properties of a particular FUE. Alternatively, a general UG-richness may suffice for FUE function, much as AU-richness contributes to intron definition in plants (10).

In terms of relative position and general sequence composition, FUEs are reminiscent of upstream sequence elements (USEs) that play a role in poly-adenylation of certain animal virus transcripts (5, 7, 41, 50). Indeed, the UAUUUGUA motif in the CaMV poly(A) signal (46) is identical to one of the sequence elements involved in mRNA 3′ end formation in hepatitis B virus (41), a mammalian pararetrovirus. To date, no animal cellular gene has been shown to possess USEs, and the experiments described above suggest that FUEs may be a general feature of Angiosperm genes. The observation that the FUE from a plant nuclear gene (the *rbcS*-E9 gene) is functionally equivalent to the FUE from a viral gene (CaMV; 31) reinforces the contention that FUEs are general features of poly(A) signals in flowering plants.

FUEs also share sequence properties with downstream sequence elements required for polyadenylation in animals, in particular a decided UG-richness. However, the UAUUUGUA element in the CaMV poly(A) signal does not function when placed downstream from the CaMV poly(A) site. It is unknown whether this sequence similarity reflects a commonality in the *trans* factors that mediate FUE function.

rbcS   5'...UGGCAUUGGGAAAACUGUUU UUCUUGUACCAUUUGUUGUG CUUGUAAUUUACUGUGUUUU
           UUAUUCGGUUUUUCGCUA..3'

CaMV   5'...AUAAUAAUGUGUGAGUAGUU CCCAGAUAAGGGAAUUAGGG UUCUUAUAGGGUUUCGCUCA
           UGUGUUGAGCAUAUAAGAAA CCCUUAGUAUGUAUUUGUAU UUGUA...3'

ocs   5'...CUAUGAUCGCAUGAUAUUUG CUUUCAAUUCUGUUGUGUCAC GUUGUAAAAAAACCUGAGCAU
           GUGUAGCUCAG...3'

zein 27 kDa   5'...CUAGCUAGUUGAGUCAUUUA GCGGCGAUGAUUGAGUAAUA AUGUGUCACGCAUCACCAUG
              GGUGGCAGUGUCAGUGUGAG CAAUGACCUGAAUGAACAAU UGA...3'

Figure 2 Comparison of the FUEs from the *rbcS*-E9, CaMV, *ocs,* and zein 27-kDa gene poly(A) signals. The motif UUGUA is doubly underlined, and UG-rich sequences are singly underlined. For the *ocs* and zein gene FUEs, only those regions clearly defined by deletion studies are indicated. The indeterminate 3′ extensions for the *ocs* and zein FUEs depicted in Figure 1 are not included here.

FUEs in plant poly(A) signals, USEs in mammalian viral signals (5, 7, 41, 50), and upstream parts of yeast poly(A) signals (1, 12, 37, 42) all share an interesting feature with the sequences that direct ρ-dependent transcription termination in *Escherichia coli*. Mutations affecting these various elements need to involve gross changes (e.g. large deletions); point mutations or linker substitution mutants do not yield defective phenotypes (38). Moreover, heterogeneous 3' ends arise from single transcription units in many of these cases. These are interesting similarities because yeast poly(A) signals are likely to be termination signals as well (36, 43). These parallels raise the possibility that, as may be the case in yeast, plant poly(A) signals may also be termination signals.

Near-upstream Elements

As noted above, a majority of plant genes do not possess an AAUAAA motif near the 3' ends of their corresponding mRNAs. Some mutagenesis studies have shown that the sequence AAUAAA, or other sequences located 6–40 nts upstream from poly(A) sites, are essential for efficient 3' end formation in plants. These sequences (NUEs) define a second class of *cis* element involved in mRNA 3' end formation in plants. In genes with multiple poly(A) sites, each site is controlled by a different NUE (27, 31).

Mutagenesis studies have defined NUEs for seven poly(A) sites in four plant genes (27, 30, 31, 45, 46, 53). An NUE can consist of the motif AAUAAA (in the CaMV poly(A) signal; 30, 45, 46), AAUGAA (in the zein 27-kDa gene; 53), or the sequence AAUGGAAUGGA (the NUE for site 1 in the *rbcS*-E9 gene; Q Li & AG Hunt, unpublished results). AAUAAA-related motifs occur in other defined NUEs (Figure 3), but the verification of their roles in mRNA 3' end formation by directed mutational analysis has not been reported. These different NUEs are probably functionally related—they are located in similar positions with respect to their poly(A) sites and can be identified by single linker-scanning mutations (in contrast to the FUEs). They are also distinct elements in that a particular NUE (or group of NUEs) can function in concert with different FUEs (31, 46).

The temptation to draw a direct functional analogy between NUEs and the mammalian poly(A) signal AAUAAA is strong. There are, however, subtle distinctions that must be made. Systematic point mutation analysis indicates that much variation in the AAUAAA motif in the CaMV poly(A) signal can be tolerated (H Rothnie & T Hohn, personal communication), more so than in mammalian systems (48). Also, naturally occurring AAUAAA-like NUEs in plant genes are much more variable than is the AAUAAA motif in mammals. Importantly, there are many plant genes that have no obvious AAUAAA-like motif near their poly(A) sites. These observations indicate that, at the very least, the range of sequences that can serve as a NUE in plant poly(A) signals is much greater than is seen with the AAUAAA motif in mammals. It is possible that plant NUEs may act through a substantially different mechanism than does the AAUAAA motif.

Although no detailed analyses have been reported, some idea of the spacing requirements between FUEs and NUEs can be gleaned from several of the published studies. No maximum allowable separation between FUEs and NUEs has been observed; FUEs can be situated at least as far as 60 nts, and perhaps more than 100 nts, from an associated NUE and still function normally (31, 46). No minimum spacing constraint has been determined, but it is probably less than 13 nts, the distance between the proximal UAUUUGUA and the AAUAAA motifs in the CaMV poly(A) signal (46).

The Cleavage/Polyadenylation Site

As is the case in mammals (48) and yeast (1, 12), the actual poly(A) site itself is a *cis* element that plays a role in 3' end formation in plants (11, 27, 31, 46, 53). Deletions, linker substitutions, or point mutations that alter a particular cleavage site change the position of poly(A) (if a suitable context is present within 30 nts of the corresponding NUE). In most cases these mutations do not affect overall poly(A) efficiency. In one instance, a particular cleavage site (CS) was able to function with two different NUEs (31), suggesting that the CS can be considered an independent *cis* element in plant poly(A) signals. Although there is no strictly conserved poly(A) site consensus in plant genes, the average base composition near these sites is distinctive (20). The defining feature for most poly(A) sites in plant genes is probably the occurrence of a YA dinucleotide in a U-rich region.

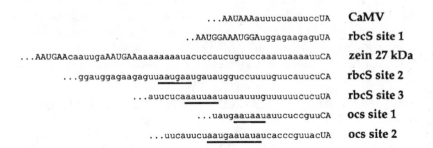

Figure 3   Comparison of the NUEs from the *rbcS*-E9, CaMV, *ocs,* and zein 27-kDa gene poly(A) signals. For the CaMV, *rbcS*-E9 site 1, and zein 27-kDa NUEs, motifs identified by site-directed mutagenesis are shown in capital letters, as are the associated cleavage sites (at the far right of each sequence). For the other four NUEs, the extents of regions defined by linker scanning analysis are shown, their associated cleavage sites noted with capital letters, and possible AAUAAA-like motifs underlined.

There seem to be rigid spatial constraints on the arrangement of NUEs and CSs. In both the *rbcS*-E9 (31) and *ocs* (27) poly(A) signals, mutation of a particular NUE leads to loss of function of a site, indicating that other NUEs located in these signals cannot function in concert with the affected CSs. Moreover, one NUE in the *rbcS*-E9 poly(A) signal is able to function with a different CS, but only if the CS is brought to an appropriate position by deletion (31). Based on known locations of NUEs and CSs in the CaMV, *rbcS*-E9, and *ocs* poly(A) signals, it appears that NUEs and CSs must lie between 20 and 30 nts of each other to permit efficient function. One notable exception is the poly(A) signal from the maize zein 27-kDa gene (53). In this case, the NUE defined by directed mutagenesis lies between 36 and 54 nts upstream from its associated site. The possible significance of this is discussed below.

Downstream Sequences

An unresolved issue is that of the involvement of downstream sequences in polyadenylation in plant genes. Sequences downstream from poly(A) sites are not needed for efficient functioning of the CaMV (11, 45, 46) and 27-kDa zein gene poly(A) signals (53), but may be important for the *rbcS*-E9 (15), *ocs* (27), and wheat histone H3 signals (34). The significance of the wheat histone H3 studies is unclear, since they were conducted in tobacco protoplasts. Because one monocot poly(A) signal has been shown to function inefficiently in tobacco (21), and the wheat histone H3 signal was inefficiently utilized (34), the downstream elements mapped in this gene may not have a function in cells where this signal is utilized efficiently. The *rbcS*-E9 and *ocs* studies involved gross deletion mutagenesis and necessarily involved the removal of several poly(A) sites in their entirety. Given the complex structure of most plant poly(A) signals, the possibility cannot be ruled out that efficient functioning of all sites in a gene requires clusters of FUEs, NUEs, CSs, or combinations of these, and that gross deletion of groups of sites can adversely affect 3′ end formation in the remaining sites. Further work is needed to resolve this issue.

DIFFERENTIAL UTILIZATION OF POLYADENYLATION SIGNALS IN PLANTS

Virtually all plant genes have multiple poly(A) sites, and multiple independent *cis* elements are involved in utilization of these sites. It is thus reasonable to ask whether a subset of the sites of any given gene is utilized preferentially under different growth or environmental conditions. This matter has been addressed for several pea *rbcS* genes in experiments in which expression of these genes was assessed by 3′ end mapping of RNAs (9, 22–24, 26, 33, 39). In these studies, no differential regulation of poly(A) site selection was seen for the pea *rbcS*-3A, -3C, and -E9 genes. The 3' end profiles were the same in light- and dark-grown pea plants; in etiolated and mature pea leaves; in plants treated with blue, red, or far-red light; and in transgenic petunia and tobacco. Moreover, the *rbcS*-3C and -E9 3' end profiles were not affected by the nature of the promoter used to drive expression of test genes containing these 3' regions, nor by the sequence of the reporter gene contained in the vectors. The 3' end profile arising from the *rbcS*-E9 3' region was not affected by leaf age (22). Therefore, in these instances, there seems to be no differential regulation of use of the multiple sites. Of course, whether this conclusion holds true for other plant poly(A) signals remains to be determined.

Williamson et al (52) found that a truncated zein poly(A) signal (in which all downstream sequences had been removed) was not utilized in transgenic tobacco, except during the early stages of seed development. At these stages, the zein poly(A) signal was apparently utilized more efficiently than a downstream nopaline synthase gene poly(A) signal. This conclusion is tentative, because no precise 3' end mapping was done. Nevertheless, there may be a strong preference for the zein poly(A) signal during seed development in tobacco. The observation that seed storage protein genes are much more likely than plant genes in general to possess AAUAAA near their 3' termini supports this hypothesis; 95% of seed storage protein genes have AAUAAA near their 3' termini, compared with 35% of plant genes in general (AG Hunt, unpublished observations). Although these observations are circumstantial, they invite new and potentially fruitful research into the possible roles of NUEs in developmentally-regulated gene expression.

One clear example of differential poly(A) site choice in plants involves the CaMV poly(A) signal. The genome-length RNA (or 35S RNA) that serves as a template for reverse transcriptase is terminally redundant. It has identical 180 nt sequences at its 5' and 3' ends; thus, the same sequences that serve as the poly(A) signal at the 3' end of this RNA are present, but are not processed, at its 5' end. In infected cells, there are small but detectable quantities of the RNA corresponding to cleavage and poly(A) addition at the 5'-proximal poly(A) site (45, 47), but most of the steady-state CaMV-encoded RNAs terminate at the second, downstream poly(A) site. This suggests a control of 3' end processing such that the majority of RNAs initiated at the 35S promoter do not terminate at the 5'-proximal site. One level of control seems to result from the proximity of the poly(A) signal to the 35S promoter. When either the CaMV or nopaline synthase poly(A) signals were moved closer than between 250–420 nts from the transcription initiation site in a gene, these signals functioned inefficiently (45). Another level of control appears to be host specific; the relative proportion of the short-stop RNAs correlated with susceptibility of a particular plant to CaMV (47).

There have been reports suggesting that the 3' regions of plant genes, including poly(A) signals, can have a significant effect on the expression of reporter genes in transient assays in protoplasts and in transgenic plants. Inglebrecht et al (18) noted as great as a 130-fold difference in the activity of the neomycin phosphotransferase II gene in transformed tobacco when the reporter gene was followed by different poly(A) signals. An et al (2) also noted a difference in chloramphenicol acetyltransferase (*cat*) gene expression levels with genes flanked by different poly(A) signals. These effects could be attributed to differences in 3' end formation mediated by the different poly(A) signals or to the relative stabilities of RNAs carrying different reporter genes and 3' regions.

In contrast, Hernandez et al (13) found no such effect of the 3' region on the relative expression levels of the *cat* gene. Others have noted that replacing wild-type with mutant poly(A) signals has little effect on the steady-state levels of mRNAs arising from test genes, because these mRNAs were polyadenylated at weak poly(A) sites situated downstream in the transformation vectors (15, 27, 46). The reasons are unclear for the discrepancies between these reports and those of Inglebrecht et al and An et al. Neighboring plant sequences (or the T-DNA border sequences themselves) might have had an effect on 3' end formation and termination of transcription or on mRNA stability in some contexts but not others. This is just conjecture, however, and the reconciliation between these observations awaits further experimentation.

DIRECTIONS FOR FUTURE RESEARCH

An interesting characteristic of the different *cis* elements involved in 3' end formation in plants is their lack of strict sequence conservation. FUEs, NUEs, and CSs can vary substantially from one gene to another. Indeed, sequences that can probably serve as any of these occur quite frequently in mRNAs, especially in 5'- and 3'-noncoding regions. This raises the issue of fidelity in 3' end choice in plant transcription units. Presumably, a discrete set of poly(A) sites are utilized, and others not, because of required interactions between the different *cis* elements and/or the *trans*-acting factors. The spatial requirements noted above may reflect such interactions. There may also be additional controlling mechanisms that govern 3' end choice in plant genes. There have been at least three reports describing a negative control of cryptic poly(A) site utilization in plants (27, 30, 53). In these cases, deletion of specific portions of 3' regions activated, or unmasked, cryptic sites lying within or outside the 3' region of interest. Detailed mechanistic explanations for these observations are lacking, but they do suggest a level of control beyond the interaction of FUEs, NUEs, and CSs in the generation of mRNA 3' ends.

Keith & Chua (21) have shown that a monocot poly(A) signal is improperly utilized in tobacco, which implies some differences between monocot and dicot poly(A) signals. It is interesting that the zein 27-kDa poly(A) signal lacks a motif (UUGUA) that occurs in three FUEs from dicot genes, a motif known in one instance to be part of a sequence that can confer FUE function (46). Moreover, the NUE-CS spacing in this signal is somewhat different from that seen in dicot signals (see above). These differences may explain the observations of Keith & Chua (21). More analysis of the structure of monocot poly(A) signals, including functional analyses in monocot cells, efficiency studies in heterologous systems, and characterizations of hybrid monocot/dicot signals, is needed to provide further insight into this phenomenon.

Several circumstantial observations suggest a link between mRNA 3' end formation and transcription termination in plants. These include intriguing sequence similarities (discussed above) as well as cases of relief of transcriptional interference in plants using a poly(A) signal (17). The effect of promoter proximity on poly(A) site function may reflect a transcription termination-linked phenomenon (44, 45). Although the link between 3' end formation and termination in plants is tenuous at best, a better understanding of transcription in plants should in turn lead to further insight into mRNA 3' end formation.

There presently is no biochemical rationale for the distinctive organization of plant poly(A) signals. It is unclear whether the occurrence of multiple sites (and their associated signals) reflects inherent properties of the processing apparatus (e.g. does the action of some factor require cooperative interactions) or of other features of mRNA 3' end formation (e.g. a transcription-linked mechanism). It is unclear whether the variation in the sequences that can serve as FUEs and NUEs in plants reflects the degree of tolerance in the sequence variation that a single set of factors can accommodate, or the existence of families of FUE- and NUE-recognizing factors, each member of which may have a discrete, unique sequence preference.

These and other aspects of plant poly(A) signals would be best addressed by in vitro assays. Unfortunately, little has been reported concerning the biochemistry of 3' end formation in plants. A polyadenylate polymerase has been purified to near homogeneity from developing mung bean cotyledons (49), and a poly(A) polymerase from wheat embryos (25) has been identified. Studies involving processing and modification of pre-mRNA precursors have not been reported with these enzymes. Other possible components of the 3' end processing machinery (e.g. RNA binding factors, site-specific nuclease activities) have not been reported. When these and other polyadenylation-related activities have been identified, purified, and the relevant genes cloned, greater insight into the rationale underlying the structure of plant poly(A) signals will be possible.

ACKNOWLEDGMENTS

I am grateful to Helen Rothnie, Thomas Hohn, Joachim Messing, François Guerineau, and Takao Minanikawa for communicating unpublished results for this review. Work done in my lab was supported by grants from the USDA Competitive Grants Program and National Research Initiative.

Literature Cited

1. Abe A, Hiraoka Y, Fusakawa T. 1990. Signal sequence for generation of mRNA 3' end formation in the *Saccharomyces cerevisiae GAL7* gene. *EMBO J.* 9:3691–97
2. An G, Mitra A, Choi HK, Costa MA, An K, et al. 1989. Functional analysis of the 3' control region of the potato wound-inducible proteinase inhibitor II gene. *Plant Cell* 1:115–22
3. Barta A, Sommergruber K, Thompson D, Hartmuth K, Matzke MA, Matzke AJM. 1986. The expression of a nopaline-synthase-human growth hormone chimaeric gene in transformed tobacco and sunflower tissue. *Plant Mol. Biol.* 6:347–57
4. Cao G-J, Sarkar N. 1992. Poly(A) RNA in *Escherichia coli*: nucleotide sequence at the junction of the lpp transcript and the polyadenylate moiety. *Proc. Natl. Acad. Sci. USA* 89:7546–50
5. Carswell S, Alwine JC. 1989. Efficiency of utilization of the simian virus late polyadenylation site: effect of upstream sequences. *Mol. Cell. Biol.* 9:4248–58
6. Dean C, Tamaki S, Dunsmuir P, Favreau M, Katayama C, et al. 1986. mRNA transcripts of several higher plant genes are polyadenylated at multiple sites *in vivo. Nucleic Acids Res.* 14:2229–40
7. DeZazzo JD, Imperiale MJ. 1989. Sequences upstream of AAUAAA influence poly(A) site selection in a complex transcription unit. *Mol. Cell. Biol.* 9:4951–61
8. Dhaese P, De Greve H, Gielen J, Seurinck J, Van Montagu M, Schell J. 1983. Identification of sequences involved in the polyadenylation of higher plant nuclear transcripts using *Agrobacterium* T-DNA genes as models. *EMBO J.* 2:419–26
9. Fluhr R, Chua N-H. 1986. Developmental regulation of two genes encoding ribulose-bisphosphate carboxylase small subunit in pea and transgenic petunia plants: phytochrome response and blue-light induction. *Proc. Natl. Acad. Sci. USA* 83:2358–62
10. Goodall GJ, Filipowicz W. 1989. The AU-rich sequences present in the introns of plant nuclear pre-mRNAs are required for splicing. *Cell* 58:473–83
11. Guerineau F, Brooks L, Mullineaux P. 1991. Effects of deletions in the cauliflower mosaic virus polyadenylation sequence on the choice of polyadenylation sites in tobacco protoplasts. *Mol. Gen. Genet.* 226:141–44
12. Heidmann S, Obermaier B, Vogel K, Domdey H. 1992 Identification of pre-mRNA polyadenylation sites in *Saccharomyces cerevisiae. Mol. Cell. Biol.* 12: 4215–29
13. Hernandez G, Cannon F, Cannon M. 1989. The effect of presumptive polyadenylation signals on the expression of the CAT gene in transgenic tobacco. *Plant Cell Rep.* 8:195–98
14. Hernandez-Lucas C, Royo J, Paz-Ares J, Ponz F, Garcia-Olmeds F, Carbonero P. 1986. Polyadenylation site heterogeneity in mRNA encoding the precursor of the barley toxin β-hordothionin. *FEBS Lett.* 200:103–6
15. Hunt AG, MacDonald M. 1989. Deletion analysis of the polyadenylation signal of a pea ribulose-1,5-bisphosphate carboxylase small subunit gene. *Plant Mol. Biol.* 13: 125–38
16. Hunt AG, Chu NM, Odell JT, Nagy F, Chua N-H. 1987. Plant cells do not properly recognize animal gene polyadenylation signals. *Plant Mol. Biol.* 7:23–35
17. Ingelbrecht I, Breyne P, Vancompernolle K, Jacobs A, Van Montagu M, Depicker A. 1991. Transcriptional interference in transgenic plants. *Gene* 109:239–42
18. Ingelbrecht ILW, Herman LMF, Dekeyser RA, Van Montagu MC, Depicker AG. 1989. Different 3' end regions strongly influence the level of gene expression in plant cells. *Plant Cell* 1:671–80
19. Irniger S, Sanfaçon H, Egli CM, Braus GH. 1992. Different sequence elements are required for function of the cauliflower mo-

saic virus polyadenylation site in *Saccharomyces cerevisiae* compared with plants. *Mol. Cell. Biol.* 12:2322–30

20. Joshi CP. 1987. Putative polyadenylation signals in nuclear genes of higher plants: a compilation and analysis. *Nucleic Acids Res.* 15:9627–40

21. Keith B, Chua N-H. 1986. Monocot and dicot pre-mRNAs are processed with different efficiencies in transgenic tobacco. *EMBO J.* 5:2419–25

22. Kuhlemeier C, Cuozzo M, Green PJ, Goyvaerts E, Ward K, Chua N-H. 1988. Localization and conditional redundancy of regulatory elements in rbcS-3A, a pea gene encoding the small subunit of ribulose-bisphosphate carboxylase. *Proc. Natl. Acad. Sci. USA* 85:4662–66

23. Kuhlemeier C, Fluhr R, Chua N-H. 1988. Upstream sequences determine the difference in transcript abundance of pea rbcS genes. *Mol. Gen. Genet.* 212:405–11

24. Kuhlemeier C, Fluhr R, Green PJ, Chua N-H. 1987. Sequences in the pea rbcS-3A gene have homology to constitutive mammalian enhancers but function as negative regulatory elements. *Genes Dev.* 1:247–55

25. Lakhani S, Kapoor R, Verma N, Sachar RC. 1989. Evidence for the *de novo* synthesis of poly(A) polymerase in germinated wheat embryos. *Phytochemistry* 28:1031–35

26. Lam E, Chua N-H. 1990. GT-1 binding site confers light responsive expression in transgenic tobacco. *Science* 248:471–74

27. MacDonald MH, Mogen BD, Hunt AG. 1991. Characterization of the polyadenylation signal of the T-DNA-encoded octopine synthase gene. *Nucleic Acids Res.* 19: 5575–81

28. Martinez-Zapater JM, Finkelstein R, Somerville CR. 1988. *Drosophila* P-element transcripts are incorrectly processed in tobacco. *Plant Mol. Biol.* 11:601–7

29. Messing J, Geraghty D, Heidecker H, Hu NT, Kridl J, Rubenstein I. 1983. Plant gene structure. In *Genetic Engineering of Plants: An Agricultural Perspective*, ed. T Kosuge, P Meredith, A Hollaender, pp. 211–27. New York: Plenum

30. Mogen BD, MacDonald MH, Graybosch R, Hunt AG. 1990. Upstream sequences other than AAUAAA are required for efficient messenger RNA 3′ end formation in plants. *Plant Cell* 2:1261–72

31. Mogen BD, MacDonald MH, Leggewie G, Hunt AG. 1992. Several distinct types of sequence elements are required for efficient mRNA 3′ end formation in a pea *rbcS* gene. *Mol. Cell. Biol.* 12:5406–14

32. Montoliu L, Rigau J, Puigdomenech P. 1990. Multiple polyadenylation sites are active in the α1-tubulin gene from *Zea mays*. *FEBS Lett.* 277:29–32

33. Nagy F, Morelli G, Fraley RT, Rogers SG,

Chua N-H. 1985. Photoregulated expression of a pea rbcS gene in leaves of transgenic plants. *EMBO J.* 4:3063–68

34. Nakayama T, Ohtsubo N, Mikami K, Kawata T, Tabata T, et al. 1989. Cis-acting sequences that modulate transcription of wheat histone H3 gene and 3′ processing of H3 premature mRNA. *Plant Cell Physiol.* 30:825–32

35. Ojala D, Montoya J, Attardi G. 1981. tRNA punctuation model of RNA processing in human mitochondria. *Nature* 290:470–74

36. Osbourne BI, Guarente L. 1988. Transcription by RNA polymerase II induces changes in DNA topology in yeast. *Genes Dev.* 2:766–72

37. Osbourne BI, Guarente L. 1989. Mutational analysis of a yeast transcriptional terminator. *Proc. Natl. Acad. Sci. USA* 86: 4097–4101

38. Platt T. 1986. Transcription termination and the regulation of gene expression. *Annu. Rev. Biochem.* 55:339–72

39. Poulsen C, Fluhr R, Kauffman JM, Boutry M, Chua N-H. 1986. Characterization of an *rbcS* gene from *Nicotiana plumbaginifolia* and expression of an *rbcS*-CAT chimeric gene in homologous and heterologous nuclear backgrounds. *Mol. Gen. Genet.* 205: 193–200

40. Proudfoot NJ. 1991. Poly[A] signals. *Cell* 64:671–74

41. Russnak R, Ganem D. 1990. Sequences 5′ to the polyadenylation signal mediate differential poly[A] site use in hepatitis B viruses. *Genes Dev.* 4:764–76

42. Russo P, Li W-Z, Hampsey DM, Zaret KS, Sherman F. 1991. Distinct *cis*-acting signals enhance 3′ endpoint formation of CYC1 mRNA in the yeast *Saccharomyces cerevisiae*. *EMBO J.* 10:563–71

43. Russo P, Sherman F. 1989. Transcription terminates near the poly(A) site in the *CYC1* gene of the yeast *Saccharomyces cerevisiae*. *Proc. Natl. Acad. Sci. USA* 86: 8348–52

44. Sanfaçon H. 1992. Regulation of mRNA formation in plants: lessons from the cauliflower mosaic virus transcription signals. *Can. J. Bot.* 70:885–99

45. Sanfaçon H, Hohn T. 1990. Proximity to the promoter inhibits recognition of cauliflower mosaic virus polyadenylation signal. *Nature* 346:81–84

46. Sanfaçon H, Brodmann P, Hohn T. 1991. A dissection of the cauliflower mosaic virus polyadenylation signal. *Genes Dev.* 5:141–49

47. Sanfaçon H, Wieczorek A. 1992. Analysis of cauliflower mosaic virus RNAs in Brassica species showing a range of susceptibility to infection. *Virology* 190:30–39

48. Sheets MD, Ogg SC, Wickens MP. 1990. Point mutations in AAUAAA and the

poly(A) addition site: effects on the accuracy and efficiency of cleavage and polyadenylation in vitro. *Nucleic Acids Res.* 18:5799–805

49. Tarui Y, Minamikawa T. 1989. Poly(A) polymerase from *Vigna unguicu lata* seedlings: a bifunctional enzyme responsible for both poly(A)-polymerizing and poly-(A)-hydrolyzing activities. *Eur. J. Biochem.* 186:591–96

50. Valsamakis A, Zeichner S, Carswell S, Alwine JC. 1991. The human immunodeficiency virus type I polyadenylation signal: a 3′ long terminal repeat element upstream from the AAUAAA necessary for efficient polyadenylation. *Proc. Natl. Acad. Sci. USA* 88:2108–12

51. Wahle E, Keller W. 1992. The biochemistry of 3′-end cleavage and polyadenylation of messenger RNA precursors. *Annu. Rev. Biochem.* 61:419–40

52. Williamson JD, Hirsch-Wyncott ME, Larkins BA, Gelvin SB. 1989. Differential accumulation of a transcript driven by the CaMV 35S promoter in transgenic tobacco. *Plant Physiol.* 90:1570–76

53. Wu L, Ueda T, Messing J. 1993. 3′-end processing of the maize 27 kDa zein mRNA. *Plant J.* In press

Annu. Rev. Plant Physiol. Plant Mol. Biol. 1994. 45:61–78

THE PLANT MITOCHONDRIAL GENOME: Physical Structure, Information Content, RNA Editing, and Gene Migration to the Nucleus

Wolfgang Schuster and Axel Brennicke

Institut für Genbiologische Forschung, Ihnestraße 63, D-14195 Berlin, Germany

KEY WORDS: Angiosperms, gene expression, gene transfer, genome structure, plant mitochondria, RNA editing

CONTENTS

INTRODUCTION 61
PHYSICAL STRUCTURE OF MITOCHONDRIAL GENOMES 62
INFORMATON CONTENT OF PLANT MITOCHONDRIAL GENOMES 63
   Introns—Origin and Fate 64
   Genes Encoded in Plant Mitochondrial Genomes 65
GENE EXPRESSION 67
   Transcription Initiation 67
   Cis- and trans-*splicing* 68
   RNA Editing 69
INTERORGANELLAR FLUX OF NUCLEIC ACIDS INFORMATION EXCHANGE 71
   Import of Genetic Information into Mitochondria 71
   Export of Genetic Information from Mitochondria to the Nucleus 72
CONCLUDING REMARKS 73

INTRODUCTION

Renewed interest in the molecular biology of plant mitochondria has generated new information about this compartment and is reflected in the decreasing intervals between reviews on the topic in this series. The first comprehensive review, published in 1982 (61), was updated after six years in 1988 (76); the third review appeared only four years later in 1992 (39); and the topic is now being covered just two years later.

With further expansion of the field, however, a review requires a more specialized focus, with some topics necessarily excluded. This review focuses on the information content of plant mitochondrial genomes, which has been advanced greatly by the recent determination of the entire genomic sequence of the mitochondrial DNA of the liverwort *Marchantia polymorpha* (80). Regarding expression of the mitochondrial information, emphasis is put on novel insights into the origin and splicing of introns and the phenomenon of RNA editing, a recently discovered and still baffling phenomenon. This review also covers gene transfer to the nucleus and how mitochondria-specific requirements of gene expression are altered after the transposition process.

Most of the topics in this review have also been covered recently elsewhere: RNA editing (16, 18, 38, 39, 90, 101, 116), mitochondrial genome structure (6, 69), transfer RNAs (73), and the transfer of genetic information to the nucleus (19, 103). Other aspects such as plant mitochondrial biogenesis (29), evolution (36, 37), cytoplasmic male sterility (64, 65, 91), and biochemistry (28) have also been reviewed recently. An overview of plant mitochondria was the subject of a recent collective publication (21).

PHYSICAL STRUCTURE OF MITOCHONDRIAL GENOMES

The in vivo structure of mitochondrial genomes has been comprehensively summarized by Bendich (6). Earlier reviews have covered the derivation of circular structures from physical mapping of overlapping restriction fragments; from these maps circular genome structures were predicted for the majority of mitochondrial genomes in eukaryotic cells (5, 69, 93). Exceptions are found as genome-sized linear molecules in the alga *Chlamydomonas reinhardtii* and the protozoa *Tetrahymena pyriformis* and *Paramecium aurelia* (see references in 6).

The small, circular DNA molecules of animal cells of about 16 kb (1, 9) have been visualized by electron microscopy and can be isolated as supercoiled molecule populations by CsCl density gradient centrifugation. Extrapolating from these molecules and the overlapping restriction fragment maps, analogous circles have been postulated for fungal and plant mitochondrial genomes (69, 85, 93; see references in 6). The in vivo structure of the larger mitochondrial DNAs of yeast and plants, however, may be different from the structures predicted by fragment mapping. Pulse field gel electrophoresis shows one fraction of molecules to be locked into highly complex arrangements that migrate with an apparent size of more than one megabase. Most fungal and plant mitochondrial DNA, however, behaves like linear molecules ranging from 50 to 100 kb in a heterogeneous population visible as a smear in the gels (7). There is also some mitochondrial DNA that does not enter pulsed-field gels. The conformation of this apparently large DNA remains unclear: it may represent relaxed circles, replication and/or recombination intermediates, or DNA somehow attached to membrane or matrix fragments.

From these observations, linear DNA molecules have been proposed as the major in vivo form of mitochondrial genomes in fungi and higher plants (6). This interpretation could explain the difficulties in identifying large circular molecules (which were at the most about 100 kb and hence smaller than the physical maps) from higher plant mitochondria by electron microscopy. In cultured plant cells circular molecules are more common, but most of these circles are subgenomic pieces of the main mitochondrial genome ranging in size from 6 kb to several hundred kb (2, 66). In *Oenothera berteriana* mito-chondria the physical presence of such molecules has been verified (20). These subgenomic circles are proposed to arise from recombination sites, defined as small nucleotide repeats of various sequences (72). The circularizations seem to occur without respect to the units of genetic information and have, for example, been found to disrupt coding regions (72, 117). Highly asymmetric concentrations of divergent genome arrangements were first found in maize, where substoichiometric amounts of variant DNA molecules have been found at one or fewer copies per cell (108, 109).

In contrast to the heterogeneity typical of most mitochondrial DNA (mtDNA), there is no detectable internal recombination in *Brassica hirta* (84) and *M. polymorpha* (79). *M. polymorpha* mtDNA was found to consist of mainly circular molecules of the size (186 kb) predicted from physical map-ping and complete sequence analysis (79, 80). These molecules have also been verified by electron microscopy, confirming the in vivo presence of such circular molecules in mitochondria (80). It is still unclear why *M. polymorpha* mtDNA is so differently organized in its physical structure compared to higher plants or fungi. Effects of DNA anchoring at membrane integral proteins and specific mechanisms of replication such as those leading to the catenated network of trypanosome mtDNA (88, 105) may be involved in determining the variation in linear and circular structures in mitochondria.

INFORMATION CONTENT OF PLANT MITOCHONDRIAL GENOMES

The complete sequence analysis of the *M. polymorpha* mitochondrial genome (80) was the major recent advance in our understanding of the information content of plant mitochondrial genomes. In particular, there are many more genes in plant mtDNA than in mammalian mtDNA. There are 94 substantial open reading frames in *M. polymorpha* mtDNA, which is about seven times the number in animal mtDNA (13 open reading frames). Gray et al's argument (39) that a large proportion of the higher plant mitochondrial genomes are noncoding sequences, which in effect distances the genes from each other, is also confirmed.

The residual size difference between the highly compact mammalian mtDNA and *M. polymorpha* mtDNA is mostly the result of introns and duplicated sequences in the liverwort (80, 81). This makes the latter mtDNA look like a nuclear genome with a considerable amount of junk sequences that apparently are non-functional. Many of these sequences are duplications of other regions, ranging from a few up to several hundred nucleotides. Both coding and noncoding regions are present in these repeats. In higher plant mitochondria, some of these duplications can be several kb long and can participate in homologous recombination by as yet unclear mechanisms (117). These recombinations between direct or inverted repeats result in the complex molecule arrays observed in numerous higher plant species.

The recombinations and sequence amplifications are mostly responsible for the size differences between liverwort and higher plant genomes, which differ by factors of between 1 and 15 (115). The other contributions to the genome expansion in higher plants are integrated chloroplast and nuclear sequences and still larger spacers of unknown (possibly nuclear) origin between individual genes. The number of genes seems to be comparable. Only the *nad7* locus and several species-specific cytoplasmic male sterility (cms)–related reading frames, which have been identified in higher plants, are not present in *M. polymorpha* mtDNA.

Introns—Origin and Fate

The mitochondrial genome of *M. polymorpha* contains 32 identified introns, contributing about 20% of the genome complexity (80). These introns have been classified as 7 members of group I and 25 members of group II. Six of the group I introns are located in the *coxI* gene, coding for subunit I of the cytochrome oxidase. Four of these introns and one of the group II introns in this gene are located at positions identical to the respective fungal introns (81, 83). This observation has been interpreted as evidence for vertical transmission of these introns from common ancestors.

In *coxI* genes of higher plants these introns are absent. All the higher plant *coxI* genes identified to date are continuous coding regions, indicating that these intervening sequences have been lost during the evolution of Angiosperm species or that they have been gained in *M. polymorpha*. Loss of introns probably occurs via reverse transcription of mature transcripts and their reinsertion. In *M. polymorpha*, a duplicated mitochondrial genomic sequence has been identified for the *cob* gene coding for cytochrome *b*, which has lost the introns present in the intact copy of this gene (80). The first *coxII* intron in higher plants (31) has been lost in the Rosidae lineage, exemplified by pea and *O. berteriana* (48) and independently in a pseudogene in *Petunia hybrida* (120). The ancestral *nad4* gene of Angiosperms had three introns, two of which were lost in specific lineages of dicots such as lettuce (33).

Introns are probably gained through duplication of existing introns and their integration into other sites. This process is deduced from the similarity between introns of different genes, for example, between the second intron of *coxI* and the first intron of *coxIII* and between the second intron of *coxIII* and *rrn26* in *M. polymorpha* mitochondria (80, 81, 83). In higher plants, such similarities have been found between the second intron of *nad1* and the first intron of *nad2* (67, 119). In this instance, similarity between adjacent exon sequences further supports an ancient duplication event into a compatible site and thus a common origin of these intervening sequences.

No group I introns have been found in Angiosperm mitochondria; they may be completely absent from these genomes. The number of introns in higher plant mitochondria is unclear, but probably does not exceed the number of introns identified in *M. polymorpha* because many of the known genes in Angiosperm mitochondria are continuous and do not contain intervening sequences.

Genes Encoded in Plant Mitochondrial Genomes

M. polymorpha and Angiosperm mtDNA contain genes for three ribosomal RNAs (rRNAs), with the 18S-5S rRNA genes linked in all species. In addition to these genes for structural ribosomal RNAs, 16 genes for ribosomal proteins have been identified in *M. polymorpha* (80). Ten of these have also been found in Angiosperm mitochondria, although the presence of individual genes varies between different plant species (Table 1 and detailed below). The organization of the ribosomal protein genes in *M. polymorpha* shows significant similarity to the respective bacterial cistrons, further evidence for the prokaryotic origin of mitochondria. In higher plants, however, only parts of these operons are conserved because the clusters, and sometimes even the genes, have been disrupted by frequent recombination events. The remaining intact ribosomal protein genes in these higher plants are now scattered throughout the genome. Some are organized in transcription units with other protein-coding genes. The functional importance of ribosomal protein genes becomes apparent in the lethal phenotype of mutants, in which some of these genes are disrupted (30, 54).

The number of transfer RNAs (tRNAs) encoded in plant mitochondria also differs between plant species. It is never sufficient to sustain a complete decoding system, although the set of genuine mitochondrial tRNA genes is supplemented by genes of chloroplast origin integrated into the mtDNA (55,

Table 1   Genes identified in plant mitochondrial genomes[a]

Translational apparatus	
ribosomal RNA	*rrn5, rrn18, rrn26*
ribosomal protein[b]	
small subunit	*rps1, rps3, rps7, rps12, rsp13 rps14, rps19*
large subunit	*rp12, rp15, rp116*
transfer RNA	at least 16
Subunits of respiratory chain complexes	
NADH-dehydrogenase	*nad1, nad2, nad3, nad4, nad4l, nad5, nad6, nad7, nad9*
cytochrome *b*	*cob*
cytochrome *c* oxidase	*coxI, coxII, coxIII*
ATP synthase	*atp1, atp6, atp9*
Cytochrome *c* biogenesis	at least 4 genes[c]
Conserved open reading frames[d]	at least 10 known

[a]The actual gene content of any plant species may vary; one or more mitochondrial genes may be transferred to the nucleus. [b] In *Marchantia polymorpha*, 16 genes for ribosomal proteins have been detected and at least a similar number is expected for higher plants. [c] Nomenclature of these genes is still to be decided. [d] Conservation of these reading frames between different plants indicates their importance, although the functions of the encoded proteins are still unclear.

73). Thus additional tRNAs have to be imported into plant mitochondria from the cytoplasm (74; see also 73).

Almost all the genes for subunits of protein complexes in the respiratory chain encoded by the animal mitochondrial genome (1, 9), with the possible exception of *atp8*, have now been identified in the mitochondrial genomes of higher plants and *M. polymorpha*. These genes code for seven subunits of complex I (*nad1-nad6* and *nad4L*), cytochrome *b*, three subunits of complex IV (*coxI-coxIII*), and subunits 6 and 9 of the ATPase (*atp6* and *atp9*; Table 1). In addition to these, the alpha subunit of the ATPase and several subunits of complex I, the NADH:ubiquinone oxidoreductase, are specified by the plant mitochondrial genomes. The *nad7* gene (17), which is also present in *P. aurelia* (92) and trypanosome (60) mtDNAs, encodes a subunit of the peripheral part of the complex, while all other mitochondrially located genes code for hydrophobic proteins in the membrane arm.

Successful purification of the NADH:ubiquinone-oxidoreductase (complex I) has recently been reported for broad bean (63), potato (46), and beetroot (96). NH_2-terminal sequence has been obtained for several subunits, allowing identification of the corresponding genes. The polypeptide sequence of the NAD7 subunit indicated that NH_2-terminal processing occurs, removing either a longer, nonconserved presequence or only the first methionine, depending on which AUG start codon is used for translation initiation (32).

A completely new set of mitochondrial genes has been identified by their similarity to bacterial reading frames that are essential for cytochrome *c* biogenesis (35, 100). The organization as two gene clusters is partially conserved in *M. polymorpha* (80), but like the ribosomal proteins, is completely rearranged and dispersed in Angiosperm mitochondria. Although some cytochrome *c* biogenesis proteins probably form a heme transporter, the functions of the other proteins are unclear. The exact properties of individual proteins involved in cytochrome *c* biogenesis will probably be defined by analysis in bacteria and extrapolation to the plant mitochondrial system.

One open reading frame conserved between higher plants and *M. polymorpha* shows significant similarity with two overlapping open reading frames in *E. coli* (113). The biochemical function of these products is unclear, but the high similarity indicates an essential, evolutionary conserved function for the encoded proteins.

About 30 additional open reading frames have been identified in the *M. polymorpha* mitochondrial genome, most of which probably encode functional proteins (80). Some of these are also present in Angiosperms among the genes in plant mtDNA for which no function or homologs can be assigned. For example, the *orfB* gene, first identified in *O. berteriana* (50) as an open reading frame cotranscribed with the *coxIII* gene, is conserved throughout the plant kingdom and the presumptive protein product has been identified in wheat (41). The present task is to define the functions of each protein-encoding gene to determine the information contained in plant mitochondrial genomes.

GENE EXPRESSION

Gray et al (39) provided an excellent summary of transcription and processing in plant mitochondria, detailing the progress made in the analysis of promoter identity and efficiency, in the processing of tRNAs in in vitro assays, and the phenomenon of RNA editing. We now focus on the determination of transcription initiation sites in dicotyledoneous plants, the identification of additional *trans*-splicing genes, and the progress made in analysis of the RNA editing process.

Transcription Initiation

Since the in vivo and in vitro identification of promoter consensus sequences in the monocot plants wheat and maize (44, 95), promoter consensus sequences have also been identified in dicots (11, 22), notably in *O. berteriana* (11). These derived consensus motifs appear to differ significantly between the various plant lineages, indicating the flexibility of promoter primary structures in plant mitochondria.

The minimal number of promoters per plant mitochondrial genome has been estimated by counting the number of transcripts with primary 5' ends. For example, 18 transcription initiation sites were identified in *O. berteriana* using the capping assay (11). The number of promoters in plant mitochondria assessed by this method exceeds those of fungi. Nevertheless, it is clear that the majority of the approximately 100 protein-coding genes of plant mtDNA are probably transcribed as multicistronic RNAs (71).

In most cases, similar promoter motifs are observed for mRNAs, rRNAs, and tRNAs, indicating that these promoters are interchangeable and not restricted to certain types of genes. The identification of a tRNA promoter similar to the mRNA transcription initiation sites (10) indicates that tRNAs can be synthesized by the same RNA polymerase that transcribes longer precursors. Several of the identified mRNA transcription initiation sites, however, show no similarity with this consensus and may thus be transcribed by different factors and/or RNA polymerases. Two different types of RNA polymerases operate in chloroplasts although their nature and origin are unclear (47, 62).

The faithful in vitro transcription initiation system developed initially in wheat mitochondria (44) has been established successfully for maize mitochondria (95). Promoter dissection by selected deletions of upstream sequences and by in vitro mutagenesis of the maize *atp1* transcription initiation site has improved our understanding of plant mitochondrial promoters. In particular, the difference between plant promoters and those in other kingdoms are more clearly underscored (94, 95). Thorough in vitro linker scanning and point mutational analysis of the *atp1* promoter from maize indicated wide tolerance of nucleotide substitutions in this promoter. Two crucial regions were identified in this study: a central domain between nucleotides −7 and +5 (5'-TAACGTATTAAA-3') and an upstream domain around nucleotides −11/−12 (consisting of a run of As relative to the first transcribed nucleotide) (94). The core nucleotide positions may be crucial for the basic activity, while the more complex array of upstream sequences appears to be involved in modulating the activity. These analyses have shown that nucleotide context rather than individual identities modulate the tuning of promoter activity in plant mitochondria.

Cis- *and* trans-*splicing*

Several group II introns in plant mitochondrial genes have been split by recombination. Thus, adjacent exons have been separated from their former *cis*-arrangements and are transcribed into separate RNA molecules, which now have to be connected by a process called *trans*-splicing.

In addition to the multiple *trans*-spliced genes *nad1* and *nad5*, *nad2* requires one *trans*-splicing event. The arrangements of *nad1* (24, 118), *nad5* (59, 87) and *nad2* (13, 86) are conserved in both monocots and dicots, indicating that the genomic interruption of these genes occurred before the split of the two lineages. The differences in *cis*- and *trans*-arrangements of the last exon of *nad1* in individual plant species demonstrates how comparatively recent changes in splicing patterns can be established genomically. The requirements of the *trans*-splicing fidelity should be analyzed in vitro. Of particular interest is the role of the base pairing in the interrupted stem of domain IV. In the *trans*-splicing intron of *nad2*, two RNA editing events have been observed that improve base-pairing in this stem (13, 67). Such editing events indicate an important functional role of these base-pairings in the stem of domain IV, but will require rigorous experimental support for verification.

RNA Editing

RNA editing in plant mitochondria involves post-transcriptional nucleotide identity switches, in most instances from C to U, but occasionally from U to C (25, 42, 43, 52, 102). The phenomenon has been reviewed recently (18, 38, 39, 101, 116) and we concentrate here on several novel deductions.

DEDUCTIONS ON THE BIOCHEMISTRY AND SPECIFICITY OF THE PROCESS
The comparative analysis of numerous genomic and cDNA sequences has resulted in the detection of several hundred editing sites in some species, notably in *O. berteriana* (116) and wheat (18). The total number of editing sites per genome is extrapolated to exceed 1000. For each edit, there must be one unique aspect to maintain precision and fidelity. The increase in the number of identified editing sites has not, however, yielded further information about the specificity determinants of RNA editing in plant mitochondria. About the only result from statistical analysis is that edited nucleotides are rarely preceded by Gs (25, 90, 99). The most favored model of specificity determinants involves *trans*-acting guide sequences analogous to those involved in trypanosome mitochondria (72, 14, 106, 107, 112), but experimental evidence is lacking. Small RNA molecules can be isolated from plant mitochondria, associated specifically with mRNA molecules of edited coding regions. The potential involvement of these small RNAs in editing awaits further work (W Schuster, unpublished information).

In this context, observations of RNA editing in tRNAs are intriguing, because these sites have been observed only in base-pairing stems, although perfect base-pairing matches are not restored in all cases (72a; S Binder & A Brennicke, manuscript in preparation). Some of the editing events found in intron sequences are likewise located in *cis*-pairing stem regions and restore A-U pairings from A:C mismatches (13, 67). These instances could indicate that nucleotides to be edited are identified at the mismatch with the other pairing strand acting as a *cis*-guiding molecule.

The RNA editing events observed in intron sequences appear in some cases to be required for correct folding of the secondary structure of the intron (13, 67, 118). They may thus have regulatory functions in establishing translatable pools of mRNAs similar to editing events that create ATG initiation codons (24).

The development of in vitro RNA editing systems is still the pressing technical problem from which clarification of the underlying biochemical process of these reactions is expected. The difficulties encountered in adapting the in vitro editing system developed for wheat mitochondria (4) to other plant species indicate loose yet complex interactions for activity. Efforts have been made to purify the participating components (3), which will allow better characterization of the enzymatic activities.

RNA EDITING MAKES FUNCTIONAL PROTEINS   Comparisons of the polypeptide sequences predicted from edited and unedited nucleotide sequences indicate that editing is required for the synthesis of functionally competent protein products. This has been tested experimentally by introducing edited and unedited *atp9* and *cob* genes with appropriate target sequences into the nuclear genome of tobacco (45). While plants with the edited proteins grew normally and exhibited fertile pollen, the unedited genes resulted in sterile pollen. The other tissues apparently were not affected, because the plants were phenotypically normal. These results indicate that—as deduced from analyses of other types of cms—normal pollen development is disturbed fatally by the presence of deviant polypeptides (45).

EVOLUTION OF PLANT MITOCHONDRIAL RNA EDITING   RNA editing in plant mitochondria initially was reported for representatives of both monocots and dicots (25, 42, 52). RNA editing does not appear to be required in *M. polymorpha* mitochondria, and an investigation has shown consequent colinearity of genomic and cDNAs in this plant (83). Analysis of *coxI* coding sequences in the Gymnosperm *Thuja plicata* (34) indicates that RNA editing occurs in this line of land plants, but our knowledge of RNA editing is still limited to seed plants.

A systematic global survey for the occurrence of mitochondrial RNA editing in different lineages of land plants has yielded a more detailed picture of the distribution of this process (49, 51). RNA editing seems to be absent from all Bryophytes, including the liverworts and the leafy mosses. Extensive RNA editing was found in all Pteridophytes and all other land plants investigated including *Lycopodium squarrosum, Ginkgo biloba, Cycas revoluta, Psilotum nudum,* and *Equisetum arvense* (49).

Because no RNA editing was detected in green algae of the Charophyceae lineage, the emergence (or loss) of RNA editing can now be timed to the separation of the Bryophytes and the other land plants. The absence of RNA editing from green algae indicates a de novo introduction and evolution of this process after plants became established on the land. Otherwise, two independent losses of this process must be postulated in algae and Bryophytes—unless present systematic genealogies are revised and the mosses are found to have originated from a separate group of terrestrial plants (51).

Observations in true ferns indicate that mitochondrial RNA editing evolved differently in the individual plant lines. In the Pteridophytes, the U to C transitions seem to be more frequent than in Gymnosperms or Angiosperms, and are required to eliminate translational stop codons (49). The editing process in these ferns may be different to allow this reverse reaction.

It is unclear whether the evolution of mitochondrial RNA editing is connected to the evolution of chloroplast RNA editing (53), which is observed in Angiosperms, but appears to be absent from *M. polymorpha* (82). The similarity of the observed alterations in both organelles indicates a common origin and possibly even common factors involved in the biochemical reactions.

INTERORGANELLAR FLUX OF NUCLEIC ACIDS INFORMATION EXCHANGE

The exchange of nucleic acids between different organelles of plant cells has been found on two different time scales. The continuous import of functional tRNAs from the cytoplasm into mitochondria occurs for several tRNA species (74). This subject was reviewed recently (73). The transfer of genetic information on a more evolutionary time scale can ultimately result in the loss of a gene from one compartment and its gain in another organelle. Several examples of such translocations recently identified in plants provide insights into the pathway of this process.

Import of Genetic Information into Mitochondria

Chloroplast sequences have been identified in the mitochondrial genomes of several higher plants (97, 98, 110, 111). Several plastid derived tRNA genes are now essential genes in plant mitochondria (55). The complete sequence of the *M. polymorpha* mitochondrial genome showed that no chloroplast sequences have been integrated into the mtDNA of this plant (80). This absence of integrated sequences may be related to the apparent lack of active recombination in this genome, to some constraint on DNA size, to rather separate organellar compartments, between which little movement of nucleic acid sequences is possible, or to the lack of some enzymatic activities required for the transfer/integration process. Nuclear sequences have been identified in the mitochondrial genome of only one plant, *O. berteriana* (97). Although such events are probably more frequent, they are difficult to confirm.

Export of Genetic Information from Mitochondria to the Nucleus

Movement of genetic information from mitochondria to the nucleus has now been documented for several higher plants. Although some of these involve pieces of only a few hundred nucleotides of noncoding regions, other events have transferred entire genes.

With respect to small sequence fragments, the transfer of intron sequences has been surveyed (56, 57, 58). Several mitochondrial intron fragments were found near nuclear genes in adjacent upstream or downstream regions. In tobacco, such mitochondrial intron pieces were identified near basic and acid glucanase genes, although a functional role for these sequences remains unlikely (57). The intron fragment inserted upstream of a lectin gene in an African bean indicates that the mitochondrial intron fragment may now play a role in conferring tissue-specific expression for this gene in seeds (56). This example shows how a duplicated sequence can by chance specify functions that are completely alien to the original context of the respective sequence. The potential influences of other transferred mitochondrial intron segments are unclear.

Several examples of gene transfer to the nucleus have been documented in the last few years. The most extensively analyzed functional gene transfer has occurred in the legume family, where the *coxII* gene has been moved to the nuclear compartment and different species of legumes display intermediate stages of the transfer (26, 77, 78). An active mitochondrial *coxII* gene is found only in pea. There is a complete, but inactive gene in soybean mitochondria; a nuclear gene copy provides the protein product for the mitochondria in this plant (26). The cowpea mitochondrial gene is completely lost and the essential COXII protein is synthesized from a nuclear gene (77). These different stages of activation of a new nuclear gene are probably derived from a single sequence translocation in one ancestral species during the evolutionary history of the legume family.

In an unrelated gene transfer event, the *rps12* gene of *O. berteriana,* a fragment of this mitochondrial gene has remained in the organelle (40). A complete gene for this mitochondrial ribosomal protein is encoded in the nuclear genome. The mitochondrial target sequence directly extends the open reading frame as a presequence, while in the legumes, an intron separates the target sequence of the *coxII* reading frame from the actual coding region.

Several other gene transfers are expected by analogy, where only gene fragments, or disrupted or untranscribed reading frames are observed in the mitochondrial genomes of different species. These include *nad7* in *M. poly-*

morpha (80); *rps13, rps19* and *rps14* in *Arabidopsis thaliana* (W Schuster, M Unseld, P Brandt, unpublished information); *coxIII* in *Selaginella elegans* (51); *rps19* in *O. berteriana* (W Schuster, unpublished information); and *rps13* in wheat (15).

Activation and correct expression of such integrated mitochondrial genes requires several changes (103). The nuclear sequence has to correspond to the edited mitochondrial sequence, and any introns originating from the mitochondrial compartment must be eliminated to provide a functional protein product. This can be accomplished most easily by starting from mature mRNA molecules in the organelle, where splicing and editing have been completed. Upon integration of the reverse transcribed nucleic acid in the nuclear genome, appropriate presequences must be added to the open reading frame to target the protein product back into the organelle. Correct expression signals for transcription and translation also are required to provide appropriate quantitative regulation. The flexibility of signals at the RNA or protein levels indicates that these signals can be added and evolve by chance to account for the observed examples.

CONCLUDING REMARKS

Recent advances in the molecular biology of plant mitochondria have yielded some unpredicted insights. The recognition of a set of new genes involved in cytochrome *c* biogenesis may be an indicator of further surprises to come concerning plant mitochondrial genes.

The assays now available for in vitro analysis of promoters, together with tissue-specific analyses of mitochondrial gene expression (114), will allow definition of plant mitochondrial gene expression regulation and its interaction with the other cellular compartments.

Dissection of the biochemical and specificity mechanisms of RNA editing has been slower than hoped for and still awaits the extensive exploitation of faithful in vitro systems. The presence of RNA editing in almost all land plants and in chloroplasts, but not in algae and mosses, casts new credence to the proposal that this process evolved from preexisting enzymatic activities rather than being a remnant of the RNA world that was inherited through evolution and modified in different lineages (27). RNA editing is widespread. It is found in trypanosomes (72, 107), measles virus (23), the mammalian nuclear gene for apolipoprotein B (89, 104), *Physarum polycephalum* (70, 75), *Acanthamoeba castellanii* (68), and plants. Although there are similar phenotypic consequences of editing and a common terminology, evolutionary and biochemical processes are distinct in each lineage.

Observations of intermediate stages of sequence transfers from the mitochondrion to the nucleus may provide insight into the endosymbiotic relationship and into mechanisms and probabilities governing the coexistence and coevolution of the multiple genomes in plant cells.

ACKNOWLEDGMENTS

We thank Charles André for his kind comments on the manuscript. Work in the authors' laboratory is supported by grants from the Bundesministerium für Forschung und Technologie and the Deutsche Forschungsgemeinschaft.

Any *Annual Review* chapter, as well as any article cited in an *Annual Review* chapter, may be purchased from the Annual Reviews Preprints and Reprints service.
1-800-347-8007; 415-259-5017; email: arpr@class.org

Literature Cited

1. Anderson S, Bankier AT, Barrell BG, de Bruijn MHL, Coulson AR, et al. 1981. Sequence and organization of the human mitochondrial genome. *Nature* 290:457–65
2. André C, Levy A, Walbot V. 1992. Small repeated sequences and the structure of plant mitochondrial genomes. *Trends Genet.* 8:128–31
3. Araya A, Begu D, Graves PV, Hernould M, Litvak S, et al. 1993. Of RNA editing and cytoplasmic male sterility in plants. See Ref. 21, pp. 83–91
4. Araya A, Domec C, Begu D, Litvak S. 1992. An in vitro system for the editing of ATP synthase subunit 9 mRNA using wheat mitochondrial extracts. *Proc. Natl. Acad. Sci. USA* 89:1040–44
5. Bendich AJ. 1985. Plant mitochondrial DNA: unusual variation on a common theme. In *Genetic Flux in Plants,* ed. B Hohn, ES Dennis, pp. 111–38. Vienna: Springer-Verlag
6. Bendich AJ. 1993. Reaching for the ring: the study of mitochondrial genome structure. *Curr. Genet.* 24:279–90
7. Bendich AJ, Loretzl CJ, Monnat RJ. 1993. The structure of the plant mitochondrial genome. See Ref. 21, pp. 171–80
8. Benne R. 1990. RNA editing in trypanosomes: Is there a message? *Trends Genet.* 6:177–81
9. Bibb MJ, Van Etten RA, Wright CT, Walberg MW, Clayton DA. 1981. Sequence and gene organization of the mouse mitochondrial DNA. *Cell* 26:167–80
10. Binder S, Brennicke A. 1993. A tRNA gene transcription initiation site is similar to mRNA and rRNA promoters in plant mitochondria. *Nucleic Acids Res.* 21:5012–19
11. Binder S, Brennicke A. 1993. Transcription initiation sites in Oenothera mitochondria. *J. Biol. Chem.* 268:7849–55
12. Deleted in proof
13. Binder S, Marchfelder A, Brennicke A, Wassinger B. 1992. RNA editing in intron sequences may be required for trans-splicing of nad2 transcripts in Oenothera mitochondria. *J. Biol. Chem.* 267:7615–23
14. Blum B, Bakalara N, Simpson M. 1990. A model for RNA editing in kinetoplast mitochondria: "guide" RNA molecules transcribed from maxicircle DNA provide the edited information. *Cell* 57:355–66
15. Bonen L. 1987. The mitochondrial S13 ribosomal protein gene is silent in wheat embryos and seedlings. *Nucleic Acids Res.* 15:10393–404
16. Bonen L. 1991. The mitochondrial genome: so simple yet so complex. *Curr. Opin. Genet. Dev.* 1:515–22
17. Bonen L, Williams K, Bird S, Wood C. 1994. The NADH dehydrogenase subunit 7 gene is interrupted by four group II introns in the wheat mitochondrial genome. *Mol. Gen. Genet.* In press
18. Bonnard G, Gualberto JM, Lamattina L, Grienenberger JM. 1992. RNA editing in plant mitochondria. *Crit. Rev. Plant Sci.* 10:503–44
19. Brennicke A. 1992. Gene translocation between organelles. *Curr. Biol.* 2:46–47
20. Brennicke A, Blanz P. 1982. Circular mitochondrial DNA species from Oenothera with unique sequences. *Mol. Gen. Genet.* 187:461–66
21. Brennicke A, Kück U. 1993. *Plant Mitochondria.* Weinheim: VCH Chemie
22. Brown GG, Auchincloss AH, Covello PS, Gray MW, Menassa R, Singh M. 1991. Characterization of transcription initiation sites on the soybean mitochondrial genome allows identification of a transcription-associated sequence motif. *Mol. Gen. Genet.* 228:345–55
23. Cattaneo R. 1991. Different types of mes-

senger RNA editing. *Annu. Rev. Genet.* 25: 71–88

24. Chapdelaine Y, Bonen L. 1991. The wheat mitochondrial gene for subunit I of the NADH dehydrogenase complex: a trans-splicing model for this gene-in-pieces. *Cell* 65:465–72

25. Covello PS, Gray MW. 1989. RNA editing in plant mitochondria. *Nature* 341:662–66

26. Covello PS, Gray MW. 1992. Silent mitochondrial and active nuclear genes for subunit 2 of cytochrome c oxidase (cox2) in soybean: evidence for RNA-mediated gene transfer. *EMBO J.* 11:3815–20

27. Covello PS, Gray MW. 1993. On the evolution of RNA editing. *Trends Genet.* 9: 265–68

28. Douce R. 1985. *Mitochondria in Higher Plants.* Orlando, FL: Academic

29. Douce R, Neuburger M. 1989. The uniqueness of plant mitochondria. *Annu. Rev. Plant Physiol. Plant Mol. Biol.* 40:371–414

30. Feiler HS, Newton KJ. 1987. Altered mitochondrial gene expression in the nonchromosomal stripe 2 mutant of maize. *EMBO J.* 6:1535–39

31. Fox TD, Leaver CJ. 1981. The Zea mays mitochondrial gene coding cytochrome oxidase subunit II has an intervening sequence and does not contain TGA codons. *Cell* 26:315–23

32. Gäbler L, Herz U, Liddell A, Leaver CJ, Schröder W, et al. 1994. The 42.5 kDa subunit of the NADH:ubiquinone oxidoreductase (complex I) is in higher plants encoded by the mitochondrial *nad7* gene. *Mol. Gen. Genet.* In press

33. Gass DA, Makaroff CA, Palmer JD. 1992. Variable intron content of the NADH dehydrogenase subunit 4 gene of plant mitochondria. *Curr. Genet.* 21:423–30

34. Glaubitz JC, Carlson JE. 1992. RNA editing in the mitochondria of a conifer. *Curr. Genet.* 22:163–65

35. Gonzalez DH, Bonnard G, Grienenberger J-M. 1993. A gene involved in the biogenesis of c-type cytochromes is co-transcribed with a ribosomal protein gene in wheat mitochondria. *Curr. Genet.* 24:248–55

36. Gray MW. 1989. Origin and evolution of mitochondrial DNA. *Annu. Rev. Cell Biol.* 5:25–30

37. Gray MW. 1989. The evolutionary origins of organelles. *Trends Genet.* 5:294–99

38. Gray MW, Covello PS. 1993. RNA editing in plant mitochondria and chloroplasts. *FASEB J.* 7:64–71

39. Gray MW, Hanic-Joyce PJ, Covello PS. 1992. Transcription, processing and editing in plant mitochondria. *Annu. Rev. Plant Physiol. Plant Mol. Biol.* 43:145–75

40. Grohmann L, Brennicke A, Schuster W. 1992. The gene for mitochondrial ribosomal protein S12 has been transferred to the nuclear genome in Oenothera. *Nucleic Acids Res.* 20:5641–46

41. Gualberto JM, Bonnard G, Lamattina L, Grienenberger J-M. 1991. Expression of the wheat mitochondrial *nad3-rps12* transcription unit: correlation between editing and mRNA maturation. *Plant Cell* 3:1109–20

42. Gualberto JM, Lamattina L, Bonnard G, Weil J-H, Grienenberger J-M. 1989. RNA editing in wheat mitochondria results in the conservation of protein sequences. *Nature* 341:660–62

43. Gualberto JM, Weil J-H, Grienenberger J-M. 1990. Editing of the wheat coxIII transcript: evidence for twelve C to U and one U to C conversions and for sequence similarities around editing sites. *Nucleic Acids Res.* 18:3771–76

44. Hanic-Joyce PJ, Gray MW. 1991. Accurate transcription of a plant mitochondrial gene in vitro. *Mol. Cell. Biol.* 11:2035–39

45. Hernould M, Suharsono S, Litvak S, Araya A, Mouras A. 1993. Male-sterility induction in transgenic tobacco plants with an unedited *atp9* mitochondrial gene from wheat. *Proc. Natl. Acad. Sci. USA* 90: 2370–74

46. Herz U, Schröder W, Liddell A, Leaver C, Brennicke A, Grohmann L. 1994. Purification of the NADH:ubiquinone oxido-reductase (complex I) from the inner mitochondrial membrane of *Solanum tuberosum*. *J. Biol. Chem.* 259:In press

47. Hess WR, Prombona A, Fieder B, Subramanian AR, Börner T. 1993. Chloroplast *rps15* and the *rpoB/C1/C2* gene cluster are strongly transcribed in ribosome-deficient plastids: evidence for a functioning nonchloroplast-encoded RNA polymerase. *EMBO J.* 12:563–71

48. Hiesel R, Brennicke A. 1983. Cytochrome oxidase subunit II gene in mitochondria of Oenothera has no intron. *EMBO J.* 2:2173–78

49. Hiesel R, Combettes B, Brennicke A. 1994. Evidence for RNA editing in mitochondria of all major groups of land plants except the bryophyta. *Proc. Natl. Acad. Sci. USA* In press

50. Hiesel R, Schobel W, Schuster W, Brennicke A. 1987. Cytochrome oxidase subunit I and III genes in Oenothera mitochondria are transcribed from identical promoter sequences. *EMBO J.* 6: 29–34

51. Hiesel R, von Haeseler A, Brennicke A. 1994. Plant mitochondrial nucleic acid sequences as a tool for phylogenetic analysis. *Proc. Natl. Acad. Sci. USA* In press

52. Hiesel R, Wissinger B, Schuster W,

Brennicke A. 1989. RNA editing in plant mitochondria. *Science* 246:1632–34

53. Hoch B, Maier RM, Appel K, Igloi GL, Kössel H. 1991. Editing of a chloroplast mRNA by creation of an initiation codon. *Nature* 353:178–80

54. Hunt MD, Newton KJ. 1991. The NCS3 mutation: genetic evidence for the expression of ribosomal protein genes in *Zea mays* mitochondria. *EMBO J.* 10:1045–52

55. Joyce PBM, Gray MW. 1989. Chloroplast-like transfer RNA genes expressed in wheat mitochondria. *Nucleic Acids Res.* 17:5461–76

56. Knoop V, Brennicke A. 1991. A mitochondrial intron sequence in the 5′-flanking region of a plant nuclear lectin gene. *Curr. Genet.* 20:423–25

57. Knoop V, Brennicke A. 1994. Promiscuous mitochondrial group II intron sequences in plant nuclear genomes. *J. Mol. Evol.* In press

58. Knoop V, Brennicke A. 1993. Group II introns in plant mitochondria—trans-splicing, RNA editing, evolution, and promiscuity. See Ref. 21, pp. 221–32

59. Knoop V, Schuster W, Wissinger B, Brennicke A. 1991. Trans-splicing integrates an exon of 22 nucleotides into the nad5 mRNA in higher plant mitochondria. *EMBO J.* 10:3483–93

60. Koslowsky DJ, Bhat GJ, Perrollaz AL, Feagin JE, Stuart K. 1990. The MURF3 gene of T. brucei contains multiple domains of extensive editing and is homologous to a subunit of NADH dehydrogenase. *Cell* 62:901–11

61. Leaver CJ, Gray MW. 1982. Mitochondrial genome organization and expression in higher plants. *Annu. Rev. Plant Physiol.* 33:373–402

62. Lerbs-Mache S. 1993. The 110-kDa polypeptide of spinach plastid DNA-dependent RNA polymerase: single-subunit enzyme or catalytic core of multimeric enzyme complexes? *Proc. Natl. Acad. Sci. USA* 90:5509–13

63. Leterme S, Boutry M. 1993. Purification and preliminary characterization of mitochondrial complex I (NADH:ubiquinone reductase) from broad bean (*Vicia faba* L.). *Plant Physiol.* 102:435–43

64. Levings CS III. 1990. The Texas cytoplasm of maize: cytoplasmic male sterility and disease susceptibility. *Science* 250:942–47

65. Levings CS III, Brown GG. 1989. Molecular biology of plant mitochondria. *Cell* 56:171–79

66. Levy AA, André CP, Walbot V. 1991. Analysis of a 120-kilobase mitochondrial chromosome in maize. *Genetics* 128:417–24

67. Lippok B, Brennicke A, Wissinger B. 1993.

Differential RNA editing in closely related introns in Oenothera mitochondria. *Mol. Gen. Genet.* In press

68. Lonergan KM, Gray MW. 1993. Editing of transfer RNAs in *Acanthamoeba castellanii* mitochondria. *Science* 259:812–16

69. Lonsdale DM. 1984. A review of the structure and organisation of the mitochondrial genome of higher plants. *Plant Mol. Biol.* 3:201–6

70. Mahendran R, Spottswood MR, Miller DL. 1991. RNA editing by cytidine insertion in mitochondria of *Physarum polycephalum*. *Nature* 349:434–38

71. Makaroff CA, Palmer JD. 1987. Extensive mitochondrial specific transcription of the *Brassica campestris* mitochondrial genome. *Nucleic Acids Res.* 15:5141–56

72. Manna E, Brennicke A. 1986. Site-specific circularisation in Oenothera mitochondria. *Mol. Gen. Genet.* 203:377–81

72a. Maréchal-Drouard L, Ramamonjisoa D, Cosset A, Weil JH, Dietrich A. 1993. Editing corrects mispairings in the acceptor stem of bean and potato mitochondrial phenylalanine transfer RNAs. *Nucleic Acids Res.* 21:4909–14

73. Maréchal-Drouard L, Weil J-H, Dietrich A. 1993. Transfer RNAs and transfer RNA genes in plants. *Annu. Rev. Plant Physiol. Plant Mol. Biol.* 44:13–32

74. Maréchal-Drouard L, Weil J-H, Guillemaut P. 1988. Import of several tRNAs from the cytoplasm into the mitochondria in bean *Phaseolus vulgaris*. *Nucleic Acids Res.* 16:4777–88

75. Miller DL, Ling M-L, Wang S, Yang N, Costandy H. 1993. RNA editing in mitochondria of *Physarum polycephalum*. See Ref. 21, pp. 53–61

76. Newton KJ. 1988. Plant mitochondrial genomes: organization, expression and variation. *Annu. Rev. Plant Physiol. Plant Mol. Biol.* 39:503–32

77. Nugent JM, Palmer JD. 1991. RNA-mediated transfer of the gene *coxII* from the mitochondrion to the nucleus during flowering plant evolution. *Cell* 66:473–81

78. Nugent JM, Palmer JD. 1993. Evolution of gene content and gene organization in flowering plant mitochondrial DNA: a general survey and further studies on *coxII* gene transfer to the nucleus. See Ref. 21, pp. 163–70

79. Oda K, Kohchi T, Ohyama K. 1992. Mitochondrial DNA of *Marchantia polymorpha* as a single circular form with no incorporation of foreign DNA. *Biosci. Biotech. Biochem.* 56:132–35

80. Oda K, Yamato K, Ohta E, Nakamura Y, Takemura M, et al. 1992. Gene organization deduced from the complete sequence of liverwort *Marchantia polymorpha* mitochondrial DNA: a primitive form of plant

mitochondrial genome. *J. Mol. Biol.* 223: 1–7

81. Ohta E, Oda K, Yamato K, Nakamura Y, Takemura M, et al. 1993. Group I introns in the liverwort mitochondrial genome: the gene coding for subunit 1 of cytochrome oxidase shares five intron positions with its fungal counterparts. *Nucleic Acids Res.* 21: 1297–305

82. Ohyama K, Fukuzawa H, Kochi T, Shira H, Sano T, et al. 1986. Chloroplast gene organization deduced from the complete sequence of liverwort *Marchantia polymorpha* chloroplast DNA. *Nature* 322: 572–74

83. Ohyama K, Oda K, Ohta E, Takemura M. 1993. Gene organization and evolution of introns of a liverwort, *Marchantia polymorpha*, mitochondrial genome. See Ref. 21, pp. 115–29

84. Palmer JD, Herbon LA. 1987. Unicircular structure of the *Brassica hirta* mitochondrial genome. *Curr. Genet.* 11:565–70

85. Palmer JD, Shields CR. 1984. Tripartite structure of the *Brassica campestris* mitochondrial genome. *Nature* 307:411–15

86. Patell VM. 1993. *Structure and expression of the gene coding for the sub-unit 2 of NADH dehydrogenase in wheat mitochondria: cis-, trans-splicing and editing of the messenger RNA.* Thesis. Univ. Strasbourg

87. Pereira de Souza A, Jubier M-F, Delcher E, Lancellin D, Lejeune B. 1991. A trans-splicing model for the expression of the tripartite *nad5* gene in wheat and maize mitochondria. *Plant Cell* 3:1363–78

88. Perez-Morga DL, Englund PT. 1993. The attachment of minicircles to kinetoplast DNA networks during replication. *Cell* 74: 703–11

89. Powell LM, Wallis SC, Pease RJ, Edwards YH, Knott TJ, Scott J. 1987. A novel form of tissue-specific RNA processing produces apolipoprotein-B48 in intestine. *Cell* 50:831–40

90. Pring D, Brennicke A, Schuster W. 1993. RNA editing gives a new meaning to the genetic information in mitochondria and chloroplasts. *Plant Mol. Biol.* 21:1163–70

91. Pring DR, Lonsdale DM. 1985. Molecular biology of higher plant mitochondrial DNA. *Int. Rev. Cytol.* 97:1–46

92. Pritchard AE, Seilhamer JJ, Mahalingam R, Sable CL, Venuti SE, Cummings DW. 1990. Nucleotide sequence of the mitochondrial genome of Paramecium. *Nucleic Acids Res.* 18:173–80

93. Quetier F, Lejeune B, Delorme S, Falconet D. 1985. Molecular organization and expression of the mitochondrial genome of higher plants. In *Higher Plant Cell Respiration,* ed. R Douce, DA Day, 18:25–36. Berlin: Springer-Verlag

94. Rapp WD, Lupold DS, Mack S, Stern DB.

1993. Architecture of the maize mitochondrial *atp1* promoter as determined by linker-scanning and point mutagenesis. *Mol. Cell. Biol.* 13:7232–38

95. Rapp WD, Stern DB. 1992. A conserved 11 nucleotide sequence contains an essential promoter element of the maize mitochondrial *atp1* gene. *EMBO J.* 11:533–43

96. Rasmussen AG, Mendel-Hartvig J, Moller IM, Wiskich JT. 1994. Isolation of the rotenone-sensitive NADH-ubiquinone reductase (complex I) from plant mitochondria. *Physiol. Plant.* In press

97. Schuster W, Brennicke A. 1987. Plastid, nuclear and reverse transcriptase sequences in the mitochondrial genome of Oenothera: Is genetic information transferred between organelles via RNA? *EMBO J.* 6:2857–63

98. Schuster W, Brennicke A. 1988. Interorganellar sequence transfer: plant mitochondrial DNA is nuclear, is plastid, is mitochondrial. *Plant Sci.* 54:1–10

99. Schuster W, Brennicke A. 1993. RNA editing in Oenothera mitochondria. See Ref. 21, pp. 63–70

100. Schuster W, Combettes B, Flieger K, Brennicke A. 1993. A plant mitochondrial gene encodes a protein involved in cytochrome c biogenesis. *Mol. Gen. Genet.* 239:49–57

101. Schuster W, Hiesel R, Brennicke A. 1993. RNA editing in plant mitochondria. *Semin. Cell Biol.* 4:279–84

102. Schuster W, Hiesel R, Wassinger B, Brennicke A. 1990. RNA editing in the cytochrome b locus of the higher plant Oenothera includes a U to C transition. *Mol. Cell Biol.* 10:2428–31

103. Schuster W, Knoop V, Hiesel R, Grohmann L, Brennicke A. 1993. The mitochondrial genome on its way to the nucleus: different stages of gene transfer in plants. *FEBS Lett.* 325:140–44

104. Scott J. 1990. Editing and modification of messenger RNA. In *Nucleic Acids and Molecular Biology,* ed F Eckstein, DMJ Lilley, 4:258–73. Berlin: Springer-Verlag

105. Simpson L. 1986. Kinetoplast DNA in trypanosomid flagellates. *Int. Rev. Cytol.* 99: 119–79

106. Sloof P, Arts GJ, Leegwater P, Speijer D, van den Burg J, et al. 1993. Properties of gRNAs involved in RNA editing in mitochondria of the insect trypanosome *Crithidia fasciculata.* See Ref. 21, pp. 37–51

107. Sloof P, Benne R. 1993. RNA editing in trypanosome mitochondria: guidelines for models. *FEBS Lett.* 325:146–51

108. Small ID, Isaac PG, Leaver CJ. 1987. Stoichiometric differences in DNA molecules containing the atpA gene suggest mechanisms for the generation of mitochondrial

genome diversity in maize. *EMBO J.* 6: 865–69

109. Small ID, Suffolk R, Leaver CJ. 1989. Evolution of plant mitochondrial genomes via sub-stoichiometric intermediates. *Cell* 58: 69–76

110. Stern DB, Lonsdale DM. 1982. Mitochondrial and chloroplast genomes of maize have a 12-kilobase DNA sequence in common. *Nature* 229:698–702

111. Stern DB, Palmer JD. 1984. Extensive and widespread homologies between mitochondrial DNA and chloroplast DNA in plants. *Proc. Natl. Acad. Sci. USA* 81: 1946–50

112. Stuart K, Corell RA, Göringer U, Koslowsky DJ, Myler PJ, et al. 1993. RNA editing in *Trypanosoma brucei*: gRNA diversity and redundancy. See Ref. 21, pp. 23–36

113. Sünkel S, Brennicke A, Knoop V. 1993. RNA editing of a conserved reading frame in plant mitochondria increases its similarity to two overlapping reading frames in *Escherichia coli*. *Mol. Gen. Genet.* In press

114. Topping JF, Leaver CJ. 1990. Mitochondrial gene expression during wheat leaf development. *Planta* 182:399–407

115. Ward BL, Anderson RS, Bendich AJ. 1981. The mitochondrial genome is large and variable in a family of plants (Cucurbitaceae). *Cell* 25:793–803

116. Wissinger B, Brennicke A, Schuster W. 1992. Regenerating good sense: RNA editing and trans-splicing in plant mitochondria. *Trends Genet.* 8:322–28

117. Wissinger B, Hiesel R, Schobel W, Unseld M, Brennicke A, Schuster W. 1991. Duplicated sequence elements and their function in plant mitochondria. *Z. Naturforsch. Teil C* 46:709–16

118. Wissinger B, Schuster W, Brennicke A. 1991. *Trans* splicing in Oenothera mitochondria: *nad1* mRNAs are edited in exon and trans-splicing group II intron sequences. *Cell* 65:473–82

119. Wolstenholme DR, Macfarlane JL, Beagley T, Thomson MC, Okada NA, Fauron CM-R. 1993. Maize mitochondrial DNA: the *nad1* gene-*mat-r* gene complex, a maturase-related pseudogene linked to a *nad2* exon, and *nad* gene intron interrelationships. See Ref. 21, pp. 151–61

120. Young EG, Hanson MR. 1987. A fused mitochondrial gene associated with cytoplasmic male sterility is developmentally regulated. *Cell* 50:41–49

Annu. Rev. Plant Physiol. Plant Mol. Biol. 1994. 45:79–112

GEMINIVIRUSES AND THEIR USES AS EXTRACHROMOSOMAL REPLICONS

Marja C. P. Timmermans, O. Prem Das, and Joachim Messing

Waksman Institute, Rutgers University, Piscataway, New Jersey 08855-0759

KEY WORDS: viral replication, viral movement, extrachromosomal replicons, gene expression,
viral evolution

CONTENTS

INTRODUCTION 80
CHARACTERISTICS OF GEMINIVIRUSES.. 81
   Genomic Organization, Host Range, and Insect Vectors 81
   Viral Protein Functions 86
   Viral Gene Transcription.. 89
GEMINIVIRUS LIFECYCLE.. 91
   Transmission 91
   Replication 92
   Movement, Symptom Development, and Host Range... 94
GEMINIVIRUSES AS VECTORS 96
   Advantages.. 96
   Disadvantages... 96
   Vector Design 96
   Vector Construction.. 97
   Vector Delivery... 99
APPLICATIONS OF GEMINIVIRUSES AND THEIR VECTORS 99
   Gene Expression 99
   Engineering Geminivirus Resistance 101
   Geminiviruses as Markers for Gene Transfer... 102
   Analysis of Transposition Mechanisms 102
   DNA Methylation.. 103
   Potential Applications 104

INTRODUCTION

Viruses have contributed to our understanding of cellular processes, and viral vectors have become indispensable tools in molecular biology. To date, however, viral vectors have not been used as extensively in plants as in bacterial and mammalian systems. This may be partly because of the efficiency of plant transformation using *Agrobacterium tumefaciens* or direct gene transfer technologies. In addition, viral vectors have not been shown to generate stable transformants via integration or transmission through the germline. Nonetheless, vectors based on plant viruses are particularly suited for some applications, and they provide a different perspective on cellular processes than does transgenic technology. For instance, vectors allow gene expression to be studied without the variation resulting from chromosomal location or DNA methylation that is encountered frequently in stable transformants. Viral vectors also permit the characterization of genes in differentiated tissues without the need for tissue culture or somatic embryogenesis, using the ability of viruses to spread systemically through the plant. Systemic spread provides an added level of amplification of the target gene, and together with high copy numbers, can result in very high expression levels. Furthermore, in mammalian cells, extrachromosomal replicons have facilitated studies of basic cellular processes such as homologous and site-specific recombination, chromatin structure, and DNA methylation (140, 80).

More than 90% of known plant viruses have RNA genomes. Because cDNAs of RNA viruses are not infectious, the development of vectors from these viruses requires in vitro transcription methods to generate infectious RNAs. The high error rate in RNA replication has been considered to limit the applications of such vectors (169), although experimental studies have not supported this view (139). A more serious drawback is that these viruses replicate in the cytosol, ruling out analysis of nuclear processes. Among RNA viruses, brome mosaic virus and tobacco mosaic virus have been used successfully to express foreign genes in protoplasts or whole plants (26, 46, 160, 161). The usefulness of these viruses as vectors should increase when key steps of the life cycle, such as replication and translation, are better understood.

Plant DNA viruses can be classified into the double-stranded (ds) caulimoviruses and the single-stranded (ss) geminiviruses. Of the former, cauliflower mosaic virus (CaMV) is the best characterized. The mechanically transmissible ds genome of this virus indicates that it is a good candidate for the design of expression vectors. However, efforts at developing CaMV into a vector are complicated by its RNA-mediated replication and polycistronic gene organization. Its restricted host range has also limited the applications of this virus in plant genetic engineering. Nevertheless, stable CaMV variants expressing foreign genes have been produced (16, 29, 102).

The simple genomic organization, nuclear replication, and broad host range of geminiviruses make them attractive candidates for vector development. Perhaps as a result, several excellent reviews on geminiviruses have appeared (6, 24, 25, 65, 95, 98, 115, 141). This review focuses on the biology of these viruses because the development of vectors depends on a clear understanding of the viral life cycle. We describe the uses of geminiviral vectors, and discuss their potential for other applications.

CHARACTERISTICS OF GEMINIVIRUSES

Genomic Organization, Host Range, and Insect Vectors

Geminiviruses have small ss circular DNA genomes of 2.5–3.0 kb and unique geminate (twinned) capsid morphology (55, 66). They infect many important crop plants including both monocots and dicots. Traditionally, geminiviruses have been classified into two subgroups, based on the tight correlation between genomic organization, host range, and insect vector for each subgroup. Recent molecular characterization of more geminiviruses has, however, identified strains that do not fit this classification, raising interesting issues about geminivirus evolution.

BIPARTITE GEMINIVIRUSES   Viruses within this subgroup have a genome comprised of two components, designated DNA A and B. Each component is encapsidated in a separate geminate particle, requiring a double inoculation for successful infection (56). Bipartite geminiviruses infect dicotyledonous hosts, and are transmitted by a single whitefly species, *Bemisia tabaci* (Table 1). This subgroup includes African cassava mosaic virus (ACMV; 143), tomato golden mosaic virus (TGMV; 61), and abutilon mosaic virus (AbMV; 50). The genomes of several bipartite geminiviruses have been cloned and sequenced (Table 1). Both DNA A and B contain a single intergenic region of approximately 200 nucleotides, called the common region. Although common regions are not conserved between viruses, they are highly conserved (> 95%) between the A and B components of each virus (95, 98). Because of the high homology, this region may contain recognition signals essential for processes common to both genomes such as replication, transcription initiation, and packaging. Indeed, a hairpin structure including the sequence motif TAATATTAC is located within this region (95). This motif is conserved among all geminiviruses, and is hypothesized to be involved in viral replication (see Replication below). In addition, potential promoter sequences for bidirectional transcription have been identified within the common region. Six open reading frames (ORFs), four on DNA A and two on B, are organized similarly on different viruses and probably encode similar functions (Figure 1) because they show extensive homology. In

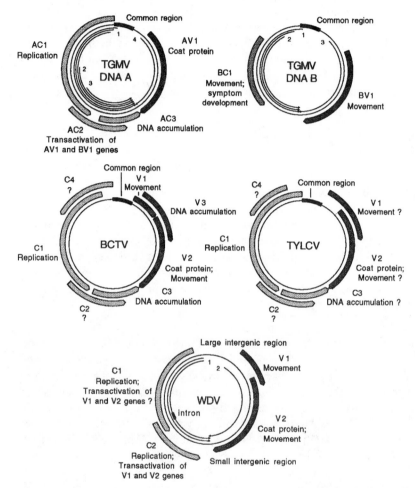

Figure 1   Geminivirus genomes. Maps of the two genomes of a typical bipartite geminivirus (TGMV DNA A and TGMV DNA B), the single genome of a typical monopartite virus (WDV), and the single genomes of two atypical viruses (BCTV and TYLCV) are shown. In all cases, lightly stippled segments represent ORFs encoded by the complementary strand, and heavily stippled segments represent ORFs encoded by the viral strand. "C" indicates a complementary strand ORF, "V" indicates a viral strand ORF, and "A" and "B" indicate which of the two genomes of bipartite viruses encodes the ORF. The functions of the encoded protein product are indicated. Other features are also indicated such as the intron and intergenic regions of the monopartite genome, and the common region that includes the origin of replication. The numbered lines on the inside represent the transcripts obtained from each genome, and are numbered sequentially. Arrows mark the 3′ end of each transcript. Transcript data is not available for BCTV and TYLCV. For TGMV DNA A, note the heterogeneity of the complementary strand transcripts at the 5′ end. Genomes are not represented to scale, but for each genome, transcripts and ORFs are drawn approximately to scale. (See 6, 95, 98, 115 for details on transcript and protein sizes.)

addition, certain viruses contain non-conserved ORFs, which may encode functions unique to each virus.

MONOPARTITE GEMINIVIRUSES This subgroup includes viruses that have a monopartite genome, infect monocotyledonous plants, and are transmitted by leafhoppers (Table 1). Among its members are maize streak virus (MSV; 79, 96, 116), wheat dwarf virus (WDV; 105, 175), and *Digitaria* streak virus (DSV; 30). Their genomes contain four ORFs and two noncoding intergenic regions. These ORFs diverge from the larger intergenic region and converge at the smaller intergenic region (Figure 1). ORF C2 of most monopartite viruses does not contain an ATG initiation codon (5, 13, 30, 81, 105, 175). This, together with the fact that consensus splice donor and acceptor sites surround the C1-C2 overlap, indicates that ORF C2 might be part of a larger protein encoded by a spliced transcript. Such splicing has been confirmed for WDV (28, 135) and DSV (2, 117; see Figure 1 for intron location). As expected from the locations of the ORFs, typical transcription initiation and termination sequences are located in the large and small intergenic regions, respectively. The large intergenic region also contains a hairpin structure, which includes the conserved sequence motif TAATATTAC that is found in the common region of bipartite viruses.

ATYPICAL GEMINIVIRUSES Several geminiviruses have properties intermediate between the two subgroups. Beet curly top virus (BCTV; 146), the first intermediate type to be characterized; tobacco yellow dwarf virus (TobYDV; 112); and tomato leaf curl virus (TLCV; 33) have monopartite genomes, but infect dicotyledonous hosts (Table 1). BCTV and TobYDV are leafhopper-transmitted, while TLCV is whitefly-transmitted. Another example, tomato yellow leaf curl virus (TYLCV; 87, 118, 129), is whitefly-transmitted, infects a dicotyledonous host, and interestingly, its genome appears to be either monopartite or bipartite depending on the isolate. Isolates of this virus from Israel (118) and Sardinia (87) appear to have a single DNA component, while the isolate from Thailand (129) contains two DNA components. The latter isolate has not been sequenced, so it is not known if its organization resembles typical bipartite geminiviruses.

The genomic organization of TobYDV is similar to a monopartite geminivirus, with features such as two intergenic regions, four ORFs and the intron (112). In contrast, the genomes of BCTV, TLCV, and TYLCV resemble both monopartite and bipartite viruses (33, 87, 118, 129, 146). From the representations of BCTV and TYLCV in Figure 1, it is clear that the left half of each genome, indicating the coding potential of the complementary strand, resembles DNA A of a bipartite geminivirus, while the right half (viral strand) resembles a monopartite virus. The single intergenic region resembles bipartite

Table 1 Geminivirus origins, genomic organizations, hosts and vectors

Virus	Abbreviation	Origin	Genomic Organization	Host	Vector[a]	Ref[b]
Abutilon mosaic	AbMV	Brazil	bipartite	Malvaceae	*Bemisia tabaci* (W)	50
Cassava mosaic[c]	ACMV	Kenya	bipartite	Euphorbiaceae, Solanaceae	*Bemisia tabaci* (W)	143
Beet curly top	BCTV	U.S.A., Mediterranean	monopartite	44 dicot families	*Circulifer tenellus* (L)	146
Bean dwarf mosaic	BDMV	Colombia	bipartite	Leguminosae, Malvaceae	*Bemisia tabaci* (W)	74
Bean golden mosaic	BGMV-BZ	Brazil	bipartite	Leguminosae	*Bemisia tabaci* (W)	53
	BGMV-DR	Dominican Rep.	bipartite	Leguminosae	*Bemisia tabaci* (W)	53
	BGMV-GA	Guatamala	bipartite	Leguminosae	*Bemisia tabaci* (W)	53
	BGMV-PR	Puerto Rico	bipartite	Leguminosae, Malvaceae	*Bemisia tabaci* (W)	77
Chloris striate mosaic	CSMV	Australia	monopartite	Gramineae	*Nesoclutha pallida* (L)	5
Digitaria streak	DSV	Vanuatu	monopartite	Gramineae	*Nesoclutha declivata* (L)	30
Miscanthus streak	MiSV	Japan	monopartite	Gramineae	unknown	20
Maize streak	MSV-K	Kenya	monopartite	Gramineae	*Cicadulina mbila* (L)	79
	MSV-N	Nigeria	monopartite	Gramineae	*Cicadulina mbila* (L)	116
	MSV-S	South Africa	monopartite	Gramineae	*Cicadulina mbila* (L)	96
Panicum streak	PSV	Kenya	monopartite	Gramineae	*Cicadulina mbila* (L)	13
Potato yellow mosaic	PYMV	Venezuela	bipartite	Solanaceae	*Bemisia tabaci* (W)	22
Squash leaf curl	SqLCV	U.S.A.	bipartite	Cucurbitaceae	*Bemisia tabaci* (W)	99
Sugarcane streak	SSV	South Africa	monopartite	Gramineae	*Cicadulina mbila* (L)	81
Tomato golden mosaic	TGMV	Brazil	bipartite	Solanaceae	*Bemisia tabaci* (W)	61
Tomato leaf curl	TLCV	Australia	monopartite	Solanaceae	*Bemisia tabaci* (W)	33
Tomato mottle	TMoV	U.S.A.	bipartite	Solanaceae	*Bemisia tabaci* (W)	1
Tobacco yellow dwarf	TobYDV	Australia	monopartite	Solanaceae, Leguminosae	*Orosius argentatus* (L)	112
Tomato yellow leaf curl	TYLCV-I	Israel	monopartite	Solanaceae	*Bemisia tabaci* (W)	118
	TYLCV-S	Sardinia	monopartite	Solanaceae	*Bemisia tabaci* (W)	87
	TYLCV-T	Thailand	bipartite	Solanaceae	*Bemisia tabaci* (W)	129
Wheat dwarf	WDV-CJI	Czechoslovakia	monopartite	Gramineae	*Psammotetix alienus* (L)	175
	WDV-S	Sweden	monopartite	Gramineae	*Psammotetix alienus* (L)	105

[a]L, leafhopper; W, whitefly
[b]The reference for the viral sequence cloning is listed
[c]Previously called cassava latent virus

viruses, and also contains the conserved motif TAATATTAC. The overlap between ORFs that is typical of geminiviruses is particularly striking for BCTV and TYLCV (Figure 1).

SEQUENCE COMPARISONS The evolutionary relationships between geminiviruses have been studied at the levels of amino acid sequence (78; MCP Timmermans & J Messing, unpublished data), nucleotide sequence (132), serological cross-reactivity (124, 125, 131), and restriction site polymorphisms (82). A composite phylogenetic tree based on the amino acid sequences of the coat proteins and replication-associated proteins is shown in Figure 2. There is greater sequence divergence among monopartite viruses compared to bipartite viruses, consistent with the observation that bipartite viruses are related serologically whereas monopartite viruses show limited cross-reactivity (65, 141). Therefore, monopartite, monocot-infecting geminiviruses may be ancestral to the bipartite class. The transition from the typical monopartite to the bipartite class requires at least three steps: infection of dicots, acquisition of whitefly transmission, and acquisition of a bipartite genome. A tentative order of these events in geminivirus evolution can be deduced by phylogenetic analysis of ORF sequences (MCP Timmermans & J Messing, unpublished data). Intermediates in the transition from the monopartite to the bipartite type may have resembled viruses of the atypical class such as TobYDV, BCTV, and possibly TLCV and TYLCV. The homology between ORFs AV1 and BV1 on the A and

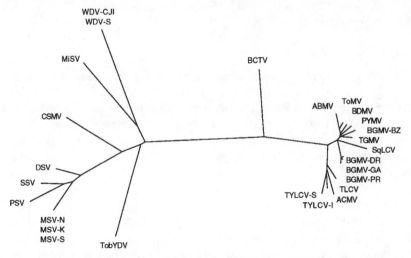

Figure 2 A phylogenetic tree of geminiviruses. Amino acid sequences of the coat protein and the replication-associated protein were aligned and distances were determined using the Dayhoff PAM matrix. The tree was constructed using the neighbor-joining program of the PHYLIP package. Line lengths are proportional to the mutation distance.

B genomes indicates that the bipartite viral genome most likely evolved from duplication of the A component (88). This transition from one to two genomes may be exemplified by TYLCV, if the bipartite genome of the isolate from Thailand turns out to be similar to other bipartite viruses.

FACTORS AFFECTING GEMINIVIRUS EVOLUTION   Geographical isolation may have played a role in geminivirus evolution (78). Thus, ACMV, TYLCV, and TLCV are viruses of the Old World, and form their own branch of the phylogenetic tree, separate from the bipartite viruses of the New World (Figure 2). In addition to geographical factors, biological constraints may have affected sequence divergence (78). For instance, divergence of the coat protein would be affected by the requirement of an insect vector for transmission. All bipartite geminiviruses are transmitted by a single whitefly species, and this may explain the reduced evolutionary distance among bipartite viruses as compared to monopartite viruses, which are transmitted by several insect vectors. Similarly, divergence among host species may also affect evolutionary interpretations. Moreover, unlike monopartite viruses, bipartite viruses require co-inoculation of both genomes for successful infection; therefore, their two genomes may be considered to coevolve. The high degree of conservation of the common regions of the two genomes for any one virus could be an example of coevolution. Recombination between these highly homologous regions of the bipartite genome is possible, though no product of such an event is known. Recombination may also have played a role in the origin of some atypical viruses, whose genomic organization resembles a monopartite genome in one half and a bipartite genome in the other half.

Viral Protein Functions

The development of viral vectors also requires an understanding of the functions of geminivirus-encoded proteins in the viral life cycle. Analysis and complementation of mutations in the various ORFs has been an effective general strategy to determine protein functions. Complementation studies have been aided by viral ORF expression in transgenic plants; the transgene approach can also identify functions that are not readily apparent from mutational analysis.

REPLICATION-ASSOCIATED PROTEIN   Mutational analysis has shown that ORF AC1 of bipartite geminiviruses (Figure 1) is necessary for viral replication (34, 40), and ACI expression in transgenic plants has shown it to be sufficient (64, 67). For monopartite viruses, the corresponding 41-kDa C1-C2 protein encoded by the spliced complementary strand transcript has been shown to be necessary and sufficient for replication (2, 135, 167). These proteins may function as site-specific topoisomerases, because geminiviruses replicate via a rolling-circle mechanism (see Replication below). Like the replication-associated proteins of bacteriophage and eubacterial plasmids, to which they bear significant homology (92), these proteins may possess nicking, helicase, and ligase activities. The replication-associated protein from TGMV binds specifically to a site located just upstream from the hairpin structure in the common region, which is the putative viral strand origin (45). Neither the binding site nor its flanking regions are conserved between viruses, and this may explain why the replication-associated proteins of related viruses are not interchangeable (101). This region is conserved between the A and B components, however, consistent with the observation that only AC1 is required for replication of both genomes (64, 67, 130, 166). Not unexpectedly, the C1-C2 protein of WDV has also been shown to act in *trans* on a cotransfected replication-deficient construct (85, 163).

COAT PROTEIN   The most abundant protein in infected tissues is the coat protein, encoded by ORF AV1 of bipartite geminiviruses (83, 143, 165) and ORF V2 of monopartite viruses (113). The coat protein of bipartite viruses is not essential for viral DNA replication, systemic spread, or symptom development because deletion mutants of this gene in ACMV and TGMV are infectious (18, 51, 147). Coat protein mutants of TGMV show some symptom attenuation and delay, indicating a role in these processes (18, 51). In contrast, the coat protein of monopartite viruses is required for viral movement and symptom development, because mutations prevent systemic infection of the host without affecting replication in protoplasts (12, 100, 176). For both classes, the coat protein protects viral DNA during transmission by insect vectors, and may contribute to vector specificity (14; see Transmission below).

MOVEMENT PROTEINS   The appearance of typical viral symptoms in transgenic tobacco expressing the ORF BC1 of SqLCV provides strong evidence for its role in symptom development in bipartite viruses (120). Mutational analysis indicates that this protein is involved in cell-to-cell spread for TGMV as well (67). Moreover, it contains amino acid sequences conserved between the cell-to-cell movement proteins of various plant viruses (93). These observations, and the localization of the BC1 protein of ACMV and SqLCV to the cell wall and plasma membrane (120, 170), indicate that this protein may alter plasmodesmata, as shown for the 30-kDa movement protein of tobacco mosaic virus (108, 174).

    The role of BV1 in viral movement is less clear. A recent study (170) indicates a possible role for BV1 in local movement because a mutation of this gene in ACMV reduced the level of viral DNA within the inoculated leaf. In contrast, in a similar experiment, mutation of the corresponding gene in SqLCV did not affect viral DNA levels in infected leaf disks (97, 120), and mutation of BV1 in TGMV did not affect local spread (67). This protein may be involved in long-distance movement, because BV1 mutants of both ACMV and SqLCV did not permit systemic infection (97, 120, 170). A similar role has also been proposed for the BV1 protein of AbMV. After agroinfection, DNA A of this virus can spread through the vascular system without DNA B, but the presence of a BV1 mutant strongly reduces spread in a dominant-negative fashion (42). Thus, BV1 has been implicated in both types of movement. Some of its functions in these processes may be mediated by interaction with viral DNA because BV1 shows significant homology (43% similarity for ACMV; 88) to the coat protein. The use of transgenic plants expressing these movement proteins should help elucidate the mechanism of viral spread.

Mutational analysis of the monopartite virus MSV has shown that ORF V1 is required for infectivity, and is probably involved in viral movement and symptom development (12, 100, 114). Because ORF V1 overlaps a possible regulatory element of the coat protein gene (see Viral Gene Transcription below), it has been argued that mutations in this ORF may affect infectivity by altered expression of the coat protein (100). Such an indirect effect is unlikely, however, because some of the ORF V1 mutants produce viral particles in maize protoplasts (11). The amino acid sequence of the V1 gene product predicts a transmembrane domain, indicating a possible membrane location, as for the BV1 and BC1 proteins of bipartite viruses.

TRANSCRIPTIONAL ACTIVATORS   The AC2 gene of bipartite gemini-viruses encodes a transcriptional activator that regulates expression of the AV1 and BV1 genes (60, 154, 155). This finding explains the pleiotropic nature of AC2 mutants. These mutants affect infectivity by preventing systemic viral movement in plants, produce no capsid protein, and accumulate reduced amounts of ss DNA in transient assays (34, 40, 63, 67, 159). The predicted amino acid sequence of AC2 shows features typical of DNA-binding proteins, including a basic domain, a zinc finger motif, and an acidic domain (155). The C1-C2 protein of monopartite viruses may function analogously as a transcriptional activator because a WDV mutant with a nonfunctional C2 ORF is not only incapable of replication, but also shows reduced activity of the viral strand promoter (75). This protein has some homology to the DNA-binding domain of myb-like plant transcription factors (75).

OTHER ORFs   Because the unspliced precursor to the C1-C2 transcript of monopartite viruses has been isolated from infected tissues (2, 28, 117, 135), a 30-kDa C1 product (Figure 1) could be expressed. However, a protein of this size has not been detected in infected plants, nor has a possible function been defined. The non-essential AC3 gene of bipartite viruses is involved in replication because mutants accumulate 50-fold less DNA (40, 111, 159). Although such mutants can produce viral particles, they exhibit attenuated symptoms (34,

67, 111), probably because the disease severity of ACMV and TGMV correlates with their B component DNA level (111, 127, 142). How the AC3 protein affects replication is unknown.

OTHER GEMINIVIRUSES For the atypical geminivirus BCTV, some gene functions have been determined by mutational analysis. As in monopartite viruses, two viral strand–encoded proteins, ORF V1 and the coat protein, have been implicated in systemic spread and symptom development (15, 76, 145). Recently, a third ORF, V3, has been identified on the viral strand (76). Mutations in this ORF appear not to affect the infectivity of BCTV on *Nicotiana benthamiana*, but infections are asymptomatic, and mutants accumulate about 8-fold more ds DNA and 9-fold less ss DNA. The reduction in ss DNA, and thus in viral titer, may account for the lack of symptom development. The altered DNA accumulation pattern implicates a role for V3 in the interconversion between ss and ds forms, possibly by stabilizing ss DNA, or by controlling a step in virion assembly (76). Like bipartite viruses, the complementary strand ORF C1 of BCTV encodes the replication-associated protein, and ORF C3 affects the efficiency of replication (145). The remaining ORFs have not been fully characterized. ORF C2 mutants remain infectious and cause normal symptoms, indicating that its putative function of transactivation (in analogy to bipartite viruses) is dispensable (145). A puzzling observation is that the introduction of stop codons at the C-terminus of ORF C4 affects symptoms, whereas a similar mutation upstream does not (144, 145).

Cognate functions clearly are encoded by roughly comparable regions of the genomes of various geminiviruses, as expected from their evolutionary relationships. Thus, related functions are encoded on the viral and complementary strands for all viruses. Similarly, the function of the AC2 ORF of bipartite genomes may be encoded by the corresponding region of the monopartite genome, i.e. the C1-C2 gene (60, 75, 154, 155). On the other hand, differences in the genomic organization of geminivirus subgroups are reflected in differences in the properties of each group. Thus, the presence of a B component correlates with the absence of phloem limitation and with the requirement for coat protein function in viral movement (141).

Viral Gene Transcription

BIDIRECTIONAL TRANSCRIPTION Bidirectional transcription from the common region or the large intergenic region has been shown for bipartite (47, 122, 153, 156, 165) and monopartite viruses (2, 28, 113, 117), respectively. The sizes and maps of the major transcripts are shown in Figure 1; in general, transcripts from each strand are 3' co-terminal, but heterogeneous at the 5' ends. Several types of transcripts are found. One type, resembling a typical eukaryotic mRNA, is flanked by a conventional TATA box and polyadenylation signal, and encodes a single protein (e.g. viral strand transcripts of bipartite viruses). A second type is also flanked by conventional expression signals, but is polycistronic (e.g. complementary strand transcripts of DNA A of bipartite genomes). Some transcripts are not flanked by appropriate 5′ expression signals and may arise from cleavage of a longer precursor (e.g. transcript 2 of DNA B of bipartite viruses). The only characterized example of RNA splicing in geminiviruses is for the complementary strand transcript of monopartite genomes (2, 28, 117, 135). Although transcription maps of the different viruses within a subgroup resemble each other, some differences exist. For instance, unlike ACMV, both TGMV (153, 156) and AbMV (47) produce a 1 kb transcript from the complementary strand of DNA A. As another example, only a single dicistronic transcript is produced from the viral strand of WDV (28), but MSV (113) and DSV (2) produce a second shorter transcript encoding the coat protein, probably from cleavage of the larger one.

VIRAL PROMOTERS   Transient expression assays have demonstrated that the AC1 gene promoter of ACMV is relatively strong, and that the promoter of the AC2/AC3 genes is also of moderate strength (177). In contrast, other genes whose products are required later in infection have weak promoters. In some cases, these may be activated by virus-encoded proteins (see below). One promoter that has been studied in detail is the MSV viral sense promoter. A 122-bp upstream activating sequence (UAS) has been identified that activates a minimal promoter in an orientation-independent, but distance-dependent manner in transient assays (43). This element bears some homology to the distal region previously identified in the CaMV 35S promoter (119), and the two elements can functionally replace each other (43). Further characterization of the UAS has identified a "rightward promoter element," which includes two tandem GC-rich repeats that interact with nuclear factors from maize (44). Similar GC-motifs are also present in the large intergenic region of the other monopartite geminiviruses, and in the common region of the bipartite viruses. Interestingly, the GC-boxes resemble the mammalian Sp1 target site, which plays a role in the bidirectional transcription of SV40 (52).

REGULATION OF GENE EXPRESSION   Temporal regulation of gene expression is a characteristic feature of many viruses, and there is evidence for such regulation in geminiviruses as well. For example, the promoters of bipartite virus genes that are required late in infection are weak in transient assays (177), but are activated in the infection cycle by an "early" gene product encoded by AC2 (60, 154, 155). In addition, the replication-associated protein AC1 of ACMV and TGMV may repress its own expression (60, 158). The binding site of this protein in the common region includes the AC1 transcription start site (45). If repression is mediated by this interaction, transcription of the B genome should be affected similarly because it shares the common region with A. For ACMV, such a reduction of BC1 transcription has been shown (60). However, for TGMV, this does not seem to be the case, possibly because another downstream transcription start site is used (Figure 1; 158). Finally, the viral and complementary sense mRNAs of bipartite geminiviruses overlap in a small region, so differential temporal regulation of transcription in the two orientations is needed (47, 122, 153, 156).

Monopartite geminiviruses also exhibit temporal control. For instance, the polyadenylation sites of the complementary and viral strands map within a few nucleotides of each other, indicative of a small transcription overlap (Figure 1; 2, 28, 113, 117). Also, expression of the viral strand genes is controlled partly by an "early" gene product, the C1-C2 protein (75). Differential splicing may also regulate viral gene expression (2, 28, 117, 135). Both the spliced transcript and the unspliced precursor of C1-C2 are present in infected tissues, and inefficient splicing is probably not the cause because the intron is typical of monocots in its splice-site junctions and A+T content (54).

GEMINIVIRUS LIFECYCLE

Transmission

Transmission of geminiviruses is thought to rely entirely on insect vectors, and it occurs in a circulative manner (65, 141). The required acquisition time, latent period and persistence in the vector, and inoculation time vary for different virus-vector combinations. In general, whitefly-transmitted geminiviruses are acquired and inoculated less rapidly than the leafhopper-transmitted viruses, which could be a result of differences in insect feeding behavior (65). Most viruses require a latent period of four hours or more, and persist in the vector for many days, possibly even for life. There is no evidence for viral replication within the insect vector, reflecting the observed dependence of transmission on the acquisition period and on the viral concentration in the plant sap (10).

Transmission of circulative viruses involves the passage of viral particles from the gut into the hemocoel (the circulatory system of the insect), then into the salivary gland, and ultimately through saliva into another plant. The specificity of leafhopper transmission of MSV is determined mostly by the passage across the gut-hemocoel barrier, because nonacquiring leafhoppers can obtain the ability to transmit by puncturing the gut or by injection of the virus directly into the hemocoel (106, 152). In contrast, vector specificity of the whitefly-transmitted geminiviruses may be determined at the level of transport from the hemocoel into the salivary gland, based on findings that a nontransmitting whiteliy species acquired SqLCV in the hemocoel (21). The virus-vector specificity seems to be determined both by the virus and by the insect. Insect specificity is demonstrated by the observation that in *Cicadulina mbila,* the ability to transmit MSV appears to be controlled by a dominant sex-linked gene (151). Among the viral-encoded products, the coat protein may play a role in determining vector specificity (128). Supporting evidence has been obtained by replacing the coat protein coding region of ACMV with BCTV; the chimeric virus is transmitted by the BCTV leafhopper vector, *Circulifer tenellus,* but unmodified ACMV is not (14).

Replication

SITE OF REPLICATION   Geminivirus replication seems to be confined to the cell nucleus. In infected plants, viral particles accumulate almost exclusively in the nuclei, either as irregular aggregates, or as hexagonal crystalline arrays (25). Moreover, in the dicot-specific geminiviruses, infection results in major cyto-pathological changes in the nucleus (4, 89), including repositioning of chromatin to the vicinity of the nuclear membrane, hypertrophy of the nucleoli, and segregation of nucleolar components into separate granular and fibrillar regions. These changes precede the accumulation of viral particles, and the nucleolar changes may reflect the shift from host to viral transcript synthesis (89). Infection is also accompanied by the appearance of fibrillar rings in the nucleus (38, 89). Because these rings consist primarily of DNA and protein, and their appearance seems to coincide with the appearance of viral particles, they may be the site of viral DNA synthesis and viral assembly (89).

REPLICATION MECHANISM   The mechanism of viral DNA replication also distinguishes geminiviruses from all other plant viruses. The ss DNA genomes and the presence of ds DNA in infected tissues indicates a replication mechanism similar to bacteriophage such as øX174 and M13. In these phage, replication consists of conversion of ss viral DNA into ds replicative form (RF), replication of RF, and production of ss viral DNA. These three steps are accomplished by two processes. One is the synthesis of the complementary strand, which occurs upon viral entry and also in conjunction with RF replication; this depends only on host functions. The second process is viral strand synthesis, which occurs via a rolling-circle mechanism, and requires a single phage-encoded protein. Evidence consistent with a rolling-circle replication mechanism has been reported for BCTV (149) and ACMV (133), and more recently for WDV (73).

COMPLEMENTARY STRAND SYNTHESIS   For monopartite geminiviruses, a complementary strand DNA fragment of approximately 80 nucleotides is tightly associated with virion DNA (32, 70, 79, 81), which may function as a primer for complementary strand synthesis. It is homologous to the small intergenic region, which has been shown to contain *cis*-acting signals for replication (85, 103, 167). This fragment has a defined 5' end and a heterogeneous 3' end, and the few ribonucleotides present at its 5' end may be the residue of an RNA molecule used to prime its synthesis (32, 70, 79, 81). It is not known if RNA polymerase (as in M13) or a primase complex (as in øX174) is involved in this priming step. The heterogeneity at the 3' end may result from interruption of complementary strand synthesis caused by viral encapsidation (32).

In contrast, no such complementary strand DNA fragment has been found associated with virions of bipartite viruses. The origin for complementary strand synthesis in these viruses is probably located within the common region, based on mutational analysis and characterization of naturally occurring subgenomic particles (49, 101, 104, 147, 150). As for monopartite viruses, complementary strand synthesis probably requires an RNA primer that is synthesized from the uncoated ss viral DNA under the direction of secondary structure determinants at the common region (134).

VIRAL STRAND SYNTHESIS   Mutational analysis has identified the hairpin structure, which contains the conserved nonanucleotide sequence, as part of the viral strand replication origin (75, 127, 136). The site of nicking of the ds RF DNA by the putative site-specific endonuclease activity of the AC1 or C1-C2 protein has not been determined. The nicking site has been mapped to a 20-bp fragment including the nonameric sequence in experiments using tandem duplications of two different BCTV strains (149). Here, initiation of replication at one origin and termination at the other releases a hybrid monomeric viral molecule. A similar study on WDV has also identified this sequence as the viral strand origin (73). Consistent with these observations, the AC1 protein from TGMV binds specifically to a site within the common region just upstream of the hairpin structure (45). It is interesting that the conserved nonamer sequence, TAATATTAC, is similar to the nicking site of øX174, TG*ATATTAT, which is also located within a hairpin structure (168).

In addition to an endonuclease activity, synthesis of viral DNA via the rolling-circle mechanism requires helicase and ligase activities. The presence of a ligase activity has not been addressed for geminiviruses, but a helicase activity has been ascribed to the AC1 protein; its binding to the common region is 4-fold stronger with ss DNA than with ds DNA, a property common to many DNA helicases (162). Protein sequence comparisons have also identified an NTP-binding site as part of a putative helicase domain at the C-terminus of AC1 and in ORF C2 of the C1-C2 protein (57). Therefore, geminiviruses may encode their own helicase activity, unlike prokaryotic phage, which rely on a host helicase for unwinding of the ds DNA template.

REGULATION OF REPLICATION   The coat protein has been suggested to regulate the transition from RF synthesis to viral ss DNA production (133, 148, 176) such that the absence of coat protein early in infection would result in continuing RF synthesis, and its presence later in the infection cycle would lead to encapsidation. However, the observation that ds DNA levels are not increased in coat protein mutants (12, 15, 18, 51, 100, 176) does not support this hypothesis. In contrast, ORF V3 mutants of BCTV do display the phenotype of increased levels of ds DNA and decreased levels of ss DNA (76). Therefore, this recently identified ORF may regulate the transition from RF to ss DNA synthesis, and similar functions may be encoded by other geminiviruses.

Because geminiviruses require host functions for DNA synthesis, viral replication is thought to be limited to the S phase of the cell cycle. This notion has been substantiated by some protoplast transfection experiments, in which cell division was found to be a prerequisite for the appearance of replicative forms (11, 15, 107, 110, 166). In other transfection experiments, however, viral replication did not require protoplast division (17, 19, 163). These results are difficult to reconcile, and subtle variables such as the form and method of introduction of viral DNA or differences in tissue culture conditions cannot be ruled out. In another study (3), nuclei at different phases of the cell cycle were separated by flow cytometry from DSV-infected plants. Replicative ds forms were present primarily in S-phase nuclei, whereas ss viral DNA could be isolated from all nuclei, also indicating that replication is confined to the S phase.

Movement, Symptom Development, and Host Range

Viral movement can be viewed as two processes: (*a*) long-distance spread through the vascular system, and (*b*) local spread, which is the movement of virus or viral DNA between cells. In addition, because geminiviruses accumulate in the nucleus of the infected cell and not in the cytoplasm, another transport mechanism may be required to facilitate movement in and out of the nucleus.

LOCAL MOVEMENT   Both monocot- and dicot-infecting monopartite geminiviruses are limited to phloem tissue, with the exception of CSMV and MSV, which in their natural hosts have been observed in mesophyll cells. In contrast, bipartite geminiviruses are found in most cell types including cortex, mesophyll, and epidermis (141). Interestingly, only bipartite viruses are readily transmissible mechanically, which may reflect their ability to infect epidermal cells. Because the two related viruses, TGMV (bipartite) and TYLCV (monopartite), differ in their ability to be mechanically transmissible in common hosts (141), a viral function correlating with the presence of the B component may be responsible for cell-to-cell spread. The observation that the A component of

ACMV (91) and AbMV (42) shows limited spread through the vascular system in the absence of DNA B when agroinfected, but not when mechanically inoculated, also supports a role for DNA B in local movement. As mentioned above, it is not clear which (or if both) of the two proteins encoded by DNA B is involved in local spread.

LONG-DISTANCE MOVEMENT Monopartite viruses are phloem-limited, and depend largely on the nutrient flow for spread. The requirement for coat protein in this process (12, 100, 176) suggests that virions mediate movement. The product of ORF V1 may also be involved (12, 100, 114), but its role is unclear. There is indirect evidence that encapsidated forms are involved in the long-distance spread of bipartite geminiviruses as well. Derivatives of ACMV and TGMV DNA A carrying inserts revert to wild-type size upon systemic movement (see Vector Design below; 36, 68, 148), which could reflect constraints induced by a packaging step. Also, systemic movement of DNA A in the absence of DNA B (42, 91) may be mediated by virions; however, mutation of the coat protein of ACMV and TGMV has little effect on long-distance spread, despite some symptom attenuation and delay (18, 51, 148). Another puzzling observation is that although the B genome is not required for spread of DNA A, mutations of BV1 block long-distance spread (97, 120, 170). It is difficult to reconcile these observations with a simple explanation, and any model would likely require the postulation of some level of interaction between the coat protein and the BV1 protein, and possibly even the BC1 protein. These interactions may be synergistic, and may involve overlapping functions such as binding to viral DNA. Note that BV1 and the coat protein share significant homology (88).

HOST RANGE AND SYMPTOM DEVELOPMENT These properties have been studied mostly by the characterization of the genomic differences between related strains that differ in their host range or symptom development. Such studies have shown that both viral replication and viral movement are involved. Thus, host range determinants of SqLCV have been mapped in one case to the common region, implicating replication, and in a second case to the BV1 ORF, implicating movement (97, 120). Similarly, detailed analysis of two variants of MSV has identified a point mutation in ORF V1 (movement) as a determinant of streak width, and a point mutation in the potential promoter of the C1-C2 gene (replication) as a determinant of severity of chlorosis, streak length, latency, and host range (9). Likewise, symptom development differences between two TGMV strains have been mapped to a fragment of the B component encompassing part of the common region and the BC1 movement protein (171). Processes controlling the level of viral accumulation in the plant should contribute to symptom severity and host range. Further investigations into these issues, perhaps using mutagenesis, may uncover additional details of the interactions between geminiviruses and their hosts.

GEMINIVIRUSES AS VECTORS

Advantages

Vectors based on geminiviruses accumulate to very high copy numbers in the inoculated cell, which can lead to high gene expression levels. In fact, a shuttle vector derived from WDV was shown to replicate to 30,000 copies per cell in maize endosperm protoplasts (163). Furthermore, the common strategy of coat protein gene replacement for vector construction has the advantage that the strong coat protein promoter drives expression of the new gene. The extrachromosomal replication of these vectors eliminates the variation in gene expression caused by position effects in integrated constructs, enabling accurate measurements of promoter strengths (see Applications below). In addition, nuclear localization may facilitate studies of DNA structure and chromatin and of RNA processing and stability. Biological containment problems for these vectors are alleviated by the lack of seed transmission and by the coat protein requirement for insect transmission. The systemic infection caused by these viruses can be used to study gene expression in differentiated tissues without creating transgenic plants. A better understanding of viral movement and symptom development, which can be expected soon based on the current pace of research in this field, may allow the design of vectors in which the plant remains symptomless, and the extent of spread can be controlled.

Disadvantages

Geminiviruses cannot be transmitted through seed, and have not been shown to integrate into host chromosomes. These characteristics may be disadvantageous for some applications. Monopartite viruses are phloem-limited, although this constraint may not be at the replication level because WDV constructs can replicate in various cell types (107, 135, 163, 164). Monopartite viruses also lack mechanical transmission and they require coat protein for movement. For both types of viruses, systemic spread can result in symptom development, and may cause vector instability (see below).

Vector Design

BIPARTITE GEMINIVIRUSES   DNA A of bipartite viruses encodes all the viral functions required for replication and encapsidation of viral DNA, while the B component is required for symptom development and viral movement (18, 39, 110, 130, 157, 166). Thus, the A genomes of both ACMV and TGMV have been used as autonomously replicating vectors. Because the coat protein of these viruses is not required for replication (18, 51, 148), this gene can be replaced with foreign sequences. Such replacement vectors with either a transcriptional or translational fusion of the introduced sequences have been constructed to express both viral and reporter genes (14, 36, 39, 62, 68, 71, 109, 154, 155, 172). When DNA B is cotransfected, or when plants carrying genomic copies of the B component are used, these vectors can spread systemically.

Such systemic movement, however, results in instability of TGMV (36, 68) and ACMV (148) constructs that are larger than wild-type (i.e. leading to deletion of cloned sequences and conversion to wild-type size). Surprisingly, increases in size to wild-type have also been seen for ACMV coat protein deletion derivatives (41, 90), but not for TGMV deletion mutants (51). Such unstable constructs replicate normally in protoplasts or leaf disks, indicating that size selection results from spread rather than replication. Therefore, rare events in the originally infected cells probably lead to wild-type–sized DNA molecules that preferentially proliferate because of their ability to spread. Systemically infected plants generally carry only one type of size revertant (36, 68).

One way to avoid vector instability and still obtain freely replicating plasmids in a large number of cells is by stably integrating multimeric copies of the viral construct in the plant genome (68, 71). In every cell type that expresses the replication-associated protein, monomeric copies can excise from the genome and be amplified extrachromosomally. The copy number obtained using this approach for TGMV was only 6-fold less than that obtained in a full systemic infection (71). An additional advantage is that viral constructs are stably maintained, and can be subjected to genetic manipulations.

MONOPARTITE GEMINIVIRUSES   All ORFs of these viruses are required for the viral life cycle, but only the intergenic regions and the complementary strand ORFs are necessary for replication (85, 100, 103, 167). Vectors based on the monopartite viruses are therefore of the replacement type where either ORF V1 or the coat protein is replaced by a marker gene (75, 100, 107, 117, 163, 164, 167). These vectors cannot spread systemically; however, mutations in both coat protein and ORF V1 of MSV can be complemented in *trans* (12), indicating that transgenic plants expressing these proteins can sustain movement of deficient vectors. Because the coat protein is involved in systemic spread for monopartite viruses (12, 100, 176), size reversion is probably a problem for these vectors as well. The previously mentioned strategy of integrating viral multimers in the genome may be applicable here too, although this has not been demonstrated.

Vector Construction

Replication-competent vector molecules can be obtained from cloned geminivirus sequences in several ways. For instance, viral copies can be cloned as a tandem duplication into a plasmid or T-DNA (17, 27, 35–37, 62, 68, 71, 75, 86, 107, 109, 154, 155, 164, 172). Upon delivery into the plant cell, monomeric viral DNA is released either through recombination between repeated sequences or, if the viral strand origin is duplicated, through rolling-circle replication (35). Alternately, direct repeats of a non-viral sequence flanking a functional viral copy can release a replicating molecule by recombination (100, 101). A third approach is to transfect a monomeric, linear viral DNA released from an *E. coli* plasmid by restriction enzymes; a cellular ligase activity probably produces replicating vectors by circularization (14, 39, 94, 107, 173). Mere linearization of the bacterial plasmid can also result in replication of ACMV constructs, possibly by recombination between linear DNAs in the cell (148). This latter strategy has the major disadvantage that the site of recombination, and thus the exact nature of the construct, is hard to control.

To circumvent the necessity of such in vivo or in vitro manipulation of constructs, a shuttle vector capable of replication in both *E. coli* and plant cells has been developed from WDV (Figure 3; 84, 163, 167). In this vector, the coat protein coding region is replaced by the selectable marker neomycin phosphotransferase type II (NPT II) and the CaMV 35S terminator, allowing kanamycin selection in plants. The presence of an *E. coli*-like promoter in the WDV genome just upstream of the coat protein initiation codon confers kanamycin resistance in *E. coli* as well. The construct also carries the p15A plasmid origin for replication in *E. coli,* and unique restriction sites facilitate cloning of expression cassettes. Shuttle vectors have the additional advantage of plasmid rescue in *E. coli,* which may have several uses. For example, it can be used to isolate products of chromosomal integration, transposition, and recombination events (163, 167). Sequences resembling *E. coli* promoter ele-

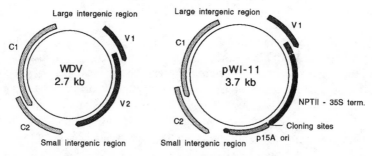

Figure 3   A WDV-based shuttle vector. Maps of the WDV genome and of the shuttle vector are shown. A translational fusion between the coat protein and the NPT II gene was created by replacing most of the V2 ORF with a construct carrying the NPT II gene linked to the 35S terminator and the p15A origin of replication (163, 167).

ments are present in the coat protein promoter region of bipartite viruses as well (121), enabling a similar vector design.

Vector Delivery

Three general methods have been used for the delivery of geminiviral vectors into plants. One method is mechanical transmission, which is possible only for bipartite geminiviruses, and occurs at low efficiency (35, 141). This method has been used successfully to infect plants using the sap of infected plants or vector DNA (14, 39, 142, 148, 172, 173). Another useful method is agroinfection, which involves T-DNA mediated delivery of viral genomes into plant cells. Repeated copies of the viral DNA are inserted between T-DNA borders; monomers are excised either by recombination or by replication. This technique was first developed for the transfer of cloned CaMV DNA to turnip plants (58), and has since been used extensively for geminiviruses (27, 36, 62, 68, 71, 101, 109). Agroinfection is indispensable for the monopartite geminiviruses, because these viruses are not amenable to mechanical transmission (100, 137, 138). Direct gene transfer technologies such as electroporation (163, 167), particle bombardment (23, 86), and PEG treatment (17, 37, 75, 94, 107, 154, 155) have also been used for transfection of protoplasts or callus tissues. In addition, imbibition of dried embryos results in the uptake of WDV DNA (164).

APPLICATIONS OF GEMINIVIRUSES AND THEIR VECTORS

Gene Expression

EXPRESSION OF FOREIGN GENES IN WHOLE PLANTS   Coat protein replacement vectors of bipartite viruses have been used to express several bacterial reporter genes in whole plants (68, 71, 173). One of these studies demonstrates quantitatively the increases in gene expression resulting from replication and systemic spread (71). The NPT II gene was fused to the coat protein promoter of TGMV, and integrated into the genome either as a multimer or as a single copy. The multimer can give rise to extrachromosomal replicating copies, which, in the presence of DNA B, can spread systemically. Replication in the absence and presence of viral spread resulted in 10-fold and 60-fold increases, respectively, in NPT II activity compared to the integrated monomer; the increase in enzyme activity correlated with the observed increases in vector copy number (71). Similarly, systemic spread of constructs with the chloramphenicol acetyltransferase (*CAT*) gene expressed from the coat protein promoter of ACMV results in high levels of *CAT* activity stable for a period of 4 weeks (173). Both these constructs are comparable in size to the viral DNA A and are therefore maintained during systemic spread. In addition, similar TGMV constructs expressing the β-glucuronidase (*GUS*) gene have been used to substantiate the absence of position effects in geminiviral vectors (68). Excision of replicating copies from 10 independent transformants carrying a partial dimer gave rise to less than 2-fold variation in total GUS activity, and only a 25% range when corrected for replicon copy number (68). This shows that promoter strengths may be measured accurately using geminiviral constructs.

EXPRESSION OF FOREIGN GENES IN LEAF DISKS AND PROTOPLASTS   Agroinfection of leaf disks provides another convenient measure of promoter strengths; transcripts of reporter genes are detectable as early as two days, and persist for several weeks (62). Using this assay, the consistency of gene expression from geminiviral vectors has been confirmed by showing that the variation in transcript levels for four independent agroinfections was less than 10%. In addition, RNA splicing studies using TGMV and agroinfection of leaf disks have shown that a dicot intron is spliced at > 95% efficiency and that monocot introns are spliced at slightly lower efficiencies (109).

Coat protein replacement constructs of WDV have been used to develop two similar assay systems in monocots, one using imbibition of dried embryos (164), and another using transfection of protoplasts (107). The shuttle vector based on WDV replicates in maize endosperm protoplasts (163, 167), and a 12-fold increase in expression has been observed for a 35S-GUS fusion cassette carried on this vector compared to a replication-deficient construct. A reduction in copy number from 30,000 to 2500 per cell is associated with this 2.9-kb insertion (163). Constructs in which the replication-associated protein gene is replaced by a 35S-GUS fusion could replicate when an intact gene was provided in trans, suggesting that vectors carrying just the intergenic regions can replicate extrachromosomally (163). Using this shuttle vector, an endogenous maize storage protein gene has been shown to be expressed in endosperm cells at high levels and with appropriate regulatory controls (MCP Timmermans & J Messing, unpublished data).

EXPRESSION OF VIRAL GENES   Geminiviral vectors have also been used to study various aspects of viral gene regulation. For example, substitution of the coat protein coding region with a reporter gene has allowed characterization of the coat protein promoter and its regulation by the AC2 or C1-C2 gene product (75, 154, 155). Substitution of this region with other viral ORFs has enabled the expression and characterization of these gene products. This approach has been used to establish a role for the coat protein in determining insect vector specificity (14), while similar experiments have been used to study the function of viral movement proteins (39, 172).

Engineering Geminivirus Resistance

Viral infections affect crop yields, and the selection of resistant varieties through classical breeding or genetic engineering has therefore been a focal point in crop improvement. Most of the effort has focused on RNA viruses, and three genetic engineering approaches have been taken (see 72 for review). One approach exploits the ability of viral satellite RNA to interfere with viral replication. The second strategy is based on cross-protection, in which infection with a mild viral strain or expression of viral proteins in transgenic plants confers protection to related and more virulent viruses. The third method uses antisense RNA expression, but this has had limited success.

Similar approaches have been used for geminiviruses. Subgenomic geminiviral DNAs occur in infected tissues, and like satellite RNAs these may affect replication of wild-type virus (48). For instance, transformation of *N. benthamiana* with a subgenomic DNA B of ACMV, and subsequent challenge with wild-type virus showed ameliorated symptoms and reduced viral DNA levels for two isolates of ACMV, and a 60% reduction in full-length DNA B (142). Symptom amelioration may result from reduced expression of the BV1 and BC1 movement proteins that are not encoded by the subgenomic DNAs. Interestingly, when virus was transmitted between transformants, extremely mild symptoms were observed, perhaps because of further reduction in levels of DNA B (142). Subgenomic DNAs are associated with many geminiviruses (49, 85, 103, 104, 150), so this method may be generally applicable. It would be effective only for different isolates of the same virus, however, because of the specificity of the replication-associated protein (142).

Natural cross-protection between geminiviral strains has not been observed, but the reason for this is unknown (65). There is one report of a phenomenon resembling cross-protection: expression of the BC1 movement protein of TGMV from the coat protein promoter of ACMV in a replacement construct resulted in replication and symptom induction on the inoculated leaves, but no systemic infection occurred (172). Surprisingly, expression of the endogenous ACMV BC1 gene in a similar construct produces normal systemic infection (39).

The antisense RNA approach has been successful for engineering protection against TGMV (27). Transgenic tobacco plants expressing antisense RNA of the TGMV AL1 ORF showed a drastic reduction in symptom development upon agroinfection with TGMV. In the most resistant line, only mild symptoms were seen on 9% of the plants. Viral replication was inhibited in these plants, and antisense RNA abundance correlated with viral DNA levels and resistance. Because the AL1 gene is conserved among geminiviruses, these plants have also been tested for suppression of replication of ACMV and BCTV, which bear 64% and 63% overall nucleotide sequence homology to

TGMV, respectively (ER Bejarano and CP Lichtenstein, submitted for publication). Using agroinfection of leaf disks, under conditions where TGMV replication was reduced 10-fold, BCTV replication was reduced 4-fold, and ACMV replication was unaffected. This may result from the fact that BCTV displays high homology to TGMV in specific stretches of the AL1 sequence, while the homology between ACMV and TGMV is more evenly dispersed (ER Bejarano and CP Lichtenstein, submitted for publication). Whether the 4-fold reduction is adequate to confer tolerance to BCTV is unknown. Antisense methods may, in general, be more successful for geminiviruses that have a single-stranded DNA genome located in the nucleus than for other viruses with life cycles restricted to the cytoplasm.

Geminiviruses as Markers for Gene Transfer

The first example of maize agroinfection by MSV (59), and the subsequent agroinfection of DSV (31) and WDV (69, 175), have demonstrated that cloned viral DNAs are biologically active, and that these viral genomes are indeed monopartite. Furthermore, these experiments have also demonstrated for the first time that T-DNA transfer to cereals is possible, providing a way to study the interaction between *Agrobacterium* and members of the Gramineae (7, see 8 for a review). Efficiency of transfer depends on the cereal host, which include barley, wheat, oat, millet, and rye. It also depends on the strain of *A. rhizogenes* or *A. tumefaciens* used. The dependence on bacterial strains has been mapped to the opine type of the Ti- or Ri-plasmid. In general, agropine strains of *A. rhizogenes* are the most efficient; for *A. tumefaciens,* strains harboring nopaline-type Ti-plasmids are able to transfer DNA but octopine strains are not. This difference between Ti-plasmids appears to map to the *virA* gene (126). Geminiviruses have also been used to assess the transformation of Gramineae by direct gene transfer. WDV has been used to show that DNA could be introduced into dry embryos of various monocot species by imbibition (164), and that microspore-derived embryos of barley can be transfected by microprojectile bombardment (23). Thus, although plant transformation technology has played an important role in the characterization of plant viruses, the introduction of viruses via agroinfection or direct gene transfer has clearly added to our understanding of factors involved in transformation of cereals.

Analysis of Transposition Mechanisms

Geminiviruses have been used to study the well-known *Activator* (*Ac*) and *Dissociation* (*Ds*) transposons of maize (94, 137, 138). The first indication that geminiviruses could be used as replicons to study transposition came from the observation that *Ac* could excise from WDV-derived constructs (94). Excision appeared to be linked to replication, because no excision product could be detected from replication-deficient plasmids. *Ds* elements were capable of transposition only when the *Ac* transposase was provided in *trans*; excision products showed footprints typical of *Ac*-mediated excision (94). Together, these observations strongly indicate that similar excision mechanisms act on these extrachromosomal replicons as on chromosomal copies of *Ds*. *Ac*-dependent transposition and typical transposition footprints have also been obtained when MSV derivatives carrying *Ds* in the small intergenic region are introduced into maize seedlings via agroinfection (138). Here, *Ds* insertion does not affect viral replication, but abolishes symptom development; therefore, excision can be monitored by the appearance of symptoms, which usually occurs within 20 days after inoculation. This strategy provides a simple, rapid, and reproducible method to obtain transposition events, and the characterization of a relatively large number of footprints has begun (137). Similar approaches should be applicable to other transposons.

DNA Methylation

Methylation levels often correlate with changes in gene expression and chromatin structure, and have been linked to phenomena such as imprinting, cosuppression, and paramutation in plants. Upon DNA replication, the methylation pattern of the parental strand is copied onto the newly synthesized DNA strand, resulting in the maintenance of methylation. The small, extrachromosomally replicating molecules provided by geminiviral vectors may offer a simple assay in which methylation and its connection to other cellular processes can be studied, analogous to SV40-based vectors in animal cells (80). Several studies have shown, however, that geminiviral vectors methylated in vitro lose their methylation upon replication in plant cells (17, 37). This applies both to constructs methylated on both strands (obtained using methylases such as Hpa II and Sss I), and to hemimethylated constructs (obtained by primer extension of ss DNA using 5-methyl-dCTP). These results indicate that replication from these methylated vectors circumvents the action of the maintenance methylase, and is consistent with the fact that no cytosine methylation has been reported for geminiviruses (17). Loss of methylation may occur because these replicons lack an appropriate chromatin structure. Preliminary studies of chromatin structure of AbMV RF forms have shown, however, that they cosediment with host chromatin, and may possess some nucleosomal organization (123). Alternately, the rolling-circle mode of replication may not allow sufficiently long-lived hemimethylated intermediates for the maintenance methylase to act on. In either case, geminiviral vectors do not seem to be suited for studies of DNA methylation in plants.

Potential Applications

Several other potential applications can be envisioned based on the high copy numbers that can be achieved. The high levels of primary transcripts obtained may facilitate measurements of the efficiency or rates of post-transcriptional processes involved in gene expression, such as mRNA capping, intron processing, polyadenylation, 3' cleavage, and RNA stability. The inefficiency of such processes, or of transcription, may contribute to the disproportionately low gene expression level observed relative to the high gene copy numbers (163). Certain applications may be less affected by these constraints. For example, the generation of high levels of antisense RNA in the nucleus may depend largely on transcription rate, and less on RNA processing. Alternately, other applications can exploit this phenomenon. Thus, a transcription factor could be depleted from cells by amplification of its binding sequence carried on a viral vector, permitting identification of the role played by the factor. It may also be possible to isolate DNA-binding factors from cell lysates as a complex with vector DNA.

The extrachromosomal replication of these vectors is another useful feature. For example, extrachromosomal recombination can be studied, analogous to experiments using SV40 vectors in mammalian cells (80). Also, geminiviruses may facilitate gene targeting, i.e. recombination between a locus on the chromosome and an extrachromosomal homologous sequence on the vector. This process could be stimulated by producing a ss nick in an integrated chromosomal copy of the geminiviral origin by means of the replication-associated protein. Similarly, transposition mechanisms can be studied profitably using geminiviral vectors (137). Questions such as the requirement for replication in transposition, the role of hemimethylated DNA and of specific flanking sequences, the structure of transposition intermediates, footprints, and dosage effects can be addressed. Transposons borne on geminiviral vectors may also aid gene tagging. For example, excision products may integrate into the chromosome, generating random mutagenesis. Alternately, non-autonomous elements (like *Ds*) integrated into the chromosome can be stimulated to transpose by a transposase carried on a geminiviral vector, obviating the need for removal of the autonomous element (like *Ac*) by segregation. The ds replicative forms may also provide a way to study chromatin structure (123). Another question of interest is the maintenance of these ds forms in daughter cells following mitosis in the absence of systemic spread. The observation that expression of the movement protein encoded by ORF BC1 of SqLCV leads to symptom development (120) indicates that intercellular communication mediated by plasmodesmata may be studied by using this protein to manipulate plasmodesmatal characteristics.

Geminiviruses offer several possibilities for vector design. Applications that do not require systemic infection can use coat protein replacement vectors without a DNA B component in the case of bipartite viruses. For such constructions, size constraints are marginal, and shuttle vectors facilitate cloning, transfection, and plasmid rescue. However, if systemic infection is required, the size constraints can be met by transfer of essential viral functions to a cotransfected helper virus. Alternately, the host chromosome can be engineered to express viral functions to support replication or movement of deficient vectors. Theoretically, the vector only needs to contain the *cis* sequences required for replication, which can be as little as a few hundred base pairs (101, 163). As discussed earlier, viral copies can be integrated into the host chromosome, leading to production of extrachromosomal replicons upon expression of the replication-associated protein (68, 71). This approach allows the replication-associated protein to be expressed in an inducible or tissue-specific manner using appropriate promoters (i.e. the viral life cycle should be inducible in specific cell types, at specific developmental stages, or in response to specific signals).

ACKNOWLEDGMENTS

We thank our many colleagues for communicating preprints and unpublished data. Dr. Sanjay Swarup is acknowledged for comments on the manuscript and valuable discussions. Our research has been supported in part by a grant from the National Institutes of Health (GM 43261).

Literature Cited

1. Abouzid AM, Polston JE, Hiebert E. 1992. The nucleotide sequence of tomato mottle virus, a new geminivirus isolated from tomatoes in Florida. *J. Gen. Virol.* 73:3225–29

2. Accotto GP, Donson J, Mullineaux PM. 1989. Mapping of *Digitaria* streak virus transcripts reveals different RNA species from the same transcription unit. *EMBO J.* 8:1033–39

3. Accotto GP, Mullineaux PM, Brown SC, Marie D. 1993. *Digitaria* streak geminivirus replicative forms are abundant in S-phase nuclei of infected cells. *Virology* 195:257–59

4. Adejare GO, Coutts RHA. 1982. Ultrastructural studies on *Nicotiana benthami-*

ana tissue following infection with a virus transmitted from mosaic-diseased Nigerian cassava. *Phytopathol. Z.* 103:87–92

5. Andersen MT, Richardson KA, Harbison S, Morris BAM. 1988. Nucleotide sequence of the geminivirus *Chloris* striate mosaic virus. *Virology* 164:443–49

6. Bisaro DM, Sunter G, Revington GN, Brough CL, Hormuzdi SG, Hartitz M. 1990. Molecular genetics of tomato golden mosaic virus replication: progress toward defining gene functions, transcription units and the origin of DNA replication. In *Viral Genes and Plant Pathogenesis*, ed. TP Pirone, JG Shaw, pp. 89–104. New York/Berlin: Springer-Verlag. 215 pp.

7. Boulton MI, Buchholz WG, Marks MS,

Markham PG, Davies JW. 1989. Specificity of *Agrobacterium*-mediated delivery of maize streak virus DNA to members of the gramineae. *Plant Mol. Biol.* 12:31–40

8. Boulton MI, Davies JW. 1990. Monopartite geminiviruses: markers for gene transfer to cereals. *Aspects Appl. Biol.* 24:79–86

9. Boulton MI, King DI, Donson J, Davies JW. 1991. Point substitutions in a promoter-like region and the V1 gene affect the host range and symptoms of maize streak virus. *Virology* 183:114–21

10. Boulton MI, Markham PG. 1986. The use of squash-blotting to detect plant pathogens in insect vectors. In *Developments in Applied Biology*, ed. RAC Jones, L Torrance, 1:55–69. Wellesbourne, Warwick: Assoc. Appl. Biol. 300 pp.

11. Boulton MI, Pallaghy CK, Chatani M, MacFarlane S, Davies JW. 1993. Replication of maize streak virus mutants in maize protoplasts: evidence for a movement protein. *Virology* 192:85–93

12. Boulton MI, Steinkellner H, Donson J, Markham PG, King DI, Davies JW. 1989. Mutational analysis of the virion-sense genes of maize streak virus. *J. Gen. Virol.* 70:2309–23

13. Briddon RW, Lunness P, Chamberlin LCL, Pinner MS, Brundish H, Markham PG. 1992. The nucleotide sequence of an infectious insect-transmissible clone of the geminivirus *Panicum* streak virus. *J. Gen. Virol.* 73:1041–47

14. Briddon RW, Pinner MS, Stanley J, Markham PG. 1990. Geminivirus coat protein gene replacement alters insect specificity. *Virology* 177:85–94

15. Briddon RW, Watts J, Markham PG, Stanley J. 1989. The coat protein of beet curly top virus is essential for infectivity. *Virology* 172:628–33

16. Brisson N, Paszkowski J, Penswick JR, Gronenborn B, Potrykus I, Hohn T. 1984. Expression of a bacterial gene in plants by using a viral vector. *Nature* 310:511–14

17. Brough CL, Gardiner WE, Inamdar NM, Zhang X-Y, Ehrlich M, Bisaro DM. 1992. DNA methylation inhibits propagation of tomato golden mosaic virus DNA in transfected protoplasts. *Plant Mol. Biol.* 18:703–12

18. Brough CL, Hayes RJ, Morgan AJ, Coutts RHA, Buck KW. 1988. Effects of mutagenesis in vitro on the ability of cloned tomato golden mosaic virus DNA to infect *Nicotiana benthamiana* plants. *J. Gen. Virol.* 69:503–14

19. Brough CL, Sunter G, Gardiner WE, Bisaro DM. 1992. Kinetics of tomato golden mosaic virus DNA replication and coat protein promoter activity in *Nicotiana tabacum* protoplasts. *Virology* 187:1–9

20. Chatani M, Matsumoto Y, Mizuta H,

Ikegami M, Boulton MI, Davies JW. 1991. The nucleotide sequence and genome structure of the geminivirus *Miscanthus* streak virus. *J. Gen. Virol.* 72:2325–31

21. Cohen SJ, Duffus JE, Liu HY. 1989. Acquisition, interference, and retention of curcurbit leaf curl viruses in whiteflies. *Phytopathology* 79:109–13

22. Coutts RHA, Coffin RS, Roberts EJF, Hamilton WDO. 1991. The nucleotide sequence of the infectious cloned DNA components of potato yellow mosaic virus. *J. Gen. Virol.* 72:1515–20

23. Creissen G, Smith C, Francis R, Reynolds H, Mullineaux P. 1990. *Agrobacterium*- and projectile-mediated viral DNA delivery into barley microspore-derived cultures. *Plant Cell Rep.* 8:680–83

24. Davies JW. 1987. Geminivirus genomes. *Microbiol. Sci.* 4:18–23

25. Davies JW, Stanley J. 1989. Geminivirus genes and vectors. *Trends Genet.* 5:77–81

26. Dawson WO, Lewandowski DJ, Hilf ME, Bubrick P, Raffo AJ, et al. 1989. A tobacco mosaic virus-hybrid expresses and loses an added gene. *Virology* 172:285–92

27. Day AG, Bejarano ER, Buck KW, Burrell M, Lichtenstein CP. 1991. Expression of an antisense viral gene in transgenic tobacco confers resistance to the DNA virus tomato golden mosaic virus. *Proc. Natl. Acad. Sci. USA* 88:6721–25

28. Dekker EL, Woolston CJ, Xue Y, Cox B, Mullineaux PM. 1991. Transcript mapping reveals different expression strategies for the bicistronic RNAs of the geminivirus wheat dwarf virus. *Nucleic Acids Res.* 19:4075–81

29. De Zoeten GA, Penswick JR, Horisberger MA, Ahl P, Schultze M, Hohn T. 1989. The expression localization and effect of human interferon in plants. *Virology* 172:213–22

30. Donson J, Accotto GP, Boulton MI, Mullineaux PM, Davies JW. 1987. The nucleotide sequence of a geminivirus from *Digitaria sanguinalis*. *Virology* 161:160–69

31. Donson J, Gunn HV, Woolston CJ, Pinner MS, Boulton MI, et al. 1988. *Agrobacterium*-mediated infectivity of cloned *Digitaria* streak virus DNA. *Virology* 162:248–50

32. Donson J, Morris-Krsinich BAM, Mullineaux PM, Boulton MI, Davies JW. 1984. A putative primer for second-strand DNA synthesis of maize streak virus is virion associated. *EMBO J.* 3:3069–73

33. Dry IB, Rigden JE, Krake LR, Mullineaux PM, Rezaian MA. 1993. Nucleotide sequence and genome organization of tomato leaf curl geminivirus. *J. Gen. Virol.* 74:147–51

34. Elmer JS, Brand L, Sunter G, Gardiner WE, Bisaro DM, Rogers SG. 1988. Genetic analysis of the tomato golden mosaic virus II. The product of the AL1 coding sequence

is required for replication. *Nucleic Acids Res.* 16:7043–60

35. Elmer JS, Sunter G, Gardiner WE, Brand L, Browning CK, et al. 1988. *Agrobacterium*-mediated inoculation of plants with tomato golden mosaic virus DNAs. *Plant Mol. Biol.* 10:225–34

36. Elmer S, Rogers SG. 1990. Selection for wild type size derivatives of tomato golden mosaic virus during systemic infection. *Nucleic Acids Res.* 18:2001–6

37. Ermak G, Paszkowski U, Wohlmuth M, Mittelsten-Scheid O, Paszkowski J. 1993. Cytosine methylation inhibits replication of African cassava mosaic virus by two distinct mechanisms. *Nucleic Acids Res.* 21: 3445–50

38. Esau K. 1977. Virus-like particles in nuclei of phloem cells in spinach leaves infected with the curly top virus. *J. Ultrastruct. Res.* 61:78–88

39. Etessami P, Callis R, Ellwood S, Stanley J. 1988. Delimitation of essential genes of cassava latent virus DNA 2. *Nucleic Acids Res.* 16:4811–29

40. Etessami P, Saunders K, Watts J, Stanley J. 1991. Mutational analysis of complementary-sense genes of African cassava mosaic virus DNA A. *J. Gen. Virol.* 72: 1005–12

41. Etessami P, Watts J, Stanley J. 1989. Size reversion of African cassava mosaic virus coat protein gene deletion mutants during infection of *Nicotiana benthamiana*. *J. Gen. Virol.* 70:277–89

42. Evans D, Jeske H. 1993. DNA B facilitates, but is not essential for, the spread of abutilon mosaic virus in agroinoculated *Nicotiana benthamiana*. *Virology* 194:752–57

43. Fenoll C, Black DM, Howell SH. 1988. The intergenic region of maize streak virus contains promoter elements involved in rightward transcription of the viral genome. *EMBO J.* 7:1589–96

44. Fenoll C, Schwarz JJ, Black DM, Schneider M, Howell SH. 1990. The intergenic region of maize streak virus contains a GC-rich element that activates rightward transcription and binds maize nuclear factors. *Plant Mol. Biol.* 15:865–77

45. Fontes EPB, Luckow VA, Hanley-Bowdoin L. 1992. A geminivirus replication protein is a sequence-specific DNA binding protein. *Plant Cell* 4:597–608

46. French R, Janda M, Ahlquist P. 1986. Bacterial gene inserted in an engineered RNA virus: efficient expression in monocotyledonous plant cells. *Science* 231:1294–97

47. Frischmuth S, Frischmuth T, Jeske H. 1991. Transcript mapping of abutilon mosaic virus, a geminivirus. *Virology* 185: 596–604

48. Frischmuth T, Stanley J. 1991. African cassava mosaic virus DI DNA interferes with the replication of both genomic components. *Virology* 183:539–44

49. Frischmuth T, Stanley J. 1992. Characterization of beet curly top virus subgenomic DNA localizes sequences required for replication. *Virology* 189:808–11

50. Frischmuth T, Zimmat G, Jeske H. 1990. The nucleotide sequence of abutilon mosaic virus reveals prokaryotic as well as eukaryotic features. *Virology* 178:461–68

51. Gardiner WE, Sunter G, Brand L, Elmer JS, Rogers SG, Bisaro DM. 1988. Genetic analysis of tomato golden mosaic virus: the coat protein is not required for systemic spread or symptom development. *EMBO J.* 7:899–904

52. Gidoni D, Kadonaga JT, Barrera-Saldane H, Takahashi K, Chambon P, Tjian R. 1985. Bidirectional SV40 transcription mediated by tandem Sp1 binding interactions. *Science* 230:511–17

53. Gilbertson RL, Faria JC, Hanson SF, Morales FJ, Ahlquist P, et al. 1991. Cloning of the complete DNA genomes of four bean-infecting geminiviruses and determining their infectivity by electric discharge particle acceleration. *Phytopathology* 81:980–85

54. Goodall G, Filipowicz W. 1991. Different effects of intron nucleotide composition and secondary structure on pre-mRNA splicing in monocot and dicot plants. *EMBO J.* 10:2635–44

55. Goodman RM. 1977. Single-stranded DNA genome in a whitefly-transmitted plant virus. *Virology* 83:171–79

56. Goodman RM, Shock TL, Haber S, Browning KS, Bowers GR. 1980. The composition of bean golden mosaic virus and its single-stranded DNA genome. *Virology* 106:168–72

57. Gorbalenya AE, Koonin EV, Wolf YI. 1990. A new superfamily of putative NTP-binding domains encoded by genomes of small DNA and RNA viruses. *FEBS Lett.* 262:145–48

58. Grimsley N, Hohn B, Hohn T, Walden R. 1986. "Agroinfection", an alternative route for viral infection of plants by using the Ti plasmid. *Proc. Natl. Acad. Sci. USA* 83: 3282–86

59. Grimsley N, Hohn T, Davies JW, Hohn B. 1987. *Agrobacterium*-mediated delivery of infectious maize streak virus into maize plants. *Nature* 325:177–79

60. Haley A, Zhan X, Richardson K, Head K, Morris D. 1992. Regulation of the activities of African cassava mosaic virus promoters by the AC1, AC2, and AC3 gene products. *Virology* 188:905–9

61. Hamilton WDO, Stein VE, Coutts RHA, Buck KW. 1984. Complete nucleotide sequence of the infectious cloned DNA components of tomato golden mosaic virus:

potential coding regions and regulatory sequences. *EMBO J.* 3:2197–205

62. Hanley-Bowdoin L, Elmer JS, Rogers SG. 1988. Transient expression of heterologous RNAs using tomato golden mosaic virus. *Nucleic Acids Res.* 16:10511–28

63. Hanley-Bowdoin L, Elmer JS, Rogers SG. 1989. Functional expression of the leftward open reading frames of the A component of tomato golden mosaic virus in transgenic tobacco plants. *Plant Cell* 1:1057–67

64. Hanley-Bowdoin L, Elmer JS, Rogers SG. 1990. Expression of functional replication protein from tomato golden mosaic virus in transgenic tobacco plants. *Proc. Natl. Acad. Sci. USA* 87:1446–50

65. Harrison BD. 1985. Advances in geminivirus research. *Annu. Rev. Phytopathol.* 23: 55–82

66. Harrison BD, Barker H, Bock KR, Guthrie EJ, Meredith G, Atkinson M. 1977. Plant viruses with circular single-stranded DNA. *Nature* 270:760–62

67. Hayes RJ, Buck KW. 1989. Replication of tomato golden mosaic virus DNA B in transgenic plants expressing open reading frames (ORFs) of DNA A: requirement of ORF AL2 for production of single-stranded DNA. *Nucleic Acids Res.* 17:10213–22

68. Hayes RJ, Coutts RHA, Buck KW. 1989. Stability and expression of bacterial genes in replicating geminivirus vectors in plants. *Nucleic Acids Res.* 17:2391–403

69. Hayes RJ, MacDonald H, Coutts RHA, Buck KW. 1988. Agroinfection of *Triticum aestivum* with cloned DNA of wheat dwarf virus. *J. Gen. Virol.* 69:891–96

70. Hayes RJ, MacDonald H, Coutts RHA, Buck KW. 1988. Priming of complementary DNA synthesis in vitro by small DNA molecules tightly bound to virion DNA of wheat dwarf virus. *J. Gen. Virol.* 69:1345–50

71. Hayes RJ, Petty ITD, Coutts RHA, Buck KW. 1988. Gene amplification and expression in plants by a replicating geminivirus vector. *Nature* 334:179–82

72. Hemenway C, Haley L, Kaniewski WK, Lawson EC, O'Connell KM, et al. 1990. Genetically engineered resistance: transgenic plants. In *Plant Viruses: Pathology,* ed. CL Mandahar, 2:347–63. Boca Raton, FL: CRC Press. 371 pp.

73. Heyraud F, Matzeit V, Schaefer S, Schell J, Gronenborn B. 1993. The conserved nonanucleotide motif of the geminivirus stem-loop sequence promotes replicational release of virus molecules from redundant copies. *Biochimie* 75:605–15

74. Hidayat SH, Gilbertson RL, Hanson SF, Morales FJ, Ahlquist P, et al. 1993. Complete nucleotide sequence of the infectious cloned DNAs of bean dwarf mosaic geminivirus. *Phytopathology* 83:181–87

75. Hofer JMI, Dekker EL, Reynolds HV, Woolston CJ, Cox BS, Mullineaux PM. 1992. Coordinate regulation of replication and virion sense gene expression in wheat dwarf virus. *Plant Cell* 4:213–23

76. Hormuzdi SG, Bisaro DM. 1993. Genetic analysis of beet curly top virus: evidence for three virion sense genes involved in movement and regulation of single- and double-stranded DNA levels. *Virology* 193: 900–9

77. Howarth AJ, Caton J, Bossert M, Goodman RM. 1985. Nucleotide sequence of bean golden mosaic virus and a model for gene regulation in geminiviruses. *Proc. Natl. Acad. Sci. USA* 82:3572–76

78. Howarth AJ, Van de Mark GJ. 1989. Phylogeny of geminiviruses. *J. Gen. Virol.* 70: 2717–27

79. Howell SH. 1984. Physical structure and genetic organization of the genome of maize streak virus (Kenyan isolate). *Nucleic Acids Res.* 12:7359–75

80. Hsieh C-L, Lieber MR. 1992. CpG methylated minichromosomes become inaccessible for V(D)J recombination after undergoing replication. *EMBO J.* 11:315–25

81. Hughes FL, Rybicki EP, Kirby R. 1993. Complete nucleotide sequence of sugarcane streak monogeminivirus. *Arch. Virol.* 132:171–82

82. Hughes FL, Rybicki EP, Von Wechmar MB. 1992. Genome typing of southern Africa subgroup 1 geminiviruses. *J. Gen. Virol.* 73:1031–40

83. Kallender H, Petty ITD, Stein VE, Panico M, Blench IP, et al. 1988. Identification of the coat protein gene of tomato golden mosaic virus. *J. Gen. Virol.* 69:1351–57

84. Kammann M, Matzeit V, Schmidt B, Schell J, Walden R, Gronenborn B. 1991. Geminivirus-based shuttle vectors capable of replication in *Escherichia coli* and monocotyledonous plant cells. *Gene* 104:247–52

85. Kammann M, Schalk H-J, Matzeit V, Schaefer S, Schell J, Gronenborn B. 1991. DNA replication of wheat dwarf virus, a geminivirus, requires two cis-acting signals. *Virology* 184:786–90

86. Kanevski IF, Thakur S, Cosowsky L, Sunter G, Brough C, et al. 1992. Tobacco lines with high copy number of replicating recombinant geminivirus vectors after biolistic DNA delivery. *Plant J.* 2:457–63

87. Kheyr-Pour A, Bendahmane M, Matzeit V, Accotto GP, Crespi S, Gronenborn B. 1992. Tomato yellow leaf curl virus from Sardinia is a whitefly-transmitted monopartite geminivirus. *Nucleic Acids Res.* 19:6763–69

88. Kikuno R, Toh H, Hayashida H, Miyata T. 1984. Sequence similarity between putative gene products of geminivirus DNAs. *Nature* 308:562

89. Kim KS, Shock TL, Goodman RM. 1978. Infection of *Phaseolus vulgaris* by bean golden mosaic virus: ultrastructural aspects. *Virology* 89:22–33

90. Klinkenberg FA, Ellwood S, Stanley J. 1989. Fate of African cassava mosaic virus coat protein deletion mutants after agroinfection. *J. Gen. Virol.* 70:1837–44

91. Klinkenberg FA, Stanley J. 1990. Encapsidation and spread of African cassava mosaic virus DNA A in the absence of DNA B when agroinoculated to *Nicotiana benthamiana. J. Gen. Virol.* 71:1409–12

92. Koonin EV, Ilyina TV. 1992. Geminivirus replication proteins are related to prokaryotic plasmid rolling circle DNA replication initiator proteins. *J. Gen. Virol.* 73:2763–66

93. Koonin EV, Mushegian AR, Ryabov EV, Dolja VV. 1991. Diverse groups of plant RNA and DNA viruses share related movement proteins that may possess chaperone-like activity. *J. Gen. Virol.* 72:2895–903

94. Laufs J, Wirtz U, Kammann M, Matzeit V, Schaefer S, et al. 1990. Wheat dwarf virus *Ac/Ds* vectors: expression and excision of transposable elements introduced into various cereals by a viral replicon. *Proc. Natl. Acad. Sci. USA* 87:7752–56

95. Lazarowitz SG. 1987. The molecular characterization of geminiviruses. *Plant Mol. Biol. Rep.* 4:177–92

96. Lazarowitz SG. 1988. Infectivity and complete nucleotide sequence of the genome of a South African isolate of maize streak virus. *Nucleic Acids Res.* 16:229–49

97. Lazarowitz SG. 1991. Molecular characterization of two bipartite geminiviruses causing squash leaf curl disease: role of viral replication and movement functions in determining host range. *Virology* 180: 70–80

98. Lazarowitz SG. 1992. Geminiviruses: genome structure and gene function. *Crit. Rev. Plant Sci.* 11:327–49

99. Lazarowitz SG, Lazdins IB. 1991. Infectivity and complete nucleotide sequence of the cloned genomic components of a bipartite squash leaf curl geminivirus with a broad host range phenotype. *Virology* 180:58–69

100. Lazarowitz SG, Pinder AJ, Damsteegt VD, Rogers SG. 1989. Maize streak virus genes essential for systemic spread and symptom development. *EMBO J.* 8:1023–32

101. Lazarowitz SG, Wu LC, Rogers SG, Elmer JS. 1992. Sequence-specific interaction with the viral AL1 protein identifies a geminivirus DNA replication origin. *Plant Cell* 4:799–809

102. Lefebvre DD, Miki BL, Laliberte J-F. 1987. Mammalian metallothionein functions in plants. *Bio/Technology* 5:1053–56

103. MacDonald H, Coutts RHA, Buck KW. 1988. Characterization of a subgenomic DNA isolated from *Triticum aestivum* plants infected with wheat dwarf virus. *J. Gen. Virol.* 69:1339–44

104. MacDowell SW, Coutts RHA, Buck KW. 1986. Molecular characterization of subgenomic single-stranded and double-stranded DNA forms isolated from plants infected with tomato golden mosaic virus. *Nucleic Acids Res.* 14:7967–84

105. MacDowell SW, MacDonald H, Hamilton WDO, Coutts RHA, Buck KW. 1985. The nucleotide sequence of cloned wheat dwarf virus DNA. *EMBO J.* 4:2173–80

106. Markham PG, Pinner MS, Boulton MI. 1984. The transmission of maize streak virus by leafhoppers, a new look at host adaptation. *Bull. Soc. Entomol. Suisse* 57: 431–32

107. Matzeit V, Schaefer S, Kammann M, Schalk H-J, Schell J, Gronenborn B. 1991. Wheat dwarf virus vectors replicate and express foreign genes in cells of monocotyledonous plants. *Plant Cell* 3:247–58

108. Maule AJ. 1991. Virus movement in infected plants. *Crit. Rev. Plant Sci.* 9:457–73

109. McCullough AJ, Lou H, Schuler MA. 1991. In vivo analysis of plant pre-mRNA splicing using an autonomously replicating vector. *Nucleic Acids Res.* 19:3001–9

110. Meyer P, Niedenhof I, Heidmann I, Saedler H. 1989. Extrachromosomal forms of CLV DNA1 in transgenic plants are inherited by symptom-free progeny. *Plant Sci.* 65:207–16

111. Morris B, Richardson K, Eddy P, Zhan X, Haley A, Gardner R. 1991. Mutagenesis of the AC3 open reading frame of African cassava mosaic virus DNA A reduces DNA B replication and ameliorates disease symptoms. *J. Gen. Virol.* 72:1205–13

112. Morris BAM, Richardson KA, Haley A, Zhan X, Thomas JE. 1992. The nucleotide sequence of the infectious cloned DNA component of tobacco yellow dwarf virus reveals features of geminiviruses infecting monocotyledonous plants. *Virology* 187: 633–42

113. Morris-Krsinich BAM, Mullineaux PM, Donson J, Boulton MI, Markham PG, et al. 1985. Bidirectional transcription of maize streak virus DNA and identification of the coat protein gene. *Nucleic Acids Res.* 13: 7237–56

114. Mullineaux PM, Boulton MI, Bowyer P, van der Vlugt R, Marks M, et al. 1988. Detection of a non-structural protein of Mr 11000 encoded by the virion DNA of maize streak virus. *Plant Mol. Biol.* 11:57–66

115. Mullineaux PM, Creissen G, Reynolds HV, Accotto GP, Woolston CJ. 1991. Organization of the genomes of geminiviruses which infect the Gramineae. In *Proceedings of the Phytochemical Society of Europe: Biochemistry and Molecular Biology of Plant-Pathogen Interactions,* ed. CJ Smith, 32:

50–75. Oxford/New York: Oxford Univ. Press. 291 pp.

116. Mullineaux PM, Donson J, Morris-Krsinich BAM, Boulton MI, Davies JW. 1984. The nucleotide sequence of maize streak virus DNA. *EMBO J.* 3:3063–68

117. Mullineaux PM, Guerineau F, Accotto GP. 1990. Processing of complementary sense RNAs of *Digitaria* streak virus in its host and in transgenic tobacco. *Nucleic Acids Res.* 18:7259–65

118. Navot N, Pichersky E, Zeidan M, Zamir D, Czosnek H. 1991. Tomato yellow leaf curl virus: a whitefly-transmitted geminivirus with a single genomic component. *Virology* 185:151–61

119. Ow DW, Jacobs JD, Howell SH. 1987. Functional regions of the cauliflower mosaic virus 35S promoter determined by the firefly luciferase gene as a reporter of promoter activity. *Proc. Natl. Acad. Sci. USA* 84:4870–74

120. Pascal E, Goodlove PE, Wu LC, Lazarowitz SG. 1993. Transgenic tobacco plants expressing the geminivirus BL1 protein exhibit symptoms of viral disease. *Plant Cell* 5:795–807

121. Petty ITD, Coutts RHA, Buck KW. 1986. Geminivirus coat protein gene promoter sequences can function in *Escherichia coli*. *Nucleic Acids Res.* 14:5113

122. Petty ITD, Coutts RHA, Buck KW. 1988. Transcriptional mapping of the coat protein gene of tomato golden mosaic virus. *J. Gen. Virol.* 69:1359–65

123. Pilartz M, Jeske H. 1992. Abutilon mosaic geminivirus double-stranded DNA is packed into minichromosomes. *Virology* 189:800–2

124. Pinner MS, Markham PG. 1990. Serotyping and strain identification of maize streak virus isolates. *J. Gen. Virol.* 71:1635–40

125. Pinner MS, Markham PG, Rybicki EP, Greber RS. 1992. Serological relationships of geminivirus isolates from Gramineae in Australia. *Plant Pathol.* 41:618–25

126. Raineri DM, Boulton MI, Davies JW, Nester EW. 1993. *VirA,* the plant-signal receptor, is responsible for the Ti plasmid-specific transfer of DNA to maize by *Agrobacterium. Proc. Natl. Acad. Sci. USA* 90:3549–53

127. Revington GN, Sunter G, Bisaro DM. 1989. DNA sequences essential for replication of the B genome component of tomato golden mosaic virus. *Plant Cell* 1:985–92

128. Roberts IM, Robinson DJ, Harrison BD. 1984. Serological relationships and genome homologies among geminiviruses. *J. Gen. Virol.* 65:1723–30

129. Rochester DE, Kositratana W, Beachy RN. 1990. Systemic movement and symptom production following agroinoculation with a single DNA of tomato yellow leaf curl geminivirus (Thailand). *Virology* 178:520–26

130. Rogers SG, Bisaro DM, Horsch RB, Fraley RT, Hoffmann NL, et al. 1986. Tomato golden mosaic virus A component DNA replicates autonomously in transgenic plants. *Cell* 45:593–600

131. Rybicki EP. 1991. The use of serological differentiation indices for the phylogenetic analysis of plant virus relationships. *Arch. Virol.* 119:83–93

132. Rybicki EP, Hughes FL. 1990. Detection and typing of maize streak virus and other distantly related geminiviruses of grasses by polymerase chain reaction amplification of a conserved viral sequence. *J. Gen. Virol.* 71:2519–26

133. Saunders K, Lucy A, Stanley J. 1991. DNA forms of the geminivirus African cassava mosaic virus consistent with a rolling circle mechanism of replication. *Nucleic Acids Res.* 19:2325–30

134. Saunders K, Lucy A, Stanley J. 1992. RNA-primed complementary-sense DNA synthesis of the geminivirus African cassava mosaic virus. *Nucleic Acids Res.* 20:6311–15

135. Schalk H-J, Matzeit V, Schiller B, Schell J, Gronenborn B. 1989. Wheat dwarf virus, a geminivirus of graminaceous plants needs splicing for replication. *EMBO J.* 8:359–64

136. Schneider M, Jarchow E, Hohn B. 1992. Mutational analysis of the "conserved region" of maize streak virus suggests its involvement in replication. *Plant Mol. Biol.* 19:601–10

137. Shen W-H, Das S, Hohn B. 1992. Mechanism of *Ds1* excision from the genome of maize streak virus. *Mol. Gen. Genet.* 233:388–94

138. Shen W-H, Hohn B. 1992. Excision of a transposable element from a viral vector introduced into maize plants by agroinfection. *Plant J.* 2:35–42

139. Siegel A. 1985. Plant-virus-based vectors for gene transfer may be of considerable use despite a presumed high error frequency during RNA synthesis. *Plant Mol. Biol.* 4:327–29

140. Smith S, Stillman B. 1991. Stepwise assembly of chromatin during DNA replication in vitro. *EMBO J.* 10:971–80

141. Stanley J. 1985. The molecular biology of geminiviruses. *Adv. Virus Res.* 30:139–77

142. Stanley J, Frischmuth T, Ellwood S. 1990. Defective viral DNA ameliorates symptoms of geminivirus infection in transgenic plants. *Proc. Natl. Acad. Sci. USA* 87:6291–95

143. Stanley J, Gay MR. 1983. Nucleotide sequence of cassava latent virus DNA. *Nature* 301:260–62

144. Stanley J, Latham JR. 1992. A symptom variant of beet curly top geminivirus pro-

duced by mutation of open reading frame C4. *Virology* 190:506–9

145. Stanley J, Latham JR, Pinner MS, Bedford I, Markham PG. 1992. Mutational analysis of the monopartite geminivirus beet curly top virus. *Virology* 191:396–405

146. Stanley J, Markham PG, Callis RJ, Pinner MS. 1986. The nucleotide sequence of an infectious clone of the geminivirus beet curly top virus. *EMBO J.* 5:1761–67

147. Stanley J, Townsend R. 1985. Characterization of DNA forms associated with cassava latent virus infection. *Nucleic Acids Res.* 13:2189–206

148. Stanley J, Townsend R. 1986. Infectious mutants of cassava latent virus generated in vivo from intact recombinant DNA clones containing single copies of the genome. *Nucleic Acids Res.* 14:5981–98

149. Stenger DC, Revington GN, Stevenson MC, Bisaro DM. 1991. Replicational release of geminivirus genomes from tandemly repeated copies: evidence for rolling-circle replication of a plant viral DNA. *Proc. Natl. Acad. Sci. USA* 88:8029–33

150. Stenger DC, Stevenson MC, Hormuzdi SG, Bisaro DM. 1992. A number of subgenomic DNAs are produced following agroinoculation of plants with beet curly top virus. *J. Gen. Virol.* 73:237–42

151. Storey HH. 1931. The inheritance by a leafhopper of the ability to transmit a plant virus. *Nature* 127:928

152. Storey HH. 1933. Investigations of the mechanism of the transmission of plant viruses by insect vectors, I. *Proc. R. Soc. London Ser. B* 133:463–85

153. Sunter G, Bisaro DM. 1989. Transcription map of the B genome component of tomato golden mosaic virus and comparison with A component transcripts. *Virology* 173:647–55

154. Sunter G, Bisaro DM. 1991. Transactivation in a geminivirus: AL2 gene product is needed for coat protein expression. *Virology* 180:416–19

155. Sunter G, Bisaro DM. 1992. Transactivation of geminivirus AR1 and BR1 gene expression by the viral AL2 gene product occurs at the level of transcription. *Plant Cell* 4:1321–31

156. Sunter G, Gardiner WE, Bisaro DM. 1989. Identification of tomato golden mosaic virus-specific RNAs in infected plants. *Virology* 170:243–50

157. Sunter G, Gardiner WE, Rushing AE, Rogers SG, Bisaro DM. 1987. Independent encapsidation of tomato golden mosaic virus A component DNA in transgenic plants. *Plant Mol. Biol.* 8:477–84

158. Sunter G, Hartitz MD, Bisaro DM. 1993. Tomato golden mosaic virus leftward gene expression: autoregulation of geminivirus replication protein. *Virology* 195:275–80

159. Sunter G, Hartitz MD, Hormuzdi SG, Brough CL, Bisaro DM. 1990. Genetic analysis of tomato golden mosaic virus: ORF AL2 is required for coat protein accumulation while ORF AL3 is necessary for efficient DNA replication. *Virology* 179: 69–77

160. Takamatsu N, Ishikawa M, Meshi T, Okada Y. 1987. Expression of bacterial chloramphenicol acetyltransferase gene in tobacco plants mediated by TMV-RNA. *EMBO J.* 6:307–11

161. Takamatsu N, Watanabe Y, Yanagi H, Meshi T, Shiba T, Okada Y. 1990. Production of enkaphalin in tobacco protoplasts using tobacco mosaic virus RNA vector. *FEBS Lett.* 269:73–76

162. Thommes P, Osman TAM, Hayes RJ, Buck KW. 1993. TGMV replication protein AL1 preferentially binds to single-stranded DNA from the common region. *FEBS Lett.* 319:95–99

163. Timmermans MCP, Das OP, Messing J. 1992. Trans replication and high copy numbers of wheat dwarf virus vectors in maize cells. *Nucleic Acids Res.* 20:4047–54

164. Topfer R, Gronenborn B, Schell J, Steinbiss H-H. 1989. Uptake and transient expression of chimeric genes in seed-derived embryos. *Plant Cell* 1:133–39

165. Townsend R, Stanley J, Curson SJ, Short MN. 1985. Major polyadenylated transcripts of cassava latent virus and location of the gene encoding coat protein. *EMBO J.* 4:33–37

166. Townsend R, Watts J, Stanley J. 1986. Synthesis of viral DNA froms in *Nicotiana plumbaginifolia* protoplasts inoculated with cassava latent virus (CLV); Evidence for the independent replication of one component of the CLV genome. *Nucleic Acids Res.* 14:1253–65

167. Ugaki M, Ueda T, Timmermans MCP, Vieira J, Elliston KO, Messing J. 1991. Replication of a geminivirus derived shuttle vector in maize endosperm cells. *Nucleic Acids Res.* 19:371–77

168. Van Mansfeld ADM, Langeveld SA, Weisbeek PJ, Baas PD, Van Arkel GA, Jansz HS. 1979. Cleavage site of øX and G4 RFI DNA. *Cold Spring Habor Symp. Quant. Biol.* 43:331–34

169. Van Vloten-Doting L, Bol JF, Cornelissen B. 1985. Plant virus-based vectors for gene transfer will be of limited use because of the high error frequency during viral RNA synthesis. *Plant Mol. Biol.* 4: 323–26

170. Von Arnim A, Frischmuth T, Stanley J. 1993. Detection and possible functions of African cassava mosaic virus DNA B gene products. *Virology* 192:264–72

171. Von Arnim A, Stanley J. 1992. Determinants of tomato golden mosaic virus symp-

tom development located on DNA B. *Virology* 186:286–93
172. Von Arnim A, Stanley J. 1992. Inhibition of African cassava mosaic virus systemic infection by a movement protein from the related geminivirus tomato golden mosaic virus. *Virology* 187:555–64
173. Ward A, Etessami P, Stanley J. 1988. Expression of a bacterial gene in plants mediated by infectious geminivirus DNA. *EMBO J.* 7:1583–87
174. Wolf S, Deom CM, Beachy R, Lucas WJ. 1991. Plasmodesmatal function is probed using transgenic tobacco plants that express a virus movement protein. *Plant Cell* 3:593–604

175. Woolston CJ, Barker R, Gunn H, Boulton MI, Mullineaux PM. 1988. Agroinfection and nucleotide sequence of cloned wheat dwarf virus DNA. *Plant Mol. Biol.* 11:35–43
176. Woolston CJ, Reynolds HV, Stacey NJ, Mullineaux PM. 1989. Replication of wheat dwarf virus DNA in protoplasts and analysis of coat protein mutants in protoplasts and plants. *Nucleic Acids Res.* 17:6029–41
177. Zhan X, Haley A, Richardson K, Morris B. 1991. Analysis of the potential promoter sequences of African cassava mosaic virus by transient expression of the β-glucuronidase gene. *J. Gen. Virol.* 72:2849–52

Annu. Rev. Plant Physiol. Plant Mol. Biol. 1994. 45:113–141

GENE EXPRESSION REGULATED BY ABSCISIC ACID AND ITS RELATION TO STRESS TOLERANCE

Peter M. Chandler and Masumi Robertson

Division of Plant Industry, CSIRO, GPO 1600, Canberra, ACT 2601, Australia

KEY WORDS: abscisic acid, gene expression, stress tolerance

CONTENTS

INTRODUCTION 113
ABA REGULATION OF GENE EXPRESSION 114
   Positive Regulation of Gene Expression 117
   Negative Regulation of Gene Expression 123
GENE PRODUCTS AND STRESS TOLERANCE 124
   Determination of Functions of ABA-Regulated Gene Products 125
   ABA-Regulated Genes and Cold Tolerance 130
   ABA-Regulated Genes and Desiccation Tolerance 133
CONCLUDING REMARKS 134

INTRODUCTION

Abscisic acid (ABA) modulates growth and development of plants, particularly during seed formation and during response to environmental stresses involving loss of water. ABA and its role as a hormone have been the subject of considerable research, with reviews covering metabolism and physiology (95, 147, 148), cell localization (49), mechanism of action at the cellular level (52), and the control of gene expression (10, 109, 124). Useful information on ABA, including mutants, gene expression, and hormone responses, can also be found in some general reviews on plant hormones (66, 96, 111, 121), and many aspects of ABA physiology and biochemistry are covered in a conference proceedings (26).

0066-4294/94/0601-0113$05.00

113

Considerable progress has been made recently in many areas of ABA research, including biosynthesis, where evidence continues to favor the carotenoid pathway, the use of monoclonal antibodies for ABA assay, structure/activity relations, mutant characterization and utilization, and gene expression regulation. The reader is referred to the reviews above for further details. Less progress has been made in other areas, most notably in identification or characterization of the ABA receptor (assuming one exists!). Several questions remain regarding the role of ABA as a stress hormone: What is the nature of the sensory pathway that leads to elevated ABA levels in a plant experiencing a lack of water, and a decline in ABA levels upon rewatering? Are the considerable changes in gene expression (and therefore protein complement) that occur in response to elevated ABA levels following stress related to stress tolerance? Given that ABA-regulated proteins have usually been defined in laboratory experiments, how relevant will they be in the analysis of, or eventually, modification of stress in the field (110)?

This review has three aims: (a) to summarize recent studies of ABA-regulated gene expression, assessing the evidence for an in vivo role of ABA in regulating gene expression in either stress or developmental situations, (b) to review progress in understanding the mechanism(s) by which ABA influences gene expression, and (c) to discuss approaches for determining the physiological roles of ABA-regulated gene products, particularly in cold and desiccation tolerance. The effects of ABA on gene expression have been known for more than a decade, and in the last five years many studies have described ABA-regulated genes, which are expressed in plants under stress. Treating plants with ABA frequently hardens them against stress; thus, ABA-regulated gene products may play a role in stress tolerance. The genes are of interest in attempts to understand the underlying physiology of stress tolerance and in efforts to genetically engineer stress tolerance in plants. Although we are still unable to detail a single example demonstrating that an ABA-regulated protein plays a role in stress tolerance, promising experimental systems have been developed to the extent that such a role, if it exists, should be demonstrable in the near future.

ABA REGULATION OF GENE EXPRESSION

In seeds, ABA influences and perhaps regulates several processes including reserve material synthesis (107), dormancy onset and maintenance (63), water loss (67), and desiccation tolerance acquisition (67). ABA levels often peak in the developing seed around the time of maximum fresh weight (65). This phase of seed development has therefore been most intensively studied to define ABA effects on both seed development and gene expression. Many proteins have been identified that are positively regulated by ABA in seeds, and that may play important physiological roles during developmental arrest or seed maturation. ABA levels decline as development proceeds past maximum fresh weight, but estimated ABA concentrations in the seed may actually rise because water loss frequently proceeds more rapidly than does ABA loss. Therefore, ABA can potentially affect gene expression even while ABA content (per seed) is falling.

ABA-regulated genes also have been studied in plants under stress (124), where the elevated levels of endogenous ABA may be involved in gene expression changes. ABA is often referred to as a generalized stress hormone, which may be a consequence of different stresses (e.g. freezing, high osmoticum, dehydration) all affecting water status in some part of the plant, thereby resulting in increased ABA levels (27). Alternatively, different environmental stresses (or even developmental cues) could lead to elevated ABA levels by a variety of mechanisms, some of which may not involve changes in water status. Further research in defining the effects of different stresses on ABA metabolism (particularly biosynthesis) will be required before this matter can be resolved.

EVIDENCE FOR ABA REGULATION   Initial evidence for ABA-regulated gene expression frequently comes from corresponding changes in endogenous ABA levels and in the levels of a particular gene product, often after imposition of a stress such as dehydration. Gene product levels have been measured using a variety of assays (based on antibodies, enzymatic activity, or hybridization); but ABA levels are now most commonly measured using monoclonal antibodies and ELISA-based procedures (reviewed in 138). The available monoclonal antibodies show high specificity for ABA, and metabolites or analogs are not generally recognized (57). However, because phaseic acid, the first stable metabolite of ABA, elicits some "ABA" responses, the use of ABA-specific antibodies could result in failure to assay an endogenous regulatory compound. Whether assaying gene product levels, or ABA levels, large numbers of samples are usually required to ensure adequate experimental replication, frequent times of sampling, and analysis of different plant components. We feel that the most benefit will come when ABA levels are determined on the same samples that are used for RNA or protein analysis (20, 25, 41, 104, 113, 114). Somewhat ironically, the availability of rapid and reliable ABA assays highlights general difficulties in interpreting data on hormone levels: Which component of the cell senses ABA, and which pool of ABA is responsible for changes in gene expression? How uniform are different cells within a tissue in terms of their ABA content, their inherent sensitivity to ABA, and their responses to ABA? These are challenging problems for future research.

Once preliminary evidence is obtained that ABA might regulate expression of a particular gene, several follow-up approaches may be taken, including studies using applied ABA, inhibitors of ABA synthesis, and mutants. Examples of these approaches will be discussed below. An important consideration is the extent to which ABA may, at one extreme, be the sole regulator of expression of a particular gene, versus the other extreme (for the purposes of this review), where ABA may only be a minor influence on expression of a gene that is regulated mostly by some other factor(s). Several gene families are regulated by ABA, as well as by stresses such as desiccation, osmotic pressure, and cold in a manner that is apparently independent of ABA (31, 39, 93, 146). For convenience we refer to ABA regulation of gene expression, but this does not mean that other factors may not also be important.

NOMENCLATURE AND RELATIONSHIPS OF ABA-REGULATED GENES   There are currently over 70 reports of different genes or gene products that are positively regulated by ABA. In many cases there is also information available on whether stresses such as dehydration, temperature extreme, salinity, high osmoticum, or wounding influence gene expression. This large number of studies allows researchers to define relationships between similar gene products from different plant species. The naming of genes and their products has led to some confusion because investigators approached related sets of genes and proteins in different systems and with different interests (86). Products of most of these genes have no known function, and they have been named using different criteria, such as developmental characteristics (late embryogenesis abundant, *lea*), regulation of expression (responsive to ABA, *rab*; dehydration induced, *dhn*), or amino acid labeling pattern (early methionine labeled, *Em*). Genes or proteins that are closely related sometimes have different names; for example, *rab17* (134) and *dhn* (17) from corn are almost identical in sequence and in their regulation of expression. In other cases, genes with similar notation are not related in sequence (e.g. *rab17* and *rab28*) (102, 134). Proposals to develop uniform nomenclature (86) may succeed in the extreme situations of close sequence relationship or where the sequences are unrelated. In many cases, however, intermediate degrees of relationship are likely to prevent establishment of a uniform nomenclature. For example, one family of ABA-regulated protein, dehydrin, was characterized originally by two conserved amino acid sequence motifs, the lysine-rich block (KIKEKLPG) and a stretch of serine residues. These conserved sequences are found in dehydrins from dicotyledonous and monocotyledonous plants. The recent characterization of a closely related protein in pea (113) indicates a similar organization of lysine-rich blocks, but the stretch of serine residues is absent. To develop uniform nomenclature based on sequence relationships, investigators must decide whether the presence of the lysine-rich block is more important than the absence of the serine stretch. Protein function, if any is found, ultimately may prove the best basis for classification. In the meantime, it seems more practical to specify the relationship of any new sequence to existing sequences rather than to attempt to standardize nomenclature for such a diverse set of gene products.

Positive Regulation of Gene Expression

INDUCTION OF GENE EXPRESSION BY APPLIED ABA   Because of the important role played by ABA in seed development, abundant transcripts from developing seeds were some of the earliest examples of cloned genes potentially regulated by ABA. The large number of cloned transcripts from developing cotton seeds is the most intensively studied example (35, 36, 55). Late embryogenesis abundant (LEA) transcripts are highly expressed during the latter part of embryogenesis, and the timing of expression is associated with elevated ABA levels. Also, precocious induction of LEA transcripts occurs following ABA treatment of cultured embryos from developing seeds. Representative *lea* genes in cotton (55) and related sequences from other plant species, both monocot and dicot, are induced following treatment with ABA (47, 50, 72, 85, 87, 140, 142). Although potentially regulated by ABA, the diversity of expression patterns observed for different LEA transcripts raises questions about the precise role of ABA as an in vivo regulator of *lea* gene expression (35, 55).

High levels of some LEA transcripts persist in seeds at maturity. This behavior contrasts with that of storage protein transcripts, which accumulate to high levels at about the same time as LEA transcripts, but which decline to very low or undetectable levels in mature seeds (35). The differences in expression between these two groups of genes may relate to functional differences between the gene products, especially with respect to the potential roles of LEA proteins in helping seeds tolerate desiccation at maturity and following imbibition and germination. Some storage proteins are positively regulated by ABA, e.g. cruciferin and napin from *Brassica napus* (22, 23, 28, 33), and the β subunit of β-conglycinin from soybean (11). However storage proteins in other species do not appear to be regulated by ABA (3). These findings highlight differences between species, and indicate that different mechanisms may regulate different groups of genes (LEA and storage protein) expressed at the same time in seed development.

In recent years there have been many reports of ABA-induced genes that have been characterized in vegetative parts of plants under stress. In most cases the initial evidence for ABA regulation of expression was an association between ABA levels and the levels of a particular transcript. Further evidence has come from induction of the appropriate transcript by application of ABA to unstressed plants. Such observations are consistent with, but do not prove, the hypothesis that stress-induced ABA is responsible for gene expression. There are also cases in which genes regulated by applied ABA are not markedly induced following stress treatments that elevate ABA levels. For example, in three ABA-induced genes in tomato, the degree of induction did not always relate to ABA levels. Stress conditions used included dehydration, high osmoticum (by PEG), NaCl, low temperature, and heat shock. Dehydration stress induced all three genes, and this was associated with a 4-fold increase in ABA levels (20, 104). PEG-treatment increased leaf ABA content to the same level as for the dehydrated leaves; however, transcript levels detected with clones pLE4 and pLE25 were barely elevated, whereas transcript levels detected with pLE16 were elevated to a much greater extent. NaCl treatment did not change ABA levels in roots, yet transcript levels detected by pLE4 and pLE25 were markedly increased. Although expression of all three genes was increased by applied ABA, the inconsistent relationship observed between mRNA levels and ABA levels in different stress situations indicates that other factors also must be involved in modulating gene expression.

Hormone application studies have also indicated that ABA-induced gene expression might be enhanced by additional stress factors present only in stressed plants. Transcript levels are sometimes greater in dehydration-stressed plants than in unstressed seedlings treated with ABA (19, 104, 113). The unstressed seedling may lack a stress factor that is able to interact with ABA. This factor would be present in the stressed seedlings and might interact synergistically with dehydration-induced ABA to generate an enhanced response. Other interpretations of this result are also possible. It is difficult to compare ABA levels from stressed plants with those from plants that have been exposed to considerable quantities of exogenous ABA. How similarly is the ABA compartmentalized in a treated versus a stressed plant? Differences in ABA metabolism may lead to differences in active ABA pools. Unstressed plants may have higher levels of factors that antagonize the action of ABA than do stressed plants. In many physiological responses, gibberellin (GA) is likely to antagonize the effects of ABA. Stressed plants, which have stopped growing, may have lower levels of active GAs than do non-stressed plants at the time of ABA application. These two hormones exert opposing effects at the level of gene expression, and gibberellic acid (GA3) reverses the ABA induction of aldose reductase (4), *rab/lea* genes (54, 55), and α-amylase inhibitor (88). These results highlight the potential importance of other factors when assessing the role of ABA in gene regulation under physiological conditions. Similar conclusions relating to the importance of other regulators also have been reached in a study of the effects of ABA on stomatal closure in stressed versus non-stressed plants (131).

USE OF MUTANTS   In some cases, such as the tomato system described above, mutants with reduced levels of endogenous ABA have been used to provide further support for the hypothesis that endogenous ABA regulates gene expression. As summarized above, these genes were expressed in response to dehydration stress in association with elevated levels of ABA. Application of ABA induced the same set of genes in unstressed plants. In studies using the ABA-deficient mutant, *flacca* (19), the genes were not induced by dehydration, which correlated with low ABA levels in the mutant (9). Application of ABA to the mutant plants led to induction of transcripts, indicating that the mutant plants would respond to ABA when the hormone was present at sufficient levels. Lower concentrations of applied ABA were required to induce ABA-regulated genes in *flacca* than in normal plants (19). Consequences of the ABA deficiency, such as increased transpiration, or reduced rates of ABA oxidation, might lead to applied ABA being more effective in *flacca,* and therefore to higher levels of gene expression.

The use of mutants also has shown organ-specific and developmental stage–specific aspects of ABA-regulated gene expression. Several genes appear to show ABA-regulated expression following dehydration stress in vegetative tissues, yet the same (or very closely related) genes do not seem to require ABA for expression in seeds, where they are presumably under a different developmental program that is independent of ABA. Viviparous mutants of corn (*vp2, vp5,* and *vp7*) have low levels of ABA in embryos during seed development and in dehydration-stressed seedlings (92). Two *rab* genes (*rab17* and *rab28*) were not induced in dehydration-stressed *vp* seedlings (101, 102), yet they were expressed at normal or near normal levels in developing embryos, indicating that ABA was not essential for expression of these genes during embryo development. In contrast, *Em* expression appears to be regulated by ABA in embryos since the level of *Em* expression was much reduced in *vp5* (140).

Another example of organ-specificity associated with ABA-regulated genes was found in the expression of a proteinase inhibitor in ABA-deficient mutants of tomato (*sitiens*; 129) and potato (*droopy*; 106). In leaves of normal plants, proteinase inhibitor II (PI-II) was absent but could be induced by wounding (42) or ABA treatment (97). Wounding experiments using the ABA deficient mutants in tomato and potato failed to induce PI-II; however, ABA application to these mutants induced the gene, demonstrating that the mutants were capable of PI-II gene expression if the appropriate signal (ABA) was received (97). Surprisingly, PI-II expression in potato tubers and in flowers of both tomato and potato occurred in both normal and mutant genotypes, indicating that developmental regulation is apparently independent of ABA in some parts of the plant (98), although it was not reported whether the isolated flowers of the mutants also showed the expected low levels of endogenous ABA.

INHIBITOR STUDIES Mutants with low endogenous levels of ABA have been characterized in only a few plant species. ABA biosynthesis inhibitors can be used to manipulate ABA levels for the majority of species where mutants are unavailable. Fluridone, which inhibits carotenoid biosynthesis, has been used to lower the expression of genes thought to be regulated by ABA (14, 50, 55, 93). The ideal inhibitor (or mutant) would be one affecting the last step of ABA biosynthesis, because this would presumably minimize some of the secondary effects that might result from a deficiency of earlier intermediates in the biosynthetic pathway. In this respect, fluridone and some *vp* mutants of corn are far from ideal because they block ABA biosynthesis prior to carotenoid formation, which has serious consequences for the plant because of chlorophyll photo-oxidation. Despite this limitation, the use of fluridone has contributed significantly to the study of ABA responses in plants, particularly when its effect on endogenous ABA levels is confirmed and when the effect can be reversed by the simultaneous application of ABA or other compounds that elevate ABA levels (62).

MECHANISM OF GENE EXPRESSION REGULATION   The mechanism(s) by which ABA regulates gene expression might involve transcriptional events, and/or a collection of post-transcriptional events such as transcript processing, mRNA stability, translational control, and protein activity and turnover. The first evidence for ABA regulation was changes in band or spot intensity shown by one- or two-dimensional gel electrophoresis of proteins. Most studies now rely on RNA blots to detail changes in gene expression because of the specificity that the hybridization probe provides. The levels of a particular mRNA (or frequently, mRNA sequence family) are assessed in different RNA samples relative to total RNA, which is assumed to stay constant. Because ABA has been found to stimulate overall transcription, including that of rRNA (80), changes in cellular RNA content may occur. Other methods of assaying gene expression with comparable specificity include immunodetection and enzyme activity measurements. Instead of dealing only with the mRNA levels that hybridization studies show, assays for ABA-regulated proteins allow for further controls over gene expression to be shown. For example, the expression of osmotin mRNA in tobacco is strongly induced by ABA in cultured cells and in young leaves of plants, yet there is no effect of ABA treatment on accumulation of osmotin protein. This may indicate different effects of ABA on osmotin gene expression, one operating on mRNA levels, and another affecting osmotin translation or stability (75). Assays of gene expression at the protein level are also desirable because such effects have to be demonstrated before the influence of ABA on gene expression can be related to physiological responses.

Many studies have shown that ABA can alter the amount of a particular mRNA relative to total RNA. Changes at the mRNA level are often thought to reflect changes in the amount of transcription, but this is not always the case: run-on transcription experiments are needed to assess whether ABA is regulating transcription of the gene of interest. Transcription experiments using nuclei isolated from immature embryos of wheat or *B. napus* showed an increase in the relative amount of transcription of ABA-regulated genes (28, 80). The differences observed in steady-state mRNA levels may not coincide exactly with differences in transcription rates measured after a particular treatment. Comparison of cruciferin and napin showed that ABA regulated expression of these two genes differently. ABA application increased the level of cruciferin mRNA 2.5-fold and napin mRNA 5-fold, while the transcription rates of these genes were increased 2- and 3-fold, respectively (28). Differences were also found among four cold/ABA-regulated genes isolated from *Arabidopsis* (45). mRNA levels increased 9- to 26-fold following cold treatment. However, cold treatment increased the relative amount of transcription of one particular gene 10-fold over the control, in contrast to the other three genes, which showed increases of less than 2.4-fold over controls. Cold treatment must therefore influence gene expression at the post-transcriptional level, presumably affecting the stability of different transcripts in different ways. It was not reported whether ABA induction might involve similar mechanisms.

Recent interest in understanding the molecular basis of gene regulation by ABA has focused on transcriptional activation and a general model in which a *trans*-acting protein factor interacts with an element in the promoter region of a gene (105). There are three essential steps in this analysis: (*a*) identification of an ABA-responsive *cis*-acting element within the promoter of an ABA-regulated gene, (*b*) characterization of the *trans*-acting factor that interacts with the defined element, and (*c*) functional analysis of the factor and the element to test whether ABA-regulated gene expression is conferred by their interaction in vivo. Several classes of ABA-regulated genes, *Em, rab,* and *Dc3* have been examined using this approach (81, 82, 90, 103, 135, 137, 145). A *cis*-acting ABA-response element (ABRE) was identified as one of the sequence motifs conserved between several ABA-regulated genes. The ABRE is a minimum 11 base sequence, and is sufficient to confer ABA-responsiveness to a reporter gene when tested in transient assay systems (44, 125). The characterization of ABRE has used homologous and heterologous systems, and has included transient expression assays as well as stable transformation. It was possible to define an ABRE for the rice *rab16* gene in both homologous and heterologous monocot assay systems (90, 125); however, the same promoter in the dicot species tobacco was not ABA responsive in vegetative tissue (145). The importance of subtle differences between *rab* promoters is highlighted by the observation that, in contrast to the rice *rab16*, the promoter from the closely related *rab17* gene from corn functioned in tobacco in response to ABA (135). These results indicate the desirability of using homologous assay systems in order to minimize the complexities that might arise from species differences in regulatory components, rather than from experimental manipulations.

Trans-acting factors (48) are nuclear proteins that bind to *cis*-acting regulatory elements in DNA. Several nuclear proteins specifically bind to ABRE (44, 90, 135). One such factor, EmBP-1, was cloned from wheat (44) and found to belong to the bZip protein family of transcription factors (64, 136). Quatrano et al recently reviewed the mechanism of Em gene regulation (109). Although EmBP-1 interacts specifically with an eight base sequence within the ABRE containing a G-box core, it remains to be shown through functional analysis that the factor acts as a transcriptional activator regulated by ABA. Studies in transgenic tobacco plants illustrate some of the subtleties in the functional analysis of DNA binding factors. A nuclear factor, TAF-1 (also a bZip protein), was isolated based on binding to the ABRE from the rice *rab16* gene (94). Functional analysis used tobacco plants that were stably transformed with a *rab16* ABRE promoter/GUS reporter gene. In control plants, GUS activity was at background levels, presumably because of the very low level of TAF-1 detected in untreated vegetative tissue. However, TAF-1 did not appear to act in ABA-regulation of the *rab16* gene, since applied ABA did not increase the reporter gene activity. The effect of ABA application on levels of TAF-1 was not reported. When TAF-1 was expressed transiently following transformation, GUS activity increased, indicating that TAF-1 could function as a transcriptional activator via ABRE. In a follow-up study, TAF-1 was found to have higher binding affinity to a sequence related to ABRE, a perfect palindromic G-box sequence (PA). The level of TAF-1 expression in leaves and roots was closely related to the activity of a reporter gene driven by the PA promoter sequence but not by ABRE (117). This indicates that TAF-1 may normally regulate the expression of genes with PA promoter sequences, and emphasizes the importance of functional analysis in vivo to establish the role of DNA binding proteins in *trans*-activation of gene expression.

As the number of characterized regulatory sequences in plant genes has increased it has become clear that G-box-like motifs are associated with many genes that are regulated by diverse factors (44, 117, 141). The different regulatory functions for G-box-like motifs (141), G-box interacting proteins (bZip class), and the relatively limited number of nuclear factor families (48) indicates that minor variations in *cis*-acting elements and *trans*-acting factors are probably important in transcriptional regulation of a gene. In tobacco, small differences in *cis*-acting sequences containing a G-box core resulted in different temporal and spatial promoter activity patterns (117). A two amino acid substitution in *trans*-acting factors, such as TAF-1, resulted in specific binding to different DNA sequences (128). Also, complex regulatory systems may be built up from various combinations of several binding domains (*cis*-elements) and *trans*-acting factors by cooperativity of factor binding, resulting in multiple element models for gene regulation (6, 13, 77). One such regulatory factor associated with ABA-regulated gene expression is the *vp1* gene product described by McCarty et al (83).

Negative Regulation of Gene Expression

The studies described above indicate that ABA positively regulates the expression of certain genes following a stress such as dehydration, or following ABA application. In stress situations it is presumably equally important that certain genes be negatively regulated, for example, genes whose products may be inappropriate for the new physiological situation. In one example, photosynthetic activity in plants under water stress is reduced through coordinated structural and biochemical changes, including closure of stomata (49), an increase in ABA levels, and a decrease in the activities of photosynthetic enzymes (122). Among the genes required for photosynthesis, a structural gene, *cab,* and a Rubisco small subunit gene, *rbcS,* were down-regulated by ABA in dehydration-stressed tomato leaves (5). Evidence for the role of ABA in mediating down-regulation of these genes included the association of increased endogenous ABA levels with reduced transcript levels in either stressed plants or plants treated with ABA, and studies on the ABA-deficient mutant, *sitiens.* Although ABA could indirectly down-regulate *cab* and *rbcS* through stomatal closure and subsequent reduced growth or increased sucrose supply, a direct effect of applied ABA on *rbcS* expression has been shown in developing embryos (108) where these influences are presumably absent.

In other cases, ABA antagonizes GA3-induced gene expression. The most extensively studied interaction between ABA and GA3 is α-amylase expression in cereal aleurone. In isolated aleurone layers, ABA antagonized GA3-induced α-amylase production. This effect was found at the mRNA level (15, 53, 116) and later it was shown that ABA was able to suppress the GA3-induced increase in transcription of the α-amylase gene (59, 149). More recently, promoter analyses of several α-amylase genes from wheat and barley have shown that this interaction between GA3 and ABA probably occurs through sequence elements that are either close together in the α-amylase promoter, or that may even be the same sequence (56, 60, 118). Further work (125) using a 21-mer oligonucleotide showed that this sequence contained elements responsive to GA3 (up-regulation) and ABA (down-regulation). Because the sequences identified for positive- and negative-regulation by ABA are different (44, 125), there are presumably different *trans*-acting factors involved. The inability to separate ABA down-regulating sequences from the up-regulating GA responsive elements indicates an alternative explanation for how ABA antagonizes GA3-induced transcription of α-amylase genes. Instead of acting via sequences in the α-amylase promoter, ABA may act earlier in the GA signal transduction pathway by reducing either the level or activity of a positive regulatory factor that binds to GA responsive elements (125).

Gibberellins play an important role in controlling shoot growth, and applied ABA antagonizes GA-stimulated growth. Although little is known about the molecular mechanisms responsible for these hormone effects, Shi et al (123) recently have identified a tomato gene, expressed in shoots, that is induced by GA3 and down-regulated by ABA.

GENE PRODUCTS AND STRESS TOLERANCE

GENERAL CONSIDERATIONS   ABA has been postulated to play a role in stress tolerance in part because ABA treatments harden plants against stress. These treatments also result in expression of ABA-regulated genes, so their products might be involved in the hardening process, in stress tolerance itself, or in recovery from the stress. It is difficult to distinguish between these different aspects of the stress response. Determining whether these proteins play a role in stress tolerance is therefore important for advancing our knowledge of stress biology and for potential applications in biotechnology. Our discussion includes not only ABA up-regulated proteins, which appear following stress imposition, but also ABA down-regulated proteins, because these could correspond to the stress-sensitive component that limits survival or growth under stress conditions.

The physiological roles of ABA-regulated proteins remain largely unknown, but they probably range widely in terms of how directly they relate to stress tolerance. At one extreme they could be the limiting factor that determines the degree of stress tolerance; at the other extreme they might have nothing to do with stress tolerance. Genes might be expressed in response to ABA under stress conditions because ABA normally regulates their expression at some particular stage of development or growth; expression at other times when ABA levels happen to be high might be coincidental. Most of the gene expression changes that result from stress-induced ABA are probably only indirectly involved in tolerating or adapting to stress; for example, they may represent adjustments that reflect the altered physiological state of the plant. Potential examples include proteins that change in abundance as elongation rates change (21), or vegetative storage proteins (127). Because most stresses lead to a reduced growth rate, many of the changes observed in levels of particular proteins are likely to be adjustments to slower growth.

Although current attention is focused on the role of ABA-regulated proteins in relation to ABA-induced stress tolerance, other aspects of ABA action, not involving gene expression changes, may also be involved. The rapid effects of ABA treatment on K^+ and Ca^{2+} ion transport in guard cells probably occur before any influence of ABA on gene expression, yet the resulting stomatal closure can lead to considerable changes in physiological processes. ABA may cause other changes that do not involve regulation of gene expression, and that may lead to altered stress tolerance. This possibility may seem unlikely, but consider the example of heat shock proteins (hsps), which in many ways are analogous to stress/ABA-induced proteins. Many biochemical roles have been postulated for hsps, and attractive hypotheses have been proposed to account for how they might confer the thermotolerance that is observed following a heat treatment. Yet two recent studies have demonstrated acquisition of thermotolerance in the absence of hsp synthesis, raising the question of whether hsp production is required for thermotolerance (1, 126, discussed in more detail below).

Determination of Functions of ABA-Regulated Gene Products

Most of the genes whose expression is regulated by ABA have no known biochemical function associated with the gene product, which makes it difficult to define the physiological roles that the gene products might play in stress tolerance. Determining the functions of the gene products is therefore of high priority. At present, the only simple approach is to compare DNA or protein sequences with those in the data base, but this often fails to yield significant similarity to any protein with a known biochemical activity. Some ABA-regulated proteins have distinctive amino acid compositions and sequences, resulting in their heat stability (61), and indicating physicochemical properties that might be functionally important in a cell losing water (29a). Future progress in determining activities of ABA-regulated proteins is likely to come from a variety of sources. As sequences of proteins with known biochemical activities become available, serendipitous relationships to sequences of ABA-regulated proteins will provide clues about their functions just as they did in the elucidation of heat shock protein functions. Studies on stress tolerance in animal (24) and microbial systems are also likely to contribute. Two notable examples include the identification of dehydrin-like proteins in cyanobacteria (18) and a protein involved in salt tolerance in yeast with plant homologs that are regulated by salt stress and ABA (37). Finally, biochemical investigations on stress tolerance in plants should continue to identify areas of metabolism that should be investigated. Recent progress includes the importance of sugars in desiccation tolerance (7) and the cloning of enzymes in production of compatible solutes (133, 139).

The possibility that a protein might possess a particular biochemical activity, usually inferred from sequence comparisons, provides clues to its potential physiological role, but this needs to be confirmed. In some cases first impressions are misleading. For example, some of the crystallin proteins of the vertebrate lens are able to form a transparent crystal-like material that allows the transmission of light, but surprisingly, these proteins also happen to be common metabolic enzymes, performing an entirely different function in other cells of the organism (143). Thus, products of the same gene, or two very closely related genes, may be used in more than one way. Two examples from plants further illustrate this point. An antisense approach was used in tomato to identify clones encoding ACC oxidase, an enzyme involved in ethylene bio-synthesis. Transformants expressing the antisense construct failed to produce ethylene during fruit ripening (46). A different study (99), also in tomato, examined antisense transformants of a gene (*E8*) that showed 53% nucleotide sequence identity to ACC oxidase. When progeny of two independent trans-formants were examined, levels of E8 protein were reduced (as expected) but ethylene production was actually enhanced. So despite considerable sequence similarity between these two proteins, reducing their levels generated opposite phenotypes. Another example involves a seed protein originally named PAPI (probable α-amylase/protease inhibitor) based on sequence relationship to other inhibitors, although no inhibitory activity was observed (89). Subse-quently, a similar close sequence relationship to lipid transfer proteins was observed, and this activity has now been confirmed (12).

Although biochemical functions proposed on the basis of sequence similar-ity may not always reliably indicate physiological roles, they provide a good starting point. For example, a cold- and ABA-induced transcript was cloned from *Arabidopsis thaliana* and the deduced protein sequence showed similar-ity to an anti-freeze protein from arctic flounder (71). Although anti-freeze activity of the protein has not yet been demonstrated, and the sequence similar-ity on which the proposed function is based is marginal, specific experimental predictions can be made and tested. Another study of cold-regulated proteins showed that one such protein, cloned from *A. thaliana* and produced by in vitro transcription/translation reactions, is highly active in protecting a freez-ing-sensitive enzyme (lactate dehydrogenase) from freezing damage. On a molar basis, the cryoprotective activity of this protein was several orders of magnitude greater than other proteins and many orders of magnitude greater than sucrose (78).

Activities such as those described above are amenable to further experi-mentation because they can be studied using in vitro systems. In many other cases it may be difficult to test possible roles of stress proteins. Our limited knowledge of the biochemistry and biophysics of traits such as cold tolerance or desiccation tolerance makes it difficult to predict which activities might be worth examining. Also, a protein may need an appropriate molecular frame-work before activity can be demonstrated, limiting the extent to which in vitro systems can be used. Promising advances in procedures that allow protein interactions to be defined in vitro should be helpful here (43). Fortunately, transformation approaches (discussed below) allow assessment of the role a protein plays in stress tolerance even if its function is unknown. However, before transformation approaches are undertaken, there is usually correlative evidence supporting a role for the protein in stress tolerance and justifying further study.

CORRELATIVE STUDIES   When the function of a gene product is unknown, the association between its level and the degree of stress tolerance may provide indirect evidence for or against its role in stress tolerance. However, many other stress-induced changes are likely to occur in addition to the change being observed: some might be critical to the response being measured; others may not affect the response. An observed positive association between the level of gene product and stress tolerance requires the hypothesis to be tested in other ways. As the number of independent lines of correlative evidence increases, confidence in an observed association increases. Further evidence might come from inhibitor studies or from mutants. Another approach is to use naturally occuring variants that exhibit different levels of stress tolerance. For example, the possible role of ABA-regulated genes in desiccation tolerance is being examined in resurrection plants and in recalcitrant seeds. Vegetative parts of resurrection plants survive extreme dehydration, and photosynthetic activity is quickly restored after rehydration. In the dicotyledonous resurrection plant, *Craterostigma plantagineum,* three transcripts cloned from desiccated tissues had not been described previously from conventional plants (100). It is unknown whether the corresponding proteins play a role in the extreme desiccation tolerance of *C. plantagineum.* In contrast to the dehydration tolerance of resurrection plants, recalcitrant seeds are unable to survive the desiccation that normally occurs with seed development and maturation. Dehydrin-like proteins were found in seeds of a recalcitrant plant, *Zizania palustris* (8), and although this might indicate that dehydrins are not involved in desiccation tolerance, recent studies have indicated that *Z. palustris* seeds are only conditionally recalcitrant (70). Further studies are required to establish the relationship between the presence of dehydrin-like proteins and the ability of seeds to withstand desiccation.

One of the most extensively studied correlations has been that of hsps and thermotolerance in animal and microbial systems. Many lines of evidence have indicated that hsps are involved in the acquisition of thermotolerance that follows a heat shock. For example, stresses that induce hsp synthesis also induce thermotolerance (reviewed in 79). Similarly, a correlation is found between synthesis and degradation of hsps and the acquisition and loss of thermotolerance (73). These correlative studies extend to work with mutants. Heat-tolerant variants and heat-sensitive variants have elevated and reduced levels (respectively) of certain hsps (76). There are even what might be considered direct demonstrations of hsp involvement in thermotolerance such as the abolition of thermotolerance following microinjection of antibodies to HSP70 (112), and a yeast mutant in the HSP104 gene that lacks thermotolerance

(119). Despite evidence for the role of hsps in thermotolerance, the results of two recent studies contradict this concept. In one study, reversible inhibitors of DNA topoisomerase I and of RNA polymerase II were used to prevent the accumulation of hsps and mRNAs in NIH3T3 cells (1). Conditions that prevented hsp accumulation in response to thermoinduction did not prevent the acquisition of thermotolerance. The other study (126) used a yeast mutant that was deficient in the heat shock transcription factor, a protein required for activation of heat shock genes. When this mutant was heat shocked the major hsps were not induced, yet thermotolerance was normal, indicating that the enhanced production of hsps was not required for thermotolerance. These authors suggested that the lack of thermotolerance in the *hsp104* mutant might indicate that a certain (constitutive?) level of this particular hsp is necessary for the acquisition of thermotolerance, but that its expression at high levels following thermoinduction is not required. However, the issue is not fully resolved because a more recent study on yeast *hsp104* mutants demonstrated a relationship between the level of HSP104 protein and the degree of thermotolerance (120). These studies illustrate some of the complexities in trying to establish functions for gene products in physiological responses.

TRANSFORMATION APPROACHES   In plant species with efficient transformation and regeneration procedures, it may be possible to demonstrate whether a cloned gene plays a role in stress tolerance by over-expression of the gene, or by interfering with its expression using an antisense/ribozyme approach. If the gene encodes a protein that is thought to play a role in stress tolerance, a suitable experimental strategy might allow a direct test of this role even if the biochemical or biophysical activity of the polypeptide is unknown. Any positive indication of a role in stress tolerance is likely to provide sufficient incentive for further efforts to determine the function of the gene product.

In the antisense/ribozyme approach, transformation with a suitable construct might eliminate accumulation of the gene product in transgenic plants under stress conditions, leading to identifiable changes in stress tolerance. We are not aware of any examples in which this approach has been successful with ABA-regulated genes. One of its main limitations stems from the fact that many ABA-regulated genes belong to related gene families. It may not be possible with a single antisense or ribozyme sequence to eliminate production of all members of the gene family. This problem will become more severe as sequence differences increase. If residual activity of the gene product persists under stress conditions there may be no observable change in stress tolerance. However, in situations where accumulation of the gene product can be prevented, this approach should allow assessment of a role in stress tolerance. A second limitation has been the lack of reliable transformation and regeneration procedures for some of the species in which ABA-regulated gene expression has been studied most intensively, especially in cereals. This limitation will become less important as new transformation procedures are developed.

An alternative strategy is to over-express the cloned gene either in the same species or in a different species following transformation, and then assess the stress tolerance of the transformed plant. This approach avoids the problems associated with gene families, but serious problems remain: (*a*) the product of the introduced gene may not be stable under non-stress conditions, so the amount of gene product accumulated may not be sufficient for stress tolerance to be shown; (*b*) there may be background problems of stress tolerance associated with expression by the host plant of gene products homologous to those of the introduced gene when stress is imposed; (*c*) the product of an introduced gene may not function in the foreign cellular environment of a different species, especially if its effect on stress tolerance requires it to interact with other components (e.g. sugar alcohols, other proteins) in the cell; and (*d*) the introduced gene product may function correctly, but it may be non-limiting in terms of stress tolerance. In this case it would be classified incorrectly as non-essential for stress tolerance. An example of the over-expression approach involves the desiccation/ABA-regulated cDNAs from *Craterostigma plantagineum* after transformation into tobacco. Although transformed plants expressed the respective proteins at high levels, there was no detectable effect on the response of the transgenic plant to osmotic stress (58).

Despite the difficulties inherent in over-expression studies, this technology has been used successfully on gene products with known activities that have modified the stress tolerance of recipient plants. Although there is no apparent relationship of these particular gene products to ABA, the studies are described here because of the principle they demonstrate. Murata et al (91) examined the chilling sensitivity of tobacco plants that had been modified genetically to alter the degree of fatty acid unsaturation in the phosphatidylglycerol of chloroplast membranes. Earlier work had established that a high proportion of *cis*-unsaturated fatty acids is associated with chilling tolerance. When tobacco was transformed with the glycerol-3-phosphate acyltransferase gene from squash (chilling sensitive), the degree of phosphatidylglycerol unsaturation was reduced and there was an increased chilling sensitivity. Transformation with the corresponding gene from *A. thaliana* (chilling tolerant) led to an increase in the degree of unsaturation and an increase in cold tolerance. This is an important result because it indicates that there are single gene products that can be used to modify whole plant characters such as chilling sensitivity. In a complementary study, Wolter et al (144) were able to increase the amount of saturated fatty acids in plastid membrane lipids of *A. thaliana* by expressing the *Escherichia coli* gene for glycerol-3-phosphate acyltransferase in transformed plants. This led to an increase in chilling sensitivity.

Tarczynski et al (130) introduced a bacterial gene for mannitol 1-phosphate dehydrogenase into tobacco plants, resulting in mannitol production in the transgenic plants. Mannitol is a compatible solute, one of a group of compounds that accumulate in plants under stress and that are thought to play a role in stress tolerance. When transgenic plants and controls were stressed by treatment with sodium chloride solution, the transformants producing mannitol showed significantly better growth than the controls, including production of new roots, new leaves, and flowers. The authors concluded that mannitol was acting either to adjust the cytosolic osmotic potential or to predispose the transgenic plants to stress tolerance.

Having discussed some of the general aspects involved in establishing a role for proteins in stress tolerance, we now turn to a more detailed discussion of two systems that appear promising for future demonstrations that ABA-regulated proteins may be involved in stress tolerance.

ABA-Regulated Genes and Cold Tolerance

An early study that indicated a role for ABA in cold acclimation investigated two potato species, *Solanum tuberosum*, whose leaves failed to cold acclimate, and *Solanum commersonii*, with leaves that hardened from a lethal temperature of −5°C to −12°C over a 15-day period at 2°C (16). During hardening the leaf ABA content of *S. commersonii* showed a transient rise (approximately 2.5-fold). Although this peak was observed on only a single time-point, it was specific to the cold acclimating species, and it was not observed in control plants maintained at 15–20°C. Application of ABA to *S. commersonii* plants growing either at 25°C or at 2°C resulted in cold acclimation even at the higher growth temperature. The effect of ABA treatment on the non-hardening species was not reported. Finally, evidence was presented that protein synthesis was required for cold acclimation, based on the effects of cycloheximide, a protein synthesis inhibitor. In this study, and many others since, the lack of identified biochemical functions for the proteins expressed during acclimation weakens mechanistic interpretations of the role of ABA in cold hardening. Little progress in understanding this role has been made. Recently, the expression of cold- and ABA-regulated genes has been associated with cold acclimation in both wild-type and mutant *A. thaliana* plants. Because these studies use mutants with complex phenotypic characteristics, we describe them in some detail before describing the results. The same mutants were used in the desiccation tolerance studies that we review later.

Two types of mutant that Koornneef and colleagues (68, 69) isolated have been used to study the role of ABA in cold acclimation and desiccation tolerance: (*a*) *aba*, denoting mutants at a single genetic locus that are low in endogenous ABA levels, probably because of a block in the ABA biosynthetic pathway (29, 115) and (*b*) *abi* (abscisic acid insensitive), denoting mutants at any of three different loci, selected as not responding to the inhibitory action of ABA on seed germination and early seedling growth. Mutants at the *aba* locus show strongly reduced seed dormancy, lack the normal light requirement for seed germination, and show a higher tendency to exhibit withering, but they respond to applied ABA. Mutants at the three *abi* loci also show a strong reduction in seed dormancy, and may wilt more readily under water stress than do normal plants. Such mutants are potentially altered in a component that recognizes ABA, such as a receptor (insensitive mutants), or in the chain of events that leads from the receptor to the measured response. The observation that several other characteristics associated with ABA (such as dormancy, desiccation tolerance, or proline accumulation) are affected in the *abi* mutants indicates that different responses to ABA share components that are altered in the mutants (32). However, since mutants at each of the three *abi* loci apparently retain some responses to ABA, they are not truly insensitive. Firn (34) has discussed some of the complexities inherent in the concept of hormone sensitivity. These are equally relevant to hormone insensitivity. The *abi3* mutant recently has been shown (40) to affect a protein that has significant similarity to the product of the *vp1* locus in corn, which functions as an ABA-regulated transcriptional activator during seed development (83).

Wild-type *A. thaliana* cold hardens at 4°C so that its freezing resistance increases from approximately −3°C to approximately −8°C (38, 74). Treatment of *A. thaliana* plants with ABA results in cold hardening occuring at 20°C (74). An alternative assay for freezing tolerance, leaf electrolyte leakage, showed a similar range over which cold tolerance increased in response to ABA. A summary of the ability of *aba* and *abi* mutants to cold acclimate is shown in Table 1. The *aba-1* mutant showed no cold acclimation (51) or reduced cold acclimation (39) relative to the wild-type. Treatment of *aba-1* plants with ABA and growth at 20°C resulted in cold acclimation (51). Therefore it can be concluded that the low levels of ABA in the mutant plants result in an impaired ability or failure to develop cold tolerance, which can be corrected by applied ABA. The extent to which the poor cold adaptation might be a direct result of the ABA deficiency, or whether the pleiotropic effects of the ABA deficiency might also be involved, is uncertain (39). The abi mutants (*abi1*, *abi2*, and *abi3*) showed cold acclimation at 4°C similar to the wild-type (39, and reported in 51). At least two conclusions could be drawn from this result: ABA responses may not be necessary for the development of cold tolerance (compare with the result using the *aba* mutant described above), or, as discussed earlier, the apparent insensitivity of the mutants to some effects of ABA does not extend to other possible effects of ABA, such as the development of cold tolerance.

Several groups have reported the isolation of cDNA clones for cold-regulated mRNAs in *A. thaliana*, which are also regulated by ABA and by drought

Table 1   Summary of cold acclimation and gene expression in *Arabidopsis thaliana*.[a]

	Genotype				
	wild-type	*aba-1*	*abi1*	abi2	abi3
Cold acclimation:					
4°C	Yes	Poor[b]	Yes	Yes	Yes
ABA/20°C	Yes	Yes	NR[c]	NR	NR
Induction of COR/LTI gene expression:					
4°C	Yes	Yes	Yes	Yes	Yes
ABA/20°C	Yes	Yes	No	Yes	Yes

[a]See text for discussion of some of the simplifying assumptions made in summarizing the data for tabular form. Data summarized from references 39, 51, 93.
[b]Differences between the results of the two studies were reported. See text for details.
[c]Not reported.

stress. These clones allow investigators to examine gene expression in the wild-type and mutant lines and may help them to determine whether the gene products might be involved in cold acclimation. Gilmour & Thomashow (39) examined mRNA levels for three of the four *cor* (cold-regulated) genes that they had previously characterized in *A. thaliana*. In wild-type, *aba-1, abi1, abi2,* and *abi3* plants mRNA levels increased markedly in response to cold treatment (Table 1). ABA treatment at 20°C also stimulated mRNA levels, except in the *abi1* mutant, which did not elevate levels of the three cold-regulated transcripts, but which did have normal levels of a control mRNA not regulated by ABA or cold treatments. Note that the *abi2* and *abi3* mutants, although selected as showing ABA-insensitive germination and early seedling growth, both responded to applied ABA by elevating levels of COR transcripts. This implies that there are at least some differences between the components involved in these two responses to ABA. Nordin et al presented a similar set of results (93) based on a different low temperature–induced clone, *lti140*. In response to low temperature, *lti140* mRNA accumulated in wild-type *A. thaliana* and in the *aba-1* and *abi1* mutants. *Lti140* transcripts also accumulated in wild-type *A. thaliana* treated with fluridone. The implication of the fluridone treatment (although its effects on the endogenous ABA levels were not reported), and of the response of the *aba-1* mutant, is that ABA may not be involved in cold induction of *lti140*. In response to ABA treatment at 20°C, *lti140* mRNA levels increased in wild-type, *aba-1*, and *abi3*, but not in *abi1*. Therefore, the gene is inducible by ABA, but ABA is not necessary for cold-induced expression.

Despite some reservations and qualifications (described below), several conclusions can be drawn from the pattern of expression of cold/ABA-regulated genes in the *aba* and *abi* mutants: 1. Expression of the cold-regulated genes does not always correlate with cold tolerance. For example, the *aba-1* mutant showed impaired cold acclimation in response to low temperature, yet levels of COR/LTI transcripts were elevated. Further tests of this correlation could be made using the *abi1* mutant. It appears that cold hardening by *abi1* plants in response to ABA at 20°C has not been studied; if hardening did occur, the fact that *cor/lti* genes are not induced under these conditions would constitute evidence that their products may not be required for cold tolerance. 2. Low temperature induction and ABA induction of *cor/lti* genes may involve different control mechanisms, since the *abi1* mutant elevated transcript levels in response to low temperature treatment, but not in response to ABA treatment. 3. The different responses of the *abi1*, *abi2*, and *abi3* mutants to ABA treatment in terms of COR/LTI transcript accumulation help define differences between the mutants and highlight the problems of interpreting a hormone insensitive phenotype.

These studies well illustrate the complexities inherent in attempting to unravel whole plant physiological responses at the molecular level. It is important to keep in mind some of the limitations of the conclusions described above. First, the effects of cold and ABA treatment on gene expression refer only to the three or four different clones that have been described. They cannot be taken as general responses until many more clones have been examined. Second, it is assumed that other aspects of the mutants are not influencing the outcomes. For example, a temperature sensitive component to the mutant phenotype; other effects on uptake, distribution, or metabolism of ABA (39); or pleiotropic effects that are characteristic of hormone mutants could all contribute to the final response shown by the plant. Third, it is assumed that the mRNA assays reflect the abundance of the corresponding polypeptides. Fourth, until the systems are described more thoroughly, interpretations will oversimplify the real situation, especially in summaries of published data such as Table 1.

ABA-Regulated Genes and Desiccation Tolerance

The same set of *A. thaliana* mutations is of value in describing the roles played by ABA during seed development, particularly in the induction of desiccation tolerance. Although the analysis of changes in gene expression that might be associated with this trait has not proceeded as far as it has for cold acclimation, this area of research is likely to make rapid progress. For these studies, double mutants were constructed involving the *aba* and *abi3* loci, since the combination of mutations affecting both ABA levels and ABA sensitivity generates a more extreme "ABA null" phenotype than either mutation alone. The *aba,abi3* seeds that developed on an *aba,abi3* plant lacked desiccation tolerance (< 10% of the seeds were desiccation tolerant), but application of ABA, or particularly of an ABA analog (LAB173711) conferred desiccation tolerance on up to 60% of the seeds. The ability of *aba,abi3* plants, homozygous for a mutation conferring ABA insensitivity, to respond to ABA or ABA analogs provides another example of how different responses to ABA presumably involve different cellular components (67).

Given these effects of the *aba,abi3* genotype on seed desiccation tolerance, and the postulated roles for LEA proteins in desiccation tolerance, it was relevant to examine gene expression in the *aba,abi3* genetic background. SDS gels indicated that the *aba,abi3* double mutant failed to accumulate the 12S and 2S storage proteins, in contrast to *aba* or *abi3* single mutants (67). Further study (84) showed that synthesis of a set of heat-stable polypeptides, expected to include LEA-type proteins, was absent specifically from the *aba,abi3* double mutant, but was restored by treatment with the ABA analog, as was desiccation tolerance. Treatment of seeds with ABA/sucrose during development effectively conferred desiccation tolerance (100% of seeds).

What conclusions can be drawn from these results? The *aba,abi3* double mutant is deficient in production of a suite of seed proteins, and shows numerous aberrations of seed development (e.g. lack of dormancy, failure to lose water, lack of desiccation tolerance). Sensitivity to desiccation (and perhaps other aberrations) can be overcome by simple treatments that also restore production of a set of seed proteins that were present in normal seeds. It may be possible to identify proteins responsible for desiccation tolerance if an appropriate transformation strategy can be developed using *aba,abi3* plants. Selection for desiccation tolerant seeds in a transformed population could provide a powerful means to isolate gene products involved in dessication tolerance.

Studies on cold acclimation and desiccation tolerance in mutants of *A. thaliana* provide a graphic example of the interplay between physiology, molecular biology, and mutational studies. Physiological and molecular responses in the mutants help further define their phenotypes, and in turn this provides information on the involvement of ABA in the responses. These studies are likely to serve as a model for future investigations.

CONCLUDING REMARKS

The analysis of genes regulated by ABA is well under way. Progress in the characterization of gene products will be rapid because there are established techniques for cloning differentially expressed mRNAs. Considerable work is being done on the mechanism of regulation, particularly in characterizing the components required for transcriptional regulation: *cis*-acting DNA elements and the proteins that bind to them. Although these events may be many steps removed from the initial perception of ABA (and perhaps other regulatory molecules), the information provided by such studies is essential for the use of ABA-regulated promoters in either experimental or practical applications.

Currently, we can specify in great detail the effects of ABA on levels of many mRNAs and proteins, and how mutations in ABA-response elements alter transcriptional activity. This situation contrasts greatly with our lack of knowledge about earlier events in ABA regulation, including stress-mediated alterations in ABA levels, proposed receptor interactions, and signal transduction. ABA biosynthetic and non-responsive mutants will be important in the analysis of these earlier events, especially if generation of the mutants can be coupled with isolation of the mutant gene, either by chromosome "walking" or by transposon "tagging." The use of transgenic plants with visible, selectable, or counter-selectable reporter genes under the control of ABA-regulated promoters should make screening for mutants easier than is currently possible.

Much of the interest in ABA-regulated proteins stems from their possible roles in stress tolerance. In principle, the biochemical or structural function of a protein need not be known for a role in stress tolerance to be demonstrated. Transformation approaches allow this role to be tested, and exciting results have recently been obtained in modifying the cold tolerance or salt tolerance of transgenic plants. We are not aware of any case where an ABA-regulated protein has been shown to influence stress tolerance in plants, although promising systems with potential candidate proteins are emerging. However, it is unknown what proportion of the proteins regulated by stress or ABA will actually be able to influence stress tolerance. Once a role in stress tolerance is demonstrated, a major problem will be determining the mechanisms by which stress/ABA-induced proteins influence stress tolerance. The principal limitations to progress in this area stem from our lack of knowledge about the functions of stress/ABA-regulated gene products and about the biochemical and structural complexities of stress tolerance.

The study of ABA-regulated genes and their relation to stress tolerance is still at the experimental stage. We hope to see considerable progress in using ABA-regulated gene products to modify the stress tolerance of transgenic plants grown under controlled conditions. A more difficult task will be the application of these findings to improve either the stress tolerance or the productivity under stress of crop plants in field situations. We do not know whether the gene products identified as limiting in laboratory stress situations will be the same as those that limit growth or survival in field stress situations. Determining the functions of ABA-regulated gene products, and their roles in plants under stress, remain challenging areas for future research.

ACKNOWLEDGMENTS

We are grateful to our colleagues Jake Jacobsen, Rod King, and Rana Munns for their comments on the manuscript. Work in our laboratory has been generously supported by a grant from the late Ken and Yasuko Myer. Masumi Robertson was the recipient of a Yasuko Hiraoka Myer Research Fellowship.

Any *Annual Review* chapter, as well as any article cited in an *Annual Review* chapter, may be purchased from the Annual Reviews Preprints and Reprints service.
1-800-347-8007; 415-259-5017; email: arpr@class.org

Literature Cited

1. Bader SB, Price BD, Mannheim-Rodman LA, Calderwood SK. 1992. Inhibition of heat shock gene expression does not block the development of thermotolerance. *J. Cell. Physiol.* 151:56–62
2. Deleted in proof
3. Barratt DHP, Whitford PN, Cook SK, Butcher G, Wang TL. 1989. An analysis of seed development in *Pisum sativum. J. Exp. Bot.* 40:1009–14
4. Bartels D, Engelhardt K, Roncarati R, Schneider K, Rotter M, Salamini F. 1991. An ABA and GA modulated gene expressed in the barley embryo encodes an aldose reductase related protein. *EMBO J.* 10:1037–43
5. Bartholomew DM, Bartley GE, Scolnik PA. 1991. Abscisic acid control of rbcS and cab transcription in tomato leaves. *Plant Physiol.* 96:291–96
6. Benfey PN, Chua N-H. 1990. The Cauliflower Mosaic Virus 35S promoter: combinatorial regulation of transcription in plants. *Science* 250:959–66
7. Blackman SA, Obendorf RL, Leopold AC. 1992. Maturation proteins and sugars in desiccation tolerance of developing soybean seeds. *Plant Physiol.* 100:225–30
8. Bradford KJ, Chandler PM. 1992. Expression of "dehydrin-like" proteins in embryos and seedlings of *Zizania palustris* and *Oryza sativa* during dehydration. *Plant Physiol.* 99:488–94
9. Bray EA. 1988. Drought and ABA-induced changes in polypeptide and mRNA accumulation in tomato leaves. *Plant Physiol.* 88:1210–14
10. Bray EA. 1991. Regulation of gene expression by endogenous ABA during drought stress. See Ref. 26, pp. 79–97
11. Bray EA, Beachy RN. 1985. Regulation by ABA of β-conglycinin expression in cultured developing soybean cotyledons. *Plant Physiol.* 79:746–50
12. Breu V, Guerbette F, Kader J, Kannangara CG, Svensson B, von Wettstein-Knowles P. 1989. A 10 kD barley basic protein transfers phosphatidylcholine from liposomes to mitochondria. *Carlsberg Res. Commun.* 54: 81–84

13. Bustos MM, Begum D, Kalkan FA, Battraw MJ, Hall TC. 1991. Positive and negative *cis*-acting DNA domains are required for spatial and temporal regulation of gene expression by a seed storage protein promoter. *EMBO J.* 10:1469–79
14. Cammue BPA, Broekaert WF, Kellens JTC, Raikhel NV, Peumans WJ. 1989. Stress-induced accumulation of wheat germ agglutinin and abscisic acid in roots of wheat seedlings. *Plant Physiol.* 91: 1432–35
15. Chandler PM, Zwar JA, Jacobsen JV, Higgins TJV, Inglis AS. 1984. The effects of gibberellic acid and abscisic acid on α-amylase mRNA levels in barley aleurone layers: studies using an α-amylase cDNA clone. *Plant Mol. Biol.* 3:407–18
16. Chen H-H, Li PH, Brenner ML. 1983. Involvement of abscisic acid in potato cold acclimation. *Plant Physiol.* 71:362–65
17. Close TJ, Kortt AA, Chandler PM. 1989. A cDNA-based comparison of dehydration-induced proteins (dehydrins) in barley and corn. *Plant Mol. Biol.* 13:95–108
18. Close TJ, Lammers PJ. 1993. An osmotic stress protein of cyanobacteria is immunologically related to plant dehydrins. *Plant Physiol.* 101:773–79
19. Cohen A, Bray EA. 1990. Characterization of three mRNAs that accumulate in wilted tomato leaves in response to elevated levels of endogenous abscisic acid. *Planta* 182: 27–33
20. Cohen A, Plant AL, Moses MS, Bray EA. 1991. Organ-specific and environmentally regulated expression of two abscisic acid-induced genes of tomato. *Plant Physiol.* 97:1367–74
21. Creelman RA, Mullet JE. 1991. Water deficit modulates gene expression in growing zone of soybean seedlings. Analysis of differentially expressed cDNAs, a new β-tubulin gene, and expression of genes encoding cell wall proteins. *Plant Mol. Biol.* 17:591–608
22. Crouch ML, Sussex IM. 1981. Development and storage-protein synthesis in *Brassica napus* (L.) embryos in vivo and in vitro. *Planta* 153:64–74

23. Crouch ML, Tenbarge K, Simon A, Finkelstein R, Scofield S, Solberg L. 1984. Storage protein mRNA levels can be regulated by abscisic acid in *Brassica* embryos. In *Molecular Form and Function of the Plant Genome*, ed. L van Vloten Doting, GSP Groot, TC Hall, pp. 555–66. New York: Plenum

24. Crowe JH, Crowe LM, Carpenter JF, Wistrom CA. 1987. Stabilization of dry phospholipid bilayers and proteins by sugars. *Biochem. J.* 242:1–10

25. Curry J, Morris CF, Walker-Simmons MK. 1991. Sequence analysis of a cDNA encoding a group 3 LEA mRNA inducible by ABA or dehydration stress in wheat. *Plant Mol. Biol.* 16:1073–76

26. Davies WJ, Jones HG. 1991. *Abscisic Acid: Physiology and Biochemistry*. Oxford: Bios

27. Davies WJ, Mansfield TA. 1983. The role of abscisic acid in drought avoidance. In *Abscisic Acid*, ed. FT Addicott, pp. 237–68. New York: Praeger

28. DeLisle AJ, Crouch ML. 1989. Seed storage protein transcription and mRNA levels in *Brassica napus* during development and in response to exogenous abscisic acid. *Plant Physiol.* 91:617–23

29. Duckham SC, Linforth RST, Taylor IB. 1991. Abscisic acid-deficient mutants at the *aba* gene locus of *Arabidopsis thaliana* are impaired in the epoxidation of zeaxanthin. *Plant Cell Environ.* 14:631–36

29a. Dure L. 1993. A repeating 11-mer amino acid motif and plant dessication. *Plant J.* 3:363–69

30. Dure L, Crouch M, Harada J, Ho TD, Mundy J, et al. 1989. Common amino acid sequence domains among the LEA proteins of higher plants. *Plant Mol. Biol.* 12:475–86

31. Espelund M, Saeboe-Larssen S, Hughes DW, Galau GA, Larsen F, Jakobsen KS. 1992. Late embryogenesis-abundant genes encoding proteins with different numbers of hydrophilic repeats are regulated differentially by abscisic acid and osmotic stress. *Plant J.* 2:241–52

32. Finkelstein RR, Somerville CR. 1990. Three classes of abscisic acid (ABA)-insensitive mutations of *Arabidopsis* define genes that control overlapping subsets of ABA responses. *Plant Physiol.* 94:1172–79

33. Finkelstein RR, Tenbarge KM, Shumway JE, Crouch ML. 1985. Role of ABA in maturation of rapeseed embryos. *Plant Physiol.* 78:630–36

34. Firn RD. 1986. Growth substance sensitivity: the need for clearer ideas, precise terms and purposeful experiments. *Physiol. Plant.* 67:267–72

35. Galau GA, Bijaisoradat N, Hughes DW.

1987. Accumulation kinetics of cotton late embryogenesis-abundant mRNAs and storage protein mRNAs: coordinate regulation during embryogenesis and the role of abscisic acid. *Dev. Biol.* 123:198–212

36. Galau GA, Hughes DW, Dure L. 1986. Abscisic acid induction of cloned cotton late embryogenesis-abundant (*Lea*) mRNAs. *Plant Mol. Biol.* 7:155–70

37. Gaxiola R, de Larrinoa IF, Villalba JM, Serrano R. 1992. A novel and conserved salt-induced protein is an important determinant of salt tolerance in yeast. *EMBO J.* 11:3157–64

38. Gilmour SJ, Hajela RK, Thomashow MF. 1988. Cold acclimation in *Arabidopsis thaliana*. *Plant Physiol.* 87:745–50

39. Gilmour SJ, Thomashow MF. 1991. Cold acclimation and cold-regulated gene expression in ABA mutants of *Arabidopsis thaliana*. *Plant Mol. Biol.* 17:1233–40

40. Giraudat J, Hauge BM, Valon C, Smalle J, Parcy F, Goodman HM. 1992. Isolation of the Arabidopsis *ABI3* gene by positional cloning. *Plant Cell* 4:1251–61

41. Gomez J, Sanchez-Martinez D, Stiefel V, Rigau J, Puigdomenech P, Pages M. 1988. A gene induced by the plant hormone abscisic acid in response to water stress encodes a glycine-rich protein. *Nature* 334: 262–64

42. Green TR, Ryan CA. 1972. Wound-induced proteinase inhibitor in plant leaves: a possible defense mechanism against insects. *Science* 175:776–77

43. Guarente L. 1993. Strategies for the identification of interacting proteins. *Proc. Natl. Acad. Sci. USA* 90:1639–41

44. Guiltinan MJ, Marcotte WR, Quatrano RS. 1990. A plant leucine zipper protein that recognizes an abscisic acid response element. *Science* 250:267–71

45. Hajela RK, Horvath DP, Gilmour SJ, Thomashow MF. 1990. Molecular cloning and expression of cor (cold-regulated) genes in *Arabidopsis thaliana*. *Plant Physiol.* 93:1246–52

46. Hamilton AJ, Lycett GW, Grierson D. 1990. Antisense gene that inhibits synthesis of the hormone ethylene in transgenic plants. *Nature* 346:284–87

47. Harada JJ, De Lisle AJ, Baden CS, Crouch ML. 1989. Unusual sequence of an abscisic acid-inducible mRNA which accumulates late in *Brassica napus* seed development. *Plant Mol. Biol.* 12:395–401

48. Harrison SC. 1991. A structural taxonomy of DNA-binding domains. *Nature* 353: 715–19

49. Hartung W, Slovik S. 1991. Physicochemical properties of plant growth regulators and plant tissues determine their distribution and redistribution: stomatal regulation

by abscisic acid in leaves. *New Phytol.* 119:361–82

50. Hatzopoulos P, Fong F, Sung ZR. 1990. Abscisic acid regulation of DC8, a carrot embryonic gene. *Plant Physiol.* 94:690–95

51. Heino P, Sandman G, Lang V, Nordin K, Palva ET. 1990. Abscisic acid deficiency prevents development of freezing tolerance in *Arabidopsis thaliana* (L.) Heynh. *Theoret. Appl. Genet.* 79:801–6

52. Hetherington AM, Quatrano RS. 1991. Mechanisms of action of abscisic acid at the cellular level. *New Phytol.* 119:9–32

53. Higgins TJ, Jacobsen JV, Zwar JA. 1982. Gibberellic acid and abscisic acid modulate protein synthesis and mRNA levels in barley aleurone layers. *Plant Mol. Biol.* 1:191–215

54. Hong B, Uknes SJ, Ho TD. 1988. Cloning and characterization of a cDNA encoding a mRNA rapidly-induced by ABA in barley aleurone layers. *Plant Mol. Biol.* 11:495–506

55. Hughes DW, Galau GA. 1991. Developmental and environmental induction of *Lea* and *LeaA* mRNAs and the postabscission program during embryo culture. *Plant Cell* 3:605–18

56. Huttly AK, Baulcombe DC. 1989. A wheat α-*Amy2* promoter is regulated by gibberellin in transformed oat aleurone protoplasts. *EMBO J.* 8:1907–13

57. Idetek. 1991. *Phytodetek-ABA.* Technical brochure. Idetek, San Bruno, CA

58. Iturriaga G, Schneider K, Salamini F, Bartels D. 1992. Expression of desiccation-related proteins from the resurrection plant *Craterostigma plantagineum* in transgenic tobacco. *Plant Mol. Biol.* 20:555–58

59. Jacobsen JV, Beach LR. 1985. Control of transcription of α-amylase and rRNA genes in barley aleurone protoplasts by gibberellin and abscisic acid. *Nature* 316:275–77

60. Jacobsen JV, Close TJ. 1991. Control of transient expression of chimaeric genes by gibberellic acid and abscisic acid in protoplasts prepared from mature barley aleurone layers. *Plant Mol. Biol.* 16:713–24

61. Jacobsen JV, Shaw DC. 1989. Heat-stable proteins and abscisic acid action in barley aleurone cells. *Plant Physiol.* 91:1520–26

62. Johnson-Flanagan AM, Huiwen Z, Thiagarajah MR, Saini HS. 1991. Role of abscisic acid in the induction of freezing tolerance in *Brassica napus* suspension-cultured cells. *Plant Physiol.* 95:1044–48

63. Karssen CM, Brinkhorst-van der Swan DLC, Breekland AE, Koornneef M. 1983. Induction of dormancy during seed development by endogenous abscisic acid: studies on abscisic acid deficient genotypes of *Arabidopsis thaliana* (L.) Heynh. *Planta* 157:158–65

64. Katagiri F, Chua N. 1992. Plant transcription factors: present knowledge and future challenges. *Trends Genet.* 8:22–27

65. King RW. 1982. Abscisic acid in seed development. In *The Physiology and Biochemistry of Seed Development, Dormancy and Germination,* ed. AA Khan, pp. 157–81. Amsterdam: Elsevier

66. Klee H, Estelle M. 1991. Molecular genetic approaches to plant hormone biology. *Annu. Rev. Plant Physiol. Plant Mol. Biol.* 42:529–51

67. Koornneef M, Hanhart CJ, Hilhorst HWM, Karssen CM. 1989. In vivo inhibition of seed development and reserve protein accumulation in recombinants of abscisic acid biosynthesis and responsiveness mutants in *Arabidopsis thaliana. Plant Physiol.* 90:463–69

68. Koornneef M, Jorna ML, Brinkhorst-van der Swan DLC, Karssen CM. 1982. The isolation of abscisic acid (ABA) deficient mutants by selection of induced revertants in non-germinating gibberellin sensitive lines of *Arabidopsis thaliana* (L.) Heynh. *Theoret. Appl. Genet.* 61:385–93

69. Koornneef M, Reuling G, Karssen CM. 1984. The isolation and characterization of abscisic acid-insensitive mutants of *Arabidopsis thaliana. Physiol. Plant.* 61:377–83

70. Kovach DA, Bradford KJ. 1992. Imbibitional damage and desiccation tolerance of wild rice *Zizania palustris* seeds. *J. Exp. Bot.* 43:747–57

71. Kurkela S, Franck M. 1990. Cloning and characterization of a cold- and ABA-inducible *Arabidopsis* gene. *Plant Mol. Biol.* 15:137–44

72. Kusano T, Aguan K, Abe M, Sugawara K. 1992. Nucleotide sequence of a rice *rab16* homologue gene. *Plant Mol. Biol.* 18:127–29

73. Landry J, Bernier D, Chretien P, Nicole LM, Tanguay RM, Marceau N. 1982. Synthesis and degradation of heat shock proteins during development and decay of thermotolerance. *Cancer Res.* 42:2457–61

74. Lang V, Heino P, Palva ET. 1989. Low temperature acclimation and treatment with exogenous abscisic acid induce common polypeptides in *Arabidopsis thaliana* (L.) Heynh. *Theoret. Appl. Genet.* 77:729–34

75. LaRosa PC, Chen Z, Nelson DE, Singh NK, Hasegawa PM, Bressan RA. 1992. Osmotin gene expression is posttranscriptionally regulated. *Plant Physiol.* 100:409–15

76. Laszlo A, Li GC. 1985. Heat-resistant variants of Chinese hamster fibroblasts altered in expression of heat shock protein. *Proc. Natl. Acad. Sci. USA* 82:8029–33

77. Lessard PA, Allen RD, Bernier F, Crispino

JD, Fujiwara T, Beachy RN. 1991. Multiple nuclear factors interact with upstream sequences of differentially regulated β-conglycinin genes. *Plant Mol. Biol.* 16:397–413

78. Lin C, Thomashow MF. 1992. A cold-regulated *Arabidopsis* gene encodes a polypeptide having potent cryoprotective activity. *Biochem. Biophys. Res. Commun.* 183:1103–8

79. Lindquist S, Craig EA. 1988. The heat-shock proteins. *Annu. Rev. Genet.* 22:631–77

80. Mansfield MA, Raikhel NV. 1990. Abscisic acid enhances the transcription of wheat-germ agglutinin mRNA without altering its tissue-specific expression. *Planta* 180:548–54

81. Marcotte WR, Bayley CC, Quatrano RS. 1988. Regulation of a wheat promoter by abscisic acid in rice protoplasts. *Nature* 335:454–57

82. Marcotte WR, Russell SH, Quatrano RS. 1989. Abscisic acid-responsive sequence from the Em gene of wheat. *Plant Cell* 1:969–76

83. McCarty DR, Hattori T, Carson CB, Vasil V, Lazar M, et al. 1991. The *viviparous-1* developmental gene of maize encodes a novel transcriptional activator. *Cell* 66:895–905

84. Meurs C, Basra AS, Karssen CM, van Loon LC. 1992. Role of abscisic acid in the induction of desiccation tolerance in developing seeds of *Arabidopsis thaliana*. *Plant Physiol.* 98:1484–93

85. Morris CF, Anderberg RJ, Goldmark PJ, Walker-Simmons MK. 1991. Molecular cloning and expression of abscisic acid-responsive genes in embryos of dormant wheat seeds. *Plant Physiol.* 95:814–21

86. Mundy J. 1989. Developing nomenclature for genes of unknown function: a case study of ABA-responsive genes. *Plant Mol. Biol. Rep.* 7:276–83

87. Mundy J, Chua N-H. 1988. Abscisic acid and water-stress induce the expression of a novel rice gene. *EMBO J.* 7:2279–84

88. Mundy J, Hejgaard J, Hansen A, Hallgren L, Jorgensen KG, Munck L. 1986. Differential synthesis *in vitro* of barley aleurone and starchy endosperm proteins. *Plant Physiol.* 81:630–36

89. Mundy J, Rogers JC. 1986. Selective expression of a probable amylase/protease inhibitor in barley aleurone cells: comparison to the barley amylase/subtilisin inhibitor. *Planta* 169:51–63

90. Mundy J, Yamaguchi-Shinozaki K, Chua N-H. 1990. Nuclear proteins bind conserved elements in the abscisic acid-responsive promoter of a rice rab gene. *Proc. Natl. Acad. Sci. USA* 87:1406–10

91. Murata N, Ishizaki-Nishizawa O, Higashi S, Hayashi H, Tasaka Y, Nishida I. 1992. Genetically engineered alteration in the chilling sensitivity of plants. *Nature* 356:710–13

92. Neill SJ, Horgan R, Parry AD. 1986. The carotenoid and abscisic acid content of viviparous kernels and seedlings of *Zea mays* L. *Planta* 169:87–96

93. Nordin K, Heino P, Palva ET. 1991. Separate signal pathways regulate the expression of a low-temperature-induced gene in *Arabidopsis thaliana* (L.) Heynh. *Plant Mol. Biol.* 16:1061–71

94. Oeda K, Salinas J, Chua N-H. 1991. A tobacco bZip transcription activator (TAF-1) binds to a G-box-like motif conserved in plant genes. *EMBO J.* 10:1793–802

95. Parry AD, Horgan R. 1991. Carotenoids and abscisic acid (ABA) biosynthesis in higher plants. *Physiol. Plant.* 82:320–26

96. Parthier B. 1989. Hormone-induced alterations in plant gene expression. *Biochem. Physiol. Pflanzen* 185:289–314

97. Pena-Cortes H, Sanchez-Serrano JJ, Mertens R, Willmitzer L, Prat S. 1989. Abscisic acid is involved in the wound-induced expression of the proteinase inhibitor II gene in potato and tomato. *Proc. Natl. Acad. Sci. USA* 86:9851–55

98. Pena-Cortes H, Willmitzer L, Sanchez-Serrano JJ. 1991. Abscisic acid mediates wound induction but not developmental-specific expression of the proteinase inhibitor II gene family. *Plant Cell* 3:963–72

99. Penarrubia L, Aguilar M, Margossian L, Fischer RL. 1992. An antisense gene stimulates ethylene hormone production during tomato fruit ripening. *Plant Cell* 4:681–87

100. Piatkowski D, Schneider K, Salamini F, Bartels D. 1990. Characterization of five abscisic acid–responsive cDNA clones isolated from the desiccation-tolerant plant *Craterostigma plantagineum* and their relationship to other water-stress genes. *Plant Physiol.* 94:1682–88

101. Pla M, Goday A, Vilardell J, Gomez J, Pages M. 1989. Differential regulation of ABA-induced 23–25 kDa proteins in embryo and vegetative tissues of the viviparous mutants of maize. *Plant Mol. Biol.* 13:385–94

102. Pla M, Gomez J, Goday A, Pages M. 1991. Regulation of the abscisic acid-responsive gene rab28 in maize *viviparous* mutants. *Mol. Gen. Genet.* 230:394–400

103. Pla M, Vilardell J, Guiltinan MJ, Marcotte WR, Niogret MF, et al. 1993. The *cis*-regulatory element CCACGTGG is involved in ABA and water-stress responses of the maize gene rab28. *Plant Mol. Biol.* 21:259–66

104. Plant AL, Cohen A, Moses MS, Bray EA.

1991. Nucleotide sequence and spatial expression pattern of a drought- and abscisic acid–induced gene of tomato. *Plant Physiol.* 97:900–6

105. Ptashne M. 1988. How eukaryotic transcriptional activators work. *Nature* 335: 683–89

106. Quarrie SA. 1982. Droopy: a wilty mutant of potato deficient in abscisic acid. *Plant Cell Environ.* 5:23–26

107. Quatrano RS. 1986. Regulation of gene expression by abscisic acid during angiosperm embryo development. *Oxford Surv. Plant Mol. Cell Biol.* 3:467–76

108. Quatrano RS, Ballo BL, Williamson JD, Hamblin MT, Mansfield M. 1983. ABA controlled expression of embryo-specific genes during wheat grain development. In *Plant Molecular Biology,* ed. R Goldberg, pp. 343–53. New York: Liss

109. Quatrano RS, Guiltinan MJ, Marcotte WR. 1992. Regulation of gene expression by abscisic acid. In *Control of Plant Gene Expression,* ed. D-PS Verma, pp. 69–90. Caldwell: Telford

110. Radin JW. 1994. Water relations in controlled environments and the field. In *NATO Advanced Workshop on Interacting Stresses on Plants in a Changing Climate,* ed. M Jackson, C Black. Berlin: Springer-Verlag. In press

111. Reid JB. 1990. Phytohormone mutants in plant research. *J. Plant Growth Regul.* 9: 97–111

112. Riabowol KT, Mizzen LA, Welch WJ. 1988. Heat shock is lethal to fibroblasts microinjected with antibodies against hsp70. *Science* 242:433–36

113. Robertson M, Chandler PM. 1992. Pea dehydrins: identification, characterisation and expression. *Plant Mol. Biol.* 19:1031–44

114. Robertson M, Walker-Simmons M, Munro D, Hill RD. 1989. Induction of α-amylase inhibitor synthesis in barley embryos and young seedlings by abscisic acid and dehydration stress. *Plant Physiol.* 91:415–20

115. Rock CD, Zeevaart JAD. 1991. The aba mutant of *Arabidopsis thaliana* is impaired in epoxy-carotenoid biosynthesis. *Proc. Natl. Acad. Sci. USA* 88:7496–99

116. Rogers JC, Milliman C. 1983. Isolation and sequence analysis of a barley amylase cDNA clone. *J. Biol. Chem.* 258:8169–74

117. Salinas J, Oeda K, Chua N-H. 1992. Two G-box-related sequences confer different expression patterns in transgenic tobacco. *Plant Cell* 4:1485–93

118. Salmenkallio M, Hannus R, Teeri TH, Kauppinen V. 1990. Regulation of α-amylase promoter by gibberellic acid and abscisic acid in barley protoplasts transformed by electroporation. *Plant Cell Rep.* 9:352–55

119. Sanchez Y, Lindquist SL. 1990. HSP104 required for induced thermotolerance. *Science* 248:1112–15

120. Sanchez Y, Taulien J, Borkovich KA, Lindquist S. 1992. Hsp104 is required for tolerance to many forms of stress. *EMBO J.* 11:2357–64

121. Scott IM. 1990. Plant hormone response mutants. *Physiol. Plant.* 78:147–52

122. Seemann JR, Sharkey TD. 1987. The effect of abscisic acid and other inhibitors on photosynthetic capacity and the biochemistry of CO_2 assimilation. *Plant Physiol.* 84: 696–700

123. Shi L, Gast RT, Gopalraj M, Olszewski NE. 1992. Characterization of a shoot-specific, GA_3- and ABA-regulated gene from tomato. *Plant J.* 2:1–7

124. Skriver K, Mundy J. 1990. Gene expression in response to abscisic acid and osmotic stress. *Plant Cell* 2:503–12

125. Skriver K, Olsen FL, Rogers JC, Mundy J. 1991. *Cis*-acting DNA elements responsive to gibberellin and its antagonist abscisic acid. *Proc. Natl. Acad. Sci. USA* 88:7266–70

126. Smith BJ, Yaffe MP. 1991. Uncoupling thermotolerance from the induction of heat shock proteins. *Proc. Natl. Acad. Sci. USA* 88:11091–94

127. Staswick PE. 1990. Novel regulation of vegetative storage protein genes. *Plant Cell* 2:1–6

128. Suckow M, von Wilcken-Bergmann B, Muller-Hill B. 1993. Identification of three residues in the basic regions of the bZIP proteins GCN4, C/EBP and TAF-1 that are involved in specific DNA binding. *EMBO J.* 12:1193–200

129. Tal M, Nevo Y. 1973. Abnormal stomatal behavior and root resistance, and hormonal imbalance in three wilty mutants of tomato. *Biochem. Genet.* 8:291–300

130. Tarczynski MC, Jensen RG, Bohnert HJ. 1993. Stress protection of transgenic tobacco by production of the osmolyte mannitol. *Science* 259:508–10

131. Tardieu F, Davies WJ. 1992. Stomatal response to abscisic acid is a function of current plant water status. *Plant Physiol.* 98:540–45

132. Torres-Schumann S, Godoy JA, Pintor-Toro JA. 1992. A probable lipid transfer protein gene is induced by NaCl in stems of tomato plants. *Plant Mol. Biol.* 18:749–57

133. Vernon DM, Bohnert HJ. 1992. A novel methyl transferase induced by osmotic stress in the facultative halophyte *Mesembryanthemum crystallinum. EMBO J.* 11: 2077–85

134. Vilardell J, Goday A, Freire MA, Torrent M, Martinez MC, et al. 1990. Gene sequence, developmental expression, and

protein phosphorylation of RAB-17 in maize. *Plant Mol. Biol.* 14:423–32

135. Vilardell J, Mundy J, Stilling B, Leroux B, Pla M, et al. 1991. Regulation of the maize *rab17* gene promoter in transgenic heterologous systems. *Plant Mol. Biol.* 17:985–93

136. Vinson CR, Sigler PB, McKnight SL. 1989. Scissors-grip model for DNA recognition by a family of leucine zipper proteins. *Science* 246:911–16

137. Vivekananda J, Drew MC, Thomas TL. 1992. Hormonal and environmental regulation of the carrot *lea*-class gene *Dc3. Plant Physiol.* 100:576–81

138. Walker-Simmons MK, Abrams SR. 1991. Use of ABA immunoassays. See Ref. 26, pp. 53–61

139. Weretilnyk EA, Hanson AD. 1990. Molecular cloning of a plant betaine-aldehyde dehydrogenase, an enzyme implicated in adaptation to salinity and drought. *Proc. Natl. Acad. Sci. USA* 87:2745–49

140. Williams B, Tsang A. 1991. A maize gene expressed during embryogenesis is abscisic acid-inducible and highly conserved. *Plant Mol. Biol.* 16:919–23

141. Williams ME, Foster R, Chua N-H. 1992. Sequences flanking the hexameric G-box core CACGTG affect the specificity of protein binding. *Plant Cell* 4:485–96

142. Williamson JD, Quatrano RS. 1988. ABA-regulation of two classes of embryo-specific sequences in mature wheat embryos. *Plant Physiol.* 86:208–15

143. Wistow GJ, Piatigorsky J. 1988. Lens crystallins: the evolution and expression of proteins for a highly specialized tissue. *Annu. Rev. Biochem.* 57:479–504

144. Wolter FP, Schmidt R, Heinz E. 1992. Chilling sensitivity of *Arabidopsis thaliana* with genetically engineered membrane lipids. *EMBO J.* 11:4685–92

145. Yamaguchi-Shinozaki K, Mino M, Mundy J, Chua N. 1990. Analysis of an ABA-responsive rice gene promoter in transgenic tobacco. *Plant Mol. Biol.* 15:905–12

146. Yamaguchi-Shinozaki K, Shinozaki K. 1993. Characterization of the expression of a desiccation-responsive *rd29* gene of *Arabidopsis thaliana* and analysis of its promoter in transgenic plants. *Mol. Gen. Genet.* 236:331–40

147. Zeevaart JAD, Creelman RA. 1988. Metabolism and physiology of abscisic acid. *Annu. Rev. Plant Physiol. Plant Mol. Biol.* 39:439–73

148. Zeevaart JAD, Rock CD, Fantauzzo F, Heath TG, Gage DA. 1991. Metabolism of ABA and its physiological implications. See Ref. 26, pp. 39–52

149. Zwar JA, Hooley R. 1986. Hormonal regulation of α-amylase gene transcription in wild oat (*Avena fatua* L.) aleurone protoplasts. *Plant Physiol.* 80:459–63

Annu. Rev. Plant Physiol. Plant Mol. Biol. 1994. 45:143–71
Copyright © 1994 by Annual Reviews Inc. All rights reserved

THE TRANSDUCTION OF BLUE LIGHT SIGNALS IN HIGHER PLANTS

Timothy W. Short

Plant Gene Expression Center, Agricultural Research Service, USDA, 800 Buchanan Street, Albany, California 94710

Winslow R. Briggs

Department of Plant Biology, Carnegie Institution of Washington, 290 Panama Street, Stanford, California 94305

KEY WORDS: blue light, *b*-type cytochrome, flavin, pterin, gene regulation, G-protein, light-induced absorbance change (LIAC[1]), protein phosphorylation, phytochrome, stomata

CONTENTS

INTRODUCTION 144
LIGHT-INDUCED PHOSPHORYLATION OF A PLASMA MEMBRANE PROTEIN 145
   Preliminary Characterization and Membrane Localization... 145
   Correlation of Light-induced Phosphorylation with Phototropism................................. 145
   Biochemical Properties of the Reaction... 146
LIGHT-INDUCED REDUCTION OF A *b*-TYPE CYTOCHROME—THE LIGHT-
   INDUCED ABSORBANCE CHANGE (LIAC).. 149
GTP-BINDING PROTEINS... 151
   G-Proteins Associated with the Plasma Membrane of Pea Apical Buds 151
PHOTOPERCEPTION: GENERAL CONSIDERATIONS 151
   Physiological Clues to the Nature of B Photoreceptors 151
B REGULATION OF GENE EXPRESSION 154
   Measurements of Transcription and Transcript Abundance 154
   Cis- and *Trans-Acting Elements in B-Mediated Gene Expression* 158
   Genetic Mutants In B Signal Transduction.. 159

[1]
   Abbreviations: B, blue light; LIAC, light-induced absorbance change; MB, methylene blue; R, red light; FR, far-red light; UV-A, ultraviolet A (320–400nm); UV-B, ultraviolet B (280–320nm); W, white light.

OTHER EFFECTS OF B... 161
   Correlations of B Effects on Cell Growth and Electrical Potentials............................... 161
   Cytoskeletal Effects of B.. 162
   Stomatal Regulation.. 163
CONCLUSIONS... 164

INTRODUCTION

Red, far-red, blue, ultraviolet-A, and ultraviolet-B light together have dramatic effects on the development of higher plants through a wide range of transduction chains that in sum define the phenomenon of photomorphogenesis. The phytochromes, well-known mediators of photomorphogenesis in the R and FR regions of the electromagnetic spectrum, have been studied in detail at the biochemical, genetic, and molecular genetic levels (45, 113). Despite considerable research during the past two decades (63, 131, 132, 133) the B- and UV-absorbing photosensory systems have remained elusive. There has been no definitive isolation and characterization of a higher plant B photoreceptor. In many cases [e.g. phototropism (65)] the approach has been mostly inferential, with responses of the whole organ or organism used to deduce properties of the photoreceptor and the signal transduction chain. Only in the last few years have meaningful inroads been made into understanding a few of the possible intracellular signal transduction chains mediating B-activated photomorphogenic processes in higher plants.

In this review we evaluate B- and UV-A–dependent photosensory systems in the context of recent studies in which some understanding has been achieved, and in which some progress has been made in defining systems in vitro. These systems include light-induced phosphorylation of a plasma membrane protein, light-induced activation of GTPase activity in the plasma membrane, and light-mediated changes in gene expression. We also consider the current status of efforts to identify definitively a photoreceptor moiety; touch briefly on some studies with guard cells, on light-induced reduction of non-mitochondrial membrane-associated *b*-type cytochromes, and on other systems relevant to the above issues; and review the promising use of mutants in B studies. We restrict ourselves primarily to genetic, cellular, biochemical, and molecular studies in higher plants where it seems that real progress has been made or is perhaps imminent. Several recent short reviews (20, 69, 72, 90, 142, 143) deal with various aspects of the above material. Plant responses to UV-B have been reviewed elsewhere recently (8, 23, 153) and will not be considered here.

LIGHT-INDUCED PHOSPHORYLATION OF A PLASMA MEMBRANE PROTEIN

Preliminary Characterization and Membrane Localization

Gallagher et al (48) first noted that irradiation of elongating stem tissue of etiolated pea seedlings (*Pisum sativum* L.) reduced the amount of phosphorylation of an approximately 120-kDa protein that could be obtained upon subsequent treatment of isolated microsomal membranes with [γ-^{32}P]ATP. The reaction was activated by B and not R, and the phosphorylated protein was not released easily from a pelletable fraction by high salt, mild detergent treatments, or osmotic shock. The pea protein was localized to the plasma membrane (48, 144), as was a homologous protein from maize coleoptiles (59, 109).

In maize and pea the effect of light both in vivo and in vitro was the potentiation of the phosphorylation. Short & Briggs (141) showed that phosphorylation of pea microsomal membranes from dark-control tissue could be enhanced strongly by irradiating the membranes before [γ-^{32}P]ATP addition. Pre-illumination of the tissue with B before membrane isolation strongly reduced the effectiveness of in vitro irradiation. Furthermore, when etiolated tissues were preloaded with ^{32}P$_i$ to permit them to synthesize endogenous radiolabeled ATP and subsequently irradiated, a single large protein was labeled [(144; pea) and (59, 109; maize)]. When tissues were preirradiated in vivo, the reduction in phosphorylation of membrane fractions in vitro simply indicated that most of the potential sites for phosphorylation were already filled with endogenously derived phosphate as a result of in vivo irradiation (144).

Correlation of Light-induced Phosphorylation with Phototropism

Short & Briggs (141) established several photobiological correlations between the light-inducible phosphorylation change and phototropism in pea. These included similar tissue distribution, adherence to the reciprocity law, and rapidity of the response (see below). In addition, the range from threshold to saturation of fluences given in vivo to reduce the subsequent capacity for phosphorylation in vitro (141) matched reasonably well the range of fluences from threshold to saturation for first positive curvature in pea (9). Short et al (144) demonstrated subsequently the same fluence-response relationship by direct measurement of in vivo phosphorylation. Likewise, the kinetic for return of phosphorylation to the dark condition in vivo following irradiation was similar to that for recovery of phototropic sensitivity in darkness following a light pulse (141, cf 16). Similar correlations for tissue distribution in maize coleoptiles and dark recovery in maize were observed subsequently (59, 108). There was also a close match between the action spectrum for potentiation of phosphorylation of maize membranes by irradiation either in vitro (108) or in vivo (59) and that for phototropism (10) in maize. Finally, in both pea (141) and maize (59), phosphorylation is complete within seconds and hence is rapid enough to be involved early in the phototropic process.

Reymond et al (116) have shown that the phosphorylation response in vivo and in vitro is found in all species examined to date. These include elongating hypocotyl tissue of sunflower (*Helianthus annuus* L.), tomato (*Lycopersicon esculentum* Mill.), *Arabidopsis thaliana* (L.) Heynh., and zucchini (*Cucurbita pepo* L.), and coleoptile tips of wheat (*Triticum aestivum* L.), barley (*Hordeum vulgare* L.), sorghum (*Sorghum vulgare* Pers.), and oat (*Avena sativa* L.). The apparent molecular weights of the phosphorylated proteins ranged from 114,000 (all the coleoptiles) to 130,000 (sunflower). In all cases, preirradiation of the tissue with saturating B dramatically reduced the amount of phosphorylation inducible with B in microsomal membranes isolated immediately after in vivo irradiation [see also (60, 109; maize); (117; *A. thaliana*)].

Although the above photobiological properties of phosphorylation are remarkably similar to those of phototropism, this apparent association of the biochemical reaction with the physiological response remains strictly correlative. Reymond et al (117) have shown that an *A. thaliana* mutant strain, JK224, reported to be roughly 30-fold reduced from wild-type in its sensitivity for first positive phototropic curvature (74), and proposed as a putative photoreceptor mutant (74, 77), also showed a dramatic deficiency in its phosphorylation response. Thus, the correlative photobiological and physiological evidence associating light-induced phosphorylation with first positive phototropism is now supported by strong genetic evidence.

Biochemical Properties of the Reaction

THE NATURE OF THE LIGHT REACTION   Significant light enhancement is still observed when an in vitro saturating light treatment is followed by a dark period, at room temperature, of up to 20 min before addition of $[\gamma\text{-}^{32}P]ATP$ (59, 109, 144). This active state apparently can be maintained indefinitely at low temperature. When samples are irradiated, immediately frozen, and thawed several days later, they still show the effects of light by increased phosphorylation (144).

Hager et al (60) showed that reducing agents such as NADH, NADPH, ascorbate, glutathione, or dithiothreitol all enhanced the light-induced increase in subsequent phosphorylation when added to microsomal membrane preparations from maize that had been frozen and thawed, but not with fresh membranes. They observed the same effect of reducing agents in H_2O_2-pretreated fresh membranes. Because the reducing agents had to be present before the light treatment to be effective, light-driven redox changes involving SH groups probably are an essential part of the reaction. In a similar vein, W. Rüdiger (personal communication) showed that sulfhydryl agents such as iodoacetate, N-ethyl maleimide, and N-phenyl maleimide inhibited light-potentiated phosphorylation, and the more hydrophobic agents were more effective. Because these reagents still inhibited subsequent phosphorylation when added after light exposure, SH groups may also be involved in reactions downstream from the initial photoexcitation.

Reymond et al (116) have shown that, at a minimum, light was activating the kinase function rather than altering the phosphorylation substrate conformation to facilitate phosphorylation. Irradiation of detergent-treated membranes from one species phosphorylated the homologous protein in nonirradiated detergent-treated membranes from another species, where the two affected proteins were of different molecular size and could be distinguished unambiguously. These latter experiments show a significant degree of functional homology in the light-inducible potentiation of phosphorylation between widely divergent species (e.g. pea and maize). They also indicate that the functional unit could be a dimer or multimer, and the reaction a cross phosphorylation rather than solely an autophosphorylation.

PROPERTIES OF THE PHOSPHORYLATION REACTION   The time course for phosphorylation in vitro is similar in pea (144, 146), A. thaliana (117), and maize (107, 109), showing a rapid rise to a maximum that lasted 2–5 min and a decline to near dark levels within 20 min regardless of treatments with B, detergents, or chase with cold ATP. The decline occurred despite the presence of either phosphatase (109, 145) or protease (109) inhibitors. The A. thaliana mutant strain JK-224 shows the same kinetics, indicating that the genetic lesion in this mutant is probably not related to a difference from wild-type in the kinetics of phosphorylation or subsequent decay (117). The basis for the decline is not known. Hager & Brich (59), in a single exception, failed to find a decline in measurable phosphorylation with time in their studies with maize. A difference in maize cultivars could account for this discrepancy.

Hager & Brich (59) have shown that the approximate K_m for ATP for maize microsomal membranes is near 20 mM for light-inducible phosphorylation. The reactions in pea (140, 145) and maize (15, 59) show broad pH optima near 7.5. In pea, where significant phosphorylation near 120 kDa is observed in membranes from dark-control tissues, the shape of the pH vs activity curve for dark-control phosphorylation is identical to that for light-activated phosphorylation (145). By contrast, the pH optimum for all other membrane-associated kinase activities detectable in maize microsomal membranes was near pH 6.0 (59). The kinase reaction in pea showed high specificity for ATP over other nucleotide triphosphates (145) and neither in pea (140, 145) nor in maize (15, 59, 109) was there a requirement for Ca^{2+}, although both systems required millimolar magnesium. The isoelectric point for the phosphorylated pea protein was reported to be near pH 7.8 (167). In maize the phosphorylation also was exquisitely sensitive to the inhibitor staurosporine (K_i near 4 nM), a known inhibitor of protein kinase C in other systems (35, 154).

Short et al (146) showed that by freeze-thawing purified right-side-out plasma membrane vesicles several times, the low level of phosphorylation detected in the fresh vesicles without detergent treatment could be enhanced significantly. Because repeated freezing and thawing of right-side-out vesicles is reported to lead to a 50/50 mixture of right-side-out and inside-out vesicles (110) this is the expected result if the reaction takes place at the inner face of the plasma membrane.

PROPERTIES OF THE PHOSPHORYLATED PROTEIN   A Coommassie-stained protein co-migrates with the radiolabeled band in gels of membranes from etiolated pea stems, but apparently is absent on gels of those from irradiated tissue (141). This observation is probably the result of two phenomena: (a) reduction in Coommassie stainability and (b) decrease in mobility in gel electrophoresis upon phosphorylation, changes known to occur upon protein phosphorylation at multiple sites in other systems (12, 58). Western blotting of plasma membrane proteins from in vivo–irradiated and dark-control pea stem tissue showed that the protein is still present in the membrane fraction in its phosphorylated form but that its mobility has been reduced (146). Careful examination of Coommassie-stained gels showed the same mobility shift (146). A similar mobility shift on phosphorylation has not been observed for the maize protein (109). Proteolytic digestion of the phosphorylated pea protein with two different proteases yielded multiple phosphorylated polypeptides on subsequent gel electrophoresis, indicative of multiple-site phosphorylation with no evidence of hierarchical phosphorylation (145). Finally, phosphoamino acid analysis of the protein in maize (109) and pea (145) showed that the predominant amino acid phosphorylated is serine, with only traces of phosphothreonine and no detectable phosphotyrosine.

Although not conclusive, several lines of evidence indicate that the photoreceptor, kinase, and substrate protein may be embodied in a single polypeptide or in a stable complex. First, the protein that becomes phosphorylated contains an ATP-binding site (146). An ATP-binding site is a necessary (though not sufficient) condition for the substrate protein itself to be a kinase. Second, the quantum efficiency for the light reaction is unaffected by triton treatment (146). Hence, individual triton micelles must contain all three components. If there are multiple proteins, they must be extremely tightly associated so that the coupling between the light reaction and the activation of the kinase function is wholly unaffected by the detergent treatment. Third, if triton-treated plasma membranes are electrophoresed on a native (triton) gel before irradiation, and the gel strips are irradiated and then treated with [γ-^{32}P]ATP, a light potentiation of phosphorylation is still observed (167). Again, if several proteins are involved, the complex must be sufficiently tight to survive electrophoresis under these conditions.

LIGHT-INDUCED REDUCTION OF A *b*-TYPE CYTOCHROME—THE LIGHT-INDUCED ABSORBANCE CHANGE (LIAC)

In the 1970s, several workers reported B-induced absorbance changes in several fungi in vivo, consistent with the reduction of *b*-type cytochrome and concomitant reduction (bleaching) of a flavin (5, 84). The action spectrum implicated a flavin as the possible photoreceptor (99). Brain et al (14) reported a LIAC from a comparable fraction of membranes obtained by differential centrifugation of maize coleoptile extracts. The difference peaks were also consistent with reduction of a cytochrome-flavin complex (84). Britz et al (21), citing several studies with fungi indicating that the dye methylene blue (MB) could confer R sensitivity to phenomena normally induced only by B/UV-A, showed that in the presence of MB, R led to reduction of a membrane-associated cytochrome. In this case, however, the difference spectrum lacked the minimum attributed to flavin bleaching. Britz et al (21) have extensively characterized the reaction (21, 176). Goldsmith et al (56) then worked out conditions needed to obtain the response in fractions from maize coleoptile membranes using the endogenous photoreceptor rather than MB. The reaction could be enhanced by added soluble flavins (riboflavin, FMN, FAD), but the endogenous photoreceptor appeared tightly bound and could not be removed by washing. The light-minus-dark difference spectra—obtained whether the endogenous chromophore, exogenous flavin, or MB were serving as photoreceptor—were identical and readily distinguishable from the bulk of dithionite-reducible cytochromes present in the fraction. Subsequently, a series of studies with membranes from maize, cauliflower, and oats showed that the LIAC co-purified with putative plasma membrane markers (178). The LIAC is not necessarily a definitive marker for plasma membrane because signals can also be observed in endoplasmic reticulum (ER) preparations (177, 179).

In a plasma membrane–enriched fraction from maize, the light-minus-dark and dithionite reduced-minus-oxidized spectra were identical, indicating a lack of contaminating cytochromes from other membrane systems or organelles (85). Triton solubilization caused little change in the quantum efficiency of the reaction, indicating that the cytochrome and flavin moieties might be on the same protein. The active fractions yielded flavin fluorescence when photoexcited at liquid nitrogen temperature (excitation spectrum characteristic of a flavoprotein with a single broad peak near 360 nm and a maximum at 470 with shoulders at 450 and 480 nm) (87). Another fluorescing component that was present (excitation maximum near 360 nm, emission maximum near 430 nm) was suggested to be a pterin (see 70). Light-minus-dark spectra at liquid nitrogen temperature indicated a b-type cytochrome with a single α-band at 555 nm. The midpoint potential E'_0 was estimated to be -65 mV. Widell & Sundqvist (180) carried out further fluorescence studies with purified plasma membrane fractions from cauliflower and oats to investigate the effects of the flavin antagonists phenylacetic acid (PAA) (62) and azide and iodide (61) on the presumptive flavin fluorescence, pterin fluorescence, and on the LIAC. The results showed that there are several fluorescent components in the plasma membrane with very different inhibitor sensitivities. It is premature to relate any of these components to the LIAC.

A photoactive flavin (or pterin) involved in efficient photochemical interaction with other redox carriers might show reduced fluorescence. Salicylhydroxamic acid (SHAM), which inhibits the LIAC (24, 25), led to a strong increase in the putative flavin fluorescence with the LIAC correspondingly inhibited (180), suggesting a relationship between the fluorescing flavin and the LIAC. The fluorescence yield is maximal only when the LIAC photochemistry is inhibited. By contrast, SHAM did not affect putative pterin fluoresence, indicating that this pigment was probably not involved in the LIAC.

Caubergs' laboratory has extensively characterized a cauliflower LIAC that differed in several ways from the maize system. However, like the maize system (and the light-inducible phosphorylation discussed above), triton solubilization had no effect on the quantum efficiency (24).

Citing evidence that the diphenyl ether herbicide acifluorfen blocked electron transport required for action and was thought to have some action at the plasma membrane, Leong & Briggs (86) investigated the effect of the compound both on phototropism and on the LIAC. Because acifluorfen inhibited dark reoxidation of the b-type cytochrome and also appeared to enhance phototropic sensitivity without affecting gravitropism, the authors suggested that the LIAC might play a role in phototropism. Subsequent work has not supported this hypothesis (47).

Efforts to relate the LIAC to any specific plant response have been unsuccessful. Schmidt & Butler (126) demonstrated light-inducible reduction of horse heart cytochrome c in solution in the presence of riboflavin, clearly a LIAC without physiological significance. Thus, the LIAC still remains an enigmatic membrane-associated reaction in search of a physiological role.

Related topics such as LIAC activity in higher plants (5), b-type cytochromes in the plant plasma membrane (6), and plasma membrane redox activity (123) have been reviewed in more detail elsewhere.

GTP-BINDING PROTEINS

G-Proteins Associated with the Plasma Membrane of Pea Apical Buds

GTP-binding regulatory proteins (G-proteins) play a critical role as transducers and amplifiers of receptor signals in a variety of organisms (55). Warpeha et al (168) recently reported a transient doubling of GTPase activity in plasma membrane–enriched sucrose gradient fractions from etiolated pea apical buds within 30–60 s after a B pulse, and a roughly 10-fold increase over the basal level within 3 min. The initial rate of GTP hydrolysis was fluence-dependent, but regardless of the B fluence, the rate of GTP hydrolysis decreased to basal levels after approximately 3 min. Nonhydrolyzable competitive inhibitors GTP-γ-S and guanylyl imidodiphosphate (GppNHp) completely abolished the B-mediated GTPase activity, while B did not affect [γ-^{32}P]ATP hydrolysis in this assay, indicating the specificity of the reaction for GTP. R was essentially ineffective. Kinetics of GTP-γ-[^{35}S] binding with plasma membrane–enriched preparations in response to B were similar to those for GTPase activity, suggesting that both responses are a function of the same process.

Warpeha et al (168) also attempted to identify G-proteins in the plasma membrane–enriched fractions as possible agents for the observed B-dependent GTPase and GTP-binding activities. Western blots probed with antibodies against the transducin $\beta\gamma$ (T$\beta\gamma$) complex recognized at least four polypeptides, while antibodies against the transducin α subunit (Tα) or against synthetic peptides from $G_o\alpha$ and $G_{i3}\alpha$ subunits recognized polypeptides of approximately 30 and 40 kDa (Tα) or 40 kDa ($G_o\alpha$/$G_{i3}\alpha$). ADP-ribosylation by pertussis toxin also yielded a single 40-kDa polypeptide in the inactive state (i.e. lacking either GTP or B). Conversely, cholera toxin only ADP-ribosylated a 40-kDa protein in the presence of both B and GTP. Finally, a polypeptide of similar apparent molecular weight was photoaffinity labeled with the GTP analog, P^3-(4-azidoanilido)-P^1 5′-[α-^{32}P]GTP, but only after preirradiation with B. The authors concluded that a 40-kDa polypeptide may be the α subunit of a typical heterotrimeric G-protein involved in B signal transduction of an unidentified low fluence response.

PHOTOPERCEPTION: GENERAL CONSIDERATIONS

Physiological Clues to the Nature of B Photoreceptors

There has been no definitive identification of a B photoreceptor in flowering plants, and the nature of the chromophore(s) is still the subject of considerable debate. For the majority of responses in higher plants, the most likely candidate for B photoreceptor pigments is a flavin or flavoprotein (support reviewed in 36, 51, 64, 147–149), although a role for carotenoids cannot be ruled out (see 17, 18, 130, 155). Conjugated or unconjugated pterins and their derivatives have been reviewed as potential primary or coreceptor pigments in some systems (50). Briggs & Iino (19) provide a detailed review of the evidence supporting flavins, carotenoids, and some possible alternative molecules in B photoperception. Later reviews attempt to integrate the most recent studies into the broader context of signal transduction (63).

Despite the lack of a chemically characterized blue light photoreceptor, some general considerations should be stressed. First, B responses are evident over a diverse range of species (131–133) and, at least among different groups of organisms, there are at least three distinct classes of B-absorbing photoreceptors separable by action spectroscopy (19). Therefore, one must exercise caution when generalizing from studies of lower plants, fungi, or microorganisms to photoreceptors of flowering plants. Second, there is growing evidence that different B responses are initiated through distinct photoreceptors within the same organism. Phototropism and the inhibition of elongation by B (both rapid and delayed) differ dramatically in their kinetics (34) and their sensitivity to B (9, 83, 169). Also, these responses are genetically separable (see below). Third, a single photomorphogenic response may be mediated by more than one photoreceptor. The absorption of B by phytochrome may elicit photomorphogenic responses strictly attributable to phytochrome or may activate pathways requiring coordinate excitation of both phytochrome and a specific B photoreceptor. Phytochromes act in place of (66, 81, 111) or in concert with (26, 185) B-specific photoreceptors in mediating phototropism. R can also moderate the effects of activating B-specific photoreceptors (for review see 97).

There is considerable evidence that multiple B photoreceptors mediate the same physiological response simultaneously. Based on the complex fluence-response relationships for oat coleoptiles, Zimmerman & Briggs (184) suggested the presence of multiple photoreceptors for detecting phototropic stimuli. A series of fluence-response curves generated at several different wavelengths and after different preirradiations for phototropism in A. thaliana provides strong photobiological evidence for at least two different photoreceptors with different absorption maxima and quantum efficiencies (77, 78). In addition, the existence of separate photoreceptors (or possibly distinct chromophores within the same photoreceptor) in mediating B and UV-A responses is supported by studies with A. thaliana mutants deficient in B-mediated processes. The inhibition of the hypocotyl-elongation response to B is defective in hy4 (79) and blu1 (88), but both show normal growth inhibition in response to UV-A (79, 181). Because the phytochrome-deficient mutant hy6 also shows a normal UV-A response, phytochrome is ruled out as the UV-A photoreceptor in these cases (181).

The *HY4* gene from *A. thaliana* has been sequenced (4) and may be a B photoreceptor. The amino-terminal half of the gene product has high homology to microbial photolyases, especially to the chromophore-binding domains where 70-80% amino acid sequence similarity has been reported between *HY4* and members of the photolyase family. These domains are at opposite ends of the photolyase-like region, and include a putative binding site for reduced flavin (FADH2) at the carboxy end and a possible site for attachment of a deazaflavin derivative (or perhaps a pterin) at the amino terminus. The carboxy-terminal region of HY4 contains some similarity to tropomyosin from rat smooth muscle and three of the five mutant alleles that have been sequenced have lesions in this portion of the gene. The other two alleles have single mutations in the putative flavin-binding domain. The cDNA encodes a protein with an expected size of 75.8 kDa and lacking any apparent membrane-spanning domains, confirming preliminary biochemical data indicating that the protein is soluble. If the *HY4* gene product is a photoreceptor, it seems unlikely that it reduces hypocotyl elongation through a signal transduction chain activated by DNA photolyase activity. It lacks one of the two tryptophan residues thought to be essential for substrate binding in all active DNA lyases characterized to date (A Cashmore, personal communication). Likewise, none of the known DNA photolyases have the carboxy-terminal tropomyosin-like domain. It seems more likely that the function of this photoactive protein was modified through evolution to assume a new function perhaps unrelated to its presumptive original activity.

Biochemical characterizations described above indicate that the photoreceptor involved in B-induced phosphorylation of a plasma membrane protein (108, 117, 141) is closely associated with the phosphorylated substrate (146, 167). This work has targeted the substrate in a biochemical approach to isolate the photoreceptor. Also, as mentioned above, the phototropism mutant JK224 is thought to be deficient in the photoreceptor (74, 77) and exhibits decreased light-induced phosphorylation (117). In addition, the mutant lacks detectable amounts of the normally phosphorylated 120-kDa protein on SDS-PAGE (E Liscum, personal communication); therefore, isolation of the gene responsible for the mutant phenotype may yield a putative photoreceptor involved in phototropism. The retention of the UV-A response in *hy4* and *blu1* mutants indicates that separate photoreceptors may be involved in the UV-A and B wavelengths. But the presence of both the B and UV-A peaks in the in vitro action spectrum for phosphorylation (108) is consistent with either a single photoreceptor absorbing in both regions or different photoreceptors that remain active in vitro and feed into a common pathway.

Fluorescence studies in pea yielded correlative evidence for a plasma membrane–associated flavin rather than a pterin as a primary chromophore candidate for the GTPase activation (171). Fluorescence emission of free riboflavin or pterin in aqueous buffer solutions was strongly quenched by both KI and phenylacetic acid (PAA). Therefore, inhibition of B responses by KI or PAA, either in vivo (127) or in vitro (144, 171), cannot differentiate reliably between these two proposed photoreceptors. Although membranes subjected to boiling in 0.1% SDS demonstrated fluorescence peaks consistent with those of free pterins, the emissions were scarcely affected by addition of KI or PAA. By contrast, fluorescence emission characteristic of free flavins appeared upon denaturation with boiling and SDS, and was inhibited strongly by addition of the quenching agents (171). This indicates that in the nondenatured state flavins are associated closely with molecules (e.g. polypeptides) able to quench the photoexcited flavins. Similar concentrations of flavin antagonists quench the fluorescence detectable in denatured plasma membrane proteins and inhibit B-induced GTP-γ-S binding to plasmalemma, correlating the G-protein activity with a reactive flavin moiety. The shape and maxima of the emission spectra indicate that more than one flavin species is present in these preparations, and it has been pointed out (49) that this study cannot rule out the role of pterin derivatives such as sepiapterin.

The indications from experiments with quenchers of activated flavins both in vivo (127, 166) and in vitro (144) are consistent with the active singlet state of a flavin as a photoreceptor in perception of phototropic stimuli. In this case the flavin is apparently not readily removed or exchanged. This association contrasts with the photoperception in the LIAC (85) or the B-mediated redox activity in the guard cell plasmalemma (163), both of which are sensitized by exogenous flavins or analogs and are therefore probably unrelated to the putative flavoproteins involved in phototropism. Unlike the photoreceptor for phototropism and phosphorylation, the concentrations of inhibitors affecting fluorescence and GTP-γ-S binding (171) are in a range reported to quench flavin triplet states (166). In addition to this differential sensitivity to flavin antagonists, the light-induced G-protein and phosphorylation activities exhibit different tissue localizations and light sensitivity, signifying that they are probably not involved in the same transduction chains.

B REGULATION OF GENE EXPRESSION

Measurements of Transcription and Transcript Abundance

Many aspects of B-mediated photomorphogenesis require the precise regulation of expression of specific genes. Although several investigators recognized that the abundance of specific trancripts or enzyme activities was regulated by light (see 68, 82, 157–159, 173), only recently has B-specific transcriptional control of individual genes been observed. Interpretation of these data is complicated by several important differences in the methods used to address questions of B regulation (e.g. plant species, tissue, age, and growth conditions; quality and quantity of inductive light treatments; time between light induction and gene assays; and the method of measuring gene regulation). Therefore, many of the experiments are not directly comparable, but some trends have begun to emerge from these studies.

Thompson et al (157) examined B-specific regulation of individual genes and gene families. They reported on the light-dependent accumulation of mRNA in pea apical buds and showed that transcript accumulation in plants grown for 7 d in R (see 19) was altered by a pulse of B given 24 h prior to mRNA isolation in nine of the thirteen gene families examined. Relative abundance of the *RBCS* gene family message increased 6-fold over that in comparably grown plants not exposed to a B pulse. Examination of individual members of the *RBCS* family in pea with gene-specific probes showed that the relative responsiveness of the *RBCS*-3A and -3C transcripts to R and B depends on the developmental stage of the plant (43, 44). In both the native pea system (33, 43) and in transgenic petunia expressing the pea *RBCS* genes (43), no effect specific to 24 h of continuous B was detected for either the *RBCS*-3A or -3C transcript in etiolated plant leaves. However, in dark-adapted mature leaves, B significantly increased the accumulation of both transcripts, with only slight regulation by phytochrome (43). The authors concluded that this phenomenon reflects a developmental switch from gene regulation by phytochrome to regulation by B although phytochrome activation is also required. Clugston et al (33) found no definitive effect of phytochrome in a similar dark-adapted pea leaf system, and the primary photoreceptor mediating fluence rate–dependent *RBCS* gene transcription in *Phaseolus vulgaris* primary leaves appears to be a B photoreceptor separate from phytochrome or photosynthetic pigments (124). Likewise, the plastidic (but not the nuclear) ferredoxin-dependent glutamate synthase in young seedlings of *Pinus sylvestris* (40) is induced strongly by either B or R, but in more mature seedlings (10 d old) the light sensitivity to R is lost.

Using cDNA clones as probes for specific nuclear and plastid-encoded gene products, Richter and colleagues also examined the gradual light-induced increases in steady state transcript levels of chloroplast-localized gene products from *Nicotiana tabacum* (119) and *Chenopodium rubrum* (120) cell cultures. They determined that in the cell suspensions plastid differentiation and transcriptional light regulation were induced primarily by B, and that R induced a relatively small effect alone but had a strongly inhibitory effect on the B-mediated increase on *RBCS* and *rbc*L (122). When observed over a shorter time course and in intact etiolated *N. tabacum* and *Lycopersicon esculentum,* however, the expression of *CAB* genes for both photosystems I and II showed transient increases within the first 16 h after onset of continuous irradiation with R, FR, B, and UV-B. These changes were shown by run-on transcription to be the result of transcription-rate increases (174). B was most effective in inducing *CAB* genes, based on fluence rate–response measurements, with phytochrome exhibiting only a small effect on the transcript accumulation.

Genes for the four major flavonoid biosynthesis enzymes [phenylalanine ammonia lyase (*PAL*), chalcone synthase (*CHS*), chalcone isomerase (*CHI*), and dihydroflavonol reductase (*DFR*)] in R-grown *A. thaliana* seedlings show induction by both UV-B and B acting through specific photoreceptors separate from phytochrome (80). This result is consistent with previous findings that *CHS* gene expression is coordinately regulated by distinct receptors of UV-B, B, and R in parsley (*Petroselinum crispum*) cell cultures (22, 39, 106). In *Sinapis alba, CHS* gene expression is also regulated by multiple photoreceptors, with R primarily inducing accumulation of *CHS* message in cotyledons and B/UV more effective at later developmental stages (11). In transgenic *A. thaliana* expressing the β-glucuronidase (*GUS*) reporter gene fused to the native *CHS* promoter, B or UV-B strongly induced accumulation of *GUS* mRNA, and R yielded only a minor increase in the steady state transcript level (42). This result matches closely the result for accumulation of endogenous *CHS* mRNA in the same plants. The response was also found to be transient following a 15 min pulse of B in 9-day old R-grown plants. Thus, the relative degree of B regulation of gene expression is strongly dependent on the individual genes or promoters examined and the species, age, and prior history of the plant material in question.

Several studies followed the accumulation of B-mediated transcripts including CAB and three unidentified cDNA clones designated pEA25, pEA207, and pEA215 (see 156) in 7-day old R-grown pea seedlings receiving a B pulse 24 h prior to harvest (93, 169, 172). Examinations of the time course, fluence response, and reciprocity characteristics of transcript abundance for both *CAB* and pEA215 (probably another *CAB* cDNA clone) yielded rapidly increasing transcript levels at low fluences, but the transcript levels decreased at higher fluences and reached control levels above 10^4 µmol · m^{-2} (172). The pEA25 transcript was negatively regulated, with a threshold below 10^{-1} µmol · m^{-2}, while pEA207 [which shows homology to a pea bud lectin (38)] had a threshold above 10^2 µmol · m^{-2}. Control of all four transcripts obeyed the reciprocity law at 10^1 and 10^4 µmol · m^{-2}. In dark-grown plants (170) the decreased response of *CAB* and pEA215 to high fluence B was not detected and instead the accumulation of these transcripts continued to increase with higher fluences. Furthermore, at higher fluences, reciprocity for *CAB* mRNA accumulation was lost, indicating possible activation of a high irradiance response. The regulation of these transcripts is at the transcriptional level (93) and is dependent on both R and B in concert. Finally, the transcription rates for both *CAB* and pEA207 (but not pEA25) is regulated by B in the absence of cycloheximide-sensitive de novo protein synthesis (94). Thus, all the necessary factors for light-dependent regulation of these genes are already present. The pEA25 transcript regulation is similar to that of the *psbC-psbD* transcript-accumulation response to B (52) in that it requires cytoplasmic protein synthesis to affect message abundance.

Another B high-irradiance response has been reported for the early light-induced proteins (ELIPs) in mature pea plants (3). These proteins are related to the *CAB* gene family (57) and their mRNAs are transiently expressed rapidly after the transition from the etiolated to light-grown regimes (95). In mature plants the mRNAs are induced specifically by high fluence rates of B (> 400 μmol $\cdot$ m$^{-2} \cdot$ s^{-1}), but not by either R or FR alone. In fact, low fluence rates of R superimposed on the B repressed the mRNA accumulation.

There is less apparent interaction between phytochrome and B photoreceptor systems on individual transcript accumulation for the plastidic *psbC-psbD* family of messages in etiolated *Hordeum vulgare* (52). R does not significantly induce the accumulation of two transcripts that are strongly accumulated in B, and an inhibitor of phytochrome synthesis [4-amino-5-fluoropentanoic acid (AFPA)] (53) does not alter *psbC-psbD* transcript accumulation while strongly inhibiting R induction of *CAB* accumulation. FR does modulate transcript abundance to a small degree, indicating that B may be necessary for initiating the response but that phytochrome may still affect the level of expression.

The down-regulation of leaf thionins was also studied in barley seedlings during the transition to light-triggered development (115). Both thionin mRNA and thionin protein accumulation decreased following 4.5 h exposure to R or B light. At least in W the decrease began immediately upon illumination, and continuous FR did not affect message abundance. Because the maximum level of B repression was greater than the level of R repression, a specific B receptor function is implied, although the relative importance of the two photosystems has not been determined.

In studies on cultured cells (120, 122) B and R have partially antagonistic effects on gene expression. Chloroplast differentiation and the concomitant accumulation of nuclear and plastid genes are coordinately induced by B, but abolished by R. This gradual induction in continuous B of *CAB, RBCS, rbc*L, *psb*A, *atp*B, and *atp*E messages is in marked contrast to the observations of gene induction by both R and B in whole plants and tissues (see 96), and indicates that in cultured cells, activation of the B/UV-A photoreceptor can substitute for phytochrome in regulating gene expression. This finding implies that the framework for B-mediated gene expression is present in plant cells, although it may be partially subverted in intact plants in favor of phytochrome regulation.

Numerous examples of co-action among the various photoreceptors have been described (97) and it has become increasingly evident that no single mechanism of co-action can explain the vast range of gene-regulation responses observed. In *Sorghum vulgare,* the chloroplast NADP-dependent gly-ceraldehyde-3-phosphate dehydrogenase (GPD) activity apparently requires activation of phytochrome and a specific B receptor (104). Both for the GPD enzyme (104) and for anthocyanin accumulation (98, 105) the authors argue that B and UV light make the plants competent to respond to the active form of phytochrome, which in turn activates individual genes. In the tomato mutant *aurea,* which is deficient in immunochemically detectable PHYA polypeptide and spectrally active phytochrome (112), transcripts for chlorophyll *a/b* binding proteins of photosystems I and II, plastocyanin, and subunit II of photosystem I—all of which are B-regulated in wild-type plants—are no longer induced by R or B (103). Anthocyanin induction is also impeded in *aurea* (2), although B-induced inhibition of stem elongation is unaffected in this mutant (1). These phenomena indicate strongly that B-mediated gene regulation requires active phytochrome in tomato. Alternatively, similar mutants in *A. thaliana* with reduced chromophore biosynthesis do not show a similar co-action of R and B in gene expression in green tissues (28, 31).

The presence of multiple phytochromes in higher plants (136) and the activation of phytochromes by B suggest a variety of hypotheses for R/B interactions in gene regulation. Some B effects require active phytochrome for full activity (46, 47, 103); one or more phytochromes may require B-photoreceptor activation to enhance phytochrome function (40, 104, 105, 125); or branched signal transduction pathways may exist and neither photoreceptor alone is able to elicit the effect without co-excitation of the other photoreceptor.

Cis- *and* Trans-*Acting Elements in B-Mediated Gene Expression*

The promoters of B-regulated genes have been investigated as a means to elucidate *cis* regulatory elements and identify possible *trans*-acting factors involved in B-controlled transcription. Feinbaum et al (42) studied light effects on a series of chimeric *CHS:GUS* constructs with variable 5′ promoter truncations in transgenic *A. thaliana.* Lines containing 523 bp of the *CHS* promoter fused to *GUS* exhibited a 50% increase in *GUS* activity after 16 h of light and 3- to 4-fold more activity than in dark control plants after a 48 h light treatment. Truncations to 186 bp of the promoter resulted in loss of B-specific *CHS* induction, although responsiveness to high intensity W was retained. The 5′ region of the *CHS* gene from *Sinapis alba* contains an element present in all other *CHS* and most *RBCS* promoters and is thought to be involved in light regulation (11), but there is no evidence to link this sequence to B-specific expression. Deletions in the pea *RBCS*-3A and -3C promoters linked to a reporter have also been used to search for light-specific *cis* elements, but no clearly defined elements were detected (43).

Recent comparisons of *psb*D-*psb*C upstream sequences of five cereals and three dicots provided initial support for a conserved light-responsive element in the complex transcription unit found in chloroplasts of divergent organisms (32). Indirect evidence supports this hypothesis, but definite conclusions about the character of the novel light-responsive element await further study.

Genetic Mutants in B Signal Transduction

Analysis of mutants exhibiting anomolous photoresponses has become a prevalent method of dissecting signal transduction pathways for B responses (for reviews see 28, 90). In 1980, Koornneef et al (79) reported their isolation of a series of EMS- and fast neutron-mutagenized *A. thaliana* plants impaired in W-induced inhibition of hypocotyl elongation. Designated *hy1-hy5*, some of these plants are deficient in one or more active phytochromes (*hy1, 2,* and *3*) while others are affected in specifically B-mediated growth (*hy4*) or in their response to R, FR, and to some degree B (*hy5*) (45).

The growth of *hy4* mutants is still partially inhibited under W, but a more recently isolated class of B-insensitive mutants (*blu1, blu2,* and *blu3*) grow normally under W, although they are as tall as dark-grown seedlings under B (88). As discussed previously, the *HY4* gene has been sequenced and its product identified as a putative chromoprotein related to DNA photolyases, implying that it may be a photoreceptor for some B responses. Just as B inhibits hypocotyl cell expansion but promotes cotyledon cell expansion in wild-type plants, in *hy4* the B-mediated hypocotyl growth inhibition and cotyledon growth promotion are decreased (69). In all three *blu* mutants, cotyledon expansion is reduced by approximately 60% under B, but in W only *blu1* and *blu2* show a reduction in cotyledon area of 30%, while *blu3* cotyledon expansion is similar to that of wild-type (88).

Mutants in *HY4* show apparently normal chloroplast development (27) and both *hy4* (27, 69) and *blu* (90, 91) mutants exhibit normal phototropic curvature and stomatal function. A second class of mutants show altered phototropism (74), but exhibit normal inhibition of hypocotyl elongation (91). Strain JK224 shows reduced sensitivity to first positive phototropic stimuli, but normal second positive phototropism and gravitropism. In addition, the mutation in this strain affects the very rapid phosphorylation response thought to be an early transduction step for phototropism (117, 141). Therefore, strain JK224 may be defective in either the photoreceptor mediating phototropism or in an early, limiting signal transduction step (74, 77, 117). JK218 exhibits a normal gravitropic response but is unresponsive to fluences of B expected to elicit either first positive or second positive curvature, indicating that phototropism (both first and second positive) and gravitropism are separable genetically.

There are several other mutants, including JK229, which shows no first positive phototropism but normal second positive and gravitropic curvatures, and JK345, which lacks first positive curvature and shows reduced second positive phototropism and gravitropism. In addition, the Poff group (75) has isolated a series of intermediate phototropic mutants (ZR8, ZR19) displaying a reduced amplitude of phototropic behavior. The isolation of separate mutants in phototropic and hypocotyl elongation responses is consistent with the physiological data from *Cucumis sativus* (34) and *Sinapis alba* (118), indicating that phototropism and inhibition of hypocotyl elongation involve separate pathways (91).

The *hy4* and *blu* mutants have been used in conjunction with other *hy* mutants to examine the roles of these genetic lesions in altering light-regulated gene expression. In the *blu1* mutants, both chlorophyll accumulation and the chlorophyll *a/b* ratio were normal, indicating that B-induced chloroplast development (121) operates through a B-mediated pathway different from that involved in growth inhibition (88). In addition, the *hy4* mutants do not show altered *RBCS* or *CAB* expression compared with those of wild-type *A. thaliana,* although under the same growth conditions *RBCS* induction was primarily a B rather than a R response (69, 90). In another study (27), reduction of both anthocyanin accumulation and *chs* mRNA abundance was observed in the *hy4* mutant. On the other hand, the accumulation of anthocyanin in W varied among the three *blu* mutants; *blu1* expressed 3-fold higher levels of pigments than did wild-type, *blu2* contained 3-fold lower levels, and *blu3* showed wild-type amounts of anthocyanins (88).

The difference in anthocyanin accumulation was the only characteristic observed in mature plants that distinguished *blu* mutants from wild-type, leading to the conclusion that the *BLU* genes operate primarily in young seedlings. This is in contrast to results of gene expression studies (11, 43) in which B becomes the dominant photoreceptor in more mature tissues. At the seedling stage the *blu* mutants and *hy4* show reduced apical hook opening in response to B, as does *hy5* (89). Homozygous double mutants between the phytochrome mutant *hy6* (31) and *blu1* lack the B, R, or FR hook opening response and appear etiolated even after 5 d in W, although eventually the hooks open and development proceeds toward an attenuated seedset (88, 89). Comparisons of hook opening in wild-type, *hy6, blu1,* and double mutants indicates that the response to R, FR, and low fluence B is mediated through one or more phytochromes, while a specific B receptor mediates the B high irradiance response for hook opening. The *blu2* and *hy4* mutants exhibit decreased sensitivity to R, but the reason for the decrease is unclear. The reduced hook opening under all light conditions in *hy5* (89) is consistent with its reduced inhibition of hypocotyl elongation by B, R, and FR (79).

Efforts to isolate additional B response mutants have resulted in identification of other mutants with reduced inhibition of hypocotyl elongation in B (69, 88). Because the characteristics of B-mediated anthocyanin induction, cotyledon expansion, and suppression of hypocotyl elongation in these other mutants vary from the previously described mutants, some of these new lines may represent lesions in distinct genes.

A series of mutants have been isolated that exhibit light-regulated phenotypes in the dark. These include the *det1* (30), *det2* (29), *cop1* (37), and *cop9* (175) mutant alleles. Although these mutants have been characterized primarily with respect to constitutive phytochrome-mediated responses, several lines of evidence indicate that they may be involved in common pathways with certain B-mediated responses including hypocotyl elongation, leaf and cotyledon expansion, anthocyanin production, and gene expression. Both *det* alleles are epistatic to *hy1, 2, 3, 4,* and *6,* indicating that both R and B signals influence the function of the *DET* genes (27). Furthermore, *det1* and *det2* have slightly different phenotypes and are additive, indicating that they may act in parallel rather than sequentially. Another indication that these genes may be involved in B signal transduction comes from *hy5,* which shows a partial loss of the B- and R-mediated growth inhibition responses (79) and hook opening responses (89) where *HY5* is a partial suppressor of *det1* and, to a lesser extent, *det2* (27, 28).

Because anthocyanin production is thought to be primarily under B control in *A. thaliana* seedlings (80), the mutant levels of these pigments is especially interesting. The *cop1* mutant contains higher *CHS* mRNA levels than either W- or dark-grown wild-type seedlings (37), and the anthocyanin level in *det1* is 6-fold higher than in wild-type. In *hy4det1* double mutants, anthocyanins accumulate to levels 40-fold higher than in wild-type plants (7-fold higher than *det1* alone) even though the *CHS* mRNA abundance is approximately the same as that for *det1*. Yet, the *det2hy4* double mutants do not exhibit excessive anthocyanin accumulation over those in *det2* alone (27). From these data it is unclear how the *DET, COP,* and *HY* gene products interact to regulate anthocyanin production and/or turnover, but the process is apparently quite complex. The picture provided by these different mutants gives further evidence for the integration of signaling pathways involved in the B photoreceptor- and phytochrome-mediated responses.

OTHER EFFECTS OF B

Correlations of B Effects on Cell Growth and Electrical Potentials

Spalding & Cosgrove (150, 151) recently demonstrated a 2–3 min transient depolarization of the plasma membrane of epidermal cells preceding the rapid B-mediated growth inhibition in *Cucumis sativus* L. hypocotyls in response to B pulses or continuous B irradiations. The lag periods and magnitudes of the change in potential and in growth rate were dependent on the fluence in a 7 s B pulse. A difference in potential also was measured with surface electrodes at the base and apex of the seedlings, yielding a response with similar time course but opposite polarity to that measured across the plasmalemma. The response apparently is not a classical action potential, because the depolarization is not propagated to unirradiated regions of the stem and it exhibits a refractory period of approximately 10 min, similar to the recovery time for resumption of growth following a B pulse. The K^+-channel blocker tetraethylammonium did not affect the response, although proton-ATPase inhibitors abolished the depolarization (151).

Because B causes a greater change in potential than can be accounted for by inhibition of the ATPase alone, the depolarization may involve a second mechanism in addition to the proton pump. This hypothesis is supported by the observation that both Ca^{2+}-channel blockers and Ca^{2+}-chelating agents slowed the repolarization process slightly, but did not affect the rate or magnitude of depolarization. However, Ca^{2+} is required for growth inhibition by B in *C. sativus* (139), indicating that the cation acts at a subsequent step in slowing elongation as well.

The opposite effect of B on cell expansion has been described in *P. vulgaris* leaves. Both R and B stimulate growth via a phytochrome and a specific B photoreceptor (162) and this response may involve light-induced proton pumping (161). Detailed examination of the kinetics of acidification and growth indicate that R- and B-stimulated cell wall acidification is insufficient to explain the full response and that other processes must be involved as well (13). A similar response measured in pea has been shown to require extracellular K^+ and is accompanied by a depolarization and subsequent hyperpolarization of the epidermal membrane potential; both the hyperpolarization and H^+ extrusion were inhibited by Ba^{2+}, indicating a role for Ca^{2+} in both responses (152).

Cytoskeletal Effects of B

Some of the growth responses to B may result from changes in the orientation of microtubules (MT). In the outer epidermal wall of maize coleoptiles, immunocytochemical visualization of microtubule orientation yielded a random MT distribution in darkness, but within 10 min after onset of B irradiation the MTs were stably realigned in a longitudinal direction (182). Continuous R, apparently acting through phytochrome, had the opposite effect, resulting in a gradual reorientation toward the transverse direction. This R effect was less stable and a 1 min B pulse given before, during, or after a continuous R irradiation took precedence and resulted in a rapid reorientation to a longitudinal MT arrangement. In contrast, auxin overcame the effects of B and resulted in a transverse orientation regardless of light conditions (182). The effects of B on MT orientation are not entirely clear, however, because B phototropic stimuli result in transverse orientation of MTs on the shaded side of maize coleoptiles (101). The causal role of MT orientation in growth responses to B is also unclear, because bending (100; maize) and inhibition of growth (83; pea) can occur without MT reorientation.

Stomatal Regulation

Blue light induces stomatal opening primarily via K^+ uptake driven by proton gradients across the plasma membrane (for review see 128). These gradients are measured by acidification of the external medium (138) or by outward electrical currents detectable in patch clamp experiments (7). There is considerable controversy about the principal mechanism responsible for H^+ extrusion (for reviews see 73, 114, 134, 183). The relative involvement of a plasma membrane redox system (163, 164) and of a specific plasma membrane ATPase (137) remains unclear. Models have been proposed suggesting that both systems act in parallel, with B acting primarily on the redox system (54, 114) or the redox process somehow regulating the ATPase (73).

The nature of the B-specific photoreceptor mediating stomatal aperture is unknown. Action spectra (71, 135) and experiments with flavin antagonists (160) and exogenous flavins (163) indicate involvement of a flavin as a possible photoreceptor, although a redox system might use flavins for electron transport separate from the primary photoreceptor. The B-mediated component of stomatal opening obeys reciprocity (71) and has been suggested to involve a two-component system similar to that described for phototropism or a photoreceptor with two interconvertible forms controlled by light activation and subsequent reversion (67, 183).

Experiments with inhibitors and activators that affect components from known signal transduction pathways suggest the involvement of Ca^{2+} and a calmodulin-dependent myosin light chain kinase-like protein (CaM-MLCK) in proton extrusion by *Vicia faba* guard cell protoplasts or stomatal opening in *Commelina benghalensis* epidermal peels (138). Inhibitors of cyclic nucleotide-dependent protein kinases and Ca^{2+}/calmodulin-dependent protein kinase II had no effect, and protein kinase C inhibitors caused only a slight inhibition of B-induced acidification of the medium. Inhibitors of either calmodulin or CaM-MLCK strongly suppressed H^+ pumping. Patch clamp studies on *Vicia faba* show that elevated Ca^{2+} levels can block K^+ channels into the guard cells (129), indicating multiple, antagonistic roles for Ca^{2+} in regulating stomatal conductance. This inhibitory role of Ca^{2+} on inwardly-rectifying K^+ channels may act through protein phosphatase 2B-like protein, because specific inhibitors of this enzyme family block the effect of Ca^{2+} on K^+ fluxes (92).

G-proteins may also play a role in regulating stomatal opening. The G-protein inactivator, GTPβS, activates K^+ influx while the analogous activator,

GTPγS, inhibits inward K$^+$ current (41). Furthermore, the effect of GTPγS is eliminated in the absence of free Ca^{2+}, indicating that an activated G-protein inhibits stomatal opening and that increased cytosolic Ca^{2+} acts in parallel with or downstream of this G-protein. Because both Ca^{2+} and the G-protein oppose the effects of B, it is unlikely that either is involved directly in B signal transduction, although B could release a constitutive G-protein-mediated inhibition. Apparently this G-protein is distinct from that described by Warpeha et al (168), because B activates the latter system, while B activation of the stomatal system would be predicted to close stomatal pores, exactly opposite to the observed effect of B.

The B-mediated response of pulvinus motor cells in *P. vulgaris* L. and *Samanea saman* has been hypothesized to involve a mechanism similar to that of stomates (102), although very little is known about the signal transduction pathway leading to leaf movements. The available data on phenomenology and stimulus-response coupling have been reviewed in depth elsewhere (76).

CONCLUSIONS

The next few years should see some exciting developments from several ongoing investigations of blue light–mediated photomorphogenesis. Research on two different systems—phototropism and hypocotyl growth inhibition—shows great promise in leading to one or more of the long elusive photoreceptors. The association of B-mediated signal transduction pathways with known elements in other signal transduction chains (e.g. phosphorylation, proton pumping, and GTPase activation) show considerable potential for leading to an understanding at the cellular, biochemical, or molecular level of some of the intracellular events activated by blue light. It is becoming increasingly clear that there may be several different, and in some cases completely independent, signal transduction chains (and photoreceptors) leading to the same physiological response. Unraveling these chains and characterizing their photoreceptors is a major challenge for the future.

ACKNOWLEDGMENTS

We are grateful to Drs. Emmanuel Liscum, Paul Oeller, and Stephen Daniel for their careful review of the manuscript and to Jane Edwards for her extensive help in preparing it. The work on protein phosphorylation was supported by NSF Grants DCB 8819137, DCB 9118392, and MCB 9219256 to W. R. B, and an NSF predoctoral fellowship to T. W. S. We very much appreciate this support.

Literature Cited

1. Adamse P, Jaspers PAP, Bakker JA, Wesselius JC, Heeringa GH, et al. 1988. Photophysiology of a tomato mutant deficient in labile phytochrome. *J. Plant Physiol.* 133:436–40

2. Adamse P, Peters JL, Jaspers PAP, van Tuinen A, Koornneef M, Kendrick RE. 1989. Photocontrol of anthocyanin synthesis in tomato seedlings: a genetic approach. *Photochem. Photobiol.* 50:107–11

3. Adamska I, Ohad I, Kloppstech K. 1992. Synthesis of the early light–inducible protein is controlled by blue light and related to light stress. *Proc. Natl. Acad. Sci. USA* 89:2610–13

4. Ahmad M, Cashmore AR. 1993. *HY4* gene of *A. thaliana* encodes a protein with characteristics of a blue-light photoreceptor. *Nature* 366:162–66

5. Asard H, Caubergs R. 1991. LIAC activity in higher plants. In *Biophysics of Photoreceptors and Photomovements in Microorganisms,* ed. F Lenci, F Ghetti, G Columbetti, D-P Häder, P-S Song, pp. 181–89. New York: Plenum

6. Asard H, Venken M, Caubergs RJ, Reijnders W, Oltmann FL, De Greef JA. 1989. *b*-Type cytochromes in higher plant plasma membranes. *Plant Physiol.* 90: 1077–83

7. Assmann SM, Simoncini L, Schröder JI. 1985. Blue light activates electrogenic ion pumping in guard cell protoplasts of *Vicia faba. Nature* 318:285–87

8. Ballaré CL, Scopel AL, Sánchez RA, Radosevich SR. 1992. Photomorphogenic processes in the agricultural environment. *Photochem. Photobiol.* 56:777–88

9. Baskin TI. 1986. Redistribution of growth during phototropism and nutation in the pea epicotyl. *Planta* 169:406–14

10. Baskin TI, Iino M. 1987. An action spectrum in the blue and ultraviolet for phototropism in alfalfa. *Photochem. Photobiol.* 46:127–36

11. Batschauer A, Ehmann B, Schäfer E. 1991. Cloning and characterization of a chalcone synthase gene from mustard and its light-dependent expression. *Plant Mol. Biol.* 16: 175–85

12. Beebe SJ, Corbin JD. 1986. Cyclic nucleotide-dependent protein kinases. In *The Enzymes,* ed. PD Boyer, EG Krebs, 17(A):44–100. New York: Academic

13. Blum DE, Elzenga JTM, Linnemeyer PA, Van Volkenburgh E. 1992. Stimulation of growth and ion uptake in bean leaves by red and blue light. *Plant Physiol.* 100: 1968–75

14. Brain RD, Freeberg JA, Weiss CV, Briggs WR. 1977. Blue light-induced absorbance changes in membrane fractions from corn and *Neurospora. Plant Physiol.* 59:948–52

15. Brich M. 1991. *Blaulichtinduzierte Schnelle Phosphorylierung eines 100-kDa Proteins im Plasmalemma aus Koleoptilespitzen von* Zea mays *L.* PhD thesis. Univ. Tübingen, Germany

16. Briggs WR. 1960. Light dosage and phototropic responses in corn and oat coleoptiles. *Plant Physiol.* 35:961–62

17. Briggs WR. 1963. The phototropic responses of higher plants. *Annu. Rev. Plant Physiol.* 14:311–52

18. Briggs WR. 1964. Phototropism in higher plants. In *Photophysiology,* ed. AC Giese, pp. 223–71. New York: Academic

19. Briggs WR, Iino M. 1983. Blue light-absorbing photoreceptors in plants. *Philos. Trans. R. Soc. Edinburgh* 303:347–59

20. Briggs WR, Short TW. 1991. The transduction of light signals in plants: responses to blue light. In *Proc. NATO Advanced Study Workshop on Phytochrome: Phytochrome Properties and Biological Action,* ed. B Thomas, CB Johnson, pp. 228–301. Berlin: Springer-Verlag

21. Britz SJ, Schrott E, Widell S, Briggs WR. 1979. Red light-induced reduction of a particle-associated *b*-type cytochrome from corn in the presence of methylene blue. *Photochem. Photobiol.* 29:359–65

22. Bruns B, Hahlbrock K, Schäfer E. 1986. Fluence dependence of the ultraviolet-light-induced accumulation of chalcone synthase mRNA and effects of blue and far-red light in cultured parsley cells. *Planta* 169:19393–98

23. Caldwell MM, Teramura AH, Tevini M. 1989. The changing solar ultraviolet climate and the ecological consequences for higher plants. *Trends Ecol. Evol.* 4:363–67

24. Caubergs RJ, De Greef JA. 1982. Light inducible cytochrome reduction in cauliflower (*Brassica oleracea*) inflorescences. *Photochem. Photobiol.* 35:405–10

25. Caubergs RJ, Goldsmith MHM, Briggs WR. 1978. Effects of inhibitors on light-induced absorbance changes in corn membrane fractions. *Carnegie Inst. Washington Yearb.* 77:356–57

26. Chon HP, Briggs WR. 1966. Effect of red light on the phototropic sensitivity of corn coleoptiles. *Plant Physiol.* 41:1715–24

27. Chory J. 1992. A genetic model for light-regulated seedling development in *Arabidopsis. Development* 115:337–54

28. Chory J. 1993. Out of darkness: mutants reveal pathways controlling light-regulated development in plants. *Trends Genet.* 9: 167–72

29. Chory J, Nagpal P, Peto C. 1991. Pheno-

typic and genetic characterization of *det2*, a new mutant that affects light-regulated seedling development in *Arabidopsis*. *Plant Cell* 3:445–59

30. Chory J, Peto C, Feinbaum R, Pratt L, Ausubel F. 1989. *Arabidopsis thaliana* mutant that develops as a light-grown plant in the absence of light. *Cell* 58:991–99

31. Chory J, Peto CA, Ashbaugh M, Saganich R, Pratt L, Ausubel F. 1989. Different roles for phytochrome in etiolated and green plants deduced from characterization of *Arabidopsis thaliana* mutants. *Plant Cell* 1:867–80

32. Christopher DA, Kim M, Mullet JE. 1992. A novel light-regulated promoter is conserved in cereal and dicot chloroplasts. *Plant Cell* 4:785–98

33. Clugston C, Barnett L, Urwin N, Jenkins G. 1990. Photoreceptors controlling transcription of *rbcS* genes in green leaf tissue of *Pisum sativum*. *Photochem. Photobiol.* 52:23–28

34. Cosgrove D. 1985. Kinetic separation of phototropism from blue-light inhibition of stem elongation. *Photochem. Photobiol.* 42:745–51

35. Davis PD, Hill CH, Keech E, Lawton G, Nixon JS, et al. 1989. Potent selective inhibitors of protein kinase C. *FEBS Lett.* 259:61–63

36. DeFabo E. 1980. On the nature of the blue light photoreceptor: still an open question. See Ref. 131, pp. 187–97

37. Deng X-W, Caspar T, Quail PH. 1991. *cop1*: A regulatory locus involved in light-controlled development and gene expression in *Arabidopsis*. *Genes Dev.* 5:1172–82

38. Dobres MS, Thompson WF. 1989. A developmentally regulated bud-specific transcript in pea has sequence similarity to seed lectins. *Plant Physiol.* 89:833–38

39. Duell-Pfaff N, Wellmann E. 1982. Involvement of phytochrome and a blue light photoreceptor in UV-B induced flavonoid synthesis in parsley (*Petroselinum hortense* Hoffm.) cell suspension cultures. *Planta* 156:213–17

40. Elmlinger MW, Mohr H. 1991. Coaction of blue/ultraviolet-A light and light absorbed by phytochrome in controlling the appearance of ferredoxin-dependent glutamate synthase in the Scots pine (*Pinus sylvestris* L.) seedling. *Planta* 183:374–80

41. Fairley-Grenot KA, Assmann SM. 1991. Evidence for G-protein regulation of inward K+ channel current in guard cells of fava bean. *Plant Cell* 3:1037–44

42. Feinbaum RL, Storz G, Ausubel FM. 1991. High intensity and blue light regulated expression of chimeric chalcone synthase genes in transgenic *Arabidopsis thaliana* plants. *Mol. Gen. Genet.* 226:449–56

43. Fluhr R, Chua N-H. 1986. Developmental

regulation of two genes encoding ribulose-bisphosphate carboxylase small subunit in pea and transgenic petunia plants: phytochrome response and blue-light induction. *Proc. Natl. Acad. Sci. USA* 83:2358–62

44. Fluhr R, Moses P, Morelli G, Coruzzi G, Chua N-H. 1986. Expression dynamics of the pea *rbcS* multigene family and organ distribution of the transcripts. *EMBO J.* 5:2063–71

45. Furuya M. 1993. Phytochromes: their molecular species, gene families, and functions. *Annu. Rev. Plant Physiol. Plant Mol. Biol.* 44:617–45

46. Gaba V, Black M, Attridge TH. 1984. Photocontrol of hypcotyl elongation in de-etiolated *Cucumis sativus* L. *Plant Physiol.* 74:897–900

47. Gaba V, Gressel J. 1987. Acifluorfen enhancement of cryptochrome-modulated sporulation following an inductive light pulse. *Plant Physiol.* 83:225–27

48. Gallagher S, Short TW, Ray PM, Pratt LH, Briggs WR. 1988. Light-mediated changes in two proteins found associated with plasma membrane fractions from pea stem sections. *Proc. Natl. Acad. Sci. USA* 85:8003–7

49. Galland P. 1992. Forty years of blue-light research and no anniversary. *Photochem. Photobiol.* 56:847–53

50. Galland P, Senger H. 1988. The role of pterins in the photoreception and metabolism of plants. *Photochem. Photobiol.* 48:811–20

51. Galston AW. 1977. Riboflavin retrospective or déja-vu in blue. *Photochem. Photobiol.* 25:503–4

52. Gamble PE, Mullet JE. 1989. Blue light regulates the accumulation of two *psbD-psbC* transcripts in barley chloroplasts. *EMBO J.* 8:2785–94

53. Gardner G, Gorton HL, Brown SA. 1988. Inhibition of phytochrome synthesis by the transaminase inhibitor, 4-amino-5-fluoropentanoic acid. *Plant Physiol.* 87:8–10

54. Gautier H, Vavasseur AGL, Boudet AM. 1992. Redox processes in the blue light response of guard cell protoplasts of *Commelina communis* L. *Plant Physiol.* 98:34–38

55. Gilman GA. 1987. G proteins: transducers of receptor-generated signals. *Annu. Rev. Biochem.* 56:615–49

56. Goldsmith MHM, Cauburgs RJ, Briggs WR. 1980. Light-inducible cytochrome reduction in membrane preparations from corn coleoptiles. I. Stabilization and spectral characterization of the reaction. *Plant Physiol.* 66:1067–73

57. Grimm B, Kruse E, Kloppstech K. 1989. Transiently expressed early light-inducible thylakoid proteins share transmembrane domains with light-harvesting chlorophyll

binding proteins. *Plant Mol. Biol.* 13:583–93
58. Guilfoyle TJ, Dietrich MA, Prenger JP, Hagen G. 1990. Phosphorylation/dephosphorylation of the carboxy terminal domain of the largest subunit of RNA polymerase II. In *Current Topics in Plant Biochemistry and Physiology,* ed. DD Randall, DG Blevins, 9:299–312. Columbia, MO: Univ. Missouri Press
59. Hager A, Brich M. 1993. Blue-light-induced phosphorylation of a plasma-membrane protein from phototropically sensitive tips of maize coleoptiles. *Planta* 189: 567–76
60. Hager A, Brich M, Bazlen I. 1993. Redox dependence of the blue-light-induced phosphorylation of a 100-kDa protein on isolated plasma membranes from tips of coleoptiles. *Planta* 190:120–26
61. Heelis PF, Parsons BJ, Phillips GO, McKellar JF. 1978. A laser flash photolysis study of the nature of flavin mononucleotide triplet states and the reactions of the neutral form with amino acids. *Photochem. Photobiol.* 28:169–73
62. Hemmerich P, Massey V, Weber G. 1967. Photo-induced benzyl substitution of flavins by phenylacetate: a possible model for flavin catalysis. *Nature* 213:728–30
63. Horwitz BA. 1994. Properties and transduction chains of the UV and blue light photoreceptors. In *Photomorphogenesis in Plants,* ed. RE Kendrick, GHM Kronenberg, pp. 327–50. Dordrecht: Kluwer Academic. 2nd ed.
64. Horwitz BA, Gressel JB. 1986. Properties and working mechanisms of the photoreceptors. In *Photomorphogenesis in Plants,* ed. RE Kendrick, GHM Kronenberg, pp. 159–83. Dordrecht: Nijhoff
65. Iino M. 1990. Phototropism: mechanisms and ecological implications. *Plant Cell Environ.* 13:633–50
66. Iino M, Briggs WR, Schäfer E. 1984. Phytochrome-mediated phototropism in maize seedling shoots. *Planta* 160:41–51
67. Iino M, Ogawa T, Zeiger E. 1985. Kinetic properties of the blue-light response of stomata. *Proc. Natl. Acad. Sci. USA* 82:8019–23
68. Jenkins G. 1988. Photoregulation of gene expression in plants. *Photochem. Photobiol.* 48:821–32
69. Jenkins GI, Jackson JA, Shaw MJ, Urwin NAR. 1994. A genetic approach to understanding responses to UV-A/blue light. In *Plant Photoreceptors and Photoperception,* ed. MG Holmes, CB Johnson. Cambridge: Cambridge Univ. Press. In press
70. Johnson JL, Hainline BE, Rajagopalan KV. 1980. Characterization of the molybdenum cofactor of sulfite oxidase, xanthine oxidase, and nitrate reductase. *J. Biol. Chem.* 255:1783–86
71. Karlsson PE. 1986. Blue light regulation of stomata in wheat seedlings. II. Action spectrum and search for action dichroism. *Physiol. Plant.* 66:207–10
72. Kaufman LS. 1993. Transduction of blue-light signals. *Plant Physiol.* 102:333–37
73. Kearns EV, Assmann SM. 1993. The guard cell-environment connection. *Plant Physiol.* 102:711–15
74. Khurana JP, Poff KL. 1989. Mutants of *Arabidopsis thaliana* with altered phototropism. *Planta* 178:400–6
75. Khurana JP, Ren Z, Steinitz B, Parks B, Best TR, Poff KL. 1989. Mutants of *Arabidopsis thaliana* with decreased amplitude in their phototropic response. *Plant Physiol.* 91:685–89
76. Koller D. 1990. Light-driven leaf movements. *Plant Cell Environ.* 13:615–32
77. Konjevic R, Khurana JP, Poff KL. 1992. Analysis of multiple photoreceptor pigments for phototropism in a mutant of *Arabidopsis thaliana. Photochem. Photobiol.* 55:789–92
78. Konjevic R, Steinitz B, Poff KL. 1989. Dependence of the phototropic response of *Arabidopsis thaliana* on fluence rate and wavelength. *Proc. Natl. Acad. Sci. USA* 86:9876–80
79. Koornneef M, Rolff E, Spruit CJP. 1980. Genetic control of light-inhibited hypocotyl elongation in *Arabidopsis thaliana* (L.) Heynh. *Z. Pflanzenphysiol.* 100:147–60
80. Kubasek WL, Shirley BW, McKillop A, Goodman HM, Briggs WR, Ausubel FM. 1992. Regulation of flavonoid biosynthetic genes in germinating *Arabidopsis* seedlings. *Plant Cell* 4:1229–36
81. Kunzelmann P, Schäfer E. 1985. Phytochrome-mediated phototropism in maize mesocotyls. Relation between light and Pfr gradients, light growth response and phototropism. *Planta* 165:424–29
82. Lamb CJ, Lawton MA. 1983. Photocontrol of gene expression. In *Encyclopedia of Plant Physiology,* (NS), ed. W Shropshire, H Mohr, 16A:213–57. Berlin: Springer-Verlag
83. Laskowski MJ, Briggs WR. 1989. Regulation of pea epicotyl elongation by blue light: fluence response relationships and growth distribution. *Plant Physiol.* 89: 293–98
84. Lemberg R, Barrett J. 1973. *Cytochromes.* London: Academic
85. Leong T-Y, Briggs WR. 1981. Partial purification and characterization of a blue light-sensitive cytochrome-flavin complex from corn membranes. *Plant Physiol.* 67:1042–46
86. Leong T-Y, Briggs WR. 1982. Evidence

from studies with acifluorfen for participation of a flavin-cytochrome complex in blue light photoreception for phototropism of oat coleoptiles. *Plant Physiol.* 70:875–81

87. Leong T-Y, Vierstra RD, Briggs WR. 1981. A blue light-sensitive cytochrome-flavin complex from coleoptiles: further characterization. *Photochem. Photobiol.* 34:697–703

88. Liscum E, Hangarter RP. 1991. *Arabidopsis* mutants lacking blue light-dependent inhibition of hypocotyl elongation. *Plant Cell* 3:685–94

89. Liscum E, Hangarter RP. 1993. Photomorphogenic mutants of *Arabidopsis thaliana* reveal activities of multiple photosensory systems during light-stimulated apical-hook opening. *Planta* 191:214–21

90. Liscum E, Hangarter RP. 1994. Tuning into the blues: mutational analysis of blue light sensing in *Arabidopsis*. *Plant Cell Environ.* In press

91. Liscum E, Young JC, Poff KL, Hangarter RP. 1992. Genetic separation of phototropism and blue light inhibition of stem elongation. *Plant Physiol.* 100:267–71

92. Luan S, Li W, Rusnak F, Assmann SM, Schreiber SL. 1993. Immunosuppressants implicate protein phosphatase regulation of K$^+$ channels in guard cells. *Proc. Natl. Acad. Sci. USA* 90:2202–6

93. Marrs KA, Kaufman LS. 1989. Blue light regulation of transcription for nuclear genes in pea. *Proc. Natl. Acad. Sci. USA* 86:4492–95

94. Marrs KA, Kaufman LS. 1991. Rapid transcriptional regulation of the *Cab* and pEA207 gene families in peas by blue light in the absence of cytoplasmic protein synthesis. *Planta* 183:327–33

95. Meyer G, Kloppstech K. 1984. A rapidly light-induced chloroplast protein with a high turnover coded for by pea nuclear DNA. *Eur. J. Biochem.* 138:201–7

96. Mohr H. 1977. Phytochrome and chloroplast development. *Endeavor* NS 1:107–14

97. Mohr H. 1986. Coaction between pigment systems. In *Photomorphogenesis in Plants*, ed. RE Kendrick, GHM Kronenberg, pp. 547–64. Dordrecht: Nijhoff

98. Mohr H, Drumm-Herrel H. 1983. Coaction between phytochrome and blue/UV light in anthocyanin synthesis in seedlings. *Physiol. Plant.* 58:408–14

99. Muñoz V, Butler WL. 1975. Photoreceptor pigment for blue light in *Neurospora crassa. Plant Physiol.* 55:421–26

100. Nick P, Furuya M, Schäfer E. 1991. Do microtubules control growth in tropism? *Plant Cell Physiol.* 32:999–1006

101. Nick P, Schäfer E, Furuya M. 1992. Auxin redistribution during first positive phototropism in corn coleoptiles. Microtubule

reorientation and the Cholodny-Went theory. *Plant Physiol.* 99:1302–8

102. Nishizaki Y. 1988. Blue light pulse–induced transient changes of electric potential and turgor pressure in the motor cells of *Phaseolus vulgaris* L. *Plant Cell Physiol.* 29:1041–46

103. Oelmüller R, Kendrick RE, Briggs WR. 1989. Blue-light mediated accumulation of nuclear-encoded transcripts coding for proteins of the thylakoid membrane is absent in the phytochrome-deficient aurea mutant of tomato. *Plant Mol. Biol.* 13:223–32

104. Oelmüller R, Mohr H. 1984. Responsivity amplification by light in phytochrome-mediated induction of chloroplast glyceraldehyde-3-phosphate dehydrogenase (NADP-dependent, EC 1.2.1.13) in the shoot of milo (*Sorghum vulgare* Pers.). *Plant Cell Environ.* 7:29–37

105. Oelmüller R, Mohr H. 1985. Mode of coaction between blue/UV light and light absorbed by phytochrome in light-mediated anthocyanin formation in the milo (*Sorghum vulgare* Pers.) seedlings. *Proc. Natl. Acad. Sci. USA* 82:6124–28

106. Ohl S, Hahlbrock K, Schäfer E. 1989. A stable blue-light derived signal modulates ultraviolet-light-induced activation of the chalcone-synthase gene in cultured parsley cells. *Planta* 177:228–36

107. Palmer JM, Reymond PR, Short TW, Briggs WR. 1992. Phosphorylation of a plasma membrane protein in maize coleoptiles may be an early step in the signal transduction chain for phototropism. *Photochem. Photobiol.* 55:S96–97

108. Palmer JM, Short TW, Briggs WR. 1993. Correlation of blue light-induced phosphorylation to phototropism in *Zea mays* L. *Plant Physiol.* 102:1219–25

109. Palmer JM, Short TW, Gallagher S, Briggs WR. 1993. Blue light-induced phosphorylation of a plasma membrane-associated protein in *Zea mays* L. *Plant Physiol.* 102:1211–18

110. Palmgren MG, Askerlund P, Fredrikson K, Widell S, Sommarin M, Larsson C. 1990. Sealed inside-out and right-side-out plasma membrane vesicles. Optimal conditions for formation and separation. *Plant Physiol.* 92:871–80

111. Parker K, Baskin TI, Briggs WR. 1989. Evidence for a phytochrome-mediated phototropism in etiolated pea seedlings. *Plant Physiol.* 89:493–97

112. Parks BM, Jones AM, Adamse P, Koornneef M, Kendrick RE, Quail PH. 1987. The aurea mutant of tomato is deficient in spectrophotometrically and immunochemically detectable phytochrome. *Plant Mol. Biol.* 9:97–107

113. Quail PH. 1991. Phytochrome: a light-activated molecular switch that regulates plant

gene expression. *Annu. Rev. Genet.* 25: 389–409

114. Raghavendra AS. 1990. Blue light effects on stomata are mediated by the guard cell plasma membrane redox system distinct from the proton translocating ATPase. *Plant Cell Environ.* 13:105–10

115. Reimann-Philipp U, Behnke S, Batschauer A, Schäfer E, Apel K. 1989. The effect of light on the biosynthesis of leaf-specific thionins in barley. *Eur. J. Biochem.* 182: 283–89

116. Reymond P, Short TW, Briggs WR. 1992. Blue light activates a specific protein kinase in higher plants. *Plant Physiol.* 100: 655–61

117. Reymond P, Short TW, Briggs WR, Poff KL. 1992. Light-induced phosphorylation of a membrane protein plays an early role in signal transduction for phototropism in *Arabidopsis thaliana. Proc. Natl. Acad. Sci. USA* 89:4718–21

118. Rich TCG, Whitelam GC, Smith H. 1985. Phototropism and axis extension in light-grown mustard (*Sinapsis alba* L.) seedlings. *Photochem. Photobiol.* 42:789–92

119. Richter G. 1984. Blue light control of the level of two plastid mRNAs in cultured plant cells. *Plant Mol. Biol.* 3:271–76

120. Richter G, Dudel A, Einspanier R, Dannhauer I, Hüsemann W. 1987. Blue-light control of mRNA level and transcription during chloroplast differentiation in photomixotrophic and photoautotrophic cell cultures (*Chenopodium rubrum* L.). *Planta* 172:79–87

121. Richter G, Ottersbach N. 1990. Blue light-dependent chloroplast differentiation in cultured plant cells: evidence for transcriptional control of plastid genes. *Bot. Acta* 103:168–73

122. Richter G, Wessel K. 1985. Red light inhibits blue light-induced chloroplast development in cultured plant cells at the mRNA level. *Plant Mol. Biol.* 5:175–82

123. Rubinstein B, Luster DG. 1993. Plasma membrane redox activity: components and role in plant processes. *Annu. Rev. Plant Physiol. Plant Mol. Biol.* 44:131–55

124. Sawbridge TI, López-Juez E, Knight MR, Jenkins GI. 1994. A blue light photoreceptor mediates the fluence rate-dependent expression of genes encoding the small subunit of ribulose 1,5-bisphosphate carboxylase/oxygenase in light-grown *Phaseolus vulgaris* primary leaves. *Planta* In press

125. Schmidt S, Drumm-Herrel H, Oelmüller R, Mohr H. 1987. Time course of competence in phytochrome-controlled appearance of nuclear-encoded plastidic proteins. *Planta* 170:400–7

126. Schmidt W, Butler WL. 1976. Flavin-mediated photoreactions in artificial systems: a possible model for the blue-light photoreceptor pigment in living systems. *Photochem. Photobiol.* 24:71–75

127. Schmidt W, Hart JW, Filner P, Poff KL. 1977. Specific inhibition of phototropism in corn seedlings. *Plant Physiol.* 60:736–38

128. Schroeder JI. 1993. Perspective on the physiology and structure of inward-rectifying K^+ channels in higher plants: biophysical implications for K^+ uptake. *Annu. Rev. Biophys. Biomol. Struct.* 23:In press

129. Schroeder JI, Hagiwara S. 1989. Cytosolic calcium regulates ion channels in the plasma membrane of *Vicia faba* guard cells. *Nature* 338:427–30

130. Shropshire W. 1980. Carotenoids as primary photoreceptors in bluelight responses. See Ref. 131, pp. 172–86

131. Senger H, ed. 1980. *The Blue Light Syndrome.* Berlin: Springer-Verlag

132. Senger H, ed. 1984. *Blue Light Effects in Biological Systems.* Berlin: Springer-Verlag

133. Senger H, ed. 1987. *Blue Light Responses: Phenomena and Occurrence in Plants and Microorganisms.* Boca Raton, FL: CRC

134. Serrano E, Zeiger E. 1989. Sensory transduction and electrical signaling in guard cells. *Plant Physiol.* 91:795–99

135. Sharkey TD, Ogawa T. 1987. Stomatal response to light. In *Stomatal Function,* ed. E Zeiger, GD Farquhar, IR Cowan, pp. 195–208. Stanford, CA: Stanford Univ. Press

136. Sharrock RA, Quail PH. 1989. Novel phytochrome sequences in *Arabidopsis thaliana*: structure, evolution, and differential expression of a plant regulatory photoreceptor family. *Genes Dev.* 3:1745–57

137. Shimazaki K, Iino M, Zeiger E. 1986. Blue light-dependent proton extrusion by guard-cell protoplasts of *Vicia faba. Nature* 319: 324–26

138. Shimazaki K, Kinoshita T, Nishimura M. 1992. Involvement of calmodulin and calmodulin-dependent myosin light chain kinase in blue light-dependent H^+ pumping by guard cell protoplasts from *Vicia faba* L. *Plant Physiol.* 99:1416–21

139. Shinkle JR, Jones RL. 1988. Inhibition of stem elongation in *Cucumis* seedlings by blue light requires calcium. *Plant Physiol.* 86:960–66

140. Short TW. 1991. *Blue light-mediated phosphorylation of a plasma membrane protein in higher plants.* PhD thesis. Stanford Univ., Stanford, CA

141. Short TW, Briggs WR. 1990. Characterization of a rapid, blue light-mediated change in detectable phosphorylation of a plasma membrane protein from etiolated pea (*Pisum sativum* L.) seedlings. *Plant Physiol.* 92:179–85

142. Short TW, Gallagher S, Briggs WR. 1990. Protein phosphorylation as a possible signal transduction step for blue light-mediated phototropism in pea (*Pisum sativum* L.) epicotyls. In *Current Topics in Plant Biochemistry and Physiology*, ed. DD Randall, DG Blevins, 9:232–48. Columbia, MO: Univ. Missouri Press

143. Short TW, Porst M, Briggs WR. 1991. On the trail of the photoreceptor for phototropism in higher plants. In *Biophysics of Photoreceptors and Photomovement in Microorganisms*, ed. F Lenci, G Columbetti, D-P Häder, P-S Song, pp. 173–80. New York: Plenum

144. Short TW, Porst M, Briggs WR. 1992. A photoreceptor system regulating in vivo and in vitro phosphorylation of a pea plasma membrane protein. *Photochem. Photobiol.* 55:773–81

145. Short TW, Porst M, Palmer JM, Fernbach E, Briggs WR. 1994. Blue light induces phosphorylation at multiple sites on a pea plasma membrane protein. *Plant Physiol.* In press

146. Short TW, Reymond P, Briggs WR. 1993. A pea plasma membrane protein exhibiting blue light-induced phosphorylation retains photosensitivity following triton solubilization. *Plant Physiol.* 101:647–55

147. Song P-S. 1980. Spectroscopic and photochemical characterization of flavoproteins and carotenoproteins as blue light photoreceptors. See Ref. 131, pp. 157–71

148. Song P-S. 1984. Photophysical aspects of blue light receptors: the old question (flavins versus carotenoids) re-examined. See Ref. 132, pp. 75–80

149. Song P-S. 1987. Possible primary photoreceptors. See Ref. 133, pp. 3–17

150. Spalding EP, Cosgrove DJ. 1989. Large plasma-membrane depolarization precedes rapid blue-light-induced growth inhibition in cucumber. *Planta* 178:407–10

151. Spalding EP, Cosgrove DJ. 1992. Mechanism of blue-light-induced plasma-membrane depolarization in etiolated cucumber hypocotyls. *Planta* 188:199–205

152. Staal M, Elzenga JTM, Prins HBA, van Volkenburgh E. 1993. Light-stimulated ion flux in epidermis and mesophyll of growing pea leaves. *Plant Physiol.* Suppl. 102:22

153. Stapleton AE. 1992. Ultraviolet radiation and plants: burning questions. *Plant Cell* 4:1353–58

154. Tamaoki T, Nomoto H, Takahashi I, Kato Y, Moromoto M, Tomita F. 1986. Staurosporine, a potent inhibitor of phospho-lipid/Ca^{++} dependent protein kinase. *Biochem. Biophys. Res. Commun.* 135:397–402

155. Thimann KV, Curry GM. 1961. Phototropism. In *Light and Life*, ed. WD McElroy, pp. 646–72. Baltimore: Johns Hopkins Press

156. Thompson WF, Everett M, Polans NO, Jorgensen RA, Palmer JD. 1983. Phytochrome control of RNA levels in developing pea and mung-bean leaves. *Planta* 158:487

157. Thompson WF, Kaufman LS, Watson JC. 1985. Induction of plant gene expression by light. *BioEssays* 3:153–59

158. Thompson WF, White MJ. 1991. Physiological and molecular studies of light-regulated nuclear genes in higher plants. *Annu. Rev. Plant Physiol. Plant Mol. Biol.* 42:423–66

159. Tobin EM, Silverthorne J. 1985. Light regulation of gene expression in higher plants. *Annu. Rev. Plant Physiol.* 36:569–93

160. Travis AJ, Mansfield TA. 1981. Light saturation of stomatal opening on the adaxial and abaxial epidermis of *Commelina communis*. *J. Exp. Bot.* 32:1169–79

161. Van Volkenburgh E, Cleland RE. 1980. Proton excretion and cell expansion in bean leaves. *Planta* 148:273–78

162. Van Volkenburgh E, Cleland RE, Watanabe M. 1990. Light-stimulated cell expansion in bean (*Phaseolus vulgaris* L.) leaves. II. Quantity and quality of light required. *Planta* 182:77–80

163. Vani T, Raghavendra AS. 1989. Tetrazolium reduction by guard cells in abaxial epidermis of *Vicia faba*: blue light stimulation of a plasmalemma redox system. *Plant Physiol.* 90:59–62

164. Vani T, Raghavendra AS. 1992. Plasma membrane redox system in guard cell protoplasts of pea (*Pisum sativum* L.). *J. Exp. Bot.* 43:291–97

165. Deleted in proof

166. Vierstra RD, Poff KL, Walker EB, Song P-S. 1981. Effect of xenon on the excited states of phototropic receptor flavin in corn seedlings. *Plant Physiol.* 67:996–98

167. Warpeha KMF, Briggs WR. 1993. Blue light-induced phosphorylation of a plasma membrane protein in pea: a step in the signal transduction chain for phototropism. *Aust. J. Plant Physiol.* 20:393–403

168. Warpeha KMF, Hamm HE, Rasenick MM, Kaufman LS. 1991. A blue-light-activated GTP-binding protein in the plasma membranes of etiolated peas. *Proc. Natl. Acad. Sci. USA* 88:8925–29

169. Warpeha KMF, Kaufman LS. 1989. Blue light regulation of epicotyl elongation in *Pisum sativum*. *Plant Physiol.* 89:544–48

170. Warpeha KMF, Kaufman LS. 1990. Two distinct blue-light responses regulate the levels of transcripts of specific nuclear-coded genes in pea. *Planta* 182:553–58

171. Warpeha KMF, Kaufman LS, Briggs WR. 1992. A flavoprotein may mediate the blue light-activated binding of guanosine 5′-triphosphate to isolated plasma membranes of

Pisum sativum L. *Photochem. Photobiol.*
55:595–603

172. Warpeha KMF, Marrs KA, Kaufman LS.
1989. Blue-light regulation of specific transcript levels in *Pisum sativum. Plant Physiol.* 91:1030–35

173. Watson JC. 1989. Photoregulation of gene expression in plants. In *Plant Biotechnology,* ed. S-D Kung, CJ Arntzen, pp. 161–205. Boston: Butterworths

174. Wehmeyer B, Cashmore AR, Schäfer E.
1990. Photocontrol of the expression of genes encoding chlorophyll *a/b* binding proteins and small subunit of ribulose-1,5-bisphosphate carboxylase in etiolated seedlings of *Lycopersicon esculentum* (L.) and *Nicotiana tabacum* (L.). *Plant Physiol.* 93:990–97

175. Wei N, Deng X-W. 1992. *Cop9:* a new genetic locus involved in light-regulated development and gene expression in *Arabidopsis. Plant Cell* 4:1507–18

176. Widell S, Britz SJ, Briggs WR. 1980. Characterization of red light induced reduction of a particle associated *b*-type cytochrome from corn in the presence of methylene blue. *Photochem. Photobiol.* 32:669–77

177. Widell S, Larsson C. 1983. Distribution of cytochrome *b* photoreductions mediated by endogenous photosensitizer or methylene blue in fractions from corn and cauliflower. *Physiol. Plant.* 57:196–202

178. Widell S, Larsson C. 1990. A critical evaluation of markers used in plasma membrane purification. In *The Plant Plasma Membrane,* ed. C Larsson, IM Müller, pp. 16–43. Berlin: Springer-Verlag

179. Widell S, Sommarin M. 1991. Purification of endoplasmic reticulum from wheat roots and shoots (*Triticum aestivum*). Comparison of their blue light sensitive redox components with those of highly purified plasma membrane. *Physiol. Plant.* 82:9–18

180. Widell S, Sundqvist C. 1987. Fluorescence properties of plasma membranes from oats and cauliflower in relation to blue light physiology. *Physiol. Plant.* 70:27–34

181. Young JC, Liscum E, Hangarter RP. 1992. Spectral-dependence of light-inhibited hypocotyl elongation in photomorphogenic mutants of *Arabidopsis:* evidence for a UV photosensor. *Planta* 188:106–14

182. Zandomeni K, Schopfer P. 1993. Reorientation of microtubules at the outer epidermal wall of maize coleoptiles by phytochrome, blue-light photoreceptor, and auxin. *Protoplasma* 173:103–12

183. Zeiger E. 1990. Light perception in guard cells. *Plant Cell Environ.* 13:739–47

184. Zimmerman BK, Briggs WR. 1963. A kinetic model for phototropic responses of oat coleoptiles. *Plant Physiol.* 38:253–61

185. Zimmerman BK, Briggs WR. 1963. Phototropic dose-response curves for oat coleoptiles. *Plant Physiol.* 38:248–53

Annu. Rev. Plant Physiol. Plant Mol. Biol. 1994. 45:173–96

CYTOKININ ACCUMULATION AND ACTION: Biochemical, Genetic, and Molecular Approaches

Andrew N. Binns

Plant Science Institute, Department of Biology, University of Pennsylvania, Philadelphia, Pennsylvania 19104-6018

KEY WORDS: cytokinin, plant hormone, transgenic plants, hormone mutants, plant molecular genetics

CONTENTS

INTRODUCTION 173
BIOCHEMICAL APPROACHES.. 174
   Cytokinin Biosynthesis.. 175
   Cytokinin Metabolism.. 176
   Cytokinin Receptors... 178
GENETIC APPROACHES 180
   Cytokinin Accumulation Mutants 180
   Cytokinin Response Mutants 182
MOLECULAR APPROACHES.. 185
   Cytokinin Control of Gene Expression... 186
   Transgenic Plants Expressing the ipt *Gene from* Agrobacterium tumefaciens 187
CONCLUDING REMARKS... 190

INTRODUCTION

Skoog and Miller's classic work in the mid 1950s demonstrated that cytokinins, N^6-substituted adenine derivatives, had critical biological activities in plant systems. The cytokinins acted in synergy with auxin to stimulate continuous cell division in cultured tobacco pith tissues (88). In contrast, they acted antagonistically with auxin in the control of shoot and root initiation in tissue cultures and in the process of apical dominance (105, 115). During the next 20–30 years, work in cytokinin biology focused on: (*a*) characterization of the naturally occurring cytokinins and their metabolites and (b) detailed examination of cytokinin involvement in a spectrum of plant activities, generally through analysis of responses to exogenously supplied hormone. These studies have yielded a wealth of information on the chemical nature and distribution of metabolites and have helped in the development of methods for manipulating plant growth and development (for reviews on synthesis and metabolism, see 67, 79; for reviews on biological activities, see 27, 90).

Fundamental questions concerning the synthesis, metabolism, and activities of cytokinin remain unanswered. For example, the enzymes involved in cytokinin biosynthesis and metabolism have not been purified or characterized in detail, nor have the genes encoding such enzymes been isolated. Little is known about the mechanism of cytokinin action (i.e. cytokinin receptors or signal transduction pathways). Finally, the activities of the various plant hormones depend on their interactions with other plant hormones, but the mechanisms by which interactions control developmental and physiological processes are unknown. This, some might say, pessimistic description of the field comes from the difficulty of carrying out genetic studies in the area (64). For example, the ability to isolate mutations in gibberellin synthesis and ethylene response has led to important advances in understanding how these growth regulators are synthesized and/or function (106, 110). The relatively underdeveloped cytokinin genetics probably reflects the crucial role played by cytokinin in the life of the plant. Mutations that disrupt its synthesis or activity may be lethal. Thus, innovative approaches to develop the genetics in this area are needed.

This review covers the progress made using biochemical, genetic, and molecular approaches to identify the genes and gene products involved in cytokinin control of growth and development. In particular, it focuses on various aspects of cytokinin biology that may provide inroads into genetic analysis: how can genes involved in cytokinin activity—synthesis, metabolism, or response—be obtained? Once such genes are identified, both classical and molecular genetic methods can be used to experimentally manipulate the plant so that basic mechanisms underlying cytokinin action can be studied.

BIOCHEMICAL APPROACHES

One clearly successful method for obtaining genes involved in particular biological responses has been through biochemical analysis of the proteins encoded by the gene. If such proteins can be purified sufficiently for use in either sequence analysis or antibody generation, then methods for isolating and manipulating the corresponding genes become available. For example, antibodies can be used to screen cDNA expression libraries and protein sequence information can be used to derive and synthesize DNA sequences for molecular

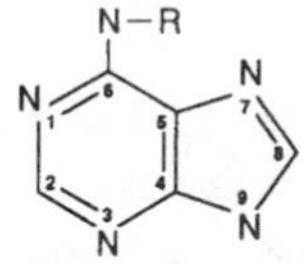

R	Trivial Name	Abbreviation
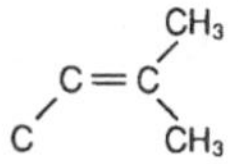 $C = C$ with CH_3 and CH_3	N^6-(Δ^2-isopentenyl) adenine	iP
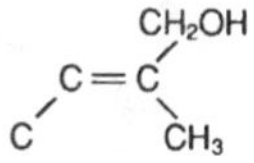 $C = C$ with CH_2OH and CH_3	*trans*-zeatin	*t*-Z
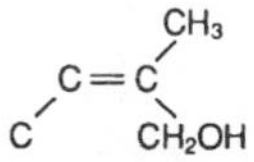 $C = C$ with CH_3 and CH_2OH	*cis*-zeatin	*c*-Z
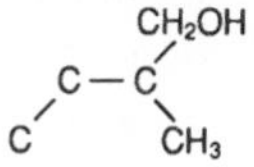 $C - C$ with CH_2OH and CH_3	dihydrozeatin	(diH)Z

Figure 1   The structure of some naturally occurring cytokinins.

probes. While this approach has not been used extensively in cytokinin biology, several systems have either potential or demonstrated success.

Cytokinin Biosynthesis

With the exception of *Agrobacterium tumefaciens*–induced crown gall tumors (see below), the biosynthetic pathway of cytokinins has not been demonstrated clearly. The currently favored model for cytokinin biosynthesis predicts the addition of dimethylallylpyrophosphate (DMAPP) to the N^6 position of AMP by an isopentenyl transferase (IPT), yielding the cytokinin ribotide, $N^6\Delta2$-isopentenylAMP ([9R-5′P]iP)[1]. This can then be modified in a series of reactions to yield other types of cytokinin ribotides, ribosides, and free bases, including zeatin and dihydrozeatin (Figure 1). Besides this de novo route, cytokinins, in particular the less active *cis*-zeatin, have been found in certain transfer RNAs as the base adjacent to the 3′-end of the anticodon (74). The physiological significance of tRNA-derived cytokinins is uncertain, particularly because the rate of tRNA turnover that would be necessary for cytokinin

---

[1]

I follow the conventions of Letham & Palni (67) to describe the cytokinins.

accumulation has not been observed (67, 79, but see below). Chen and co-workers have provided the best evidence that plants normally use the de novo pathway. They extracted and partially purified IPT from tobacco cell lines that were capable of growing in the absence of exogenous cytokinin (22, 23). Another source of IPT has been the slime mold *Dictyostelium discoideum* where iP serves as a precursor to the germination inhibitor discadenine (60). In this case, the IPT was partially purified and found to have similar, but not identical, substrate specificity as the plant enzyme. Unfortunately, in neither case was purified enzyme obtained in quantities sufficient to provide material for either antibody generation or microsequencing.

In 1974 Miller (87) provided unequivocal evidence that crown gall tumors contained elevated levels of cytokinins. Later studies showed that this increase in cytokinin content reflects activity of the *ipt* gene of the transferred DNA (T-DNA) from *Agrobacterium tumefaciens* (for reviews, see 11, 133). This enzyme is produced by plant cells only when they express the *ipt* locus of the T-DNA, and, as with the endogenous plant IPT, it catalyzes the formation of [9R-5'P]iP from 5'AMP and DMAPP (1, 5, 18). Despite this similarity, there have been no reports of plant DNA sequences homologous to the bacterial *ipt* gene. Biochemical analysis has not been extensive: enzyme characterization has been achieved mainly through examination of the protein overexpressed in bacteria (1, 5, 18). It was difficult to prove that the activity observed in crown gall tumors came from the T-DNA *ipt* gene (1, 18), though this was probably the case. Given the proposed importance of IPT in cytokinin biosynthesis, it is surprising that more detailed biochemical analysis of this T-DNA encoded protein has not been carried out. A variety of different bacteria have been shown to contain *ipt* or *ipt*-like type genes (25, 93). The characterization of regions of homology between these enzymes, the definition of their active site, and their interaction with substrates could lead to the development of strategies for gene isolation or the synthesis of antagonists that might be used to block cytokinin synthesis.

Cytokinin Metabolism

An important way to regulate the availability of active cytokinins in plant tissues is through metabolic interconversions. I define cytokinin metabolism as the conversion of [9R-5'P]iP to any other N^6-substituted adenine derivative cytokinin. The major conversions are the dephosphorylation and deribosylation yielding the riboside and free base, respectively; the hydroxylation of the side chain to yield the *trans*-zeatin (*t*-Z) derivatives ([9R-5'P] *t*-Z, [9R] *t*-Z, and *t*-Z), and the reduction of the side chain to yield dihydrozeatin derivatives ([9R-5'P] (diH)Z, [9R] (diH)Z, and (diH)Z). These molecules can also be further metabolized, often by O-glycosylation of Z or (diH)Z, and N-glycosylation at the 7 and 9 positions of the adenine base. As Turner et al demonstrated, sugars other than glucose and ribose (e.g. xylose) may be added to the cytokinin molecule (127). Recently, Auer & Cohen (4) identified metabolites of N^6 benzyl adenosine ([9R] BA) in which the ribose was glycosylated at its 3'OH. Finally, cytokinins can be degraded through the activity of a cytokinin oxidase that removes the isopentenyl side chain, yielding the adenine or adenosine from the cytokinin base or nucleoside, respectively (20, 67).

The diversity of metabolites leads to the question: Which molecules have biological activity, which are used for temporary storage, and which are used to essentially remove cytokinin from the mobilizable pool? Based on studies showing that the N-glycosides are very stable and have low biological activity in vitro, these forms appear to withdraw cytokinin from the potential pool (67). In contrast, the O-glycosides are deglycosylated readily and may be involved in homeostatic control of active cytokinin levels (67, 79). Numerous studies have documented a correlation between various steady state levels of specific cytokinin metabolites and specific states and/or responses of the plant or plant tissues. For example, tissues from *Phaseolus* spp. that differed in their response to exogenous cytokinin also differed in their cytokinin metabolism (89). These species also showed interesting differences in the cytokinin metabolism in the embryo (66, also see below). Similarly, a detailed examination of the genetics of sex determination in *Mercuralis annua* has yielded a series of near isogenic lines that differ in sex based on the form of three different segregating genes. *t*-Z was abundant in the female plants, whereas iP was the predominant form in the male plants. Treatment of male plants with *t*-Z resulted in female flowers (34, 35, 70), indicating that regulation of iP hydroxylation to yield *t*-Z is critical in sex determination of this plant.

The active cytokinins have not been identified clearly, mainly because mutants or inhibitors that block particular metabolic steps are not available. However, the isolation, purification, and characterization of metabolic enzymes and their use in deriving molecular probes for gene isolation and manipulation should provide new opportunities. There are reports of crude activities for virtually all of the major metabolic steps (67, 79); however, relatively few of these enzymes have been purified enough to be used in antibody production and/or protein sequencing. Partial purification and characterization of cytokinin oxidase (20), zeatin reductase (78), and a *cis-trans* zeatin isomerase (6) have been reported. The latter seems particularly interesting in that it converts *cis*-zeatin to *trans*-zeatin, and could bring tRNA-derived molecules into the pool of active cytokinins. The most advanced molecular analysis of a cytokinin metabolic enzyme has been the purification and characterization of the zeatin O-glucosyl and O-xylosyl transferases from embryos of *Phaseolus vulgaris* (32, 127). A monoclonal antibody raised against PAGE-purified O-xylosyl transferase (75) has been used (*a*) in immunoelectron microscopy to show that the enzyme is found in both the cytoplasm and the nucleus (76) and (*b*) to identify a λ gt22 cDNA clone containing an insert proposed to code for the enzyme (77). Preliminary analysis of this sequence indicates that it encodes a 51-kDa protein, agreeing with the estimated size of the purified enzyme. As these types of genes become available, transgenic methodologies could be used to alter cytokinin metabolism. Enzyme activity could be increased by overexpression or decreased via antisense methods (99), thereby testing models of the role metabolism may play in particular cytokinin activities.

An example of how altered cytokinin metabolism could affect plants comes from the analysis of the *rolC* gene, one of several genes of the T-DNA from *Agrobacterium rhizogenes* involved in the evocation of the hairy root disease (for review, see 134). Transgenic plants overexpressing *rolC* exhibited a variety of phenotypes, some of which (e.g. dwarfing, reduced apical dominance) were suggestive of cytokinin effects (42, 98, 112). Analysis of the RolC protein derived by overexpression of *rolC* in *Escherichia coli* showed that it had a β-glucosidase activity that could cleave 7- and 9-linked glucose residues from zeatin and other cytokinins (42). Given that accumulation of the N-linked glucosides is thought to reduce the overall active cytokinin pool, this glucosidase activity could increase the available cytokinin pools and account for the abnormal phenotypes seen in the plants overexpressing *rolC*. However, extensive analysis of such plants has failed to establish that they contain substantially increased levels of cytokinins. One report (112) indicated that such plants have a 2- to 4-fold increase in both *t*-Z and iP derivatives, whereas another report (98) showed no significant increase in *t*-Z or (diH)Z and a 5-fold decrease in iP-type cytokinins with no alteration in the levels of 9-glucosylated zeatin. These conflicting results prevent the conclusion that the *rolC* gene product affects plant development by altering cytokinin metabolism. The RolC protein may exert its effects through gibberellin metabolism (98, 112), and further characterization of the protein, including estimations of the substrate specificity and kinetics of the glucosidase activity, could help resolve this issue.

Cytokinin Receptors

The activities of any signal molecule must be mediated by the interaction of that signal with some perceiving element, which then affects response elements leading to molecular, biochemical, or physiological events. In most well characterized systems, small molecule signals such as cytokinin bind to protein receptors that may, for example, affect systems controlling ion flux, phoshorylation of regulatory proteins, transcription, translation, and/or secretion (e.g. 13, 24, 50, 69). Characterization of cytokinin receptors and the early stages of signal transduction is in its infancy. Several reports of cytokinin binding proteins have appeared over the past 10–15 years. Some of the earliest reports concerned cytokinin binding proteins from wheat germ (41, 107) and the best characterized of these (CBF-1) has amino acid similarity to vicilin-type seed storage proteins (17). This protein has reasonably high affinity ($K_d = 10^{-7}$ M) for active cytokinins, but much lower affinity for nonactive cytokinins or other plant metabolites. Photoaffininty labeling using a BA derivative showed that a single histidine residue of CBF-1 was labeled, indicating a very specific interaction of cytokinin with the protein (17). Although the physiological significance of storage proteins as cytokinin binding proteins is unclear, they may serve as a cytokinin storage compartment allowing rapid release of cytokinins upon germination. This hypothesis has not been rigorously tested.

Several searches for cytokinin receptors in various vegetative tissues have yielded partially purified cytokinin binding proteins. Momotani & Tsuji (92) purified a water soluble cytokinin binding protein from adult tobacco leaves by successive chromatography on BA-linked and zeatin-linked Sepharose beads. The predominant protein in this fraction had an apparent molecular weight of 31 kDa. Using equilibrium dialysis, the K_d for BA was ~ 10^{-7} M and binding of BA to this protein was blocked by other cytokinins, but not by adenine or ATP. Another recent approach was based on the increased activity, on a molar basis, of many synthetic diphenylurea derivatives over naturally occurring or even synthetic N^6-adenine type cytokinins (91, 95). Nagata et al (95) used tritiated versions of diphenylurea derivatives to study cytokinin binding proteins from etiolated mung bean seedlings. A partially purified protein that was isolated (~ 21 kDa) had a K_d of ~ 10^{-8}–10^{-9} M for cytokinins such as zeatin or BA. Both free diphenylurea-type molecules and the N^6 adenyl cytokinin bases competed efficiently in binding assays, but the cytokinin ribosides did not. This difference between the binding of the bases and ribosides was not observed in several other cytokinin binding proteins (92, 107, 109).

Another biochemical approach to cytokinin receptor analysis was to use cytokinin antagonists (anticytokinins). The most potent and specific of these appear to be substituted pyrollo [2,3-d] pyrimidines and 7-substituted-3methylpyrazolo [4,3-d] pyrimidines. At reasonably low doses (~ 1 μM) these molecules inhibit cytokinin-dependent growth in tobacco tissue cultures, but the inhibition can be reversed by increasing the concentration of cytokinin in the medium (52, 53, 116, 117). Hamaguchi et al (53) used a fluorescent analog of a pyrollo [2,3-d] pyrimidine as a probe for cytokinin binding proteins, and found several candidates from cytokinin-dependent tobacco cultures. One of these had a reasonably high affinity (K_d ~ 10^{-7} M) for authentic cytokinins, though the specificity of this binding was not characterized extensively. These results help in the search for cytokinin binding proteins, and they indicate that anticytokinins may be useful in genetic screens for cytokinin receptor mutants.

Clearly, cytokinin receptors have not been identified, much less experimentally manipulated. An issue raised about many of the reports of cytokinin binding proteins is that the K_d for the cytokinin is often orders of magnitude higher than either the endogenous level found in the tissue, or the levels supplied in bioassay. This concern, however, rests on the assumption that the reported concentrations of cytokinin in plant tissues, often in the nanomolar range, are the concentrations available to potential receptors. Endogenous hormone levels usually are presented on a fresh weight basis. This is useful in terms of relative comparisons, but it does not provide the actual concentration at the site of cytokinin action. One question that can be raised is whether cytokinins (or their metabolites) are found in vacuoles, which often account for a large percentage of the fresh weight of a tissue. Morris and coworkers (31) found that developing corn embryos contain high cytokinin levels (~ 0.2 μM). Furthermore, immunocytochemical localization has shown that levels of cytokinin can vary dramatically between cell and tissue types (38, 122). Until the well-developed hormone measuring methods are used to characterize the actual concentration of cytokinins present and available in specific cell types, it is premature to speculate on the biochemical properties of cytokinin receptors.

GENETIC APPROACHES

The use of genetic methods for elucidating mechanisms by which plant hormones are synthesized, metabolized, and/or utilized has been most clearly documented in gibberellin (GA), abscisic acid (ABA), and ethylene (for reviews, see 63, 64, 113). For example, selection for dwarfism in corn led to a series of mutants that affect GA biosynthesis and metabolism. These mutants have been invaluable in determining which is the "active" GA and in characterizing the biochemical nature of the synthetic pathway (106). Another striking example of a successful genetic approach has been with the ethylene response system of *Arabidopsis thaliana* seedlings. In this case, mutants have been isolated that are insensitive to exogenously supplied ethylene, or that act as though ethylene is always present at high levels. Several of these *A. thaliana* genes have been identified and cloned, and molecular manipulation of the system is now possible (14, 62, 110).

Cytokinin Accumulation Mutants

The key to any genetic analysis is the design of selection criteria: What is the expected phenotype caused by the desired mutation? Cytokinin overaccumulation would seem easy to identify. Based on studies of transgenic plants that overexpress the *ipt* gene from *Agrobacterium* (see below), one might expect that at least some mutants would have significantly altered apical dominance, dwarfing, or delayed senescence.

The most dramatic example of cytokinin overproducing mutants are from the moss *Physcomitrella patens*. When protonema of the wild-type, haploid gametophyte are exposed to exogenous cytokinin, they form buds that develop into gametophores (2). Ashton et al (3) isolated a series of mutants (OVE) that overproduced sterile gametophores when cultured in the absence of cytokinin. This indicated they were overaccumulating cytokinin or had a constitutively activated response pathway. Physiological analysis of these mutants showed that they had exceptionally high endogenous cytokinin levels (> 100 times those of wild-type), and that the pool available from tRNAs was not likely to be the source (104, 129). However, biochemical analysis necessary to determine whether these mutations affect cytokinin biosynthetic or metabolic pathways has not yet been achieved. Genetic analysis of these mutants used protoplast fusion methods (44). Somatic diploid hybrids were constructed using the OVE mutants, which were wild-type for amino acid prototrophy, and auxotrophic strains that were wild-type for gametophore production. Of the 15 independently isolated mutants, 14 formed somatic hybrids with wild-type strains that gave a wild-type phenotype for gametophore production, documenting the recessive nature of the OVE alleles. Seven of these recessive mutant strains were shown to belong to three different complementation groups. One OVE strain appeared to result from a dominant mutation because its somatic hybrid with a wild-type auxotroph yielded a gametophore-overproducing phenotype. These studies have not led to the identification of the genes involved in the cytokinin overaccumulation phenotype. Given that protoplast regeneration and culture of the mosses is straightforward, the development of methods to deliver and integrate DNA into them is the next critical technical step that must be accomplished.

Two strategies have been used to select for cytokinin accumulation mutants in *A. thaliana*. Town and coworkers used x-irradiation to induce tumors that were capable of growth in vitro on hormone-free medium. *A. thaliana* stem tissues usually require auxin and cytokinin for continuous growth. Interestingly, none of the hormone-independent tumors induced by x-irradiation contained high levels of cytokinin (19). Thus, they probably resulted from genetic alterations in the cytokinin response system(s) (see below). Chaudhury and coworkers pursued a different strategy (21). They screened mutagenized *A. thaliana* for seedlings that had aberrant cotyledon formation. One such mutant, *amp1* (altered meristem program), produced seedlings with multiple cotyledons and developed into an adult plant with altered patterns of leaf emergence, severely reduced apical dominance, altered floral morphology, and an increased potential to regenerate shoot buds in culture. These phenotypes indicated that this recessive mutation affected cytokinin levels in the plant. Indeed, analysis of endogenous cytokinins showed that Z and (diH)Z were 4- to 8-fold higher in the mutant than in the wild-type. Further physiological and biochemical investigations will be needed to determine whether the increased cytokinin level is the result of increased synthesis or altered metabolism and whether these aberrations are restricted to particular cell or tissue types.

The tumor-prone hybrids of *Nicotiana* spp. are an example of how genetic imbalance may affect cytokinin accumulation. Hybrids between various species of *Nicotiana* spontaneously produce tumors, particularly at wound sites. Smith (120, 121) has extensively studied the genetic basis of tumor formation in these hybrids and their physiology has been reviewed recently (59). Näf (94) originally proposed that *Nicotiana* spp. fall into two groups—plus and minus—and that crosses between them result in tumor-prone hybrids. Further genetic characterization indicated that the plus group plants carry one dominant gene (*I*) involved in tumor induction and the minus group plants carry a recessive gene (or group of genes, *ee*) that influences tumor expression (see 120, 121 for discussion). Cytokinin accumulation—in particular the conversion of cytokinin nucleotides to free bases—has been hypothesized to be critical to cytokinin autotrophic growth of the tumors (47, 96, 97), but detailed biochemical characterization of this has not been achieved. One intriguing report indicated that genes homologous to the *rolB* and *rolC* genes of *A. rhizogenes* found in *Nicotiana glauca* were expressed in tumors of its interspecific hybrid with *Nicotiana langsdorffii*, but not in either parent (58). The genetic aspects of the tumor-prone hybrid system are complicated, but these tumors allow for the examination of the hormonal physiology related to uncontrolled cell proliferation in a system that does not depend on insertion of *Agrobacterium* genes. In principle, they appear to be a source of both biosynthetic enzymes and metabolic enzymes critical to cytokinin accumulation.

In contrast to mutants that accumulate excessive quantities of cytokinin, the lateral suppresser mutant of Craigella tomato appears to contain less cytokinin than does the wild-type. This mutant is characterized by a lack of axillary meristems. ELISA and immunocytochemical analysis indicated that the terminal bud of the mutant contains significantly less cytokinin than the wild-type (73, 122). These results indicate that the *in planta* levels of cytokinin are critical for axillary bud formation by the apical meristem. Interestingly, the cytokinin levels in the root tips of the mutant and wild-type were equivalent (122), indicating that the mutation does not disrupt the cytokinin biosynthetic pathway, but rather affects cell-type specific control of accumulation.

Cytokinin Response Mutants

Alterations in the levels of cytokinins resulting from mutations in synthesis or metabolism are not the only type of mutation that could exhibit phenotypes indicative of changes in the cytokinin physiology of the plant. Another class of hormone mutants are those that survive exposure to inhibitory or toxic concentrations of the hormone. Such mutants are expected to carry alterations in some component of the signal transduction pathway that leads to the toxic response, including the hormone receptor(s).

Cytokinin-resistant mutants have been isolated in *Nicotiana* spp., *Arabidopsis* spp., and *Physcomitrella* spp.. *N. plumbaginifolia* mutants were selected by their ability to survive and exhibit a somewhat normal phenotype in the presence of 20 μM BA (15). This recessive mutation resulted in numerous phenotypes, including germination that was less sensitive to cytokinin, and leaf development that occurred at cytokinin concentrations inhibitory to this process in wild-type. The mutant plants also had a wilty phenotype. Physiological analysis showed that the mutant could not convert ABA-aldehyde to ABA, the water stress–induced signal that controls stomatal closure (102). It is unclear whether the cytokinin resistance causes this ABA metabolic defect or vice versa. It is interesting to note (see below) that mRNA accumulation from cytokinin-inducible genes can also be caused by a variety of stresses (e.g. 46, 126). The relationship between cytokinin biology and stress biology needs further investigation.

Su & Howell (123) developed an innovative selection scheme for cytokinin resistant mutants in *A. thaliana*. EMS-mutagenized M2 seeds were germinated in the presence of a dose of cytokinin (2.5 μM) that would inhibit root growth, a classic cytokinin activity. Seedlings with normal appearing roots were rescued by transfer to cytokinin-free conditions and characterized phenotypically and genetically. Surprisingly, all five cytokinin resistant (*ckr*) mutants isolated in this fashion were in the same complementation group. These *ckr1* mutants were resistant to other cytokinins in the root inhibition assay, but not to auxin, which, at high concentrations, could also inhibit root growth. Grown in the absence of cytokinin, *ckr1* plants exhibited relatively few differences compared to wild-type, except that the rosette leaves were "more cup shaped and yellow" and the roots had altered root hair morphology (123). Further characterization of the relationship between this mutation and cytokinin activity should be intriguing.

Ashton and co-workers (2) isolated several recessive cytokinin-resistant mutants from *Physcomitrella patens*. Auxin treatment reversed many of the mutant phenotypes (e.g. aberrant morphology, little gametophore production) observed in the absence of exogenous cytokinin. These results illustrated, at the genetic level, the critical nature of the interaction between the plant hormones. Mutants selected as resistant to one hormone are often resistant to others. For example, the auxin resistant *A. thaliana* mutant, *axr2,* is also resistant to ethylene, ABA, and cytokinin (64, 130).

Perhaps one of the most intriguing examples of cytokinin response mutants are not, or at least need not be, mutants at all. Cells that normally require an exogenous supply of cytokinin for growth in culture can lose this requirement either spontaneously or as the result of specific inductive treatment, and are called cytokinin habituated (for a review, see 81). In a series of studies on cytokinin habituation in *Nicotiana tabacum* cv Havana 425 (H425 tobacco), Meins and coworkers demonstrated the habituated state in H425 tobacco pith cells can be readily reversed: single cell cytokinin-habituated clones responded to exogenous cytokinin by forming complete, fertile plants. When pith tissue was excised from such regenerated plants and cultured, it again required cytokinin for continuous growth (7, 82, 84). Thus, the habituated phenotype was not the result of a classic mutation, but represented an alternate, stable epigenetic state (for a further discussion, see 81).

Most examples of cytokinin habituation in H425 tobacco have been epigenetic in origin, but two cases of a genetic origin have been described. Contrasting with wild-type tobacco, leaves of plants regenerated from certain cytokinin-habituated lines grew in culture in the absence of cytokinin (83, 85). These studies identified two different habituated leaf (*Hl*) loci as monogenic and either partially (*Hl-1*) or fully (*Hl-2*) dominant. When *Hl-1* plants were inoculated with *Agrobacterium tumefaciens* mutants that lacked the *ipt* gene, fully autonomous tumors developed (54), indicating that the aberrant locus has oncogenic functions similar to the wild-type *ipt* gene. Although extensive genetic and physiological analysis of the *Hl-1* and *Hl-2* mutant plants has been possible, the genes have not been cloned.

The physiological basis of cytokinin habituation is unknown. Several cytokinin habituated lines have been shown to contain elevated (or at least detectable) levels of cytokinin (37, 56) or cytokinin biosynthesis (23, 40), but several others have not (55, 56, 86, 131). For example, crown gall tumors of H425 tobacco induced by wild-type *Agrobacterium* had significant quantities of both iP and *t*-Z derivatives, whereas the cytokinin-habituated tobacco cells (either genetic or epigenetic in nature) did not contain detectable levels of these cytokinins (55, 56, 86). Such results indicated that the cytokinin response system in the cytokinin-habituated lines was constitutively active. This interpretation, however, is not compatible with results showing that expression of the cytokinin responsive gene β-1,3 glucanase was not down-regulated in either cytokinin-habituated or cytokinin-requiring lines unless exogenous cytokinin was present (39, also see below). A similar picture evolves from analysis of bud initiation in cytokinin-habituated lines: stimulation of adventitious shoots from them required the same cytokinin dose that was necessary in the cytokinin-requiring lines (7).

The studies described above indicate that neither cytokinin overproduction nor a general increase in cytokinin responsiveness is responsible for the cytokinin-habituated phenotype. What then is its biochemical basis? Broadly stated, there are two possibilities. First, steps in a cytokinin-mediated cell cycle control system may be constitutively activated while other cytokinin activities are unaffected (Figure 2). Alternatively, controls necessary to advance through the cell cycle that can be activated by cytokinin (49) may also be activated by other environmental cues, and these latter systems are actually involved in the cytokinin habituation process. For example, the oligosaccharide "Nod" factors produced by *Rhizobium* spp. in response to its hosts cause localized cell division in the absence of exogenous cytokinin (30). Perhaps molecules such as these may play a role in habituation. Another possibility is based on the finding that cytokinin-habitutated tobacco lines contained high levels of the dehydrodiconiferyl alcohol glucosides (DCGs) (9). The DCGs are naturally occurring, glucosylated phenolic dimers that can replace cytokinin in tobacco cell division bioassays but not tobacco shoot initiation assays (124). It is unknown whether the DCGs are involved in cytokinin-mediated control of cell division, or in some other control pathway (e.g. wound cell division) that can function in the place of cytokinin. Molecular and genetic manipulation of DCG synthesis or metabolism would provide a critical test of their involvement in habituation, specifically, and plant growth and development, in general.

MOLECULAR APPROACHES

The use of molecular biology to explore cytokinin action is in its earliest stages. The two major approaches have been to (*a*) identify genes whose

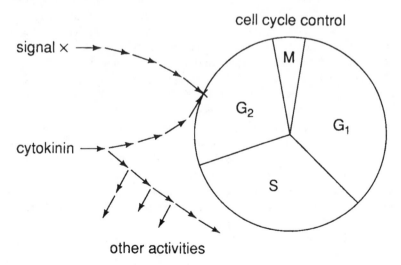

Figure 2 Model illustrating alternative signal transduction pathways for cytokinin control of the cell cycle and other activities (see text).

expression is affected by cytokinin and characterize this control and (*b*) use the *ipt* gene from the T-DNA of *Agrobacterium* in the production of transgenic plants that would (or could) have altered cytokinin levels.

Cytokinin Control of Gene Expression

Given the complexity of cytokinin action at the biological level, complexity at the molecular level is not surprising. As is the case with most plant hormones, activities ascribed to cytokinin are often effected by activities of other signals. One classic example is the role of cytokinin in senescence. In this case effects of light, nutritional status, and other hormones (e.g. ethylene) can modulate the response of particular tissues to cytokinin (51). Similarly, various types of stress responses show complex interactions with cytokinin. For example, salt stress and cytokinin stress independently induced accumulation of mRNA encoding phosphoenolpyruvate carboxylase (126). Several different types of signals are known to affect the expression of the basic form of β-1,3-glucanase in tobacco, all of which seem to affect the abundance of its mRNA. Ethylene induced accumulation of this mRNA, as did wounding, whereas the combination of auxin and cytokinin (but neither alone) repressed expression (29, 45, 46, 89). Chitinase mRNA accumulation showed nearly an identical pattern of hormonal control (114). In tobacco tissue cultures, auxin and cytokinin repressed β-1,3-glucanase expression, but at the same time caused an increase in the amounts of ethylene produced by the tissue (46). These results indicate that the repression of β-1,3 glucanase expression by auxin and cytokinin overrides any inductive effects of ethylene. Another example is nitrate reductase, in which a combination of light, cytokinin, and nitrogen status are involved in transcriptional regulation of gene expression (71, 72, 100). Finally, cytokinin is known to interact with light in the greening of nonpigmented tissues (103). In particular, cytokinin has been shown to induce accumulation of the light harvesting complex (LHCP) mRNA (28a, 48). Nuclear run-on experiments indicated that in *Lemna gibba* the major cytokinin effects on the accumulation of LHCP mRNA are post-transcriptional (48).

Many of the studies described above monitored effects over long periods of time (i.e. days). Thus, these cytokinin-stimulated genes are not likely to represent the primary molecular responses to cytokinin but rather the results of a series of signaling activities. Recent studies have begun to characterize cDNA libraries derived from cells that respond rapidly to cytokinin (26, 33). In the case of cultured soybean cells (26), cDNAs were prepared from mRNA of cells that had been starved for cytokinin for 3 days and then exposed to cytokinin (in the presence of auxin) for 4 h, and cytokinin-inducible cDNA clones were identified by differential hybridization. Using these cDNAs as probes, Crowell et al (26) showed that the levels of the corresponding cytokinin-inducible mRNAs increased 2- to 20-fold within 4 h of cytokinin treatment and preceded the resultant greening and initiation of cell growth. Although two of these clones were identified as ribosomal proteins, most did not show sequence similarity to other known genes. Analysis of mRNA from cells starved for auxin (in the presence of cytokinin) for 3 days followed by auxin treatment showed that many of the cytokinin-inducible genes were also auxin-inducible (26).

The studies described above indicate that genes involved in growth control can be regulated by both auxin and cytokinin and their induction is specific to the process of growth rather than to the hormone. Similarly, cytokinin treatment of *Sebania rostrata* roots induces expression of *SrEnod2* (28). Transcription of this gene was first observed at an early stage of nodule development after interaction of *S. rostrata* roots with symbiotic bacterium, *Azorhizobium caulinodans* (28). It is unknown whether cytokinin, or the oligosaccharide Nod factors (30), or simply growth activation, are the critical signal(s) inducing expression of this gene in the developing nodule.

Dominov et al (33) constructed a cDNA library from BA stimulated *N. plumbaginifolia* cells and characterized in detail one of the cDNAs (pLS216). This cDNA has sequence similarity to auxin-induced and stress-induced genes from other plants. The transcript homologous to it is rapidly, but transiently, up-regulated by either BA or the auxin 2,4-dichlorophenoxyacetic acid (2,4-D), each hormone yielding a different time course of induction. Studies on the transient nature of this induction showed that 24 h after 2,4-D treatment, the mRNA levels had dropped back to near control levels. Addition of fresh 2,4-D to cells that had been exposed to it for 24 h would not re-stimulate transcription of this gene, although BA treatment of 2,4-D treated cells did. These results indicate that the cells can become temporarily desensitized—but only to the hormone originally used to induce them. These kinds of experiments, as well as those described above, indicate the complexity of hormonal interactions at a molecular level and are reminiscent of the complexities of hormonal interactions described in genetic, physiological, and cellular studies. Determining how the hormones interact—either cooperatively, independently, or antagonistically—to affect gene expression and how this leads to or controls the processes in question will be a crucial goal over the next several years.

Transgenic Plants Expressing the ipt Gene from Agrobacterium tumefaciens

The ability to generate transgenic plants represents one of the most powerful genetic tools developed over the past decade (for reviews, see 65, 108). The availability of the *ipt* sequence of *A. tumefaciens* has allowed several investigators to insert it into plant cells, either under the control of its own promoter or behind other types of plant promoters. *N. tabacum* and *Solanum tuberosum* tissues transformed with the *ipt* gene spontaneously produced shoots, but the transformed shoots could not form roots as expected for cytokinin overproducing cells (36, 61, 111). A somewhat surprising result was observed when *N. glutinosa* cells were transformed with the *ipt* gene: the transformed tissue was unorganized and capable of growing in the absence of either auxin or cytokinin (8). This was unexpected because exogenous cytokinin could not support the growth of nontransformed *N. glutinosa* cells in the absence of auxin. These results indicate that cells producing cytokinin may respond differently to its signals than do cells not producing this hormone, and provide even more reason to manipulate the levels of endogenous cytokinin to determine how it functions in plant development.

The first transgenic plants carrying the *ipt* gene were actually shoots derived from *A. tumefaciens* strain T37–transformed tobacco teratomas. These shoots could not form roots and were therefore grafted onto healthy host root systems (16). They exhibited many of the phenotypes associated with cytokinin overproduction such as reduced apical dominance and spontaneous shoot formation at wound sites. Otherwise they were remarkably normal: the grafted shoot systems, although slightly thicker than usual, produced normal leaves and were fertile. When the leaf tissues were excised and cultured, they grew as typical teratomatic tumors (16). Thus, the intact regenerated shoots were more capable of normal growth in the presence of excess cytokinin than were the cell cultures. These studies, and others like it (36) had the drawback that expression of the *ipt* gene was from its own promoter, resulting in constitutive expression and the abnormalities described above. Yusibov et al (132), who obtained complete, rooted plants carrying the *ipt* gene, reported reduced expression of the *ipt* gene in the regenerated plants and significantly lower levels of cytokinins than had been reported previously in both tumors and grafted *ipt*-containing shoots. The low expression and low cytokinin levels apparently allowed root formation to occur. In experiments with an *ipt* promoter fused to the β-glucuronidase (GUS) coding sequence, Dymcock et al showed that expression was strongly influenced by tissue type and plant culture conditions (36).

The effects of *ipt* on normal plant tissues have been studied using regulatable promoters to control its expression. For example, the *ipt* coding sequence has been fused to promoters of heat shock genes (*hsp*) in several instances (80, 111, 118, 119). Plants carrying such constructs formed normal roots, but were generally smaller and more highly branched, had darker green leaves, and contained higher cytokinin levels than did control plants. Variation in these features apparently reflected the expression of the *ipt* gene in non-heat shock conditions. When heat shock was applied, the *ipt* expression and cytokinin levels increased, but the response to this treatment varied. For example, Smigocki (119) noted that heat shock caused increased lateral branching. In contrast, Medford et al (80) saw no change in the phenotype of *hsp-ipt* plants heat shocked at weekly intervals over a twelve week period even though cytokinin concentration increased dramatically. These results indicate that the absolute concentration of cytokinin in the plants is not the crucial factor controlling their development (64). One tempting speculation is that the *hsp-ipt* plants have become desensitized to cytokinin as a result of the higher cytokinin concentrations found in the non–heat shocked condition, resulting in little effect from further increased cytokinin levels. An alternate model has been presented of the effects of increased cytokinin levels on cytokinin sensitivity, based on analysis of transgenic potato plants carrying the intact *ipt* gene (101).

In the studies described above, entire plants were heat shocked. To analyze the effects of a more localized increase in cytokinins, Smart et al (118) heat shocked only a small part of a single leaf and then monitored cytokinin levels and phenotypic effects thereafter. Again, the non–heat shocked plants were shorter, darker green, and had more released axillary shoots than did controls. As expected, localized heat shock caused the treated area to produce more cytokinin. Interestingly, the nontreated areas of the same leaf had only slightly higher levels, indicating that the cytokinin does not readily move from the site of production. The heat shocked areas stayed greener than did the rest of the leaf and this treatment accelerated senescence of leaves higher up the stem. These experiments provide strong evidence that local changes in cytokinin level can have global consequences for the plant, perhaps by mobilizing resources to the site of cytokinin production (118).

Problems encountered with the use of *hsp* promoters are that (*a*) heat shock could itself affect plant responses and (*b*) the promoters appear leaky—low levels of expression from uninduced plants were enough to have profound effects on the plant. To avoid heat shock as the means of increasing cytokinin level, Li et al (68) fused the *ipt* coding sequence to one end of the bi-directional small-auxin-up-regulated (SAUR) promoter and GUS to the other end and introduced it into tobacco plants. This study indicated a correlation between the tissue specific expression of the SAUR-GUS construct and cytokinin levels in the tissue. It further showed that the transformed plants had a variety of phenotypes typically associated with increased cytokinin levels such as dwarfing, reduced apical dominance, and reduced root initiation. Although certain phenotypes indicated a local response to locally synthesized cytokinins (e.g. numerous adventitious buds forming over the petioles and leaf veins), the expression from the SAUR promoter was found in tissues throughout the plant. Thus, the contribution of local cytokinin products in generating the observed whole plant phenotypes was unclear. Determination of a clearer relationship between local cytokinin production and its effect was attempted by placing a transposable element (*Ac*) between a 35S cauliflower mosaic virus (CAMV) promoter and the *ipt* coding sequence (43). The *ipt* gene could only be transcribed in the transgenic material when the *Ac* element transposed away from the construct. In this case, adventitious buds arose spontaneously from leaves of transgenic plants that appeared normal otherwise. Most of these buds expressed the *ipt* gene. These results indicate that transposition of *Ac* away from the 35S-*Ac-ipt* construct causes expression of *ipt* and cytokinin production, which could then induce bud formation in cells that had not been activated by *Ac* transposition (43). This confirmed earlier clonal analyses of crown gall tumors, which showed that hormones produced by *Agrobacterium*-transformed cells can cause growth and shoot formation in nearby non-transformed cells (10, 12, 128).

CONCLUDING REMARKS

The use of genetics and molecular genetics to elucidate mechanisms of cytokinin action in plants lags behind the experimental examination of similar problems for other plant hormones. Because cytokinin has numerous complex activities essential to plant growth and development, mutants affecting cytokinin synthesis, metabolism, and response are difficult to identify. The material reviewed here indicates, however, that inroads toward genetic manipulation of cytokinin activities are being made. The isolation of a cDNA that encodes a cytokinin metabolic enzyme has been accomplished in one case and should be achievable soon in several other cases. It is only a matter of time before such genes are expressed in transgenic plants under the control of regulatable promoters. Because such a situation should be cell autonomous (it is unlikely that metabolic enzymes would move out of the cell), this type of experiment should provide information on how localized depletion of cytokinin affects the particular tissue or cell type expressing the gene. The outlook concerning the effects of controlled cytokinin synthesis is also hopeful. Use of the *ipt* gene to provide regulatable cytokinin production has been achieved and further refinements of this strategy are likely to provide more detailed information. The control of cytokinin biosynthesis by plant-encoded enzymes is still unresolved. Perhaps cytokinin overproducing mutants will provide a way to identify the genes involved in this crucial process.

The issue of cytokinin response is less clear. cDNAs representing cytokinin responsive genes, generally of unknown function, have been isolated, and analysis of the cytokinin-resistant mutants is just beginning to provide information on the relationship between cytokinin and other signaling systems such as stress. Clear advances in this area have yet to be made. Perhaps more attention should be given to cell culture systems. Hayashi et al constructed a T-DNA vector that, when inserted into the plant genome, would cause over-expression of the adjacent gene (57). Tobacco leaf mesophyll cells were co-cultivated with Agrobacteria carrying this construct and then selected for auxin-autonomous growth. This method proved successful and eleven genes involved in auxin control of growth are now being characterized. It is likely that such mutants could not have been selected at the whole plant level, because the intact plant is already auxin autonomous. Plants regenerated from these cells were normal. Such results indicate the complexity of systems that the plant, in contrast to cultured cells, uses to control cell division. Another experimental cell culture system is cytokinin habituation. As described above, at least some of these cell lines survive and grow in the absence of cytokinin and have neither an increased endogenous cytokinin content nor a general increase in cytokinin sensitivity. This indicates that these habituated cells, in contrast to intact plants, require neither cytokinin nor at least certain cytokinin responses for survival. It should be possible to devise selection strategies that use such cells to uncover mutations in cytokinin response systems. These types of novel genetic strategies using cultured cells should provide useful alternatives to studies carried out on whole plants in the efforts to elucidate mechanisms of cytokinin action.

ACKNOWLEDGMENTS

I would like to thank the numerous colleagues who sent preprints of in press material and who took the time to discuss various issues raised in this review. I also would like to thank Dr. Michael Savka for reading an earlier version of this manuscript. Finally, I thank the National Science Foundation and the National Institutes of Health for providing support for my laboratory to study many of the issues discussed here.

Literature Cited

1. Akiyoshi DE, Klee H, Amasino RM, Nester EW, Gordon MP. 1984. T-DNA of *Agrobacterium tumefaciens* encodes an enzyme of cytokinin biosynthesis. *Proc. Natl. Acad. Sci. USA* 81:5994–98
2. Ashton NW, Cove DJ, Featherstone DR. 1979. The isolation and physiological analysis of mutants of the moss, *Physcomitrella patens,* which over-produce gametophores. *Planta* 144:437–42
3. Ashton NW, Grimsley NH, Cove DJ. 1979. Analysis of gametophytic development in the moss, *Physcomitrella patens,* using auxin and cytokinin resistant mutants. *Planta* 144:427–35
4. Auer CA, Cohen JD. 1993. Identification of a benzyladenine disaccharide conjugate produced during shoot organogenesis in *Petunia* leaf explants. *Plant Physiol.* 102: 541–45
5. Barry GF, Rogers SG, Fraley RT, Brand L. 1984. Identification of a cloned cytokinin biosynthetic gene. *Proc. Natl. Acad. Sci. USA* 81:4776–80
6. Bassil NV, Mok DWS, Mok MC. 1993. Partial purification of a *cis-trans*-isomerase of zeatin from immature seed of *Phaseolus vulgaris* L. *Plant Physiol.* 102:867–72
7. Binns A, Meins F Jr. 1973. Evidence that habituation of tobacco pith cells for cell division factors is heritable and potentially reversible. *Proc. Natl. Acad. Sci. USA* 70: 2660–62

8. Binns AN, Labriola J, Black RC. 1987. Initiation of auxin autonomy in *Nicotiana glutinosa* cells by the cytokinin-biosynthesis gene from *Agrobacterium tumefaciens*. *Planta* 171:539–48

9. Binns AN, Roman G, Teutonico RA. 1992. Accumulation of dehydrodiconiferyl alcohol glucosides in cytokinin habituated cell lines. In *Physiology and Biochemistry of Cytokinins in Plants*, ed. M Kamínek, DWS Mok, E Zazímalová, pp. 59–64. The Hague: SPB Academic

10. Binns AN, Sciaky D, Wood HN. 1982. Variation in hormone autonomy and regeneration potential of cells transformed by strain A66 of *Agrobacterium tumefaciens*. *Cell* 31:605–12

11. Binns AN, Thomashow MF. 1988. Cell biology of *Agrobacterium* infection and transformation of plants. *Annu. Rev. Microbiol.* 42:575–606

12. Binns AN, Wood HN, Braun AC. 1981. Suppression of the tumorous state in crown gall teratomas of tobacco: a clonal analysis. *Differentiation* 19:97–102

13. Blatt MR, Thiel G. 1993. Hormonal control of ion channel gating. *Annu. Rev. Plant Physiol. Plant Mol. Biol.* 44:543–67

14. Bleecker AB, Estelle MA, Somerville C, Kende H. 1988. Insensitivity to ethylene conferred by a dominant mutation in *Arabidopsis thaliana*. *Science* 241:1086–89

15. Blonstein AD, Parry AD, Horgan RH, King PJ. 1991. A cytokinin-resistant mutant of *Nicotiana plumbaginifolia* is wilty. *Planta* 183:244–50

16. Braun AC, Wood HN. 1976. Suppression of the neoplastic state with the acquisition of specialized functions in cells, tissues, and organs of crown gall teratomas of tobacco. *Proc. Natl. Acad. Sci. USA* 73:496–500

17. Brinegar AC, Cooper G, Stevens A, Hauer CR, Shabanowitz J, et al. 1988. Characterization of a benzyladenine binding-site peptide isolated from a wheat cytokinin-binding protein: sequence analysis and identification of a single affinity-labeled residue by mass spectrometry. *Proc. Natl. Acad. Sci. USA* 85:5927–31

18. Buchmann I, Marner F-J, Schröder G, Waffenschmidt S, Schröder J. 1985. Tumour genes in plants: T-DNA encoded cytokinin biosynthesis. *EMBO J.* 4:853–59

19. Campell BR, Town CD. 1991. Physiology of hormone autonomous tissue lines derived from radiation-induced tumors of *Arabidopsis thaliana*. *Plant Physiol.* 97:1166–73

20. Chatfield JM, Armstrong DJ. 1986. Regulation of cytokinin oxidase activity in callus tissues of *Phaseolus vulgaris* L. cv Great Northern. *Plant Physiol.* 80:493–99

21. Chaudhury AM, Letham S, Craig S, Dennis ES. 1994. *amp 1*-a mutant with high cytokinin levels and altered embryonic pattern, faster vegetative growth, constitutive photomorphogenesis, and precocious flowering. *Plant J.* 4:In press

22. Chen C. 1982. Cytokinin biosynthesis in cell free systems. In *Plant Growth Substances*, ed. PF Wareing, pp. 155–64. New York: Academic

23. Chen C, Melitz DK. 1979. Cytokinin biosynthesis in a cell-free system from cytokinin-autotrophic tobacco tissue cultures. *FEBS Lett.* 107:15–20

24. Chrispeels MJ. 1991. Sorting of proteins in the secretory system. *Annu. Rev. Plant Physiol. Plant Mol. Biol.* 42:21–53

25. Crespi M, Messens E, Caplan AB, Van Montagu M, Desomer J. 1992. Fasciation induction by the phytopathogen *Rhodococcus fascians* depends upon a linear plasmid encoding a cytokinin synthase gene. *EMBO J.* 11:795–804

26. Crowell DN, Kadlecek AT, Manorama CJ, Amasino RM. 1990. Cytokinin-induced mRNAs in cultured soybean cells. *Proc. Natl. Acad. Sci. USA* 87:8815–19

27. Davies PJ, ed. 1987. *Plant Hormones and their Role in Plant Growth and Development*. Dordrecht: Nijhoff. 681 pp.

28. Dehio C, de Bruijn FJ. 1992. The early nodulin gene *SrEnod2* from *Sesbania rostrata* is inducible by cytokinin. *Plant J.* 2:117–28

28a. de la Serve BT, Axelos M, Péaud-Lenoel C. 1985. Cytokinins modulate the expression of genes encoding the protein of the light-harvesting chlorophyll α/β complex. *Plant Mol. Biol.* 5:155–63

29. De Loose M, Alliotte T, Gheysen G, Genetello C, Gielen J, et al. 1988. Primary structure of a hormonally regulated β-glucanase of *Nicotiana plumbaginifolia*. *Gene* 70:13–23

30. Dénarié J, Debellé F, Rosenberg C. 1992. Signaling and host range variation in nodulation. *Annu. Rev. Microbiol.* 46:497–531

31. Dietrich JT, Morris RO, Kaminek M, Blevins DG, Reinbott TM. 1993. Cytokinins and grain development in maize. *Plant Physiol.* 102:23 (Suppl.)

32. Dixon SC, Martin RC, Mok MC, Shaw G, Mok DWS. 1989. Zeatin glycosylation enzymes in *Phaseolus*. *Plant Physiol.* 90:1316–21

33. Dominov JA, Stenzler L, Lee S, Schwarz JJ, Leisner S, et al. 1992. Cytokinins and auxins control the expression of a gene in *Nicotiana plumbaginifolia* cells by feedback regulation. *Plant Cell* 4:451–61

34. Durand B, Durand R. 1991. Male sterility and restored fertility in annual mercuries, relations with sex differentiation. *Plant Sci.* 80:107–18

35. Durand B, Durand R. 1991. Sex determina-

tion and reproductive organ differentiation in *Mercurialis. Plant Sci.* 80:49–65

36. Dymock D, Risiott R, de Pater S, Lancaster J, Tillson P, et al. 1991. Regulation of *Agrobacterium tumefaciens* T-*cyt* gene expression in leaves of transgenic potato (*Solanum tuberosum* L. cv. Désirée) is strongly influenced by plant culture conditions. *Plant Mol. Biol.* 17:711–25

37. Dyson WH, Hall RH. 1972. $N^6(\Delta^2$-isopentenyl)adenosine: its occurrence as a free nucleoside in a cytokinin autonomous strain of tobacco tissue. *Plant Physiol.* 50: 616–21

38. Eberle J, Wang TL, Cook S, Wells B, Weiler EW. 1987. Immunoassay and ultrastructural localization of isopentenyladenine and related cytokinins using monoclonal antibodies. *Planta* 172:289–97

39. Eichholz R, Harper J, Felix G, Meins F Jr. 1983. Evidence for an abundant 33,000-dalton polypeptide regulated by cytokinins in cultured tobacco tissues. *Planta* 158: 410–15

40. Einset JW, Skoog F. 1973. Biosynthesis of cytokinin in cytokinin autotrophic callus. *Proc. Natl. Acad. Sci. USA* 70:658–60

41. Erion JL, Fox JE. 1981. Purification and properties of a protein which binds cytokinin-active 6-substituted purines. *Plant Physiol.* 67:156–62

42. Estruch JJ, Chriqui D, Grossmann K, Schell J, Spena A. 1991. The plant oncogene *rolC* is responsible for the release of cytokinins from glucoside conjugates. *EMBO J.* 10:2889–95

43. Estruch JJ, Prinsen E, Onckelen HV, Schell J, Spena A. 1991. Viviparous leaves produced by somatic activation of an inactive cytokinin-synthesizing gene. *Science* 254: 1364–67

44. Featherstone DR, Cove DJ, Ashton NW. 1990. Genetic analysis by somatic hybridization of cytokinin overproducing developmental mutants of the moss, *Physcomitrella patens. Mol. Gen. Genet.* 222:217–24

45. Felix G, Meins F Jr. 1986. Developmental and hormonal regulation of β-1,3-glucanase in tobacco. *Planta* 167:206–11

46. Felix G, Meins F Jr. 1987. Ethylene regulation of β-1,3-glucanase in tobacco. *Planta* 172:386–92

47. Feng X-H, Dube SK, Bottino PJ, Kung S. 1990. Restoration of shooty morphology of a nontumorous mutant of *Nicotiana glauca* × *N. langsdorffii* by cytokinin and the isopentenyltransferase gene. *Plant Mol. Biol.* 15:407–20

48. Flores S, Tobin EM. 1988. Cytokinin modulation of LHCP mRNA levels: the involvement of post-transcriptional regulation. *Plant Mol. Biol.* 11:409–15

49. Fosket DE, Short KC. 1973. The role of cytokinin in the regulation of growth, DNA synthesis and cell proliferation in cultured soybean tissues (*Glycine max* var. Biloxi). *Physiol. Plant.* 28:14–23

50. Gallie DR. 1993. Posttranscriptional regulation of gene expression in plants. *Annu. Rev. Plant Physiol. Plant Mol. Biol.* 44:77–105

51. Goldthwaite JJ. 1987. Hormones in plant senescence. See Ref. 27, pp. 553–73

52. Gregorini G, Laloue M. 1980. Biological effects of cytokinin antagonists 7-(pentylamino) and 7-(benzylamino)- 3-methyl-pyrazolo(4,3-*d*)pyrimidines on suspension-cultured tobacco cells. *Plant Physiol.* 65:363–67

53. Hamaguchi N, Iwamura J, Fujita T. 1985. Fluorescent anticytokinins as a probe for binding. *Eur. J. Biochem.* 153:565–72

54. Hansen CE, Meins F Jr. 1986. Evidence for a cellular gene with potential oncogenic activity in plants. *Proc. Natl. Acad. Sci. USA* 83:2492–95

55. Hansen CE, Meins F Jr, Aebi R. 1987. Hormonal regulation of zeatin-riboside accumulation by cultured tobacco cells. *Planta* 172:520–25

56. Hansen CE, Meins F Jr, Milani A. 1985. Clonal and physiological variation in the cytokinin content of tobacco cell lines differing in cytokinin requirement and capacity for neoplastic growth. *Differentiation* 29:1–6

57. Hayashi H, Czaja I, Lubenow H, Schell J, Walden R. 1992. Activation of a plant gene by T-DNA tagging: auxin-independent growth in vitro. *Science* 258:1350–53

58. Ichikawa T, Ozeki Y, Syono K. 1990. Evidence for the expression of the *rol* genes of *Nicotiana glauca* in genetic tumors of *N. glauca* × *N. langsdorffii. Mol. Gen. Genet.* 220:177–80

59. Ichikawa T, Syono K. 1991. Tobacco genetic tumors. *Plant Cell Physiol.* 32:1123–28

60. Ihara M, Taya Y, Nishimura S, Tanaka Y. 1984. Purification and some properties of Δ^2-Isopentenylpyrophosphate:5'AMP Δ^2-Isopentenyltransferase from the cellular slime mold *Dictyostelium discoideum. Arch. Biochem. Biophys.* 230:652–60

61. Inzé D, Follin A, Van Lijsbettens M, Simoens C, Genetello C, et al. 1984. Genetic analysis of individual genes of *Agrobacterium tumefaciens*: further evidence that two genes are involved in indole-3-acetic acid synthesis. *Mol. Gen. Genet.* 194:265–74

62. Kieber JJ, Rothenberg M, Roman G, Feldman KA, Ecker JR. 1993. *CTR1*, a negative regulator of the ethylene response pathway in Arabidopsis, encodes a member of the Raf family of protein kinases. *Cell* 72:427–41

63. King PJ. 1988. Plant hormone mutants. *Trends Genet.* 4:157–62
64. Klee H, Estelle M. 1991. Molecular genetic approaches to plant hormone biology. *Annu. Rev. Plant Physiol. Plant Mol. Biol.* 42:529–51
65. Klee HJ, Horsch R, Rogers S. 1987. Agrobacterium mediated plant transformation and its further applications to plant biology. *Annu. Rev. Plant Physiol.* 38:467–86
66. Lee Y, Mok MC, Mok DWS, Griffin DA, Shaw G. 1985. Cytokinin metabolism in *Phaeolus* embryos—genetic difference and the occurrence of novel zeatin metabolites. *Plant Physiol.* 77:635–41
67. Letham DS, Palni LMS. 1983. The biosynthesis and metabolism of cytokinins. *Annu. Rev. Plant Physiol.* 34:163–97
68. Li Y, Hagen G, Guilfoyle TJ. 1992. Altered morphology in transgenic tobacco plants that overproduce cytokinins in specific tissues and organs. *Dev. Biol.* 153:386–95
69. Libbenga KR, Mennes AM. 1987. Hormone binding and its role in hormone action. See Ref. 27, pp. 194–221
70. Louis J-P, Augur C, Teller G. 1990. Cytokinins and differentiation processes in *Mercurialis annua. Plant Physiol.* 94:1535–41
71. Lu J, Ertl JR, Chen C. 1990. Cytokinin enhancement of the light induction of nitrate reductase transcript levels in etiolated barley leaves. *Plant Mol. Biol.* 14:585–94
72. Lu J, Ertl JR, Chen C. 1992. Transcriptional regulation of nitrate reductase mRNA levels by cytokinin-abscisic acid interactions in etiolated barley leaves. *Plant Physiol.* 98:1255–60
73. Maldiney R, Pelese F, Pilate G, Sotta B, Sossountzov L, et al. 1986. Endogenous levels of abscissic acid, indole-3-acetic acid, zeatin and zeatin riboside during the course of adventitious root formation in cuttings of Craigella and Craigella lateral suppressor tomatoes. *Physiol. Plant.* 68:426–30
74. Maréchal-Drouard L, Weil JH, Dietrich A. 1993. Transfer RNAs and transfer RNA genes in plants. *Annu. Rev. Plant Physiol. Plant Mol. Biol.* 44:13–32
75. Martin RC, Martin RR, Mok MC, Mok DWS. 1990. A monoclonal antibody specific to zeatin O-glycosyltransferases of *Phaeolus. Plant Physiol.* 94:1290–94
76. Martin RC, Mok MC, Mok DWS. 1993. Cytolocalization of zeatin O-xylosyltransferase in *Phaseolus. Proc. Natl. Acad. Sci. USA* 90:953–57
77. Martin RC, Mok MC, Mok DWS. 1993. Localization and cloning of a cytokinin metabolic enzyme, zeatin O-xylosyltransferase. *Plant Physiol.* 102:24 (Suppl.)
78. Martin RC, Mok MC, Shaw G, Mok DWS. 1989. An enzyme mediating the conversion

of zeatin to dihydrozeatin in *Phaseolus* embryos. *Plant Physiol.* 90:1630–35
79. McGaw BA. 1987. Cytokinin biosynthesis and metabolism. See Ref. 27, pp. 76–93
80. Medford JI, Horgan R, El-Sawi Z, Klee HJ. 1989. Alterations of endogenous cytokinins in transgenic plants using a chimeric isopentenyl transferase gene. *Plant Cell* 1: 403–13
81. Meins F Jr. 1989. Habituation: heritable variation in the requirement of cultured plant cells for hormones. *Annu. Rev. Genet.* 23:395–408
82. Meins F Jr, Binns AN. 1982. Rapid reversion of cell-division-factor habituated cells in culture. *Differentiation* 23:10–12
83. Meins F Jr, Foster R. 1986. A cytokinin mutant derived from cultured tobacco cells. *Dev. Genet.* 7:159–65
84. Meins F Jr, Foster R. 1986. Trans-determination of plant cells. *Differentiation* 30:188–89
85. Meins F Jr, Foster R, Lutz J. 1983. Evidence for a Mendelian factor controlling the cytokinin requirement of cultured tobacco cells. *Dev. Genet.* 4:129–41
86. Meins F Jr, Hansen CE. 1992. Elevated *trans*-zeatin riboside content is not required for the neoplastic growth of Ti plasmid-transformed tobacco cells. In *Physiology and Biochemistry of Cytokinins in Plants,* ed. M Kamínek, DWS Mok, E Zazimalová, pp. 83–85. The Hague: SPB Academic
87. Miller CO. 1974. Ribosyl-*trans*-zeatin, a major cytokinin produced by crown gall tumor tissue. *Proc. Natl. Acad. Sci. USA* 71:334–38
88. Miller CO, Skoog F, Von Saltza MH, Strong F. 1955. Kinetin, a cell division factor from deoxyribonucleic acid. *J. Am. Chem. Soc.* 77:1392–93
89. Mohnen D, Shinshi H, Felix G, Meins F Jr. 1985. Hormonal regulation of β 1,3-glucanase messenger RNA levels in cultured tobacco tissues. *EMBO J.* 4:1631–35
90. Mok DWS, ed. 1994. *Cytokinins—Chemistry, Activity and Function.* Boca Raton, FL: CRC. In press
91. Mok MC, Mok DWS, Armstrong DJ, Shudo K, Tsogai Y, et al. 1982. Cytokinin activity of N-phenyl-N'-1,2,3-thiadiazol-5-ylurea (thidiazuron). *Phytochemistry* 21: 1509–11
92. Momotani E, Tsuji H. 1992. Isolation and characterization of a cytokinin-binding protein from the water-soluble fraction of tobacco leaves. *Plant Cell Physiol.* 33: 407–12
93. Morris RO. 1986. Genes specifying auxin and cytokinin biosynthesis in phytopathogens. *Annu. Rev. Plant Physiol.* 37:509–38
94. Näf U. 1958. Studies in tumor formation in *Nicotiana* hybrids I. The classification of

parents into two etiologically significant groups. *Growth* 22:167–80

95. Nagata R, Kawachi E, Hashimoto Y, Shudo K. 1993. Cytokinin-specific binding protein in etiolated mung bean seedlings. *Biochem. Biophys. Res. Commun.* 191:543–49

96. Nandi SK, de Klerk GJM, Parker CW, Palni LMS. 1990. Endogenous cytokinin levels and metabolism of zeatin riboside in genetic tumour tissues and non-tumourous tissues of tobacco. *Physiol. Plant.* 78:197–204

97. Nandi SK, Palni LMS, Parker CW. 1990. Dynamics of endogenous cytokinins during the growth cycle of a hormone-autotrophic genetic tumor line of tobacco. *Plant Physiol.* 94:1084–89

98. Nilsson O, Moritz T, Imbault N, Sandberg G, Olsson O. 1993. Hormonal characterization of transgenic tobacco plants expressing the *rolC* gene of *Agrobacterium rhizogenes* T$_L$-DNA. *Plant Physiol.* 102:363–71

99. Oeller PW, Min-Wang L, Pike DA, Theologis A. 1991. Reversible inhibition of tomato fruit senescence. *Science* 254:437–39

100. Oelmüller R, Briggs WR. 1990. Intact plastids are required for nitrate- and light-induced accumulation of nitrate reduced activity and mRNA in squash cotyledons. *Plant Physiol.* 92:434–39

101. Ooms G, Risiott R, Kendall A, Keys A, Lawlor D, et al. 1991. Phenotypic changes in T-*cyt*-transformed potato plants are consistent with enhanced sensitivity of specific cell types to normal regulation by root-derived cytokinin. *Plant Mol. Biol.* 17:727–43

102. Parry AD, Blonstein AD, Babiano MJ, King PJ, Horgan R. 1991. Abscisic-acid metabolism in a wilty mutant of *Nicotiana plumbaginifolia*. *Planta* 183:237–43

103. Parthier B. 1979. The role of phytohormones (cytokinins) in chloroplast development. *Biochem. Physiol. Pflanzen* 174:173–214

104. Perry KC, Cove DJ. 1986. Transfer RNA pool sizes and half lives in wild-type and cytokinin over-producing strains of the moss *Physcomitrella patens*. *Physiol. Plant.* 67:680–84

105. Phillips IDJ. 1975. Apical dominance. *Annu. Rev. Plant Physiol.* 26:341–67

106. Phinney BO, Spray CR. 1982. Chemical genetics and the gibberellin pathway in *Zea mays* L. In *Plant Growth Substances*, ed. PF Wareing, pp. 101–10. New York: Academic

107. Polya GM, Davis AW. 1978. Properties of a high-affinity cytokinin-binding protein from wheat germ. *Planta* 139:139–47

108. Potrykus I. 1991. Gene transfer to plants: assessment of published approaches and results. *Annu. Rev. Plant Physiol. Plant Mol. Biol.* 42:205–25

109. Romanov GA, Taran VY, Venis MA. 1990. Cytokinin-binding protein from maize shoots. *J. Plant Physiol.* 136:208–12

110. Rothenberg M, Ecker J. 1993. Mutant analysis as an experimental approach towards understanding plant hormone action. *Dev. Biol.* 4:3–13

111. Schmülling T, Beinsberger S, De Greef J, Schell J, Van Onckelen H, et al. 1989. Construction of a heat-inducible chimaeric gene to increase the cytokinin content in transgenic plant tissue. *FEBS Lett.* 249:401–6

112. Schmülling T, Fladung M, Grossmann K, Schell J. 1993. Hormonal content and sensitivity of transgenic tobacco and potato plants expressing single *rol* genes of *Agrobacterium rhizogenes* T-DNA. *Plant J.* 3:371–82

113. Scott IM. 1990. Plant hormone response mutants. *Physiol. Plant.* 78:147–52

114. Shinshi H, Mohnen D, Meins F Jr. 1987. Regulation of a plant pathogenesis-related enzyme: inhibition of chitinase and chitinase mRNA accumulation in cultured tobacco tissues by auxin and cytokinin. *Proc. Natl. Acad. Sci. USA* 84:89–93

115. Skoog F, Miller CO. 1957. Chemical regulation of growth and organ formation in plant tissues cultured in vitro. *Symp. Soc. Exp. Biol.* 11:118–31

116. Skoog F, Schmitz RY, Bock RM, Hecht SM. 1973. Cytokinin antagonists: synthesis and physiological effects of 7-substituted 3-methylpyrazolo[4,3-*d*]pyrimidines. *Phytochemistry* 12:25–37

117. Skoog F, Schmitz RY, Hecht SM, Frye RB. 1975. Anticytokinin activity of substituted pyrrolo[2,3-*d*]pyrimidines. *Proc. Natl. Acad. Sci. USA* 72:3508–12

118. Smart CM, Scofield SR, Bevan MW, Dyer TA. 1991. Delayed leaf senescence in tobacco plants transformed with *tmr*, a gene for cytokinin production in *Agrobacterium*. *Plant Cell* 3:647–56

119. Smigocki AC. 1991. Cytokinin content and tissue distribution in plants transformed by a reconstructed isopentenyl transferase gene. *Plant Mol. Biol.* 16:105–15

120. Smith HH. 1972. Plant genetic tumors. *Prog. Exp. Tumor Res.* 15:138–59

121. Smith HH. 1988. The inheritance of genetic tumors in *Nicotiana* hybrids. *J. Hered.* 79: 277–83

122. Sossountzov L, Maldiney R, Sotta B, Sabbagh I, Habricot Y, et al. 1988. Immunocytochemical localization of cytokinins in Craigella tomato and a sideshootless mutant. *Planta* 175:291–304

123. Su W, Howell SH. 1992. A single genetic locus, Ckr1, defines Arabidopsis mutants in which root growth is resistant to low con-

centrations of cytokinin. *Plant Physiol.* 99: 1569–74

124. Teutonico RA, Dudley MW, Orr JD, Lynn DG, Binns AN. 1991. Activity and accumulation of cell division-promoting phenolics in tobacco tissue cultures. *Plant Physiol.* 97:288–97

125. Deleted in proof

126. Thomas JC, McElwain EF, Bohnert HJ. 1992. Convergent induction of osmotic stress-responses: abscisic acid, cytokinin, and the effects of NaCl. *Plant Physiol.* 100: 416–23

127. Turner JE, Mok DWS, Mok MC, Shaw G. 1987. Isolation and partial purification of the enzyme catalyzing the formation of novel zeatin metabolites in *Phaseolus vulgaris* embryo. *Proc. Natl. Acad. Sci. USA* 84:3714–17

128. Van Slogteren GMS, Hoge JHC, Hooykaas PJJ, Schilperoort RA. 1983. Clonal analysis of heterogeneous crown gall tumor tissues induced by wild type and shooter mutant strains of *Agrobacterium tumefaciens* expression of T-DNA genes. *Plant Mol. Biol.* 2:321–33

129. Wang TL, Buetelmann P, Cove DJ. 1981. Cytokinin biosynthesis in mutants of the moss *Physcomitrella patens. Plant Physiol.* 68:739–44

130. Wilson AK, Pickett FB, Turner JC, Estelle M. 1990. A dominant mutation in *Arabidopsis* confers resistance to auxin, ethylene and abscisic acid. *Mol. Gen. Genet.* 222: 377–83

131. Wyndaele R, Christiansen J, Horseele R, Rüdelsheim P, Onckelen HV. 1988. Functional correlation between endogenous phytohormone levels and hormone autotrophy of transformed and habituated soybean cell lines. *Plant Cell Physiol.* 29:1095–101

132. Yusibov VM, Il PC, Andrianov VM, Piruzian ES. 1991. Phenotypically normal transgenic T-*cyt* tobacco plants as a model for the investigation of plant gene expression in response to phytohormonal stress. *Plant Mol. Biol.* 17:825–36

133. Zambryski PC. 1992. Chronicles from the *Agrobacterium* plant cell DNA transfer story. *Annu. Rev. Plant Physiol. Plant Mol. Biol.* 43:465–90

134. Zambryski PC, Tempé J, Schell J. 1989. Transfer and function of T-DNA genes from *Agrobacterium* Ti and Ri plasmids. *Cell* 56:193–201

Annu. Rev. Plant Physiol. Plant Mol. Biol. 1994. 45:197–209

PPfMs AND OTHER COVERT CONTAMINANTS: IS THERE MORE TO PLANT PHYSIOLOGY THAN JUST PLANT?

Mark A. Holland

Department of Biology, Salisbury State University, Salisbury, Maryland 21801

Joseph C. Polacco

Department of Biochemistry and Interdisciplinary Plant Group, University of Missouri, Columbia, Missouri 65211

KEY WORDS: tissue culture, bacteria, contamination, *Methylobacterium*

CONTENTS

INTRODUCTION 197
PLANTS AND MICROBES 198
THE PPfMs 200
PLANTS OR MICROBES? 204
LIFE WITHOUT CONTAMINATION 206
PROSPECTS 207

INTRODUCTION

You can observe a lot just by watching.—Yogi Berra

This review focuses on covert bacterial contamination—that is, undetected contamination of cell cultures and plant tissues used in the laboratory, which

0066-4294/94/0601-0197$05.00

197

may affect the results of an experiment. Reports of such contaminants appear infrequently in the literature for two reasons: First, when contamination is noticed, it is usually the reason results are discarded, and only rarely (e.g. 9, 47) the reason they are published. Second, but more significantly, when contaminants remain covert, their activities masquerade as plant activities and may never be recognized as microbial in origin. Covert contaminants represent a vast majority of microorganisms usually overlooked because their activities do not cause "problems," but which may be very important to plant biology. In particular, we are interested in contaminants that are normally associated with plants in vivo.

Two examples of microbial contamination illustrate these points. Hill & Rogers (29) reported a bacterial origin for L-serine dehydratase in French beans. They measured the enzyme activity in freshly prepared extracts, then by chance made additional measurements over a period of several days. Enzyme activity increased over the course of the experiment. This increase could be attributed to the growth of microbes in the extract, not to activity native to the plant.

Horsch & King (35) isolated an auxotrophic mutant cell line from *Datura innoxia* suspension cultures. Their initial characterization described the mutation as leaky. After they discovered that the culture contained *Hyphomicrobium* spp. (what they called a covert contaminant) and cured it with a combination of antibiotics, the leaky expression disappeared and the mutant became "tight."

In both examples, a microbial contaminant was recognized by chance. It is difficult to say how often significant microbial contaminants remain undiscovered in our laboratories or what kinds of activities they may cause. Leifert and colleagues (41, 43) presented a substantial list of bacterial, fungal, and insect contaminants found in cell culture. Many of these are regularly associated with plant tissues. Many are difficult to detect. What are they doing in those cultures and how many of them have fooled us in the laboratory?

PLANTS AND MICROBES

Plant/microbe relationships like the soybean/*Bradyrhizobium japonicum* symbiosis in nitrogen-fixing root nodules or the genetic transformation of plant tissues by *Agrobacterium* spp. are physiologically meaningful to the plant, but less obvious relationships may also be significant. If understanding the relationships between plants and microbes in nature indicates the kinds of interactions likely to be found between them in the laboratory, the reverse is also true. Insights from the laboratory indicate previously unappreciated relationships between microbes and plants in nature.

In some rain forest ecosystems, microbial life on the surfaces of leaves may form a layer 50 μm thick. It would be surprising were the plant insensitive to so substantial a microbial population or if the activities of so large a population had no effect on the plant. Many of the microbes living on the phylloplane probably lead a saprophytic lifestyle, feeding on materials leached from the leaf. A variety of organic and inorganic substances has been found in leaf leachates (69) including all essential elements, free sugars and sugar alcohols, all amino acids, many organic acids, growth regulating substances, vitamins, alkaloids, and phenolics. The author speculated that at least under some conditions, all plant substances could be leached from the plant. Moreover, the amount of material leached from a leaf can be substantial. In one study, up to 6% of the dry weight of *Phaseolus vulgaris* leaves (mainly in the form of carbohydrates) could be leached in a 24 h period (69). The availability of what can be a plentiful food resource in plant leachates gives the impression that perhaps the microbial inhabitants of the phylloplane are opportunistic, but accidental visitors. Certainly this is true for some, but some are permanent residents. Do any of them give back to plants as well as take? Phylloplane bacteria produce B vitamins, auxins, and cytokinins, among other products (4, 37, 62).

Among pathogenic bacteria, disease states are obvious interactions with plants, but latent or chronic infections are covert relationships that may escape detection and that can confound experimental work. Hayward (27) published a review on latent infections of plants by bacteria. Yet, despite the work reviewed in that paper and what has come out in the 20 years since it was published, there is little appreciation that plant tissues, inside and out, support a substantial microbial population. Grimm & Baumann (24) tested methods of detection for *Erwinia carotovora* var. atroseptica in potato cultures. They noted that the bacterium can persist for many years in tissue cultures without producing any obvious sign of contamination. But neither ELISA nor tests using nutrient media were reliable indicators of the status of infection of the cultures in that study. This difficulty in assessing the presence of covert contaminants is a recurrent theme in the literature. Many authors (e.g. 43) stress the importance of finding appropriate media for use in detection of contaminants, but even this may not produce reliable results. In *Erwinia* (24), the bacteria did not grow well in indicator media, presumably because they were somehow attached to the plant cell walls. The authors suggested that an enzymatic pretreatment of the plant tissue might help to release the bacterial cells into the culture medium. Tripepi & George (68) identified *Pseudomonas syringae* pv. syringae and *Curtobacterium flaccumfaciens* ssp. as seed-borne contaminants of Mung bean (see also 15). These contaminants are resistant to surface sterilization with NaOCl and can interfere with rooting bioassays for which the beans are subsequently used. Cooke et al (9) tested the effects of

Erwinia caratovora, Pseudomonas syringae, and *Xanthomonas campestris* on tissue cultures of *Aster, Cheiranthus, Delphinium, Iris,* and *Rosa* by inoculating the cultures with the bacteria. Interestingly, some bacteria survived in the tissue cultures as covert infections, were maintained in culture, and did not produce disease symptoms until after plants regenerated from the cultures were weaned onto soil.

Endophytic bacteria or fungi seem especially likely to escape detection in vivo and in vitro. The presence of bacteria within seemingly healthy plant tissues has been recognized for some time (e.g. 33) and removal of systemic contaminants is a considerable problem for some plant species (e.g. 14, 70). Significantly, microbes can affect the biology of the experimental system. Consider the report from Stierle et al (65) that not only the bark of the Pacific yew, but also an endophytic fungus, *Taxomyces andreanae,* isolated from the tree are sources of the anti-cancer drug taxol. Are other plant-derived drugs also produced by plant-associated microbes? Should plant-associated microbes be a target for conservation efforts? Tor et al (66) demonstrated that covert infection by endophytic *Curtobacterium* was responsible for false positive results in β-glucuronidase (GUS) assays for plant transformation in *Dioscorea* spp. Although others have noted some background GUS activity in higher plants (30, 36), this report was the first to show that some background activity can be of bacterial origin. Because GUS is popular as a reporter gene in plant tranformation studies, an appreciation of the potential to produce detectable activity by covert contaminants is essential.

During the last few years, the relationship between plants and one covert contaminant has been unfolding in our laboratory. *Methylobacterium* spp., pink-pigmented facultative methyltrophs (PPFMs), provide a useful model for the unappreciated kinds of interactions between plants and bacteria that take place routinely on lab benches and in culture dishes (2, 22, 23, 55).

THE PPFMs

In particular, PPFM refers to the species *Methylobacterium mesophilicum,* a species that is seed-transmitted and is regularly associated with plants (4, 15, 16). Although the bacterium has been known to microbiologists for some time, its relationship to plants is relatively unknown among plant scientists. This is somewhat surprising because the bacterium is probably distributed ubiquitously, having been isolated from more than 70 species of plants including monocots and dicots, angiosperms and gymnosperms, and vascular and nonvascular plants (10, 11, 16). Significantly, all plant tissue analyzed to date has been colonized with the bacteria (12, 16, 17, 31). We first became aware of the PPFMs during a study of the urease isozymes of soybean. Briefly, a soybean mutant at the urease structural gene locus *Eu4* (67) unexpectedly retained substantial background urease activity (15–40% of wild-type activity) in callus cultures and in the unifoliate leaves of seedlings, but not in other tissues that normally contain the ubiquitous urease isozyme encoded at *Eu4* (57). Moreover, the observed activity did not resemble the ubiquitous urease by several biochemical criteria (31). We partially cured the plant of the bacteria and demonstrated a concomitant reduction in the levels of background urease activity, thus showing that the background urease in cell cultures and leaves was the product of bacterial activity in those tissues (31).

Several difficulties encountered during this work illustrate why contaminants like *Methylobacterium* are often overlooked. First, their presence is not obvious. We have found cultures that contain as many as 10^8 colony forming units per gram fresh weight, although they do not appear to be contaminated. Also, isolated PPFMs do not grow well on plant tissue culture media. These observations also have been made for other common culture contaminants (13, 44). Because they do not grow on the culture medium, they remain confined within the mass of cultured cells. In old cultures the bacteria occasionally appear as a puddle around the senescing callus. This raises a question of cause and effect. Are the bacteria responsible for the decline of such cultures (as they are often blamed to be) or do they simply desert the culture as it senesces? A second problem in determining the biological significance of PPFM contamination is that quantifying PPFM populations requires a selective medium. The PPFMs are slow growers (10, 12) and if a selective medium is not used, they are overgrown rapidly by other bacteria or fungi present on plant tissues. On non-selective (rich) media, the PPFMs may be easily overlooked. Of course, there may be other contaminants of which we are unaware. A recent survey of contaminants in plant cell culture cataloged more than 300 isolates of contaminants without mention of the PPFMs (43). These authors also discussed the difficulties of detecting contamination without knowledge of appropriate culture media for the contaminants present. In the case of the PPFMs, some other unrecognized microbial contaminant could also contribute to background urease activity in soybean cultures. However, this seems unlikely because the background urease resembles the PPFM urease biochemically and because PPFMs are the only bacteria we found in every tissue sample examined.

The real biological significance of the PPFMs is in their intimate relationship with the plant. A class of pleiotropic mutations in soybean (mutations at either of the loci *Eu2* or *Eu3*) that affect expression of all soybean-encoded ureases (50) also affect the urease of the PPFMs living on the leaves of the mutant plants (31). PPFMs living on these urease-null plant mutants are themselves urease-null for as long as they are associated with the plant. In free-living culture, they become urease-positive. The effect occurs because nickel is unavailable to the PPFMs in the mutant plants (urease is a nickel-containing metalloenzyme). We know the effect is not the result of altered nickel uptake or transport in the mutant (31). The molecular basis of the phenotype is unclear.

In on-going experiments, additional points of interaction between the PPFMs and plants have been discovered in nitrogen metabolism (Figure 1) (32, 64). It is apparent that the PPFMs have access to plant metabolites and that they are able to degrade both ureides (a major transport form for nitrogen in nodulated soybeans) and urea (31, 56). PPFMs probably use ureides while living on soybean because they either do not take up arginine (the other logical source of metabolic urea) or do not have an arginase (56). When embryos lacking both plant and bacterial urease activities (*eu3-e1/eu3-e1*) develop on a plant of the same genotype, urea accumulates in the seeds. If the bacterial urease alone is active in those embryos, however, urea does not accumulate (56, 63). Similarly, during soybean germination, urea accumulation in the embryonic axis is mitigated by the activity of the bacterial urease (56, 63).

Basile et al (5) first described what they later nicknamed PPFMs as a contaminant of tissue cultures of the leafy liverwort *Scapania nemorosa*. In their early work, they showed that the bacteria were capable of stimulating the growth and reproductive development of the liverwort cultures, even when they were inoculated on the medium next to the explant. They also demonstrated that the bacteria normally are plant surface dwellers. Later, they showed that the PPFMs produce vitamin B12 and that addition of the vitamin to cultures of the bryophytes *Jungermannia leiantha* and *Gymnocolea inflata* could stimulate the plant's growth and development (4). While vitamin B12 has not been demonstrated to be essential for higher plants, there have been reports of cobalamin-dependent enzymes from higher plants (58, 59). Corpe & Basile (11) also reported that the PPFMs stimulated plantlet regeneration of cell cultures of *Streptocarpus prolixis* (gracilis) Burtt. Pieces of callus were

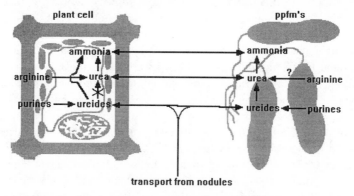

Figure 1 Interactions between soybean and the PPFMs. The figure illustrates points of plant and bacterial interaction for which we have found evidence (32, 56, 64). Significantly, the plant both supplies the bacteria with metabolites and receives metabolites from them.

transferred to a medium without exogenous hormones. When such cultures were inoculated with PPFMs, plantlets were regenerated within 15 days. Untreated controls did not regenerate plantlets after 30 days. Presumably, this effect is different from the vitamin effect seen with *J. leiantha* and *G. inflata*. These results are more compatible with the hypothesis that PPFMs affect cytokinin levels in the plant tissue. This is not without precedent. *Agrobacterium* spp., among other familiar plant-associated bacteria, is known to produce cytokinins. Currently, we are testing the PPFMs to see whether they do produce substances with plant hormone activity.

Another interesting effect of PPFMs is on seed germination. Such effects have been described previously. For example, Klincare et al (37) showed a correlation between lowered populations of seed microflora and decline in germination rates in a variety of species. To study the effect of PPFMs on seed germination, we first heat-treated soybean seeds (50°C, 48 h) to reduce the populations of PPFMs (31). We then set up five different treatments using both heated and unheated seeds: imbibition in sterile water, imbibition in PPFM culture medium (salts only), imbibition in spent PPFM culture medium (i.e. complete PPFM medium in which PPFMs were grown to stationary phase, filtered through a 0.45 μm filter to remove bacterial cells), imbibition in PPFM culture medium (salts only) plus cytokinins (0.5 mg/L BA and 0.5 mg/L zeatin), and imbibition in a suspension of washed PPFM cells. After imbibition, the seeds were allowed to germinate in sterile germination paper. After four days, germination frequency was computed (Table 1). There appears to be a PPFM-mediated effect on germination that can be replaced either by spent medium or by the addition of hormones to the medium in which the seeds are imbibed. Is it conceiveable that one of the factors contributing to the loss of germinability of old seed is a decline in the population of bacteria resident within?

Corpe & Rheem (12) suggested that the facultative methylotrophy of the PPFMs was partly responsible for the maintenance of their relationship with plants. Corpe & Rheem showed that cuticle from the leaves of white clover contains a substantial quantity of methanol, especially while the plants are growing. Growing tissues also contain the highest populations of PPFMs. By exploiting the unusual food resource, methanol, PPFMs may remove that toxin from the plant tissue and occupy a niche in the leaf environment available to few other microorganisms. The degradation of methylated pectin is one likely source of methanol in the plant. Such an activity is expected in growing tissues. Methanol production by plants in the field (48) and in culture (3) as well as a stimulatory effect of small quantities of methanol on plants (53) have been discussed recently. It will be interesting to see whether the PPFMs are involved in any of these evolving stories. Does the application of methanol to

Table 1   Effects of PPFM bacteria on germination of soybean seeds. Heated seeds were treated at 50°C, 48 h to reduce the numbers of viable PPFMs in them (31). Seeds were imbibed in each of the media below. Data are percents of germination. Each entry represents the mean of two trials, each of which included 50 seeds.

Treatment	Unheated	Heated
Water	96%	70%
PPFM Medium[a]	95	70
Spent Medium[b]	95	79
Medium + BA + Zn[c]	100	87
Washed PPFMs[d]	98	86

[a]Ammonium Mineral Salts (AMS) medium (31) without carbon or nitrogen sources added.

[b]Complete AMS medium in which PPFMs have grown to stationary phase.

[c]AMS medium without carbon or nitrogen, but containing 0.5 mg/L benzyl adenine and 0.5 mg/L zeatin.

[d]PPFMs harvested from spent medium above by centrifugation, resuspended in sterile water, recentrifuged, and resuspended in sterile water (final volume equivalent to volume of initial culture).

plant leaves stimulate the activity of the PPFMs or other metylotrophs living there?

PLANTS OR MICROBES?

Bacterial production of vitamins, their participation in nitrogen metabolism as shown by plant mutants, and their effect on germinability of seeds all point to an evolved and intimate association between the bacteria and the plant. Other possibilities also come to mind. Perhaps the bacteria function as waste managers for the plant. The fate of metabolic wastes in plants has not received a great deal of attention. Because plants are generally not motile, either their excreted wastes must be volatile (like methanol) or they must be removed from the plant's immediate environment by physical (e.g. leaching) or biotic (e.g. microbial) processes. Production of methanol by the plant was mentioned earlier in connection with the PPFMs. Few other plant waste products have been identified as such in the literature. One intriguing idea is that plant allelopathic or defense compounds originate as metabolic wastes. Muller (52) recounted of these compounds: "…elimination may involve volatilization of terpenes in arid climates, leaching of phenols in humid climates, isolation in deciduous organs, or being rendered innocuous by chemical bonding as in the formation of glycosides." Although some of these plant compounds have anti-microbial properties, others are metabolized by bacteria and/or fungi. Another possibility is that plant-associated microbes might participate in the production of some of these secondary metabolites.

The production of triterpenoids, for example, is a microbial trick done by *Methylobacterium organophilum* (a PPFM relative) (7, 61, 72). Taxol, mentioned earlier, is produced by an endophytic fungus associated with the Pacific yew (65). These products and many others, which were thought to be solely the products of plant metabolism, may be at least partly, if not wholly, of microbial origin.

Of related interest is the production of plant growth regulators by bacteria. It was mentioned earlier that PPFMs produce what appear in culture to be cytokinin effects. The relationship between *Ardisia crispa* and its leaf-resident bacteria is the most outstanding example in support of a cytokinin contribution. *A. crispa* and related Myrsinaceae family members, as well as *Psychotria* spp. and relatives (Rubiaceae), are characterized by bacteria-containing nodules on the margins of their leaves (20, 45). The bacteria are seed-transmitted (51). Occasionally, seeds are produced that germinate but fail to develop beyond the seedling stage. These so-called cripples are thought to lack the bacterial leaf symbiont because cripples can be generated at will by heating the seeds before germination to kill the bacteria resident in them. A small proportion of crippled seedlings eventually recover and resume normal growth. Although bacteria are not isolated from the cripples, those that recover do harbor bacterial populations. Others never grow, their meristems dedifferentiate somewhat to resemble callus tissue, their leaves drop, and eventually they die. The identity of the bacterial symbiont, if indeed only one bacterial species participates in the symbiosis, is uncertain. Several bacteria have been isolated and identified (6, 34), but successful re-establishment of the symbiosis in cured plants, required by Koch's postulates, has not been accomplished (20). Results from work on two different species indicate that the crippled phenotype results from a lack of cytokinin and the inability of the plants to produce the hormone. LaMotte & Lersten (39) demonstrated a requirement for cytokinins in cell cultures of *Psychotria* spp. and a tolerance to high levels of the hormone. Rodrigues Pereira et al (62) isolated two different bacteria from *A. crispa* and showed that both produce cytokinins.

It is interesting to speculate on the kinds of plant activities described in the literature for which covert contaminants may be responsible. In tissue culture, covert contaminants could be responsible for some examples of habituation or accomodation phenomena (49) often attributed to epigenetic changes in the plant tissue. Such phenomena are slow, adaptive responses to changes in the culture medium. Often, they also are rapidly reversible. Examples are loss of dependence on exogenous growth substances and changes in the levels of assimilatory enzymes in response to an altered nitrogen source. Covert contaminants might alter the capacity of a cell culture to regenerate plants or even to grow in culture at all. In work with whole plants, contaminants might produce any number of measureable enzyme activities. Are some invariant or non-segregating isozymes really microbial products?

LIFE WITHOUT CONTAMINANTS

Three general methods have been reported for removing microbial contaminants from plant material: physical methods (like heating), chemical disinfectants (like NaOCl), and antibiotics. The best results generally are obtained by using a combination of the three strategies. Heating dry soybean seeds in a laboratory oven at 50°C for 48 h was successful in our work with the PPFMs (31) and was also suggested by Rodrigues Pereira et al (62). Once seeds were treated with dry heat, the PPFM populations associated with the seedlings developed from them remained at 3–5% of the normal level for more than one month during which time the plants were maintained in the growth chamber and greenhouse without regard for sterility. Seeds that subsequently developed on those plants, however, carried normal levels of the bacterium. Sinclair and coworkers used a hot (90°C) vegetable oil treatment to control fungal contaminants in soybean (60, 71). Hankin & Sands used microwaves to disinfect tobacco seeds (26), and we have used their method successfully both with tobacco and with *Arabidopsis thaliana*. Seed in a paper envelope, held above the floor of the oven and microwaved for 20 min at full power was unaffected in germination and was clean enough to be used in tissue culture applications, although it still harbored PPFM populations (MA Holland & JC Polacco, unpublished observation).

Chemical disinfection is a routine step in tissue culture. Generally, protocols use commercial bleach, detergents, and alcohols, although chlorine gas, mercurials, and other agents have been used. These are effective in removing many contaminants from the tissue; however, they are not 100% effective. PPFMs, for example, are not removed by surface sterilization. Many investigators have found that certain contaminants are removed only at the expense of killing the tissue being cultured (e.g. 19).

Several different antibiotic regimes have been tested for their ability to control microbial contaminants in cell culture. The most successful regimes combine a number of antibiotics, each of which has a different mode of action. For example, streptomycin and carbenicillin were used in combination on *Datura innoxia* (35). The use of antibiotics in cell culture, both singly and in combination, has been studied extensively (e.g. 18, 21, 38, 42) and was reviewed recently (43).

PROSPECTS

It ain't over till it's over. —Yogi Berra

Covert contaminants present problems in the laboratory, but they also offer opportunities. There is much to learn about microbes whose activites are so tuned to the plants they inhabit that their activities are indistinguishable. Some of these organisms may be useful targets for manipulation by biotechnologists. Methylotrophic bacteria like the PPFMs are known to produce a variety of useful products from simple substrates (25, 40, 46). Might plant-associated bacteria or fungi be manipulated to confer disease or pest resistance on their hosts? Do they already do this? Might engineering make them enhance plant quality? In culture, can plant/microbe associations be exploited to make products that result from a joint metabolism? Forced associations in vitro and in vivo continue to be established (1, 8, 28, 54) with varying degrees of success.

Literature Cited

1. Aloysius SKD, Paton AM. 1984. Artificially induced symbiotic associations of L-form bacteria and plants. *J. Appl. Bacteriol.* 56:465–77
2. Austin B, Goodfellow M. 1979. *Pseudomonas mesophilica,* a new species of pink bacteria isolated from leaf surfaces. *Int. J. Syst. Bacteriol.* 29:373–78
3. Aveni MT, Gorecki RJ, Amable RA, Koch JL, Obendorf RL. 1989. Methanol accumulates during in vitro growth of zygotic embryos. *Plant Physiol.* 89(4):1028A
4. Basile DV, Basile MR, Li QY, Corpe WA. 1985. Vitamin B12-stimulated growth and development of *Jungermannia leiantha* Grolle and *Gymnocolea inflata* (Huds.) Dum. (Hepaticae). *Bryologist* 88(2):77–81
5. Basile DV, Slade LL, Corpe WA. 1969. An association between a bacterium and a liverwort, *Scapania nemorosa.* *Bull. Torrey Bot. Club* 96(6):711–14
6. Bettelheim KA, Gordon JF, Taylor J. 1968. The detection of a strain of *Chromobacterium lividum* in the tissues of certain leaf-nodulated plants by the immunofluorescence technique. *J. Gen. Microbiol.* 54:177–84
7. Bisseret P, Zundel M, Rohmer M. 1985. Prokaryotic triterpenoids 2. 2-beta-methylhopanoids from *Methylobacterium or-*

ganophilum and *Nostoc muscorum,* a new species of prokaryotic triterpenoids. *Eur. J. Biochem.* 150:29–34
8. Carlson PS, Chaleff RS. 1974. Forced association between higher plant and bacterial cells. *Nature* 252:393–95
9. Cooke DL, Waites WM, Leifert C. 1992. Effects of *Agrobacterium tumefaciens, Erwinia caratovora, Pseudomonas syringae* and *Xanthomonas campestris* on plant tissue cultures of *Aster, Cheiranthus, Delphinium, Iris,* and *Rosa*; disease development in vivo as a result of latent infection in vitro. *Z. Pflanzenkr. Pflanzenschutz/J. Plant Dis. Protect.* 99(5):469–81
10. Corpe WA. 1985. A method for detecting methylotrophic bacteria on solid surfaces. *J. Microbiol. Methods* 3:215–21
11. Corpe WA, Basile DV. 1982. Methanol-utilizing bacteria associated with green plants. *Dev. Indust. Microbiol.* 23:483–93
12. Corpe WA, Rheem S. 1989. Ecology of the methylotrophic bacteria on living leaf surfaces. *FEMS Microbiol. Ecol.* 62:243–50
13. Debergh PC, Vanderschaeghe AM. 1974. Some symptoms indicating the presence of bacterial contaminants in plant tissue cultures. *Acta Hortic.* 225:77–81
14. DeBoer SH, Copeman RJ. 1974. Endophytic bacterial flora in *Solanum tuber-*

osum and its significance in bacterial ring rot diagnosis. *Can. J. Plant Sci.* 54:115–22

15. Dunleavy JM. 1988. *Curtobacterium plantarum* sp. nov. is ubiquitous in plant leaves and is seed transmitted in soybean and corn. *Int. J. Syst. Bacteriol.* 39:240–49

16. Dunleavy JM. 1988. *In vitro expression of the cellulose gene in* Methylobacterium mesophilicum, *a seed-transmitted bacterium ubiquitous in soybean.* Presented at 2nd Biennial Conf. Mol. Cell. Biol. Soybean, Ames, Iowa, July 25–27

17. Dunleavy JM. 1990. *Urease production by* Methylobacterium mesophilicum, *a seed-transmitted bacterium ubiquitous in soybean.* Presented at 3rd Biennial Conf. Mol. Cell. Biol. Soybean, Ames, Iowa, July 23–25

18. Falkiner FR. 1988. Strategy for the selection of antibiotics for use against common bacterial pathogens and endophytes of plants. *Acta Hortic.* 225:53–56

19. Finer JJ, Saxton RW, Norris BL, Steele JA, Rahnema S. 1991. Growth and disinfestation of 6 different bacteria in embryogenic suspension cultures of cotton. *Plant Cell Rep.* 10:380–83

20. Fletcher LM. 1976. Bacterial symbioses in the leaf nodules of Myrsinaceae and Rubiaceae. In *Microbiology of Aerial Plant Surfaces,* ed. CH Dickinson, TF Preece, pp. 465–85. New York: Academic

21. Gilbert JE, Shohet S, Caligari PDS. 1991. The use of antibiotics to eliminate latent bacterial contamination in potato tissue cultures. *Ann. Appl. Bot.* 119(1):113–20

22. Green PN, Bousfield IJ. 1982. A taxonomic study of some gram-negative facultatively methylotrophic bacteria. *J. Gen. Microbiol.* 128:623–38

23. Green PN, Bousfield IJ. 1983. Emendation of *Methylobacterium* (Patt, Cole, and Hanson 1976); *Methylobacterium rhodinum* (Heumann 1962) comb. nov. corrig.; *Methylobacterium radiotolerans* (Ito and Iizula 1971) comb. nov. corrig.; and *Methylobacterium mesophilicum* (Austin and Goodfellow 1979) comb. nov. *Int. J. Syst. Bacteriol.* 33:875–77

24. Grimm F, Baumann B. 1991. Untersuchungen zum nachweis latenter kontaminationen von *Erwinia carotovora* var. *atroseptica* (van Hall) Dye an in vitropflanzen von kartoffeln. *Potato Res.* 34: 47–55

25. Haber CL, Allen LN, Zhao S, Hanson RS. 1983. Methylotrophic bacteria: biochemical diversity and genetics. *Science* 221: 1147–53

26. Hankin L, Sands DC. 1977. Microwave treatment of tobacco seed to eliminate bacteria on the seed surface. *Phytopathology* 67:794–95

27. Hayward AC. 1974. Latent infections by bacteria. *Annu. Rev. Phytopathol.* 12:87–97

28. Herman EB. 1990. Non-axenic plant tissue culture: possibilities and opportunities. *Acta Hortic.* 280:233–38

29. Hill HM, Rogers LJ. 1972. Bacterial origin of L-serine dehydratase in French beans. *Phytochemistry* 11:9–18

30. Hodal L, Bochardt A, Neilsen JE, Mattsson O, Okkels FT. 1992. Detection, expression and specific elimination of endogenous beta-glucuronidase activity in transgenic and non-transgenic plants. *Plant Sci.* 87(1): 115–22

31. Holland MA, Polacco JC. 1992. Urease-null and hydrogenase-null phenotypes of a phylloplane bacterium reveal altered nickel metabolism in two soybean mutants. *Plant Physiol.* 98:942–48

32. Holland MA, Stebbins NE, Polacco JC. 1992. Evidence of interactions in nitrogen metabolism between soybean and a phylloplane bacterium. *Plant Physiol.* 99(1): 30A

33. Hollis JP. 1951. Bacteria in healthy potato tissue. *Phytopathology* 41:350–66

34. Horner HT Jr, Lersten NR. 1972. Nomenclature of bacteria in leaf nodules of the families Myrsinaceae and Rubiaceae. *Int. J. Syst. Bact.* 22(2):117–22

35. Horsch RB, King J. 1983. A covert contaminant of cultured plant cells: elimination of a *Hyphomicrobium* spp. from cultures of *Datura innoxia* (Mill.) *Plant Cell Tissue Organ. Cult.* 2:21–28

36. Hu CY, Chee PP, Chesny RH, Zhou JH, Miller PD, O'Brien WT. 1990. Intrinsic GUS-like activity in seed plants. *Plant Cell Rep.* 9:1–5

37. Klincare AA, Kreslina DJ, Mishke IV. 1971. Composition and activity of the epiphytic microflora of some agricultural plants. In *Ecology of Leaf Surface Microorganisms,* ed. TF Preece, CH Dickinson, pp. 191–201. New York: Academic

38. Kneifel W, Leonhardt W. 1992. Testing of different antibiotics against gram-positive and gram-negative bacteria isolated from plant tissue culture. *Plant Cell Tissue Organ. Cult.* 29(2):139–44

39. LaMotte CE, Lersten NR. 1972. Attempts to obtain bacteria-free plants of *Psychotria punctata* (Rubiaceae): growth and root formation in callus cultures. *Am. J. Bot.* 59: 89–96

40. Large PJ, Bamforth CW. 1988. *Methylotrophy and Biotechnology.* New York: Longman. 303 pp.

41. Legatt IV, Waites WM, Leifert C, Nicholas J. 1988. Characterisation of micro-organisms isolated from plants during micropropagation. *Acta Hortic.* 225:93–102

42. Leifert C, Canotta H, Wright SM, Waites B, Cheyne VA, Waites WM. 1991. Elimina-

tion of *Lactobacillus plantarum, Corynebacterium* spp., *Staphylococcus saprophyticus* and *Pseudomonas paucimobilus* from micropropagated *Hemerocallis, Choisya,* and *Delphinium* cultures using antibiotics. *J. Appl. Bacteriol.* 71:307–30

43. Leifert C, Ritchie JY, Waites WM. 1991. Contaminants of plant-tissue and cell cultures. *World J. Microbiol. Biotechnol.* 7: 452–69

44. Leifert C, Waites WM. 1992. Bacterial growth in plant tissue culture media. *J. Appl. Bacteriol.* 72(6):460–66

45. Lersten NR, Horner HT Jr. 1976. Bacterial leaf nodule symbiosis in angiosperms with emphasis on Rubiaceae and Myrsinaceae. *Bot. Rev.* 42(2):145–214

46. Lidstrom ME, Stirling DI. 1990. Methylotrophs: genetics and commercial applications. *Annu. Rev. Microbiol.* 44:27–58

47. Long RD, Curtin TF, Cassells AC. 1988. An investigation of the effects of bacterial contaminants on potato nodal cultures. *Acta Hortic.* 225:83–91

48. MacDonald RC, Fall R. 1992. Methanol emission from plants. *Plant Physiol.* 99(1): 427A

49. Meins F Jr. 1989. Habituation: heritable variation in the requirement of cultured plant cells for hormones. *Annu. Rev. Genet.* 23:395–408

50. Meyer-Bothling LE, Polacco JC, Cianzio SR. 1987. Pleiotropic soybean mutants defective in both urease isozymes. *Mol. Gen. Genet.* 209:432–38

51. Miller IM, Donnelly AE. 1987. Location and distribution of symbiotic bacteria during floral development in *Ardisia crispa. Plant Cell Environ.* 10:715–24

52. Muller CH. 1969. The "co-" in coevolution. *Science* 164:197–98

53. Nomura AM, Benson AA. 1992. The path of carbon in photosynthesis: improved crop yields with methanol. *Proc. Natl. Acad. Sci. USA* 89:9794–98

54. Paton AM, Innes CMJ. 1991. Methods for the establishment of intracellular associations of L-forms with higher plants. *J. Appl. Bacteriol.* 71:59–64

55. Patt TE, Cole GC, Hanson RS. 1976. Methylobacterium, a new genus of facultatively metylotrophic bacteria. *Int. J. Syst. Bacteriol.* 26(2):226–29

56. Polacco JC, Holland MA. 1993. Roles of urease in plant cells. *Int. Rev. Cytol.* 145: 65–103

57. Polacco JC, Judd AK, Dybing JK, Cianzio SR. 1989. A new mutant class of soybean lacks urease in leaves but not in leaf-derived callus or in roots. *Mol. Gen. Genet.* 217:257–62

58. Poston JM. 1977. Leucine 2,3-aminomutase: a cobalamin-dependent enzyme present in bean seedlings. *Science* 195:301–2

59. Poston JM. 1978. Coenzyme B12-dependent enzymes in potatoes: leucine 2,3-aminomutase and methylmalonyl-CoA mutase. *Phytochemistry* 17:401–2

60. Pyndji MM, Sinclair JB, Singh T. 1987. Soybean seed thermotherapy with heated vegetable oils. *Phytopathology* 71(3):213–16

61. Renoux JM, Rohmer M. 1985. Prokaryotic triterpenoids: new bacteriohopanetetrol cyclitol ethers from the methylotrophic bacterium *Methylobacterium organophilum. Eur. J. Biochem.* 151:405–10

62. Rodrigues Pereira AS, Houwen PJW, Deurenberg-Vos HWJ, Pey EBF. 1972. Cytokinins and the bacterial symbiosis of *Ardisia* species. *Z. Pflanzenphysiol.* 68:170–77

63. Stebbins N, Holland MA, Cianzio SR, Polacco JC. 1991. Genetic tests of the roles of the embryonic ureases of soybean. *Plant Physiol.* 97:1004–10

64. Stebbins NE, Holland MA, Polacco JC. 1992. Metabolic origins of urea in soybean. *Plant Physiol.* 99(9):33A

65. Stierle A, Strobel G, Stierle D. 1993. Taxol and taxane production by *Taxomyces andreanae,* an endophytic fungus of Pacific yew. *Science* 260:214

66. Tor M, Mantell SH, Ainsworth C. 1992. Endophytic bacteria expressing beta-glucuronidase cause false positives in transformation of *Dioscorea* species. *Plant Cell Rep.* 11(9):452–56

67. Torisky RS, Griffin JD, Yenofsky RA, Polacco JC. 1993. Evidence that a single gene (*Eu4*) encodes the ubiquitous urease of soybean. *Mol. Gen. Genet.* In press

68. Tripepi RR, George MW. 1991. Identification of bacteria infecting seedlings of Mung bean used in rooting assays. *J. Am. Soc. Hortic. Sci.* 116(1):80–84

69. Tukey HB Jr. 1971. Leaching of substances from plants. In *Ecology of Leaf Surface Microorganisms,* ed. TF Preece, CH Dickinson, pp. 67–80. New York: Academic

70. Wilson ZA, Power JB. 1989. Elimination of systemic contamination in explant and protoplast cultures of rubber (*Hevea brasiliensis* Muell. Arg.). *Plant Cell Rep.* 7:622–25

71. Zinnen TM, Sinclair JB. 1982. Thermotherapy of soybean seeds to control seedborne fungi. *Phytopathology* 72(7): 831–34

72. Zundel M, Rohmer M. 1985. Prokaryotic triterpenoids. 3. The biosynthesis of 2 beta-methylhopanoids and 3 beta-methylhopanoids of *Methylobacterium organophilum* and *Acetobacter pasteurianus* ssp. pasteurianus. *Eur. J. Biochem.* 150:35–39

Annu. Rev. Plant Physiol. Plant Mol. Biol. 1994. 45:211–34
Copyright © 1994 by Annual Reviews Inc. All rights reserved

MOLECULAR ANALYSIS OF PROTEINS IN THE PLANT PLASMA MEMBRANE

M. R. Sussman

Cell and Molecular Biology Program and Department of Horticulture, University of Wisconsin, Madison, Wisconsin 53706

KEY WORDS: plasma membrane, proton pump, H$^+$-ATPase, channels, carriers, protein kinases

CONTENTS

INTRODUCTION 211
PROTON PUMP (H$^+$-ATPase) 213
　Pre-DNA Cloning Years 213
　A Gene Family is Born.. 215
　The Next Generation.. 220
CARRIERS AND CHANNELS 222
PROTEIN KINASES.. 226
FUTURE PROSPECTS 229

INTRODUCTION

The water-insoluble nature of membrane proteins makes them difficult to study using biochemical techniques. Fortunately, new techniques for cloning and sequencing DNA have provided more facile strategies for elucidating protein structure and function. DNA-based strategies have thus made a great impact in the molecular study of "buttery" membranous proteins, particularly those in the plasma membrane of higher plants. In comparison to membrane proteins in the chloroplasts and mitochondria, routine large-scale protein chemistry techniques have been applied less frequently to the study of proteins in the plant plasma membrane. For many years investigators optimized methods to obtain the most pure and homogeneous preparations of plasma membrane vesicles from crude plant extracts. However, purity is only part of the picture. Success in purifying and solubilizing proteins from these plasma membrane vesicles has required that equal attention be paid to optimizing catalytic stability, for example, by minimizing post-homogenization proteolytic and lipolytic activities.

The proton pump (H^+-ATPase) was the first higher plant plasma membrane protein to be purified and characterized at the molecular level. This protein is at least 10-fold more abundant in the plasma membrane of fungi such as yeast and *Neurospora crassa*; therefore, it is not surprising that a protocol for its solubilization and purification without denaturation was first established using fungal membranes (74,75). In 1989, the proton pump became the first plant plasma membrane enzyme whose gene was cloned and sequenced (30, 50). In the past four years new DNA-based strategies have also been used to clone and characterize genes encoding other proteins in the plasma membrane of higher plants.

Previous reviews have discussed in depth the molecular structure of the plasma membrane proton pump (H^+-ATPase), from both plants and fungi (9, 26, 56, 76, 77, 79, 85, 86). Cloning and sequencing of proton pump gene isoforms from many species (93), as well as site-directed mutagenesis studies in yeast (57) are providing important information on essential amino acids and how they may participate in the catalytic cycle. This review covers recent studies concerned with the biological function of individual members of the gene family encoding this important plant enzyme. Recent discoveries on the molecular structure of several other plant plasma membrane proteins are also reviewed. The genes encoding three of these proteins (potassium channel, sucrose carrier, amino acid carrier) were all cloned by a new and powerful genetic strategy, the rescue and genetic complementation of yeast transport mutants using plant cDNA. A fourth protein, the nitrate carrier, was cloned by gene tagging (insertional mutagenesis) techniques using *Arabidopsis thaliana*.

Plasma membrane transporters show rapid changes in catalytic activity in response to changes in the environment. In other eukaryotic organisms, phosphorylation of the transporter polypeptides at serine (ser), threonine (thr), or tyrosine (tyr) residues plays an important role in regulating their activity. Of course, plasma membrane transporters can be phosphorylated by membrane-bound or cytoplasmic protein kinases. Although a burgeoning number of such protein kinases are being reported in the plant kingdom, there is no evidence of which ones, if any, play a role in regulating proteins at the plasma membrane.

Plant genes encoding two different classes of ser/thr-type plasma membrane protein kinases were cloned and sequenced recently: calcium-dependent protein kinases (CDPKs) and receptor protein kinases (RPKs). A new class of prokaryotic-like receptor protein kinases that may phosphorylate histidine or aspartyl residues was identified recently via a chromosome walk to the ethylene-resistance (*etr*) mutant locus of *Arabidopsis thaliana*. Although the intracellular location has not been established conclusively for these three classes of protein kinases, they currently represent our best candidates as regulators of plasma membrane function in higher plants.

THE PROTON PUMP (H^+-ATPase)

Pre-DNA Cloning Years

In animal cells, a sodium pump (Na^+,K^+-ATPase) is the primary active transport system at the plasma membrane. This enzyme transports three sodium ions out of the cell and takes up two potassium ions every time an ATP molecule is cleaved. The unequal stoichiometry of monovalent cations transported during each cycle of catalysis generates an electric potential (negative inside). The electric potential and sodium gradient are the two components of a sodium motive force that provides the energy to propel solutes across the plasma membrane, via ion channels and sodium-coupled carriers. Other reviews provide a more detailed discussion of the distinction between pumps, carriers, and channels (10, 42, 85). For our purposes, channels are proteinaceous pores that allow ions and other solutes to equilibrate with the prevailing electrochemical gradients. Carriers perform the same job, but they must couple the flow of one solute (e.g. sugar) with that of another (e.g. protons). In this way, a carrier can couple the inward transport of sugar to both the electrical and chemical gradient of sodium ions or protons. Pumps couple the transport of a solute such as sodium ions or protons directly to the breakage of a chemical bond (e.g. ATP).

The plasma membranes of plants and fungi do not appear to have a Na^+,K^+-ATPase. Instead, the primary active transport system is a H^+-ATPase. The enzyme generates two products, an electric potential (negative inside), and a chemical gradient of protons (e.g. a slightly basic pH of 7.1–7.5 in the cytoplasm, and a considerably lower acidic pH of 5–6 in the cell wall). The plant and fungal plasma membrane proton pump (H^+-ATPase) contains a catalytic polypeptide with an approximate M_r of 100 kDa that is phosphorylated and dephosphorylated at an essential aspartyl residue during each turnover of the catalytic cycle. This latter property has led to the classification of the plasma membrane enzyme as a P-type ATPase, and distinguishes it from two other well-characterized types of proton translocating ATPases: the F_0F_1-type H^+-ATPase of mitochondria, chloroplasts, and bacteria, and the V-type H^+-ATPase in the vacuolar membrane.

Channel proteins are fast, readily moving several million ions per second. In contrast, pumps are fairly sluggish, and only catalyze the transport of several hundred ions per second per protein molecule. Thus, in cells transporting a lot of ions, the H^+-ATPase is an abundant protein in the plasma membrane. From electrophysiological measurements in such cells one can calculate that the H^+-ATPase is a major consumer of cellular ATP (23). Recent immunocytochemical studies confirm this interpretation because they indicate that guard cells, phloem cells, and root epidermal cells show the greatest immunoreactivity with an antibody directed against domains present in known plant plasma membrane H^+-ATPase genes (51, 66, 91).

Because the ATPase biochemical activity is easy to assay (phosphate release from ATP), it is not surprising that the 100-kDa H^+-ATPase was the first plasma membrane protein to be purified to homogeneity and, via Edman degradation of internal tryptic peptides, was the first for which amino acid sequence was obtained (71). This endeavor required the identification of a tissue from which large amounts of the ATPase could be extracted in a non-proteolyzed intact form. Roots from 6-day old etiolated seedlings of *Avena sativa* cv. Stout were one of the few tissues satisfying this condition. Standard rate-zonal and equilibrium sucrose density gradient centrifugation techniques were modified to allow one person, in one day, to process large amounts of this tissue into purified plasma membrane vesicles of high specific activity H^+-ATPase. A balance sheet of vanadate-sensitive, but nitrate-, molybdate-, and azide-insensitive, ATPase present in membrane vesicles at various stages of purification showed that the plasma membrane H^+-ATPase specific activity increased from 0.064 units (μmoles Pi released/min/mg protein) in the tissue homogenate, to 1.74 units in the purified vesicles (37). A similar specific activity and SDS-PAGE profile was obtained with plasma membrane vesicles purified by a two phase dextran procedure (68), as compared to sucrose density gradient centrifugation. An approximately 30-fold enrichment of the H^+-ATPase specific activity in the purified vesicles relative to the crude homogenate indicated that plasma membrane protein represented, at most, approximately 3% of the total cellular protein.

On SDS-PAGE, these purified plasma membrane vesicles display a prominent Coomassie-stained polypeptide at 100 kDa (83). Using Bio-Gel A-1.5m column chromatography in the presence of 1% (w/v) lithium dodecyl sulfate to perform molecular sieving separation at 4°C on a scale not readily feasible with gel electrophoresis, this 100-kDa polypeptide was purified, alkylated with iodoacetamide, and subjected to tryptic hydrolysis. Eight random tryptic peptides were then purified by reverse phase HPLC, and subjected to Edman degradations with an automated protein sequenator. Approximately 110 amino acid residues were identified from this procedure and many of these internal sequences were readily matched to known conserved regions of P-type ATPases whose genes were cloned and sequenced from animals, fungi, and bacteria. Subsequent cloning and sequencing of an *A. thaliana* gene family encoding the plasma membrane H^+-ATPase has shown extensive sequence identity to the eight *A. sativa* random tryptic peptides.

This large-scale plasma membrane protein purification showed that the H^+-ATPase concentration in the plant plasma membrane is fairly low compared to animal and fungal membranes. From a crude extract of 3.6 kg fresh weight of oat roots, we obtained approximately 200 mg of purified plasma membrane vesicles, of which only 0.5 mg of the 100-kDa H^+-ATPase polypeptide was obtained, indicating that the ATPase represents less than 1% of the plasma membrane protein (83). The experimentally observed maximal turnover number of purified P-type ATPases is 100–300 µmoles Pi released/min/mg protein. Thus, theory confirms that membrane vesicles with a specific activity of 1.7 µmoles Pi released/min/mg protein should contain 1% or less of their protein as the ATPase. In summary, these early experiments with oat roots indicated that the H^+-ATPase is 1% or less of the plasma membrane protein, and the plasma membrane is overall 3% or less of the total cellular protein. Thus, the H^+-ATPase, which is one of the more abundant plant plasma membranes, represents 0.03% or less of the total cellular protein. Our discovery of one H^+-ATPase cDNA clone out of a population of several thousand random cDNA clones supports this value. These observations demonstrate that, unlike more abundant cytoplasmic or organellar proteins, those in the plasma membrane may be particularly rare (e.g. ion channels or protein kinases, which may be at least 100-fold less abundant than the H^+-ATPase). Although several other plant plasma membrane proteins have been characterized using DNA-based strategies, the H^+-ATPase remains the only plant plasma membrane protein for which amino acid sequence has been derived directly from the native protein purified from plants.

In a typical animal cell, the 100-kDa Na^+,K^+-ATPase is likewise relatively low in abundance, and most biochemical studies use organs with specialized transport functions because they contain an especially high abundance of the ATPase (e.g. mammalian kidneys or the electric organ of the electric eel). From these tissues, it is relatively easy to obtain milligram amounts of the purified Na^+,K^+-ATPase, with high catalytic activity. The low natural abundance of the plasma membrane H^+-ATPase in plant extracts along with its propensity for denaturation during detergent solubilization and purification has made it difficult to conduct careful biochemical studies similar to those performed with the animal enzyme. Recent success in expression of cDNA encoding the plant plasma membrane H^+-ATPase in yeast offers an attractive alternative for future biochemical studies of the plant enzyme (48, 49, 92).

A Gene Family Is Born

Using the oat protein–derived amino acid sequence, as well as knowledge of a highly conserved amino acid sequence found in all P-type ATPases, a single-stranded DNA oligonucleotide was synthesized and used as a radioactive probe for filter hybridization with an oat cDNA library. A cross-hybridizing, partial-length cDNA oat clone was isolated, purified, and sequenced and was found to encode a protein with extensive amino acid sequence identity to the conserved P-type ATPase domains. The oat clone was then used to probe *A. thaliana* cDNA and genomic libraries, from which two different gene clones were isolated, AHA-1 and AHA-2, for *Arabidopsis* H⁺-ATPase. A complete sequence for the protein encoded by each was published (29, 30). At the same time, complete sequence of a third *A. thaliana* isoform (AHA-3) was reported (50), as well as the sequence of genes encoding similar enzymes from tobacco (6), tomato (19), and rice (94).

Subsequent studies in *A. thaliana,* tobacco, and tomato showed that the H⁺-ATPase is encoded by a surprisingly large gene family. Northern blots of mRNA extracted from various plant tissues demonstrated that expression of the three known *A. thaliana* and four known tobacco gene isoforms were regulated differentially (29, 45, 52). To make sense out of this diversity of gene structure and expression, the gene family in one plant species must be analyzed exhaustively. *A. thaliana* is well suited for this analysis because in comparison to other plants, it has a small genome and, thus, is less likely to maintain redundant genes. In 1990 we embarked on an analysis of the AHA gene family, using both PCR and low stringency hybridizations. Because *A. thaliana* genes typically contain few and short introns, genomic DNA may be used as a template for PCR reactions, thus circumventing problems of detection caused by low abundance of expression in a small number of cells. Experiments using the PCR were the most informative, as discussed below (JF Harper, L Manney, MR Sussman, manuscript submitted).

In all known prokaryotic and eukaryotic P-type ion translocating ATPases, there are two short domains that show the highest degree of amino acid sequence conservation. Both domains lie within the central large hydrophilic loop where ATP binding and hydrolysis occurs. One domain surrounds the phosphorylated aspartyl residue and the second is located 300–350 amino acids toward the carboxy terminus, in a region known to define the ATP binding site. Approximately midway between the two is a third highly conserved domain, which has also been used for PCR experiments to identify new P-type ATPases from *Drosophila melanogaster* DNA (90). Using a series of primers corresponding to the domains separated by 300 amino acids, with varying degrees of degeneracy, under a variety of annealing temperatures and magnesium concentrations, we amplified, cloned, and sequenced 1.2–1.6 kilobase pair fragments corresponding to nine new genomic clones highly similar to, but still distinct from the known isoforms AHA-1, 2, and 3. The clones contained a third conserved domain that is the hallmark of P-type ATPases, but only seven of the nine showed high sequence identity to AHA1-3. The other two clones encoded much more highly divergent proteins and were given the abbreviation AXA to signify the possibility that they represent P-type ATPases that transport cations other than protons. A recent report (35) describes such an AXA clone, which is thought to be a Ca^{+2}-ATPase and has only 22% identity to AHA-1.

Gene-specific probes were designed from intron sequences of the seven AHA-like fragments and Southern blots confirmed that all seven represent fragments of bona fide *A. thaliana* genes. These have been used to probe genomic libraries, to isolate full-length clones. The complete sequence for the protein encoded by one such clone, AHA-10, was obtained by a combination of genomic and cDNA sequence (JF Harper, L Manney, MR Sussman, manuscript submitted) and a complete full-length sequence for AHA-9 was reported recently (33).

Are all 10 AHA genes truly plasma membrane proton pumps? Three pieces of evidence favor this interpretation. First, the presence of extensive sequence identity between the 10 AHA isoforms and greatly reduced identity to AXAs indicate that the 10 AHA genes represent similar but non-identical isoforms of the plasma membrane proton pump (H^+-ATPase) (see Figure 1 and discussion below). Second, there is extensive amino acid sequence identity between the proteins predicted from the DNA sequence and the amino acid sequence of random tryptic peptides derived from the 100-kDa H^+-ATPase catalytic polypeptide purified from oat root plasma membrane vesicles. Third, cDNAs corresponding to the first three AHA isoforms have been expressed in yeast, and the purified transgenic plant polypeptides display catalytic properties similar to *pmal,* the single essential yeast gene encoding the H^+-ATPase (49, 92). Furthermore, when cDNA encoding an AHA-2 polypeptide is mutated via deletion of carboxy terminal residues, the plant enzyme translocates to the plasma membrane and is able to support yeast growth in the absence of *pmal* (48). Obviously, sequence identity is an important criterion, but the acid test is expression of each of the seven new AHA cDNAs in yeast and direct measurement of their ion specificity. It is also necessary to determine whether each gene isoform directs expression of the protein to the plasma membrane of *A. thaliana,* rather than to an internal membrane. These studies will require isoform-specific protein probes, and a novel approach to this problem is discussed below.

Why are there so many isoforms encoding the *A. thaliana* H^+-ATPase? Although Northern blots indicated that the three *A. thaliana* and four tobacco genes were expressed differentially in various tissues, these studies do not provide the resolution needed to determine cell-specific locations. For this purpose, we prepared fusions of 5′ upstream sequences from the AHA genes with the reporter gene, β-glucuronidase (GUS). Histochemical localization of GUS activity under the light microscope demonstrated that (*a*) AHA-3 is

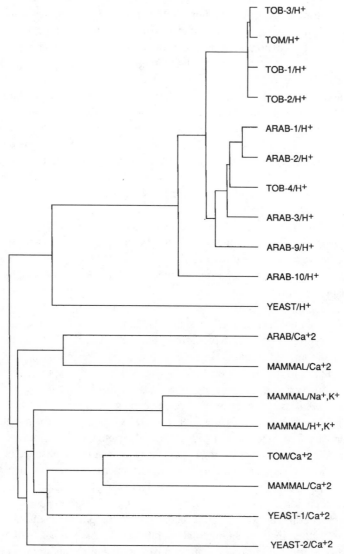

Figure 1 A graphical representation of amino acid sequence relatedness among P-type cation translocating ATPases derived from mammals, yeast, and plants. From bottom to top, the following 19 proteins are compared: (1) and (2) are two yeast genes encoding endomembrane Ca^{+2}-ATPases, (3) is a mammalian endoplasmic/sarcoplasmic reticulum Ca^{+2}-ATPase, (4) is a tomato endoplasmic reticulum Ca^{+2}-ATPase, (5) is a mammalian plasma membrane H^+,K^+-ATPase, (6) is a mammalian plasma membrane Na^+,K^+-ATPase, (7) is a mammalian plasma membrane Ca^{+2}-ATPase, (8) is an *A. thaliana* chloroplast envelope Ca^{+2}-ATPase, (9) is a yeast plasma membrane H^+-ATPase and the remaining 10 are all plasma membrane H^+-ATPase isoforms from tobacco, *A. thaliana,* and tomato. All protein sequences were obtained from the GENBANK and EMBL databases, and were analyzed using University of Wisconsin Genetics Computer Group software on a VAX computer.

expressed in the phloem of vegetative tissue (14), (*b*) AHA-2 is expressed in root hairs and other surface tissues of the root (JF Harper, personal communication), and (*c*) AHA-10 is expressed only in the seed (J Harper, L Manney, MR Sussman, manuscript submitted). At the same time, Northern blots showed that AHA-9 mRNA is expressed mainly in anther tissue (33). Taken together, these results have led us to hypothesize that each of the AHA isoforms is expressed in different specialized transport tissues of higher plants. According to this model, AHA-3 is the gene that encodes an H^+-ATPase that generates the proton-motive force used by sucrose carriers in the phloem, AHA-2 encodes the one that drives mineral absorption from the soil into roots, AHA-10 is involved in the flow of nutrients that nourish the embryo, and AHA-9 provides the proton-motive force involved in anther function (and perhaps in pollen tube growth). Based on this hypothesis, it is reasonable to predict that stomatal guard cells use one of the remaining AHA genes (i.e. AHA-1, -4, -5, -6, -7 or -8) for which cell-specific expression has not yet been measured. Future studies will determine whether this prediction is correct.

There are two plausible reasons why different cells use a specific H^+-ATPase gene isoform. First, the small amount of amino acid sequence differences may be important for generating proton pumps with unique catalytic properties tailored to the different cells' transport functions. Second, each of the ATPases promoters may be specially designed to allow just the right amount of expression at the right time in that cell type. A recent report comparing the catalytic properties of AHA-1, -2 and -3 in yeast demonstrate that there are substantial differences in the K_m for ATP, pH optimum, and inhibitor sensitivities of the various isoforms (49). For example, AHA-3 showed a several fold lower affinity for ATP and vanadate, compared to AHA-2 and AHA-1. Also noted were differences in distance migrated on SDS-PAGE despite nearly identical molecular weights, and a difference in stimulation by exogenously added lysolecithin, a lipid activator of the plant enzyme. These experiments demonstrate conclusively that even in the same lipid matrix (i.e. the yeast endoplasmic reticulum), three *A. thaliana* H^+-ATPase isoforms have surprisingly different chemical, catalytic, and regulatory properties .

These two possibilities (i.e. promoter function and differences in catalytic activity) are not mutually exclusive. Both may have contributed to the evolution, over hundreds of millions of years, of this large gene family. Fick's law of diffusion states that over distances beyond 1 mm, simple chemical diffusion is too slow to support the flow of nutrients needed to nourish a large multicellular structure. For this reason, mammals have evolved specialized structures such as the blood-based circulatory system, kidneys, and intestines. Independently, over a similarly long period of time, higher plants have evolved specialized transport structures such as root hairs, stomata, phloem, and xylem.

Figure 1 is a graphical representation of the amino acid sequence relatedness of the various P-type plant ATPase proteins whose complete gene sequence has been determined. In this analysis, the percent identity of all possible pair-wise combinations was calculated and those clones with the greatest similarity are graphically presented with a close, common root. It is striking to note that all of the AHA sequences align more closely with homologs from tobacco, tomato, and yeast, than with the two plant ATPases (*A. thaliana* and tomato), which are more similar to the Ca^{+2}-ATPase family. AHA-10 is the most diverged of the plant H^+-type sequences but even here it is clearly more aligned with the other AHA sequences than with the Ca^{+2}-types. It is also interesting to note that, as reported earlier (45), the first three tobacco sequences define a class significantly different from the known AHA clones. In contrast, tobacco isoform 4 is closely related to the known AHA-1, -2, and -3 group. These results imply that the uncharacterized AHA gene isoforms -4, -5, -6, -7, and -8 may represent isoforms more closely related to the first three tobacco sequences. This prediction will be tested once complete sequences are determined for these AHA isoforms.

The Next Generation

Our model, in which each of the AHA genes is expressed in specific cell types with unique transport functions, is an hypothesis based on observations with a limited number of isoforms. This model may be overly simplistic in that overlap of expression of specific isoforms in different tissues might occur. Furthermore, the bacterial reporter gene procedure is a facile, but indirect means of microscopically determining the likely cellular location of AHA gene expression. It is thus important to confirm or refute the AHA promoter analyses by performing direct microscopic in situ hybridization using gene specific mRNA probes. To be certain that these probes are absolutely gene specific, it seems wise to wait until the 5′ and 3′ untranslated sequences of all of the AHAs are determined because these are the most likely sources of transcribed, gene specific sequences. However, completely gene-specific probes may be impossible to design because at least two of the isoforms (AHA-1 and AHA-2) contain significant regions of identity in the 3′ untranslated ends.

An alternative route is to generate monoclonal antibodies specific for the protein encoded by each isoform. However, because several of the isoforms have high amino acid sequence identity (e.g. AHA-1 and AHA-2 are 95% identical), it is questionable whether a truly isoform-specific antibody can be obtained. Also, there may be additional isoforms not yet identified that contain sequence similarities to the known isoforms. This criticism applies equally to mRNA (in situ hybridizations) or protein (monoclonal antibody) probes.

A new approach, not subject to the criticism discussed above, is to tag each of the isoforms with a foreign sequence that is not present in the A. thaliana genome. This approach, called epitope tagging, is becoming commonly used in animal and yeast studies, and involves inserting a short sequence of foreign amino acids into the plant protein, which is then recognized by a preexisting available monoclonal antibody (24, 41). This approach has been applied successfully in a follow-up to the analysis of the AHA-3 gene promoter (ND DeWitt, MR Sussman, manuscript in preparation). In these experiments, an 11–amino acid long sequence found in the animal oncogene protein, c-myc, was placed in the ATPase protein near the carboxy terminus by ligating a synthetic 33-bp oligonucleotide into the appropriate location within a 10,000-bp genomic DNA clone containing the entire AHA-3 gene and its putative promoter. Transgenic A. thaliana plants were then generated in which this modified AHA-3 gene was inserted ectopically at one of several locations in the nuclear genome. Western blots using a commercially available c-myc monoclonal antibody immunodecorated a 100-kDa polypeptide in membrane protein isolated from several independent transgenic plants but not in membranes from control, nontransformed plants. Immunocytochemical studies are under way to determine the cellular and subcellular location of the epitope-tagged AHA-3 protein.

In addition to providing knowledge about where the AHA-3 protein is expressed, these transgenic plants also provide a new type of experimental material (i.e. phloem plasma membrane that has been specifically tagged by a foreign epitope). Previous biochemical studies using plant plasma membrane vesicles were limited by the inability to isolate or even identify plasma membrane vesicles derived from a specific cell type. Using conventional immunological techniques, it should now be possible to physically separate c-myc labeled phloem plasma membrane vesicles in crude extracts of transgenic plants. Because the same DNA manipulation can be performed with all the AHA genes, it should be possible to generate transgenic plants in which the plasma membrane of many different specialized cells are labeled. Cell-specific A. thaliana promoters usually are expressed with similar cell-specific fidelity in tobacco. Thus, if a larger plant species with more readily available bulk tissue is desired, tobacco or another related dicot species could be transformed readily with the engineered AHA clones.

In summary, the identification of at least ten different gene isoforms expressing similar, but distinct, forms of the proton pump was an unexpected finding. It is no longer possible to talk about the plasma membrane H^+-ATPase. Instead, one must specify which isoform is being discussed. Two new DNA-based techniques allow elucidation of the catalytic functions and cellular locations of these polypeptides. Expression of the AHA cDNA clones in yeast provides ample amounts of the proteins to perform comparative biochemical studies of kinetic parameters and inhibitor sensitivities. If appropriate unique sequences can be located, these yeast-expressed plant proteins may also provide suitable antigen to elicit gene-specific monoclonal antibodies for further studies of the pumps in planta. Alternatively, epitope tagging techniques provide a new and elegant means of specifically labeling each AHA isoform. Thus, although the door seems closed on simplistic studies of plant extracts containing a mixture of ATPase polypeptides, DNA-based strategies provide us with a window opening out on new ways of studying the plant plasma membrane.

CARRIERS AND CHANNELS

It has long been known that the influx of potassium across the plasma membrane of plants and fungi is coupled tightly to metabolic energy and in particular, to the actions of an ATPase. Earlier biochemical observations of potassium-stimulated ATPase activity in plasma membrane vesicles led investigators to conclude that potassium moved through this ATPase, analogous to the K^+-ATPase of the *E. coli* inner membrane, or the H^+,K^+-ATPase in stomach plasma membrane. It is now more widely accepted that the plant and fungal plasma membrane ATPase is a pure and simple proton pump (H^+-ATPase), and that potassium moves through a separate protein. This movement is coupled indirectly to the electric potential or the pH gradient generated by the proton pump.

In diverse organisms, from *E. coli* to *Drosophila melanogaster,* the molecular characterization of proteins involved in potassium transport has been greatly aided by classical genetic techniques. In microbes, potassium transport mutants were isolated by selecting for cells that require more potassium for growth. In this manner, several different transport systems with differing affinities for potassium were identified in *E. coli*. These different affinities allow the cells to grow at a multitude of external potassium levels (17). Although the genetic evidence is clear that these proteins are required to move potassium into the cells to satisfy nutritional needs, it does not prove whether the proteins are the actual transporters, or simply plasma membrane proteins that help to regulate the transporters. Sometimes the amino acid sequence is informative, as in the case of *KdpB,* the gene whose protein is induced by potassium starvation and which, when mutated, prevents cells from growing at micromolar levels of potassium (1, 46). The amino acid sequence of the protein encoded by *KdpB* is clearly a P-type ATPase, with the highly conserved phosphorylated aspartate residue and other conserved sequences found only in this family of ion pumps.

Similar genetic experiments in yeast have identified two homologous genes, *trk1* and *trk2*, that when mutated, cause yeast to require higher than normal potassium concentrations in the nutrient medium. Although these genes encode a protein with many transmembrane domains, there is no homology to known transporters; therefore, in the absence of any other data, it is uncertain whether this protein is the transporter, or a regulator of the real transporter (26).

The first success in identifying an amino acid sequence encoded by a potassium channel protein was achieved with mutants of *D. melanogaster* known as *shaker, ether a go-go,* and *slowpoke.* These mutants contained dysfunctional potassium channels in nerve cells that resulted in phenotypic changes of behavior that were easily scored. DNA sequence of the mutant genes showed a family of proteins with a similar topology of six membrane spanning domains and sequence identity (36). Patch clamp electrophysiological studies of the gene products after heterologous expression in *Xenopus laevis* oocytes demonstrated that these were the actual genes for potassium channels. Furthermore, site-directed mutagenesis experiments coupled to channel recordings in *X. laevis* showed the existence of a voltage-gating domain. The electrophysiological data had indicated that these potassium channels were outward rectifying (i.e. they catalyzed potassium efflux out of the cells) and the electric potential at which they opened could now be ascribed to the function of a small number of amino acids. After these *shaker* clones became available, low-stringency hybridizations were performed with cDNA and genomic DNA from a wide variety of organisms, and *shaker* homologs were discovered in the genomes of mammals and other multicellular eukaryotes with nervous systems.

Despite observations of ion channel recordings with membranes from plants, yeast, and bacteria, the amino acid sequence for these more primitive ion channels remained elusive until 1992. The breakthrough came from two independent laboratories performing experiments with yeast potassium transport mutants that require higher than normal concentrations of potassium in the growth medium (2, 73, 84). These investigators transformed the mutant yeast cells, called *trk* minus, with *A. thaliana* cDNA and screened for plant clones that would allow the mutants to grow on medium containing a nonpermissive low potassium concentration. These experiments demonstrated the existence of two *A. thaliana* genes encoding proteins with a remarkable similarity to the *D. melanogaster* and mammalian *shaker* genes. It is important to recognize that the yeast wild-type *trk* gene, which also rescues these *trk* minus mutants, bears no sequence homology to *shaker* or, for that matter, to any protein in the database. Furthermore, none of the electrophysiological data reported to date indicates that the *trk* gene product is a potassium channel. In addition, no one has reported a *shaker* homolog in the yeast genome. It is thus ironic that a bona fide *shaker* potassium channel homolog was found in the plant genome using a yeast mutant with a defect in a gene that lacks recognizable channel sequences. Of course, the *trk* gene product may be a true potassium channel and if so, it could represent a new class of channel sequences yet to be identified in more advanced multicellular eukaryotes.

The conclusion that these *A. thaliana shaker*-like genes are responsible for potassium transport in higher plants is supported by several lines of evidence. First, in patch clamp studies with guard cell protoplasts, the predominant plasma membrane potassium channel observed was an inward-rectifying channel that opened at hyperpolarizing electric potentials, and was thus probably the transporter responsible for potassium uptake in these cells (72). At the time, this result was surprising because although potassium channels clearly were responsible for potassium efflux in other eukaryotes, there had been no inward-rectifying channel genes cloned and it was uncertain whether a channel could account for the thermodynamic considerations involved in potassium uptake. The conclusion that *A. thaliana shaker*-like cDNA clones expressed channel proteins that were responsible for uptake was later confirmed by patch clamp electrophysiological studies after expression in *X. laevis* oocytes (67). Ultimate proof of this conclusion must await the isolation of plants in which these specific genes have been altered.

An approach to isolate *A. thaliana* potassium transport mutants was reported in 1993 (80). These investigators observed that low concentrations of cesium, a potassium analog, block the growth of germinated seedlings. This toxic effect is strictly competitive with potassium supplied in the medium. Another monovalent cation, lithium, is also inhibitory, but unlike cesium its action is unaffected by the potassium concentration. Radioactive tracer uptake studies with *A. thaliana* germlings indicated that cesium is also a potent competitive inhibitor of potassium uptake, which confirms electrophysiological observations of cesium inhibition of potassium channel activity in *A. thaliana* cells. These results indicated that it may be possible to isolate potassium transport mutants by selecting for cesium-insensitivity. One cesium-insensitive mutant has been isolated that behaves as a single point recessive mutation, mapped to the top of chromosome IV. The molecular identity of this gene, and an electrophysiological analysis of the mutant plant, are under investigation.

Recent results from two laboratories indicate that a second type of potassium transport system is present in the plant plasma membrane (38, 43). This additional system is a high-affinity system that would only operate at lower external potassium concentrations. Measurements of the electric potential indicate that the electrochemical gradient of potassium is insufficient to account for the internal levels of potassium observed at low external potassium concentrations. Because channels only use the potassium electrochemical concentration (i.e. the electric potential plus potassium chemical gradient), the second transport system must be either a potassium carrier using the pH gradient, or a potassium pump, directly coupled to metabolic energy via ATP or another high energy compound. Electrophysiological studies in *Neurospora crassa* also indicate the operation, at low external potassium concentrations, of a potassium-proton symport carrier (64). Although these electrophysiological arguments are persuasive, it is interesting to note that the rescued yeast *trk* minus mutants, utilizing the *A. thaliana shaker*-like clones, can grow and divide even at low micromolar levels of external potassium. One can account for this observation only if the electric potential in yeast is more negative than in plants, which is a distinct possibility. Measurements of the electric potential in large cells such as those found in *N. crassa* or giant algae are routinely performed by microelectrode insertion techniques. Unfortunately, because yeast and plant cells are much smaller, it is technically more difficult to measure this electric potential accurately.

By continued screening, it should be possible to use the *trk* minus rescue strategy to identify plant cDNA clones encoding high-affinity potassium transporters, but only the *shaker*-like channel clones have been obtained to date. The yeast strategy assumes that the proteins encoded by the plant clones are synthesized, folded properly, and then transported to the fungal plasma membrane. Although this is the case for the *A. thaliana shaker* clones and for plant genes encoding plasma membrane proton-coupled symporters for sugars and amino acids (25, 34, 60), this may not be the case for all plant transporters. When the plant plasma membrane proton pump genes AHA-1, -2, or -3 are expressed in yeast, the protein is synthesized and folded into a catalytically active form, but it remains stuck in the endoplasmic reticulum and is not translocated to the plasma membrane. Alternatively, if a mutated form of the AHA-2 gene lacking a portion of the carboxy terminus is expressed in yeast, the plant protein appears in the plasma membrane and functions sufficiently well to genetically rescue yeast mutants containing a defective fungal proton pump gene. Because the plant genes encoding potassium channels, amino acid carriers, and sugar carriers seem to express proteins that function well in yeast, this problem of membrane protein trafficking could represent a situation unique to the plant proton pump.

The nitrate carrier is the only other plant plasma membrane transporter for which an amino acid sequence has been obtained. The molecular identity of this protein was determined through an elegant use of *A. thaliana* genetics. In 1978, Doddema et al reported on the identification of a specific chlorate-resistant *A. thaliana* mutant locus, *chl1*, which appeared to be defective in its uptake of chlorate and nitrate (15, 16). Other chlorate resistant mutant loci are known, but these all encompass defects in nitrate/chlorate reduction rather than transport.

To isolate and clone the CHL1 gene, an allele was found in which insertion of T-DNA caused the mutation. This mutant plant was found by screening Feldmann's large population of transgenic *A. thaliana* plants, which contain mutations caused by T-DNA insertion (89). A second allele was identified in which this same gene was disrupted by an endogenous *A. thaliana* transposon (88). The coding sequence of the undisrupted wild-type CHL1 gene predicts a hydrophobic protein with 12 putative transmembrane domains that shows no sequence homology with other transporters or any other protein in the database. Expression of the CHL1 gene in *X. laevis* oocytes proved that the CHL1 protein catalyzes the uptake of nitrate. Inward movement of nitrate was associated with a pH-dependent depolarization of the electric potential, indicating that the transporter was a nitrate/proton symporter with a stoichiometry of at least two protons transported per nitrate ion. This elegant combination of classical and molecular genetics and electrophysiological characterization of the gene product after heterologous expression is a benchmark by which future efforts will be judged.

PROTEIN KINASES

Protein kinases are essential components in signal transduction pathways involving the plasma membrane in all eukaryotes. The first plant protein kinase to be purified to homogeneity from a plant extract was a calcium-dependent protein kinase [also called calmodulin domain protein kinase (CDPK)] (55, 61). Using amino acid sequence derived by Edman degradations from tryptic peptides of the purified kinase, a cDNA encoding CDPK from soybean was isolated and sequenced (31). Although the sequence predicted for the protein encoded by this first CDPK gene lacked a transmembrane domain or other known membrane-targeting sequences, biochemical fractionation studies indicated that a protein immunoreactive with CDPK monoclonal antibodies was associated with the oat root plasma membrane (69). Immunocytochemical studies demonstrate that at least some of the CDPK in plant cells is associated with the cytoskeleton (54) and may be involved in regulating cytoplasmic streaming (44). CDPK associated with the membrane may be the result of the enzyme binding to cytoskeletal filaments attached to the membrane. More recent studies indicate that the CDPK gene family is large, with at least six different gene isoforms expressed in *A. thaliana* (E Hrabak, personal communication). Several of these isoforms contain an N-terminal myristylation consensus sequence that may be involved in targeting to the cytoskeleton and/or the plasma membrane (28).

In vitro experiments have indicated a role for CDPK in regulating plasma membrane transport functions. In these experiments, purified oat root plasma membrane vesicles were incubated with $[\gamma\text{-}^{32}P]ATP$ and time-dependent incorporation of radioactivity into serine and threonine residues in the H^+-ATPase and other proteins was observed. However, as mentioned earlier, the difficulties of isolating and purifying sufficient amounts of the plant plasma membrane H^+-ATPase have hampered efforts to determine directly whether phosphorylation regulates catalytic activity. Recent biochemical and genetic experiments with the yeast plasma membrane proton pump have established that kinase-mediated phosphorylation probably does play a biological role. Chang & Slayman (11) demonstrated that the fungal pump is phosphorylated in vivo at several different amino acids, and only one or two of these phosphorylation events is involved in the rapid activation of the pump that occurs when glucose-starved cells are treated with glucose. In the same experiments, these investigators found that as the pump moves from its site of synthesis on the endoplasmic reticulum to the plasma membrane, it picks up additional serine and threonine phosphates. Thus, in yeast, kinase-mediated phosphorylation of the proton pump may be involved in its biogenesis and/or protein trafficking, as well as modulating its catalytic activity.

In contrast with plants, there is little or no calcium-stimulated protein kinase activity associated with plasma membrane vesicles isolated from exponentially growing yeast cells (70). A protein kinase that resembles animal casein kinases has been found tightly associated with the ATPase in plasma membrane vesicles (40), and a gene encoding a membrane associated member of this family has been reported to be involved in salt tolerance and cell growth polarity (62, 63). The exact site(s) of serine/threonine phosphorylations in the fungal proton pump has not been identified, but site-directed mutagenesis of a threonine in the carboxy terminus eliminates some of the stimulatory effects of glucose (78). As with many integral membrane proteins, lipids are required for maximal activity of the proton pump, and changes in essential lipids may be occuring in concert with kinase-mediated regulatory events.

A puzzling aspect of biochemical studies with CDPK is that the detergent-solubilized plasma membrane, as well as the polypeptide produced by expression of A. thaliana CDPK genes in E. coli, is stimulated by specific lipids (28). The specificity of lipid stimulation is unlike mammalian protein kinase C, where activation by diacylglycerol and other lipids fits well into known transduction pathways involving phospholipases. Gene clones resembling protein kinase C have not been reported in plants, although a yeast homolog has been found. Alternatively, CDPK has not been reported in yeast or eukaryotes other than the malaria parasite, Plasmodium (98). Further studies are required to establish whether lipids play a role in regulating CDPK action in situ, analogous to lipid effects on protein kinase C.

In vitro and in vivo studies have implicated plasma membrane protein kinases in diverse signal transduction events, from pathogen resistance (20, 22) to blue light photoreception (47, 58, 59, 81). Receptor protein kinases

(RPKs) are the only other plasma membrane protein kinases with known molecular structures. The first plant RPK clone was obtained from a maize cDNA library and contained a carboxy-terminal kinase domain separated from an N-terminal extracellular receptor domain by a single membrane spanning region. In mammalian cells, most of these types of protein kinases are tyrosine kinases, but the plant clones all seem to bear closest sequence identity to the few serine/threonine types of receptor kinases. Shortly after the maize clone was reported, a family of RPKs was found in the genome of *A. thaliana*. Most of these clones can be considered orphan ligand receptor kinases because the ligand that binds to the extracellular domain is unidentified, and the substrates for the kinase are unknown.

The *A. thaliana* RPK gene family encodes three types of proteins, based on the amino terminal receptor domain: (*a*) S-receptor kinases, which have homology to the S-locus glycoproteins involved in the self-incompatibility reaction of *Brassica* pollen (7, 18, 82, 87, 96, 97); (*b*) leucine-rich repeat RPKs, which contain the leucine-rich repeat sequence thought to be involved in protein-protein interactions (13); and (*c*) epidermal growth factor RPKs, which contain repeat sequences seen previously in the epidermal growth factor receptor of animal cells (39). The biological role of only the S-receptor kinases is certain because mutations in the genes encoding these proteins have been implicated in incompatibility reactions.

The latest plant protein kinase to be discovered with a possible role in plasma membrane functions is the biggest surprise of all in the field of eukaryotic signal transduction. A chromosome walk to the *A. thaliana* ethylene resistance mutant (*etr*) locus uncovered a new type of receptor protein kinase, previously seen only in prokaryotes (12). In bacteria, this kinase is called a sensor-regulator and is an essential step in the signal transduction of diverse phenomena, from pathogenicity and virulence to chemotaxis and turgor regulation. In many cases, there are several protein domains involved. The first resembles mammalian RPKs because it contains an extracellular hydrophilic sequence that acts as the receptor for external ligands. This is followed by a membrane spanning segment and an internal hydrophilic domain that contains an essential histidine autophosphorylating activity. An additional protein domain contains an aspartyl residue to which the histidine phosphate is transferred, and which then becomes activated. For example, the D and E genes of the *E. coli kdp* operon gene encode sensor-regulator proteins that sense changes in turgor pressure, and activate the transcription of genes A, B, and C, which encode subunits of a plasma membrane P-type K^+-ATPase that is a high-affinity potassium uptake system (1, 46, 53, 95).

Although the intracellular location of the *etr* locus is uncertain, its presence in the plant genome indicates that serine/threonine/tyrosine phosphorylation may not be the only means by which plants use protein kinases to modify the catalytic activity of proteins. Both plants and yeast rapidly alter gene expression after changes in turgor pressure (8); thus, it may be worthwhile to examine whether genes resembling *kdpD*/E and *etr* are involved.

FUTURE PROSPECTS

In the past five years we have witnessed the first cloning and sequencing of genes encoding plasma membrane transporters and protein kinases. Finding the amino acid sequences of these proteins is an essential first step in generating the tools and probes used to study their biological roles, but the sequences alone may provide little information about these roles. Heterologous expression in yeast or *X. laevis* and epitope tagging to determine the cellular and intracellular distribution of expression are important strategies for elucidating potential roles for the gene product, but in many cases the biological role for a cloned gene is unclear. For example, an auxin-binding protein in the lumen of the endoplasmic reticulum has been purified and from its amino acid sequence, its structural gene has been cloned. Electrophysiological studies imply that this protein is present in the plasma membrane and may act as the receptor for auxin (65). The obvious next step to evaluate directly the hypothesized biological role of this protein is to create mutant plants in which expression of this gene is reduced or elevated.

A key for further progress is the creation of knockout plants (i.e. mutant plants in which particular cloned plasma membrane genes are disrupted). There are various ways to conduct this type of reverse genetics. One method is to screen for insertional mutations from large populations of *A. thaliana* plants containing T-DNA inserted at random locations. There are approximately 15,000 genes in the *A. thaliana* genome; with the approximately 10,000 different T-DNA disruptions now available, many of these genes may be disrupted somewhere in this population. The polymerase chain reaction or Southern blots with pooled populations of transgenic plants can be used to identify these plants. Most disruptions are expected to be recessive mutations; thus, heterozygote plants carrying one disrupted allele should be present.

A second approach is to use transposon tagging to isolate mutant plants. In this procedure, the cloned gene is mapped to a chromosomal locus in the *A. thaliana* genome, and a movable transposon located nearby is used to generate disrupted alleles. For example, investigators have placed *Ds,* a corn transposable element that lacks its own transposase, at many different locations on the five chromosomes of *A. thaliana.* When these transgenic plants are crossed with a plant containing the *Ac* element (which cannot move, but which contains a transposase gene), the *Ds* element is induced to leave its current location, and insert in regions nearby (3–5). Thus, by choosing a plant with *Ds* near the gene of interest, it should be possible to screen for disrupted alleles in the progeny from a cross with *Ac* plants. Again, the polymerase chain reaction or Southern blots with a pooled population of plants can be used to identify the plants carrying a disruption in the gene of interest.

The above approaches are alternatives to antisense expression, which may be impossible to use on one gene isoform within a large family of similar genes. When there is not a gene family, this approach may well be worth taking. The recent development of inducible promoters (21, 27, 32) also allows the creation of transgenic plants in which antisense expression is under control of the investigator. This is an important point because it may be impossible to obtain transgenic plants from tissue culture if the mutant phenotype includes altered growth and development.

The ability to generate gene knockouts is revolutionizing the study of signal transduction pathways in mice and other model eukaryotic systems. Unfortunately, homologous recombination is not yet a reality with *A. thaliana* or any other plant; thus, we must be content with one of the disruption approaches outlined above. Of course, the classical genetic approach using point mutations to identify interesting phenotypes will continue to provide plants in which map-based strategies are used to elucidate the gene sequence. Either way, it is no longer satisfactory to isolate a clone encoding a plasma membrane protein and determine its sequence. The biological function of the plasma membrane protein encoded by these clones must be determined, and the recently devised DNA-based strategies provide new ways to achieve this goal.

Literature Cited

1. Altendorf K, Siebers A, Epstein W. 1992. The *KDP* ATPase of *Escherichia coli. Ann. NY Acad. Sci.* 671:228–43
2. Anderson JA, Huprikar SS, Kochian LV, Lucas WJ, Gaber RF. 1992. Functional expression of a probable *Arabidopsis thaliana* potassium channel in *Saccharomyces cerevisiae. Proc. Natl. Acad. Sci. USA* 89: 3736–40
3. Bancroft I, Bhatt AM, Sjodin C, Scofield S, Jones JDG, Dean C. 1992. Development of an efficient two-element transposon tagging system in *Arabidopsis thaliana. Mol. Gen. Genet.* 233:449–61
4. Bancroft I, Jones JDG, Dean C. 1993. Heterologous transposon tagging of the *DRL1* locus in *Arabidopsis. Plant Cell* 5:631–38
5. Bhatt AM, Dean C. 1992. Development of tagging systems in plants using heterologous transposons. *Curr. Opin. Biotechnol.* 3:152–58
6. Boutry M, Michelet B, Goffeau A. 1989. Molecular cloning of a family of plant genes encoding a protein homologous to plasma membrane H^+-translocating ATPases. *Biochem. Biophys. Res. Commun.* 162:567–74
7. Boyes DC, Nasrallah JB. 1993. Physical linkage of the SLG and SRK genes at the self-incompatibility locus of *Brassica oleracea. Mol. Gen. Genet.* 236:369–73
8. Brewster JL, De Valoir T, Dwyer ND, Winter E, Gustin MC. 1993. An osmosensing signal transduction pathway in yeast. *Science* 259:1760–63
9. Briskin DP. 1990. The plasma membrane H^+-ATPase of higher plant cells: biochemistry and transport function.

Biochim. Biophys. Acta Rev. Biomembr. 1019:95–109

10. Bush DR. 1993. Proton-coupled sugar and amino acid transporters in plants. *Annu. Rev. Plant Physiol. Plant Mol. Biol.* 44: 513–42

11. Chang A, Slayman CW. 1991. Maturation of the yeast plasma membrane [H^+]ATPase involves phosphorylation during intracellular transport. *J. Cell Biol.* 115:289–95

12. Chang C, Kwok SF, Bleecker AB, Meyerowitz EM. 1993. *Arabidopsis* ethylene-response gene *ETR1*: similarity of product to two-component regulators. *Science* 262: 539–44

13. Chang C, Schaller GE, Patterson SE, Kwok SF, Meyerowitz EM, Bleecker AB. 1992. The *TMK1* gene from *Arabidopsis* codes for a protein with structural and biochemical characteristics of a receptor protein kinase. *Plant Cell* 4:1263–71

14. DeWitt ND, Harper JF, Sussman MR. 1991. Evidence for a plasma membrane proton pump in phloem cells of higher plants. *Plant J.* 1:121–28

15. Doddema H, Hofstra JJ, Feenstra WJ. 1979. Uptake of nitrate by mutants of *Arabidopsis thaliana*, disturbed in uptake or reduction of nitrate. III. Regulation. *Physiol. Plant.* 45:339–43

16. Doddema H, Otten H. 1979. Uptake of nitrate by mutants of *Arabidopsis thaliana*, disturbed in uptake or reduction of nitrate. II. Kinetics. *Physiol. Plant.* 45:332–39

17. Dosch DC, Helmer GL, Sutton SH, Salvacion FF, Epstein W. 1991. Genetic analysis of potassium transport loci in *Escherichia coli*: evidence for three constitutive systems mediating uptake of potassium. *J. Bacteriol.* 173:687–96

18. Dwyer KG, Lalonde BA, Nasrallah JB, Nasrallah ME. 1992. Structure and expression of *AtS1*, an *Arabidopsis thaliana* gene homologous to the S-locus related genes of *Brassica*. *Mol. Gen. Genet.* 231: 442–48

19. Ewing NN, Wimmers LE, Meyer DJ, Chetelat RT, Bennett AB. 1990. Molecular cloning of tomato plasma membrane proton-ATPase. *Plant Physiol.* 94:1874–81

20. Farmer EE, Moloshok TD, Saxton MJ, Ryan CA. 1991. Oligosaccharide signaling in plants: specificity of oligouronide-enhanced plasma membrane protein phosphorylation. *J. Biol. Chem.* 266:3140–45

21. Faryar K, Gatz C. 1992. Construction of a tetracycline-inducible promoter in *Schizosaccharomyces pombe*. *Curr. Genet.* 21: 345–49

22. Felix G, Grosskopf DG, Regenass M, Boller T. 1991. Rapid changes of protein phosphorylation are involved in transduction of the elicitor signal in plant cells. *Proc. Natl. Acad. Sci. USA* 88:8831–34

23. Felle HH. 1982. Effects of fusicoccin upon membrane potential, resistance, and current-voltage characteristics in root hairs of *Sinapis alba*. *Plant Sci. Lett.* 25:219–25

24. Finley D, Bartel B, Varshavsky A. 1989. The tails of ubiquitin precursors are ribosomal proteins whose fusion to ubiquitin facilitates ribosome biogenesis. *Nature* 338:394–400

25. Frommer WB, Hummel S, Riesmeier JW. 1993. Expression cloning in yeast of a cDNA encoding a broad specificity amino acid permease from *Arabidopsis thaliana*. *Proc. Natl. Acad. Sci. USA* 90:5944–48

26. Gaber RF. 1992. Molecular genetics of yeast ion transport. *Int. Rev. Cytol.* 137A: 299–353

27. Gatz C, Kaiser A, Wendenburg R. 1991. Regulation of a modified CaMV 35S promoter by the Tn10-encoded Tet repressor in transgenic tobacco. *Mol. Gen. Genet.* 227: 229–37

28. Harper JF, Binder BM, Sussman MR. 1993. Calcium and lipid regulation of an *Arabidopsis* protein kinase expressed in *Escherichia coli*. *Biochemistry* 32:3282–90

29. Harper JF, Manney L, DeWitt ND, Yoo MH, Sussman MR. 1990. The *Arabidopsis thaliana* plasma membrane H^+-ATPase multigene family. Genomic sequence and expression of a third isoform. *J. Biol. Chem.* 265:13601–8

30. Harper JF, Surowy TK, Sussman MR. 1989. Molecular cloning and sequence of cDNA encoding the plasma membrane proton pump (H^+-ATPase) of *Arabidopsis thaliana*. *Proc. Natl. Acad. Sci. USA* 86: 1234–38

31. Harper JF, Sussman MR, Schaller GE, Putnam-Evans C, Charbonneau H, Harmon AC. 1991. A calcium-dependent protein kinase with a regulatory domain similar to calmodulin. *Science* 252:951–54

32. Hershey HP, Stoner TD. 1991. Isolation and characterization of cDNA clones for RNA species induced by substituted benzenesulfonamides in corn. *Plant Mol. Biol. Int. J. Mol. Biol. Biochem. Genet. Eng.* 17:679–90

33. Houlne G, Boutry M. 1994. Identification of an *Arabidopsis thaliana* gene encoding a plasma membrane H^+-ATPase whose expression is restricted to anther tissues. *Plant J.* In press

34. Hsu L-C, Chiou T-J, Chen L, Bush DR. 1993. Cloning a plant amino acid transporter by functional complementation of a yeast amino acid transport mutant. *Proc. Natl. Acad. Sci. USA* 90:7441–45

35. Huang L, Berkelman T, Franklin AE, Hofman NE. 1994. Characterization of a gene encoding a Ca^{+2}-ATPase-like protein in the plastid envelope. *Proc. Natl. Acad. Sci. USA*. In press

36. Jan LY, Jan YN. 1992. Tracing the roots of ion channels. *Cell* 69:715–18
37. Katz D. 1989. *Studies on the plasma membrane H⁺-ATPase of oat roots-preparation and assay, cytological localization and sulfhydryl chemistry.* PhD thesis. Univ. Wisconsin, Madison
38. Kochian LV, Lucas WJ. 1993. Can K⁺ channels do it all? *Plant Cell* 5:720–21
39. Kohorn BD, Lane S, Smith TA. 1992. An *Arabidopsis* serine/threonine kinase homologue with an epidermal growth factor repeat selected in yeast for its specificity for a thylakoid membrane protein. *Proc. Natl. Acad. Sci. USA* 89:10989–92
40. Kolarov J, Kulpa J, Baijot M, Goffeau A. 1988. Characterization of a protein serine kinase from yeast plasma membrane. *J. Biol. Chem.* 263:10613–19
41. Kuchler K, Dohlman HG, Thorner J. 1993. The a-factor transporter (STE6 gene product) and cell polarity in the yeast, *Saccharomyces cerevisiae. J. Cell Biol.* 120:1203–15
42. Maathuis FJM, Sanders D. 1992. Plant membrane transport. *Curr. Opin. Cell Biol.* 4:661–69
43. Maathuis FJM, Sanders D. 1993. Energization of potassium uptake in *Arabidopsis thaliana. Planta.* 191:302–7
44. McCurdy DW, Harmon AC. 1992. Phosphorylation of a putative myosin light chain in Chara by calcium-dependent protein kinase. *Protoplasma* 171:85–88
45. Moriau L, Bogaerts P, Jonniaux JL, Boutry M. 1993. Identification and characterization of a second plasma membrane H⁺-ATPase gene subfamily in *Nicotiana plumbaginifolia. Plant. Mol. Biol. Int. J. Mol. Biol. Biochem. Genet. Eng.* 21:955–63
46. Naprstek J, Walderhaug MO, Epstein W. 1992. Purification and kinetic characterization of the *Kdp*-ATPase. *Ann. NY Acad. Sci.* 671:481–83
47. Palmer JM, Short TW, Gallagher S, Briggs WR. 1993. Blue light-induced phosphorylation of a plasma membrane-associated protein in *Zea mays* L. *Plant Physiol.* 102: 1211–18
48. Palmgren MG, Christensen G. 1993. Complementation in situ of the yeast plasma membrane H⁺-ATPase gene *pma1* by an H⁺-ATPase gene from a heterologous species. *FEBS Lett.* 317:216–22
49. Palmgren MG, Christensen G. 1994. Functional comparisons between plant plasma membrane H⁺-ATPase isoforms expressed in yeast. *J. Biol. Chem.* In press
50. Pardo JM, Serrano R. 1989. Structure of a plasma membrane H⁺-ATPase gene from the plant *Arabidopsis thaliana. J. Biol. Chem.* 264:8557–62
51. Parets-Soler A, Pardo JM, Serrano R. 1990.
Immunocytolocalization of plasma membrane H⁺-ATPase. *Plant. Physiol.* 93: 1654–58
52. Perez C, Michelet B, Ferrant V, Bogaerts P, Boutry M. 1992. Differential expression within a three-gene subfamily encoding a plasma membrane proton ATPase in *Nicotiana plumbaginifolia. J. Biol. Chem.* 267: 1204–11
53. Polarek JW, Williams G, Epstein W. 1992. The products of the kdpDE operon are required for expression of the *Kdp* ATPase of *Escherichia coli. J. Bacteriol.* 174:2145–51
54. Putnam-Evans C, Harmon AC, Palevitz BA, Fechheimer M, Cormier MJ. 1989. Calcium-dependent protein kinase is localized with F-actin in plant cells. *Cell Motil. Cytoskel.* 12:12–22
55. Putnam-Evans CL, Harmon AC, Cormier MJ. 1990. Purification and characterization of a novel calcium-dependent protein kinase from soybean. *Biochemistry* 29:2488–95
56. Rao R, Nakamoto RK, Verjovski-Almeida S, Slayman CW. 1992. Structure and function of the yeast plasma membrane H⁺-ATPase. *Ann. NY Acad. Sci.* 671:195–203
57. Rao R, Slayman CW. 1993. Mutagenesis of conserved residues in the phosphorylation domain of the yeast plasma membrane H⁺-ATPase. Effects on structure and function. *J. Biol. Chem.* 268:6708–13
58. Reymond P, Short TW, Briggs WR. 1992. Blue light activates a specific protein kinase in higher plants. *Plant Physiol.* 100: 655–61
59. Reymond P, Short TW, Briggs WR, Poff KL. 1992. Light-induced phosphorylation of a membrane protein plays an early role in signal transduction for phototropism in *Arabidopsis thaliana. Proc. Natl. Acad. Sci. USA* 89:4718–21
60. Riesmeier JW, Willmitzer L, Frommer WB. 1993. Isolation and characterization of a sucrose carrier cDNA from spinach by functional expression in yeast. *EMBO J.* 11:4705–13
61. Roberts DM, Harmon AC. 1992. Calcium-modulated proteins: targets of intracellular calcium signals in higher plants. *Annu. Rev. Plant. Physiol. Plant. Mol. Biol.* 43:375–414
62. Robinson LC, Hubbard EJA, Graves PR, Depaoli RAA, Roach PJ. 1992. Yeast casein kinase I homologues: an essential gene pair. *Proc. Natl. Acad. Sci. USA* 89: 28–32
63. Robinson LC, Menold MM, Garrett S, Culbertson MR. 1993. Casein kinase I-like protein kinases encoded by *YCK1* and *YCK2* are required for yeast morphogenesis. *Mol. Cell. Biol.* 13:2870–81

64. Rodriguez-Navarro A, Blatt MR, Slayman CL. 1986. A potassium proton symport in *Neurospora crassa. J. Gen. Physiol.* 87: 649–74

65. Rück A, Palme K, Venis MA, Napier RM, Felle HH. 1993. Patch-clamp analysis establishes a role for an auxin binding protein in the auxin stimulation of plasma membrane current in *Zea mays* protoplasts. *Plant J.* 4:41–46

66. Samuels AL, Fernando M, Glass ADM. 1992. Immunofluorescence localization of plasma membrane H$^+$-ATPase in barley roots and effects of K nutrition. *Plant Physiol.* 99:1509–14

67. Schachtman DP, Schroeder JI, Lucas WJ, Anderson JA, Gaber RF. 1992. Expression of an inward-rectifying potassium channel by the *Arabidopsis KAT1* cDNA. *Science* 258:1654–58

68. Schaller GE. 1990. *Studies on the structure and kinase-mediated phosphorylation of the H$^+$-ATPase from oat plasma membrane.* PhD thesis. Univ Wisconsin, Madison

69. Schaller GE, Harmon AC, Sussman MR. 1992. Characterization of a calcium- and lipid-dependent protein kinase associated with the plasma membrane of oat. *Biochemistry* 31:1721–27

70. Schaller GE, Sussman MR. 1988. Phosphorylation of the plasma membrane H$^+$-ATPase of oat roots by a calcium-stimulated protein kinase. *Planta* 173:509–18

71. Schaller GE, Sussman MR. 1988. Isolation and sequence of tryptic peptides from the proton-pumping ATPase of the oat plasma membrane. *Plant Physiol.* 86:512–16

72. Schroeder JI, Fang HH. 1991. Inward-rectifying K$^+$ channels in guard cells provide a mechanism for low-affinity K$^+$ uptake. *Proc. Natl. Acad. Sci. USA* 88:11583–87

73. Sentenac H, Bonneaud N, Minet M, Lacroute F, Salmon JM. 1992. Cloning and expression in yeast of a plant potassium ion transport system. *Science* 256: 663–65

74. Serrano R. 1984. Purification of the proton pumping ATPase from plant plasma membranes. *Biochem. Biophys. Res. Commun.* 121:735–40

75. Serrano R. 1985. *Plasma Membrane ATPase of Plants and Fungi.* Boca Raton, FL: CRC

76. Serrano R. 1989. Structure and function of the plasma membrane ATPase. *Annu. Rev. Plant Physiol. Plant Mol. Biol.* 40:61–94

77. Serrano R. 1993. Structure, function and regulation of plasma membrane H$^+$-ATPase. *FEBS Lett.* 325:108–11

78. Serrano R, Portillo F, Monk BC, Palmgren MG. 1992. The regulatory domain of fungal and plant plasma membrane H$^+$-ATPase. *Acta Physiol. Scand.* 146 (Suppl. 607):131–36

79. Serrano R, Villalba JM, Palmgren MG, Portillo F, Parets-Soler A. 1992. Studies of the plasma membrane H$^+$-ATPase of yeast and plants. *Biochem. Soc. Trans.* 20:562–66

80. Sheahan JJ, Ribeiro-Neto L, Sussman MR. 1993. Cesium-insensitive mutants of *Arabidopsis thaliana. Plant J.* 3:647–56

81. Short TW, Reymond P, Briggs WR. 1993. A pea plasma membrane protein exhibiting blue light-induced phosphorylation retains photosensitivity following triton solubilization. *Plant Physiol.* 101:647–55

82. Stein JC, Howlett B, Boyes DC, Nasrallah ME, Nasrallah JB. 1991. Molecular cloning of a putative receptor protein kinase gene encoded at the self-incompatibility locus of *Brassica oleracea. Proc. Natl. Acad. Sci. USA* 88:8816–20

83. Surowy TK, Sussman MR. 1987. Molecular cloning of the plant plasma membrane H$^+$-ATPase. In *Genetic Aspects of Plant Mineral Nutrition,* ed. HW Gabelman, BC Loughman, pp. 579–90. Dordrecht: Martinus Nijhoff

84. Sussman MR. 1992. Shaking *Arabidopsis thaliana. Science* 256:619

85. Sussman MR, Harper JF. 1989. Molecular biology of the plasma membrane of higher plants. *Plant Cell* 1:953–60

86. Sussman MR, Surowy TK. 1987. Physiology and molecular biology of membrane ATPases. *Oxford Surv. Plant. Mol. Cell Biol.* 4:47–70

87. Tobias CM, Howlett B, Nasrallah JB. 1992. An *Arabidopsis thaliana* gene with sequence similarity to the S-locus receptor kinase of *Brassica oleracea*—sequence and expression. *Plant Physiol.* 99:284–90

88. Tsay YF, Frank MJ, Page T, Dean C, Crawford NM. 1993. Identification of a mobile endogenous transposon in *Arabidopsis thaliana. Science* 260:342–44

89. Tsay YF, Schroeder JI, Feldmann KA, Crawford NM. 1993. The herbicide sensitivity gene CHL1 of *Arabidopsis* encodes a nitrate-inducible nitrate transporter. *Cell* 72:705–13

90. Varada A, Gilmore-Heber M, Benz AJ. 1993. Amplification of the phosphorylation site-ATP binding site cDNA fragment of the Na$^+$,K$^+$-ATPase and the Ca^{+2}-ATPase of *Drosophila melanogaster* by polymerase chain reaction. *FEBS Lett.* 258: 203–7

91. Villalba JM, Lutzelschwab M, Serrano R. 1991. Immunocytolocalization of plasma-membrane H$^+$-ATPase in maize coleoptiles and enclosed leaves. *Planta* 185:458–61

92. Villalba JM, Palmgren MG, Berberián GE, Ferguson C, Serrano R. 1992. Functional expression of plant plasma membrane H$^+$-ATPase in yeast endoplasmic reticulum. *J. Biol. Chem.* 267:12341–49

93. Wach A, Schlesser A, Goffeau A. 1992. An alignment of 17 deduced protein sequences from plant, fungi, and ciliate H⁺-ATPase genes. *J. Bioenerg. Biomembr.* 24:309–17
94. Wada M, Takano M, Kasamo K. 1992. Nucleotide sequence of a complementary DNA encoding plasma membrane H⁺-ATPase from rice (*Oryza sativa* L.). *Plant Physiol.* 99:794–95
95. Walderhaug MO, Polarek JW, Voelkner P, Daniel JM, Hesse JE. 1992. *KdpD* and *KdpE*, proteins that control expression of the kdpABC operon, are members of the two-component sensor-effector class of regulators. *J. Bacteriol.* 174:2152–59
96. Walker JC. 1993. Receptor-like protein kinase genes of *Arabidopsis thaliana. Plant. J.* 3:451–56
97. Walker JC, Zhang R. 1990. Relationship of a putative receptor protein kinase from maize to the S-locus glycoproteins of *Brassica. Nature* 345:743–46
98. Zhao Y, Kappes B, Franklin RM. 1993. Gene structure and expression of an unusual protein kinase from *Plasmodium falciparum* homologous at its carboxyl terminus with the EF hand calcium-binding proteins. *J. Biol. Chem.* 268:4347–54

Annu. Rev. Plant Physiol. Plant Mol. Biol. 1994. 45:235–56

DIURNAL REGULATION OF PHOTOSYNTHETIC CARBON METABOLISM IN C3 PLANTS

Donald R. Geiger and Jerome C. Servaites

Department of Biology, University of Dayton, Dayton, Ohio 45469-2320

KEY WORDS: Calvin cycle, carbon allocation, end product inhibition, assimilatory segment, light activation

CONTENTS

INTRODUCTION 235
ORGANIZATION AND REGULATION OF THE CALVIN CYCLE 236
   Organization of the Cycle 236
   The Calvin Cycle—A Responsive Carbon Conveyor 237
   Levels of Regulation 239
   Biochemical Conductance and Potential 240
ACCLIMATION DURING THE DIURNAL LIGHT REGIME 241
INTEGRATION OF ASSIMILATION AND CARBON ALLOCATION 243
   Carbon Allocation in the Calvin Cycle 243
   Diurnal Regulation of Carbon Allocation and Export 248
   End Product Inhibition of Photosynthesis 250
CONCLUDING REMARKS 251

INTRODUCTION

A measure of successful diurnal regulation of photosynthesis is the ability of the plant to maintain internal balance among the various component processes, while responding to current environmental conditions. Balance must be preserved among the component physical, photochemical, and biochemical processes, the supply of carbon and energy to these processes and the allocation of products of carbon assimilation. Achieving this balance is difficult because the processes being regulated and integrated have durations ranging from 10^{-15} to

235

10^{+4} s, a span of time covering 20 orders of magnitude (49). This review focuses on diurnal regulation of C_3 photosynthesis with emphasis on how the component processes interact.[1]

ORGANIZATION AND REGULATION OF THE CALVIN CYCLE

Organization of the Cycle

The Calvin cycle (Figure 1) uses the products of the light reactions (ATP and NADPH) to convert CO_2 into pivotal intermediates that supply carbon to various synthetic pathways inside as well as outside the chloroplast. Integration of the Calvin cycle into carbon metabolism can be understood by considering the working of its two operational parts: an unbranched, assimilatory segment and a system of branched pathways from which carbon is allocated to various uses (31, 81).

The assimilatory segment consists of the sequence of five enzymes (PRK, Rubisco, PGK, Gal3PDH and TPI). Loading this segment with carbon from CO_2, phosphate from ATP, and electrons from NADPH transforms Ru5P and CO_2 to 2 molecules of triose-P (Figure 1). The sum of these inputs produces a large change in free energy, which favors carbon assimilation (4, 5). Lacking branching, the carbon is conveyed through each of the enzymes in the segment according to a fixed relationship.

By contrast, the system of branched pathways allocates newly assimilated carbon to RuBP regeneration or one of several synthetic pathways outside the cycle. The only input is the product of the assimilatory segment, Gal3P, and its availability sets the pace. Except for FBPase and SBPase, the reactions of this portion of the cycle are all nearly reversible. The internal branching produces a variety of intermediates that are common to both the regeneration of RuBP and

---

[1]
  Abbreviations: ADPG: ADP-glucose; ADPGPP: ADP-glucose pyrophosphorylase (EC 2.7.7.27); BPGA: 1,3-bisphosphoglycerate; CA1P: 2-carboxyarabinitol-1-phosphate; C_i: leaf internal CO_2 concentration; DAHP: 3-deoxy-D-arabino-heptulosonate-7-phosphate; DHAP: dihydroxyacetone phosphate; E4P: erythrose 4-phosphate; E_t: enzyme concentration; EPSP: 5-enolpyruvylshikimic acid-3-phosphate; F6P: fructose 6-phosphate; F2,6BP: fructose 2,6-bisphosphate; FBP: fructose 1,6-bisphosphate; FBPase: fructose bisphosphatase (EC 3.1.3.11); Fd/Td: ferredoxin/thioredoxin system; G1P: glucose 1-phosphate; G6P: glucose 6-phosphate; Gal3P: glyceraldehyde 3-phosphate; Gal3PDH: $NADP^+$-glyceraldehyde 3-phosphate dehydrogenase (EC 1.2.1.13); [Gal3PDH]: the assemblage of enzymes involved in conversion of PGA to DHAP; PCO: photorespiratory carbon oxidation; PGA: 3-phosphoglyceric acid; PGI: phosphoglucoisomerase (EC 5.3.1.9); PGK: 3-phosphoglycerate kinase (EC 2.7.2.3); PGM: phosphoglucomutase (EC 2.7.5.1); PRK: phosphoribulokinase (EC 2.7.1.19); R5P: ribose 5-phosphate; RA: Rubisco activase; Rubisco: ribulose 1,5-bisphosphate carboxylase/oxygenase (EC 4.1.1.39); RuBP: ribulose 1,5-bisphosphate; Ru5P: ribulose 5-phosphate; S7P: sedoheptulose 7-phosphate; SBP: sedoheptulose 1,7-bisphosphate; SBPase: sedoheptulose 1,7- bisphosphatase (EC 3.1.3.37); SPS: sucrose phosphate synthase (EC 2.4.1.14); TPI: triose-P isomerase (EC 5.3.1.1); Xu5P: xylulose 5-phosphate

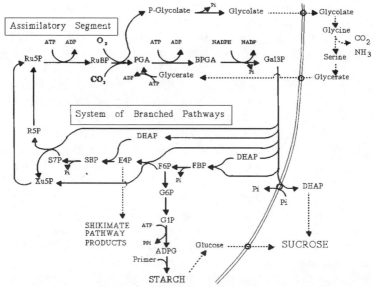

Figure 1   Diagram of the Calvin cycle showing inputs, the paths of carbon flow and allocation.

the synthetic pathways that supply carbon to make essential products. For instance, the chloroplastic FBPase reaction provides carbon both for regeneration of RuBP and starch synthesis (70). While the carbon allocated to various uses does not follow a fixed stoichiometry, the amount leaving the cycle from the branched portion must not exceed that required to support the assimilatory segment. If exit of carbon from the cycle were unregulated, the loss of carbon from the cycle could outstrip the rate of carbon entry and quickly deplete the cycle. For example, if an excessive amount of E4P were used by the shikimate pathway, insufficient carbon would be available to regenerate RuBP.

The Calvin Cycle—A Responsive Carbon Conveyor

Comparing the Calvin cycle to an assembly line conveyor can provide some unique insights and can enable us to envision how carbon flow through the various enzyme catalyzed steps is regulated. The analogy also may help provide insights for understanding integration of the cycle with CO_2 supply, the light reactions, and carbon allocation.

As with the sequence of operations that occur along a conveyor, the product of each reaction in the Calvin cycle is the starting material for the next. If flow through any point fails to keep pace, then starting materials accumulate and products deplete downstream. Because the sequential reactions comprise a closed loop, flux through each reaction must be in balance with the rest, except for brief periods of transition.

Because flux is rapid and the pools of cycle intermediates are small and turn over within seconds, maintaining stability could be difficult. For example, in a photosynthesizing leaf, flux through the assimilatory segment may be on the order of 25 μmol CO_2 m^{-2} s^{-1}, while the concentration of RuBP in the stroma may be as low as 50 μmol m^{-2}. Clearly, if the regeneration rate of RuBP continued to lag behind the carboxylation rate, RuBP would be depleted below its effective concentration in a matter of seconds and the Calvin cycle would be disrupted (29).

Fortunately, such disruptions and instability generally do not occur. Unlike an ordinary conveyor, the Calvin cycle is self-correcting, that is, imbalance sets in motion its own correction. Changes in concentrations of reactants and products that result from imbalance alter the potential that drives the reaction, adjust flux and, thereby, correct the imbalance (for details see 31). A factor that constrains a process such as regeneration of RuBP will be the primary cause in the slowing of the cycle only during the brief transition to the new state. While this factor may be singled out as being limiting, it is clearly now only one among a number of factors that determine the new state.

Flux within the cycle must also be adjusted to establish balance compatible with cycle inputs and outputs. The result is just-in-time availability of CO_2, ATP and NADPH and delivery of Calvin cycle metabolites to produce starch, sucrose and other products in correct proportions. In this regard, control mechanisms are present to dissipate excess light energy to match the use of light energy with the rate of carbon assimilation (15, 53, 62). Likewise, utilization of the triose-P produced by the cycle may regulate carbon assimilation (26, 36, 41, 83, 100).

Table 1   Regulation of individual Calvin cycle enzymes at different levels of physiological organization.

Level of regulation	Acclimation achieved	Regulation at enzyme level	Regulation at system level
Transcription and translation; E_t, k_{cat}	Adjustment to conditions that prevail during development.	Total amount or activity adjusted to conditions prevailing over days.	Relative enzyme amounts of activities adjusted to determine system capacity.
Enzyme activation state RA, CA1P, Fd/Td	General adjustment to diurnal conditions.	Activation state level adjusted to diurnal factors.	Activation state balances photon flux with pathway conductance.
Catalysis K_m, K_i	Fine adjustment to moment to moment changes.	Levels of substrates, products, and effectors control rate of catalysis.	Self-regulation establishes balance among enzymes in a sequence or pathway.

Levels of Regulation

Metabolic regulation of photosynthetic carbon metabolism enables a plant to respond to both environmental and internal factors over a wide range of time spans. Ultimately, such regulation is based on the regulation of individual enzymes at different levels of physiological and structural organization (Table 1). Regulation of gene expression determines the potential maximum enzymatic capacity in response to general growth conditions that prevail over a span of days. The levels of individual enzymes also appear to be regulated in relation to one another. For instance, a reduction in the level of Rubisco results in an adjustment in the levels of other Calvin cycle enzymes (67). The molecular mechanisms involved in balancing copy number of enzymes in relation to each other are not yet clear.

Some enzymes that comprise a metabolic pathway and perform a specific synthetic reaction are arranged into multienzyme complexes (43). Immunoelectron microscopy has shown that PRK, Gal3PDH, SBPase, and ferredoxin-NADP$^+$ reductase are bound to stroma-faced thylakoid membranes, whereas Rubisco and RA are distributed randomly throughout the stroma (94).

Maximal enzymatic capacity, which depends on the total amount of enzyme present, can be attenuated by lowering activation state, among other ways. In the Calvin cycle, the activities of five enzymes are modulated by light driven changes in activation state (10, 11). Deactivation at night prevents futile cycling (10, 11), but, more importantly, changes in activation state maintain a general correspondence between the level of enzyme activity and the flux of light driven electron transport during the diurnal light period (81). For PRK, FBPase, SBPase and Gal3PDH, activation involves reduction of sulfhydryl groups via the Fd/Td system (10, 11). Rubisco activation, by contrast, involves carbamylation of a specific lysine residue (60), mediated via RA (73, 74). Regulation of the sulfhydryl-activated enzymes is complex because the levels of substrates or effector molecules also influence the activation state of these enzymes (75).

The activation states of the various light regulated enzymes of the Calvin cycle appear to change separately and at different rates in response to light (39, 52, 73, 75, 81). To balance flux throughout the Calvin cycle under these conditions requires additional control based on emergent properties of the system of control metabolites interacting with a series of responsive enzymes (31). The resulting self-regulation involves interaction of metabolites not only with a single enzyme but also with other enzymes in the pathway. The increasing levels of complexity in regulation are needed so carbon assimilation can be matched closely with the rate of ATP and NADPH synthesis by the light reactions during the diurnal cycle.

Table 2 Factors involved in regulating biochemical conductance and potential of enzymes of the assimilatory segment of the Calvin cycle.

Enzyme	Biochemical conductance	Biochemical potential
PRK	E_t, k_{cat}, Fd/Td	$\dfrac{[ATP]}{[ADP][Pi]}$ and $\dfrac{[RuBp]}{[R5P]}$
Rubisco	E_t, k_{cat}, CA1PRA	$\dfrac{[CO_2]}{[O_2]}$ and $\dfrac{[RuBp]}{[PGA]}$
[Gal3PDH]	E_t, k_{cat}, Fd/Td	$\dfrac{[ATP]}{[ADP][Pi]}$, $\dfrac{[PGA]}{[DHAP]}$ and $\dfrac{[NADPH][H^+]}{[NADP]}$

Biochemical Conductance and Potential

Regulatory mechanisms vary flux through an enzyme catalyzed reaction by altering biochemical conductance, potential, or both (Table 2). Biochemical conductance may be changed in the long term by altering the amount of enzyme and, more rapidly, by changing activation state or the concentration of effectors or inhibitors (Table 1). The simultaneous increases in light and temperature that usually occur throughout the day allow enzyme activity to keep pace with increasing availability of light energy. The rate of enzyme-catalyzed reactions generally increases with temperature within the temperature range in which the enzyme is stable and retains full activity. In view of the Q_{10} for Rubisco of 2.21 (20) and the likelihood that other reactions in the cycle have similar Q_{10} values (70), increasing temperature should allow photosynthesis rate to follow irradiance. Increasing temperature also raises the level for light saturation of photosynthesis (20), allowing the leaf to utilize light energy at midday.

Light-mediated changes in the activation states of PRK, Rubisco, and Gal3PDH and the levels of control metabolites regulate biochemical conductance, that is, they affect "resistance to flux through the pathway" (38). Woodrow & Berry (100) have developed equations for describing Rubisco regulation and Dietz & Heber (17) have done likewise for [Gal3PDH]. Changing biochemical potential by altering the concentration ratio of substrates and products is crucial for matching the activities of a particular enzyme with those of other enzymes in the assimilatory segment and with photon flux. In the assimilatory segment, only PRK and [Gal3PDH] use ATP as substrate and thereby are dependent upon phosphorylation potentials while only [Gal3PDH] is sensitive to changes in reduction potential. These potentials generally change little with irradiance level and are not used as a primary means of regulating diurnal photosynthesis rate (see below).

ACCLIMATION DURING THE DIURNAL LIGHT REGIME

The amount of sunlight striking a unit area of the earth at any time in the course of a day is a direct function of the sine of the angle that the sun makes with the earth's surface. The light available to leaves for photosynthesis depends both on the time course of diurnal irradiance and on factors, such as leaf orientation, that affect the interception of the incident light. Evolution of photosynthesizing organisms under cyclic diurnal irradiance has equipped plants to regulate the photosynthetic process in ways that allow carbon to be assimilated efficiently over the wide range of diurnal cycle irradiance. Because of this adaptation, it is advantageous to study regulation of photosynthesis and carbon metabolism in the context of the diurnal light cycle. In particular, timing aspects of carbon assimilation regulation can be analyzed more effectively because the gradual changes in irradiance allow us to observe the step by step progress of the daily acclimation process. A number of researchers used this approach to elucidate mechanisms involved in the regulation of photosynthesis in nature (25, 33, 69, 78, 79).

The rate of net CO_2 assimilation generally follows the time course of irradiance during the first several hours (25, 33) but the amount of carbon fixed per incident photon decreases at midday (53). The changes in apparent quantum efficiency throughout the day require that the photosynthetic apparatus acclimates to the change in the proportion of incident light being used for carbon fixation (15, 53). To maintain the critical balance between energy absorption and use, and so maintain photosynthetic efficiency (62), carbon flux through each enzymatic reaction in the Calvin cycle is regulated to keep pace with changes in light level during the diurnal cycle. Light regulation of carbon assimilation enzymes involves changes in both activation state (39) and levels of control metabolites (78, 79). A key issue in understanding diurnal regulation of the Calvin cycle is ascertaining the roles and relative importance of changes in biochemical potential and conductance.

Changes in phosphorylation and reduction potentials generally are small compared to the large diurnal range of irradiance levels and carbon fluxes (9, 17, 18, 39, 81). Leaf ATP and NADPH levels increase somewhat during the dark/light transition but generally change little during the ensuing day (81, 91), evidence that the rate of use remains close to the rate of synthesis (9, 39, 81). Because these metabolites are maintained at relatively steady levels in photosynthesizing cells (18), their turnover rates rather than their concentrations are good indicators of carbon or electron fluxes of photosynthesis. The fact that the levels of these metabolites do not follow closely the time course of photosynthesis (81) indicates that their concentrations are not major factors in the regulation of photosynthesis rate (39). Furthermore, it does not seem feasible to regulate the activity of one enzyme by changing the level of these metabolites, because this would unavoidably alter the activity of other enzymes for which they are a substrate.

Rather, the match between photosynthesis rate and irradiance occurs by changes in biochemical conductance of the assimilatory segment enzymes (17, 18, 38). The relatively small changes in reduction potential that occur during the course of the dark/light transition and subsequent diurnal irradiance cycle are sufficient to induce the observed changes in the Fd/Td activated enzymes. Large changes in irradiance such as sun flecks produce a mismatch between the rates of synthesis and use of ATP and NADPH (64) and result in short term oscillations in the ratios of $[ATP]/[ADP][P_i]$ and $[NADPH][H^+]/[NADP^+]$ (84, 90, 99).

The enzymes PRK, Gal3PDH, and coupling factor reach full activation at rather low light levels (39, 73), while the activation of FBPase and SBPase, like Rubisco, require higher light levels (81). Hutchison & Ort (47) have shown that PRK and coupling factor, which attain full activation at low light, have lower midpoint redox potentials, as determined by in vitro titration. The response of these two enzymes to light level likely is significant in the initiation of the Calvin cycle during the dark/light transition. The rather rapid activation of the Fd/Td activated enzymes with increasing light and their rather slow deactivation with decreasing light (52) is evidence for a sizable pool of reductants within the chloroplast stroma that buffers change in potential. Metabolite levels may also affect the process of light mediated enzyme activation (75). For example, the presence of FBP considerably reduces the redox potential needed to activate FBPase (75).

Differences in the time course of light activation of the various Calvin cycle enzymes can induce a momentary imbalance in carbon flux at points in the pathway and this induces self-regulation by metabolites. By the time the activation state of Rubisco begins to increase during the diurnal light regime, the activation states of the two flanking enzymes, PRK and Gal3PDH, are maximal (81). Sequential activation of enzymes also appears to help prepare Calvin cycle control systems and thus enables carboxylation rate to follow the increasing light driven electron flux. Imbalance in conductance between enzymes is translated into changes in levels of RuBP and PGA, which in turn modulate enzyme activity through self-regulation (31, 81). The early light activation of coupling factor and PRK provides a way to generate ATP and RuBP. The increase in the level of RuBP during the early part of the light period is necessary for the RA-mediated activation of Rubisco (73) and the down-regulation of the highly activated PRK.

The amount of active Rubisco is regulated by various combinations of RA-mediated carbamylation of the enzyme and release of CA1P, depending on the species of plant (63, 73, 76, 77). Some of the RuBP that is made early in the light period is bound to Rubisco to form an inactive complex (8). This complex functions as the substrate for activation of Rubisco by RA (65, 73). Another portion of Rubisco is present in the carbamylated state, but is inactive because of the binding of the inhibitor CA1P. Finally, a certain portion of the Rubisco remains catalytically active even in the dark. Diurnal increase in Rubisco activity is mediated by carbamylation of the enzyme and release of CA1P, both mediated by RA (65). It is not clear how increases in irradiance level are translated into a higher Rubisco activation state. ATP is required for RA-mediated activation of Rubisco, but, because ATP level does not change markedly with the course of diurnal irradiance, it cannot be the agent that explains the time course of Rubisco activation (9). RA is activated by light, but at much lower levels than Rubisco (57), indicating some additional involvement of irradiance in activating Rubisco. Campbell & Ogren (12) concluded that stimulation of light activation of Rubisco by RA required electron transport through photosystem I and that the needed light was not acting simply to supply ATP for the RA reaction.

INTEGRATION OF ASSIMILATION AND CARBON ALLOCATION

Carbon Allocation in the Calvin Cycle

Triose-P produced by the assimilatory segment of the Calvin cycle is distributed among a number of pathways. Most of the carbon remains in the chloroplast and enters reactions of the system of branched pathways at five points (Figure 1). The rest enters the cytosol via the phosphate translocator. The analogy to an assembly line conveyor helps us envision the need for regulating distribution of triose-P by a responsive and versatile system. Calvin cycle regulatory mechanisms serve three major functions in relation to entry into these pathways. First, entry of triose-P into the pathways at the five points needs to be regulated in response to current conditions. Control of the entry of triose-P into these pathways for alternative uses integrates photosynthetic carbon metabolism with the plant's physiology and development. Second, restricting entry of triose-P into each of these synthetic pathways helps assure that a sufficient reserve of newly assimilated carbon is available for RuBP regeneration within the Calvin cycle. Carbon flux for regeneration of RuBP must be preserved by regulating exit into alternative pathways, because even one open exit would quickly deplete the Calvin cycle of carbon and, thus, be detrimental to carbon assimilation (29, 34, 82, 85). Finally, regulating triose-P allocation enables the plant to acclimate when a partially closed pathway restricts triose-P utilization. For example, decreased ability to synthesize sucrose because of sink limitation or other causes of limited export may reduce the rate of regeneration of P_i needed for photophosphorylation (41, 83). Feedback limitation of triose-P utilization is characterized by the ability of photosynthesis to respond to an increase in CO_2 and a decrease in O_2 concentrations (83, 84). Alternative pathways for utilizing triose-P allow carbon to be diverted to other products, thereby allowing turnover of newly assimilated carbon and associated P_i to continue (54).

Regulation of triose-P allocation provides for its transport into the cytosol to maintain adequate carbon export to sinks, while retaining an adequate amount in the chloroplast for regeneration of RuBP and synthesis of other products of the Calvin cycle. Transport of triose-P into the cytosol by the phosphate transporter in strict counter exchange for P_i or PGA is thought to be regulated primarily by cytosolic processes, particularly by regulation of sucrose synthesis (92). It is the rate of regeneration of P_i in the cytosol rather than the level of triose-P in the stroma that paces the exit of triose-P. When mannose was fed to leaves, the level of P_i in the cytosol was lowered, thus slowing exit of triose-P from the chloroplast and entry of P_i into the chloroplast. As a consequence, allocation of newly assimilated carbon to starch increased and the lower level of P_i in the chloroplast slowed photosynthesis (42, 98).

RuBP REGENERATION   An adequate pool of RuBP to support carboxylation and oxygenation is maintained by entry into the pathway for regeneration of RuBP of 5 of the 6 triose-P produced per 3 CO_2 fixed. Two aspects of regulation help assure that the rate of RuBP regeneration is adequate to support rates of carboxylation and oxygenation. First, in the absence of marked stress, carbon assimilation produces more than enough triose-P to regenerate the RuBP needed to accept CO_2 at the current fixation rate. Second, PRK catalyzes the committed step that takes carbon from the system of branched pathways and produces RuBP. It is very unlikely that PRK and Rubisco could outpace one or the other for very long since the activities of the two are closely balanced through self-regulation. However, stresses may cause a slowing of RuBP regeneration, resulting in a slowing of photosynthesis (20, 83). For instance, diversion of carbon from RuBP synthesis to other Calvin cycle products (85) or inhibition of key enzymes (74a) may cause this slowing.

Regulation of RuBP regeneration is an important process for explaining how photosynthesis rate responds to a change in internal leaf CO_2 level (20, 72, 83). Under high light at CO_2 concentrations somewhat below atmospheric level, carbon assimilation rate is considered to be limited by Rubisco activity, whereas above atmospheric concentration range, CO_2 levels have much less effect on photosynthesis rate and carbon fixation is said to be limited by RuBP regeneration. Light level affects both phosphorylation and reduction potentials and the activation state of a number of enzymes, whereas changing CO_2 level, as far as we know, affects the activity of only one Calvin cycle enzyme,

Rubisco. If this is the case, changing CO_2 level will have its earliest direct effects on the Rubisco reaction and on the levels of the other substrate (RuBP) and the product (PGA) (71, 72, 96). Not only are these metabolites directly involved in the reaction but they also interact with the level of Rubisco activity in a self-regulating system (81). Subsequent regulatory responses, principally a change in the activation state of Rubisco (71, 72, 96), appear likely to be initiated directly or indirectly by these metabolite changes.

When the C_i level in a leaf photosynthesizing in air is increased, the immediate CO_2-induced increase in Rubisco activity increases photosynthesis and causes RuBP level to begin decreasing and PGA to begin increasing, subject to self-regulation. To the extent that the CO_2-induced increase in flux through Rubisco can be supported by the current phosphorylation and reduction potentials, the increase in carbon assimilation rate will be maintained. The decrease in RuBP and the increase in PGA are a necessary part of regulation that limits the increase in carboxylation rate to equal the rates of light-driven phosphorylation and reduction in the assimilatory segment. Likewise, these metabolite changes increase the rates of the enzymes flanking Rubisco to match carboxylation rate. If the increase in carbon assimilation is to be sustained, RuBP regeneration also must increase correspondingly. In the RuBP regeneration segment, CO_2 does not appear to increase the activation state of FBPase and SBPase, so their activities must be increased in other ways, probably by changes in metabolite levels. As a result of these self-regulated changes in PGA, RuBP and, perhaps, other metabolites, the rate of regeneration of RuBP increases to a level that equals the rate of its use. Clearly, the lowering of RuBP is not the result of a limitation of RuBP regeneration (20, 83). It is a necessary part of the regulatory process that restores balance between carboxylation and the activities of the flanking enzymes and matches all of these to light flux. It is the self-regulated lowering of the RuBP level and not the inability to regenerate it faster that is a major factor in restoring and maintaining metabolic balance. With time, Rubisco activation state may be lowered and the metabolite levels restored to near the earlier levels (71, 72), possibly in response to the changes in metabolite levels themselves (75).

When C_i is lowered from atmospheric level in a photosynthesizing leaf, the immediate CO_2-induced decrease in Rubisco activity causes RuBP levels to increase and PGA to decrease. Because the degree of restriction of light level on assimilation rate is relaxed somewhat, the change in assimilation rate per unit decrease in CO_2 at a point below atmospheric CO_2 level will be larger than the change per unit increase in CO_2 above atmospheric level. The activation state of Rubisco does not change initially but its activity decreases as C_i is lowered. The reaction rates of the assimilatory segment enzymes flanking Rubisco are corrected by the self-regulated increase in RuBP and decrease in PGA levels to match Rubisco activity. Under lower CO_2 level, less light energy is used for carbon assimilation and so presumably must be dissipated (15, 53). RuBP regeneration also decreases correspondingly, presumably as a result of the decrease in the rate of triose-P production. Metabolites attain levels where the flux of carbon through Rubisco equals the rate through the other enzymes of the assimilatory segment.

PHOTORESPIRATORY CARBON OXIDATION CYCLE   Under atmospheric levels of CO_2 and O_2, as much as 50% of the recently fixed carbon enters the PCO pathway as P-glycolate (55, 61). About three fourths of this carbon returns to the chloroplast as glyceric acid and reenters the Calvin cycle as PGA. The existence of the PCO cycle enables the initial product, P-glycolate, which cannot be metabolized in the Calvin cycle, to be converted to PGA, thereby returning a substantial amount of carbon to the Calvin cycle (Figure 1). In this sense, the PCO cycle restricts loss of carbon from the Calvin cycle caused by oxygenation of RuBP. Chemically inhibiting specific enzymes of the PCO cycle results in carbon accumulation in photorespiratory intermediates, for example, glycolate or glycine (80). Loss of carbon from the Calvin cycle and severe inhibition of photosynthesis follow. Mutant plants lacking specific enzymes of the PCO cycle likewise accumulate various photorespiratory intermediates and cannot photosynthesize efficiently or grow under normal atmospheres (87). When increasing amounts of carbon drain into PCO intermediates, less and less carbon is incorporated into starch and, eventually, the amount allocated to sucrose also declines.

SHIKIMATE PATHWAY   The shikimate pathway leads to the synthesis of the three aromatic amino acids, several plant growth regulators, and a number of secondary products including lignin and flavanoids. Jensen (48) concluded that regardless of the diversity of secondary metabolites originating from the shikimate pathway in plants, a massive output of products derived from shikimate can be expected. Up to 60% of the ultimate plant dry mass is represented by molecules that once traversed the shikimate pathway. Entry of carbon into the shikimate pathway is restricted by feedback regulation of DAHP synthase. This first enzyme of the shikimate pathway is regulated by arogenate, a precursor of tyrosine and phenylalanine (48). The effectiveness of glyphosate (N-[phosphonomethyl] glycine) as a herbicide is based on the formation of a tight-binding inhibitor of EPSP synthase (1), an enzyme that lies between DAHP synthase and the enzymes involved in synthesis of arogenate and further products.

The importance of restricting entry of carbon into the shikimate pathway is illustrated by the results of studies in which glyphosate was applied to a photosynthesizing sugar beet leaf (34, 82, 85). Within minutes after application, the RuBP level began to decrease, while shikimate-P, a substrate for EPSP synthase, accumulated (29). Amrhein et al (1) demonstrated that this accumulation occurs in the chloroplast. Disruption of feedback regulation of DAHP synthase allows unregulated entry of E4P into the initial portion of the shikimate pathway (29). At first, flexibility of leaf carbon metabolism may allow carbon assimilation to acclimate to the decreasing RuBP level and in this case the photosynthesis rate is maintained. After RuBP concentration decreases to about twice the concentration of RuBP binding sites on Rubisco, starch accumulation and photosynthesis rate begin to decline (2, 82). When the products of the light reactions are not used as fast as produced, photoinhibition results (53). The effect on photosynthesis becomes particularly pronounced at higher light levels, evidence that photoinhibition is an important factor in the herbicidal effect (85).

Starch accumulation is also inhibited by glyphosate-induced disruption of the shikimate pathway (34, 82), further illustrating that unregulated entry of carbon into the shikimate pathway depletes the triose-P used for RuBP regeneration and starch synthesis. Sucrose synthesis continues unabated long after these other processes have been inhibited (37), which supports the view that triose-P entry into the cytosol by the phosphate translocator is regulated primarily by cytosolic metabolism (92). The continued synthesis of sucrose maintains the translocation of both sucrose and glyphosate to sinks, an important factor in its effectiveness as a systemic herbicide (29).

STARCH   As much as half of the triose-P produced by net photosynthesis may remain in the chloroplast and be converted to transitory or metabolic starch for supporting respiration and export of carbon. Regulation of sucrose synthesis in the cytosol is a critical factor in determining the supply of triose-P available for allocation to starch (54, 92). This synthesis is controlled at the ADPGPP step (66, 89) by the levels of PGA and P_i (66). Allocation of triose-P for starch synthesis must be regulated without restricting excessively the synthesis of sucrose for export, and it must provide an adequate reserve of carbon for subsequent metabolism and export (22, 25). Stark et al (89) observed that constitutive expression of a modified gene for ADPGPP that lacks the usual metabolite regulation found in higher plants was detrimental to growth and development of potato plants. When the bacterial form of ADPGPP was expressed in leaves, regulation of carbon entry into starch was lost. The plants required an exogenous sucrose supply for growth, presumably because carbon was diverted from synthesis of sucrose into unregulated synthesis of starch.

In addition to regulation at the ADPGPP step, other sites such as PGI (54) and PGM (13) may regulate allocation of carbon to starch. Results from experimental application of control analysis to regulation of carbon allocation to sucrose and starch (54) demonstrated that even enzymes catalyzing reactions that are close to equilibrium are not present in excess, as was formerly thought. The enzymes actually are found in amounts that are only sufficient to maintain reactants close to equilibrium. Although a pathway may appear to be regulated by only one limiting enzyme, other enzymes that seem to be present in excess also exert measurable control of flux in the pathway.

SUCROSE   To provide a steady supply of carbon for export from leaves, sucrose synthesis must achieve a rate that assures a supply of carbon available for export throughout both light and dark periods (22, 25, 78, 79). Factors that enhance the rate of sucrose synthesis will decrease retention of assimilated carbon in the chloroplast (92). On the other hand, export of triose-P from the chloroplast must not be so fast that it hinders RuBP regeneration. The need to maintain a relatively constant rate of sucrose synthesis in spite of changes in the rate of diurnal photosynthesis (25, 78), different stages of development (21), different levels of sink strength (22), and other internal and environmental factors demands a corresponding complexity in the system for regulation. Several control mechanisms are involved in regulating sucrose synthesis (46), including metabolite regulation of cytosolic FBPase by F2,6BP (26, 50) and of SPS by P_i and G6P (19). Furthermore, SPS activation state is regulated by reversible protein phosphorylation (46) that occurs in response to light (86), circadian regulation (51), nitrogen metabolism (68), and increased leaf carbohydrate content (26). The variation in modes of sucrose synthesis regulation reflects the diversity of carbohydrate metabolism among plants and the responsiveness of sucrose metabolism to different conditions.

Worrell et al (101) expressed maize SPS in tomato leaves under control of the promoter for a gene encoding the light-regulated and leaf-specific Rubisco small subunit from tobacco. Expression of maize SPS in tomato leaves resulted in SPS levels varying from the normal tomato level to six times that level. The maize enzyme does not appear to be regulated in tomato leaves in response to physiological conditions. During the light period, tomato leaves ordinarily accumulate a high level of starch, but only a low level of sucrose. Transgenic plants with an enhanced SPS level had leaf sucrose levels up to twice that of wild-type plants, while starch level was reduced accordingly. Enhanced SPS activity in the photosynthetic tissues increased the shoot to root ratio in favor of shoots (28). Photosynthesis rate of the transgenic plants in air was not changed, but the light- and CO_2-saturated photosynthesis rate was increased by 20%. The ability to increase sucrose synthesis rate may allow higher photosynthesis rates under certain conditions.

Diurnal Regulation of Carbon Allocation and Export

Effective regulation of carbon metabolism in sink tissues requires a relatively steady supply of sucrose and other organic compounds to sustain development of growing regions (22–24). Diurnal regulation of source leaf metabolism generally maintains sucrose synthesis rate throughout both night and day periods, thereby providing relatively steady export to sinks (22–25). In sugar beet leaves, sucrose export is closely controlled throughout the light period by regulation of the proportion of carbon going to sucrose (22). Changing the sink-source ratio has little effect on the diurnal pattern of carbon allocation between sucrose and starch and photosynthesis rate (22). Rather, acclimation to changes in source-sink ratio appears to occur mainly by a change in partitioning of imported carbon among sinks.

Starch synthesis also is closely regulated and follows a characteristic diurnal pattern in photosynthesizing sugar beet leaves. Under the gradually cycling irradiance similar to what occurs outdoors during a natural day period, starch accumulation in exporting leaves follows a sinusoidal time course (25). Starch synthesis begins only after 1 to 2 h and begins to slow about 2 h before the end of the day period. Even under light regimes in which irradiance begins rapidly and remains at a constant intensity throughout the entire day period, starch accumulation approximates the sinusoidal time course of a natural day (24). When irradiance is kept high and photosynthesis rate is well above that needed to maintain starch synthesis, starch accumulation still begins to decrease at the usual time near the end of the day (25).

Circadian regulation of carbon allocation between sucrose and starch was evident when plants were studied under a light regime in which irradiance was kept high during the usual night period (59). Starch accumulation began to slow at the usual time near the day's end and ceased by 4 h into the extended light period. The decrease in starch accumulation resulted in an equal increase in the rate of sucrose synthesis, and because export from sugar beet leaves is set by sucrose synthesis, a near doubling of export was observed by the end of the transition period. This pattern of carbon allocation continued for the remainder of the extended light period. Under these conditions, photosynthesis rate was maintained during the usual day period and began to decrease gradually only near the beginning of the extended portion of the light period (59). Transition in carbon allocation, which normally occurs at the end of the day, appears to be a regular circadian event that prepares the leaf for the accustomed night-time metabolism. Endogenous regulation of carbon allocation is part of the diurnal regulation that responds to photosynthetic duration and length of the night (14, 22, 35). The gradual diurnal shift from sucrose derived from starch to sucrose derived from newly-fixed carbon (25) and back again appears to involve regulation of two partially separate pathways that supply carbon for sucrose synthesis. Beck (6) has proposed that there are different mechanisms for transport of carbon from the chloroplast to the cytosol for sucrose synthesis. At night or under low photosynthesis, a precursor of sucrose other than triose-P (25, 78) appears to exit from the chloroplast by a path that does not involve the phosphate translocator (6). Sucrose is synthesized from triose-P during the day whereas at night this part of the pathway may be down-regulated at cytosolic FBPase by high levels of F2,6BP. At that time, synthesis of sucrose supplied from starch degradation and that from newly assimilated carbon are regulated in a complementary fashion, thereby producing a fairly steady rate of synthesis throughout day and night (25).

End Product Inhibition of Photosynthesis

Source-sink imbalance resulting from impaired export affects diurnal carbon allocation, carbon metabolism, and photosynthesis, but the responses and mechanisms involved are not clear (30, 32). When the rate of carbon use in sinks slows, the level of sucrose in the free space appears to increase and sugars may accumulate in the mesophyll, resulting in an eventual effect on source leaf metabolism (30). The gradual onset of the reduction in carbon use leads to a gradual accumulation of sugars in the source leaf, allowing time for acclimation and restoration of balance. In some cases, photosynthesis may adjust to a new level within one to several days following impaired export (3). Depending on the severity of the imbalance and the species of plant, a change in allocation or inhibition of photosynthesis may occur, but this acclimation is a measured response, often requiring several days (3, 44). The type of carbohydrate stored, sucrose or starch, and the presence of invertase and other enzymes, like fructokinase, appear to be factors affecting the severity of the response (36, 45).

Physical manipulation and, more recently, the study of transgenic plants, in which export has been biochemically inhibited by the expression of a yeast invertase in the leaf cell walls (16, 40, 88, 97), have shown the effects of inhibited export on source metabolism. Girdling of leaf petioles to inhibit export is a rapid but drastic procedure that may produce secondary effects. Following girdling, export is drastically reduced and free space sugars are likely to increase rapidly, although inhibition is not seen immediately. Inhibition of export by transgenic manipulation is generally present throughout the life of the plant. Export normally begins gradually as sink leaves undergo transition to source leaves (21) and the induced blockage of export occurs when a particular section of the leaf converts from a sink to a source. The transport species, sucrose, is converted in the free space to hexoses, which are not readily loaded into the phloem, and consequently export is halted in leaves of plants having apoplastic loading. Depending on the level of transgenic expression, blockage would be expected to vary from incomplete to nearly complete. Both of these have some aspects in common with slowed import that occasionally occurs in ordinary plants with changes in sink strength, but the effects of the two manipulations are much more drastic. The effect on photosynthesis will vary, largely depending on metabolic characteristics of the particular species of plant (36, 45).

As a result of studies involving manipulations of export, several investigators have proposed possible mechanisms to describe how carbon fixation responds to impaired export. Foyer et al (27) hypothesized that inhibition of export leads to increased hydrolysis of sucrose in the vacuole and phosphorylation of the resulting hexoses in the cytosol. Sequestration of cytosolic P_i would increase the net synthesis of F2,6BP and decrease sucrose synthesis. Overall these events would significantly increase the ratio of triose-P to P_i in the cytosol, thus depriving the chloroplasts of P_i and decreasing photosynthesis. However, depletion of cytosolic P_i may be difficult to achieve because cytoplasmic P_i is maintained constant at the expense of large fluctuations in vacuolar P_i (7, 58, 95).

Goldschmidt & Huber (36) proposed that strong end-product inhibition of photosynthesis resulting from impaired export was dependent on the presence of high levels of acid-invertase in leaves. Under these conditions, sucrose hydrolysis within the vacuoles of these plants leads to cycling of sucrose and hexose sugars. When sucrose hydrolysis occurs faster than does phosphorylation of the released hexoses, the latter accumulate to a significant extent. In some species (e.g. soybean), hexose accumulation may be transient. The appearance of free hexoses, even transiently, may initiate down-regulation of the Calvin cycle, possibly as a result of effects on gene expression.

Leaves of transgenic tobacco plants expressing yeast-derived invertase in their cell walls developed green and bleached areas as they matured (93). The green areas contained low levels of soluble sugars and starch, which turned over daily whereas the pale areas accumulated carbohydrate and showed lower levels of photosynthesis and higher levels of respiration. Inhibition of photosynthesis was accompanied by decreased levels of Rubisco and other Calvin cycle enzymes, while enhanced respiration was accompanied by increased levels of glycolytic enzymes. These changes in enzyme levels appeared to be important for mid- to long-term adjustment to high leaf carbohydrate levels. Lambers (56) has suggested that high levels of carbohydrate lead to wasteful respiration (i.e. respiration that depletes assimilated carbon without producing a corresponding amount of ATP) possibly by the alternative oxidase pathway. Indeed, abnormally high rates of respiration were observed in starchless mutants (13). When sucrose accumulates in leaves, short-term regulation mediated by changes in P_i and other metabolites may allow sucrose synthesis to decrease while starch synthesis increases without any loss of photosynthetic rate (93).

CONCLUDING REMARKS

Regulation at three distinct levels of physiological organization provides flexibility that enables plants to acclimate effectively to changing diurnal irradiance. The interactions of these mechanisms are complex and provide flexibility for regulation (31). Studying photosynthesis in intact plants under a diurnal cycle reveals important emergent properties of the intact system that would not be seen from the study of individual components. One of these emergent mechanisms is self-regulation, which integrates these complex interactions and thus refines regulation. Extreme environmental conditions or drastic manipulations may compromise the plant's ability to acclimate. Changes in the global environment, such as increased atmospheric CO_2 and temperature fluctuations, which are products of human activity, may drive regulatory mechanisms to instability. Likewise, attempts to increase productivity and improve yield by genetic manipulations, whether by plant breeding or molecular technology, instead may inadvertently alter the ability of the plant to restore and maintain source-sink balance. Although the altered plant may function well under optimal conditions, it may fail to acclimate well under stress. For example, although enhanced expression of SPS in photosynthetic tissues of transgenic plants may have the potential to increase photosynthetic rates under favorable conditions, it lowers root to shoot ratio (28), a feature that probably is a detriment under water stress. If the potential advantages of recombinant DNA technology are to be realized, manipulations should not disrupt the ability of the plant to regulate photosynthesis and carbon metabolism and to establish and maintain source-sink balance.

ACKNOWLEDGMENTS

We thank Drs. Raymond Chollet and Steven C. Huber for their helpful criticism of the manuscript and Dr. Wen-J. Shieh for constructing the figure. Research in the authors' laboratory has been supported by National Science Foundation, the Competitive Research Grants Office of the United States Department of Agriculture and the Monsanto Agriculture Company.

Literature Cited

1. Amrhein N, Deus B, Gehrke P, Steinrücken HC. 1980. The site of inhibition of the shikimate pathway by glyphosate. II. Interference of glyphosate with chorismate formation in vivo and in vitro. *Plant Physiol.* 66:830–34

2. Badger MR, Sharkey TD, von Cammerer S. 1984. The relationship between steady-state gas exchange of bean leaves and the

levels of carbon-reduction-cycle intermediates. *Planta* 160:305–13

3. Bagnall DJ, King RW, Farquhar GD. 1988. Temperature-dependent feedback inhibition of photosynthesis in peanut. *Planta* 175:348–54

4. Bassham JA, Buchanan BB. 1982. Carbon dioxide fixation pathways in plants and bacteria. In *Photosynthesis*, ed. Govindjee,

2:141–89. New York: Academic
5. Bassham JA, Krause GH. 1969. Free energy changes and metabolic regulation in steady state photosynthetic carbon reduction. *Biochim. Biophys. Acta* 189:207–21
6. Beck E. 1985. The degradation of transitory starch granules in chloroplasts. In *Regulation of Carbon Partitioning in Photosynthetic Tissue*, ed. RL Heath, J Preiss, pp. 27–44. Rockville, MD: Am. Soc. Plant Physiol.
7. Bligny R, Gardeström P, Roby C, Douce R. 1990. ^{31}P NMR studies of spinach leaves and their chloroplasts. *J. Biol. Chem.* 265: 1319–26
8. Brooks A, Portis AR Jr. 1988. Protein-bound ribulose bisphosphate correlates with deactivation of ribulose bisphosphate carboxylase in leaves. *Plant Physiol.* 87: 244–49
9. Brooks A, Portis AR Jr, Sharkey TD. 1988. Effects of irradiance and methyl viologen treatment on ATP, ADP, and activation of ribulose bisphosphate carboxylase in spinach leaves. *Plant Physiol.* 89:735–39
10. Buchanan BB. 1980. Role of light in regulation of chloroplast enzymes. *Annu. Rev. Plant Physiol.* 31:341–74
11. Buchanan BB. 1991. Regulation of CO_2 assimilation in oxygenic photosynthesis: the ferridoxin/thioredoxin system. Perspective on its discovery, present status, and future development. *Arch. Biochem. Biophys.* 288:1–9
12. Campbell WJ, Ogren WL. 1990. Electron transport through photosystem I stimulates light activation of ribulose bisphosphate carboxylase/oxygenase (Rubisco) by rubisco activase. *Plant Physiol.* 94:479–84
13. Caspar T, Huber SC, Somerville C. 1985. Alterations in growth, photosynthesis and respiration in a starchless mutant of *Arabidopsis thaliana* (L.) Heynh, deficient in chloroplast phosphoglucomutase activity. *Plant Physiol.* 79:11–17
14. Chatterton NJ, Silvius JE. 1979. Photosynthate partitioning into starch in soybean leaves. I. Effects of photoperiod versus photosynthesis duration. *Plant Physiol.* 64: 749–53
15. Demmig-Adams B, Adams WW. 1992. Photoprotection and other responses of plants to high light stress. *Annu. Rev. Plant Physiol. Plant Mol. Biol.* 43:599–626
16. Dickinson CD, Altabella T, Crispeels MJ. 1991. Slow-growth phenotype of transgenic tomato expressing apoplastic invertase. *Plant Physiol.* 95:420–25
17. Dietz K-J, Heber U. 1984. Rate-limiting factors in leaf photosynthesis. I. Carbon fluxes in the Calvin cycle. *Biochem. Biophys. Acta* 767:432–43
18. Dietz K-J, Heber U. 1986. Light and CO_2 limitations of photosynthesis and states of

the reactions regenerating ribulose 1,5-bisphosphate or reducing 3-phosphoglycerate. *Biochem. Biophys. Acta* 848: 392–401
19. Doehlert DC, Huber SC. 1983. Regulation of spinach leaf sucrose-phosphate synthase by glucose-6-phosphate, inorganic phosphate and pH. *Plant Physiol.* 73:989–94
20. Farquhar GD, von Caemmerer S. 1982. Modelling of photosynthetic response to environmental conditions. In *Physiological Plant Ecology. II. Water Relations and Carbon Assimilation*, ed. OL Lange, PS Nobel, CB Osmond, H Ziegler, *Encycl. Plant Physiol.* (NS), 12B:549–87. New York: Springer
21. Fellows RJ, Geiger DR. 1974. Structural and physiological changes in sugar beet leaves during sink to source conversion. *Plant Physiol.* 54:877–85
22. Fondy BR, Geiger DR. 1980. Effect of rapid changes in sink-source ratio on export and distribution of products of photosynthesis in leaves of *Beta vulgaris* L. and *Phaseolus vulgaris* L. *Plant Physiol.* 66: 945–49
23. Fondy BR, Geiger DR. 1982. Diurnal pattern of translocation and carbohydrate metabolism in source leaves of *Beta vulgaris* L. *Plant Physiol.* 70:671–76
24. Fondy BR, Geiger DR. 1985. Diurnal changes in allocation of newly fixed carbon in exporting leaves of sugar beet. *Plant Physiol.* 78:753–57
25. Fondy BR, Geiger DR, Servaites JC. 1989. Photosynthesis, carbohydrate metabolism and export in *Beta vulgaris* L. and *Phaseolus vulgaris* L. during a sinusoidal light regime. *Plant Physiol.* 89:396–402
26. Foyer C. 1988. Feedback inhibition of photosynthesis through source-sink regulation in leaves. *Plant Physiol. Biochem.* 26:483–92
27. Foyer C, Furbank RT, Walker DA. 1987. Interactions between ribulose-1,5-bisphosphate carboxylase and stromal metabolites. I. Modulation of enzyme activity by Benson-Calvin cycle intermediates. *Biochim. Biophys. Acta* 894:157–64
28. Galtier N, Foyer CH, Huber J, Voelker TA, Huber SC. 1993. Effects of elevated sucrose-phosphate synthase activity on photosynthesis, assimilate partitioning, and growth in tomato (*Lycopersicon esculentum* var UC82B). *Plant Physiol.* 101:535–43
29. Geiger DR, Bestman H. 1990. Self-limitation of herbicide mobility by phytotoxic action. *Weed Sci.* 38:324–29
30. Geiger DR, Fondy BR. 1991. Regulation of carbon allocation and partitioning: status and research agenda. In *Recent Advances in Phloem Transport and Assimilate Compartmentation*, ed. JL Bonnemain, S Del-

rot, WJ Lucas, J Dainty, pp. 1–9. Nantes, France: Ouest Editions, Presses Acad.

31. Geiger DR, Servaites JC. 1994. Dynamics of self-regulation of photosynthetic carbon metabolism. *Plant Physiol. Biochem.* In press

32. Geiger DR, Servaites JC, Shieh W-J. 1992. Balance among parts of the source-sink system: a factor in crop productivity. In *Crop Photosynthesis: Spatial and Temporal Determinants*, ed. N Baker, H Thomas. *Top. Photosynth.* 12:155–92. Amsterdam: Elsevier

33. Geiger DR, Shieh W-J, Lu LS, Servaites JC. 1991. Carbon assimilation and leaf water status in sugar beet leaves during a simulated natural light regime. *Plant Physiol.* 97:1103–8

34. Geiger DR, Tucci MA, Servaites JC. 1987. Glyphosate effects on carbon assimilation and gas exchange in sugar beet leaves. *Plant Physiol.* 85:365–69

35. Gerhardt R, Stitt M, Heldt HW. 1987. Subcellular metabolite levels in spinach leaves. *Plant Physiol.* 83:399–407

36. Goldschmidt EE, Huber SC. 1992. Regulation of photosynthesis by end-product accumulation in leaves of plants storing starch, sucrose, and hexose sugars. *Plant Physiol.* 99:1443–48

37. Gougler JA, Geiger DR. 1984. Carbon partitioning and herbicide transport in glyphosate-treated sugarbeet (*Beta vulgaris*). *Weed Sci.* 32:546–51

38. Heber U, Niemanis S, Dietz K-J, Viil J. 1986. Assimilatory power as a driving force in photosynthesis. *Biochim. Biophys. Acta* 852:144–55

39. Heber U, Takahama U, Niemanis S, Shimizu-Takahama M. 1982. Transport as the basis of the Kok effect. Levels of some photosynthetic intermediates and activation of light-regulated during photosynthesis and green leaf protoplasts. *Biochim. Biophys. Acta* 679:287–99

40. Heineke D, Sonnewald U, Büssis D, Günter G, Leidreiter K, et al. 1992. Apoplastic expression of yeast-derived invertase in potato. Effects on photosynthesis, leaf solute composition, water relations, and tuber composition. *Plant Physiol.* 100: 301–8

41. Herold A. 1980. Regulation of photosynthesis by sink activity—the missing link. *New Phytol.* 86:131–44

42. Herold A, Lewis DH. 1977. Mannose and green plants: occurrence, physiology and metabolism, and use as a tool to study the role of orthophosphate. *New Phytol.* 79:1–40

43. Hrazdina G, Jensen RA. 1992. Spatial organization of enzymes in plant metabolic pathways. *Annu. Rev. Plant Physiol. Plant Mol. Biol.* 43:241–67

44. Huang LK, Wong SC, Terashima I, Zhang X, Lin DX, Osmond CB. 1989. Chilling injury in mature leaves of rice. I. Varietal differences in the effect of chilling on canopy photosynthesis under simulated 'dry cold dew wind' conditions experienced in south-east China. *Aust. J. Plant Physiol.* 16:321–37

45. Huber SC. 1989. Biochemical mechanism for regulation of sucrose accumulation in leaves during photosynthesis. *Plant Physiol.* 91:656–62

46. Huber SC, Huber JL. 1992. Role of sucrose-phosphate synthase in sucrose metabolism in leaves. *Plant Physiol.* 99:1275–78

47. Hutchison RS, Ort DR. 1991. Investigation of the reductive activation of tomato stromal fructose bisphosphatase following light chilling. *Plant Physiol.* 96S:12

48. Jensen RA. 1986. The shikimate/arogenate pathway: link between carbohydrate metabolism and secondary metabolism. *Physiol. Plant.* 66:164–68

49. Kamen MD. 1963. *Primary Processes in Photosynthesis*, pp. 1–21. New York: Academic. 183 pp.

50. Kerr PS, Huber SC. 1987. Coordinate control of sucrose formation in soybean leaves by sucrose-phosphate synthase and fructose-2,6-bisphosphate. *Planta* 170:197–204

51. Kerr PS, Rufty TW Jr, Huber SC. 1985. Endogenous rhythm in photosynthesis, sucrose-phosphate synthase activity and stomatal resistance in leaves of soybean *Glycine max* (L.) Merr. *Plant Physiol.* 77:275–80

52. Kobza J, Edwards GE. 1987. The photosynthetic induction response in wheat leaves: net CO_2 uptake, enzyme activation, and leaf metabolites. *Planta* 171:549–59

53. Krause GH. 1988. Photoinhibition of photosynthesis. An evaluation of damaging and protective mechanisms. *Physiol. Plant.* 74: 566–74

54. Kruckeberg AL, Neuhaus HE, Feil R, Gottlieb LD, Stitt M. 1989. Decreased-activity mutants of phosphoglucose isomerase in the cytosol and chloroplasts of *Clarkia xantiana*. *Biochem. J.* 261:457–67

55. Laing WA, Ogren WL, Hageman RH. 1974. Regulation of soybean net photosynthetic CO_2 fixation by the interaction of CO_2, O_2, and ribulose 1,5-diphosphate carboxylase. *Plant Physiol.* 54:678–85

56. Lambers H. 1990. Oxidation of mitochondrial NADH and the synthesis of ATP. In *Plant Physiology, Biochemistry and Molecular Biology*, ed. DT Dennis, DH Turpin, pp. 124–43. New York: Longman

57. Lan Y, Woodrow IE, Mott KA. 1992. Light-dependent changes in ribulose bisphos-

phate carboxylase activase activity in leaves. *Plant Physiol.* 99:304–9
58. Lauer MJ, Blevins DG, Sierzputowska-Gracz H. 1989. ^{31}P-Nuclear magnetic resonance determination of phosphate compartmentation in leaves of reproductive soybeans as affected by phosphate nutrition. *Plant Physiol.* 89:1331–36
59. Li B, Geiger DR, Shieh WJ. 1992. Evidence for circadian regulation of starch and sucrose synthesis in sugar beet leaves. *Plant Physiol.* 99:1393–99
60. Lorimer GH, Miziorko HM. 1980. Carbamate formation on the ε-amino group of a lysyl residue as the basis for the activation of ribulosebisphosphate carboxylase by CO_2 and Mg^{2+}. *Biochemistry* 19:5321–28
61. Ogren WL. 1984. Photorespiration: pathways, regulation and modification. *Annu. Rev. Plant Physiol.* 35:415–42
62. Osmond CB. 1981. Photorespiration, photoinhibition: some implications for the energetics of photosynthesis. *Biochim. Biophys. Acta* 639:77–98
63. Parry MAJ, Delgado E, Vadell J, Keys AJ, Lawlor D, Medrano H. 1993. Water stress and the diurnal activity of ribulose-1,5-bisphosphate carboxylase in field grown *Nicotiana tabacum* genotypes selected for survival at low CO_2 concentrations. *Plant Physiol. Biochem.* 31:113–20
64. Pearcy RW. 1990. Sunflecks and photosynthesis in plant canopies. *Annu. Rev. Plant Physiol. Plant Mol. Biol.* 41:421–53
65. Portis AR Jr. 1992. Regulation of ribulose bisphosphate carboxylase/oxygenase activity. *Annu. Rev. Plant Physiol. Plant Mol. Biol.* 43:415–37
66. Preiss J. 1988. Biosynthesis of starch and its regulation. In *The Biochemistry of Plants,* ed. J Preiss, 14:181–254. Orlando: Academic
67. Quick WP, Schurr U, Scheibe R, Schulze E-D, Rodermel SR, et al. 1991. Decreased ribulose-1,5-bisphosphate carboxylase-oxygenase in transgenic tobacco transformed with "antisense" rbcS. I. Impact on photosynthesis in ambient growth conditions. *Planta* 183:542–54
68. Quy LV, Foyer C, Champigny M-L. 1991. Effect of light and NO_3^- on wheat leaf phosphoenolpyruvate carboxylase activity. Evidence for covalent modulation of the C_3 enzyme. *Plant Physiol.* 97:1476–82
69. Raschke K, Resemann A. 1986. The midday depression of CO_2 assimilation in leaves of *Arbutus unedo* L.: diurnal changes in photosynthetic capacity related to changes in temperature and humidity. *Planta* 168:546–58
70. Robinson SP, Walker DA. 1981. Photosynthetic carbon reduction cycle. In *Biochemistry of Plants,* ed. MD Hatch, NK Boardman, 8:193–239. New York: Academic

71. Sage RF, Sharkey TD, Seemann JR. 1988. The in vivo response of the ribulose-1,5-bisphosphate carboxylase activation state and the pool sizes of photosynthetic metabolites to elevated CO_2 in *Phaseolus vulgaris* L. *Planta* 174:407–16
72. Sage RF, Sharkey TD, Seemann JR. 1990. Regulation of ribulose-1,5-bisphosphate carboxylase activity in response to light intensity and CO_2 in the C_3 annuals *Chenopodium album* L. and *Phaseolus vulgaris* L. *Plant Physiol.* 94:1735–42
73. Salvucci ME. 1989. Regulation of Rubisco activity in vivo. *Physiol. Plant.* 77:164–71
74. Salvucci ME, Portis AR Jr, Ogren WL. 1985. A soluble chloroplast protein catalyzes ribulosebisphosphate carboxylase/oxygenase activation in vivo. *Photosynth. Res.* 7:193–201
74a. Sassenrath GF, Ort DR, Portis AR Jr. 1990. Impaired reductive activation of stromal bisphosphatases in tomato leaves following low-temperature exposure at high light. *Arch. Biochem. Biophys.* 282:302–8
75. Scheibe R. 1987. $NADP^+$-malate dehydrogenase in C_3 plants: regulation and role of a light-activated enzyme. *Physiol. Plant.* 71:393–400
76. Seemann JR, Kobza J, Moore BD. 1990. Metabolism of 2-carboxyarabinitol 1-phosphate and regulation of ribulose-1,5-bisphosphate carboxylase activity. *Photosynth. Res.* 23:119–30
77. Servaites JC. 1990. Inhibition of ribulose 1,5-bisphosphate carboxylase/oxygenase by 2-carboxyarabinitol-1-phosphate. *Plant Physiol.* 92:867–70
78. Servaites JC, Fondy BR, Li B, Geiger DR. 1989. Sources of carbon for export from spinach leaves throughout the day. *Plant Physiol.* 90:1168–74
79. Servaites JC, Geiger DR, Tucci MA, Fondy BR. 1989. Leaf carbon metabolism and metabolite levels during a period of sinusoidal light. *Plant Physiol.* 89:403–8
80. Servaites JC, Ogren WL. 1977. Chemical inhibition of the glycolate pathway in soybean leaf cells. *Plant Physiol.* 60:461–66
81. Servaites JC, Shieh W-J, Geiger DR. 1991. Regulation of photosynthetic carbon reduction cycle by ribulose bisphosphate and phosphoglyceric acid. *Plant Physiol.* 97:1115–21
82. Servaites JC, Tucci MA, Geiger DR. 1987. Glyphosate effects on carbon assimilation, ribulose bisphosphate carboxylase activity, and metabolite levels in sugar beet leaves. *Plant Physiol.* 85:370–74
83. Sharkey TD. 1985. Photosynthesis in intact leaves of C_3 plants: physics, physiology and rate limitations. *Bot. Rev.* 51:53–105
84. Sharkey TD, Stitt M, Heineke D, Gerhardt R, Raschke K, Heldt HW. 1986. Limitation of photosynthesis by carbon metabolism.

II. O$_2$-insensitive CO$_2$ uptake results from limitation of triose phosphate utilization. *Plant Physiol.* 81:1123–29

85. Shieh W-J, Geiger DR, Servaites JC. 1991. Effect of N-(phosphono)methyl glycine on carbon assimilation and metabolism during a simulated natural day. *Plant Physiol.* 97: 1109–14

86. Sicher RC, Kremer DF. 1985. Possible control of maize leaf sucrose-phosphate synthase activity by light modulation. *Plant Physiol.* 79:695–98

87. Somerville C. 1986. Analysis of photosynthesis with mutants of higher plants and algae. *Annu. Rev. Plant Physiol.* 37:467–507

88. Sonnewald U, Brauer M, von Schaewen A, Stitt M, Willmitzer L. 1991. Transgenic tobacco plants expressing yeast-derived invertase either in the cytosol, vacuole or apoplast: a powerful tool for studying sucrose metabolism and sink/source interactions. *Plant J.* 1:95–106

89. Stark DM, Timmerman KP, Barry GF, Preiss J, Kishore G. 1992. Regulation of the amount of starch in plant tissues by ADP-glucose pyrophosphorylase. *Science* 258: 287–92

90. Stitt M. 1986. Limitation of photosynthesis by carbon metabolism. I. Evidence for excess electron transport capacity in leaves carrying out photosynthesis in saturating light and CO$_2$. *Plant Physiol.* 81:1115–22

91. Stitt M, Lilley R McC, Heldt HW. 1982. Adenine nucleotide levels in the cytosol, chloroplasts and mitochondria of wheat leaf protoplasts. *Plant Physiol.* 70:971–77

92. Stitt M, Quick WP. 1989. Photosynthetic carbon partitioning: its regulation and possibilities for manipulation. *Physiol. Plant.* 77:633–41

93. Stitt M, von Schaewen A, Willmitzer L. 1990. "Sink" regulation of photosynthetic metabolism in transgenic tobacco plants expressing yeast invertase in their cell wall involves a decrease of the Calvin cycle enzymes and an increase of glycolytic enzymes. *Planta* 183:40–50

94. Süss K-H, Arkona C, Mannteuffel R, Adler K. 1993. Calvin cycle multienzyme complexes are bound to chloroplast thylakoid membranes of higher plants in situ. *Proc. Natl. Acad. Sci. USA* 90:5514–18

95. Theodorou ME, Plaxton WC. 1993. Metabolic adaptations of plant respiration to nutritional phosphate deprivation. *Plant Physiol.* 101:339–44

96. von Caemmerer S, Edmondson DL. 1986. Relationship between steady-state gas exchange, in vivo ribulose bisphosphate carboxylase activity and some carbon reduction cycle intermediates in *Raphanus sativus. Aust. J. Plant Physiol.* 13:669–88

97. von Schaewen A, Stitt M, Schmidt R, Sonnewald U, Willmitzer L. 1990. Expression of yeast-derived invertase in the cell wall of tobacco and *Arabidopsis* plants leads to accumulation of carbohydrate and inhibition of photosynthesis and strongly influences growth and phenotype of transgenic tobacco plants. *EMBO J.* 9:3033–44

98. Walker DA, Herold A. 1977. Can the chloroplast support photosynthesis unaided? In *Photosynthetic Organelles: Structure and Function*, ed. S Miyachi, S Katoh, K Shibata, pp. 295–310. Spec. Issue *Plant Cell Physiol.*, Jpn. Soc. Plant Physiol.

99. Walker DA, Osmond CB. 1986. Measurement of photosynthesis in vivo with a leaf disc electrode: correlations between light dependence of steady-state photosynthetic O$_2$ evolution and chlorophyll *a* fluorescence transients. *Phil. Trans. R. Soc. London Ser. B* 227:267–80

100. Woodrow IE, Berry JA. 1988. Enzymatic regulation of photosynthetic CO$_2$ fixation in C$_3$ plants. *Annu. Rev. Plant Physiol. Plant Mol. Biol.* 39:533–94

101. Worrell AC, Bruneau J-M, Summerfelt K, Boersig M, Voelker TA. 1991. Expression of a maize sucrose phosphate synthase in tomato alters leaf carbohydrate partitioning. *Plant Cell* 3:1121–30

Annu. Rev. Plant Physiol. Plant Mol. Biol. 1994. 45:257–85

ALKALOID BIOGENESIS: Molecular Aspects

T. Hashimoto and Y. Yamada

Nara Institute of Science and Technology, Takayama-chyo 8916-5, Ikoma 630-01, Nara, Japan

KEY WORDS: alkaloid biosynthesis, compartmentation, gene expression, mutants, biotechnology

CONTENTS

INTRODUCTION 257
BIOSYNTHETIC ENZYMES AND GENES 258
   Nicotine and Tropane Alkaloids 258
   Isoquinoline Alkaloids 261
   Indole Alkaloids 264
SUBCELLULAR COMPARTMENTATION 268
   Alkaloid Vesicles 268
   Other Compartments 270
REGULATION 271
   Tissue- and Cell-Specificity, and Light Regulation 271
   Responses to Elicitors and Growth Regulators 272
   Promoter Analysis 273
MUTANTS AND CELL VARIANTS 274
   Nicotine Mutants 274
   Other Mutants 276
   Cell Variants 277
BIOTECHNOLOGICAL APPLICATIONS 278
PROSPECTS 279

INTRODUCTION

Alkaloids are defined as secondary metabolites that contain secondary, tertiary, or quaternary nitrogen atoms in their molecules. The chemical structures of about 100,000 alkaloids are currently known (22). Some of these alkaloids function as phytoalexins or in plant-insect interactions, and others have long

0066-4294/94/0601-0257$05.00

been used by man as stimulants, drugs, narcotics and poisons. Based on their core nitrogen-containing (often cyclic) skeletons, alkaloids are classified into several biogenically related groups. Since these groups are formed by quite different biosynthetic routes, this review focuses on only a few groups for which the enzymes and genes involved in biosynthesis have been characterized. Accordingly, we emphasize the molecular and genetic aspects of alkaloid biogenesis. The biosynthesis of particular groups of alkaloids has been reviewed previously (15, 47, 48, 65, 66, 85).

BIOSYNTHETIC ENZYMES AND GENES

Nicotine and Tropane Alkaloids

Nicotine and tropane alkaloids share a common early biosynthetic pathway that originates from ornithine and/or arginine by the actions of ornithine decarboxylase (OrnDC) and arginine decarboxylase (ArgDC) (Figure 1). This early pathway involves the symmetric diamine putrescine, which is found ubiquitously in plants and usually is metabolized to conjugated forms or to the polyamines, spermidine and spermine. Although both ornithine and arginine can be metabolized to putrescine, evidence suggests that arginine supplies most of the putrescine for alkaloid biosynthesis (47). All living organisms possess OrnDC. OrnDC genes have been cloned from several vertebrates and yeast, but not yet from plants. In contrast, all mammalian cells and many lower eukaryotes lack ArgDC. ArgDC genes have been cloned from *Escherichia coli* (89) and barley (6), and a partial cDNA clone encoding a polypeptide homologous to barley ArgDC was recently isolated from a tomato cDNA library (28). These cDNA clones can be used to obtain ArgDC cDNAs or genes from alkaloid-producing species. Putrescine *N*-methyltransferase (PMT) is the first committed enzyme, which drives the flow of nitrogen away from polyamine biosynthesis to alkaloid biosynthesis. Partially purified PMT from cultured roots of *Hyoscyamus albus* has been characterized in terms of substrate specificity and inhibition by amines, and the resulting model of its active site is remarkably similar to the model reported for spermidine synthase (52). Cloning of PMT cDNA has provided molecular evidence of the similarity of these two enzymes that act on the same substrate (putrescine): the deduced amino acid sequence of tobacco PMT is 49% identical to the sequence of human spermidine synthase (N Hibi, unpublished information). Expression of cloned tobacco cDNA in spermidine synthase–deficient *E. coli* has demonstrated that tobacco PMT shows PMT activity but has no spermidine synthase activity. Thus, PMT has probably evolved recently from spermidine synthase in higher plants. In addition to the portion that is homologous to spermidine synthase, tobacco PMT has at its N-terminus an 82–amino acid extension that is not

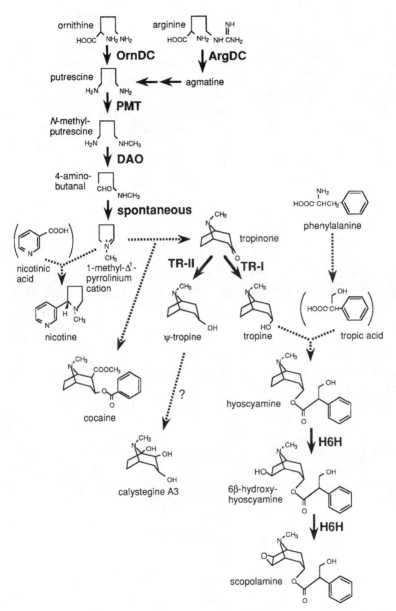

Figure 1 Biosynthetic pathways of nicotine and tropane alkaloids. ArgDC, arginine decarboxylase; OrnDC, ornithine decarboxylase; PMT, putrescine *N*-methyltransferase; DAO, diamine oxidase; TR, tropinone reductase; H6H, hyoscyamine 6β-hydroxylase.

present in spermidine synthase. This N-terminal portion is highly hydrophilic and contains six putative glycosylation signals. N-Methylputrescine (MP) is then oxidatively deaminated to 4-aminobutanal, which spontaneously cyclizes to the 1-methyl-Δ^1-pyrrolinium cation. A diamine oxidase (DAO) catalyzes this oxidative deamination. DAOs are widespread in nature and are particularly active in leguminous plants. Symmetric diamines (e.g. putrescine and cadaverine) are good substrates for most DAOs, but the DAOs in alkaloid-producing plants highly prefer N-methylated diamines over symmetric diamines (43). For example, the specificity constant of *H. niger* DAO for MP is 9-fold higher than the constant for putrescine. Considering that the free MP content in *H. albus* root cultures is 5-fold greater than the putrescine content at the onset of alkaloid formation (49), putrescine is probably metabolized predominantly to MP and then to the 1-methyl-Δ^1-pyrrolinium cation, rather than being deaminated directly by DAO. The 1-methyl-Δ^1-pyrrolinium cation is a reactive compound. Its imino bond condenses, at least in vitro, with acetoacetic acid to give hygrine, a putative precursor of the tropane ring, whereas its coupling with nicotinic acid is presumed to form nicotine. Evolution of PMT from spermidine synthase may have opened up the alkaloid pathway to this reactive cation and, in some plant genera, the original short pathway might have been extended to synthesize alkaloids with more complex structures. Most tropane alkaloids (especially of the hyoscyamine-type) are restricted in Solanaceae but many tropane-type alkaloids are found in several taxonomically unrelated genera (105). This sporadic occurrence indicates that the evolutionary transition from spermidine synthase to functional PMT may have occurred independently in many different genera. This hypothesis should be testable when PMT genes are cloned from several taxonomically distant species.

The first intermediate with a tropane ring is the ketone tropinone, which is located at the branching point in the biosynthetic pathway. Two different enzymes that reduce the keto group of tropinone have been found: tropinone reductase-I (TR-I) forms tropine, which has a 3α-hydroxyl group, whereas tropinone reductase-II (TR-II) gives ψ-tropine, which has a 3β-hydroxyl group (e.g. 44). Thus, the two alcohol products are diastereomers of each other. Tropine is metabolized to hyoscyamine and scopolamine. Although the metabolism of ψ-tropine is not well understood, calystegins may be derived from this alcohol (22). Analysis of the cDNA clones for both reductases has shown that TR-I and TR-II of *Datura stramonium* belong to the short-chain, nonmetal dehydrogenase family, and are 64% identical to each other in 260 amino acid residues (93). We dissected the two TRs into five peptide fragments and assembled various chimeric reductases that contain part of TR-I and part of TR-II. The stereospecificities of tropinone reduction displayed by these chimeric constructs indicated that the N-terminal half of each enzyme contains a common NADPH-binding site and that a peptide fragment of about 70 amino acid residues at the C-terminal half (of which about 30 amino acid residues are different between the two wild-type TRs) is critical in positioning tropinone for stereospecific reduction (K Nakajima, unpublished information). Thus, a relatively small number of amino acid substitutions is sufficient to alter the stereospecificity of enzyme reactions.

Conversion of hyoscyamine to its epoxide, scopolamine, proceeds in two oxidative steps: hydroxylation at the 6β-position of the tropane ring (45), followed by intramolecular epoxide formation by removal of the 7β-hydrogen (41). Expression of a hyoscyamine 6β-hydroxylase (H6H) cDNA clone in *E. coli* has demonstrated that H6H catalyzes both reactions, and that the hydroxylase activity is about 40-fold stronger than the epoxidase activity (42). This bifunctional enzyme requires 2-oxoglutarate, ferrous ion, ascorbate, and molecular oxygen for catalysis (46), and therefore belongs to the 2-oxoacid-dependent oxygenase family (102). The amino acid sequence of *H. niger* H6H is homologous, among others, to ethylene-forming enzymes (EFE; also called ACC oxidase) (78), and the positions of three introns found in the H6H gene and the tomato EFE gene are conserved strictly (T Kanegae, unpublished information). Because the H6H gene has been found only in scopolamine-producing solanaceous species, while the EFE gene exists in most (probably all) higher plants, the H6H gene must have evolved from a ubiquitous plant 2-oxoacid-dependent oxygenase gene, such as the EFE gene, during diversification of the Solanaceae.

Isoquinoline Alkaloids

The 13 enzymes necessary for the synthesis of berberine have been discovered, but aligning these enzymes into a complete picture has not been easy. The pathway involves several methyltransferases and oxidases that accept more than one potential intermediate. Moreover, contamination by interfering activities may complicate enzyme assays and the obtained results. Figure 2 shows what is currently thought to be a major (but probably not exclusive) pathway that can be found in *Coptis* spp. (91) and *Thalictrum* spp. (32), and perhaps in *Berberis* spp. (MH Zenk, personal communication).

The early pathway from tyrosine to (*S*)-norcoclaurine is based on various feeding experiments and partial characterization of the enzymes involved (e.g. 111), but these studies have also indicated that pathway intermediates may be somewhat interconvertible, thus forming a metabolic lattice in the pathway. For example, there may be a branch pathway by way of 3,4-dihydroxyphenylalanine (DOPA) to dopamine. Three cDNA clones and four genomic clones encoding tyrosine decarboxylase (TyrDC) have been isolated from parsley (57), and these can be used to obtain TyrDC genes from plant species that produce isoquinoline alkaloids.

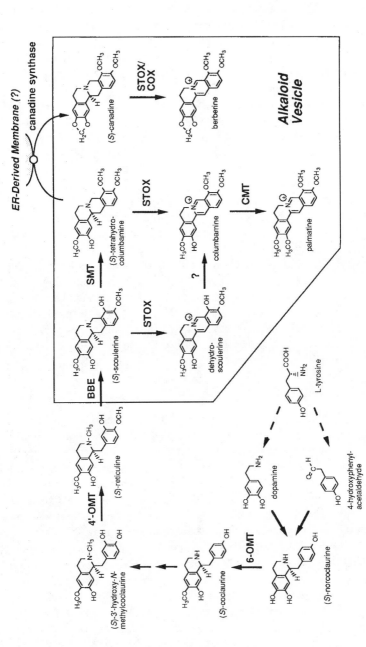

Figure 2 Biosynthetic pathways of berberine and related alkaloids. 6-OMT, norcoclaurine 6-*O*-methyltransferase; 4′-OMT, 3′-hydroxy-*N*-methylcoclaurine 4′-*O*-methyltransferase; BBE, berberine bridge enzyme; SMT, scoulerine 9-*O*-methyltransferase; CMT, columbamine *O*-methyltransferase; STOX, (*S*)-tetrahydroprotoberberine oxidase; COX, canadine oxidase.

(S)-Norcoclaurine is converted to (S)-reticuline by two O-methyltransferases, one N-methyltransferase, and one phenolase. During purification of 3'-hydroxy-N-methylcoclaurine 4'-O-methyltransferase (4'-OMT) from *Berberis koetineana* cell cultures, norcoclaurine 6-O-methyltransferase (6-OMT) was concomitantly and unintentionally purified: the 380-fold enriched 4'-OMT preparation at 5% yield, in which only two protein bands were detected, contained a 125-fold purified 6-OMT at 2% yield (30). This indicates either that these O-methyltransferases have similar physical properties, or that they are physically associated during purification. Interestingly, a highly purified 6-OMT preparation from *Coptis japonica* cell cultures also contained a 40-kDa and a 41-kDa protein. When a cDNA clone encoding the 41-kDa protein was expressed in *E. coli,* it showed 4'-OMT activity but not 6-OMT activity, suggesting again that 6-OMT and 4'-OMT were co-purified unintentionally (F Sato and T Tsujita, unpublished information). 4'-OMT and the 40-kDa protein, which presumably corresponds to 6-OMT, showed 52% identity in amino acid sequences. This level of amino acid identity is expected to give O-methyltransferases with different positional specificities, but does not appear to be high enough to explain the almost identical chromatographic properties of the two enzymes during purification. Simultaneous expression of both enzymes in *E. coli* and co-precipitation experiments with specific antibodies may give us a better understanding of this unusual phenomenon.

(S)-Reticuline is thought to enter a specific vesicle where most, if not all, of the subsequent enzymes reside (see discussion of alkaloid vesicles). The oxidative cyclization of the N-methyl group of (S)-reticuline forms the berberine bridge of (S)-scoulerine. The berberine bridge enzyme (BBE) consumes 1 mole of O_2 and produces 1 mole each of scoulerine and H_2O_2 for each mole of reticuline (124). The primary structure of BBE was deduced from a BBE cDNA isolated from *Eschscholtzia californica* cell cultures, and contained at the amino terminus a putative signal peptide of 22 amino acids (21). Three potential glycosylation sites were found in the sequence. Amino acid sequence analysis of purified BBE confirmed that one of the sites contained a modified asparagine residue. BBE shows 25% overall homology to 6-hydroxy-D-nicotine oxidase of *Arthrobacter oxidans,* which includes the consensus sequence for a covalently flavinylated histidine. This homology, together with the finding that chelating agents inhibit this enzyme activity (124) and a ^{3}H NMR mechanistic study of BBE (29), suggests that metal and flavin are involved in channeling two electrons from reticuline to oxygen.

(S)-Scoulerine is then methylated to (S)-tetrahydrocolumbamine, in which a methylenedioxy bridge is formed to yield (S)-canadine. The enzyme activity that was initially thought to represent formation of the methylenedioxy bridge (110) was actually caused by the demethylation activity of a peroxidase within the vesicle (5). A cytochrome P-450 enzyme from microsomes of *Berberis* spp., *Thalictrum* spp., and *Coptis* spp. formed the methylenedioxy bridge in (S)-tetrahydrocolumbamine but not in the quaternary alkaloid columbamine (M Rueffer, personal communication). Because of this substrate specificity of canadine synthase, the berberine pathway by way of columbamine that was once proposed to operate in *Berberis* spp. (139) must be abandoned. At the final step of this pathway, (S)-canadine is oxidized to the quaternary alkaloid berberine. The oxidase from *Berberis* spp. cell cultures converts several (S)-tetrahydroprotoberberines, including (S)-canadine and (S)-1-benzylisoquinoline alkaloids to corresponding quaternary alkaloids (2), and is designated (S)-tetrahydroprotoberberine oxidase (STOX) because of its broad substrate specificity. STOX contains covalently bound flavin. In the presence of oxygen, alkaloid substrate is first oxidized by STOX to an iminium intermediate with simultaneous production of 1 mole each of H_2O_2 and H_2O. With a tetrahydroprotoberberine as a substrate, the iminium intermediate oxidizes spontaneously to the protoberberine, although this spontaneous oxidation does not occur with the 1,2-dehydroiminium ion produced from 1-benzylisoquinoline. Such a reaction mechanism and the presence of flavin in STOX are reminiscent of BBE, but a critical difference is that, unlike BBE, STOX is not inhibited by EDTA or 1,10-phenanthroline. The *Coptis* spp. and *Thalictrum* spp. oxidases that convert (S)-canadine to berberine show biochemical properties that are so different from STOX in *Berberis* spp. that these enzymes deserve a different name: i.e. (S)-canadine oxidase (COX). COX, which probably contains iron but not flavin (96); produces 2 moles of H_2O_2 and 1 mole of berberine in the presence of oxygen (96), and acts primarily on (S)-canadine, but not on other tetrahydroprotoberberines or on norreticuline, for which STOX is highly active (32). Molecular cloning of STOX and COX should show whether (and if so, how) these oxidases are related structurally to each other and to BBE.

In the vesicles, there are side pathways other than the main pathway leading to berberine. In particular, STOX converts scoulerine and tetrahydrocolumbamine to the corresponding quaternary alkaloids (2), which are then methylated by the intra-vesicle methyltransferases that act specifically on quaternary alkaloids (108; M Rueffer, personal communication).

Indole Alkaloids

The pathways that lead to complex terpenoid indole alkaloids, such as vinblastine and vincristine, may involve well over 20 biosynthetic enzymes, only a few of which have been isolated and characterized biochemically, and even fewer of which have been characterized at the molecular level (Figure 3). Tryptophan decarboxylase (TryDC) converts tryptophan to tryptamine, consists of two identical subunits, contains one mole of pyridoxal phosphate and probably one mole of pyrroloquinoline quinone per subunit (100), and is localized in the cytoplasm (17). TryDC appears to exist in vivo in an equilibrium between a dimeric active form, which is stabilized by Mg^{2+}, and a monomeric form, which is rapidly degraded by proteolysis (27). TryDC cDNAs have been isolated from *Catharanthus roseus* (19) and *Camptotheca acuminata* (C Nessler, personal communication). The deduced primary sequence of *C. roseus* TryDC exhibits 57% and 39% of amino acid identities with parsley TyrDC and with dopa decarboxylase of *Drosophila melanogaster,* respectively. A signal peptide sequence has not been found at the N-terminus, which confirms the cytoplasmic localization of TryDC. Although a homogeneous TryDC preparation from *C. roseus* cell cultures contained an N-terminus that started from the fourteenth amino acid residue of the cDNA sequence

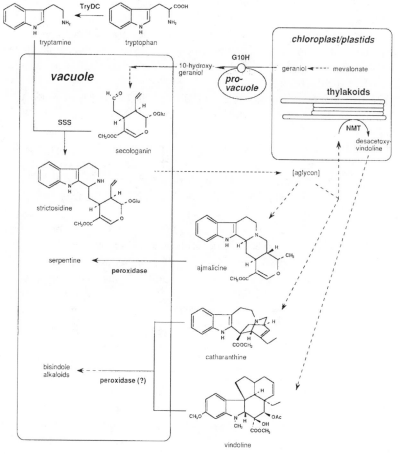

Figure 3 Biosynthetic pathways of terpenoid indole alkaloids. TryDC, tryptophan decarboxylase; G10H, geraniol 10-hydroxylase; SSS, strictosidine synthase; NMT, 11-methoxy-2,16-dihydro-16-hydroxytabersonine *N*-methyltransferase.

(33), the molecular weight of this purified TryDC is considerably smaller than that of other *C. roseus* TryDC preparations. This observation indicates that the missing N-terminal peptide is the result of an accidental protease cleavage during purification procedures.

In addition to tryptamine, the formation of strictosidine requires secologanin. The mevalonic acid pathway in the plastids yields geranyl phosphate from two isoprene units, which is then transformed to secologanin via geraniol by several steps. The hydroxylation of geraniol at the C-10 position is catalyzed by geraniol 10-hydroxylase (G10H or monoterpene hydroxylase), which also converts nerol, the *cis*-isomer of geraniol, to 10-hydroxynerol (83). G10H was one of the first cytochrome P-450 monooxygenases to be characterized in plants (74, 86). The hydroxylase is microsome-bound, requires molecular oxygen and NADPH, and is inhibited by carbon monoxide, although the inhibition can be reversed selectively by light with a maximal intensity between 420 and 450 nm. Molecular cloning of G10H has been attempted using several different approaches; however, there is no definite proof to date that the G10H gene has been cloned.

The first plant cytochrome P-450 cDNA was isolated (10) from ripening avocado fruit that contained high G10H activity (36). Expression of this P-450 gene (CYP71A1) in yeast showed that it was not a *trans*-cinnamic acid hydroxylase; however, G10H activity has not been examined in yeast preparation (9). Differential screening of *C. roseus* cDNA libraries showed two highly homologous P-450 genes (Cros1 and Cros2) that were induced in alkaloid production medium (131). The Cros1 and Cros2 proteins showed only 26% amino acid identity to the avocado CYP71A1 protein. When expressed in yeast (134), as well as in tobacco and *Arabidopsis* spp. (J Schröder, personal communication), they showed neither G10H activity nor other P-450 activities that are present in *C. roseus*. Based on the N-terminal amino acid sequence obtained from a highly purified, but not homogeneous, preparation of *C. roseus* G10H, an oligonucleotide primer was synthesized and used to amplify a 1.8-kb PCR product from a *C. roseus* cDNA library. Screening of the same cDNA library with the PCR fragment resulted in isolation of two closely related P-450 clones, which turned out to be identical to the Cros1 and Cros2 genes (81). In line with the transgenic expression of the Cros genes, tobacco plants that overexpressed one of these P-450 genes did not show G10H activity. Moreover, when these genes were introduced in the sense or antisense orientation to *C. roseus* tumors, G10H activity in the transgenic tumors did not change as expected (81). These results substantiated the belief that neither of the cDNA clones encodes G10H. PCR amplification using a set of degenerate primers that specified the conserved heme-binding region of cytochrome P-450 protein resulted in isolation of 16 different P-450 cDNA sequences from *C. roseus* (84). It is unknown whether any of these clones corresponds to

G10H. The large number of isolated P-450 sequences indicates that many P-450 proteins are expressed in a given plant tissue.

Electrons from NADPH are transferred to cytochrome P-450 mono-oxygenases by a common flavoprotein, NADPH-cytochrome P-450 reductase. This reductase has been highly purified from *C. roseus* seedlings, and reconstitutes G10H activity in the presence of partially purified G10H and *C. roseus* crude lipid (73). Fluorimetric analysis of the purified reductase has shown the presence of noncovalently bound FAD and FMN as prosthetic groups. The primary structure of the *C. roseus* reductase was deduced from a partial cDNA clone and a genomic clone, and was shown to have conserved domains that may be involved in the binding of FMN, FAD, and NADPH (82). Plant P-450 reductases have also been cloned from mung bean (119) and *Arabidopsis* spp. (cDNA sequences are registered in Gen Bank by C Mignote et al). In fact, two reductase genes have been found in *Arabidopsis* spp., which raises the interesting possibility that different reductases might interact with certain groups of P-450 proteins in higher plants. The mung bean reductase can substitute for the rat reductase in the reconstitution of mammalian P-450-catalyzed hydroxylation reactions (119). A rabbit liver P-450, when expressed in tobacco, seems to interact with a tobacco reductase, causing marked phenotypic changes in the transgenic tobacco plants (112). These results indicate that the plant reductase and the animal reductase are basically interchangeable as electron carriers to P-450, although a reductase isolated from bovine liver did not substitute for plant reductases in the reconstitution of bisbenzylisoquinoline alkaloid coupling reactions (123).

The condensation of the glucoiridoid secologanin (derived from 10-hydroxygeraniol by several enzymatic steps) with the indole tryptamine is catalyzed by strictosidine synthase (SSS). SSS does not require cofactors, is a glycoprotein, and exists in more than four charge isoforms in *C. roseus* (101) and *Cinchona robusta* (126), but as a single enzyme in *Rauvolfia serpentina* (39). SSS cDNAs have been cloned from *R. serpentina* (64) and *C. roseus* (80, 99). The deduced amino acid sequences of these SSS cDNAs show no apparent homology to the proteins in the data base and, respectively, contain N-terminal signal peptides of 27 and 29 amino acid residues, and one and two putative N-linked glycosylation site(s). In both *R. serpentina* (11) and *C. roseus* (99), *SSS* is a single-copy gene in the haploid genome; therefore, the multiple isoforms found in *C. roseus* must be caused by differences in the post-translational modifications of one gene product. Interestingly, when genomic DNA blots prepared from many plant species were probed with the *Rauvolfia* SSS cDNA, a strong hybridization was observed with genomic DNA from soybean that has no SSS activity or indole alkaloids (60). It is unknown whether the hybridizing DNA encodes a protein related to SSS.

Strictosidine is deglucosylated and then converted to ajmalicine, catharanthine, vindoline, and other indole alkaloids. The biosynthesis of vindoline involves desacetoxyvindoline 4-hydroxylase, a 2-oxoglutarate-dependent dioxygenase. Amino acid sequences of the 4-hydroxylase purified from *C. roseus* plants showed distinct homology to the amino acid sequence of H6H, another dioxygenase involved in tropane alkaloid biosynthesis (14).

SUBCELLULAR COMPARTMENTATION

Efficient synthesis of complex alkaloids may benefit from physical association of a series of complex enzymes or from subcellular compartmentation of part of the pathway so that pathway intermediates are made more readily available to enzymes than if the intermediates were allowed to diffuse throughout the cytoplasm. Although an increasing amount of compelling data supports metabolite channeling for interpreting some unpredicted results of metabolic studies in numerous biological systems (55, 97), no such data has been reported in alkaloid biosynthesis. Microscopic and biochemical studies do, however, support subcellular compartmentation of alkaloid metabolism.

Alkaloid Vesicles

Laticifers of opium poppy were shown to contain abundant, irregularly shaped vesicles that appeared to be derived from elongated endoplasmic reticulum (ER) (95), and that were thought to not only contain morphine, but also to synthesize and metabolize it (25). Only after careful biochemical studies was it clearly shown that part of the isoquinoline biosynthetic pathway is localized within distinct vesicles. Several enzymes in the biosynthetic pathway from reticuline to berberine and related quaternary alkaloids (Figure 2) are contained in a vesicle with a specific gravity of $r = 1.14$ g cm^{-3}. The vesicle enzymes include BBE (3), STOX (3), COX (31), columbamine *O*-methyltransferase (109), and dehydroscoulerine *O*-methyltransferase (M Rueffer, personal communication). There has been some disagreement over whether scoulerine 9-*O*-methyltransferase is cytosolic or inside the vesicle (32, 90, 139). SMT was purified to homogeneity from *C. japonica* cell cultures (113), and partial cDNA clones encoding SMT were obtained (H Fujiwara, N Takeshita, and F Sato, unpublished information). The complete primary structure of SMT would indicate the presence or absence of an amino terminal signal peptide, which would be required for targeting the vesicle.

Electron micrographs of *Berberis* spp. cells have shown that smooth spherical vesicles, which appear to correspond to the isolated alkaloid vesicles, frequently are located in clusters inside what appear to be small vacuoles (3). Incubation of purified vesicles with reticuline and *S*-adenosyl-L-methionine produced scoulerine and dehydroscoulerine. In addition, fortification of this reaction mixture with SMT also produced tetrahydrocolumbamine and columbamine (3). These feeding experiments support the notion that the purified vesicles contain both BBE and STOX, but little, if any, SMT. The alkaloid vesicles isolated from *Berberis wilsoniae* (3) and *C. japonica* (31) contain protoberberine alkaloids (jatrorrhizine, palmatine, coptisine, berberine, and columbamine) in the same compositions that are found in intact cells or isolated vacuoles. If we assume that the alkaloid vesicles, like the vacuoles isolated from *Thalictrum glaucum* (107), are permeable to tertiary protoberberines but not to quaternary alkaloids, then all of the quaternary alkaloids must be formed in the vesicles by STOX/COX and the resident methyltransferases. These quaternary alkaloids may be deposited into a central vacuole by direct fusion of the vesicle, or by first being taken up by a small vacuole which fuses subsequently with a central vacuole.

When quaternary benzylisoquinoline alkaloids were added exogenously to the culture medium, these alkaloids were taken up rapidly, in a kinetic pattern similar to an enzymatic reaction following the Michaelis-Menten equation, into the vacuoles of *C. japonica* and *T. flavum,* but not into the vacuoles of *Datura innoxia,* which does not produce these alkaloids (114, 115). According to the above model, exogenous quaternary alkaloids might be reduced in the cytosol to the corresponding tertiary alkaloids, which are then taken up by the vesicles and reoxidized back to the quaternary forms in the vesicles. Along these lines, an NADPH-dependent reductase that converts dehydroscoulerine to (*S*)-scoulerine was recently discovered in *Berberis* spp., *Coptis* spp., and *Thalictrum* spp. cell cultures (108).

It is still unclear how the alkaloid vesicle is formed. Based on the vesicle-mediated protein sorting mechanisms that have been clarified by a combination of biochemistry in animal cell-free systems and genetics in yeast (103, 106), a working model might begin with an analogy to animal endosomes and yeast vesicle intermediates that are bound to lysosomes or vacuoles from the *trans*-Golgi network. Correct enzyme sorting into the alkaloid vesicles probably involves the function of many proteins, including small GTP-binding proteins and membrane-associated signal transduction complexes, some of which may be specific to alkaloid biogenesis. Vacuolar proteins of yeast and probably of plants are recognized as such by short peptide segments on the sequences, but not by glycosylation, as with animal lysosomal proteins. The vesicular enzyme BBE is glycosylated and has an N-terminal signal peptide that is probably required to enter into the lumen of ER (21). When BBE is expressed with the signal peptide in army worm cells, almost all of the recombinant BBE is excreted into the culture medium as a mature enzyme without the signal sequence (62). Because the host insect cells do not have alkaloid vesicles or vacuoles, the BBE may be sorted from ER to the *trans*-Golgi network, from which it must be removed from the cells by a default transport mechanism. When the same BBE construct was expressed in yeast, however, the enzyme was recovered in a microsomal fraction, as analyzed by sucrose density gradient centrifugation, while the BBE expressed without the signal peptide stayed in the cytosol (T Kutchan, personal communication). Such heterologous expression experiments are currently at a preliminary stage, but should become important in elucidating fundamental similarities and differences between mechanisms that sort molecules into specialized alkaloid vesicles and general cellular compartments, such as vacuoles. Cultured *Thalictrum* spp. cells that excrete indigenous berberine into the medium may be a valuable material for studying the function of alkaloid vesicles, and will be discussed later.

Other Compartments

Interesting aspects of P-450 enzymes and interacting reductases are their subcellular localization and their topology on the membranes. In *C. roseus,* G10H was found in a membrane fraction that was different from ER but that contained mostly *provacuoles,* i.e. vesicles that were identified as small vacuoles by electron microscopy and by the presence of the vacuolar carotenoid lutein (75). Some NADPH-cytochrome P-450 reductase activity was also recovered in the same membrane fraction. Electron microscopy showed that provacuoles resembling purified G10H-containing vesicles coalesced to the large central vacuole. These results indicate that both G10H and the reductase are probably located on the membranes of provacuoles, but this possibility must be confirmed by immunogold electronmicroscopy. If these enzymes are located on the membranes, how can a particular P-450/reductase pair be targeted to a specific membrane, instead of to the usual destination of the ER membrane? Only one P-450 reductase gene has been detected in the genome of *C. roseus* by genomic DNA blot hybridization (82). G10H may be localized initially on a restricted region of the ER membrane, which buds off, together with the nearby reductase, to form a vesicle intermediate with potential vacuolar characteristics. Vesicles derived from dilation of ER are well documented in laticifers of opium poppy (95). This model requires cytoplasmic components that specifically interact and recognize either G10H or the particular region of ER membrane where G10H might accumulate. Molecular characterization of G10H and high-resolution immunogold electronmicroscopy are necessary to obtain a better understanding of vesicle biogenesis.

Plant vacuoles also play a role in alkaloid metabolism. Basic peroxidases in the vacuoles of *C. roseus* cells are responsible for the oxidation of ajmalicine to the charged quaternary form, serpentine (8), and perhaps also for the oxidative coupling of vindoline and catharanthine, leading to bisindole alkaloids (24). Ultrastructural immunolocalization (79) and subcellular fractionation (125) have demonstrated that SSS in the indole alkaloid pathway is a vacuolar protein. The N-terminal cleavable peptide of SSS functions as the signal peptide for ER entry. When the SSS cDNA encoding the precursor form was expressed in insect cells, a processed form without the signal peptide was secreted into the culture medium (62). In contrast, expression of the precursor form in tobacco cells targeted the enzyme (probably as a mature form) into the tobacco vacuole (79). Correct targeting of proteins into plant vacuoles requires, in addition to the signal peptide, vacuolar sorting signals in the mature protein, or in the cleavable N-terminal or C-terminal propeptides (12, 94). Comparison of the N-terminal amino acid sequences in purified SSS from *C. roseus* cells (99) to those in SSS protein expressed in insect cells (62) indicates that SSS does not contain an N-terminal propeptide for vacuolar targeting. Strictosidine, the reaction product of SSS, is deglucosylated by a β-glucosidase that is probably located on the outside of the tonoplast membrane (125).

The methyltransferase that functions at the third to the last step in vindoline biosynthesis is associated with thylakoid membranes (16). Thus, vindoline biosynthesis involves, at least, chloroplast, a vacuole, and an alkaloid vesicle, in addition to cytoplasm (Figure 3).

REGULATION

Like many other plant genes, expression of the genes encoding biosynthetic enzymes for alkaloids are regulated spatially and temporally, as well as by internal and external stimuli such as plant hormones, light, and stress. Although fundamental molecular mechanisms leading to transcriptional activation or repression of alkaloid genes may be, in many respects, the same as those of other plant genes, some molecular components in the regulatory network might function specifically for alkaloid metabolism.

Tissue- and Cell-Specificity, and Light Regulation

Classical reciprocal grafting experiments between alkaloid-producing and nonproducing plants, and alkaloid formation in plant organ cultures, indicate that nicotine and the tropane alkaloids hyoscyamine and scopolamine are synthesized primarily in the root and translocated through the xylem to the aerial parts of the plants (reviewed in 47). Accordingly, the transcripts for PMT, TR-I (K Nakajima, unpublished information) and H6H (78) were found almost exclusively in the root, and particularly strong expression of these genes was observed in cultured roots. Immunohistochemistry showed that the H6H protein was located in the pericycle cells of the young root, which are parenchyma cells at the periphery of the vascular cylinder present in developmentally young roots without secondary growth (40). Stronger expression of H6H in cultured roots than in the roots of mature plants reflects the much higher proportion of pericycle cells in the cultured root tissue.

The enzyme activities of 16-methoxy-2, 3-dihydro-3-hydroxytabersonine N-methyltransferase, desacetoxyvindoline 4-hydroxylase, and deacetylvindoline 17-O-acetyltransferase that together constitute part of the vindoline biosynthetic pathway from strictosidine are detected only in hypocotyls and cotyledons of C. roseus seedlings (18), and the activities of the last two enzymes are induced by light, probably through a signal transduction pathway mediated by phytochrome (1).

Responses to Elicitors and Growth Regulators

Pathogenic attack of plants induces several defense-related genes that code for enzymes with antimicrobial activities and enzymes involved in the synthesis of flavonoids, lignins, and antimicrobial phytoalexins. Several alkaloids possess antimicrobial activities and accumulate after pathogen attack. Alkaloid formation in several cultured cells is also induced or enhanced markedly by treating with cell wall preparations of yeast or fungi (e.g. 118). Activities or transcript levels of several enzymes necessary for the biosynthesis of antimicrobial alkaloids have been induced by yeast or fungal elicitors (e.g. 4, 21, 23, 99). A fungal elicitor also induces transcript levels of S-adenosylmethionine synthetase and S-adenosylhomocysteine hydrolase, indicating an increased turnover of activated methyl groups that may be used for synthesis of antifungal alkaloids (58). Inducability seems to depend on the physiological conditions of cell culture and the type of final alkaloid product: a low alkaloid-producing strain of Eschscholtzia californica was induced to synthesize benzo-[c]phenanthridine alkaloids derived from (S)-scoulerine, but high-alkaloid producing strains could not be further stimulated by treatment with the elicitor (118). Formation of bisbenzylisoquinoline alkaloids, which are also derived from (S)-scoulerine, was not enhanced in cell cultures of several Berberis species and Stephania japonica (118).

In several defined systems, jasmonic acid mediates elicitor-induced plant gene activation. When cell cultures of Rauvolfia canescens and E. californica were challenged with a yeast elicitor, endogenous jasmonic acid accumulated rapidly and transiently, followed by a steady increase in alkaloid accumulation (35). Exogenous application of methyl jasmonate to E. californica cell cultures, in turn, induced the formation of alkaloids, with concomitant induction of BBE expression. 12-Oxo-phytodienoic acid, the first pentacyclic intermediate in the jasmonate biosynthetic pathway, also strongly induced alkaloid biosynthesis in E. californica cell cultures (61). Jasmonate induced not only several different alkaloids, but also other classes of secondary metabolites, including flavonoids, guaianolides, and anthraquinones (35).

Negative effects of auxin on alkaloid biosynthesis have been known for some time, and it was shown recently that auxin rapidly down-regulates several genes involved in alkaloid biosynthesis. Addition of auxin to C. roseus cell cultures that had been grown in an auxin-free medium repressed transcription of the *TryDC* and *SSS* genes within 3 h (34, 99). Removal of flower heads and several young leaves (topping), which is a common practice among tobacco growers, temporarily eliminates apical dominance and increases nicotine content in the tobacco leaves. Topping of hydroponically grown tobacco plants increased the PMT activity in the root more than 10-fold within one day, but the presence of 50 mM indoleacetic acid in the nutrient solution at the time of topping completely blocked this increase (88). Auxin may be transported from the shoot to the root where it suppresses the genes for nicotine biosynthesis.

Conversion of nicotine to nornicotine in tobacco varieties that have a nicotine converter gene occurs primarily during curing when the fresh harvested leaves are hung in barns under regulated conditions of temperature and humidity for several days to several weeks. Expression of the converter gene appears to be induced by ethylene formed during the curing process because treatment of tobacco seedlings with ethephon or ethrel, which releases ethylene, induces conversion of nicotine to nornicotine (e.g. 116).

Promoter Analysis

The greatest concern in the functional analysis of alkaloid biosynthesis genes using transgenic plants is whether the heterologous plant hosts regulate faithfully the introduced promoters. Given the evolution of the complex series of biosynthetic enzymes and the intricate compartmentation of pathways including specific vesicles, transcriptional factors may be found that are unique to certain groups of alkaloid pathways.

The promoter of the *C. roseus TryDC* gene has been analyzed in tobacco. In the transient expression system using tobacco mesophyl protoplasts, auxin did not down-regulate the promoter (33). When assayed in transgenic plants, the *TryDC* promoter is not expressed strongly in the tobacco root (33), unlike the indigenous *Catharanthus* gene, but it is inducible by elicitors (PBF Ouwerkerk and J Memelink, personal communication). Likewise, the promoter of the *C. roseus SSS* gene is inducible by elicitors in transgenic tobacco plants (G Pasquali and J Memelink, personal communication). Thus, elicitor induction of alkaloid biosynthesis genes seems to be mediated by a signal transduction mechanism of general stress responses in plants. Pericycle-specific expression of *H6H* has been analyzed in several transgenic hosts (T Kanegae and H Kajiya, unpublished information). The 5'-upstream region of the *H. niger H6H* gene was fused to the coding region of the β-glucuronidase (GUS) gene, and introduced into *H. niger, Atropa belladonna, Nicotiana tabacum,* and *N. rustica* by a binary vector system based on wild-type *Agrobacterium rhizogenes. A. belladonna* has an endogenous *H6H* gene which is expressed weakly in the root, while the two *Nicotiana* species do not have an *H6H* gene.

In the transgenic hairy roots of the *Nicotiana* species, GUS activity stained only the root apex; however, in *H. niger* hairy roots, both the pericycle and the root apex were stained. Many clones of belladonna hairy roots showed GUS activity only at the root apex, but several clones were stained at the pericycle, as well as the root apex. Histochemical GUS staining of regenerated transgenic plants of tobacco and belladonna showed no GUS activity in the aerial parts. These results indicate that expression of the *H6H* promoter at the pericycle requires *trans*-acting factors that are present in the roots of scopolamine-producing plants. Thus far, deletion studies have been unable to separate the DNA sequences necessary for expression at the pericycle and at the root apex.

MUTANTS AND CELL VARIANTS

Well-defined mutants that are defective in one or more steps of alkaloid biosynthesis are not expected to be lethal, and should be useful for understanding ecological roles of plant alkaloids and regulation of their synthesis. Regulatory mutants have been especially important in characterizing and cloning, by transposon tagging, the regulatory genes of flavonoid metabolism (130). In contrast to flavonoid pigments that can be scored easily with the naked eye, changes in alkaloid contents and compositions must usually be analyzed by time-consuming chromatographic procedures. This disadvantage is reflected in the fact that most available alkaloid mutants are found in economically important crops species. Cultured plant cells often show heterogeneous alkaloid contents and compositions. Variant cell lines that produce no, little, or abundant alkaloids may be screened from such heterogeneous cell populations. Because of the high frequency at which variants are obtained and the frequent instability of the phenotypes, such variant cell lines are probably not caused by true mutations, and are often called epigenetic cell variants. Some aspects of variant cell lines with altered alkaloid metabolism are discussed below.

Nicotine Mutants

The predominant alkaloid in the cured leaves of most tobacco varieties is nicotine, but certain varieties possess a gene that demethylates nicotine to nornicotine during curing. Because the transfer of nornicotine into tobacco smoke is less than one fourth of the transfer for nicotine, the nicotine converter gene was once considered to be a means of lowering the nicotine content of tobacco. However, this approach was discouraged later because of the poor smoking quality of tobacco products with high nornicotine contents. The modern descendants of the probable progenitors of amphidiploid *Nicotiana tabacum* (e.g. *N. sylvestris* and *N. tomentosiformis*) convert nicotine to nornicotine (77). The converter gene in *N. tomentosiformis* is active in the green leaves so that the nornicotine level is high even before curing, whereas the converter gene in *N. sylvestris* is activated only during curing (132). During the recent evolution of *N. tabacum* the two converter loci from the progenitors appear to have mutated. When the artificially synthesized amphidiploid between *N. sylvestris* and *N. tomentosiformis* was maintained by self-pollination for about 20 generations, one of the converter loci was inactivated (76). One scenario proposes that primitive man began to use an early tobacco mutant with a defective converter locus from *N. tomentosiformis,* by using tobacco as a green leaf product or by rapidly drying the leaf with heat treatment, thereby minimizing the remaining converter locus from *N. sylvestris* (132). In *N. tabacum,* the inactivated converter allele from *N. tomentosiformis* changes to an activated state at a remarkably high frequency of 0.8% per generation (133). This backmutation probably occurs in both reproductive and somatic cells. The converter alleles in *N. tabacum* and *N. tomentosiformis,* however, do not mutate readily to the inactivated alleles. This instability of the mutated *N. tomentosiformis* allele is reminiscent of a transposon insertion. Since the converter gene in *N. tomentosiformis* and the gene in tobacco nornicotine varieties are active before curing and during curing, respectively, regulation of the promoter must have changed in the process of mutation and backmutation. Mutations caused by insertion and excision of a transposon at a critical regulatory region of the promoter might account for such a change, as has been well documented for the genes of flavonoid metabolism (13).

Cuban cigar tobacco varieties of very low nicotine content were used in the 1930s to develop cigarette varieties of low nicotine content. By introducing low alkaloid genes from the Cuban varieties into Burley 21 through a series of backcrosses (129), LA Burley 21 was developed and registered in 1969 as a genetically stable breeding line with an extremely low alkaloid content (69). LA Burley 21 does not differ from Burley 21 in days to flower, number of leaves, leaf size and plant height, but it is extremely susceptible to insect damage, probably because of its very low nicotine content. Careful genetic studies (67, 68) have demonstrated convincingly that the nicotine content is controlled by two semi-dominant genes, A and B. These two genes show additive effects, and the A locus has an approximately 2.4-fold stronger effect than the B locus. At least two of the enzyme activities involved in nicotine biosynthesis, PMT and quinolinic acid phosphoribosyltransferase, in root tissues of hydroponically grown plants of the four genotypes—AABB (wild-type Burley 21), AAbb, aaBB, and aabb (LA Burley 21)—were proportional to the nicotine levels in the leaves (117). Mutants with low nicotine levels incorporated proportionally less nicotine precursors in both leaves and roots (cited in 87), and accumulated proportionally more polyamines (mostly putrescine and spermidine) in leaves and cultured roots than the wild-type, probably because of the block at PMT (72; T Hashimoto and S Higashiguchi, unpublished information). Differential screening between the cultured roots of Burley 21 and LA Burley 21 resulted in isolation of two cDNA clones whose expression in the root were both repressed proportionally to the nicotine contents of the four genotypes (N Hibi, unpublished information). One clone encoded PMT, while the other clone encoded a protein homologous to the alfalfa isoflavone reductase (IFR)(98). This IFR-like protein, however, did not reduce 2′-hydroxyformononetin, a substrate for IFR, when expressed in *E. coli,* and might be an NADPH-dependent reductase that functions in nicotine metabolism. RFLP has not been detected between Burley 21 and LA Burley 21, when these two cDNA clones were used as probes (T Hashimoto, unpublished information). Both the PMT gene and the IFR-like gene are expressed specifically in the root, the site of nicotine biosynthesis, and are regulated synergetically by the two loci, A and B, indicating that the two loci might be regulatory genes for nicotine biosynthesis.

Other Mutants

The most successful example of the use of alkaloid mutants in agriculture is arguably the lupin. During the late 1920s, von Sengbusch screened some 1.5 million plants of the Mediterranean *Lupinus* species and found a few "sweet," low-alkaloid genotypes that had less than one twentieth of the lupin alkaloids in seed as compared to wild-type lupins (54). Several additional low-alkaloid mutants have since been discovered in other *Lupinus* species, and at least five mutant alleles are at different loci. Phytochemical analyses of these sweet mutants have shown that mutations to these loci significantly reduce alkaloid contents but do not completely eliminate any one component alkaloid (e.g. 134). Some of the sweet mutants might be defective in regulatory genes involved in lupin alkaloid metabolism, but the polyploid nature of the *Lupinus* genus and the scarcity of pure isogenic lines are hampering the use of these mutants in molecular genetic studies on lupin alkaloids.

Many barley cultivars contain the toxic indole alkaloid gramine, which is synthesized from tryptophan via 3-aminomethylindole (AMI). The last two consecutive methylation steps leading from AMI to gramine are probably catalyzed by a single methyltransferase (71). Some barley cultivars (e.g. Proctor) lack the ability to synthesize AMI, while in others (e.g. Morex) both AMI synthesis and methyltransferase activity are defective (70). The methyltransferase protein itself is absent in the latter cultivars. The two metabolic defects in Morex are tightly linked, and reciprocal crosses between Proctor and Morex did not recover the ability to synthesize AMI in F1 hybrids. A regulatory mutation might explain the metabolic defects in Morex, but more detailed genetic and biochemical studies are needed. Although mutants with altered alkaloid contents have also been found for other alkaloid types, such as opium alkaloids, they are often associated with morphological changes.

Compartmentation of biosynthetic enzymes in specialized alkaloid vesicles and probable fusion of the vesicles with the vacuole should involve functions of genes specific to alkaloid biogenesis. Mutations in such genes may disrupt alkaloid biosynthesis or proper targeting of the alkaloids into the vacuole. Berberine secretion by *Thalictrum minus* is an intriguing example. Cell suspension cultures derived from the leaf of *T. minus* var. *hypoleucum*, a perennial herb native to Japan, release most of the berberine produced into the culture medium (92). Cell suspension cultures derived from 17 different plants of this *T. minus* variety all released berberine (127), but cell suspension cultures of *T. minus* var. *minus*, a variety native to Europe, accumulated berberine within cells rather than releasing it into the medium (128). The secretory ability of the Japanese variety may be caused by a mutation. Analysis of F1 hybrids between the Japanese variety and the European variety of *T. minus*, and their progeny is in progress (M Tabata, personal communication). The secreting cell strain of *T. minus* does not take up berberine supplied to the culture medium, whereas cultured cells of *T. flavum* readily take up berberine and related alkaloids (136). Electron microscopy has shown that the secreting cells have numerous spherical vesicles in the cytoplasm that closely resemble the alkaloid vesicles that are well characterized in *Berberis* spp. and *Coptis* spp. (135). It should be interesting to see whether the same enzymes (COX, in particular) that are found in the alkaloid vesicles of other cell strains are also contained in the vesicles of the *T. minus* cells. Alternatively, a defect in some components of the fusion machinery between the vesicle and the vacuole might result in mistargeting of the vesicle to the plasma membrane.

Cell Variants

Among initial populations of cultured cells induced recently from explant tissues, heterogeneity was sometimes noticed in the cellular contents of colored alkaloids. Even seemingly stable cell lines segregate occasionally into alkaloid-producing and nonproducing strains during subculture. For example, a white strain that does not produce any protoberberines or tetrahydro-derivatives was derived spontaneously from a yellow strain of *Thalictrum glaucum* that produces considerable amounts of berberine and columbamine (32). Pairwise comparison of the activities of ten enzymes involved in berberine biosynthesis showed that the white strain had very low activities of all four of the methyltransferases, but activities of the other six enzymes were similar to those found in the yellow strain (Figure 2). The enzyme activity of one methyltransferase, SMT, was only 5-fold lower in the white strain than in the yellow strain, and was considerably higher than those of the other three methyltransferases in the white strain. When (*S*)-reticuline and (*R,S*)-coclaurine were fed independently to the white strain, visibly detectable amounts of berberine were formed from reticuline, but not from coclaurine. Therefore, the very low activities of the three methyltransferases that lead from coclaurine to reticuline are the main causes of the white phenotype. Likewise, enzyme activities of at least two O-methyltransferases (4′-OMT and 6-OMT) in a low-producing strain of *Coptis japonica* are lower than those in a high-producing strain (F Sato, unpublished information). Future molecular studies will address how the activities of the methyltransferase genes are regulated in coordination with, and independently of, the other biosynthesis genes in cell cultures.

BIOTECHNOLOGICAL APPLICATIONS

One or a few genes encoding biosynthetic enzymes in a given pathway may be overexpressed in an enzymatically active form in a microbe that can be used as a bioconversion catalyst. Recombinant *E. coli* that expressed SSS converted exogenously provided tryptamine plus secologanin into strictosidine (42). Similarly, *E. coli* cells expressing H6H converted hyoscyamine that had been added to the culture medium into scopolamine (59). These product alkaloids diffused freely into the medium where they accumulated. This excretion property of the *E. coli* biotransformation system will simplify greatly the downstream product recovery process. It is unknown whether recombinant yeast stores product alkaloids in the vacuole or secretes them into the medium.

A large quantity of purified plant enzyme that has been produced in a heterologous organism may be used as a biocatalyst or for probing reaction mechanisms. Enzymatically active SSS was overexpressed in *E. coli,* yeast, and insect cells as a proenzyme with the signal peptide, as a mature form, or as a fusion protein to ubiquitin (59, 62, 63). One of the most efficient expression systems uses a baculovirus-transformed insect cell culture, which yields 4 mg of purified, active SSS per liter cell culture after a simple purification procedure. The insect cell culture system also produced active BBE in a similarly high yield (62). When SSS was overexpressed as a mature form in *E. coli,* most of the expressed enzyme was recovered as insoluble bodies, but solubilization of the insoluble enzyme in 8 M urea and refolding in 0.1 M Tris-HCl recovered active SSS in a high yield (104).

Engineering alkaloid metabolism by introducing cloned genes into plants or cell cultures promises to be a great challenge for commercial exploitation of plant biotechnology. However, the rate at which an alkaloid is synthesized is generally controlled at multiple steps by several enzymes, subsets of which are regulated both spatially and temporally. Thus, overexpression of only one enzyme in a multi-step pathway does not usually result in a significant increase in the amount of end-product of the pathway. For example, constitutive, enhanced expression of one decarboxylase (OrnDC, TryDC, or lysine decarboxylase) in tobacco, *C. roseus,* or *Peganum harmala* increased the concentration of the enzyme's reaction product or its immediate metabolite, but only marginally, if at all, affected the accumulation of the final, desired alkaloids (7, 20, 26, 33, 37, 38, 51, 120). In one experiment in which TryDC was over-expressed in tobacco, accumulation of tyramine was observed in many trans-formants (121). The molecular basis for this puzzling metabolic disturbance is unknown since TryDC does not convert tyrosine to tyramine.

A surprisingly successful case of metabolic engineering has been reported for scopolamine production in *A. belladonna* (137). Wild-type belladonna accumulates primarily hyoscyamine, and the content of scopolamine, which is much preferred to hyoscyamine as an anticholinergic medicine, is low because of the low H6H activity in the root. The H6H cDNA of *H. niger* was intro-duced into belladonna, in which the enzyme was strongly and constitutively expressed in all parts of the plant. The primary transformant, and its progeny that inherited the transgene, contained scopolamine almost exclusively in the leaf and the stem, while the conversion of hyoscyamine to scopolamine in the root was not as efficient. Feeding of hyoscyamine or 6β-hydroxyhyoscyamine to the culture medium of a tobacco plantlet that had been transformed with the same H6H transgene suggested that hyoscyamine was converted efficiently to scopolamine during translocation from the root to the aerial parts (138). When transgenic belladonna hairy roots were used for the overexpression of H6H, the conversion efficiency of hyoscyamine to scopolamine was considerably lower than in the transgenic plants (50).

The transgenic approach that involves fortifying rate-limiting enzyme ac-tivities can be applied successfully to only a few metabolic steps. The future success of biotechnological metabolic engineering will depend on the molecu-lar cloning of elusive regulatory genes that control the expression of a series of alkaloid biosynthesis genes. Enhanced and constitutive expression of such regulatory genes in transgenic medicinal plants is expected to increase signifi-cantly the total amount of useful alkaloids produced per plant.

PROSPECTS

Studies of alkaloids and their metabolism were originally prompted by the unique pharmacological properties and the bewildering phytochemical diver-sity of alkaloids. Scientists with backgrounds in organic chemistry, plant cell culture, or biochemistry have studied every group of alkaloids. Although progress in this field has been slow, several interesting aspects of alkaloid biogenesis and its control are now emerging. Two major subjects are expected to attract considerable attention in the near future. The first subject concerns the genes that regulate alkaloid metabolism. There is still no direct evidence of such genes, but there is reason to believe that the control of multi-step path-ways by specific regulatory genes is not unique to flavonoid biosynthesis. Molecular cloning of putative alkaloid-specific master genes may require some ingenuity and time, but it should be rewarding considering the biotech-nological applications. The second subject concerns the intricate compartmentation of alkaloid biosynthesis. The biogenesis of specific alkaloid vesicles that house selected sets of enzymes must be broken down into individual steps of vesicle differentiation, and each step should be compared to the well-characterized vesicle transport systems in yeast and mammals. Advances in our understanding of the biogenesis of plant vacuoles are essential. The unique and fundamental principles that are to be gained from studies of a few model alkaloids are expected to apply to many other alkaloids with different biosynthetic origins, and may provide a stimulating incentive to plant biology in general.

ACKNOWLEDGMENTS

We thank our colleagues who generously supplied the preprints and unpublished information referred to in this review—particularly V de Luca, K Hahlbrock, JHC Hoge, TM Kutchan, A Meijer, CL Nessler, R Robins, F Sato, J Schröder, M Tabata, R Verpoorte, and MH Zenk.

> Any *Annual Review* chapter, as well as any article cited in an *Annual Review* chapter, may be purchased from the Annual Reviews Preprints and Reprints service. 1-800-347-8007; 415-259-5017; email: arpr@class.org

Literature Cited

1. Aerts RJ, de Luca V. 1992. Phytochrome is involved in the light-regulation of vindoline biosynthesis in *Catharanthus roseus*. *Plant Physiol.* 100:1029–32
2. Amann M, Nagakura N, Zenk MH. 1988. Purification and properties of (*S*)-tetrahydroprotoberberine oxidase from suspension-cultured cells of *Berberis wilsoniae*. *Eur. J. Biochem.* 175:17–25
3. Amann M, Wanner G, Zenk MH. 1986. Intracellular compartmentation of two enzymes of berberine biosynthesis in plant cell cultures. *Planta* 167:310–20
4. Arakawa H, Clark WG, Psenak M, Coscia CJ. 1992. Purification and characterization of dihydrobenzophenanthridine oxidase from elicited *Sanguinaria canadensis* cell cultures. *Arch. Biochem. Biophys.* 299:1–7
5. Bauer W, Stadler R, Zenk MH. 1992. Peroxidase catalyzed dimerization and demethylation of protoberberine alkaloids. *Bot. Acta* 105:370–74
6. Bell E, Malmberg RL. 1990. Analysis of a cDNA encoding arginine decarboxylase from oat reveals similarity to the *Escherichia coli* arginine decarboxylase and evidence of protein processing. *Mol. Gen. Genet.* 224:431–36

7. Berlin J, Rugenhagen C, Dietze P, Fecker LF, Goddijn OJM, Hoge JHC. 1993. Increased production of serotonin by suspension and root cultures of *Peganum harmala* transformed with a tryptophan decarboxylase cDNA clone from *Catharanthus roseus*. *Trans. Res.* In press
8. Blom TJM, Sierra M, van Vliet TB, Franke-van Dijk MEI, de Koning P, et al. 1991. Uptake and accumulation of ajmalicine into isolated vacuoles of cultured cells of *Catharanthus roseus* (L.) G. Don. and its conversion into serpentine. *Planta* 183:170–77
9. Bozak KR, O'Keefe DP, Christoffersen RE. 1992. Expression of a ripening-related avocado (*Persea americana*) cytochrome P450 in yeast. *Plant Physiol.* 100:1976–81
10. Bozak KR, Yu H, Sirevag R, Christoffersen RE. 1990. Sequence analysis of ripening-related cytochrome P-450 cDNAs from avocado fruit. *Proc. Natl. Acad. Sci. USA* 87:3904–8
11. Bracher D, Kutchan TM. 1992. Strictosidine synthase from *Rauvolfia serpentina*: analysis of a gene involved in indole alkaloid biosynthesis. *Arch. Biochem. Biophys.* 294:717–23

12. Chrispeels MJ, Raikhel NV. 1992. Short peptide domains target proteins to plant vacuoles. *Cell* 68:613–16

13. Coen ES, Carpenter R, Martin C. 1986. Transposable elements generate novel spatial patterns of gene expression in *Antirrhinum majus. Cell* 47:285–96

14. de Carolis E, de Luca V. 1993. Purification, characterization, and kinetic analysis of a 2-oxoglutarate-dependent dioxygenase involved in vindoline biosynthesis from *Catharanthus roseus. J. Biol. Chem.* 268: 5504–11

15. de Luca V. 1993. Enzymology of indole alkaloid biosynthesis. In *Methods in Plant Biochemistry,* ed. PM Dey, JB Harborne, 9:345–68. London: Academic

16. de Luca V, Balsevich J, Tyler RT, Kurz WGW. 1987. Characterization of a novel *N*-methyltransferase (NMT) from *Catharanthus roseus* plants. *Plant Cell Rep.* 6: 458–61

17. de Luca V, Cutler AJ. 1987. Subcellular localization of enzymes involved in indole alkaloid biosynthesis in *Catharanthus roseus. Plant Physiol.* 85:1099–1102

18. de Luca V, Fernandes JA, Campbell D, Kurz WGW. 1988. Developmental regulation of enzymes of indole alkaloid biosynthesis in *Catharanthus roseus. Plant Physiol.* 86:447–50

19. de Luca V, Marineau C, Brisson N. 1989. Molecular cloning and analysis of cDNA encoding a plant tryptophan decarboxylase: comparison with animal dopa decarboxylase. *Proc. Natl. Acad. Sci. USA* 86: 2582–86

20. DeScenzo RA, Minocha SC. 1993. Modulation of cellular polyamines in tobacco by transfer and expression of mouse ornithine decarboxylase cDNA. *Plant Mol. Biol.* 22: 113–27

21. Dittrich H, Kutchan TM. 1991. Molecular cloning, expression, and induction of berberine bridge enzyme, an enzyme essential to the formation of benzophenanthridine alkaloids in the response of plants to pathogenic attack. *Proc. Natl. Acad. Sci. USA* 88:9969–73

22. Dräger B, Funck C, Höhler A, Mrachatz G, Portsteffen A, et al. 1993. Calystegines as a new group of tropane alkaloids in Solanaceae. *Plant Cell Tissue Cult.* In press

23. Eilert U, de Luca V, Constabel F, Kurz WGW. 1987. Elicitor-mediated induction of tryptophan decarboxylase and strictosidine synthase activities in cell suspension cultures of *Catharanthus roseus. Arch. Biochem. Biophys.* 254:491–97

24. Endo T, Goodbody A, Vukovic J, Misawa M. 1988. Enzymes from *Catharanthus roseus* cell suspension cultures that couple vindoline and catharanthine to form 3′,4′-anhydrovinblastine. *Phytochemistry* 27: 2147–49

25. Fairbairn JW, Hakim F, Kheir YE. 1974. Alkaloidal storage, metabolism and translocation in the vesicles of *Papaver somniferum* latex. *Phytochemistry* 13: 1133–39

26. Fecker LF, Hillebrandt S, Rugenhagen C, Herminghaus S, Landsmann J, Berlin J. 1992. Metabolic effects of a bacterial lysine decarboxylase gene expressed in a hairy root culture of *Nicotiana glauca. Biotech. Lett.* 14:1035–40

27. Fernandes JA, Kurz WGW, de Luca V. 1989. Conformation-dependent inactivation of tryptophan decarboxylase from *Catharanthus roseus. Biochem. Cell Biol.* 67:730–34

28. Fleming AJ, Mandel T, Roth I, Kuhlemeier C. 1993. The patterns of gene expression in the tomato shoot apical meristem. *Plant Cell* 5:297–309

29. Frenzel T, Beale JM, Kobayashi M, Zenk MH, Floss HG. 1988. Stereochemistry of enzymatic formation of the berberine bridge in protoberberine alkaloids. *J. Am. Chem. Soc.* 110:7878–80

30. Frenzel T, Zenk MH. 1990. S-Adenosyl-L-methionine: 3′-hydroxy-*N*-methyl-(*S*)-coclaurine- 4′-*O*-methyltransferase, a regioand stereoselective enzyme of the (*S*)-reticuline pathway. *Phytochemistry* 29:3505–11

31. Galneder E, Rueffer M, Wanner G, Tabata M, Zenk MH. 1988. Alternative final steps in berberine biosynthesis in *Coptis japonica* cell cultures. *Plant Cell Rep.* 7:1–4

32. Galneder E, Zenk MH. 1990. Enzymology of alkaloid production in plant cell cultures. In *Progress in Plant Cellular and Molecular Biology,* ed. HJJ Nijkamp, LHW van der Plas, J van Aartrijk, pp. 754–62. Dordrecht: Kluwer

33. Goddijn OJM. 1992. *Regulation of terpenoid indole alkaloid biosynthesis in* Catharanthus roseus: *the tryptophan decarboxylase gene.* PhD thesis. Leiden Univ., The Netherlands

34. Goddijn OJM, de Kam RJ, Zanetti A, Schilperoort RA, Hoge JHC. 1992. Auxin rapidly down-regulates transcription of the tryptophan decarboxylase gene from *Catharanthus roseus. Plant Mol. Biol.* 18:1113–20

35. Gundlach H, Muacller MJ, Kutchan TM, Zenk MH. 1992. Jasmonic acid is a signal transducer in elicitor-induced plant cell cultures. *Proc. Natl. Acad. Sci. USA* 89:2389–93

36. Hallahan DL, Nugent JHA, Hallahan BJ, Dawson GW, Smiley DW, et al. 1992. Interactions of avocado (*Persea americana*) cytochrome P-450 with monoterpenoids. *Plant Physiol.* 98:1290–97

37. Hamill JD, Robins RJ, Parr AJ, Evans DM, Furze JM, et al. 1990. The effects of over-expressing the yeast ornithine decarboxylase gene upon the nicotine and polyamine levels in transformed roots of *Nicotiana rustica*. In *Progress in Plant Cellular and Molecular Biology*, ed. HJJ Nijkamp, LHW van der Plas, J van Aartrijk, pp. 732–37. Dordrecht: Kluwer

38. Hamill JD, Robins RJ, Parr AJ, Evans DM, Furze JM, Rhodes MJC. 1990. Over-expressing a yeast ornithine decarboxylase gene in transgenic roots of *Nicotiana rustica* can lead to enhanced nicotine accumulation. *Plant Mol. Biol.* 15:27–38

39. Hampp N, Zenk MH. 1988. Homogeneous strictosidine synthase from cell suspension cultures of *Rauvolfia serpentina*. *Phytochemistry* 27:3811–15

40. Hashimoto T, Hayashi A, Amano Y, Kohno J, Iwanari S, et al. 1991. Hyoscyamine 6β-hydroxylase, an enzyme involved in tropane alkaloid biosynthesis, is localized at the pericycle of the root. *J. Biol. Chem.* 266:4648–53

41. Hashimoto T, Kohno J, Yamada Y. 1989. 6β-Hydroxyhyoscyamine epoxidase from cultured roots of *Hyoscyamus niger*. *Phytochemistry* 28:1077–82

42. Hashimoto T, Matsuda J, Yamada Y. 1993. Two-step epoxidation of hyoscyamine to scopolamine is catalyzed by bifunctional hyoscyamine 6β-hydroxylase. *FEBS Lett.* 329:35–39

43. Hashimoto T, Mitani A, Yamada Y. 1990. Diamine oxidase from cultured roots of *Hyoscyamus niger*: its function in tropane alkaloid biosynthesis. *Plant Physiol.* 93:216–21

44. Hashimoto T, Nakajima K, Ongena G, Yamada Y. 1992. Two tropinone reductases with distinct stereospecificities from cultured roots of *Hyoscyamus niger*. *Plant Physiol.* 100:836–45

45. Hashimoto T, Yamada Y. 1986. Hyoscyamine 6β-hydroxylase, a 2-oxoglutarate-dependent dioxygenase, in alkaloid-producing root cultures. *Plant Physiol.* 81:619–25

46. Hashimoto T, Yamada Y. 1987. Purification and characterization of hyoscyamine 6β-hydroxylase from root cultures of *Hyoscyamus niger* L. *Eur. J. Biochem.* 164:277–85

47. Hashimoto T, Yamada Y. 1992. Tropane alkaloid biosynthesis: regulation and application. In *Proc. 7th Annu. Penn. State Symp. Plant Physiol.*, pp. 122–34. Rockville: Am. Soc. Plant Physiol. Press

48. Hashimoto T, Yamada Y. 1993. Nicotine and tropane alkaloids. See Ref. 15, pp. 369–79

49. Hashimoto T, Yukimune Y, Yamada Y. 1989. Putrescine and putrescine *N*-methyltransferase in the biosynthesis of tropane alkaloids in cultured roots of *Hyoscyamus albus*. I. Biochemical studies. *Planta* 178:123–30

50. Hashimoto T, Yun D-J, Yamada Y. 1993. Production of tropane alkaloids in genetically engineered root cultures. *Phytochemistry* 32:713–18

51. Herminghaus S, Schreier PH, McCarthy JEG, Landsmann J, Botterman J, Berlin J. 1991. Expression of a bacterial lysine decarboxylase gene and transport of the protein into chloroplasts of transgenic tobacco. *Plant Mol. Biol.* 17:475–86

52. Hibi N, Fujita T, Hatano M, Hashimoto T, Yamada Y. 1992. Putrescine *N*-methyltransferase in cultured roots of *Hyoscyamus albus*. *Plant Physiol.* 100:826–35

53. Deleted in proof

54. Hondelmann W. 1984. The lupin—ancient and modern crop plant. *Theor. Appl. Genet.* 68:1–9

55. Hrazdina G, Jensen RA. 1992. Spatial organization of enzymes in plant metabolic pathways. *Annu. Rev. Plant Physiol. Plant Mol. Biol.* 43:241–67

56. Deleted in proof

57. Kawalleck P, Keller H, Hahlbrock K, Scheel D, Somssich IE. 1993. A pathogen-responsive gene of parsley encodes tyrosine decarboxylase. *J. Biol. Chem.* 268:2189–94

58. Kawalleck P, Plesch G, Hahlbrock K, Somssich IE. 1992. Induction by fungal elicitor of *S*-adenosyl-L-methionine synthase and *S*-adenosyl-L-homocysteine hydrolase mRNA in cultured cells and leaves of *Petroselinum crispum*. *Proc. Natl. Acad. Sci. USA* 89:4713–17

59. Kutchan TM. 1989. Expression of enzymatically active cloned strictosidine synthase from the higher plant *Rauvolfia serpentina* in *Escherichia coli*. *FEBS Lett.* 257:127–30

60. Kutchan TM. 1993. Strictosidine: from alkaloid to enzyme to gene. *Phytochemistry* 32:493–506

61. Kutchan TM. 1994. 12-Oxo-phytodienoic acid induces accumulation of berberine bridge enzyme transcript in a manner analogous to methyl jasmonate. *J. Plant Physiol.* In press

62. Kutchan TM, Bock A, Dittrich H. 1994. Heterologous expression of the plant proteins strictosidine synthase and berberine bridge enzyme in insect cell culture. *Phytochemistry* In press

63. Kutchan TM, Dittrich H, Bracher D, Zenk MH. 1991. Enzymology and molecular biology of alkaloid biosynthesis. *Tetrahedron Lett.* 47:5945–54

64. Kutchan TM, Hampp N, Lottspeich F, Beyreuther K, Zenk MH. 1988. The cDNA clone for strictosidine synthase from *Rauvolfia serpentina*: DNA sequence de-

termination and expression in *Escherichia coli. FEBS Lett.* 237:40–44

65. Kutchan TM, Zenk MH. 1993. Enzymology and molecular biology of benzophenanthridine alkaloid biosynthesis. *J. Plant Res.* Special Issue 3:165–73

66. Leete E. 1990. Recent developments in the biosynthesis of the tropane alkaloids. *Planta Med.* 56:339–52

67. Legg PD, Chaplin JF, Collins GB. 1969. Inheritance of percent total alkaloids in *Nicotiana tabacum* L. *J. Hered.* 60:213–17

68. Legg PD, Collins GB. 1971. Inheritance of per cent total alkaloids in *Nicotiana tabacum* L. II. Genetic effects of two loci in Burley 21 × LA Burley 21 populations. *Can. J. Genet. Cytol.* 13:287–91

69. Legg PD, Collins GB, Litton CC. 1970. Registration of LA Burley 21 tobacco germplasm. *Crop Sci.* 10:212

70. Leland TJ, Grumet R, Hanson AD. 1985. Biochemical, immunological and genetic characterization of natural gramine-free variants of *Hordeum vulgare* L. *Plant Sci.* 42:77–82

71. Leland TJ, Hanson AD. 1985. Induction of a specific *N*-methyltransferase enzyme by long-term heat stress during barley leaf growth. *Plant Physiol.* 79:451–57

72. Madsen JP, Bush LP, Gay SL. 1985. Effects of curing on polyamine content of leaves of *Nicotiana tabacum* L. genotypes with different alkaloid levels. *J. Agric. Food Chem.* 33:1182–85

73. Madyastha KM, Coscia CJ. 1979. Detergent-solubilized NADPH-cytochrome c (P-450) reductase from the higher plant, *Catharanthus roseus. J. Biol. Chem.* 254:2419–27

74. Madyastha KM, Meehan TD, Coscia CJ. 1976. Characterization of a cytochrome P-450 dependent monoterpene hydroxylase from the higher plant *Vinca rosea. Biochemistry* 15:1097–1102

75. Madyastha KM, Ridgway JE, Dwyer JG, Coscia CJ. 1977. Subcellular localization of a cytochrome P-450-dependent monooxygenase in vesicles of the higher plant *Catharanthus roseus. J. Cell Biol.* 72:302–13

76. Mann TJ, Weybrew JA. 1958. Inheritance of alkaloids in hybrids between flue-cured tobacco and related amphiploids. *Tob. Sci.* 2:29–34

77. Mann TJ, Weybrew JA, Matzinger DF, Hall JL. 1964. Inheritance of the conversion of nicotine to nornicotine in varieties of *Nicotiana tabacum* L. and related amphiploids. *Crop Sci.* 4:349–53

78. Matsuda J, Okabe S, Hashimoto T, Yamada Y. 1991. Molecular cloning of hyoscyamine 6β-hydroxylase, a 2-oxoglutarate-dependent dioxygenase, from cultured roots of *Hyoscyamus niger. J. Biol. Chem.* 266:9460–64

79. McNight TD, Bergey DR, Burnett RJ, Nessler CL. 1991. Expression of enzymatically active and correctly targeted strictosidine synthase in transgenic tobacco plants. *Planta* 185:148–52

80. McNight TD, Roessner CA, Devagupta R, Scott AI, Nessler CL. 1990. Nucleotide sequence of a cDNA encoding the vacuolar protein strictosidine synthase from *Catharanthus roseus. Nucleic Acids Res.* 18:4939

81. Meijer AH. 1993. Cytochrome P-450 and secondary metabolism in *Catharanthus roseus.* PhD thesis. Leiden Univ., The Netherlands

82. Meijer AH, Cardoso MIL, Voskuilen JT, de Waal A, Verpoorte R, Hoge JH. 1993. Isolation and characterization of a cDNA clone from *Catharanthus roseus* encoding NADPH:cytochrome P-450 reductase, an enzyme essential for reactions catalysed by cytochrome P-450 mono-oxygenases in plants. *Plant J.* 4:47–60

83. Meijer AH, de Waal A, Verpoorte R. 1993. Purification of the cytochrome P-450 enzyme geraniol 10-hydroxylase from cell cultures of *Catharanthus roseus. J. Chromatogr.* 635:237–49

84. Meijer AH, Souer E, Verpoorte R, Hoge JHC. 1993. Isolation of cytochrome P-450 cDNA clones from the higher plant *Catharanthus roseus* by a PCR strategy. *Plant Mol. Biol.* 22:379–83

85. Meijer AH, Verpoorte R, Hoge JH. 1993. Regulation of enzymes and genes involved in terpenoid indole alkaloid biosynthesis in *Catharanthus roseus. J. Plant Res.* Special Issue 3:145–64

86. Mihaliak CA, Karp F, Croteau R. 1993. Cytochrome P-450 terpene hydroxylases. See Ref. 15, pp. 261–79

87. Miller RD, Collins GB, Davis DL. 1983. Effects of nicotine precursors on nicotine content in callus cultures of Burley tobacco alkaloid lines. *Crop Sci.* 23:561–65

88. Mizusaki S, Tanabe Y, Noguchi M, Tamaki E. 1973. Changes in the activities of ornithine decarboxylase, putrescine *N*-methyltransferase and *N*-methylputrescine oxidase in tobacco roots in relation to nicotine biosynthesis. *Plant Cell Physiol.* 14:103–10

89. Moore RC, Boyle SM. 1990. Nucleotide sequence and analysis of the *speA* gene encoding biosynthetic arginine decarboxylase in *Escherichia coli. J. Bacteriol.* 172:4631–40

90. Muemmler S, Rueffer M, Nagakura N, Zenk MH. 1985. *S*-Adenosyl-L-methionine: (*S*)-scoulerine 9-*O*-methyltransferase, a highly stereo- and regio-specific enzyme in tetrahydroprotoberberine biosynthesis. *Plant Cell Rep.* 4:36–39

91. Müller MJ, Zenk MH. 1992. The nor-coclaurine pathway is operative in berberine biosynthesis in *Coptis japonica. Planta Med.* 58:524–27

92. Nakagawa K, Konagai A, Fukui H, Tabata M. 1984. Release and crystallization of berberine in the liquid medium of *Thalictrum minus* cell suspension cultures. *Plant Cell Rep.* 3:254–57

93. Nakajima K, Hashimoto T, Yamada Y. 1993. Two tropinone reductases with different stereospecificities are short-chain dehydrogenases evolved from a common ancestor. *Proc. Natl. Acad. Sci. USA* In press

94. Nakamura K, Matsuoka K. 1993. Protein targeting to the vacuole in plant cells. *Plant Physiol.* 101:1–5

95. Nessler CL, Mahlberg PG. 1977. Ontogeny and cytochemistry of alkaloidal vesicles in laticifers of *Papaver somniferum* L. (Papaveraeae). *Am. J. Bot.* 64:541–51

96. Okada N, Shinmyo A, Okada H, Yamada Y. 1988. Purification and characterization of (*S*)-tetrahydroberberine oxidase from cultured *Coptis japonica* cells. *Phytochemistry* 27:979–82

97. Ovadi J, Srere PA. 1992. Channel your energies. *Trends Biochem. Sci.* 17:445–47

98. Paiva NL, Edwards R, Sun YJ, Hrazdina G, Dixon RA. 1991. Stress responses in alfalfa (*Medicago sativa* L.) 11. Molecular cloning and expression of alfalfa isoflavone reductase, a key enzyme of isoflavonoid phytoalexin biosynthesis. *Plant Mol. Biol.* 17:653–67

99. Pasquali G, Goddijn OJM, de Waal A, Verpoorte R, Schilperoort RA, et al. 1992. Coordinated regulation of two indole alkaloid biosynthetic genes from *Catharanthus roseus* by auxin and elicitors. *Plant Mol. Biol.* 18:1121–31

100. Pennings EJM, Groen BW, Duine JA, Verpoorte R. 1989. Tryptophan decarboxylase from *Catharanthus roseus* is a pyridoxoquinoprotein. *FEBS Lett.* 255:97–100

101. Pfitzner U, Zenk MH. 1989. Homogeneous strictosidine synthase isoenzymes from cell suspension cultures of *Catharanthus roseus. Planta Med.* 55:525–30

102. Prescott AG. 1993. A dilemma of dioxygenases (or where biochemistry and molecular biology fail to meet). *J. Exp. Bot.* 44:849–61

103. Pryer NK, Wuestehube LJ, Schekman RW. 1992. Vesicle-mediated protein sorting. *Annu. Rev. Biochem.* 61:471–516

104. Roessner CA, Devagupta R, Hasan M, Williams HJ, Scott AI. 1992. Purification of an indole alkaloid biosynthetic enzyme, strictosidine synthase, from a recombinant strain of *Escherichia coli. Protein Expr. Purif.* 3:295–300

105. Romeike A. 1978. Tropane alkaloids—oc-currence and systematic importance in angiosperms. *Bot. Not.* 131:85–96

106. Rothman JE, Orci L. 1992. Molecular dissection of the secretory pathway. *Nature* 355:409–15

107. Rueffer M. 1985. The production of isoquinoline alkaloids by plant cell cultures. In *The Chemistry and Biology of Isoquinoline Alkaloids,* ed. JD Phillipson, MF Roberts, MH Zenk, pp. 265–80. Berlin: Springer-Verlag

108. Rueffer M. 1992. Side reactions in the protoberberine biosynthetic pathway. *Planta Med.* 58:579

109. Rueffer M, Amann M, Zenk MH. 1986. *S*-Adenosyl-L-methionine: columbamine-*O*-methyltransferase, a compartmentalized enzyme in protoberberine biosynthesis. *Plant Cell Rep.* 3:182–85

110. Rueffer M, Zenk MH. 1985. Berberine synthase, the methylenedioxy group forming enzyme in berberine synthesis. *Tetrahedron Lett.* 26:201–2

111. Rueffer M, Zenk MH. 1987. Distant precursors of benzylisoquinoline alkaloids and their enzymatic formation. *Z. Naturforsch. Teil C* 42:319–32

112. Saito K, Noji M, Ohmori S, Imai Y, Murakoshi I. 1991. Integration and expression of a rabbit liver cytochrome P-450 gene in transgenic *Nicotiana tabacum. Proc. Natl. Acad. Sci. USA* 88:7041–45

113. Sato F, Takeshita N, Fitchen J, Fujiwara H, Yamada Y. 1993. *S*-Adenosyl-L-methionine: scoulerine-9-*O*-methyltransferase from cultured roots of *Coptis japonica* cells. *Phytochemistry* 32:659–64

114. Sato H, Kobayashi Y, Fukui H, Tabata M. 1990. Specific differences in tolerance to exogenous berberine among plant cell cultures. *Plant Cell Rep.* 9:133–36

115. Sato H, Tanaka S, Tabata M. 1993. Kinetics of alkaloid uptake by cultured cells of *Coptis japonica. Phytochemistry.* In press

116. Sato M, Asaine K. 1982. Studies on the accumulation of nicotine-converting mutants in tobacco. III. A cotyledon-test method for the detection of nicotine-converting mutants. *Iwata Tob. Rep.* 14:17–21

117. Saunders JW, Bush LP. 1979. Nicotine biosynthetic enzyme activities in *Nicotiana tabacum* L. genotypes with different alkaloid levels. *Plant Physiol.* 64:236–40

118. Schumacher H-M, Gundlach H, Fiedler F, Zenk MH. 1987. Elicitation of benzophenanthridine alkaloid synthesis in *Eschscholtzia* cell cultures. *Plant Cell Rep.* 6:410–13

119. Shet MS, Sathasivan K, Arlotto MA, Mehdy MC, Estabrook RW. 1993. Purification, characterization, and cDNA cloning of an NADPH-cytochrome P450 reductase from mung bean. *Proc. Natl. Acad. Sci. USA* 90:2890–94

120. Songstad DD, de Luca V, Brisson N, Kurz WGW, Nessler CL. 1990. High levels of tryptamine accumulation in transgenic tobacco expressing tryptophan decarboxylase. *Plant Physiol.* 94:1410–13

121. Songstad DD, Kurz WGW, Nessler CL. 1991. Tyramine accumulation in *Nicotiana tabacum* transformed with a chimeric tryptophan decarboxylase gene. *Phytochemistry* 30:3245–46

122. Southon IW, Buckingham J, eds. 1989. *Dictionary of Alkaloids.* London: Chapman & Hall

123. Stadler R, Zenk MM. 1993. The purification and characterization of a unique cytochrome P-450 enzyme from *Berberis stolonifera* plant cell cultures. *J. Biol. Chem.* 268:823–31

124. Steffens P, Nagakura N, Zenk M. 1985. Purification and characterization of the berberine bridge enzyme from *Berberis beaniana* cell cultures. *Phytochemistry* 24: 2577–83

125. Stevens LH, Blom TJM, Verpoorte R. 1994. Subcellular localization of tryptophan decarboxylase, strictosidine synthase and strictosidine glucosidase in suspension cultured cells of *Catharanthus roseus* and *Tabernaemontana divaricata. Plant Cell Rep.* 12:573–76

126. Stevens LH, Giroud C, Pennings EJM, Verpoorte R. 1993. Purification and characterization of strictosidine synthase from a suspension culture of *Cinchona robusta. Phytochemistry* 33:99–106

127. Suzuki M, Nakagawa K, Fukui H, Tabata M. 1987. Relationship of berberine-producing capability between *Thalictrum* plants and their tissue cultures. *Plant Cell Rep.* 6:260–63

128. Tabata M. 1991. Transport and secretion of natural products in plant cell cultures. *Planta Med.* 57:21–26

129. Valleau WD. 1949. Breeding low-nicotine tobacco. *J. Agric. Res.* 78:171–81

130. van Tunen AJ, Mol JNM. 1991. Control of flavonoid synthesis and manipulation of flower colour. In *Developmental Regulation of Plant Gene Expression,* ed. D Grierson, pp. 94–130. Glasgow: Blackie

131. Vetter H-P, Mangold U, Schröder G, Marner F-J, Werck-Reichhart D, Schröder J. 1992. Molecular analysis and heterologous expression of an inducible cytochrome P-450 protein from periwinkle (*Catharanthus roseus* L.). *Plant Physiol.* 100:998–1007

132. Wernsman EA, Matzinger DF. 1968. Time and site of nicotine conversion in tobacco. *Tob. Sci.* 12:226–28

133. Wernsman EA, Matzinger DF. 1970. Relative stability of alleles at the nicotine conversion locus of tobacco. *Tob. Sci.* 14:34–36

134. Williams W, Harrison JE, Jayasekera S. 1984. Genetical control of alkaloid production in *Lupinus mutabilis* and the effect of a mutant allele mutal isolated following chemical mutagenesis. *Euphytica* 33:811–17

135. Yamamoto H, Nakagawa K, Fukui H, Tabata M. 1986. Cytological changes associated with alkaloid production in cultured cells of *Coptis japonica* and *Thalictrum minus. Plant Cell Rep.* 5:65–68

136. Yamamoto H, Suzuki M, Kitamura T, Fukui H, Tabata M. 1989. Energy-requiring uptake of protoberberine alkaloids by cultured cells of *Thalictrum flavum. Plant Cell Rep.* 8:361–64

137. Yun D-J, Hashimoto T, Yamada Y. 1992. Metabolic engineering of medicinal plants: transgenic *Atropa belladonna* with an improved alkaloid composition. *Proc. Natl. Acad. Sci. USA* 89:11799–803

138. Yun D-J, Hashimoto T, Yamada Y. 1993. Transgenic tobacco plants with two consecutive oxidation reactions catalyzed by hyoscyamine 6β-hydroxylase. *Biosci. Biotech. Biochem.* 57:502–3

139. Zenk MH, Rueffer M, Kutchan TM, Galneder E. 1988. Future trends of exploration for the biotechnological production of isoquinoline alkaloids. In *Applications of Plant Cell and Tissue Culture,* ed. Y Yamada, pp. 213–23. Chichester: Wiley

Annu. Rev. Plant Physiol. Plant Mol. Biol. 1994. 45:287–301

MOLECULAR BIOLOGY OF CAROTENOID BIOSYNTHESIS IN PLANTS

Glenn E. Bartley and Pablo A. Scolnik

DuPont Central Research, P.O. Box 80402, Wilmington, Delaware 19880-0402

Giovanni Giuliano

Ente per Le Nuove Tecnologie, L'Energia E L'Ambiente, Casaccia Research Center, P.O. Box 2400, Rome 00100 AD, Italy

KEY WORDS: carotenoid(s), chloroplast, chromoplast, *Arabidopsis*, maize, tomato, photooxidation

CONTENTS

INTRODUCTION 288
GENE CLONING 288
   Geranylgeranyl Pyrophosphate Synthase (GGPS) 288
   Phytoene Synthase (PSY) 291
   Phytoene Desaturase (PDS) 292
   Other Genes 292
GENETIC CONTROL 293
   Arabidopsis thaliana 293
   Lycopersicon esculentum 294
   Zea mays 295
   Other Species 295
REGULATION OF EXPRESSION 295
   Light 295
   Photooxidative Stress 296
   Development 296
FINAL COMMENTS 297

INTRODUCTION

Carotenoids, the most diverse and widespread group of pigments found in nature, are synthesized de novo by all photosynthetic and many non-photosynthetic organisms. Our current knowledge of the molecular biology of carotenoids derives primarily from the study of the pathway in specific organisms, including the photosynthetic prokaryotes *Rhodobacter* and *Synechococcus*, bacteria of the genus *Erwinia*, the fungi *Neurospora* and *Phycomyces*, and the higher plants *Zea mays* (corn), *Lycopersicon esculentum* (tomato), *Narcissus pseudonarcissus* (daffodil), and *Capsicum annuum* (pepper). In green tissues of higher plants carotenoids provide essential photoprotective functions, blocking the formation of reactive oxygen species. These pigments also accumulate in chromoplasts, providing the yellow, orange, and red colors of many flowers, fruits, and storage roots. The photoprotective functions of carotenoids have been discussed extensively in other publications (2, 24, 41). There is no firm evidence for a functional role of carotenoids in flowers and fruits, although they may play a role in pollen formation (reviewed in 70) and in the attraction of pollinating insects or animals involved in seed dispersal. The distribution of carotenoid pigments in flowers and fruits has been discussed elsewhere (32, 33). Although the chemical structures of most plant carotenoids were elucidated decades ago, the cloning of plant genes for the biosynthetic enzymes and the study of their expression is a relatively recent development. In this article we review current information on the molecular aspects of carotenoid biosynthesis in plants, and we discuss possible future directions for the field.

GENE CLONING

Plant genes for carotenoid enzymes have been cloned by four approaches: heterologous hybridization, differential cDNA screening, transposon tagging, and screening of expression cDNA libraries with antibodies raised against purified enzymes. In this section we describe the gene identification and cloning strategies, the chemical reactions catalyzed by the products of the cloned genes, and the structural features of the corresponding proteins. Numbered steps in a simplified version of the biosynthetic pathway are shown in Figure 1, and a list of cloned genes is shown in Table 1.

Geranylgeranyl Pyrophosphate Synthase (GGPS)

GGPPS of higher plants is a multifunctional enzyme that catalyzes steps 2, 3, and 4 (25). Antibodies against GGPPS purified from *Capsicum annuum* (pepper) chromoplasts have been used to clone the corresponding cDNA from an expression library made from ripening fruit mRNA (42). The identity of this

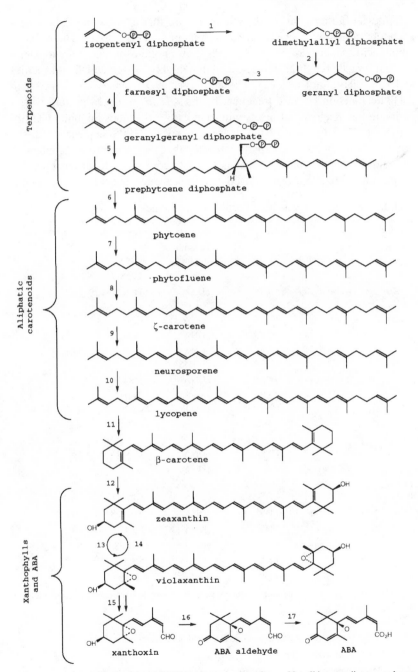

Figure 1   Carotenoid–abscisic acid (ABA) biosynthesis pathway. Not all intermediates are shown. Specific steps are described in the text.

cDNA was further confirmed by determining that expression of the corresponding protein in *Escherichia coli* results in GGPPS activity. The deduced amino acid sequence of pepper GGPPS shows regions of homology to bacterial and fungal prenyltransferases, including highly conserved DD (diaspartic) and RR (diarginine) residues. The conserved DD residues are involved in catalytic function because treatment of the *E. coli*–produced protein with a carbodiimide reagent that reacts with free carboxyl groups inhibited GGPPS activity (42). Alignment of the deduced pepper GGPPS sequence with the bacterial and fungal counterparts indicates that the plant protein contains a 60-residue amino terminal extension, likely to correspond to a transit peptide for plastid localization. The deduced peptide sequence, but not the correspond-

Table 1   Cloned genes for plant carotenoid biosynthesis enzymes.

Enzyme	Organism	Gene	Clone type	Accession #	Reference
GGPP synthase	*A. thaliana*	GGPS	genomic fragment	L22347	—
		GGPS1	cDNA	L25813	—[a]
	C. annuum	GGPS	cDNA	P80042	42[b]
Phytoene synthase	*A. thaliana*	PSY	cDNA	L25812	—
	C. annuum	PSY	cDNA	X68017	61a
	L. esculentum	PSY1	cDNA (TOM5)	Y00521	54
		PSY1	cDNA	M84744	7
		PSY1	genomic (GTOM5)	X60441	55
		PSY2	genomic (GTOMF)	X60440	55[c]
		PSY2	cDNA	L23424	6
	Z. mays	Psy	cDNA	—	—[d]
Phytoene desturase	*A. thaliana*	PDS	cDNA	L16237	61b
	C. annuum	PDS	cDNA	X68058	34
	G. max	PDS	cDNA	M64704	8
	L. esculentum	PDS	cDNA	M88683	30
		PDS	cDNA	X59948	51[e]
		PDS	genomic	—	—[f]

[a]Sequence divergence from genomic fragment suggests that this cDNA corresponds to a second copy of the gene.

[b]Deduced amino acid sequency only.

[c]GTOMF may actually be a PSY2 genomic DNA and not a pseudogene (6).

[d]B Buckner, manuscript in preparation.

[e]This cDNA contains and intron and is truncated at the 3′ end.

[f]B Aracri, GE Bartley, PA Scolnik, & G Giuliano, manuscript in preparation.

ing DNA sequence of the cDNA, has been published (42). Based on this information Bartley and Scolnik cloned and reported to Genbank the sequence of an *Arabidopis thaliana GGPS* gene fragment (Table 1).

Phytoene Synthase (PSY)

PSY converts two molecules of geranylgeranyl pyrophosphate into phytoene (steps 5 and 6), and it is the first dedicated enzyme of the pathway. Clone TOM5, one of a collection of cDNAs for genes induced during tomato fruit ripening (22, 46) shows significant homology to the *crtB* genes of *Erwinia* and *Rhodobacter,* which encode PSY (3). The TOM5 cDNA encodes a 47-kDa protein carrying a 120–amino acid transit peptide absent in the corresponding bacterial proteins (7, 54). Transgenic tomato plants expressing a 5' segment of TOM5 in the antisense orientation under the control of the CaMV 35S promoter show significantly decreased levels of carotenoids in flowers and fruits (13). Extensive biochemical analysis of these antisense transgenic plants showed that in fruits the block is between geranylgeranyl pyrophosphate and phytoene. Furthermore, the cDNA complements *Rhodobacter capsulatus crtB* mutants (7). This finding, together with the phenotype of TOM5 antisense fruits, which accumulate GGPP, indicates that this gene encodes PSY.

Fruits of TOM5 antisense plants show up to 97% reduction in carotenoid levels, but leaf carotenoids are not affected (14). This indicates that a second *PSY* gene, with lower homology to TOM5 in the 5' region used in the antisense construct, may be active in somatic tissues (7). An mRNA encoding an active PSY with characteristics similar to the one encoded by TOM5 has been identified in tomato leaves (6). These authors renamed the corresponding gene, previously reported as a pseudogene, *PSY2*. Therefore, we refer to TOM5 as *PSY1*, and to the second gene as *PSY2*. Both PSY proteins show extensive conservation with human and fungal squalene synthases, enzymes that also catalyze prenyl substrate condensations (59). Two genomic regions homologous to TOM5 have been cloned and sequenced (55). The *PSY1* genomic clone (*GTOM5*) contains 5 introns and 6 exons in a 3.6-kb sequence. In the second clone (clone F, or *PSY2*), the first exon of *PSY1* is split into 3 exons.

PSY genes also have been cloned from maize and pepper plants. The pepper gene encodes a protein with a high degree of homology to tomato PSY (61a). This enzyme catalyzes reactions 5 and 6, confirming that the plant PSY is a bifunctional enzyme (26). A putative carotenoid regulatory locus from maize, *Yellow1* (*Y1*) was tagged using the *Mu* transposable element. Subsequent cloning and sequencing of a *Y1* cDNA indicated that there was high homology to tomato and bacterial *PSY* genes, indicating that *Y1* encodes phytoene synthase in maize (18).

Phytoene Desaturase (PDS)

The yellow, red, and orange colors of most common carotenoids result from the absorption of light by chromophores of seven or more conjugated double bonds. PDS creates additional double bonds in the C40 backbone of phytoene, thus transforming this colorless carotenoid into a colored compound. Different PDS enzymes, able to catalyze 2, 3, or 4 successive desaturation reactions, are present in different organisms (5, 43). The plant PDS enzyme only catalyzes the two desaturations that convert phytoene into zeta-carotene (steps 7 and 8).

A soybean cDNA encoding PDS was cloned (8) using a heterologous probe derived from a *Synechococcus* gene able to confer resistance to the bleaching herbicide norflurazon, a PDS inhibitor (19). The tomato homolog was cloned with a similar approach (51). The identity of the cDNAs was further confirmed by expression in *Rhodobacter* and *E. coli* strains accumulating phytoene. In both cases, expression of the PDS cDNA resulted in the formation of zeta-carotene (8, 51). The tomato genomic region encoding PDS has been sequenced. The transcribed region contains 15 exons and 14 introns and extends over approximately 7.7 kb of genomic DNA (B Aracri, GE Bartley, PA Scolnik, G Giuliano, manuscript in preparation). The pepper PDS was purified from fruit chromoplasts and cDNA clones were isolated using antibodies raised to the purified protein. The resulting cDNA shows high homology to soybean and tomato PDS, and its overexpression in *E. coli* results in the synthesis of an active PDS protein (34).

The plant *PDS* clones isolated so far encode proteins of approximately 64 kDa deduced molecular weight that show high homology to cyanobacterial, but not to *Rhodobacter* or *Neurospora* PDS proteins (8, 34, 51). In agreement with the observation that purified PDS contains bound FAD (34), a highly conserved FAD-binding motif has been observed in the amino terminal regions of all PDS enzymes characterized to date (5, 8, 34, 51). Plant PDS also have transit peptides of approximately 110 amino acids (8).

Other Genes

Ongoing research in several laboratories will lead to the cloning of genes for additional steps of the pathway. A cDNA possibly encoding capsanthin/ capsorubin synthase, one of the late steps in pepper fruit xanthophyll biosynthesis, has been isolated recently using antibodies raised to the purified enzyme (M Kuntz, personal communication). Cyanobacterial genes encoding zeta-carotene desaturase (ZDS; 44) and lycopene cyclase (LCY; 20) have been isolated using, respectively, an *E. coli*–based complementation assay and genetic transfer of a herbicide-resistance trait into cyanobacteria. These genes could possibly be used as probes for cloning the corresponding higher plant genes.

GENETIC CONTROL

Genetic blocks in the carotenoid biosynthesis pathway can affect early or late stages of plant development. Mutations affecting carotenoid biosynthesis in green tissues may have pleiotropic phenotypes, including photobleaching and severe alterations in leaf development, that are also observed in mutants not affected in this pathway. In contrast, mutations affecting late stages such as chromoplast development in maize endosperm and tomato fruit provide a more unambiguous phenotype. We mention only those mutants for which there is strong biochemical or molecular evidence for primary lesions in carotenoid biosynthesis. Other mutants affected in chloroplast biogenesis may fall in this category, but the available evidence is inconclusive. The mutants selected and the carotenoid intermediates they accumulate are listed in Table 2.

Arabidopsis thaliana

Many *A. thaliana* mutants affected in pigmentation have been described (39). The first mutants shown to be arrested in some step of the carotenoid pathway were originally isolated as abscisic acid–deficient variants, and three such

Table 2 Plant mutants arrested in carotenoid biosynthesis.

Species	Mutant	Step	Expression	Reference
Z. mays	y1	5, 6	endosperm, leaf[a,b,c]	17, 18, 57
	vp2	7	all tissues	57, 65
	vp5	7	all tissues	57, 58
	w3	7	all tissues	2, 57
	vp9	9	all tissues	57
	ps	11	all tissues	57
	y9	?	endosperm, leaf	57, 58
L. esculentum	r	5, 6	all tissues	29, 35
	gh	7	all tissues	56, 62
	not	15	all tissues	49
A. thaliana	aba	13	all tissues	27, 61
H. vulgare	tig-b	9	all tissues	47
	tig-o	11	all tissues[d]	47
H. annuus	nd-1	5, 6?	all tissues[e]	28

[a]Phenotype present only in some alleles.
[b]Gene cloned.
[c]Putatitve transposon mutants available.
[d]Viviparous or non-dormant mutant.
[e]Variegated mutant.

isolates were found to be allelic (40). Plants with any of these three alleles accumulate zeaxanthin, indicating a block in the epoxidation of this xanthophyll to violaxanthin (Figure 1, step 13; see 27, 61). These mutants are altered in chloroplast development and chlorophyll fluorescence yields (60), in accordance with the proposed role of zeaxanthin in thermal dispersion of excess excitation energy (24). Two mutant lines that accumulate phytoene were identified during a screen of a T-DNA mutagenized collection (4). Also, the variegated *immutans* line accumulates phytoene in its white sectors, and the defect maps to a locus different from *PDS* (69). Thus, there are some interesting parallels between *A. thaliana immutans* and tomato *ghost*. Both lines have a variegated, epigenetic phenotype; accumulate phytoene in white sectors; and do not appear to be affected in the structural genes for phytoene desaturase.

Lycopersicon esculentum

The tomato fruit pericarp has been used widely as a system for genetic, biochemical, and molecular studies of carotenoid biosynthesis. Goodwin (31) and Tanskley & Mutschler (66) have listed a series of loci affecting carotenoid content in tomato. Most of these loci only affect pigmentation in the fruits and flowers; only one locus (*ghost, gh*) controls pigmentation in all somatic tissues (56). Mutations in this locus give rise to a somatically unstable albino phenotype resulting from a block in the desaturation of phytoene (Figure 1, steps 7 and 8; see 62). The *PDS* gene from tomato has been mapped recently to the end of chromosome 3; it is a single copy gene (30; B Aracri, GE Bartley, PA Scolnik, G Giuliano, manuscript in preparation). The *gh* locus maps to chromosome 11. This observation indicates that *GHOST* encodes a function different from PDS, but necessary for phytoene desaturation.

Mutants of the *YELLOW FLESH* (*R*) locus change the color of the fruit pericarp from red to yellow, severely reducing the carotenoid content of this tissue. Some alleles also affect flower pigmentation (21, 35). This locus has been shown to correspond to *PSY1* (GTOM5) by DNA sequencing of mutant alleles and genetic complementation with the TOM5 cDNA (29). Inhibition and ectopic production of carotenoids were observed in some transgenic plants overexpressing the TOM5 cDNA (29). The inhibition may reflect co-suppression of *PSY* expression by the introduction of homologous DNA sequences. One of the *R* mutant alleles is caused by the insertion of a highly repeated genetic element, reminiscent of transposable elements. At least two genomic loci showing homology to *PSY* are present, respectively, on chromosomes 2 and 3 of tomato (37). The chromosome 3 locus corresponds to *PSY1/R* while the chromosome 2 gene probably corresponds to *PSY2,* which has been mapped to this linkage group by using single chromosome addition lines (6).

Zea mays

Robertson (57) has surveyed a series of maize carotenoid mutants. Several mutants (*vp2, vp5, vp9, ps, w3*) have altered carotenoid content in both endosperm and leaves. Leaves of some mutants show a variable extent of chlorophyll photobleaching (2) depending on the severity of the allele. This phenotype is often coupled to a more or less pronounced viviparous phenotype, resulting from lack of abscisic acid (ABA), a by-product of the carotenoid pathway (Figure 1). The *Y1* and *Y9* loci control pigmentation predominantly in the endosperm, although some alleles have a slightly pale leaf color. In some cases, *y9* mutant alleles confer vivipary.

The *Y1* locus was cloned by transposon tagging with the *Mu* transposable element: *Y1/Y1* plants carrying the *Mu* element were crossed to *y1/y1* plants and the rare *y1/y1-mum* kernels caused by *Mu* insertion in the *Y1* gene were recovered from the progeny. Leaf color of some *y1-mum* alleles was lighter (pastel alleles), which is a temperature sensitive phenomenon observed previously, most likely from photobleaching of chlorophyll. Mutability of the various *y1-mum* pastel alleles resulted in small green sectors within the pale green background of homozygous *y1-mum* leaves (17). One of the multiple bands detected by a *Mu3* probe in DNA blots of mutant plants was found to be genetically linked to one *y1-mum* allele and was subsequently cloned. The DNA sequence of the corresponding cDNA shows high homology to *PSY* from tomato and bacteria (18). It is not known whether a second *PSY* gene is active in maize leaves. Alternatively, the lack of complete leaf photobleaching in the *Y1* pastel alleles could be the result of the mutations' leaky character.

Other Species

Genetic loci affecting different steps of the carotenoid pathway have been described in barley (47), pepper, and sunflower (28) and they are listed in Table 2. The *Or* locus induces xanthophyll accumulation in potato tubers, which are normally devoid of carotenoids (16).

REGULATION OF EXPRESSION

Light

In green tissues carotenoids are synthesized in chloroplasts and are essential for the formation of the light-harvesting and photosynthetic reaction center complexes (9, 45, 52, 71; reviewed in 15, 63). As explained above, carotenoid biosynthesis enzymes are coded for by nuclear genes, translated as precursors in cytoplasmic ribosomes, and imported post-translationally into plastids. Therefore, the final concentration of carotenoids may be controlled by various mechanisms affecting enzyme levels and by the availability of binding sites in photosynthetic complexes. The light-induced formation of photosynthetic complexes is controlled by transcriptional and post-transcriptional events (68). Carotenoid levels increase several fold during deetiolation of seedlings (30; reviewed in 53). A phytochrome effect on carotenoid biosynthesis in both leaves and fruits has been reported (reviewed in 53), and some phytochrome-deficient mutants have altered leaf carotenoid composition (10). The tomato *high pigment* mutant, previously regarded as a carotenoid regulatory mutant, may be altered in phytochrome response (1 and references therein). In tomato, seedling deetiolation has little effect on *PDS* and *PSY1* transcript levels, whereas dark-adaptation has a negative effect on *PSY1*, but not on *PDS* mRNA levels (30). Thus, control of *PSY1* and *PDS* expression does not appear to play a significant role in the regulation of carotenoid content during deetiolation. It is important to note that, at least in tomato plants and with the exception of *PSY1* in fruits, *PSY* and *PDS* mRNAs are rare and difficult to detect with standard RNA blotting technology. More sensitive mRNA amplification techniques are useful for the study of gene regulation in tomato plants (6, 30).

Photooxidative Stress

Transitions from low to high light intensity in the field trigger a photooxidative stress response in which the levels of components of the xanthophyll cycle, particularly zeaxanthin, increase significantly. This carotenoid may dissipate excess light energy in the photosystems, and a model for the regulation of its synthesis from violaxanthin by the decreased luminal pH at high light intensities has been proposed (reviewed in 24).

Photooxidative stress under laboratory conditions can be produced by illuminating plants in which the biosynthesis of colored carotenoids has been blocked genetically or by treatment with bleaching herbicides such as norflurazon, a PDS inhibitor. This light stress response represses the transcription of some nuclear genes that encode photosynthetic proteins, particularly members of the *LHCP* family (reviewed in 67). However, expression of *PDS* and, to a lesser extent, *PSY1* and *PSY2*, is induced under photooxidative conditions (6, 30). These results were obtained with the tomato *ghost* mutant and with norflurazon-treated seedlings, both essentially devoid of colored carotenoids. Two mechanisms (30) have been proposed for this regulatory effect: feedback inhibition by end products, as described for *Phycomyces* (11), or induction of expression by photooxidative stress. Preliminary results obtained in our laboratories indicate that the latter is the case.

Development

Plant development controls the qualitative and quantitative composition of carotenoids. Root carotenoid levels in tomato are about 0.3% of leaf levels, and xanthophylls predominate (50). *PDS* and *PSY1* mRNA levels in roots are approximately 10-fold lower than in leaf (30). During the development of tomato flowers and fruits there is a temporal correlation between the increase in carotenoid levels (36), the transition from chloroplasts to chromoplasts, and the induction of carotenoid gene expression, particularly *PSY1* (30, 46). *PSY2* expression is not induced during fruit ripening (6). In other species, high levels of enzymes for carotenoid biosynthesis are found in petal chromoplasts (12). In tomato flowers, *PSY1* and *PDS* mRNA levels are induced 10-fold during anthesis and high levels of both transcripts are found in mature petals and anthers, but not in sepals and ovaries (30). The induction of these genes in petals may reflect an adaption to the proposed function of carotenoids in the attraction of pollinating insects. Expression in anthers is of interest because of the possible role of carotenoids in pollen development (70). Flowers from plants containing the *PSY1* antisense construct have reduced carotenoid levels (13), indicating that *PSY1* expression is necessary to achieve wild-type flower pigmentation.

Expression of several tomato genes, including *PSY1,* is induced in fruits by the ripening hormone ethylene (46), a process blocked in ripening-inhibited mutants (38), and in fruits treated with silver ions, which antagonize ethylene action (23). Because expression of the gene encoding polygalacturonase, also found to respond to ethylene, is unaffected in tomato antisense fruits blocked in the biosynthesis of this hormone (48), the putative regulation of *PSY1* by ethylene must also be reassessed. Carotenoid levels are also induced during pepper fruit ripening (64). *GGPS* and *PDS* mRNA and the levels of the corresponding enzymes were measured during this process (34, 42). During ripening, mRNA and enzyme levels are concomitantly induced for *GGPPS*; however, PDS enzymatic activity, but not the *PDS* mRNA level, is induced. This indicates that post-transcriptional events play a role in the regulation of carotenogenic enzymes during pepper fruit ripening. Pepper *PSY* mRNA levels are only slightly induced by ripening (61a), indicating that regulation of the same gene in two related solanaceous plants, tomato and pepper, may be quite different. It is important to note, however, that fruit ripening is climacteric in tomato but not in pepper.

FINAL COMMENTS

In the last few years the major emphasis in carotenoid analysis has been the cloning of genes for enzymes of the biosynthetic pathway. This effort will continue until all genes are identified. The study of gene regulation, in spite of recent significant advances, is still in its infancy. The preliminary picture that emerges from the study of *PSY* and *PDS* in tomato is that the major regulatory factors are photooxidation in leaves and the development of reproductive structures. The signal transduction pathway for the negative and positive regulation of gene expression by photooxidation remains unknown, as are the regulatory factors that control carotenoid levels during the development of flowers, fruits, and endosperm. Research opportunities also exist for the application of classical and molecular genetics to the manipulation of the quantity and quality of carotenoids in food and feed crops.

Literature Cited

1. Adamse P, Jaspers PAPM, Koornneef M, Kendrick RE. 1986. A genetic approach to the study of photomorphogenesis in higher plants. In *Proceedings of the XVI Yamada Conference—Phytochrome and Plant Photo Morphogenesis,* organized by M Furuya, Abstr. 80. Okazaki, Japan: Yamada Sci. Found.
2. Anderson IC, Robertson DS. 1960. Role of carotenoids in protecting chlorophyll from photodestruction. *Plant Physiol.* 35:531–34
3. Armstrong GA, Alberti M, Hearst JE. 1990. Conserved enzymes mediate the early reactions of carotenoid biosynthesis in nonphotosynthetic and photosynthetic procaryotes. *Proc. Natl. Acad. Sci. USA* 87:9975–79
4. Barrette T, Norris SR, DellaPenna D. 1993. *Genetic dissection of carotenoid and chlorophyll synthesis in* Arabidopsis thaliana. Presented at 5th Int. Conf. *Arabidopsis* Res., Ohio State Univ., Columbus
5. Bartley GE, Schmidhauser TJ, Yanofsky C, Scolnik PA. 1990. Carotenoid desaturases from *Rhodobacter capsulatus* and *Neurospora crassa* are structurally and functionally conserved and contain domains homologous to flavoprotein disulfide oxidoreductases. *J. Biol. Chem.* 265:16020–24
6. Bartley GE, Scolnik PA. 1993. cDNA cloning, expression during development and genome mapping of *PSY2,* a second tomato gene encoding phytoene synthase. *J. Biol. Chem.* 268:25718–21
7. Bartley GE, Viitanen PV, Bacot KO, Scolnik PA. 1992. A tomato gene expressed during fruit ripening encodes an enzyme of the carotenoid biosynthesis pathway. *J. Biol. Chem.* 267:5036–39
8. Bartley GE, Viitanen PV, Pecker I, Chamovitz D, Hirschberg J, Scolnik PA. 1991. Molecular cloning and expression in photosynthetic bacteria of a soybean cDNA coding for phytoene desaturase, an enzyme of the carotenoid biosynthesis pathway. *Proc. Natl. Acad. Sci. USA* 88:6532–36
9. Bassi R, Pineau B, Dainese P, Marquardt J. 1993. Carotenoid-binding proteins of photosystem II. *Eur. J. Biochem.* 212:297–303
10. Becker TW, Foyer C, Caboche M. 1992. Light-regulated expression of the nitrate reductase and nitrite reductase genes in tomato and in the phytochrome-deficient aurea mutant of tomato. *Planta* 188:39–47
11. Bejarano ER, Cerdá-Olmedo E. 1989. Inhibition of phytoene dehydrogenation and activation of carotenogenesis in *Phycomyces. Phytochemistry* 28:1623–26
12. Beyer P, Weiss G, Kleinig H. 1985. Solubilization and reconstitution of the membrane-bound carotenogenic enzymes from daffodil chromoplasts. *Eur. J. Biochem.* 153:341–46
13. Bird CR, Ray JA, Fletcher JD, Boniwell JM, Bird AS, et al. 1991. Using antisense RNA to study gene function: inhibition of carotenoid biosynthesis in transgenic tomatoes. *BioTechnology* 9:635–39
14. Bramley P, Teulieres C, Blain I, Bird C, Schuch W. 1992. Biochemical characterization of transgenic tomato plants in which carotenoid synthesis has been inhibited through the expression of antisense RNA to pTOM5. *Plant J.* 2:343–49
15. Britton G. 1986. Biosynthesis of chloroplast carotenoids. In *Regulation of Chloroplast Differentiation,* ed. G Akoyunoglou, H Benger, pp. 125–34. New York: Liss
16. Brown CR, Edwards CG, Yang C-P, Dean BB. 1993. Orange flesh trait in potato—inheritance and carotenoid content. *J. Am. Soc. Hortic. Sci.* 118:145–50
17. Buckner B, Kelson TL, Robertson DS. 1990. Cloning of the *y1* locus of maize, a

gene involved in the biosynthesis of carot-enoids. *Plant Cell* 2:867–76
18. Buckner B, San Miguel P, Bennetzen JL. 1993. The *Y1* gene codes for phytoene synthase. *Maize Genet. Coop. Newsl.* 67: 65
19. Chamovitz D, Pecker I, Hirschberg J. 1991. The molecular basis of resistance to the herbicide norflurazon. *Plant Mol. Biol.* 16: 967–74
20. Cunningham FX Jr, Chamovitz D, Misawa N, Gantt E, Hirschberg J. 1993. Cloning and functional expression in *Escherichia coli* of a cyanobacterial gene for lycopene cyclase, the enzyme that catalyzes the bio-synthesis of β-carotene. *FEBS Lett.* 1:130– 38
21. Darby LA, Ritchie DB, Taylor IB. 1978. Isogenic lines of tomato 'Ailsa Craig.' In *The Glasshouse Crops Research Institute 1977 Annual Report,* ed. JN Davies, pp. 168–84. Littlehampton, UK: GCRI
22. Davies KM, Grierson D. 1989. Identifica-tion of cDNA clones for tomato (*Lycopersicon esculentum* Mill.) mRNAs that accu-mulate during fruit ripening and leaf senes-cence in response to ethylene. *Planta* 179: 73–80
23. Davies KM, Hobson GE, Grierson D. 1990. Differential effect of silver ions on the accumulation of ripening-related mRNAs in tomato fruit. *J. Plant Physiol.* 135:708–13
24. Demmig-Adams B, Adams WW III. 1992. Photoprotection and other responses of plants to high light stress. *Annu. Rev. Plant Physiol. Plant Mol. Biol.* 43:599–626
25. Dogbo O, Camara B. 1987. Purification of isopentenyl pyrophosphate isomerase and geranylgeranyl pyrophosphate synthase from *Capsicum* chromoplasts by affinity chromatography. *Biochim. Biophys. Acta* 920:140–48
26. Dogbo O, Laferriere A, D'Harlingue A, Camara B. 1988. Carotenoid biosynthesis: isolation and characterization of a bifunc-tional enzyme catalyzing the synthesis of phytoene. *Proc. Natl. Acad. Sci. USA* 85: 7054–58
27. Duckham SC, Linforth RST, Taylor IB. 1991. Abscisic-acid-deficient mutants at the *aba* gene locus of *Arabidopsis thaliana* are impaired in the epoxidation of zeax-anthin. *Plant Cell Environ.* 14:601–6
28. Fambrini M, Pugliesi G, Vernieri P, Giuliano G, Baroncelli S. 1993. Character-ization of a sunflower (*Helianthus annuus* L.) mutant deficient in carotenoid synthesis and abscisic acid content induced by in vitro tissue culture. *Theor. Appl. Genet.* 87:65–69
29. Fray RG, Grierson D. 1993. Identification and genetic analysis of normal and mutant phytoene synthase genes of tomato by se-quencing, complementation and co-sup-pression. *Plant Mol. Biol.* 22:589–602
30. Giuliano G, Bartley GE, Scolnik PA. 1993. Regulation of carotenoid biosynthesis dur-ing tomato development. *Plant Cell.* 5: 379–87
31. Goodwin TW. 1971. Biosynthesis. In *Carotenoids,* ed. O Isler, pp. 577–636. Basel: Birkhauser Verlag
32. Goodwin TW. 1980. *Biochemistry of Carotenoids,* Vol. 1. London: Chapman & Hall
33. Goodwin TW, Britton G. 1988. Distribu-tion and analysis of carotenoids. In *Plant Pigments,* ed. TW Goodwin, pp. 61–131. London: Academic
34. Hugueney P, Römer S, Kuntz M, Camara B. 1992. Characterization and molecular cloning of a flavoprotein catalyzing the synthesis of phytofluene and zeta-carotene in *Capsicum* chromoplasts. *Eur. J. Biochem.* 209:399–407
35. Jenkins JA, MacKinney G. 1955. Carot-enoids of the apricot tomato and its hybrids with yellow and tangerine. *Genetics* 40: 715–20
36. Khudairi AK. 1972. The ripening of toma-toes. *Am. Sci.* 60:696–707
37. Kinzer SM, Schwager SJ, Mutschler MA. 1990. Mapping of ripening-related or -spe-cific cDNA clones of tomato (*Lycopersicon esculentum*). *Theor. Appl. Genet.* 79: 489–96
38. Knapp J, Moureau P, Schuch W, Grierson P. 1989. Organization and expression of polygalacturonase and other ripening-re-lated genes in Ailsa Craig 'Neverripe' and 'Ripening inhibitor' tomato mutants. *Plant Mol. Biol.* 12:105–16
39. Koornneef M. 1990. Linkage map of *Arabidopsis thaliana.* In *Genetic Maps,* ed. SJ O'Brien, 6:95–98. Cold Spring Harbor: Cold Spring Harbor Lab.
40. Koornneef M, Jorna ML, Brinkhorst-Van der Swan DLC, Karssen CM. 1982. The isolation of abscisic acid (ABA) deficient mutants by selection of induced revertants in non-germinating gibberellin-sensitive lines of *Arabidopsis thaliana* (L.) Heynh. *Theor. Appl. Genet.* 61:385–93
41. Krinsky NI. 1979. Carotenoid protection against photooxidation. *Pure Appl. Chem.* 51:649–60
42. Kuntz M, Römer S, Suire C, Hugueney P, Weil JH, et al. 1992. Identification of a cDNA for the plastid-located geranylger-anyl pyrophosphate synthase from *Capsicum annuum*: correlative increase in en-zyme activity and transcript level during fruit ripening. *Plant J.* 2:25–34
43. Linden H, Misawa N, Chamovitz D, Pecker I, Hirschberg J, Sandmann G. 1991. Func-tional complementation in *Escherichia coli* of different phytoene desaturase genes and

analysis of accumulated carotenes. *Z. Naturforsch. Teil C* 46:1045–51

44. Linden H, Vioque A, Sandmann G. 1993. Isolation of a carotenoid biosynthesis gene coding for zeta-carotene desaturase from *Anabaena* PCC 7120 by heterologous complementation. *FEMS Microbiol. Lett.* 106: 99–104

45. Markgraf T, Oelmüller R. 1991. Evidence that carotenoids are required for the accumulation of a functional photosystem II, but not photosystem I in the cotyledons of mustard seedlings. *Planta* 185:97–104

46. Maunders MJ, Holdsworth MJ, Slater A, Knapp JE, Bird R, et al. 1987. Ethylene stimulates the accumulation of ripening-related mRNAs in tomatoes. *Plant Cell Environ.* 10:177–84

47. Nielsen OF, Gough S. 1974. Macromolecular physiology of plastids. XI. Carotenes in etiolated *tigrina* and *xantha* mutants of barley. *Physiol. Plant.* 30:246–51

48. Oeller PW, Wong LM, Taylor LP, Pike DA, Theologis A. 1991. Reversible inhibition of tomato fruit senescence by antisense RNA. *Science* 254:437–39

49. Parry AD, Griffiths A, Horgan R. 1992. Abscisic acid biosynthesis in roots. II. The effects of water-stress in wild-type and abscisic-acid-deficient mutant (notabilis) plant of *Lycopersicon esculentum* Mill. *Planta* 187:192–97

50. Parry AD, Horgan R. 1992. Abscisic acid biosynthesis in roots. I. The identification of potential abscisic acid precursors, and other carotenoids. *Planta* 187:185–91

51. Pecker I, Chamovitz D, Linden H, Sandmann G, Hirschberg J. 1992. A single polypeptide catalyzing the conversion of phytoene to zeta-carotene is transcriptionally regulated during tomato fruit ripening. *Proc. Natl. Acad. Sci. USA* 89:4962–66

52. Peter GF, Thornber JP. 1991. Biochemical composition and organization of higher plant photosystem II light-harvesting pigment-proteins. *J. Biol. Chem.* 266:16745–54

53. Rau W. 1985. Mechanism of photoregulation of carotenoid biosynthesis in plants. *Pure Appl. Chem.* 57:777–84

54. Ray J, Bird CR, Maunders M, Grierson D, Schuch W. 1987. Sequence of pTOM5, a ripening related cDNA from tomato. *Nucleic Acids Res.* 15(24):587

55. Ray J, Moureau P, Bird C, Bird A, Grierson D, et al. 1992. Cloning and characterization of a gene involved in phytoene synthesis from tomato. *Plant Mol. Biol.* 19:401–4

56. Rick CM, Thompson AE, Brauer O. 1959. Genetics and development of an unstable chlorophyll deficiency in *Lycopersicon esculentum. Am. J. Bot.* 46:1–11

57. Robertson DS. 1975. Survey of the albino and white-endosperm mutants of maize. *J. Hered.* 66:67–74

58. Robertson DS. 1989. The source of the mutator system and the first *Mu*-induced mutants. *Maize Genet. Coop. Newsl.* 63:9–10

59. Robinson GW, Tsay YH, Kienzle BK, Smith-Monroy CA, Bishop RW. 1993. Conservation between human and fungal squalene synthetases: similarities in structure, function, and regulation. *Mol. Cell Biol.* 13:2706–17

60. Rock CD, Bowlby NR, Hoffmann-Benning S, Zeevaart JAD. 1992. The ABA mutant of *Arabidopsis thaliana* L. Heynh. has reduced chlorophyll fluorescence yields and reduced thylakoid stacking. *Plant Physiol.* 100:1796–801

61. Rock CD, Zeevaart JAD. 1991. The *aba* mutant of *Arabidopsis thaliana* is impaired in epoxy-carotenoid biosynthesis. *Proc. Natl. Acad. Sci. USA* 88:7496–99

61a. Römer S, Hugueney P, Bouvier F, Camara B, Kuntz M. 1993. Expression of the genes encoding the early carotenoid biosynthetic enzymes in *Capsicum annum. Biochim. Biophys. Res. Comm.* 196:1414–21

61b. Scolnik PA, Bartley GE. 1993. Phytoene desaturase from *Arabidopsis. Plant Physiol.* 103:1475

62. Scolnik PA, Hinton P, Greenblatt IM, Giuliano G, Delanoy MR, et al. 1987. Somatic instability of carotenoid biosynthesis in the tomato *ghost* mutant and its effect on plastid development. *Planta* 171:11–18

63. Siefermann-Harms D. 1985. Carotenoids in photosynthesis. Location in photosynthetic membranes and light-harvesting function. *Biochim. Biophys. Acta* 811:325–55

64. Simpson DJ, Rahman FMM, Buckle KA, Lee TH. 1974. Chemical regulation of plastid development. II. Effect of CPTA on the ultrastructure and carotenoid composition of chromoplasts of *Capsicum annuum* cultivars. *Aust. J. Plant Physiol.* 1:135–47

65. Stinard PS, Buckner B. 1992. *vp2–5080*, a new mutable allele of *vp2* with large revertant sectors. *Maize Genet. Coop. Newsl.* 66:9

66. Tanksley SD, Mutschler MA. 1990. Linkage map of the tomato (*Lycopersicon esculentum*). In *Genetic Maps*, ed. SJ O'Brien, 6:3–15. Cold Spring Harbor: Cold Spring Harbor Lab.

67. Taylor WC. 1989. Regulatory interactions between nuclear and plastid genomes. *Annu. Rev. Plant Physiol. Plant Mol. Biol.* 40:211–33

68. Thompson WF, White MJ. 1991. Physiological and molecular studies of light-regulated nuclear genes in higher plants. *Annu. Rev. Plant Physiol. Plant Mol. Biol.* 42:423–66

69. Wetzel CM, Jiang C-Z, Meehan LJ, Voytas DF, Rodermel SR. 1993. The *immutans* variegation mutant is plastid autonomous and impaired in carotenoid biosynthesis. Presented at 5th Int. Conf. *Arabidopsis* Res., Ohio State Univ., Columbus

70. Wiermann R, Gubatz S. 1992. Pollen wall and sporopollenin. *Int. Rev. Cytol.* 140:35–72

71. Young AJ, Britton G. 1989. The distribution of α-carotene in the photosynthetic pigment-protein complexes of higher plants. *Plant Sci.* 64:179–83

Annu. Rev. Plant Physiol. Plant Mol. Biol. 1994. 45:303–22

STORAGE PROTEINS OF VEGETATIVE PLANT TISSUES

P. E. Staswick

Department of Agronomy, University of Nebraska, Lincoln, Nebraska 68583-0915

KEY WORDS: nitrogen, jasmonic acid, reserves, soybean, gene expression

CONTENTS

INTRODUCTION 303
NITROGEN STORAGE AND REDISTRIBUTION 304
WHAT IS A STORAGE PROTEIN? 306
VEGETATIVE STORAGE PROTEINS OF SOYBEAN 307
    Regulation of the Soybean VSP Genes 310
    The Role of Nitrogen in Gene Regulation 310
    The Role of Jasmonate in Gene Regulation 311
    VSP and Stress Responses 313
    VSP is Related to a Tomato Acid Phosphatase 313
SINK REMOVAL ELEVATES OTHER SOYBEAN PROTEINS 314
PROTEINS RELATED TO SOYBEAN VSP IN OTHER SPECIES 316
PROTEIN STORAGE: VARIATIONS ON A THEME 317

INTRODUCTION

Plants accumulate nutrient elements and assimilate them into a wide variety of compounds necessary for their growth and development. These elements often are limiting in the environments in which plants live and therefore must be rationed among the many metabolic processes requiring them. When plentiful, however, nutrients can be stored for later use at a time when the assimilation capacity may be inadequate. Classic examples of stored reserves are those found in seeds: in particular, starch, proteins, and lipids. These reserves survive long periods of seed dormancy and then are catabolized after germination to nourish early seedling growth (see 13 and references therein). The annual storage of nitrogen in tree bark during the fall, and its subsequent mobilization for regrowth in the spring is another example (65).

Storage reserves not only provide a cache of nutrients for future needs, but they also can conserve resources by recovering nutrients that would otherwise be lost as tissues senesce. The redistribution of nutrients from mature organs to actively growing tissues is thus an important strategy for the efficient utilization of potentially limiting resources.

This review focuses on proteins that function as storage reserves in the vegetative tissues of soybean (*Glycine max* L.). These proteins are distinct from the seed reserve proteins in that they accumulate transiently, and are synthesized and then degraded within a single life cycle. Furthermore, the vegetative storage proteins may be synthesized and degraded simultaneously in organs that differ in their stage of development. Although this topic was reviewed previously (58), I revisit important early results and discuss more recent discoveries relating to the regulation of these genes. The soybean proteins are also compared briefly with other proteins that apparently function as storage reserves in various organs of other plant species. Only limited reference to the seed storage proteins will be found here; numerous reviews on this topic have appeared previously (e.g. see 1, 13, 30).

NITROGEN STORAGE AND REDISTRIBUTION

Although carbon skeletons and sulfur are derived from the degradation of proteins, proteinaceous storage reserves are most often considered important for their N content. Amino acids released by protein turnover can be used directly for synthesis of new proteins, or they may be metabolized for other purposes. To understand the role of storage proteins in vegetative tissues it is important to consider how N is partitioned among the various organs and tissues that compete for this essential element (for reviews see 17, 43, 52).

As is also true for carbon (67), actively growing organs are sinks for N, which they import via both the phloem and xylem to sustain their growth. As leaves mature, import gradually declines and they eventually become net exporters. Translocation out of these source leaves is primarily through the phloem and is regulated by the strength or size of sinks elsewhere on the plant, such as expanding leaves, developing fruits, and apices (52). The change in status from a sink to a source as leaves mature is depicted in Figure 1.

Although some of the biosynthetic precursors used by seeds during their development are acquired during this growth phase, a substantial portion can come from the redistribution of previous assimilates (17). Nitrogen assimilation actually declines during soybean seed fill (64) and generally is not sufficient to meet the high demand for N by developing seeds. In this species up to

90% of seed N is acquired prior to seed development (28, 77). Because developing seeds are strong sinks for mobilized nutrients, their growth initiates a dramatic change in metabolic processes throughout the plant. This is particularly true in monocarpic species for which this is the terminal phase of their development. As leaves senesce protein, chlorophyll and starch are degraded, and photosynthesis declines (74). Although ultimately leading to death, the senescent phase of monocarpic plant development is highly organized to facilitate the efficient translocation of resources from vegetative tissues to developing reproductive organs (62).

Leaves are the major source of N that is mobilized to soybean seeds, although stems, seed pods, and, to a lesser extent, roots also contribute (25). Amino acids and amides are the principal forms of N that are translocated from leaves and stems to developing seeds in soybean (39). Transport of N fixed in roots occurs primarily as ureides, which can be stored along with nitrate, to some extent, in soluble pools that are readily accessible for future metabolic needs or for export to other tissues (52).

High cellular concentrations of soluble N would lead to osmotic imbalance and eventually death. In contrast, proteins affect osmolarity to a much smaller extent and thus are a desirable form in which to store N. Millard (38) also points out that reducing nitrate for protein synthesis in young, actively grow-

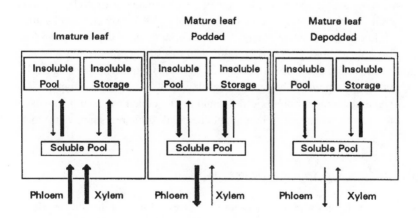

Figure 1 Model for nitrogen distribution in soybean leaves. Soluble pools include nitrate, amino acids, and amides, some of which may be stored temporarily. Insoluble pool refers to macromolecules, especially proteins, that have a primary function other than storage (e.g. enzymatic or structural). Insoluble storage includes proteins, such as VSP, used primarily for storage purposes. N flux in a young expanding leaf and in mature leaves from a plant bearing seed pods (podded) and from a depodded plant are depicted. Width of arrows indicates relative N flux to or from each pool. Adapted from Feller & Keist (17) and Simpson (52).

ing tissues, when light energy is plentiful, is preferable to storing unreduced N as nitrate until later, when the reducing capacity may be limited. Thus, although the energy cost to synthesize and then degrade protein is high, this may be an important strategy for accumulating N, particularly in species that remobilize significant quantities of N for the development of their protein-rich seeds.

Much of the N in plants occurs in protein (10). Upon their degradation, proteins contribute substantially to the pool of soluble N that is transported to other organs (51, 52, 64). Protein accounts for approximately 80% of the reduced N in wheat leaves (44) and proteolysis increases dramatically in vegetative plant parts during seed development (69). Wittenbach (74) found that soluble leaf protein declined at least 6-fold during seed development in soybean, beginning 3–4 weeks after flowering. Because of its abundance in leaves, ribulose-1,5-bisphosphate carboxylase (Rubisco) accounts for the largest proportion of protein that is degraded during this time.

In some plants, notably soybean, the shift in leaf metabolism from a sink to a source can be reversed to some degree by removal of developing seed pods. Superficially it would appear that depodding delays senescence, since leaves are retained and remain green for an extended period. However, Wittenbach (74) demonstrated that depodding does not delay the photosynthetic decline that is correlated with a loss of Rubisco. On the other hand, the chlorophyll loss normally associated with this stage of development is slowed by depodding, while starch and protein continue to accumulate for several weeks after they decline in pod-bearing plants. Thus, depodding leads to a change in leaf function from a photosynthesizing source leaf to a storage site for nitrogen and carbon that would normally be mobilized to developing seeds (see Figure 1). These leaves accumulate extremely high levels of two similar polypeptides, the vegetative storage protein (VSP) α and β. Depodded soybean plants have been useful for investigating how plants respond to altered source/sink relations.

WHAT IS A STORAGE PROTEIN?

Although the function of storage proteins in plants is evident, identifying those proteins that should be considered storage reserves is less obvious. This is because storage proteins are a heterogeneous group and there is no direct biochemical assay for their function. Historically, the seed reserve proteins have been thought to be 1. localized in storage vacuoles (protein bodies), 2. five percent or more of the total protein, 3. stored for the nutritional needs of a later developmental stage, and 4. lacking in any other metabolic or structural role (13).

The protein reserves in seeds are sequestered in organelles called *protein bodies,* which are vacuoles that are modified for storage purposes (5). Analogous storage vacuoles are found in other organs as well. Plant vacuoles have been compared to animal lysosomes because they contain numerous hydrolases, including high levels of proteinases. Nevertheless, storage proteins are stably accumulated to high levels in these organelles.

All proteins sequester amino acids, which are later released as proteins turn over. In a sense all proteins are storage reserves; however, most plant proteins are not abundant and individually they account for an insignificant fraction of the total N. A particularly striking exception is Rubisco, the most abundant of plant proteins. Some have considered Rubisco a storage protein because it can be a major source of the N translocated to seeds during their development (46, 63). But the biosynthesis and degradation of most proteins is regulated according to the need for their enzymatic or other metabolic functions rather than for storage purposes. Thus, although Rubisco is an important source of mobilized N, Wittenbach et al (76) point out that unlike true storage proteins, Rubisco and most other proteins are not laid away and biologically inactive until their degradation. Proteins that are synthesized and degraded solely according to the availability of excess N would provide added flexibility since they could be harvested freely for their N content at any time.

Another distinguishing feature of storage proteins is that they are preferentially degraded. That is, their concentration relative to the total declines when reserves are mobilized for reuse. For example, the abundant seed storage proteins of legume cotyledons are degraded while new proteins, those involved in seedling growth, are being synthesized. The vegetative storage proteins of soybean are also preferentially lost from leaves and seed pods during seed development (57). Consistent with a N storage function, storage proteins also often have a relatively high proportion of the N-rich amino acids, particularly arginine and the amides.

VEGETATIVE STORAGE PROTEINS OF SOYBEAN

A distinguishing feature of the soybean VSP is its relatively transient pattern of accumulation and loss. Unlike seed and most other plant storage proteins that supply resources for the succeeding generation or following a period of dormancy, synthesis and degradation of the soybean VSP occurs within a single growing season. Wittenbach (74) first described two glycosylated polypeptides, now called VSPα and VSPβ, that are abundant in young sink leaves (6% to 15% of the soluble protein) but decline to 1% during seed fill. A similar pattern of accumulation in young tissue was subsequently shown for seed pods (57) and seedling hypocotyls (35). N is mobilized from seed pods during seed protein synthesis and VSP accounts for about 45% of the protein lost from pods during this time (57). This pattern of accumulation and disappearance of VSPα and β in accordance with organ source/sink status is consistent with their role as temporary storage reserves for N.

Soybean VSP was discovered when leaves of depodded soybean plants were found to maintain high levels of protein even though they they had lost most of their Rubisco (74). Strikingly, in these sink-deprived plants VSPα and β accumulated to nearly 50% of the soluble protein, essentially quantitatively replacing Rubisco (75). Subsequent studies showed that the mRNA corresponding to these proteins accounted for as much as 16% of the total pA$^+$ mRNA following depodding (56). This high level of mRNA is apparently an important factor contributing to the extreme abundance of VSPα and β under these conditions.

Further evidence supporting a storage role for VSPα and β came from immunolocalization studies. These proteins are located principally in vacuoles of the paraveinal mesophyll and closely associated bundle sheath cells in leaves (21), although they can also be quite abundant in epidermal cells at some stages of development (58). The paraveinal mesophyll tissue is a specialized cell layer sandwiched between the palisade and spongy mesophyll, interconnecting the vascular bundles. This essentially nonphotosynthetic tissue apparently occurs in only a few plant species and is involved in the transfer of assimilates between the leaf vasculature and the photosynthetic mesophyll cells (19). The VSP polypeptides are associated with vascular tissues in stems and seed pods as well (27). Deposition of storage proteins and other reserves near xylem and phloem appear to facilitate their subsequent mobilization.

The two VSP polypeptides are about 80% identical, based on sequence deduced from cDNAs. They have no sequence similarity with the known seed storage proteins, although the amino acid composition is somewhat similar to legume seed storage proteins—both are deficient in sulfur-containing amino acids (55). Molecular weight estimates for the two polypeptides have varied among laboratories and although one (VSPα) has higher mobility on SDS-PAGE than the other (VSPβ) the sequence predicts the mature polypeptides to be about 25 kDa. The polypeptides assemble as either hetero- or homodimers (47, 54) and they contain characteristic signal peptides, consistent with transit through the secretory pathway to the vacuole.

VSPα and β are most abundant in above ground vegetative tissues and only minor amounts have been detected in roots and seeds (57). Although seed pods are not strictly considered vegetative organs, they function as other vegetative tissues in that they accumulate N and protein prior to seed development and then lose protein and remobilize N during seed fill.

Although only minor amounts of VSPα and β occur in seeds, they accumulate abundantly in cotyledons following seed germination, when these organs become leaf-like and photosynthetic. Figure 2 illustrates that the VSP polypeptides increased markedly in cotyledons while total protein was declining, because of the degradation of seed storage proteins. VSPα and β were not detected at 2 days after germination. By 8 days they had increased at least 70-fold as a percent of total protein, which decreased about 5-fold during this period.

It may seem paradoxical that these new VSPs are synthesized while seed storage protein is being degraded in seedling cotyledons. This may indicate that the soluble N released from the degradation of the abundant seed reserve proteins is more than can be used immediately for seedling growth. Some of this excess N is transiently partitioned into the VSPs, until it can be used later in seedling development. Why is seed protein degradation not controlled in a more timely manner, releasing N only as it is needed? The answer may be that conversion of the nutrient-storing cotyledons of seeds to photosynthetic leaf-like organs requires a dramatic cellular reorganization, involving the complete dissolution of seed protein storage vacuoles and the release of their entire contents during germination. The rapid cell expansion and requisite enlargement of cell membranes occurring during germination may depend on reuse of membranes derived from these storage vacuoles.

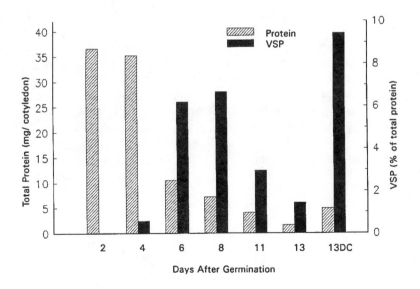

Figure 2 Changes in total protein and VSPα and β content in soybean seedling cotyledons following germination. Seedlings were grown in controlled environment chambers and VSP polypeptides were detected with antisera and quantified as described previously (57). DC refers to seedlings that were decapitated by removal of the shoot meristem above the cotyledons at 4 days after germination.

Consistent with the idea that the soybean VSPs may be needed to store excess N in cotyledons after germination, decapitation to remove the active sinks (growing shoots) from these seedlings further increased the VSPs and slowed the loss of total protein from cotyledons. This treatment also extended the life of cotyledons, which abscised at 13 to 15 days on control plants but remained viable for more than 19 days on decapitated seedlings. Although wounding caused by decapitation may have induced the VSP genes, this result is consistent with that obtained for mature plants following seed pod removal, for which clear evidence was provided that the effect of wounding was minimal (56).

Regulation of the Soybean VSP Genes

The soybean VSP genes are remarkable in the number and diversity of both developmental and external stimuli that influence their expression. These include source/sink status, phloem blockage, N availability, wounding, water deficit, light, jasmonic acid (JA), and sucrose. However, most of these appear to be directly or indirectly related to the source/sink status, and consequently to the temporary storage needs of the organs involved.

The normal developmental decline in VSP gene expression as organs convert from a sink to a source is reversed strikingly by the removal of seed pods (Figure 1). This is not primarily a wound response, since the amount of VSPα and β accumulating in leaves is inversely related to the number of seed pods (sink size) remaining on plants, even though the amount of wounding is essentially the same (56). This result also reveals that the gene regulatory system is quite finely tuned to vary the amount of VSPα and β according to the need for storage. These genes are also induced in leaves by blocking the phloem export path at the leaf petiole, either by heat girdling or by chilling the petiole and gene expression increases modestly in the dark, when leaf export declines. These observations suggest that a phloem mobile substance may be used to monitor the export status of individual leaves and to regulate expression of the VSP genes accordingly.

The Role of Nitrogen in Gene Regulation

If VSPα and β are temporary N storage reserves, then we might expect that compounds containing N would regulate these genes. VSP mRNA and protein levels are closely tied to N availability. mRNA in N-deficient plants was essentially undetectable even in young leaves, but mRNA abundance increased according to the amount of ammonium nitrate provided to plants during their growth (60). At high N concentrations VSP mRNA was abundant even in the oldest leaves, at much higher levels than in nodulated control plants. Total leaf protein was also higher in these plants. Thus when excess N is available some of it can be sequestered in these VSPs.

The regulatory picture is more complicated, however, as N apparently does not control the VSP genes directly. Various forms of mineral and organic N had no effect on VSP mRNA in either soybean suspension cell cultures or leaf explants, although N-deficient plants restored to N-fertility increased VSP gene expression after a few days (60). Because there is a decreased need to mobilize N from mature leaves when N is plentiful, the effect of N availability during plant growth seems related to altered source/sink relations, rather than having a direct regulatory role.

The Role of Jasmonate in Gene Regulation

In contrast to N, methyl jasmonate (MeJA) induces soybean VSP gene expression rapidly in whole plants, leaf explants, and in suspension cell cultures (2, 35, 60). Furthermore, jasmonate elevates VSP mRNA levels even in leaves from N-deficient plants, albeit to a lower level than in plants receiving adequate N. This suggests that jasmonate plays a more direct role than does soluble N.

JA and MeJA recently have received considerable attention as a potentially important new class of plant growth regulator. Jasmonate affects a variety of plant responses and induces the expression of several genes, including many thought to be important in stress responses (for recent reviews see 42, 59). A variety of evidence (discussed below) supports a role for jasmonate in the regulation of the soybean VSP genes.

Jasmonate is widely distributed in plants, including soybean, and the concentrations that have been found are comparable with other plant growth regulators (37). Furthermore, developmental changes in soybean VSP gene expression correlate with general changes in jasmonate levels in plants. Highest levels of both are generally found in young plant organs and a direct relationship between the two was recently demonstrated in soybean hypocotyls (33). Although there have been few studies on the distribution of jasmonate within plant organs, one found that jasmonate was most abundant in vascular bundles of pericarp (seed pod) tissue (41), where VSP mRNA is also prevalent (27). The cell-specific pattern of VSP gene expression in leaves can be abolished by exogenous MeJA, which causes high level expression in all cell types, rather than in just the paraveinal mesophyll and bundle sheath cells (27). This could indicate that although all cells are competent to express VSP, expression is normally limited to those tissues having higher concentrations of jasmonate, or greater sensitivity to endogenous jasmonate. This hypothesis requires further testing.

Further evidence implicating endogenous jasmonate in VSP gene regulation comes from studies using inhibitors of lipoxygenase, the first enzyme in the JA biosynthetic pathway. Gene induction by wounding and petiole girdling was blocked by pretreatment with various lipoxygenase inhibitors, but the genes were still fully inducible by exogenous MeJA (60). This suggests that de novo synthesis of JA in response to these stimuli is necessary for VSP induction. In agreement with this result, wounding elevates jasmonate levels in soybean hypocotyls in a manner consistent with its involvement in the induction of VSP genes (8). The fact that some plant lipoxygenase genes are induced by jasmonate (4, 36) may suggest a jasmonate signal amplification mechanism. Modest increases in jasmonate levels could stimulate the JA biosynthetic pathway, causing further induction of jasmonate-responsive genes.

Evidence from other species also supports a gene regulatory role for jasmonate. Tomato leaf proteinase inhibitors are induced by low concentrations of airborne MeJA (15), as are the soybean VSP genes (20). Octadecanoid precursors of JA also effectively induce proteinase inhibitors, but closely related compounds that are not JA precursors do not have this effect (16). The assumption is that the precursors are converted to jasmonate, which has gene inducing activity. This also provides evidence that the exogenously applied jasmonate and its precursors do not simply cause a toxic or stress response, since closely related compounds had no effect. Jasmonate may also be involved in signaling phytoalexin biosynthesis, since fungal elicitors that induce enzymes of the phytoalexin biosynthetic pathway also elevate jasmonate in cell cultures (24). Exogenously applied jasmonate also induces enzymes of this pathway. An *Arabidopsis thaliana* mutant that is defective in its response to MeJA has been characterized (61). Such mutants should shed additional light on the role of jasmonate in signaling various plant processes, including gene expression.

In addition to jasmonate, soluble sugars also appear important for VSP gene regulation. Sucrose, fructose, or glucose interact synergistically with MeJA to induce VSP genes in soybean leaves and in suspension cell cultures (33). Endogenous levels of total soluble sugars, as well as MeJA, also correlate with VSP gene expression. This is consistent with previous evidence that VSP genes are regulated according to the source/sink status of tissues. It would be interesting to know whether jasmonate and/or soluble sugars are more prevalent in those cells expressing the highest levels of VSP (the paraveinal mesophyll, bundle sheath, and epidermal cells). Photosynthate also seems important for the wound response of VSP, since induction was greatly suppressed in leaves placed in the dark or sprayed with inhibitors of photosynthesis following wounding (33).

The VSP genes have been cloned and sequenced (47–49) and only one copy of *vspA* and two copies of *vspB* are present in the soybean genome, which indicates the promoters must be quite active to produce the high levels of expression that have been noted. *vspB* was apparently duplicated recently, since the inverted tandem repeat encompassing both gene copies is identical over the entire length of the two genes, including introns, at least 500 nucleotides of 5' flanking sequence and at least 200 nucleotides of 3' flanking sequence (47, 49). There has been only limited study of the VSP gene promoters, but recent evidence demonstrated that a G-box motif in the *vspB* promoter was important for jasmonate response in tobacco (34). A similar sequence is also found in *vspA* (48) and it presumably plays the same role in this gene. Furthermore, the jasmonate response element was separable from sequences conferring response to sucrose. A G-box element is also essential for jasmonate-response in a potato proteinase inhibitor gene (29).

VSP and Stress Responses

Although both wounding and water deficit induce the soybean VSP genes, there is no indication that the VSPα and β are directly involved in plant defense, as are the wound-inducible proteinase inhibitors of the Solanaceae (50). Nor is it clear that VSPα and β ameliorate the effects of water deficit, although such a function cannot be ruled out. However, under stress conditions the VSP genes may be responding to a decrease in the strength of plant sinks, because of stress-induced slowing of plant growth. It is questionable whether synthesis of VSPα and β for storage purposes is of benefit over short periods of stress. Over the longer term, the finely tuned signaling mechanism for detecting changes in source/sink status may aid in temporarily sequestering resources that are not needed for growth, and subsequently releasing them when conditions are more favorable. Nothing is known about the signaling mechanisms that control degradation of the VSPs.

Much of the current research on how plants react to stress is focused on rapid responses that are presumed to be directly involved in limiting the deleterious effects of a single stimulus such as wounding or pathogen attack. This approach is yielding important and useful results, but a plant's response to the natural environment is likely to be much more complex and long term. Repeated assaults by multiple stresses of varying intensity over an extended period undoubtedly cause additional more subtle alterations in plant metabolism (e.g. salvaging resources from damaged tissues and translocating them to healthy tissues, limiting plant size, or altering plant architecture). These types of responses may allow plants not only to survive stress but to subsequently grow and reproduce effectively, but they are clearly more difficult to study, undoubtedly involving modulation of the expression of many genes.

VSP is Related to a Tomato Acid Phosphatase

A tomato acid phosphatase was recently found to be about 40% identical to the VSP subunits (72). This raised the question of whether VSPα and β have phosphatase activity, and if so, whether a storage function is incidental to the enzymatic activity of these proteins. It was reported recently (14) that soybean VSPα and β have acid phosphatase activity. However, specific activities of the different VSP isomers ranged from 0.3 to 10 units per mg protein, which is roughly 2–4 orders of magnitude lower than many other plant acid phosphatases. A recent investigation found that in depodded plants, where VSPα and β comprised at least 40% of the leaf soluble protein, the VSPs accounted for no more than 0.1% of the total acid phosphatase activity (P Staswick, unpublished results). The insignificant fraction of the total activity contributed by VSPα and β supports the other evidence that the primary function of these proteins is for storage purposes. However, the VSPs may still have a unique enzymatic role, such as high activity on a presently unknown substrate, or activity in a particular cellular compartment where other acid phosphatases are not present. If this were the case, it would be interesting to know why soybean has such large quantities of this enzyme.

In contrast to soybean VSPα and β, the tomato acid phosphatase is apparently low in abundance, suggesting an enzymatic rather than a storage function. One explanation for this incongruity between the tomato and soybean proteins relates to the possible origin of storage proteins. Some may be derived from genes that originally encoded enzymes or other biologically active proteins. Duplicate copies of these genes could be free to acquire a promoter that stimulates abundant expression according to the storage needs of a specific plant organ. These genes may also have lost some or all of the biological activity associated with their previous function.

Other proteins may fit this model, including patatin, an abundant storage protein of potato tubers that also has lipid acyl hydrolase activity (3). Certain seed lectins and proteinase inhibitors are also abundant in seeds, particularly in legumes. Although some of these undoubtedly have important defensive or other roles, evidence regarding the function of most of these proteins is lacking. It is interesting that certain cultivars lack the abundant seed lectins and in wild populations of *Phaseolus* there appears to be no selective advantage associated with this agglutinating activity in seeds (45). This supports the idea that these lectins may serve a storage role rather than having an essential function such as plant defense. Abundant lectins in the bark of certain deciduous trees also function as overwinter storage reserves (22). Although difficult to prove, reasonable evidence suggests that soybean VSPα, VSPβ, and some other storage proteins were recruited from genes that originally had another metabolic function.

SINK REMOVAL ELEVATES OTHER SOYBEAN PROTEINS

Several additional soybean proteins have been found to be elevated in leaves after fruit removal, but not all genes responding to depodding necessarily encode storage proteins. The dramatic change in leaf metabolism from a source to a storage organ in response to depodding undoubtedly involves a significant modulation in the expression of a number of genes. Some of these could include enzymes involved in the biosynthesis and processing of storage macromolecules (i.e. protein and starch) or their assembly and packaging in storage vesicles.

An abundant 80-kDa polypeptide first noted by Wittenbach (74) was subsequently shown to be a lipoxygenase of about 94 kDa (66). Like VSPα and β, it is elevated by MeJA (23) and is found in vacuoles of leaf paraveinal mesophyll and bundle sheath cells. However, the regulatory pattern of the 94-kDa protein is distinct from that of VSPα and β, calling into question its primary function as a storage reserve. The protein increase in response to depodding is less striking than for VSPα and β and the protein is maintained in pod-bearing plants for at least 6 weeks after flowering, when most of VSPα and β have been degraded. Furthermore, after an initial induction at 2 weeks after flowering in response to depodding, lipoxygenase mRNA levels actually decline in depodded plants to a lower level than at anthesis. mRNA was also more abundant in older than in younger leaves (at anthesis vs 2 weeks preanthesis) (23). This contrasts sharply with the mRNA accumulation pattern for VSPα and β, which is higher in younger leaves and continues to increase in response to depodding for at least 11 weeks (56). Therefore, the regulation of the lipoxygenase genes may be for enzymatic rather than storage purposes.

A 30-kDa protein that is tightly associated with Rubisco in leaf extracts is also induced by depodding (7). Its function is not known, but we have recently cloned the cDNA and found that the gene is induced by a variety of stimuli, closely paralleling the pattern found for VSPα and β (P Staswick & S Crafts-Brandner, unpublished results). Interestingly, the protein is located in the cytoplasm rather than in vacuoles, where storage proteins normally occur (S Crafts-Brandner, personal communication). The metabolic function of this protein is unclear at this time. Other soybean proteins, possibly related to lectins, also respond to depodding (54) but little more is known about these.

Soybean leaf acid phosphatase activity increases 10-fold in response to depodding and most of this enzyme activity colocalizes with VSPα and β in paraveinal mesophyll cells, vascular tissue, and lower epidermal cells (61a). However, most of the increase is not due to VSPα and β. A single protein responsible for the majority of this activity has been purified. The specific activity of this acid phosphatase is at least 1000-fold higher than soybean VSP and the protein increases at least 5-fold in depodded plants. In contrast to VSPα and β the 51-kDa phosphatase is only about 0.2% of the total leaf protein after 9 weeks of depodding (61a). Thus it does not contribute significantly as a storage reserve. It may be involved in other activities associated with the change in leaf function from a source to a storage organ, such as phosphate metabolism, but its role and the reason for its increase in response to depodding is still being investigated.

PROTEINS RELATED TO SOYBEAN VSP IN OTHER SPECIES

We have examined whether other plants species accumulate proteins related to soybean VSPα and β using antisera to these proteins. Figure 3 shows that immunologically-related proteins similar in size to the soybean proteins are present in common bean (*Phaseolus vulgaris*), pea (*Pisum sativa*), alfalfa (*Medicago sativa*), tobacco (*Nicotiana tabacum*), and corn (*Zea mays*). Furthermore, providing high levels of N to plants increased the amount of protein detected in every case, but the signal was still quite weak compared with that of soybean (note that 400-fold less protein was loaded for soybean). Proteins of apparently low abundance in *A. thaliana* are also detected with the soybean VSP antisera and these are elevated by MeJA treatment (61).

Despite these results, it is questionable whether proteins related to soybean VSP serve as storage reserves in other species. Although sequence divergence could account partially for the weak signal when VSPα and β antisera is used as a probe for other species, sink removal in common bean, pea, or alfalfa did not elevate specific polypeptides as in soybean (P Staswick, unpublished results). Similar results were also reported for wheat (*Triticum aestivum* L.) leaves (31). The evidence suggests the proteins identified with soybean VSP antisera are not abundant and may be homologs of the low-abundance acid phosphatase from tomato, but not significant for their storage capacity.

Why would soybean be unique among these species in having highly abundant transient VSPs? One reason may be the high nitrogen requirement of the protein-rich soybean seeds. One study found the seed protein content (percent-

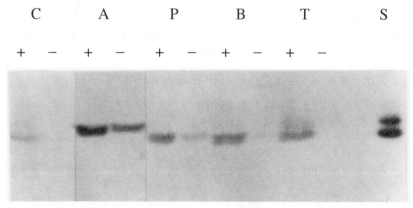

Figure 3 Effect of N in several plant species on the abundance of proteins that crossreact with soybean VSPα and β antisera. For procedures see Figure 2. Total soluble leaf protein loaded in each lane was 0.25 μg for soybean and 100 μg for all others. + and − refer to plants grown with and without supplemental N, respectively (40 mM ammonium nitrate supplied twice weekly as described in 60). C, corn; A, alfalfa; P, pea; B, common bean; T, tobacco; S, soybean.

age of dry weight) of soybean ranged from 31% to 375% higher than 24 other crop species, including several other legumes (53). Soybean seeds also had the highest N content per gram of available photosynthate. Another difference is that soybean is apparently one of the few legumes that have paraveinal mesophyll tissue. Because this tissue is implicated in the temporary storage of surplus metabolites, it may be an indicator of plant species that use abundant protein storage reserves, such as the VSPs of soybean. Two other species that have paraveinal mesophyll cells, wing bean (*Psophocarpus tetragonolobus* D.C.) and mung bean (*Vigna radiata* L.), also have dense staining vacuolar material (probably protein) that increases in these cells in response to depodding (18). The putative storage proteins in these species have not yet been examined.

PROTEIN STORAGE: VARIATIONS ON A THEME

Although it is not clear whether homologs of soybean VSPα and β are vegetative storage reserves in other species, several other proteins do have this function in other situations. Unlike VSPα and β, most of these proteins store resources from one growing season to the next, but there are interesting parallels between some of these proteins and VSPα and β, particularly relating to the effect of source/sink status.

Tubers are overwintering vegetative propagules that contain large quantities of starch and protein. Storage proteins occurring in tubers of common potato (*Solanum tuberosum*) and sweet potato (*Ipomea batatas*) include patatin and sporamin, respectively. These normally are synthesized only in the tuberous sinks during their development, where they can comprise 40% and 80% of the total soluble protein in *Solanum* and *Ipomea,* respectively (32, 46). They are then degraded preferentially during the growth of sprouts from the tuberous root.

Patatin (as well as starch) accumulates in stems and leaf petioles (but not leaves) following the removal of tubers (40). Thus while patatin bears no resemblance in sequence to soybean VSPα and β, it can serve as a storage protein in above-ground tissues when these become storage rather than source organs following removal of the normal sinks (tubers). Furthermore, the availability of photosynthate is apparently an important regulatory factor for these genes. As for soybean VSPα and β, sucrose induces both patatin (70) and sporamin (26) in leaves.

Deciduous trees also synthesize storage proteins and these play an important role in growth and development. N is cycled from foliage that will senesce in the fall to overwintering bark tissue. The process is reversed in the spring. Bark storage proteins have been characterized in several species (65) and where examined they have been found in vacuolar storage organelles in the parenchyma cells of inner bark (71). These proteins vary in size, generally ranging from about 25–40 kDa. At least one of these, a 32-kDa protein of poplar (*Populus deltoides*), is induced by short-day photoperiod (6). It is unknown if signaling the expression of this gene occurs via an altered source-sink status in response to shorter days or if a more direct signaling mechanism involving phytochrome is active. Recent evidence that the 32-kDa protein is induced under long-day light regimes by providing N to plants (68) supports a role for source-sink relations.

There may be vegetative storage proteins in poplar leaves as well. Wounding induces the *Win3* multigene family, which encodes leaf proteins with sequence homology to sweet potato sporamin. *Win3* is induced in leaves distant from a wounded leaf, but only if the leaves are directly connected through the vasculature (12). This indicates a role for signaling through the transport stream. These leaf proteins may be storage reserves, although a direct role in ameliorating effects of wounding cannot presently be ruled out. Another wound-inducible leaf protein (WIN4) is related to bark storage protein (11). Like soybean VSPα and β, this protein is induced by airborne MeJA, but not by the short-day treatments that induce the related bark storage proteins. Many wound-inducible plant genes evidently are regulated by several other developmental and external factors. As is the case for soybean VSPα and β, these wound-induced proteins may not be immediately involved in overcoming the effects of a particular stress. Some may have a more general role, aimed at altering plant metabolism for long-term survival.

A 24-kDa glycoprotein comprising up to 20% of the soluble protein in trifoliate orange (*Poncus trifoliata*) leaves also has characteristics of a storage reserve (73). It is abundant in young expanding leaves but declines preferentially as leaves mature. Cold hardening plants accelerated the loss of the leaf protein and leaf reserves were thought to be remobilized in preparation for overwinter storage elsewhere, perhaps in bark tissues. Consistent with this scenario, cold treatment stimulates synthesis rather than degradation of bark storage proteins in poplar (68).

Another site of storage protein accumulation is in the overwintering root system of certain herbaceous perennials. Although morphologically and functionally distinct from tubers and stolons, before dormancy these organs also sequester nutrients that are subsequently remobilized for regrowth. In leafy spurge (*Euphorbia esula*) roots, the accumulation of a predominant 26-kDa protein correlates with the decline in sink activity in above ground tissues during the fall, and the protein is degraded preferentially during regrowth in the spring (9). The ability of leafy spurge to store reserves below ground is thought to be an important factor contributing to the persistence of this pernicious weed.

Protein storage is apparently quite common in plants, occurring in a variety of organs and for somewhat different purposes. Some provide nutrients for rapid growth after a period of dormancy, whereas others, such as the soybean VSPs, play a more transient role during a single growing season. The regulatory blueprint for most of these proteins is lacking, but interesting parallels are beginning to emerge. The storage of reserves as protein is evidently an important strategy for dealing with the scarcity of mineral elements in many soils and the potentially limiting effect of adverse environments on nutrient assimilation.

Any *Annual Review* chapter, as well as any article cited in an *Annual Review* chapter, may be purchased from the Annual Reviews Preprints and Reprints service.
1-800-347-8007; 415-259-5017; email: arpr@classorg

Literature Cited

1. Altschul MA, Wilke HL. 1985. *New Protein Foods. Seed Storage Proteins*, Vol. 5. New York: Academic
2. Anderson JM. 1991. Jasmonic acid-dependant increase in vegetative storage protein in soybean tissue cultures. *J. Plant Growth Regul.* 10:5–10
3. Andrews DL, Beams B, Summers MD, Park WD. 1988. Characterization of the lipid acyl hydrolase activity of the major potato (*Solanum tuberosum*) tuber protein, patatin, by cloning and abundant expression in a baculovirus vector. *Biochem. J.* 252:199–206
4. Bell E, Mullet JE. 1991. Lipoxygenase gene expression is modulated in plants by water deficit, wounding, and methyl jasmonate. *Mol. Gen. Genet.* 230:456–62
5. Chrispeels MJ. 1984. Biosynthesis, processing and transport of storage proteins and lectins in cotyledons of developing legume seeds. *Philos. Trans. R. Soc. London Ser. B* 304:309–22
6. Coleman GD, Chen TH, Ernst SG, Fuchigami L. 1991. Photoperiod control of poplar bark storage protein accumulation. *Plant Physiol.* 96:686–92
7. Crafts-Brandner SJ, Salvucci ME, Egli DB. 1991. Fruit removal in soybean induces the formation of an insoluble form of ribulose-1,5-bisphosphate carboxylase/oxygenase in leaf extracts. *Planta* 183:300–6
8. Creelman RA, Tierney ML, Mullet JE. 1992. Jasmonic acid/methyl jasmonate accumulate in wounded soybean hypocotyls and modulate wound gene expression. *Proc. Natl. Acad. Sci. USA* 89:4938–41
9. Cyr ER, Bewley JD. 1990. Seasonal variation in nitrogen storage reserves in the roots of leafy spurge (*Euphorbia esula*) and re-sponses to decapitation and defoliation. *Physiol. Plant.* 78:361–66
10. Dalling MJ. 1987. Proteolytic enzymes and leaf senescence. In *Plant Senescence: Its Biochemistry and Physiology*, ed. WW Thompson, EA Nothnagel, RC Huffaker, pp. 54–60. Rockville, MD: Am. Soc. Plant Physiol.
11. Davis JM, Coleman GD, Chen TH, Gordon MP. 1993. Win4, a family of wound-induced genes in poplar trees. *Plant Physiol.* 102S:15
12. Davis JM, Gordon MP, Smit BA. 1991. Assimilate movement dictates remote sites of wound-induced gene expression in poplar leaves. *Proc. Natl. Acad. Sci. USA* 88: 2393–96
13. Derbyshire E, Wright DJ, Boulter D. 1976. Legumin and vicilin, storage proteins of legume seeds. *Phytochemistry* 15:3–24
14. DeWald DB, Mason HS, Mullet JE. 1992. The soybean vegetative storage proteins VSPα and VSPβ are acid phosphatases active on polyphosphates. *J. Biol. Chem.* 267:15958–64
15. Farmer EE, Ryan CA. 1990. Interplant communication: airborne methyl jasmonate induces synthesis of proteinase inhibitors in plant leaves. *Proc. Natl. Acad. Sci. USA* 87:7713–16
16. Farmer EE, Ryan CA. 1992. Octadecanoid precursors of jasmonic acid activate the synthesis of wound-inducible proteinase inhibitors. *Plant Cell* 4:129–34
17. Feller U, Keist M. 1986. Senescence and nitrogen metabolism in annual plants. In *Fundamental, Ecological and Agricultural Aspects of Nitrogen Metabolism*, ed. H Lambers, JJ Neetson, I Stulen, pp. 219–34. Dordrecht, Netherlands: Nijhoff

18. Franceschi VR, Giaquinta RT. 1983. Specialized cellular arrangements in legume leaves in relation to assimilate transport and compartmentation: comparison of the paraveinal mesophyll. *Planta* 159:415–22

19. Franceschi VR, Giaquinta RT. 1983. The paraveinal mesophyll of soybean leaves in relation to assimilate transfer and compartmentation. II. Structural, metabolic and compartmental changes during reproductive growth. *Planta* 157:422–31

20. Franceschi VR, Grimes HD. 1991. Induction of soybean vegetative storage proteins and anthocyanins by low level atmospheric methy jasmonate. *Proc. Natl. Acad. Sci. USA* 88:6745–49

21. Franceschi VR, Wittenbach VA, Giaquinta RT. 1983. Paraveinal mesophyll of soybean leaves in relation to assimilate transfer and compartmentation. III. Immunohistochemical localization of specific glycopeptides in the vacuole after depodding. *Plant Physiol.* 72:586–89

22. Greenwood JS, Stinissen HM, Peumans WJ, Chrispeels MJ. 1986. *Sambucus nigra* agglutinin is located in protein bodies in the phloem parenchyma of the bark. *Planta* 167:275–78

23. Grimes HD, Koetje DS, Franceschi VR. 1992. Expression, activity and cellular accumulation of methyl jasmonate-responsive lipoxygenase in soybean seedlings. *Plant Physiol.* 100:433–43

24. Gundlach H, Muller MJ, Kutchan TM, Zenk MH. 1992. Jasmonic acid is a signal transducer in elicitor-induced plant cell cultures. *Proc. Natl. Acad. Sci. USA* 89:2389–93

25. Hanway JJ, Weber CR. 1971. Accumulation of N, P, and K by soybean (*Glycine max* (L.) Merrill plants.) *Agron. J.* 63:406–8

26. Hattori T, Fukumoto H, Nakagawa S, Nakamura K. 1991. Sucrose-induced expression of genes coding for the tuberous root storage protein, sporamin, of sweet potato in leaves and petioles. *Plant Cell Physiol.* 32:79–86

27. Huang J-F, Bantroch DJ, Greenwood JS, Staswick PE. 1991. Methyl jasmonate treatment eliminates cell-specific expression of vegetative storage protein genes in soybean leaves. *Plant Physiol.* 97:1512–20

28. Jeppson RG, Johnson RR, Hadley HH. 1978. Variation in mobilization of plant nitrogen to grain in nodulating and non-nodulating soybean genotypes. *Crop Sci.* 18:1058–62

29. Kim S, Choi J, Costa MA, An G. 1992. Identification of G-Box sequences as an essential element of methyl jasmonate response of potato proteinase inhibitor II promoter. *Plant Physiol.* 99:627–31

30. Kreis M, Shewry PR, Forde BG, Miflin BJ.

1985. Structure and evolution of seed storage proteins and their genes with particular reference to those of wheat, barley and rye. In *Oxford Surveys of Plant Molecular and Cell Biology*, ed. BJ Miflin, pp. 253–317. New York: Oxford Univ. Press

31. MacKown CT, Van Sanford DA, Zhang N. 1992. Wheat vegetative nitrogen compositional changes in response to reduced reproductive sink strength. *Plant Physiol.* 99: 1469–74

32. Maeshima M, Sasaki T, Asahi Y. 1985. Characterization of major proteins in sweet potato tuberous roots. *Phytochemistry* 24: 1899–1902

33. Mason HS, DeWald DB, Creelman RA, Mullet JE. 1992. Coregulation of soybean vegetative storage protein gene expression by methyljasmonate and soluble sugars. *Plant Physiol.* 98:859–67

34. Mason HS, DeWald DB, Mullet JE. 1993. Identification of a methyl jasmonate-responsive domain in the soybean *vspB* promoter. *Plant Cell* 5:241–51

35. Mason HS, Mullet JE. 1990. Expression of two vegetative storage protein genes during development and in response to water deficit, wounding and jasmonic acid. *Plant Cell* 2:569–79

36. Melan MA, Dong X, Endara ME, Davis KR, Ausubel FM, Peterman TK. 1993. An *Arabidopsis thaliana* lipoxygenase gene can be induced by pathogens, abscisic acid, and methyl jasmonate. *Plant Physiol.* 101: 441–50

37. Meyer A, Miersch O, Buttner C, Dathe W, Sembdner G. 1984. Occurrence of the plant growth regulator jasmonic acid in plants. *J. Plant Growth Regul.* 3:1–8

38. Millard P. 1988. The accumulation and storage of nitrogen by herbaceous plants. *Plant Cell Environ.* 11:1–8

39. Ohyama T. 1984. Comparative studies on the distribution of nitrogen in soybean plants supplied with NO_2 and NO_3^- at the pod filling stage. II. Assimilation and transport of nitrogenous constituents. *Soil Sci. Plant Nutr.* 30:219–29

40. Paiva E, Lister RM, Park WD. 1983. Induction and accumulation of major tuber proteins of potato in stems and petioles. *Plant Physiol.* 71:161–68

41. Parthier B. 1990. Jasmonates: hormonal regulators of stress factors in leaf senescence? *J. Plant Growth Regul.* 9:57–63

42. Parthier B. 1991. Jasmonates, new regulators of plant growth and development: many facts and few hypotheses on their actions. *Bot. Acta* 104:405–64

43. Pate JS. 1980. Transport and partitioning of nitrogenous solutes. *Annu. Rev. Plant Physiol.* 31:313–40

44. Peoples MB, Dalling MJ. 1988. The interplay between proteolysis and amino acid

metabolism during senescence and nitrogen reallocation. In *Senescence and Aging in Plants*, ed. LD Nooden, AC Leopold, pp. 181–217. New York: Academic

45. Pusztai A, Croy RRD, Grant G, Stewart JC. 1983. Seed lectins: distribution, location and biological role. In *Seed Proteins*, ed. J Daussant, JV Mosse, pp. 53–80. London: Academic

46. Racusen D, Foote M. 1965. Protein synthesis in dark-grown soybean leaves. *Can. J. Bot.* 43:817–24

47. Rapp WD, Lilley GG, Nielsen NC. 1990. Characterization of soybean vegetative storage proteins and genes. *Theor. Appl. Genet.* 79:785–92

48. Rhee Y, Staswick PE. 1992. Nucleotide sequence of a soybean vegetative storage protein *vspA* gene. *Plant Physiol.* 98:792–93

49. Rhee Y, Staswick PE. 1992. Nucleotide sequence of a soybean vegetative storage protein *vspB* gene. *Plant Physiol.* 98:795–96

50. Ryan CA. 1992. The search for the proteinase inhibitor-inducing factor, PIIF. *Plant Mol. Biol.* 19:123–33

51. Sesay A, Shibles R. 1980. Mineral depletion and leaf senescence in soybean as influenced by foliar nutrient application during seed filling. *Ann. Bot.* 45:47–55

52. Simpson RJ. 1986. Translocation and metabolism of nitrogen: whole plant aspects. See Ref. 17, pp. 71–96

53. Sinclair TR, DeWit CT. 1975. Photosynthate and nitrogen requirements for seed production by various crops. *Science* 189:565–67

54. Spilatro SR, Anderson JM. 1989. Characterization of a soybean leaf protein that is related to the seed lectin and is increased with pod removal. *Plant Physiol.* 90:1387–93

55. Staswick PE. 1988. Soybean vegetative storage protein structure and gene expression. *Plant Physiol.* 87:250–54

56. Staswick PE. 1989. Developmental regulation and the influence of plant sinks on vegetative storage protein gene expression in soybean leaves. *Plant Physiol.* 89:309–15

57. Staswick PE. 1989. Preferential loss of an abundant storage protein from soybean pods during seed development. *Plant Physiol.* 90:1252–55

58. Staswick PE. 1990. Novel regulation of vegetative storage protein genes. *Plant Cell* 2:1–6

59. Staswick PE. 1992. Jasmonate, genes, and fragrant signals. *Plant Physiol.* 99:804–7

60. Staswick PE, Huang J, Rhee Y. 1991. Nitrogen and methyl jasmonate induction of soybean vegetative storage protein genes. *Plant Physiol.* 96:130–36

61. Staswick PE, Su W, Howell SH. 1992. Methyl jasmonate inhibition of root growth and induction of a leaf protein are decreased in an *Arabidopsis thaliana* mutant. *Proc. Natl. Acad. Sci. USA* 89:6837–40

61a. Staswick PE, Papa C, Huang J-F, Rhee Y. 1994. Purification of a major leaf acid phosphatase that is increased by seed pod removal. *Plant Physiol.* In press

62. Stoddart JL, Thomas H. 1982. Leaf senescence. In *Encyclopedia of Plant Physiology*, New Ser. 14A, ed. D Boulter, B Parthier, pp. 592–636. Berlin: Springer-Verlag

63. Thayer SS, Choe HT, Tang A, Huffaker RC. 1987. Protein turnover during senescence. See Ref. 10, pp. 71–80

64. Thibodeau PS, Jaworski EG. 1975. Patterns of nitrogen utilization in the soybean. *Planta* 127:133–47

65. Titus JS, Kang SM. 1982. Nitrogen metabolism, translocation, and recycling in apple trees. *Hortic. Rev.* 4:204–6

66. Tranbarger TJ, Franceschi VR, Hildebrand DF, Grimes HD. 1991. The soybean 94-kilodalton vegetative storage protein is a lipoxygenase that is localized in paraveinal mesophyll cell vacuoles. *Plant Cell* 3:973–87

67. Turgeon R. 1989. The sink-source transition in leaves. *Annu. Rev. Plant Physiol. Plant Mol. Biol.* 40:119–38

68. van Cleve B, Apel K. 1993. Induction by nitrogen and low temperature of storage-protein synthesis in poplar trees exposed to long days. *Planta* 189:157–60

69. Waters SP, Peoples MB, Simpson RJ, Dalling MJ. 1980. Nitrogen redistribution during grain growth in wheat (*Triticum aestivum* L.) I. Peptide hydrolase activity and protein breakdown in the flag leaf, glume and stem. *Planta* 148:422–28

70. Wenzler HC, Mignery GA, Fisher LM, Park WD. 1989. Analysis of a chimeric class-I patatin-GUS gene in transgenic potato plants: high level expression in tubers and sucrose-inducible expression in cultured leaf and stem explants. *Plant Mol. Biol.* 12:41–50

71. Wetzel S, Demmers C, Greenwood JS. 1989. Seasonally fluctuating bark proteins are a potential form of nitrogen storage in three temperate hardwoods. *Planta* 178:275–81

72. Williamson VM, Colwell G. 1991. Acid phosphatase-1 from nematode resistant tomato. *Plant Physiol.* 97:139–46

73. Witt W, Mauk CS, Yelenosky G. 1989. Effect of leaf development and temperature on a foliar storage glycoprotein from *Poncirus trifoliata* (L.) Raf. *Plant Sci.* 62:157–63

74. Wittenbach VA. 1982. Effect of pod removal on leaf senescence in soybeans. *Plant Physiol.* 70:1544–48

75. Wittenbach VA. 1983. Purification and characterization of a soybean leaf storage glycoprotein. *Plant Physiol.* 73:125–29

76. Wittenbach VA, Franceschi VR, Giaquinta RT. 1984. Soybean leaf storage proteins. *Curr. Top. Plant Biochem. Physiol.* 3:19–30

77. Zehir C, Egli DB, Leggett JE, Reicosky DA. 1982. Cultivar differences in N redistribution in soybeans. *Agron. J.* 74:375–79

Annu. Rev. Plant Physiol. Plant Mol. Biol. 1994. 45:323–37

THE GLYCINE DECARBOXYLASE COMPLEX FROM PLANT MITOCHONDRIA

David J. Oliver

Department of Microbiology, Molecular Biology, and Biochemistry, University of Idaho, Moscow, Idaho 83843

KEY WORDS: photorespiration, respiration, gene expression, light induction, protein structure

CONTENTS

INTRODUCTION 323
GLYCINE DECARBOXYLASE IN THE PHOTORESPIRATORY PATHWAY 324
THE GLYCINE DECARBOXYLASE REACTION 325
PROTEIN CHEMISTRY 327
    P-protein 327
    H-protein 327
    T-protein 328
    L-protein 328
THE STRUCTURE OF THE GLYCINE DECARBOXYLASE COMPLEX 329
METABOLIC CONTROL OF THE GLYCINE DECARBOXYLASE REACTION AND
    THE ROLE OF SUBSTRATE TRANSPORTERS 330
SERINE HYDROXYMETHYLTRANSFERASE 331
DEVELOPMENTAL EXPRESSION OF GLYCINE DECARBOXYLASE 331
    Light-dependent Expression 331
    H-Protein and the Small Subunit of Rubisco 332
    Tissue-specific Expression 333
CONCLUSIONS 334

INTRODUCTION

Most mitochondria can convert glycine to serine. The sequential reactions of two enzymes, the glycine decarboxylase multienzyme complex (GDC—sometimes referred to as the glycine cleavage system or serine synthase) and serine hydroxymethyltransferase (SHMT), convert two molecules of glycine to one each of serine, CO_2, and NH_3 with the reduction of NAD^+ to NADH. Glycine decarboxylase, which is found in plants (48), animals (16, 19), and bacteria (22, 64), oxidizes glycine to CO_2, NH_3, and a one-carbon fragment in the form of N^5, N^{10}-methylene tetrahydrofolate (THF). The N^5, N^{10}-methylene THF reacts with a second molecule of glycine in the reaction catalyzed by SHMT to yield serine.

Glycine Decarboxylase Reaction
  Glycine + NAD^+ + THF $\rightarrow$ methylene–THF + CO_2 + NH_3 + NADH

Serine Hydroxymethyltransferase Reaction
  Glycine + methylene–THF + H_2O $\rightarrow$ Serine + THF

Overall Reaction
  2 Glycine + NAD^+ $\rightarrow$ Serine + CO_2 + NH_3

The role of GDC in all organisms is to interconnect the metabolism of one-, two-, and three-carbon compounds (19, 36, 37). In mammalian tissues, GDC converts glycine derived from protein catabolism into one-carbon fragments that can be used in the anabolic functions of the mitochondria and serine that can enter three-carbon metabolism and glycolysis. In humans, mutations in any of the three unique enzymes of the glycine decarboxylase complex result in the fatal genetic disease, nonketotic hyperglycinemia (25). GDC is postulated to have been recruited during plant evolution to fill a key step in photosynthesis and photorespiration. In C_3 plants, GDC mutants are lethal only under photorespiratory conditions. They have no phenotype when grown at elevated CO_2 concentrations (3, 52).

GLYCINE DECARBOXYLASE IN THE PHOTORESPIRATORY PATHWAY

The photorespiratory C_2 cycle diverges from the photosynthetic carbon fixation reactions with the oxygenation of RuBP by Rubisco (17, 26). Phosphoglycolate, a product of the reaction, serves as the first metabolite of the C_2 cycle or glycolate pathway. It is hydrolyzed to glycolate, which is converted to glycine in the peroxisome. Glycine enters the mitochondria where the combined reactions of GDC and SHMT convert two moles of glycine to one each of serine and photorespiratory NH_3 and CO_2. The three-carbon fragment in serine eventually reenters the C_3 cycle (17, 26).

In C_3 plants the flux of carbon through the C_2 pathway is about equal on a molar basis to the flux through the photosynthetic C_3 cycle thus making it a major metabolic pathway in leaves. Glycine decarboxylase has a pivotal role in this cycle as the site of photorespiratory CO_2 and NH_3 release. The CO_2 released by this pathway is about 25% of the amount fixed by Rubisco, and thus this pathway decreases the carbon that is available to support plant growth and crop yield. The rate of NH_3 release is nearly 10 times higher than the rate of primary NH_3 fixation, and the released NH_3 must be recycled with near perfect efficiency to prevent unacceptable levels of N loss (17, 26).

THE GLYCINE DECARBOXYLASE REACTION

Glycine decarboxylase is a multienzyme complex that requires the concerted efforts of four different component enzymes to complete its reactions (5, 68) (Figure 1). These proteins are P-protein, the pyridoxal 5-phosphate (PLP)-dependent amino acid decarboxylase proper; H-protein, a carrier protein with a bound lipoamide cofactor; T-protein, a tetrahydrofolate (THF) transferase; and L-protein, the flavoprotein dihydrolipoamide dehydrogenase. The reaction begins as the Schiff base between the carbonyl of PLP and lys[792] of the P-protein exchanges for a Schiff base between PLP and the α-amino group of glycine

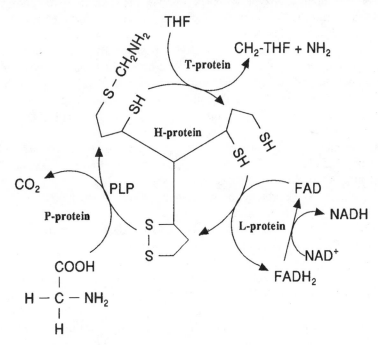

Figure 1   The reaction mechanism of the glycine decarboxylase multienzyme complex. The GDC reaction requires all four component proteins of the multienzyme complex. The lipoamide cofactor on the H-protein carries reaction intermediates between the reactive sites of the P-protein, T-protein, and L-protein.

(24, 62). Three nearly simultaneous reactions follow. An α-elimination reaction results in the decarboxylation of glycine and the loss of the α-carboxyl carbon as CO_2. As the carbon skeleton is oxidized, electrons are passed to the lipoamide of the H-protein, reducing it to dihydrolipoamide. Simultaneously, the methylamine that remains after glycine decarboxylation is transferred to one of the sulfhydryl groups of the dihydrolipoamide. The methylamine dihydrolipoamide intermediate of the H-protein next leaves the active site of the P-protein and moves to the active site of the T-protein. The methylene carbon of glycine (the α-carbon) is transferred to THF to make N^5, N^{10}-methylene THF and the α-amino group of glycine is released as NH_3. The dihydrolipoamide now leaves the active site of the T-protein and migrates to the active site of the L-protein. This lipoamide dehydrogenase catalyzes the oxidation of dihydrolipoamide to lipoamide with the reduction of FAD to $FADH_2$, which ultimately is reoxidized with the reduction of NAD^+ to NADH. In this swinging arm model of the enzyme, the long flexible arm of the lipoamide attached to the ε-amino group of the H-protein lys^{63} moves reaction intermediates and reducing equivalents between the reactive sites located on the other three subunits of the multienzyme complex (19, 68). Thus, the reactive intermediates remain enzyme-bound, which increases the reaction rates by decreasing the distances intermediates would need to diffuse (38).

This reaction mechanism was originally postulated for the enzyme from liver mitochondria (19), but substantial experimental data have shown that it applies equally well to the enzyme from pea mitochondria (5, 37, 38, 68). PLP, NAD^+, and THF are all essential cofactors for the reactions (5, 31, 48, 68). Reagents targeted to PLP [isonicotinic acid hydrazide (43) and carboxymethoxylamine (49)] and lipoamide [arsenite (67)] inhibit the reaction. CO_2, not bicarbonate, is the product of the decarboxylation step (48). The three predicted forms of the H-protein (lipoamide, dihydrolipoamide, and methylamine dihydrolipoamide) have all been isolated (32, 68). Finally, the complex, or component proteins of the complex, catalyze the partial reactions predicted from the mechanism.

The overall reaction can be measured as the release of $^{14}CO_2$ from [1-^{14}C] glycine (48, 68) or the glycine-dependent reduction of NAD^+ to NADH (5, 31). Both reactions require all four subunits of the complex and excess NAD^+ and THF as the terminal electron and methylene acceptors, respectively (68). The P-protein and H-protein catalyze the exchange of $^{14}CO_2$ into the carboxyl carbon of glycine, the reversal of the first step of the reaction pathway (48, 51, 68). The ^{14}C-labeled methylene carbon on the methylamine dihydrolipoamide form of the H-protein is transferred to THF in a reaction dependent on the T-protein (10, 32, 68). The L-protein will transfer electrons from NADH to the H-protein, forming a dihydrolipoamide cofactor, in the reversal of the mechanism's last step. The dihydrolipoamide can be detected and reoxidized by reaction with DTNB (30, 32, 38). The L-protein passes electrons from dihydrolipoic acid to NAD^+ (5). Finally, the entire reaction can be reversed with the synthesis of glycine from $^{14}CO_2$, NH_3, NADH, and methylene THF (usually formed from serine by SHMT) (48, 50). In addition to providing partial reactions for the individual component enzymes of the complex, these reactions strongly support the proposed mechanism.

PROTEIN CHEMISTRY

All of the component proteins of GDC are nuclear-encoded and synthesized with N-terminal presequences that are cleaved after targeting the enzymes to the mitochondrial matrix (42, 54). Based on the ready assembly of the intact complex in vitro, the components are probably capable of spontaneous assembly within the matrix to form the multienzyme complex (38). Despite earlier suggestions to the contrary (29, 48), GDC is clearly a soluble matrix enzyme with no strong association with the mitochondrial membrane.

P-protein

The P-protein of the complex is the actual amino acid decarboxylase and, in the animal literature, is often referred to as glycine decarboxylase, which, unfortunately, is the name the plant literature uses to describe the entire multienzyme complex. The mature P-protein from peas is comprised of 971 amino acids with a predicted MW of 105,000. It is synthesized with an N-terminal extension of 86 amino acids (62) that is removed following uptake into the matrix (21). This is the longest mitochondrial targeting sequence described to date and appears to contain no information other than the matrix targeting specifications filled by the smaller presequences of other mitochondrial proteins. Comparisons with a peptide from the P-protein of chicken liver specifically labeled with PLP and NaB^3H_4, indicate that lys^{792} is the site of PLP binding (11). Substantial sequence similarities in this region with the human and chicken enzymes include a glycine-rich region that has been implicated in the function of this and other PLP enzymes (24, 28, 62).

H-protein

In peas (20, 27) and *Arabidopsis* (55), the mature H-protein of the glycine decarboxylase complex is 131 amino acids, with a mass of 13.9 kDa. The mitochondrial targeting sequence for both proteins is 34 amino acids long and can target this protein to mitochondria in vitro (54). The sequences of the plant proteins are highly conserved with both the chicken and human sequences particularly around lys^{63}, the site of lipoic acid attachment (21, 27, 65). The bovine H-protein has been demonstrated to be taken up by isolated mitochondria as the apoprotein; lipoylation occurs within the matrix after removal of the presequence (12, 13). Although equivalent work has not been done with plant enzyme, it is likely that a similar maturation process occurs in higher plants.

T-protein

Although the T-protein isolated from the matrix of plant mitochondria has an apparent mass of about 45 kDa as estimated by SDS-PAGE (5, 68), recent cloning of the gene from peas shows that the protein has an actual molecular weight of 40,961 (J Bourguignon, P Vauclare, V Merand, E Forest, M Neuburger, & R Douce, unpublished information). T-protein is synthesized as a 438 amino acid precursor that is processed to a 408 amino acid mature protein following the cleavage of the 30 amino acid N-terminal presequence. Comparison of the pea sequence with that for the T-protein from chicken (33) and bovine (34) liver, shows about 50% identity to the amino acid with great similarities in predicted secondary structure. As with the animal enzymes, there are no obvious similarities between the pea T-protein and other THF enzymes.

The dihydrolipoamide acyltransferase function of other respiratory multienzyme complexes [the pyruvate dehydrogenase complex (PDC), the α-ketoglutarate dehydrogenase complex (α-KGDC), and the branched ketoacid dehydrogenase complex] has been divided between the H-protein and the T-protein of GDC. In the acyltransferases there are 1–3 lipoyl domains on the N-terminal region of the protein separated by conserved hinge regions that connect them to the carboxyl terminal catalytic region (46). The hinge regions are apparently important for increased flexibility of the lipoamide domain. In the proposed mechanism, the 28Å reach of the swinging lipoamide arm must extend to cover the distance between the active sites of the multienzyme complex. The lipoyl and transferase catalytic domains in GDC have been separated onto two different polypeptide chains. The H-protein contains the lipoyl binding site and, according to the model presented below, comprises the core of the complex. The transferase domain of the dihydrolipoamide acyltransferases resides in the T-protein of the GDC. T-protein is a tetrahydrofolate-based methyl transferase instead of the CoASH-dependent acyl transferase that is used by the other complexes.

L-protein

Turner et al (63) and Bourguignon et al (4) both cloned the L-protein of GDC. The mature pea protein is 470 amino acids with a calculated mass of 50 kDa (substantially less than the 60 kDa estimated by SDS-PAGE) (5, 68). The presequence is 31 amino acids. Sequence comparison with the yeast enzyme shows a high degree of conservation and identifies sites associated with flavin and pyridine nucleotide binding as well as the active site and sites for dimerization (63). In addition to its role in GDC, the L-protein also serves the same function in other multienzyme complexes, including PDC and α-KGDC. Early work with the GDC from pea mitochondria showed that monoclonal antibodies raised against the L-protein inhibited both its activity and that of mitochondrial PDC (68). Molecular studies verify that there is only one gene producing mitochondrial lipoamide dehydrogenase and, therefore, that this subunit is shared in common by all the multienzyme complexes (4, 63).

THE STRUCTURE OF THE GLYCINE DECARBOXYLASE COMPLEX

The structure of the glycine decarboxylase complex may never be fully solved. While stability is not a problem with the components of this multienzyme complex, the assembled structure is highly unstable (38, 48). At high concentrations of the component proteins, the complex is stable and is retained on a 300-kDa cutoff ultrafiltration membrane without losing any enzyme activity or component proteins. When the protein concentration is diluted below 0.25 mg ml^{-1}, the complex dissociates (38) and can then reform if concentrated subsequently. The ready dissociation of the complex into its component proteins has precluded purifying it directly from the mitochondrial matrix fraction. The only method that yields pure complex is the isolation and purification of each of the four individual component proteins, followed by reconstitution of the functional complex in vitro (5, 38, 68).

The ratio of the subunits of the complex has been determined immunochemically on the crude matrix extract (38). Given the limitations of this assay, particularly that the L-protein is also associated with other multienzyme complexes, the monomeric subunit ratio is 4P:27H:9T:2L. The P-protein and the L-protein are dimers. The T-protein is a monomer. The H-protein can be either a monomer or a dimer. Binary complexes of the P-protein and H-protein and of the T-protein and H-protein have been observed (68). Based on these observations, our model of the complex is a core of H-protein dimers with one dimer for each monomer of the larger component proteins decorating the surface of the H-protein cluster. The complex is held together by interactions between the H-proteins and between the H-proteins and the larger subunits. No direct interactions between the P-, T-, and L-proteins have been measured, and none are included in the model.

In this model, the H-protein is pivotal in determining the structure of the intact complex. When the complex has dissociated, H-protein serves as a substrate for each of the three reactions of the complex. In the P-protein reaction it is a co-substrate with glycine, in the T-protein reaction it is a co-substrate with THF, and in the L-protein reactions it is a co-substrate with NAD^+. Under the dissociated conditions, the apparent K_m for H-protein is 2.4 µM and it often limits the reaction rate, possibly accounting for the difficulty originally associated in identifying this enzyme in solubilized plant mitochondria (38, 68). In the undissociated complex, the H-protein acts as a catalyst and not as a substrate and the rate of the reaction is not stimulated by the addition of extra H-protein.

METABOLIC CONTROL OF THE GLYCINE DECARBOXYLASE REACTION AND THE ROLE OF SUBSTRATE TRANSPORTERS

Glycine decarboxylase is inhibited by its two major products, NADH and serine. NADH is competitive with NAD^+ for binding to the L-protein (5, 31). The K_m for NAD^+ is 75 μM and the K_i for NADH is 15 μM. Since the L-protein of GDC is shared with PDC and α-KGDC, the NAD^+/NADH ratio coordinately regulates the metabolism of glycine, pyruvate, and α-ketoglutarate (41). The inhibition by serine is competitive with glycine, suggesting that they both are interacting with the P-protein of the complex (K_m for glycine = 6 mM, K_i for serine = 4 mM) (39). Interestingly, in both cases the inhibition is prevented by unique substrate transporters in plant mitochondria (Figure 2). Reducing equivalents are shuttled out of the mitochondria by a combination of

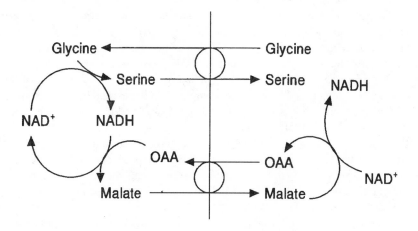

Figure 2 Shuttle for removing reducing equivalents generated by the glycine decarboxylase reaction from the mitochondria. Glycine entry and serine exit are mediated by one or more substrate transporters. Reducing equivalents, including those generated by glycine oxidation, are rapidly shuttled out of the mitochondrial matrix by a unique OAA transporter and malate dehydrogenase activity in the matrix and cytosol.

malate dehydrogenase activity in the matrix and cytosol and an active oxaloac-etate/malate transporter in the inner mitochondrial membrane (9, 40). Gly-cine/serine exchange proteins in the inner mitochondrial membrane prevent inhibition by serine (35, 66). Working together, these two transport systems prevent feedback inhibition by the products of glycine oxidations and allow the rapid rates of glycine oxidation needed for the rates of photorespiration occurring in the leaves of C_3 plants (Figure 2).

SERINE HYDROXYMETHYLTRANSFERASE

Serine hydroxymethyltransferase is a homotetramer (5). In peas, each subunit is comprised of 487 amino acids, with a calculated mass of 54 kDa, and each subunit binds one molecule of PLP. The protein's cDNA, which Turner et al (61) have cloned, indicates the existence of a mitochondrial targeting sequence of 27 or 31 amino acids. SHMT activity is closely tied to glycine decarboxyl-ase activity because the former must recycle the methylene THF produced by the latter. SHMT is not part of the glycine decarboxylase complex but rather is connected through a soluble THF pool (5). There appear to be two forms of the protein in the mitochondrial matrix and an isoform in the plastid (61).

DEVELOPMENTAL EXPRESSION OF GLYCINE DECARBOXYLASE

While some glycine decarboxylase activity is probably present in all cells within a plant, the level of expression is tightly coupled to the photosynthetic activity of the tissues in C_3 plants.

Light-dependent Expression

The greening of etiolated tissues is one of the most carefully analyzed pro-cesses for investigating the light-dependent expression of photosynthetic en-zymes. Mitochondria isolated from etiolated pea shoots, like those isolated from potato tubers and soybean and cucumber cotyledons (1), oxidize glycine poorly. Mitochondria isolated from green leaves or greened cotyledons of C_3 plants, however, show rapid rates of glycine oxidation. Glycine is the preferred substrate for these green leaf mitochondria in vitro in much the same way that it is the predominate substrate in vivo (7, 8, 67).

Walker & Oliver (69) initially showed that the increase in glycine oxidation activity during greening of etiolated peas resulted from an increase in the levels of all four of the component proteins of the complex. The major differ-ence between the mitochondria isolated from green and etiolated pea leaves are the four component proteins of GDC and the subunits of SHMT (6, 54).

Several photosynthetic enzymes increase in activity during the greening of etiolated tissue. Rubisco is the most studied example. The accumulation of glycine decarboxylase activity closely parallels the increase in Rubisco activity. Following illumination of the etiolated peas there was a 3–5 hr lag before either activity increased measurably. The two activities then increased together over the course of the next 8 hr of illumination (53, 54, 69). The P-, H-, and T-proteins increased 5- to 10-fold while the L-protein only increased about 2- to 3-fold because of the larger amount of L-protein in non-green tissues where this subunit was a component of PDC and α-KGDC. Rawsthorne's group (60) has followed the time course for the accumulation of the glycine decarboxylase proteins along with SHMT during greening of etiolated pea plants. The T- and H-proteins and SHMT accumulated at similar rates after an inital lag. The P-protein showed a shorter lag than the other three and it reached a plateau while the other three continued to accumulate. L-protein levels increased little during the greening process.

cDNA clones for the H-protein (20, 27), P-protein (21, 62), T-protein (60), L-protein (4, 63), and SHMT (61) have been isolated and characterized. The mRNAs for the H-, P-, and T-proteins were present at low but detectable levels in etiolated pea tissues (20, 21, 27, 60, 62, 63). Following transfers to the light there was a 4 hr lag before the level of H-protein mRNA increased (20, 53, 54). Similar kinetics were seen for SHMT and T-protein mRNAs (60, 61). In contrast, the L-protein mRNA was present at higher concentrations in etiolated tissues and its level changed little when the plants were illuminated (4, 63). P-protein mRNA expression appeared to have a shorter lag and reached a steady state level sooner (21, 62).

The light-dependent increase in mRNA for the H-protein and P-protein are controlled at the transcriptional level. Run-on transcription with nuclei isolated from etiolated and green pea leaves demonstrated that about 5–7 times more H-protein and P-protein mRNA was transcribed in the nuclei of green leaves than in the nuclei from etiolated leaves (54). This bias was about the same magnitude of change measured in mRNA steady state pool levels, which implies that most of the mRNA increase during greening results from increased transcription.

H-Protein and the Small Subunit of Rubisco

There are obvious parallels between the H-protein and the small subunit of Rubisco (SSU). Both proteins are about the same size and accumulate to millimolar concentrations within the soluble phase of their respective organelles; both are nuclear-encoded and intimately involved in photosynthesis; and both are light-regulated at the transcriptional level. These observations suggest strongly that similar mechanisms control the transcription of these two genes. Kinetic analysis of the time course for the transcriptional activation of both *rbcS* and *gdcH* when pea plants were transferred from the dark to light showed that the mRNA levels for these genes responded in synchrony (20, 21, 53, 54). Both mRNAs were present at low levels in etiolated tissues and remained low for about 4 hr following illumination. Although there was several times more SSU than H-protein mRNA, the amounts of each increased proportionally (54).

With *rbcS,* the lag before the level of its mRNA begins to increase is associated with the time it takes for chloroplast maturation (23, 56, 59). The lag time can be shortened by exposing light-grown plants to 48–72 hr of darkness and then following the kinetics of photosynthetic gene expression after the plants are returned to light. During the dark period mRNA abundance of both the SSU and H-protein drop to very low levels (even lower than in etiolated leaves). Upon returning to light, the lag time before the mRNA for both SSU and H-protein begins to accumulate decreases to about 1 hr (53, 54). Mature chloroplasts are present in these tissues and immediately signal for the activation of nuclear genes once the plant is exposed to light. Again, the levels of H-protein and SSU mRNA increase in a coordinated manner. While the nature of this mature chloroplast signal is unknown, it can be disrupted by either using a genotype where chloroplast development is disrupted or by using the herbicide, Norflurazon, which blocks carotenoid synthesis and allows photobleaching and chloroplast disruption in the light (56, 57). Norflurazon-treated pea plants still undergo initial phytochrome-dependent changes such as shorter stature and leaf expansion. They do not green, however, which blocks the maturation of the chloroplasts and prevents the chloroplast signal from activating the nuclear-encoded photosynthetic genes, thereby preventing the increased transcription of both SSU and H-protein.

The promoter region from the *Arabidopsis* H-protein gene was fused to the β-glucuronidase (GUS) reporter gene and expressed in tobacco (53; R Srinivasan & DJ Oliver, unpublished information). GUS expression was photoregulated with about 7 times more GUS activity in light-grown than dark-grown plants. The DNA sequence of the promoter region revealed a number of *cis* elements that are normally associated with light-dependent expression of the well-characterized photosynthetic genes, *cab* and *rbcS*. Progressive 5′ deletions of the promoter region reduced both total promoter activity and light-dependent GUS expression.

Tissue-specific Expression

Transgenic tobacco plants expressing GUS under control of the *Arabidopsis* H-protein promoter have shown that the expression of the reporter gene in different tissues closely correlated with the amount of Rubisco activity (53). This tissue specificity was lost as the promoter region was shortened (R Srinivasan & DJ Oliver, unpublished information).

GDC expression has been studied in wheat leaves, which contain about 4–5 times more P-, H-, and T-proteins in green tissue than in etiolated leaves (47, 58). Monocots have age and developmental gradients through the growing leaves, with the oldest, most mature tissue at the leaf tips and the youngest tissue at the base. Glycine decarboxylase activity increases from the base to tip along with the other photorespiratory enzymes (58). Measurements of the levels of individual proteins suggest that the P-, H-, and T-proteins increase together as the tissue matures. L-protein is present at higher levels in the juvenile tissue and does not increase proportionally during tissue maturation.

Possibly the most interesting studies concerning spatial distribution of GDC comes from Rawsthorne's laboratory, which has provided a description of the immunolocalization the GDC enzymes in C_3-C_4 intermediates (18, 44, 45). These plants have photosynthetic gas exchange characteristics (CO_2 compensation point and O_2 sensitivity of CO_2 fixation) that are intermediate between the C_3 and C_4 plants. In *Moricandia arvensis* leaves, the photosynthetic, as well as most of the photorespiratory enzymes, are distributed uniformly between the mesophyll and bundle sheath tissues. In contrast, GDC and SHMT are localized to the bundle sheath. Photorespiratory CO_2 release would occur in bundle sheath mitochondria and the longer diffusion path could account for the increased portion of photorespired CO_2 that is refixed before escaping from the leaf. This mechanism would require substantial fluxes of glycine and serine between the two cell types. Surprisingly, mesophyll mitochondria of *M. arvensis* only lack the P-protein of GDC and still contain some complement of the other component proteins of the complex. Because the lack of any subunit would block the function of the complex, this does not affect the arguments presented but might suggest that in this case the tissue distribution of GDC activity is a fairly recent evolutionary change.

CONCLUSIONS

Glycine decarboxylase is a light-induced multienzyme complex that accumulates to over 130 mg protein per ml in the matrix of mitochondria from the leaves of C_3 plants. The mechanism of the enzyme complex is fairly well understood, and there has been progress in cloning the genes involved and studying their light- and tissue-dependent expression patterns. Several major problems are still unresolved. Among these problems are potential metabolic control of the complex, the coordination of gene expression for the component proteins of the complex, and understanding the three-dimensional structure of this unusual protein.

ACKNOWLEDGMENTS

This research was funded by the USDA/SEA National Research Initiatives Competitive Grants Program and Idaho Agricultural Experiment Station.

Literature Cited

1. Arron GP, Edwards GE. 1980. Light-induced development of glycine oxidation by mitochondria from sunflower cotyledons. *Plant Sci. Lett.* 18:229–35
2. Besson V, Rebeille F, Neuburger M. 1993. Effects of tetrahydrofolate polyglutamates on the kinetic parameters of serine hydroxymethyltransferase and glycine decarboxylase from pea leaf mitochondria. *Biochem. J.* 292:425–30
3. Blackwell D, Murray AJS, Lea PJ. 1990. Photorespiratory mutants of the mitochondrial conversion of glycine to serine. *Plant Physiol.* 94:1316–22
4. Bourguignon J, Macherel D, Neuburger M, Douce R. 1992. Isolation, characterization, and sequence analysis of a cDNA clone encoding L-protein, the dihydrolipoamide dehydrogenase component of the glycine cleavage system from pea-leaf mitochondria. *Eur. J. Biochem.* 204:865–73
5. Bourguignon J, Neuburger M, Douce R. 1988. Resolution and characterization of the glycine-cleavage reaction in pea leaf mitochondria. *Biochem. J.* 255:169–78
6. Day DA, Neuburger M, Douce R. 1985. Biochemical characterization of chlorophyll-free mitochondria from pea leaves. *Aust. J. Plant Physiol.* 12:219–28
7. Day DA, Neuburger M, Douce R. 1985. Interactions between glycine decarboxylase, the tricarboxylic acid cycle, and the respiratory chain in pea leaf mitochondria. *Aust. J. Plant Physiol.* 12:119–30
8. Dry IB, Wiskich JT. 1985. Characteristics of glycine and malate oxidation by pea leaf mitochondria: evidence of differential access to NAD and respiratory chains. *Aust. J. Plant Physiol.* 12:329–39
9. Ebbighausen H, Jia C, Heldt HW. 1985. Oxaloacetate translocator in plant mitochondria. *Biochem. Biophys. Acta* 47:184–99
10. Fujiwara K, Okamura-Ikeda K, Motokawa Y. 1984. Mechanism of the glycine cleavage reaction: further characterization of the intermediate attached to H-protein and of

the reaction catalyzed by T-protein. *J. Biol. Chem.* 259:10664–68
11. Fujiwara K, Okamura-Ikeda K, Motokawa Y. 1987. Amino acid sequence of the phosphopyridoxyl peptide from P-protein of the chicken liver glycine cleavage system. *Biochem. Biophys. Res. Commun.* 149:621–27
12. Fujiwara K, Okamura-Ikeda K, Motokawa Y. 1990. cDNA sequence, in vitro synthesis, and intramitochondrial lipoylation of H-protein of the glycine cleavage system. *J. Biol. Chem.* 265:17463–67
13. Fujiwara K, Okamura-Ikeda K, Motokawa Y. 1992. Expression of mature bovine H-protein of the glycine cleavage system in *Escherichia coli* and *in vitro* lipoylation of the apoform. *J. Biol. Chem.* 267:20011–16
14. Gardeström P, Bergman A, Ericson I. 1980. Oxidation of glycine via the respiratory chain of mitochondria prepared from different parts of spinach. *Plant Physiol.* 65:389–91
15. Gilmartin PM, Sarokin L, Memelink J, Chua N-H. 1990. Molecular light switches for plant genes. *Plant Cell* 2:369–78
16. Hiraga K, Kikuchi G. 1980. The mitochondrial glycine cleavage system: purification and properties of glycine decarboxylase from chicken liver mitochondria. *J. Biol. Chem.* 255:11671–76
17. Husic DW, Husic HD, Tolbert NE. 1987. The oxidative photosynthetic carbon cycle. *Crit. Rev. Plant Sci.* 5:45–100
18. Hylton CM, Rawsthorne S, Smith AM, Jones DA, Woolhouse HW. 1988. Glycine decarboxylase is confined to the bundle-sheath cells of leaves of C_3-C_4 intermediate species. *Planta* 175:452–59
19. Kikuchi G. 1973. The glycine cleavage system: composition, reaction mechanism, and physiological significance. *Mol. Cell. Biochem.* 1:169–87
20. Kim Y, Oliver DJ. 1990. Molecular cloning, transcriptional characterization, and sequencing of the cDNA encoding the H-protein of the mitochondrial glycine decar-

boxylase complex in peas. *J. Biol. Chem.* 265:848–53

21. Kim Y, Shah K, Oliver DJ. 1991. Cloning and light-dependent expression of the gene coding for the P-protein of the glycine decarboxylase complex from peas. *Physiol. Plant.* 81:501–6

22. Klein SM, Sagers RD. 1966. Glycine metabolism: 1. Properties of the system catalyzing the exchange of bicarbonate with the carboxyl group of glycine in *Peptococcus glycinophilus. J. Biol. Chem.* 241:197–205

23. Kuhlemeier C, Green PJ, Chua N-H. 1987. Regulation of gene expression in higher plants. *Annu. Rev. Plant Physiol.* 38:221–57

24. Kume A, Koyata H, Sakakibara T, Ishiguro Y, Kure S, Hiraga K. 1991. The glycine cleavage system, molecular cloning of the chicken and human glycine decarboxylase cDNAs and some characteristics involved in the deduced protein structures. *J. Biol. Chem.* 266:3323–29

25. Kume A, Kure S, Tada K, Hiraga K. 1988. The impaired expression of glycine decarboxylase in patients with hyperglycinemias. *Biochem. Biophys. Res. Commun.* 154:292–97

26. Lorimer GH, Andrews TJ. 1981. The C_2 chemo- and photorespiratory carbon oxidation cycle. In *The Biochemistry of Plants, A Comprehensive Treatise,* ed. MD Hatch, NK Boardman, 8:329–74. New York: Academic

27. Macherel D, Lebrun M, Gagnon J, Neuburger M, Douce R. 1990. Primary structure and expression of H-protein, a component of the glycine cleavage system of pea leaf mitochondria. *Biochem. J.* 268:783–89

28. Marceau M, Lewis SD, Shafer JA. 1988. The glycine-rich region of *Escherichia coli* D-serine dehydratase. *J. Biol. Chem.* 263:16934–41

29. Moore AL, Jackson C, Halliwell B, Dench JE, Hall DO. 1977. Intramitochondrial localisation of glycine decarboxylase in spinach leaves. *Biochem. Biophys. Res. Commun.* 78:483–91

30. Motokawa Y, Kikuchi G. 1969. Glycine metabolism by rat liver mitochondria. IV. Isolation and characterization of hydrogen carrier protein, an essential factor for glycine metabolism. *Arch. Biochem. Biophys.* 135:402–9

31. Neuburger M, Bourguignon J, Douce R. 1986. Isolation of a large complex from the matrix of pea leaf mitochondria involved in the rapid transformation of glycine into serine. *FEBS Lett.* 207:18–22

32. Neuburger M, Jourdain A, Douce R. 1991. Isolation of H-protein loaded with methylamine as a transient species in glycine decarboxylase reactions. *Biochem. J.* 278:765–69

33. Okamura-Ikeda K, Fujiwara K, Motokawa Y. 1982. Purification and characterization of chicken liver T-protein, a component of the glycine cleavage system. *J. Biol. Chem.* 257:135–39

34. Okamura-Ikeda K, Fujiwara K, Yamamoto M, Hiraga K, Motokawa Y. 1991. Isolation and sequence determination of cDNA encoding T-protein of the glycine cleavage system. *J. Biol. Chem.* 266:4917–21

35. Oliver DJ. 1987. Glycine uptake by pea leaf mitochondria: a proposed model for the mechanism of glycine—serine exchange. In *Plant Mitochondria: Structural, Functional, and Physiological Aspects,* ed. AL Moore, RB Beechey, pp. 219–22. New York: Plenum

36. Oliver DJ, Kim Y. 1990. Biochemistry of the photorespiratory C-2 cycle. In *Biochemical and Genetic Regulation of Photosynthesis,* ed. I Zelitch, pp. 253–69. New York: Liss

37. Oliver DJ, Neuburger M, Bourguignon J, Douce R. 1990. Glycine metabolism by plant mitochondrion. *Physiol. Plant.* 80:487–91

38. Oliver DJ, Neuburger M, Bourguignon J, Douce R. 1990. Interaction between the component enzymes of the glycine decarboxylase multienzyme complex. *Plant Physiol.* 94:833–39

39. Oliver DJ, Sarojini G. 1987. Regulation of glycine decarboxylase by serine. In *Progress in Photosynthesis Research,* ed. J Biggins, 3:573–76. The Hague: Nijhoff

40. Oliver DJ, Walker GH. 1984. Characterization of the transport of oxaloacetate by pea leaf mitochondria. *Plant Physiol.* 76:409–13

41. Pascal N, Dumas R, Douce R. 1990. Comparison of the kinetic behavior toward pyridine nucleotides of NAD^+-linked dehydrogenases from plant mitochondria. *Plant Physiol.* 94:189–93

42. Pfanner N, Neupert W. 1990. The mitochondrial protein import apparatus. *Annu. Rev. Biochem.* 59:331–53

43. Pritchard GG, Griffin WH, Whittingham CP. 1962. The effect of isonicotinyl hydrazide on the photosynthetic incorporation of radioactive CO_2 into ethanol-soluble compounds of *Chlorella. J. Exp. Bot.* 14:281–89

44. Rawsthorne S. 1992. C_3-C_4 intermediate photosynthesis: linking physiology to gene expression. *Plant J.* 2:267–74

45. Rawsthorne S, Hylton CM, Smith AM, Woolhouse HW. 1988. Photorespiratory metabolism and immunogold localization of photorespiratory enzymes of C_3 and C_3-C_4 intermediate species of *Moricandia. Planta* 173:298–308

46. Reed LJ, Hackert ML. 1990. Structure-function relationships in dihydrolipoamide

acyltransferases. *J. Biol. Chem.* 265:8971–74

47. Rogers WJ, Jordan BR, Rawsthorne S, Tobin AK. 1991. Changes in the stoichiometry of glycine decarboxylase subunits during wheat and pea leaf development. *Plant Physiol.* 96:952–56

48. Sarojini G, Oliver DJ. 1983. Isolation and partial characterization of glycine decarboxylase from pea leaf mitochondria. *Plant Physiol.* 72:194–99

49. Sarojini G, Oliver DJ. 1985. Inhibition of glycine oxidation by carboxymethoxylamine, methoxylamine, and acethydrazide. *Plant Physiol.* 77:786–89

50. Sato T, Kochi H, Motokawa Y, Kawasaki H, Kikuchi G. 1969. Glycine metabolism by rat liver mitochondria: I. Synthesis of two molecules of glycine from one molecule each of serine, bicarbonate, and ammonia. *J. Biochem.* 65:63–70

51. Sato T, Kochi H, Motokawa Y, Sato N, Kikuchi G. 1969. Glycine metabolism by rat liver mitochondria: III. The glycine cleavage and the exchange of carboxyl carbon of glycine with bicarbonate. *J. Biochem.* 65:77–83

52. Somerville CR, Ogren WL. 1982. Mutants of the cruciferous plant *Arabidopsis thaliana* lacking glycine decarboxylase activity. *Biochem. J.* 202:373–80

53. Srinivasan R, Berndt WA, Oliver DJ. 1993. Coordinated expression of photosynthetic and photorespiratory genes. In *Plant Mitochondria,* ed. A Brennicke, U Kuck, pp. 160–69. Weinheim, Germany: VCH

54. Srinivasan R, Kraus C, Oliver DJ. 1992. Developmental expression of the glycine decarboxylase multienzyme complex in greening pea leaves. In *Molecular, Biochemical, and Physiological Aspects of Plant Respiration,* ed. H Lambers, LHW van der Plas, pp. 323–34. The Hague: SPB Academic

55. Srinivasan R, Oliver DJ. 1992. H-protein of the glycine decarboxylase multienzyme complex, complementary DNA encoding the protein from *Arabidopsis thaliana.* *Plant Physiol.* 98:1518–19

56. Susek RE, Chory J. 1992. A tale of two genomes: role of a chloroplast signal in coordinating nuclear and plastid genome expression. *Aust. J. Plant Physiol.* 19:387–99

57. Taylor WC. 1989. Regulatory interactions between nuclear and plastid genomes. *Annu. Rev. Plant Physiol. Plant Mol. Biol.* 40:211–33

58. Tobin AK, Thorpe JR, Hylton CM, Rawsthorne S. 1989. Spatial and temporal influences on the cell-specific distribution of glycine decarboxylase in leaves of wheat and peas. *Plant Physiol.* 91:1219–25

59. Tobin EM, Silverthorne J. 1985. Light regulation of gene expression in higher plants. *Annu. Rev. Plant Physiol.* 36:569–93

60. Turner SR, Hellen R, Ireland R, Ellis N, Rawsthorne S. 1993. The organisation and expression of the gene encoding the mitochondrial glycine decarboxylase complex and serine hydroxymethyltransferase in pea *(Pisum sativum).* *Mol. Gen. Genet.* 236:402–8

61. Turner SR, Ireland R, Morgan C, Rawsthorne S. 1992. Identification and localization of multiple forms of serine hydroxymethyltransferase in pea *(Pisum sativum)* and characterization of a cDNA encoding a mitochondrial isoform. *J. Biol. Chem.* 267:13528–34

62. Turner SR, Ireland R, Rawsthorne S. 1992. Cloning and characterization of the P subunit of glycine decarboxylase from pea. *J. Biol. Chem.* 267:5355–60

63. Turner SR, Ireland R, Rawsthorne S. 1992. Purification and primary amino acid sequence of the L subunit of glycine decarboxylase. *J. Biol. Chem.* 267:7745–50

64. Vanden Boom TJ, Reed KE, Cronan JE. 1991. Lipoic acid metabolism in *E. coli. J. Bact.* 173:6411–20

65. Yamamoto M, Koyata H, Matsui C, Hiraga K. 1991. The glycine cleavage system, occurrence of two types of chicken H-protein mRNAs presumably formed by the alternative use of the polyadenylation consensus sequences in a single exon. *J. Biol. Chem.* 266:3317–22

66. Walker GH, Oliver DJ. 1982. Identification of a glycine transporter from pea leaf mitochondria. *Biochem. Biophys. Res. Commun.* 107:856–61

67. Walker GH, Oliver DJ, Sarojini G. 1982. Simultaneous oxidation of glycine and malate by pea leaf mitochondria. *Plant Physiol.* 70:1465–69

68. Walker JL, Oliver DJ. 1986. Glycine decarboxylase multienzyme complex, purification and partial characterization from pea leaf mitochondria. *J. Biol. Chem.* 261:2214–21

69. Walker JL, Oliver DJ. 1986. Light-induced increases in the glycine decarboxylase multienzyme complex from pea leaf mitochondria. *Arch. Biochem. Biophys.* 248:626–38

Annu. Rev. Plant Physiol. Plant Mol. Biol. 1994. 45:339–67
Copyright © 1993 by Annual Reviews Inc. All rights reserved

INHIBITORS OF PHOTOSYNTHETIC ENZYMES/CARRIERS AND METABOLISM

Leszek A. Kleczkowski

Plant Molecular Biology Laboratory, Agricultural Research Council of Norway, P.O. Box 5051, 1432 Aas, Norway

KEY WORDS: ammonia assimilation, fatty acid biosynthesis, herbicides, inhibitors, photorespiration, photosynthesis, starch biosynthesis, sucrose biosynthesis

CONTENTS

INTRODUCTION 339
PRIMARY TARGETS FOR INHIBITORS DURING PHOTOSYNTHESIS 340
   Light Reactions 340
   C_3 *Cycle* 343
   C_4 *Cycle* 345
   C_2 *Cycle* 348
   Starch Biosynthesis... 350
   Nitrogen Assimilation 351
   Fatty Acid Biosynthesis.. 353
INHIBITORS OF ASSIMILATE EXPORT FROM CHLOROPLASTS 355
INHIBITORS OF SUCROSE BIOSYNTHESIS 356
PHOTOSYNTHETIC INHIBITORS AND HERBICIDE RESEARCH 358
CONCLUSIONS.. 359

INTRODUCTION

Inhibitors have played an important part in advancing our knowledge of biochemistry of photosynthesis-related processes, frequently being instrumental in the elucidation of metabolic pathways and regulatory steps in vivo. For instance, inhibitors of the light reactions of photosynthesis were essential in uncovering the identities and functions of the components of the electron transport system, helping to establish the arrangement of redox components in the *Z-scheme*. Furthermore, inhibitors have been valuable tools in the study of kinetic mechanisms of selected enzymes, and they helped to identify active groups involved in catalysis and regulation. Many of these substances have successfully been used as herbicides to eliminate weeds. Inhibitors of photosynthetic events are either naturally occurring regulatory metabolites or chemicals used as nonphysiological analogs of substrates and regulators or active site modifiers. This review considers both of these groups of compounds. Particular emphasis is given to inhibitors likely to affect activities of their protein targets under physiological conditions. This criterion limits the compounds discussed to those characterized by K_i values lower than, or similar to, K_m values for substrate(s) of a given enzyme/carrier, and/or to those that act as irreversible inhibitors. Recent developments in metabolite or chemical inhibition of photosynthesis are presented.

PRIMARY TARGETS FOR INHIBITORS DURING PHOTOSYNTHESIS

Light Reactions

Photosynthetic assimilatory metabolism is dependent on the transduction of light to chemical energy (ATP) and the release of electrons for chemical reduction by the thylakoid membranes, which contain the electron and proton translocating components. The thylakoid membrane contains both photosystem I (PSI) and photosystem II (PSII) polypeptide complexes as well as the coupling factor complex responsible for ATP synthesis. These components of the electron and proton transport systems of photosynthesis are the primary targets of about half of the commercial compounds used as herbicides.

ATPase Photosynthetic ATP synthesis in green plants is catalyzed by the oligomeric ATPase complex. This enzyme complex consists of two separable moieties, a hydrophobic membrane component (CF_0) and a water-soluble peripheral component (CF_1). The CF_0 component is involved in proton translocation across the energy-conserving membrane and the CF_1 component contains the catalytic site(s) (129, 165). Photophosphorylation can be inhibited by a direct inhibition of chloroplastic ATP synthase by energy-transfer inhibitors without any direct effects on electron flow and the establishment of a proton electrochemical potential difference across the thylakoid membrane. Alternatively, uncoupling of photosynthetic electron transport from ATP synthesis leads to the dissipation of the proton gradient necessary to drive phosphorylation of ADP (129, 141). Uncouplers such as phenylhydrazones, carbanilates, diphenylamines, and ethane amines also inhibit photosynthetic electron transport

(141). Nitrophen and related chlorinated *p*-nitro diphenyl ethers may inhibit by binding directly to the coupling factor and preventing the exchange of ADP at CF_1 (71, 141). Non-herbicidal compounds that behave as energy transfer inhibitors include the antibiotic Dio-9 and phloridzin (119).

A variety of modified adenine nucleotides have been used in studies on the action of ATPase. 2'- and 3'-methylated or -deoxidized ADP analogs can fully replace the parental nucleotide in photophosphorylation and the corresponding ATP analogs can serve as substrates in the reverse reaction. On the other hand, 2' and 3'-acylated ADP derivatives are inhibitors of photophosphorylation, mostly because they are competitive inhibitors of the binding of endogenous ADP to the active site on ATP synthase. These compounds are useful tools in enzymatic studies of nucleotide-dependent reactions because they allow the investigation of the nucleotide binding process without a subsequent enzymatic conversion (125, 165, 170). Another approach involves the use of irreversibly-binding active-site modifiers, especially those reacting with positively charged lysine and arginine residues, which are believed to interact with the phosphate groups of adenine nucleotides and inorganic phosphate (P_i). Modification of these amino acids causes inhibition of ADP phosphorylation and of ATP hydrolysis by isolated CF_1 (reviewed in 165).

PHOTOSYNTHETIC ELECTRON TRANSPORT SYSTEM   The most widely used inhibitors of photosynthetic electron transport are urea and triazine herbicides. For all the members of these groups, the site and mode of action are the same, but their effectiveness depends on various lipophilic side chains. The most important members of these two herbicide types are diuron (DCMU) and atrazine (2-chloro-4-ethylamino-6-isopropylamino-s-triazine). Both of these herbicides inhibit PSII activity by displacing a bound plastoquinone from the Q_B^- binding site in PSII (Figure 1). Binding studies have shown that atrazine displaces the oxidized form of plastoquinone and occupies the specific binding site for the quinone acceptor, which is thought to lie on the Q_B protein (also known as D1 protein or herbicide binding 32-kDa protein). The herbicide is not able to accept electrons, so the electron is unable to leave Q_A, the first quinone acceptor. The binding of the herbicide, thus, effectively blocks electron flow and inhibits photosynthesis (114, 119, 141, 146, 169). Maize and sorghum are tolerant to triazines (but not to the urea derivatives) because they contain isozymes of glutathione transferase that conjugate these compounds, and thus detoxify them (149, 166).

Phenol-type herbicides are another large group of chemicals that affect PSII activity. They include 2,4-dinitro-6-s-butylphenol (DNOC, also called dinoseb), 3,5-dibromo-4-hydroxybenzonitrile (bromoxynil), and 3,5-diiodo-4-hydroxy-benzonitrile (ioxynil). All these compounds are believed to interact at the same region of PSII as ureas and triazines, although their mechanisms of action are different. Ureas interfere noncompetitively with the specific binding site of the phenolic herbicide iso-butyl-dinoseb, whereas the phenol-type herbicides are competitive inhibitors (122). In PSII there are apparently two binding sites for the ureas and triazines, and one site for phenolic compounds. The urea and triazine herbicides have one binding site in common (146, 169).

Electron flow immediately after the PSII components can be inhibited by benzoquinone analogs [e.g. 2,5-di-bromo-3-methyl-6-isopropyl-p-benzoquinone (DBMIB)], as well as diphenyl ethers, which interfere with oxidation of plastoquinone, whereas plastocyanin function is affected by potassium cyanide and mercuric chloride. These compounds probably interfere with the cytochrome b_6/f complex (129, 141) (Figure 1).

Another group of compounds inhibiting electron transport are those affecting PSI activity (Figure 1). The most important inhibitors to interact with PSI are paraquat and diquat, which belong to the family of bipyridyl compounds (31). Paraquat, also known as methyl viologen, intercepts electrons between the bound ferredoxin acceptors and NADP and then reduces oxygen to superoxide (O_2^-). Superoxide is a free radical that reacts nonspecifically with a wide range of molecules in the chloroplast, leading to rapid loss of chloroplast activity. Lipid molecules in cell membranes are especially sensitive. The lack of $NADP^+$ reduction results in the rapid cessation of CO_2 incorporation. The production of O_2^- via electron transport from ferredoxin to oxygen in chloroplasts [or through different mechanisms in other cell compartments (28)] may occur to a limited extent in vivo depending on the content of reduced ferredoxin that has not been reoxidized by ferredoxin-$NADP^+$ reductase (16, 31). Superoxide radicals generated both as a by-product of normal physiological activities and through the action of bipyridyl compounds are detoxified to

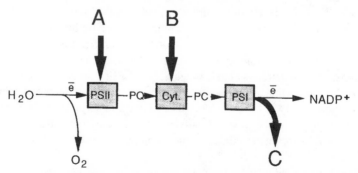

Figure 1   A simplified scheme showing photosynthetic electron flow within the chloroplast from water to the natural acceptor $NADP^+$. (A) site of inhibition by ureas, triazines, and phenol-type herbicides, (B) inhibition of the electron flow through the cytochrome b_6/f complex by diphenyl ethers, (C) site of diversion of electron flow by bipyridyls such as paraquat and diquat. (PQ) plastoquinone, (PC) plastocyanin.

hydrogen peroxide by a chloroplastic isoenzyme of superoxide dismutase. Increased activity of this and other enzymes of oxygen metabolism in transgenic plants have often been correlated with the mechanism of paraquat survival (6, 16, 28, 52).

C_3 Cycle

RIBULOSE-1,5-BISPHOSPHATE CARBOXYLASE/OXYGENASE (RUBISCO)   Rubisco catalyzes the first reactions of the C_3 and C_2 cycles, using ribulose-1,5-bisphosphate (RuBP) and CO_2 or O_2 as substrates. Activity of the enzyme is regulated by several factors, including CO_2 and O_2 concentrations, light, temperature, pH, magnesium, and chloroplastic metabolites. One of the most important metabolite control mechanisms appears to be exerted by a recently discovered tight-binding inhibitor identified as 2-carboxyarabinitol-1-P (CA1P), a naturally occurring analog of a transition state intermediate of the Rubisco reaction (12, 54, 130, 148). CA1P is identical with D-hamamelonic acid-2-phosphate, a member of a naturally occurring family of branched-chain monosaccharides (11). In the darkened leaves, CA1P binds tightly to Rubisco with a dissociation constant of 32 nM (12). Upon illumination of leaves, CA1P is metabolized and its concentration declines. An enzyme present in the leaves of tobacco initially was reported to destroy CA1P by a reaction dependent on NADPH (140). This enzyme later was identified as a phosphatase that hydrolyzed CA1P to carboxyarabinitol (53, 61, 88). CA1P phosphatase is stimulated by NADPH, RuBP, and fructose-1,6-bisphosphate, whereas P_i acts as an inhibitor (62, 139). Carboxyarabinitol, a product of the phosphatase, is present both in species with high and low levels of CA1P (118). Both CA1P and carboxyarabinitol appear to participate in a metabolic substrate cycle in vivo (117).

The removal of the tightly bound CA1P from Rubisco is facilitated by Rubisco activase, a chloroplast regulatory protein, which mediates activation of Rubisco (134, 139, 158). Rubisco activase promotes dissociation of CA1P and other sugar-phosphates [e.g. 6-phosphogluconate, 3-phosphoglycerate (3-PGA)] from the active site of Rubisco, converting the enzyme into a form that has a high affinity for carbamylation (reviewed in 130). Carbamylation of a specific lysyl residue makes Rubisco competent for catalysis. In plants such as bean or potato, at least some part of the regulation of Rubisco appears to be achieved via CA1P, whereas in plants like spinach, which do not produce CA1P, the control is generally by changes in activation (117, 148). Alternatively, given ubiquitous occurence of the activase in plants, the role of CA1P may be limited to regulating metabolite binding to Rubisco in the dark rather than modulating Rubisco activity in the light (reviewed in 139). In plants lacking CA1P (e.g. spinach), there could be other as yet unidentified CA1P analogs, which may have eluded detection because of weaker binding to

Rubisco or because of their lability during isolation. It has been noted that an effective Rubisco inhibitor needs only to bind tightly enough to avoid displacement by RuBP (148).

The discovery of CA1P as the nocturnal inhibitor of Rubisco was preceded by studies on the mechanism of the Rubisco reaction with two related compounds, 2-carboxyarabinitol-1,5-bisphosphate and its keto derivative (5, 84, 111, 145). Crystal structure analysis of the quaternary complex of spinach Rubisco with the former compound (3) provided the evidence of the binding of substrate analogs to the active site of the activated enzyme.

The activation of purified Rubisco (25) and the process of light-dependent activation of Rubisco in intact, lysed, or reconstituted chloroplasts (21) are inhibited by glyoxylate, a metabolite of the C_2 pathway. The inhibitory effect of glyoxylate on Rubisco activation may be of importance under conditions of nitrogen deficiency (21). It is unclear, however, whether this regulation is of physiological significance. The major question is whether glyoxylate is available in the chloroplasts in vivo (47, 93).

Because Rubisco can use either O_2 or CO_2 as a substrate, with both compounds competing for the active site on the enzyme (5, 72), the carboxylation reaction (essential for the operation of the C_3 cycle), can be inhibited by an increase in O_2 concentration. The inhibition of photosynthesis by elevated $[O_2]$ has frequently been used as an indicator of photorespiration. However, the inhibitory effect of O_2 on photosynthesis can also be partially attributed to a generation of oxygen free-radicals within the chloroplasts and to other processes (37, 72). Thus, inhibition of photosynthesis by O_2 may not be a reliable quantitative index of photorespiration.

The possibility of enhancing photosynthesis by modifying Rubisco, so that it has a reduced ratio of RuBP oxygenase to RuBP carboxylase activity, has been discussed frequently (e.g. 123). Enzymes from various plants have different ratios of these two activities (123), which indicates that this parameter is not immutably fixed by the inherent constraints of the chemistry of Rubisco reaction. However, there is substantial uncertainty about whether the enzyme can be modified by genetic engineering methods so that it has a reduced ratio of RuBP oxygenase to RuBP carboxylase activity without any decrease in RuBP carboxylase activity. Many compounds have been used in attempts to alter the ratio of carboxylase to oxygenase activities, but they have either had no effect or inhibited both activities (17). Interpretations of data from some experiments are complex because the alterations often are actually effects on the stability or on the activation state of the Rubisco enzyme (17, 72).

OTHER ENZYMES   The C_3 cycle, which comprises several soluble enzymes, all localized in the chloroplast stroma, can be considered as an autocatalytic pathway where for every three turns of the cycle one molecule of triose-P is generated from three molecules of CO_2. Triose-P could then be used either to regenerate the CO_2 acceptor, RuBP, or to provide carbon skeletons for starch or sucrose synthesis. The activity of the C_3 cycle is thus under several forms of control, including short-term (metabolite regulators) and longer-term (light activation) mechanisms (37).

The most important metabolites, apart from CA1P, that may inhibit photosynthesis by inhibiting components of the C_3 cycle are 2-P-glycolate [target-triose-P-isomerase (1)], glycerate [chloroplastic fructose-1,6-bisphosphatase (FBPase) and sedoheptulose-1,7-bisphosphatase (144)], sedoheptulose-7-P [sedoheptulose-1,7-bisphosphatase (144)], RuBP [sedoheptulose-1,7-bisphosphatase and phosphoribulose kinase (42, 144)], 6-phosphogluconate [phosphoribulose kinase (2, 42)], and ADP [3-PGA kinase and phosphoribulose kinase (37, 42, 126)]. Among nonphysiological compounds, there is a strong irreversible inhibition of phosphoribulose kinase by iodoacetol phosphate, an analog of triose-P (168); and inhibition of phosphoribulose kinase by DL-glyceraldehyde (10, 155), which is one of few compounds actually shown to inhibit specifically dark reactions of photosynthesis (10, 38, 164). The latter compound, besides its effect on phosphoribulokinase, may also block other sites in the conversion of triose-P to RuBP (10, 155). The regenerative phase of photosynthesis is also inhibited by glycolaldehyde (38, 150).

C_4 Cycle

PHOSPHOENOLPYRUVATE CARBOXYLASE (PEPCase)   PEPCase, the primary carboxylation step of the C_4 pathway, catalyzes principally the reaction of carboxylation of phosphoenolpyruvate (PEP) with HCO_3^-. A variety of mechanisms have been postulated for controlling the activity of the enzyme, including allosteric effectors, phosphorylation, and oligomerization (4, 83, 90). In terms of allosteric regulation, plant PEPCase is known to be activated by glucose-6-P and glycine and inhibited by L-malate, oxaloacetate, and aspartate (69, 124). Malate inhibition of the enzyme is complex, with both competitive and noncompetitive components. Chemical modification studies have shown that there is a distinct site responsible for noncompetitive malate inhibition of *Crassula argentea* enzyme (137). Noncompetitive malate inhibition might be associated with a specific aggregation state of the enzyme (116). Depending on storage conditions for isolated PEPCase, the enzyme may exist in distinct aggregation/conformation states, which substantially differ in their kinetic and regulatory properties (90).

Several traditional amino-acid–specific modifying reagents have been tested with respect to their effect on activity of PEPCase (reviewed in 4, 49).

For instance, the enzyme can be inactivated by sulfhydryl reagents, including bromopyruvate (50), p-(chloromercuri)-benzoate (161) or 2-oxo-3-butenoate (175). This last compound probably binds to a different sulfhydryl group than is reactive with bromopyruvate (175). O'Leary and coworkers have synthesized and tested a number of possible *suicide substrates* for PEPCase, including bromophosphoenolpyruvate, 1-carboxyallenyl phosphate, phosphoenolthiopyruvate, and 3-chlorophosphoenolpyruvate. Incubation of maize PEPCase with any of these compounds resulted in a slow inactivation of the enzyme ($t_{1/2}$ = 20 min–5 h) (30, 109, 152). The slow inhibition was caused by accumulated products rather than PEP analogs. Interestingly, analogs of PEP that can serve as substrates for PEPCase reaction undergo hydrolysis rather than carboxylation. The carboxylation reaction occurs only with PEP, which serves as a physiological substrate, principally (109). Given the variety of substrates that have already been tested and for which this abnormal reaction occurs, it is likely that both steric and electronic effects are important in diverting the reaction to the hydrolysis pathway (30). Analogs of PEP have also been synthesized from phosphonomethacrylates (115). Derivatives containing two halogen substituents in the gamma position have been found to be the most potent inhibitors of the enzyme (115).

PEPCase is also very sensitive to inhibition by dodecylbenzenesulfonate, an anionic detergent (174). The inhibition (K_i of about 14 µM) is noncompetitive against both PEP and HCO_3^-. Perhaps the most useful inhibitor used in studies on functional aspects of PEP carboxylase is 3,3-dichloro-2(dihydroxyphosphinoylmethyl)propenoate (DCDP), which inhibits both PEPCase activity (as a competitive inhibitor) and photosynthetic rates, especially with respect to C_4 species (76). In these plants, DCDP produces about 80-100% inhibition of photosynthetic CO_2 assimilation (76, 78, 79). The inhibition could be prevented by high $[CO_2]$ (76), presumably by overcoming the bundle sheath cells' diffusive resistance to CO_2. Inhibition of PEPCase by DCDP limits the operation of the C_4 pathway, thus, effectively preventing the supply of CO_2 to the bundle sheath cells. Although secondary effects of DCDP cannot be ruled out unequivocally, the compound's lack of effects on other PEP-utilizing enzymes (79) or photosynthesis by isolated bundle sheath strands suggests a specific effect on PEPCase. The inhibitory effect of DCDP on C_3 species was usually observed after a long lag period, suggesting secondary sites of inhibition (76). DCDP has been proposed as a convenient tool to assess the contribution of C_4 photosynthesis to net CO_2 uptake among C_3-C_4 intermediates (18).

NADP- AND NAD-MALIC ENZYME   The NADP-malic enzyme, localized in chloroplasts of bundle sheath cells, catalyzes decarboxylation of malate in the so called NADP-malic enzyme C_4 species (36, 56). In addition to NADP-dependent activity, the enzyme can also use NAD as a substrate. This latter activity is potently (K_i of 1 μM) inhibited by NADP or NADPH (57). The reaction product, CO_2, inhibits the enzyme at concentrations likely to occur in bundle sheath cells (about 0.5 mM) and this may provide negative feedback of physiological relevance (77). Structural analogs of NADP have been used successfully to study regions of the $NADP^+$ molecule that are important for binding to the NADP-malic enzyme. The results have suggested that the 2'-phosphate group of $NADP^+$ is crucial for the nucleotide binding to the enzyme, whereas the charge density in the C_4 atom of the pyridine ring is the major factor that governs the coenzyme activity (36, 159).

NAD-malic enzyme is a primary regulatory enzyme for the metabolism of malate in plant mitochondria. The enzyme serves an anaplerotic function for the production of pyruvate, but in certain C_4 and Crassulacean acid metabolism plants it provides CO_2 for refixation in the Calvin cycle, playing a key role in photosynthesis (8, 56, 173). The enzyme is regulated by several compounds, including chloride, citrate, nitrate, NADH, and NADPH, which act as inhibitors; and CoA, fructose-1,6-P_2, and sulfate, which are activators (reviewed in 8, 56, 173). NAD-malic enzyme displays substrate- and ligand-induced changes in oligomeric state that have major kinetic consequences (51). It is not yet clear how some effectors regulate the malic enzyme, as the enzyme may exist in several different forms with respect to its affinity for substrates. For instance, the enzyme from bacteria is inhibited by CoA, while plant NAD malic enzyme is activated by this compound (173).

ADENYLATE KINASE   Adenylate kinase catalyzes the magnesium-dependent conversion of ATP and AMP to ADP, and thus equilibrates adenylate pool(s) in cells (97). The enzyme from C_4 plants is believed to be directly involved in photosynthesis (52) by having its activity coupled to that of pyruvate, P_i dikinase, which produces PEP, the primary substrate of C_4-type of photosynthesis (37, 56). In C_3 plants, adenylate kinase probably is not directly involved in the photosynthetic metabolism (94). The enzyme from maize leaves is potently inhibited by nanomolar concentrations of P^1,P^5-di(adenosine-5')pentaphosphate (97), a compound resembling a transition state analog of the adenylate kinase reaction. The inhibitor binds only to the free enzyme form of adenylate kinase, as evidenced by competitive patterns with all substrates of the reaction. Its K_i values of 11–80 nM (depending on varied substrate) are about 2–3 orders of magnitude lower than the corresponding K_m values for substrates of the enzyme (97). There has been considerable interest in the physiological function of P^1,P^5-di(adenosine-5')pentaphosphate and related dinucleotides that, at least in non-plant tissues, might be involved in DNA replication, and could play an important role in response of organisms to various forms of environmental stresses (9). Unfortunately, no physiological studies on the effect of this inhibitor on plant metabolism have been reported.

C_2 Cycle

GLYCOLATE OXIDASE   Glycolate oxidase, the peroxisomal enzyme, is a flavo-protein catalyzing oxidation of glycolate by O_2 to form glyoxylate and hydrogen peroxide. The enzyme can also use a hydrated form of glyoxylate as a substrate, producing oxalate (72). Oxalate formation might be an unavoidable side reaction of the photorespiratory metabolism (74). Glycolate oxidase is strongly inhibited by butyl-2-hydroxy-3-butynoate, which irreversibly inactivates the enzyme by modifying its FMN prosthetic group (82). Hydroxysulphonates, especially α-hydroxy-2-pyridinemethane sulfonic acid (HPMS), have often been described as inhibitors of glycolate oxidase (72, 178). Other inhibitors of the enzyme include oxalate, mesooxalate, oxaloacetate, malonate, iodoacetate, hydroxylamine, and linoleic acid (cited in 72).

Inhibitors of glycolate oxidase have often been used in attempts to inhibit the C_2 cycle (reviewed in 72, 93). In most cases, both C_2 and C_3 pathways were inhibited. HPMS had been suggested to effectively stimulate rates of photosynthesis in leaf discs (178), but subsequent reports from other laboratories have indicated that this compound acts as a strong inhibitor of photosynthesis and may exert multiple nonspecific effects on plant metabolism (e.g. 72, 87, 110). At low doses, HPMS exerts its effect first on glycolate oxidation in the peroxisomes (inhibiting photorespiration) and then, after a lag phase, it inhibits CO_2-uptake-related processes. With increasing concentrations of the effector, the lag phase shortens or disappears and photosynthesis becomes even more inhibited than photorespiration (110). Indeed, continued loss of carbon from the Calvin-Benson cycle, caused by blockage of the C_2 pathway, would be expected eventually to decrease the concentration of cycle intermediates and to inhibit CO_2 fixation.

CATALASE   Activity of catalase, both in vivo and in vitro, is inhibited by aminotriazole (40, 59). Isozymes of catalase may substantially differ in their sensitivity to this inhibitor, the sensitive form(s) having the K_i for aminotriazole of about 0.1 mM. The aminotriazole-resistant activity is probably caused by the form of catalase with enhanced peroxidatic activity, which has impaired inhibitor-binding site (59).

Under conditions of rapid photosynthesis by intact leaves, at high temperatures and low CO_2 (or high O_2), if there is insufficient catalase activity to react with the hydrogen peroxide, excess peroxide may rapidly decarboxylate keto-acids, such as glyoxylate or hydroxypyruvate, to generate CO_2 (180). Although this reaction may effectively increase rates of photorespiratory CO_2

release and contribute to lower photosynthetic rates, under normal physiological conditions the decarboxylation of glyoxylate is believed to occur only to a minimal extent in vivo (e.g. 72, 93).

AMINOTRANSFERASES   Conversion of glyoxylate to glycine in the C_2 pathway requires a coordinate action of two peroxisomal glyoxylate-dependent aminotransferases [serine:glyoxylate and glutamate(alanine):glyoxylate aminotransferases]. Both enzymes apparently have pyridoxal phosphate as a coenzyme and are inhibited by several compounds, including structural analogs of the preferred amino acid donor substrate. Ammonia and amino acids usually serve as competitive inhibitors of the aminotransferases (reviewed in 44, 72). Potent inhibition of aminotransferase activities is observed with aminooxyacetate and hydroxylamine, both compounds known to inhibit enzymes that require pyridoxal phosphate as a coenzyme (19, 72, 80). The two glyoxylate-dependent aminotransferases are irreversibly inhibited by carbinolamine, a serine analog (58). Glycidate inhibits glutamate:glyoxylate aminotransferase (179), and vinylglycine strongly inhibits serine:glyoxylate aminotransferase (22).

GLYCINE DECARBOXYLASE AND SERINE HYDROXYMETHYLTRANSFERASE   The interconversion of glycine and serine in plant mitochondria involves the enzymatic system of the glycine decarboxylase complex and serine hydroxymethyltransferase. These reactions account for the loss of CO_2 in the C_2 cycle (72, 93). The CO_2 release is specifically blocked by aminoacetonitrile (26, 167). Aminoacetonitrile was, at one time, thought to effectively stimulate the rate of photosynthesis in intact leaves (176). Subsequent studies, however, demonstrated that aminoacetonitrile is a competitive inhibitor of glycine oxidation by mitochondria, but has no effect on photosynthesis of isolated wheat chloroplasts (167). In wheat leaf protoplasts, aminoacetonitrile decreased CO_2 assimilation under photorespiratory conditions (low bicarbonate, 0.3 mM) but not under nonphotorespiring conditions (high bicarbonate, 10 mM). Thus, specific blockage of the glycolate pathway under photorespiring conditions leads to an inhibition of photosynthesis. Failure to recycle glycolate-derived carbon to the chloroplast in the form of glycerate, or feedback inhibition of photosynthesis by product accumulation in the glycolate path may account for the inhibitory effect (167). These results are consistent with other studies in which the blockage of the glycolate pathway, either by inhibitors or by a mutation, caused inhibition of photosynthesis under photorespiratory conditions (23, 72, 156).

The glycine to serine conversion in isolated mitochondrial fractions is also inhibited by aminooxyacetate, methoxylamine, acethydrazide, isonicotinyl-hydrazide, hydroxylamine, glycine hydroxamate, vinylglycine, and other glycine and serine analogs (22, 34, 43, 142). Depending on plant species and tissue, these compounds affect both or only one of the components of the glycine to serine conversion in mitochondria (i.e. glycine decarboxylase and serine hydroxymethyltransferase). Among physiological effectors, inhibition by NADH (K_i of 10 µM) is of significance for the regulation of glycine decarboxylase activity in vivo (15).

GLYOXYLATE AND HYDROXYPYRUVATE REDUCTASES   The reduction of glyoxylate and hydroxypyruvate in leaves is now known to be carried out by at least three reductases having different specificities for cofactor and for glyoxylate and/or hydroxypyruvate. Activity of NADPH-preferring hydroxypyruvate reductase (cytosolic) is strongly inhibited by oxalate, an end product of carbon metabolism in plants, with micromolar K_i values (92, 96). The enzyme is also selectively inhibited by tartronate and phosphohydroxypyruvate, with K_i values of less than 0.4 mM (91, 96). Activity of NADPH-preferring glyoxylate reductase (cytosolic) is inhibited by acetohydroxamate (K_i of 0.3 mM) and other hydroxamates (95) and by glycidate (95, 179). It is also nonspecifically inhibited by aminooxyacetate (95) similarly to the peroxisomal NADH-preferring hydroxypyruvate reductase (80, 95), which is thought to represent the major hydroxypyruvate-reducing activity during photorespiration in leaves. Glyoxylate, besides serving as a substrate in the transamination and reduction reactions, can also be used in several other pathways, e.g. leading to malate, tartrate, or formate synthesis (72, 73). The magnitude of these processes may depend on plant species, tissue, and type of metabolism, but they probably occur only to a limited extent in leaves.

Starch Biosynthesis

The major regulatory reaction of starch biosynthesis is that of ADP-glucose pyrophosphorylase (AGPase), a two-gene encoded enzyme, catalyzing the first committed step of starch production in all tissues. AGPase from both photosynthetic and nonphotosynthetic plant tissues (and from cyanobacteria) is potently inhibited by P_i and activated by 3-PGA, which is the common product of the C_3 and C_2 pathways (98, 131). Interestingly, for AGPase from leaves of *Xerosicyos dangui* and *Hoya carnosa,* a low concentration of 3-PGA was required for maximal inhibition with P_i (153), while for AGPase from barley leaves, P_i was required for maximal activation with 3-PGA (100). In most cases, binding of 3-PGA caused an increase in the interactions between inhibitor sites, and binding of P_i increased the interactions between activator sites. The binding sites for 3-PGA are localized close to the C-termini for both subunit-types of spinach leaf AGPase, but their amino acid sequences may differ (131). The sites responsible for P_i binding have yet to be identified, but they probably overlap with those responsible for 3-PGA binding, as shown by a recent study of the P_i-insensitive AGPase from barley seed endosperm (99).

A low sensitivity, if any, to P_i inhibition for this enzyme was accompanied by insensitivity to 3-PGA activation. The K_m and V_{max} values of the endosperm AGPase, when assayed in the absence of 3-PGA, were comparable to those of the 3-PGA-activated enzyme from different species and tissues. The K_i values with P_i were at least an order of magnitude higher for barley endosperm AGPase when compared to those of the enzyme from barley leaves (99, 100). Although the molecular basis for this apparent insensitivity of barley seed AGPase to the 3-PGA/P_i regulation is unknown at present, it is probably related to an apparently extraamyloplastic localization of this enzyme, as found by immunocytochemical approaches (LA Kleczkowski, unpublished data).

Physiological evidence supporting the importance of P_i in regulation of starch synthesis comes from a series of experiments from Walker's laboratory (e.g. 37, 154) in which P_i was sequestered in the form of a mannose derivative. In leaves, mannose is converted in the cytosol to a phosphate ester, which cannot be further metabolized in the cell. Feeding of mannose to leaf discs resulted in up to a 15-fold increase in the starch content. In related studies, plants growing in a P_i-deficient medium showed up to 10-times higher starch content than control plants. These experiments clearly indicate the major role for P_i in regulating starch formation in vivo, supporting the notion that AGPase is an important regulatory step in starch formation in leaves.

Pyrophosphate, the product of AGPase reaction, is another inhibitor that may be of physiological significance for starch production (100, 131). In leaves, there is apparently an appreciable pool of pyrophosphate, which—in addition to its regulatory role—is also a storage form of P_i and an energy source (13). Plant AGPase is also strongly inhibited (K_i of 20 μM) by chromium-ATP, a general competitive inhibitor of ATP-utilizing enzymes (100).

Nitrogen Assimilation

Nitrogen assimilation constitutes, besides CO_2 fixation, a major function of a leaf cell. There is now strong evidence that the major pathway of assimilation for the ammonia produced in leaf tissues by nitrite reductase and by glycine decarboxylase is its conversion into glutamine, followed by glutamate formation. Biosynthesis and metabolism of these and some other amino acids (glycine, serine, alanine, aspartate, and asparagine) in plants are closely associated with photosynthetic processes (46, 86, 104).

Nitrogen assimilation in leaves is now believed to be carried out by the closely coupled reactions of glutamine synthetase and ferredoxin-dependent glutamate synthase, resulting in the incorporation of ammonia into glutamate (46, 86). Glutamine synthetase, the primary target for inhibitors of nitrogen assimilation, is inhibited at micromolar levels by methionine sulphoximine.

Inactivation of the enzyme triggers rapid metabolic changes, including accumulation of free ammonia and a decrease in contents of most, but not all, amino acids (105, 133, 160). In excised wheat leaves, methionine sulphoximine stimulated the deamination and decarboxylation of glutamate to succinate, malate, and CO_2; and it decreased the incorporation of $^{14}CO_2$ into glycine, serine, alanine, glutamate, and aspartate, presumably because of the lack of available amino donors (171, 172). The methionine sulphoximine–induced accumulation of ammonia is usually O_2 concentration-dependent and can be blocked by inhibitors of the C_2 pathway (113, 172), indicating the key role of photorespiration in ammonia production (46, 86). In some fungi or green algae grown on high concentrations of ammonia, methionine sulphoximine has no considerable effect owing to operation of glutamate dehydrogenase (105).

Glutamine synthetase is also potently inhibited by L-phosphinothricin [2-amino-(methylphosphinyl)-butanoic acid] (glufosinate) (reviewed in 105, 112), an analog of L-glutamic acid, which was originally discovered as a component of bialaphos, an antibiotic produced by *Streptomyces* species. Bialaphos is a tripeptide comprised of phosphinothricin and two residues of L-alanine. Inhibition of glutamine synthetase by phosphinothricin causes a rapid buildup of intracellular ammonia levels and an associated disruption of chloroplast structure, resulting in the inhibition of photosynthesis and plant cell death (143). Although ammonia toxicity appears to be the primary mode of action of phosphinothricin, ammonium may not be the main cause of cell death in the presence of the inhibitor (e.g. 103, 127). In addition to ammonia toxicity, the inhibitory effect of phosphinothricin on photosynthesis may be explained by inhibition of protein synthesis (especially Q_B protein involved in electron transport), toxic glyoxylate accumulation due to shortage of amino donors, or insufficient regeneration of intermediates of the C_3 cycle (103, 143).

Phytotoxic characteristics of plants treated with inhibitors of glutamine synthetase are remarkably similar to those of glutamine synthetase-deficient mutant plants exposed to air (14, 105, 156). Plants treated with the inhibitor are unable to fix CO_2 at normal rates in air and rapidly show severe signs of stress and, eventually, necrosis. Strategies for developing phosphinothricin-resistant plants involve isolation of a phosphinothricin-resistant gene from herbicide-resistant cells, an expression of an enzyme that detoxifies phosphinothricin, or overexpression of glutamine synthetase protein by amplification of its gene (reviewed in 89, 103, 114, 146).

Ferredoxin-dependent glutamate synthase, the other key activity of the photosynthetic nitrogen assimilation cycle (46, 86), is potently inhibited by azaserine (O-diazoacetylserine) (103, 160). This compound might have potential as a herbicide, since inhibition of glutamate synthase has an even more toxic effect on plants than blockage of the glutamine synthetase step (103).

Mutants deficient in glutamate synthase accumulate glutamine, analogously to the massive buildup of ammonia concentration in glutamine synthetase-deficient plants (14, 85, 156, 157).

Fatty Acid Biosynthesis

Plastids contain acetyl-CoA synthetase, acetyl-CoA carboxylase, and fatty-acid synthetase, key components of fatty-acid synthesis in plants. Inhibitors that affect any of these enzymatic steps (Figure 2) can block glycerolipid and phospholipid synthesis, which results in inhibition of membrane formation. Most affected are young developing leaves and meristematic tissues, which depend on an efficient fatty acid supply (106, 108).

Acetyl-CoA synthetase is a target for allicin, a naturally occurring antibiotic from garlic (41). Allicin also affects other SH-group containing en-

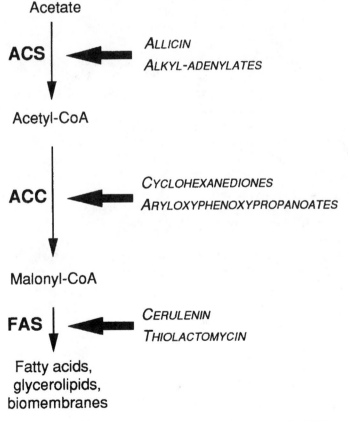

Figure 2 Key steps in fatty acid metabolism that are the primary targets for inhibitors. Arrows indicate sites of inhibitor action. (ACS) acetyl-CoA synthetase, (ACC) acetyl-CoA carboxylase, (FAS) fatty acid synthetase cycle.

zymes (107). Several alkyl-adenylates recently have been reported as potent (K_is of few micromolar or less) inhibitors of acetyl-CoA synthetase from radish seedlings (48). They are competitive against ATP and noncompetitive with respect to acetate. Acetyl-adenylate has been proposed as an intermediate in the enzymic formation of acetyl-CoA from acetate by acetyl-CoA synthetase (48).

Acetyl-CoA carboxylase is the major target of inhibitors affecting fatty acid biosynthesis in plants. The enzyme produces malonyl-CoA, which is a key intermediate in both fatty acid and flavonoid (gibberellin, abscisic acid, carotenoids, and other isoprenoids) biosynthesis. In plants, acetyl-CoA, the substrate for the enzyme, is derived directly from acetate via the action of acetyl-CoA synthetase or from pyruvate via pyruvate dehydrogenase. In plastids, both of these mechanisms take place (45, 177). In the absence of any other means of synthesizing malonyl-CoA, inhibition of acetyl-CoA carboxylase would deprive the plant of a key intermediate essential to both lipid and flavonoid biosynthesis and would lead to phytotoxic effects. Acetyl-CoA carboxylase is inhibited by several substituted 1,3-cyclohexanediones (e.g. alloxydim, sethoxydim, and clethodim) and various derivatized phenoxypropionic acids (diclofop, haloxyfop, and trifop) (Figure 2). Both groups of inhibitors show selectivity for grass species (monocots), whereas broadleaf species (dicots) are tolerant to the inhibitors. Cyclohexanediones and phenoxypropionic acids probably bind to the same region of the target enzyme but they occupy different binding sites (106). These compounds inhibit lipid and/or flavonoid biosynthesis in susceptible species (24, 64, 65), reflecting the key role of acetyl-CoA carboxylase in these processes (20, 101, 132, 147). In some plants, more than one isoenzyme of acetyl-CoA carboxylase are present, localized within and outside of chloroplasts, with different sensitivities to the inhibitors (39). The differences in sensitivity of particular grasses toward diclofop and cycloxydim may be the result of a domain in the acetyl-CoA carboxylase-peptide that is modified to prevent herbicide binding (107). Mutant plants of maize with an altered herbicide-resistant acetyl-CoA carboxylase activity have been isolated by selection of regenerable maize cells showing tolerance to sethoxydim application (128).

The fatty-acid synthetase cycle consists of several proteins/enzymes with the acyl carrier protein and the condensing enzymes as essential constituents. Two antibiotics, cerulenin and thiolactomycin, and their derivatives are known to inhibit enzymes of the cycle (reviewed in 41). Given the variety of protein components of the fatty acid synthetase cycle, it appears inevitable that more inhibitors resembling substrates/products of the partial reactions will be uncovered.

As pointed out by Lichtenthaler (106), intact plastids might be used as an efficient test system to study effects of inhibitors of fatty acid biosynthesis.

The K_i values found for inhibitors of acetyl-CoA carboxylase using this partly in vivo system were about 5–10 fold lower than those obtained with the isolated enzyme. The use of plastids to study inhibitory characteristics seems the most reliable system to identify biotypes with different sensitivities or tolerance to the inhibitors (106).

INHIBITORS OF ASSIMILATE EXPORT FROM CHLOROPLASTS

Studies of metabolite transport inhibition in and out of chloroplasts have focused on the phosphate translocator of the inner chloroplast envelope. The phosphate translocator facilitates an exchange of triose phosphates made in chloroplasts with cytoplasmic P_i, 3-PGA, and (in C4 plants) PEP (60). The metabolites transported on the phosphate translocator show competitive interactions, which is consistent with a common binding site for these compounds (60). The translocator is potently (micromolar levels) and irreversibly inhibited by 4,4'-diisothiocyanatostilbene-2,2'-disulfonic acid (DIDS). When applied to crabgrass mesophyll chloroplasts, DIDS inhibited 3-PGA-dependent O_2 evolution (135). The inhibition could be prevented by preincubation of chloroplasts with 3-PGA, PEP, or P_i before addition of DIDS, indicating that this compound binds to the substrate binding site on the phosphate translocator protein. DIDS was found to react specifically with a 29-kDa polypeptide in the envelope of C3 and C4 mesophyll chloroplasts (136). DIDS was also found to be a potent inhibitor of starch synthesis in isolated cauliflower plastids, probably reflecting dependence of plastidic AGPase on activation by 3-PGA, which is imported to nongreen plastids from the cytosol (121).

Chloroplast phosphate translocator is also inhibited by pyridoxal phosphate (136), but not by iodoacetol phosphate, an analog of triose phosphate (168). Glyphosate, an inhibitor of the shikimate pathway in chloroplasts (162), recently was found to be a potent inhibitor of P_i transport into protoplasts of bean leaves (29). Because these data may reflect the activity of the plasma membrane phosphate translocator rather than the chloroplast phosphate carrier, it is still unclear whether the latter can also be inhibited by glyphosate application.

Activity of phosphate translocator and, consequently, photosynthetic rates depend on provision of external P_i (37, 154). Without sufficient P_i supply, photosynthesis is inhibited both in isolated chloroplasts and in intact plants (60, 75, 154). Feeding P_i to leaf tissue can increase the rate of photosynthesis (154), indicating that the supply of P_i to chloroplasts could be a limiting factor under normal conditions. The P_i-dependent inhibition of photosynthesis may also reflect an important regulatory role of this compound in several metabolic pathways in leaves (70, 75, 89, 131).

Another transport system extensively studied using inhibitors is the glycolate/glycerate exchange across the chloroplast membranes during C_2 pathway. Studies by Howitz & McCarty (e.g. 66, 67) have indicated that this transporter is distinct from phosphate and dicarboxylate translocators operating during photosynthesis. N-ethylmaleimide strongly inhibits glycolate or glycerate transport in pea chloroplasts in a substrate protectable manner (67). The translocator is also inhibited by L-phenylglycolic acid, lactate, and glyoxylate. These compounds may inhibit transport by diminishing a proton gradient across the chloroplast membranes (66, 67). The transport is believed to be driven by either a proton symport or a hydroxyl antiport.

INHIBITORS OF SUCROSE BIOSYNTHESIS

Sucrose (formed in the cytosol) is the other major product of photosynthesis in most plants. Both starch and sucrose are produced from photosynthetically generated triose-P via pathways that are in some respects similar to the gluconeogenesis pathway of animal cells. Flux control is shared among key enzymes of the pathway, including especially FBPase and sucrose phosphate synthase (27, 55, 163).

The cytosolic FBPase occupies a strategic site in metabolism, catalyzing the first irreversible reaction during the conversion of triose-P to sucrose in the cytosol. The enzyme is potently inhibited by fructose-2,6-bisphosphate (Fru-2,6-P$_2$) (Figure 3), which is effective at micromolar concentrations (68, 163). Physiological concentration of Fru-2,6-P$_2$ is usually in the 1–10 µM range,

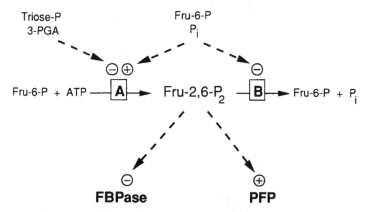

Figure 3   Fru-2,6-P$_2$ as the key regulator of carbon partitioning, an outline. (A) Fru-6-P,2 kinase, (B) Fru-2,6-bisphosphatase, (PFP) pyrophosphate-dependent phosphofructokinase. Regulatory effects of Fru-2,6-P and photosynthetic metabolites on enzymatic activities are denoted as (−) for inhibition or (+) for activation.

although this could be an overestimate owing to binding of the effector to target proteins (163). Levels of Fru-2,6-P$_2$ may fluctuate considerably depending on light/dark conditions and tissue type (13). The inhibitor increases K$_m$ values of FBPase (about 100-fold) and induces sigmoidal substrate saturation kinetics. It also increases the enzyme's sensitivity to other inhibitors, such as AMP and P$_i$. Because of its potent inhibition of cytosolic FBPase, Fru-2,6-P$_2$ will inhibit carbon flow of CO$_2$ fixed by Rubisco toward sucrose synthesis. Conversely, a decrease in Fru-2,6-P$_2$ concentration will promote sucrose synthesis from triose-P (Figure 3).

Fru-2,6-P$_2$ is formed from fructose-6-P (Fru-6-P) and ATP by a specific kinase, and is degraded by a phosphatase activity (Figure 3) (27, 163). The kinase, which forms Fru-2,6-P$_2$, as well as phosphatase, which hydrolyzes the phosphate group in the 2- position, have a number of regulatory properties. Plant Fru-6-P,2 kinase is activated by Fru-6-P and P$_i$ and inhibited by 3-PGA and dihydroxyacetone phosphate; whereas Fru-2,6-bisphosphatase is inhibited by P$_i$, Fru-6-P, and Mg^{2+}. Thus, an extensive export of triose-P from chloroplasts to the cytosol would lower the Fru-2,6-P$_2$ concentration (by inhibiting Fru-6-P,2 kinase), promoting the use of photosynthate for sucrose synthesis by cytosolic FBPase. On the other hand, increased concentrations of P$_i$ or Fru-6-P (e.g. at high sucrose levels) would raise the Fru-2,6-P$_2$ concentration, and thus inhibit sucrose synthesis and/or stimulate starch production (102, 163). In such a way, Fru-2,6-P$_2$ can integrate the carbon metabolism of leaves, regulating carbon flow between the chloroplasts and the cytosol. The regulatory role of Fru-2,6-P$_2$ appears to be of crucial importance for control of carbon partitioning for starch (chloroplasts) and sucrose (cytosol) synthesis (163). Fru-2,6-P$_2$ is also a potent activator of pyrophosphate-dependent phosphofructokinase (13, 27, 138).

In addition to the FBPase reaction controlled by Fru-2,6-P$_2$, the sucrose synthesis is regulated at the sucrose phosphate synthase site (27, 55, 70). This enzyme is under both coarse (covalent modification) and fine control (70). The latter consists of regulation by the ratio of glucose-6-P (activator) and P$_i$ (inhibitor). Glucose-6-P increases V$_{max}$ of sucrose phosphate synthase and its affinity for Fru-6-P. The activation of the enzyme by glucose-6-P is countered by P$_i$. Both glucose-6-P and P$_i$ probably compete for a regulatory site on the enzyme, and this control accounts for most of the inhibitory effect of P$_i$ (33, 70). The regulatory site apparently contains a sulfhydryl group, which is inactivated in the absence of reductant, resulting in a loss of regulatory properties but not of catalytic activity (33). The metabolic control of sucrose phosphate synthase activity provides a mechanism where accumulation of hexose phosphates, which is indicative of an increased activity of the cytosolic FBPase, can increase the rate of sucrose synthesis.

PHOTOSYNTHETIC INHIBITORS AND HERBICIDE RESEARCH

Inhibitors of plant metabolism frequently have been used commercially as herbicides to kill weeds selectively. Most herbicides are directed against the photosynthetic process, which is essential for autotrophic growth. Other primary targets for herbicide action are plant cell growth and development processes (32), and the biosynthesis of lipids (101), amino acids (81, 89, 162), and pigments (35, 120). The critically important feature of herbicides is their ability to distinguish between crop and weed species. Selectivity of the action of herbicides is usually based on differential herbicide uptake between weed and crop, or controlled timing and site of application, or detoxification of the herbicide by the crop plant. Genetic modification of any of these steps may affect the selectivity of a given herbicide, making it less or more effective (63, 114, 146). In some cases, application of a second chemical that reduces the toxicity of the herbicide to the crop has helped attain selectivity of the herbicide effect (114). Screening of plants and other organisms is likely to reveal some new enzymes that can be useful in either catabolizing or conjugating herbicides to reduce their toxic effects. Introduction of genes encoding these proteins to crop species is one of the most important tasks for current and future agriculture research (114, 146).

In recent years, several naturally occurring weed biotypes have developed herbicide resistance via changes in the properties of target proteins (7, 63, 114). For instance, atrazine-resistant weeds have a much reduced, typically 1000-fold lower, binding capacity for atrazine, mostly because of a single mutation in the chloroplastic gene encoding D1 protein (63, 114). In the nonmutants, this polypeptide binds both Q_B and triazines; the mutant D1 still binds Q_B so that electron transport can proceed, but it no longer binds triazines or substituted ureas. Mechanisms of resistance to herbicides other than atrazine are rather poorly understood, even though this knowledge may assist substantially in the identification of genes that confer enhanced resistance and in their introduction to crop species. Rather than understanding herbicide resistance in weeds, most of the relevant research is concerned with the search for new herbicide-resistant crop cultivars (63).

Most of the compounds used as herbicides resulted from random screening procedures (151). Several thousand chemicals may have to be examined to find a new herbicide that merits commercialization. This random screening approach could be improved with an understanding of the biochemical basis for herbicide action. Information on how inhibitors interact with their target sites, both in vitro and under physiological conditions, should be useful in a rational design of new specific herbicides. Some of these compounds could owe their selectivity to differences in major metabolic processes between weeds and crops. For instance, one of the challenges for future herbicide research may be the design of an effective inhibitor of the C_4 metabolic pathway. Because many weeds are C_4 plants (37), inhibitors affecting the C_4 pathway might be invaluable to agricultural weed control. Another major task may be the design of a chemical that could affect Rubisco inactivation in species producing CA1P (148). Such a compound, which could inhibit breakdown of CA1P (or stimulate its synthesis) in leaves, would be expected to impair photosynthesis and plant growth. Recent developments in the biochemical and kinetic characterization of the C_4 photosynthetic pathway and Rubisco should provide necessary background for such studies. Acceleration of new herbicide development through better understanding of biochemistry and molecular biology of photosynthesis-related processes will certainly have a decisive impact on future applications of herbicide research to agriculture.

CONCLUSIONS

During the past 5–10 years, research on action of inhibitors of photosynthetic enzymes has yielded much detailed information about catalytic mechanisms and regulation of target proteins and has helped to evaluate metabolite control. Furthermore, our understanding of the photosynthetic process has been greatly enhanced by the discovery of naturally occurring inhibitors, such as CA1P and Fru-2,6-P_2, and by the elucidation of their role as primary regulators of photosynthetic carbon metabolism. Inhibitors were also invaluable in our understanding of functional groups involved in the catalytic activity of a given enzyme (e.g. work on PEPCase). Apart from purely academic interest in elucidating photosynthetic mechanisms, chemical inhibitors are now frequently used to modulate physiological processes for agricultural benefit.

An important question regarding the use of inhibitors, both in basic and applied research, concerns their specificity of action. There is now increased evidence of isozymes that may show selectivity toward a given inhibitor [e.g. glyoxylate-utilizing leaf reductases (47), or isozymes of acetyl-CoA carboxylase (39) and catalase (59)]. The question of whether known inhibitors of specific enzymes actually inhibit the target protein in vivo is also inherently difficult to answer, but the issue is important both in terms of metabolic regulation and for herbicide research. In fact, the usefulness of a very effective inhibitor may be nullified by defects in its uptake and translocation, or by the presence of enzymes able to metabolize the inhibitor, rendering it biologically inactive.

One of the most important reasons for searching for specific inhibitors is to evaluate the role of a given enzyme in a pathway or to rationalize the importance of a pathway in a physiological response of a plant. Recent developments in isolating mutants lacking a given protein or using antisense DNA technology to inhibit expression of a given gene have, to some extent, diminished the value of inhibitors. However, because of major differences in photosynthetic metabolism between plant species (e.g. C_3, C_4, and Crassulacean acid metabolism plants), a specific inhibitor still represents a powerful (and cheap!) tool to study protein function. In C_4 plants, for instance, inhibitors affecting the C_4 pathway could be useful in elucidating the complexity of the morphological and biochemical adaptations associated with the syndrome. Specific inhibitors of enzymes of the C_2 pathway for which no mutants have been isolated, e.g. glycolate oxidase or glycerate kinase, should still be instrumental in verifying the role of these enzymes in photosynthetic metabolism. Inhibitors could also be used to verify results obtained with mutant plants (and vice versa), as demonstrated for key activities of the C_2 and ammonia assimilation pathways (14, 72, 91, 156).

Inhibitors affecting photosynthetic processes include a diversity of chemical classes from very simple, like oxygen or ammonia, to more complex compounds, like P^1,P^5-di(adenosine-5')pentaphosphate or HPMS. Most inhibitors are analogs of substrates/products of a given protein target, but for many (e.g. P_i inhibition of AGPase, or HPMS inhibition of glycolate oxidase) there is no obvious rationale for their effectiveness. More studies are needed, particularly those using biochemical and molecular biology approaches, to examine metabolite regulation (e.g. chemical and/or site-directed mutagenesis modifications of amino acid sequences responsible for effector-binding), to comprehend the modes of regulation and mechanisms underlying effects of inhibitors of photosynthetic reactions. These basic research studies should be of direct benefit for rational design of commercial herbicides and other compounds in efforts to control photosynthetic events for agricultural, environmental, and industrial purposes.

ACKNOWLEDGMENTS

The author would like to thank Drs. N. R. Baker, G. E. Edwards, P. J. Lea, and M. E. Salvucci for helpful comments and suggestions during preparation of the manuscript.

Literature Cited

1. Anderson LE. 1971. Chloroplast and cytoplasmic enzymes. II. Pea leaf triose phosphate isomerases. *Biochim. Biophys. Acta* 235:237–44
2. Anderson LE. 1973. Regulation of pea leaf ribulose-5-phosphate kinase activity. *Biochim. Biophys. Acta* 321:484–88
3. Andersson I, Knight S, Schneider G, Lindqvist Y, Lundqvist T, et al. 1989. Crystal structure of the active site of ribulose-

bisphosphate carboxylase. *Nature* 337: 229–34

4. Andreo CS, Gonzales DH, Iglesias AA. 1987. Higher plant phosphoenolpyruvate carboxylase. Structure and regulation. *FEBS Lett.* 213:1–8

5. Andrews TJ, Lorimer GH. 1987. Rubisco: structure, mechanisms, and prospects for improvement. In *The Biochemistry of Plants,* ed. MD Hatch, NK Boardman, 10: 131–218. New York: Academic

6. Aono M, Kubo A, Saji H, Tanaka K, Kondo N. 1993. Enhanced tolerance to photooxidative stress of transgenic *Nicotiana tabacum* with high chloroplastic glutathione reductase activity. *Plant Cell Physiol.* 34: 129–35

7. Arntzen CJ, Ditto CL, Brewer PE. 1979. Chloroplast membrane alterations in triazine-resistant *Amaranthus retroflexus* biotypes. *Proc. Natl. Acad. Sci. USA* 76: 278–82

8. Artus NN, Edwards GE. 1985. NAD-malic enzyme from plants. *FEBS Lett.* 182:225–33

9. Baker JC, Jacobson MK. 1986. Alteration of adenyl dinucleotide metabolism by environmental stress. *Proc. Natl. Acad. Sci. USA* 83:2350–52

10. Bamberger ES, Avron M. 1975. Site of action of inhibitors of carbon dioxide assimilation by whole lettuce chloroplasts. *Plant Physiol.* 56:481–85

11. Beck E, Scheibe R, Reiner J. 1989. An assessment of the Rubisco inhibitor. 2-Carboxyarabinitol-1-phosphate and D-hamamelonic acid 2-phosphate are identical compounds. *Plant Physiol.* 90:13–16

12. Berry JA, Lorimer GH, Pierce J, Seeman, JR, Meek J, Freas S. 1987. Isolation, identification, and synthesis of 2-carboxyarabinitol 1-phosphate, a diurnal regulator of ribulose bisphosphate carboxylase activity. *Proc. Natl. Acad. Sci. USA* 84:734–38

13. Black CC, Carnal NW, Paz N. 1985. Role of pyrophosphate and fructose-2,6-bisphosphate in regulating plant sugar metabolism. In *Regulation of Carbon Partitioning in Photosynthetic Tissue,* ed. RL Heath, J Preiss, pp. 45–62. Baltimore: Waverly

14. Blackwell RD, Murray AJS, Lea PJ, Kendall AC, Hall NP, et al. 1988. The value of mutants unable to carry out photorespiration. *Photosynth. Res.* 16:155–76

15. Bourguignon J, Neuburger M, Douce R. 1988. Resolution and characterization of the glycine-cleavage reaction in pea leaf mitochondria. *Biochem. J.* 255:169–78

16. Bowler C, Van Montagu M, Inzé D. 1992. Superoxide dismutase and stress tolerance. *Annu. Rev. Plant Physiol. Plant Mol. Biol.* 43:83–116

17. Boyle FA, Keys AJ. 1984. Oxygenase and carboxylase activities of RuBP carboxylase from wheat leaves. In *Advances in Photosynthesis Research,* ed. C Sybesma, 3:771–74. The Hague: Nijhoff/Junk

18. Brown HR, Byrd GT, Black CC. 1991. Assessing the degree of C_4 photosynthesis in C_3-C_4 species using an inhibitor of phosphoenolpyruvate carboxylase. *Plant Physiol.* 97:985–89

19. Brunk DG, Rhodes D. 1988. Amino acid metabolism of *Lemna minor* L. III. Responses to aminooxyacetate. *Plant Physiol.* 87:447–53

20. Burton JD, Gronwald JW, Somers DA, Connelly JA, Gengenbach BG, Wyse DL. 1987. Inhibition of plant acetyl-coenzyme A carboxylase by the herbicides sethoxydim and haloxyfop. *Biochem. Biophys. Res. Commun.* 148:1039–44

21. Campbell WJ, Ogren WL. 1990. Glyoxylate inhibition of ribulosebisphosphate carboxylase/oxygenase activation in intact, lysed, and reconstituted chloroplasts. *Photosynth. Res.* 23:257–68

22. Cho C, Ishii R, Hyeon S-B, Suzuki A. 1987. Inhibition of serine:glyoxylate aminotransferase and mitochondrial glycine oxidation in the photorespiratory glycolate pathway by vinylglycine. *Agric. Biol. Chem.* 51: 2597–98

23. Cho C, Sugimoto Y, Kim J-M, Usuda H, Ishii R, et al. 1983. Search for photorespiration inhibitors: glycine and serine derivatives. *Agric. Biol. Chem.* 47:2685–87

24. Cho H-Y, Widholm JM, Slife FW. 1986. Effects of haloxyfop on corn and soybean cell suspension cultures. *Weed Sci.* 34:496–501

25. Cook CM, Mulligan RM, Tolbert NE. 1985. Inhibition and stimulation of ribulose-1,5-bisphosphate carboxylase/oxygenase by glyoxylate. *Arch. Biochem. Biophys.* 240: 392–401

26. Creach E, Stewart CR. 1982. Effects of aminoacetonitrile on net photosynthesis, ribulose-1,5-bisphosphate levels, and glycolate pathway intermediates. *Plant Physiol.* 70:1444–48

27. Cseke C, Buchanan BB. 1986. Regulation of the formation and utilization of photosynthate in leaves. *Biochim. Biophys. Acta* 853:43–63

28. del Rio LA, Sandalio LM, Palma JM. 1990. A new cellular function for peroxisomes related to oxygen free radicals. *Experientia* 46:989–92

29. Denis M-H, Delrot S. 1993. Carrier-mediated uptake of glyphosate in broad bean (*Vicia faba*) via a phosphate translocator. *Physiol. Plant.* 87:569–75

30. Diaz E, O'Laughlin JT, O'Leary MH. 1988. Reaction of phosphoenolpyruvate carboxylase with (Z)-3-bromophos-

phoenolpyruvate and (Z)-3-fluorophos-phoenolpyruvate. *Biochemistry* 27:1336–41

31. Dodge AD. 1989. Herbicides interacting with photosystem I. See Ref. 31a, pp. 37–50

31a. Dodge AD, ed. 1989. *Herbicides and Plant Metabolism* Cambridge: Cambridge Univ. Press

32. Dodge AD. 1990. The mode of action and metabolism of herbicides. In *Weed Control Handbook: Principles*, ed. RJ Hance, K Holly, pp. 201–15. Oxford: Blackwell Scientific

33. Doehlert DC, Huber SC. 1984. Phosphate inhibition of spinach leaf sucrose phosphate synthetase as affected by glucose-6-phosphate and phosphoglucose isomerase. *Plant Physiol.* 76:250–53

34. Dry IB, Wiskich JT. 1986. Comparative aspects of aminooxyacetate inhibition of glycine oxidation and aminotransferase activity of pea leaf mitochondria. *Plant Sci.* 44:23–28

35. Duke SO, Lydon J, Becerril JM, Sherman TD, Lehnen LP, Matsumoto H. 1991. Protoporphyrinogen oxidase-inhibiting herbicides. *Weed Sci.* 39:465–73

36. Edwards GE, Andreo CS. 1992. NADP-malic enzyme in plants. *Phytochemistry* 31:1845–57

37. Edwards GE, Walker DA. 1983. C_3, C_4: *Mechanisms, and Cellular and Environmental Regulation, of Photosynthesis*. Oxford/London: Blackwell Scientific

38. Edwards GE, Walker DA. 1984. Influence of glycerate on photosynthesis by wheat chloroplasts. *Arch. Biochem. Biophys.* 231:124–35

39. Egli MA, Gengenbach BG, Gronwald JW, Somers DA, Wyse DL. 1993. Characterization of maize acetyl-coenzyme A carboxylase. *Plant Physiol.* 101:499–506

40. Feierabend J, Schubert B. 1978. Comparative investigation of the action of several chlorosis-inducing herbicides on the biogenesis of chloroplasts and leaf microbodies. *Plant Physiol.* 61:1017–22

41. Focke M, Feld A, Lichtenthaler HK. 1991. Inhibition of early steps of *de novo* fatty-acid biosynthesis by different xenobiotica. *Physiol. Plant.* 81:251–55

42. Gardemann A, Stitt M, Heldt HW. 1983. Control of CO_2 fixation. Regulation of spinach ribulose-5-phosphate kinase by stromal metabolites. *Biochim. Biophys. Acta* 722:51–60

43. Gardeström P, Bergman A, Ericson I. 1981. Inhibition of the conversion of glycine to serine in spinach leaf mitochondria. *Physiol. Plant.* 53:439–44

44. Givan CV. 1980. Aminotransferases in higher plants. In *The Biochemistry of Plants*, ed. BJ Miflin, 5:365–406. New York: Academic

45. Givan CV. 1983. The source of acetyl-CoA in chloroplasts of higher plants. *Physiol. Plant.* 57:311–16

46. Givan CV, Joy KW, Kleczkowski LA. 1988. A decade of photorespiratory nitrogen cycling. *Trends Biochem. Sci.* 13:433–37

47. Givan CV, Kleczkowski LA. 1992. The enzymic reduction of glyoxylate and hydroxypyruvate in leaves of higher plants. *Plant Physiol.* 100:552–56

48. Golz A, Lichtenthaler HK. 1992. Inhibition of plant acetyl-CoA synthetase by alkyl-adenylates. *Z. Naturforsch. Teil C* 47:845–50

49. Gonzalez DH, Andreo CS. 1988. Identification of 2-enolbutyrate as the product of the reaction of maize leaf phosphoenolpyruvate carboxylase with (Z) and (E)-2-phosphoenolbutyrate: evidence from NMR and kinetic measurements. *Biochemistry* 27:177–83

50. Gonzalez DH, Iglesias AA, Andreo CS. 1986. Active-site-directed inhibition of phosphoenolpyruvate carboxylase from maize leaves by bromopyruvate. *Arch. Biochem. Biophys.* 245:179–89

51. Grover SD, Wedding RT. 1984. Modulation of the activity of NAD malic enzyme from *Solanum tuberosum* by changes in oligomeric state. *Arch. Biochem. Biophys.* 234:418–25

52. Gupta AS, Heinen JL, Holaday AS, Burke JJ, Allen RD. 1993. Increased resistance to oxidative stress in transgenic plants that overexpress chloroplastic Cu/Zn superoxide dismutase. *Proc. Natl. Acad. Sci. USA* 90:1629–33

53. Gutteridge S, Julien B. 1989. A phosphatase from chloroplast stroma of *Nicotiana tabacum* hydrolyzes 2′-carboxyarabinitol-1-phosphate, the natural inhibitor of Rubisco, to 2′-carboxyarabinitol. *FEBS Lett.* 254:225–30

54. Gutteridge S, Parry MAJ, Burton S, Keys AJ, Mudd A, et al. 1986. A nocturnal inhibitor of carboxylation in leaves. *Nature* 324:274–76

55. Harbron S, Foyer C, Walker D. 1981. The purification and properties of sucrose-phosphate synthase from spinach leaves: the involvement of this enzyme and fructose bisphosphatase in the regulation of sucrose biosynthesis. *Arch. Biochem. Biophys.* 212:237–46

56. Hatch MD. 1987. C_4 photosynthesis: a unique blend of modified biochemistry, anatomy and ultrastructure. *Biochim. Biophys. Acta* 895:81–106

57. Hatch MD, Mau SL. 1977. Association of NADP and NAD-linked malic enzyme activities in *Zea mays*: relation to C_4 pathway

of photosynthesis. *Arch. Biochem. Biophys.* 179:361–69

58. Havir EA. 1986. Inactivation of serine:glyoxylate and glutamate:glyoxylate aminotransferases from tobacco leaves by glyoxylate in the presence of ammonium ion. *Plant Physiol.* 80:473–78

59. Havir EA. 1992. The *in vivo* and *in vitro* inhibition of catalase from leaves of *Nicotiana sylvestris* by 3-amino-1,2,4-triazole. *Plant Physiol.* 99:533–37

60. Heldt HW, Flügge U-I, Borchert S. 1991. Diversity of specificity and function of phosphate translocators in various plastids. *Plant Physiol.* 95:341–43

61. Holbrook GP, Bowes G, Salvucci ME. 1989. Degradation of 2-carboxyarabinitol 1-phosphate by a specific chloroplast phosphatase. *Plant Physiol.* 90:673–78

62. Holbrook GP, Galasinski SC, Salvucci ME. 1991. Regulation of 2-carboxyarabinitol 1-phosphatase. *Plant Physiol.* 97:894–99

63. Holt JS, Powles SB, Holtum JAM. 1993. Mechanisms and agronomic aspects of herbicide resistance. *Annu. Rev. Plant Physiol. Plant Mol. Biol.* 44:203–29

64. Hoppe HH. 1985. Differential effect of diclofop-methyl on fatty acid synthesis in leaves of sensitive and tolerant plant species. *Pestic. Biochem. Physiol.* 23:297–308

65. Hoppe HH, Zacher H. 1986. Inhibition of fatty acid biosynthesis in isolated bean and maize chloroplasts by herbicidal phenoxy-phenoxypropionic acid derivatives and structurally related compounds. *Pestic. Biochem. Physiol.* 24:298–305

66. Howitz KT, McCarty RE. 1985. Substrate specificity of the pea chloroplast glycolate transporter. *Biochemistry* 24:3645–50

67. Howitz KT, McCarty RE. 1987. Tentative identification of the pea chloroplast envelope glycolate transporter. In *Progress in Photosynthesis Research*, ed. J Biggins, 3: 593–96. Dordrecht: Nijhoff

68. Huber SC. 1986. Fructose-2,6-bisphosphate. *Annu. Rev. Plant Physiol.* 37:233–46

69. Huber SC, Edwards GE. 1975. Inhibition of phosphoenolpyruvate carboxylase from C_4 plants by malate and aspartate. *Can. J. Bot.* 53:1925–33

70. Huber SC, Huber JL. 1992. Role of sucrose-phosphate synthase in sucrose metabolism in leaves. *Plant Physiol.* 99:1275–78

71. Huchzermeyer B, Loehr A. 1983. Effects of nitrophen on chloroplast coupling factor-dependent reactions. *Biochim. Biophys. Acta* 724:224–29

72. Husic DW, Husic HD, Tolbert NE. 1987. The oxidative photosynthetic carbon cycle or C_2 cycle. *CRC Crit. Rev. Plant Sci.* 5:45–100

73. Igamberdiev AU. 1989. Pathways of gly-

colate conversion in plants. *Biol. Rundsch.* 27:137–44

74. Igamberdiev AU, Zemlyanukhin AA, Rodionova LG. 1989. Glycolate oxidase from wheat and sugarbeet leaves: catalytic properties and role in oxalate biosynthesis. *Biochemistry* 53:1738–44

75. Iglesias AA, Plaxton WC, Podesta FE. 1993. The role of inorganic phosphate in the regulation of C_4 photosynthesis. *Photosynth. Res.* 35:205–11

76. Jenkins CLD. 1989. Effects of the phosphoenolpyruvate carboxylase inhibitor 3,3-dichloro-2-(dihydroxyphosphinoylme thyl)propenoate on photosynthesis. *Plant Physiol.* 89:1231–37

77. Jenkins CLD, Burnell JN, Hatch MD. 1987. Form of inorganic carbon involved as a product and as an inhibitor of C_4 acid decarboxylases operating in C_4 photosynthesis. *Plant Physiol.* 85:952–57

78. Jenkins CLD, Furbank RT, Hatch MD. 1989. Inorganic carbon diffusion between C_4 mesophyll and bundle sheath cells. Direct bundle sheath CO_2 assimilation in intact leaves in the presence of an inhibitor of the C_4 pathway. *Plant Physiol.* 91:1356–63

79. Jenkins CLD, Harris RLN, McFadden HG. 1987. 3,3-Dichloro-2-dihydroxyphosphi-noylmethyl-2-propenoate, a new specific inhibitor of phosphoenolpyruvate carboxylase. *Biochem. Int.* 14:219–26

80. Jenkins CLD, Rogers LJ, Kerr MW. 1983. Inhibition of glycolate metabolism by aminooxyacetate: consequences for photosynthesis. *Phytochemistry* 22:19–23

81. Jensen RA. 1986. The shikimate/arogenate pathway: link between carbohydrate metabolism and secondary metabolism. *Plant Physiol.* 66:164–68

82. Jewess PJ, Kerr MW, Whitaker DP. 1975. Inhibition of glycolate oxidase from pea leaves. *FEBS Lett.* 53:292–94

83. Jiao JA, Chollet R. 1991. Posttranslational regulation of phosphoenolpyruvate carboxylase in C_4 and Crassulacean acid metabolism plants. *Plant Physiol.* 95:981–85

84. Johal S, Partridge BE, Chollet R. 1985. Structural characterization and the determination of negative cooperativity in the tight binding of 2-carboxyarabinitol bisphosphate to higher plant Rubisco. *J. Biol. Chem.* 260:9894–9904

85. Joy KW, Blackwell RD, Lea PJ. 1992. Assimilation of nitrogen in mutants lacking enzymes of the glutamate synthase cycle. *J. Exp. Bot.* 43:139–45

86. Keys AJ, Bird IF, Cornelius MJ, Lea PJ, Wallsgrove RM, Miflin BJ. 1978. Photorespiratory nitrogen cycle. *Nature* 275: 741–42

87. Khavari-Nejad RA. 1977. Effects of α-hydroxy-2-pyridinemethanesulfonic acid

on photosynthetic carbon dioxide uptake and stomatal movements in excised tomato leaves. *Plant Physiol.* 60:44–46

88. Kingston-Smith AH, Major I, Parry MAJ, Keys AJ. 1992. Purification and properties of a phosphatase in French bean (*Phaseolus vulgaris* L.) leaves that hydrolyses 2'-carboxy-D-arabinitol 1-phosphate. *Biochem. J.* 287:821–25

89. Kishore GM, Shah DM. 1988. Amino acid biosynthesis inhibitors as herbicides. *Annu. Rev. Biochem.* 57:627–63

90. Kleczkowski LA, Edwards GE. 1991. A low temperature-induced reversible transition between different kinetic forms of maize leaf phosphoenolpyruvate carboxylase. *Plant Physiol. Biochem.* 29:9–17

91. Kleczkowski LA, Edwards GE, Blackwell RD, Lea PJ, Givan CV. 1990. Enzymology of the reduction of hydroxypyruvate and glyoxylate in a mutant of barley lacking peroxisomal hydroxypyruvate reductase. *Plant Physiol.* 94:819–25

92. Kleczkowski LA, Edwards GE, Randall DD. 1992. Effects of oxalate on the reduction of hydroxypyruvate and glyoxylate in leaves. *Phytochemistry* 31:51–54

93. Kleczkowski LA, Givan CV. 1988. Serine formation in leaves by mechanisms other than the glycolate pathway. *J. Plant Physiol.* 132:641–52

94. Kleczkowski LA, Randall DD. 1988. Development of adenylate kinase activity upon greening of etiolated seedlings of C_3 and C_4 species. *Photosynthetica* 22:112–15

95. Kleczkowski LA, Randall DD, Blevins DG. 1987. Inhibition of spinach leaf NADPH(NADH)-glyoxylate reductase by acetohydroxamate, aminooxyacetate and glycidate. *Plant Physiol.* 84:619–23

96. Kleczkowski LA, Randall DD, Edwards GE. 1991. Oxalate as a potent and selective inhibitor of spinach (*Spinacia oleracea*) leaf NADPH-dependent hydroxypyruvate reductase. *Biochem. J.* 276:125–27

97. Kleczkowski LA, Randall DD, Zahler WL. 1990. Adenylate kinase from maize leaves: true substrates, inhibition by P^1, P^5-di-(adenosine-5')pentaphosphate, and kinetic mechanism. *Z. Naturforsch. Teil C* 45:607–13

98. Kleczkowski LA, Villand P, Lönneborg A, Olsen O-A, Lüthi E. 1991. Plant ADP-glucose pyrophosphorylase—recent advances and biotechnological perspectives. *Z. Naturforsch. Teil C* 46:605–12

99. Kleczkowski LA, Villand P, Lüthi E, Olsen O-A, Preiss J. 1993. Insensitivity of barley endosperm ADP-glucose pyrophosphorylase to 3-phosphoglycerate and orthophosphate regulation. *Plant Physiol.* 101:179–86

100. Kleczkowski LA, Villand P, Preiss J, Olsen O-A. 1993. Kinetic mechanism and regulation of ADP-glucose pyrophosphorylase from barley (*Hordeum vulgare*) leaves. *J. Biol. Chem.* 268:6228–33

101. Kobek K, Focke M, Lichtenthaler HK. 1988. Fatty-acid biosynthesis and acetyl-CoA carboxylase as a target of diclofop, fenoxafop and other aryloxy-phenoxy-propionic acid herbicides. *Z. Naturforsch. Teil C* 43:47–54

102. Kruckeberg AL, Neuhaus HE, Feil R, Gottlieb LD, Stitt M. 1989. Reduced activity mutants of phosphoglucose isomerase in the chloroplast and cytosol of *Clarkia xantiana*. *Biochem. J.* 261:457–67

103. Lea PJ. 1991. The inhibition of ammonia assimilation: a mechanism of herbicide action. In *Herbicides,* ed. NR Baker, MP Percival, pp. 267–97. New York/Amsterdam: Elsevier

104. Lea P, Blackwell RD. 1992. The role of amino acid metabolism in photosynthesis. In *Biosynthesis and Molecular Regulation of Amino Acids in Plants,* ed. BK Singh, JC Shannon, H Flores, pp. 98–110. Rockville, MD: Am. Soc. Plant Physiol.

105. Lea PJ, Ridley SM. 1989. Glutamine synthetase and its inhibition. See Ref. 31a, pp. 137–70

106. Lichtenthaler HK. 1990. Mode of action of herbicides affecting acetyl-CoA carboxylase and fatty acid biosynthesis. *Z. Naturforsch. Teil C* 45:521–28

107. Lichtenthaler HK, Focke M, Golz A, Hoffmann S, Kobek K, Motel A. 1992. Investigation on the starting enzymes of plant fatty acid biosynthesis and particular inhibitors. In *Metabolism, Structure and Utilization of Plant Lipids,* ed. A Cherif, pp. 103–12. Tunis: Centre Natl. Pedagogique

108. Lichtenthaler HK, Meier D. 1984. Inhibition by sethoxydim of chloroplast biogenesis, development and replication in barley seedlings. *Z. Naturforsch. Teil C* 39:115–22

109. Liu J, Peliska J, O'Leary MH. 1990. Synthesis and study of (Z)-3-chlorophosphoenolpyruvate. *Arch. Biochem. Biophys.* 277:143–48

110. Loboda T, Kleczkowski LA, Tarłowski J, Nalborczyk E. 1988. A transient stimulation of net photosynthesis of rye leaves by α-hydroxy-2-pyridinemethane sulfonic acid (α-HPMS) due to inhibition of photorespiratory CO_2 release. *J. Exp. Bot.* 39: 1765–70

111. Lorimer GH, Andrews TJ, Pierce J, Schloss JV. 1986. 2'-Carboxy-3-keto-D-arabinitol 1,5-bisphosphate, the six carbon intermediate of the Rubisco reaction. *Philos. Trans. R. Soc. London Ser. B* 313:397–407

112. Manderscheid R, Wild A. 1986. Studies on the mechanism of inhibition by phosphinothricin of glutamine synthetase iso-

lated from *Triticum aestivum* L. *J. Plant Physiol.* 123:135–42

113. Martin F, Winspear MJ, McFarlane JD, Oaks A. 1983. Effect of methionine sulphoximine on the accumulation of ammonia in C3 and C4 leaves. *Plant Physiol.* 71:177–81

114. Mazur BJ, Falco SC. 1989. The development of herbicide resistant crops. *Annu. Rev. Plant Physiol. Plant Mol. Biol.* 40: 441–70

115. McFadden HG, Harris RLN, Jenkins CLD. 1989. Potential inhibitors of phosphoenolpyruvate carboxylase. II. Phosphonic acid substrate analogues derived from reaction of trialkyl phosphites with halomethacrylates. *Aust. J. Chem.* 42:301–14

116. Meyer CR, Willeford KO, Wedding RT. 1991. Regulation of phosphoenolpyruvate carboxylase from *Crassula argentea*: effect of incubation with ligands and dilution on oligomeric state, activity, and allosteric properties. *Arch. Biochem. Biophys.* 288: 343–49

117. Moore BD, Seemann JR. 1992. Metabolism of 2'-carboxyarabinitol in leaves. *Plant Physiol.* 99:1551–55

118. Moore BD, Sharkey TD, Kobza J, Seemann JR. 1992. Identification and levels of 2'-carboxyarabinitol in leaves. *Plant Physiol.* 99:1546–50

119. Moreland DE. 1980. Mechanisms of action of herbicides. *Annu. Rev. Plant Physiol.* 31:597–638

120. Mostowska A. 1992. Photodynamic herbicides: their influence on the development and structure of photosynthetic apparatus. *Photosynthetica* 26:19–31

121. Neuhaus HE, Henrichs G, Scheibe R, 1993. Characterization of glucose-6-phosphate incorporation into starch by isolated intact cauliflower-bud plastids. *Plant Physiol.* 101:573–78

122. Oettmeier W, Masson K. 1980. Synthesis and thylakoid membrane binding of the radioactively labeled herbicide dinoseb. *Pestic. Biochem. Physiol.* 14:86–97

123. Ogren WL. 1984. Photorespiration: pathways, regulation and modification. *Annu. Rev. Plant Physiol.* 35:415–42

124. O'Leary MH. 1982. Phosphoenolpyruvate carboxylase: an enzymologist's view. *Annu. Rev. Plant Physiol.* 33:297–315

125. Onur G, Schafer G, Strotmann H. 1983. Synthesis of 2'-/3'-O-acylated adenine nucleotide analogs and their interactions in photophosphorylation. *Z. Naturforsch. Teil C* 38:49–59

126. Pacold I, Anderson LE. 1975. Chloroplast and cytoplasmic enzymes. VI. Pea leaf 3-phosphoglycerate kinases. *Plant Physiol.* 55:168–71

127. Palmer CE, Oelck M, 1993. The relationship of phosphinothricin to growth and metabolism in cell cultures of *Brassica napus* L. *J. Plant Physiol.* 141:105–10

128. Parker WB, Marshall LC, Burton JD, Somers DA, Wyse DL, et al. 1990. Dominant mutations causing alterations in acetyl-coenzyme A carboxylase confer tolerance to cyclohexanodione and aryloxyphenoxypropionate herbicides in maize. *Proc. Natl. Acad. Sci. USA* 87:7175–79

129. Percival MP, Baker NR. 1991. Herbicides and photosynthesis. In *Herbicides*, ed. NR Baker, MP Percival. New York/Amsterdam: Elsevier

130. Portis AR. 1992. Regulation of ribulose-1,5-bisphosphate carboxylase/oxygenase activity. *Annu. Rev. Plant Physiol. Plant Mol. Biol.* 43:415–37

131. Preiss J. 1991. Biology and molecular biology of starch synthesis and its regulation. In *Oxford Surveys of Cellular and Molecular Biology*, ed. BJ Miflin, 7:59–114. Oxford: Oxford Univ. Press

132. Rendina AR, Felts JM, Beaudoin JD, Craig-Kennard AC, Look LL, et al. 1988. Kinetic characterization, stereoselectivity, and species selectivity of the inhibition of plant acetyl-CoA carboxylase by the aryloxyphenoxypropionic acid grass herbicides. *Arch. Biochem. Biophys.* 265:219–25

133. Rhodes D, Deal L, Haworth P, Jamieson GC, Reuter CC, Ericson MC. 1986. Amino acid metabolism of *Lemna minor* L. I. Responses to methionine sulphoximine. *Plant Physiol.* 82:1057–62

134. Robinson SP, Portis AR. 1988. Release of the nocturnal inhibitor, carboxyarabinitol-1-phosphate, from ribulose bisphosphate carboxylase/oxygenase by rubisco activase. *FEBS Lett.* 233:413–16

135. Rumpho ME, Edwards GE. 1985. Characterization of 4,4'-diisothiocyano-2,2'-disulfonic acid stilbene inhibition of 3-phosphoglycerate-dependent O2 evolution in isolated chloroplasts. Evidence for a common binding site of the C4 phosphate translocator for 3-phosphoglycerate, phosphoenolpyruvate, and inorganic phosphate. *Plant Physiol.* 78:537–44

136. Rumpho ME, Edwards GE, Yousif AE, Keegstra K. 1988. Specific labeling of the phosphate translocator in C3 and C4 mesophyll chloroplasts by tritiated dihydro-DIDS (1,2-ditritio-1,2-[disulfo-4,4'-diisothiocyano]diphenylethane). *Plant Physiol.* 86:1193–98

137. Rustin P, Meyer CR, Wedding RT. 1988. Identification of substrate and effector binding sites of phosphoenolpyruvate carboxylase from *Crassula argentea*. A possible role of phosphoenolpyruvate as substrate and activator. *J. Biol. Chem.* 263: 17611–14

138. Sabularse DC, Anderson RL. 1981. D-

Fructose 2,6-bisphosphate: a naturally occurring activator for inorganic pyrophosphate: D-fructose 6-phosphate 1-phospho-
transferase in plants. *Biochem. Biophys.
Res. Commun.* 103:848–55

139. Salvucci ME. 1989. Regulation of Rubisco
activity *in vivo. Physiol. Plant.* 77:164–71

140. Salvucci ME, Holbrook GP, Anderson JC,
Bowes G. 1988. NADPH-dependent metabolism of the ribulose bisphosphate car-
boxylase/oxygenase inhibitor 2-carboxy-
arabinitol-1-phosphate by a chloroplast
protein. *FEBS Lett.* 231:197–201

141. Sandmann G, Böger P. 1986. Sites of herbicide inhibition at the photosynthetic apparatus. In *Encyclopedia of Plant Physiology,* ed. LA Staehelin, CJ Arntzen, 19:596–
602. Berlin: Springer-Verlag

142. Sarojini G, Oliver DJ. 1985. Inhibition of
glycine oxidation by carboxymethoxylamine, methoxylamine and acethydrazide.
Plant Physiol. 77:786–89

143. Sauer H, Wild A, Rühle W. 1987. The effect
of phosphinothricin (glufosinate) on photosynthesis. II. The causes of inhibition of
photosynthesis. *Z. Naturforsch. Teil C* 42:
270–78

144. Schimkat D, Heineke D, Heldt HW. 1990.
Regulation of sedoheptulose-1,7-bisphos-
phatase by sedoheptulose-7-phosphate and
glycerate, and of fructose-1,6-bisphospha-
tase by glycerate in spinach chloroplasts.
Planta 181:97–103

145. Schneider G, Lindqvist Y, Branden C-I.
1992. Rubisco: structure and mechanism.
Annu. Rev. Biophys. Biomol. Struct. 21:
119–43

146. Schultz A, Wengenmayer F, Goodman HM.
1990. Genetic engineering of herbicide resistance in higher plants. *CRC Crit. Rev.
Plant Sci.* 9:1–15

147. Secor J, Cseke C. 1988. Inhibition of acetyl-CoA carboxylase activity by haloxyfop
and tralkoxydim. *Plant Physiol.* 86:10–12

148. Servaites JC. 1990. Inhibition of ribulose-
1,5-bisphosphate carboxylase/oxygenase
by 2-carboxyarabinitol-1-phosphate. *Plant
Physiol.* 92:867–70

149. Shimabukuro RH. 1985. Detoxification of
herbicides. In *Weed Physiology,* ed. SO
Duke, 2:215–40. Boca Raton, FL: CRC
Press

150. Sicher RC. 1984. Glycolaldehyde inhibi-
tion of photosynthetic carbon assimilation
by isolated chloroplasts and protoplasts. In
Advances in Photosynthesis Research, ed.
C Sybesma, 3:413–16. The Hague:
Nijhoff/Junk

151. Siehl DL. 1992. Considerations in selecting
a target site for herbicide design. In *Biosyn-
thesis and Molecular Regulation of Amino
Acids in Plants,* ed. BK Singh, HE Flores,
JC Shannon, pp. 146–62. Rockville, MD:
Am. Soc. Plant Physiol.

152. Sikkema KD, O'Leary MH. 1988. Synthesis and study of phosphoenolthiopyruvate.
Biochemistry 27:1342–47

153. Singh B, Greenberg E, Preiss J. 1984.
ADPglucose pyrophosphorylase from the
CAM plants *Hoya carnosa* and *Xerosicyos
danguyi. Plant Physiol.* 74:711–16

154. Sivak MN, Walker DA. 1986. Photosynthesis *in vivo* can be limited by phosphate
supply. *New Phytol.* 102:499–512

155. Slabas AR, Walker DA. 1976. Inhibition of
spinach phosphoribulokinase by DL-glyceraldehyde. *Biochem. J.* 153:613–19

156. Somerville CR. 1986. Analysis of photosynthesis with mutants of higher plants and
algae. *Annu. Rev. Plant Physiol.* 37:467–
507

157. Somerville CR, Ogren WL. 1980. The inhibition of photosynthesis in *Arabidopsis*
mutants lacking glutamate synthase activity. *Nature* 286:257–59

158. Somerville SC, Portis AR, Ogren WL.
1982. A mutant of *Arabidopsis thaliana*
which lacks activation of RuBP carboxylase *in vivo. Plant Physiol.* 70:381–87

159. Spampinato CP, Paneth P, O'Leary MH,
Andreo CS. 1991. Analogues of NADP+ as
inhibitors and coenzymes for NADP+
malic enzyme from maize leaves. *Photosynth. Res.* 28:69–76

160. Stewart GR, Rhodes D. 1976. Evidence for
the assimilation of ammonia via the glutamine pathway in nitrate-grown *Lemna
minor* L. *FEBS Lett.* 64:296–99

161. Stiborova M, Leblova S. 1986. Structure of
phosphoenolpyruvate carboxylase from
maize leaves. *FEBS Lett.* 205:32–34

162. Stidham MA. 1991. Herbicidal inhibitors
of branched chain amino acid biosynthesis.
In *Herbicides,* ed. NR Baker, MP Percival,
pp. 247–66. New York/Amsterdam: Elsevier

163. Stitt M. 1990. Fructose-2,6-bisphosphate
as a regulatory metabolite in plants. *Annu.
Rev. Plant Physiol. Plant Mol. Biol.* 41:
153–85

164. Stokes DM, Walker DA. 1972. Photosynthesis by isolated chloroplasts. Inhibition
by DL-glyceraldehyde of carbon dioxide
assimilation. *Biochem. J.* 128:1147–57

165. Strotmann H, Bickel-Sandkötter S. 1984.
Structure, function, and regulation of chloroplast ATPase. *Annu. Rev. Plant Physiol.*
35:97–120

166. Timmerman KP. 1989. Molecular characterization of corn glutathione S-transferase
isozymes involved in herbicide detoxica-
tion. *Physiol. Plant.* 77:465–71

167. Usuda H, Arron GP, Edwards GE. 1980.
Inhibition of glycine decarboxylation by
aminoacetonitrile and its effect on photosynthesis in wheat. *J. Exp. Bot.* 31:1477–83

168. Usuda H, Edwards GE. 1981. Inhibition of
photosynthetic carbon metabolism in iso-

lated chloroplasts by iodoacetol phosphate. *Plant Physiol.* 67:854–58

169. Van Rensen JJS. 1989. Herbicides interacting with photosystem II. See Ref. 31a, pp. 21–36

170. Wagner R, Ponse G, Strotmann H. 1986. Binding of 2'(3')-*O*-(2,4,6-trinitrophenyl)-adenosine 5'-diphosphate opens the pathway for protons through the chloroplast ATPase complex. *Eur. J. Biochem.* 161: 205–9

171. Walker KA, Givan CV, Keys AJ. 1984. Glutamic acid metabolism and the photorespiratory nitrogen cycle in wheat leaves. *Plant Physiol.* 75:60–66

172. Walker KA, Keys AJ, Givan CV. 1984. Effect of L-methionine sulphoximine on the products of photosynthesis in wheat (*Triticum aestivum*) leaves. *J. Exp. Bot.* 35:1800–10

173. Wedding RT. 1989. Malic enzymes of higher plants. *Plant Physiol.* 90:367–71

174. Weidner M, Burchartz N. 1978. Inhibition of phosphoenolpyruvate carboxylase by formulated herbicides and anionic detergents. *Biochem. Physiol. Pflanzen.* 173: 381–89

175. Wirsching P, O'Leary MH. 1988. (Z)-3-(Fluoromethyl)phosphoenolpyruvate: synthesis and enzymatic studies. *Biochemistry* 27:1348–55

176. Yun S, Ishii R, Hyeon S, Suzuki A, Murata Y, Tamura S. 1979. Effects of some chemicals on photorespiration and photosynthesis in the excised rice leaves. *Agric. Biol. Chem.* 43:2207–209

177. Zeiher CA, Randall DD. 1991. Spinach leaf acetyl-coenzyme A synthetase: purification and characterization. *Plant Physiol.* 96: 382–89

178. Zelitch I. 1966. Increased rate of net photosynthetic carbon dioxide uptake caused by the inhibition of glycolate oxidase. *Plant Physiol.* 41:1623–31

179. Zelitch I. 1978. Effect of glycidate, an inhibitor of glycolate synthesis in leaves, on the activity of some enzymes of the glycolate pathway. *Plant Physiol.* 61:236–41

180. Zelitch I. 1992. Control of plant productivity by regulation of photorespiration. *BioScience* 42:510–16

Annu. Rev. Plant Physiol. Plant Mol. Biol. 1994. 45:369–92

THE ROLE OF CARBONIC ANHYDRASE IN PHOTOSYNTHESIS

Murray R. Badger and G. Dean Price

Plant Environmental Biology Group, Research School Biological Sciences, Australian National University, Canberra 2601, ACT, Australia

KEY WORDS: CO_2 concentrating mechanisms, aquatic macrophytes, cyanobacteria, algae, carboxysome, pyrenoid

CONTENTS

INTRODUCTION 369
THE REQUIREMENT FOR CARBONIC ANHYDRASES IN DIFFERENT
    ORGANISMS 370
CYANOBACTERIA 371
    Carboxysomal Carbonic Anhydrase 373
    Is the Ci Pump a Membrane Carbonic Anhydrase? 374
MICROALGAE... 374
    Chloroplast Carbonic Anhydrase... 375
    Periplasmic Carbonic Anhydrase.. 376
    Cytosolic Carbonic Anhydrase.. 377
    Induction of Carbonic Anhydrase at Low-CO_2.. 378
AQUATIC MACROALGAE AND ANGIOSPERMS 379
    External Carbonic Anhydrase.. 379
    Internal Carbonic Anhydrase 380
C_4 PLANTS 381
CAM PLANTS 382
C_3 PLANTS 382
EVOLUTION OF PHOTOSYNTHETIC CARBONIC ANHYDRASES 385
CONCLUSIONS... 387

INTRODUCTION

Carbonic anhydrase (CA, EC 4.2.1.1) is a zinc-containing metalloenzyme that catalyzes the interconversion of CO_2 and HCO_3^- in many organisms (Equation 1).

$$CO_2 + H_2O \rightarrow HCO_3^- + H^+ \qquad\qquad 1.$$

0066-4294/94/0601-0369$05.00

369

In the hydration reaction, CO_2 reacts with a Zn-OH intermediate at the active site of the enzyme. In the reverse dehydration direction, HCO_3^- reacts with $Zn-H_2O$. This reaction has been studied extensively, particularly with CA enzymes from animal sources. Recent reviews cover extensively the mechanistic functioning and evolutionary diversity of this enzyme (20, 94, 107).

Carbonic anhydrase is required in biological systems because the uncatalyzed interconversion between CO_2 and HCO_3^- is slow when compared to the flux rates between these species, which may be required in living cells. For example, in a higher plant chloroplast, the rate of photosynthesis may approach 300 μmol mg Chl^{-1} h^{-1}. This is about 2.8 mM s^{-1}, assuming a chloroplast volume of 30 μl mg Chl^{-1}. The uncatalyzed first-order rate constant for the conversion of CO_2 to HCO_3^- is about 0.05 s^{-1} at 25 °C and pH 8 (57). Thus at present atmospheric CO_2 levels (350 ppm, leading to a chloroplast [CO_2] of about 8 μM) the uncatalyzed rate of interconversion of CO_2 to HCO_3^- will be about 0.5×10^{-3} mM s^{-1}. This is about 10^4 times slower than the biological flux that takes place via CO_2 fixation by the primary CO_2-fixing enzyme, ribulose bisphosphate carboxylase/oxygenase (Rubisco). If the interconversion of CO_2 and HCO_3^- were important for the supply of CO_2 to the active site of Rubisco, then CA would have to be present to enable such high flux rates to occur in the chloroplast stroma.

For many years, carbonic anhydrase has been known to exist in many photosynthetic organisms and on the basis of the foregoing argument, it has been assumed to play an important role in the photosynthetic process. However, only recently has work with various photosynthetic organisms started to define the functional roles of CA and how these roles may differ between photosynthetic groups. Some organisms may have multiple types of CAs in various cellular locations and each type of CA may play a different role in the photosynthetic process. Several reviews have documented the evolution in our understanding of CA function in a range of photosynthetic organisms, including terrestrial and aquatic higher plants, micro and macro algae from marine and freshwater habitats, and cyanobacteria (1, 34, 73, 88, 103, 109). This review summarizes recent advances in our understanding of CA function, highlighting the diverse ways that this enzyme is used in the efficient functioning of photosynthetic CO_2 fixation by Rubisco.

THE REQUIREMENT FOR CARBONIC ANHYDRASE IN DIFFERENT ORGANISMS

Figure 1 shows a comparison of the photosynthesis rate that would be expected at current atmospheric levels of CO_2 both in the presence and absence of CA in various organisms. The requirement for CA varies significantly

Requirement for CA in Photosynthesis

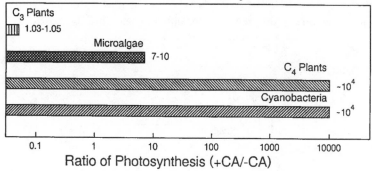

Figure 1   The relative requirement for CA for photosynthesis at ambient CO_2 levels (350 ppm) in different photosynthetic organisms. Data is shown as an estimated ratio of photosynthesis in the presence and in the absence of CA, for cyanobacteria (76, 89, 90), green microalgae (97), C_4 higher plants (40), and C_3 higher plants (19 and this review). The values were derived from data presented in the cited references.

among photosynthetic organisms. For cyanobacteria and C_4 higher plants, CA is required if any photosynthesis is to occur and almost all the CO_2 that is fixed must pass through the CA-catalyzed reaction. For green microalgae, a lesser but still significant involvement of CA is predicted. Rubisco may fix 10–15% of CO_2 directly, but the majority must still interact with CA. At the lower end of the spectrum, C_3 plants require minimal CA. In these plants, CA probably stimulates photosynthesis by 3–5% at ambient CO_2, and most of the CO_2 flux can occur without CA. The relative importance of CA in these organisms depends on the role that CA plays in each. In cyanobacteria and microalgae, CA is required to convert an accumulated pool of HCO_3^- to CO_2, which is the substrate for Rubisco (5). In C_4 plants, CA is needed to convert CO_2 to HCO_3^-, which is the substrate for PEP carboxylase (40). In both cases, CA acts as a primary enzyme in the CO_2 fixing pathway and little fixation can occur in its absence. For C_3 plants, CA helps the facilitated diffusion of CO_2 (as both CO_2 and HCO_3^-) across the chloroplast stroma. Without CA, CO_2 can still diffuse at a reasonable rate and photosynthesis is affected only marginally (19, 80).

Cyanobacteria

Cyanobacteria possess an active CO_2 concentrating mechanism (CCM) that elevates CO_2 around the active site of Rubisco. This CCM greatly improves the efficiency of photosynthesis in inorganic carbon (Ci) acquisition over what would occur based solely on the kinetic properties of cyanobacterial Rubisco (for review, see 2, 5, 47, 62). Figure 2 presents a schematic diagram of this CCM and highlights the central role that CA plays. An active Ci transport system, most likely located on the plasma membrane, uses photosynthetic energy to accumulate HCO_3^- to levels as high as 50 mM within the cytosol of the cell. This Ci transport system seems able to use both CO_2 and HCO_3^- as

CYANOBACTERIA

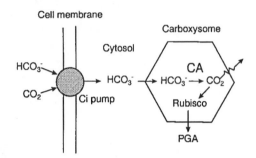

MICROALGAE

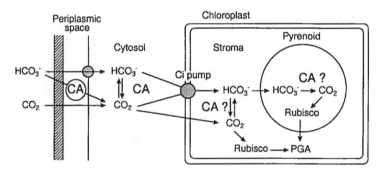

AQUATIC MACROPHYTES

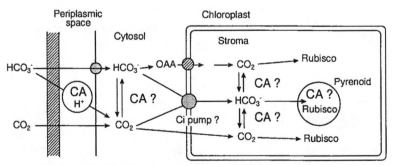

Figure 2   Models showing the roles of carbonic anhydrase during photosynthesis in cyanobacteria, microalgae, and aquatic macrophytes.

external substrates and may represent the activity of more than one pumping system (5, 62).

Carboxysomal Carbonic Anhydrase

For the internal HCO_3^- pool to be available for fixation by Rubisco, it first must be converted to CO_2. In a unique structural association, both Rubisco and CA are localized within the carboxysomes in the interior of the cell (5, 47, 56, 79). These small polyhedral bodies appear most analogous to virus-like particles. They are surrounded by a protein sheath and are found in many prokaryotic organisms containing Rubisco (15). The carboxysomal CA dehydrates the HCO_3^-, which apparently is able to diffuse in from the cytosol. Although there are only low levels of carboxysomal CA in the cyanobacteria that have been examined (notably *Synechococcus* PCC7942), there is sufficient activity to support its proposed role (79). The CO_2 produced within the carboxysome exhibits restricted diffusion from its generation site, which can lead to a localized elevation of CO_2. Various models have been proposed for how CO_2 diffusion is restricted (89, 90), but the mechanism is poorly understood. In the simplest case, the carboxysome shell could be somewhat impermeable to CO_2 but relatively permeable to HCO_3^-; however, this is largely speculative. In another more complex hypothesis, CA and Rubisco could form discrete associations within the carboxysome. Models have shown that this could lead to localized elevation of CO_2 at the active site of Rubisco without the need of an outer permeability barrier (89). In cyanobacteria, CA can be present only in the carboxysome. Any CA in the cytosol would short circuit the CCM and allow the futile efflux of HCO_3^- from the cytosol via its conversion to CO_2. This has been shown by the transgenic expression of a human carbonic anhydrase protein in the cytosol of *Synechococcus* PCC7942 (77).

The gene and protein product for carboxysomal CA from *Synechococcus* PCC7942 has been isolated and characterized recently (31, 113) using high-CO_2–requiring mutants. It is most similar to CA from *Escherichia coli* and from the chloroplast of higher plants (20–30% identity at the amino acid level) and there may be a common ancestry for prokaryotic-type CAs (31, 36). Carboxysomal CA has a 60–70 amino acid extension at the carboxyl end of the protein (31), giving a monomer size of 31 kDa, compared to 24 kDa for the chloroplast enzyme of higher plants. This sequence may be important for targeting CA to the carboxysome and/or for establishing some specific association with Rubisco. This hypothesis remains to be tested by direct genetic manipulation of CA and Rubisco genes using transformable cyanobacterial models such as *Synechococcus* PCC7942.

Is the Ci Pump a Membrane Carbonic Anhydrase?

As described above, the Ci transport system appears to take either CO_2 or HCO_3^- from the external medium and transport both Ci species to the cytosol, delivering them as HCO_3^- (5). In CO_2 transport, this process must involve the hydration of CO_2 as it crosses the plasma membrane and some workers have hypothesised that a CA-like reaction may be involved. Because neither the Ci transport proteins nor their genes have been isolated, this theory is not supported by direct evidence. Several observations are, however, consistent with such a notion. First, a lipid soluble CA inhibitor, ethoxyzolamide, can specifically inhibit both CO_2 and HCO_3^- transport activities in cyanobacterial cells (75). However, the I_{50} for inhibition is quite high (100 μM compared to about 5 μM for the carboxysomal CA) (79). Second, another type of CA activity has been identified in cyanobacterial cell extracts, with about half the activity of the carboxysomal CA, a high I_{50} for ethoxyzolamide, and a possible plasma membrane location (6, 79). Third, cells with a mutation in the carboxysomal CA gene are still able to catalyze a significant light dependent exchange of Ci species with water. This may represent a cycling flux of Ci species through the Ci transport system and a resulting loss of label to H_2O (4, 76).

Isolation and analysis of the Ci transport proteins and their genes is needed before any further mechanistic relationships between a CA-like reaction mechanism and Ci transport can be established. In vitro assay techniques for both native and reconstituted Ci transport systems should also be developed. Membrane-associated CAs have been isolated and characterized in animal systems (114), but these CAs bear little relationship to the Ci transport system described here. The animal CAs are certainly anchored in the membrane, but the active site of the protein is located on the globular part of the molecule located in the liquid phase. These CAs do not participate in the unidirectional transport of Ci species across a lipid membrane. Thus, if the Ci transport system in cyanobacteria does employ a CA-like reaction mechanism, it will represent an entirely new class of CA-like proteins not previously described.

MICROALGAE

Although CA activity is intimately involved in photosynthetic CO_2 fixation by microalgae, a relatively simple picture of the mechanism is not available because of the presence of multiple CA isozymes located in the chloroplast, cytosol, and the external periplasmic space of a range of microalgal cells (1, 2, 5, 103, 109). CA activities have been examined in several microalgal species, but *Chlamydomonas reinhardtii* has been studied the most extensively and much of our understanding of CA functioning has been derived from this organism. Thus the scheme depicted in Figure 2, showing the possible roles of

CA in microalgae, is derived largely from studies with *C. reinhardtii.* Microalgae have a CCM, but because of the eukaryotic nature of the cells, the mechanism may not be entirely the same as that found in cyanobacteria. For example, in microalgae, Ci transport systems may be located both on the plasma membrane and the chloroplast envelope (2, 5), and there are multiple CAs.

Chloroplast Carbonic Anhydrase

Chloroplastic CA seems to be the most important CA in the photosynthetic CO_2 flux of microalgal cells. This hypothesis comes primarily from studies of a high-CO_2–requiring *C. reinhardtii* mutant (*ca-1*) that appears to be defective in an internal CA activity (96). Unfortunately, the internal CA of *C. reinhardtii* and other microalgae is quite low in detectable activity and it has been difficult to quantify and isolate the various internal isozymes. However, at least two groups of investigators appear to have shown that this particular mutant is devoid of chloroplastic CA and that in the wild-type this CA may be pelletable in nature (48, 66; and DF Sültemeyer, personal communication).

The role of chloroplastic CA can be hypothesized based on experimental evidence and by analogy of the algal chloroplast to the cyanobacterial model (Figure 2). Studies of isolated chloroplasts have shown that the primary Ci transport system is probably located on the chloroplast envelope (65, 66, 101). In addition, studies of the *ca-1* mutant, and the use of ethoxyzolamide as an inhibitor of internal CA, have indicated that inhibition of internal CA actually leads to an overaccumulation of internal Ci (3, 96). These two observations indicate that the Ci transport system located on the chloroplast envelope delivers HCO_3^- to the chloroplast stroma and that chloroplast CA is required to convert the HCO_3^- to CO_2 in much the same way that carboxysomal CA is required in cyanobacteria.

A structural equivalent to the carboxysome is found in the microalgal chloroplast in the form of the pyrenoid body, which contains most if not all of the Rubisco in several algal species that have been studied closely. The pyrenoid is found in many algal species. It appears as an electron-dense protein inclusion body located in the chloroplast and may be surrounded by a starch sheath with pores (50, 51). In some algal species, unstacked thylakoid membranes also have been shown to traverse the pyrenoid (21, 54), forming several species-specific patterns. Recent work has focused on the protein components of the pyrenoid structure, in an attempt to define its role in the algal chloroplast. Immuno-gold labeling has shown that Rubisco is localized predominantly in this body (53, 54), as is Rubisco activase (55), which presumably is required for continued activation of Rubisco. However, the evidence for CA co-localization is conflicting. Immuno-cytochemical studies of the red alga *Porphyridium cruentum* (112) showed that homologous CA antibody labeling was found predominantly in the chloroplast, although the pyrenoid labeling was only at background levels. In the diatom, *Phaeodactylum tricornutum* (54), spinach CA antibodies also appeared to label the chloroplast, although the pyrenoid again had little label. In contrast to these results, studies with *C. reinhardtii*, using dansyl amide as an active site fluorescent probe for CA (52), demonstrated specific pyrenoid staining. The pelletable nature of the *C. reinhardtii* chloroplastic CA is also consistent with the notion that CA may be present in the pyrenoid (99). More work is needed to establish the specific location of CA in algal chloroplasts; however, it is reasonable at this stage to draw the model shown in Figure 2, placing CA in the pyrenoid.

Further work must be done to refine the role of CA in the algal chloroplast. The isolation of both the chloroplast CA protein and its gene will aid progress, providing more specific probes for detection and localization. In addition, direct genetic manipulation of *C. reinhardtii* may be promising, either through chloroplast or nuclear transformation. For example, it would be interesting to express a foreign CA protein in the chloroplast stroma to see if this caused a loss of a chloroplast HCO_3^- pool and the creation of a high-CO_2–requiring mutant. In addition, manipulation of the Rubisco large and small subunit sequences may be fruitful in detection of sequences that target the proteins to the pyrenoid and in creating mutants in which Rubisco is present in the chloroplast stroma. Finally, selection in *C. reinhardtii* of chemically induced mutants lacking pyrenoids may provide a simple, yet powerful, insight into CA and pyrenoid function.

Some microalgae do not contain pyrenoids and a study of these species may also contribute to an understanding of chloroplast CA. Studies have been conducted recently with a green algae, *Coccomyxa* PA, which lacks pyrenoids. This microalga can exist in a free-living state or as a photobiont in lichen associations. These studies indicate that *Coccomyxa* may lack a CCM when compared to species such as *C. reinhardtii* (69, 70). *Coccomyxa* also appears to contain rather high levels of internal CA, some 100-fold greater than *C. reinhardtii* (K Palmqvist, personal communication). It is intriguing to speculate that pyrenoid-less algae may be like C_3 higher plants (see Figure 3), with both Rubisco and high levels of CA distributed throughout the stroma, and lacking an active Ci transport system.

Periplasmic Carbonic Anhydrase

An external CA has been identified in several microalgae including *C. reinhardtii, Chlorella* species, *Dunaliella tertiolecta,* and *Scenedesmus obliquus* (1, 2, 103, 109). This CA is located in the periplasmic space and its activity can be assayed in intact cells because of its access to the external medium. The activity of this CA increases dramatically when the CCM activity of cells is induced by growing them at low CO_2 levels (1, 2, 16, 103, 109) and the absolute level of activity of the periplasmic CA may greatly exceed that of the internal CA of the cell. However, the occurrence of periplasmic CA varies significantly between microalgae. *C. reinhardtii* has high levels of periplasmic CA while *C. moewusii* appears to have little if any (33). Differences also occur among *Chlorella* (63) and *Dunaliella* (33) species. These species do have active CCMs but it appears as though periplasmic CA may be an inducible but not essential part of this mechanism.

The role of the periplasmic CA is not entirely clear, but Figure 2 presents a reasonable explanation of its probable function. The external CA probably converts external HCO_3^- to CO_2, which can be taken up through the plasma membrane and then used for photosynthesis. In support of this notion, the ability of microalgal cells to use external HCO_3^- for photosynthesis has been correlated with the presence of periplasmic CA (2, 63). However, periplasmic CA is not absolutely necessary for HCO_3^- utilization because cells of *C. reinhardtii* and *S. obliquus* seem able to transport HCO_3^- across the plasma membrane without needing to convert it to CO_2 (71, 100, 102, 108). Nevertheless, in the presence of external CA inhibitors, the efficiency with which external Ci is used for photosynthesis is decreased significantly in the absence of periplasmic CA (64, 71). Periplasmic CA probably increases the efficiency with which cells can access external Ci. This includes both the supply of CO_2 for diffusion across the plasma membrane and the supply of HCO_3^- to the plasma membrane HCO_3^- transport system.

The periplasmic CA of *C. reinhardtii* has been characterized extensively because it is abundant and easy to purify. The cDNA and genomic sequences of two members of a gene family, which are 92% homologous, have been identified and termed *CAH1* and *CAH2* (29, 30, 106). These code for 41.6-kDa polypeptides that undergo proteolytic cleavage and glycosylation to yield a 4.2-kDa small subunit and a 38-kDa large subunit (16). Investigators have suggested that these subunits form a heterotetramer L_2S_2 holoenzyme with the large subunits joined by disulfide bonds (41, 46). This periplasmic CA shows more homology with animal CAs than with higher plant chloroplast CA (30). The overall amino acid homology is about 20%, but there is greater conservation of sequences around the zinc-binding active site of the large subunit. The similarity of periplasmic CA to animal CAs is correlated with its high sensitivity to sulphonamide inhibitors such as ethoxyzolamide and acetazolamide, while higher plant chloroplast CAs and cyanobacterial carboxysomal CA appear to be relatively insensitive. More detailed information on this periplasmic CA can be found in recent reviews (16, 103).

Cytosolic Carbonic Anhydrase

In *C. reinhardtii,* recent studies using immunological and mass spectrometric assays have detected the presence of both cytosolic and chloroplastic CA activities (42, 67, 99). The cytosolic CA appears to be quite different from the chloroplastic CA. The cytosolic CA is larger in size (110 kDa vs 45 kDa), is very sensitive to sulphonamide inhibitors, and is antigenically distinct and soluble in nature (42, 99). Similar observations also have been made with other algal species. Multiple intracellular CA activities have been detected in *Chlorella* spp. (18, 82, 83) and *Dunaliella* spp. (33, 86). Whether the cytosolic CA is distinct from the periplasmic CA is unknown. The size of the protein and its sensitivity to sulphonamides indicates strongly that it could be a form of the periplasmic enzyme that is accumulated in the cytosol. This hypothesis could be tested with specific antibodies and micro-sequencing of purified protein.

The function of the cytosolic CA is speculative and depends to an extent on the location and specificity of the Ci transporters associated with microalgae. If there is a Ci transporter located on the chloroplast envelope and it has a particular preference for either CO_2 or HCO_3^-, then CA may be required to catalyze the transfer of this species from the plasma membrane to the chloroplast. This may involve simple facilitated diffusion of CO_2 across the cytosol or a more specific conversion of either CO_2 to HCO_3^- or vice versa.

Induction of Carbonic Anhydrase at Low CO_2

All forms of CA associated with *C. reinhardtii* are inducible when the cells are transferred from growth at high CO_2 to air. For periplasmic CA, the activity increases from almost nothing to quite high levels (16, 17, 30), while for cytosolic and chloroplast forms, there is a 5- to 20-fold increase (99). This induction is obviously an adaptation to allow more efficient acquisition of CO_2 and low external Ci levels. The increase in periplasmic CA will allow external HCO_3^- to be used more effectively, while the increased internal CA presumably allows the accumulated inorganic carbon within the cell to generate higher CO_2 levels around Rubisco, possibly with lower levels of accumulated Ci. In particular, this may be the case for the chloroplastic CA, if the stroma is the primary site for active Ci accumulation.

There has been some attempt to understand the mechanisms underlying the induction of periplasmic CA and to identify how the cell senses limiting CO_2 and translates this into modified gene expression (see 16). The expression of *CAH1* and *CAH2* genes in *C. reinhardtii* are under strong transcriptional control; however, these two genes appear to be regulated in an opposing manner (29, 30). The *CAH1* mRNA transcript levels increase when cells are transferred to low CO_2, while *CAH2* mRNA is decreased. The reverse is true when cells are transferred from air to high CO_2. *CAH1* transcription appears to be stimulated by light, while *CAH2* mRNA is increased in the dark. The changes in transcription are much greater for *CAH1* than for *CAH2* and this is the most highly expressed member of the gene family. The functional relevance of these two types of periplasmic CA and their differential expression remains to be resolved.

AQUATIC MACROALGAE AND ANGIOSPERMS

The occurrence of CA in various aquatic macrophytes from both marine and freshwater environments is well documented (2, 32, 103). Many of these organisms appear to have a CCM that is functionally similar to the mechanism found in eukaryotic microalgae and many must use external HCO_3^- for photosynthesis (2, 103). CA functions in this context in these organisms. There is evidence that both intracellular and external (periplasmic) CAs are present, but species and growth environments vary greatly, so it is difficult to specify roles for CA that can apply to all macrophytes. Despite this difficulty, evidence indicates that external and internal CA in separate cell compartments may be important to photosynthesis, in much the same way as has already been described for microalgae. A model of possible roles is given in Figure 2.

External Carbonic Anhydrase

The evidence for external CA comes from measurements of CA activities associated with whole cells as well as the effects on photosynthesis of impermeable sulphonamide inhibitors such as acetazolamide and dextran-bound sulphonamide (2, 8, 9, 25, 37, 38, 104). In many organisms, these inhibitors decrease the affinity of photosynthesis for external Ci, particularly at alkaline pHs where HCO_3^- is the dominant form of external Ci. Such experiments have led to the conclusion that external CA, when present, is involved in the conversion of HCO_3^- to CO_2, which then diffuses across the plasma membrane into the cell. However, the extent to which this involvement of CA explains all the HCO_3^- utilization of a macrophyte depends on the individual organism. In some marine macroalgae, such as *Gracilaria tenuistipitata, Fucus serratus, Laminaria saccharina,* and *Ulva rigida* (8, 37, 38), most of the HCO_3^- uptake is the result of indirect conversion to CO_2. However, in other macrophytes there appear to be additional pathways for HCO_3^-, such as the direct active uptake of HCO_3^- that has been described for the marine algal macrophytes *Ulva fasciata, Ulva lactuca,* and *Gracilaria conferta* (7, 22, 43). Alternatively, CA in external acid regions may aid in the conversion of HCO_3^- to CO_2, as has been described for freshwater angiosperms such as *Potamogeton* and *Elodea* spp., and the giant alga *Chara corallina,* (2, 78, 81). Thus external CA, when present, fulfills a role that is similar to the role proposed in Figure 2 for microalgae. Little work has been done on characterizing or purifying the external CA from any macrophyte. It would be interesting to see how this enzyme relates to periplasmic CAs from microalgae such as *C. reinhardtii,* and whether it is related to the chloroplastic CA of the same organism.

Internal Carbonic Anhydrase

The presence of internal CA in macrophytes has been indicated by the measurement of CA activities in whole cell homogenates and through the effects on photosynthesis of the permeable CA inhibitor ethoxyzolamide (8, 9, 32, 37, 104). Levels of CA in different classes of macrophytes vary considerably and little effort has been made to localize the activity between the chloroplast and the cytosol. However, there is evidence of a chloroplast association with some CA in the marine macroalgae *Ulva rigida* (8), *Gracilaria tenustipitata* (37), *Codium decorticatum,* and *Udotea flabellum* (92). In *Ulva* and *Gracillaria* spp., both soluble and pelletable forms of internal CA were indicated (8, 37). In most macrophytes examined, ethoxyzolamide appears to reduce the affinity of photosynthesis for external Ci at both acidic and alkaline pHs, indicating an important role for internal CA in photosynthesis.

The hypothesized role of internal CA is based on our knowledge of microalgae and C_3 photosynthesis. The location and nature of any active Ci transport systems remains to be determined for any macrophyte species, but plasma membrane and chloroplast envelope systems are a possibility and there is probably species to species variation. A cytosolic CA could facilitate the diffusion of Ci to the chloroplast envelope or convert HCO_3^- to CO_2 if HCO_3^- was accumulated actively in the cytosol by plasma membrane Ci transport. A chloroplast CA could catalyze a simple facilitated diffusion of CO_2 to the active site of Rubisco, as in C_3 plants, or it could have a special association with Rubisco, as in the cyanobacterial carboxysome or the pyrenoid, using a chloroplast HCO_3^- pool to elevate CO_2 for fixation by Rubisco.

Chloroplast pyrenoids are abundant in macroalgae, but they show a considerable degree of diversity in structure (24, 35, 49). It will become increasingly important to establish the importance of these pyrenoids in macroalgal photosynthesis and in CCMs. Advances in this area are unlikely to be made until the function of pyrenoids in microalgae has been established more clearly. Heterologous DNA and antibody probes derived from microalgal and higher plant studies may be useful in this effort if sufficient homology exists.

Some macrophytes appear to have a type of CCM that relies more on fixation of C_4 acids, and subsequent decarboxylation in the chloroplast, than on active accumulation of Ci. They may perform a C_4-type photosynthesis within the one cell (10, 91). In at least one macrophyte, *Hydrilla verticllata* (10), a cytosolic PEP carboxylase appears to be involved in the formation of malate, which would require HCO_3^- as a substrate. A cytosolic CA would be required if CO_2 diffusing through the plasma membrane were to be fixed rapidly, as in C_4 plants. For other macrophytes, such as *Udotea flabellum,* where PEP-CK seems to be involved as a carboxylase (91), its preference for CO_2 as a substrate would negate the need for cytosolic CA, unless HCO_3^- was actively transported into this compartment. In macrophytes performing a C_4-type of photosynthesis, chloroplastic CA might not be needed if CO_2 is produced by decarboxylation and if leakage of CO_2 is to be minimized.

C_4 PLANTS

CA plays a crucial role in the photosynthesis of C_4 plants, acting as the first enzyme in a pathway catalyzing the fixation of CO_2 into C_4 acids (40). Although the initial carboxylation reaction of C_4 photosynthesis is catalyzed by PEP carboxylase, this reaction uses HCO_3^- rather than CO_2 as a substrate. To provide substrate for this reaction, the CO_2 entering the mesophyll cells from the atmosphere must be converted rapidly to HCO_3^-, which is mediated by CA (Figure 3). Supporting this role is the fact that the CA of C_4 leaves is almost exclusively confined to the cytosol of the mesophyll cells (13), together with PEP carboxylase (see 39). The function of CA in C_4 plants has been modeled using measured activities and kinetic parameters (40, 45). This analysis indicates that although high levels of CA activity are observed in C_4 plants, the calculated activity of this enzyme at in vivo CO_2 concentrations may be just enough to support the observed rates of C_4 photosynthesis. Some studies indicate that the activities of both PEP carboxylase and CA may be regulated tightly in the mesophyll of C_4 plants such as maize (14, 98). This regulation may maintain a balance of activities that is just sufficient for photosynthesis under various environmental conditions such as different growth irradiances and nitrogen nutrition.

CA activity is absent from the bundle sheath tissue of C_4 plants (13, 45). In all types of C_4 plants, the C_4 acids that are transported into the bundle sheath are decarboxylated to yield CO_2 (39). Models have shown that if CA were present in this tissue, a significant leak rate could be established through the conversion of CO_2 to HCO_3^-, followed by leakage of the HCO_3^- to the mesophyll via the plasmodesmata (45). The diffusion of CO_2 from the bundle sheath is minimized in two ways: (a) by the presence of a direct diffusion barrier in the form of the bundle sheath and (b) by the absence of CA. These leakage barriers allow CO_2 to be elevated some 10–50 times over the ambient concentration and they minimize the leakage and recycling of bundle sheath CO_2 to less than 50%. Although this cannot be proven yet, it is a probable explanation for the apparent absence of CA from the bundle sheath. A good test of this model may be to express a foreign CA in the bundle sheath of a C_4 plant using a general non-specific leaf promoter, and to assess its effects on physiology.

It is generally accepted that C_4 plants have evolved from C_3 ancestors (39); thus, it is reasonable to expect that C_4 CA has probably evolved from a pre-existing C_3 CA and its function and cellular location have been modified to meet its new role. At present there are no published DNA sequences available from C_4 sources for the relevant comparisons to be made. Immunological studies of CAs from a variety of sources show a degree of cross-reactivity between both C_3 and C_4 species of monocot and dicot origin. Cross-reactivity varies considerably, but in most cases a single protein band of 24–26 kDa was detected (11). This is similar to the size of the mature C_3 chloroplast CA (12, 27, 58, 59, 85, 93). The kinetic properties of C_4 CA appear akin to C_3 enzymes, with similar affinities for CO_2 and sensitivity to inhibitors (40).

CAM PLANTS

CAM plants are unique in that they perform elements of both C_4 and C_3 photosynthesis within the same cell, with the two phases separated in time rather than space (95). C_4 acid synthesis, via PEP carboxylase, occurs in the dark, while the decarboxylation and CO_2 refixation by Rubisco proceeds in the following light period. In addition to this strict CAM fixation, there are periods when CO_2 may be fixed directly by Rubisco in the light, thus avoiding the need for PEP carboxylase. There have been few studies of CA in CAM plants and as a result there is limited evidence for defining the function of CA in these plants. Based on our understanding of C_4 and C_3 plants, the role of CA in CAM photosynthesis may be speculated on and a possible model for its function is shown in Figure 3. A cytosolic CA should be required to convert CO_2 to HCO_3^-, the substrate for PEP carboxylase, in the same manner as was described for C_4 plants. In addition, a chloroplastic CA may be required to allow efficient C_3 photosynthesis during periods of direct CO_2 fixation by Rubisco (see 95). If CAM plants were fixing CO_2 via a strict C_4 acid fixation cycle, then little or no chloroplastic CA would be required. Facultative CAM plants that can spend long periods performing only C_3 photosynthesis (95) would not require the cytosolic CA.

Measured total CA activity in CAM plants seems to be similar to that in both C_3 and C_4 plants (111). However, there is some confusion about its cellular location. In some species examined, a cytosolic location appeared dominant while in others CA was most active in the chloroplast fraction (111). The location of CA probably varies depending on the amount of CAM and C_3 photosynthesis that a plant is performing. It would be interesting to examine the induction of both types of activity under growth conditions that entrain CAM plants in either their CAM or C_3 photosynthetic modes.

C_3 PLANTS

CA has long been recognized as an abundant protein in C_3 plants, representing some 0.5–2% of total soluble leaf protein (19, 68). The predominant CA in C_3

leaves is located in the stroma of chloroplasts (44, 110), but some investigators
have suggested the existence of cytosolic forms (34, 73, 88). In recent years,
the chloroplast CA genes have been isolated and characterized from spinach

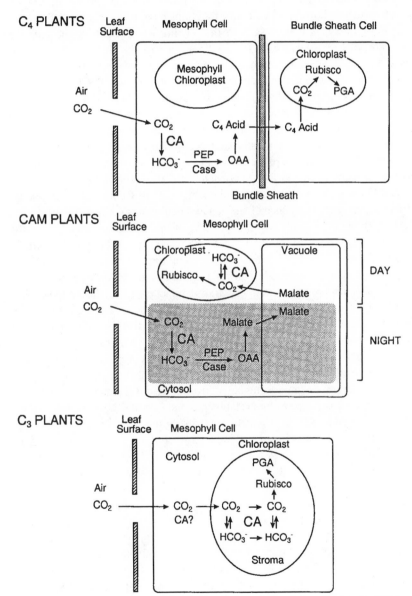

Figure 3   Models showing the roles of carbonic anhydrase during photosynthesis in C₄, CAM and
C₃ higher plants.

(12, 27), pea (58, 93), *Arabidopsis thaliana* (85), and tobacco (59). These are nuclear encoded proteins that have a transit peptide for import into the chloroplast and that show a high degree of homology within the mature polypeptide.

Little progress has been made in fully defining the role of CA in C_3 photosynthesis; however, two major functions have been proposed. The first, and most likely, is that the CA in the alkaline chloroplast stroma aids the diffusion of CO_2 species (CO_2 and HCO_3^-) across the chloroplast by facilitated diffusion, thus minimizing the gradient in CO_2 concentration between the chloroplast envelope and the active site of Rubisco (Figure 3). CO_2 is assumed to be the only Ci species that moves across the chloroplast envelope. Cytosolic CA has little function in such a model because HCO_3^- levels are low in the more acidic cytosol and the diffusion distance between the cell wall and the chloroplast envelope is generally small (19). The second proposed function involves CA and Rubisco participating in a special association, perhaps similar to that already described for cyanobacterial carboxysomes. This association would allow CO_2 to be elevated at the active site of Rubisco. Unfortunately, direct experimental evidence of such an association has not been found.

In defining the role of CA, attempts have been made to specifically alter the activity of chloroplast CA and examine the resultant effect on photosynthesis and other processes. CA inhibitors, such as ethoxyzolamide and acetazolamide, have been used with both intact plants and isolated chloroplasts (44, 105). These inhibitors have a problem penetrating intact plants, so these experiments have not been fruitful. Studies with intact chloroplasts have indicated that there is only a small effect of CA inhibition on CO_2 fixation (44). Modulation of CA activity in intact plants by controlling either zinc or nitrogen nutrition has also been used (23, 87). The results of these experiments have suggested that CA may be present in excess and that considerable reduction in its activity has little effect on the photosynthetic performance of C_3 plants.

The most direct approach to establishing the role of C_3 CA is probably via the isolation and analysis of genetic mutants. Some attempts have been made to do this by isolating chemically induced *A. thaliana* mutants that are deficient in CA, but even though large numbers of plants were screened, a suitable mutant was not recovered (N Artus and C Somerville, personal communication). However, chloroplastic CA in tobacco plants was specifically reduced recently using antisense RNA techniques (80). This genetic manipulation resulted in plants with a range of CA activities down to 1% of wild-type levels. The most extreme reduction of CA resulted in plants with no detectable changes in photosynthesis per unit leaf area at ambient levels of CO_2 and similar amounts of Rubisco. The only measurable parameters that showed significant changes were a slightly increased stomatal conductance (0.4 vs 0.45 mol m^{-2} s^{-1}) and a reduction (about 1‰) in ^{13}C isotope discrimination of leaf dry matter. This reduction in discrimination can be equated to a 15–20 µbar decrease in the average chloroplastic CO_2 concentration. These results are consistent with a facilitated diffusion model for the action of CA (19), which predicts that C_3 plants may improve their photosynthetic rate in air by about 5% in the presence of chloroplast CA. CA probably functions in C_3 photosynthesis as depicted in Figure 3. It would be useful to study mutants that completely lacked chloroplast CA, to determine if small amounts of CA, such as those in the transgenic tobacco, can enhance photosynthesis in some other manner.

C3 chloroplasts appear to contain a relatively large excess of CA, but CA protein levels seem to be regulated in a similar fashion to that described for Rubisco. This conclusion is supported by studies of nitrogen nutrition and growth at different CO_2 levels, in which amounts of both Rubisco and CA levels vary in a similar manner (61, 72, 74, 85). One possible reason for this result is that a certain ratio of CA to Rubisco is optimal for efficient photosynthesis and nitrogen use efficiency (19). However, there is some variation in the levels of CA in different C_3 plant species. The most notable examples are some wheat cultivars, which can have up to 30-fold less CA than do other monocot and dicot species (40, 61). The limitation placed on photosynthesis in C_3 plants may be species-specific and the CO_2 transfer resistance to the chloroplast may be lessened in some species by increased chloroplast CA. The studies of antisense-CA tobacco plants (80) and growth of plants at different zinc levels (23, 87) show that coordination between CA and Rubisco expression is not obligatory and can be broken easily without significant effects on photosynthesis. Thus the importance and mechanistic basis for the apparent coordination between CA and Rubisco are still unknown.

EVOLUTION OF PHOTOSYNTHETIC CARBONIC ANHYDRASES

Characterization of animal-type CA enzymes has progressed considerably in the last few years with the determination of the crystal structure of the human CA2 protein and the use of site directed mutagenesis of an increasing number of cloned enzymes (26, 94, 107). In contrast, there has been relatively little work on the characterization of purified CAs from photosynthetic sources. However, this will improve through current work on the crystal structure of pea chloroplast CA (Y Lunqvist, personal communication) and the increasing availablitiy of cloned and expressed enzymes (84). Thus, any comments made now on the evolution of photosynthetically related CAs must be tentative and based on limited evidence.

The CA proteins and genes from higher plant chloroplasts (12, 27, 58, 59, 85, 93), cyanobacteria (31, 113), and *E. coli* (36) have essentially no homology with animal CAs (31, 94), although they have homology between themselves. Indeed, there may have been convergent evolution of CA proteins from separate prokaryotic and eukaryotic ancestors (31). This makes it difficult to extrapolate from any knowledge of the structure and function of animal CAs to the properties and possible diversity of CAs in photosynthetic organisms, with the exception of the periplasmic CA in *C. reinhardtii,* which has considerable homology with eukaryotic CAs (31). The reaction mechanisms of higher plant chloroplastic CAs have evolved their reaction mechanism to a state of perfection that approaches that of the HCA2 enzyme (the k_{cat}/K_m term for the enzyme appears to be in the order of 10^7–10^8 M s^{-1}) (94). This indicates that at limiting substrate concentrations of either CO_2 or HCO_3^-, the reaction is diffusion- or encounter-limited, and there may be little scope for improving the catalytic performance of the enzyme. Whether this is also the case for cyanobacterial and algal chloroplastic CAs remains to be determined.

The $K_m(CO_2)$ and $K_m(HCO_3^-)$ for the higher plant chloroplast and C_4 mesophyll CA enzymes are on the order of 1–3 mM and 30–50 mM, respectively (34, 40, 103). This means that in most photosynthetic systems, the enzyme may be operating far below saturation. This is particularly true in higher plant chloroplasts and in C_4 mesophyll, where CO_2 is about 8 and 4 µM, respectively, and HCO_3^- does not exceed 0.8 mM. However, for cyanobacteria and algae with CCMs, this may not be the case. In cyanobacteria, the carboxysomal [CO_2] and [HCO_3^-] are likely to be around 0.5–1.0 mM and 20–50 mM, respectively (76, 89, 90), which means that carboxysomal CA may be operating around its half–saturation point. It is interesting that the levels of CA found in photosynthetic organisms seem inversely proportional to the inorganic carbon concentrations of the environments in which the enzyme operates. Thus the highest levels of CA are found in higher plants and the lowest levels are found in cyanobacteria and the chloroplasts of algae (40, 79, 80, 99). Even the production of high levels of periplasmic CA in algae at low external CO_2 is consistent with this notion. These findings indicate that a major adaptation in photosynthetically related CAs may have occurred in the regulation of their quantity rather than their kinetic properties. This would make sense if the enzyme from all sources is close to evolutionary perfection. Thus it is the product of [substrate] × [enzyme] (which determines flux rate) that is important in determining the required amount of CA that any photosynthetic organism will have to synthesize to meet its photosynthetic needs. The ratio of CA/Rubisco activities within photosynthetic organisms (ignoring external CA in aquatic species) may be a measure of the extent to which CO_2 is elevated at the site of both CA and Rubisco. On a normalized scale, the ratio of CA to Rubisco would be around 1–5 for *Synechococcus* PCC7942 (79), 4–16 for *C. reinhardtii* (99) and 500–2000 for C_4 and C_3 plants (40, 80). C_4 species are obviously anomolous in that they do have a CCM and a high ratio, but in this case, CA is localized in the low-CO_2 environment of the mesophyll and Rubisco is in the high-CO_2 bundle sheath.

CONCLUSIONS

Carbonic anhydrase plays a key role in photosynthetic CO_2 fixation in a wide range of photosynthetic organisms, but there is obvious diversity in the functional roles it has been required to perform. CA functions in three primary modes in photosynthetic systems: (*a*) to convert HCO_3^- to CO_2, for fixation by Rubisco; (*b*) to convert CO_2 to HCO_3^- for fixation by PEP carboxylase; and (*c*) to provide rapid equilibration between CO_2 and HCO_3^- so that facilitated diffusion of CO_2 can be enhanced. In its most novel role, CA and Rubisco can be associated in the carboxysomes of cyanobacteria and possibly the pyrenoids of algae. In these associations, the CA dehydrates an accumulated HCO_3^- pool, which results in the localized elevation of CO_2 around the active site of Rubisco. This role of CA can only be supported through the operation of active inorganic carbon transport systems that sustain elevated internal HCO_3^- pools. Rubisco and CA associations may be a common feature of many aquatic phototrophs that appear to have active CCMs. CA also plays a specialized role as an extracellular or periplasmic enzyme that is produced by many aquatic phototrophs to aid in the conversion of external HCO_3^- to CO_2, for subsequent uptake by the cell. In C4 and CAM higher plants, cytosolic CA plays a vital role in the conversion of CO_2 to HCO_3^-, which can then serve as the substrate for PEP carboxylase, the primary carboxylation enzyme in these systems. Finally, in its least specialized role, CA may simply aid in the facilitated diffusion of CO_2 through the cytosol and the chloroplast stroma. This is the case for the chloroplastic CA of C3 higher plants and possibly also the cytosolic CA of number of aquatic phototrophs.

Future research on photosynthetic CAs may take several directions. By far the most intriguing will be to discover how both the cyanobacterial carboxysome and the algal pyrenoid can form specific associations between CA and Rubisco, which are then used to locally elevate CO_2. Advances will probably be made at the protein structure and function level using molecular genetics and protein chemistry, but may include more sophisticated techniques for in situ protein visualization. Results obtained in these areas may help explain why such specific associations seem to have been used successfully by some aquatic organisms but appear to be absent in higher plants. A more detailed study of pyrenoids from different species is needed, considering the diversity of pyrenoid structures that exist in nature. Does CA play a role in all these structures or are there significant differences in the extent to which pyrenoids contribute to CCMs? The role of cytosolic CA in photosynthetic species that do not use PEP carboxylase needs to be defined. The existence of a cytosolic

CA that is distinct from chloroplastic or periplasmic CAs needs to be established as does its contribution to overall photosynthetic CO_2 fixation. In this context, further studies of cytosolic CA from C_4 plants may help explain how such enzymes have evolved (e.g. by modification of pre-existing chloroplastic enzymes). Finally, it will be interesting to see the development of structure-functions studies of CA enzymes purified from photosynthetic sources. Such studies could give a clear picture of how prokaryotic-type CAs have developed their CA reaction mechanism in what seems to be a process of convergent evolution when compared to CAs from animal or eukaryotic sources. Whether investigators find the same or different structural elements as those of animal CAs will be intriguing and illuminating and will help us understand how CAs have adapted to their various roles in photosynthesis.

Literature Cited

1. Aizawa K, Miyachi S. 1986. Carbonic anhydrase and CO_2 concentrating mechanisms in microalgae and cyanobacteria. *FEMS Microbiol. Rev.* 39:215–33
2. Badger MR. 1987. The CO_2 concentrating mechanism in aquatic phototrophs. In *The Biochemistry of Plants: A Comprehensive Treatise.* Vol. 10: *Photosynthesis,* ed. MD Hatch, NK Boardman, pp. 219–74. New York: Academic
3. Badger MR, Kaplan A, Berry JA. 1980. Internal inorganic carbon pool of *Chlamydomonas reinhardtii*: evidence for a carbon dioxide concentrating mechanism. *Plant Physiol.* 66:407–13
4. Badger MR, Price GD. 1989. Carbonic anhydrase associated with the cyanobacterium *Synechococcus* PCC7942. *Plant Physiol.* 89:51–60
5. Badger MR, Price GD. 1992. The CO_2 concentrating mechanism in cyanobacteria and green algae. *Physiol. Plant.* 84:606–15
6. Bedu S, Laurent B, Joset F. 1992. Membranous and soluble carbonic anhydrase activities in a cyanobacterium, *Synechocystis* PCC6803. In *Research in Photosynthesis,* ed. N. Murata, 3:819–22. Dordrecht: Kluwer Academic
7. Beer S, Israel A. 1990. Photosynthesis of *Ulva fasciata.* IV pH, carbonic anhydrase and inorganic carbon conversion in the unstirred layer. *Plant Cell Environ.* 13:555–60
8. Bjork M, Haglund K, Ramazanov Z, Garcia-Reina G, Pedersén M. 1992. Inorganic carbon assimilation in the green seaweed *Ulva rigida* C. Ag. (Chlorophyta). *Planta* 187:152–56
9. Bjork M, Haglund K, Ramazanov Z, Pedersén M. 1993. Inducible mechanisms for HCO_3^- utilization and repression of photorespiration in protoplasts and thalli of three species of *Ulva* (Chlorophyta). *J. Phycol.* 29:166–73
10. Bowes G, Salvucci ME. 1989. Plasticity in the photosynthetic carbon metabolism of submerged aquatic macrophytes. *Aquatic Bot.* 34:233–66
11. Burnell JN. 1990. Immunological study of carbonic anhydrase in C_3 and C_4 plants using antibodies to maize cytosolic and spinach chloroplastic carbonic anhydrase. *Plant Cell Physiol.* 31:423–27
12. Burnell JN, Gibbs MJ, Hatch MD. 1990. Spinach chloroplastic carbonic anhydrase. Nucleotide sequence analysis of cDNA. *Plant Physiol.* 92:37–40
13. Burnell JN, Hatch MD. 1988. Low bundle sheath carbonic anhydrase is apparently essential for effective C_4 pathway operation. *Plant Physiol.* 86:1252–56
14. Burnell JN, Suzuki I, Sugiyama T. 1990. Light induction of the effect of nitrogen

status upon the activity of carbonic anhydrase in maize leaves. *Plant Physiol.* 94:384–87

15. Codd GA, Marsden WJN. 1984. The carboxysomes (polyhedral bodies) of autotrophic prokaryotes. *Biol. Rev.* 59:389–422

16. Coleman JR. 1991. The molecular and biochemical analyses of CO_2 concentrating mechanisms in cyanobacteria and green algae. *Plant Cell Environ.* 14:861–67

17. Coleman JR, Grossman AR. 1984. Biosynthesis of carbonic anhydrase in *Chlamydomonas reinhardtii* during adaptation to low CO_2. *Proc. Natl. Acad. Sci. USA* 81:6049–53

18. Coleman JR, Rotatore C, Williams T, Colman B. 1991. Identification and localisation of carbonic anhydrase in two *Chlorella* species. *Plant Physiol.* 95:331–34

19. Cowan IR. 1986. Economics of carbon fixation in higher plants. In *On the Economy of Plant Form and Function,* ed. TJ Givinish, pp. 130–70. London: Cambridge Univ. Press

20. Deutsch HF. 1987. Carbonic anhydrases. *Int. J. Biochem.* 19:101–13

21. Dodge JD. 1973. *The Fine Structure of Algal Cells.* New York: Academic

22. Drechsler Z, Beer S. 1991. Utilization of inorganic carbon by *Ulva lactuca. Plant Physiol.* 97:1439–44

23. Edwards GE, Mohamed AK. 1973. Reduction in carbonic anhydrase activity in zinc deficient leaves of *Phaseolus vulgaris* L. *Crop. Sci.* 13:351–54

24. Ekman P, Lignell Å. Pedersén M. 1989. Localization of ribulose-1,5-bisphosphate carboxylase/oxygenase in *Gracilaria secundata* (Rhodophyta) and its role as a nitrogen storage pool. *Bot. Marina* 32:527–34

25. Elzenga JTM, Prins HBA. 1988. Adaptation of *Elodea* and *Potamogeton* to different inorganic carbon levels and the mechanism for photosynthetic bicarbonate utilization. *Aust. J. Plant Physiol.* 15:727–35

26. Eriksson AE, Jones TA, Liljas A. 1988. Refined structure of human carbonic anhydrase II at 2.0 Å resolution. *Proteins: Struct. Funct. Genet.* 4:274–82

27. Fawcett TW, Browse JA, Volokita M, Bartlett SG. 1990. Spinach carbonic anhydrase primary structure deduced from the sequence of a cDNA clone. *J. Biol. Chem.* 265:5414–17

28. Deleted in proof

29. Fujiwara S, Fukuzawa H, Tachiki A, Miyachi S. 1990. Structure and differential expression of two genes encoding carbonic anhydrase in *Chlamydomonas reinhardtii. Proc. Natl. Acad. Sci. USA* 87:9779–83

30. Fukuzawa H, Fujiwara S, Yamamoto Y, Dionisio-Sese ML, Miyachi S. 1990. cDNA cloning, sequence, and expression

of carbonic anhydrase in *Chlamydomonas reinhardtii*: regulation by environmental CO_2 concentration. *Proc. Natl. Acad. Sci. USA* 87:4383–87

31. Fukuzawa H, Suzuki E, Komukai Y, Miyachi S. 1992. A gene homologous to chloroplast carbonic anhydrase (*icfA*) is essential to photosynthetic carbon fixation in the cyanobacterium *Synechococcus* PCC7942. *Proc. Natl. Acad. Sci. USA* 89: 4437–41

32. Giordano M, Maberly SC. 1989. Distribution of carbonic anhydrase in British marine macroalgae. *Oecologia* 81:534–39

33. Goyal A, Shiraiwa Y, Husic HD, Tolbert NE. 1992. External and internal carbonic anhydrases in *Dunaliella* species. *Marine Biol.* 113:349–55

34. Graham D, Reed ML, Patterson BD, Hockley DG, Dwyer MR. 1984. Chemical properties, distribution and physiology of plant and algal carbonic anhydrases. *Ann. NY Acad. Sci.* 429:222–37

35. Griffiths DJ. 1970. The pyrenoid. *Bot. Rev.* 36:29–58

36. Guilloton MB, Korte JJ, Lamblin AF, Fuchs JA, Anderson PM. 1992. Carbonic anhydrase in *Escherichia coli*. A product of the cyn operon. *J. Biol. Chem.* 267:3731–34

37. Haglund K, Bjork M, Ramazanov Z, Garcia-Reina G, Pedersén M. 1992. Role of carbonic anhydrase in photosynthesis and inorganic carbon assimilation in the red alga *Gracillaria tenuistipitata. Planta* 187: 275–81

38. Haglund K, Ramazanov Z, Mtolera M, Pedersén M. 1992. Role of carbonic anhydrase in light-dependent alkalization by *Fucus serratus* L. and *Laminaria saccharina* (L.) Lamour. (Phaeophyta). *Planta* 188:1–6

39. Hatch MD. 1987. C_4 photosynthesis: a unique blend of modified biochemistry, anatomy and ultrastructure. *Biochim. Biophys. Acta* 895:81–106

40. Hatch MD, Burnell JN. 1990. Carbonic anhydrase activity in leaves and its role in the first step of C_4 photosynthesis. *Plant Physiol.* 93:380–83

41. Husic HD, Hsieh SC, Berrier AL. 1991. Effects of dithiothreitol on the catalytic activity, quaternary structure, and sulphonamide binding properties of an extracellular carbonic anhydrase from *Chlamydomonas reinhardtii. Biochim. Biophys. Acta* 1078: 35–42

42. Husic HD, Kitayama M, Togasaki RK, Moroney JV, Morris KL, Tolbert NE. 1989. Identification of intracellular carbonic anhydrase in *Chlamydomonas reinhardtii* which is distinct from the periplasmic form of the enzyme. *Plant Physiol.* 89:904–9

43. Israel A, Beer S. 1992. Photosynthetic car-

bon acquisition in the red alga *Gracilaria conferta*. II. Rubisco carboxylase kinetics, carbonic anhydrase and HCO_3^- uptake. *Marine Biol.* 112:697–700

44. Jacobsen BS, Fong F, Heath RL. 1975. Carbonic anhydrase of spinach. Studies on its location, inhibition and physiological function. *Plant Physiol.* 55:468–74

45. Jenkins CLD, Furbank RT, Hatch MD. 1989. Mechanism of C_4 photosynthesis. A model describing the inorganic carbon pool in bundle sheaths. *Plant Physiol.* 91:1372–81

46. Kamo T, Shimogawara K, Fukuzawa H, Muto S, Miyachi S. 1990. Subunit constitution of carbonic anhydrase from *Chlamydomonas reinhardtii*. *Eur. J. Biochem.* 192: 557–62

47. Kaplan A, Schwarz R, Lieman-Hurwitz J, Reinhold L. 1991. Physiological and molecular aspects of the inorganic carbon concentrating mechanism in cyanobacteria. *Plant Physiol.* 97:851–55

48. Katzman GL, Marcus Y, Moroney JV, Carlson SJ, Togasaki RK. 1989. Carbonic anhydrase activity in isolated chloroplasts of wild-type and high CO_2 dependent mutants of *Chlamydomonas reinhardtii* as studied by a new assay. *Plant Physiol.* 89:173

49. Kerby NW, Evans LV. 1981. Pyrenoid protein from the brown alga *Pilayella littoralis*. *Planta* 151:469–75

50. Kuchitsu K, Tsuzuki M, Miyachi S. 1988. Changes in starch localization within the chloroplast induced by changes in CO_2 concentration during growth of *Chlamydomonas reinhardtii*: independent regulation of pyrenoid starch and stromal starch. *Plant Cell Physiol.* 29:1269–78

51. Kuchitsu K, Tsuzuki M, Miyachi S. 1988. Characterization of the pyrenoids isolated from the unicellular green algae *Chlamydomonas reinhardtii*: particulate form of Rubisco protein. *Protoplasma* 144:17–24

52. Kuchitsu K, Tsuzuki M, Miyachi S. 1991. Polypeptide composition and enzyme activities of the pyrenoid and its regulation by CO_2 concentration in unicellular green algae. *Can. J. Bot.* 69:1062–69

53. McKay RML, Gibbs SP. 1989. Immunocytochemical localization of ribulose 1,5-bisphosphate carboxylase/oxygenase in light-limited and light-saturated cells of *Chlorella pyrenoidosa*. *Protoplasma* 149: 31–37

54. McKay RML, Gibbs SP. 1991. Composition and function of pyrenoids: cytochemical and immunocytochemical approaches. *Can. J. Bot.* 69:1040–52

55. McKay RML, Gibbs SP, Vaughn KC. 1991. Rubisco activase is present in the pyrenoid of green algae. *Protoplasma* 162:38–45

56. McKay RML, Gibbs SP, Espie GS. 1993. Effect of dissolved inorganic carbon on the expression of carboxysomes, localization of Rubisco and the mode of inorganic carbon transport in cells of the cyanobacterium *Synechococcus* UTEX 625. *Arch. Microbiol.* 159:21–29

57. Magid E, Turbeck BO. 1968. The rates of spontaneous hydration of CO_2 and the reciprocal reaction in neutral aqueous solutions between 0° and 38°. *Biochim. Biophys. Acta* 165:515–24

58. Majeau N, Coleman JR. 1991. Isolation and characterization of a cDNA coding for pea chloroplastic carbonic anhydrase. *Plant Physiol.* 95:264–68

59. Majeau N, Coleman JR. 1992. Nucleotide sequence of a complementary DNA encoding tobacco chloroplastic carbonic anhydrase. *Plant Physiol.* 100:1077–78

60. Deleted in proof

61. Makino A, Sakashita H, Hidema J, Mae T, Ojima K, Osmond B. 1992. Distinctive responses of ribulose-1,5-bisphosphate carboxylase and carbonic anhydrase in wheat leaves to nitrogen nutrition and their possible relationships to CO_2-transfer resistance. *Plant Physiol.* 100:1737–43

62. Miller AG, Espie GS, Canvin DT. 1990. Physiological aspects of CO_2 and HCO_3^- transport by cyanobacteria: a review. *Can. J. Bot.* 68:1291–302

63. Miyachi S, Tsuzuki M, Avramova ST. 1983. Utilization modes of inorganic carbon for photosynthesis in various species of *Chlorella*. *Plant Cell Physiol.* 24:441–51

64. Moroney JV, Husic HD, Tolbert NE. 1985. Effects of carbonic anhydrase inhibitors on inorganic carbon accumulation by *Chlamydomonas reinhardtii*. *Plant Physiol.* 79: 177–83

65. Moroney JV, Kitayama M, Togasaki RK, Tolbert NE. 1987. Evidence for inorganic carbon transport by intact chloroplasts of *Chlamydomonas reinhardtii*. *Plant Physiol.* 83:460–63

66. Moroney JV, Mason CB. 1991. The role of the chloroplast in inorganic carbon acquisition by *Chlamydomonas reinhardtii*. *Can. J. Bot.* 69:1017–24

67. Moroney JV, Togasaki RK, Husic HD, Tolbert NE. 1987. Evidence that an internal carbonic anhydrase is present in 5% CO_2 grown and air-grown *Chlamydomonas*. *Plant Physiol.* 84:757–61

68. Okabe K, Yang S, Tsuzuki M, Miyachi S. 1984. Carbonic anhydrase: its content in spinach leaves and its taxonomic diversity studied with anti-spinach leaf carbonic anhydrase antibody. *Plant Sci. Lett.* 33:145–53

69. Palmqvist K. 1993. Photosynthetic CO_2 use efficiency in lichens and their isolated photobionts: possible role of a CO_2 concen-

trating mechanism in cyanobacterial lichens. *Planta* 191:48–56

70. Palmqvist K, Ogren E, Lernmark U. 1993. The CO_2-concentrating mechanism is absent in the green alga *Coccomyxa*. A comparative study of photosynthetic CO_2 and light responses of *Coccomyxa, Chlamydomonas reinhardtii* and barley protoplasts. *Plant Cell Environ.* 17:In press

71. Palmqvist K, Yu J-W, Badger MR. 1993. Carbonic anhydrase activity and inorganic carbon fluxes in low- and high-Ci cells of *Chlamydomonas reinhardtii* and *Scenedesmus obliquus*. *Physiol. Plant.* In press

72. Peet MM, Huber SC, Patterson DT. 1986. Acclimation to high CO_2 in monoecious cucumber. II. Carbon exchange rates, enzyme activities, and starch and nutrient conditions. *Plant Physiol.* 80:63–67

73. Poincelot RP. 1979. Carbonic anhydrase. In *Encyclopedia of Plant Physiology (NS),* ed. M Gibbs, E Latzko, 6(2):230–38. Berlin: Springer-Verlag

74. Porter MA, Grodzinski B. 1984. Acclimation to high CO_2 in bean: carbonic anhydrase and ribulose bisphosphate carboxylase. *Plant Physiol.* 74:413–16

75. Price GD, Badger MR. 1989. Ethoxyzolamide inhibition of CO_2 uptake in the cyanobacterium *Synechococcus* PCC7942 without apparent inhibition of internal carbonic anhydrase activity. *Plant Physiol.* 89:37–43

76. Price GD, Badger MR. 1989. Isolation and characterization of high-CO_2 requiring mutants of the cyanobacterium *Synechococcus* PCC7942: two phenotypes that accumulate inorganic carbon but are unable to generate CO_2 within the carboxysome. *Plant Physiol.* 91:514–25

77. Price GD, Badger MR. 1989. Expression of human carbonic anhydrase in the cyanobacterium *Synechococcus* PCC7942 creates a high CO_2-requiring phenotype. *Plant Physiol.* 91:505–13

78. Price GD, Badger MR, Bassett ME, Whitecross MI. 1985. Involvement of plasmalemmasomes and carbonic anhydrase in photosynthetic utilization of bicarbonate in *Chara corallina*. *Aust. J. Plant Physiol.* 12:241–56

79. Price GD, Coleman JR, Badger MR. 1992. Association of carbonic anhydrase activity with carboxysomes isolated from the cyanobacterium *Synechococcus* PCC7942. *Plant Physiol.* 100:784–93

80. Price GD, von Caemmerer S, Evans JR, Yu J-W, Lloyd J, et al. 1993. Specific reduction of chloroplast carbonic anhydrase activity by antisense RNA in transgenic tobacco plants has a minor effect on photosynthetic CO_2 assimilation. *Planta.* In press

81. Prins HBA, Elzenga JTM. 1989. Bicarbon-

ate utilization: function and mechanism. *Aquatic Bot.* 34:59–83

82. Pronina NA, Semenenko VE. 1984. Localization of membrane bound and soluble forms of carbonic anhydrase in the *Chlorella* cell. *Fiziol. Rast.* 31:241–51

83. Pronina NA, Semenenko VE. 1990. Membrane-bound carbonic anhydrase takes part in CO_2 concentration in algae cells. In *Current Research in Photosynthesis,* ed. M Baltscheffsky, 4:489–92. Dordrecht: Kluwer Academic

84. Provart NJ, Majeau N, Coleman JR. 1993. Characterization of pea chloroplastic carbonic anhydrase. Expression in *Escherichia coli* and site-directed mutagenesis. *Plant Mol. Biol.* 22:937–42

85. Raines CA, Horsnell PR, Holder C, Lloyd JC. 1992. *Arabidopsis thaliana* carbonic anhydrase: cDNA sequence and effect of CO_2 on mRNA levels. *Plant Mol. Biol.* 20:1143–48

86. Ramazanov Z, Cardenas J. 1992. Inorganic carbon transport across cell compartments of the halotolerant alga *Dunaliella salina*. *Physiol. Plant.* 85:121–28

87. Randall PJ, Bouma D. 1973. Zinc deficiency, carbonic anhydrase, and photosynthesis in leaves of spinach. *Plant Physiol.* 57:229–32

88. Reed ML, Graham D. 1981. Carbonic anhydrase in plants: distribution, properties and possible physiological roles. *Progr. Phytochem.* 7:47–94

89. Reinhold L, Kosloff R, Kaplan A. 1991. A model for inorganic carbon fluxes and photosynthesis in cyanobacterial carboxysomes. *Can. J. Bot.* 69:984–88

90. Reinhold L, Zviman M, Kaplan A. 1989. A quantitative model for inorganic carbon fluxes and photosynthesis in cyanobacteria. *Plant Physiol. Biochem.* 27:945–54

91. Reiskind JB, Beer S, Bowes G. 1989. Photosynthesis, photorespiration and ecophysiological interactions in marine macroalgae. *Aquatic Bot.* 34:131–52

92. Reiskind JB, Seamon PT, Bowes G. 1988. Alternative methods of photosynthetic carbon assimilation in marine macroalgae. *Plant Physiol.* 87:686–92

93. Roeske CA, Ogren WL. 1990. Nucleotide sequence of pea cDNA encoding chloroplast carbonic anhydrase. *Nucleic Acids Res.* 18:3413

94. Silverman DN. 1991. The catalytic mechanism of carbonic anhydrase. *Can. J. Bot.* 69:1070–78

95. Smith JAC, Bryce JH. 1992. Metabolite compartmentation and transport in CAM plants. *Society for Experimental Biology Seminar Ser.: Plant Organelles,* ed. AK Tobin, 50:141–67. London: Cambridge Univ. Press

96. Spalding MH, Spreitzer RJ, Ogren WL.

1983. Carbonic anhydrase deficient mutant of *Chlamydomonas* requires elevated carbon dioxide concentration for photoautotrophic growth. *Plant Physiol.* 73:268–72

97. Spalding MH, Portis AR. 1985. A model of carbon dioxide assimilation in *Chlamydomonas reinhardtii. Planta* 164:308–20

98. Sugiharto B, Burnell JN, Sugiyama T. 1992. Cytokinin is requires to induce the nitrogen-dependent accumulation of mRNAs for phosphoenolpyruvate carboxylase and carbonic anhydrase in detached maize leaves. *Plant Physiol.* 100: 153–56

99. Sültemeyer DF, Fock HP, Canvin DT. 1990. Mass spectrometric measurements of intracellular carbonic anhydrase activity in high and low Ci cells of *Chlamydomonas. Plant Physiol.* 94:1250–57

100. Sültemeyer DF, Fock HP, Canvin DT. 1991. Active uptake of inorganic carbon by *Chlamydomonas reinhardtii*: evidence for simultaneous transport of HCO_3^- and CO_2 and characterization of active CO_2 transport. *Can. J. Bot.* 69:995–1002

101. Sültemeyer DF, Klock G, Kreutzberg K, Fock HP. 1988. Photosynthesis and apparent affinity for dissolved inorganic carbon by cells and chloroplasts of *Chlamydomonas reinhardtii* grown at high and low CO_2 concentrations. *Planta* 176:256–60

102. Sültemeyer DF, Miller AG, Espie GS, Fock HP, Canvin DT. 1989. Active CO_2 transport by the green alga *Chlamydomonas reinhardtii. Plant Physiol.* 89:1213–19

103. Sültemeyer DF, Schmidt C, Fock HP. 1993. Carbonic anhydrases in higher plants and aquatic microorganisms. *Physiol. Plant.* 88:179–90

104. Surif MB, Raven JA. 1989. Exogenous inorganic carbon sources for photosynthesis in seawater by members of the *Fucales* and the *Laminariales* (Phaeophyta): ecological and taxonomic implications. *Oecologia* 78: 97–105

105. Swader JA, Jacobson BS. 1972. Acetazolamide inhibition of photosystem II in isolated spinach chloroplasts. *Phytochemistry* 11:65–70

106. Tachiki A, Fukuzawa H, Miyachi S. 1992. Characterization of carbonic anhydrase isozyme CA2, which is the *CAH2* gene product, in *Chlamydomonas reinhardtii. Biosci. Biotechnol. Biochem.* 56:794–98

107. Tashian RE. 1989. The carbonic anhydrases: widening perspectives on their evolution, expression and function. *BioEssays* 10:186–92

108. Theilmann J, Tolbert NE, Goyal A, Senger H. 1990. Two systems for concentrating CO_2 and bicarbonate during photosynthesis by *Scenedesmus. Plant Physiol.* 92: 622–29

109. Tsuzuki M, Miyachi S. 1989. The function of carbonic anhydrase in aquatic photosynthesis. *Aquatic Bot.* 34:85–104

110. Tsuzuki M, Miyachi S, Edwards G. 1985. Localization of carbonic anhydrase in mesophyll cells of terrestrial C_3 plants in relation to CO_2 assimilation. *Plant Cell Physiol.* 26:881–91

111. Tsuzuki M, Miyachi S, Winter K, Edwards G. 1982. Localization of carbonic anhydrase in crassulacean acid metabolism plants. *Plant Sci. Lett.* 24:211–18

112. Yagawa Y, Muto S, Miyachi S. 1987. Carbonic anhydrase of a unicellular red alga *Porphyridium cruentum* R-1. II. Distribution and role in photosynthesis. *Plant Cell Physiol.* 28:1509–16

113. Yu J-W, Price GD, Song L, Badger MR. 1992. Isolation of a putative carboxysomal carbonic anhydrase gene from the cyanobacterium *Synechococcus* PCC7942. *Plant Physiol.* 100:794–800

114. Zhu SL, Sly WS. 1990. Carbonic anhydrase IV from human lung. *J. Biol. Chem.* 265: 8795–801

Annu. Rev. Plant Physiol. Plant Mol. Biol. 1994. 45:393–420

AUXIN-BINDING PROTEINS

Alan M. Jones

Department of Biology, University of North Carolina, Chapel Hill, North Carolina
27599-3280

KEY WORDS: plant hormone-binding proteins, auxin-binding proteins, auxin receptors, plant
growth regulators, plant morphogen

CONTENTS

INTRODUCTION 393
AUXIN-BINDING PROTEINS IN THE ENDOMEMBRANE SYSTEM 394
    ABP1 Identification and Purification.. 394
    ABP1 Structure 398
    Auxin-binding Protein Localization in Organs, Tissues, and Cells................................ 401
    ABP1 Function.. 403
AUXIN-BINDING PROTEINS IN THE PLASMA MEMBRANE.. 406
AUXIN-BINDING PROTEINS IN THE NUCLEUS... 408
SOLUBLE AUXIN-BINDING PROTEINS THAT ARE NOT YET LOCALIZED.............. 410
    Enzymes Labeled by 5-Azidoindole-3-acetic acid.. 410
    Soluble Auxin-binding Sites.. 411
CONCLUDING REMARKS.. 412

INTRODUCTION

"Where shall I begin?"
"Well, it's always best to begin in the beginning."—(*Possible answer from the
Good Witch Glinda to Dorothy Gale's question on studying signal transduc-
tion chains.*)

When it comes to signal transduction, it is not really clear where the beginning
is. It would appear more useful to consider signal transduction chains not as
linear sets of reactions but rather as signal transduction networks to accommodate our understanding of signal feedbacks, the cellular influences of various
elements of hormone signaling, and the contributions of multiple receptors,
multiple inputs toward single responses. Where shall I begin? Molecular

393

changes in the number or activities of receptors typically have the largest effect on development compared to changes in the level of downstream effectors. In network theory parlance, the highest control value for such a series of molecular reactions leading to a hormone response is most often ligand activation of the hormone receptor. Thus, for the study of receptors to be an effective beginning for understanding the molecular mechanisms of a specific hormone-mediated process, the traps of linear thinking must be avoided. We must acknowledge that receptors, ligand-modulated enzymes, and other ligand-regulated proteins act simultaneously. It is the context of these simultaneous activities, not the action of a single receptor, that leads the cell down a specific developmental pathway, for example, toward differentiation vs elongation.

This review is limited to the growth-promoting class of auxins. The most has been published on this class, although there have been recent breakthroughs in sibling fields. For example, there is new evidence on the identification of 1. a salicylate-binding protein (10, 10a), 2. gibberellin-binding proteins (36, 77), 3. a naphthylphthalamic acid-binding protein (133), 4. cytokinin-binding proteins (56), and 5. a fusicoccin-binding protein (17, 74). Progress has been made toward identifying elicitor receptors (11). An overview of this broad area of plant hormone receptors can be found in several recent reviews and books (e.g. 54, 58, 72a, 80, 127). This review covers recent information on the identification and characterization of auxin-binding proteins (ABPs), for the most part omitting work describing only auxin-binding. I refer readers to previous reviews in this series for a historical perspective (52, 103). For reviews and opinions on the role of ABPs, see references 13, 14, 42, 48, 55, 82, 83, 92, 108, 116.

AUXIN-BINDING PROTEINS IN THE ENDOMEMBRANE SYSTEM

ABP1 Identification and Purification

Work during the 1970s on characterizing the auxin-binding activity in maize shoots (reviewed in 103, 127) provided the groundwork for purifying the proteins responsible for this activity. Ligand specificity, optimum binding conditions, and reagents that modified auxin binding were determined for an auxin-binding site, designated Site I, that was found to comigrate on sucrose gradients with endoplasmic reticulum (ER) enzyme markers. In 1985, Dieter Klambt's group at the University of Bonn used a creative approach to identify an ABP in maize with characteristics of Site I auxin binding (62). They found that auxin-binding activity in crude extracts of acetone solubilized membrane fractions could be retained by diiodobenzoic acid Sepharose chromatography (ether linked at the second position), permitting them to separate two pools of protein, one pool containing auxin-binding activity and one pool lacking it. These two pools of protein were used to generate antisera and the purified IgG fractions were used to prepare immunoaffinity columns designated anti-ABP and anti-nonABP. Crude extracts were then processed by diiodobenzoic acid-Sepharose, anti-nonABP, and anti-ABP chromatographies to yield a 600-fold enrichment of auxin activity and a substantially-enriched protein of reported molecular mass of 20 kDa. Auxin-binding activity in this purified fraction was heat labile, had a sharp pH optimum of 5.5, and a dissociation constant for $[^{14}C]$ naphthalene-1-acetic acid (1-NAA) of $3-6 \times 10^{-8}$ M (measured directly). The binding protein appears to be dimeric (49, 115, 121), binds 1–2 moles of NAA per mole of dimer and contains sugar adjuncts (29, 64). Designation of this protein is generally ABP1, but in the literature it has been referred as Zm-ERabp1, axr1, and 22-kDa ABP.

Other groups used conventional approaches to purify ABP1. Shimomura et al (115) used DEAE-Sephacel, 1-NAA-Sepharose (coupled via the naphthalene carboxyl group), and gel filtration chromatographies to purify from maize shoots a 21 kDa-subunit ABP that behaves as a dimer. The pH optimum for binding is approximately 5.0, yet ABP1 is unstable at a pH below ~ 5.5. The dissociation constants for seven auxins and analogs agree with dissociation constants determined for solubilized auxin-binding activity. Napier et al (85) prepared polyclonal and monoclonal antibodies to ABP1, partially purified by successive anion exchange chromatographies and then to adequate purity for antibody production by preparative electrophoresis. The standard procedure in most labs is a combination of ion exchange and PAA affinity chromatography (96, 130), which appears to give the highest yields and purity to date. Mike Venis' early discovery that auxin-binding activity in the microsomal fraction can be solubilized by acetone (126) has been central in all the purification protocols because this step affords substantial purification on its own and avoids detergents that complicate purification and analyses.

The first direct demonstration that the 20–22-kDa subunit protein, and not a contaminant, was responsible for the purified auxin-binding activity (49) came using a non-equilibrium binding technique of photoaffinity labeling for ABP that was created by Nelson Leonard in the late 1970s at the University of Illinois (46, 47, 49, 71, 72). Jones & Venis (49) prepared a sample for photoaffinity labeling using Napier et al's method (85) to enrich auxin-binding activity in maize shoots. $[7-^{3}H]$, 5-Azidoindole-3-acetic acid ($[^{3}H]5N_3IAA$) was added at concentrations at least 10 times below the expected K_d of auxin binding. Of the many bands apparent in the Coomassie-stained gel, only three proteins with molecular masses of ~ 22, 24, and 43 kDa specifically incorporated the photoaffinity label after UV photolysis (see Figure 1). Photolabeling was blocked by micromolar concentrations of indole-3-acetic acid (IAA) and labeling did not occur without UV activation. Antibodies prepared against the

22-kDa subunit ABP were used in conjunction with photoaffinity labeling to show that both the 22- and 24-kDa ABPs appeared to be dimeric.

Much of the identification of ABP has depended on photoaffinity labeling (see Table 2). Thus, a brief description of the advantages and limitations of this method is worthwhile. Photoaffinity labeling is a powerful technique for identifying receptors and characterizing structure because it provides direct proof of physical interaction between the hormone analog and the binding site.

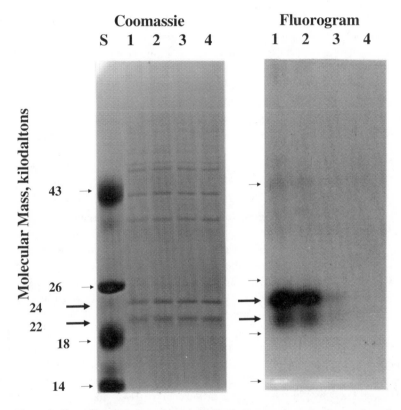

Figure 1 Photoaffinity labeling of three solubilized microsomal proteins from maize coleoptile using [³H]5-azidoindole-3-acetic acid ([³H]5N₃IAA). Auxin-binding activity was enriched in maize coleoptile extracts by the method of Napier et al (85). Proteins were incubated with [³H]5N₃IAA plus four increasing amounts of indole-3-acetic acid (IAA). The four samples then were photolyzed as described in Jones & Venis (49). The left panel is the Coomassie stained gel of these four reactions and the right panel is the corresponding fluorogram. Lane 1 shows the control reaction, i.e. no competing IAA was included; lane 2, 0.16 μM IAA; lane 3, 1.6 μM IAA; lane 4, 16 μM IAA. S, molecular weight standards. Note that incorporation of [³H]5N₃IAA by the three proteins with molecular masses of 22, 24, and 43 kDa is blocked by IAA, indicating specific incorporation. The 22-kDa protein is ABP1 as described in the text. The identities of the 24- and 43-kDa proteins are not certain, but ABPs of these sizes are also described in the text. This figure came from reference 49.

The limitation of the technique is in the interpretation of this interaction. Nonspecific interactions before and after the UV-induced nitrene generation can occur and result in photolytic tagging under certain conditions (105). Validation of binding site labeling should be determined by four experiments (94) designed to demonstrate that 1. only specific proteins in a complex are labeled, 2. the photolabeling is a saturable reaction over a physiologically meaningful concentration, 3. labeling is protected by compounds known to competitively bind the protein(s) of interest, and 4. photolabeling is dependent on activation. The latter experiment addresses the problem of enzymatic processes and the possibility of generation of nitrenes that have exceptionally long half-lives. These long-lived compounds are not suitable for binding site labeling unless the off rate for the photogenerated product is much lower than the nitrene half-life, which is not likely for auxins. The former set of experiments address the problem of specificity and physiological significance. Once specificity is established, the conclusion that the photoaffinity labeling agent is interacting at or near the binding site is valid. However, proximal to the binding site does not unequivocally mean on the protein containing the binding site entirely. Photoaffinity labeling can be temperature dependent; therefore, it should be performed at more than one temperature (25°C, 4°C, and −196°C are used typically). Labeling at −196°C often lowers nonspecific labeling, but labeling at this temperature is not always possible. Despite other limitations that will be discussed in the individual cases to follow, photoaffinity labeling has had a great impact on identifying ABPs.

Table 1   A comparison of ABP1 and Site I auxin binding.

Character	ABP1	Site I	References
Apparent size	~ 40 kDa	~ 40 kDa	49, 126
Binding pH optimum	5.5	5.5	62, 101
Some binding affinities	$K_D^{IAA} = 3.9\ \mu M$	$K_D^{IAA} = 3.9\ \mu M$	62, 99
	$K_D^{NAA} = 0.3\ \mu M$	$K_D^{NAA} = 0.8\ \mu M$	
	$K_D^{2,4D} = 63\ \mu M$	$K_D^{2,4D} = 100\ \mu M$	
Red light regulation	abundance down 50% in 12 h	activity down 50% in 12 h	43, 102
Subcellular localization	predominantly endoplasmic reticulum	predominantly endoplasmic reticulum	45, 98
Tissue localization	growing zones of maize shoots	growing zones of maize shoots	45, 102

Because of a correlation between ABP1 characteristics and characteristics of Site I auxin-binding activity in maize extracts, auxin binding to ABP1 is accepted generally as genuine Site I auxin binding. Table 1 lists some of these similarities. By making the connection between Site I binding and ABP1, we can deduce more about the characteristics of ABP1. For example, most estimates on the abundance of Site I are 10–100 pmoles/g fresh weight of maize coleoptiles, one of the richest sources for ABP1 known. This indicates that ABP1 is ~ 0.01–0.1% of the total cell protein in this organ. Other characteristics of Site I binding have enabled researchers to make hypotheses on ABP1 binding site structure.

ABP1 Structure

Soon after ABP1 was identified, three different groups cloned the ABP1 cDNA and other related cDNAs from maize seedling cDNA libraries either by screening with degenerate oligonucleotides designed from the amino-terminal sequence of ABP1 (29, 37) or by screening with anti-ABP1 serum (122). The cDNA sequences of ABP1 homologs from tobacco, *Arabidopsis thaliana* (91), and strawberry have been identified and the appropriate cDNAs were used to determine that the ABP1 gene is located on chromosome 3 in maize (61, 131) and chromosome 4 in *A. thaliana* (91, 115).

The deduced primary structures of ABP1 (Figure 2), together with biochemical data obtained using purified maize ABP1, provide a general picture of ABP1 structure. The maize cDNA encodes a unique 201-residue preprotein containing a 38-residue signal sequence. There is one glycosylation site containing a high-mannose type sugar adjunct (29). This type of oligosaccharide, found on many ER proteins, is consistent with the presence on ABP1 of a carboxy-terminal tetrapeptide sequence, KDEL, a putative ER lumenal residence signal, and with previous work demonstrating that 90–95% of the intracellular pool of ABP1 (i.e. the pool that is associated with microsomes) comigrates on sucrose gradients with the ER enzyme marker, cytochrome c reductase (45, 79, 114). The ER localization is perplexing and yet the most interesting characteristic of ABP1 because this would make it unique among hormone receptors in plants and animals, and it challenges us to hypothesize what role, if any, a hormone receptor would have within the ER. More about the intra- and extracellular localization of ABP1 and its possible biological significance is discussed below.

ABP1 is a soluble protein that survives organic solvent extraction of the microsomes from which it can be isolated, indicating that ABP1 is not an integral membrane protein. Structure predictive algorithms also do not predict membrane spanning domains in the mature protein. Neither circular dichroism spectral analyses nor structure predictive algorithms support a significant amount of helical secondary structure. ABP1 is probably composed mostly of

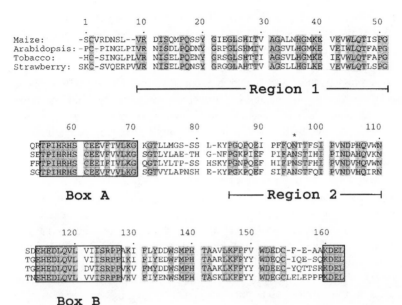

Figure 2 Comparison of four protein sequences deduced from cDNA encoding maize, *Arabidopsis*, tobacco, and strawberry ABP1 homologs. The maize sequence comes from references 29, 37, 122; the *Arabidopsis* sequence from reference 91; the tobacco sequence from the gene bank (accession number X70903); and the strawberry sequence from Drs. Heather Macdonald and Colin Lazarus, University of Bristol. Identical residues are shaded. The asterisk indicates the glycosylation site. Not shown are other ABP isoforms found in maize (111a). These have approximately 90% sequence identities. All have the carboxy-terminal KDEL.

β sheets and turns. Proline and histidine residues occur more frequently than average in ABP1. ABP1 behaves as a globular dimer in solution, although a rod-like monomeric gross structure is not ruled out by the existing data. The apparent quaternary structure is not altered by β-mercaptoethanol, indicating that if intermolecular disulfide bonds exist, they do not provide the mutually exclusive force for dimerization (49).

Shimomura & Watanabe (117) used circular dichroism spectral analysis to demonstrate hormone-induced conformational changes in ABP1. Binding of naphthalene-1-acetic acid (NAA) and IAA to ABP1 causes changes in the 240–290 nm region of the spectrum. The antiauxin naphthalene-2-acetic acid causes different and generally larger changes in the spectrum including the region at 231 nm. Hormone-induced conformational changes have been detected for ABP1 using the monoclonal antibody designated MAC256 for which the epitope was mapped to the carboxy terminus (79). Napier & Venis (81) have shown that MAC256 recognition of ABP1 by enzyme-linked immunosorbent assay (ELISA) is inhibited by 1-NAA between 1 μM and 1 mM.

The level of inhibition generally correlates, but not without exception, with the affinity of the inhibiting ligand to ABP1, indicating that the ELISA assay is reporting native conformational changes.

The deduced amino acid sequences of the mature ABP1 from three dicots and one monocot show a high degree of conservation. There is 63% sequence identity between the maize and *A. thaliana* sequences, the two most divergent sequences available. ABP1 has two highly conserved contiguous stretches of sequences indicated as Boxes A and B in Figure 2. Box A is located between residues T54 and G70 (maize sequence numbering) and differs by a single methyl group in the dicot comparison and by a substitution of a methyl group for a hydroxyl in the monocot/dicot comparison (i.e. V vs T). Box B is located between E113 and P127 and differs by a methyl group except for the tobacco sequence, which contains a sterically similar but charged aspartic acid side chain instead of valine. Both boxes are rich in charged residues and Box A contains the residues cysteine, histidine, and arginine, which have been implicated in auxin binding (86, 101, 125). This result prompted Venis et al (128) to propose that Box A contains the structural requisites for the auxin-binding site. Figure 2 also illustrates two regions designated 1 and 2 that are less conserved overall but that have invariant residue pairs and clusters throughout. This depiction of poor conservation for regions 1 and 2 may be somewhat misleading because the differences between the regions are actually very small structurally. ABP1 is too conserved to draw conclusions about specific domains.

Region 2 is especially rich in invariant proline residues and also contains the single potential glycosylation site (N95-S/T-T) for a high-mannose type oligosaccharide. Napier & Venis (84) have mapped to this region some of the epitopes for antibodies directed against purified and recombinant ABP1, indicating that this region is antigenic. One possible interpretation of this antigenicity is that region 2 is on the surface of the protein. The carboxy-terminal portion of ABP1 is not conserved except for the tetrapeptide KDEL at the terminus and DE clusters nearby, which are putative retention sequences for ER lumenal proteins (93).

Antibodies that were directed against a synthetic Box A induced an in vitro auxin response in the absence of auxin (128). This result supports the idea that Box A contains structure of the auxin binding site because antibody recognition of this site may have induced conformational changes similar to auxin. Theoretical studies on the binding site have resulted in an active site model (20) whereby the site on ABP1 is composed of a hydrophobic table that accommodates the auxin ring system, a hydrophobic transition region, and a carboxylic acid binding group. Specificity and a substantial change in free energy (~ 2.4 kcal/mol) occurs by a hydrogen bond or ion-pair interaction between the astriding carboxylic acid oxygen and a positive point charge or hydrogen bond possibly provided by an arginine amino group. This model of the auxin-binding site of ABP1 based on auxin binding is similar to a model proposed based on auxin structure and auxin-induced growth relationships (97). The model also draws on the pharmacophore concept and introduces a new term, phytophore, which describes the conformational space of an idealized plant ligand.

Auxin-binding Protein Localization in Organs, Tissues, and Cells

There are several isoforms of ABP1. Hesse et al (29) reported the amino-terminal sequence of three ABP1-like sequences and noted that one of these encodes a protein larger than 22 kDa. There are at least three expressed members in the ABP gene family (111a). Antibodies directed against ABP1 detected low levels of other bands on Western blots. Tillmann et al (122) noted two bands on Westerns probed with antiABP1 serum and acknowledged that one interpretation was the existence of multiple forms. Jones & Venis (49) used photoaffinity labeling to identify two ABPs of similar subunit molecular masses. Feldwisch et al (24) also used photoaffinity labeling to identify an ABP that is slightly larger than ABP1 and provided evidence that it is localized to the plasma membrane. Napier & Venis used Western analysis to compare the subunit sizes and abundance of proteins recognized by anti-ABP1 antiserum and found size heterogeneity and multiple species in different plants (84). These observations indicate clearly that in plants there are several isoforms of ABP1 that are antigenically similar but that may have different functions and/or different subcellular locations. This conclusion limits the interpretation of results on localization of ABP1 in organs, tissues, and cells until isoform-specific reagents are available.

ABP1 abundance correlates roughly with the growing regions of the plant. For example, in the etiolated maize seedling, ABP1 levels are highest on a fresh weight basis in the apical region of the mesocotyl (45), the basal region of the coleoptile, and the rolled young leaves (D Harden & A Jones, unpublished results). These are all rapidly growing regions of the plant. The abundance of ABP1 is regulated by red light but the red-induced decrease in ABP1 in the mesocotyl is not rapid enough to be a primary mode of action for red-light inhibited growth in this seedling tissue (43). In the mature maize plant, ABP1 mRNA is relatively abundant in ears and styles and lower in roots (29). ABP1 mRNA is about twofold more abundant in maize ears than in tassels, whereas ABP4 mRNA is about twofold less abundant in ears than in tassels (29a). ABP4, another member of this gene family, shares 86% sequence identity to ABP1. ABP1 can be detected in leaves, coleoptiles, mesocotyls and roots of red-light grown maize seedlings although the relative abundance cannot be determined from the immunoprecipitation data (114). Based on the binding activity, Shimomura et al concluded that ABP1 levels are about 7 times higher in the coleoptile epidermis than in the residual coleoptile tissue. Lobler & Klambt (63) reached the same conclusion using an immuno-histochemical technique.

Using antibodies directed against ABP1, Radermacher & Klambt (96) detected a putative ABP in purified extracts of maize roots. Based on the slightly smaller size of this root ABP, its binding affinity for 1-NAA, and its relative abundance, they concluded that this ABP could be a root isoform of the coleoptile ABP and that it is present in roughly equal amounts. As discussed above, organ-specific isoforms may lead to variations in auxin action in different organs.

Isopycnic centrifugation was used to determine the subcellular localization of ABP1 and in all cases, ABP1 that was identified immunochemically (45, 79, 114) and by binding activity (98) predominantly comigrated with the enzyme marker cytochrome c reductase. This indicates that more than 90% of the microsomal population is located in the ER. Because previous localization studies were limited to membrane-associated proteins, thus excluding cytoso-lic and extracytoplasmic populations, Jones & Herman (44) used immunohis-tochemical techniques in conjunction with electron microscopy to determine the overall subcellular localization of ABP1 and its homologs. Using different monospecific antibodies to ABP1, they observed that ABPs are found mostly in the endomembrane system but not in the nucleus, vacuole, or plastids, or on the corresponding organellar membranes (Figure 3; 44). Some ABP1 was found in the Golgi apparatus, appressed to the plasma membrane, and within the cell wall space, indicating that some ABP1 is secreted by the normal pathway (Figure 3; 44). Because morphometric analyses were not performed, a quantitative distribution is unknown, although the relative amounts of ABP1 in the endomembrane system clearly varied either between adjacent cells or as a position within individual cells. These analyses are not quantitative in sub-cellular distribution because of differences in epitope presentation within different compartments and because of the problem of assessing organellar vol-umes. ABP1 found in the cell wall space is probably a result of a low level secretion in the normal pathway, perhaps best viewed as a leakage. This conclusion supports the extracytoplasmic location of at least some of the auxin action sites, evident from the growth promoting activities of membrane im-permeant auxins (12, 129).

ABP1 is not predicted at the plasma membrane and within the cell wall space because ABP1 and its known isoforms having the canonical ER reten-tion sequence (111a), should be retained in the plant ER lumen and plants clearly use these sequences (79). Nonetheless, there is evidence that ABP1 or a homolog acts at the outer face of the plasma membrane. This evidence is consistent with an extracytoplasmic location for at least some of the ABP1 pool. How much ABP1 is extracytoplasmic is unknown but crude estimates

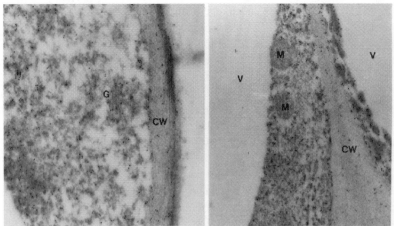

Figure 3   Subcellular localization of ABP in maize determined by immunochemical localization. Monospecific antibodies to maize ABP (made monospecific using the purified 22-kDa maize ABP) were used to localize ABP in maize coleoptile epidermal (outer) cells (left panel) and black Mexican sweet maize cells from culture (right panel). Secondary anti-rabbit Ig antibodies conjugated to 10 nm gold were used for the secondary label. Cells and tissue were minimally fixed in 2% glutaraldehyde to preserve antigenicity. Note the gold particles profusely labeling the endomembrane including a Golgi apparatus (G), but not the mitochondria (M) or vacuole (V). Also note that some gold is also found within the cell wall space (CW). These micrographs were provided by Dr. Eliot Herman, USDA, Beltsville, MD.

are that about 1% of the total cellular signal on Western blots is detected in the medium of maize cell cultures. This percentage is well within the unaccounted fraction reported in the earlier membrane fractionation studies.

The narrow pH optimum of 5.5 for auxin binding to ABP1 and its lability at this low pH have been used to draw separate conclusions. Optimal binding at pH 5.5 indicates that ABP1 must perceive auxin in acidic cellular compartments such as the cell wall space where it acts instantaneously before deactivation. A short half-life for activity or protein stability is an expected feature of a regulatory protein. One would expect that most of the ABP1 in the ER should not be occupied by auxin because this compartment is not considered acidic. Alternatively, Shimomura et al (117) have placed more importance on the pH lability and argue that the acid sensitivity of ABP1 indicates that it must operate in neutral pH compartments even though receptor occupancy at this pH is only a fraction of the maximal value. The observation that ABP1 or an immunochemically similar protein is found in both the endomembrane system and in the cell wall space still leaves open both interpretations (44).

ABP1 Function

By pharmacological standards, ABP1 is an auxin receptor, but many who study auxin binding, including me, are reluctant to define ABP1 as such just yet. ABP1 binds a series of auxins with affinities that, for the most part, correlate with the efficacy of the compound to stimulate cell elongation (62, 99). Furthermore, ABP1 is a low abundance protein found in the expected auxin-targeted tissues. These are the criteria to define a ligand receptor, so why the reluctance to label ABP1? The reluctance lies in the uniqueness of ABP1 as a hormone receptor, begging for a more rigorous set of standards to be applied. If, it is shown that ABP1, as a result of binding auxin, initiates the set of reactions signaling cells to elongate or some other auxin response, then ABP1 will be the first growth hormone receptor found to reside predominantly in the ER in plants and animals.

What set of reactions signals cells to elongate and what should we look for to test receptor function? This is also unclear, but many groups have focused on reactions that are probably involved in auxin-induced proton secretion into the cell wall space. Auxin-induced proton secretion is probably a molecular step involved in the cell wall loosening process (112) and it may be a regulatory step playing a causal role. Auxins cause the plasma membrane to hyperpolarize (21) through the increased activity of plasma membrane outward driven H^+ATPase pumps, either directly or indirectly, for example, by rapidly recruiting more pumps to the plasma membrane (27). Jean Guern's group at the CNRS laboratory has documented auxin-induced hyperpolarization using whole tobacco mesophyll protoplasts (3–7, 21). The auxin induction is small, ranging from 14 mV (3) to 5 mV (4), but the change is statistically significant. Antibodies directed against ABP1 as well as antibodies against the H^+ATPase block the auxin-induced hyperpolarization while nonimmune antibodies have no effect on auxin induction (3, 4). Both auxin induction and antibody inactivation are dosage dependent. Maize ABP1 added to the medium bathing the tobacco protoplasts changes the dosage dependency. For example, added ABP1 over the range of 1 pM to 1 nM changes the optimal concentration of 1-NAA to induce the maximal hyperpolarization of 5 mV from micromolar to nanomolar concentration (4). Conversely, anti-maize ABP antibodies in the picomolar range cause the optimal concentration of NAA for maximum hyperpolarization to shift more than 10-fold to a higher concentration (4). What the anti-maize ABP1 antibodies recognize on the tobacco mesophyll plasma membrane has not been shown; however, based on the deduced amino acid sequence, the tobacco homolog shares considerable sequence identity with maize ABP1 (see Figure 2) and considerable cross-reactivity is expected.

The hyperpolarization phenomenon was also examined using protoplasts derived from auxin-insensitive tobacco mutants and lotus and tobacco plants transformed with Ri plasmid of the pathogenic bacterium *Agrobacterium rhizogenes* (7, 70). Shen et al (113) concluded that transformed plants are more sensitive to auxin because the growth of root segments from transformed lotus are inhibited by 1-NAA at concentrations up to 100 times lower than for the wild-type control. Auxin-induced hyperpolarization in Ri transformed lotus root protoplasts and tobacco mesophyll protoplasts is similarly more sensitive, i.e. the concentration of auxin needed to obtain the maximal hyperpolarization is up to 100 times lower than for the nontransformed controls. On the other hand, hyperpolarization of mesophyll protoplasts from a 1-NAA-tolerant tobacco mutant requires more auxin for the maximal effect. The effectiveness of anti-maize ABP1 antibodies in the over-sensitive (transformed) and under-sensitive (mutants) sets of plants suggests that sensitivity to 1-NAA is related to the amount of ABP1 or a functionally equivalent and immunochemically similar homolog at the outer face of the plasma membrane (6).

A different antiserum was also used in the same types of experiments described above. Antibodies against a synthetic Box A of ABP1 (Figure 2) induce hyperpolarization of tobacco protoplast plasma membrane in the absence of auxin (128). The antibodies recognize several proteins in purified samples but not crude extracts. One protein is approximately the expected size; thus the antibodies in this serum (designated D16) presumably recognize the tobacco ABP1 homolog and act agonistically. The authors conclude that ABP1 mediates hyperpolarization and that Box A is the hormone-binding site (128). These results have been confirmed using an independent serum directed against Box A (P Bourgade, B Knauth, A Walther, & D Klambt, unpublished results).

Almost all the evidence for a receptor function rests on data obtained from the hyperpolarization assay described above. Recently, however, the patch-clamp technique has been used to investigate the effect of anti-ABP1 antibodies and auxin on whole cell current in maize protoplasts. The results seem consistent with conclusions based on the hyperpolarization assay. Hubert Felle's group at the University of Giessen has measured the activity of $H^+ATPase$ directly and shown that auxins activate this outward proton pump (104). The antibodies against maize ABP1 that were used in the previous hyperpolarization assays affected the current either by blocking the auxin activation or, in the case of the antiserum against Box A (D16), by stimulating this current in the absence of auxin.

Despite the small effect that auxin has on hyperpolarization, the unknown biological meaning of auxin-induced hyperpolarization, and the uncertain immunoreaction of the antibodies to tobacco plasma membrane proteins, data from the hyperpolarization and the patch-clamp studies build a testable hypothesis on ABP1 action. The idea is that ABP1 perceives auxin on the outer surface of the plasma membrane or within the cell wall space and stimulates membrane hyperpolarization either directly by regulating the $H^+ATPase$, indirectly by interaction with some other signal coupling protein, or both (4, 55). This theory is similar to the prokaryotic two-component systems such as the ATP-binding cassette family of proteins of which some specialized channel proteins are members (87). John Cross (14) has proposed an alternative theory that ABP1 acts both intra- and extracellularlly in a cycling fashion and in some way mediates the rapid exocytosis (27) that is involved in cell wall loosening and growth (100). Gunter Scherer's group at the University of Bonn has been studying auxin activation of membrane associated proteins such as the plasma membrane ATPase and enzymes in the phospholipid metabolism such as phospholipase A. They found that phospholipase A activity in zucchini (1) and soybean (110a) microsomal vesicles, determined by the release of lysophos-photidylcholine, was inhibited by anti-ABP1 IgG fraction, whereas the same concentration of preimune IgG had only a slight effect (1). They argue that auxin action at the plasma membrane involves changes in lysophospholipids and is mediated by ABP1. Clearly, the possibilities of ABP1 action are still numerous and speculative at this point.

AUXIN-BINDING PROTEINS IN THE PLASMA MEMBRANE

Terri Lomax's group at Oregon State University has focused on the plasma membrane as sites for auxin receptors and auxin uptake and efflux carriers. Highly purified, right-side-out vesicles of plasma membrane origin were pre-pared by the phase partition method and membrane proteins solubilized by Triton X-100 or unsolubilized were subjected to photoaffinity labeling with [^{3}H]5-N$_3$IAA (30). To reduce nonspecific interactions between membrane proteins and the photogenerated nitrene, they photolyzed samples at $-196°C$ because they reasoned that for weak ligand receptor interactions, low tempera-tures would lower the ligand off rate without affecting the photochemical reactions. As shown in Figure 4, fluorography revealed two proteins, with apparent molecular masses of 40 and 42 kDa, that appeared to specifically incorporate the photolabel. Triton X-100 solubilization enhanced the competi-tion by 1 mM IAA. The pattern of competition of the photolabeling by auxins and structurally similar compounds mimics competition of [^{14}C]IAA uptake into zucchini vesicles (39, 65, 68) and not the pattern for auxin efflux or auxin-induced growth, indicating that these ABPs may be the auxin uptake carrier. The 40/42-kDa ABPs that appear to be dimeric units of an oligomeric complex (32) partition into the hydrophilic phase of Triton X-114 phase ex-traction similar to other membrane channel proteins. The 40/42-kDa ABPs are found in divergent plant species including gymnosperms, dicots, and cereals such as maize (67).

Hicks et al applied the same approach to the auxin-insensitive mutant of tomato designated *diageotropica* (*dgt*). This recessive mutant has normal lev-els of auxin but is insensitive to exogenous auxin (51), making it a potential auxin receptor mutant. In wild-type tomato shoots and, to a lesser extent, roots, the same 40/42-kDa doublet of proteins could be photolabeled, but in the *dgt* mutant only in the root microsomal proteins could the 40/42-kDa proteins be photolabeled (31). Mutant shoots either lacked these ABPs or contained non-functional ABPs. The *dgt* phenotype indicates an altered auxin transport system. Although auxin transport rates in *dgt* do not appear to be affected by the genetic lesion, the flux of auxin through *dgt* tissues is increased (16; D Rayle & T Lomax, unpublished data). The nature of the genetic lesion is unknown, however, other hormones such as ethylene and cytokinins may be involved (66). Under certain environmental conditions, it is possible to photolabel the

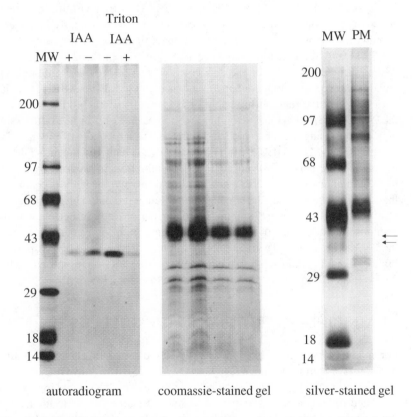

autoradiogram   coomassie-stained gel   silver-stained gel

Figure 4   40/42-kDa ABPs located on the plasma membrane from zucchini are photolabeled by 0.5µM [³H]5,azidoindole-3-acetic acid. Photolabeling was performed as described by Hicks et al (30). Fluorogram (left panel) shows photolabeling in the absence (–) or presence (+) of 1 mM competing IAA. Labeling is shown with (+Triton) and without (–Triton) 0.1% Triton X-100. The corresponding Coomassie-stained gel (center panel) and the silver-stained gels (right panel) of pure plasma membrane (pm) proteins illustrate that of the many proteins present, only the low abundance 40/42-kDa proteins incorporate the label. This figure was taken from reference 30 and modified with permission.

40/42-kDa ABPs in the *dgt* mutant (D Rayle, R Hopkins, S Verhey & T Lomax, unpublished data). Also, the *dgt* phenotype can be phenocopied in wild-type by exogenous cytokinins; thus, there appears to be interplay between hormones and evidence that the genetic lesion is not in the gene(s) encoding the 40/42-kDa ABPs.

Klaus Palme's laboratory at the Max Plank Institute also used the photoaffinity labeling technique to detect ABPs in purified preparations of plasma membrane. They found that maize coleoptile plasma membranes are enriched with four ABPs of molecular masses 60, 58, 24 and 23 kDa, but they did not detect the 40/42-kDa ABPs in maize that Lomax & Hicks observed (67). However, like the 40/42-kDa ABPs, the 24/23-kDa ABPs partitioned into the aqueous phase of Triton X-114 phase extraction, which similarly indicates a membrane channel. Antibodies to recombinant maize ABP1 cross-react with the 23-kDa ABP indicating sequence homology. Both the 23- and 24-kDa ABPs were sequenced, but the sequences were not provided although the authors claimed that both contained sequence similarities to ABP1. It is also not clear what the relationship could be between these 23/24-kDa ABPs and the 24-kDa ABP that Jones & Venis reported earlier (49) in nearly the same preparations using the same method (Figure 1). Because photoaffinity labeling of the 23/24-kDa ABPs by $[^3H]5$-N_3IAA (24) and by 5-azido-1-naphthylphthalamic acid (133), an auxin efflux inhibitor, is blocked by 1 mM naphthylphthalamic acid (NPA) and IAA, respectively, the authors concluded that the two labeling reagents are tagging the same protein, which is the auxin efflux carrier. Jacobi et al (38) found a protein, peripherally associated with the symbiosome membrane of soybean root nodules, with some characteristics of the 23/24-kDa maize plasma membrane protein. Because auxin transport inhibitors like NPA induce pseudonodules (33), a role has been proposed for this 23/24-kDa ABP for auxin transport in nodule formation (33, 38).

AUXIN-BINDING PROTEINS IN THE NUCLEUS

For many years, Kees Libbenga's group has pursued in plants the animal paradigm for steroid hormone receptors (58, 59, 73). They have reasoned that such ligand-regulated transcription factors must exist in plants and used auxin-regulated transcription as the system to study. Soluble factors in the nuclear preparations of cultured tobacco cells were enriched sufficiently to detect high affinity auxin binding but the identity of these factors is still uncertain. Libbenga's group has focused on a 50-kDa subunit protein because it is retained by IAA-Sepharose affinity chromatography and is phosphorylated, two characteristics that they predict a nuclear auxin receptor to have (2, 59, 73, 123, 124). Addition of these crude preparations with auxin-binding activity to isolated tobacco nuclei stimulates RNA polymerase II activity compared to the addition of samples without auxin-binding activity, but the effect on specific transcription rates of known auxin-regulated mRNA is not known. The group ascribes poor reproducibility of this system to protein modification, possibly phosphorylation of these putative auxin receptors (59, 73).

Shingo Saikai's laboratory, now at the University of Tsukuba, has purified two proteins from mung bean hypocotyls that appear to have auxin-binding activity (106, 107, 109) and may function in the nucleus. Their designations ABP-I and ABP-II might be confused with the ER-localized ABP1 identified in maize, but these proteins appear to be different based on subunit and quaternary structure. ABP-I has subunits (or proteolytic fragments) of ~ 50 and 15 kDa and behaves as 390,000 kDa in size exclusion chromatography (SEC) while ABP-II has a subunit mass of ~ 50 kDa and behaves as 190,000 kDa in SEC. The relationship between the two proteins is unclear and the reported molecular and binding characteristics are not dissimilar enough to distinguish them yet. Sakai argues that these ABPs have a nuclear function because addition of these ABPs to mung bean nuclei during in vitro run-on experiments causes an increase in [^{3}H]UTP incorporation in an α amanitin-sensitive fashion. Unfortunately, this effect is independent of IAA addition, even in nuclei that the authors attempted to deplete of endogenous auxin. However, adding IAA with ABP-I or II seems to cause subtle qualitative changes among the newly-induced transcripts. Transcripts that incorporated [γ-S]-GTP were purified from the in vitro run-on reactions by Hg-Sepharose chromatography and the translated products were examined by 2-D gel electrophoresis (53). Small differences in the translation patterns led the authors to conclude that ABP-I and II alone cause quantitative differences in polymerase II activity and that the addition of IAA to these ABPs causes qualitative changes in polymerase selectivity during reinitiation, possibly by physically interacting with RNA polymerase II (108, 110). Mennes et al also concluded that auxin (2,4-D) causes reinitiation of RNA polymerase II on specific genes (73a). Because the changes are small and the in vitro reinitiation experiments are extremely difficult to perform, these exciting results must be viewed with caution for now.

Prasad & Jones (95) used anti-idiotypic antibodies to detect an ABP that probably is enriched in nuclei. Primary rabbit antibodies were directed against IAA linked via the indole nitrogen to BSA. These IAA-specific antibodies were purified on IAA-affigel columns and were used to prepare a secondary set of anti-idiotypic (anti-ID) antibodies. The anti-ID antibodies recognized a 65-kDa band in common to 8 divergent plant species but not in yeast or mice. Recognition of this 65-kDa protein by the anti-ID antibodies could be blocked by the addition of 2 μM 2,4-D or 10 μM IAA, indicating that the antibodies recognized an epitope structurally similar, but not necessarily functionally similar, to an auxin-binding site. However, the anti-ID antibodies retain auxin-binding activity on immunocolumns, supporting the conclusion that the 65-kDa protein is an ABP. Direct photolabeling using [^{3}H]5N$_3$IAA has not been accomplished nor have equilibrium techniques demonstrated that this 65-kDa protein binds auxins. Highly-purified nuclei contain this 65-kDa ABP, which supports the idea that this protein may act in the nucleus, possibly in analogy to other known nuclear ligand receptors. The protein has been purified and a preliminary amino-terminal sequence has been obtained (P Prasad, S Trauco, M-J Wu, & A Jones, unpublished data).

SOLUBLE AUXIN-BINDING PROTEINS THAT ARE NOT YET LOCALIZED

Enzymes Labeled by 5-Azidoindole-3-acetic Acid

Soluble auxin-binding activities, presumed to be cytosolic, have been studied since the early 1980s, but identification of soluble ABPs is relatively recent. Pat King's group at the Friedrich Miescher Institute used photoaffinity labeling to detect ABPs in the soluble fraction of wild-type *Hyoscyamus muticus* cell cultures (8, 69). These cells were chosen as starting material because of the potential to investigate differences in labeling patterns in auxin-insensitive mutant lines of *Hyoscyamus* spp. (89). Photoaffinity labeling with [^{3}H]5N^{3}IAA was performed at both 4°C and −196°C using different ammonium sulfate cuts of the soluble proteins. In the low salt cut, a 31- and 24-kDa protein incorporated the photolabel at 4°C but not at −196°C, and in the higher salt cut fraction, a 25-kDa protein was labeled only at −196°C. The 25-kDa ABP has a narrow pH optimum for labeling and photolabeling was competed by auxins but not structurally similar compounds. In contrast, the 31/24-kDa ABP photolabeled over a broad pH range and photolabeling was competed by the non-auxins, indole and tryptophan, but not other nonindole auxins. Macdonald et al (69) and Bilang et al (8) identified the 31/24- and 25-kDa ABPs as the basic form of β1,3-glucanase and glutathione S-transferase (GST), respectively. Zettle et al recently confirmed the results on GST photolabeling using *A. thaliana* (134). Interestingly, auxin modulates the expression of the genes for both these enzymes (18, 19, 25, 75, 120). Auxin binding to these enzymes may be an artifact of the photolabeling technique discussed above. The high cellular abundance of both enzymes, the very low pH optimum for photolabeling GST (at least for *Hyocyamus* spp. GST), and the known affinity of GST for radicals seem to diminish the biological significance of β1,3-glucanase and GST being auxin receptors, but we should not dismiss these observations because there is precedent at least for ligand regulation of GST. For example, several groups reported that steroids bind to noncatalytic sites of GST and proposed that this interaction is involved in GST molecular trafficking and/or nuclear uptake of steroid hormones (e.g. 15, 34, 35). Interestingly, indole-3-carbinol blocks steroid hormone photolabeling of the murine GST in vitro (15). One possible mode of GST action noted by Bilang et al (8) is to alter the redox potential in cells, which has been correlated with a change in cell proliferation in culture (19a). There is new evidence that gene expression in plants also may be controlled by the redox state of the cell. Chen et al (10a) have shown that the plant hormone salicylic acid (SA) binds catalase non-competitively to inhibit its peroxide reduction activity. This is a novel signal transduction pathway. Because GSTs are related functionally to catalases, auxin may modulate a specific class of GSTs, resulting in altered redox potential to alter gene expression and cellular development.

Some enzymes are predicted to bind auxin (i.e. those enzymes for which auxin is a substrate or product). Using photoaffinity labeling, Palme's group has identified a 60-kDa hydrolase in maize (9). High levels of auxin analogs were required to block photolabeling (indicating a low affinity for auxin) and there is little hydrolase activity with any of the obvious IAA-ester conjugates such as IAA-myoinositol, the major IAA conjugate in maize. This protein is different from the 60-kDa protein on the plasma membrane identified by the same group using the same technique (24, 76). The lack of substrate specificity for IAA conjugates and the sequence similarity to a possible cytokinin hydrolase (22) prompted Palme's group to explore other phytohormone substrates for the photolabeled 60-kDa protein. They recently found that this protein has high activity and specificity for O-linked cytokinin glucosides, a temporary storage form for this hormone. Moreover, as predicted, the abundance of the protein was highest in the root where cytokinins are produced. As with GST, we are confronted with the meaning of auxin binding to cytokinin-O-glucosidase. Auxin to cytokinin ratios are important in regulating plant development and this auxin interaction with a cytokinin-releasing enzyme may be central to the interplay between the two hormones. The expression of prokaryotic phytohormone hydrolases has profound effects on plant development (22, 23), although with less than the expected effect on tissue levels of these hormones (88, 111), demonstrating that the hormone interaction is complex and will be difficult to elucidate.

Soluble Auxin-binding Sites

A low affinity auxin-binding activity recently was purified from the growing shoots of peach trees (90). This activity corresponds to a 20-kDa glycoprotein that behaves as 100 kDa in SEC. Antibodies to maize ABP1 do not cross-react to this peach protein. The function of this protein is unknown.

Many more ABPs will probably be found, because other groups are currently studying auxin-binding activities in the context of a biological response such as somatic embryogenesis (26) and rooting (78). Nakamura et al (78) have solubilized a membrane auxin-binding activity and claim that it is different from ABP1 based on the binding parameters. There are now several sets of antisera to various ABPs (43, 48, 62, 85, 114, 128), which could be used to test the relationships of novel binding activities to known ABPs and which should be useful in purification of these and other soluble ABPs (28, 40, 41, 50, 132).

CONCLUDING REMARKS

Table 2 and Figure 5 summarize this review. The individual ABPs are listed in Table 2 and grouped according to our first approximation of their cellular locations. The subcellular locations have provided the best hints on the putative functions of these ABPs. The individual roles of each are unknown, although some ideas for function are provided in Figure 5 and discussed below. It is clear that there is not enough information to label any single ABP as "the auxin receptor," a mediator of all the cellular events induced by auxin.

The ABPs found predominantly in the endoplasmic reticulum are unique. Their location indicates that they may play a role in mediating auxin effects on exocytosis, one of the rapid events in auxin-stimulated cell elongation. I have presented evidence that this ABP, designated ABP1, or an immunochemically similar (and functionally the same) ABP acts at low levels in the cell wall space. This ABP1 may interact with a plasma membrane protein to initiate a set of reactions (exocytosis, perhaps) by modulating channel or pump activities; initiate a chain of reactions coupled by secondary messengers such as GTP, calcium, lysophospolipids, or phosphoinositides; or could be involved in auxin transport in analogy to small molecule transport in bacteria (87). Alternately, this ABP may act exclusively in the endoplasmic reticulum and its

Table 2   Auxin-binding proteins.

Cell location	Subunit size	Identification	References
Endomembrane	22 kDa	immunoaffinity	62
		affinity	85, 115
	22/24/43 kDa	photoaffinity	49
	22	cDNA cloning	29, 37, 122
Plasma membrane	40/42 kDa	photoaffinity	30
	60/58 kDa, 24/23 kDa	photoaffinity	24
Nuclear	65 kDa	anti-idiotypic antibodies	48
	47/15 kDa	affinity	107
Soluble	60-kDa hydrolase	photoaffinity	8a
	31-kDa β glucanse	photoaffinity	69
	25-kDa GST	photoaffinity	8

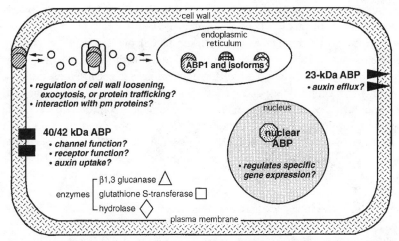

Figure 5 Summary of the different ABPs that have been identified and speculation about their possible roles in plant growth and development. There is probably more than one ABP located predominantly in the endomembrane system. These may be involved in trafficking of proteins, membranes, and cell wall materials. Alternatively, a small population may interact with other proteins at the plasma membrane. Plasma membrane ABPs could be involved in auxin perception or transport. Nuclear ABPs may control auxin-regulated gene transcription. Some enzymes incorporate the photoaffinity labeling reagent [^{3}H]5-azidoindole-3-acetic acid. Auxin-modulation of these enzyme activities may be one aspect of the auxin mechanism of action.

localization in post-ER compartments merely reflects its lower efficiency in ER retention and not its function. There are at least two ABPs on the plasma membrane, a classic location for membrane receptors. These may be involved in auxin uptake, efflux, or perception. The nucleus contains a 65-kDa ABP or possibly other ABPs that may mediate auxin effects on gene transcription. Finally, Figure 5 shows ABPs with known enzymatic activities that bind auxin artifactually, catalytically, or possibly at regulatory sites. Some of these and more to come will be shown to be involved in maintaining the auxin economy and concentration gradients while others may have novel roles in cell development.

Much of our effort to formulate working models depends on precedent from other systems and the paradigms that emerge. This applies to auxin receptors. For example, the nuclear ABP immediately evokes the paradigms of steroid hormone receptor action but the ER ABP1, lacking in precedent, leaves us perplexed. Paradigms can be useful but they can also lead us astray and paradigms are sometimes chosen by what we call the signal and how we imagine that the signal functions. If an auxin was discovered today, would we call it a hormone? Probably not, because it has all the characteristics of a morphogen, not a hormone (118). There are spatial gradients of auxins in organs and tissues, different levels of auxins induce different positive re-

*auxin   why   not   na
receptor   ~*

sponses, auxins instruct multiple responses, and all cells are pluripotent toward the application of auxins. It does not matter what we call auxin (hormone, morphogen, growth regulator, or growth substance), but it does matter how we think auxins work. We would expect to find multiple binding proteins, multiple receptor types that bind with different affinities and specificities, multiple isoforms of receptor types, and interactions between receptor types. The limited knowledge we have of auxin receptors is enough to borrow aspects of the paradigm for the morphogen, retinoic acid (RA; 57, 60). Gradients of RA are established by cytoplasmic RA-binding proteins (CRABP, two types) and retinol-binding proteins (CRBP) that require retinol binding in the ER for folding and secretion. RA regulates the expression of genes encoding CRABP and CRBP. These binding proteins are not receptors; they serve as RA reservoirs and sinks. However, there are three receptor types (RAR) that bind the *trans* isomers of RA, three types (RXR) that bind 9-*cis* RA, and each type has multiple isoforms. It is also known that different RAR:RXR heterodimers as well as homodimers form to interact differentially with response elements in a plethora of RA-inducible promoters. Obviously, the initial perception of RA and its mode of action when fully unraveled will represent a complex network of reactions, and not a linear chain of reactions that come to mind when thinking of some hormone transduction chains such as for insulin and acetylcholine. Will auxin perception mimic the retinoic acid paradigm? Probably not exactly, but the element of complexity is there and complexity in signal perception in plants has already been established with the phytochrome family of red-light receptors (119). The point is not that the details of auxin perception and retinoic acid perception are the same, but rather that we should expect multiple auxin receptors that interact. I predict that it will be the context of this interaction that directs a specific auxin response and not the consequence of auxin activating only one receptor, the concomitant step of a single transduction chain. If this is true then we are challenged to design novel experiments that sort the functions of individual auxin receptor types and isoforms, and to design novel genetic screens that yield genuine auxin receptor mutants.

ACKNOWLEDGMENTS

I thank Dr. Terri Lomax, Oregon State University; Dr. Dieter Klambt, University of Bonn; and members of my laboratory for critically reading draft versions of this review. I thank Drs. Heather Macdonald and Colin Lazarus, University of Bristol, for providing the strawberry cDNA sequence to ABP1 prior to submission to the genebanks and all others who provided me with unpublished data. Work in my laboratory is supported by the US Department of Agriculture (NRICGO) and the National Science Foundation (Cell Biology Program).

Any *Annual Review* chapter, as well as any article cited in an *Annual Review* chapter, may be purchased from the Annual Reviews Preprints and Reprints service.
1-800-347-8007; 415-259-5017; email: arpr@class.org

Literature Cited

1. Andre B, Scherer GFE. 1991. Stimulation by auxin of phospholipase A in membrane vesicles from an auxin-sensitive tissue is mediated by an auxin receptor. *Planta* 185: 209–14
2. Bailey HM, Barker RDJ, Libbenga KR, van der Linde PCG, Mennes AM, Elliott MC. 1985. Auxin binding sites in tobacco cells. *Biol. Plant.* 27:105–9
3. Barbier-Brygoo H, Ephritikhine G, Klambt D, Ghislain M, Guern J. 1989. Functional evidence for an auxin receptor at the plasmalemma of tobacco mesophyll protoplasts. *Proc. Natl. Acad. Sci. USA* 86:891–95
4. Barbier-Brygoo H, Ephritikhine G, Klambt D, Maurel C, Palme K, et al. 1991. Perception of the auxin signal at the plasma membrane of tobacco mesophyll protoplasts. *Plant J.* 1:83–93
5. Barbier-Brygoo H, Ephritikhine G, Maurel C, Guern J. 1992. Perception of the auxin signal at the plasma membrane of tobacco mesophyll protoplasts. *Biochem. Soc. Trans.* 20:59–63
6. Barbier-Brygoo H, Ephritikhine G, Shen WH, Delbarre D, Klambt D, Guern J. 1989. Characterization and modulation of the sensitivity of plant protoplasts to auxin. In *Transducing Pathways: Activation and Desensitization,* ed. TM Konjin, MD Houslay, P Van Haastert, *NATO ASI Ser.,* H44:231–44. Berlin: Springer-Verlag
7. Barbier-Brygoo H, Guern J, Ephritikhine G, Shen WH, Maurel C, Klambt D. 1989. The sensitivity of plant protoplasts to auxins: modulation of receptors at the plasmalemma. In *Plant Gene Transfer. UCLA Symp. Mol. Cell. Biol. (NS),* ed. C Lamb, R Beachy, 129:165–73. New York: Liss
8. Bilang J, Macdonald H, King PJ, Sturm A. 1993. A soluble auxin-binding protein from *Hyoscyamus muticus* is a glutathione S-transferase. *Plant Physiol.* 102:29–34
8a. Brzobohaty B, Moore I, Kristoffersen P, Bako L, Campos N, et al. 1993. Release of active cytokinin by a β-glucosidase localized to the maize root meristem. *Science* 262:1051–54
9. Campos N, Bako L, Feldwisch J, Schell J, Palme K. 1992. A protein from maize labeled with azido-IAA has novel β-glucosidase activity. *Plant J.* 2:675–84
10. Chen Z, Ricigliano J, Klessig DF. 1993. Purification and characterization of a soluble salicylic acid binding-protein from tobacco. *Proc. Natl. Acad. Sci. USA* 90: 9533–37
10a. Chen Z, Silva H, Klessig DF. 1993. Active oxygen species in the induction of plant systemic acquired resistance by salicylic acid. *Science* 262:1883–86
11. Cheong J-J, Hahn MG. 1991. A specific, high-affinity binding site for the hepta-β-glucoside elicitor exists in soybean membranes. *Plant Cell* 3:137–47
12. Cohen JD, Baldi BG, Bialek K. 1985. Strongly acidic auxin indole-3-methanesulfonic acid. *Plant Physiol.* 77:195–99
13. Cross JW. 1985. Auxin action: the search for the receptor. *Plant Cell Environ.* 8:351–59
14. Cross JW. 1991. Cycling of auxin-binding protein through the plant cell: pathways in auxin signal transduction. *New Biol.* 8: 813–19
15. Danger DP, Baldwin WS, Leblanc GA. 1992. Photoaffinity labelling steroid-hormone-binding glutathione S-transferase with [³H]methyltrienolone. *Biochem. J.* 288:361–67
16. Daniel SG, Rayle DL, Cleland RE. 1989. Auxin physiology of the tomato mutant *diageotropica. Plant Physiol.* 91:804–7
17. deBoer AH, Watson BA, Cleland RE. 1989. Purification and identification of the fusicoccin binding protein from oat root plasma membrane. *Plant Physiol.* 89:250–59
18. Dominov JA, Stenzler L, Lee S, Schwarz JJ, Leisner S, Howell SH. 1992. Cytokinins and auxins control the expression of a gene in *Nicotiana plumbaginifolia* cells by feedback regulation. *Plant Cell* 4:451–61
19. Droog FNJ, Hooykaas PJJ, Libbenga KR, van der Zaal EJ. 1993. Proteins encoded by an auxin-regulated gene family of tobacco share limited but significant homology with glutathione S-transferase and one member indeed shows in vitro GST activity. *Plant Mol. Biol.* 21:965–72
19a. Earnshaw BA, Johnson MA. 1985. The effect of glutathione on development in wild carrot suspension cultures. *Biochem. Biophys. Res. Commun.* 133:988–93
20. Edgerton MD, Tropsha A, Jones AM. 1994. Modelling the auxin-binding site of auxin-binding protein 1 of maize. *Phytochemistry* 35:In press
21. Ephritikhine G, Barbier-Brygoo H, Muller

J-F, Guern J. 1987. Auxin effect on the transmembrane potential difference of wild-type and mutant protoplasts exhibiting a differential sensitivity to auxin. *Plant Physiol.* 83:801–4

22. Estruch JJ, Chriqui D, Grossman K, Schell J, Spena A. 1991. The plant oncogene *rolC* is responsible for the release of cytokinins from glucoside conjugates. *EMBO J.* 10:2889–95

23. Estruch JJ, Schell J, Spena A. 1991. The protein encoded by the *rolB* plant oncogene hydrolyses indole glucosides. *EMBO J.* 10:3125–28

24. Feldwisch J, Zettl R, Hesse F, Schell J, Palme K. 1992. An auxin-binding protein is localized to the plasma membrane of maize coleoptile cells: identification by photoaffinity labeling and purification of a 23-kDa polypeptide. *Proc. Natl. Acad. Sci. USA* 89:475–79

25. Felix G, Meins F. 1985. Purification, immunoassay and characterization of an abundant, cytokinin-regulated polypeptide in cultured tobacco tissues. *Planta* 164:423–28

26. Filippini F, Terzi M, Vallone D, Lo Schiavo F. 1992. Modulation of auxin-binding proteins in cell suspensions. II. Isolation and initial characterization of carrot cell variants impaired in somatic embryogenesis. *Theor. Appl. Genet.* 84:430–34

27. Hager A, Debus G, Edel H-G, Stransky H, Serrano R. 1991. Auxin induces exocytosis and the rapid synthesis of a high-turnover pool of plasma-membrane H⁺-ATPase. *Planta* 185:527–37

28. Herber B, Ulbrich B, Jacobsen H-J. 1988. Modulation of soluble auxin-binding proteins in soybean cell suspensions. *Plant Cell Rep.* 7:178–81

29. Hesse T, Feldswisch J, Balshusemann D, Bauw G, Puype M, et al. 1989. Molecular cloning and structural analysis of a gene from *Zea mays* (L.) coding for a putative receptor for the plant hormone auxin. *EMBO J.* 8:2453–61

29a. Hesse T, Garbers C, Brzobohaty B, Kreimer G, Söll D, et al. 1993. Two members of the ER *abp* gene family are expressed differently in reproductive organs but to similar levels in the coleoptile of maize. *Plant Mol. Biol.* 23:57–66

30. Hicks GR, Rayle DL, Jones AM, Lomax TL. 1989. Specific photoaffinity labeling of two plasma membrane polypeptides with an azido auxin. *Proc. Natl. Acad. Sci. USA* 86:4948–52

31. Hicks GR, Rayle DL, Lomax TL. 1989. The *diageotropica* mutant of tomato lacks high specific activity auxin binding sites. *Science* 245:52–54

32. Hicks GR, Rice MS, Lomax TL. 1993. Characterization of auxin-binding proteins from zucchini plasma membrane. *Planta* 189:83–90

33. Hirsch AM, Bhuvaneswari TV, Torrey JG, Bisseling T. 1989. Early nodulin genes are induced in alfalfa root outgrowths elicited by auxin transport inhibitors. *Proc. Natl. Acad. Sci. USA* 86:1244–48

34. Homma H, Listowsky I. 1985. Identification of Y_b-glutathione S-transferase as a major rat liver protein labeled with dexamethasone 21-methanesulfonate. *Proc. Natl. Acad. Sci. USA* 82:7165

35. Homma H, Maruyama H, Niitsu Y, Listowsky I. 1986. A subclass of glutathione S-transferases as intracellular high-capacity and high affinity steroid binding proteins. *Biochem. J.* 235:763–68

36. Hooley R, Smith SJ, Beale MH, Walker RP. 1993. In vivo photoaffinity labelling of gibberellin-binding proteins in *Avena fatua* aleurone. *Aust. J. Plant Physiol.* 20:573–84

37. Inohara N, Shimomura S, Fukui T, Futai M. 1989. Auxin-binding protein located in the endoplasmic reticulum of maize shoots: molecular cloning and complete primary structure. *Proc. Natl. Acad. Sci. USA* 86:3564–68

38. Jacobi A, Zettl R, Palme K, Werner D. 1993. An auxin binding protein is localized in the symbiosome membrane of soybean nodules. *Z. Naturforsch. Teil C* 48:35–40

39. Jacobs M, Lomax T, Hertel R. 1984. A comparison of the auxin specificity of medium acidification and elongation in maize coleoptiles. *Plant Sci. Lett.* 34:35–41

40. Jacobsen H-J, Hajek K. 1985. Genotype-specific soluble auxin-binding in etiolated pea epicotyls. *Biol. Plant* 27:110–13

41. Jacobsen H-J, Hajek K, Mayerbacher R, Herber B. 1987. Soluble auxin-binding: Is there a correlation between growth-stage dependent high affinity auxin binding and auxin competence? In *Plant Hormone Receptors*, ed. D Klambt, pp. 63–70. Berlin: Springer-Verlag

42. Jones AM. 1990. Do we have the auxin receptor yet? *Physiol. Plant.* 80:154–58

43. Jones AM, Cochran DS, Lamerson PL, Cohen J, Evans M. 1991. Red-light induced changes in auxin, an auxin-binding protein, and auxin transport in maize mesocotyl. *Plant Physiol.* 97:352–58

44. Jones AM, Herman E. 1993. KDEL-containing auxin-binding protein is secreted to the plasma membrane and cell wall. *Plant Physiol.* 101:595–606

45. Jones AM, Lamerson P, Venis MA. 1989. Comparison of site I auxin binding and a 22-kilodalton auxin-binding protein in maize. *Planta* 179:409–14

46. Jones AM, Melhado LL, Ho T-HD, Leonard NJ. 1984. Azido auxins: quantitative binding data in maize. *Plant Physiol.* 74:294–302

47. Jones AM, Melhado LL, Ho T-HD, Pearce CJ, Leonard NJ. 1984. Azido auxins: photoaffinity labelling of auxin-binding proteins in maize coleoptile with tritiated 5-azidoindole-3-acetic acid. *Plant Physiol.* 75:1111–16

48. Jones AM, Prasad PV. 1991. Auxin-binding proteins and their possible roles in auxin-mediated plant growth. *BioEssays* 14:43–48

49. Jones AM, Venis MA. 1989. Photoaffinity labelling of auxin-binding proteins in maize. *Proc. Natl. Acad. Sci. USA* 86:6153–56

50. Kaur H, Kapoor HC. 1989. A membrane-associated binding site from chickpea (*Cicer arietum* L.) epicotyls. *Plant Sci.* 65:135–41

51. Kelly MO, Bradford KJ. 1986. Insensitivity of the *diageotropica* tomato mutant to auxin. *Plant Physiol.* 82:713–17

52. Kende H, Gardner GM. 1976. Hormone binding in plants. *Annu. Rev. Plant Physiol.* 27:267–90

53. Kikichi M, Imaseki H, Sakai S. 1989. Modulation of gene expression in isolated nuclei by auxin-binding proteins. *Plant Cell Physiol.* 30:765–73

54. Klambt D. 1986. *Plant Hormone Receptors.* Berlin: Springer-Verlag. 320 pp.

55. Klambt D. 1990. A view about the function of auxin-binding proteins at plasma membranes. *Plant Mol. Biol.* 14:1045–50

56. Kulaeva ON, Karavaiko NN, Moshkov IE, Selivankina SY, Novikova GV. 1990. Isolation of a protein with cytokinin-receptor properties by means of anti-idiotypic antibodies. *FEBS Lett.* 261:410–12

57. Leid M, Kastner P, Chambon P. 1992. Multiplicity generates diversity in the retinoic acid signalling pathways. *Trends Biochem. Sci.* 17:427–33

58. Libbenga KR, Mennes AM. 1987. Hormone binding and its role in hormone action. In *Plant Hormones and Their Role in Plant Growth and Development,* ed. PJ Davies, pp. 194–221. Boston: Nijhoff

59. Libbenga KR, van Telgen HJ, Mennes AM, van der Linde PCG, van der Zaal EJ. 1987. Characterization and function analysis of a high-affinity cytoplasmic auxin-binding protein. In *Molecular Biology of Plant Growth Control,* ed. JE Fox, M Jacobs, pp. 229–43. New York: Liss

60. Livrea MA, Packer L, eds. 1993. *Retinoids.* New York: Dekker. 672 pp.

61. Lobler M, Hirsch AM. 1990. RFLP mapping of the *abp1* locus in maize (*Zea mays* L.). *Plant Mol. Biol.* 15:513–16

62. Lobler M, Klambt D. 1985. Auxin-binding proteins of corn (*Zea mays* L.) I. Purification by immunological methods and characterization. *J. Biol. Chem.* 260:9848–53

63. Lobler M, Klambt D. 1985. Auxin-binding proteins of corn (*Zea mays* L.) II. Localization of a putative receptor. *J. Biol. Chem.* 260:9854–59

64. Lobler M, Simon K, Hesse T, Klambt D. 1987. Auxin receptors in target tissue. In *Molecular Biology of Plant Growth Control,* ed. JE Fox, M Jacobs, pp. 279–88. New York: Liss

65. Lomax TL. 1986. Active auxin uptake by specific plasma membrane carriers. In *Plant Growth Substances 1985,* ed. M Bopp, pp. 209–13. Berlin: Springer-Verlag

66. Lomax TL, Coenen C, Gaiser JC, Hopkins R, Rayle DL, Rice MS. 1993. Auxin perception and the regulation of tomato growth and development. In *Molecular Biology of Tomato,* ed. J Yoder, pp. 129–38. Lancaster, PA: Technomic

67. Lomax TL, Hicks GR. 1992. Specific auxin-binding proteins in the plasma membrane: receptors or transporters? *Biochem. Soc. Trans.* 20:64–69

68. Lomax TL, Melhorn RJ, Briggs WR. 1985. Active auxin uptake by zucchini membrane vesicles: quantitation using ESR volume and ΔpH determinations. *Proc. Natl. Acad. Sci. USA* 82:6541–45

69. Macdonald H, Jones AM, King PJ. 1991. Photoaffinity labelling of soluble auxin-binding proteins. *J. Biol. Chem.* 266:7393–99

70. Maurel C, Barbier-Brygoo H, Spena A, Tempe J, Guern J. 1991. Single *rol* genes from the *Agrobacterium rhizogenes* T$_L$-DNA alter some of the cellular responses to auxin in *Nicotiana tabacum. Plant Physiol.* 97:212–16

71. Melhado LL, Jones AM, Ho T-HD, Leonard NJ. 1984. Azido auxins: photolysis in solution and covalent binding to soybean. *Plant Physiol.* 74:289–94

72. Melhado LL, Jones AM, Leonard NJ, Vanderhoef LN. 1981. Azido auxins: synthesis and biological activity of fluorescent labelling agents. *Plant Physiol.* 68:469–75

72a. Mennes AM, Maan A, Hall MA. 1991. Plant hormone receptors. In *Cell to Cell Signals in Plants and Animals,* ed. V Neuhoff, J Friend. *NATO ASI Ser.,* H44:301–13. Berlin: Springer-Verlag

73. Mennes AM, Nakamura C, van der Linde PCG, van der Zall EJ, van Telgen H-J, et al. 1987. Cytosolic and membrane-bound high-affinity auxin-binding proteins in tobacco. In *Plant Hormone Receptors,* ed. D Klambt, pp. 51–62. Berlin: Springer-Verlag

73a. Mennes AM, Quint A, Gribnau JH, Boot CJM, van der Zaal EJ, et al. 1992. Specific transcription and reinitiation of 2,4-D-induced genes in tobacco nuclei. *Plant Mol. Biol.* 18:109–17

74. Meyer C, Feyerabend M, Weiler EW. 1989. Fusicoccin-binding proteins in *Arabidop-*

sis thaliana (L.) Heynh. *Plant Physiol.* 89: 692–99

75. Mohnen D, Shinshi H, Felix G, Meins F Jr. 1985. Hormonal regulation of β1,3-glucanse messenger RNA levels in cultured tobacco tissues. *EMBO J.* 4:1631–35

76. Moore I, Feldwisch J, Campos N, Zettl R, Brzobohaty B, et al. 1992. Auxin-binding proteins of *Zea mays* identified by photoaffinity labeling. *Biochem. Soc. Trans.* 20:70–73

77. Nakajima M, Sakai S, Kanazawa K, Kizawa S, Yamaguchi I, et al. 1993. Partial purification of a soluble gibberellin-binding protein from mung bean hypocotyl. *Plant Cell Physiol.* 34:289–96

78. Nakamura C, van Telgen H-J, Mennes AM, Ono H, Libbenga KR. 1988. Correlation between auxin resistance and the lack of a membrane-bound auxin binding protein and a root-specific peroxidase in *Nicotiana tabacum. Plant Physiol.* 88:845–49

79. Napier RM, Fowke LC, Hawes C, Lewis M, Pelham HRB. 1992. Immunological evidence that plants use both HDEL and KDEL for targetting proteins to the endoplasmic reticulum. *J. Cell Sci.* 102:261–71

80. Napier RM, Venis MA. 1990. Receptors for plant growth regulators: recent advances. *J. Plant Growth Regul.* 9:113–26

81. Napier RM, Venis MA. 1990. Monoclonal antibodies detect an auxin-induced conformational change in the maize auxin-binding protein. *Planta* 182:313–18

82. Napier RM, Venis MA. 1991. Auxin receptors: recent developments. *Plant Growth Regul.* 10:329–40

83. Napier RM, Venis MA. 1991. From auxin-binding protein to plant hormone receptor? *Trends Biochem. Sci.* 16:72–75

84. Napier RM, Venis MA. 1992. Epitope mapping reveals conserved regions of an auxin-binding protein. *Biochem. J.* 284:841–45

85. Napier RM, Venis MA, Bolton MA, Richardson LI, Butcher GW. 1988. Preparation and characterization of monoclonal and polyclonal antibodies to maize membrane auxin-binding proteins. *Planta* 176:519–26

86. Nave J-F, Benveniste P. 1984. Inactivation by phenylgloxal of the specific binding of 1-naphthylacetic acid with membrane-bound auxin binding sites from maize coleoptiles. *Plant Physiol.* 74:1035–40

87. Nikaido H, Saier MH Jr. 1992. Transport proteins in bacteria: common themes in their design. *Science* 258:936–42

88. Nilsson O, Crozier A, Schmulling T, Sandberg G, Olsson O. 1993. Indole-3-acetic acid homeostasis in transgenic tobacco plants expressing the *Agrobacterium rhizogenes rol B* gene. *Plant J.* 3:681–89

89. Oetiker J, Gebhardt C, King PJ. 1990. A temperature-sensitive auxin auxotroph not deficient in indole-3-acetic acid. *Planta* 180:220–28

90. Ohmiya A, Kikuchi M, Sakai S, Hayashi T. 1993. Purification and properties of an auxin-binding protein from the shoot apex of peach tree. *Plant Cell Physiol.* 34:177–83

91. Palme K, Hesse T, Campos N, Garbers C, Yanofsky MF, Schell J. 1992. Molecular analysis of an auxin binding protein gene located on chromosome 4 of *Arabidopsis. Plant Cell* 4:193–201

92. Palme K, Hesse T, Moore I, Campos N, Feldwisch J, et al. 1991. Hormonal modulation of plant growth: the role of auxin perception. *Mech. Dev.* 33:97–106

93. Pelham HRB. 1989. Control of protein exit from the endoplasmic reticulum. *Annu. Rev. Cell Biol.* 5:1–23

94. Potter R, Haley BE. 1982. Photoaffinity labeling of nucleotide binding sites with 8-azidopurine analogs. *Methods Enzymol.* 91:613–33

95. Prasad PV, Jones AM. 1991. A putative receptor for the plant growth hormone auxin identified and charactertized by anti-idiotypic antibodies. *Proc. Natl. Acad. Sci. USA* 88:5479–83

96. Radermacher E, Klambt D. 1993. Auxin dependent growth and auxin-binding proteins in primary roots and root hairs of corn (*Zea mays* L.). *J. Plant Physiol.* 141:689–703

97. Rakhaminova AB, Khavkin EE, Yaguzhinskii LS. 1978. Construction of a model of the auxin receptor. *Biokhimiya* 43:806–23. Transl. in *Biochem. USSR* 43:639–53

98. Ray PM. 1977. Auxin-binding sites of maize coleoptiles are localized on membranes of the endoplasmic reticulum. *Plant Physiol.* 59:594–99

99. Ray PM. 1977. Specificity of auxin-binding sites on maize coleoptile membranes as possible receptor sites for auxin action. *Plant Physiol.* 60:585–91

100. Ray PM. 1985. Auxin and fusicoccin enhancement of β-glucan synthase in peas. *Plant Physiol.* 78:466–72

101. Ray PM, Dohrman U, Hertel R. 1977. Characterization of naphthaleneacetic acid binding to receptor sites on cellular membranes of maize coleoptile tissue. *Plant Physiol.* 59:357–64

102. Ray PM, Walton JD. 1981. Evidence for receptor function of auxin binding sites in maize. *Plant Physiol.* 68:1334–38

103. Rubery PH. 1981. Auxin receptors. *Annu. Rev. Plant Physiol.* 32:569–96

104. Ruck A, Palme K, Venis MA, Napier RM, Felle HH. 1993. Patch-clamp analysis establishes a role for an auxin binding protein in the auxin stimulation of plasma membrane current in *Zea mays* protoplasts. *Plant J.* 4:41–46

105. Ruoho AE, Kiefer H, Roeder PE, Singer SJ. 1973. The mechanism of photoaffinity labeling. *Proc. Natl. Acad. Sci. USA* 70: 2567–71

106. Sakai S. 1984. Characterization of 2,4-D binding to the auxin-binding protein purified from etiolated mung bean seedlings. *Agric. Biol. Chem.* 48:257–59

107. Sakai S. 1985. Auxin-binding protein in etiolated mung bean seedlings: purification and properties of auxin-binding protein II. *Plant Cell Physiol.* 26:185–92

108. Sakai S. 1992. Regulatory functions of soluble auxin-binding proteins. *Int. Rev. Cytol.* 135:239–67

109. Sakai S, Hanagata T. 1983. Purification of an auxin-binding protein from etiolated mung bean seedlings by affinity chromatography. *Plant Cell Physiol.* 24:685–93

110. Sakai S, Kikuchi M, Nakajima N. 1992. Interaction between auxin-binding protein-I and RNA polymerase II. *Biosci. Biotech. Biochem.* 56:1225–29

110a. Scherer GFE, Andre B. 1993. Stimulation of phospholipase A_2 by auxin in microsomes from suspension-cultured soybean cells is receptor-mediated and influenced by nucleotides. *Planta* 191:515–23

111. Schmülling T, Fladung M, Grossmann K, Schell J. 1993. Hormonal content and sensitivity of transgenic tobacco and potato plants expressing single *rol* genes of *Agrobacterium rhizogenes* T-DNA. *Plant J.* 3: 371–82

111a. Schwob E, Choi S-Y, Simmons C, Migliaccio F, Ilag L, et al. 1993. Molecular analysis of three maize 22 kDa auxin-binding protein genes—transient promoter expression and regulatory regions. *Plant J.* 4:423–32

112. Senn AP, Goldsmith MHM. 1988. Regulation of electrogenic proton pumping by auxin and fusicoccin as related to the growth of *Avena* coleoptiles. *Plant Physiol.* 88:131–38

113. Shen WH, Davioud E, David C, Barbier-Brygoo H, Tempe J, Guern J. 1990. High sensitivity to auxin is a common feature of hairy root. *Plant Physiol.* 94:554–60

114. Shimomura S, Inohara N, Fukui T, Futai M. 1988. Different properties of two types of auxin-binding sites in membranes from maize coleoptiles. *Planta* 175:558–66

115. Shimomura S, Liu W, Inohara N, Watanabe S, Futai M. 1993. Structure of the gene for an auxin-binding protein and a gene for 7SL RNA from *Arabidopsis thaliana*. *Plant Cell Physiol.* 34:633–37

116. Shimomura S, Sotobayashi S, Futai M, Fukui T. 1986. Purification and properties of an auxin-binding protein from maize shoot membranes. *J. Biochem.* 99:1513–24

117. Shimomura S, Watanabe S. 1993. Auxin receptors in plasmalemma. *Membrane* 18: 34–42

118. Slack JMW. 1991. *From Egg to Embryo.* Cambridge: Cambridge Univ. Press. 328 pp. 2nd ed.

119. Smith H, Whitelam GC. 1990. Phytochrome, a family of photoreceptors with multiple physiological roles. *Plant Cell Environ.* 13:695–707

120. Takahashi Y, Nagata T. 1992. *parB*: an auxin-regulated gene encoding glutathione S-transferase. *Proc. Natl. Acad. Sci. USA* 89:56–59

121. Tappeser B, Wellnitz D, Klambt D. 1981. Auxin affinity proteins prepared by affinity chromatography. *Z. Pflanzenphysiol.* 101: 295–302

122. Tillmann U, Viola G, Kayser B, Siemester G, Hesse T, et al. 1989. cDNA clones of the auxin-binding protein from corn coleoptiles (*Zea mays* L.): isolation and characterization by immunological methods. *EMBO J.* 8:2463–67

123. van der Linde PCG, Bouman H, Mennes AM, Libbenga KR. 1984. A soluble auxin-binding protein from cultured tobacco tissues stimulates RNA synthesis in vitro. *Planta* 160:102–8

124. van der Linde PCG, Mennes AM. 1990. Modulation, purification and function of a soluble auxin receptor. In *Plant Aging: Basic and Applied Approaches. NATO ASI Ser.*, ed. R Rodriquez, R Sanchez Tames, DJ Durzaan, 186:263–67. New York: Plenum

125. Venis MA. 1977. Affinity labels for auxin binding sites in corn coleoptile membranes. *Planta* 134:145–49

126. Venis MA. 1977. Solubilization and partial purification of auxin-binding sites of corn membranes. *Nature* 266:268–69

127. Venis MA. 1985. *Hormone Binding Sites in Plants.* London: Longman. 191 pp.

128. Venis MA, Napier RM, Barbier-Brygoo H, Maurel C, Perrot-Rechenmann C, Guern J. 1992. Antibodies to a peptide from the maize auxin-binding protein have auxin agonist activity. *Proc. Natl. Acad. Sci. USA* 89:7208–12

129. Venis MA, Thomas EW, Barbier-Brygoo H, Ephritikhine G, Guern J. 1990. Impermeant auxin analogues have auxin activity. *Planta* 182:232–35

130. Viola G. 1991. *Proteine aus mais (Zea mays L.) und Lowenmaulchen (Antirrhinum majus L.): Untersuchungen zur Isolierung und Charakterisierung auf Protein- und Nucleinsaureebene.* Diss. thesis. Univ. Bonn

131. Yu L-X, Lazarus CM. 1991. Structure and sequence of an auxin-binding protein gene from maize (*Zea mays* L.). *Plant Mol. Biol.* 16:925–30

132. Zaina S, Bertani A, Lambardi L, Mapelli S,

Torti G. 1989. Membrane-associated binding sites for indoleacetic acid in the rice coleoptile. *Planta* 179:222–27

133. Zettl R, Feldwisch J, Boland W, Schell J, Palme K. 1992. 5'-Azido-[3,6³H]-1-naphthylphthalamic acid, a photoactivatable probe for naphthylphthalamic acid receptor proteins from higher plants: identification of a 23-kDa protein from maize coleoptile plasma membranes. *Proc. Natl. Acad. Sci. USA* 89:480–84

134. Zettl R, Schell J, Palme K. 1993. Photoaffinity labeling of *Arabidopsis thaliana* plasma membrane vesicles by 5-axido-[7-³H]-indole-3-acetic acid: Identification of a glutathione-S-transferase. *Proc. Natl. Acad. Sci. USA* In press

Annu. Rev. Plant Physiol. Plant Mol. Biol. 1994. 45:421–45

THE RIBONUCLEASES OF HIGHER PLANTS

Pamela J. Green

MSU-DOE Plant Research Laboratory and Department of Biochemistry, Michigan State University, East Lansing, Michigan 48824-1312

KEY WORDS: RNA degradation, S-RNases, S-like RNases, phosphate starvation, senescence

CONTENTS

INTRODUCTION 421
MAIN CLASSES OF RNASES IN HIGHER PLANTS 422
    Biochemical Classification 422
    Sequence Similarity among Plant RNases from Diverse Species 423
LOCALIZATION 428
    Secretory RNases 428
    RNases in Chloroplasts and Mitochondria 429
    Other Intracellular RNases 430
REGULATION OF PLANT RNASES 431
    Senescence 432
    Phosphate Starvation 433
    Plant Disease 433
    Light and Other Developmental Cues 434
    Genetic Determinants 436
FUNCTIONAL ROLES OF PLANT RNASES 437
    Self-Incompatibility 437
    Phosphate Remobilization 438
    Plant Defense 438
    mRNA Decay 439
CONCLUSIONS AND FUTURE PROSPECTS 440

INTRODUCTION

RNA-degrading enzymes, better known as ribonucleases (RNases), have been studied in eukaryotic cells for many years. As the first enzyme to be chemically sequenced, bovine pancreatic RNase A has been an excellent model

E C number an 1 myright!

enzyme for protein structure/function studies (18). More recently it has become apparent that pancreatic RNase is a member of a superfamily of enzymes with roles that extend beyond the digestion of dietary RNA. Members of this superfamily include enzymes that exhibit angiogenic, neurotoxic, antitumor, or immunosuppressive activities (34). Although plants do not appear to contain RNases in this superfamily, the theory that ribonucleases can have diverse functions in plants has also been developing rapidly. Perhaps the major stimulus behind this was McClure et al's discovery (85) that the S-glycoproteins associated with gametophytic self-incompatibility (SI) in the Solanaceae had ribonuclease activity. These S-glycoproteins, now called S-RNases, are secreted into the style mucilage where they are thought to abort the growth of pollen bearing the same S-allele (reviewed in 31, 64, 104, 113). Since the discovery of these S-RNases, several S-like RNases from self-compatible plants have been identified that are related to, but distinguishable from, the S-RNases (e.g. 110). The S-like RNases may participate in starvation rescue and senescence as well as general RNA turnover.

Plant ribonucleases were studied actively before the 1980s, climaxing with the publication of two excellent reviews by Farkas (40) and Wilson (123). Interest began to wane subsequently because little correlation was found between total RNase activity and RNA content in most plant samples and because experiments to determine the function of individual enzymes were not feasible at the time. Fortunately, this situation was relatively short-lived as the identification of the S- and S-like RNases and their regulatory properties has led to renewed enthusiasm for the study of plant RNases. This review emphasizes the most recent findings and highlights previous work in the areas of RNase biochemistry, localization, and regulation that are most relevant to RNase function. I also discuss how new technologies such as plant transformation and molecular genetic approaches are providing new ways to study the roles of RNases in many plant processes. Two specialized types of RNases, ribosome inactivating proteins and plant viral ribozymes, have been the subjects of other recent reviews (108 and 117, respectively) and, therefore, are not covered in detail here.

MAIN CLASSES OF RNASES IN HIGHER PLANTS

Biochemical Classification

The RNA-degrading enzymes of higher plants have classically been grouped into four main enzyme families (40, 123): 1. *RNase I enzymes* are soluble endoribonucleases with a pH optimum of 5.0–5.5 and a molecular weight of 20–25 kDa. They produce 3′ nucleotide products (E.C. 3.1.27.1), and are not sensitive to EDTA. 2. *RNase II enzymes* are similar to the RNase I class: they are endoribonucleases that produce 3' nucleotide products (E.C. 3.1.27.1) with a low sensitivity to EDTA. They are distinguished from RNase I enzymes by their lower molecular weight (17–21 kDa), higher pH optimum (6.0–7.0), and microsomal location. Both RNase I and RNase II enzymes prefer to cleave after purine residues. 3. *Nuclease I enzymes* are endonucleases that degrade both ssDNA and RNA and have a pH optimum of 5.0–6.5. They have a molecular weight of 31–35 kDa, produce 5' nucleotides as products (E.C. 3.1.30.1), and are highly sensitive to EDTA. 4. *Exonuclease I enzymes* are large (molecular weight ≥ 100 kDa) soluble exonucleases that degrade both ssDNA and RNA to 5' nucleotides (E.C. 3.1.15.1), and have a pH optimum of 7.0–9.0 and a high sensitivity to EDTA.

These four types of enzymes have been found in a variety of plants (40, 123). Recent examples include RNase I–type enzymes from tomato (3), and barley (69, 95); an RNase II–type enzyme from barley (95); nuclease I–types from barley (25) and zinnia (112) and an exonuclease I–type from petunia (96). Many plant RNA-degrading enzymes cannot easily be classified into one of these four main classes. *Arabidopsis thaliana* appears to contain representatives of the RNase I, RNase II, and nuclease I classes, but most activities in the *A. thaliana* RNase profile do not fit precisely into a given category (131). A further complication is that the distinction between the classes is not always clear. This is partly because of the variety of enzyme preparations and characterization procedures used in various laboratories. For example, in some cases RNases are identified that have not been tested for DNase activity, while in other cases the microsomal preparations used to distinguish between RNase I and II enzymes are crude at best. Even when clear distinctions have been made on the basis of biochemical properties, such as for RNase I and RNase II enzymes from barley, immunological evidence can indicate strong structural similarities (95).

Sequence Similarity among Plant RNases from Diverse Species

An alternative approach to classifying plant RNases has been to compare their amino acid sequences. The first plant RNases to be sequenced were the S-RNases of *Nicotiana alata*, but they were not known to be RNases at the time (8). The cDNAs for these enzymes were isolated because their products, small basic glycoproteins, cosegregated with the S-locus, known to control SI in *N. alata* (31). Amino acid sequence comparisons led Sakyama to recognize that the S-glycoproteins were homologous to a set of fungal RNases typified by RNase T2 (66, 85). Enzymatic assays performed by Clarke and coworkers then demonstrated that purified S-glycoproteins did indeed have RNase activity (85). These findings have influenced greatly the development of mechanistic models for SI. Following the work in *N. alata*, S-RNases were cloned from a variety of other Solanaceous species that exhibit SI (31, 64). Many of these sequences are shown in Figure 1.

To address whether related RNases exist in self-compatible species, a polymerase chain reaction (PCR) approach was used to search for possible homologs from *A. thaliana*. PCR primers corresponding to the most highly conserved regions of the S- and T2-type RNases were synthesized and used to amplify *A. thaliana* cDNA (111). Three different PCR products were isolated, each containing all the amino acids that were invariant among the fungal and *N. alata* S-RNases within the amplified region. The genes corresponding to these PCR products were designated *RNS1, RNS2*, and *RNS3* (111). Expression of the *RNS* cDNA clones in yeast demonstrated that each gene encodes an active RNase (110; PA Bariola, CJ Howard, CB Taylor, MT Verburg, VD Jaglan, PJ Green, submitted). The identification of S-RNase homologs (S-like RNases) in self-compatible plants indicates that RNases in the T2/S superfamily are not restricted to self-incompatible species and, therefore, must also play a more fundamental role in RNA decay in plants.

Amino acid sequences deduced from the *RNS* cDNA clones are aligned with other plant RNases in Figure 1. The sequences of three additional plant RNases have been determined by directly sequencing the isolated proteins. RNases LX and LE were isolated from cultured cells of *Lycopersicon esculentum* (63, 78), a self-compatible tomato species. RNase MC is from the seeds of the bitter gourd *Momordica charantia* (55), which is also self-compatible. As shown in the alignment in Figure 1, RNases LE, LX, and MC are all S-like RNases because they exhibit sequence homology to the S- and the fungal T2-type RNases. The partial sequence of tryptic peptides of RNase LV-3 of *L. esculentum* indicates that it too is an S-like RNase (77). Although the number of RNase sequences is still rather limited, particularly from self-compatible species, the fact that all of the sequences fit into the same superfamily indicates that this type of enzyme represents a major class of RNase in the Angiosperms and probably in the plant kingdom.

The most highly conserved regions among the plant RNases that are boxed in Figure 1 have been designated C1-C5 (58). Two of these, C2 and C3, are also found in the fungal enzymes (see asterisks in Figure 1) and contain two of the three active site histidine residues that are required for the catalytic activity of RNase Rh of *Rhizopus niveus* (91). The third required histidine residue is also conserved among the plant enzymes (position 103 in Figure 1), as are the pairs of cysteine residues at positions 58 and 111 and positions 177 and 217 that form disulfide bonds in RNase T2 (66). Glutamic acid 104, the acidic residue participating in catalysis (92) is invariant among the S-like RNases and highly conserved among the S-RNases. All of the RNases in Figure 1 for which DNA sequences are known contain a typical signal sequence for entry into the secretory pathway (signal sequences are not shown in the alignment).

Although the similarity among the sequences shown in Figure 1 extends throughout their length, the S- and S-like RNases are distinct subclasses within the RNase superfamily. Each subclass contains highly conserved residues that are not conserved among members of the other subclass (110). The residues that distinguish the S-like RNases from the S-RNases are located primarily between the active-site histidines, where most of the residues common to the fungal enzymes are located, and at the N-terminus. A cluster of amino acids that are conserved only in the S-RNases is located between positions 143 and 156 in Figure 1. These residues may be important for the specialized function of the S-RNases, a hypothesis that can now be tested (see below). Genealogies generated from the genus aligned in Figure 1 and previous alignments indicate that some members of this RNase superfamily diverged before speciation. This is true for the divergence of S- and S-like subclasses (110) and for members of the individual subclasses (57; PA Bariola, CJ Howard, CB Taylor, MT Verburg, VD Jaglan, PJ Green, submitted). Finally, in contrast to the RNase superfamily typified by pancreatic RNase A, which has been found only in some vertebrates, the T_2/S-RNase superfamily in Figure 1 includes members from species as diverse as bacteria (41) and mammals (59). The crystal structure of a fungal member of this family, RNase Rh, was determined recently at a resolution of 2.5 Å (73), and a plant member, RNase MC, was recently crystallized (37). The fungal and seed RNases appear to have somewhat different preferences for hydrolysis of specific bases (61), so it will be interesting to compare their 3-dimensional structures.

Figure 1 (overleaf)   Alignment of S- and S-like RNase amino acid sequences. S-like RNase sequences are LE of *Lycopersicon esculentum* (63), RNS1 and RNS3 of *Arabidopsis thaliana* (PA Bariola, CJ Howard, CB Taylor, MT Verburg, VD Jaglan, PJ Green, submitted), LX of *L. esculentum* (78), RNS2 of *A. thaliana* (110), and MC of *Momordica charantia* (55). S-RNase sequences are 1Stu, r1Stu, and 2Stu of *Solanum tuberosum* (65); 2Pet, 3Pet, and 1Pet of *Petunia inflata* (6); Ps2A, Ps3A, and Ps1B of *Petunia hybrida* (30); 5Lyc of *Lycopersicon peruvianum* (116); a, Z, F11, and 1Nic of *Nicotiana alata* (70); XPet and OPet of self-compatible *P. hybrida* (7); 3Nic, 6Nic, and 2Nic of *N. alata* (9); and 3Sol and 2Sol of *Solanum chacoense* (128). X2 of *P. inflata* (76) is not an S-RNase but was placed with this group because it is more closely related to the S-RNases than the S-like RNases (76). The sequences are aligned from predicted mature N-termini of the *N. alata* S-RNases (see 49). Heavy-bordered boxes enclose conserved regions C1-C5 (58). Light shading indicates residues that are identical or functionally identical in at least 22 of the 28 sequences. Light-bordered boxes enclose residues that are identical or functionally identical in at least 5 of the 6 S-like RNase sequences but not highly conserved among the S-RNases. Dark shading indicates residues that are identical or functionally identical in at least 18 of the 22 S-RNase sequences but not highly conserved among the S-like RNases. Asterisks above the sequences denote residues in the fungal RNases T_2 (66), Rh (52), M (118a), and Trv (56) that are identical or functionally identical to the corresponding boxed or shaded residues. Functionally identical residues are grouped as follows: A, S, T; I, L, M, V; H, K, R; F, W, Y; D, E; Q, N.

Multiple sequence alignment (amino acid positions 1–120). Conserved regions are boxed and labeled C1, C2, and C3. Asterisks (*) above the alignment mark conserved positions.

```
              1        10         20         30         40         50         60         70         80         90        100        110        120
              |         |          |          |          |          |          |          |          |          |          |          |          |
                 *                *                              *******                    ***                *     ** *** **  *    ***       ******        *
         ┌──────────────────────┐                                    ┌──────────┐                                                        ┌─────────┐
LE    D  │FDFFYFVQQWP│ GSYC.DTKQS.C...CYPTTGKPAA  .PIPN. │DFGIHGLWP│ NNNDGTYB. SNCDPNGP. YDQSQISDLISSMQQNWPTLAC. PS..GSG.STFWSHEWE │KHGTCA│ ESVLTNQHA
RNS1  D  │FDFFYFVQQWP│ GSYC.DTQKK.C...CYPNSGKPAA  .PIPN. │DFGIHGLWP│ NYKDGTYB. SNCDASKP. FDSSTISDLLTSMKKSWPTLAC. PS..SSG.LKFWSHEWI │KHGTCS│ ESVL.DQHE
LX    D  │FDFFYFVQQWP│ ASYC.DTRRS.C...CYPDTGKPDE  .KSN.  │DFSIHGLWP│ NYKDGKWB. QNCDRESS. LDBSEFSDLLSTMEKNKWPSLAC. PS..SDG.LKFWSHEWI │KHGTCS│ .AL.NQHA
RNS3  D  │FDFFYFMLQWP│ GAYC.DSRHS.C...CYPQTGKPAA  .PIPN. │DFGIHGLWP│ NYKTGGWB. QNCNPDSR. FDDLRVSDLMSDLQREWPTLSC. PS..NDG.MKFWTHEWE │KHGTCA│ ESEL.DQHD
RNS2  E  │FDYFALSLQWP│ GTYCRGTRHC.CSKNACCRGSDAPT  YDYNDGSWB. SCCYR.SD. FKEKEISTLMDGLEKYWPSLSCGEPSSCNGGKGSFWGHEWE │KHGTCS│ SPVFHDEYN
MC    .  │FDSFWFEVQQWP│ PAVCSFQKSGSGC...LBGSG.LR  QFTIHGLWP  Q.GSGT.SLTNCPQQSP. HDITKLSHLQSQLNTLWPNVLRANNQQ... FWSHEWT │KHGTCS│ EST.FNQAA

                 *                *                              *******                    ***                *     ** *** **  *    ***       ******        *
         ┌─────────────────────┐                                    ┌──────────┐
1Stu  D  │FELLELVSTWP│ ATFCYAYG...CSKR....PIPN.  │NFTIHGLWP│ DNKSTV. LNFCNLAHEDEYIPTDHKILTED.KRWPQLRY. DYLYGIRQYLWKNEFI │KHGTCS│ INR.YKQPA
r1Stu N  │FELLELVSTWP│ ATFCYAYG...CSR.....PIPN.  │NFTIHGLWP│ DNKSVI. LNDCKVVNKEGYVKITDPKQITED.KRWPQLRY. EKLYGIDRQYLWKNEFI │KHGSCS│ INR.YKQBA
2Pet  N  │FDYFQLVLTWP│ ASFCY.PKNF.CKR.....KSN.   │NFTIHGLWP│ EKKHFR. LEFCTGDKYSRF..KEDNIINVLERHWIQMRF. DEKYASTKQPLWEHEYN │RHGICC│ KNL.YDQBA
Ps2A  N  │FDYFQLVLTWP│ ASFCY.PKNK.CQR.....RSN.   │NFTIIHGLWP│ EKKHFR. LEFCPGDKFSRF..KEDNINVLERHWIQMRF. DEDYANAKQPLWQHEYN │RHGICC│ KNL.YDQKA
Ps3A  N  │FDYFQLVLTWP│ ASFCY.PKNK.CQR.....RSN.   │NFTIIHGLWP│ EKKRF.  LEFCTGDKYKRF..LEEDNINVLERHWIQMRF. DETYANTKQPLWEHEYN │KHGICC│ KNL.YDQKA
3Pet  N  │FDYIQLVLTWP│ ASFCYRPKNI.CRR.....IPN.   │NFTIHGLWP│ EKEHFR. LEFCDGDKFVSF.SLKD..RIVNDLERHWVQMKF. DEKFAKIKQPLWTHEYN │KHGICS│ SNL.YDQRA
1Pet  N  │FEYIQLVLTWP│ ASFCFRPKNI.CKR.....PAK.   │NFTIHGLWP│ EITGFR. LEFCTGDFKYT..FKDNNIVDYLERHHVQMKF. DENYAKYHQPLWSYEYF │KHGMQC│ SKI.YNQKA
2Stu  D  │FDYMQLVLTWP│ RSFCYPYGF..CNR.....IPPN.  │NFTIHGLWP│ DKKPMRGQLQFCTSD..DYIKFTPGSVLDALDHWIQLKF. EREIGIRDQPLWKDQYK │KHGTCC│ LPR.YNQLQ
5Lyc  D  │FDYMQLVLTWP│ RSFCYPRRF..CNR.....IPPN.  │NFTIHGLWP│ DKMGIPGHLQFCTSEK..YEIFEPGSVLDALDQHWIQLKF. ERETGLRNQPLWRDQYH │KHGTCC│ LQR.YNQLQ
              a
Z     D  │FDYMQLVLFWP│                    IPK.   │NFTIHGLWP│ DEQHGM. LNDC.GE.T.FTKLREPREKKELDDRWPDLKR. SRSDAQDVESFWEYEYN │KHGTCC│ TEL.YDQAA
XPet  D  │FDFCYPKNF..CSRI....APK.│            │NFTIHGLWP│ DKVRGR. LQFCTSEKYVNF..AQDSPILDDLDHHWMELKY. HRDFGLENQPLWRGQYC │KHGTCC│ IPR.YNQMQ
F11   D  │FEYLQLVLTWP│ DCYRPRYL.CRI.....TAPN.   │NFTIHGLWP│ DNEQRR. LQFCT.ST.EYSLF.DGDILDDLDRHWIQLKF. DKETGMQLDRWHEQFF │KHGTCC│ ENR.YKQMP
Ps1B  D  │FEYLQLVLTWP│ ASFCY.ANH..CERI....APN.  │NFTIHGLWP│ DNVKTR. LHNCKPKPTYSY..F.TGKMLNDLDKHWMQLKF. EQDYGRTEQPSWKYQYI │KHGSCC│ QKR.YNQNT
OPet  S  │FDHWQLVLIWP│ AGYCVKG...CPRV....IPN.   │DFTIHGLWP│ DSISVI. MYNC..DPTKTFAITIEIKQLTELKEKWPELTS. TAQFALTSQSFWRYQYE │KHGTCC│ FPV.YSQSA
3Nic  A  │FDHWQLVIIWP│ AGYCKIKG..CPRT....VFPD.   │NFTIHGLWP│ DSVSVM. MYNC..DPPTRFNKIRETNIKNELEKRWPELTS. TAQFALKSQSFWKTQYE │KHGTCC│ LPF.YSQSA
6Nic  A  │AFCHTTPSP.CKR.....IPN.│              │NFTIHGLWP│ DNVSTM. LNYC.DREDEYEKLDDDKKKKDLDRWPDLTI. ARADCIEHQFWKHEYN │KHGTCC│ SKS.YNLTQ
X2    A  │TAFCHTT.P.CKN.....IPS.│              │NFTIHGLWP│ DNVSTT. LNFC.GKEDEYNIIMDGPERNGLYVRWPDLIR. EKADCMKTQNFWRREYI │KHGTCC│ SEI.YNQVQ
1Nic  Y  │YEYMQLVLQWP│ TAFCHASPT..CK.....VTPN.   │NFTIHGLWP│ DNVSTE. LNYCKSKTG.KYNNIKDPTIKNELYKRWPDLTT. SETDCLGNQNFWKREYN │KHGTCC│ SGR.YNLQQ
2Nic  A  │FEYMQLVLTWP│ TAFCNVM..N.CER.....TPT.   │NFTIHGLWP│ DNVSTE. LNYC.DRQKKFKLFEDDKKQNDLDDRWPDLTL. DRDDCKNGQGFWSYEYR │KHGTCC│ LPS.YNLQQ
3Sol  A  │FEYMQLVLTWP│ ITFCRIK..H.CER.....TPT.   │NFTIHGLWP│ DNHTTM. LNYC.DRSKPYNMFTDGKKKNDLDERWPDLTK. TKFDSLDKQAFWKDQYV │KHGTCC│ SDK.FDREQ
2Sol  T  │FEHLQLVLTWP│ TISFCHKE..R.CIR.....SSS.  │NFTIHGLWP│ DNSTSTR. LNYC.KIVKYNKLIEDEHKIDALEYGWPNLTT. TEAVSKEDQVFWGKQYI │KHGSCC│ TDL.YDKDA
2Sol  T  │FEYMKLVLQWP│ PMYCNKF...CER.....IPR.    │NFTVHGLWP│ DNKKYL. LNNC..RSYAYNALTNVREQSKLDDRWPDLTS. NKSMTMKDQKFWEYEYN │KHGTCC│ EKL.YNQAQ

         C1                                             C2                                                               C3
```

```
              130       140       150       160       170       180       190       200       210       220       230       240
              |         |         |         |         |         |         |         |         |         |         |         |
     * *   **                   *         *          *         *          *           *                    *
LE   YFKKALD L...KNQIDLL SILQGADIHPDGE.SYDIVNIRNAIKSAI.GYTPWIQCNVDQSGN.S QLYQVYICV DGSGDSLIECPIPFG..GK.CGTSIE.FPTF
NS1  YFQTALN L...KQKTNLL GALTKAGINPDGK.SYSLESIRDSIKESI.GHTPWVECNRDSQGN.S QLYQVYLCV DRSGSGLIECPVFPH..GK.CGAEIE.FPSF
LX   YFQTALD F...KTKSNLL QNLNNAGIKPRNGDYYGVES.KKAIEKGV.GHTPFIECNVDSQGN.H QLYQVYLCV DSSASKFIDCPIFPHG.GK.CGSKIE.FPSFSTNDDHDEF
NS3  YFEAGLK L...KQKANLL HALTNAGIKPDDK.PYEMKDIENTIKQVV.GHAPGIECNHDSSHN.S QLYQIYLCV DTSASNFINCPVMPH..GR.CDDSRVQ.FPK
NS2  YFLTTLN LYLKHNVTDVL ..YQAGYVASNSE.KYPLGGIVTAIQNAF.HITPEVVCKRDAI..G D.EIRICF YKDFKPRD.CVGSQDLTSRKSC.PKYVSLPEYTPLDGEAMVLKMPTEREAL
MC   YFKLAVD ...DII GALRPHAAGPNGR.TKSRQAIKGFLKAKF.GFFPGLRCRTDPQTKVS YLUQVVACF AQDGSTLIDC....T..LRDTCGANFIF

Stu  YFDLAMK I...KDKFDLL GTLRNHGINPGST.YELDDIERAIMTVSIEV.PSLKCIQK.PLGNV ELNEIGICL DPEAKYMVPCPRTGS....CHNMG.HKIKFR
Stu  YFDLAMK I...KDKFDLL GTLRNHGINPGST.YELDDIERAIMTVSIEV.PSLKCIQK.PLGNV ELNEIGICL DPEAKYMVPCPRTGS....CHNMG.HKIKFR
Pet  YFLLAIR L...KDKLDLL TTLRTHGITPGTK.HTFGEIQKAIKTVTNNKDPDLKCVEN.IKGVK ELNEIGICF NPAADSFHDCRHSKT....CDETDSTQTLFRR
s2A  YFLLAMR L...KDKLDLL TTLRTHGITPGTK.HTFGEIQKAIKTVTNNKDPDLKCVEN.IKGVK ELNEIGICF NPAADSFHDCRHSVT....CDETDSTQTLFRR
s3A  YFLLAMR L...KDKLDLL TTLRTHGITPGTK.HTPGEIQKAIKTVTSNNDPDLKCVEN.IKGVM ELNEIGICF TPAADRFDRCRHSNT....CDETSSTKLFRG
Pet  YFLLAMR L...KDKFDLL TTLRTHGITPGTK.HTPGDIQKAIKTVTNQVDPDLKCVEH.IKGVR ELKEVGICF TPAADSFHDCRHSNT....CDETDSTKLLFR
Pet  YFLLATR V...KEKFDLL TTLRTHGITPGTK.HTPGDIQKAIKTVTNQVDPDLKCVEH.IKGVQ ELNEIGICF NPAADNFYPCHHST....CDETDSKMLLFR
Stu  YFLLAMR L...KEKFDLL TTLRTHGITPGTK.HTPKKIQDAIKTVT.QEVPDLKCVQY.IQGVI ELYEGICF TPEADSLFPCRQSKS....CHPTENPLIFRL
Lyc  YFLLAMR L...KEKFDLL TTLRTHGITPGTK.HTYKKIQDAIKTVT.QEVPDLKCVQY.TRGVI ELYEGICF TPEADSPSLCRQSNS....CHPTENPLIFR
a    YFDLAKN L...KDKFDLL RNLKNEGIIPGST.YTVDECEKQSEAVT.QAYPNLNCVGDPQK.II ELSEIGICF DRGATKVITCRRTT....CNPINKKESFPLN
Z    YFLLAMR L...KDKFDLL ATLRTHGITPGTK.HTFNETQKAIKTVTNQVDPDLKCVEH.IKGVH ELSEIGICF CDETGITKLIFRR
Pet  YFLLAMR L...KNKFDLL TTLRTHGIIPGTK.HTFDEIQKAIKTVTNQVDPDLKCVQH.IQGVE ELNEIGICF TPAADRFFPCPQSKS....CDKTA.RVLFRR
F11  YFGLAIR L...KDKTDLL SIIRSQGVTPGST.YTGERINSSIASVT.RVKPNLKCIYYRGK..I ELTEIGICF TPNADSMRPCPQSDI....CPKTGT.KLLFR
s1B  YFDFAIK L...KDKTDLL TIIRNQGVTPGST.YTGEKLNSSIASVT.RVAPNLKCIYY..QGKI ELTEIGICF DRITVAMMSCPRISTS...CKFGTNARIIFRQ
Pet  YFDFAIK L...KDKFDLL TSLRHGIIPGNS.YTVQKINSTKAIT.QGYPNLSC....TKRQM ELLEIGICF NRTIVAMMSCPRISTS...CKFGTNAGITFRQ
Nic  YFDLAMA L...KDKFDLL TSLKNHGIIRGYK.YTVQKINNTIKTVT.KGYPNLSC....TKGQ ELLEIGICF DSKYKNVIDCPHPKT....CKPMGNRGIKFP
Nic  YFPRLAMA L...KDKFDLL TSLTNHGIIPGSN.YNVOKINSTIKTIT.RGYPNLSC....TEE.M ELLEIGICF DSTVKNVIDCPHPKT....CKTASNQGIMFP
X2   YFPHLAMA L...KDKFDLL KSFRNHGIIPTKS.YTVQKYNNTVKAIT.KGFPNLTC....NK.QM ELMEIGICF CNPT...IIKFP
Nic  YFDLAMA L...RDKFDLL SSIRNHGISRFS.YTVQNLNNTKAIT.GGFPNLTC....SR.LR ELQEIGICF DQKVKNVIDCPRPKT....CKATR.NGIIFP
Nic  YFDLAMT L...RDKFDLL KILAHGIIPGTS.HHTSNIONAVKSVT.QGVPHVTCPNNRFKGTS ELXEIGICF DETVKNVIDCPNPKT....CKPTN.KGIMFP
Sol  YFDLAMN L...KDRFDLL KILAHGIIPGTS.HHTSNIQNAVKSVT.QGVPHVTCPNNRFKGTS ELLELALCF DPQAQNVIHCPRPKT....CNSKGTKGITFP
Sol  YFNLTMN L...KDKFDLL RIIRNHGIVPGE.LALLSNSGRPLRQLTNKVFPSLKCID.NNGIM ELLEVGICF DPAATKVIPCHRPWI....CHADENTRIELVK

        C4                                                          C5
```

LOCALIZATION

Secretory RNases

RNases that enter the secretory system may be targeted to the vacuole or the extracellular space, or they may be retained in one of the compartments of the secretory system such as the endoplasmic reticulum (ER). Approximately 70–80% of the RNase activity in plant cells is located in the vacuole (2, 23); therefore, most of the soluble RNase activity in plant homogenates is probably derived from vacuolar enzymes rather than enzymes located in the nucleus or cytoplasm. Specific examples include three RNase I–type enzymes (LV-1, LV-2, and LV-3) from vacuoles of cultured tomato cells (3, 77), and an acid RNase (presumably an RNase I type) that has been localized immunologically to the vacuoles of morning glory flowers (16). It has been proposed that RNS2 is a vacuolar enzyme (110) because it contains a 20 amino acid C-terminal extension that has the features of a vacuolar targeting sequence (29). Matile has reviewed some earlier reports of RNase activity associated with the vacuoles of plants (82). Glund and coworkers (1) have reported that 3% of the total RNA in cultured tomato cells is found in isolated vacuoles, in the form of fragments less than 80 nucleotides in length. The presence of this RNA in the vacuole indicates that vacuolar RNases may be involved in the overall turnover of RNA in living cells (1); however, direct evidence for this hypothesis has not been reported.

One prominent class of extracellular RNases are the S-RNases of Solanaceous plants. Immunocytochemical studies have shown that the S-RNases of *N. alata* are localized to the surface of the stigmatic papillae and the extracellular matrix in the transmitting tract of the style (9). They are also found in the walls of cells that comprise the inner epidermis of the ovary (9). Like the S-RNases, most of the S-like RNases shown in Figure 1 have a signal sequence (110; PA Bariola, CJ Howard, CB Taylor, MT Verburg, VD Jaglan, PJ Green, submitted) and, therefore, enter the secretory pathway. RNase LE is secreted into the medium of cultured tomato cells, and its high structural similarity (75–79% similarity) with RNS1 and RNS3 indicates that the latter two enzymes are also extracellular. Other RNA-degrading enzymes that are secreted include a nuclease I and two RNases from barley aleurones (25), and an RNase and a nuclease found in the walls of potato tuber cells (89). RNase LX of tomato may be localized to the ER because the sequence HDEF, present at the C-terminus (see Figure 1) of the enzyme, may function as an ER retention signal (78). This would be consistent with the observation that LX is an intracellular enzyme that does not copurify with vacuoles (77). Retention of RNase LX in the ER may provide a means to sequester its activity away from cytoplasmic RNA, but it is difficult to predict what biological role the enzyme may have in this compartment. Another possibility is that the HDEF sequence may be removed by proteolytic cleavage, allowing the protein to be targeted elsewhere.

RNases in Chloroplasts and Mitochondria

Recent efforts to study the processing and stability of chloroplast transcripts in vitro and in vivo have pointed to the importance of both exonuclease and endonuclease activities in plastid mRNA metabolism. In vitro and in vivo experiments (reviewed in 48) indicate that a 3′ to 5′ exoribonuclease and RNA-binding proteins are involved in 3′-end formation in chloroplasts. Potential stem-loop structures present at the 3′ ends of mature chloroplast mRNAs may block progression of the exoribonuclease, thereby contributing to mRNA stability. Similar mechanisms are known to stabilize transcripts in bacteria (50). The situation is complicated further because the presence of these stem-loops is not always sufficient to confer mRNA stability (38), and the stem-loops may also mediate recognition of the transcript by endoribonucleases (53).

Evidence that at least two endonucleases exist in spinach chloroplasts was obtained recently from a study of *petD* RNA processing in vitro (53). The first endonuclease, designated EndoC1, appears to recognize a mutant form of the *petD* (*petD2*) stem-loop RNA. This activity has not been separated from the exoribonuclease activities present in the extract so its identification remains tentative. The second endonuclease, EndoC2, is biochemically separable from exoribonucleases and appears to cleave the *petD* transcript at the stop codon and at its mature 3′ end. In vitro RNA decay experiments indicate that cleavage at the stop codon can lead to rapid degradation of the upstream RNA. Based on these observations, EndoC2 may contribute to the control of mRNA stability and RNA maturation in vivo (53). Interestingly, EndoC2 appears to co-fractionate with an RNA-binding protein that binds to sites adjacent to or overlapping EndoC2 cleavage sites (53, 54). The nature of the association between these two activities remains unknown.

Less is known about the RNases of mitochondria. RNase MRP, an enzyme implicated in the processing of RNA primers during mitochondrial DNA replication in several eukaryotes, has been reported to exist both in the nucleolus and in mitochondria of mammalian cells (32). A nuclear-encoded RNA (called MRP or 7-2 RNA) is an essential component of the enzyme in several organisms (103). The mitochondrial location of the enzyme was recently called into question with the isolation and characterization of plant MRP RNAs. These RNAs were undetectable in mitochondria purified from tobacco leaves and suspension cultures (72). A re-examination of HeLa cell mitochondria for MRP RNA concluded that little or no MRP RNA is present (71). Despite the controversy over the existence of significant levels of RNase MRP in the mitochondria (71, 114), its nucleolar localization has been established firmly, indicating a role for the enzyme in rRNA processing (see 72 and references therein).

Like the MRP RNAs from other eukaryotes (103), *A. thaliana* and tobacco MRP RNAs are related structurally to RNase P RNA, the RNA component of a tRNA processing activity (35). RNase P is expected to exist in all cells or organelles that synthesize tRNA (35). In spinach chloroplasts, an enzyme with the cleavage characteristics of RNase P has been characterized partially and has a lower density than its bacterial counterpart (118). The enzyme is also resistant to micrococcal nuclease (118). These observations and the apparent lack of RNA in partially purified preparations, have prompted the suggestion that the chloroplast RNase P may not contain the RNA component that is characteristic of the enzyme from other sources (118). Because nuclease resistance and low density were also characteristics of an archaebacterial RNase P that copurified with an RNA (36), further study will be necessary to establish whether or not the spinach activity lacks an RNA subunit (35).

Other Intracellular RNases

Knowledge about the RNases that reside in the cytosol is extremely limited, not only in plants but also in other eukaryotes. This is unfortunate because the cytosol is presumably the major site of mRNA degradation. In yeast, one well characterized ribonuclease is XRN1, a 5' to 3' exoribonuclease that has been highly purified from ribosomes (reviewed in 107). Although XRN1 has been implicated in mRNA decay and is presumed to be cytoplasmic, there is considerable molecular genetic evidence that the enzyme also has roles in other processes, some of which are nuclear (67, 107). XRN1 is unlikely to be a member of the soluble exonuclease family of enzyme found in higher plants (discussed above) because enzymes in the exonuclease class have high activity toward DNA and RNA substrates, whereas the XRN1 activity is much more active toward RNA than DNA (107). A cytoplasmic enzyme that has been well characterized and for which the gene was recently cloned from mammalian cells is the 2-5A-dependent RNase that is implicated in interferon action and mRNA degradation (132). Efforts to identify a plant homolog of this activity have not been successful (27a; B Silverman, M Sullivan, PJ Green, unpublished results).

Many of the RNase activities found in plant cells have been reported to be particle bound (40). Unfortunately, the nature or purity of the various particulate fractions is rarely investigated. One well-characterized particle-bound enzyme is an RNase II–type enzyme from maize that is associated with crude microsomal preparations (120). It is not known whether RNase II enzymes are associated with the ER in vivo, or if they just adsorb to the particles during isolation. The endonuclease class of activities in many plants has also been reported to be associated with particulate fractions (reviewed in 40).

In addition to the activities associated with microsomes, RNase activity has been shown to co-purify with nuclei (79), chromatin (106), and ribosomes (4, 98) from plants. Although the RNase activities associated with plant ribosomes (or polyribosomes) have not been studied actively for several years, recent developments indicate that polyribosome-associated enzymes warrant further study. In the appropriate in vitro assays, RNase activities associated with polyribosomes have been shown to degrade mammalian transcripts in a fashion that reflects the in vivo degradation of the same transcripts (99). Initial efforts to develop polyribosome-based in vitro mRNA decay systems in plants have also been encouraging. Ribonuclease activities associated with oat polyribosomes degrade several polysomal transcripts in vitro at rates similar to those measured in vivo (27). In cell-free RNA-decay assays containing either petunia or soybean polysomes, *SRS4* transcripts (encoding the small subunit of ribulose bisphosphate carboxylase) that were endogenous to the polysomes or added exogenously were degraded to form products corresponding to those observed in vivo (MM Tanzer & RB Meagher, submitted). The ribonucleases responsible for the formation of these degradation products remain to be characterized, but an endoribonuclease may initiate the decay process (MM Tanzer & RB Meagher, submitted).

REGULATION OF PLANT RNASES

Plants exhibit changes in RNase levels in response to a variety of different endogenous and exogenous stimuli. In many of the earlier studies, only changes in total RNase content were measured (40). The use of RNase activity gels is now the preferred method for investigating the effects of regulatory stimuli on RNases that can be separated electrophoretically. The details of this procedure, which involves a negative stain for RNase activity, have been discussed in several reports (e.g. 20, 22, 131). Because a number of genes for RNases have been isolated, nucleic acid probes have become useful for studying the control of RNase mRNA levels. Gene-specific probes have begun to provide a powerful means to dissect the regulatory properties of genes that encode nearly identical polypeptides. In some cases, the information provided by comparing the sequences of similar genes has indicated strategies to generate anti-sera specific for unique domains on closely related proteins. To date, each of these approaches has furthered our knowledge of how RNases are controlled in higher plants. In particular, our understanding of the control of RNases during senescence, phosphate starvation, and development have advanced considerably. These and other areas that provide much potential for future study are discussed below.

Senescence

Senescence is among the most prominent processes that influence RNase content. Historically, the literature indicates that senescence is correlated to an increase in RNase activity in plants. However, many of the studies leading to this conclusion were performed in excised leaves, and wounding alone is known to lead to increased RNase activity (40). Based on studies performed on the senescing corolla of morning glory, senescence appears to precede the increase in RNase activity, rather than the other way around (15, 68). In studies performed on attached leaves, the effects are less clear cut. Depending on the plant, RNase activity has been reported to increase, decrease, or increase and then decrease as reviewed previously (40). Although some of this variation undoubtedly results from differences in plant material and experimental conditions, the remainder may reflect the varying contributions to individual enzymes on total RNase activity. For example, in attached oat leaves, the activity of an RNase I–type enzyme decreases while the activity of an endonuclease increases during senescence. When the oat leaves were excised, it was the RNase I enzyme that increased (40, 127).

In wheat, the activity of a single-strand preferring nuclease (probably a nuclease I–type enzyme) increased dramatically during senescence of flag leaves (19, 21). Wheat flag leaves were also used for detailed analysis of three biochemically distinct RNases (WL$_A$, WL$_B$, and WL$_C$). By varying the assay conditions, the activity of the individual enzymes could be monitored in solution assays and on activity gels (21). In these experiments, all three RNases increased during senescence, with the effects on RNase WL$_B$ > WL$_C$ > WL$_A$. The increases were most pronounced late in senescence, i.e. 30–37 days post-anthesis (21). In contrast, the nuclease began to increase at an earlier stage [20 days post-anthesis (19)] and the magnitude of its induction exceeded that noted for the three RNases. The nuclease and RNases WL$_B$ and WL$_C$ appear to be composed of multiple isozymes judging from their mobility on activity gels (21). It will be interesting to determine whether each set of isozymes consists of modified versions of the same gene product or of different gene products.

Studies of the expression of the gene for RNS2, an S-like RNase of *A. thaliana,* indicate that RNS2 is also a senescence-induced RNase. Northern blot experiments with a gene-specific probe demonstrated that the *RNS2* transcript increases in senescing petals of intact *A. thaliana* plants (110). Other experiments have shown that *RNS2, RNS3,* and, to a lesser degree, *RNS1* are also induced during leaf senescence in *A. thaliana* (110; PA Bariola, CJ Howard, CB Taylor, MT Verburg, VD Jaglan, PJ Green, submitted).

Phosphate Starvation

Several recent reports indicate that phosphate (P_i) starvation can induce a number of RNases and RNase genes. The first RNase shown to be controlled by P_i starvation was the extracellular S-like RNase LE of tomato (90). Because it is extracellular rather than intracellular P_i concentration that induces RNase LE (44), the signal may be transduced at the plasma membrane (45). The four other S-like RNases identified from tomato suspensions were also found to be induced during P_i limitation. Three of these were the vacuolar enzymes, RNase LV-1, RNase LV-2 and RNase LV-3, and one is the putative ER ribonuclease, RNase LX (77). The kinetics of their induction indicate that these four RNases are regulated coordinately with RNase LE. The response seems to be specific for Pi because starvation for nitrate or the application of other stresses did not induce RNase LE activity (44).

Based on the work done in tomato, the S-like RNase genes that have been cloned have also been investigated for their response to P_i limitation. For *RNS2*, P_i starvation of *A. thaliana* seedlings resulted in a clearly elevated level of mRNA compared to seedlings incubated in P_i-containing medium (110). P_i limitation induces the *RNS1* mRNA even more dramatically than the *RNS2* transcript (PA Bariola, CJ Howard, CB Taylor, MT Verburg, VD Jaglan, PJ Green, submitted). Interestingly, P_i starvation had very little effect on expression of *RNS3*. Because all the *RNS* genes were induced by senescence, the lack of induction of *RNS3* by P_i starvation may be indicative of a fundamental difference in the mechanism of induction mediated by the two stimuli. This hypothesis is supported further by the observation that *RNS1* was only modestly induced by senescence, but had the highest level of induction among the *RNS* genes in response to P_i starvation.

Plant Disease

Many reports document increases in RNase activity in diseased plants (121). Such increases may not represent direct responses to attack by a pathogen but rather to a secondary effect of wounding or senescence (40). The most common observation is that total RNase activity in tissue homogenates or intercellular wash fluids increases after infection. When RNase activity gels are used to monitor the response, several bands of RNase activity appear to increase, and changes in pH optima also have been observed (13). The increases in RNase activities are often more pronounced in compatible rather than incompatible reactions. For example, in rust-infected wheat, induction of the late RNase is specific for compatible interactions (28). In some cases, new enzymes appear, but it is often unknown whether they are new plant enzymes or are produced by the pathogen (40).

Another characteristic of the work on diseased plants is that only a limited effort has been made to relate the enzymes in question to the majority of biochemical knowledge of plant RNases that exists outside the plant pathology field. One exception is a study on the effect of the host-specific toxin of the fungus *Helminthosporium maydis* race T in maize roots (124). All the maize RNase activities that were well characterized previously increased following toxin treatment, but only after growth was reduced drastically. This timing indicates that the changes were a secondary effect of toxin treatment (124). Similarly, a modest and delayed induction of the RNS2 transcript has been observed in *A. thaliana* following infection with *Pseudomonas* spp., again characteristic of a secondary effect (PA Bariola, J Tsuji, X Dong, F Ausubel, S Somerville, PJ Green, unpublished results).

Light and Other Developmental Cues

Light and darkness both induce specific RNase activities. For example, in primary leaves of wheat, light deprivation appears to induce a model senescing process (dark-induced senescence) (126). RNases WL_B and WL_C, which are induced during natural senescence in wheat (see above), are also induced in plants incubated in darkness (21). After a 2-day lag (RNase WL_B) or decline (RNase WL_C), both activities increased linearly for the remainder of the 7-day dark treatment. The senescence-induced nuclease I activity was also induced following dark treatment (19). In contrast to what was observed during natural senescence in flag leaves, RNase WL_A exhibited no detectable increase in activity in darkness (21). Upon reillumination to reverse the dark-induced senescence, nuclease I and RNases WL_B and WL_C declined in activity while RNase WL_A increased and then decreased. It is unknown whether the atypical response of WL_A is the result of inherent differences between flag leaves and primary leaves, or between natural and dark-induced senescence. RNases reported to increase in response to light include a guanine-preferring RNase of oat (127) and a ribosome-associated RNase of lupin (4). The latter appears to be synthesized de novo, and it is under phytochrome control (5).

Other developmental processes that lead to general changes in RNase content include seed development; seed germination; root, fruit, and flower development; and xylogenesis. RNase activity usually increases during seed development except during the later stages, when decreases can occur (40, 125). Increases in RNase activity also occur during seed germination (12, 125) and during cell elongation and maturation in roots (115). Barley aleurones synthesize and secrete a nuclease in response to gibberellic acid, which is thought to participate with other hydrolases in the breakdown of the endosperm (25). In mammalian cells undergoing programmed cell death, a common characteristic is an induction of nucleolytic activity (33). Aspects of programmed cell death may also occur in cultured zinnia cells when they are hormonally induced (with auxin and cytokinin) to differentiate into xylem. This process leads to the induction of a nuclease with N-terminal sequence homology to nuclease I of barley (26, 112). Several RNase activities were also induced during xylogenesis, but unfortunately these have not been studied in much detail. A 17-kDa RNase activity is particularly intriguing because it is induced transiently (112) and, therefore, could provide a clue to this interesting example of cell differentiation leading to cell death.

Two RNases have been characterized recently from developing tomato fruit (87). These enzymes, designated Tf1 and Tf2, are both glycoproteins and have apparent molecular weights of 59 and 29 kDa, respectively. When Tf1 is reduced and denatured, two subunits (30 and 29 kDa) are observed, only one of which has RNase activity upon renaturation. The active subunit of Tf1 is related immunologically to Tf2. Both enzymes resemble the S-RNases in terms of their preference for homopolymers, pH optima, and apparent size (87). The accumulation of Tf1 and Tf2 in tomato fruits is maximal from 5–10 days after anthesis when total RNase activity (assayed at neutral pH) is also maximal (87). Both enzyme activities decline during fruit maturation and then increase slightly from mature green through the breaker stage. Because these RNase activities are most abundant when cell division and cell differentiation are occurring, it is unlikely that their primary roles are to mediate cell death and breakdown (87).

The accumulation patterns of the S-RNases of *N. alata* and their transcripts have been investigated in great detail via in situ hybridization and EM immunocytochemistry (8, 9, 46, 86). These studies demonstrated that the S-RNase genes are expressed at high levels in the transmitting tract (particularly in the upper portion) of the style and in the epidermal cells of the placenta. The proteins are present in the intercellular matrix but not in the cytoplasm of transmitting tract cells. In the placental epidermis, the S-RNases are again found outside the cells in the innermost regions of the walls adjacent to the plasma membrane. Thus the expression of the S-genes and the accumulation of their products correlates exactly with the path that the pollen tube takes as it grows toward the ovules. Similar conclusions have been reached from RNA blot analysis of the *Petunia hybrida* S-RNase genes; however, the petunia genes are also expressed at a low level in petals (30).

Some aspects of the expression patterns of the *A. thaliana RNS* genes are similar to those of the S-RNase genes, most notably that *RNS1*, *RNS2* and *RNS3* are all highly expressed in flowers. In dissected flowers, the *RNS2* transcript is found in both pistils (stigma and style) and petals (at a slightly higher level in the latter) (110). However, *RNS2* is also expressed in leaves, stems and roots of *A. thaliana* plants and the *RNS3* transcript is present in roots and stems at a significant level (110; PA Bariola, CJ Howard, CB Taylor, MT Verburg, VD Jaglan, PJ Green, submitted). As mentioned above, *RNS1* can be highly induced in seedlings during P_i starvation. Therefore, unlike the S-RNases, the S-like RNases of *A. thaliana* are not flower-specific.

An RNase called X2 was identified recently from *Petunia inflata,* a species exhibiting self-incompatibility (76). X2 is not linked to the S-locus, is not polymorphic, and has a different preference for homopolymers than the S-RNases of *P. inflata.* Although not associated with self-incompatibility, X2 is pistil-specific and an extracellular enzyme as are the S-RNases. Its structure resembles that of the S-RNases much more than that of the S-like RNases (see Figure 1) (76). X2 and the S proteins may have diverged from a common ancestral RNase or X2 may have diverged later in evolution from the S-RNases (76).

Genetic Determinants

Several genetic determinants have been reported to control the levels of RNase activities in plants. In maize, the *opaque*-2 mutant has been reported to overproduce RNase I–type activity (122). The *Opaque*-2 gene encodes a transcriptional control protein that regulates several other genes (102); thus, the gene(s) that encode maize RNase I may also be targets of control by the Opaque-2 protein. In studies of the variegated mutant *im* of *A. thaliana,* an increase in the amount of RNase active in the acidic pH range was observed (97). This increase, which was attributed primarily to the white sectors of *im* plants, is probably not a primary effect of the *im* mutation because the lesion is highly pleiotropic (97). Instead, the effect may be the result of a secondary effect of photooxidation that may occur in white sectors. Treatment of wild-type plants with norflurazon (which results in photooxidation of plastids and the formation of white leaves or sectors), gives rise to the same pattern of increased RNase activity as does the *im* mutation (CM Wetzel & S Rodermel, unpublished results).

More recently, an effort has been made to use the activity gel assay to screen mutagenized *A. thaliana* plants (M2s) for those that alter the typical profile of RNases. Although tedious, this screen has led to the isolation of several plants with heritable genetic lesions that affect specific RNase activities (M Abler & PJ Green, unpublished results). Some of these mutants may be regulatory mutants because they increase the activity of one or more RNases, while others that lack a band of activity may affect the gene for the RNase in question. Accordingly, the analysis of these mutants could provide insight into both RNase regulation and RNase function.

To analyze DNase activities, reciprocal chromosome substitution lines between wheat cultivars Wichita and Cheyenne have been used to show that Wichita chromosome 2D enhances the activities of two DNA-degrading enzymes (130). These two DNA-degrading activities have electrophoretic

mobilities that are nearly identical to wheat RNases (129), so it will be interesting to test whether Wichita chromosome 2D also controls these activities. In
earlier studies, the contribution of genetic background to the RNase isozyme
pattern in wheat and maize was also investigated to a limited extent (reviewed
in 40).

FUNCTIONAL ROLES OF PLANT RNASES

Self-Incompatibility

Over the past several years, evidence has been accumulating in strong support
of a functional role for the S-RNases in gametophytic SI. In particular, the
S-genes were found to be highly polymorphic and to always co-segregate with
a specific S-genotype. Moreover, the S-RNases were produced in the appropriate cells and at the appropriate developmental stage to abort pollen tube
growth in incompatible reactions. Despite the abundance of highly correlative
data, until recently, there was no proof that the S-RNases were required for SI.
Working with *P. inflata,* Kao and coworkers (75) transformed an S_2S_3 plant
(which will reject either S_2 or S_3 pollen) with an S_3 antisense gene, resulting in
several transgenic plants that were self-fertile. The fertility characteristics of
these plants correlated well with decreased expression of the corresponding
S-RNase genes at both the RNA and protein levels. Plants with drastically
reduced levels of S_3 protein fail to reject S_3 pollen and plants with drastically
reduced levels of both S_2 and S_3 proteins lose the ability to reject both S_2 and
S_3 pollen (75). Moreover, when an S_3 transgene was expressed in S_1S_2 plants,
these transgenic plants gained the ability to reject S_3 pollen. These experiments represent the first demonstration that expression of the S-RNase genes
is required and sufficient for the pistil to reject self-pollen in the Solanaceae.
This work should make it possible to transform antisense plants with a new
S-gene to change their incompatibility phenotypes. In addition, expression of
mutated S-RNase genes in transgenic plants should allow identification of
regions of the proteins that are critical for function.

One of the most important questions that can be addressed by the aforemen-
tioned mutagenesis studies is whether RNase activity per se is required for the
S-gene products to function in SI. Again correlative evidence indicates that it
is. During incompatible interactions, pollen rRNA is degraded (84), although it
remains intact during compatible interactions. This degradation is thought to
be the cause rather than the effect of aborted pollen tube growth (31). Other
experiments have shown that S-RNases are able to enter (46) and inhibit the
growth (62) of pollen tubes germinated in vitro, although allele specificity has
not been demonstrated in these cases. Genetic evidence indicates that the
S-locus functions in the pollen and the style, but neither S-RNases nor their

transcripts have been found in pollen (31, 104). Therefore, most mechanistic models of gametophytic SI include a specificity role for the as yet unidentified pollen component.

In one favored model, the pollen component mediates the selective uptake of S-RNases of the same genotype, resulting in the degradation of pollen RNA in incompatible interactions. Alternatively, the pollen component may act as an RNase inhibitor that inactivates S-RNases with different genotypes but does not inactivate S-RNases derived from the same genotype. In this model, S-RNases enter pollen tubes of any genotype, similar to the in vitro situation, but only S-RNases that are derived from the same genotype as the pollen will cause RNA degradation and arrest of pollen tube growth. Future study of the S-RNases may help elucidate SI mechanisms and provide insights about the T_2/S RNase superfamily in general.

Phosphate Remobilization

The induction of S-like RNases and their genes during senescence and P_i limitation indicates that P_i remobilization may be one functional role of S-like RNases in higher plants. Because plants often exist under Pi-limiting conditions (42), it seems logical that they would have RNases such as LE, LX, and the *RNS* gene products that could help scavenge the P_i from dying organs for the benefit of growing or reproductive structures. RNases that are secreted could help degrade the RNA from senescing cells that have been damaged or that have lysed. Those secreted from roots could also help degrade any RNA that might be present in the rhizosphere. RNases in the vacuole could help facilitate the degradation of the RNA reported to exist in this compartment (1) or cytoplasmic RNA if the vacuoles lyse (82). Uptake of RNA into the vacuole may occur via autophagy (83), a process known to increase during senescence (82). Based on work done on the S-RNases, it should be possible to inhibit the expression of S-like RNase genes with antisense constructs and then to investigate how the transgenic plants behave during senescence and phosphate starvation.

Plant Defense

RNases may also help defend the plant against pathogen attack. In particular, RNases produced in the pistil may function in this manner (e.g. X2 in *P. inflata* and RNS2 in *A. thaliana*) (76). The pistil is a rich source of nutrients that could provide a favorable environment for plant pests, although it is rarely colonized. This could be explained by the combined action of RNases, proteinase inhibitors (10), and other plant defense proteins in the pistil, which could render the organ less amenable for invasion by various plant pathogens. An RNase functioning in this manner may have been recruited during the evolution of SI to protect Solanaceous plants from invasion by pollen tubes with the same genotype (76).

mRNA Decay

One process in which RNases are considered to play a prominent role is mRNA degradation. Unfortunately, our knowledge of which RNases participate in mRNA decay in plants and other higher eukaryotes is extremely limited (47, 107). The half-lives of different transcripts vary over a wide range in plants, so the decay process must be highly regulated (43, 47, 109). Two types of sequences recently were shown to target reporter transcripts for rapid degradation in tobacco (88, 94). The mechanisms by which these mRNA instability sequences function are unknown. The degradation of most or all mRNAs may be carried out by general RNases whose specificity is dictated by other effector molecules. These effectors could include RNA binding proteins, RNA localization factors, or RNase inhibitors. Recognition of instability sequences by effector molecules could make the corresponding transcripts more effective substrates for general RNases. In other aspects of RNA metabolism, such as polyadenylation, there is already precedent for specificity factors that can provide nonspecific enzymes with the ability to discriminate between transcripts that contain or lack specific sequences (119). Therefore, RNases that are capable of digesting bulk RNA in vitro may still play a primary role in specific mRNA decay pathways in vivo if their access to the right substrates is controlled tightly.

Alternatively, most cytoplasmic transcripts could be protected from rapid decay by the 5' cap and 3' poly(A) tail. Removal of the cap by a nucleotide pyrophosphatase (14), the poly(A) tail by a poly(A) nuclease (101), or generation of an internal cleavage by a site-specific endoribonuclease (24) could then serve as the rate-limiting step for mRNA decay, with nonspecific RNases acting only after this initial event. One RNase implicated in mRNA decay in yeast is a poly(A) nuclease (PAN), whose activity is dependent on the poly(A)-binding protein (PAB) (101). If a similar activity exists in plants, it may be highly regulated because of the multiplicity and differential control of plant PABs (17, 51).

To differentiate plant RNases that participate in mRNA decay from those with other roles, further investigation of the RNases in the plant cytoplasm is essential. Detailed analysis of the RNases associated with polyribosomes may be informative because there is considerable evidence that translation and mRNA decay can be coupled tightly in plants (109) and other eukaryotic systems (100). Another attractive avenue is to use RNase-deficient mutants to address the contribution of individual RNases to plant mRNA degradation pathways, an approach that has been useful in bacteria (39).

CONCLUSIONS AND FUTURE PROSPECTS

During the past several years our knowledge of plant RNases, particularly at the molecular level, has grown considerably. We have learned that the S-glycoproteins that are required for gametophytic SI are active RNases and are part of an RNase superfamily that is probably represented in all plants. Data indicate that members of this superfamily and possibly other types of RNases participate in diverse physiological processes in higher plants, ranging from senescence and xylogenesis to seed germination and fruit development. These functions are in addition to their general housekeeping roles in RNA processing and turnover in the nucleus, cytoplasm, and organelles. It is difficult to estimate how many different RNA-degrading enzymes are encoded by the average plant genome; however, it is unlikely that plants have a simpler RNase complement than *Escherichia coli*, in which at least 20 RNases have been identified (39). Continued efforts to separate the different gene products that contribute to previously characterized RNase activities will be essential.

Our increased knowledge of individual RNases and their genes prompts us to address new questions and to revisit old ones. One new question is whether plants synthesize one or more proteinaceous inhibitors specific for this class of enzyme, analogous to the well-studied RNase inhibitor that binds to enzymes in the RNase A superfamily of animal cells (74). S-RNases appear to be insensitive to the mammalian RNase inhibitor (105). If plants do produce inhibitors for members of the T_2/S superfamily, they could be involved intimately in SI mechanisms depending on their specificities. Another new pursuit is to investigate the role of RNases in processes such as xylogenesis, where cell death is part of the developmental program. With respect to old questions, studies of the roles of RNases in plant defense, senescence, and organ development will benefit tremendously from future efforts to combine the molecular, biochemical, and genetic approaches that are now available. In addition, responses to other stimuli, such as water stress and wounding, that were studied to the limits of available technology several years ago (40), should be reinvestigated. Beyond its contribution to basic knowledge, work on plant RNases may indicate new ways to improve crops by genetic engineering. If RNases with high anti-viral, anti-bacterial, or anti-fungal activity are found, it may be possible to engineer pathogen-resistant plants that secrete such an RNase in large amounts. In closing, it is interesting to note that male-sterile transgenic plants and restorer lines already have been engineered using genes for a bacterial RNase (Barnase) and its specific inhibitor (Barnstar), respectively (80, 81). The success of this approach indicates that novel applied uses also may be found for plant RNase genes.

ACKNOWLEDGMENTS

I am grateful to Pauline Bariola for preparing Figure 1; to many colleagues who kindly provided reprints and preprints; to Pauline Bariola and Christie Howard, and Drs. Michael Abler, Crispin Taylor, Teh-hui Kao, and Tom McKeon for comments on the manuscript; and to Karen Bird for editorial assistance. Grant support from the DOE (FG02-90ER-20021) and NSF (DCB9105968) is also gratefully acknowledged.

Any *Annual Review* chapter, as well as any article cited in an *Annual Review* chapter, may be purchased from the Annual Reviews Preprints and Reprints service.
1-800-347-8007; 415-259-5017; email: arpr@class.org

Literature Cited

1. Abel S, Blume B, Glund K. 1990. Evidence for RNA-oligonucleotides in plant vacuoles isolated from cultured tomato cells. *Plant Physiol.* 94:1163–71
2. Abel S, Glund K. 1986. Localization of RNA-degrading enzyme activity within vacuoles of cultured tomato cells. *Physiol. Plant.* 66:79–86
3. Abel S, Glund K. 1987. Ribonuclease in plant vacuoles: purification and molecular properties of the enzyme from cultured tomato cells. *Planta* 172:71–78
4. Acton GJ. 1974. Phytochrome controlled acid RNase: an "attached" protein of ribosomes. *Phytochemistry* 13:1303–10
5. Acton GJ, Schopfer P. 1974. Phytochrome induced synthesis of ribonuclease de novo in lupin hypocotyl sections. *Biochem J.* 142:449–55
6. Ai Y, Singh A, Coleman CE, Ioerger TR, Kheyr-Pour A, Kao T-h. 1990. Self-incompatibility in *Petunia inflata*: isolation and characterization of cDNAs encoding three S-allele-associated proteins. *Sex. Plant Reprod.* 3:130–38
7. Ai Y, Tsai D-S, Kao T-h. 1992. Cloning and sequencing of cDNAs encoding two S proteins of a self-compatible cultivar of *Petunia hybrida. Plant Mol. Biol.* 19:523–28
8. Anderson MA, Cornish EC, Mau S-L, Williams EG, Hoggart R, et al. 1986. Cloning of cDNA for a stylar glycoprotein associated with expression of self-incompatibility in *Nicotiana alata. Nature* 321:38–44
9. Anderson MA, McFadden GI, Bernatzky R, Atkinson A, Orpin T, et al. 1989. Sequence variability of three alleles of the self-incompatibility gene of *Nicotiana alata. Plant Cell* 1:483–91
10. Atkinson AH, Heath RL, Simpson RJ, Clarke AE, Anderson MA. 1993. Proteinase inhibitors in *Nicotiana alata* stigmas are derived from a precursor protein

which is processed into five homologous inhibitors. *Plant Cell* 5:203–13
11. Deleted in proof
12. Barker GR, Bray CM, Walter TJ. 1974. The development of ribonuclease and acid phosphatase during germination of *Pisum arvense. Biochem. J.* 142:211–19
13. Barna B, Ibenthal WD, Heitefuss R. 1989. Extracellular RNase activity in healthy and rust infected wheat leaves. *Physiol. Mol. Plant Pathol.* 35:151–60
14. Bartkiewicz M, Sierakowska H, Shugar D. 1984. Nucleotide pyrophosphatase from potato tubers. Purification and properties. *Eur. J. Biochem.* 143:419–26
15. Baumgartner B, Kende H, Matile P. 1975. Ribonuclease in senescing morning glory. Purification and demonstration of de novo synthesis. *Plant Physiol.* 55:734–37
16. Baumgartner B, Matile P. 1976. Immuno-chemical localization of acid ribonuclease in morning glory flower tissue. *Biochem. Physiol. Pflanz.* 170:279–859
17. Belostotsky DA, Meagher RB. 1993. Differential organ-specific expression of three poly(A)-binding-protein genes from *Arabidopsis thaliana. Proc. Natl. Acad. Sci. USA* 90:6686–90
18. Blackburn P, Moore S. 1982. Pancreatic ribonuclease. *Enzymes* XV:317–433
19. Blank A, McKeon TA. 1989. Single-strand-preferring nuclease activity in wheat leaves is increased in senescence and is negatively photoregulated. *Proc. Natl. Acad. Sci. USA* 86:3169–73
20. Blank A, McKeon TA. 1991. Three RNases in senescent and nonsenescent wheat leaves. *Plant Physiol.* 97:1402–8
21. Blank A, McKeon TA. 1991. Expression of three RNase activities during natural and dark-induced senescence of wheat leaves. *Plant Physiol.* 97:1409–13
22. Blank A, Sugiyama RH, Dekker CA. 1982.

Activity staining of nucleolytic enzymes after sodium dodecyl sulfate-polyacrylamide gel electrophoresis: use of aqueous isopropanol to remove detergent from gels. *Anal. Biochem.* 120:267–75

23. Boller T, Kende H. 1979. Hydrolytic enzymes in the central vacuole of plant cells. *Plant Physiol.* 63:1123–32

24. Brown BD, Zipkin ID, Harland RM. 1993. Sequence-specific endonucleolytic cleavage and protection of mRNA in *Xenopus* and *Drosophila. Genes Dev.* 7:1620–31

25. Brown PH, Ho T-HD. 1986. Barley aleurone layers secrete a nuclease in response to gibberellic acid. *Plant Physiol.* 82:801–6

26. Brown PH, Ho T-HD. 1987. Biochemical properties and hormonal regulation of barley nuclease. *Eur. J. Biochem.* 168:357–64

27. Byrne DH, Seeley KA, Colbert JT. 1993. Half-lives of oat mRNAs *in vivo* and in a polysome-based *in vitro* system. *Planta* 189:249–56

27a. Cayley PJ, White RF, Antoniw JF, Walesby NJ, Kerr IM. 1982. Distribution of the ppp(A2'p)nA-binding protein and interferon-related enzymes in animals, plants, and lower organisms. *Biochem. Biophys. Res. Commun.* 108:1243–50

28. Chakravorty AK, Shaw M, Scrubb LA. 1974. Ribonuclease activity of wheat leaves and rust infection. *Nature* 247:577–80

29. Chrispeels MJ, Raikhel NV. 1992. Short peptide domains target proteins to plant vacuoles. *Cell* 68:613–16

30. Clark KR, Okuley JJ, Collins PD, Sims TL. 1990. Sequence variability and developmental expression of S-alleles in self-incompatible and pseudo-self-compatible petunia. *Plant Cell* 2:815–26

31. Clarke AE, Newbigin E. 1993. Molecular aspects of self-incompatibility in flowering plants. *Annu. Rev. Genet.* 27:257–79

32. Clayton DA. 1991. Nuclear gadgets in mitochondrial DNA replication and transcription. *Trends Biochem. Sci.* 16:107–11

33. Collins MKL, Rivas AL. 1993. The control of apoptosis in mammalian cells. *Trends Biochem. Sci.* 18:307–9

34. D'Alessio G. 1993. New and cryptic biological messages from RNases. *Trends Cell Biol.* 3:106–9

35. Darr SC, Brown JW, Pace NR. 1992. The varieties of ribonuclease P. *Trends Biochem. Sci.* 17:178–82

36. Darr SC, Pace B, Pace NR. 1990. Characterization of ribonuclease P from the archaebacterium *Sulfolobus solfataricus. J. Biol. Chem.* 266:5689–95

37. De A, Funatsu G. 1992. Crystallization and preliminary X-ray diffraction analysis of a plant ribonuclease from the seeds of the bitter gourd *Momordica charantia. J. Mol. Biol.* 228:1271–73

38. Deng XW, Gruissem W. 1987. Control of plastid gene expression during development: the limited role of transcriptional regulation. *Cell* 49:379–87

39. Deutscher MP. 1993. Ribonuclease mutiplicity, diversity, and complexity. *J. Biol. Chem.* 268:13011–14

40. Farkas GL. 1982. Ribonucleases and ribonucleic acid breakdown. In *Encyclopedia of Plant Physiology,* ed. B Parthier, D Boulter, 14B:224–62. Berlin: Springer-Verlag. 774 pp.

41. Favre D, Ngai PK, Timmis KN. 1993. Relatedness of a periplasmic, broad-specificity RNase from *Aeromonas hydrophila* to RNase I of *Escherichia coli* and to a family of eukaryotic RNases. *J. Bacteriol.* 175:3710–22

42. Fried M, Brosehart H. 1967. *The Soil-Plant System in Relation to Organic Mineral Nutrition.* New York: Academic

43. Gallie DR. 1993. Posttranscriptional regulation of gene expression in plants. *Annu. Rev. Plant Physiol. Plant Mol. Biol.* 44:77–105

44. Glund K, Goldstein AH. 1993. Regulation, synthesis, and excretion of a phosphate starvation inducible RNase by plant cells. In *Control of Plant Gene Expression,* ed. DPS Verma, pp. 311–23. Boca Raton, FL: CRC

45. Glund K, Nürnberger T, Abel S, Jost W, Preisser J, Komor E. 1990. Intracellular Pi compartmentalization during phosphate starvation-triggered induction of an intracellular ribonuclease in tomato cell culture. In *Progress in Plant Cellular and Molecular Biology,* ed. HJJ Nijkamp, WH Van Der Plas, J Van Aartrijk, pp. 338–42. Dordrecht: Kluwer Academic

46. Gray JE, McClure BA, Bönig I, Anderson MA, Clarke AE. 1991. Action of the style product of the self-incompatibility gene of *Nicotiana alata* (S-RNase) on *in vitro*-grown pollen tubes. *Plant Cell* 3:271–83

47. Green PJ. 1993. Control of mRNA stability in higher plants. *Plant Physiol.* 102:1065–70

48. Gruissem W, Schuster G. 1993. Control of mRNA degradation in organelles. In *Control of Messenger RNA Stability,* ed. J Belasco, G Brawerman, pp. 329–65. San Diego: Academic. 517 pp.

49. Haring V, Gray JE, McClure BA, Anderson MA, Clarke AE. 1990. Self-incompatibility: a self-recognition system in plants. *Science* 250:937–41

50. Higgins CF, Causton HC, Dance GSC, Mudd E. 1993. The role of the 3' end in mRNA stability and decay. In *Control of Messenger RNA Stability,* ed. J Belasco, G Brawerman, pp. 329–65. San Diego: Academic. 517 pp.

51. Hilson P, Carroll KL, Masson PH. 1993. Molecular characterization of PAB2, a member of the multigene family coding for poly(A)-binding proteins in *Arabidopsis thaliana*. *Plant Physiol.* 103:525–33

52. Horiuchi H, Yanai K, Takagi M, Yano K, Wakabayashi E, et al. 1988. Primary structure of a base non-specific ribonuclease from *Rhizopus niveus. J. Biochem. (Tokyo)* 103:408–18

53. Hsu-Ching C, Stern DB. 1991. Specific ribonuclease activities in spinach chloroplasts promote mRNA maturation and degradation. *J. Biol. Chem.* 266: 24205–11

54. Hsu-Ching C, Stern DB. 1991. Specific binding of chloroplast proteins in vitro to the 3′ untranslated region of spinach chloroplast *petD* mRNA. *Mol. Cell. Biol.* 11: 4380–88

55. Ide H, Kimura M, Arai M, Funatsu G. 1991. The complete amino acid sequence of ribonuclease from the seeds of bitter gourd (*Momordica charantia*). *FEBS Lett.* 284: 161–64

56. Inada Y, Watanabe H, Ohgi K, Irie M. 1991. Isolation, characterization, and primary structure of a base non-specific and adenylic acid preferential ribonuclease with higher specific activity from *Trichoderma viride. J. Biochem. (Tokyo)* 110:896–904

57. Ioerger TR, Clark AG, Kao T-h. 1990. Polymorphism at the self-incompatibility locus in Solanaceae predates speciation. *Proc. Natl. Acad. Sci. USA* 87:9732–35

58. Ioerger TR, Gohlke JR, Xu B, Kao T-h. 1991. Primary structural features of the self-incompatibility protein in solanaceae. *Sex. Plant Reprod.* 4:81–87

59. Irie M. 1993. Structure-function relationship and distribution of RNase T2 family enzymes. *Ribonucleases: Chemistry, Biology, Biotechnology, 3rd Int. Meet.,* Capri, L22. Abstr.

60. Deleted in proof

61. Irie M, Watanabe H, Ohgi K, Minami Y, Yamada H, Funatsu G. 1993. Base specificity of two plant seed ribonucleases from *Momoridica charantia* and *Luffa cylindrica. Biosci. Biotechnol. Biochem.* 57:497–98

62. Jahnen W, Lush WM, Clarke AE. 1989. Inhibition of *in vitro* pollen tube growth by isolated S-glycoproteins of *Nicotiana alata. Plant Cell* 1:501–10

63. Jost W, Bak H, Glund K, Terpstra P, Beintema JJ. 1991. Amino acid sequence of an extracellular, phosphate-starvation-induced ribonuclease from cultured tomato (*Lycopersicon esculentum*) cells. *Eur. J. Biochem.* 198:1–6

64. Kao T-h. 1993. Self-incompatibility: a self/nonself recognition mechanism of sexual reproduction. In *Plant Signals in Interaction with Others,* ed. J Schultz, I Raskin. Rockville, MD: Am. Soc. Plant Physiol. In press

65. Kaufmann H, Salamini F, Thompson RD. 1991. Sequence variability and gene structure at the self-incompatibility locus of *Solanum tuberosum. Mol. Gen. Genet.* 226: 457–66

66. Kawata Y, Sakiyama F, Tamaoki H. 1988. Amino-acid sequence of ribonuclease T_2 from *Aspergillus oryzae. Eur. J. Biochem.* 176:683–97

67. Kearsey S, Kipling D. 1991. Recombination and RNA processing: a common strand. *Trends Cell Biol.* 1:110–12

68. Kende H, Baumgartner B. 1974. Regulation of aging in flowers of *Ipomoea tricolor* by ethylene. *Planta* 116:279–89

69. Kenefick DG, Blake TK. 1986. Low ribonuclease I activity prior to cold acclimation in freeze selected winter barley. *Crop Sci.* 26:1099–1103

70. Kheyr-Pour A, Bintrim SB, Ioerger TR, Remy R, Hammond SA, Kao T-h. 1990. Sequence diversity of pistil S-proteins associated with gametophytic self-incompatibility in *Nicotiana alata. Sex. Plant Reprod.* 3:88–97

71. Kiss T, Filipowicz W. 1992. Evidence against a mitochondrial location of the 7-2/MRP RNA in mammalian cells. *Cell* 70: 11–16

72. Kiss T, Marshallsay C, Filipowicz W. 1992. 7-2 MRP RNAs in plant and mammalian cells: association with higher order structures in the nucleolus. *EMBO J.* 11:3737–46

73. Kurihara H, Mitsui Y, Ohgi K, Irie M, Mizuno H, Nakamura KT. 1992. Crystal and molecular structure of RNase Rh, a new class of microbial ribonuclease from *Rhizopus niveus. FEBS Lett.* 306: 189–92

74. Lee FS, Vallee BL. 1993. Structure and action of mammalian ribonuclease (angiogenin) inhibitor. *Prog. Nucleic Acid Res. Mol. Biol.* 44:1–30

75. Lee H-S, Huang S, Kao T-h. 1993. S-proteins control rejection of incompatible pollen in *Petunia inflata. Nature* 367:560–63

76. Lee H-S, Singh A, Kao T-H. 1992. RNase X2, a pistil-specific ribonuclease from *Petunia inflata,* shares sequence similarity with solanaceous S proteins. *Plant Mol. Biol.* 20:1131–41

77. Löffler A, Abel S, Jost W, Beintema JJ, Glund K. 1992. Phosphate-regulated induction of intracellular ribonucleases in cultured tomato (*Lycopersicon esculentum*) cells. *Plant Physiol.* 98:1472–78

78. Löffler A, Glund K, Irie M. 1993. Amino acid sequence of an intracellular, phosphate-starvation-induced ribonu-

clease from cultured tomato (*Lycopersicon esculentum*) cells. *Eur. J. Biochem.* 214: 627–33

79. Lyndon RF. 1966. Intracellular distribution of ribonuclease activity in pea roots. *Biochim. Biophys. Acta* 113:110–19

80. Mariani C, De Beuckeleer M, Truettner J, Leemans J, Goldberg RB. 1990. Induction of male sterility in plants by a chimaeric ribonuclease gene. *Nature* 347:737–41

81. Mariani C, Gossele V, De Beuckeleer M, De Block M, Goldberg RB, et al. 1992. A chimaeric ribonuclease-inhibitor gene restores fertility to male sterile plants. *Nature* 357:384–87

82. Matile P. 1975. *The Lytic Compartment of Plant Cells.* Vienna: Springer-Verlag

83. Matile P. 1978. Biochemistry and function of vacuoles. *Annu. Rev. Plant Physiol.* 29: 193–213

84. McClure BA, Gray JE, Anderson MA, Clarke AE. 1990. Self-incompatibility in *Nicotiana alata* involves degradation of pollen rRNA. *Nature* 347:757–60

85. McClure BA, Haring V, Ebert PR, Anderson MA, Simpson RJ, et al. 1989. Style self-incompatibility gene products of *Nicotiana alata* are ribonucleases. *Nature* 342: 955–58

86. McFadden GI, Anderson MA, Bönig I, Gray JE, Clarke AE. 1992. Self-incompatibility: insights through microscopy. *J. Microscopy* 166:137–48

87. McKeon TA, Lyman ML, Prestamo G. 1991. Purification and characterization of two ribonucleases from developing tomato fruit. *Arch. Biochem. Biophys.* 290:303–11

88. Newman TC, Ohme-Takagi M, Taylor CB, Green PJ. 1993. DST sequences, highly conserved among plant *SAUR* genes, target reporter transcripts for rapid decay in tobacco. *Plant Cell* 5:701–14

89. Nguyen TT, Palcic MM, Hadziyev D. 1988. Characterization of cell-wall-bound nuclease and ribonuclease from potato tuber. *Agric. Biol. Chem.* 52:957–65

90. Nürnberger T, Abel S, Jost W, Glund K. 1990. Induction of an extracellular ribonuclease in cultured tomato cells upon phosphate starvation. *Plant Physiol.* 92: 970–76

91. Ohgi K, Horiuchi H, Watanabe H, Iwama M, Takagi M, Irie M. 1992. Evidence that three histidine residues of a base non-specific and adenylic acid preferential ribonuclease from *Rhizopus niveus* are involved in the catalytic function. *J. Biochem. (Tokyo)* 112:132–38

92. Ohgi K, Horiuchi H, Watanabe H, Iwama M, Takagi M, Irie M. 1993. Role of Asp51 and Glu105 in the enzymatic activity of a ribonuclease from *Rhizopus niveus. J. Biochem. (Tokyo)* 113:219–24

93. Deleted in proof

94. Ohme-Takagi M, Taylor CB, Newman TC, Green PJ. 1993. The effect of sequences with high AU content on mRNA stability in tobacco. *Proc. Natl. Acad. Sci. USA* 90: 11811–15

95. Pietrzak M, Cudny H, Maluszynski M. 1980. Purification and properties of two ribonucleases and a nuclease from barley seeds. *Biochim. Biophys. Acta* 614:102–12

96. Plischke W, Hess D. 1980. Nucleases from *Petunia hybrida*: a sugar-unspecific nuclease isolated from leaves. *Biochem. Physiol. Pflanz.* 175:629–36

97. Redei GP. 1967. Biochemical aspects of a genetically determined variegation in *Arabidopsis. Genetics* 56:431–43

98. Rijven AHGC. 1978. Ribosomal wash ribonucleases from fenugreek (*Trigonella foenum graecum* L.) and soybean (*Glycine max* (L) Merr) cotyledons and their interactions with poly(A) and some modified nucleosides. *Plant Sci. Lett.* 11:293–303

99. Ross J, Kobs G. 1986. H4 histone messenger RNA decay in cell-free extracts intiates at or near the 3′ to 5′ terminus and proceeds 3′ to 5′. *J. Mol. Biol.* 188:579–93

100. Sachs AB. 1993. Messenger RNA degradation in eukaryotes. *Cell* 74:413–21

101. Sachs AB, Deardorff JA. 1992. Translation initiation requires the PAB-dependent poly(A) ribonuclease in yeast. *Cell* 70: 961–73

102. Schmidt RJ. 1993. Opaque-2 and zein gene expression. In *Control of Plant Gene Expression,* ed. DPS Verma, pp. 337–55. Boca Raton, FL: CRC. 579 pp.

103. Schmitt ME, Bennett JL, Dairaghi DJ, Clayton DA. 1993. Secondary structure of RNase MRP RNA as predicted by phylogenetic comparison. *FASEB J.* 7:208–13

104. Sims TL. 1993. Genetic regulation of self-incompatibility. *Crit. Rev. Plant Sci.* 12: 129–67

105. Singh A, Ai Y, Kao T-h. 1991. Characterization of ribonuclease activity of three S-allele-associated proteins of *Petunia inflata. Plant Physiol.* 96:61–68

106. Srivastava BIS, Matsumoto H, Chadha KC. 1971. Studies on chromatin-associated nuclease from barley leaves. *Plant Cell Physiol.* 12:609–18

107. Stevens A. 1993. Eukaryotic nucleases and mRNA turnover. In *Control of Messenger RNA Stability,* ed. JG Belasco, G Brawerman, pp. 449–71. San Diego: Academic. 517 pp.

108. Stirpe F, Barbieri L, Battelli MG, Soria M, Lappi DA. 1992. Ribosome-inactivating proteins from plants: present status and future prospects. *BioTechnology* 10:405–12

109. Sullivan ML, Green PJ. 1993. Mechanisms controlling the stability and translation of nuclear encoded transcripts in higher plants. *Plant Mol. Biol.* 23:1091–1104

110. Taylor CB, Bariola PA, delCardayré SB, Raines RT, Green PJ. 1993. RNS2: a senescence-associated RNase of *Arabidopsis* that diverged from the S-RNases before speciation. *Proc. Natl. Acad. Sci. USA* 90: 5118–22

111. Taylor CB, Green PJ. 1991. Genes with homology to fungal and S-gene RNases are expressed in *Arabidopsis thaliana. Plant Physiol.* 96:980–84

112. Thelen MP, Northcote DH. 1989. Identification and purification of a nuclease from *Zinnia elegans* L.: a potential molecular marker for xylogenesis. *Planta* 179:181–95

113. Thompson RD, Kirch H-H. 1992. The S locus of flowering plants: when self-rejection is self-interest. *Trends Genet.* 8:381–87

114. Topper JN, Bennett JL, Clayton DA. 1992. A role for RNAase MRP in mitochondrial RNA processing. *Cell* 70:16–20

115. Trebal JP, Beopoulos N, Esnault R. 1979. Ribonuclease activities in bean roots. *Phytochemistry* 18:1635–37

116. Tsai D-S, Lee H-S, Post LC, Kreiling KM, Kao T-h. 1992. Sequence of an S-protein of *Lycopersicon peruvianum* and comparison with other solanaceous S-proteins. *Sex. Plant Reprod.* 5:256–63

117. von Ahsen U, Schroeder R. 1993. RNA as a catalyst: natural and designed ribozymes. *BioEssays* 15:299–306

118. Wang MJ, Davis NW, Gegenheimer P. 1988. Novel mechanisms for maturation of chloroplast transfer RNA precursors. *EMBO J.* 7:1567–74

118a. Watanabe H, Naitoh A, Suyama Y, Inokuchi N, Shimada H, et al. 1990. Primary structure of a base non-specific and adenylic acid preferential ribonuclease from *Aspergillus saitoi. J. Biochem. (Tokyo)* 108: 303–10

119. Wickens M. 1990. How the messenger got its tail: addition of poly(A) in the nucleus. *Trends Biochem. Sci.* 15:277–81

120. Wilson CM. 1968. Plant nucleases. II. Properties of corn ribonucleases I and II and corn nuclease I. *Plant Physiol.* 43: 1339–46

121. Wilson CM. 1975. Plant nucleases. *Annu. Rev. Plant Physiol.* 26:187–208

122. Wilson CM. 1980. Plant nucleases VI. Genetic and developmental variability in ribonuclease activity in inbred and hybrid corn endosperms. *Plant Physiol.* 66:119–25

123. Wilson CM. 1982. Plant nucleases: biochemistry and development of multiple molecular forms. *Isozymes: Curr. Top. Biol. Med. Res.* 6:33–54

124. Wilson CM, Apel GA. 1975. Effect of *Helminthosporium maydis,* race T, pathotoxin on growth and ribonuclease levels of corn roots. *Crop Sci.* 15:385–89

125. Winchcombe CF, Bewley JD. 1992. Changes in ribonucleic acid and polyribosomes as related to ribonuclease activities during development and germination of the castor bean seed. *Phytochemistry* 31:2591–97

126. Wittenbach VA. 1977. Induced senescence of intact wheat seedlings and its reversibility. *Plant Physiol.* 59:1039–42

127. Wyen NV, Erdei S, Farkas GL. 1971. Isolation from *Avena* leaf tissues of a nuclease with the same type of specificity towards RNA and DNA. Accumulation of the enzyme during leaf senescence. *Biochim. Biophys. Acta* 232:472–83

128. Xu B, Mu J, Nevins DL, Grun P, Kao T-h. 1990. Cloning and sequencing of cDNAs encoding two self-incompatibility associated proteins in *Solanum chacoense. Mol. Gen. Genet.* 224:341–46

129. Yen Y, Baenziger PS. 1993. Identification, characterization, and comparison of RNA-degrading enzymes of wheat and barley. *Biochem. Genet.* 31:133–45

130. Yen Y, Baenziger PS. 1993. Wheat chromosome 2D carries genes controlling the activity of two DNA-degrading enzymes. *Theor. Appl. Genet.* In press

131. Yen Y, Green PJ. 1991. Identification and properties of the major ribonucleases of *Arabidopsis thaliana. Plant Physiol.* 97: 1487–93

132. Zhou A, Hassel BA, Silverman RH. 1993. Expression cloning of 2–5A-dependent RNAase: a uniquely regulated mediator of interferon action. *Cell* 72:753–65

Annu. Rev. Plant Physiol. Plant Mol. Biol. 1994. 45:447–67

MALATE COMPARTMENTATION—RESPONSES TO A COMPLEX METABOLISM

Enrico Martinoia[1] *and Doris Rentsch*

Institute of Plant Sciences, Swiss Federal Institute of Technology, CH-8092 Zürich, Switzerland

KEY WORDS: malate, compartmentation, transport

CONTENTS

INTRODUCTION 448
MALATE CONCENTRATIONS IN PLANT TISSUES... 450
    Leaves... 450
    Roots.. 451
    Xylem and Phloem.. 452
    Apoplast and Exudates... 453
INTRACELLULAR COMPARTMENTATION 453
MALATE TRANSPORT SYSTEMS 455
    Chloroplasts... 455
    Mitochondria.. 456
    Vacuoles... 458
    Peroxisomes.. 459
    Symbiosomes... 459
    Plasma Membrane Anion Channel.. 460
REGULATION OF MALATE DISTRIBUTION... 460
    Chloroplasts... 460
    Mitochondria.. 461
    Vacuoles... 461
FINAL COMMENTS 462

---

[1] E. Martinoia's address through December 1995: Université de Poitiers, Station Biologique de Beau Site, 86000 Poitiers, France.

0066-4294/94/0601-0447$05.00

INTRODUCTION

Malic acid is a dicarboxylic acid with pK values of approximately 3.3 and 5.1, depending on the salt concentration in the medium. This means that at physiological pH, malic acid is present in the dianionic form (malate); however, $Hmal^-$ or H_2mal also may be present in plants, if localized in the acidic vacuoles. Enzymes using malate as substrate are located in the cytosol, chloroplasts, mitochondria, peroxisomes, and glyoxysomes (see Table 1) and are involved in various metabolic pathways (62). The complex enzymology of malate metabolism reflects its versatility:

1. Malate serves as an intermediate in the TCA cycle, and malate accumulated during the day may be used as respiratory energy during the night (25, 38).
2. Assimilation of HCO_3^- by PEP carboxylase (PEPC) yields oxaloacetic acid, which is readily converted to malate, the store for both CO_2 and reduction equivalents in CAM (66, 105, 114) and NADP–malic enzyme (ME) type C_4 plants (27, 28). Temporary (CAM-plants) or spatial (C_4-plants) separation of CO_2 fixation from the Calvin cycle permits a high water use efficiency.
3. Transport of reduction equivalents to the cytosol or peroxisomes is usually mediated by an oxaloacetate-malate shuttle operating at the inner membrane of chloroplasts and mitochondria, respectively (25, 48, 89).
4. Malate can be used as an osmoticum (42, 85, 86, 90, 118). The best known example are guard cells, in which malate synthesis and degradation play an important role in stomatal movement.
5. Malic acid synthesis and degradation are components of the pH state mechanism (1, 61, 97). Malic acid synthesis supplies the cytosol with H^+; its degradation produces OH^-.
6. Unequal cation or anion uptake by roots can be balanced by malate synthesis in response to cation excess, or by its degradation in the case of anion excess (59, 106).
7. Some plant roots excrete acids to increase phosphate availability in the soil (52, 80). Malate is frequently an important component of the exudate.
8. A malate-oxaloacetate shuttle is hypothesized to function in the generation of apoplastic NADH, which is used in a complex reaction to generate apoplastic H_2O_2. Hydroxycinnamyl alcohols then may be polymerized by a peroxidase to give lignin (40).
9. Recently, malate has been shown to modulate the voltage-dependence of the stomatal anion channel, and malate may be part of the CO_2 sensor mechanism (46).

Compartmentation of a central metabolite, such as malate, that is shared by several organelles, cells, or organs involves complex metabolic interactions

Table 1 Enzymes directly involved in malate metabolism

Enzyme	Reaction	Cofactors	Regulation/Properties	Compartment	References
malate dehydrogenase (MDH)	OAA → mal	NADH		cytosol mitochondria peroxisomes	
		NADPH	light activated (thioredoxin m) NADP/NADPH modulated	glyoxysomes chloroplasts	29, 53, 92
malic enzyme (ME)	mal → pyr + CO_2	NAD/Mn^{2+}	activated by acetyl CoA, CoA, F1, 6P_2	mitochondria	25, 27, 28
		NADP/Mg^{2+} or $Mn2+$	inhibited by NADPH and pyruvate	chloroplasts (C_4, C_3)	27, 28, 30
malate synthase (MS)	gly + AcCoA → mal	NADP/Mg^{2+} or Mn^{2+}		cytosol (C_3, CAM) glyoxysomes	18, 114
fumarase (FUM)	fum + H_2O → mal		stereospecific *trans* addition of H and OH	mitochondria	17, 39
PEP carboxylase (PEPC)	PEP + HCO_3^- → OAA	Mg^{2+} or Mn^{2+}	ativated by phosphorylation (CAM, C_4; not C_3), modulated by different factors (pH, mal, G6P)	cytosol	4, 19, 55, 105

PEPC is listed in the table since most OAA synthesized by this enzyme is thought to be readily converted to malate in NADP-ME C^4 and CAM plants. mal: malate; OAA: oxaloacetate; pyr: pyruvate; gly: glyoxylate; AcCoA: AcetylCoA; F1, 6P^2: fructose-1,6-bisphosphate; fum: fumarate; PEP: phosphoenol pyruvate; G6P: glucose 6-phosphate.

and cannot be considered independently from these processes. Nevertheless, this review focuses primarily on malate compartmentation per se, and on the transport mechanisms involved in malate fluxes. We trust that the references will guide the reader to additional information on the various aspects of malate metabolism.

MALATE CONCENTRATIONS IN PLANT TISSUES

Leaves

Plants exhibiting CAM metabolism use malic acid as a store of readily available CO_2 during the night and, as a result, accumulate the highest levels of malic acid known to occur in plant tissues [up to 350 mM (66)]. Decarboxylation during the day may decrease the malic acid level by 200 mM. In contrast to CAM plants, C_3 and C_4 plants accumulate malic acid as its salt (i.e. malate) even when its concentration is very high [100 mM (116)]. Malate levels vary between species, cultivar, or even between individual leaves of a single plant (116; E Martinoia, unpublished results). When grown under identical conditions, pea accumulates far more malate (74 ± 10 mM) than spinach (3.6 ± 0.6 mM) (99). The malate content increases in leaves during light-dependent nitrate reduction. This phenomenon is especially apparent in detached leaves, in which the NO_3^- supply is stopped and malate export to the roots interrupted (57). Malic acid synthesis has a double function during nitrate reduction: production of H^+, which can buffer the OH^- generated during nitrate reduction, and replacement of the NO_3^- anion for charge balancing. Inhibition of nitrate reductase activity by mild wilting or transfer to CO_2-free air is correlated with inhibition of malate synthesis, indicating that nitrate reduction and malic acid synthesis are closely linked processes (57).

In C_3 plants, malic acid levels decrease during the night, probably by decarboxylation to pyruvate, which can be used as an energy source in the TCA cycle. Synthesis of malate during the day and its consumption during the night also result in a diurnal rhythm in malate content in C_3 plants; however, in contrast to CAM plants, maximal malate levels are found at the end of the day (38; G Kaiser & E Martinoia, unpublished data; WM Kaiser, personal communication). Substituting NH_4^+ for NO_3^- results in reduced malate accumulation (116), and diurnal malate fluctuation is no longer observed (WM Kaiser, personal communication).

Malate levels usually decrease when plants are exposed to NaCl stress (77, 99), probably because of decreased NO_3^- uptake. However, some plants may change their metabolic status, switching from C_3 metabolism to CAM under salt stress conditions, and they accumulate large amounts of malic acid as a result of adaptation to high salt concentrations (105, 114). This switch has been investigated in the most detail in *Mesembryanthemum crystallinum* (19, 114). Water stress enhances the expression of CAM metabolism in many CAM plants (66, 105, 114), while both increases and decreases in malate levels have been observed in C_3 plants (33).

Investigations of the intercellular distribution of malate in leaves of *Zea mays* have not produced conclusive results. In two contributions, most of the malate was found in the mesophyll (64, 100), whereas a recent study found most of the malate in the bundle sheath (111). In the latter study, the overall malate concentration in the leaves was about three times higher than in the material used in the first study by the same group, and it may be that the intercellular distribution is a function of the total malate content. Because most of the malate is confined to the vacuole (see below), no predictions can be made about whether malate concentration gradients direct a mass flow of malate from the mesophyll to the bundle sheath.

Unequal distribution of malate within a leaf has been observed frequently. Concentrations generally are lower in the epidermis than in the mesophyll (115; MJ Schramm, KJ Dietz, & E Martinoia, unpublished data). No significant diurnal changes in malic acid concentrations were observed in the epidermis of *M. crystallinum* that exhibited CAM metabolism (115).

As mentioned above, guard cells accumulate large quantities of malate during the opening of stomata (85). In many plants, such as *Tulipa* spp. or *Vicia faba*, malate is the most important counter ion to potassium and may be present at 100–200 mM concentration (85). Isolated guard cells of other plants [e.g. *Allium cepa* (onion)] have been reported to accumulate chloride rather than malate. Unpublished results of K Raschke and his group (personal communication) have shown that onion guard cells contain remarkably high activities of PEPC and NADP-malate dehydrogenase (MDH), which together can support malate synthesis. Because malate can be found in epidermal strips of illuminated onion leaves (94), the energy and carbon for malate synthesis in guard cells must be supplied directly by mesophyll cells.

Roots

When plants absorb various ions at different rates, the concentrations of cation and anion may be unbalanced. Excess cation uptake has been reported for plants exposed to K_2SO_4 or $(NH_4)_2SO_4$ (59, 106). The excess was associated with a corresponding release of H^+ into the medium to maintain electroneutrality. The resulting increase in cellular OH^- concentration has to be buffered by the synthesis of organic acids, equivalent to the excess of cation taken up. An increase of the malate concentration from 3 mM to more than 6 mM was observed in maize root tips in response to K_2SO_4 treatment (13, 14), and high malate concentrations were found in the vicinity of the root tips in soybean plants grown in the absence of a nitrogen source (102). This result may indicate that in these plants, a cation excess was taken up and malate had to be synthesized to produce H^+.

In the presence of NO_3^-, anion uptake exceeds cation uptake and OH^- (or HCO_3^-) is released to the rhizosphere (59, 106). OH^- may arise directly as a result of NO_3^- reduction in the roots. At low external NO_3^- concentrations, nitrate reduction occurs predominantly in the root, but some plants transfer nitrate to the shoot even if its external concentration is low (3). As external nitrate concentration increases (in agricultural soils, from 1 to 20 mM), shoot nitrate reduction becomes increasingly important in all species (3) because the OH^- produced in the roots is not sufficient to counterbalance the excess in anion uptake. It was postulated that under these conditions malate is transported from the shoots to the roots where it is decarboxylated by malic enzyme, yielding pyruvate and HCO_3^- (59). Recent experiments with soybean showed that interruption of the phloem flow by the girdling technique caused an inhibition of NO_3^- uptake. Addition of malate to the transpiration flow increased the malate concentration in the phloem and enhanced NO_3^- uptake capacity (106). These results indicate that NO_3^- uptake by roots depends on the availibility of malate produced in the shoot and its translocation in the phloem.

An exotic example may demonstrate the diversity of roots. It has been reported that roots of the shootless, epiphytic orchid *Chiloschista usneoides* are able to exhibit CAM metabolism and hence accumulate malate in a diurnal rhythm in their chlorophyll-containing aerial roots, which do not possess stomata (16).

Xylem and Phloem

Together with citrate, malate is usually the predominant organic acid both in xylem and phloem sap (119); however, the concentrations reported are usually fairly low. They are affected by the plant's nutritional status and thus reflect the metabolism of the root's uptake zone. Most of the relevant investigations have been performed with *Ricinus* spp., in which phloem and xylem sap can be obtained easily (1, 108). Under conditions of excess cation absorption by the roots (e.g. NH_4^+ nutrition), malate concentrations are similar in the xylem and phloem of *Ricinus* spp. NO_3^- nutrition causes increased malate concentration in the phloem (at least 2-fold compared to NH_4^+ fed plants), indicating that under these conditions an excess of anions is taken up, which must be balanced by OH^- or HCO_3^- extrusion (see above discussion of roots). In the xylem, malate concentrations are higher under NH_4^+ nutrition than under NO_3^- nutrition.

Apoplast and Exudates

Malate has been recovered recently from the apoplast of pea leaves (99). The observation that apoplastic malate may act as a modulator of the guard cell anion channel (46, see below) raised the question of whether the levels of apoplastic malate are subject to modulation. Indeed, low CO_2 concentrations diminished and high CO_2 concentration doubled the apoplastic malate concentration in *V. faba* (R Hedrich, personal communication).

In several plants, such as rape or chickpea, excretion of carboxylic acids is stimulated under phosphate deficiency (52, 80). Citric acid is usually the main carboxylic acid excreted; however, malate often plays an important role, too. Because the amount and composition of carboxylic acids exuded from chickpea roots seems to be correlated with those in the leaves rather than with those in the roots, it has been suggested that excreted carboxylic acids are synthesized in and translocated from the shoot (80).

INTRACELLULAR COMPARTMENTATION

Considering the high malate content often observed in plants, the majority of malate has been postulated to be localized within the vacuole. Indeed, Buser & Matile (11) were the first to show that isolated *Bryophyllum* spp. vacuoles contained most of the cellular malate. Further studies using isolated vacuoles confirmed this result (37, 93). But plants containing low malate levels do not accumulate malate within their vacuoles, and vacuolar concentrations may be equal or even lower than cytosolic concentrations (95, 99). When tracer experiments are performed, by feeding protoplasts with $^{14}CO_2$, and vacuoles are isolated quickly, it appears that newly synthesized malate (or total organic acids) (10, 56) is transferred rapidly from the cytosol to the vacuole. Thus the cytosol is not exposed to large fluctuations in malate levels (38, 56). Because of the large volume of the vacuole and the high vacuolar malate content, estimations of cytosolic malate concentrations tend to be inaccurate. Also, vacuole isolation often is accompanied by a loss of low molecular weight metabolites from these organelles.

Heldt and coworkers developed an alternative approach to estimating intracellular malate concentrations (37, 38). They used a non-aqueous fractionation procedure to determine solute contents in different subcellular fractions. Using this approach, they were able to show that the bulk of malate was localized in the vacuole, and they could also estimate the cytosolic and chloroplastic malate concentrations. Cytosolic malate concentrations of spinach leaves were estimated to be less than 1 mM during the night and about 2.5 mM during the day. Chloroplastic concentrations were about 5 mM during the day and less than 1.2 mM in the dark. In recent work by the same group (48), stromal concentrations were estimated to be 3 mM in the day and 1 mM in the night, whereas the cytosolic concentration remained unchanged (1mM). A limiting factor of the non-aqueous fractionation method is the inaccurate estimation of the volume of the cellular compartments. Also, the leaves are assumed to consist of uniform cells, which is not the case. As a first attempt to improve the reliability of this method, the exact volumes of the different compartments have been determined using stereological techniques (113); however, this study did not investigate malate distribution.

Using oat protoplasts and a fast differential filtration technique, Hampp and coworkers (41) have observed a higher malate concentration in chloroplasts of illuminated protoplasts than in the dark. The malate concentration in chloroplasts was 1.9 mM in the light but 1 mM in the dark. Mitochondrial malate concentrations rose slightly in the light from 1 mM to 1.2 mM. One must consider, however, that compared to leaf cells, protoplasts are in a medium with reduced oxygen tension. This may affect respiration and hence malate metabolism.

Kaiser and coworkers have reported malate concentrations in rapidly, aqueously isolated spinach (2 mM) and pea (3 mM) chloroplasts (99). These concentrations are in the same range as those reported by Heldt's group. Salt stress decreased the overall malate content and the chloroplastic malate content from 3 mM to 1 mM in pea and from 2 mM to 1 mM in spinach (99). In earlier reports from the same laboratory, malate concentrations in spinach chloroplast were somewhat higher (6–12 mM) (95).

Roberts and coworkers used ^{13}C-NMR to determine the cytosolic malate concentration in intact maize root tips (13, 14). Root tips were perfused with $KH^{13}CO_3$, which was incorporated in malate. The different pH values of the vacuole and the cytosol resulted in a shift of the ^{13}C-signal, allowing the determination of the vacuolar and cytosolic proportion of malate. Cytosolic malate concentration was estimated initially to be 5–6 mM. In a later, more elaborate investigation, malate synthesis was modulated by the addition of K_2SO_4 to the medium and the cytosolic concentration was determined for different malate contents in the tissue. Cytoplasmic malate (3.5 mM) increased temporarily to 7.5 mM in response to increased malate synthesis and then dropped to a steady state value of about 6 mM.

To our knowledge, malate concentrations in peroxisomes have not been reported. The peroxisomal membrane is permeable to small solutes (see discussion of transport systems); therefore, the peroxisomal malate concentration must reflect the cytosolic malate concentration in the dark. In the light, the high metabolic activity in leaf peroxisomes probably decreases the malate concentration compared to the cytosolic malate concentration.

A special situation of intracellular malate distribution is found in root nodules, where carboxylic acids (mainly succinate, malonate, and malate) are crucial as energy supply for the bacteria. These acids have been found in the cytosol and in the bacteroids of nodules (34, 101). Because of the unknown volumes of the compartments involved, it is difficult to estimate the concentration gradients across the peribacteroidal membrane. The results presented support the hypothesis that the concentration of malate and the other dicarboxylates is higher in the cytosol than in the bacteroids, indicating that export of carboxylates from the cytosol may be regulated at the peribacteroidal membrane.

MALATE TRANSPORT SYSTEMS

Chloroplasts

Initially, various dicarboxylates, including 2-oxoglutarate, malate, succinate, glutamate, and aspartate, were found to cross the inner envelope of spinach chloroplasts by carrier-mediated transport (31, 65), each dicarboxylate inhibiting the transport of the other. The possibility remained that more than one translocator (or binding site) with overlapping specificities might be involved in the translocation of the dicarboxylates. Similar results were obtained with pea chloroplasts (84). For mesophyll chloroplasts of C_4 plants, dicarboxylate transport occurred at considerably faster rates (20). The K_m value for malate was similar to that observed in spinach (0.5 mM for maize, 0.4 mM for spinach). In contrast to the phosphate translocator, which catalyzes a strict counter exchange, the dicarboxylate translocator may also operate as an uniport, even though at consistently lower rates. Further detailed studies by Heldt and coworkers (32, 117) showed that malate is transported by two translocators playing an important role in NH_3 assimilation. Uptake of 2-oxoglutarate into and release of glutamate from chloroplasts is required for the glutamine synthetase/glutamate synthetase pathway (see Figure 1). These two dicarboxylates are, however, not exchanged directly. Uptake of 2-oxoglutarate is accompanied by an efflux of malate, while the release of glutamate is accompanied by malate uptake. The overall process results in a net oxoglutarate/glutamate exchange. The 2-oxoglutarate/malate translocator from spinach has been purified recently (78), and its identity has been established by functional reconstitution. The purified translocator has an apparent molecular mass of 45 kDa and accounts for approximately 3% of the envelope protein. It accepts 2-oxoglutarate, malate, succinate, fumarate, and glutarate, but glutamate only poorly. The K_m value for malate (0.21 mM) corresponds to previously published values. 2-Oxoglutarate inhibited the transport of malate competitively ($K_i = 0.95$ mM), while competitive inhibition by glutamate was weak ($K_i = 15$ mM). In a mutant of *Arabidopsis thaliana,* which is not viable under photorespiratory conditions, a polypeptide was identified that might be identical to the glutamate/malate translocator (98).

In the mesophyll chloroplasts of C_4 plants, oxaloacetate synthesized by PEPC is reduced to malate by the light-activated and NADPH/NADP-modulated MDH. Early polarographic measurements demonstrated oxaloacetate reduction in illuminated chloroplasts (2). The affinity of the dicarboxylate transport system is similar for malate and oxaloacetate, and MDH equilibrum favors malate. A separate transport system for oxaloacetate must therefore exist. Indeed, a highly specific transport system was identified that efficiently transports oxaloacetate into the chloroplast (K_m = 45 µM for maize, 9 µM for spinach) (43). This carrier has a much lower affinity for malate (K_i = 7.5 mM for maize, 1.4 mM for spinach). In C_3 plants the chloroplastic oxaloacetate/malate shuttle is also present and supplies the cytosol and the peroxisomes with reduction equivalents. It is still unclear which transporter is responsible for the export of malate. Another question is whether malate is transferred to the bundle-sheath chloroplasts of NADP-type C_4 plants by one of the existing translocators or by a special, bundle-sheath specific carrier.

Mitochondria

Mitochondrial transport systems have been characterized mainly in mammalian systems (63). A tricarboxylate-malate exchange across the inner mitochondrial membrane has been demonstrated repeatedly in plants (22, 76 and refer-

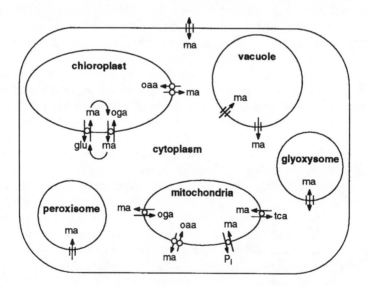

Figure 1   Carriers and channels involved in malate transport. The putative transport direction under physiological conditions is indicated. For mitochondria, the direction may change depending on the metabolic state. (antiport carrier ⌀; uniport carrier ⌀; channels ; glu ‖‖ ‖‖; glutamate; ma: malate; oaa: oxaloacetate; oga: oxoglutarate; tca: tricarboxylate)

ences therein). Recently this translocator was partially purified from pea mitochondria and functionally reconstituted in liposomes. The translocator is inhibited strongly by 1-2-3-benzene-tricarboxylate. The plant carrier exhibits some striking differences from the mammalian tricarboxylate transporter: it does not exchange PEP for citrate and in pea, isocitrate is not exchanged for citrate. The transporters from rat and bovine liver have been purified [30 kDa for rat liver, 37–38 kDa for bovine liver (6, 15)]. The plant citrate^{3-}/malate^{2-} exchanger is thought to supply the cytosol with citrate, which serves as a precursor for 2-oxoglutarate, which in turn is required for NH_3 assimilation. Because the matrix of the mitochondria has a negative membrane potential with respect to the cytosol, citrate^{3-}/malate^{2-} exchange may function as a unidirectional transporter.

The dicarboxylate carrier exchanges malate for phosphate, malonate, succinate, and arsenate, but not 2-oxoglutarate (22, 63). Because both substrates carry two negative charges, the carrier does not transfer net charges at physiological pH. The K_m values are approximately 0.2 mM for malate and 1.5 mM for phosphate. The plant translocator has been identified using monoclonal antibodies and has an apparent molecular mass of about 26 kDa (110). A molecular mass of 28 kDa was reported for rat liver, whereas in bovine liver, two polypeptides copurified with molecular masses of 34 kDa and 36 kDa, respectively (7, 104).

The 2-oxoglutarate carrier exchanges 2-oxoglutarate for malate, malonate, succinate, and oxaloacetate, but not for phosphate or arsenate (22, 63). The K_m for 2-oxoglutarate is 0.05 mM; the values for malate and oxaloacetate are 0.1–0.3 mM. Phthalonate, a classical inhibitor of the mammalian 2-oxoglutarate translocator, inhibits the 2-oxoglutarate transport in mitochondria from some but not all plant sources (see references in 36). The 2-oxoglutarate carrier from maize mitochondria has been partially purified and reconstituted (36). The mammalian counterpart has been cloned and sequenced recently (91). This protein has 314 amino acids and is structurally related to other mitochondrial transport proteins such as the ADP/ATP translocator and the phosphate carrier (91).

The exchange of oxaloacetate and malate is a plant-specific transport, and is actually not an exchange process. It is probably catalyzed by two electrogenic uniporters that may be linked to each other for charge compensation (21, 24, 25, 26, 81, 120). The transport has a high affinity for oxaloacetate ($K_m = 3$–15 μM) and is inhibited strongly by phthalonate but only weakly by malate (K_i of malate is 150-fold greater than K_m of oxaloacetate). These properties enable mitochondria to take up oxaloacetate from the cytosol even though the equilibrum of the MDH reaction greatly favors malate formation ($K_{eq} = 3 \times 10^{-5}$). Inhibition of malate transport by low concentrations of oxaloacetate (120) indicates that oxaloacetate in the mitochondrial matrix controls transport of malate to the cytosol, allowing the maintenance of a redox gradient between the NADH/NAD systems in the matrix and the cytosol. Krömer & Heldt (60) recently presented evidence that the oxaloacetate/malate shuttle operates under simulated physiological conditions; however, investigations of the NAD-ME–type C_4 plant *Panicum miliaceum* showed that possibly not all plant mitochondria contain such an oxaloacetate transporter (35). Anion uniport is mediated by a Mg^{2+}-insensitive inner membrane channel, which is also permeable to different di- and tricarboxylates (5). It is unknown whether this channel is involved in the oxaloacetate/malate shuttle.

Vacuoles

A permease for malate was first described for vacuoles of the CAM plant *Kalanchoe daigremontiana* (12). The apparent K_m was 1 mM; the uptake was not specific for this dianion because other dicarboxylates were inhibitory. Quite unexpectedly, an energized uptake, which had been postulated for thermodynamic reasons (68), could not be detected. Mg ATP–dependent malate uptake was observed in barley mesophyll vacuoles (72). Determination of the vacuolar malate concentration showed that the ATP-stimulated uptake occurred against a concentration gradient and was not the result of enhanced malate exchange. Energization of the uptake is driven by the vacuolar proton pumps (87, 103), as is true for most of the vacuolar transport processes (71). This conclusion is based on the observation that inhibitors of these pumps also inhibit the active component of malate uptake (72, 73, 112). As for *K. daigremontiana,* uptake of malate into barley mesophyll vacuoles is not specific for L-malate. Calculation of the K_m values at different external pH values indicates that the malate dianion is the transported form (88). Similar results were found in *K. daigremontiana, Catharanthus roseus,* and lettuce (8, 69, 70, 71, 112). In barley and *K. daigremontiana,* both a histidine and a lysine group appear to be involved in malate binding (5a, 74, 75). The presence of a histidine group is also thought to occur in *C. roseus* (23). The malate transporter of barley has been partially purified and functionally reconstituted in liposomes (75); however, the molecular nature of this carrier is still unknown. Patch clamp experiments with vacuoles have revealed a channel with properties similar to those of the uptake system described with tracer flux analysis (54; JAC Smith & AJ Pennington, personal communication). The channel is voltage dependent and opens only at positive membrane potentials. This channel probably is identical to the malate carrier.

A further voltage-dependent and malate-permeable channel, which is also permeable for other anions and monovalent cations, has been observed in the vacuolar membrane (SV-channel) (44, 45, 47). This channel opens at high Ca^{2+} concentrations (> 0.3 μM) and at negative voltages (negative inside the vacuole). Typical single channel conductance is 60–80 pS in 50–100 mM salt solutions. A consistently higher conductivity was observed in guard cell vacuoles, indicating that this channel is involved in the release of malate during stomata closure (44).

Peroxisomes

Reduction equivalents needed to drive the photorespiratory carbon cycle are supplied by a malate-oxaloacetate shuttle rather than by direct transfer of NADH (89); however, no specific transporters have been identified in the peroxisomal membrane (79). For peroxisomes of rat liver, researchers proposed that a pore with an exclusion limit for molecules of 700–800 Da is responsible for the transfer of small solutes (109). From this preparation, a pore was reconstituted in liposomes, and researchers suggested that a 22-kDa protein was responsible for pore formation. These results need to be confirmed.

Extensive studies on spinach peroxisome metabolism led to the conclusion that the observed compartmentation of peroxisomal metabolism is not primarily the result of the boundary membrane being a permeability barrier. Rather it is a function of the structural arrangement of the multi-enzyme complex in the peroxisomal matrix allowing metabolite channeling (50, 51, 89).

Symbiosomes

Transport of compounds across the peribacteroid membrane (i.e. the membrane separating the bacteroids from the cytosol) provides fixed nitrogen to the plant in exchange for carbon compounds for the symbionts. The main energy source for the bacteroids are carboxylic acids. Transport of malate and succinate across the peribacteroid membrane of soybean nodules is probably mediated by the same translocator because reciprocal competitive inhibition has been observed (82, 107). Malonate, which often occurs in high concentrations in nodules, is transported slowly. The transport system of the peribacteroid membrane can be distinguished from that of the bacteroids by its higher K_m towards malate (approximately 160 μM, compared to 9 μM) and by its inhibition by vanadate. Similar data have been reported for nodules in French bean (49). ATP slightly stimulated malate accumulation. Pretreatment of symbiosomes with alkaline phosphatase substantially inhibited malate uptake, and this inhibition was completely reversed by the addition of ATP. The rate of malate uptake across the peribacteroid membrane may be controled by phosphorylation. In in vitro phosphorylation assays the major phosphorylated polypeptide appears to be nodulin 26. Because the phosphorylation status of the putative nodulin 26 is correlated with the rate of malate uptake, Ouyang et al (83) proposed that nodulin 26 might be the dicarboxylate carrier of the peribacteroid membrane; however, this needs to be confirmed.

Plasma Membrane Anion Channel

Closing of stomata is accompanied by an efflux of K^+ and Cl^- from the guard cells and by degradation of malate (85, 118). Patch-clamp studies indicate that an anion channel that is slightly permeable for malate (permeability: NO_3^- > I^- > Br^- > Cl^- >> malate) is localized in the plasma membrane of guard cells. This anion channel is subject to modulation by extra- and intracellular signals (46). Calcium and nucleotides increase the open probability, while auxins and malate direct the working gate and thus the working range of the anion channel toward the resting potential of the cell. A malate concentration corresponding to its K_m (0.4 mM) elicits a 38 mV shift. The binding site responsible for shifting the gate is on the extracellular face of the channel. Because malate has been detected in the apoplast, the stomatal anion channel may be part of the plant's CO_2 sensor (46). This is an exciting hypothesis that needs to be investigated further.

REGULATION OF MALATE DISTRIBUTION

Chloroplasts

Two components controlling redox fluxes between the cytosol and chloroplasts have been identified. First, the chloroplastic NADP-dependent MDH is a light-activated enzyme, which also is modulated by the NADPH/NADP ratio. It has been compared to a valve allowing the release of excessive redox equivalents, in the form of malate, from the chloroplast (92). In fact, calculation of the hypothetical equilibrum ratio of NADPH/NADP in the chloroplast of illuminated spinach leaves using the equation

$$\frac{NADPH}{NADP} = \frac{[glutamate] \times [malate] \times K_{GOT} \times K_{MDH}}{[2-oxoglutarate] \times [aspartate] \times [H^+]}$$

assuming a stroma pH of 8, yields a value that is 15 times lower than the measured stromal NADPH/NADP ratio (48). Because the transaminase reaction can be assumed to be in equilibrum, the oxaloacetate concentration must be higher than it would be in equilibrum with the stromal NADPH/NADP. For leaves in darkness, when the stromal MDH is inactive, the calculated ratio is 88 times less than the measured value. Second, the ratio [glutamate] × [malate]/[2-oxoglutarate] × [aspartate] was found to be 12 times higher in the stroma of illuminated leaves than in the cytosol. This gradient disappears in the dark. Metabolite transport across the envelope, therefore, must contribute to the maintenance of the redox gradient. It is unknown how this gradient is maintained or whether malate export is a regulated process under these conditions.

Mitochondria

Different NADH/NAD ratios can also be observed between the cytosol and the mitochondrial matrix. Electrogenic malate transport across the inner mitochondrial membrane is inhibited by relatively low oxaloacetate concentrations (120). Inhibition of malate transport by oxaloacetate may be a regulated step allowing maintenance of a NADH/NAD gradient between the mitochondria and the cytosol. Interestingly, it was also shown that effective decarboxylation of aspartate in *Panicum miliaceum,* a NAD-ME–type C4 plant, was dependent on malate supply (35). Malate probably enters the mitochondria via the dicarboxylate transporter in exchange for P_i and sustains oxoglutarate import by an exchange. Varying the malate concentration in the cytosol of bundle sheath cells has been hypothesized to provide a regulatory link between the metabolic reactions in the mesophyll that synthesize C4 acids, on the one hand, and C4 acid decarboxylation in the bundle sheath, on the other hand. Stimulation of malate decarboxylation by aspartate in bundle sheath cell chloroplasts has been shown in NADP-ME–type C4 plants (9).

Vacuoles

Two observations indicate that regulatory mechanisms may be involved in malate compartmentation between the cytosol and the vacuole: 1. malate concentration changes observed in the cytosol are relatively small compared to those observed in the vacuole (38) and 2. in tissues with low malate content, the malate concentration in vacuoles does not exceed that in the cytosol. The vacuolar pH in C3 plants is usually 5.5–6. Under these conditions, malate is present primarily as mal^{2-}. Malate accumulation is driven by the $\Delta\psi$, which is generated by one of the vacuolar proton pumps (87, 103). Maximal accumulation is given by the Nernst equation. Because of its dianionic nature, malate can be accumulated to higher extent within the vacuole as a monoanion. Furthermore, small changes in the transtonoplast $\Delta\psi$ result in large differences in the potential of malate accumulation. For example, a 5 mV increase in $\Delta\psi$, from 30 to 35 mV, changes the accumulation capacity of the vacuole from 10- to 15-fold. Modulation of the tonoplast proton pumps in response to the dark/light regime has been postulated from indirect experiments (96 and references therein). It is therefore tempting to speculate that small changes in the proton pump activities may contribute to the regulation of malate distribution between the cytosol and the vacuole.

Exclusion of malate from the vacuole implies that the transport of malate to the vacuole is inhibited. Iwasaki et al (54) and JAC Smith & AJ Pennington (personal communication) have shown that a malate channel with properties similar to the malate transport system opens only at membrane potentials over 30 mV (positive inside the vacuole). Because other anions have to be retained within the vacuole, the $\Delta\psi$ may not be reduced. Some cytosolic factors may affect the gating of the channel, allowing malate transport only when excess malate is synthesized. The SV-channel described first by Hedrich et al (45) is permeable for malate and might be involved in malate export. The much higher permeability observed for the SV-channel of guard cell vacuoles, where malate fluxes have to be larger than in the mesophyll, supports this hypothesis.

The situation is somewhat different for CAM plants, which accumulate malic acid rather than its salt, malate. The accumulated malic acid strongly acidifies the vacuolar sap. During acidification, the dianion, monoanion, and free acid are present in the vacuole in changing proportions. Siedow and coworkers (58) have examined fluxes of malate in the dark period after $^{13}CO_2$ fixation in the crassulacean *Kalanchoe tubiflora*. Malate fluxes through the mitochondria were greatest at the onset and at the end of the dark period. The $[4\text{-}^{13}C]$ malate label was randomized even several hours after $^{13}CO_2$ was applied, indicating that vacuolar malate is released to the cytosol even in the dark. These results show that malate turnover takes place continuously during the vacuole loading phase. The enhanced randomization at the end of the dark period may reflect reduced uptake of malate into the vacuole, resulting from energy limitation or already enhanced release of malate to the cytosol. Thermodynamic studies have indicated that malic acid is released to the cytosol in the H_2mal form by diffusion (67). It has also been shown that the tonoplast permeability increases in response to increased environmental temperatures. However, as soon as the pH rises above 3.5, most of the malic acid will be present as the monoanion. Because Kalt et al (58) observed randomization of $[4\text{-}^{13}C]$ malate to $[1\text{-}^{13}C]$ malate in the early stage of dark fixation, when vacuolar pH is between 4 and 5, there must be a transporter or channel that allows the release of the monoanionic form of malate. The question remains whether, during the day, efflux is enhanced, malate uptake inhibited, or both events combined. While malate is decarboxylated rapidly during the light phase, a malate concentration gradient over the tonoplast is generated continously, which may further enhance the malate release from the vacuole.

FINAL COMMENTS

Malate has a pivotal function in most organelles, and the vacuole is a store for an excess of malate. The biochemistry of malate metabolism has been investigated intensively, and the kinetics of many transport proteins have been described. The 2-oxoglutarate translocator of the chloroplast (78) is the first translocator involved in malate transport to have been purified. Kalt et al's experiments (58) indicate that fluxes of malate between the different organelles are fast and that compartmentation cannot be regarded as a semi-static condition. In the future, we will need to understand this aspect of malate metabolism and compartmentation, and we must extend our knowledge of the molecular nature of the carriers involved in malate transport. The regulation of malate transfer can then be investigated, and a picture of the concerted action of the malate translocators in a cell can emerge.

ACKNOWLEDGMENTS

We thank N. Amrhein, H. W. Heldt, and W. M. Kaiser for critical discussion and many colleagues for providing manuscripts and unpublished information.

Literature Cited

1. Allen S, Raven JA. 1987. Intracellular pH regulation in *Ricinus communis* grown with ammonium or nitrate as N source: the role of long distance transport. *J. Exp. Botany* 38:580–96

2. Anderson JW, House CM. 1979. Polarographic study of dicarboxylic-acid-dependent export of reducing equivalents from illuminated chloroplasts. *Plant Physiol.* 64: 1064–69

3. Andrews M. 1986. The partitioning of nitrate assimilation between root and shoot of higher plants. *Plant Cell Environ.* 9:511–19

4. Bakrim N, Prioul JL, Deleens E, Rocher JP, Arrio-Dupont M, et al. 1993. Regulatory phosphorylation of C_4 phosphoenolpyruvate carboxylase. *Plant Physiol.* 101: 891–97

5. Beavis AD, Vercesi AE. 1992. Anion uniport in plant mitochondria is mediated by a Mg^{2+}-insensitive inner membrane anion channel. *J. Biol. Chem.* 267:3079–87

5a. Bettey M, Smith JAC. 1994. Dicarboxylate transport at the vacuolar membrane of the CAM plant *Kalanchoe daigremontianis*: sensitivity to protein-modifying and sulphhydryl reagents. *Biochim. Biophys. Acta* In press

6. Bisaccia F, De Palma A, Palmieri F. 1989. Identification and purification of the tricarboxylate carrier from rat liver mitochondria. *Biochim. Biophys. Acta* 977:171–76

7. Bisaccia F, Indivieri C, Palmieri F. 1988. Purification and reconstitution of two anion carriers from rat liver mitochondria: the dicarboxylate and the 2-oxoglutarate carrier. *Biochim. Biophys. Acta* 933:229–40

8. Blom-Zandstra M, Koot HTM, Hattum J, Borstlap AC. 1990. Interactions of uptake of malate and nitrate into isolated vacuoles from lettuce leaves. *Planta* 183:10–16

9. Boag S, Jenkins CLD. 1986. The involvement of aspartate and glutamate in the decarboxylation of malate by isolated bundle sheath chloroplasts from *Zea mays*. *Plant Physiol.* 81:115–19

10. Boller T, Alibert G. 1983. Photosynthesis in protoplasts from *Meliliotus alba*: distribution of products between vacuole and cytosol. *Z. Pflanzenphysiol.* 110:231–38

11. Buser C, Matile P. 1977. Malic acid in vacuoles isolated from *Bryophyllum* leaf cells. *Z. Pflanzenphysiol.* 82:462–66

12. Buser-Sutter C, Wiemken A, Matile P. 1982. A malic acid permease in isolated vacuoles of a crassulacean acid metabolism plant. *Plant Physiol.* 69:456–59

13. Chang K, Roberts JKM. 1989. Observation of cytoplasmic and vacuolar malate in maize root tips by ^{13}C-NMR spectroscopy. *Plant Physiol.* 89:197–203

14. Chang K, Roberts JKM. 1991. Cytoplasmic malate levels in maize root tips during K^+ ion uptake determined by ^{13}C-NMR spectroscopy. *Biochim. Biophys. Acta* 1092:29–34

15. Claeys D, Azzi A. 1989. Tricarboxylate carrier of bovine liver mitochondria. *J. Biol. Chem.* 264:14627–630

16. Cockburn W, Goh CJ, Avadhani PN. 1985. Photosynthetic carbon assimilation in a shootless orchid, *Chiloschita usneoides* (DON) LDL. *Plant Physiol.* 77:83–86

17. Comai L, Dietrich RA, Maslyar DJ, Baden CS, Harada JJ. 1989. Coordinate expression of transcriptionally regulated isocitrate lyase and malate synthase genes in *Brassica napus* L. *Plant Cell* 1: 293–300

18. Cushman JC. 1992. Characterization and expression of a NADP-malic enzyme cDNA induced by salt stress from the fac-

ultative crassulacean acid metabolism plant, *Mesembryanthemum crystallinum. Eur. J. Biochem.* 208:259–66

19. Cushman JC, Meyer G, Michalowski CB, Schmitt JM, Bohnert HJ. 1989. Salt stress leads to differential expression of two isogenes of phosphoenolpyruvate carboxylase during crassulacean acid metabolism induction in the common ice plant. *Plant Cell* 1:715–25

20. Day DA, Hatch MD. 1981. Dicarboxylate transport in maize mesophyll chloroplasts. *Arch. Biochem. Biophys.* 211:738–42

21. Day DA, Wiskich JT. 1981. Glycine metabolism and oxalacetate transport by pea leaf mitochondria. *Plant Physiol.* 68:425–29

22. Day DA, Wiskich JT. 1984. Transport processes in isolated plant mitochondria. *Physiol. Vég.* 22:241–61

23. Dietz KJ, Canut H, Marigo G. 1992. Identification of an essential histidine residue at the active site of the tonoplast malate carrier in *Catharanthus roseus* cells. *J. Membr. Biol.* 129:137–43

24. Douce R, Bonner WD. 1972. Oxaloacetate control of Krebs cycle oxidation in purified plant mitochondria. *Biochem. Biophys. Res. Commun.* 275:619–24

25. Douce R, Neuburger M. 1989. The uniqueness of plant mitochondria. *Annu. Rev. Plant Physiol. Plant Mol. Biol.* 40:371–414

26. Ebbighausen H, Chen J, Heldt HW. 1985. Oxaloacetate translocator in plant mitochondria. *Biochim. Biophys. Acta* 810: 184–99

27. Edwards G, Walker DA. 1983. *C3C4: Mechanisms and Cellular and Environmental Regulation of Photosynthesis.* Oxford: Blackwell. 542 pp.

28. Edwards GE, Huber SC. 1981. The C_4 pathway. In *The Biochemistry of Plants: A Comprehensive Treatise*, ed. MD Hatch, NK Boardman, 8:237–81. New York: Academic. 521 pp.

29. Edwards GE, Nakamoto H, Burnell JN, Hatch MD. 1985. Pyruvate,P_i dikinase and NADP-malate dehydrogenase in C_4 photosynthesis: properties and mechanism of light/dark regulation. *Annu. Rev. Plant Physiol.* 36:255–86

30. El-Shora HM, ap Rees T. 1991. Intracellular location of $NADP^+$-linked malic enzyme in C_3 plants. *Planta* 185:362–67

31. Flügge UI, Heldt HW. 1991. Metabolite translocators of the chloroplast envelope. *Annu. Rev. Plant Physiol. Plant Mol. Biol.* 42:129–44

32. Flügge UI, Woo KC, Heldt HW. 1988. Characteristics of 2-oxoglutarate and glutamate transport in spinach chloroplasts. *Planta* 174:534–41

33. Ford CW. 1984. Accumulation of low molecular weight solutes in water-stressed tropical legumes. *Phytochemistry* 23: 1007–15

34. Fougère F, Le Rudulier D, Streeter JG. 1991. Effect of salt stress on amino acid, organic acid and carbohydrate composition of roots, bacteroids, and cytosol of alfalfa (*Medicago sativa* L.). *Plant Physiol.* 96: 1228–36

35. Furbank RT, Agostino A, Hatch MD. 1990. C_4 acid decarboxylation and photosynthesis in bundle sheath cells of NAD-malic enzyme-type C_4 plants: mechanism and the role of malate and orthophosphate. *Arch. Biochem. Biophys.* 276:374–81

36. Genchi G, De Santis A, Ponzone C, Palmieri F. 1991. Partial purification and reconstitution of the α-ketoglutarate carrier from corn (*Zea mays* L.) mitochondria. *Plant Physiol.* 96:1003–7

37. Gerhardt R, Heldt HW. 1984. Measurement of subcellular metabolite levels in leaves by fractionation of freeze-stopped material in nonaqueous media. *Plant Physiol.* 75:542–47

38. Gerhardt R, Stitt M, Heldt HW. 1987. Subcellular metabolite levels in spinach leaves. *Plant Physiol.* 83:399–407

39. Graham IA, Leaver CJ, Smith SM. 1992. Induction of malate synthase gene expression in senescent and detached organs of cucumber. *Plant Cell* 4:349–57

40. Gross GG, Jause C, Elstner EF. 1977. Involvement of malate, monophenols and the superoxide radical in hydrogen peroxide formation by isolated cell walls from horse radish. *Planta* 136:271–76

41. Hampp R, Goller M, Füllgraf H. 1984. Determination of compartmented metabolite pools by a combination of rapid fractionation of oat mesophyll protoplasts and enzymic cycling. *Plant Physiol.* 75:1017–21

42. Haschke HP, Lüttge U. 1975. Stochiometric correlation of malate accumulation with auxin-dependent K^+-H^+ exchange and growth in *Avena* coleoptile segments. *Plant Physiol.* 56:696–98

43. Hatch MD, Dröscher L, Flügge UI, Heldt HW. 1984. A specific translocator for oxaloacetate transport in chloroplasts. *FEBS Lett.* 178:15–19

44. Hedrich R, Barbier-Brygoo H, Felle H, Flügge UI, Lüttge U, et al. 1988. General mechanisms for solute transport across the tonoplast of plant vacuoles: a patch-clamp survey of ion channels and proton pumps. *Bot. Acta* 101:7–13

45. Hedrich R, Flügge UI, Fernandez JM. 1986. Patch-clamp studies of ion transport in isolated plant vacuoles. *FEBS Lett.* 204: 228–32

46. Hedrich R, Marten I. 1993. Malate-induced feedback regulation of plasma membrane

anion channels could provide a CO_2 sensor to guard cells. *EMBO J.* 12:897–901

47. Hedrich R, Schroeder JI. 1989. The physiology of ion channels and electrogenic pumps in higher plants. *Annu. Rev. Plant Physiol.* 40:539–69

48. Heineke D, Riens B, Grosse H, Hoferichter P, Flügge UI, Heldt HW. 1991. Redox transfer across the inner chloroplast envelope. *Plant Physiol.* 95:1131–37

49. Herrada G, Puppo A, Rigaud J. 1989. Uptake of metabolites by bacteroid-containing vesicles and by free bacteroids from French bean nodules. *J. Gen. Microbiol.* 135: 3165–71

50. Heupel R, Heldt HW. 1994. Protein organization in the matrix of leaf peroxisomes: a proposed multi-enzyme complex involved in photorespiratory metabolism. *Eur. J. Biochem.* In press

51. Heupel R, Markgraf T, Robinson DG, Heldt HW. 1991. Compartmentation studies on spinach leaf peroxisomes. *Plant Physiol.* 96:971–79

52. Hoffland E, Van de Boogard R, Nelemans J, Findenegg G. 1992. Biosynthesis and root exudation of citric and malic acids in phosphate-starved rape plants. *New Phytol.* 122:675–80

53. Issakidis E, Miginiac-Maslow M, Decottignies P, Jacquot JP, Cretin C, Gadal P. 1992. Site-directed mutagenesis reveals the involvement of an additional thioredoxin-dependent regulatory site in the activation of recombinant *Sorghum* leaf NADP-malate dehydrogenase. *J. Biol. Chem.* 267: 21577–83

54. Iwasaki I, Arata H, Kijima H, Nishimura M. 1992. Two types of channels involved in the malate ion transport across the tonoplast of a crassulacean acid metabolism plant. *Plant Physiol.* 98:1494–97

55. Jiao J, Chollet R. 1991. Posttranslational regulation of phosphoenolpyruvate carboxylase in C_4 and crassulacean acid metabolism plants. *Plant Physiol.* 95:981–85

56. Kaiser G, Martinoia E, Wiemken A. 1981. Rapid appearance of photosynthetic products in the vacuoles isolated from barley mesophyll protoplasts by a new fast method. *Z. Pflanzenphysiol.* 107:103–13

57. Kaiser WM, Förster J. 1989. Low CO_2 prevents nitrate reduction in leaves. *Plant Physiol.* 91:970–74

58. Kalt W, Osmond CB, Siedow JN. 1990. Malate metabolism in the dark after $^{13}CO_2$ fixation in the crassulacean plant *Kalanchoe tubiflora*. *Plant Physiol.* 94:826–32

59. Kirkby EA, Knight AH. 1977. Influence of the level of nitrate nutrition on ion uptake and assimilation, organic acid accumulation, and cation-anion balance in whole tomato plants. *Plant Physiol.* 60:349–53

60. Krömer S, Heldt HW. 1991. Respiration of

pea leaf mitochondria and redox transfer between the mitochondrial and extramitochondrial compartment. *Biochim. Biophys. Acta* 1057:42–50

61. Kurkdjian A, Guern J. 1989. Intracellular pH: measurement and importance in cell activity. *Annu. Rev. Plant Physiol. Plant Mol. Biol.* 40:271–303

62. Lance C, Rustin P. 1984. The central role of malate in plant metabolism. *Physiol. Vég.* 22:625–41

63. LaNoue JF, Schoolwerth AC. 1979. Metabolite transport in mitochondria. *Annu. Rev. Biochem.* 48:871–922

64. Leegood RC. 1985. The intercellular compartmentation of metabolites in leaves of *Zea mays* L. *Planta* 164:163–71

65. Lehner K, Heldt HW. 1978. Dicarboxylate transport across the inner membrane of the chloroplast envelope. *Biochim. Biophys. Acta* 501:531–44

66. Lüttge U. 1987. Carbon dioxide and water demand: crassulacean acid metabolism (CAM), a versatile ecological adaption exemplifying the need for integration in ecophysiological work. *New Phytol.* 106:593–629

67. Lüttge U, Smith JAC. 1984. Mechanism of passive malic-acid efflux from vacuoles of the CAM plant *Kalanchoe daigremontiana*. *J. Membr. Biol.* 81:149–58

68. Lüttge U, Smith JAC, Marigo G, Osmond CB. 1981. Energetics of malate accumulation in the vacuoles of *Kalanchoe tubiflora* cells. *FEBS Lett.* 126:81–84

69. Marigo G, Bouyssou H, Laborie D. 1988. Evidence for malate transport into vacuoles isolated from *Catharanthus roseus* cells. *Bot. Acta* 101:187–91

70. Marquardt-Jarczyk G, Lüttge U. 1990. Anion transport at the tonoplast of mesophyll cells of the CAM plant *Kalanchoe daigremontiana*. *J. Plant Physiol.* 136: 129–36

71. Martinoia E. 1982. Transport processes in vacuoles of higher plants. *Bot. Acta* 105: 232–45

72. Martinoia E, Flügge UI, Kaiser G, Heber U, Heldt HW. 1985. Energy-dependent uptake of malate into vacuoles isolated from barley mesophyll protoplasts. *Biochim. Biophys. Acta* 806:311–19

73. Martinoia E, Grill E, Tommasini R, Kreuz K, Amrhein N. 1993. ATP-dependent glutathione S-conjugate 'export' pump in the vacuolar membrane of plants. *Nature* 364: 247–49

74. Martinoia E, Vogt E, Amrhein N. 1990. Transport of malate and chloride into barley mesophyll vacuoles. Different carriers are involved. *FEBS Lett.* 261:109–11

75. Martinoia E, Vogt E, Rentsch D, Amrhein N. 1991. Functional reconstitution of the malate carrier of barley mesophyll vacu-

oles in liposomes. *Biochim. Biophys. Acta* 1062:271–78

76. McIntosh CA, Oliver DJ. 1992. Isolation and characterization of the tricarboxylate transporter from pea mitochondria. *Plant Physiol.* 100:2030–34

77. McNulty IB. 1985. Rapid osmotic adjustment by a succulent halophyte to saline shock. *Plant Physiol.* 78:100–3

78. Menzlaff E, Flügge UI. 1993. Purification and functional reconstitution of the 2-oxoglutarate/malate translocator from spinach chloroplasts. *Biochim. Biophys. Acta* 1147:13–18

79. Mettler IJ, Beevers H. 1980. Oxidation of NADH in glyoxisomes by a malate-aspartate shuttle. *Plant Physiol.* 66:555–60

80. Ohwaki Y, Hirata H. 1992. Differences in carboxylic acid exudation among P-starved leguminous crops in relation to carboxylic acid contents in plant tissues and phospholipid level in roots. *Soil Sci. Plant Nutr.* 38:235–43

81. Oliver DJ, Walker GH. 1984. Characterization of the transport of oxaloacetate by pea leaf mitochondria. *Plant Physiol.* 76:409–13

82. Ouyang LJ, Udvardi MK, Day DA. 1990. Specificity and regulation of the dicarboxylate carrier on the peribacteroid membrane of soybean nodules. *Planta* 182:437–44

83. Ouyang LJ, Whelan J, Weaver DC, Roberts DM, Day DA. 1991. Protein phosphorylation stimulates the rate of malate uptake across the peribacteroid membrane of soybean nodules. *FEBS Lett.* 293:188–90

84. Proudlove MO, Thurman DA, Salisbury J. 1984. Kinetic studies on the transport of 2-oxoglutarate and L-malate in isolated pea chloroplasts. *New Phytol.* 96:1–8

85. Raschke K. 1979. Movements of stomata. In *Encyclopedia of Plant Physiology*. Vol. 7: *Physiology of Movements*, ed. W Haupt, E Feinleib, pp. 57–100. Berlin: Springer-Verlag. 584 pp

86. Raschke K, Hedrich R, Reckmann U, Schroeder J. 1988. Exploring biophysical and biochemical components of the osmotic motor that drives stomatal movement. *Bot. Acta* 101:283–94

87. Rea PA, Sanders D. 1987. Tonoplast energization: two H^+ pumps, one membrane. *Physiol. Plant.* 71:131–41

88. Rentsch D, Martinoia E. 1991. Citrate transport into barley mesophyll vacuoles—comparison with malate uptake activity. *Planta* 184:532–37

89. Reumann S, Heupel R, Heldt HW. 1994. Compartmentation studies on spinach leaf peroxisomes. II. Evidence for the transfer of reductant from the cytosol to the peroxisomal compartment via malate-oxaloacetate shuttle. *Planta* In press

90. Rieger A, Lutz A, Hampp R. 1992. Compartmentation of soluble carbohydrates, of starch and of malate in motor organs (pulvini) and other parts of *Phaseolus coccineus* L. leaves. *Planta* 187:95–102

91. Runswick MJ, Walker JE, Bisaccia F, Iacobazzi V, Palmieri F. 1990. Sequence of the bovine 2-oxoglutarate/malate carrier protein: structural relationship to other mitochondrial transport proteins. *Biochemistry* 29:11033–40

92. Scheibe R. 1987. $NADP^+$-malate dehydrogenase in C_3 plants: regulation and role of a light-activated enzyme. *Physiol. Plant.* 71:393–400

93. Schnabl H, Kottmeier C. 1984. Determination of malate levels during the swelling of vacuoles isolated from guard-cell protoplasts. *Planta* 161:27–31

94. Schnabl H, Raschke K. 1980. Potassium chloride as stomatal osmoticum in *Allium cepa* L., a species devoid of starch in guard cells. *Plant Physiol.* 65:88–93

95. Schröppel-Meier G, Kaiser WM. 1988. Ion homeostasis in chloroplasts under salinity and mineral deficiency. *Plant Physiol.* 87:822–27

96. Siebke K, Yin ZH, Raghavendra AS, Heber U. 1992. Vacuolar pH oscillations in mesophyll cells accompany oscillations of photosynthesis in leaves: interdepedence of cellular compartments, and regulation of electron flow in photosynthesis. *Planta* 186:526–31

97. Smith FA, Raven JA. 1979. Intracellular pH and its regulation. *Annu. Rev. Plant Physiol.* 30:289–311

98. Somerville SC, Somerville CR. 1985. A mutant of *Arabidopsis* deficient in chloroplast dicarboxylate transport is missing an envelope protein. *Plant Sci. Lett.* 37:217–20

99. Speer M, Kaiser WM. 1991. Ion relations of symplastic and apoplastic space in leaves from *Spinacia oleracea* L. and *Pisum sativum* L. under salinity. *Plant Physiol.* 97:990–97

100. Stitt M, Heldt HW. 1985. Generation and maintenance of concentration gradients between the mesophyll and bundle sheath in maize leaves. *Biochim. Biophys. Acta* 808:400–14

101. Streeter JG. 1987. Carbohydrate, organic acid, and amino acid composition of bacteroids and cytosol from soybean nodules. *Plant Physiol.* 85:768–73

102. Stumpf DK, Burris RH. 1981. Organic acid contents of soybean: age and source of nitrogen. *Plant Physiol.* 68:989–91

103. Sze H. 1985. H^+-translocating ATPases: advances using membrane vesicles. *Annu. Rev. Plant Physiol.* 36:175–208

104. Szewczyk A, Nalcez MJ, Broger C, Wojtczak L, Azzi A. 1987. Purification by affinity chromatography of the dicarboxylate

carrier from bovine heart mitochondria. *Biochim. Biophys. Acta* 894:252–60

105. Ting IP. 1985. Crassulacean acid metabolism. *Annu. Rev. Plant Physiol.* 36:595–622

106. Touraine B, Muller B, Grignon C. 1992. Effect of phloem-translocated malate on NO_3^- uptake by roots of intact soybean plants. *Plant Physiol.* 99:1118–23

107. Udvardi MK, Price CD, Gresshof PM, Day DA. 1988. A dicarboxylate transporter on the peribacteroid membrane of soybean nodules. *FEBS Lett.* 231:36–40

108. Van Beusichem ML, Kirby EA, Baas R. 1988. Influence of nitrate and ammonium nutrition on the uptake, assimilation, and distribution of nutrients in *Ricinus communis*. *Plant Physiol.* 86:914–21

109. Van Veldhoven PP, Just WW, Mannaerts GP. 1987. Permeability of the peroxisomal membrane to cofactors of β-oxydation. *J. Biol. Chem.* 262:4310–18

110. Vivekananda J, Beck CF, Oliver DJ. 1988. Monoclonal antibodies as tools in membrane biochemistry. *J. Biol. Chem.* 263: 4782–88

111. Weiner H, Heldt HW. 1992. Inter- and intracellular distribution of amino acids and other metabolites in maize (*Zea mays* L.) leaves. *Planta* 187:242–46

112. White PJ, Smith JA. 1989. Proton and anion transport at the tonoplast in crassulacean acid metabolism plants: specificity of the malate-influx system in *Kalanchoe daigremontiana*. *Planta* 179:265–74

113. Winter H, Robinson DG, Heldt HW. 1993. Subcellular volumes and metabolite concentrations in barley leaves. *Planta* 191:180–90

114. Winter K. 1985. Crassulacean metabolisms. In *Photosynthetic Mechanisms and the Environment*, ed. J Barber, NR Baker, pp. 329–87. Amsterdam: Elsevier

115. Winter K, Edwards GE, Holtum JAM. 1981. Nocturnal accumulation of malic acid occurs in mesophyll tissue without proton transport to epidermal tissue in the inducible crassulacean acid metabolism plant *Mesembryanthemum crystallinum*. *Plant Physiol.* 68:355–57

116. Winter K, Usuda H, Tsuzuki M, Schmitt M, Edwards GE, et al. 1982. Influence of nitrate and ammonia on photosynthetic characteristics and leaf anatomy of *Moricandia arvensis*. *Plant Physiol.* 70:616–25

117. Woo KC, Flügge UI, Heldt HW. 1987. A two-translocator model for the transport of 2-oxoglutarate and glutamate in chloroplasts during ammonia assimilation in the light. *Plant Physiol.* 84:624–32

118. Zeiger E, Farquhar GD. 1987. *Stomatal Function*. Stanford: Stanford Univ. Press. 503 pp.

119. Ziegler H. 1975. Nature of substances in phloem. In *Encyclopedia of Plant Physiology*. Vol. 1: *Transport in Plants*, ed. MH Zimmermann, JA Milburn, pp. 57–100. Berlin: Springer-Verlag. 535 pp.

120. Zoglowek C, Krömer S, Heldt HW. 1988. Oxaloacetate and malate transport by plant mitochondria. *Plant Physiol.* 87:109–15

Annu. Rev. Plant Physiol. Plant Mol. Biol. 1994. 45:469–91

STRUCTURAL AND FUNCTIONAL ASPECTS OF CHAPERONIN-MEDIATED PROTEIN FOLDING

Anthony A. Gatenby and Paul V. Viitanen

Central Research and Development, DuPont, Wilmington, Delaware 19880-0402

KEY WORDS: GroE, heat shock proteins, cpn60, molecular chaperones, TCP1

CONTENTS

INTRODUCTION 469
CHAPERONIN MOLECULES.. 470
   Chaperonin 60 471
   Chaperonin 10 472
   Cytosolic T-Complex Polypeptide-1 Related Chaperonins.................................... 473
MECHANISM OF CHAPERONIN ACTION IN FACILITATING PROTEIN
   FOLDING.. 475
   ATPase Activity... 475
   ATP-Dependent Association of cpn10 with cpn60.. 479
   Binding of Polypeptides to Chaperonins... 480
   Release of Polypeptides from Chaperonins.. 483

INTRODUCTION

A considerable amount of cellular energy is invested in the synthesis of correctly folded and biologically active proteins. Many of the steps in this intricate pathway from template to protein are understood in considerable detail, but the final stage, the folding of a polypeptide chain into its correct three-dimensional structure, is still obscure. The informational content in the primary amino acid sequence must be converted rapidly into a correct structure under

0066-4294/94/0601-0469$05.00

conditions of high protein concentrations in the cell, in the presence of other rapidly folding species that could present vulnerable interactive surfaces (30), and at physiological temperatures that destabilize aggregation-prone folding intermediates (17, 51). Although earlier data, based on spontaneous refolding of several denatured proteins in vitro, led to the view that protein folding in cells was a simple consequence of terminated chains being released from ribosomes, it is now generally accepted that protein folding in cells is more complex. The folding process in vivo is regulated by the interaction of incompletely folded polypeptides with a group of proteins known as molecular chaperones. This interaction partitions the polypeptides toward productive folding pathways by suppressing "off-pathway" reactions, such as aggregation (21, 29, 32, 44, 51, 62).

Molecular chaperones are a class of abundant cellular proteins that modulate the folding of numerous other polypeptides during progression to their native states. This review focuses on one group of molecular chaperones, the chaperonins, which possess an intriguing molecular architecture, and mechanistically, are perhaps the best understood of these exceptional proteins. Two types of chaperonins are now recognized (20): the GroE form found in eubacteria, chloroplasts, and mitochondria, and the recently identified TCP1 form found in archaebacteria and the cytosol of eukaryotic cells. Both types will be reviewed here, especially their structural and functional aspects. Several recent reviews cover the genetics of chaperonins, their role in the heat shock response, and possible applications in biotechnology (1, 21, 32, 33, 93, 103). The first identified chaperonins, the 10-kDa GroES and 60-kDa GroEL polypeptides from *Escherichia coli,* and related homologs are now commonly referred to as chaperonin 10 (cpn10) and chaperonin 60 (cpn60), respectively (34, 63), reflecting the molecular mass of their subunits. The t-complex polypeptide 1 and related proteins are simply referred to as TCP1, or TCP1-related.

CHAPERONIN MOLECULES

Chaperonins are the most varied and structurally complex members of the molecular chaperone group of proteins. All are oligomeric proteins and have ring-like or toroidal structures with central cavities. This simple architectural plan, however, has developed some interesting permutations. The GroE group of chaperonins, comprising cpn60 and cpn10, have ring structures with 7-fold rotational symmetry. The heptameric cpn60 toroid can be a double-stacked ring (*E. coli,* mitochondria, chloroplasts), a single ring (some mitochondria), a single subunit type (*E. coli,* mitochondria), or two different subunits (chloroplasts). Cpn10 chaperonins exist as single rings with a single 10-kDa subunit type (*E. coli,* mitochondria) or with tandem fusion giving double cpn10 subunits of 24-kDa (chloroplasts). The TCP1 group of chaperonins contain related proteins with double rings of six (plants), and eight or possibly nine subunits (mammals, archaebacteria). Furthermore, these TCP1-related rings have a more complex subunit composition than do the cpn60 chaperonins, and they contain five or six different polypeptides of about the same size (52–65-kDa). Despite these structural variations, all chaperonins facilitate protein folding. After a description of their various molecular configurations, we discuss how these chaperonins influence protein folding.

Chaperonin 60

GroEL (cpn60) and GroES (cpn10) were among the first molecular chaperones identified and characterized in detail (1, 103). Proteins related to either GroES or GroEL from *E. coli* have now been identified in numerous prokaryotic organisms, and they display a high degree of amino acid sequence homology (103). The *groEL* gene product was initially identified as being required for bacteriophage morphogenesis (1, 103), but is now also known to be essential for cell survival (23). Viewed by electron microscopy, *E. coli* cpn60 has a structure of two stacked rings with seven 60-kDa subunits in each ring (45). The diameter of the cpn60 cylinder is about 14.5 nm with a longitudinal axis of 16 nm (58). The center of the rings is occupied by a cavity, about 6 nm in diameter (10, 58), which enlarges in the presence of ADP (58, 85). Folding intermediates are bound inside the central cavities within individual cpn60 rings (10, 58, 85), and it is estimated that each cavity can accommodate a globular protein of up to 60-kDa (44) or 90-kDa (10).

Mitochondria and chloroplasts contain nuclear-encoded cpn60 molecules that are highly conserved at the level of amino acid sequence homology, both with each other and with bacterial cpn60 (43, 66, 68, 82). Electron micrographs of the cpn60 proteins from mitochondria and plastids show similar structural features to those observed with *E. coli* cpn60 (47, 68, 81). Although the two-layered heptameric ring structure is observed for chloroplast and most mitochondrial cpn60s, an exception is a single seven-membered ring isolated from the mitochondria of Chinese hamster ovary cells (96) and from moth sperm (72). In contrast to *E. coli* and mitochondria, which contain only a single type of cpn60 polypeptide (43, 82), chloroplasts contain equal amounts of two distinct cpn60 polypeptides (α and β) (66, 76). It is not known whether the α and β subunits reside in the same or different cpn60 tetradecamers. Chloroplast cpn60 was originally identified as an important component in the biosynthesis of ribulose bisphosphate carboxylase (Rubisco), where it binds to nascent large subunit polypeptides before their assembly into holoenzyme (4, 6, 83). More generally a variety of proteins imported into isolated chloroplasts can form stable complexes with the chaperonin, and are then dissociated from these complexes with ATP (31, 63). This result is similar to observations with yeast mitochondria, in which several imported proteins form stable complexes with cpn60, which is required for correct folding and assembly (16, 77). A protein synthesized within maize mitochondria will also bind to endogenous cpn60 and is released by the addition of ATP, demonstrating that the folding of nuclear and mitochondrial encoded proteins is mediated by the chaperonin (80).

Chaperonin 10

We refer to chaperonin 10 (cpn10 or GroES) as a co-chaperonin (64), because it behaves primarily as a cofactor in releasing bound target proteins from cpn60 in a form that can progress to the native state. The first identified representative of this group of proteins was the GroES protein from *E. coli* (1, 103). The *groES* gene is the first of two genes in the *groE* operon, and as with *groEL,* it was identified initially by mutations that prevent the growth of several bacteriophages. Subsequent characterization of the *groES* gene products showed that the 10-kDa polypeptides form heptameric and homooligomeric ring structures (14). Intuitively, this 7-fold rotational symmetry makes sense, because a stable complex can be formed between cpn10 and the 7-fold rotationally symmetrical cpn60 molecule in the presence of certain adenine nucleotides (14, 97). *groES* is essential for cell viability (23), and its overexpression (together with *groEL*) allows suppression of many heat-sensitive mutants (92) and enhanced assembly of foreign proteins synthesized in *E. coli* (13, 99), including Rubisco (35).

Homologs of cpn10 are present in mammalian mitochondria (40, 64) and higher plant chloroplasts (5). The bovine and rat liver mitochondrial cpn10s were identified because they are functionally compatible with *E. coli* cpn60 (64). The mitochondrial cpn10 can substitute for bacterial cpn10 in the cpn60-facilitated refolding of Rubisco, and can form a stable ATP-dependent complex with cpn60. Amino acid sequence analysis of the bovine (5) and rat (40) co-chaperonins show that both are highly homologous to prokaryotic cpn10s. Thus, the protein-folding machinery of mitochondria requires a co-chaperonin for full biological activity. Like *E. coli* cpn10, the mitochondrial protein is homooligomeric.

The chloroplast cpn10 has also been isolated from plastids, again by its ability to interact functionally with bacterial cpn60 in the chaperonin-dependent refolding of Rubisco, and the formation of a stable complex with cpn60 in the presence of ATP (5). The name cpn10 is somewhat misleading, because the functional chloroplast homolog has an actual subunit molecular mass of about 24 kDa, and each subunit has a novel structure comprising two distinct cpn10-like domains fused together in tandem (5). The spinach protein is synthesized as a precursor molecule and has a typical N-terminal chloroplast transit peptide.

Remarkably, there are only 8 residues that are completely conserved in the 16 bacterial cpn10 sequences that are currently known, and these same residues are also found in both halves of the "duplex" cpn10 molecule from chloroplasts (5). This high degree of conservation of important amino acid residues suggests that both halves of the plastid cpn10 may be able to function independently, or perhaps they perform different functions. To address this directly, the full-length duplex cpn10 molecule, or each of its two halves, were expressed separately in *E. coli* and tested for functionality. The *E. coli* strain that was used for this test has a defective *groES* gene, which prevents bacteriophage growth (22). Expression in the mutant strain of the duplex cpn10, or each of the two halves, restored bacteriophage growth, indicating that all three protein configurations are functional and can substitute for bacterial cpn10 (F Baneyx, U Bertsch, C Kalbach, SM van der Vies, J Soll, AA Gatenby, submitted for publication). This reveals that each half of the chloroplast duplex cpn10 is functional. Although its exact subunit composition has been difficult to determine because of dissociation during chromatography, spinach cpn10 is known to be homooligomeric. The highly efficient interaction of chloroplast cpn10 with bacterial cpn60, and the formation of a stable cpn10-cpn60 binary complex (5), indicates that cpn10 probably exists as a heptamer in its functional state. If so, we imagine that the ring of seven duplex subunits may have two functional surfaces, perhaps for differential interaction with the α and β subunits of chloroplast cpn60s (66, 76).

Cytosolic T-Complex Polypeptide-1 Related Chaperonins

If nascent polypeptide chains interact sequentially with different types of molecular chaperones during progression to their native states (57, 65), then representatives of each chaperone group might be localized in various cell compartments. For example, chloroplasts, mitochondria, and bacteria have hsp70, cpn60, and cpn10 proteins, or their related homologs. Because the eukaryotic cytosol is an active site for protein folding, and contains hsp70 molecular chaperones, cpn60-like components of the protein folding machinery should also be present. The cytosolic t-complex polypeptide-1 related molecular chaperones probably represent a functional equivalent of the cpn60-type proteins. Originally identified as a ubiquitous cytosolic protein in mouse cells, limited similarities between the TCP1 and cpn60 polypeptide sequences were described (19, 39). Further support that TCP1 could be a molecular chaperone comes from the isolation of a related protein that had some of the properties of chaperones and was 36–40% identical to TCP1 (90). Specifically, the thermophilic factor 55 (TF55) from the thermophilic archaebacterium, *Sulfolobus shibatae,* is a two stacked-ring complex, with eight or nine 55-kDa subunits in each ring and a weak ATPase that binds to unfolded proteins but not their native forms (90). The three-dimensional structure of a related protein from the thermophile *Pyrodictium occulatum* has been reconstructed, and shows a complex of two rings of eight subunits each, in the center of which is a large (6.7 nm diameter) cavity with corrugated walls (78). Striated rectangular side views are observed, similar to those seen for cpn60 (45, 47, 58, 68, 81).

Direct evidence that the cytosol of eukaryotic cells contains TCP1-related proteins that are fully functional as molecular chaperones has now been reported for animal (26–28, 60, 100) and plant (75) species. The protein from mammals was isolated as a large heterooligomeric complex of about 800–970 kDa, with an outer diameter of 12–16 nm and an inner diameter of 6 nm (26, 27, 60, 100). Electron microscopy shows two stacked rings, each with eight or nine subunits, and a central cavity (26, 60). Other views appear as rectangular striated structures, again reminiscent of negative stain images of cpn60 (45, 47, 58, 68, 81). The complex contains at least five or six distinct polypeptides of 52–65 kDa (26, 27), some of which are sequentially and antigenically related to TCP1 (26, 27, 60, 100). The particle has been referred to as TRiC (TCP1 ring complex) (26). At present it is not clear whether there is a single type of TCP1 complex with different subunits, or a number of related oligomeric forms with varying subunit composition. The complex is also reported to be associated with the hsp70 molecular chaperone (60). As described later, these TCP1-related complexes can interact with the nonnative conformations of several target proteins, although they interact mainly with the major cytoskeletal components. Bound proteins are released subsequently in an assembly-competent form in the presence of ATP (26, 27, 100), although there may be additional cofactor requirements for tubulin folding (28).

In contrast to the mammalian TCP1 type of chaperonin, the TCP1-related molecular chaperone isolated from etiolated oat seedlings has a different quaternary structure (75). The smaller plant protein is about 600 kDa, and the reduced size is the result of a different configuration of subunits. Although the individual subunits are about 60 kDa and are therefore similar to other chaperonin polypeptides, they are arranged in two stacked rings of six subunits each, with a central cavity. The number of subunits in the stacked rings of chaperones varies between six (oat TCP1), seven (cpn60), and eight or nine (mammalian and archaebacteria TCP1), although all retain this important central cavity. Like other TCP1-related chaperonins, the oat protein is heterooligomeric with six distinct polypeptides, five of which cross-react with anti-TCP1 (75). Sequencing of an internal protease-generated fragment gives similarities of 62% when compared to the predicted TCP1 proteins from *Arabidopsis thaliana* (74). The chaperonin copurifies with the phytochrome photoreceptor, and can stimulate refolding of denatured phytochrome to a photoactive form in the presence of ATP. This does not exclude a role for the oat chaperone in the folding of other plant proteins, especially because it appears to be an abundant protein. It may also have a role in stabilization of the two physiologically important P_r and P_{fr} conformations of phytochrome. Immuno-electron microscopy of oat cells also shows the presence of cross-reacting cytosolic proteins when incubated with antibacterial cpn60 (38).

MECHANISM OF CHAPERONIN ACTION IN FACILITATING PROTEIN FOLDING

A major objective in studying chaperonins is to understand the molecular mechanisms that enable them to assist in protein folding. During chaperonin-facilitated folding of proteins, several partial reactions in the overall folding pathway can be identified and examined in more detail using purified components. These partial reactions include (*a*) the interactions between chaperonins and adenine nucleotides, (*b*) the ATP-dependent formation of a stable binary complex between chaperonins and co-chaperonins, (*c*) the association of chaperonins with nonnative proteins, and (*d*) the release of these target proteins from chaperonins. An understanding of these partial reactions should lead to a clearer understanding of how chaperonins participate in protein folding.

ATPase Activity

The only legitimate catalytic activity of the chaperonin proteins is their ability to hydrolyze ATP to ADP and inorganic phosphate in the absence of other proteins. This wasteful hydrolysis of high energy phosphate has been called the uncoupled ATPase (97), to distinguish it from the coupled hydrolysis that is observed during chaperonin-assisted protein folding reactions. ATPase activity is an intrinsic property of all prokaryotic (14, 45, 88, 97) and eukaryotic (43, 79, 81) cpn60s that have been examined. It is also a property of the eukaryotic cytosolic TCP1-related proteins (26, 75) and the archaebacterial homolog TF55 (90). Even under optimal conditions, cpn60 ATPase activity is extremely slow in comparison to other ATPases, with turnover numbers of ~ 0.04–0.2 molecules of ATP per second per protomer. Similar turnover numbers are observed for the TCP1-related proteins from plants (75), animals (26), and archaebacteria (90).

Most of our knowledge about the uncoupled ATPase comes from studies with purified cpn60 (GroEL) from *E. coli*. The reaction is quite specific for ATP and, not surprisingly, full catalytic activity requires Mg^{2+}, although other divalent cations ($Ca^{2+} > Mn^{2+} > Zn^{2+}$) can substitute partially. In addition, bacterial cpn60 (97) and mitochondrial cpn60 (96) require certain monovalent cations, in particular, potassium. When the ATP concentration is saturating, half maximal activation occurs at 100 μM K^+, about one order of magnitude lower than that required by other K^+-activated enzymes. However, in the presence of subsaturating ATP, the requirement for K^+ can increase several orders of magnitude (89). As with other K^+-activated enzymes, low concentration of Rb^+ or NH_4^+ ions are equally efficacious in stimulating the ATPase. The mechanistic importance of these observations is highlighted by parallel studies that show similar K^+ requirements for the ATP-dependent, chaperonin-assisted reconstitution of Rubisco (97), rhodanese (70), and α-glucosidase (46). Thus, K^+ ions are required for both the coupled and uncoupled hydrolysis of ATP.

ATP hydrolysis by cpn60 is positively cooperative with respect to ATP and exhibits a Hill coefficient of 2.0 (36). Moreover, the extent of cooperativity is increased in the presence of the co-chaperonin cpn10 (GroES) (36), and is related reciprocally to the concentration of K^+ ions (89). For example, as the K^+ concentration is increased from 1 to 500 mM, the Hill coefficient for ATP decreases from 3 to 2. Interestingly, the cpn60 ATPase is also cooperative with respect to K^+, in a manner that is inversely related to the concentration of ATP. Thus, the Hill coefficient for K^+ increases from 1 to 2 as the ATP concentration is decreased from 50 μM to 2 μM. In the presence of high concentrations of K^+, the $K_{0.5}$ for ATP hydrolysis by cpn60 is about 5 μM ATP (89, 97).

Equilibrium studies with cpn60 that has been modified covalently with a fluorescent reporter group (pyrene maleimide) demonstrate that the binding of adenine nucleotides is also cooperative (49). The Hill coefficient and $K_{0.5}$ for ATP binding were found to be 4 and 10 μM, respectively. The ability of cpn60 to bind ADP is much weaker with a $K_{0.5}$ of 2.5 mM. Because of such low affinity interactions, it is not surprising that neither the binding of ATP nor ADP by cpn60 alone can be detected using spin columns under dissociating conditions (8). In contrast, under equilibrium conditions each cpn60 oligomer can bind at least 4 molecules of $[\gamma^{-32}P]$-ATP without cpn10 or ATP hydrolysis (89). It is likely that there is at least one ATP binding site per cpn60 protomer, and in the absence of K^+ ions or cpn10 (see below), the affinity of the remaining sites is so weak that complete site occupancy is precluded.

As with ATP hydrolysis (36), the cooperativity of ATP binding by cpn60 is enhanced in the presence of its co-chaperonin (cpn10) and the Hill coefficient increases from 4 to 6 (49). Moreover, in the presence of cpn10, cpn60 binds ADP with much higher affinity (8, 49, 89). This tight binding results from the formation of an extremely stable complex between the two chaperonin proteins and ADP. Using the pyrene maleimide–tagged protein (49), the affinity of cpn60 for ADP was estimated to increase more than 30,000-fold in the presence of cpn10, resulting in a $K_{0.5}$ of ~ 70 nM. Despite its remarkable stability, the cpn60-ADP-cpn10 complex forms very slowly at low concentrations of ADP, and its formation is thought to result from in situ hydrolysis of ATP. In the absence of adenine nucleotides there is no detectable interaction between cpn60 and cpn10.

The formation of a stable complex between bacterial cpn60 and cpn10 results in an inhibition of the cpn60 ATPase (14), which can be nearly complete (67, 89, 97). Both chloroplast (5) and mitochondrial (40, 64) cpn10 co-chaperonins can substitute effectively for bacterial cpn10 in this reaction. This inhibition is more complicated than originally envisioned. Jackson et al (49) reported that maximal inhibition of the ATPase by cpn10 was only 50% and appeared immediately without a lag period. Other groups also have observed only partial inhibition (14, 36). Based on single-turnover experiments, the halved rate of ATP hydrolysis observed in the presence of cpn10 was attributed to a complete inhibition of half of the active sites of $cpn60_{14}$ (49). Titration experiments with cpn10 show a tight protein-protein interaction (K_d ~ 0.5–3 nM oligomers), with maximal inhibition occurring with 1 mol of $cpn10_7$ per 1 mol of $cpn60_{14}$. This stoichiometry is in agreement with the asymmetric chaperonin complexes that have been observed with electron microscopy (84). In contrast, other groups have detected complete inhibition of the ATPase, following a brief uninhibited or partially inhibited lag period (67, 97). Recently, this inhibition has been dissected into three distinct kinetic phases (89). In phase I, both heptameric toroids of cpn60 turn over once at exactly the same rate that they do in the absence of cpn10. Phase I ends with the formation of an asymmetric chaperonin complex ($cpn60_{14}$-ADP_7-$cpn10_7$) in which 7 mols of ADP are trapped in a form that is poorly exchangeable with free ADP. During phase II, the remaining 7 sites have the potential to turn over at the same rate that they do in the absence of cpn10, such that the overall rate of hydrolysis is maximally 50%. However, the kinetic properties of these sites are altered and they do not necessarily hydrolyze all available ATP. Instead, phase II can give way to phase III, a completely inhibited state, the onset of which is dependent on the relative affinities of the remaining 7 sites for ATP (substrate) versus ADP (inhibitor).

In general, the complete inhibition of phase III is partially the result of the remaining catalytically functional toroid in the $cpn60_{14}$-ADP_7-$cpn10_7$ complex becoming sensitized to the free [ADP]/[ATP] ratio (89). When this ratio becomes sufficiently high, ATP hydrolysis shuts down. Accordingly, phase III can be prevented (or reversed) by the addition of an ATP-regenerating system that prevents the accumulation of ADP. Alternatively, phase III can be returned to phase II by increasing the concentration of either ATP or K^+ ions; the latter act by increasing the relative affinity of cpn60 for ATP.

The observation that cpn10 inhibits the uncoupled ATPase of cpn60, and yet is essential for the ATP hydrolysis-dependent functions of the chaperonins, suggests that it serves as a coupling factor (97). Thus, it was expected that the coupled chaperonin system would hydrolyze ATP only in the presence of unfolded target proteins. Indeed, nonnative rhodanese stimulates ATP hydrolysis by the inhibited $cpn60_{14}$-ADP_7-$cpn10_7$ complex with kinetics similar to that of its chaperonin-assisted refolding (67). Remarkably, 130 mols of ATP are hydrolyzed per mol of rhodanese refolded! It was argued that this ATP consumption is trivial compared to the total cost of synthesizing a protein. Similar results have been obtained using denatured lactate dehydrogenase (49). As the in vitro chaperonin system becomes better understood, it is likely that these large stoichiometries probably will come down to more realistic values. It is also probable that different proteins are likely to yield different stoichiometries, which in turn will depend on the experimental conditions.

Other interactions have been reported, most relating to the influence of adenine nucleotides on cpn60 oligomers. Using urea as a denaturant, a stable population of folded cpn60 monomers was isolated (61). These monomers were unable to reassemble into stacked ring structures unless certain adenine nucleotides were provided. ATP was the most effective, although ADP and AMP-PNP were partially active. Interestingly, both cpn10 and intact cpn60 oligomers potentiated the maximum effect of ATP. In contrast, genetic studies indicate that in vivo assembly of yeast mitochondrial cpn60 (mt-cpn60) re-quires chaperoning by preexisting functional mt-cpn60$_{14}$ (15). It is unclear whether this in vivo requirement relates to oligomerization per se or some earlier step such as monomer folding.

The oligomeric state of the chloroplast cpn60 (ch-cpn60) is also influenced by ATP (6, 42, 76), which in this case shifts the equilibrium toward dissocia-tion. This effect is enhanced at low temperatures and appears to be fully reversible. Bacterial cpn60 also dissociates in the presence of ATP under certain conditions (43). Although dissociation of ch-cpn60 occurs at physio-logical concentrations of ATP, these in vitro experiments were performed with dilute chaperonin solutions, a condition that in itself would favor dissociation (76). Thus, the ATP-dependent dissociation of ch-cpn60 is probably not phys-iologically relevant. However, it could create problems during in vitro studies on the influence of the chloroplast chaperonins on protein folding.

The results above indicate that the binding and/or hydrolysis of ATP by cpn60$_{14}$ is accompanied by significant conformational changes. Clearly, cpn60 has a toroidal architecture ideally suited to propagate conformational changes arising from any of its monomeric subunits. Each of these 14 identical subunits are in direct contact with at least 3 of its nearest neighbors. This high degree of structural organization allows ample opportunity for cooperative communication between the monomers of a given ring and also permits cross-talk between the two stacked rings. The most obvious manifestation of confor-mational change induced by ATP is the observed decrease in the affinity of cpn60$_{14}$ for its nonnative protein substrates (2, 9, 24, 37, 53, 67, 94, 95). Adenine nucleotide–dependent conformational changes in cpn60 have also been detected using certain proteases (3, 58), and some of these changes do not require ATP hydrolysis, indicating that binding of the nucleoside triphosphate is responsible for at least some of the structural rearrangements that lead to release of bound protein (3, 49). As already noted, significant fluorescent enhancements are observed upon the binding of adenine nucleotides to pyrene-labeled cpn60 (49). Finally, adenine nucleotide–induced morphological changes in cpn60 oligomers have been visualized directly through electron microscopy (58, 85).

ATP-Dependent Association of cpn10 with cpn60

A physical interaction between cpn60 and cpn10 was first reported by Chandrasekhar et al (14), who found that the two proteins co-purify in the presence of ATP, and that the latter protein inhibits the ATPase activity of the former. Biochemical data, therefore, supported earlier genetic observations that the two chaperonin proteins interact functionally in vivo (1, 103). It was shown subsequently that cpn60-cpn10 complexes can also form in the presence of ADP (61), and that, indeed, this is the adenine nucleotide that is stably associated with complexes formed in the presence of ATP (8). Upon binding to cpn60, a highly mobile and accessible loop, which is present in the N-terminal third of the cpn10 monomer, becomes immobilized and inaccessible within the asymmetric complex (56). Several mutant cpn10 proteins that are defective in their interaction with cpn60 contain lesions that map to this region.

The requirement for adenine nucleotides in the formation of the complex between chaperonin and co-chaperonin is rather unique. Most proteins that interact with cpn60 only do so in their nonnative states. Furthermore, such interactions occur spontaneously and are destabilized by ATP and other adenine nucleotides. The two chaperonins may always exist as complexes in vivo, because the levels of ATP required for complex formation (submillimolar) are well within the physiological range. Dilution during purification probably results in complex dissociation. Even so, a stable chaperonin complex has been purified from a thermophilic bacterium (87) and has been shown to assist in the in vitro refolding of several proteins. Not surprisingly, the thermophilic holo-chaperonin complex consists of two rings of $cpn60_7$, one ring of $cpn10_7$, and tightly bound ADP (48, 101). Thus, certain cpn60-cpn10 complexes are extremely stable, even in the absence of added adenine nucleotides, supporting the theory that this is the usual in vivo form of the two proteins. A mutation in *groEL* that prevents the assembly of bacteriophages or Rubisco holoenzyme in *E. coli* results in a suboptimal interaction between the mutant cpn60 and cpn10 (3). This results in an inefficient discharge reaction, which presumably interferes with the kinetics of successful protein folding and probably accounts for the slower growth rates of strains harboring this mutation.

Considering that both cpn60 and cpn10 are toroidal and have 7-fold rotational symmetry, it seems intuitive that both faces of the double ring $cpn60_{14}$ should be able to bind a ring of $cpn10_7$. However, electron micrographs of the

ATP-dependent complexes formed from the purified *E. coli* chaperonins indicate that $cpn10_7$ binds asymmetrically to $cpn60_{14}$ (58, 84, 85). Side views of these complexes are bullet-shaped, and exhibit a gross distortion thought to exist in one ring of the cpn60 oligomer. Bullets are also observed in side views of the stable chaperonin complex of *T. thermophilus* (48). Moreover, antibodies directed against the cpn10 component of this complex bind only to the rounded head of the bullet. These images presumably represent chaperonin complexes that consist of one molecule of $cpn10_7$ bound to one molecule of $cpn60_{14}$. This would agree with the stoichiometries reported for cpn60-cpn10 complexes isolated by sucrose gradient centrifugation (8) and native polyacrylamide gel electrophoresis (58); the chaperonin-assisted refolding of dihydrofolate reductase (67), rhodanese (58), and malate dehydrogenase (71); $cpn10$ inhibition of the uncoupled ATPase of cpn60 (89); and binding studies conducted with pyrene-labeled cpn60 (49).

To account for this unexpected stoichiometry, it has been suggested that cpn60 is a nonsymmetric dimer of two 7-mers, only one of which can bind cpn10 (18). In theory, conversion of the apparently symmetrical $cpn60_{14}$ to a nonsymmetric state could result from its interaction with ATP, $cpn10_7$, or both. Alternatively, it was proposed that the two rings of $cpn60_{14}$ are stacked head-to-tail rather than tail-to-tail, which could also explain the asymmetric binding (48). Recent experimental evidence supports the former hypothesis. Langer et al (58) found that the ATP-dependent binding of one ring of $cpn10_7$ protects only half of the subunits of $cpn60_{14}$ from proteolytic truncation of about 50 C-terminal residues. Moreover, when cpn10 was dissociated and removed from these partially digested asymmetric complexes, half of the remaining intact cpn60 subunits were protected from further proteolysis by a second addition of cpn10 and ATP. This clearly indicates that the co-chaperonin is initially free to interact with either end of a symmetrical cpn60 cylinder and that its random binding to one end drastically decreases the affinity of the other end for cpn10. The available evidence indicates that the two identical rings of bacterial $cpn60_{14}$ are not necessarily structurally or functionally equivalent within the stacked rings. Nevertheless, the results obtained with a mammalian mitochondrial cpn60 (96) imply that all the information necessary for recognition and binding of nonnative protein substrates, and for ATP-dependent formation of a stable complex with cpn10, resides within a single heptameric ring of cpn60.

Binding of Polypeptides to Chaperonins

The cpn60 chaperonins are extremely promiscuous and interact with a variety of proteins that are unrelated structurally in their native states. The TCP1-related chaperonins have not been as extensively characterized with regard to target protein binding. However, because interactions between TCP1 homologs and phytochrome (75), tubulin (26, 28, 100), actin (27), luciferase (26), and the Su9-dihydrofolate reductase fusion protein (26, 90) have been reported, they may also exhibit promiscuity with a wide range of unrelated proteins. These interactions with chaperonins occur spontaneously and result in the formation of stable complexes that interfere with both off-pathway and on-pathway folding reactions. This event constitutes the initial step in all chaperonin-mediated processes, and its biochemical basis resides in some common structural element or motif that is only present or accessible in incompletely folded proteins (95). The formation of complexes between cpn60 and certain "native" proteins, such as pre-β-lactamase (53) and dihydrofolate reductase (94), occurs indirectly through stable interactions with nonnative conformers known to exist in equilibrium with these proteins. In what amounts to a net unfolding reaction, all the native protein is eventually sequestered on cpn60 in an inactive form through mass action. The time required for such interactions can be significant, and depends on the height of the energy barrier that separates those conformations recognized by cpn60 from those that are not. Thus, most native proteins fail to interact with cpn60 at or below physiological temperatures (95). On the other hand, studies with α-glucosidase (46) and malate dehydrogenase (41) indicate that rather mild increases in temperature can shift the equilibrium toward conformers that can bind to cpn60.

Chemically denatured proteins interact rapidly with cpn60, when conditions that favor refolding are established. Studies with Rubisco show that the predominate species present in solution at the time of interaction with cpn60 are folding intermediates with some degree of secondary and tertiary structure (91). In contrast, it has been suggested that a more unfolded form of lactate dehydrogenase is the preferred substrate of cpn60 (2). Unfortunately, the actual identity of the reactive species remains obscure because of the likely existence of rapid equilibria between protein folding intermediates and the fully unfolded state (17). It is generally agreed, however, that cpn60 does not interact with insoluble protein aggregates. Indeed, both Rubisco (34) and rhodanese (70) lose their ability to refold in a chaperonin-assisted manner when their labile folding intermediates partition toward aggregated states. Moreover, cpn60 completely suppresses the formation of citrate synthase (12), rhodanese (67), α-glucosidase (46), and phytochrome (37) aggregates, but cannot resolubilize them after they have formed. Thus, the well documented and important ability of cpn60 to prevent aggregation during protein folding is accomplished through its interactions with labile protein folding intermediates whose formation precedes that of the aggregate.

While sequestered on cpn60, target proteins are unable to progress spontaneously to their native state (2, 24, 53, 87, 91, 94, 97). Consequently, a stable interaction with cpn60 inhibits both legitimate and illegitimate folding events. In both cases, the degree of inhibition depends on several factors including the concentrations of cpn60 and the target protein, the dissociation constant for the binary complex in question, and the relative rate constants leading to the native state or malfolded aggregated states that are no longer recognized by cpn60. Depending on the protein and the experimental conditions, folding inhibition can be nearly complete, lasting for many hours (25, 94, 97), or only partial and transient (73, 86). Some proteins may fail to interact with cpn60 at all, simply because of rapid kinetic partitioning. For binary complex formation to occur, cpn60 must successfully encounter the refolding protein before it progresses to states that it no longer recognizes. This situation is clearly different from most receptor-ligand interactions where the ligand remains a stable species that is capable of rebinding to its receptor. Nevertheless, some cpn60-target protein complexes are stable enough to survive one or more rounds of gel filtration chromatography (3, 37, 94, 95). Complexes between several target proteins and the TCP1-related chaperonins have also been detected (26–28, 90, 100).

The nonnative character of target proteins bound to cpn60 has been demonstrated repeatedly by the failure to detect catalytic activity, or by dramatically enhanced susceptibility to exogenously added proteases (2, 12, 24, 34, 46, 53, 67, 91, 94). Results of fluorescent studies with cpn60-bound dihydrofolate reductase, rhodanese (67), and α-glucosidase (46) suggest that cpn60 stabilizes conformational intermediates that resemble "molten globules" or "compact intermediates" (reviewed in 17), partially folded states that are rich in secondary structure, but lack the highly organized tertiary structure of the native state. However, because of potential interactions between cpn60 and the target protein's chromophores, changes in the intrinsic fluorescence of a protein when bound to cpn60 versus when free in solution are difficult to interpret.

Using 2-D NMR, the low affinity interactions between cpn60 and two synthetic peptides have been observed (54, 55). Both were stabilized as α-helices on cpn60, despite the fact that these were not the major species in free solution. It is not known whether this structural motif was present initially in the peptide population or if it was a conformation imposed upon binding to the chaperonin. Nevertheless, the observation that cpn60 can also interact with an all β-protein (86) indicates that recognition by cpn60 may depend more on the nature of early folding intermediates than on some specific secondary structural element. A logical candidate is the excessive surface exposure of hydrophobic amino acid side-chains, because most partially folded proteins have not completely buried their hydrophobic interiors (55). Predominantly hydrophobic interactions are also indicated by the dissociating effects of low salt and low temperature on cpn60 complexes formed with pre-β-lactamase (7) and the bacteriophage P22 tailspike protein (11), respectively. Thus, the same forces that are largely responsible for the aggregation of folding intermediates may also account for their interactions with cpn60.

It is generally agreed that each cpn60 tetradecamer can bind only one (53, 59, 69) or two (8, 94, 102) molecules of a target protein. Reported differences in stoichiometries relate to several factors, most of which could lead to underestimates. These include (*a*) titrations based on the suppression of spontaneous folding or aggregation, where rapid kinetic partitioning affects the outcome; (*b*) uncertainties in protein concentration, either of cpn60, the target protein, or both; and (*c*) partial occupancy of cpn60 binding sites by contaminating endogenous protein ligands. Binding stoichiometries may even differ for target proteins of different sizes. The observed cooperative binding of rhodanese to cpn60 (8) would not be expected with a stoichiometry of 1 mol of rhodanese per cpn60$_{14}$. In addition, scanning transmission electron microscopy of complexes formed between cpn60 and a gold cluster conjugate of dihydrofolate reductase support the theory that the chaperonin can bind two polypeptides simultaneously (10).

In summary, the ability of cpn60 to arrest both productive and nonproductive in vitro folding reactions results from its stable interactions with folding intermediates, which possess certain properties characteristic of the molten globule state. As a result of binary complex formation these transient, labile conformational intermediates are protected from intermolecular reaction that can lead to aggregation. Binding to cpn60 may be mediated in part through hydrophobic patches that are exposed on the surfaces of incompletely folded proteins. Only one or two nonnative proteins can bind to the cpn60$_{14}$ oligomer. Recent morphological evidence (10, 58) suggests that this binding takes place within the central cavity of the cpn60 cylinder. Although this cavity can only accommodate a globular protein of up to 60–90 kDa (10, 44), it is still possible for even larger proteins to form a stable complex with the chaperonin. For example, the interaction of the 124-kDa phytochrome polypeptide with cpn60 (37) may be a case where the binding of particular collapsed domains occurs within cpn60, instead of the entire polypeptide entering the cavity. Target proteins, sequestered within this cage-like, protected environment (85), probably establish multiple contacts with the monomeric subunits that comprise the stacked ring structures.

Release of Polypeptides from Chaperonins

The manner in which target proteins are discharged from chaperonins to yield biologically active proteins is not entirely understood. If the released target protein can refold spontaneously, reactivation merely requires its dissociation from the binary complex, in which case there is no need to invoke a folding event while the protein is associated with cpn60. However, the recovery of refolded protein is frequently greater in the presence of the chaperonins, which is the observation that must be explained. In an extreme case, it was reported that the spontaneous dissociation of lactate dehydrogenase (LDH) from its complex with cpn60 resulted in a 3-fold increase in yield of active enzyme, compared to the unassisted folding reaction (2). This release did not require adenine nucleotides or cpn10, although it was extremely slow. Apparently, cpn60 had served as a buffer for unstable LDH folding intermediates, a large proportion of which normally partition to aggregates at higher concentrations. In effect, the slow dissociation of cpn60-bound LDH had maintained a low concentration of free folding intermediates, thus minimizing aggregation, and allowing more of the protein to engage in productive folding. Similar mechanisms have been invoked to explain the assistance of folding that is observed with Hsp90 (98) and the small heat-shock proteins (50).

The spontaneous release of most cpn60-bound proteins is too slow to be of physiological importance. Even if the target protein momentarily escapes from the chaperonin, there is no assurance that it will partition to the native state before rebinding occurs. Fortunately, net release can be accelerated by certain factors, the most important of which is the combination of the cpn10 co-chaperonin and ATP. Even in the absence of cpn10, ATP and certain other adenine nucleotides effect large conformational changes in cpn60 (3, 49, 85), and in so doing, weaken its affinity for nonnative proteins. But it is difficult to generalize, because the magnitude of this effect, and indeed the relative abilities of different adenine nucleotides to induce it, varies from one target protein to the next.

Nevertheless, there are several examples showing that ATP alone can accelerate the net release of biologically active proteins from binary complexes formed between cpn60 and target proteins (2, 11, 24, 37, 52, 53, 67, 73, 86, 94). In some of these discharge reactions, even nonhydrolyzable ATP analogs were at least partially effective. Compared to spontaneous folding, there were no significant rate enhancements, indicating that cpn60 does not alter the transition state of folding (49, 94, 102). However, in most cases, there were significant increases in the recovery of folded protein, and these are explained adequately by the "buffer" effect. The influence of cpn10 on these chaperonin-assisted folding reactions depended on the target protein. In some cases it had no effect, while in others, it significantly increased either the rate, the yield, or both. Examination of the factors causing successful release in vitro of assembly-competent target proteins from the cytosolic TCP1 chaperonins, also indicates that in most cases hydrolysis of ATP is sufficient (26, 27, 75, 100). This does not exclude the existence of a TCP1-specific co-chaperonin for maximal activity in vivo, and it has been reported that two co-factors required for efficient release of tubulins from TCP1 are present in the cytosol (28).

These results indicate that cpn60 plays only a passive role in protein folding, and that there is no strict requirement for cpn10 or ATP hydrolysis. Through conformational changes that destabilize cpn60-target protein complexes, the binding of adenine nucleotides merely accelerates a process that could otherwise occur in their absence. As a result, the released target protein has more time to isomerize toward its native state before being recaptured by the chaperonin. Higher order off-pathway reactions are suppressed, resulting in a greater recovery of folded protein. In this scenario, the co-chaperonin is not obligatory, but plays a "fine-tuning" role, which in some cases may increase the efficiency of the discharge reaction.

These observations stand in glaring contrast to earlier results obtained with Rubisco (34, 97), where reactivation to the native state required ATP hydrolysis and both the cpn10 and cpn60 chaperonin components. Similar results were reported subsequently for the successful refolding of citrate synthase (12), rhodanese (67, 70), ornithine transcarbamylase (40, 104), and malate dehydrogenase (41, 71). These results imply that there are two distinct classes of target proteins, those that require cpn10 and those that do not. Surprisingly, studies with Rubisco (3, 95) and rhodanese (67), under conditions where spontaneous folding cannot occur, have shown that these target proteins can be released from cpn60 by the addition of ATP alone. However, in the absence of cpn10, the released proteins were not committed to the native state and were still susceptible to aggregation. In contrast, when cpn10 was present during the discharge reaction, the released Rubisco and rhodanese proceeded to the native state. Recent studies with ornithine transcarbamylase (104) indicate that its efficient conversion from an uncommitted to a committed state requires ATP hydrolysis in addition to cpn10. Clearly, more than a simple buffering effect is operating here. In addition to highlighting the importance of cpn10 for full chaperonin function, two obvious implications emerge: (*a*) certain folding events may occur while target proteins are in association with cpn60, and (*b*) the degree of foldedness of a released target protein may differ depending upon whether or not cpn10 is present.

A requirement, or lack of requirement, for cpn10 is not necessarily an intrinsic property of a target protein, but could also reflect the folding environment that the protein is in. Participation of the co-chaperonin might only be mandatory under non-permissive folding conditions, where spontaneous reactivation cannot occur. To test this theory, we have experimentally defined permissive and nonpermissive folding environments for Rubisco, citrate synthase, and malate dehydrogenase (M Schmidt, J Buchner, MT Todd, GH Lorimer, PV Viitanen, submitted for publication), three target proteins reportedly requiring both chaperonin proteins for their successful reactivation. As predicted, in the permissive environment, the addition of ATP alone was sufficient to release active proteins from cpn60. For Rubisco, spontaneous refolding at 25°C was dependent on the presence of Cl⁻ ions (62), thus defining a permissive condition. This peculiar ionic requirement was also evident in the discharge of catalytically active Rubisco from cpn60 by ATP alone. Interestingly, the recovery of folded Rubisco was completely unaffected by Cl⁻

ions when cpn10 was present. We conclude that some degree of folding must have occurred while Rubisco was associated with the cpn60-cpn10 complex.

Whether some of these folding events take place on cpn60 or entirely in free solution, perhaps in a protected environment (85), remains to be determined. It has been suggested that partial folding occurs on the surface of cpn60, with the target protein making multiple contacts with the subunits of the chaperonin (67). In this model, cpn10 coordinates the release of bound segments of the protein, in an ATP-dependent hydrolysis manner that permits the step-wise folding of domains. Others have suggested that through multiple cycles of ATP binding and hydrolysis, cpn60 may actually reverse certain off-pathway reactions that lead to aggregation and decrease folding yields (49). In this imaginative model, cpn60 oscillates between states with low and high affinity for nonnative proteins and was likened to a "selective and reversible protein denaturant."

Available evidence indicates that the products of the chaperonin discharge reaction are monomeric species, even in those cases where the biologically active target proteins are oligomeric (34, 37, 91, 104). Thus, it remains to be demonstrated that these molecular chaperones participate in higher order assembly reactions that result in quaternary structure. Future studies are clearly warranted on the influence of the chaperonin proteins on the reconstitution of oligomeric proteins, under conditions where the appearance of the native state exhibits second order kinetics. Finally, regarding the role of the chaperonins as "catalysts of protein folding," with the exception of Rubisco (62, 97), the chaperonin proteins do not accelerate the rates of refolding of most target proteins. This has led to the general conclusion that chaperonins do not catalyze protein folding in the traditional sense of the word (49, 62, 71, 94, 102). The rate enhancement observed with Rubisco is not understood, but may result from the suppression of slowly reversible, higher order off-pathway reactions that could be rate determining for spontaneous folding (62). Rate enhancements might also result from the suppression, or reversal (49), of certain intramolecular folding events that are not directly in line to the native state.

In summary, a number of potential mechanisms could account for the assistance of protein folding observed with chaperonin proteins. However, it is only the primary amino acid sequence of the target protein that drives the folding process and gives rise to the native state. Because of their known promiscuity and lack of substrate specificity, it is difficult to envision chaperonins directly facilitating on-pathway folding reactions. Thus, it is generally agreed that chaperonins function primarily in the suppression of higher order reactions that compete with productive folding and lead to aggregation. Much of this can be explained by the buffer effect of chaperonins, and the destabilization of chaperonin-target protein complexes by adenine nucleotides. However, this passive mechanism does not account for the profound effects frequently observed when cpn10 is also present. Because the requirement for cpn10 may be linked to the experimental conditions, the mechanistic significance of the co-chaperonin could be completely obscured in a permissive in vitro environment. Although chaperonins probably do not alter the transition state of folding, they do share with enzymes and catalysts the ability to undergo multiple turnovers (41, 57, 71). This observation is consistent with a general cellular requirement for chaperonins to facilitate efficient folding of a wide range of diverse proteins using a common reusable binding site, perhaps a protected niche within the chaperonin central cavity.

Literature Cited

1. Ang D, Liberek K, Skowyra D, Zylicz M, Georgopoulos C. 1991. Biological role and regulation of the universally conserved heat shock proteins. *J. Biol. Chem.* 266: 24233–36
2. Badcoe IG, Smith CJ, Wood S, Halsall DJ, Holbrook JJ, et al. 1991. Binding of a chaperonin to the folding intermediates of lactate dehydrogenase. *Biochemistry* 30: 9195–200
3. Baneyx F, Gatenby AA. 1992. A mutation in GroEL interferes with protein folding by reducing the rate of discharge of sequestered polypeptides. *J. Biol. Chem.* 267: 11637–44
4. Barraclough R, Ellis RJ. 1980. Protein synthesis in chloroplasts. IX. Assembly of newly-synthesized large subunits into ribulose bisphosphate carboxylase in isolated pea chloroplasts. *Biochim. Biophys. Acta* 608:19–31
5. Bertsch U, Soll J, Seetharam R, Viitanen PV. 1992. Identification, characterization, and DNA sequence of a functional "double" groES-like chaperonin from chloroplasts of higher plants. *Proc. Natl. Acad. Sci. USA* 89:8696–700
6. Bloom MV, Milos P, Roy H. 1983. Light-dependent assembly of ribulose 1,5-bisphosphate carboxylase. *Proc. Natl. Acad. Sci. USA* 80:1013–17
7. Bochkareva ES, Girshovich AS. 1992. A newly synthesized protein interacts with groES on the surface of chaperonin groEL. *J. Biol. Chem.* 267:25672–75
8. Bochkareva ES, Lissin NM, Flynn GC, Rothman JE, Girshovich AS. 1992. Positive cooperativity in the functioning of molecular chaperone GroEL. *J. Biol. Chem.* 267:6796–800
9. Bochkareva ES, Lissin NM, Girshovich AS. 1988. Transient association of newly synthesized unfolded proteins with the heat-shock GroEL protein. *Nature* 336: 254–57
10. Braig K, Simon M, Furuya F, Hainfeld JF, Horwich AL. 1993. A polypeptide bound by the chaperonin groEL is localized within a central cavity. *Proc. Natl. Acad. Sci. USA* 90:3978–82
11. Brunschier R, Danner M, Seckler R. 1993. Interactions of phage P22 tailspike-protein with groE molecular chaperones during refolding in vitro. *J. Biol. Chem.* 268:2767–72
12. Buchner J, Schmidt M, Fuchs M, Jaenicke R, Rudolph R, et al. 1991. GroE facilitates refolding of citrate synthase by suppressing aggregation. *Biochemistry* 30: 1586–91
13. Carrillo N, Ceccarelli EA, Krapp AR, Boggio S, Ferreyra RG, Viale AM. 1992. Assembly of plant ferredoxin-NADP$^+$ oxidoreductase in *Escherichia coli* requires GroE molecular chaperones. *J. Biol. Chem.* 267:15537–41
14. Chandrasekhar GN, Tilly K, Woolford C, Hendrix R, Georgopoulos C. 1986. Purification and properties of the groES morphogenetic protein of *Escherichia coli*. *J. Biol. Chem.* 261:12414–19
15. Cheng MY, Hartl F-U, Horwich AL. 1990. The mitochondrial chaperonin hsp60 is required for its own assembly. *Nature* 348: 455–58
16. Cheng MY, Hartl F-U, Martin J, Pollock RA, Kalousek F, et al. 1989. Mitochondrial heat-shock protein hsp60 is essential for assembly of proteins imported into yeast mitochondria. *Nature* 337:620–25

17. Creighton TE. 1990. Protein folding. *Biochem. J.* 270:1–16
18. Creighton TE. 1991. Unfolding protein folding. *Nature* 352:17–18
19. Ellis RJ. 1990. Molecular chaperones: the plant connection. *Science* 250:954–59
20. Ellis RJ. 1992. Cytosolic chaperonin confirmed. *Nature* 358:191–92
21. Ellis RJ, van der Vies SM. 1991. Molecular chaperones. *Annu. Rev. Biochem.* 60:321–47
22. Fayet O, Louarn J-M, Georgopoulos C. 1986. Suppression of the *Escherichia coli dnaA46* mutation by amplification of the *groES* and *groEL* genes. *Mol. Gen. Genet.* 202:435–45
23. Fayet O, Ziegelhoffer T, Georgopoulos C. 1989. The *groES* and *groEL* heat shock gene products of *Escherichia coli* are essential for bacterial growth at all temperatures. *J. Bacteriol.* 171:1379–85
24. Fisher MT. 1992. Promotion of the *in vitro* renaturation of dodecameric glutamine synthetase from *Escherichia coli* in the presence of GroEL (chaperonin-60) and ATP. *Biochemistry* 31:3955–63
25. Fisher MT. 1993. On the assembly of dodecameric glutamine synthetase from stable chaperonin complexes. *J. Biol. Chem.* 268: 13777–79
26. Frydman J, Nimmesgern E, Erdjument-Bromage H, Wall JS, Tempst P, Hartl F-U. 1992. Function in protein folding of TRiC, a cytosolic ring complex containing TCP-1 and structurally related subunits. *EMBO J.* 11:4767–78
27. Gao Y, Thomas JO, Chow RL, Lee G-H, Cowan NJ. 1992. A cytoplasmic chaperonin that catalyzes β-actin folding. *Cell* 69:1043–50
28. Gao Y, Vainberg IE, Chow RL, Cowan NJ. 1993. Two cofactors and cytoplasmic chaperonin are required for the folding of α- and β-tubulin. *Mol. Cell. Biol.* 13:2478–85
29. Gatenby AA. 1992. Protein folding and chaperonins. *Plant Mol. Biol.* 19:677–87
30. Gatenby AA, Ellis RJ. 1990. Chaperone function: the assembly of ribulose bisphosphate carboxylase-oxygenase. *Annu. Rev. Cell Biol.* 6:125–49
31. Gatenby AA, Lubben TH, Ahlquist P, Keegstra K. 1988. Imported large subunits of ribulose bisphosphate carboxylase/oxygenase, but not imported β-ATP synthase subunits, are assembled into holoenzyme in isolated chloroplasts. *EMBO J.* 7:1307–14
32. Gatenby AA, Viitanen PV, Lorimer GH. 1990. Chaperonin assisted polypeptide folding and assembly: implications for the production of functional proteins in bacteria. *Trends Biotechnol.* 6:95–101
33. Gatenby AA, Viitanen PV, Speth V, Grimm R. 1994. Identification, cellular localization and participation of chaperonins in protein folding. In *Molecular Processes of Photosynthesis*, ed. J Barber. Greenwich, CT: JAI. In press
34. Goloubinoff P, Christeller JT, Gatenby AA, Lorimer GH. 1989. Reconstitution of active dimeric ribulose bisphosphate carboxylase from an unfolded state depends on two chaperonin proteins and Mg-ATP. *Nature* 342:884–89
35. Goloubinoff P, Gatenby AA, Lorimer GH. 1989. GroE heat-shock proteins promote assembly of foreign prokaryotic ribulose bisphosphate carboxylase oligomers in *Escherichia coli. Nature* 337:44–47
36. Gray TE, Fersht AR. 1991. Cooperativity in ATP hydrolysis by GroEL is increased by GroES. *FEBS Lett.* 292:254–58
37. Grimm R, Donaldson G, van der Vies SM, Schäfer E, Gatenby AA. 1993. Chaperonin-mediated reconstitution of the phytochrome photoreceptor. *J. Biol. Chem.* 268: 5220–26
38. Grimm R, Speth V, Gatenby AA, Schäfer E. 1991. GroEL-related molecular chaperones are present in the cytosol of oat cells. *FEBS Lett.* 286:155–58
39. Gupta RS. 1990. Sequence and structural homology between a mouse T-complex protein TCP-1 and the 'chaperonin' family of bacterial (GroEL, 60–65 kDa heat shock antigen) and eukaryotic proteins. *Biochem. Int.* 20:833–41
40. Hartman DJ, Hoogenraad NJ, Condron R, Hoj PB. 1992. Identification of a mammalian 10-kDa heat shock protein, a mitochondrial chaperonin 10 homologue essential for assisted folding of trimeric ornithine transcarbamoylase *in vitro. Proc. Natl. Acad. Sci. USA* 89:3394–98
41. Hartman DJ, Surin BP, Dixon NE, Hoogenraad NJ, Hoj PB. 1993. Substoichiometric amounts of the molecular chaperones groEL and groES prevent thermal denaturation and aggregation of mammalian mitochondrial malate dehydrogenase in vitro. *Proc. Natl. Acad. Sci. USA* 90:2276–80
42. Hemmingsen SM, Ellis RJ. 1986. Purification and properties of ribulosebisphosphate carboxylase large subunit binding protein. *Plant Physiol.* 80:269–76
43. Hemmingsen SM, Woolford C, van der Vies SM, Tilly K, Dennis DT, et al. 1988. Homologous plant and bacterial proteins chaperone oligomeric protein assembly. *Nature* 333:330–34
44. Hendrick JP, Hartl F-U. 1993. Molecular chaperone functions of heat-shock proteins. *Annu. Rev. Biochem.* 62:349–84
45. Hendrix RW. 1979. Purification and properties of groE, a host protein involved in bacteriophage assembly. *J. Mol. Biol.* 129: 375–92
46. Höll-Neugebauer B, Rudolph R, Schmidt

M, Buchner J. 1991. Reconstitution of a heat shock effect in vitro: influence of GroE on the thermal aggregation of α-glucosidase from yeast. *Biochemistry* 30:11609–614

47. Hutchinson EG, Tichelaar W, Hofhaus G, Weiss H, Leonard KR. 1989. Identification and electron microscopic analysis of a chaperonin oligomer from *Neurospora crassa* mitochondria. *EMBO J.* 8:1485–90

48. Ishii N, Taguchi H, Sumi M, Yoshida M. 1992. Structure of holo-chaperonin studied with electron microscopy: oligomeric cpn10 on top of two layers of cpn60 rings with two stripes each. *FEBS Lett.* 299:169–74

49. Jackson GS, Staniforth RA, Halsall DJ, Atkinson T, Holbrook JJ, et al. 1993. Binding and hydrolysis of nucleotides in the chaperonin catalytic cycle: implications for the mechanism of assisted protein folding. *Biochemistry* 32:2554–63

50. Jacob U, Gaestel M, Engel K, Buchner J. 1993. Small heat shock proteins are molecular chaperones. *J. Biol. Chem.* 268:1517–20

51. Jaenicke R. 1991. Protein folding: local structures, domains, subunits and assemblies. *Biochemistry* 30:3147–61

52. Kern G, Schmidt M, Buchner J, Jaenicke R. 1992. Glycosylation inhibits the interaction of invertase with the chaperone groEL. *FEBS Lett.* 303:203–5

53. Laminet AA, Ziegelhoffer T, Georgopoulos C, Plückthun A. 1990. The *Escherichia coli* heat shock proteins GroEL and GroES modulate the folding of the β-lactamase precursor. *EMBO J.* 9:2315–19

54. Landry SJ, Gierasch LM. 1991. The chaperonin GroEL binds a polypeptide in an α-helical conformation. *Biochemistry* 30:7359–62

55. Landry SJ, Jordan R, McMacken R, Gierasch LM. 1992. Different conformations for the same polypeptide bound to chaperones DnaK and GroEL. *Nature* 355:455–57

56. Landry SJ, Zeilstra-Ryalls J, Fayet O, Georgopoulos C, Gierasch LM. 1993. Characterization of a functionally important mobile domain of GroES. *Nature* 364:255–58

57. Langer T, Lu C, Echols H, Flanagan J, Hayer MK, Hartl F-U. 1992. Successive action of DnaK, DnaJ and GroEL along the pathway of chaperone-mediated protein folding. *Nature* 356:683–89

58. Langer T, Pfeifer G, Martin J, Baumeister W, Hartl F-U. 1992. Chaperonin-mediated protein folding: GroES binds to one end of the GroEL cylinder, which accommodates the protein substrate within its central cavity. *EMBO J.* 11:4757–65

59. Lecker S, Lill R, Ziegelhoffer T, Georgopoulos C, Bassford PJ, et al. 1989. Three pure chaperone proteins of *Escherichia coli*—SecB, trigger factor and GroEL—form soluble complexes with precursor proteins in vitro. *EMBO J.* 8:2703–9

60. Lewis V, Hynes GM, Zheng D, Saibil H, Willison K. 1992. T-complex polypeptide-1 is a subunit of a heteromeric particle in the eukaryotic cytosol. *Nature* 358:249–52

61. Lissin NM, Venyaminov SY, Girshovich AS. 1990. (Mg-ATP)-dependent self-assembly of molecular chaperone GroEL. *Nature* 348:339–42

62. Lorimer GH, Todd MJ, Viitanen PV. 1993. Chaperonins and protein folding: unity and disunity of mechanisms. *Philos. Trans. R. Soc. London Ser. B* 339:297–304

63. Lubben TH, Donaldson GK, Viitanen PV, Gatenby AA. 1989. Several proteins imported into chloroplasts form stable complexes with the GroEL-related chloroplast molecular chaperone. *Plant Cell* 1:1223–30

64. Lubben TH, Gatenby AA, Donaldson GK, Lorimer GH, Viitanen PV. 1990. Identification of a groES-like chaperonin in mitochondria that facilitates protein folding. *Proc. Natl. Acad. Sci. USA* 87:7683–87

65. Manning-Krieg UC, Scherer PE, Schatz G. 1991. Sequential action of mitochondrial chaperones in protein import into the matrix. *EMBO J.* 10:3273–80

66. Martel R, Cloney LP, Pelcher LE, Hemmingsen SM. 1990. Unique composition of plastid chaperonin-60: α and β polypeptide-encoding genes are highly divergent. *Gene* 94:181–87

67. Martin J, Langer T, Boteva R, Schramel A, Horwich AL, Hartl F-U. 1991. Chaperonin-mediated protein folding at the surface of groEL through a 'molten globule'-like intermediate. *Nature* 352:36–42

68. McMullin TW, Hallberg RL. 1988. A highly evolutionarily conserved mitochondrial protein is structurally related to the protein encoded by the *Escherichia coli* groEL gene. *Mol. Cell. Biol.* 8:371–80

69. Mendoza JA, Lorimer GH, Horowitz PM. 1991. Intermediates in the chaperonin-assisted refolding of rhodanese are trapped at low temperature and show a small stoichiometry. *J. Biol. Chem.* 266:16973–76

70. Mendoza JA, Rogers E, Lorimer GH, Horowitz PM. 1991. Chaperonins facilitate the in vitro folding of monomeric mitochondrial rhodanese. *J. Biol. Chem.* 266:13044–49

71. Miller AD, Maghlaoui K, Albanese G, Kleinjan DA, Smith C. 1993. *Escherichia coli* chaperonins cpn60 (groEL) and cpn10 (groES) do not catalyse the refolding of mitochondrial malate dehydrogenase. *Biochem. J.* 291:139–44

72. Miller SG, Leclerc RF, Erdos GW. 1990.

Identification and characterization of a testis-specific isoform of a chaperonin in a moth, *Heliothis virescens. J. Mol. Biol.* 214:407–22

73. Mizobata T, Akiyama Y, Ito K, Yumoto N, Kawata Y. 1992. Effects of the chaperonin groE on the refolding of tryptophanase from *Escherichia coli. J. Biol. Chem.* 267: 17773–79

74. Mori M, Murata K, Kubota H, Yamamoto A, Matsushiro A, Motita T. 1992. Cloning of a cDNA encoding the *Tcp-1* (*t* complex polypeptide 1) homologue of *Arabidopsis thaliana. Gene* 122:381–82

75. Mummert E, Grimm R, Speth V, Eckerskorn C, Schiltz E, et al. 1993. A TCP1-related molecular chaperone from plants refolds phytochrome in its photoreversible form. *Nature* 363:644–47

76. Musgrove JE, Johnson RA, Ellis RJ. 1987. Dissociation of the ribulosebisphosphate-carboxylase large-subunit binding protein into dissimilar subunits. *Eur. J. Biochem.* 163:529–34

77. Ostermann J, Horwich AL, Neupert W, Hartl F-U. 1989. Protein folding in mitochondria requires complex formation with hsp60 and ATP hydrolysis. *Nature* 341: 125–30

78. Phipps BM, Typke D, Hegerl R, Volker S, Hoffmann A, et al. 1993. Structure of a molecular chaperone from a thermophilic archaebacterium. *Nature* 361:475–77

79. Picketts DJ, Mayanil CSK, Gupta RS. 1989. Molecular cloning of a Chinese hamster mitochondrial protein related to the "chaperonin" family of bacterial and plant proteins. *J. Biol. Chem.* 264:12001–8

80. Prasad TK, Hack E, Hallberg RL. 1990. Function of the maize mitochondrial chaperonin hsp60: specific association between hsp60 and newly synthesized F1-ATPase alpha subunits. *Mol. Cell. Biol.* 10:3979–86

81. Pushkin AV, Tsuprun VL, Solovjeva NA, Shubin VV, Evstigneeva ZG, Kretovich WL. 1982. High molecular weight pea leaf protein similar to the groE protein of *Escherichia coli. Biochim. Biophys. Acta* 704:379–84

82. Reading DS, Hallberg RL, Myers AM. 1989. Characterization of the yeast HSP60 gene coding for a mitochondrial assembly factor. *Nature* 337:655–59

83. Roy H, Bloom M, Milos P, Monroe M. 1982. Studies on the assembly of large subunits of ribulose bisphosphate carboxylase in isolated pea chloroplasts. *J. Cell Biol.* 94:20–27

84. Saibil H, Dong Z, Wood S, Auf der Mauer A. 1991. Binding of chaperonins. *Nature* 353:25–26

85. Saibil HR, Zheng D, Roseman AM, Hunter AS, Watson GM, et al. 1993. ATP induces large quaternary rearrangements in a cage-like chaperonin structure. *Curr. Biol.* 3: 265–73

86. Schmidt M, Buchner J. 1992. Interaction of GroE with an all-β-protein. *J. Biol. Chem.* 267:16829–33

87. Taguchi H, Konishi J, Ishii N, Yoshida M. 1991. A chaperonin from a thermophilic bacterium, *Thermus thermophilus,* that controls refoldings of several thermophilic enzymes. *J. Biol. Chem.* 266:22411–18

88. Terlesky KC, Tabita FR. 1991. Purification and properties of the chaperonin 10 and chaperonin 60 proteins from *Rhodobacter sphaeroides. Biochemistry* 30:8181–86

89. Todd MJ, Viitanen PV, Lorimer GH. 1993. The hydrolysis of ATP by *Escherichia coli* groEL: the effects of groES and K⁺. *Biochemistry* 32:8560–67

90. Trent JD, Nimmesgern E, Wall JS, Hartl F-U, Horwich AL. 1991. A molecular chaperone from a thermophilic archaebacterium is related to the eukaryotic protein t-complex polypeptide-1. *Nature* 354:490–93

91. van der Vies SM, Viitanen PV, Gatenby AA, Lorimer GH, Jaenicke R. 1992. Conformational states of ribulose bisphosphate carboxylase and their interaction with chaperonin 60 (GroEL). *Biochemistry* 31: 3635–44

92. Van Dyk TK, Gatenby AA, LaRossa RA. 1989. Demonstration by genetic suppression of interaction of GroE products with many proteins. *Nature* 343:451–53

93. Vierling E. 1991. The roles of heat-shock proteins in plants. *Annu. Rev. Plant Physiol. Plant Mol. Biol.* 42:579–620

94. Viitanen PV, Donaldson GK, Lorimer GH, Lubben TH, Gatenby AA. 1991. Complex interactions between the chaperonin 60 molecular chaperone and dihydrofolate reductase. *Biochemistry* 30:9716–23

95. Viitanen PV, Gatenby AA, Lorimer GH. 1992. Purified chaperonin 60 (groEL) interacts with the nonnative states of a multitude of *Escherichia coli* proteins. *Protein Sci.* 1:363–69

96. Viitanen PV, Lorimer GH, Seetharam R, Gupta RS, Oppenheim J, et al. 1992. Mammalian mitochondrial chaperonin 60 functions as a single toroidal ring. *J. Biol. Chem.* 267:695–98

97. Viitanen PV, Lubben TH, Reed J, Goloubinoff P, O'Keefe DP, Lorimer GH. 1990. Chaperonin-facilitated refolding of ribulosebisphosphate carboxylase and ATP hydrolysis by chaperonin 60 (groEL) are K⁺ dependent. *Biochemistry* 29:5665–71

98. Wiech H, Buchner J, Zimmermann R, Jakob U. 1992. Hsp90 chaperones protein folding in vitro. *Nature* 358:169–70

99. Wynn RM, Davie JR, Cox RP, Chuang DT. 1992. Chaperonins GroEL and GroES promote assembly of heterotetramers (α2β2)

of mammalian mitochondrial branched-chain α-keto acid decarboxylase in *Escherichia coli. J. Biol. Chem.* 267:12400–3

100. Yaffe MB, Farr GW, Miklos D, Horwich AL, Sternlicht ML, Sternlicht H. 1992. TCP1 complex is a molecular chaperone in tubulin biogenesis. *Nature* 358:245–48

101. Yoshida M, Ishii N, Muneyuki E, Taguchi H. 1993. A chaperonin from a thermophilic bacterium *Thermus thermphilus. Phil. Trans. R. Soc. London* 339:305–12

102. Zahn R, Plückthun A. 1992. GroE prevents the accumulation of early folding interme-

diates of pre-β-lactamase without changing the folding pathway. *Biochemistry* 31: 3249–55

103. Zeilstra-Ryalls J, Fayet O, Georgopoulos C. 1991. The universally conserved GroE (Hsp60) chaperonins. *Annu. Rev. Microbiol.* 45:301–25

104. Zheng X, Rosenberg LE, Kalousek F, Fenton WA. 1993. groEL, groES, and ATP-dependent folding and spontaneous assembly of ornithine transcarbamylase. *J. Biol. Chem.* 268:7489–93

Annu. Rev. Plant Physiol. Plant Mol. Biol. 1994. 45:493–526

THE USE AND CHARACTERISTICS OF THE PHOTOACOUSTIC METHOD IN THE STUDY OF PHOTOSYNTHESIS

Shmuel Malkin

Biochemistry Department, The Weizmann Institute of Science, Rehovot, 76100 Israel

Ora Canaani

Department of Microbiology, Jefferson University, Philadelphia, Pennsylvania 19107 and Department of Biology, Temple University, Philadelphia, Pennsylvania 19140

KEY WORDS: photoacoustics, photosynthesis, energetics, oxygen evolution and uptake

CONTENTS

INTRODUCTION 494
   Photoacoustic and Photothermal Methods and their Place in
      Photosynthesis Research 494
   Historical Notes 495
   Other Applications of the PA Method 497
PA SIGNAL GENERATION AND UTILIZATION USING THE GAS-COUPLED
   MICROPHONE ARRANGEMENT 497
   Experimental 497
   The Photothermal Signal in Periodically Modulated Light 501
   The Modulated Photobaric Signal 505
   Photothermal and Photobaric PA Signals in the Time Domain 509
PA SIGNALS IN DIRECTLY-COUPLED-DETECTOR CONFIGURATION 510
   Experimental 510
RESULTS OF PHOTOSYNTHETIC ACTIVITIES 512
   Energy Storage 512
   Photobaric Measurements 516
PHOTOSYNTHESIS PHYSIOLOGY STUDIES 516
   Photosynthesis Transients 516

Light Distribution Between the Two Photosystems: State 1–State 2
  Transitions 518
"Low Light" State 520
The Effect of Possible Stromal Sinks 520
Indications for Independent Activities of PS I In Vivo 521
Stress Physiology and Ecology 521
EPILOGUE 522

INTRODUCTION

Photoacoustic and Photothermal Methods and their Place in Photosynthesis Research

Photoacoustics is a research technique that uses the ability of periodically modulated or pulsed light to produce sound upon absorption. This effect is commonly explained by the transformation of light energy to heat, which correspondingly is also periodically modulated or pulsed. The resulting temperature modulation leads to cycles of expansion and contraction, which excite sound waves. This mechanism is therefore called *photothermal* (93).

Although the photothermal mechanism occurs universally, other mechanisms contribute to the photoacoustic (PA) signal in special cases. In photosynthetic systems there are two such mechanisms: (*a*) volume changes, resulting directly from the change in the molecular packing following photochemical reactions (2, 72a); and (*b*) gas evolution (or uptake), which produces pressure changes by rapid injection (or elimination) of gaseous material in a restricted volume. The latter mechanism, called *photobaric* (89, but cf 76), occurs in leaves and lichens, where the gas phase is adjacent to the photosynthesizing cell (6, 26), and it commonly involves oxygen evolution, although oxygen uptake through photosystem I (PS I) (Mehler reaction) also may be superimposed. The contribution of each mechanism to the photoacoustic signal can be separated out, as will be explained later. Each contribution conveys independent pieces of information about the photosynthetic process. The photothermal contribution provides information on the extent of light conversion to heat and, indirectly, on the amount of energy stored in the intermediates and products of the chemical reaction. Volume changes indicate structural changes. The photobaric signal is a sensitive and rapid analytical handle for photosynthetic activity of leaves and lichens, monitoring separately the photosystem II (PS II) activity by oxygen evolution and PS I activity by oxygen uptake. These characteristics make PA a unique, noninvasive method to study photosynthesis, particularly with special advantages for the study of leaf photosynthesis (69).

There are other means to detect the photothermal effect alone, based on measurements of temperature changes. For example, in photothermal radiometry (PTR), the modulated heat is monitored by the corresponding modulation in the amount of infrared radiation emitted from the sample, sensed by an infrared detector (65). Another example is the use of the spatial and temporal refractive index changes, caused by the heat modulation. These may be probed by a weak laser beam, using the resulting deviations in the beam direction [photothermal beam deflection (PBD) or "mirage" method] (61). These methods have been applied in photosynthesis infrequently, but they are useful for determining the photothermal contribution to the total PA signal (65).

There are several advantages of using PA (as well as the alternative photothermal methods), because the response is a modulated signal: 1. Only signals synchronized to the light modulation are detected; unwanted signals and ambient noise are eliminated, including non-modulated responses (e.g. respiration, CO_2 fixation, and photorespiration, which are presumably not modulated in the frequency range used for PA measurements). PA signals therefore reflect gross photosynthesis. 2. It is possible to superimpose an additional non-modulated (background) light excitation, which by itself cannot produce any signal, for changing the physiological conditions during measurement. 3. It is possible to obtain kinetic and energetic information on the various reaction steps, following the primary photoact, either from the modulation frequency dependence, in the case of periodically modulated excitation (frequency-domain measurements), or from the signal pulse time dependence, for pulsed excitation (time-domain measurements).

Historical Notes

The PA effect was discovered in solid samples at the end of the nineteenth century. The systematic development of PA as an ordered discipline came about thirty years later with studies of gases into which a microphone was embedded, and wherein sound generation occurred momentarily at each locus where light was absorbed, by local thermal expansion (93). Years later the photoacoustic effect was "rediscovered" in a condensed phase sample, where the sound was generated and detected in the surrounding gas phase (93). A theoretical treatment of the photothermal mechanism, in this case, considering that the modulated heat has to reach the surface before its transduction to sound, showed the existence of temperature modulation wavelets, which carry the modulated heat flux but which are exponentially attenuated along their propagation distance. Only that part of the modulated light energy, absorbed in an optical path below the surface, which is within the effective range of a thermal wave (usually only several microns), can be expressed as an acoustic signal (10, 78, 93). PA is thus essentially a surface phenomenon. Based on the assumption that light energy is converted fully and rapidly to heat, absorption spectra were constructed (1, 9, 93) by scanning PA signals vs the wavelength at a constant incident power. PA spectroscopy was found to be more effective than the conventional transmission spectroscopy for extreme cases of either nearly transparent or opaque samples, or in samples that have large surface to volume ratios, and that are highly light-scattering such as blood smears, algal layers, and leaves (1, 93). Other photothermal methods for studying such samples also were developed (93).

The PA method was introduced to study photosynthesis because it was thought (71) that the PA signal would reflect only a fractional conversion of light energy to heat, resulting from energy storage (ES), which therefore could be quantified. This required a comparison to a signal from a reference in which the conversion of the light to heat was complete. After some unsatisfactory attempts based on inhibition treatment (13), a reference signal was created reversibly by adding non-modulated photosynthetically saturating background light to the studied sample. This strategy worked well in aqueous suspensions or in water-soaked samples, leading to quantitative estimation of the energy storage (68, 74), and it was adopted by other workers. In leaves, however, the strategy failed and background light led to unexpected effects. It was soon recognized that the PA signals in leaves are more complex. In fact, an earlier study, which monitored transients in the PA signal from a leaf (63), and which considered the existence of the photothermal mechanism only, could only explain them by assuming (incorrectly) that the physical thermal parameters of the leaf vary in time. This was resolved by the discovery that in air-filled structures (leaves and lichens), the PA signal at low modulation frequencies has a significant contribution from oxygen evolution (6). Using a single pulse excitation, the two components of the PA signal in leaves could also be seen and resolved in the time domain (24, 67, 76). Using a train of a few saturating flashes, S-state phenomena could be observed (24, 76). Contributions to the PA signal having a negative sign were seen in special cases, either in the frequency or time domains, and were tentatively assigned to oxygen uptake (49, 70).

A different research line used a set-up in which a microphone was directly inserted into the aqueous suspension sample, to serve as a volume change sensor in the time domain (14). This led to new results: At a temperature where the thermal expansion coefficient is zero ($\sim 4°C$) and no photothermal expansion was expected, a negative signal, indicating shrinkage, accompanied the electron transfer from P_{870} to Q_A. (2, 15). At any other temperature, the resulting signal was resolved into the above volume change and the usual photothermal expansion, from which energy storage was calculated. Unfortunately, there was a major inconsistency, still unresolved, between these results and those obtained from delayed-luminescence measurements (2a). Using the same principle of a direct coupling between the sensor and the sample, a method known as "laser induced optoacoustic spectroscopy" (LIOAS) (3, 4) was developed using very short (ns) laser flashes, which created sound pulses within the sample. These sound pulses were detected by piezoelectric sensors attached directly to the sample, or through the cuvette walls, which allowed a time resolution of less than a μs. Time-domain PA signals were measured from suspensions of bacterial cells, chromatophores, PS I and PS II particles (81, 82), and leaves submerged in water (64). ES was calculated, assuming the photothermal mechanism alone, whereas there was no awareness that volume changes could take place. Volume changes were considered in a recent LIOAS work (72a).

This review intends to critically examine the various uses of the PA method in the study of photosynthesis. Figure 1 presents the basic experimental set-up for both PA methods discussed here. The interested reader should also consult previous reviews (3, 10, 69, 77; see especially 42). The general use of PA in chemistry and biochemistry, for energetic and kinetic aspects, is discussed in reference 87.

Other Applications of the PA Method

The PA method can be used in non-invasive plant pigment spectroscopy, considering the photothermal mechanism alone. This application, in particular, includes depth profile analysis, performed by changing the modulation frequency and thus the effective range of the thermal waves. The location of pigments down the light path can be evaluated from the change of the spectrum vs the frequency (1, 9, 10, 77, 80, 85, 93). Generally, however, the old PA spectra from leaves were mostly imprecise because they were obtained under conditions in which oxygen evolution contributed appreciably and energy storage (ES) modified the photothermal contribution, and both oxygen evolution and ES may be wavelength dependent. (With regard to the depth profile analysis, cf below for an alternative explanation for the change of the PA spectrum vs frequency.)

Gas-phase PA can be used to detect trace quantities (in the ppb range) of important compounds (e.g. ethylene, ammonia, water vapor, and CO_2) in situ (48, 84). This is a useful spectroscopic application, but it is outside the scope of this review.

PA SIGNAL GENERATION AND UTILIZATION USING THE GAS-COUPLED MICROPHONE ARRANGEMENT

Experimental

The experimental set-up includes light sources and optics, which produce beams of modulated or pulsed light and constant (background) light. Light guide optics is often used for light delivery and mixing (6, 77). Light is modulated with a chopper or electronic modulation of the power supply (67, 94). In the latter case, several modulation frequencies can be mixed, which

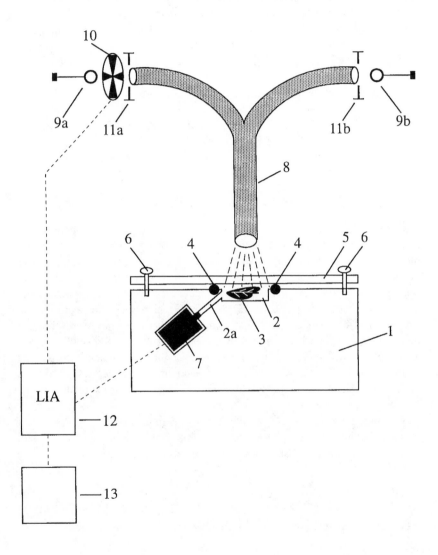

Figure 1a Schematic drawing of the PA cell with gas-coupled sensor and the basic elements of the experimental set-up. 1. cell body; 2. PA cavity; 2a. connecting channel from PA cavity to microphone; 3. sample; 4. O-ring; 5. cover window; 6. clamping screws; 7. microphone; 8. light-guide; 9a. modulated light source; 9b. background light source; 10. chopper; 11a and b. shutters; 12. lock-in amplifier; 13. recorder (or data acquisition system).

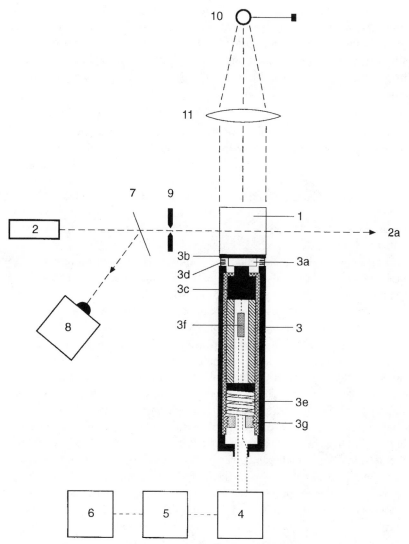

Figure 1b   Schematic drawing of the PA cell with directly coupled sensor in the LIOAS mode. 1.
spectrophotometric cell with sample; 2. laser; 2a. laser beam propagation path; 3. piezoelectric
sensor assembly; 3a. sensor; 3b. metallic foil for electrical connection with the upper side of the
sensor; 3d. metallic spring; 3e. main spring to press the sensor for a good coupling with the cuvette
wall; 3f. preamplifier; 3g. screwed base to support and press the spring; 4. amplifier; 5. digital
oscilloscope; 6. data acquisition system; 7. beam splitter (to divert part of the laser beam to the energy
meter); 8. laser energy meter; 9. diaphragm (pinhole); 10. background light source; 11. optics for
collimating the background light onto the sample. [Solid black] = metallic parts; [diagonal pattern]
= electrically insulating material; [cross-hatch] = insulating envelope.

allows a simultaneous separated detection for each frequency in the resulting signal. This is advantageous for leaf measurements because it allows simultaneous measurement of energy storage and oxygen evolution in one experiment (94). Strong pulsed light is provided from a xenon flash lamp or a laser. Weaker pulses and periodically modulated light can be obtained convenienly from light emitting diodes (LED), which also can be incorporated into the cell structure (94).

THE GAS (AIR)-COUPLED MICROPHONE PA CELL   Different PA cell contructions are described in the literature (e.g. 6, 12, 39, 77, 93, 94, 96); however, their basic principle of construction is similar. The volume of the PA cell cavity (where the sample is placed) is small (typical area ≈ 1 cm^2 and thickness ≈ 1 mm). A microphone is coupled to the cavity through a small air channel (12, 39, 93), and the cavity is tightly enclosed with a transparant window, to secure acoustic insulation. The entire PA cell must be placed so as to minimize ground vibrations (Figure 1a).

The small air volume around the sample, needed for the sensitivity, is problematic for a leaf study that requires a defined atmosphere, because CO_2 is rapidly depleted by photosynthesis. Nonetheless, the obtained PA signals indicate efficient photosynthesis, which can be measured continuously for hours. There are two tentative possible explanations: 1. The PA signals still reflect gross photosynthesis at the CO_2 compensation point. Although at this point the time averaged rates of gross photosynthesis and respiration are balanced, with no net average change in the concentrations of the participating compounds, the modulations of the photosynthetic apparatus, caused by the modulated light, persist. 2. The PA signals reflect electron transport activity involving abundant acceptors like oxygen. To counter these problems, the following approaches have been taken: (a) using a leaf piece with a periphery exposed to the outside (70), so CO_2 can be carried by lateral diffusion from the exposed part; (b) replenishing CO_2 by valve-controlled streaming through the PA cavity (91), during the time between measurements (the signal is too noisy if measured during streaming); and (c) using a CO_2 permeable membrane, supported on a sintered-glass disc for sealing the PA chamber, which allows CO_2 to enter by diffusion from continuous streaming above the chamber (41).

Flat, thin samples (e.g. leaves or algal thalli), with high surface concentration of pigments, are most suitable for this mode of PA measurement. The use of aqeous suspensions is somewhat problematic. For a good signal to noise ratio, very high concentrations (e.g. a few hundred mg chl/ml) were used in early experiments (13, 68, 74), but this caused problems of manipulation, reproducibility, and cleanliness. Better results were obtained using such samples on a solid support [e.g. layering on a filter paper (17), vacuum filtration onto a millipore filter (28, 36), or aspiration onto a thin nitrocellulose filter (30–35)].

REPRESENTATION OF THE PERIODICALLY MODULATED PA SIGNAL   Conventional measurements are made with periodically modulated light excitation. The resulting alternating electrical current from the microphone is analyzed by a lock-in amplifier (LIA), which selects signals at exactly the same frequency as that of the light modulation, rejecting irrelevant signals and random noise (10, 83, 89). Measurements refer to the fundamental mode (in a Fourier series representation) of all the periodic signals. The light modulation is represented mathematically by $i = i_0 \sin(2\pi f t)$, where i_0 is the amplitude, f is the frequency, and t is the time. For synchronization, the LIA uses a reference signal from the light modulating device and in response, an internal reference, r, is generated within the LIA, represented by $r = r_0 \sin(2\pi f t - \varepsilon)$, where ε, the phase angle, can be adjusted arbitrarily. The microphone signal, represented by $S = S_0 \sin(2\pi f t - [\delta - \varepsilon]) = S_I \sin(2\pi f t - [\delta - \varepsilon]) + S_Q \cos(2\pi f t - [\delta - \varepsilon])$, is measured by the LIA, which can give out simultaneously either the amplitude, S_0, and the phase shift, $[\delta - \varepsilon]$, with respect to the reference, or alternatively, the amplitudes of the components S_I and S_Q, termed *in-phase* and *quadrature,* respectively. These exhibited parameters may vary slowly (relative to $1/f$) on a physiological time scale. The relation between these two optional representations is $S_I = S_0 \cos(\delta - \varepsilon)$ and $S_Q = S_0 \sin(\delta - \varepsilon)$, or conversely, $S = [S_I^2 + S_Q^2]^{1/2}$ and $(\delta - \varepsilon) = \text{arctg} [S_I/S_Q]$. Mathematically, the result of the LIA measurement is a two-dimensional mathematical vector (expressed by the amplitude S_0 and direction angle $[\delta - \varepsilon]$, or by its two components $[S_I; S_Q]$. This is important when different contributions combine to give a total signal (e.g. photothermal and photobaric), which must be added as vectors (83, 89).

TIME DOMAIN MEASUREMENTS   An alternative measurement involves using a brief pulse of light. In response, a sound pulse is generated, which is recorded in the time domain (e.g. using an oscilloscope) (24, 36, 67, 76, 91). Averaging of several pulse responses might be needed for a good signal/noise ratio (36, 67, 91).

The Photothermal Signal in Periodically Modulated Light

MECHANISM   The theoretical description of the photothermal effect is as follows (10, 88, 93): Light impinges from above the surface of the sample and is absorbed according to Beer-Lambert Law. Diffusion of the produced heat to the surface establishes a partly modulated temperature field, $T = T (x,y,z,t)$, where T is the modulated temperature part and x,y,z are the spatial coordinates. Modulated heat transfer from the surface to the nearby gas layer causes its periodic expansion and contraction, exciting, in turn, a sound wave through the bulk gas. For a mathematical description, the conventional heat propagation equation is used, including the continuous distribution of heat sources, according to the rate of light-energy absorption at any particular location. The main parameter in this equation is the thermal diffusivity (D_{TH}), equal to $\kappa/(\rho C)$, where κ is the heat conductivity, ρ is the density, and C is the specific heat (subscript TH is used for thermal). When heat is produced promptly after light absorption, the modulated heat flux from the sources is given by: $\sigma N h \nu i_0 \sin(2\pi f t)$, where N is Avogadro number, h is Planck's constant, ν is the electromagnetic frequency, i_0 is expressed in molar (einstein) units, and σ is the fraction of light absorption per unit optical path. The same heat conduction equation can be used for the modulated part only, which follows:

$$\partial T/\partial t = D_{TH}\left[\partial^2 T/\partial x^2 + \partial^2 T/\partial y^2 + \partial^2 T/\partial z^2\right] + \left[(\sigma N h\nu\, i_0)/C\rho\right]\sin(2\pi f t) \quad 1$$

Although Equation 1 is over-simplified, considering the heterogeneity and the complex optics of the biological sample, there are some general characteristics of the solution: T oscillates at the frequency f, with varying amplitude and phase. It can be viewed as a superposition of elementary wavelets (thermal waves), propagating away from each local heat source element. The thermal waves have the peculiar property that their amplitude decreases exponentially as a function of the distance, d, traveled from their source (i.e. they are proportional to $\exp(-d/\mu_{TH})$, where μ_{TH} is a characteristic attenuation distance, called the *thermal diffusion length* and is given by $\mu_{TH} = (D_{TH}/\pi f)^{1/2}$. μ_{TH} expresses the range of the thermal wave's effectiveness (and is also equal to the wavelength), usually in the order of microns (e.g. $\mu_{TH} = 38$ μm at 30 Hz in water). The surficial T, hence the PA signal, is contributed by thermal waves arriving from a small depth below the surface, in the order of μ_{TH}. By calculation, the PA signal in an opaque thick sample is approximately proportional to the percent of light absorption within a distance equal to μ_{TH} (74). Varying μ_{TH}, through f, allows measurement of light absorption spectra for different depths, probing possible non-uniform pigment distribution (depth profile analysis) (10, 77, 93). PA saturation is reached in an opaque sample as f decreases, corresponding to near total light absorption and a loss in spectral resolution (see 74; 93, fig. 9.2 for a pictorial description of thermal waves).

PHOTOCALORIMETRY   The most important use of photothermal methods in photosynthesis research is photocalorimetry—i.e. comparing signals resulting from different heat emission rates and then determining energy (enthalpy) values of intermediates and products. To determine the proportionality factor between the photothermal signal and the emitted heat, a reference measurement must be obtained, for which the conversion of the light energy is known. The compared sample and reference must have exactly the same thermal and optical parameters, so that the relation between their signal vectors is the same as between the corresponding rates of modulated heat generation. Another requirement is that the sample's photochemical properties must be homogeneous throughout the sample.

The kinetics of heat release is an important consideration in photocalorimetric measurements. For all the kinetic stages with a rate constant much larger than the modulation frequency, the associated heat release is considered as prompt. This part of the heat release will closely follow the light modulation, while the heat release from the reaction intermediates of much longer lifetimes ($\gg f^{-1}$) will not be modulated and therefore would not be expressed as a PA signal. By scanning through the frequency it is therefore possible, in principle, to measure the ES in the different intermediates in succession and also find the kinetic constants of their interconversion (71). At a frequency where the above conditions are met, the PA signals from the sample and reference will have the same phase angle and their difference will be proportional to the energy, which is not converted to heat. Most of this energy is stored as chemical energy and a small part is emitted as fluorescence. This is analyzed quantitatively by using the following energy balance equation (71):

$$h\nu = H + \phi_F h\nu_F + \phi_P(\Delta E) \qquad\qquad 2$$

In Equation 2, $h\nu$ is the energy of the absorbed photon, H is the prompt heat release (per absorbed photon), ν_F is the average frequency of a fluorescence photon, ϕ_F and ϕ_P are the quantum yields of fluorescence and photochemistry, respectively, and ΔE is the stored energy (per absorbed photon). The PA signals from the sample, S, and from an ideal reference, R, are respectively proportional to H and $h\nu$. It therefore follows that:

$$(R-S)/R = \phi_F(\nu_F/\nu) + \phi_P(\Delta E/h\nu) \qquad\qquad 3$$

The experimental ratio $(R - S)/R$, termed *photochemical loss* (PL), expresses the relative deficit in the PA signal resulting from all competing processes (71). The term $\phi_P(\Delta E)/h\nu$ is the photochemical energy storage, relative to the exciting photon energy, abbreviated ES. In many reported measurements the fluorescence term in Equation 3 was neglected, leading to ES $\approx$ PL.

REFERENCING METHODS   For absolute quantification, the reference must be an inert light absorber, with exactly the same physical (optical, thermal, and thermoelastic) properties as the sample. Obtaining this condition in the small volume PA cell with gas-coupled sensor, by replacing the sample with the reference, is extremely difficult if not impossible. In one ad hoc approach, the same sample, after inactivation by DCMU, was used as a reference (28, 29). In another approach, a comparison is made between a PA spectrum (under appropriate conditions for PA spectroscopy) and ordinary percent absorption spectrum (28, 44, 45, 100). The ratio between the two was considered to represent a relative spectrum of the heat emission yield, called *thermal deactivation spectrum* (TD), proportional to (1 − PL). However, this method gives only relative values for the efficiency of heat production as a function of the wavelength. Internal referencing is the best and most general approach, which is particularly feasible in photosynthesis research. In internal referencing, the sample is brought to photosynthetic saturation with an intense non-modulated (background) light (68, 74, 30–35). Under these conditions, photochemical efficiency drops close to zero and the photothermal signal increases to its maximum, serving as a reference. This procedure is usually nondestructive and reversible. The relaxation of the photothermal signal after switching off the background light is usually biphasic: an immediate partial relaxation followed by a slower phase of several seconds (72, 96). This evidently reflects a transitory state of an initial small photochemical quantum yield, caused by adaptation to the previous intense light, which probably parallels the non-photochemical fluorescence quenching process (see below).

Sample stability, with regard to the relevant physical properties, is required for internal referencing. In leaves, the amplitude and phase of the maximum reference photothermal signal are remarkably constant for hours; however, stress treatments in leaves may cause gross structural changes, which in turn change the maximum photothermal signal [in drought (54) or photoinhibition (50, 60) treatments]. Evidently, an internal referencing measurement must be performed for any new condition.

THE EFFECT OF FLUORESCENCE   Equation 2 may be used to estimate the error in neglecting fluorescence and to account for its presence. For a photochemically active sample, Equation 2 is $h\nu = H' + \phi\Delta E + \phi_F h\nu_F$. At photosynthetic saturation, Equation 2 is $h\nu = H' + \phi_{Fm} h\nu_F$. H and H' refer to the sample and the reference, respectively. The quantum yield of fluorescence changes from ϕ_F (in the sample) to the maximum (light saturated) ϕ_{Fm} (in the reference) and ν_F is the same in the above two equations (equal emission spectra). The resulting connection between ES and PL is therefore:

$$(ES) = (PL) [1 - \phi_{Fm}(\nu_F/\nu)] + (\phi_{Fm} - \phi_F) (\nu_F/\nu) \qquad 4$$

To calculate ES, the PL should be corrected by multiplying with $1 - \phi_{Fm}(\nu_F/\nu)$ and adding $(\phi_{Fm} - \phi_F)(\nu_F/\nu)$. The effect of the first correction is somewhat compensated by the second. A general estimation of PS II fluorescence yield may be used (82) to demonstrate the extent of the correction: $\phi_{Fm} \approx 0.1$ and $\phi_F \approx 0.03$; for 685 nm excitation, where $\nu_F/\nu \approx 0.97$, and for common PL values 0.2–0.3, a systematic error in ES of about +13% is estimated.

For a more precise estimation, the actual fluorescence quantum yields must be measured for each experiment, which can be done using PA. The (reference) photothermal signal and the relative fluorescence yield, F_m, are measured simultaneously during sufficient time in the presence of an intense photosynthetically saturating background light to observe the main drop of F_m, from an initial value F_{m1} to a lower value, F_{m2}, as a result of the non-photochemical quenching process. This drop is quantified by the coefficient q_N (which equals $1 - [F_{m2}/F_{m1}]$ or $1 - [\phi_{Fm2/Fm1}]$). In both states the photochemical quantum yield is nearly zero and the only competing processes are fluorescence and heat release. Hence, the non-photochemical decrease of fluorescence must be followed by increased heat emission, depicted by a small rise from an initial photothermal reference signal, R_1, to a final value, R_2. From the proportionality of R_1 and R_2 to $(h\nu - \phi_{Fm1}h\nu)$ and $(h\nu - \phi_{Fm2} h\nu)$, respectively, it is possible to obtain the absolute quantum yield of fluorescence (Equation 5).

$$\phi_{Fm1} = [\nu F/\upsilon] \, [(R_2/R_1) - 1] \, / \, [(R_2/R_1) - 1 + q_N] \qquad\qquad 5$$

One report (52) has not detected measurable changes in R, although q_N, at about 0.9–0.95, was appreciable. With an estimated signal/noise ratio of about 2% (i.e. $R_2/R_1 \leq 1.02$), $\phi_{Fm1} \leq 0.021$. In another study (38), which used intense modulated light, and hence had an improved signal/noise ratio, $R_2/R_1 \approx 1.05$ and $q_N \approx 0.6$, from which $\phi_{Fm1} = 0.07$. From the measured ratio F_o/F_m the value of ϕ_{Fo} was also estimated at 0.03. These values reconfirm the marginal correction of the PL values to obtain ES (Equation 4). For a common PL of about 0.3, the resulting ES is about 0.32 (i.e. a systematic deviation close to +6%).

Because measurements of the absolute fluorescence yields are also important per se, it is important to stress that the use of PA for this purpose is relatively facile, particularly for optically difficult samples such as leaves.

The Modulated Photobaric Signal

The photothermal mechanism can account for the entire PA signal of organisms and sub-cellular organelles, soaked or submerged in water. This mechanism is characterized by a saturating background light effect, which increases the signal as a result of the increased heat emission. In organisms or structures where the photosynthetic membranes lie close enough to a gaseous phase, the photothermal signal is usually accompanied by the photobaric signal. The photobaric mechanism was recognized by the abnormal behavior of the intense background light at sufficiently low modulation frequencies, where it caused a marked decrease of the signal amplitude, accompanied by a large phase shift (6, 89). This effect indicated a contribution to the PA signal in addition to the photothermal one, which was eliminated at light saturation. This additional contribution, distinguished only at low modulation frequencies (typically below 100 Hz), was considered to be the result of periodically pulsed gas evolution (O_2) and/or uptake (O_2, CO_2), which tend to zero at light saturation (when the rate does not vary with changes in the light intensity). The remaining signal, at light saturation, was presumably the result of the usual photothermal contribution, which was then at a maximum. At high modulation frequencies, the gas evolution modulation is damped, since it cannot follow the rapid light modulation.

Several observations are consistent with the above interpretation: (*a*) the change of phase angle, upon application of the background light, indicating at least two distinct contributions, with different phase angles (89); (*b*) the different behavior of PTR (65, 75) or PBD (61) signals, resulting from the photothermal effect alone, which expectedly increase at light saturation; (*c*) transients in the non-photothermal contribution, during the photosynthetic induction period, which parallel the behavior of photosynthetic oxygen evolution (6, 49, 70, 96); (*d*) the absolute requirement of a gaseous phase in close proximity to the photosynthetic membranes. Elimination of the inner air space of leaves by water infiltration abolished the non-photothermal contribution (72); (*e*) the behavior of the ratio of the amplitudes (non-photothermal/photothermal), which depends on the modulation frequency according to a material diffusion model (6, 89); and (*f*) kinetic parallelism to various oxygen evolution phenomena (Emerson enhancement, S-state periodicity in time-domain measurements), as described below.

RESOLUTION AND QUANTITATION OF THE PHOTOBARIC SIGNAL   Using vectorial representation, **O** represents the photobaric signal vector ($[O_I;O_Q]$), and **T** represents the maximum photothermal vector ($[T_I;T_Q]$). With superscripts (+) and (−) denoting the presence or absence of saturating background light, the total signal vector **S** is

$$\mathbf{S}^{(-)} = (1 - PL)\mathbf{T} + \mathbf{O} \qquad \text{and} \qquad \mathbf{S}^{(+)} = \mathbf{T} \qquad\qquad 6$$

from which it is possible to obtain **O**:

$$\mathbf{O} = \mathbf{S}^{(-)} - (1 - PL)\mathbf{S}^{(+)} \qquad\qquad 7$$

An independent estimation of PL is needed to estimate **O**. This may be provided by (*a*) auxillary photothermal measurements [e.g. PTR (65) or PBD(61)]; (*b*) PA photothermal measurements of water infiltrated samples (72); or (*c*) PA photothermal measurements at high modulation frequencies, taking into account that PL varies little in the relevant frequency range (about 10–500 Hz) (61, 65, 72). A rough zero-order approximation, sufficient for indicative purposes, uses PL = 0 (i.e. the maximum photothermal signal serves as a base line for the photobaric signal).

In many comparative measurements, under conditions when no phase shift is expected, the photobaric signal vector is best quantitated by its projection in a direction perpendicular to the photothermal signal, for which the PL value is not needed. For this, the phase ε of the internal LIA reference is adjusted in the presence of the saturating background light, until one of the signal components, e.g. the quadrature, is nulled. This procedure provides a new coordinate system for the vectorial representation of the signals, in which the quadrature photothermal component is zero and the photothermal signal appears completely as an in-phase component. When the background light is removed, the phase of the photothermal signal remains nearly the same (72) and the appearing quadrature signal represents a pure component of **O** (17, 25, 83, 89, 94–96). Instead of performing the above procedure physically, the same pure component of **O** can be computed from data obtained at any arbitrary phase setting (preferably if data are accessed by a computer). This component is given by the expression $[S_Q^{(+)}S_I^{(-)} - S_I^{(+)}S_Q^{(-)}]/[S_I^{(+)2} + S_Q^{(+)2})]^{1/2}$, deduced from simple vectorial calculation.

WHICH PROCESSES CONTRIBUTE TO THE PHOTOBARIC SIGNAL? Modulated gas evolution or uptake arise from modulation in the formation or disappearance of the corresponding species. Pure dark reactions and enzymatic steps that operate at substrate saturation levels are not expected to be modulated significantly. This includes respiration, but also photosynthetic CO_2 fixation and photorespiration, since RuBP concentration is saturating for the carboxylation and oxygention reaction. It seems, therefore, that the steady-state photobaric signal comes mainly from oxygen evolution. Under certain conditions it is necessary to consider the possible effect of a Mehler type O_2 uptake in PS I, which may be modulated by and contribute to the photobaric signal. This possibility was noticed in circumstances such as when normal photosynthesis was at a low level, for example, immediately after dark adaptation (67, 91, 76) and during photosynthesis induction (49, 70), or following heat stress (57, 59). When measurements were made at very low frequencies (about 5–15 Hz) or in time-domain measurements (67, 76, 91), such an effect was discerned by the negative sense relative to the direction of the normal steady-state photobaric (i.e. oxygen evolution) signal vector, and by its elimination by levels of far-red background illumination sufficient to saturate PS I (S Malkin, unpublished information).

THE EFFECT OF DIFFUSION To generate a photobaric signal, oxygen has to diffuse to the gaseous phase. The treatment of the diffusion, with regard to the modulated flux (6, 89), is mathematically analogous to the modulated heat diffusion case (Equation 1), with oxygen concentrations, fluxes, and the diffusion constant, D_{OX}, replacing temperatures, heat fluxes, and the heat diffusivity, D_{TH}, respectively. Here too, the modulated part of the concentration is mathematically expressed by superposition of concentration wavelets—generated in the photosynthetic membrane and propagating toward the inner air phase. These waves die out exponentially, with an effective range (oxygen diffusion length), μ_{OX}, given by $(D_{OX}/\pi f)^{1/2}$. By assuming that for exerting their effect both the thermal and concentration waves propagate to the inner air phase, where presumably most of the acoustic signal is created, it was possible to obtain a simplified theoretical model and to derive the frequency dependence of the ratio A_{OX}/A_{PT} of the photobaric and photothermal signal amplitudes. As a first approximation, a unidimensional model was used, reducing the assembly of photosynthetic membranes to a single thin sheet, parallel and at a distance l to the cell boundary. It was also assumed that no sinks exist in the stromal and plasmatic spaces between the membrane and the inner air phase. A_{OX}/A_{PT} was determined by the relative exponential attenuations of the oxygen and thermal waves:

$$A_{OX}/A_{PT} = \mathbf{a} \exp\left[-\pi^{1/2}(1/D_{OX}^{1/2} - 1/D_{TH}^{1/2}) \, l \, f^{1/2}\right] \qquad 8$$

A plot of $\mathrm{Ln}\,(A_{OX}/A_{PT})$ vs $f^{1/2}$ is expected to yield a straight line, with a slope $-\pi^{1/2}(1/D_{OX}^{1/2} - 1/D_{TH}^{1/2})l$. The intercept $\mathbf{a}$ is regarded as the true A_{OX}/A_{PT}, free of diffusion effects. Many studies have verified Equation 8 (26, 51–54, 55, 72, 89). Moreover, the ratio of slopes from two plots, for a leaf and a lichen (26), was consistent with a rough estimate of the ratio of the corresponding l values. From the slope and an estimate of l, D_{OX} in vivo was determined (89) to be significantly smaller than in pure water. The value of D_{OX} limits the detectability of photobaric signals, at ordinary frequencies, to diffusion distances not larger than about a micron. No photobaric signals are therefore expected from an ordinary suspension, although photobaric signals have been measured from thin layered algae on solid support (17) as a result of the close contact with air.

THE LOCUS OF PA SIGNAL GENERATION IN A LEAF   The above assumption that the sound is generated solely or mostly in the inner air phase, from where it is transmitted to the outside, contradicts a concept used formerly in PA spectroscopy of leaves (1, 9, 80, 93). According to this concept, the leaf was viewed as a homogeneous substance through which the thermal waves traveled all the way to the outside surface, and the profound changes in the uv-visible PA spectrum as f increases resulted from an uneven distribution of pigments along the leaf width, combined with the increased damping of the thermal waves (93). An alternative explanation, consistent with the present concept, is to distinguish between thermal waves originating in the epidermal pigments, which indeed travel to the outside surface and thermal waves generated in the photosynthetic machinery, which effectively travel to the inner air phase. The ratio between the corresponding contributions to the PA signal (which determines the resulting spectrum) depends on the physical transmission of sound from the inner air phase to the outside, which is presumably frequency dependent and is governed by the leaf structure, physical flexibility, and the extent of stomatal opening. Indeed, the frequency dependence of the maximum photothermal contribution (54) indicates the insignificance of thermal wave diffusion from the photosynthetic pigments to the outside. This mode of thermal wave propagation did, however, gain more importance only in a drought stressed leaf (54), where drought treatment resulted in structural changes of the leaf anatomy and hence of overall diffusion paths and distances.

THE EFFECT OF THE REACTION KINETICS   The time lag between the primary photoact and the release of oxygen from water, or uptake through PS I, exerts additional damping. Calculations (6) predict an additional factor to Equation 8, $k/(4\pi^2 f^2 + k^2)^{1/2}$, where k is a rate constant for a limiting step in the relevant electron transport segment(s). This factor is equal to 1 at sufficiently low f and decreases significanly as f approaches k, tending finally to zero as f increases further. This effect should cause a negative deviation from linearity, at high frequency, of the Ln (A_{OX}/A_{PT}) vs $f^{1/2}$ plot derived from Equation 8. While such deviations were indeed found (26, 89), later work (72) indicated that they could be artifacts, resulting from an underestimation of the PL, which introduced a large error when A_{OX} became small at high f. Also, the magnitude of the deviations from the linear plot did not conform to the above theoretical relation (89) and occurred at different frequency ranges for a leaf and a lichen (26). More accurate data did not show such deviation (72), resulting in an estimation k >> 500 sec^{-1} for the rate limiting step in the electron tansport to water.

Photothermal and Photobaric PA Signals in the Time Domain

RESPONSE TO SATURATION FLASHES IN LEAVES   A single ms flash excitation resulted in a pulsed response lasting several ms (24, 76). The signal was proportional to the derivative of the pressure pulse inside the PA cell (24), with its overall shape reflecting the diffusion of heat and oxygen, respectively, and the microphone properties. The photobaric contribution from each successive flash in a series was determined by subtracting the signal obtained in the presence of strong background light from the signal following the first flash, which was purely photothermal. The small lag time (1 ms) until the photobaric signal appeared indicated an equal or shorter reaction time for oxygen production (24). Oscillations were seen in the resolved photobaric signals, typical of S-state oscillations (24, 76), which in a leaf cannot be observed easily in any other way. The photobaric signals, obtained with strong flashes, were quite small (24, 76) and closely saturated, compared to the photothermal signals, which become proportional to the flash intensity. A relatively strong negative photobaric signal was obtained with weaker flashes after dark adaptation (76), indicating oxygen uptake. This result was not observed with strong flashes, presumably because oxygen uptake saturates earlier than oxygen evolution.

THE RESPONSE TO WEAK FLASHES   An alternative mode of measurement used very weak, relatively long flashes (ms), obtained from LEDs (67), which required averaging of the PA signal from a large number of pulses to obtain a reasonable signal to noise ratio. In this light limiting condition, the photobaric signal pulse was comparable to or even bigger than the photothermal signal. The two signals had distinct and separated time dependencies: the photothermal contribution rose sharply during the light pulse and terminated immediately after; the photobaric contribution lagged (reflecting the oxygen diffusion), often peaking after the termination of the light pulse. The PL was calculated from the initial slopes of the signal, assigned to the photothermal contribution alone, by comparison to the initial slope of the signal in the presence of saturating background light, which eliminated the slow photobaric contribution. This allowed the time profile of the photothermal contribution in the total signal to be obtained by proper normalization. The photobaric signal was then resolved by subtracting the photothermal contribution. A more general procedure was tried (67), in which the normalized time dependence of each signal component was determined at some prefixed standard conditions: (a) at the steady state where the photobaric signal was positive and resulted from oxygen evolution or (b) after sufficiently long dark adaptation time where the photobaric signal was negative, presumably because of oxygen uptake. The amplitude of each component could then be followed by a computerized linear analysis (67). This procedure was meant as an alternative to the LIA method with modulated light. It did not work well for the uptake because its time dependence during the PA pulse was found to be uncontrollably variable (91).

PA SIGNALS IN DIRECTLY-COUPLED-DETECTOR CONFIGURATION

Experimental

THE PA SET-UP   In this configuration, the PA cell has a relatively large volume (typically, a few ml) and is best utilized for dilute suspensions. It can be assembled from an ordinary spectrophotometric cuvette and a suitable sensor. It is used mainly for relatively fast time-domain measurements. Earlier experiments used a capacitor microphone placed in the aqueous medium to detect volume changes (2, 14, 15). More recently, fast piezo electric films pressed against the outer cuvette walls were used as pressure pulse sensors (Figure 1b).

MECHANISM OF SIGNAL GENERATION Heat diffusion, which is slow, is a negligible factor here. In the case of laser excitation, for example, the sound is produced mainly from the immediate local volume change, which occurs along the light path only and produces a pressure pulse there. The pulse propagates with the velocity of sound, c, toward the sensor, positioned at a close distance s from the laser beam. The response time lag after the flash is s/c. For a sufficiently short flash, the time resolution is limited by d/c, which is the time it takes for the sound to pass through the laser beam cross-section, d, which is set by a pinhole as a compromise between having sufficient intensity and a reasonable time resolution. Another limiting factor is the response time of the sensor. Piezo electric film gets into resonance, resulting in extensive ringing at a longer time scale, which complicates the analysis of the heat evolution time dependence. In LIOAS, only the first peak of the signal, corresponding to one time-point (around 1 μs), was used (64, 81, 82). In earlier experiments, using a microphone and ordinary μs flash excitation (2, 14, 15), the time resolution was essentially limited by the high frequency cut-off of the microphone response, with no limitation on a longer time scale up to several hundred ms. In longer times the system relaxes to its resting position, by leaks.

PHOTOTHERMAL AND VOLUME CHANGES In suspension of chromatophores or bacterial reaction-centers, a light pulse induces a time-dependent microphone response, which reflects a combination of photochemical and photothermal volume changes associated with the electron transfer from P_{870} to Q_A (2, 14, 15). This conclusion was deduced from an experiment performed at about 4°C, where the thermal expansion coefficient was essentially zero. Nevertheless a (negative) signal was obtained, which therefore was assigned to a pure (negative) volume change, induced by the photochemistry. Assuming that the volume change was temperature independent, this signal was subtracted from the signal obtained in any other temperature, resulting in a pure photothermal signal. In the recent LIOAS studies (81, 82) there was no awareness of possible volume changes, so the total signal was considered to be purely photothermal. In this case, however, the photothermal effect was enhanced about 2–3-fold, by the use of 30% glycerol (which has a higher thermal expansion coefficient); in retrospect, the error arising from the neglect of the volume change was less serious. In more recent LIOAS work on *R. sphaeroides* reaction centers (72a), however, measurements were taken at a range of temperatures and the sound pulse was found to arise from a combination of photochemical volume change (measured separately and appearing as a negative signal at about 4°C) and of photothermal volume change.

In this measurement mode, in contrast to the gas-phase coupled sensor mode, an external reference may conveniently be used for both the photochemical volume change and the photothermal contributions to the signal. A dilute solution, in the same buffer as that used for the sample, of a photochemically inert absorber (a colored solution of an inorganic salt or a photochemically inert organic dye) of exactly the same light absorbance at the wavelength used may serve as a proper reference, having opitcal and thermoelastic properties sufficiently close to that of the sample (2, 14, 15, 81, 82). The absolute value of the photochemical volume change can be calibrated by comparing its value, obtained at 4°C, to the photothermal signal of an external reference measurement (at a temperature far greater than 4°C), using the values of the thermoelastic properties (thermal expansion coefficient, specific heat, and mass density) and the absorbed flash energy (2, 15).

RESULTS OF PHOTOSYNTHETIC ACTIVITIES

Energy Storage

Using the definition $ES = (\phi_P \Delta E)/h\nu = (\phi_P \Delta E)\lambda/hc$, where λ is the wavelength and c is the speed of light, ES can measure both the quantum yield, ϕ_P, and the absolute stored energy of some intermediate(s), ΔE. The simple observation of energy storage, by the reversible increase of the photothermal signal upon application of intense background light, is by itself an indication of photosynthetic activity. Most studies were done in continuously modulated light and this effect was found in all representative healthy organisms and active cell fractions: thylakoid membranes (30, 32, 35, 68), PS I core complex (86), PS II enriched sub-membranal fractions (16, 31), isolated bundle-sheath cells (88), green microalgae (20, 27), cyanobacteria (29) and heterocysts (33), purple anoxygenic bacteria and chromatophores (74), and marine macroalgae (43, 62, 66, 73). In leaves and lichens the reversible increase of the PA signal upon application of the intense background light was observed only under conditions where the photobaric signal was not expressed (6, 26, 62, 72, 85, 89, 94). However, even in the presence of the photobaric signal, it is possible to observe the same effect by using other photothermal methods (PTR or PBD) (61, 65, 75). Thermal deactivation spectra were obtained for Anacystis as an indication of photochemical activity (29), but thermal deactivation spectra were used most often in studies on model and viable systems oriented in polymer films (e.g. 44, 45, 100), where a good referencing method is not available. Time-domain pulsed PA measurements also indicated energy storage in PS I and PS II complexes (81), bacterial cells (82), and whole leaves submerged in water (64), at around 1 μs, and in a layered green alga at around 8 ms (36).

ES vs ϕ_P ES measurements can be used to follow ϕ_P changes, which may be caused by various treatments or slow physiological adaptations to changes in light intensity and quality. Changes in ϕ_P and the parallel ES changes result from changes either in the degree of openness of reaction centers or the inherent quantum yield of open reaction centers. The inherent quantum yield may be measured under strict light limiting conditions, where all reaction centers are presumably open, or by considering variable fluorescence as an indicator for PS II openness (67a). It is not always practicable to measure the inherent ES for true light limiting conditions (i.e. at sufficiently low light intensity) because of low signal to noise ratio. The problem was circumvented in chloroplast and subchloroplast preparations (30–35, 68) by using linear 1/ES vs light intensity plots (based on the traditional hyperbolic relation between rate and light intensity) and to obtain inherent ES, $(ES)_m$, by extrapolating ES to zero light intensity. Another parameter of this plot was the intensity, i_{50} at which ES decreases to $(ES)_m/2$, which is proportional to the maximum saturation rate of the involved reaction. A puzzling result was the invariability of $(ES)_m$ in thylakoid preparations following inhibition treatments [gluteraldehyde fixation (30) or DCMU addition (32)]. Although i_{50} decreased as expected, $(ES)_m$ was invariable, contradicting the well-known inhibition of ϕ_P at light limiting conditions. Perhaps this indicates a transient energy storing reaction when PS II is blocked in the Q_A^- state (e.g. cyclic electron transport around PS II; transient reduction of pheophytine).

Comparison of ES changes with fluorescence changes provides some more details on the nature of the photsynthetic reactions involved. ES and variable fluorescence $(F - F_0)$ should be antiparallel, if only changes in reaction center openness are involved. This was seen, for example, when both changed as a function of time during the induction period (66). In thylakoid membranes aspirated on solid support, this relation was observed and a similar light intensity dependence was found for both (35), which is typical of normal electron transport through PS II. In previous work (68), however, using a concentrated thylakoid suspension, variable fluorescence was saturated much earlier than was ES. It was then concluded (supported also by the increase of ES in the far-red wavelength range) that PS I was engaged in an independent activity (possibly a cyclic electron flow) when PS II was saturated—a situation characteristic of the absence of electron acceptors. A possible explanation for this discrepancy is that in the recent experiments on layered membranes (35), oxygen is readily available to act as an efficient electron acceptor.

ES vs λ It is particularly easy to obtain the photosynthetic activity vs λ, by scanning PA photothermal signals, once without and once with saturating background light, plotting the resulting ES/λ vs λ (6, 16, 25–27, 65, 73, 97). The particular advantage of PA is that such activity spectrum does not obviously require any measurement of light intensities and absorptions (because these are built into the method).

ES AS A MONITOR FOR THE ENERGETICS OF THE PHOTOSYNTHETIC PRO-CESSES   From the ES value, which is relative ($\phi_P \Delta E/h\nu$), it is possible to calculate the absolute energy, ΔE, for the intermediates of the photosynthetic reaction, if the photon energy, $h\nu$, and the quantum yield, ϕ_P, are known. In previous LIOAS time-domain measurements (in which volume changes were not considered), the following ES values were obtained for the 1–2 µs range: > 0.6 in leaves at 675 nm (64), 0.83 in PS I particles and 0.67 in PS II particles at 677 nm (81), and 0.3 in bacterial cells at 590 nm (82). Assuming a conventional value for ϕ_P (about 0.9), ΔE was accordingly estimated for $P680^+Q_A^-$, $P700^+A_1^-$ and $P870^+Q_A^-$ (1.2, 1.5, 0.6 eV, respectively). By equating ΔE with standard free energy change, the redox potentials were estimated.

There is a discrepancy between these results and those of ms time-domain measurements for bacterial systems, where volume changes were accounted for, and the resulting ΔE for $P^+Q_A^-$ was unreasonably minute (2, 2a), as already described above. Volume changes were in the order of −10 ml/einstein (2).

The more recent LIOAS measurements, which demonstrated and accounted for volume changes (72a), allowed more rigorous estimation of the energy storage in $P_{870}Q_A^-$ (about 0.83 eV for *R. sphaeroides* R-26 strain). The photo-chemical volume change tended to reach a saturation value (~ 6–8 ml/mole reaction centers) for the R-26 strain as the flash intensity increased. The initial slope (volume change/einstein) and the maximum saturation value (volume change/mole reaction centers) in the plot of volume change vs flash intensity were nearly the same and indicated a quantum yield of photochemistry near 1. Although this work (72a) and the former slower microphone measurements (2) agree roughly on the photochemical volume changes, the appreciable energy storage found for the R-26 strain stands in contrast to the very small enthalpy change (~ 0.13 eV) found earlier. Further work is required to explain this difference. Comparison of the results (72a) between different *R. sphaeroides* strains, R-26 and (M)Y210W, indicate a large range of variability in both volume change and energy storage.

The LIOAS experiments (81, 82) also monitored energy storage in a triplet state, formed by recombination in closed reaction centers. This was evident from the smaller signal, even in the presence of background light or high flash intensity at the saturation range for the primary reaction, compared to an external reference, which indicated energy storage. The signal maximized to the level of the reference only at exceedingly high intensities, when the high density of excited states tends to quench the triplet state. This effect might be further exploited to obtain energetic and kinetic information on the triplet and on processes that lead to its quenching at extreme high intensities.

Results relevant to longer times were obtained with the gas-coupled-micro-phone PA cell. For this measurement mode it is possible to prove that inherent volume changes affect the PA signal only negligibly and, therefore, need not be considered. Most $(ES)_m$ values in leaves were around 0.30–0.35 or less, for wavelengths of 650–685 nm, with a weak dependence on f between about 10 and 500 Hz (61, 72). It appears that there is significant energy dissipation between about 1 and 100 μs, but only a little is dissipated at longer times. Indeed, $(ES)_m = 0.35$, for the conventional $\phi_P = 0.5/e^-$, corresponds to a ΔE that is slightly greater than the enthalpy of glucose (per electron equivalent) (65, 72).

A change may occur in the reaction type, affecting ΔE, when scanning ES as a function of some variable, particularly λ, because contributions of reactions associated with each of the two photosystems may vary (43, 62, 73, 97). A scan of ES/λ may reflect changes in both ϕ_P and ΔE. An attempt was made to segment ES into the sum of contributions from PS I and PS II, $[ES]_I$ and $[ES]_{II}$, respectively (97, 98). In the presence of saturating background far-red light ($\lambda > 690$ nm—denoted "light-1"), selectively absorbed in PS I, $[ES]_I$ was supposedly eliminated; hence, $[ES]_{II}$ was defined as the ES value in presence of light-1 and $[ES]_I$ was defined as $[ES] - [ES]_{II}$. Accordingly, both were scanned as a function of λ (97). This approach may be criticized by noting that $[ES]_I$ and $[ES]_{II}$ do not simply reflect ΔE_I and ΔE_{II} in proportion, but also depend on the light distribution ratios of the two photosystems: $[ES]_{II}$ is actually proportional to the light distribution ratio of PS II but $[ES]_I$ is related in a more complex manner. One should note that a linear segmentation into independent, individual ES contributions, as in the above, is only possible for sufficiently short reaction times, before the photochemical products in each photosystem have reacted. This probably applies to this case, at a modulation frequency of 80 Hz or above.

A different approach to ES segmentation was attemped for layered *Chlorella* cells, in a gas-coupled microphone PA cell, using flash time-domain measurements (36). The results were applicable to a time-point 8 ms after the flash. It was assumed that 695 nm flashes excited only cyclic electron transport around PS I. The resulting ES was independent of the flashing time-interval (0.35–60 s). With 650 nm flashes at very low repetition rates (60 s), isolated electron transport reactions in the two photosystems were probed. When the repetition rate was high (0.35 s), whole chain electron transport was probed. ΔE values were estimated from the values of ES. ΔE for the cyclic electron flow around PS I was evaluated to be 0.68 eV/hv (ES = 0.4 at 695 nm), tentatively assigned to reduced Fd and oxidized cyt b_6f. A ΔE of 0.5 eV/hv was assigned to whole linear electron transport from water to NADP and a ΔE of 1.07 eV/hv was assigned to the electron transport from the PS II Mn complex to Q_B.

Photobaric Measurements

Since the photobaric signal O (regarding amplitude, or any vectorial component) is proportional to $\phi_P i_0$ and the photothermal signal T is proportional to the absorbed energy flux, $h v i_0$ (which is equal to $[hc/\lambda]i_0$), where i_0 is the photon flux amplitude, the ratio $(O/T\lambda)$ can be used as a measure proportional to ϕ_P, eliminating the need for tedious light intensity and absorbance measurements. Scanning the last ratio vs λ made it particularly easy to obtain photosynthetic activity spectra in leaves and lichens (25, 26). When changes in the oxygen diffusion parameters are expected, e.g. after a stress treatment that affects the details of the anatomical structure or the stromal viscosity, the amplitudes ratio A_{OX}/A_{PT} must be measured at several modulation frequencies, using its zero-frequency extrapolated value (Equation 8).

LIGHT SATURATION CURVES   As with the ES measurement, maximum O/T is obtained with modulated light of low (photosynthetically limiting) intensity. O/T can be measured at various average light intensities (e.g. in presence of an additional background light, varying in its intensity between zero and the maximum used for referencing). The resulting O/T is approximately proportional to the differential quantum yield, i.e. the derivative of the rate light saturation curve (89). The regular quantum yield, in relative terms, was constructed by integration (25, 89). Using an independent estimation of the maximum quantum yield, the absolute maximal rate could be calculated.

PHOTOSYNTHESIS PHYSIOLOGY STUDIES

Photosynthetic Transients

Using PA in the study of transient phenomena has the advantage of rapid response and in particular has the ability to manipulate the physiological and electron-transport state with addition of nonmodulated lights. PA provides a variety of different kinds of information (e.g. oxygen evolution, oxygen uptake, and energy storage). Compared to fluorescence, which reports mostly on PS II, PA reports on both PS II (oxygen evolution), PS I (Mehler-type oxygen uptake) and on some combination of the two, in the measurements of ES.

TRANSIENT PHOTOBARIC SIGNALS   In dark adapted leaves, the onset of periodically modulated light first produces an immediate photobaric pulse of short duration (a second or so, depending on the intensity), closely parallel to the initial fluorescence rise (49, 70, 96). This pulse presumably reflects the "oxygen gush," corresponding to the limited electron transport from water to the small pool of acceptors between the photosystems (mainly plastoquinone), previously observed by polarography in algae (64a). This effect also can be used for an absolute calibration of the photobaric signal (70), provided that the plastoquinone pool can be quantified. Denoting the area below the PA pulse, in the relative scale, by **A** (in time units, e.g. sec), the steady-state photobaric signal by O, and the pool size by **P** (electron-equivalents Chl^{-1}), the absolute oxygen evolution rate would then be [4**P**/**A**] O (in moles per unit time per chlorophyll).

An intermediate period follows the initial pulse, with complex photobaric transients of positive and negative waves, presumably reflecting waves of oxygen evolution and uptake (49, 70). Finally, there is a slow increase of the photobaric signal toward the final steady-state level. The modulated uptake contribution is damped much more than is the normal evolution signal when using higher modulation frequency, f, in the measurement, indicating a smaller rate limiting constant for the uptake. The uptake is saturated (i.e. eliminated) at relatively low background light intensities (49). Under these conditions, the normal photobaric signal remains, reflecting the ordinary modulated oxygen evolution. These observations indicate the presence of electron transport activity, involving the two photosystems, during the induction period from water to oxygen, with a low rate-limiting step at PS I. The involvement of PS I is indicated by the presence of an uptake signal with modulated light-1 (49), and also by its elimination with excess background light-1, as PS I is saturated (S Malkin, unpublished information).

In a time-domain study with repetitive LED pulses (91), a strong negative (i.e. uptake) signal occurred in transient but also in the steady-state, when 5% CO_2 was streamed through the PA cell. It was concluded that the occurrence of an uptake following dark adaption period was tied to the accumulation of (respiratory) CO_2. Consistently, the uptake disappeared momentarily with a flush of CO_2-free air. The CO_2 effect was explained by stroma acidification, which presumably increases the affinity to O_2 as an electron acceptor. The modulated uptake was associated with the first reduction step of O_2 to super-oxide-radical anion (91).

THE RELATIONSHIP BETWEEN THE PHOTOCHEMICAL YIELD AND THE NON-PHOTOCHEMICAL QUENCHING OF FLUORESCENCE   The influence of the non-photochemical quenching, q_N, on net steady-state photosynthesis was examined thoroughly by conventional methods (67a). PA, which monitors gross photosynthesis, a more relevant aspect, was used to check this relation during transient periods, taking advantage of its rapid response. Relations similar to the steady state were found between the time-dependent q_N and the inherent ϕ_P (i.e. the quantum yield of open centers), measured immediately after turning off an intense radiation, which caused the non-photochemical the quenching (95, 96). From these relations it was found that the extrapolated ϕ_P, for $q_N = 1$, was very close to zero (95). In some older reports, there was no obvious relation between fluorescence and photobaric PA transients (7, 8, 49), because the change of ϕ_P with q_N was not recognized then.

ENERGY STORAGE TRANSIENTS   The induction period is marked by a continuous slow decline of the PA photothermal signal (i.e. increase of ES), approaching steady-state (6, 72, 94, 96). Several individual stages of the photothermal signal decline were seen in water infiltrated leaves (72), but reproducibility between individual experiments was too poor for a clear identification of the stages. For the initial part, a very fast increase of photothermal signal, parallel to the initial fluorescence rise and the oxygen gush, was expected. However, this was not clearly observed (cf 72). When fluorescence reached its peak and PS II nearly closed, ES was still significant (6, 72), indicating possible PS I activity.

Light Distribution Between the Two Photosystems: State 1–State 2 Transitions

PA offers several possibilities for the quantitative study of light distribution to the two photosystems, using combinations of modulated and background lights such that the first is absorbed more favorably in one photosystem and the second is absorbed more favorably in the other photosystem. In the following discussion, we use the customary abbreviations light-1 and light-2 for the kinds of light absorbed more favorably in PS I and PS II, respectively. For example, when PS I is saturated by excessive background light-1, the resulting modulated signal arises solely from PS II (22), which is proportional to the fraction of the modulated light absorption in PS II. Another example is the use of modulated light-1, distributed almost exclusively to PS I, where addition of background light with a better distribution to PS II removes the limitation in PS I and gives rise to a change in the PS I signal (43, 73, 75), marking an increase of its photoactivity. In such experiments one may use all the different aspects of the PA signal [oxygen evolution (22), energy storage (43, 73, 75), or oxygen uptake (which has not yet been used)] to obtain different information.

Although the effect of background light addition or removal is almost immediate, consequential light adaptation effects, responding to the combined lights, or to the removal of the background light, are observed by PA as slow transients on the physiological time scale. One example is the state 1–state 2 transition, resulting from slow adaptation to an initial unequal light distribution of light-2, which in higher plants ends in a more equal distribution (43a).

LIGHT DISTRIBUTION FROM EMERSON ENHANCEMENT OF THE OXYGEN EVOLUTION SIGNAL   The procedure below refers to modulated light that is absorbed to a greater extent in PS II than in PS I (i.e. light-2). Two theoretical extreme cases are distinguished. For the case of non-interacting separated photosystems ("separate-package"), the modulated signals in the absence, O^-, and presence, O^+, of saturating light-1 are respectively proportional to $\phi_1 \alpha$ and $\phi_2 \beta$, where ϕ_1 and ϕ_2 are the maximum quantum yields in open centers for PS I and PS II, respectively, and α and β are the light distribution coefficients ($\phi_2 \beta$ > $\phi_1 \alpha$; $\alpha + \beta = 1$). Hence, the oxygen evolution signal is enhanced by the application of the background light-1 (Emerson enhancement effect). For $\phi_1 = \phi_2 = 1$, the magnitude of the enhancement, E, is given by $E = O^+/O^- = \beta/\alpha$, from which α and β are calculated. For interacting photosystems, with efficient excitation transfer from PS II closed reaction centers to PS I ("spill-over"), O^- would be independent of β, and O^+ would be proportional to β, so that $E = 2\beta$ (22). Another way in which the above two cases were distinguished used the light-1 saturation curve of the enhancement effect (17,19,22). Different values for the ratio of initial slope to the saturation value of the enhancement are predicted.

QUANTIFICATION OF LIGHT DISTRIBUTION FROM ENERGY STORAGE   ES, when measured with modulated light-1, is usually quite low (62, 75), reflecting the extremely uneven light distribution. However, ES is enhanced considerably when background light-2 is added, reflecting more opening of PS I by normal linear electron transport (43, 73, 75). This effect allows determination of the ratio α/β for the modulated light-1 (75). In red and brown marine algae (43, 73) the situation is more complex, with modulated light-1 measuring substantial ES, which was attributed to independent PS I activity (most likely cyclic electron flow, cf below). An analytical approach was developed to obtain light distribution and energy transfer parameters from the PA energy storage data (43, 73).

STATE 1–STATE 2 TRANSITIONS   By following the changes in light distribution parameters over a physiological time scale, PA studies characterized the different states of light distribution (states 1 and 2) and the transitions between them. In leaves, for example, the nearly reciprocal changes in α and β represented proof of the loss of pigments from PS II and a partial transfer of them to PS I, in the transition from state 1 to state 2 (21, 22). The effect of added phosphatase inhibitors on the PA transients demonstrated protein phosphorylation involvement (21, 27, 29). The involvement of chlorophyll b was indicated by the changes in the spectrum of Emerson enhancement (22), by the absence of the state 1–state 2 transitions in a LHC II–less mutant (22), and by the small E and faster state transitions (11) in mutants with low LHC II content. PA photobaric signals from the different sides of a leaf showed that palisade and spongy cells behave differently (90) with respect to these transitions.

Although ES is expected to increase in the transition from state 1 to state 2, as light energy becomes utilized more optimally by a better light distribution, there is a contradicting report for a maple tree leaf (98). In this case ES stays constant, although ES_{II} decreases as expected, following the decrease in β.

Observation of PA signals resulting from oxygen evolution in layered microalgae (17) and in lichens (18) was used to obtain information about the light distribution and state transitions for algal cells, either free or as symbionts, similar to the case for leaves. Otherwise, particularly in macroalgae, energy storage measurements were used (43, 73). These studies indicate that in phycobilisome-containing organisms several mechanisms may exist to balance the excitation energy between the two photosystems. In certain cases it seems that the two photosystems do not interact, each separately connected to phycobilisomes (17) with a possibility of detachment of PS I in state 1 (17, 18). In other cases light absorbed by the phycobilisomes was channeled to both photosystems, but in a different manner in the two states: in state 1, PS I receives excitation by energy transfer mainly from closed PS II reaction centers; in state 2, PS I becomes attached to the phycobilisomes and receives excitation directly from the phycobilisomes (73). A case was made for an invariability of ES, observed by a LIOAS study (78), which here is expected.

"Low Light" State

An adaptation to sufficiently low-level modulated light led to a state where the oxygen evolution photobaric signal decreased (23), although it was momentarily restored by background light-1. This state was thought to be a reflection of an extreme case of light-2 inefficiency in PS I; however, fluorescence measurements did not support this conclusion. Additional data on the possibility of oxygen uptake provided a better interpretation—that a gradual increase of oxygen uptake at the low light level opposes the oxygen evolution signal and decreases the total photobaric signal. This contribution tends to be eliminated when background light-1 is applied, as PS I becomes saturated.

The Effect of Possible Stromal Sinks

Another transient—decrease of the oxygen-evolution photobaric signal after attaining steady-state—was seen in a maple tree leaf, although it had completely different characteristics. This transient was not restored by light-1 and occured at ordinary light intensities. Intense saturating light reversed the transient. It was suggested (37) that the transient reflected the effect of oxygen sinks (respiration, photorespiration) present in the path of oxygen diffusion to the inner air phase. Such sinks cannot produce PA signals, but they can affect the already created modulated oxygen flux. Indeed, this occurs when the sink action is proportional to the total oxygen concentration (at low concentrations), but not when the oxygen concentration is saturating with respect to the sink action (a situation created in the strong light). More concrete evidence must be supplied, however, to support this concept. Part of the weak light intensity transient, described in the former section above, might also be related to the presence of sinks.

Indications for Independent Activities of PS I In Vivo

ES measurements can detect photoactivities in cases that otherwise are not easily probed in vivo, e.g. those arising from coupled or even futile electron-transport cycles. For example, relatively high ES values in modulated light-1 indicated photoactivity assigned to PS I alone in a variety of organisms (20, 27, 43, 62, 72, 73). This activity had characteristics that differed from activity at shorter wavelengths (e.g. it had a much smaller i_{50} value, was more resistant to stress or aging, and was inducible by stress conditions; 20, 27, 53). More work is needed to ascertain the type of reaction associated with this activity, although cyclic electron flow is a likely candidate (20). Based on this supposition, the correlation between changes in this activity and state 1–state 2 transitions was investigated to determine whether cyclic electron transport affects light distribution, but no effect was found (53).

Stress Physiology and Ecology

There are many aspects of PA measurements that are useful for studying stress conditions. For example, changes in the photobaric signal arise from both changes in the rate of oxygen evolution and from changes in the anatomical structural parameters of the leaf, which modify the oxygen diffusion from the chloroplasts to the inner air phase. Therefore, plots of Ln A_{OX}/A_{PT} vs $f^{1/2}$ must be examined (54, 55) to see separately the effects on the slope (diffusion parameters) and the intercept (true photobaric signal). Inhibition can be also monitored by the decline of ES. Here too, changes in the leaf anatomical parameters affect the reference maximum PA signal. Indeed, this is probably the main reason for the significant increase in the maximum photothermal signal after a long photoinhibitory treatment (50, 60). The assumption (50, 60) that the maximum photothermal signal increase at light saturation conditions signifies a corresponding proportional heat emission increase is unreasonable, considering the initial low fluorescence yield, which is presumably the only competing process under the measurement conditions. Calibrated correctly, however, a true increase of heat emission under light-limiting conditions was found (7, 8, 40) to follow photoinhibition conditions, with a particular correlation to the accumulation of zeaxanthin (40). Differential stress effects on the two photosystems may be monitored by changes in the Emerson enhancement parameters (55–57, 59). New activities may appear as the result of the stress [e.g. uptake photobaric signals after heat shock (50) or increased PS I dependent energy storage (20, 27, 59)].

PA methods have been used to monitor effects of water drought (54, 55), photoinhibition (27, 40, 50, 58), heat shock (56, 57, 59), chilling (101), and senescence (47). PA signals in lichens hung for long periods of time in air-polluted areas have been used to monitor a long-term pollution level (92). Aspects of environmental stress were investigated by PA, monitoring the effects of gaseous and dissolved SO_2 introduced to the foliage or to the soil (99). PA spectroscopic characterization of environmental stress has been made on conifer needles (79). The effects of herbicide and characterization of herbicide-resistant mutants have also been studied using PA (46, 51).

EPILOGUE

The above survey demonstrates the many research aspects that can be treated by the PA method for the study of photosynthesis. There are still many questions that remain unanswered, partly as a result of the existing limitations in the experimental gadgets and technical details, and partly because the PA method has so far remained a specialized technique. Hopefully, there will be a second phase of PA studies, in which its use will be expanded widely, based also on a more careful, rigorous approach and better design of cells and sensors.

Literature Cited

1. Adams MJ, Beadle BC, King AA, Kirkbright GF. 1976. Analytical photoacoustic spectrometery, part II. Ultra-violet and visible optoacoustic spectra of some inorganic, biochemical and phytochemical samples. *Analyst* 101:553–56
2. Arata H, Parson WW. 1981. Enthalpy and volume changes accompanying electron transfer from P-870 to quinones in *Rps. sphaeroides* reaction centers. *Biochim. Biophys. Acta* 636:70–81
2a. Arata H, Parson WW. 1981. Delayed luminescence from *Rps. sphaeroides* reaction centers. Enthalpy and free energy changes accompanying electron transfer from P870 to quinones. *Biochim. Biophys. Acta* 638:201–9
2b. Biggins J, ed. 1987. *Progress in Photosynthesis Research*, Vol. 2. Dordrecht: Nijhoff
3. Braslavsky SE. 1986. Photoacoustic and photothermal methods applied to the study of radiationless deeactivation process in biological systems and in substances of biological interest. *Photochem. Photobiol.* 43:667–75
4. Braslavsky SE, Heihoff K. 1989. Photothermal methods. In *CRC Handbook of Organic Photochemistry*, ed. JC Scaiano, 1:327–55. Boca Raton, FL: CRC Press
5. Bruce D, Salehian O. 1992. The efficiency of primary photosynthetic processes in state 1 and state 2. *Biochim. Biophys. Acta* 1100:242–50
6. Bults G, Horwitz BA, Malkin S, Cahen D. 1981. Photoacoustic measurements of photosynthetic activities in whole leaves—photochemistry and gas exchange. *Biochim. Biophys. Acta* 679:452–65

7. Buschmann C. 1987. Induction kinetics of heat emission before and after photoinhibition in cotyledons of *Raphanus sativus*. *Photosynth. Res.* 14:229–40

8. Buschmann C, Kocsányi L. 1989. Correlation between the induction kinetics of heat emission and that of chlorophyll fluorescence and its quenching mechanisms. *Photosynth. Res.* 21:129–36

9. Buschmann C, Prehn H. 1983. *In vivo* photoacoustic spectra of *Raphanus* and *Tradescantia* leaves taken at different chopping frequencies of the excitation light. *Photobiochem. Photobiophys.* 5:63–69

10. Buschmann C, Prehn H. 1990. Photoacoustic spectroscopy—photoacoustic and photothermal effects. In *Modern Methods in Plant Analysis*, ed. H-F Linskens, JF Jackson, 11:148–80. Berlin: Springer-Verlag

11. Buurmeijer WF, Wonders JHAM, Vredenberg WJ. 1990. State transitions in photomorphogenetic mutants of tomato. In *Current Research in Photosynthesis*, ed. M Baltscheffsky, 2:891–94. Dordrecht: Kluwer

12. Cahen D. 1981. Photoacoustic cell for reflection and transmission measurements. *Rev. Sci. Instr.* 52:1306–10

13. Cahen D, Malkin S, Lerner EI. 1978. Photoacoustic spectroscopy of chloroplast membranes: listening to photosynthesis. *FEBS Lett.* 91:339–42

14. Callis JB, Gouterman M, Danielson JDS. 1969. Flash calorimeter for measuring triplet yields. *Rev. Sci. Instr.* 40:1599–1605

15. Callis JB, Parson WW, Gouterman M. 1972. Fast changes of enthalpy and volume on flash excitation of *chromatium* chromatophores. *Biochim. Biophys. Acta* 267: 348–62

16. Camm EL, Popovic R, Lorrain L, Leblanc RM, Fragata M. 1988. Photoacoustic characterization of energy storage of photosystem 2 core enriched particles from barley isolated with octyl-β-D-glucopyranoside detergent. *Photosynthetica* 22: 27–32

17. Canaani O. 1986. Photoacoustic detection of oxygen evolution and state 1–state 2 transitions in cyanobacteria. *Biochim. Biophys. Acta* 852:74–80

18. Canaani O. 1987. Control of state 1–state 2 transitions in the blue-green alga *Nostoc muscorum*. See Biggins 1987, pp. 769–72

19. Canaani O. 1990. Photoacoustic studies on the dependence of state transitions on grana stacking. *Photosynth. Res.* 25:225–32

20. Canaani O. 1990. The role of cyclic electron flows around Photosystem I and excitation energy transfer distribution between the photosystems upon acclimation to high ionic stress in *Dunaliella salina*. *Photochem. Photobiol.* 52:591–99

21. Canaani O, Barber J, Malkin S. 1984. Evidence for phosphorylation and dephosphorylation *in vivo* in the mechanism for regulation of the distribution of excitation energy—photoacoustic and fluorimetric study of an intact leaf. *Proc. Natl. Acad. Sci. USA* 81:1614–18

22. Canaani O, Malkin S. 1984. Distribution of light excitation in an intact leaf between the two photosystems of photosynthesis—changes in absorption cross-sections following state 1–state 2 transitions. *Biochim. Biophys. Acta* 766:513–24

23. Canaani O, Malkin S. 1984. Physiological adaptation to low light intensity in intact leaves resulting in extreme imbalanced light distribution between the two photosystems of photosynthesis. *Biochim. Biophys. Acta* 766:525–32

24. Canaani O, Malkin S, Mauzerall D. 1988. Pulsed photoacoustic detection of flash-induced oxygen evolution from intact leaves and its oscillation. *Proc. Natl. Acad. Sci. USA* 85:4725–29

25. Canaani O, Motzan Z, Malkin S. 1985. Comparison between photosynthetic parameters of Tobacco aurea mutants Su/su and Tobacco wild type *in-vivo*. *Planta* 164: 480–86

26. Canaani O, Ronen R, Garty J, Cahen D, Malkin S, Galun M. 1984. Photoacoustic study of the green alga *Trebouxia* in the lichen *Ramalina duriaei in-vivo*. *Photosynth. Res.* 5:297–306

27. Canaani O, Schuster G, Ohad I. 1989. Photoinhibition in *Chlamydomonas reinhardtii*: effect on state transitions, intersystem energy distribution and Photosystem I cyclic electron flow. *Photosynth. Res.* 20: 129–46

28. Carpentier R, LaRue B, Leblanc RM. 1983. Photoacoustic spectroscopy of *Anacystis nidulans* I. effect of the sample thickness on the photoacoustic signal; II. characterizatrion of pigment holochroms and thermal deactivation spectrum. *Arch. Biochem. Biophys.* 222:403–15

29. Carpentier R, LaRue B, Leblanc RM. 1984. Photoacoustic spectroscopy of *Anacystis nidulans* III. Detection of photosynthetic activities. *Arch. Biochem. Biophys.* 228: 534–43

30. Carpentier R, Leblanc RM, Mimeault M. 1988. Monitoring electron transfer by photoacoustic spectroscopy in native and immobilized thylakoid membranes. *Biotechnol. Bioeng.* 32:64–67

31. Carpentier R, Leblanc RM, Mimeault M. 1989. Photoacoustic detection of photosynthetic energy storage in Photosystem II submembrane fractions. *Biochim. Biophys. Acta* 975:370–76

32. Carpentier R, Leblanc RM, Mimeault M. 1990. On the nature of the photosynthetic

energy monitored by photoacoustic spectroscopy. *Photosynth. Res.* 23:313–18

33. Carpentier R, Matthijs HCP, Leblanc RM, Hind G. 1986. Monitoring energy conversion in photosystem I of cyanobacteria heterocysts by photoacoustic spectroscopy. *Can. J. Phys.* 64:1136–38

34. Carpentier R, Nakatani H, Leblanc RM. 1985. Photoacoustic detection of energy conversion in a Photosystem II submembrane preparation from spinach. *Biochim. Biophys. Acta* 808:470–73

35. Carpentier R, Owens TG, Leblanc RM. 1991. Simultaneous saturation of variable fluorescence yield and photoacoustically monitored thermal emission in thylakoid membranes. *Photochem. Photobiol.* 53: 565–69

36. Cha Y, Mauzerall D. 1992. Energy storage of linear and cyclic electron flows in photosynthesis. *Plant Physiol.* 100:1869–77

37. Charland M, Veeranjaneyulu K, Charlebois D, Leblanc RM. 1992. Photoacoustic signal generation in leaves: are O_2-consuming processes involved? *Biochim. Biophys. Acta* 1098:261–65

38. Dau H, Hansen U-P. 1990. A study on the energy-dependent quenching of chlorophyll fluorescence by means of photoacoustic measurements. *Photosynth. Res.* 25:269–78

39. Ducharme D, Tessier A, Leblanc RM. 1979. Design and characteristics of a cell for photoacoustic spectroscopy of condensed matter. *Rev. Sci. Instr.* 50:1461–62

40. Eyletters M, Lannoye R. 1992. Photoacoustically monitored energy dissipation and xanthophyll cycle compared in two durum wheat varieties exposed to a highlight treatment. In *Springer Series in Optical Sciences: Photoacoustic and Photothermal Phenomena III*, ed. D Bicanic, 69: 62–64. Heidelberg: Springer-Verlag

41. Fork DC, Herbert SK. 1991. A gas-permeable photoacoustic cell. *Photosynth. Res.* 27:151–56

42. Fork DC, Herbert SK. 1993. The application of photoacoustic techniques to studies of photosynthesis. *Photochem. Photobiol.* 57:207–20

43. Fork DC, Herbert SK, Malkin S. 1991. Light energy distribution in the brown alga *Macrocystis pyrifera* (giant kelp). *Plant Physiol.* 95:731–39

43a. Fork DC, Satoh K. 1986. The control by state transitions of the distribution of excitation energy in photosynthesis. *Annu. Rev. Plant Physiol.* 37:335–61

44. Frackowiak D, Cegielski R, Leblanc RM. 1990. Thermal deactivation of excitation of *Rhodospirillum rubrum* cells and their fragments immobilized in polymer films. *Photosynthetica* 24:85–95

45. Frackowiak D, Hendrich W, Romanowski M, Sczepaniak A, Leblanc RM. 1987. Polarized spectra of photosystem 2 particles in isotropic and streched polyvinylalcohol films. *Photosynthetica* 21:147–55

46. Fuks B, Homble F, Van Eycken F, Figeys H, Lannoye R. 1992. Photoacoustic spectroscopy as a tool for monitoring herbicide effects on triazine-resistant and susceptible biotypes of black nightshade (*Solanum nigrum*). *Weed Sci.* 40:371–77

47. Gruszecki WI, Veeranjaneyulu K, Bogumil Z, Leblanc RM. 1991. Energy transfer processes during senescence: fluorescence and photoacoustic studies of intact pea leaves. *Biochim. Biophys. Acta* 1056:173–80

48. Harren FHM, Reuss J, Woltering EJ, Bicanic DD. 1990 Photoacoustic measurements of agriculturally interesting gases and detection of C_2H_4 below the ppb level. *Appl. Spectr.* 44:1360–68

49. Havaux M. 1988. Induction of photosynthesis in intact leaves under normal and stressing conditions followed simultaneously by transients in chlorophyll fluorescence and photoacoustically monitored O_2 evolution. *Plant Physiol. Biochem.* 26: 695–704

50. Havaux M. 1989. Increased thermal deactivation of excited pigments in pea leaves subjected to photoinhibitory treatments. *Plant Physiol.* 89:286–92

51. Havaux M. 1989. Photoacoustic characteristics of leaves of atrazine resistant weed mutants. *Photosynth. Res.* 21:51–59

52. Havaux M. 1990. "Energy"-dependent quenching of chlorophyll fluorescence and thermal energy dissipation in intact leaves during induction of photosynthesis. *Photochem. Photobiol.* 51:481–86

53. Havaux M. 1993. Photoacoustic measurements of cyclic electron flow around photosystem I in leaves adapted to light-states 1 and 2. *Plant Cell Physiol.* 33:799–803

54. Havaux M, Canaani O, Malkin S. 1986. Photosynthetic responses of leaves to water stress, expressed by photoacoustics and related methods. I. Probing the photoacoustic method as an indicator for water stress in vivo. II. The effect of rapid drought on the electron transport and the relative activities of the two photosystems. *Plant Physiol.* 82:827–39

55. Havaux M, Canaani O, Malkin S. 1987. Inhibition of photosynthetic activities under slow water stress measured *in vivo* by the photoacoustic method. *Physiol. Plant* 70:503–10

56. Havaux M, Canaani O, Malkin S. 1987. Extreme changes in the distribution of excitation energy in the photochemical apparatus of intact leaves induced by progressive heat stress. See Ref. 2b, pp. 749–52

57. Havaux M, Canaani O, Malkin S. 1987.

Oxygen uptake by tobacco leaves after heat shock. *Plant Cell Environ.* 10:677–83
58. Havaux M, Eyletters M. 1991. Is the *in vivo* photosystem-I function resistant to photoinhibition? An answer from photoacoustic and far-red absorbance measurements in intact leaves. *Z. Naturforsch. Teil C* 46:233–38
59. Havaux M, Greppin H, Strasser RJ. 1991. Functioning of photosystems I and II in pea leaves exposed to heat stress in the presence or absence of light: analysis using *in vivo* fluorescence, absorbance, oxygen and photoacoustic measurements. *Planta* 186:88–98
60. Havaux M, Gruszecki WI, Dupont I, Leblanc RM. 1991. Increased heat emission and its relationship to the xanthophyll cycle in pea leaves exposed to strong light stress. *J. Photochem. Photobiol. B* 8:361–70
61. Havaux M, Lorrain L, Leblanc RM. 1990. Photothermal beam deflection: a new method for *in vivo* measurements of thermal energy dissipation and photochemical energy conversion in intact leaves. *Photosynth. Res.* 24:63–73
62. Herbert SK, Fork DC, Malkin S. 1990. Photoacoustic measurements *in-vivo* of energy storage by cyclic electron flow in algae and higher plants. *Plant Physiol.* 94:926–34
63. Inoue Y, Watanabe A, Shibata K. 1979. Transient variation of photoacoustic signal from leaves accompanying photosythesis. *FEBS Lett.* 101:321–23
64. Jabben M, Schaffner K. 1985. Pulsed laser induced optoacoustic spectroscopy of intact leaves. *Biochim. Biophys. Acta* 809:445–51
64a. Joliot P. 1965. Cinétique des reactions liées à l'émission d'oxygén photosynthétique. *Biochim. Biophys. Acta* 102:116–34
65. Kanstad SO, Cahen D, Malkin S. 1983. Simultaneous detection of photosynthetic energy storage and oxygen evolution in leaves by photothermal radiometry and photoacoustics. *Biochim. Biophys. Acta* 722:182–89
66. Katoh S, Yamagishi A. 1984. Parallel inductive kinetics of fluorescence and photoacoustic signal in dark-adapted thalli of *Bryopsis maxima. Biochim. Biophys. Acta* 767:185–91
67. Kolbowski J, Reising H, Schreiber U. 1990. Computer-controlled pulse modulation system for analysis of photoacoustic signals in the time domain. *Photosynth. Res.* 25:309–16
67a. Krause GH, Weis E. 1991. Chlorophyll fluorescence and photosynthesis: the basics. *Annu. Rev. Plant Physiol. Plant Mol. Biol.* 42:313–49
68. Lasser-Ross N, Malkin S, Cahen D. 1980.

Photoacoustic detection of photosynthetic activities in isolated broken chloroplasts. *Biochim. Biophys. Acta* 593:330–41
69. Malkin S. 1986. Photoacoustic probing of energy storage and gas exchange in the photosynthesis of leaves. *J. Chem. Soc. Faraday Trans. 2.* 82:2233–35
70. Malkin S. 1987. Fast photoacoustic transients from dark adapted intact leaves—I. Oxygen evolution and uptake pulses. A phenomenological record. *Planta* 171:65–72
71. Malkin S, Cahen D. 1979. Photoacoustic spectroscopy and radiant energy conversion: theory of the effect with special emphasis on photosynthesis. *Photochem. Photobiol.* 29:803–13
72. Malkin S, Charland M, Leblanc RM. 1992. A photoacoustic study of water infiltrated leaves. *Photosynth. Res.* 33:37–50
72a. Malkin S, Churio MS, Shochat S, Braslavsky S. 1994. Photochemical energy storage and volume changes in the microsecond range in bacterial photosynthesis—a laser induced optoacoustic study. *Photochem. Photobiol.* In press
73. Malkin S, Herbert SK, Fork DC. 1990. Light distribution, transfer and utilization in the marine red alga *Porphyra perforata* from photoacoustic energy storage measurement. *Biochim. Biophys. Acta* 1016:177–89
74. Malkin S, Lasser-Ross N, Bults G, Cahen D. 1981. Photoacoustic spectroscopy in photosynthesis. In *Structure and Molecular Organization of the Photosynthetic Apparatus,* ed. G Akoyunoglou, 3:1031–42. Philadelphia: Balaban
75. Malkin S, Schreiber U, Jansen M, Canaani O, Shalgi E, Cahen D. 1992. The use of photothermal radiometry in assessing leaf photosynthesis: I. General properties and correlation of energy storage to P_{700} redox state. *Photosynth. Res.* 29:87–96
76. Mauzerall D. 1990. Determination of oxygen emission and uptake in leaves by pulsed, time resolved, photoacoustics. *Plant Physiol.* 94:278–83
77. Moore TA. 1983. Photoacoustic spectroscopy and related techniques applied to biological materials. In *Photochemical and Photobiological Reviews,* ed. KC Smith, 7:187–221. New York: Plenum
78. Mullineaux CW, Griebenow S, Braslavsky SE. 1991. Photosynthetic energy storage in cyanobacterial cells adapted to light-states 1 and 2. A laser-induced optoacoustic study. *Biochim. Biophys. Acta* 1060:315–18
79. Nagel E, Buschmann C, Lichtenthaler HK. 1987. Photoacoustic spectra of needles as an indicator for photosynthetic capacity of healthy and damaged conifers. *Physiol. Plant* 70:427–37
80. Nery JW, Pessoa O, Vargas H, Reis FAM,

Gabrielli AC, Miranda LCM, Vinha CA. 1987. Photoacoustic spectroscopy for depth-profile analysis and herbicide monitoring in leaves. *Analyst* 112:1487–90

81. Nitsch C, Braslavsky SE, Schatz GH. 1988. Laser-induced optoacoustic calorimetry of primary processes in isolated photosystem I and photosystem II particles. *Biochim. Biophys. Acta* 934:201–12

82. Nitsch C, Schatz GH, Braslavsky SE. 1989. Laser-induced optoacoustic calorimetry of primary processes in cells of *Rhodospirillum rubrum*. *Biochim. Biophys. Acta* 975:88–95

83. N'soukpoé-Kossi CN, Belanger R, Keilani S, Proteau H, Boivin P, Leblanc RM. 1990. Short-term acid damage to photosynthesis in corn and sugar maple leaves assessed by photoacoustic spectroscopy. *Can. J. Bot.* 68:2292–300

84. Oehler O, Blum H. 1990. Photoacoustic measurement of plant gas exchange. In *Springer Series in Optical Sciences: Photoacoustic and Photothermal Phenomena II*, ed. JC Murphy, JW Maclachlan-Spicer, LC Aamodt, BSH Royce, 62:369–71. Berlin: Springer-Verlag

85. O'Hara EP, Tom RD, Moore TA. 1983. Determination of the *in vivo* absorption and photosynthetic properties of the lichen *Acarospora schleicheri* using photoacoustic spectrum. *Photochem. Photobiol.* 38:709–15

86. Owens TG, Carpentier R, Leblanc RM. 1990. Detection of photosynthetic energy storage in a photosystem I reaction center preparation by photoacoustic spectroscopy. *Photosynth. Res.* 24:201–8

87. Peters KS, Synder GJ. 1988. Time resolved photoacoustic calorimetry: probing the energetics and dynamics of fast chemical and biochemical reactions. *Science* 241:1053–57

88. Popovic R, Leblanc RM, Beauregard M, 1988. Photoacoustic studies of bundle-sheath cell photosynthesis in *Zea mays*. *J. Plant Physiol.* 132:94–97

89. Poulet P, Cahen D, Malkin S. 1983. Photoacoustic detection of photosynthetic oxygen evolution from leaves—quantitative analysis by phase and amplitude measurements. *Biochim. Biophys. Acta* 724:433–46

90. Punnett T. 1987. Environmental control of photosynthetic enhancement in palisade mesophyll chloroplasts measured by photoacoustic spectroscopy. See Ref. 2a, pp. 753–56

91. Reising H, Schreiber U. 1992. Pulse-modulated photoacoustic measurements reveal strong gas-uptake component at high CO_2 concentration. *Photosynth. Res.* 31:227–38

92. Ronen R, Canaani O, Garty J, Cahen D, Malkin S, Galun M. 1983. The effect of air pollution and bisulfite treatment in the lichen *Ramalina duriaei* studied by photoacoustics. In *Advances in Photosynthesis Research*, ed. C Sybesma, 4:251–54. The Hague: Junk

93. Rosencwaig A. 1980. *Photoacoustics and Photoacoustic Spectroscopy*. New York: Wiley

94. Snel JFH, Polm MW, Buurmeijer WF, Vredenberg WJ. 1992. Deconvolution of photobaric and photothermal signals from spinach leaves. In *Springer Series in Optical Sciences: Photoacoustic and Photothermal Phenomena III*, ed. D Bicanic, 69:65–68. Heidelberg: Springer

95. Snel JFH, van Leperen W, Vredenberg WJ. 1990. Complete suppression of oxygen evolution in open PS2 centers by non-photochemical fluorecence quenching? In *Current Research in Photosynthesis*, ed. M Baltschevsky, 2:911–14. Dordrecht: Kluwer

96. Snel JFH, Kooijman M, Vredenberg WJ. 1990. Correlation between chlorophyll fluorescence and photoacoustic signal transients in spinach leaves. *Photosynth. Res.* 25:259–68

97. Veeranjaneyulu K, Charland M, Charlebois DCN, Leblanc RM. 1991. Photosynthetic energy storage of photosystems I and II in the spectral range of photosynthetically active radiation in intact sugar maple leaves. *Photosynth. Res.* 30:131–38

98. Veeranjaneyulu K, Charland M, Charlebois DCN, Leblanc RM. 1991. Photoacoustic study of changes in energy storage of PS I and PS II during state 1–state 2 transition. *Plant Physiol.* 97:330–34

99. Veeranjaneyulu K, N'soukpoé-Kossi CN, Leblanc RM. 1991. SO_2 effect on photosynthetic activities of intact sugar maple leaves as detected by photoacoustic spectroscopy. *Plant Physiol.* 97:50–54

100. Wróbel D, Hendrich W. 1989. Thermal deactivation and energy transfer in isolated photosystem II and light-harvesting complexes in polyvinyl alcohol film. *J. Photochem. Photobiol. B* 3:319–32

101. Yakir D, Rudich J, Bravdo BA, Malkin S. 1986. Prolonged chilling under moderate light—effect on photosynthetic activity measured with the photoacoustic method. *Plant Cell Environ.* 9:581–88

Annu. Rev. Plant Physiol. Plant Mol. Biol. 1994. 45:527–44

PLANT HORMONE-INDUCED CHANGES IN THE ORIENTATION OF CORTICAL MICROTUBULES:
Alterations in the Cross-linking Between Microtubules and the Plasma Membrane

Hiroh Shibaoka

Department of Biology, Faculty of Science, Osaka University, Toyonaka, Osaka 560, Japan

KEY WORDS: cell expansion, cross-bridge, microtubule, plasma membrane, plant hormone

CONTENTS

INTRODUCTION 528
HORMONE-INDUCED CHANGES IN THE ORIENTATION OF CORTICAL
     MICROTUBULES 528
   Auxins 528
   Gibberellins 530
   Cytokinins 533
   Ethylene 533
   Abscisic Acid 534
   Other Hormones 534
INTERACTIONS BETWEEN CORTICAL MICROTUBULES AND THE PLASMA
     MEMBRANE 535
   Dissociation of Microtubules from the Plasma Membrane 535
   Stabilization and Destabilization of Cortical Microtubules 537
   *Factors that Mediate the Association of Microtubules with the Plasma
     Membrane* 538
CONCLUDING REMARKS 540

INTRODUCTION

The morphology of an individual plant depends to a great extent on the shape of its individual cells. The shape of each plant cell depends, in turn, on the direction of cell expansion, which is regulated by the properties of the cell wall and, in particular, by the orientation of cellulose microfibrils in the wall (71). Cortical microtubules (MTs) have long been considered to control the orientation of cellulose microfibrils and, in this way, to control the direction of cell expansion (18, 20, 26, 52, 73). Plant hormones, which also influence plant morphology, change cortical MT orientation; therefore, hormone-regulated MT orientation has been postulated to be an important aspect of hormone-mediated control of plant morphology (22, 83). However, because hormone-induced MT reorientation is usually accompanied by cell expansion rate changes, which have been considered important factors in MT reorientation (19, 97, 98), hormone-induced MT reorientation is sometimes regarded as a result, rather than a cause, of hormone-controlled cell expansion (100). In this review, we focus on whether hormone-induced MT reorientation occurs as a result of hormone-induced changes in the rate of cell expansion or via a different mechanism.

Because the orientation of newly deposited cellulose microfibrils changes when plant hormones change the orientation of cortical MTs (2, 7, 49, 91), MTs are considered to regulate the orientation of cellulose microfibrils either before or after their own reorientation. Although the mechanism of this regulation is not fully understood, cortical MTs must be associated with the plasma membrane if they are to regulate the movement of the cellulose-synthesizing complexes (17); therefore, MTs must be associated with the plasma membrane either before or after the hormone-induced reorientation of MTs. Moreover, hormone-induced MT reorientation must involve MT depolymerization or dissociation from the plasma membrane, as well as MT reorganization or reassociation with the membrane. This review also examines the mechanism of this dissociation and reassociation.

HORMONE-INDUCED CHANGES IN THE ORIENTATION OF CORTICAL MICROTUBULES

Auxins

Auxins, such as indoleacetic acid (IAA), induce cell elongation by changing the mechanical properties of the cell wall (57). In azuki bean epicotyls, MTs are not involved in auxin-induced cell elongation since disruption of cortical MTs by colchicine (81) or ethyl N-phenylcarbamate (85) does not inhibit IAA-induced epicotyl elongation. In this plant system, IAA does not alter cortical MT orientation (41, 82); however, IAA causes cortical MTs to orient themselves transversely to the cell axis in tissues of several plant species [e.g. wheat coleoptiles (94), oat mesocotyls (33), maize coleoptiles (7, 67, 100), and radish hypocotyls (76)]. Despite these observations, it remains unclear whether cortical MT reorientation is involved in auxin-induced cell elongation in these tissues because disruption of cortical MTs does not reduce the auxin-induced cell elongation, except in wheat coleoptiles, in which colchicine reduces elongation by about one-third (51). Because colchicine is unable to reduce auxin-induced cell elongation in maize coleoptiles, Zandomeni & Schopfer (100) proposed that auxin-induced reorientation is not involved in auxin-induced cell elongation.

The observation that IAA changes the orientation of cortical MTs in gibberellin A_3 (GA$_3$)-pretreated azuki bean epicotyls but not in untreated epicotyls (41) may explain why auxin causes cortical MT reorientation in some plant tissues and not in others. The effect of GA$_3$ pretreatment on MT reorientation is negated by the simultaneous application of actinomycin D, which indicates that GA$_3$-induced mRNA synthesis is involved in MT reorientation (see below for further details). Auxin appears to cause the predominance of transverse MTs in cells that possess a factor identical or equivalent to the mRNA that is supposed to be synthesized in azuki bean epicotyl cells during the pretreatment with GA$_3$. The cells in which auxin causes cortical MT reorientation seem to have synthesized such a factor in response to the endogenous gibberellins (GAs).

Transverse MTs predominate under conditions favorable for cell elongation, for example, in the presence of auxin or GAs (83). Longitudinal MTs predominate under conditions that are unfavorable for cell elongation, such as during irradiation with visible light (33, 50, 100), during the later stages of aging (28), and in the presence of growth inhibitors (32, 74–76). The question of whether cortical MT reorientation under such conditions is a cause or a result of the accelerated or the suppressed cell elongation has been debated (50, 75, 100). The observation that auxin changes the arrangement of cortical MTs before it induces cell elongation in maize coleoptiles (67) and in radish hypocotyls (76) indicates that auxin is able to induce the arrangement of cortical MTs by a mechanism that does not involve the cell-elongation properties of auxin. In GA$_3$-pretreated epicotyl segments of azuki bean seedlings, cortical MT reorientation does not become apparent until auxin-induced cell elongation becomes discernible (41). It is thus unclear whether the reorientation of MTs requires the expansion of cells in azuki bean epicotyls, although the same amount of cell elongation does not cause MT reorientation in GA$_3$-unpretreated epicotyls (41), which indicates that the cell elongation alone does not cause MT reorientation (see below for further details).

It is unclear whether auxin-induced cortical MT reorientation is involved in auxin-induced cell elongation, but the reorientation clearly is not involved in short-term auxin-induced cell elongation because the elongation that involves cortical MT reorientation involves another time-consuming process—the reinforcement of the cell wall by the deposition of newly synthesized cellulose—which occurs after MTs have become reoriented. It is interesting to note that 2,6-dichlorobenzonitrile, a specific inhibitor of cellulose synthesis (30), does not inhibit auxin-induced cell elongation in maize coleoptiles (13), in which auxin causes cortical MT reorientation (7, 67, 100), or in pea epicotyls (8). Since the short-term auxin-induced cell elongation does not involve auxin-induced cortical MT reorientation, it is unlikely that MT reorientation is involved in the tropic responses of plants (67, 76). The involvement of MT reorientation in tropic bending was indicated by the apparent correlation between the growth rate and the orientation of cortical MTs on the concave and convex sides of bending plant shoots (67); however, a detailed reexamination of this phenomenon indicated that such a correlation was not always observed in bending tissues (68). The GA3-induced cell elongation in azuki bean epicotyls that involves cortical MT reorientation does not become apparent for the first one to two hours (41, 81). This observation also indicates that short-term auxin-induced elongation does not involve MT reorientation.

Colchicine inhibits the IAA-induced cell elongation by about 30% in wheat coleoptiles (51), in which IAA is known to cause cortical MT reorientation (94). This coincides with the result that colchicine does not inhibit the elongation caused by IAA alone but does inhibit the additional elongation caused by GA3 (81). IAA seems to regulate cell expansion via at least two processes (one colchicine-insensitive process and one colchicine-sensitive process) in wheat coleoptiles and in GA3-treated epicotyl segments of azuki bean seedlings. In the former process, IAA induces the loosening of cell walls and in the latter it regulates the orientation of cortical MTs. The latter process is considered to be accomplished only in the presence of specific mRNA(s), the synthesis of which is induced by endogenous or exogenous GAs. Studies examining the effects of auxins on the arrangement of cortical MTs in GA-depleted mutants or in plants in which the biosynthesis of endogenous GAs is suppressed by so-called growth retardants should tell us whether this hypothesis is valid. If it is valid, auxins do not cause the reorientation of MTs in such plants.

Gibberellins

GAs, which promote longitudinal cell expansion and suppress lateral expansion in stems (83, 84), cause a predominance of transverse cortical MTs in tissues of various plant species [e.g. azuki bean epicotyls (41, 82, 85), onion leaf-sheaths (63), dwarf maize mesocotyls (31, 61), dwarf pea epicotyls (2, 74, 75), and cucumber hypocotyls (32)]. In the internodes of deepwater rice, GA3 prevents the transverse to longitudinal reorientation of cortical MTs, which occurs with the progression of cell maturation (77). Treatment with S-3307 (uniconazol), an inhibitor of the biosynthesis of GAs, causes cortical MTs to orient themselves obliquely or longitudinally to the cell axis in onion leaf-sheath cells (64), indicating that endogenous GAs are involved in maintaining a transverse array of MTs. In azuki bean epicotyls, the GA₃-induced elongation is reversed by MT-disrupting agents, such as colchicine (81) and ethyl N-phenylcarbamate (85), and by inhibitors of cellulose synthesis, such as coumarin (29) and 2,6-dichlorobenzonitrile (30). GA₃ increases the percentage of cells with transversely arranged cellulose microfibrils on the inner surface of the cell walls in epicotyls of both azuki bean (91) and dwarf pea (2). Thus, it appears that GAs cause changes in the orientation of newly deposited cellulose microfibrils by changing the orientation of cortical MTs and, in this way, GAs regulate the direction of cell expansion (83, 84).

MTs do not seem to be involved in short-term GA-induced cell expansion. In lettuce hypocotyls, GA₃ induces cell expansion within 10–15 minutes of treatment (66a) and this GA₃-induced cell expansion is not inhibited by colchicine (12a). In this plant material, GA₃ does not cause MT reorientation (77a).

Since GA₃ causes cortical MTs to orient transversely to the cell axis without changing the rate of cell elongation in onion leaf-sheaths (63), the GA₃-induced reorientation of MTs does not seem to be a result of the GA₃-induced cell elongation. The recently reported result that GA₃ causes a predominance of cortical MTs in epicotyl cells of decapitated dwarf pea seedlings, in which GA₃ does not change the rate of cell elongation (75), also indicates that GA₃ is able to control cortical MT orientation by a mechanism that does not involve a GA₃-induced change in the rate of cell elongation.

GA₃ causes a predominance of transverse MTs only when it is applied together with auxin. GA₃ alone does not change the orientation of MTs in segments of azuki bean epicotyls (41); however, it (or GA₄) alone does cause a predominance of transverse MTs in onion leaf-sheaths (63), dwarf-maize mesocotyls (31, 61), dwarf-pea epicotyls (2, 74, 75), and cucumber hypocotyls (32) when it is applied to intact plants. The observation that GAs change cortical MT orientation in intact plants, in which cells are supplied with endogenous auxin, but not in excised segments, in which endogenous auxin is depleted, indicates that auxin is required for GA-induced cortical MT reorientation. Because cortical MTs reorient before auxin induces cell elongation in maize coleoptiles (67) and in radish hypocotyls (76), it appears that auxin, and not cell elongation, is required for MT reorientation. Detailed examination of the relationship between MT reorientation and cell elongation is needed to clarify how auxin cooperates with GAs in inducing cortical MT reorientation.

Although it has been well established that GA-induced mRNA synthesis is involved in GA-induced synthesis of hydrolytic enzymes in cereal aleurone layers (36), it has not yet been determined whether GA-induced mRNA synthesis is involved in GA-induced cell elongation. GAs require auxin if they are to cause cell elongation and auxin-induced cell elongation is inhibited by RNA or protein synthesis inhibitors (14); thus, the result that inhibitors of mRNA synthesis also inhibit GA-induced cell elongation does not necessarily imply that GA-induced mRNA synthesis is involved in GA-induced cell elongation. GA3, applied before treatment with IAA, stimulates IAA-induced cell elongation, which allows us to examine whether GA-induced mRNA synthesis is involved in GA-induced cell elongation. Pretreatment with GA3, when followed by treatment with IAA, causes elongation of segments of azuki bean epicotyls and causes a predominance of transverse MTs (41). Actinomycin D, an inhibitor of mRNA synthesis, applied during the GA3 pretreatment, reverses the promotion of GA3-induced cell elongation and inhibits GA3-induced reorientation of MTs, an indication that GA3-induced mRNA synthesis is involved in these phenomena (41). Because pretreatment with actinomycin D does not inhibit the cell elongation that occurs during the IAA-posttreatment, it is clear that the amount of actinomycin D remaining in the cells is not sufficient to inhibit the processes induced by IAA-posttreatment. GA3 stimulates cell elongation and increases the level of tubulin transcripts in *Avena* internodes, but the increase is inhibited by several drugs that inhibit cell elongation, an indication that expression of the gene for tubulin is closely associated with cell elongation (60).

Although inhibitors of protein kinases do not inhibit the IAA-induced cell elongation in azuki bean epicotyls, they prevent the additional elongation caused by GA3 and inhibit GA3-induced MT reorientation (66). In experiments in which GA3 and IAA are applied sequentially, 6-dimethylaminopurine, an inhibitor of protein kinases, does not inhibit GA3-induced cell elongation when it is applied during pretreatment with GA3, but it does inhibit GA3-induced cell elongation when it is applied during posttreatment with IAA (T Kaneta, unpublished observation). This observation indicates that the protein kinase inhibitor does not inhibit steps at which GA3 is perceived by cells and induces mRNA synthesis. The inhibitor seems to inhibit the actual reorientation of MTs.

It has not yet been determined if the depolymerization of cortical MTs is involved in the GA3-induced reorientation of cortical MTs. Recently, Wasteneys et al (95) observed the dynamics of cortical MTs in living cells using a method that included microinjection of brain tubulin with a fluorescent label into living cells (101). Such methods may be useful for determining whether GA3-induced MT reorientation involves depolymerization of MTs. Cultured cells derived from cotton ovule callus elongate in response to GA3 (11). If the cortical MTs change their orientation in response to GAs, this cell line should be useful for such examinations.

GA3 stabilizes cortical MTs when it causes them to become arranged transversely to the cell axis in onion leaf-sheath cells (63). MT-disrupting agents, such as colchicine and cremart, disrupt cortical MTs in the leaf-sheath cells and cause swelling of the basal parts of the leaf-sheaths, but they do not have this effect in GA3-pretreated seedlings. Cortical MTs in onion leaf-sheath cells are sensitive to cold and are disrupted by cold treatment. Treatment with GA3 renders MTs resistant to cold. GA3 also stabilizes cortical MTs in maize mesocotyl cells (21). Cortical MTs in protoplasts prepared from mesocotyls of GA3-treated maize seedlings are more cold-stable than those prepared from untreated seedlings (21), indicating that the cell wall is not involved in the GA3-induced stabilization of cortical MTs. In contrast, although GA3 causes the transverse arrangement of MTs in dwarf-pea epicotyl cells (2, 74), MTs in GA3-treated seedlings are cold-sensitive, whereas longitudinal MTs in untreated seedlings are cold-resistant.

Cytokinins

Cytokinins, such as kinetin, inhibit the auxin-induced elongation of stems and cause stem thickening. MT-disrupting agents, such as colchicine (82) and ethyl N-phenylcarbamate (85), partially alleviate the inhibition by kinetin of the IAA-induced elongation of azuki bean epicotyl segments. This inhibition is also alleviated by cellulose synthesis inhibitors, such as coumarin (29) and 2,6-dichlorobenzonitrile (30). These results indicate that MTs are involved in kinetin-induced inhibition of epicotyl elongation and that cellulose synthesis is required for this inhibition. These findings also indicate, together with the observation that kinetin causes cortical MTs to become oriented longitudinally to the cell axis (82, 94), that kinetin regulates the direction of cell expansion by causing this longitudinal orientation of cortical MTs, thereby causing the arrangement of newly deposited cellulose microfibrils in the same orientation.

Ethylene

Ethylene inhibits longitudinal cell expansion and promotes lateral expansion in stems. Treatment with ethylene causes a predominance of longitudinal MTs (49, 72, 90) and results in the deposition of longitudinal cellulose microfibrils in cell walls (15). As a result, ethylene has been considered to regulate the direction of cell expansion by causing cortical MTs to orient longitudinally to the cell axis, which in turn regulates the orientation of newly laid down cellulose microfibrils. The reorientation of cortical MTs by ethylene becomes evident 30 min after the start of treatment when the ethylene-induced lateral cell expansion has not yet become apparent (72). Ethylene appears to reorient cortical MTs by a mechanism that does not involve changes in the rate of cell expansion. MTs are not depolymerized during reorientation, so it appears that the depolymerization of MTs is not involved in the ethylene-induced reorientation of MTs (72).

Abscisic acid

Abscisic acid (ABA), which inhibits stem elongation, causes cortical MTs to become arranged longitudinally to the cell axis in dwarf-pea epicotyls (74, 75) and in cucumber hypocotyls (32). As described above, GA3 promotes cell elongation and causes a predominance of transverse cortical MTs in dwarf-pea epicotyls. ABA applied after treatment with GA3 decreases the rate of GA3-induced epicotyl elongation, and it eliminates GA3-induced predominance of transverse MTs and causes a predominance of longitudinal MTs (74). ABA also increases the stability of MTs to cold after GA3 has decreased such stability (74). The orientation change of cortical MTs is preceded by evidence of the ABA-induced inhibition of elongation of pea epicotyls. Therefore, the inhibition of epicotyl elongation by ABA, which appears soon after application of the hormone, is not a direct consequence of the ABA-induced change in MT orientation (74). However, because ABA causes a predominance of longitudinal cortical MTs in decapitated cuttings of dwarf pea epicotyls, in which ABA does not change the rate of epicotyl elongation (75), it is clear that the ABA-induced change of cortical MT orientation does not involve ABA-induced changes in the rate of cell elongation. Although the early effects of ABA on epicotyl elongation do not involve MTs, the predominance of longitudinal MTs caused by ABA might result in the ABA-induced cessation of elongation of epicotyls.

ABA reduces the extensibility of the cell wall in maize coleoptiles (46). The deposition of longitudinally oriented cellulose microfibrils in the wall, which occurs under the regulation of longitudinally oriented MTs, is responsible for the ABA-induced reduction in extensibility of the cell wall. Water stress, which increases the level of ABA in cells and inhibits cell elongation, increases the percentage of longitudinally or obliquely oriented MTs in cucumber hypocotyls (42). The reorientation of MTs by ABA seems to be involved in the inhibition of growth caused by water stress.

Other Hormones

Jasmonic acid and related compounds induce the formation of potato tubers (44). The levels of jasmonic acid increase (69) and cortical MTs disappear (62) in leaf-sheath cells of onion plants as the cells swell to form bulbs. As might be expected from these observations, jasmonic acid and methyl jasmonate disrupt cortical MTs in tobacco BY-2 cells (1). However, since these compounds exert their MT-disrupting effects only in cells at the S phase of the cell cycle, it is implausible that they mediate the formation of onion bulbs. Methyl jasmonate disrupts MTs in cultured potato cells but such disruption occurs only in a small fraction of treated cells (58).

Raphanusanin, a natural growth inhibitor from radish that is postulated to play an important role both in the light-induced inhibition of elongation and in phototropism in radish hypocotyls, inhibits the auxin-induced reorientation of cortical MTs (76). Although this inhibition is hypothesized to be involved in phototropism of radish hypocotyls (76), this hypothesis is untenable because, as mentioned above, short-term auxin-induced elongation does not involve cortical MT reorientation.

INTERACTIONS BETWEEN CORTICAL MICROTUBULES AND THE PLASMA MEMBRANE

Dissociation of Microtubules from the Plasma Membrane

As discussed above, cortical MTs are probably associated with the plasma membrane either before or after their hormone-induced reorientation; thus, they should be depolymerized or dissociated from the membrane before they reorient themselves. MTs are not depolymerized during ethylene-induced reorientation of MTs (72), which suggests that depolymerization is not involved in the reorientation. Seagull's observation (79) that cortical MT reorientation does not occur in the presence of taxol, an MT-stabilizing agent, seems to contradict the suggestion that depolymerization is not involved in reorientation of MTs. But if taxol stabilizes not only the MTs themselves but also the association of MTs with the plasma membrane, this observation is not necessarily inconsistent with that suggestion. An experiment that examined interactions between MTs assembled in vitro and membrane ghosts (fragments of plasma membranes from burst protoplasts) indicates that taxol may facilitate the association of cortical MTs with the plasma membrane. MTs assembled in vitro were only bound to membrane ghosts in the presence of taxol (89; see below for further details). If taxol strengthens the association of MTs with the plasma membrane, Seagull's observation (79) also indicates that the dissociation of MTs from the plasma membrane occurs during the process of reorientation of MTs.

Cortical MTs are oriented randomly in protoplasts, even when protoplasts are prepared from cells with parallel arrays of MTs (16, 24, 54), an indication that the MTs are dissociated from and then reassociated with the membrane during protoplast preparation. If MTs never became dissociated from the membrane, their parallel arrays would not be disturbed. The dissociation of MTs is indicated by the absence of MTs on membrane ghosts from freshly prepared *Mougeotia* sp. protoplasts, which contain many cortical MTs (56). MTs in protoplasts prepared in the presence of taxol are parallel to each other (59). This observation also suggests that taxol inhibits the dissociation of MTs from the plasma membrane. Taxol hastens the change in MT arrangement from random arrays to parallel arrays in tobacco BY-2 protoplasts that are cultured in a medium that facilitates cell elongation (45). Taxol seems to hasten MT reorientation by accelerating the polymerization of MTs and causing the polymerized MTs to become associated with the plasma membrane. If MTs are associated with the plasma membrane over their entire length and the number of such MTs increases, the MTs will inevitably become arranged parallel to each other.

Although taxol inhibits MT reorientation in cotton fibers (79), randomly arranged MTs in tobacco BY-2 protoplasts become reoriented in parallel arrays in the presence of taxol (45). This contrast may result from the interaction between the plasma membrane and the cell wall (45). If the plasma membrane is not associated with the cell wall, MTs may reorient without dissociating from the plasma membrane. The change in orientation of parallel MTs during the preparation of protoplasts in the presence of taxol (59) also indicates that MT reorientation can occur without the dissociation of MTs from the membrane in cells in which the plasma membrane is not associated with the cell wall. It is still possible that hormone-mediated MT reorientation involves the hormone-mediated dissociation of the plasma membrane from the cell wall. MTs in protoplasts prepared from mesocotyls of GA_3-pretreated maize seedlings form parallel arrays (21). Cold treatment, which causes the disruption of cortical MTs in protoplasts prepared from untreated seedlings, does not disrupt MTs but does disturb their parallel arrays in protoplasts prepared from GA_3-treated seedlings (21). The association of MTs with the plasma membrane appears to be cold-sensitive. Treatment with pressure, which can cause depolymerization of MTs if applied too quickly, causes MT reorientation in leaf cells of *Lolium* sp. (9). It is tempting to speculate that pressure causes MT reorientation by causing the dissociation of MTs from the plasma membrane. If so, we may further speculate that the association of MTs with the plasma membrane is sensitive to the same factors that cause the depolymerization of MTs.

Immunofluorescence microscopy (70, 88) and immunoelectron microscopy (48) have shown co-localization of actin filaments with cortical MTs. MTs around the subapex of fern protonemata disappear on treatment with cytochalasin (38), suggesting that actin filaments stabilize the cortical MTs. Cytochalasin hastens cortical MT reorientation in developing cotton fibers (79), and it allows cortical MTs to reorient under conditions in which inhibitors of protein kinases inhibit GA_3-induced reorientation (K Mizuno, unpublished observation; see below for further details). Actin filaments probably stabilize not only MTs themselves but also the association between MTs and the plasma membrane.

Stabilization and Destabilization of Cortical Microtubules

The depolymerization of MTs does not seem to be involved in ethylene-induced reorientation of MTs (72), and it is also unclear whether depolymerization is involved in IAA- or GA-induced reorientation. If cortical MTs turn over continually, the selective stabilization of MTs in a certain orientation should result in a predominance of MTs in that orientation (96). Plant hormones, such as GA_3 and ABA, change the stability of cortical MTs when they change the orientation of MTs and they may cause the predominance of a certain orientation by selectively stabilizing the MTs in that orientation. It is expected that studies on the mechanism that regulates the stability of cortical MTs will give us some clues to the mechanism that regulates the orientation of MTs.

ROLE OF THE CELL WALL   In general, MTs are cold-sensitive and, therefore, they depolymerize on exposure to low temperatures; however, the cortical MTs in the cells of some plant tissues are cold-resistant [e.g. *Azolla* sp. root tips (23), *Lolium* sp. internodes (37), dwarf-pea epicotyls (2, 74), and *Nicotiana* sp. pollen tubes (6)]. Cortical MTs in tobacco BY-2 cells are cold-resistant (4), but the MTs assembled in vitro from tubulin isolated from the same cells are cold-sensitive (3). This result indicates that the MTs within the cells are protected against cold by some mechanism. Removal of the cell wall renders the cortical MTs sensitive to cold and MTs in protoplasts are cold-sensitive. Extensin, a component of the cell wall, stabilizes MTs in BY-2 protoplasts (4). The cell wall or some component(s) thereof seem to be involved in the mechanism that stabilizes cortical MTs. Although treatment of BY-2 protoplasts with protease does not disrupt the cortical MTs in the protoplasts, MTs are absent from the membrane ghosts prepared from such protease-treated protoplasts (5); which indicates that transmembrane proteins are involved in the association of cortical MTs with the plasma membrane and, moreover, that the digestion of extracellular portions of the proteins causes the dissociation of cortical MTs from the plasma membrane. Extensin does not stabilize cortical MTs in protease-pretreated BY-2 protoplasts (5). MTs are probably associated with the plasma membrane when extracellular materials, such as extensin and the cell wall, stabilize the cortical MTs. Protease treatment of BY-2 cells renders cortical MTs cold-sensitive and often disturbs the parallel arrays of cortical MTs (5). Transmembrane proteins appear to be involved in the stabilization and organization of cortical MTs. The arrangement and the stability of cortical MTs seem to be regulated by a mechanism that involves the cell wall, transmembrane proteins, and MT-plasma membrane cross-linking systems. The disruption of cortical MTs in subapical regions of fern protonemata weakens the association of the plasma membrane with the cell wall, indicating that cortical MTs are involved in the association of the membrane with the cell wall (39). Simmonds recently discussed the role of the cell wall in maintaining ordered arrays of MTs in plant cells (86).

POSSIBLE INVOLVEMENT OF PROTEIN PHOSPHORYLATION   Treatment with ATP causes the disappearance of cortical MTs and of the preprophase band of MTs on membrane ghosts prepared from tobacco BY-2 cells at the late G_1 phase or at prophase (43). ATP also causes MTs to disappear from ghosts prepared from BY-2 cells at the logarithmic phase of growth, in which about 90% of cells are at interphase, and a cytoplasmic extract of the same cells accelerates the disappearance (87). This indicates that cytosolic factors collaborate with ATP in causing the disappearance of cortical MTs. Staurosporine, an inhibitor of protein kinases, inhibits the ATP-dependent disappearance of MTs on ghosts prepared from cells at the late G_1 phase or at prophase (43); and okadaic acid, a specific inhibitor of protein phosphatases 1 and 2A, causes the disappearance of cortical MTs (25). These results indicate protein phosphorylation involvement in MT disappearance (43). Because taxol is a potent inhibitor of MT depolymerization, taxol inhibition of ATP-dependent disappearance of MTs strongly indicates that the disappearance is caused by MT depolymerization. The taxol inhibition also indicates that the disappearance could be caused by MT dissociation from the plasma membrane, because, as described above, taxol can strengthen the association of MTs with the plasma membrane. Inhibitors of protein kinases, such as 6-dimethylaminopurine and staurosporine, render cortical MTs of tobacco BY-2 cells cold-stable (65). Inhibitors of protein kinases also inhibit GA_3-induced cell elongation and MT reorientation in azuki bean epicotyls (66). It could be that the inhibition of protein phosphorylation stabilizes cortical MTs by stabilizing the association of MTs with the plasma membrane. Dissociation of MTs from the plasma membrane seems to be an important part of the process leading to MT reorientation. Moreover, MTs associated with the membrane are more stable than those that are not. Cytochalasin can overcome protein kinase inhibition of GA_3-induced cell elongation (K Mizuno, unpublished observation). The disruption of actin microfilaments seems to weaken MT association with the plasma membrane, which allows GA_3 to cause MT reorientation. The observation that cytochalasin hastens cortical MT reorientation in cotton fibers (79) also indicates that the disruption of actin filaments weakens the association of MTs with the plasma membrane.

Factors that Mediate the Association of Microtubules with the Plasma Membrane

MICROTUBULE-ASSOCIATED PROTEINS   Cortical MTs become reassociated with the plasma membrane, probably via cross-bridges, after reorientation following dissociation. Electron microscopy has shown cross-bridges between cortical MTs and the plasma membrane (17, 23, 37, 47, 80). The observation that many MTs are present on membrane ghosts and that they are not removed by washing (4, 5, 12, 40, 43, 54, 56, 87, 92) also indicates the tight association of MTs with the plasma membrane. There have been abundant reports of the presence of systems that cross-link MTs to the plasma membrane, but no details have been reported of the biochemical nature of such systems. Because these systems bind MTs, a component of the system that attaches directly to MTs must be able to bind to MTs, a feature common to microtubule-associated proteins (MAPs). Observation of the MT reorganization process in *Spirogyra* sp. cells containing disrupted MTs indicates that a factor that regulates MT orientation is present in the cortex of the cells (27). Although such a factor is hard to displace from the cortex by centrifugation in cells with organized MTs, it is easily displaced in cells with disrupted MTs, an indication that a factor that regulates MT orientation is associated with the MTs. Components that are involved in the cross-linking systems will be identified among those MAPs that co-localize with cortical MTs. Several kinds of MAP have been isolated from plant sources such as carrot cultured cells (10), thalli of *Dichotomosiphon* sp. (55), maize cultured cells (78, 93), and tobacco BY-2 cells (34, 99). Although a 76-kDa MAP from carrot cells has been shown to co-localize with cortical MTs (10), its involvement in MT–plasma membrane cross-linking systems has not been studied.

A method that includes temperature-dependent polymerization and depolymerization of MTs, which has been used to isolate animal MAPs, cannot be used to isolate plant MAPs partly because of the deleterious effects of vacuolar components in extracts of plant cells. Tubulin in an extract prepared from evacuolated protoplasts of tobacco BY-2 cells can polymerize into MTs even in the absence of MT-stabilizing agents, and the polymerized MTs are cross-linked to one another to form bundles (35). A 65-kDa MAP was isolated from such an extract by a polymerization and depolymerization procedure (34), indicating that the use of extracts of evacuolated protoplasts may lead to the successful isolation of various kinds of plant MAP. Immunofluorescence microscopy using antibodies raised against the 65-kDa protein showed that this protein is co-localized with all types of MT array (34). Isolation and characterization of plant MAPs will help to elucidate the mechanism by which plant hormones cause the cortical MT reorientation.

ASSOCIATION OF MICROTUBULES WITH THE PLASMA MEMBRANE IN VITRO   An experiment with an extract from evacuolated protoplasts of tobacco BY-2 cells and membrane ghosts prepared from the same cells demonstrated the ability of the plasma membrane to cross-link MTs. When the ghosts were pretreated with a cold solution of Ca^{2+}ions to remove preexisting MTs and the extract was placed on the pretreated ghosts, tubulin in the extract polymerized into MTs and polymerized MTs attached themselves to the ghosts (89). The MTs on the ghosts could be removed by treatment with a 0.6 M solution of KCl or a 0.1 M solution of Na_2CO_3. KCl- or Na_2CO_3-pretreated ghosts did not bind MTs, an indication that either KCl or Na_2CO_3 removes a factor that is involved in cross-linking MTs to the plasma membrane. However, the KCl- or Na_2CO_3-treated ghosts bound MTs if the extract was preincubated with taxol and then placed on the ghosts (89). We speculate that MTs assembled in the presence of taxol are associated with a factor that can overcome the damage that treatment with KCl or Na_2CO_3 causes to the cross-linking systems. If such a factor has properties similar to those of tubulin, or if it is a member of the tubulin superfamily, then taxol would strengthen the association of such a factor with MTs, thereby strengthening the association of MTs with the plasma membrane. If tubulin-like proteins are involved in the cross-linking system between MTs and the plasma membrane, the conditions that cause the depolymerization of MTs would be expected to cause the dissociation of MTs from the membrane. As discussed above, pressure and low temperature seem to cause MT reorientation by causing MTs to dissociate from the plasma membrane (9, 21). In this regard, it is important to study the role of γ-tubulin, a member of the tubulin superfamily, in the association of MTs with the plasma membrane. Liu et al (53) recently showed that a γ-tubulin-related protein is concentrated in the cortex of the root cells of soybean and *Allium* sp. at interphase; however, it is unclear whether this protein is co-localized with cortical MTs.

CONCLUDING REMARKS

Many authors have discussed the involvement of physical factors, such as cell shape and the rate of cell expansion, in determining cortical MT orientation (e.g. 19, 97, 98). Studies on hormone-induced cortical MT reorientation indicate that biochemical events, such as hormone-induced gene expression and protein phosphorylation, are important aspects of this phenomenon. Although the results summarized in this review do not allow us to postulate a mechanism by which plant hormones cause cortical MT reorientation, we can outline the studies that are needed to define such a mechanism: 1. studies to examine the dynamics of cortical MTs in hormone-treated living cells; 2. studies to identify the mRNAs that are synthesized in hormone-treated cells in which the hormone-induced reorientation of MTs is scheduled to occur; 3. studies to identify proteins that are involved in the cross-linking between MTs and the plasma membrane; 4. studies to clarify how protein phosphorylation is involved in the association and dissociation of MTs with the plasma membrane; 5. studies on MAPs that are associated with cortical MTs; and 6. studies on the effects of plant hormones on the properties and molecular species of factors that mediate the association of MTs with the plasma membrane. We are fortunate that methods for such studies are now available.

ACKNOWLEDGMENTS

I thank Reiko Nagai for her comments on the manuscript and Koichi Mizuno and Tsuyoshi Kaneta for providing unpublished information. Research in my laboratory has been supported by grants from the Ministry of Education, Science and Culture, Japan.

Any *Annual Review* chapter, as well as any article cited in an *Annual Review* chapter, may be purchased from the Annual Reviews Preprints and Reprints service.
1-800-347-8007; 415-259-5017; email: arpr@class.org

Literature Cited

1. Abe M, Shibaoka H, Yamane H, Takahashi N. 1990. Cell cycle-dependent disruption of microtubules by methyl jasmonate in tobacco BY-2 cells. *Protoplasma* 156:1–8
2. Akashi T, Shibaoka H. 1987. Effects of gibberellin on the arrangement and the cold stability of cortical microtubules in epidermal cells of pea internodes. *Plant Cell Physiol.* 28:339–48
3. Akashi T, Izumi K, Nagano E, Enomoto M, Mizuno K, Shibaoka H. 1988. Effects of propyzamide on tobacco cell microtubules in vivo and in vitro. *Plant Cell Physiol.* 29:1053–62
4. Akashi T, Kawasaki S, Shibaoka H. 1990. Stabilization of cortical microtubules by the cell wall in cultured tobacco cells. Effects of extensin on the cold-stability of cortical microtubules. *Planta* 182:363–69
5. Akashi T, Shibaoka H. 1991. Involvement of transmembrane proteins in the association of cortical microtubules with the plasma membrane in tobacco BY-2 cells. *J. Cell Sci.* 98:169–74
6. Åström H, Virtanen I, Raudaskoski M. 1991. Cold-stability in the pollen tube cytoskeleton. *Protoplasma* 160:99–107
7. Bergfeld R, Speth V, Schopfer P. 1988. Reorientation of microfibrils and microtubules at the outer epidermal wall of maize coleoptiles during auxin-mediated growth. *Bot. Acta* 101:57–67
8. Brummell DA, Hall JL. 1985. The role of cell wall synthesis in sustained auxin-induced growth. *Physiol. Plant.* 63:406–12
9. Cleary AL, Hardham AR. 1993. Pressure induced reorientation of cortical microtubules in epidermal cells of *Lolium rigidum* leaves. *Plant Cell Physiol.* 34:1003–8
10. Cyr RJ, Palevitz BA. 1989. Microtubule-binding proteins from carrot. I. Initial characterization and microtubule bundling. *Planta* 177:245–60
11. Davidonis GH. 1990. Gibberellic acid-induced cell elongation in cotton suspension cultures. *J. Plant Growth Regul.* 9:243–46
12. Doohan ME, Palevitz BA. 1980. Microtubules and coated vesicles in guard-cell protoplasts of *Allium cepa* L. *Planta* 149:389–401
12a. Durnam DJ, Jones RL. 1982. The effects of colchicine and gibberellic acid on growth and microtubules in excised lettuce hypocotyls. *Planta* 154:204–11
13. Edelmann H, Bergfeld R, Schopfer P. 1989. Role of cell-wall biogenesis in the initiation of auxin-mediated growth in coleoptiles of *Zea mays* L. *Planta* 179:486–94
14. Edelmann H, Schopfer P. 1989. Role of protein and RNA synthesis in the initiation of auxin-mediated growth in coleoptiles of *Zea mays* L. *Planta* 179:475–85
15. Eisinger W. 1983. Regulation of pea internode expansion by ethylene. *Annu. Rev. Plant Physiol.* 34:225–40
16. Galway ME, Hardham AR. 1986. Microtubule reorganization, cell wall synthesis and establishment of the axis of elongation in regenerating protoplasts of the alga *Mougeotia*. *Protoplasma* 135:130–43
17. Giddings TH, Staehelin LA. 1988. Spatial relationship between microtubules and plasma-membrane rosettes during the deposition of primary wall microfibrils in *Closterium* sp. *Planta* 173:22–30
18. Giddings TH, Staehelin LA. 1991. Microtubule-mediated control of microfibril deposition: a re-examination of the hypothesis. See Ref. 53b, pp. 85–99
19. Green PB, Selker JML. 1991. Mutual alignments of cell walls, cellulose, and cytoskeletons: their role in meristems. See Ref. 53b, pp. 303–22
20. Gunning BES, Hardham AR. 1982. Microtubules. *Annu. Rev. Plant Physiol.* 33:651–98
21. Hamada H, Mita T, Shibaoka H. 1994. Stabilization of cortical microtubules in maize

mesocotyl cells by gibberellin A₃. *Plant Cell Physiol.* 35:In press

22. Hardham AR. 1982. Regulation of polarity in tissues and organs. See Ref. 53a, pp. 377–403

23. Hardham AR, Gunning BES. 1978. Structure of cortical microtubule arrays in plant cells. *J. Cell Biol.* 77:14–34

24. Hasezawa S, Hogetsu T, Syono K. 1988. Rearrangement of cortical microtubules in elongating cells derived from tobacco protoplasts—a time-course observation by immunofluorescence microscopy. *J. Plant Physiol.* 133:46–51

25. Hasezawa S, Nagata T. 1992. Okadaic acid as a probe to analyze the cell cycle progression in plant cells. *Bot. Acta* 105:63–69

26. Hepler PK, Palevitz BA. 1974. Microtubules and microfilaments. *Annu. Rev. Plant Physiol.* 25:309–62

27. Hogetsu T. 1987. Re-formation and ordering of wall microtubules in *Spirogyra* cells. *Plant Cell Physiol.* 28:875–83

28. Hogetsu T, Oshima Y. 1986. Immunofluorescence microscopy of microtubule arrangement in root cells of *Pisum sativum* L. var. Alaska. *Plant Cell Physiol.* 27:939–45

29. Hogetsu T, Shibaoka H, Shimokoriyama M. 1974. Involvement of cellulose synthesis in actions of gibberellin and kinetin on cell expansion. Gibberellin-coumarin and kinetin-coumarin interactions on stem elongation. *Plant Cell Physiol.* 15:265–72

30. Hogetsu T, Shibaoka H, Shimokoriyama M. 1974. Involvement of cellulose synthesis in actions of gibberellin and kinetin on cell expansion. 2,6-Dichlorobenzonitrile as a new cellulose-synthesis inhibitor. *Plant Cell Physiol.* 15:389–93

31. Ishida K, Katsumi M. 1991. Immunofluorescence microscopical observation of cortical microtubule arrangement as affected by gibberellin in d₅ mutant of *Zea mays* L. *Plant Cell Physiol.* 32:409–17

32. Ishida K, Katsumi M. 1992. Effects of gibberellin and abscisic acid on the cortical microtubule orientation in hypocotyl cells of light-grown cucumber seedlings. *Int. J. Plant Sci.* 153:155–63

33. Iwata K, Hogetsu T. 1989. The effects of light irradiation on the orientation of microtubules in seedlings of *Avena sativa* L. and *Pisum sativum* L. *Plant Cell Physiol.* 30:1011–16

34. Jiang CJ, Sonobe S. 1993. Identification and preliminary characterization of a 65 kDa higher-plant microtubule-associated protein. *J. Cell Sci.* 105:891–901

35. Jiang CJ, Sonobe S, Shibaoka H. 1992. Assembly of microtubules in a cytoplasmic extract of tobacco BY-2 miniprotoplasts in the absence of microtubule-stabilizing agents. *Plant Cell Physiol.* 33:497–501

36. Jones RL, Jacobsen JV. 1991. Regulation of synthesis and transport of secreted proteins in cereal aleurone. *Int. Rev. Cytol.* 126:49–88

37. Juniper BE, Lawton JR. 1979. The effect of caffeine, different fixation regimes and low temperature on microtubules in the cells of higher plants. Evidence for diversity in their response to chemical and physical treatments. *Planta* 145:411–16

38. Kadota A, Wada M. 1992. The circular arrangement of cortical microtubules around the subapex of tip-growing fern protonemata is sensitive to cytochalasin B. *Plant Cell Physiol.* 33:99–102

39. Kagawa T, Kadota A, Wada M. 1992. The junction between the plasma membrane and the cell wall in fern protonemal cells, as visualized after plasmolysis, and its dependence on arrays of cortical microtubules. *Protoplasma* 170:186–90

40. Kakimoto T, Shibaoka H. 1986. Calcium-sensitivity of cortical microtubules in the green alga *Mougeotia*. *Plant Cell Physiol.* 27:91–101

41. Kaneta T, Kakimoto T, Shibaoka H. 1993. Actinomycin D inhibits the GA₃-induced elongation of azuki bean epicotyls and the reorientation of cortical microtubules. *Plant Cell Physiol.* 34:1125–32

42. Katsumi M, Mori S, Ishida K. 1989. GA-induced growth of cucumber seedlings under water stress. In *Plant Water Relations and Growth under Stress,* ed. M Tazawa, M Katsumi, Y Masuda, H Okamoto, pp. 349–56. Tokyo: Myu

43. Katsuta J, Shibaoka H. 1992. Inhibition by kinase inhibitors of the development and the disappearance of the preprophase band of microtubules in tobacco BY-2 cells. *J. Cell Sci.* 103:397–405

44. Koda Y. 1992. The role of jasmonic acid and related compounds in the regulation of plant development. *Int. Rev. Cytol.* 135:155–99

45. Kuss-Wymer CL, Cyr RJ. 1992. Tobacco protoplasts differentiate into elongate cells without net microtubule depolymerization. *Protoplasma* 168:64–72

46. Kutschera U, Schopfer P. 1986. Effect of auxin and abscisic acid on cell wall extensibility in maize coleoptiles. *Planta* 167:527–35

47. Lancelle SA, Callaham DA, Hepler PK. 1986. A method for rapid freeze fixation of plant cells. *Protoplasma* 131:153–65

48. Lancelle SA, Hepler PK. 1991. Association of actin with cortical microtubules revealed by immunogold localization in *Nicotiana* pollen tubes. *Protoplasma* 165:167–72

49. Lang JM, Eisinger WR, Green PB. 1982. Effects of ethylene on the orientation of microtubules and cellulose microfibrils of

pea epicotyl cells with polylamellate cell walls. *Protoplasma* 110:5–14

50. Laskowski MJ. 1990. Microtubule orientation in pea stem cells: a change in orientation follows the initiation of growth rate decline. *Planta* 181:44–52

51. Lawson VR, Weintraub RL. 1975. Interactions of microtubule disorganizers, plant hormones, and red light in wheat coleoptile segment growth. *Plant Physiol.* 55:1062–66

52. Ledbetter MC, Porter KR. 1963. A "microtubule" in plant cell fine structure. *J. Cell Biol.* 19:239–50

53. Liu B, Marc J, Joshi HC, Palevitz BA. 1993. A γ-tubulin-related protein associated with the microtubule arrays of higher plants in a cell cycle-dependent manner. *J. Cell Sci.* 104:1217–28

53a. Lloyd CW, ed. 1982. *The Cytoskeleton in Plant Growth and Development.* London: Academic

53b. Lloyd CW, ed. 1991. *The Cytoskeletal Basis of Plant Growth and Form.* London: Academic

54. Lloyd CW, Slabas AR, Powell AJ, Lowe SB. 1980. Microtubules, protoplasts and plant cell shape. An immunofluorescent study. *Planta* 147:500–6

55. Maekawa T, Ogihara S, Murofushi H, Nagai R. 1990. Green algal microtubule-associated protein with a molecular weight of 90 kDa which bundles microtubules. *Protoplasma* 158:10–18

56. Marchant HJ, Hines ER. 1979. The role of microtubules and cell-wall deposition in elongation of regenerating protoplasts of *Mougeotia. Planta* 146:41–48

57. Masuda Y. 1990. Auxin-induced cell elongation and cell wall changes. *Bot. Mag.* 103:345–70

58. Matsuki T, Tazaki H, Fujimori T, Hogetsu T. 1992. The influences of jasmonic acid methyl ester on microtubules in potato cells and formation of potato tubers. *Biosci. Biotech. Biochem.* 56:1329–30

59. Melan MA. 1990. Taxol maintains organized microtubule patterns in protoplasts which lead to the resynthesis of organized cell wall microfibrils. *Protoplasma* 153:169–77

60. Mendu N, Silflow CD. 1993. Elevated levels of tubulin transcripts accompany the GA$_3$-induced elongation of oat internode segments. *Plant Cell Physiol.* 34:973–83

61. Mita T, Katsumi M. 1986. Gibberellin control of microtubule arrangement in the mesocotyl epidermal cells of the d$_5$ mutant of *Zea mays* L. *Plant Cell Physiol.* 27:651–59

62. Mita T, Shibaoka H. 1983. Changes in microtubules in onion leaf sheath cells during bulb development. *Plant Cell Physiol.* 24:109–17

63. Mita T, Shibaoka H. 1984. Gibberellin stabilizes microtubules in onion leaf sheath cells. *Protoplasma* 119:100–9

64. Mita T, Shibaoka H. 1984. Effects of S-3307, an inhibitor of gibberellin biosynthesis, on swelling of leaf sheath cells and on the arrangement of cortical microtubules in onion seedlings. *Plant Cell Physiol.* 25:1531–39

65. Mizuno K. 1992. Induction of cold stability of microtubules in cultured tobacco cells. *Plant Physiol.* 100:740–48

66. Mizuno K. 1993. Changes of microtubule orientation, and of tubulin isoform in relation to gibberellin-induced elongation of azuki epicotyl segments. *Plant Cell Physiol.* 34:s90 (Abstr.)

66a. Moll C, Jones RL. 1981. Short-term kinetics of elongation growth of gibberellin-responsive lettuce hypocotyl sections. *Planta* 152:442–49

67. Nick P, Bergfeld R, Schäfer E, Schopfer P. 1990. Unilateral reorientation of microtubules at the outer epidermal wall during photo- and gravitropic curvature of maize coleoptiles and sunflower hypocotyls. *Planta* 181:162–68

68. Nick P, Furuya M, Schäfer E. 1991. Do microtubules control growth in tropism? Experiments with maize coleoptiles. *Plant Cell Physiol.* 32:999–1006

69. Nojiri H, Yamane H, Seto H, Yamaguchi I, Murofushi N, et al. 1992. Qualitative and quantitative analysis of endogenous jasmonic acid in bulbing and non-bulbing onion plants. *Plant Cell Physiol.* 33:1225–31

70. Pierson ES, Kengen HMP, Derksen J. 1989. Microtubules and actin filaments colocalize in pollen tubes of *Nicotiana tabacum* L. and *Lilium longiflorum* Thunb. *Protoplasma* 150:75–77

71. Preston RD. 1974. *The Physical Biology of Plant Cell Walls.* London: Chapman & Hall. 490 pp.

72. Roberts IN, Lloyd CW, Roberts K. 1985. Ethylene-induced microtubule reorientations: mediation by helical arrays. *Planta* 164:439–47

73. Robinson DG, Quader H. 1982. The microtubule-microfibril syndrome. See Ref. 53a, pp. 109–26

74. Sakiyama M, Shibaoka H. 1990. Effects of abscisic acid on the orientation and cold stability of cortical microtubules in epicotyl cells of the dwarf pea. *Protoplasma* 157:165–71

75. Sakiyama-Sogo M, Shibaoka H. 1993. Gibberellin A$_3$ and abscisic acid cause the reorientation of cortical microtubules in epicotyl cells of the decapitated dwarf pea. *Plant Cell Physiol.* 34:431–37

76. Sakoda M, Hasegawa K, Ishizuka K. 1992. Mode of action of natural growth inhibitors in radish hypocotyl elongation—influence

of raphanusanin on auxin-mediated microtubule orientation. *Physiol. Plant.* 84:509–13

77. Sauter M, Seagull RW, Kende H. 1993. Internodal elongation and orientation of cellulose microfibrils and microtubules in deepwater rice. *Planta* 190:354–62

77a. Sawhney VK, Srivastava LM. 1975. Wall fibrils and microtubules in normal gibberellic-acid-induced growth of lettuce hypocotyl cells. *Can. J. Bot.* 53:824–35

78. Schellenbaum P, Vantard M, Peter C, Fellous A, Lambert AM. 1993. Co-assembly properties of higher plant microtubule-associated proteins with purified brain and plant tubulins. *Plant J.* 3:253–60

79. Seagull RW. 1990. The effects of microtubule and microfilament disrupting agents on cytoskeletal arrays and wall deposition in developing cotton fibers. *Protoplasma* 159:44–59

80. Seagull RW, Heath IB. 1980. The organization of cortical microtubule arrays in the radish root hair. *Protoplasma* 103:205–29

81. Shibaoka H. 1972. Gibberellin-colchicine interaction in elongation of azuki bean epicotyl sections. *Plant Cell Physiol.* 13:461–69

82. Shibaoka H. 1974. Involvement of wall microtubules in gibberellin promotion and kinetin inhibition of stem elongation. *Plant Cell Physiol.* 15:255–63

83. Shibaoka H. 1991. Microtubules and the regulation of cell morphogenesis by plant hormones. See Ref. 53b, pp. 159–68

84. Shibaoka H. 1993. Regulation by gibberellins of the orientation of cortical microtubules in plant cells. *Aust. J. Plant Physiol.* 20:461–70

85. Shibaoka H, Hogetsu T. 1977. Effects of ethyl N-phenylcarbamate on wall microtubules and on gibberellin- and kinetin-controlled cell expansion. *Bot. Mag.* 90:317–21

86. Simmonds DH. 1992. Plant cell wall removal: cause for microtubule instability and division abnormalities in protoplast cultures? *Physiol. Plant.* 85:387–90

87. Sonobe S. 1990. ATP-dependent depolymerization of cortical microtubules by an extract in tobacco BY-2 cells. *Plant Cell Physiol.* 31:1147–53

88. Sonobe S, Shibaoka H. 1989. Cortical fine actin filaments in higher plant cells visualized by rhodamine-phalloidin after pretreatment with m-maleimidobenzoyl N-hydroxysuccinimide ester. *Protoplasma* 148:80–86

89. Sonobe S, Takahashi S, Yagawa S, Shibaoka H. 1993. Interaction between cortical microtubules and the plasma membrane in tobacco BY-2 cells. *Int. Bot. Congr., 15th, Yokohama,* p. 82 (Abstr.)

90. Steen DA, Chadwick AV. 1981. Ethylene effects in pea stem tissue. Evidence for microtubule mediation. *Plant Physiol.* 67:460–66

91. Takeda K, Shibaoka H. 1981. Effects of gibberellin and colchicine on microfibril arrangement in epidermal cell walls of *Vigna angularis* Ohwi et Ohashi epicotyls. *Planta* 151:393–98

92. van der Valk P, Rennie PJ, Connolly JA, Fowke LC. 1980. Distribution of cortical microtubules in tobacco protoplasts. An immunofluorescence microscopic and ultrastructural study. *Protoplasma* 105:27–43

93. Vantard M, Schellenbaum P, Fellous A, Lambert AM. 1991. Characterization of maize microtubule-associated proteins, one of which is immunologically related to tau. *Biochemistry* 30:9334–40

94. Volfová A, Chvojka L, Haňkovská J. 1977. The orientation of cell wall microtubules in wheat coleoptile segments subjected to phytohormone treatment. *Biol. Plant.* 19:421–25

95. Wasteneys GO, Gunning BES, Hepler PK. 1993. Microinjection of fluorescent brain tubulin reveals dynamic properties of cortical microtubules in living plant cells. *Cell Motil. Cytoskelet.* 24:205–13

96. Wasteneys GO, Williamson RE. 1989. Reassembly of microtubules in *Nitella tasmanica*: quantitative analysis of assembly and orientation. *Eur. J. Cell Biol.* 50: 76–83

97. Williamson RE. 1990. Alignment of cortical microtubules by anisotropic wall stresses. *Aust. J. Plant Physiol.* 17:601–13

98. Williamson RE. 1991. Orientation of cortical microtubules in interphase plant cells. *Int. Rev. Cytol.* 129:135–206

99. Yasuhara H, Sonobe S, Shibaoka H. 1992. ATP-sensitive binding to microtubules of polypeptides extracted from isolated phragmoplasts of tobacco BY-2 cells. *Plant Cell Physiol.* 33:601–8

100. Zandomeni K, Schopfer P. 1993. Reorientation of microtubules at the outer epidermal wall of maize coleoptiles by phytochrome, blue-light photoreceptor, and auxin. *Protoplasma* 173:103–12

101. Zhang D, Wadsworth P, Hepler PK. 1990. Microtubule dynamics in living dividing plant cells: confocal imaging of microinjected fluorescent brain tubulin. *Proc. Natl. Acad. Sci. USA* 87:8820–24

Annu. Rev. Plant Physiol. Plant Mol. Biol. 1994. 45:545–75

PROTEIN IMPORT INTO PLANT MITOCHONDRIA

Anthony L. Moore, Carlton K. Wood, and Felicity Z. Watts

Biochemistry Department, University of Sussex, Falmer, Brighton BN1 9QG, United Kingdom

KEY WORDS: plant mitochondria, protein import, chaperones, processing, respiratory chain

---

CONTENTS

INTRODUCTION 545
THE RESPIRATORY CHAIN 546
BASIC PROCESSES FOR PROTEIN IMPORT 546
PROTEINS REQUIRED FOR PROTEIN IMPORT 548
    Cytosolic Presequence Binding Factors 548
    Receptor and Import Channel Proteins... 549
    Cytochrome c Heme Lyase 551
    Molecular Chaperones 551
PROCESSING OF IMPORTED PROTEINS 554
PRESEQUENCES OF NUCLEAR-ENCODED PLANT MITOCHONDRIAL
    PROTEINS 556
USE OF HETEROLOGOUS SYSTEMS... 557
ENERGETIC REQUIREMENTS FOR PROTEIN IMPORT 559
    The Role of the Membrane Potential.. 560
    The Role of ATP.. 562
    What Drives Protein Translocation? 563
DEVELOPMENTAL REGULATION.. 564
    Respiratory Activity 564
    Protein Import... 565
CONCLUDING REMARKS... 567

INTRODUCTION

In cellular energy metabolism in non-photosynthetic and photosynthetic tissues, mitochondria provide cytosolic ATP, via oxidative phosphorylation, and

0066-4294/94/0601-0545$05.00

carbon skeletons for biosynthesis. Mitochondria are composed of a double membrane, the outer and inner membranes, which enclose two aqueous compartments, the intermembrane space and the matrix. Because most mitochondrial proteins are encoded in the nucleus, mitochondrial function and biogenesis depends on the ordered and, in some cases, complex processes of protein import. The mechanisms by which proteins are imported into *Saccharomyces cerevisiae* (yeast) and *Neurospora* mitochondria (59, 61, 63, 142) have been reviewed recently. This review concentrates on recent advances in our understanding of protein targeting to plant mitochondria and reference to other systems will only be made for comparison and for purposes of clarity.

THE RESPIRATORY CHAIN

The respiratory chain of plant mitochondria possesses several multiprotein complexes comparable to those found in other eukaryotic organisms (35). The four major complexes are NADH dehydrogenase (Complex I), succinate dehydrogenase (Complex II), ubiquinol-cytochrome c reductase (the bc_1 complex; Complex III), and cytochrome c oxidase (Complex IV). Plant systems are also

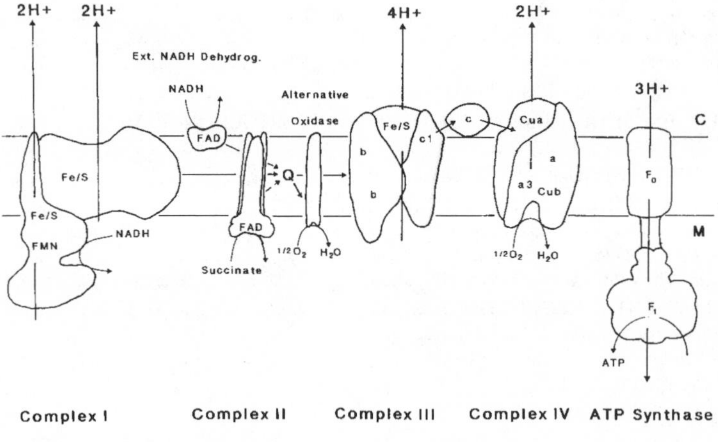

Figure 1   The respiratory chain of plant mitochondria. Abbreviations used: NADH, nicotinamide adenine dinucleotide-reduced; FMN, flavin mononucleotide; FAD, flavin adenine dinucleotide; Fe/S, iron-sulphur center; b, cytochrome b; c and c1, cytochromes c and c_1; a and a3, cytochromes a and a_3; Q, ubiquinone pool; C and M, the cytosolic and matrix facing sides of the inner membrane; F_0 and F_1, the hydrophobic and hydrophilic components of the ATP synthase.

characterized by the presence of an NAD(P)H dehydrogenase located on the outer surface of the inner membrane, a rotenone-insensitive NADH dehydrogenase located on the inner surface of the inner membrane (which may be part of Complex I), and to a varying extent, depending on the plant species, a cyanide- and antimycin-resistant alternative oxidase (115), an integral membrane protein (159).

Although nuclear DNA encodes most of the components of the respiratory chain complexes, mitochondrial DNA (mtDNA) encodes approximately 5% (127). To date, the subunits of the complexes that are encoded by plant mtDNA include eight subunits of the NADH dehydrogenase [NAD1 (21, 191); NAD2 (8, 194); NAD3 (64); NAD4 (51, 96); NAD5 (92); NAD6 (68, 130); NAD7 (see 130); NAD9 (see 130)], one subunit of Complex III [apocytochrome *b* (30, 199)], the three major subunits of cytochrome *c* oxidase [COXI (29, 72); COXII (29, 104); COXIII (72)], and four subunits of the ATP synthase [ATP6 (33); ATP8 (see 127); ATP9 (31, 34, 66); ATPA (11, 65)] (Figure 1). The alternative oxidase is nuclear-encoded (150) and although there is little information on either the structure or genetic origin of the external NADH dehydrogenase, it is also assumed to be nuclear-encoded.

BASIC PROCESSES FOR PROTEIN IMPORT

Figures 2 and 3 summarize the processes involved in targeting proteins to the different compartments and membranes in mitochondria. These mechanisms have been deduced mainly through in vitro import assays using yeast or *Neurospora* mitochondria. The initial steps are thought to involve interaction of the precursor protein, through its presequence, with outer membrane receptor proteins (3). Outer membrane proteins are inserted directly into the outer membrane (69), although it is not clear whether all outer membrane proteins need to interact with receptors before insertion into the membrane. Proteins destined for other locations are transported to import channels. Import channels are thought to correspond to contact sites, areas of transient association between the inner and outer mitochondrial membranes. After proteins destined for the matrix enter import channels, the proteins interact with mitochondrial HSP70 (mtHSP70) (154) and the presequence is cleaved off. Correct folding and assembly then requires matrix-localized Cpn60 (25). Targeting of proteins to the inner membrane can occur by one of several mechanisms depending on the protein. For example, the ADP/ATP translocator, which in yeast is not synthesized with a presequence, uses some of the components of the import machinery and is targeted to the inner membrane without entering the matrix (103). In contrast, the subunit IV of yeast cytochrome *c* oxidase is imported completely into the matrix before it is assembled in the inner membrane (79a). Another mechanism by which proteins are targeted to the inner membrane

depends on hydrophobic sequences within the protein, which stop translocation and anchor the proteins to the inner membrane (58, 128). There is still
some debate about how proteins are targeted to the intermembrane space: do
proteins destined for this space enter the matrix and then get re-exported [a
mechanism known as conservative sorting (94)] or are they prevented from
entirely entering the matrix via a stop-transfer mechanism (60) (see Figure 2)?

PROTEINS REQUIRED FOR PROTEIN IMPORT

A combination of biochemical and genetic analyses have been used to identify
yeast and *Neurospora* proteins required for protein import into mitochondria.
A list of these proteins is shown in Table 1, and their location is given
schematically in Figure 3.

Cytosolic Presequence Binding Factors

Two presequence binding factors (PBFs) have been identified in animal cells
through their ability to stimulate import of proteins into purified mitochondria
(120, 121, 132). These proteins, like the HSP70 family of chaperones (see
below) are thought to contribute to the import-competent conformation of the

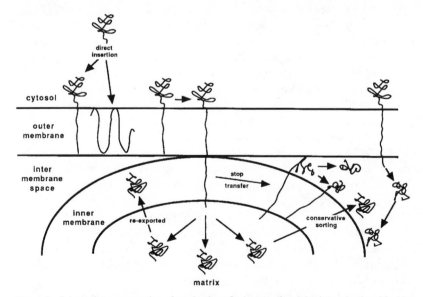

Figure 2 Schematic representation of mechanisms for import of proteins into mitochondria. Note
that targeting to the inner membrane and intermembrane space can occur by more than one process
depending upon the protein.

precursor proteins. It is unclear whether these two PBFs are identical to each other and so far no homologs have been identified in either fungi or plants.

Receptor and Import Channel Proteins

Protease studies have indicated that outer membrane protein receptors have overlapping specificities for proteins to be imported. One of these, a protein of approximately 70 kDa identified in both yeast and *Neurospora* (Mas70p and MOM72, respectively) has a hydrophobic, membrane-associated domain and a 60-kDa cytosolic domain (69, 70). This exposed region is thought to be involved in binding precursor proteins and in their subsequent delivery to the

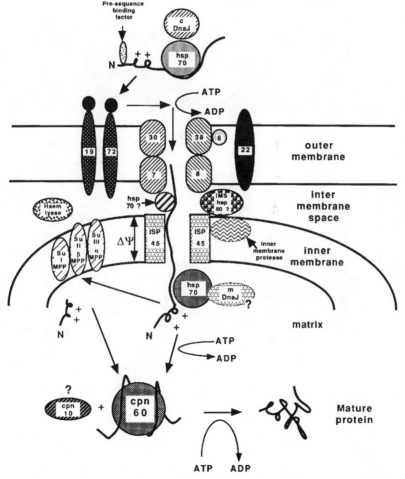

Figure 3 Components required for protein import into mitochondria. The question mark indicates proteins proposed to be involved in the import machinery.

import channel. Another receptor protein, MOM19 (161), has been shown to function independently in the absence of MOM72 (112). A third potential receptor protein, a 32-kDa integral membrane protein (136), was identified

Table 1 Components required for protein import into mitochondria.

Location	Protein	Essential for Viability	Fungi	Plants	References
cytosol	HSP70	no[a]	SSA1–4	HSP70	32
	DnaJ-like	yes[b]	MAS5		1
	presequence binding factor	?			119, 121, 32
outer membrane	receptor protein	?	MOM19		112, 161
		no	Mas70p/MOM72		75, 162
		no	p32		136
	channel protein	yes	ISP42/MOM38	PSIS42	2, 89, 175, 181
		?	MOM30		111, 163
		?	MOM22		111, 163
			MOM8		111, 163
		?	MOM7		111, 163
		no	ISP6		88
	adhesion site protein[d]				142
intermembrane space	IC chaperone[d]				142
	cytochrome c haem lyase	no	CYC3		36, 123, 166
inner membrane	IM protease I	no	IMP1		5, 157
	channel protein	yes	ISP45		76
matrix	mHSP70	yes	SSC1	PHSP1[e]	18, 87, 154, 176
	DnaJ-like	no	SCJ1		9
	matrix protease	yes	mas1p/PEP	see text	144, 192, 195
		yes	mas2p/MPP	subIIIcc[f]	71, 91, 135
	Cpn60	yes	mif4	Cpn60	101, 146

[a] redundant, since four related cytosolic proteins exist

[b] essential for growth at high temperature

[c] identified in rats

[d] hypothetical proteins

[e] also found associated with the inner membrane

[f] subunit III of cytochrome c reductase

using anti-idiotypic antibodies raised against a mitochondrial presequence peptide. This protein, which is not essential for viability in yeast, is identical to the mitochondrial phosphate transporter (143).

The import channel is thought to contain several proteins, including ISP42 [yeast (2, 175)] or MOM38 [*Neurospora* (89)]. Although this protein does not contain an obvious signal sequence or membrane-spanning domain, it is believed to be located in the outer membrane. The protein is thought to be part of a hetero-oligomeric complex that contains several other outer membrane proteins including MOM30, MOM22, MOM8, MOM7, and ISP6 (76, 88, 89, 111). To date, little is known about receptor proteins in plant mitochondria, although a potential homolog of ISP42 (PSISP42) has been identified in purified pea leaf mitochondria using anti-ISP42 antisera for Western analysis (181).

Cytochrome c Heme Lyase

Apocytochrome *c*, an intermembrane space protein, reaches its destination via direct insertion into the outer membrane (166). The driving force for its subsequent import into the intermembrane space is thought to be provided by cytochrome *c* haem lyase (CCHL), which is also required for attachment of the haem to apocytochrome *c*. Yeast mutants and *Neurospora* strains lacking CCHL fail to import apocytochrome *c* (36, 129), indicating that CCHL has a role in the import process. Similarly, site-directed mutagenesis of the cysteine residues required for heme-binding of yeast apocytochrome *c* also reduced the amount of imported apocytochrome *c*, although this was not observed in the import into mammalian mitochondria of mutagenized *Drosophila* apocytochrome *c* (67).

Molecular Chaperones

Molecular chaperone is the term given to proteins that mediate the correct assembly of other polypeptides that are not themselves components of the final structure of the protein. They comprise a wide range of proteins including some that were identified initially as heat shock proteins (HSPs). There are several recent reviews on molecular chaperones (41, 53, 73, 177), so this review covers only those aspects that apply to protein import into mitochondria.

CYTOSOLIC HSP70s   In contrast to protein import into the endoplasmic reticulum (ER), which occurs co-translationally, import into mitochondria in yeast is thought to occur post-translationally. Post-translational import requires that the proteins be maintained in a conformation that can be translocated across the membrane(s), because proteins in their native, fully folded state cannot be translocated (24, 38, 40). Such import-competent conformations are brought about through the interaction with cytosolic HSP70, and the requirement for these proteins in yeast has been shown using a mutant strain deleted for three out of the four cytosolic HSP70 (*SSA*) genes (32). More recently, a yeast *DnaJ* homolog (MAS5) has been identified (1) that is required for protein import into mitochondria and that is thought to act in concert with the cytoplasmic HSP70s.

To date, several plant HSP70 genes have been isolated that do not appear to be targeted to organelles. They do not fall into the BiP (immunoglobulin binding protein), mitochondrial, or chloroplast subsets of proteins, because they do not encode organelle targeting presequences, and are therefore likely to encode cytosolic HSP70s (151, 193). Preliminary biochemical evidence of a role for plant cytosolic HSP70s in protein import into mitochondria has been obtained by adding HSP70 protein purified from a cytosolic fraction from *Vicia faba* to in vitro import assays along with mitochondrial precursor proteins (137). Addition of the purified HSP70 increases binding of precursor proteins to mitochondria but has little effect on the overall amount of protein imported. This indicates that the cytosolic HSP70 may be needed to expose a region, possibly the presequence, to allow interaction with surface receptors. A role has been demonstrated for a maize cytosolic HSP70 in translocation into endosperm microsomes. This cytosolic HSP70 was capable of autophosphorylation, and the phosphorylated form had reduced ability to promote membrane translocation (110). A plant *DnaJ* cDNA has recently been isolated, but since the cDNA is lacking the N-terminal coding sequence and the subcellular location of the protein has not been determined, it is unclear whether it encodes a cytoplasmic or organellar plant DnaJ protein, nor is it known whether it has a role in protein targeting (7).

MITOCHONDRIAL HSP70s   Genetic and biochemical evidence has indicated that a mtHSP70 is required in protein import (87). In yeast this protein is distinct from the cytosolic HSP70s since it is encoded by a separate nuclear gene, *SSC1* (27), and then imported into mitochondria. The protein is associated with newly-imported precursor proteins and has been proposed to act in conjunction with a mitochondrial DnaJ-like protein, SCJ1, in yeast (9).

Mitochondrial HSP70 genes have been identified from at least four different plant species [pea (182), bean (176), potato and tomato (125)] and sequence comparison indicates that they have high levels of sequence identity not only with each other but also with mtHSP70s from yeast (27) and *Schizosaccharomyces pombe* (145). By analogy with the yeast SSC1 protein, the mtHSP70 genes have been proposed to have a role in protein import into mitochondria. Indirect evidence for this comes from co-import experiments of proteins into purified pea leaf mitochondria (117). In these experiments, in vitro import of proteins into plant mitochondria is enhanced by the addition to the import reaction of mtHSP70 precursor protein, which is also imported. The implications of this for the developmental regulation of protein import are discussed later. The pea, tomato, and potato mtHSP70 genes were isolated by virtue of their relatedness to *S. pombe* and *Escherichia coli* HSP70s (125, 182), but interestingly the bean mtHSP70 gene was isolated in a screen for calcium-binding protein genes (176). This bean mtHSP70 has not been shown to bind calcium, but in vitro phosphorylation experiments have shown that it is phosphorylated in a calcium-dependent manner by protein extracts from lysed mitochondria. Calcium-dependent autophosphorylation of mtHSP70 has been demonstrated for pea mtHSP70 (109), and this is unaffected by the addition of calmodulin. The involvement of calcium with animal HSP70s (all non-mitochondrial) has been suggested previously: (*a*) mammalian BiP has been described as a calcium-binding protein (101), (*b*) sequence analysis of cytosolic HSP70s has identified a calmodulin-binding domain (165), and (*c*) a mouse hsc70 has been shown to bind calmodulin in a calcium-dependent manner (165). However the significance of the presence of a potential calmodulin-binding site and the calcium-dependent phosphorylation of the mtHSP70 remains to be determined. In yeast the mtHSP70 has been shown to be associated with the inner membrane solely as a result of its binding to proteins that are in the process of being imported (105, 154). Such a transient association is unlikely to account for the observation that the bean mtHSP70 co-segregates with the inner membrane as well as the matrix (176). The protein has also been found in higher molecular weight forms, indicating it is part of a complex, and based on its homology with the *E. coli* protein DnaK, it may interact with DnaJ and have other roles within mitochondria (176).

INTERMEMBRANE SPACE HSP70s   By analogy with the multiple HSP70s found within chloroplasts (106), there may be one or more HSP70 protein located within the intermembrane space (142). Such proteins could be involved in the import of proteins destined for the intermembrane space.

MITOCHONDRIAL Cpn60s   Mitochondrial Cpn60 is essential for viability of yeast cells, while cells harboring a gene encoding a temperature-sensitive Cpn60 protein fail to fold and assemble newly imported proteins at the nonpermissive temperature (25). Several plant Cpn60 genes have been identified (147, 171) and, by analogy with yeast, are assumed to be involved in protein folding and complex assembly. A maize Cpn60 was shown to interact with mitochondrially synthesized proteins, although folding and assembly of the proteins into complexes was not observed, possibly owing to the lack of imported components required for the complex formation (146).

PROCESSING OF IMPORTED PROTEINS

After transport through the inner membrane, the precursor protein is processed to its mature form. The processing steps are believed to require a number of proteases that are involved in the cleavage of the presequence of the imported protein. Processing proteases from yeast and *Neurospora* were identified initially by screening temperature-sensitive mutants for any that accumulated uncleaved precursor proteins between the outer and inner mitochondrial membranes (196, 197). Two unlinked genes were identified originally, *MAS1* and *MAS2*; the proteins encoded by these genes (mas1p and mas2p) have been subsequently renamed β-MPP (for β-mitochondrial processing peptidase) and α-MPP (for α-mitochondrial processing peptidase), respectively (86).

Yeast α- and β-MPPs have been identified subsequently as two nonidentical subunits of a matrix-located processing protease (84, 144, 192, 198). This protease is a water soluble, metal-requiring heterodimer (see 3), the subunits of which are nuclear-encoded as precursor proteins with targeting presequences that are in turn cleaved by MPP already assembled in its mature form (192). The yeast MPP is essential for cell viability. Deletion of the *MPP* gene is lethal (144), possibly because cleavage of the presequence of imported proteins is required for correct assembly to the mature form, or because dissociation of the immature protein is not effected until the presequence is cleaved. Alternatively, the lack of processing activity may cause the import channel to become "jammed" with unprocessed proteins since inactivation or depletion of either the α- or β-subunit of MPP causes the import process to cease (197).

In *Neurospora,* as in yeast, the processing peptidase activity is the result of two interacting monomer units now called α-MPP (originally named MPP) and β-MPP (originally named PEP—processing enhancing protein)(71). β-MPP in *Neurospora* is partly associated with the inner membrane, unlike the β-MPP from yeast, which is matrix-located. However, the location of α-MPP in *Neurospora* appears to be analogous to the yeast equivalent in that it is also matrix-located. Isolated α-MPP has shown only low protease activity while isolated β-MPP has no protease activity and a complex containing both forms restores full processing activity (71).

Mammals do not appear to have an equivalent of the yeast β-MPP, but they have a matrix protease that shows some homology with the yeast α-MPP (135). It was originally named matrix protease I or P-55, but in line with the new nomenclature it is now called α-MPP (86). A second mammalian matrix protease, the mitochondrial intermediate protease (MIP) has also been characterized. Its function is to remove the C-terminal portion of the presequence from some precursor proteins (83, 85). MIP protease recognizes a consensus octapeptide sequence of FXXSXXXX where S represents Ser, Thr, Gly and the enzyme cuts adjacent to the right-hand residue (74). The full length cDNA sequence of MIP has been reported recently (83a) and sequence comparison shows some homology (47% similarity, 24% identity) to a putative MIP homolog identified during the sequencing of yeast chromosome III (131).

In addition to possessing matrix-located α-MPP and MIP, mammalian species also have a peptidase that is located in the mitochondrial inner membrane. This protein, called inner membrane peptidase (IMP), has also been identified in yeast (5, 157). The putative function of IMP is to cleave presequences of proteins targeted to the inner membrane space (e.g. cytochrome b_2).

Because the import and processing activities of yeast and *Neurospora* show similarities, it has been assumed that further evolutionary conservation would be observed in plants (22). Indeed, experiments using a yeast presequence to target a foreign protein, *E. coli* β-glucuronidase (GUS), into tobacco mitochondria gave import and processing of the chimeric protein (156). Subsequently, potato was used to characterize components that interact with presequences of imported plant mitochondrial proteins (18). Submitochondrial localization was used to fractionate processing activity, and activity was detected using antibodies directed against the matrix protease from *Neurospora*. SDS-PAGE of the immunoprecipitate showed that the processing peptidase activity co-fractionated with cytochrome *c* reductase. This was a surprising result because cytochrome *c* reductase is an integral membrane protein and is one of the respiratory chain complexes. Interestingly, direct protein sequencing and characterization of cDNA clones have shown that potato cytochrome *c* reductase subunit III is equivalent to the yeast matrix α-MPP (17, 19). Furthermore, subunits I and II of the complex share 40–50% sequence identity with the β-MPP from *Neurospora* (44, 45). Individual subunits or combinations of subunits from in vitro processing assays were inactive, indicating that the subunits need to be part of a complex for activity. The peptidase appears to be active even in the presence of high salt concentrations but does not require the presence of externally added metal ions for function (42). These points are in direct contrast to the αβ-MPP complex from *Neurospora* (71), which is dependent on divalent metal ions (e.g. Zn^{2+}, Mn^{2+}, and Co^{2+}). Both the *Neurospora* and the potato cytochrome *c* reductase β-MPP complexes, however, are inhibited by metal chelators such as EDTA and orthophenanthroline, indicating that both are metalloproteinases (10). Molecular genetic and physiological evidence, however, indicates that the "cytochrome *c* reductase/processing peptidase complex" (αβ-MPP) belongs to a new class of metalloendoproteases (46). Similar work (48, 49, 57) using the F_1-β subunit of the F_0F_1-ATP synthase of *Nicotiana plumbaginifolia* and the Rieske FeS protein of *Neurospora* has also localized a processing protease to the inner mitochondrial membrane. This protease's activity is also associated with the bc_1 respiratory complex. This is in contrast with yeast Rieske FeS protein, in which processing occurs in the matrix (74). Further work will show whether this protease studied by Glaser's group (48, 49) is similar in nature and function to that studied by Schmitz' group (17, 45).

A question so far unaddressed is what happens to the presequence that is cleaved from the imported precursor protein? In the past, these presequences were assumed simply to be degraded. However, recent work investigating the mitochondrial targeting presequence of the Rieske FeS protein has shown that the presequence is processed in a single step after insertion into the cytochrome bc_1 complex of rat liver mitochondria and thereafter is retained as a subunit in the complex (16). This work indicates that investigation of the fate of other presequences (both those cleaved in the matrix and those cleaved within the inner mitochondrial membrane) may also yield important information.

PRESEQUENCES OF NUCLEAR-ENCODED PLANT MITOCHONDRIAL PROTEINS

In general, proteins targeted to mitochondria are translated as precursor proteins containing N-terminal presequences that are responsible for determining the intramitochondrial location of the mature protein. There are exceptions to this, however, since some proteins do not contain cleavable presequences [e.g. isopropylmalate synthase (matrix) (6); the fungal (200) and mammalian (124), but not plant (43, 190), adenine nucleotide (ADP/ATP) translocator (inner membrane); and apocytochrome c (intermembrane space) (166)]. In a few cases, targeting information is contained within the mature sequence [e.g. aspartate aminotransferase (54) and cytochrome c oxidase subunit V_a (58)].

Extensive analysis of yeast and animal mitochondrial presequences has yielded information about the determinants involved in targeting precursor proteins to the matrix. Figure 4 shows a list of presequences published to date. The lengths of matrix targeting sequences are variable, ranging from 27–86 amino acids, and there is little if any sequence conservation between different genes. In some instances, however, there are conserved regions within the presequences from homologous genes from different plant species (e.g. mtHSP70s); this homology does not extend to the presequences of the equivalent yeast and animal sequences. The presequences of proteins destined for the matrix all contain a high proportion of Ser, Ala, Leu, and basic residues, with few acidic amino acids, similar to the presequences from several other organisms (178). In their survey of 36 non-plant mitochondrial matrix presequences, von Heijne et al (178) found that the majority of sequences had Arg at the −2 and −10 positions (13/36 at −2, 10/36 at −3, and 16/36 at −10). In the plant presequences shown in Figure 4, the −3 position for Arg seems to be preferred over the −2 position (12/21 vs 3/21) and Arg is never present at −10. Thus, despite the similarity in the overall amino acid content, the precise spacing of the basic residues appears to differ in these plant presequences. The significance of this remains to be determined. For example, does the spacing reflect preferred recognition site(s) for plant processing proteases?

The plant presequences for proteins destined for the inner membrane or intermembrane space are, in general, bipartite in nature, as is the case in other organisms. The presequences of the ADP/ATP translocator and cytochrome c_1 comprise an N-terminal region capable of forming an amphiphilic α-helix, followed by a hydrophobic domain. In contrast, the Reiske FeS presequence does not contain a hydrophobic domain, which is the case in FeS proteins from other organisms. In yeast the presequence undergoes two cleavage events in the matrix prior to insertion of the protein within the membrane (74). Similarly the presequences for the *Sauromatum* and *Arabidopsis* alternative oxidases do not contain hydrophobic domains, and although the *Sauromatum* enzyme is processed after import into mitochondria in vitro, there are no data as to whether this is a one- or two-step reaction.

USE OF HETEROLOGOUS SYSTEMS

Many protein import studies involve the use of heterologous systems. These can take several forms, for example, (*a*) where a yeast mitochondrial presequence is fused to a mammalian cDNA for a cytosolic protein (e.g. COXIV-DHFR) for import into yeast mitochondria or (*b*) where a yeast presequence fused to a reporter gene is used for import into plant organelles. Such studies have provided useful information about factors required for protein import, although care is needed when interpreting some of the data because DHFR has been shown to contain a cryptic matrix targeting sequence within the first 85 amino acids of the protein (80). Thus, sequences within the DHFR moiety may influence the import process. For example, in experiments to determine the minimum length of presequences required for matrix targeting, the result may be influenced by the presence of the cryptic targeting sequence in DHFR. Similarly, the use of DHFR to study targeting to locations other than the matrix (e.g. the intermembrane space) may not be very informative because there may be conflict between signals provided by the intermembrane space targeting sequences and the matrix targeting signal in DHFR. To determine the mechanism required for targeting to the intermembrane space, Koll et al (94) used fusion proteins containing 167 or 561 amino acids from cytochrome b_2 fused to DHFR; thus, any influence of the matrix targeting sequence may have been far enough removed from the N-terminus to prevent it from affecting the mechanism of import. However, there are instances where sequences within the mature portion of the protein affect the import of the protein [e.g. aspartate aminotransferase (54) and cytochrome c oxidase subunit V_a (58)]. In the case

Matrix

```
Atcpn60*                                    MYRFASNLASKARIAQN   A RQVSSRMSWSRNY
Zmcpn60                                     MYRAAASLASKARQAGNSLATRQVGSRLAWSRNY
Ccpn60.1                                    MHRFATGLASKARLARN GA NQIASRSNWRRNY
Ccpn60.2                                    MHRFASGLASKARLARN GA NQIASRSSWSRNY

PSHSP70*                    MAATLLRSLQRRNLSSSVSAFRSLTGSTKTSYATHKLASLTRPFSSRPA
PVHSP70*                    MAAVLRSLRRRDVASATFSAYRSLTGSTKPAYVAQKWSCLARPFSSRPA
StHSP70*                  MATAALLRSLRRREFATSSISAYRTLASNTKPSWCPSLVGAKWAGLARPFSSKPA

PsSHMT                                  MAMAMALRKLSSSVNKSSRPLFSASSLYYKS

Psgdc-L                                    MAMANLARRKGYSLLSSETLRYSFSLRSRAF
Psgdc-P MERARRLANRATLKRLLSEAKQNRKTESTSTTTTTPLPFSLSGSSSRYVSSVSNSILRGRGSKPDNNVSRVGFLGVGYPSQSRS
Psgdc-H                                    MALRMWASSTANALKLSSSSRLHLSPTFSISRCF
Atgdc-H                                    MALRMWASSTANALKLSSSVSKSHLSPFSFSRCF

CvmMDH                                  MKASILRSVSRVSAVSRSSSSNRLLSRSF

NpBATPase*        MASRRLLASLLRQSAQRGGGLISRSLGNSIPKSASRAS SRASPKGFLLNRAVQY
HbBATPase*        MASRRLLSSLLRSSSRSVSKSPISNINPKLSSSSPSSKSRASPYGYLLTRAAEY
ZmBATPase*        MASRRVVSSLLRSASRLRAASPAAPRPRAPPHRP        SPAGYLFNRAAAY
OsBATPASE*        MATRRALSSLVR AASRLRGASPAPRPRGPLHRP        SPSGYLFNRAAAY

IbdFlATPase                     MAMTGRARSMGFSILQKALSSAQRSNAHRSILCPTLSNSELLRNY

Ibd'FlATPase                            MFRHSSRLLARATTMGWRRPF

ZmSOD                              MALRTLASKKVLSFPFGGAGRPLAAAASARG
NpSOD                              MALRTLVSRRTLAT  GLGFRQQ     LRG
```

Outer surface inner membrane

```
Stcytc1          MSLGKKIRIGFDGFGRINRFITRGAAQRNDSKLPSRNDALKHGLDGLGSAGSKSFRALAAIGAGVSGLLSFATIAY

ZmRFS*           MLRVAGRRLSSSLSWRPAAAVARGPLAGAGVPDRDDDSARGRQSPRFSIDSPFFVASRGFSSTETVV

StAAC*         MADMNQHPTVFQKAANQLDLRSSLSQDVHARYGGVQPAIYQRHFACGNY SNAGLQRCQATQDLSLITSNASPVFVQ
ZmAAC.1*       MADQANQPTVLHKLGGQFHLRSIISEGVRARNICPSVSYERRFATRNYMTQSLWGPSMSVSGGINVPVMQTPLCAN
ZmAAC.2*       MADQANQPTVLHKLGGQFHLSSSFSEGVRARNICPSFSPYERRFATRNYMTQSLWGPSMSVSGGINVPVMPTPLFAN
```

Inner membrane

```
Sgaox*                MMSSRLVGTALCRQLSHVPVPQYLPALRPTADTASSLLHGCSAAAPAQRAGLWPPSWFSPPRHA(ST)
Ataox*                                             MDTRAPT IGGMRFA

IbcOV-I                                            M
```

Figure 4   Plant mitochondrial presequences. *Atcpn60, Zmcpn60*: *Arabidopsis* and maize cpn60 (147); *Ccpn60.1, Ccpn60.2*: pumpkin Cpn60 (171); *PsHSP70*: pea mtHSP70 (182); *PvHSP70*: bean mtHSP70 (176); *StHSP70*: potato mtHSP70 (125); *PsSHMT*: pea serine hydroxymethyltransferase (172); *Psgdc -L, -P, -H*: pea L, P, and H subunits of glycine decarboxylase (102, 173, 174); *Atgdc-H*: *Arabidopsis* H subunit of glycine decarboxylase (164); *CvmMDH*: watermelon malate dehydrogenase (56); *NpBATPase*: *Nicotiana plumbaginifolia* β subunit of F₁ ATPase (13); *HbBATPase*: *Hevea brasiliensis* β subunit of F₁ ATPase (26); *ZmBATPase*: maize β subunit of F₁ ATPase (37, 189); *OsBATPase*: rice β subunit of F₁ ATPase (153); *IbdF1ATPase* and *Ipd'F1ATPase*: sweet potato δ and δ' subunits F₁ ATPase (90, 118); *ZmSOD*: maize superoxide dismutase (188); *NpSOD*: *Nicotiana plumbaginifolia* superoxide dismutase (15); *Stcytc1*: potato cytochrome c₁ (18); *ZmRFS*: maize Rieske Fe/S (78); *StAAC*: potato ADP/ATP translocator (43); *ZmAAC.1, ZmAAC.2*: maize ADP/ATP translocator (190); *Sgaox*: *Sauromatum* alternative oxidase (150); *Ataox*: *Arabidopsis* alternative oxidase (95); *IbCOV-I*: sweet potato subunit V_c cytochrome c oxidase (note this protein is not encoded with a presequence)(122). Gaps have been introduced to maximize the alignment. * indicates that the cleavage site has not been determined, but is postulated from position in related proteins.

of cytochrome *c* oxidase subunit V_a, removal of a hydrophobic domain from the mature portion causes its mislocalization to the matrix.

Other heterologous systems include the use of yeast mitochondrial presequences, which have been demonstrated to import reporter proteins into plant mitochondria (12, 184). Such experiments cannot be used to study the import of the ADP/ATP translocator, for example, because the plant precursor contains a presequence, whereas fungal and mammalian proteins do not (43, 124, 190, 200). Because targeting of plant proteins, as opposed to yeast, requires the added ability to discriminate between mitochondria and chloroplasts, plant mitochondrial presequences may need to contain more information than yeast presequences to target the proteins specifically to mitochondria. Thus, identification of factors required for protein import into plant mitochondria should involve the use of plant genes or presequences rather than yeast sequences. Another example where higher eukaryotic mitochondrial presequences may provide more specific signals than yeast presequences is the mammalian phosphate carrier (148, 152), which is encoded as a precursor, whereas the yeast homolog is not (143).

ENERGETIC REQUIREMENTS FOR PROTEIN IMPORT

As indicated earlier, protein import is a multistep process that requires various components within the cytosol, associated with the outer and inner membranes and within the mitochondrial matrix. In addition to the components described above, the protein import machinery requires energy in the form of ATP and a transmembrane potential ($\Delta\psi$) (34, 59, 60, 62, 81, 107, 126, 140). Because proteins targeted to the outer membrane, intermembrane space, inner membrane, and mitochondrial matrix are imported by a variety of routes, the exact energetic requirements for targeting to specific locations may also vary.

The Role of the Membrane Potential

The proton-pumping activity of the respiratory chain complexes results in the generation of a proton-motive force across the inner mitochondrial membrane (see Figure 1). The proton-motive force comprises two interconvertible components, a transmembrane electrical potential ($\Delta\psi$) and a pH gradient. Evidence for an essential role of the membrane potential (or an energized membrane) in the translocation of proteins across the mitochondrial membranes is based on well-documented observations in yeast and *Neurospora* that protein import is inhibited either by the addition of a respiratory inhibitor (such as antimycin A or cyanide) or an ionophore (such as CCCP or valinomycin) (52, 81, 107, 138, 141, 155). When mitochondria are incubated with any of these compounds, protein import is arrested or prevented, resulting in a decrease in the mature form of the protein and increased accessibility of the protein to externally added protease. Because preincubation with nigericin (which collapses the pH gradient by exchanging K^+ for H^+) has little effect either on the rates of protein import or on the generation of the mature form of the protein, a membrane potential, but not a pH gradient, has been concluded to be necessary to drive the uptake of precursor proteins (155). In fact, the sensitivity of protein import to the addition of uncouplers (such as CCCP) is generally taken as a criterion that faithful import has occurred (22, 23, 39, 155).

The actual role of the membrane potential in protein import is still uncertain since it is unknown whether the membrane potential is required simply for insertion of the precursor protein into the membrane or if it is also required for translocation of the protein across the membrane. There is also confusion over the exact nature of the membrane potential (i.e. Is it a result of proton-pumping activity of the respiratory chain or does it result from membrane energization as a result of electron transport activity?).

Evidence indicates that the requirement for a membrane potential is independent of the requirement for ATP (59, 126, 140). This is based on the observation that when a precursor protein bound for the matrix is imported into mitochondria, which maintain a membrane potential but are depleted of ATP, the precursor is transported across the outer membrane. However, after insertion of the presequence into the inner membrane, translocation is arrested. The protein can then be chased into the matrix after further addition of ATP, indicating that the potential is only required for initial insertion of the protein into the inner membrane (82, 149). Early studies showed that complete translocation and processing could be driven solely by the membrane potential (138, 155); however, because these experiments were not carried out with ATP-depleted mitochondria, these conclusions are not necessarily valid.

The observed sensitivity of the import process to ionophores and respiratory inhibitors led to the conclusion that the driving force for uptake was the membrane potential component of the proton-motive force (52, 138, 155, 158). A more detailed analysis (107) of the nature of the membrane potential showed that it was an electrical potential (possibly the result of charge displacement) that was independent of the movement of protons and hence not necessarily a component of the proton-motive force. This study also demonstrated that an electrical gradient of only 35–60 mV was sufficient to drive the translocation process. Because the actual value of the potential required for translocation depended on, and varied with, the number of positively charged residues in the presequence, it has been concluded that the role of the electrical potential is to drive protein translocation across the membrane electrophoretically (107).

Similar conclusions have also been reached following studies with plant mitochondria (180). When substrates were oxidized proton-motively (via cytochrome oxidase) or non-proton-motively (via the alternative oxidase), no difference could be detected in the levels of either protein import or processing. The alternative oxidase does not pump protons (113), but an electrical potential of approximately 80 mV is generated during the oxidation of substrates via this pathway. Since protein import is observed under these conditions, the electrical gradient obviously is of sufficient magnitude to drive protein translocation. The results described above support the theory that an electrical potential gradient (negative inside), and not a proton gradient, is required for the translocation of the positively charged presequences, possibly via an electrophoretic effect. To prevent confusion over the meaning of these two gradients, perhaps $\Delta\psi$ should be restricted to mean the membrane potential component of the proton-motive force and should not be used interchangeably to describe both gradients.

Biophysical studies with artificial membranes have also shown that presequences interact strongly with membranes that possess negatively charged phospholipids, indicating that these interactions are necessary for the import process (50, 167). The ability of mitochondrial presequences to induce cardiolipin-specific interbilayer contacts, which appear to be dissociated by a transmembrane potential, has been described recently (99). The membrane potential is therefore proposed to have a dual role resulting in insertion of the presequence into the inner membrane and in inducing dissociation of the inner and outer membranes after insertion. Because only the presequence of cytochrome c oxidase subunit IV was studied, there is an obvious need to investigate the binding of other presequences to determine whether they also support this model.

The Role of ATP

Opinion on the role of ATP in protein import suggests that ATP is required for at least three of the steps (4, 77, 133, 134, 140), most of which involve molecular chaperones, with the exact energetic requirement dependent on the ultimate destination of the precursor protein.

CYTOSOLIC REQUIREMENT   The release of newly synthesized precursor proteins, which are normally bound to cytosolic HSP70, requires the hydrolysis of cytosolic ATP (32, 119). This appears to be the case for all proteins destined for the matrix. The only notable exceptions are apocytochrome c (166), cytochrome c haem lyase (100), fusion proteins attached to DHFR (81), and cytochrome c oxidase subunit V_a (110a). The reason these proteins do not require cytosolic ATP is currently unknown, although it has been suggested that perhaps they remain in an import-competent conformation without the aid of chaperones (4) or, alternatively, become unfolded upon contact with acidic phospholipids at the membrane surface (47). The fact that fusion proteins (possessing matrix targeting signals) can be imported efficiently into mitochondria in an ATP-independent manner further stresses the importance of using native proteins, which more closely resemble in vivo conditions, to resolve the energetic requirements of protein import.

If the destination of the precursor protein is either the intermembrane space or the outer surface of the inner membrane, then ATP is only required on the cytosolic side. For example, the import and processing of cytochrome c_1 (61) and the ADP/ATP translocator (103, 139), as well as the assembly of the ADP/ATP translocator, are unaffected by depletion of matrix ATP.

MATRIX REQUIREMENT   For proteins destined for the matrix, the additional ATP requirement probably reflects the ATP requirement of the matrix-located HSP70 (87). If mtHSP70 acts as the driving force for pulling the precursor protein across the inner membrane (87, 105, 154), then ATP is required for release of the bound protein from the chaperone. However, it is unknown how much ATP is required to pull proteins across the membrane or whether import requires multiple rounds of ATP hydrolysis. As previously described, the use of ATP-depleted mitochondria has demonstrated that ATP within the matrix is required for complete translocation of proteins across the inner membrane (82, 149). An additional requirement for ATP within the matrix is for refolding and oligomerization of some of the imported precursors, a process catalysed by Cpn60 (25, 105, 133).

The ATP requirements described for the above steps should be considered a minimum. There may be additional energetic requirements both for cytosolic anti-folding proteins other than the HSP70s and for molecular chaperones involved in the import machinery at locations other than those already identified (e.g. in the intermembrane space or bound to the outer surface of the outer membrane).

What Drives Protein Translocation?

Although there is general agreement on the overall energetic requirement for protein import, controversy still surrounds the question of what actually drives import. Is it a result of the precursor protein being pulled through the inner membrane by mtHSP70 after the presequence has emerged, as Neupert and colleagues (87, 134) suggest, or does the mtHSP70 interact with incoming precursor polypeptides simply because they are not completely folded (4)? Although the former theory is supported by experiments in yeast (87, 134), in which the deletion of the gene encoding the mtHSP70 results in accumulation of precursor proteins that are associated with, but not translocated across, the mitochondrial inner membrane, there is a discrepancy with respect to the phenotype of this null mutant and the requirement for ATP in vitro. For instance, the mutant appears unable to import cytochrome c_1 or the ADP/ATP translocator, both of which, as noted above, do not require matrix ATP for import. This inhibition may be a direct result of inactivation of the mtHSP70 (134). Inhibition of import could be an indirect effect of the mutation resulting in a jamming of all import channels with proteins that require mtHSP70 (179). Because import of cytochrome c_1 and the ADP/ATP translocator occurs via the import channels, inhibition of import will be observed but does not necessarily implicate a direct requirement for this chaperone in these cases. Furthermore, for mtHSP70 to pull the precursor protein across the membrane (126), the chaperone must bind tightly enough to pull the protein through the membrane and have little specificity since the segments of the translocated protein will vary considerably in hydrophobicity and charge. In this context, a model (160) based on a Brownian ratchet mechanism has suggested that protein translocation is driven by biased random thermal motion. The model proposes that Brownian motion can cause unfolded proteins to fluctuate back and forth across a membrane without any net displacement. Net translocation is considered to occur as a result of modification on one side of the membrane as a consequence of binding and dissociation of chaperones, chain-coiling induced by pH differences, and/or ionic gradients. Although the model predicts that protein translocation can be driven by several thermodynamic energy sources and that the relative role of each will vary with the translocated protein, there is relatively little experimental evidence to support it. However, the model is able to explain the lack of specificity required for a general protein import mechanism and it may explain why different laboratories have implicated so many different energy sources for protein translocation.

DEVELOPMENTAL REGULATION

The import of precursor proteins into mitochondria requires, as indicated earlier, a complex enzymatic mechanism based on results of in vitro protein import experiments using isolated organelles. In brief, precursor proteins are synthesized by a coupled in vitro transcription/translation system in the presence of labeled amino acid residues followed by incubation with mitochondria under varying metabolic conditions. After import (which is normally inhibited by the addition of valinomycin) and processing, mitochondria are re-isolated by centrifugation and resuspended in SDS sample buffer, and proteins are separated by SDS-PAGE and assessed by fluorography. The criteria indicating faithful import include dependence on an energy source, intact membranes, and insensitivity to externally added proteases.

Although protein import into plant mitochondria has been studied in less detail (12–14, 22, 55, 79, 93, 117, 122, 180, 183–187) than in other systems, the same criteria are used as indicators of efficient import. In contrast to the yeast and *Neurospora* systems, protein import into plant mitochondria in vitro has been observed, in the majority of cases, to result in the survival of a considerable proportion of the precursor form even after protease treatment. This has led to the impression that protein import is an inefficient process in plant mitochondria. It does not appear to be a reflection of the intactness or activity of the organelles used in these experiments because the mitochondria tend to be purified and show reasonable respiratory activity and coupling. Furthermore, low efficiency of import does not appear to be the result of using heterologous systems (22), in which non-plant presequences are used for targeting, since in most cases similar efficiencies are observed even when homologous systems are employed (183–185). Perhaps this low efficiency reflects the varying roles of mitochondria during plant development.

Respiratory Activity

Distinct biochemical and morphological changes occur to mitochondria as the plant cell matures and becomes photosynthetically active (98, 168–170). Wheat and barley are the best systems for studying these changes because all the cells in these tissues originate from a single meristematic region at the leaf base (98). Morphometric measurements, during the development of light-grown primary leaves of wheat, have indicated that although there is an increase in mesophyll cell volume, in contrast to the chloroplasts and cytosol, there does not appear to be any significant change in the fractional volume occupied by mitochondria (168). However, there appear to be changes in the composition of the mitochondrial population. For instance, in the basal region of the leaf, mitochondria are small and numerous, and as the cells expand, mitochondrial numbers increase (although the fractional volume they occupy remains constant), keeping pace with cell expansion (168). After cell elongation is complete and cells have reached their maximum size, mitochondrial size appears to have increased concomitant with a decrease in overall number per cell. Changes in mitochondrial enzymatic activity parallel morphometric changes. Although cytochrome c oxidase activity is relatively high in the basal meristematic cells, overall activity increases 2-fold during cell expansion and then remains relatively constant in later stages of leaf development (169). The pattern of development of glycine decarboxylase, a key mitochondrial enzyme in the photorespiratory pathway (35), appears to parallel that of other enzymes involved in photorespiration (170), all of which have negligible activity until the emergence of photosynthesis. These results indicate a coordinated development of photorespiratory enzymes during leaf growth, irrespective of the enzymes' cellular compartment. These biochemical and morphological changes probably reflect the changing metabolic roles of mitochondria as the plant cell matures and becomes photosynthetically active. In tissues with a high metabolic rate, such as regions of cell division and expansion, ATP synthesis is a key mitochondrial function and is likely to be modulated as photosynthetic activity develops, with mitochondria taking on a more biosynthetic role in the provision of carbon skeletons (114, 116). Because increases in enzymatic activity require coordinated accumulation of proteins, mitochondrial protein import is probably also regulated, for example, by the regulated synthesis of protein components of the import machinery during the development of the plant cell. This appears to be the case for the chloroplast import apparatus, which seems to be modulated in concert with the protein demands of the developing plastids (28).

Protein Import

Protein import into mitochondria from mature plant cells may be less efficient than import from cells with a high metabolic rate because of limitations in the amount of some of the components of the import machinery. Some evidence (55, 117) indicates that mtHSP70 is one of the components whose level limits protein import into mitochondria from mature tissues. Comparison of immunogold-labeling densities of pea leaf and *S. pombe* mitochondria probed with a specific anti-mtHSP70 antibody (182) has shown that the level of cross-reactivity is 10-fold lower in pea leaf mitochondria isolated from 14-day old leaves than in mitochondria from exponentially growing *S. pombe,* with significantly higher densities in mitochondria from immature pea leaf tissue (117). Similarly Western-blot analysis indicates that the intensity of cross-reactivity with this antiserum decreases with cell development. Further confirmation that the levels of mtHSP70 may be limiting the efficiency of protein import came from experiments in which unlabeled mtHSP70 synthesized in vitro was included in the import reaction mix during the import of COXIV-

DHFR into pea leaf mitochondria (117, 137). In the absence of exogenously synthesized mtHSP70 precursor, the import efficiency of COXIV-DHFR was relatively low (as evidenced by the presence of the protease-resistant precursor form) but was enhanced considerably by the presence of mtHSP70. If components such as chaperones are limiting uptake into mitochondria isolated from mature plant tissue, this may explain the low levels of protein import observed with these organelles. Interestingly, most plant import experiments to date have been carried out with organelles derived from mature tissue, which may be inherently less competent at importing proteins.

Figure 5 illustrates a possible explanation for the presence of the precursor form of the protein, even after protease treatment. After passage of the precursor protein through the outer membrane in both yeast and plant mitochondria, the protein interacts with the inner membrane import machinery (Figure 5A and C). In yeast mitochondria the precursor interacts with mtHSP70 and is imported (Figure 5B); however, in mitochondria from mature plant tissue, the precursor is arrested because of limiting amounts of mtHSP70. Translocation intermediates are formed and accumulate after traversing the outer membrane, in which the C-terminal portion of the precursor is exposed to the intermembrane

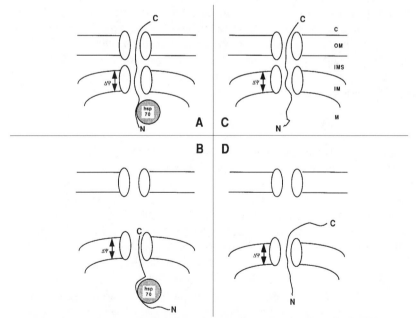

Figure 5 Diagramatic representation of protein import into yeast and plant mitochondria. A and B: stages of import into yeast mitochondria; C and D: stages of import into mitochondria from mature plant tissue. Abbreviations used: N and C refer to amino- and carboxy-termini; c, cytosol; OM, outer membrane; IMS, intermembrane space; IM, inner membrane; M, matrix; Δψ, the mitochondrial membrane potential.

space while the N-terminal portion is partly exposed to the matrix (Figure 5D). Because of the dynamic nature of the translocation machinery in the inner and outer membranes (142), the dissociation of translocation contact sites results in protection of the arrested precursor protein against protease treatment. Similar translocation intermediates have been observed previously in yeast systems (82, 141, 149). Thus the low levels of protein import that are normally observed with plant mitochondria may not be the result of an inherent inefficiency of the process per se but merely may reflect the limiting amounts of some of the components of the import machinery.

The above discussion suggests that caution should be used when performing and interpreting the results from import experiments with isolated plant mitochondria. In addition to ensuring that the required cytosolic factors are present, mitochondria should be isolated from tissues in which mitochondrial chaperones are not limiting.

CONCLUDING REMARKS

In this review we have sought to bring together recent developments in the area of protein import into plant mitochondria, and have related them to data on the import of proteins into mitochondria of other organisms. Studies on yeast and *Neurospora* have led the way in the past, but experiments using plant mitochondria are now contributing important information (e.g. on processing proteases and in the developmental regulation of import). Such studies are likely to help determine how discrimination specifically targets proteins to mitochondria rather than chloroplasts. Gene cloning and the ability to apply antisense technology will permit manipulation of components within the system.

The field of protein import in general, and plants in particular, has witnessed significant advances in the last few years and we hope this review will stimulate further research.

ACKNOWLEDGMENTS

We are grateful to all those who provided us with reprints or preprints of their work and to all our colleagues for their patience and understanding.

Literature Cited

1. Atencio DP, Yaffe MP. 1992. MAS5, a yeast homologue of DnaJ involved in mitochondrial protein import. *Mol. Cell. Biol.* 12:283–91
2. Baker KP, Schaniel A, Vestweber D, Schatz G. 1990. A yeast mitochondrial outer membrane protein essential for protein import and cell viability. *Nature* 348:605–9
3. Baker KP, Schatz G. 1991. Mitochondrial proteins essential for viability mediate protein import into yeast mitochondria. *Nature* 349:205–8
4. Beasley EM, Wachter C, Schatz G. 1992. Putting energy into mitochondrial protein import. *Curr. Opin. Cell Biol.* 4:646–51
5. Behrens M, Michaelis G, Pratje E. 1991. Mitochondrial inner membrane protease 1 of *Saccharomyces cerevisiae* shows sequence similarity to the *Escherichia coli* leader peptidase. *Mol. Gen. Genet.* 228: 167–76
6. Beltzer JP, Morris SR, Kohlhaw GB. 1988. Yeast *LEU4* encodes mitochondrial and non-mitochondrial forms of α-isopropylmalate synthase. *J. Biol. Chem.* 263:368–74
7. Bessoule J-J. 1993. Occurrence and sequence of a DnaJ protein in plant (*Allium porrum*) epidermal cells. *FEBS Lett.* 323: 51–54
8. Binder S, Marchfelder A, Brennicke A, Wissinger B. 1992. RNA editing in transsplicing intron sequences of NAD2 messenger-RNAs in *Oenothera* mitochondria. *J. Biol. Chem.* 267:7615–23
9. Blumberg H, Silver PA. 1991. A homologue of the bacterial heat-shock gene DnaJ that alters protein sorting in yeast. *Nature* 349:627–29
10. Boehni P, Gasser S, Leaver C, Schatz G. 1980. A matrix-localised mitochondrial protease processing cytosolically-made precursors to mitochondrial proteins. In *The Organisation and Expression of the Mitochondrial Genome*, ed. AM Kroon, C Saccone, pp. 423–33. Amsterdam: Elsevier
11. Boutry M, Briquet M, Goffeau A. 1983. The alpha subunit of a plant mitochondrial F_1-ATPase is translated in mitochondria. *J. Biol. Chem.* 258:8524–26
12. Boutry M, Chaumont F. 1993. Protein targeting to plant mitochondria. See Ref. 20, pp. 323–29
13. Boutry M, Chua N-H. 1985. A nuclear gene encoding the beta subunit of the mitochondrial ATP synthase in *Nicotiana plumbaginifolia. EMBO J.* 4:2159–65
14. Boutry M, Ferenc N, Chua NH. 1987. The beta subunit of a plant mitochondrial ATP synthase has a presequence involved in mitochondrial targeting. In *Plant Molecular*

Biology, ed. J Wettstein, NH Chua, NATO Ser. 140:127–34. New York: Plenum
15. Bowler C, Alliotte T, De Loose M, Van Montagu M, Inze D. 1989. A plant manganese superoxide dismutase is efficiently imported and correctly processed by yeast mitochondria. *EMBO J.* 8:31–38
16. Brandt U, Yu L, Yu CA, Trumpower BL. 1993. The mitochondrial targeting presequence of the Rieske iron-sulfur protein is processed in a single step after insertion into the cytochrome bc_1 complex in mammals and retained as a subunit in the complex. *J. Biol. Chem.* 268:8387–90
17. Braun H-P, Emmermann M, Kruft V, Schmitz UK. 1992. The general mitochondrial processing peptidase from potato is an integral part of cytochrome *c* reductase of the respiratory chain. *EMBO J.* 11:3219–27
18. Braun H-P, Emmermann M, Kruft V, Schmitz UK. 1992. Cytochrome c_1 from potato: a protein with a presequence for targeting to the mitochondrial intermembrane space. *Mol. Gen. Genet.* 231: 217–25
19. Braun H-P, Emmermann M, Schmitz UK. 1993. Cytochrome *c* reductase from potato mitochondria: a protein complex involved in respiration and protein import. See Ref. 20, pp. 307–13
20. Brennicke A, Kuck U. 1993. *Plant Mitochondria: With Emphasis on RNA Editing and Cytoplasmic Male Sterility.* Weinheim: VCH
21. Chapdelaine Y, Bonen L. 1991. The wheat mitochondrial gene for subunit-I of the NADH dehydrogenase complex—a transsplicing model for this gene-in-pieces. *Cell* 65:465–72
22. Chaumont F, O'Riordan V, Boutry M. 1990. Protein transport into mitochondria is conserved between plant and yeast species. *J. Biol. Chem.* 265:16856–62
23. Chen WJ, Douglas MG. 1987. Phosphodiester bond cleavage outside the mitochondria is required for the completion of protein import into the mitochondrial matrix. *Cell* 49:651–58
24. Chen WJ, Douglas MG. 1987. The role of protein structure in the mitochondrial import pathway. Unfolding of mitochondrially bound precursors is required for membrane translocation. *J. Biol. Chem.* 262: 15605–9
25. Cheng MY, Hartl F-U, Martin J, Pollock RA, Kalousek F, et al. 1989. Mitochondrial heat-shock protein hsp60 is essential for assembly of proteins imported into yeast mitochondria. *Nature* 337:620–25
26. Chye M-L, Tan C-T. 1992. Isolation and

nucleotide sequence of a cDNA clone encoding the beta subunit of mitochondrial ATPase from *Hevea brasiliensis*. *Plant Mol. Biol.* 18:611–12

27. Craig EA, Kramer J, Kosic Smithers J. 1987. *SSC1*, a member of the 70-kDa heat shock protein multigene family of *Saccharomyces cerevisiae* is essential for growth. *Proc. Natl. Acad. Sci. USA* 84: 4156–60

28. Dahlin C, Cline K. 1991. Developmental regulation of the plastid protein import apparatus. *Plant Cell* 3:1131–40

29. Dawson AJ, Hodge TP, Isaac PG, Leaver CJ, Lonsdale DM. 1986. Location of the genes for cytochrome oxidase subunits I and II, apocytochrome *b*, subunit of the F_1-ATPase and the ribosomal genes on the mitochondrial genome of maize. *Curr. Genet.* 10:561–64

30. Dawson AJ, Jones VP, Leaver CJ. 1984. The apocytochrome *b* gene in maize mitochondria does not contain introns and is preceded by a potential ribosome binding site. *EMBO J.* 3:2107–13

31. Dell'Orto P, Moenne A, Graves PV, Jordana X. 1993. The potato mitochondrial ATP synthase subunit 9: gene structure, RNA editing and partial protein sequence. *Plant Sci.* 88:45–53

32. Deshaies RJ, Koch BD, Werner-Washburne M, Craig EA. 1988. A subfamily of stress proteins facilitates translocation of secretory and mitochondrial precursor polypeptides. *Nature* 332:800–5

33. Dewey RE, Levings CS III, Timothy DH. 1985. Nucleotide sequence of the ATPase subunit 6 gene of maize mitochondria. *Plant Physiol.* 79:914–19

34. Dewey RE, Schuster AM, Levings CS III, Timothy DH. 1985. Nucleotide sequence of F0-ATPase proteolipid (subunit 9) gene of maize mitochondria. *Proc. Natl. Acad. Sci. USA* 82:1015–19

35. Douce R, Neuburger M. 1989. The uniqueness of plant mitochondria. *Annu. Rev. Plant Physiol. Plant Mol. Biol.* 40:371–414

36. Dumont ME, Ernst JF, Sherman F. 1988. Coupling of heme attachment to import of cytochrome *c* into yeast mitochondria. Studies with heme lyase-deficient mitochondria and altered apocytochromes *c*. *J. Biol. Chem.* 263:15928–37

37. Ehrenschaft M, Brambl R. 1990. Respiration and mitochondrial biogenesis in germinating embryos. *Plant Physiol.* 93:295–304

38. Eilers M, Hwang S, Schatz G. 1988. Unfolding and refolding of a purified precursor protein during import into isolated mitochondria. *EMBO J.* 7:1139–45

39. Eilers M, Oppliger W, Schatz G. 1987. Both ATP and an energised inner membrane are required to import a purified precursor protein into mitochondria. *EMBO J.* 6:1073–77

40. Eilers M, Schatz G. 1986. Binding of a specific ligand inhibits import of a purified precursor protein into mitochondria. *Nature* 332:228–32

41. Ellis RJ, van der Vies SM. 1991. Molecular chaperones. *Annu. Rev. Biochem.* 60:321–47

42. Emmermann M, Braun H-P, Arretz M, Schmitz UK. 1993. Characterisation of the bifunctional cytochrome *c* reductase-processing peptidase complex from potato mitochondria. *J. Biol. Chem.* 268:18936–42

43. Emmermann M, Braun H-P, Schmitz UK. 1991. The ADP/ATP translocator from potato has a long aminoterminal extension. *Curr. Genet.* 20:405–10

44. Emmermann M, Braun H-P, Schmitz UK. 1992. The processing protease from plant mitochondria. See Ref. 97, pp. 373–76

45. Emmermann M, Braun H-P, Schmitz UK. 1993. The two high molecular weight subunits of cytochrome *c* reductase from potato are immunologically related to the mitochondrial processing enhancing protein. *Biochim. Biophys. Acta* 1142:306–10

46. Emmermann M, Schmitz UK. 1993. The cytochrome *c* reductase integrated processing peptidase from potato mitochondria belongs to a new class of metalloendoproteases. *Plant Physiol.* In press

47. Endo T, Eilers M, Schatz G. 1988. Binding of a tightly folded artificial mitochondrial precursor protein to the mitochondrial outer membrane involves a lipid-mediated conformational change. *J. Biol. Chem.* 264: 2951–56

48. Eriksson A-C, Glaser E. 1992. Mitochondrial processing proteinase: a general processing proteinase of spinach leaf mitochondria is a membrane-bound enzyme. *Biochim. Biophys. Acta* 1140:208–14

49. Eriksson A-C, Sjoling S, Glaser E. 1993. A general processing proteinase of spinach leaf mitochondria is associated with the bc_1 complex of the respiratory chain. See Ref. 20, pp. 299–306

50. Frey S, Tamm LK. 1990. Membrane insertion and lateral diffusion of fluorescence-labelled cytochrome-*c*-oxidase subunit-IV signal peptide in charged and uncharged phospholipid bilayers. *Biochem. J.* 272: 713–19

51. Gass DA, Makaroff CA, Palmer JD. 1992. Variable intron content of the NADH dehydrogenase subunit-4 gene of plant mitochondria. *Curr. Genet.* 21:423–30

52. Gasser SM, Daum G, Schatz G. 1982. Import of proteins into mitochondria—energy-dependent uptake of precursors by

isolated mitochondria. *J. Biol. Chem.* 257: 13034–41

53. Gething M-J, Sambrook J. 1992. Protein folding in the cell. *Nature* 355:33–45

54. Giannattasio S, Marra E, Vacca RA, Iannace G, Quagliariello E. 1992. Import of mutant forms of mitochondrial aspartate aminotransferase into isolated mitochondria. *Arch. Biochem. Biophys.* 298: 532–37

55. Gietl C, Hock B. 1984. Uptake and processing of in vitro synthesised mitochondrial malate dehydrogenase by isolated watermelon mitochondria. *Z. Pflanzenphysiol. Bd.* 114:393–401

56. Gietl C, Lehnerer M, Olsen O. 1990. Mitochondrial malate dehydrogenase from watermelon: sequence of cDNA clones and primary structure of the higher plant precursor protein. *Plant Mol. Biol.* 14:1019–30

57. Glaser E, Eriksson A, Knorpp C, Whelan J, Hugosson M. 1992. In vitro import and processing of nuclear encoded precursor proteins with spinach leaf mitochondria. See Ref. 97, pp. 309–22

58. Glaser SM, Miller BR, Cumsky MG. 1990. Removal of a hydrophobic domain within the mature portion of a mitochondrial inner membrane protein causes its mislocalization to the matrix. *Mol. Cell. Biol.* 10:1873–81

59. Glick B, Schatz G. 1991. Import of proteins into mitochondria. *Annu. Rev. Genet.* 25: 21–44

60. Glick BS, Beasley EM, Schatz G. 1992. Protein sorting in mitochondria. *Trends Biochem. Sci.* 17:453–59

61. Glick BS, Brandt A, Cunningham K, Muller S, Hallberg RL, Schatz G. 1992. Cytochromes c_1 and b_2 are sorted to the intermembrane space of yeast mitochondria by a stop-transfer mechanism. *Cell* 69:809–22

62. Glick BS, Wachter C, Schatz G. 1992. The energetics of protein import into mitochondria. *Biochim. Biophys. Acta* 1101:249–51

63. Glover LA, Lindsay JG. 1992. Targeting proteins to mitochondria: a current overview. *Biochem. J.* 284:609–20

64. Gualberto JM, Wintz H, Weil JH, Grienenberger JM. 1988. The genes coding for subunit 3 of NADH dehydrogenase and for ribosomal protein S12 are present in the wheat and maize mitochondrial genomes and are co-transcribed. *Mol. Gen. Genet.* 215:118–27

65. Hack E, Leaver CJ. 1983. The alpha-subunit of the maize F_1-ATPase is synthesised in the mitochondrion. *EMBO J.* 2:1783–89

66. Hack E, Leaver CJ. 1984. Synthesis of a dicyclohexylcarbodiimide-binding proteolipid by cucumber (*Cucumis sativus* L.). *Curr. Genet.* 8:537–42

67. Hakvoort TBM, Sprinkle JR, Margoliash E. 1990. Reversible import of apocytochrome *c* into mitochondria. *Proc. Natl. Acad. Sci. USA* 87:4996–5000

68. Haouazine N, de Souza AP, Jubier M-F, Lancelin D, Delcher E, Lejeune B. 1992. The wheat mitochondrial genome contains an ORF showing sequence homology to the gene encoding the subunit 6 of the NADH-ubiquinone oxidoreductase. *Plant Mol. Biol.* 20:395–404

69. Hase T, Muller U, Riezman H, Schatz G. 1984. A 70-kd protein of the yeast mitochondrial outer membrane is targeted and anchored via its extreme amino terminus. *EMBO J.* 3:3157–64

70. Hase T, Riezman H, Suda K, Schatz G. 1983. Import of proteins into mitochondria; nucleotide sequence of the gene for a 70-kD protein of the yeast mitochondrial outer membrane. *EMBO J.* 2:2169–72

71. Hawlitschek G, Schneider H, Schmidt B, Tropschug M, Hartl F-U, Neupert W. 1988. Mitochondrial protein import: identification of a processing peptidase and of PEP, a processing enhancing protein. *Cell* 53: 795–806

72. Heisel R, Schobel W, Schuster W, Brennicke A. 1987. The cytochrome oxidase subunit I and subunit III genes in *Oenothera* mitochondria are transcribed from identical promoter sequences. *EMBO J.* 6:29–34

73. Hendrick JP, Hartl F-U. 1993. Molecular chaperone functions of heat shock proteins. *Annu. Rev. Biochem.* 62:349–84

74. Hendrick JP, Hodges PE, Rosenberg LE. 1989. Survey of amino-terminal proteolytic cleavage sites in mitochondrial precursor proteins: leader peptides cleaved by two matrix proteases share a three amino acid motif. *Proc. Natl. Acad. Sci. USA* 86:4056–60

75. Hines V, Brandt A, Griffiths G, Horstmann H, Brutsch H, Schatz G. 1990. Protein import into yeast mitochondria is accelerated by the outer membrane protein MAS70. *EMBO J.* 9:3191–200

76. Horst M, Jeno P, Kronidou NG, Bolliger L, Oppliger W, et al. 1993. Protein import into yeast mitochondria: the inner membrane import site protein ISP45 is the *MPI1* gene product. *EMBO J.* 12:3035–41

77. Horst M, Kronidou NG, Schatz G. 1993. Through the mitochondrial inner membrane. *Curr. Biol.* 3:175–77

78. Huang J, Struck F, Matzinger DF, Levings CJ III. 1991. Functional analysis in yeast of cDNA coding for the mitochondrial Rieske iron-sulfur protein of higher plants. *Proc. Natl. Acad. Sci. USA* 88:10716–20

79. Hugosson M, Boman HG, Glaser E. 1992. Effect of some antibacterial peptides on mitochondrial respiration and in vitro im-

port of nuclear encoded precursor protein. See Ref. 97, pp. 367–72

79a. Hurt EC, Pessold-Hurt B, Schatz G. 1984. The amino-terminal region of an imported mitochondrial precursor polypeptide can direct cytoplasmic dihydrofolate reductase into the mitochondrial matrix. *EMBO J.* 3:3149–56

80. Hurt EC, Schatz G. 1987. A cytosolic protein contains a cryptic mitochondrial targeting signal. *Nature* 325:499–503

81. Hwang ST, Schatz G. 1989. Translocation of proteins across the mitochondrial inner membrane, but not into the outer membrane, requires nucleoside triphosphates in the matrix. *Proc. Natl. Acad. Sci. USA* 86: 8432–36

82. Hwang ST, Wachter C, Schatz G. 1991. Protein import into the yeast mitochondrial matrix—a new translocation intermediate between the two mitochondrial membranes. *J. Biol. Chem.* 266:21083–89

83. Isaya G, Kalousek F, Fenton WA, Rosenberg LE. 1991. Cleavage of precursors by the mitochondrial processing peptidase requires a compatible mature protein or an intermediate octapeptide. *J. Cell Biol.* 113: 65–76

83a. Isaya G, Kalousek F, Rosenberg LE. 1992. Sequence analysis of rat mitochondrial intermediate peptidase: similarity to zinc metallopeptidases and to a putative yeast homologue. *Proc. Natl. Acad. Sci. USA* 89:8317–21

84. Jensen RE, Yaffe MP. 1988. Import of proteins into yeast mitochondria: the nuclear *MAS2* gene encodes a component of the processing protease that is homologous to the *MAS1*-encoded subunit. *EMBO J.* 7: 3863–71

85. Kalousek F, Hendrick JP, Rosenberg LE. 1988. Two mitochondrial matrix proteases act sequentially in the processing of mammalian matrix enzymes. *Proc. Natl. Acad. Sci. USA* 85:7536–40

86. Kalousek F, Neupert W, Omura T, Schatz G, Schmitz UK. 1993. Uniform nomenclature for the mitochondrial peptidases cleaving precursors of mitochondrial proteins. *Trends Biochem. Sci.* 18:249

87. Kang PJ, Ostermann J, Shilling J, Neupert W, Craig EA, Pfanner N. 1990. Requirement for hsp70 in the mitochondrial matrix for translocation and folding of precursor proteins. *Nature* 348:137–43

88. Kassenbrock CK, Cao W, Douglas MG. 1993. Genetic and biochemical characterisation of ISP6, a small mitochondrial outer membrane protein associated with the protein translocation complex. *EMBO J.* 12: 3023–34

89. Kiebler M, Pfaller R, Sollner T, Griffiths G, Horstmann H, et al. 1990. Identification of a mitochondrial receptor complex required for recognition and membrane insertion of precursor proteins. *Nature* 348:610–16

90. Kimura T, Takeda S, Asahi T, Nakamura K. 1990. Primary structure of a precursor for the δ-subunit of sweet potato mitochondrial F₁-ATPase deduced from full-length cDNA. *J. Biol. Chem.* 265:6079–85

91. Kleiber J, Kalousek F, Swaroop M, Rosenberg LE. 1990. The general mitochondrial matrix processing protease from rat liver: structural characterisation of the catalytic subunit. *Nature* 348:610–16

92. Knoop V, Schuster W, Wissinger B, Brennicke A. 1991. Transsplicing integrates an exon of 22 nucleotides into the NAD5 messenger-RNA in higher plant mitochondria. *EMBO J.* 10:3483–93

93. Knorpp C, Hugosson M, Glaser E. 1992. Temperature dependence and kinetics of plant mitochondrial protein import. See Ref. 97, pp. 361–65

94. Koll H, Guiard B, Rassow J, Ostermann J, Horwich AL, et al. 1992. Antifolding activity of hsp60 couples protein import into the mitochondrial matrix with export to the intermembrane space. *Cell* 68:1163–75

95. Kumar AM, Soll D. 1992. *Arabidopsis* alternative oxidase sustains *Escherichia coli* respiration. *Proc. Natl. Acad. Sci. USA* 89: 10842–46

96. Lamattina L, Grienenberger JM. 1991. RNA editing of the transcript coding for subunit-4 of NADH dehydrogenase in wheat mitochondria—uneven distribution of the editing sites among the 4 exons. *Nucleic Acids Res.* 19:3275–82

97. Lambers H, van der Plas LH W. 1992. *Molecular, Biochemical and Physiological Aspects of Plant Respiration.* The Hague: SPB Academic

98. Leech RM. 1984. Chloroplast development in angiosperms: current knowledge and future prospects. In *Topics in Photosynthesis.* Vol. 5: *Chloroplast Biogenesis,* ed. NR Baker, J Barber, pp. 1–21. Amsterdam: Elsevier

99. Leenhouts JM, de Gier J, de Kruijff B. 1993. A novel property of a mitochondrial presequence: its ability to induce cardiolipin-specific interbilayer contacts which are dissociated by a transmembrane potential. *FEBS Lett.* 327:172–76

100. Lill R, Stuart RA, Drygas ME, Nargang FE, Neupert W. 1992. Import of cytochrome *c* heme lyase into mitochondria: a novel pathway into the intermembrane space. *EMBO J.* 11:449–56

101. Macer DRJ, Koch GLE. 1988. Identification of a set of calcium-binding proteins in reticuloplasm, the luminal content of the endoplasmic reticulum. *J. Cell. Sci.* 91:61–70

102. Macherel D, Bourguignon J, Douce R. 1992. Cloning of the gene (*gdcH*) encoding

H-protein, a component of the glycine decarboxylase complex of pea (*Pisum sativum* L.). *Biochem. J.* 286:627–30

103. Mahkle K, Pfanner N, Martin J, Horwich AL, Hartl F-U, Neupert W. 1990. Sorting pathways of mitochondrial inner membrane proteins. *Eur. J. Biochem.* 192:551–55

104. Mann V, Ekstein I, Nissen H, Hiser C, McIntosh L, Hirschberg J. 1991. The cytochrome oxidase II gene in mitochondria of the sugar-beet *Beta vulgaris* L. *Plant Mol. Biol.* 17:559–66

105. Manning-Krieg UC, Scherer PE, Schatz G. 1991. Sequential action of mitochondrial chaperones in protein import into the matrix. *EMBO J.* 10:3273–80

106. Marshall JS, DeRocher AE, Keegstra K, Vierling E. 1990. Identification of heat shock protein HSP70 homologues in chloroplasts. *Proc. Natl. Acad Sci. USA* 87:374–78

107. Martin J, Mahkle K, Pfanner N. 1991. Role of an energised inner membrane in mitochondrial protein import. *J. Biol. Chem.* 266:18047–51

108. McMullin TW, Hallberg RL. 1988. A highly evolutionarily conserved protein is structurally related to the protein encoded by the *Escherichia coli* groEL gene. *Mol. Cell. Biol.* 8:371–80

109. Miernyk JA, Duck NB, David NR, Randall DD. 1992. Autophosphorylation of the pea mitochondrial heat-shock protein homolog. *Plant Physiol.* 100:965–69

110. Miernyk JA, Duck NB, Shatters RG, Folk WR. 1992. The 70-kilodalton heat shock cognate can act as a molecular chaperone during the membrane translocation of a plant secretory protein precursor. *Plant Cell* 4:821–29

110a. Miller BR, Cumsky MG. 1993. Intramitochondrial sorting of the precursor to yeast cytochrome *c* oxidase subunit V$_a$. *J. Cell. Biol.* 121:1021–29

111. Moczko M, Dietmeier K, Sollner T, Segui B, Steger HF, et al. 1992. Identification of the mitochondrial receptor complex in *Saccharomyces cerevisiae*. *FEBS Lett.* 310:265–68

112. Moczko M, Gartner F, Pfanner N. 1993. The protein import receptor MOM19 of yeast mitochondria. *FEBS Lett.* 326:251–54

113. Moore AL, Bonner WD Jr, Rich PR. 1978. The determination of the proton-motive force during cyanide-insensitive respiration in plant mitochondria. *Arch. Biochem. Biophys.* 186:298–306

114. Moore AL, Leach G, Whitehouse DG. 1993. The regulation of oxidative phosphorylation in plant mitochondria: the roles of the quinone oxidising and reducing pathways. *Biochem. Soc. Trans.* 21:765–69

115. Moore AL, Siedow JN. 1991. The regulation and nature of the cyanide-resistant alternative oxidase of plant mitochondria. *Biochim. Biophys. Acta* 1059:121–40

116. Moore AL, Siedow JN, Fricaud AC, Vojnikov V, Walters AJ, Whitehouse DG. 1992. Regulation of mitochondrial respiratory activity in photosynthetic systems. In *Plant Organelles: Compartmentation of Metabolism in Photosynthetic Tissue, Society for Experimental Biology Seminar*, ed. AK Tobin, 50:189–210. Cambridge: Cambridge Univ. Press

117. Moore AL, Walters AJ, Lennon AM, Wattts FZ. 1993. Developmental regulation of the protein import apparatus of plant mitochondria. See Ref. 20, pp. 291–98

118. Morikami A, Aiso K, Asahi T, Nakamura KJ. 1992. The δ-subunit of higher plant six-subunit mitochondrial F$_1$-ATPase is homologous to the δ-subunit of animal mitochondrial F$_1$-ATPase. *J. Biol. Chem.* 267:72–76

119. Murakami H, Pain D, Blobel G. 1988. 70-kD heat shock-related protein is one of at least two distinct cytosolic factors stimulating protein import into mitochondria. *J. Cell Biol.* 107:2051–57

120. Murakami K, Mori M. 1990. Purified presequence binding factor (PBF) forms an import-competent complex with a purified mitochondrial precursor protein. *EMBO J.* 9:3201–8

121. Murakami K, Tokunaga F, Iwanaga S, Mori M. 1990. Presequence does not prevent folding of a purified mitochondrial precursor protein and is essential for association with a reticulocyte cytosolic factor(s). *J. Biochem.* 108:207–14

122. Nakagawa T, Maeshima M, Nakamura K, Asahi T. 1990. Molecular cloning of a cDNA for the smallest nuclear-encoded subunit of sweet potato cytochrome *c* oxidase: analysis with the cDNA of the structure and import into mitochondria of the subunit. *Eur. J. Biochem.* 191:557–61

123. Nargang FE, Drygas ME, Kwong PL, Nicholson DW, Neupert W. 1988. A mutant of Neurospora crassa deficient in cytochrome *c* heme lyase activity cannot import cytochrome *c* into mitochondria. *J. Biol. Chem.* 263:9388–94

124. Neckermann N, Li K, Wade RP, Shuster R, Wallace DC. 1987. cDNA sequence of a human skeletal muscle ADP/ATP translocator: lack of a leader peptide, divergence from a fibroblast translocator cDNA, and coevolution with mitochondrial DNA genes. *Proc. Natl. Acad. Sci. USA* 84:7580–84

125. Neumann D, Emmermann M, Thierfelder J-M, zur Nieden U, Clericus M, et al. 1993. HSP68—a DnaK-like heat-stress protein of plant mitochondria. *Planta* 190:32–43

126. Neupert W, Hartl F-U, Craig EA, Pfanner N. 1990. How do polypeptides cross the mitochondrial membranes? *Cell* 63:447–50

127. Newton KJ. 1988. Plant mitochondrial genomes: organisation, expression and variation. *Annu. Rev. Plant Physiol. Plant Mol. Biol.* 39:503–32

128. Nguyen M, Bell AW, Shore GC. 1988. Protein sorting between mitochondrial membranes specified by position of the stop-transfer domain. *J. Cell. Biol.* 106:1499–506

129. Nicholson DW, Hergersberg C, Neupert W. 1988. Role of cytochrome-*c* heme lyase in the import of cytochrome-*c* into mitochondria. *J. Biol. Chem.* 263:19034–42

130. Nugent JM, Palmer JD. 1993. Characterisation of the *Brassica campestris* mitochondrial gene for subunit six of NADH dehydrogenase: nad6 is present in the mitochondrion of a wide range of flowering plants. *Curr. Genet.* 23:148–53

131. Oliver SG, van der Art JM, Agostoni-Carbone ML, Aigle M, Alberghina L, et al. 1992. The complete DNA sequence of yeast chromosome III. *Nature* 357:38–46

132. Ono H, Tuboi S. 1990. Purification and identification of a cytosolic factor required for import of precursors into mitochondria. *Arch. Biochem. Biophys.* 280:299–304

133. Ostermann J, Horwich AL, Neupert W, Hartl F-U. 1989. Protein folding in mitochondria requires complex formation with hsp60 and ATP hydrolysis. *Nature* 341:125–30

134. Ostermann J, Voos W, Kang PJ, Craig EA, Neupert W, Pfanner N. 1990. Precursor proteins in transit through the mitochondrial contact sites interact with hsp70 in the matrix. *FEBS Lett.* 277:281–84

135. Ou WJ, Ito A, Okazaki H, Omura T. 1989. Purification and characterisation of a processing protease from rat liver mitochondria. *EMBO J.* 8:2605–12

136. Pain D, Murakami H, Blobel G. 1990. Identification of a receptor for protein import into mitochondria. *Nature* 347:444–49

137. Perryman RA, Mooney B, Lennon A, Moore AL, Harmey MA. 1993. HSP70 and import of precursor proteins in plant mitochondria. See Ref. 20, pp. 315–22

138. Pfanner N, Neupert W. 1986. Transport of F$_1$-ATPase subunit β into mitochondria depends on both a membrane potential and nucleoside triphosphates. *FEBS Lett.* 209:152–56

139. Pfanner N, Neupert W. 1987. Distinct steps in the import of ADP/ATP carrier into mitochondria. *J. Biol. Chem.* 262:7528–36

140. Pfanner N, Neupert W. 1990. The mitochondrial import apparatus. *Annu. Rev. Biochem.* 59:331–54

141. Pfanner N, Rassow J, Guiard B, Sollner T, Hartl F-U, Neupert W. 1990. Energy requirements for unfolding and membrane translocation of precursor proteins during import into mitochondria. *J. Biol. Chem.* 265:16324–29

142. Pfanner N, Rassow J, van der Klei IJ, Neupert W. 1992. A dynamic model of the mitochondrial protein import machinery. *Cell* 68:999–1002

143. Phelps A, Schobert CT, Wohlrab H, 1991. Cloning and characterisation of the mitochondrial phosphate transporter gene from the yeast *Saccharomyces cerevisiae*. *Biochemistry* 30:248–52

144. Pollock RA, Hartl F-U, Cheng MY, Ostermann J, Horwich A, Neupert W. 1988. The processing peptidase of yeast mitochondria: the two cooperating components MPP and PEP are structurally related. *EMBO J.* 7:3493–500

145. Powell MJ, Watts FZ. 1990. Isolation of a gene for a mitochondrial HSP70 of *Schizosaccharomyces pombe*. *Gene* 95:105–11

146. Prasad TK, Hack E, Hallberg RL. 1990. Function of the maize mitochondrial chaperonin hsp60: specific association between hsp60 and newly synthesised F$_1$-ATPase alpha subunits. *Mol. Cell. Biol.* 10:3979–86

147. Prasad TK, Stewart CR. 1992. cDNA clones encoding *Arabidopsis thaliana* and *Zea mays* mitochondrial chaperonin HSP60 and gene expression during seed germination and heat shock. *Plant Mol. Biol.* 18:873–85

148. Pratt RD, Ferraira GC, Pederson PL. 1991. Mitochondrial phosphate transport—import of the H$^+$/Pi symporter and role of the presequence. *J. Biol. Chem.* 266:1276–80

149. Rassow J, Pfanner N. 1991. Mitochondrial preproteins en route from the outer membrane to the inner membrane are exposed to the intermembrane space. *FEBS Lett.* 293:85–88

150. Rhoads DM, McIntosh L. 1991. Isolation of a cDNA clone encoding an alternative oxidase protein of *Sauromatum guttatum* (Schott). *Proc. Natl. Acad. Sci. USA* 88:2122–26

151. Rochester DE, Winer JA, Shah DM. 1986. The structure and expression of maize genes encoding the major heat shock protein, HSP70. *EMBO J.* 5:451–58

152. Runswick MJ, Powell SJ, Nyren P, Walker JE. 1987. Sequence of the bovine mitochondrial phosphate carrier protein—structural relationship to ADP/ATP translocase and the brown fat mitochondria uncoupling protein. *EMBO J.* 6:1367–73

153. Sakamoto M, Shimada H, Fujimura T. 1992. Nucleotide sequence of a cDNA encoding a β-subunit of the mitochondrial

ATPase from rice (*Oryza sativa*). *Plant Mol. Biol.* 20:171–74

154. Scherer PE, Krieg UC, Hwang ST, Vestweber D, Schatz G. 1990. A precursor protein partly translocated into yeast mitochondria is bound to a 70kd mitochondrial stress protein. *EMBO J.* 9:4315–22

155. Schleyer M, Schmidt B, Neupert W. 1982. Requirement of a membrane potential for the posttranslational transfer of proteins into mitochondria. *Eur. J. Biochem.* 125: 109–16

156. Schmitz UK, Lonsdale DM. 1989. A yeast mitochondrial presequence functions as a signal for targeting to plant mitochondria in vivo. *Plant Cell* 1:783–91

157. Schneider A, Behrens M, Scherer P, Pratje E, Michaelis G, Schatz G. 1991. Inner membrane protease I, an enzyme mediating intramitochondrial protein sorting in yeast. *EMBO J.* 10:247–54

158. Segui-Real B, Kispal G, Lill R, Neupert W. 1993. Functional independence of the protein translocation machineries in mitochondrial outer and inner membranes: passage of preproteins through the intermembrane space. *EMBO J.* 12:2211–18

159. Siedow JN, Whelan J, Wiskich JT, Day DA. 1992. Topology of the alternative oxidase in soybean mitochondria. See Ref. 97, pp. 19–27

160. Simon SM, Peskin CS, Oster GF. 1992. What drives the translocation of proteins? *Proc. Natl. Acad. Sci. USA* 89:3770–74

161. Sollner T, Griffiths G, Pfaller R, Neupert W. 1989. MOM19, an import receptor for mitochondrial precursor proteins. *Cell* 59: 1061–70

162. Sollner T, Pfaller R, Griffiths G, Pfanner N, Neupert W. 1990. A mitochondrial import receptor for the ADP/ATP carrier. *Cell* 62: 107–15

163. Sollner T, Rassow J, Wiedmann M, Schlossmann J, Keil P, et al. 1992. Mapping of the protein import machinery in the mitochondrial outer membrane by crosslinking of translocation intermediates. *Nature* 355:84–87

164. Srinivasan R, Oliver DJ. 1992. H-protein of the glycine decarboxylase multienzyme complex. *Plant Physiol.* 98:1518–19

165. Stevenson MA, Calderwood SK. 1990. Members of the 70-kilodalton heat shock protein family contain a highly conserved calmodulin-binding domain. *Mol. Cell. Biol.* 10:1234–38

166. Stuart RA, Neupert W. 1990. Apocytochrome *c*: an exceptional mitochondrial precursor protein using an exceptional import pathway. *Biochimie* 72:115–21

167. Swanson ST, Roise D. 1992. Binding of a mitochondrial presequence to natural and artificial membranes: role of surface potential. *Biochemistry* 31:5746–51

168. Tobin AK, Rogers WJ. 1992. Metabolic interactions of organelles during wheat leaf development. In *Plant Organelles: Compartmentation of Metabolism in Photosynthetic Tissue. Society for Experimental Biology Seminar,* ed. AK Tobin, 50:293–323. Cambridge: Cambridge Univ. Press

169. Tobin AK, Sumar N, Patel M, Moore AL, Stewart GR. 1988. Development of photorespiration during chloroplast biogenesis in wheat leaves. *J. Exp. Bot.* 39:833–43

170. Tobin AK, Thorpe JR, Hylton CM, Rawsthorne S. 1989. Spatial and temporal influences on the cell-specific distribution of glycine decarboxylase in leaves of wheat (*Triticum aestivum* L.) and pea (*Pisum sativum* L.). *Plant Physiol.* 91:1219–25

171. Tsugeki R, Mori H, Nishimura M. 1992. Purification, cDNA cloning and northern-blot analysis of mitochondrial chaperonin 60 from pumpkin cotyledons. *Eur. J. Biochem.* 209:453–58

172. Turner S, Ireland R, Morgan C, Rawsthorne S. 1992. Identification and localisation of multiple forms of serine hydroxymethyltransferase in pea (*Pisum sativum*) and characterisation of a cDNA encoding a mitochondrial isoform. *J. Biol. Chem.* 267: 13528–34

173. Turner S, Ireland R, Rawsthorne S. 1992. Cloning and characterisation of the P subunit of glycine decarboxylase from pea (*Pisum sativum*). *J. Biol. Chem.* 267:5355–60

174. Turner S, Ireland R, Rawsthorne S. 1992. Purification and primary amino acid sequence of the L subunit of glycine decarboxylase. *J. Biol. Chem.* 267:7745–50

175. Vestweber D, Brunner J, Baker A, Schatz G. 1989. A 42K outer-membrane protein is a component of the yeast mitochondrial protein import site. *Nature* 341:205–9

176. Vidal V, Ranty B, Dillenschneider M, Charpenteau M, Ranjeva R. 1993. Molecular characterisation of a 70 kDa heat-shock protein of bean mitochondria. *Plant J.* 3:143–50

177. Vierling E. 1991. The roles of heat shock proteins in plants. *Annu. Rev. Plant Physiol. Plant Mol. Biol.* 42:579–620

178. von Heijne G, Steppuhn J, Herrmann RG. 1989. Domain structure of mitochondrial and chloroplast targeting peptides. *Eur. J. Biochem.* 180:535–45

179. Wachter C, Schatz G, Glick BS. 1992. The role of ATP in the intramitochondrial sorting of cytochrome c_1 and the adenine nucleotide translocator. *EMBO J.* 11:4787–94

180. Walters AJ, Watts FZ, Moore AL. 1992. Bioenergetic aspects of protein targeting to plant mitochondria. See Ref. 97, pp. 407–12

181. Watts FZ, Lennon AM, Walters AJ, Moore AL. 1992. Identification and characterisa-

tion of two components, PHSP1 and PSISP42, of the mitochondrial import mechanisms of *Pisum sativum*. See Ref. 97, pp. 399–405

182. Watts FZ, Walters AJ, Moore AL. 1992. Characterisation of PHSP1, a cDNA encoding a mitochondrial HSP70 from *Pisum sativum*. *Plant Mol. Biol.* 18:23–32

183. Whelan J, Dolan L, Harmey MA. 1988. Import of precursor proteins into *Vicia faba* mitochondria. *FEBS Lett.* 236:217–20

184. Whelan J, Knorpp C, Glaser E. 1990. Sorting of precursor proteins between isolated spinach leaf mitochondria and chloroplasts. *Plant Mol. Biol.* 14:977–82

185. Whelan J, Knorpp C, Harmey MA, Glaser E. 1991. Specificity of leaf mitochondrial and chloroplast processing systems for nuclear-encoded precursor proteins. *Plant Mol. Biol.* 16:283–92

186. Whelan J, O'Mahony P, Harmey MA. 1990. Processing of precursor proteins by plant mitochondria. *Arch. Biochem. Biophys.* 279:281–85

187. White JA, Scandalios JG. 1987. In vitro synthesis, importation and processing of Mn-superoxide dismutase (SOD-3) into maize mitochondria. *Biochim. Biophys. Acta* 926:16–25

188. White JA, Scandalios JG. 1988. Isolation and characterisation of a cDNA for mitochondrial manganese superoxide dismutase (SOD-3) of maize and its relation to other manganese superoxide dismutases. *Biochim. Biophys. Acta* 951:61–70

189. Winning BM, Bathgate B, Purdue PE, Leaver CJ. 1990. Nucleotide sequence of a cDNA encoding the beta subunit of the mitochondrial ATP synthase from *Zea mays*. *Nucleic Acids Res.* 18:5885

190. Winning BM, Day CD, Sarah CJ, Leaver CJ. 1991. Nucleotide sequence of two cDNAs encoding the adenine nucleotide translocator from *Zea mays* L. *Plant Mol. Biol.* 17:305–7

191. Wissinger B, Schuster W, Brennicke A. 1991. Transsplicing in *Oenothera* mitochondria—NAD1-messenger RNAs are edited in exon and transsplicing group-II intron sequences. *Cell* 65:473–82

192. Witte C, Jensen RE, Yaffe MP, Schatz G. 1988. *MAS1*, a gene essential for yeast mitochondrial assembly, encodes a subunit of the mitochondrial processing protease. *EMBO J.* 7:1439–47

193. Wu CH, Caspar T, Browse J, Lindquist S, Somerville C. 1988. Characterization of an HSP70 cognate gene family in *Arabidopsis*. *Plant Physiol.* 88:731–40

194. Xue Y, Davies DR, Thomas CM. 1990. Sugar-beet mitochondria contain an open reading frame showing extensive sequence homology to the subunit 2 gene of the NADH-ubiquinone reductase complex. *Mol. Gen. Genet.* 221:195–98

195. Yaffe MP, Ohta S, Schatz G. 1985. A yeast mutant temperature-sensitive for mitochondrial assembly is deficient in a mitochondrial protease activity that cleaves imported precursor polypeptides. *EMBO J.* 4:2069–74

196. Yaffe MP, Schatz G. 1984. The future of mitochondrial research. *Trends Biochem. Sci.* 9:179–81

197. Yaffe MP, Schatz G. 1984. Two nuclear mutations that block mitochondrial protein import in yeast. *Proc. Natl. Acad. Sci. USA* 81:4819–23

198. Yang M, Jensen RE, Yaffe MP, Schatz G. 1988. Import of proteins into yeast mitochondria: the purified matrix processing protease contains two subunits which are encoded by the nuclear *MAS1* and *MAS2* genes. *EMBO J.* 7:3857–62

199. Zanlungo S, Litvak S, Jordana X. 1991. Isolation and nucleotide sequence of the potato mitochondrial gene for apocytochrome *b*. *Plant Mol. Biol.* 17:527–30

200. Zimmermann R, Paluch U, Sprinzl M, Neupert W. 1979. Cell-free synthesis of the mitochondrial ADP/ATP carrier protein of *Neurospora crassa*. *Eur. J. Biochem.* 99:247–52

Annu. Rev. Plant Physiol. Plant Mol. Biol. 1994. 45:577–607
Copyright © 1994 by Annual Reviews Inc. All rights reserved

INTEGRATION OF CARBON AND NITROGEN METABOLISM IN PLANT AND ALGAL CELLS

H. C. Huppe and D. H. Turpin

Biology Department, Queen's University, Kingston, Ontario, Canada K7L 3N6

KEY WORDS: metabolic regulation, carbon and nitrogen interactions, activation of respiration, nitrogen assimilation, carbon partitioning

CONTENTS

INTRODUCTION 578
AN APPROACH TO STUDY.. 579
PHYSIOLOGICAL EVIDENCE OF INTERACTION 580
CARBON REQUIREMENTS FOR NITROGEN ASSIMILATION 581
   Photosynthetic Carbon Fixation.. 583
   Anapleurotic Carbon Fixation 585
   The Source of 2-Oxoglutarate for Glutamate Synthesis 585
REGULATION OF CARBON FLOW TO THE TCA CYCLE 586
   Phosphoenolpyruvate Carboxylase.. 586
   Pyruvate Kinase... 586
   Pyruvate Dehydrogenase Complex 587
   Phosphofructokinase and Pyrophosphate Fructose-6-phosphate
      1-Phosphotransferase... 587
   Fructose-1,6-bisphosphatase.. 588
   Sucrose Phosphate Synthetase 589
   Starch Metabolism... 589
A MODEL FOR ACTIVATION OF RESPIRATORY CARBON FLOW TO
   NITROGEN ASSIMILATION... 590
ENERGY CONSIDERATIONS... 591
   The Enzymes of Nitrate Reduction 591
   Cytosolic Reductant Supply... 592
   Energy Supply in the Plastid 593
   ATP Supply 595
REGULATION IN DIFFERENT SYSTEMS.. 596

0066-4294/94/0601-0577$05.00

The Algae 596
The Photosynthetic Higher Plant Cell 597
Heterotrophic Tissues 598
CONCLUDING REMARKS 599

INTRODUCTION

The metabolic pathways for carbon and nitrogen consume the greater part of the photosynthetic cell's carbon and energy resources. Up to 55% of net plant carbon is committed to nitrogen assimilation and metabolism in some tissues. Carbon and nitrogen metabolism are linked because they must share organic carbon and energy supplied directly from photosynthetic electron transport and CO_2 fixation, or from respiration of fixed carbon via glycolysis, the tricarboxylic acid (TCA) cycle and the mitochondrial electron transport chain (mitETC). The integration of these two important metabolic processes must involve extensive regulation between the two pathways.

This review focuses on the involvement of carbon assimilation and respiration in the primary assimilation of nitrogen and, conversely, the effect of nitrogen assimilation on the rate and route of carbon metabolism in the cell.[1] Many controversies surrounding either pathway will be summarized to show how our current understandings fit with the requirements of metabolic integration. We have restricted this review to the assimilation of new nitrogen into organic matter in photosynthetic eukaryotes.

The primary assimilation of inorganic nitrogen into amino acids requires carbon skeletons in the form of ketoacids and energy in the form of ATP and reductant (Figure 1a). In both photosynthetic and heterotrophic cells, the keto-acids are intermediates of respiratory metabolism (76). Respiration of stored and/or translocated photosynthate provides energy in heterotrophic tissues whereas photosynthesis usually supplies energy in photosynthetic cells (3, 6, 28, 97). In all cells, an increased rate of primary nitrogen assimilation requires

---

[1]

Abbreviations: 2-OG: 2-oxoglutarate; 3-PGA: 3-phosphoglycerate; Asp: aspartate; DCMU: 3-(3,4-dichlorophenyl)-1,1-dimethylurea; DHAP: dihydroxyacetone phosphate; F26P$_2$: fructose 2,6-bisphosphate; F6P: fructose 6-phosphate; FBP: fructose 1,6-bisphosphate; FBPase: fructose 1,6-bisphosphatase; Fd: ferredoxin; FNR: Fd-NADP oxidoreductase; G6P: glucose 6-phosphate; G6PDH: glucose 6-phosphate dehydrogenase; GAPDH: glyceraldehyde 3-phosphate dehydrogenase; Gln: glutamine; Glu: glutamate; GOGAT: glutamine 2-oxoglutarate aminotransferase; GS: glutamine synthetase; IDH: isocitrate dehydrogenase; MDH: malate dehydrogenase; mitETC: mitochondrial electron transport chain; NH_4^+: ammonium; NiR: nitrite reductase; NO_3^-: nitrate; NR: nitrate reductase; OAA: oxaloacetate; OPP: oxidative pentose phosphate; PDC: pyruvate dehydrogenase complex; PEP: phosphoenolpyruvate; PEPcase: phosphoenolpyruvate carboxylase; PFK: phosphofructokinase; PFP: pyrophosphate fructose-6-phosphate 1-phosphotransferase; Pi: inorganic phosphate; PK: pyruvate kinase; Pyr: pyruvate; Rubisco: ribulose 1,5-bisphosphate carboxylase; RuBP: ribulose 1,5-bisphosphate; SPS: sucrose phosphate synthetase; TCA: tricarboxylic acid; TP: triose phosphate.

some increased flow of carbon through respiratory pathways, which affects the metabolism of the whole cell (8, 161, 165).

AN APPROACH TO STUDY

The pioneering studies of carbon and nitrogen interactions were carried out by Syrett (158–161), Bassham, Kanazawa and co-workers (8, 80, 81) primarily using unicellular algae. An early observation that is central to the understanding and research of this problem was that nitrogen-sufficient algae required light to assimilate nitrogen, but nitrogen-limited cells rapidly assimilated nitrogen in the dark (2, 158, 163). Intracellular carbohydrate stores decreased during dark assimilation and assimilation ceased when these stores were depleted (158, 160). Exogenously supplied carbohydrate could also overcome dark inhibition (139, 158). The ability to induce nitrogen assimilation in the dark allowed physiological studies of this process in the absence of the complexities of photosynthetic gas exchange. Further, it indicated that carbohydrate respiration is important to the assimilation of nitrogen.

The use of unicellular algae remains central to our understanding of how respiration, photosynthesis, and nitrogen assimilation interact. Algae can be grown as a homogeneous cell population in chemostats (168) where nutrient availability sets growth rates so that long-term and short-term effects of differing medium composition can be compared. The rapid, uniform response of algal cells to changes such as increased availability of inorganic nitrogen allows determination of short-term fluxes that highlight the enzymes and regulatory factors controlling the response. Specific enzymes regulate metabolic pathways. The inhibition or activation of these regulatory enzymes radically change the flux in the whole pathway. Determination of substrate/product ratios can be used to identify regulatory enzymes. Combining in vitro studies of individual enzymes with examination of how metabolites and enzyme effectors change during the activation of a metabolic pathway provides an indication of what factors sustain the activity of that pathway.

Multicellularity and organ differentiation adds complexity to the study of carbon and nitrogen interactions in higher plants (8, 113). Many basic questions about nitrogen metabolism in higher plants are still answered by complex descriptions. Plants can assimilate both nitrate (NO_3^-) or ammonium (NH_4^+). The site of assimilation and type of nitrogen used varies diurnally, seasonally, ecologically, developmentally, and with species (3, 112). Plants absorb nitrogen though their roots, and most, if not all, NH_4^+ absorbed is assimilated in the roots (144), whereas much of the NO_3^- is transported to and assimilated in the shoots (3, 145). Some plants assimilate nitrogen in the dark in both roots and shoots (3, 6, 126, 133). Most plants store large amounts of NO_3^- in leaf vacuoles, so the nitrogen available for assimilation in a cell can exceed the amount available for uptake from the environment (145). Since nitrogen is a mobile nutrient, plants are able to move nitrogen into new growth by senescing older tissue. Finally, high C/N ratios equate to nitrogen metabolism affecting carbon metabolism less in higher plants than in algae (165). The nutrient status of a higher plant changes gradually, which makes it difficult to determine the order in which enzymes become activated (fine control), instead emphasizing adjustments of enzyme content (coarse control). Long-term differences in metabolite levels as affected by nutrient status can facilitate identification of regulatory sites important in the simultaneous metabolism of carbon and nitrogen in both higher plants and algae

PHYSIOLOGICAL EVIDENCE OF INTERACTION

Gas exchange measurements reflect the effect of nitrogen on carbon flow. Addition of a nitrogen source to N-limited or starved algal cells in the dark increases respiratory CO_2 release, demonstrating the activation of respiratory carbon flow (158, 177, 179). The higher reductant demand of NO_3^- assimilation requires that more carbon be mobilized to assimilate NO_3^- vs NH_4^+. This causes higher relative rates of CO_2 release during NO_3^- assimilation than when the NH_4^+ is the nitrogen source (179).

The differing reductant demands of NH_4^+ and NO_3^- also lead to differential coupling between the TCA cycle and the mitETC. During NH_4^+ assimilation in algae, O_2 consumption and CO_2 evolution increase proportionally, showing that respiratory carbon flow for amino acid synthesis is coupled to the mitETC and O_2 is the respiratory electron acceptor (177, 179). In contrast, O_2 uptake is only slightly higher during NO_3^- assimilation despite the larger increase in CO_2 release. This result indicates that some respiratory electrons flow to NO_3^- rather than to the mitETC (179). In barley roots, the respiratory quotient increases in response to NO_3^-, but not to NH_4^+, which reflects a large drop in O_2 uptake and little change in CO_2 release (13). Overall, this response shows that the requirements for NH_4^+ and NO_3^- assimilation differ similarly in plants and algae, although the energy partitioning may vary somewhat. Nitrate assimilation by barley shoots also results in an increased respiratory quotient, which shows that a significant portion of either respiratory or photosynthetic electrons is coupled to NO_3^- reduction in higher plants (12). The role of NO_3^- as a sink for respiratory electrons is also indicated by the lower inhibition of both respiratory carbon flow and nitrogen assimilation observed when NO_3^- vs NH_4^+ is the nitrogen source under anaerobic conditions (179).

Studies in N-limited algae during the activation of respiration for amino acid synthesis do not support a role for the alternative oxidase in supplying carbon for biosynthesis (111). The cytochrome oxidase pathway must couple the nitrogen-stimulated respiratory carbon flow to the mitETC in *Selenastrum minutum* because this alga lacks alternate pathway activity (178). In *Chlamydomonas reinhardtii*, despite its high alternate pathway capacity, $^{18}O_2$ discrimination demonstrated no alternative oxidase activity during NH_4^+ assimilation (176).

Respiratory CO_2 release and O_2 consumption can be differentiated during photosynthesis using a membrane inlet mass spectrometer and media containing $^{18}O_2$ and $^{13}CO_2$ (115). In algal cells, the respiratory response is similar during photosynthetic and dark nitrogen assimilation (177, 179). Despite stimulating a greater release of CO_2 than NH_4^+ assimilation, NO_3^- assimilation causes only a slight increase in respiratory O_2 consumption (179). In N-limited *S. minutum,* respiratory carbon flow provides carbon for amino acid synthesis during photosynthesis as well as some of the reductant for NO_3^- assimilation (179).

Nitrogen-sufficient algae do not accumulate large carbohydrate reserves (e.g. starch); therefore, the supply of carbon to amino acid synthesis is dependent on recent photosynthate (2, 161). Withholding either light or CO_2 inhibits nitrogen assimilation (2, 92, 162, 163, 169). Inhibitors of photosynthetic electron transport and carbon fixation [e.g. 3-(3,4-dichlorophenyl)-1,1-dimethylurea (DCMU) or DL-glyceraldehyde, respectively] mimic this effect (2, 163). If cells are provided with a metabolizable source of organic carbon (e.g. glucose, acetate, or internal carbohydrate) (139, 161), nitrogen assimilation becomes independent of recent photosynthate (2, 161).

Nitrogen assimilation by higher plants and algae is often accompanied by a stimulation of O_2 evolution because nitrogen metabolism provides another sink for the energy harvested in photosynthesis (34, 93, 145). Despite the increased O_2 evolution, the rate of CO_2 fixation is sometimes suppressed during nitrogen assimilation, especially at non-saturating light intensities (42, 53, 94, 123). In some N-limited algae, this photosynthetic suppression is dramatic even at saturating light (44) and starch is mobilized to provide carbon (102, 146).

CARBON REQUIREMENTS FOR NITROGEN ASSIMILATION

Carbohydrates formed via photosynthetic CO_2 fixation are either transported from the site of assimilation to other parts of the plant or used in the chloroplast to supply energy production, biosynthesis, or storage. Much work in carbon metabolism has focused on understanding the allocation of carbon within the cell, and between the source and sink tissues of plants (32, 151). A major demand on the carbon assimilated in the chloroplasts is the synthesis of amino acids from inorganic nitrogen. When the rate of nitrogen metabolism is reduced by nitrogen limitation or by decreasing the amount of nitrate reductase

(NR), carbon storage compounds accumulate and photosynthetic efficiency may be reduced (51, 52, 132). Activation of nitrogen assimilation consumes TCA cycle intermediates, which must be replaced by anapleurotic CO_2 fixa-

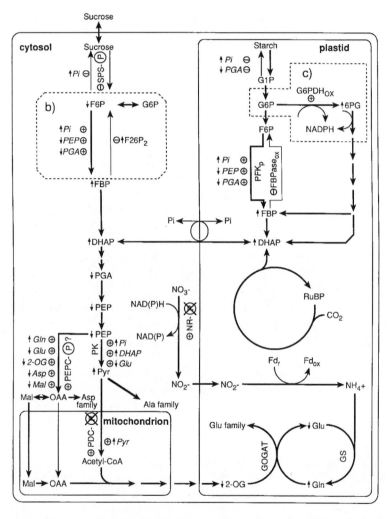

Figure 1a   The control of carbon partitioning to primary nitrogen assimilation. This figure indicates the general features associated with the provision of carbon from starch, sucrose, and recent photosynthate to the synthesis of amino acids. The significance of each source depends on the tissue in question and its developmental and physiological status. The reactions boxed by dashed lines and designated *b* and *c* are detailed in Figures 1b and c. Small arrows preceding a metabolite indicate whether it is an increase ($\uparrow$) or decrease ($\downarrow$) in the level of the metabolite. Abbreviations/symbols are as in text, except Acetyl-CoA: acetyl-coenzyme A; Mal: malate; activate, $\oplus$; deactivate $\ominus$; phosphorylated, $\textcircled{P}$; not phosphorylated; $\textcircled{R}$.

tion (Figure 1a). The regulation of carbon flow as effected by nitrogen assimilation provides insight into how carbon demands are balanced in the cell.

Photosynthetic Carbon Fixation

Recent photosynthate is used in amino acid synthesis (2, 8). If carbohydrate stores are available, however, nitrogen can be assimilated by plant and algal cells in the absence of CO_2 (2, 7). In fact, CO_2 fixation is suppressed in some cells during nitrogen assimilation. Although reported in some cases for higher plants (51, 123), the most dramatic and well-studied examples of this drop in CO_2 fixation are in algae where the proximate cause is the limitation of ribulose 1,5-bisphosphate carboxylase (Rubisco) by ribulose 1,5-bisphosphate (RuBP) (42, 44). The regulatory mechanisms are unknown, but the decline is correlated with the onset of starch mobilization (146) and may reflect incompatibility between carbon reduction and respiration in the chloroplast. The suppression in photosynthetic metabolism is transient, lasting until the nitrogen assimilation rates decline or the available starch reserves are depleted (44, 160, 161).

Whether recent or stored photosynthate provides carbon for amino acid synthesis depends on the capacity of photosynthesis and the availability of stored carbohydrate (110). For example, given the photosynthetic CO_2 fixation rate and C/N ratio of protein in N-limited *S. minutum,* carbon from photosynthesis would sustain a rate of NH_4^+ assimilation of ~ 50 mol mg^{-1} Chl h^{-1}.

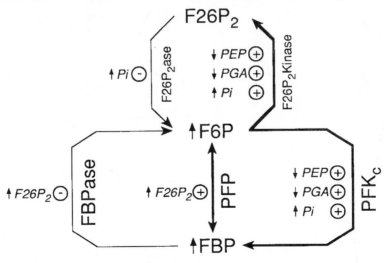

Figure 1b   Inhibition of cytosolic gluconeogenesis and activation of glycolysis at the fructose 6-phosphate (F6P)/fructose 1,6-bisphosphate (FBP) interconversion during nitrogen assimilation. Abbreviations are as in the text, except F26P₂K: fructose-2,6-bisphosphate kinase; F26P₂ase: fructose 2,6-bisphosphatase.

Without nitrogen, illuminated cells accumulate starch. If the NH_4^+ assimilation rate increases, the rate of starch accumulation decreases. When the demands for carbon by nitrogen assimilation exceed the rate of photosynthetic CO_2 fixation, starch degradation is observed in the light (165). Evidence for the integration of carbon and nitrogen assimilation has also been reported in higher plants. Soybean plants grown at three nitrogen levels (3, 9, and 18 μM) partitioned less photosynthate to starch as the nitrogen supply increased (129). The excess carbon accumulated in the first day that soybean plants were deprived of all nitrogen equaled the amount of carbon necessary to support the daily rate of nitrogen assimilation in those plants (132).

In the dark, stored carbohydrate must provide carbon for amino acid synthesis (28, 102). It has been demonstrated in *S. minutum* that the stoichiometry between the rate of starch degradation and the rate of NH_4^+ assimilation is consistent with the C/N ratio in this alga. Cells mobilize starch at a higher rate during NO_3^- assimilation. This larger carbon demand is reflected by the higher rate of respiratory CO_2 release when NO_3^- assimilation rather than NH_4^+ is the nitrogen source (174).

The capacity of wheat roots to assimilate nitrogen is correlated directly with their carbohydrate status, and the carbon skeletons for nitrogen assimilation are derived from the stored carbon (28). The provision of exogenous carbon to leaves, roots, and algae in the dark significantly stimulates the rate of nitrogen assimilation (7, 139). The greater availability of carbohydrate in the light accounts, in part, for the diurnal cycle of nitrogen assimilation in both heterotrophic and photosynthetic tissues (133). This indicates a general relationship

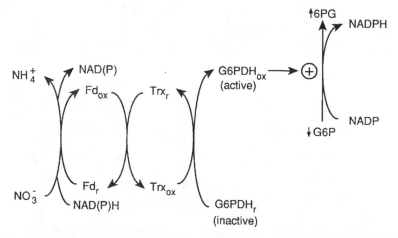

Figure 1c   A diagrammatic representation of the cascade involved in the thioredoxin activation of the oxidative pentose phosphate pathway to provide reducing power for NO_3^- reduction. Abbreviations are as in the text, except Trx: thioredoxin; r: reduced; ox: oxidized.

between carbohydrate status and the differing dark assimilation rates reported in plants.

Anapleurotic Carbon Fixation

Studies of dark CO_2 fixation in plants show that anapleurotic carbon is required to replace TCA cycle intermediates consumed in biosynthesis. The onset of nitrogen assimilation in both higher plant tissues (8, 95, 185) and algal cells (8, 159, 171) results in a large stimulation of dark CO_2 fixation. There is a linear relationship between the rate of nitrogen assimilation and anapleurotic carbon fixation (95, 171). In *S. minutum*, ~ 0.3 mol carbon is fixed per mol nitrogen assimilated, a stoichiometry corresponding well to the required anapleurotic HCO_3^- fixation calculated from the amino acid composition of total cellular protein (171). If anapleurotic reactions are blocked by stringent CO_2/HCO_3^- removal, dark NH_4^+ assimilation stops (2). Guy et al (59) have demonstrated the activation of anapleurotic carbon fixation by phosphoenolpyruvate carboxylase (PEPcase) during photosynthetic nitrogen assimilation using the differential discrimination by Rubisco and PEPcase against ^{13}C (relative to ^{12}C).

The Source of 2-Oxoglutarate for Glutamate Synthesis

Amino acids are synthesized from the transamination of ketoacids with glutamate (Glu) and other amino donors. The ketoacids are all intermediates of respiratory pathways (76). The complexities of amino acid synthesis are detailed elsewhere (22), but of key importance to this analysis is the provision of 2-oxoglutarate (2-OG) for Glu synthesis. Three potential pathways for the production of 2-OG have been described. It has been thought that citrate is produced and oxidized to 2-OG by the sequential action of aconitase and NAD-isocitrate dehydrogenase (NAD-IDH) within the mitochondria. The 2-OG is then exported to the cytosol and chloroplast for Glu synthesis via glutamine synthetase and glutamine 2-oxoglutarate aminotransferase (GS/GOGAT) (183). The identification of cytosolic aconitase (21) and NADP-IDH (21, 30, 31, 60) means that citrate exported from the mitochondria could be converted to 2-OG in the cytosol before transport to the chloroplast. A chloroplastic isozyme of NADP-IDH has also been found (29, 41, 122).

Chen & Gadal (31) have proposed that the cytosol is the major source of 2-OG for amino acid synthesis based on the relative activities of the cytosolic and mitochondrial IDH isozymes, the high activity of the mitochondrial 2-oxoglutarate dehydrogenase, and the high capacity 2-OG transporter in the chloroplastic envelope (121). The 2–3-fold higher export rate of citrate over 2-OG by isolated spinach mitochondria oxidizing malate is consistent with the cytosol rather than the mitochondria being the more significant source of 2-OG (60). The chloroplastic enzyme is not thought to be a significant contributor to

2-OG production. No source of isocitrate for the chloroplastic NADP-IDH is known because neither a chloroplastic aconitase (41) nor a chloroplastic transport system for isocitrate has been identified (30).

REGULATION OF CARBON FLOW TO THE TCA CYCLE

The demand for TCA cycle intermediates increases with the rate of nitrogen assimilation. To understand how this increased demand for carbon from the TCA cycle is balanced with the storage and translocation of carbohydrates in the cell, we must examine the regulation of enzymes that direct carbon during nitrogen assimilation. In general, the effectors discussed here have been found to act at physiologically relevant concentrations on the isolated enzyme.

Phosphoenolpyruvate Carboxylase

Phosphoenolpyruvate carboxylase is the major enzyme involved in replacing the TCA intermediates removed for biosynthesis in higher plants and algae (4, 124, 140). We know less about PEPcase regulation in C_3 plants than in C_4 plants (77, 95, 124, 140, 142).

Key inhibitors of the S. minutum PEPcase are the amino acids Glu and aspartate (Asp) and the TCA cycle intermediates 2-OG and malate (140). Malate inhibits PEPcase from Euglena gracilis (114), R. communis cotyledons (FE Podestá, manuscript in preparation), and nodule enzymes from soybean (142) and lupin (100). Asp and Glu inhibition has been reported for all but the Euglena PEPcase, whereas 2-OG affects the enzymes from nodules. Other intermediates, including citrate, isocitrate, succinate, and pyruvate (Pyr), have been inhibitory to some but not every PEPcase examined.

The only activators of anapleurotic PEPcase demonstrated to date are glutamine (Gln) and dihydroxyacetone phosphate (DHAP) in S. minutum (140), and DHAP in soybean nodule PEPcase (142). The response to effectors can vary dramatically with pH (142; FE Podestá, manuscript in preparation) and the presence of glycerol (142). These preliminary findings indicate a central role for Asp, Glu, and the TCA cycle intermediates in the feedback inhibition of anapleurotic PEPcase. The greater phosphorylation of wheat PEPcase in illuminated vs darkened leaves has led to speculation that phosphorylation also may be involved in the regulation of C_3 PEPcase activity (27).

Pyruvate Kinase

Initiation of de novo amino acid synthesis requires the provision of carbon to the TCA cycle via pyruvate kinase (PK). There are at least two isozymes of PK in a given plant tissue, the cytosolic form (PKc) and the plastidic form (PKp), which show markedly different regulatory kinetics (98, 117, 118). Our under-

standing of PK regulation is limited despite the evidence of its importance in
modulating plant metabolism under a variety of physiological conditions.
Most work on the regulatory properties of plant PK has demonstrated
inhibition by ATP or citrate (9, 117, 164) and activation by AMP (9, 164).
Only four studies have related PK regulation with its role in providing carbon
for amino acid synthesis (9, 98, 119, 186), and these studies have shown that
amino acids are also central regulators of PKc in tissues active in protein
synthesis. Glutamate inhibited spinach leaf PKII (no localization determined),
whereas Asp reversed Glu inhibition, indicating a role for the Glu/Asp ratio in
controlling this enzyme (9). Glu was also a potent inhibitor of PKc from *S.
minutum* and *C. reinhardtii* (98, 186). DHAP was an activator and Asp weakly
activated algal PKc, but inhibited the plastidic isozyme (98).

In *R. communis* cotyledons, Glu, Asp, and arginine regulation was dependent on pH; at pH 6.9 they inhibited and at pH 7.5 they activated PK (FE
Podestá, manuscript in preparation). Cytosolic PK from developing seed endosperm of *R. communis* showed no response to amino acids at either pH (117,
118). The regulatory differences in *R. communis* PKc isozymes are consistent
with the role of the endosperm as a substrate exporter and the cotyledons as a
tissue active in biosynthesis and nitrogen assimilation.

Pyruvate Dehydrogenase Complex

The pyruvate dehydrogenase complex (PDC) controls entry of carbon into the
TCA cycle. Although this multi-enzyme complex has been purified from
plants (122a), Randall and co-workers have elucidated its regulation mainly
from impure preparations. Light-induced inhibition of PDC by phosphorylation is affected by the relative activity of both PDC kinase and PDC phospho-
rylase, enzymes that form part of the complex (25). High ATP/ADP ratios also
favor inactivation of PDC, which indicates energetic control of the TCA cycle
and supports the hypothesis that phosphorylation plays a role in the control of
PDC (25, 26). Pyruvate acts as an effective PDC activator by overriding ATP
inhibition of PDC kinase (141).

Phosphofructokinase and Pyrophosphate Fructose-6-phosphate 1-Phosphotransferase

The interconversion of fructose 6-phosphate (F6P) and fructose 1,6-bisphos-
phate (FBP) is a key feature in plant respiratory metabolism (Figure 1b).
Typically, phosphofructokinase (PFK) converts F6P to FBP while fructose
1,6-bisphosphatase (FBPase) catalyzes the reverse reaction in both plastids
and the cytosol (35). A cytosolic pyrophosphate fructose-6-phosphate 1-phos-
photransferase (PFP) that operates in either direction has been purified from a
wide range of higher plants (147). The regulation of carbon flow through these
pathways is crucial for providing F6P to glycolysis and the TCA cycle (36).

All cells in higher plants have a cytosolic and plastidic PFK (PFK$_c$ and PFK$_p$, respectively); however, these isozymes may vary within the same plant between tissue type and developmental stage (35, 36, 38). In green algae it appears that there is a single, plastidic enzyme (17, 82, 83, 87). PFK$_c$ and PFK$_p$ are distinguishable immunologically, kinetically, and by their stability during purification (16, 35, 82, 86). Both isozymes show compex behavior toward their substrates, ATP and F6P, and this behavior can change depending on the pH used or the concentration of inorganic phosphate (Pi) present (16, 17, 35, 55, 86, 182). Phosphofructosekinase can undergo changes in its aggregation state (35), and in carrot root, these changes alter the regulatory and kinetic properties of PFK and are affected by the levels of ATP and/or F6P (182). Inorganic phosphate activates both cytosolic and plastidic PFK at low concentrations (5 mM), whereas high Pi concentrations (> 20 mM) inhibit the plastidic isozyme (16, 35, 82). PFK$_c$ and PFK$_p$ are allosterically inhibited by phosphoenolpyruvate (PEP) and 3-phosphoglycerate (3-PGA) (55). Activation by low levels of Pi can override the inhibition by PEP and PGA, and the Pi/PGA and Pi/PEP ratios are thought to be key in regulating PFK$_c$ and PFK$_p$. The regulatory properties of PFK are consistent with the enzyme's significance in converting F6P to FBP to provide carbon for nitrogen assimilation (16, 17, 35, 82, 86). PFK$_p$ should act in the provision of carbon from starch, whereas PFK$_c$ should be important in tissues that import sucrose versus those that depend on triose phosphate (TP) export from the plastid.

The role of the PFP has been debated since its discovery in plant tissues (35). PFP is fully reversible (127) and is activated in both directions by fructose 2,6-bisphosphate (F26P$_2$) (88, 154). Many functions have been proposed for this enzyme including the production of pyrophosphate for sucrose synthesis (5) and as a glycolytic bypass for PFK during Pi deficiency (38a). Study of PFP in tissues active in nitrogen assimilation is required to determine what, if any, role this enzyme has in providing carbon for amino acid synthesis.

Fructose-1,6-bisphosphatase

Both cytosolic and chloroplastic FBPase have been purified and characterized from several plants and algae (18, 143, 188). The presence of plastidic FBPase has not been universally reported in heterotrophic plant tissue, and there appears to be a correlation between the type of storage carbohydrate and the presence of this enzyme (47).

The cytosolic FBPase is regulated by two major factors, F26P$_2$ and AMP, which are both inhibitors of the enzyme (35, 147). These metabolites have a key role in controlling sucrose synthesis and degradation. The synthesis and degradation of F26P$_2$ itself is controlled by metabolites. F26P$_2$-kinase activity is inhibited by 3-PGA, PEP, and TP and is activated by F6P and Pi, while the F26P2-phosphatase is inhibited by Pi and F6P (32, 147, 149).

Chloroplastic FBPase is active only at alkaline pH and if the protein is reduced via the ferredoxin (Fd)-thioredoxin system (54). These two chloroplastic changes occur with illumination (23). The activated enzyme requires a divalent cation (Mg^{+2}, Ca^{+2}) as a cofactor (54, 143). Redox regulation of the FBPase and other enzymes of the reductive pentose phosphate pathway ensures against futile cycling of carbon between the chloroplastic reductive and oxidative pentose phosphate (OPP) pathways during photosynthesis (24). Little is known about how these pathways are regulated if starch is mobilized during photosynthesis; however, during photosynthetic NO_3^- assimilation in N-limited algae there is an oxidative change in the activation state of key enzymes of these pathways (HC Huppe, manuscript in preparation). This may indicate that the Fd-thioredoxin system can be involved in regulating enzyme activity in response to metabolic demands even in saturating light environments.

Sucrose Phosphate Synthetase

Sucrose phosphate synthetase (SPS), the major enzyme of sucrose biosynthesis, is activated upon transition from dark to light (70, 72). Activation decreases throughout the day as the enzyme's sensitivity to Pi inhibition increases (70). In contrast, the ability of glucose 6-phosphate (G6P) to override SPS inhibition also increases during the day. Although the deactivation of SPS is correlated in leaves with an accumulation of sucrose, there is no direct effect of sucrose on the enzyme (72, 153). Instead, high sucrose levels may inhibit TP transport from the chloroplast leading to a rise in the SPS inhibitor, Pi.

The regulation of SPS relates to the phosphorylation status of the enzyme; phosphorylated SPS is more sensitive to Pi inhibition (69, 70). The SPS-phosphatase is light activated by a mechanism involving protein synthesis (180). Inorganic phosphate and G6P are effectors of the phosphorylation mechanism as well as direct allosteric effectors of SPS. Inorganic phosphate inhibits SPS-phosphatase and G6P inhibits the kinase. These changes are consistent with a dephosphorylative activation of SPS upon transition from dark to light (180).

Starch Metabolism

Inorganic phosphate inhibition and PGA activation of ADP-glucose pyrophosphorylase are key to the regulation of starch synthesis (120). Little is known about the regulation of starch degradation (152). Starch may be degraded continually in the light, albeit at a very low rate relative to its biosynthesis (150). Although TP is usually presumed to be the product of starch degradation exported from the plastid, maltose or other hexoses might also be produced and directly exported (148). An understanding of how the regulation of starch degradation is integrated into nitrogen metabolism will require fundamental advances in our knowledge of starch degradation.

A MODEL FOR ACTIVATION OF RESPIRATORY CARBON FLOW TO NITROGEN ASSIMILATION

The demand created by an increase in nitrogen assimilation resets some priorities for cellular resources. The coordinated response of many enzymes that is required to meet these demands must be mediated by regulatory signals shared among those enzymes.

If little nitrogen is being assimilated, the amination of Glu by GS would be slow, Gln pools would be low, and Glu and TCA cycle intermediates would be high. Pyruvate kinase and PEPcase activity would be inhibited. If NH_4^+ becomes available, the levels of GS and GOGAT substrates, Glu and 2-OG, will drop (Figure 1a). Increased activity of GS/GOGAT establishes a new balance in Glu and Gln by which the demand for carbon can be transduced to both PK and PEPcase to increase glycolytic and anapleurotic carbon flow, respectively. The drop in Glu relieves inhibition on both enzymes whereas Gln may activate PEPcase in some systems. Any decrease in the TCA intermediates would also help to activate both PEPcase and PK. The Pyr produced can reverse inhibition of the PDC by ATP/ADP ratios. The combined activation of PK, PEPcase, and PDC would provide carbon to the TCA cycle, depleting levels of PEP. The disappearance of PEP would relieve inhibition of PFK. In both starch and sucrose-dependent tissue, an active PFK facilitates the flow of carbon toward glycolysis rather than sucrose synthesis. Triose phosphate export from the chloroplast in exchange for Pi would increase plastidic Pi, thus supporting plastidic PFK activation and, along with the decrease in PGA, inhibiting starch synthesis, thereby diverting carbon to respiratory metabolism.

An inhibited cytosolic FBPase is required to decrease the drain on either TP exported from chloroplasts or to minimize futile cycling of imported sucrose. The level of $F26P_2$, an inhibitor of cytosolic FBPase, should be elevated because decreased PEP and PGA would activate the $F26P_2$-kinase, while a high level of cytosolic Pi would inhibit the $F26P_2$-phosphatase. A high level of cytosolic Pi would also inhibit SPS in two ways: a direct allosteric effect and an increase in its phosphorylation state via inhibition of the SPS phosphatase.

The results of metabolite measurements during activation of NH_4^+ assimilation in a variety of tissues are consistent with many aspects of this scenario (see below). Most of the metabolite fluxes induced by NO_3^- are similar but slower than with NH_4^+(170). There are also changes in pyridine nucleotides and OPP pathway components during the onset of NO_3^- assimilation that reflect the requirement to reduce NO_3^- to NH_4^+ (75, 170).

ENERGY CONSIDERATIONS

The carbon requirements for amino acid synthesis are independent of the form of inorganic nitrogen assimilated. The major differences between the assimilation of inorganic NO_3^- and NH_4^+ are the energy costs associated with the reduction of NO_3^- to NH_4^+. Compartmentalization of NO_3^- assimilatory pathway enzymes increases the complexity of integrating and controlling nitrogen and carbon metabolism during assimilation. How energy requirements are met depends on the type of tissue and its physiological circumstances. Ultimately, photosynthetic light reactions supply all energy in the plant cell. Although the presence of light is required for carbon assimilation, nitrogen assimilation can occur without the direct input of light energy (1), relying instead on carbon and energy allocated initially to carbon assimilation and stored in carbon bonds.

The Enzymes of Nitrate Reduction

The enzymes of the nitrate assimilatory pathway have been reviewed extensively (11, 84, 96, 144). The NR is a cytosolic enzyme that reduces NO_3^- to NO_2^- by the addition of two electrons from NADH (46, 50, 172). An additional NR that uses either NADPH and NADH has been identified in several organisms and is confined to roots in most higher plants that have been examined (33, 84, 97, 125). Nitrite reductase (NiR), a plastid enzyme in algae, leaves, and roots, reduces NO_2^- to NH_4^+ with 6 electrons donated from Fd (84). GS/GOGAT, rather than glutamate dehydrogenase, is accepted as the pathway responsible for primary ammonia assimilation (101, 130). Multiple forms of GS, including both plastidic and cytosolic isozymes, have been identified in photosynthetic cells and root extracts (50, 96, 173). Molecular tagging indicates that GS isozymes are expressed in separate tissues of the plant, and thus may have different functions (40). GS is the first step in nitrogen assimilation that requires ATP. Both a Fd-dependent and NADP-dependent GOGAT have been found in leaves, roots, and green algae (96, 156). The chloroplastic, Fd-dependent species is the predominant form and its activity is higher in green than in heterotrophic tissues (155).

Of these enzymes, only the regulation of NR has been studied in detail. NR is thought to limit the rate of NO_3^- assimilation because any decrease in NR activity results in a corresponding decrease in NO_3^- assimilation (11, 52). Transgenic tobacco plants that overexpress NR, however, do not assimilate substantially more nitrate (52), indicating that NR activity may be optimized to the general plant metabolism.

A diurnal biosynthesis/degradation cycle regulates NR, decreasing its potential activity throughout the day (72). Superimposed on the coarse control of enzyme quantity is a rapid increase of sensitivity to Mg^{+2} inhibition upon darkening (68, 78, 79, 128). Sensitivity to Mg^{+2} disappears within an hour of exposing leaves to light, but then increases slowly during the day (72). Several lines of evidence link light activation to dephosphorylation of the leaf enzyme decreasing its sensitivity to Mg^{+2} (68, 79). In vivo application of okadic acid, a phosphatase inhibitor, prevents this dephosphorylation and completely blocks light activation of NR (71).

Treatment of leaves with cycloheximide indicates that NR phosphatase itself is light activated by a process dependent on protein synthesis and that the enzyme must be synthesized continually (71). The NR kinase is neither turned over rapidly nor affected by light, but it is inhibited by G6P. NR activation, therefore, depends on regulation and synthesis of the NR phosphatase. Addition of cycloheximide to illuminated leaves leads to a slow decrease in NR activation (71), indicating that the rapid dark deactivation of the phosphatase involves more than a halt in protein synthesis.

Cytosolic Reductant Supply

Possible sources of cytosolic reductant for NR include cytosolic dehydrogenases, mitochondrial TCA activity, and/or photosynthesis. Whether observations of the induction of NO_3^- assimilation by light, CO_2, and the provision of TP and/or respiratory intermediates in the dark is used to argue for a particular route depends on interpretation (2, 97, 99, 106, 110, 123, 135, 184). For example, the light dependency of NO_3^- reduction could implicate a requirement for photosynthetic energy, for photosynthate to be respired or oxidized, or for light inhibition of the mitETC to make reductant available to NR (97).

In the plant cell the low NADPH/NADP and NADH/NAD ratios in the cytosol vs the chloroplast or mitochondria, respectively, favor the organelles as the source for cytosolic reductant (61, 62, 89). Neither NADPH nor NADH can be exported directly from chloroplasts or mitochondria so the activity of the shuttles between these compartments must be regulated to maintain these redox differentials.

The chloroplast has two transporters for reductant (61). TP transported from the plastid via a DHAP/3-PGA shuttle can supply NADH and ATP via the cytosolic NAD-glyceraldehyde 3-phosphate dehydrogenase (GAPDH) and 3-PGA-kinase. NADH can also be exported as malate by the malate-oxaloacetate (OAA) shuttle, which is regulated by NADP-malate dehydrogenase (MDH). The dual requirement of reduced thioredoxin and high NADPH/NADP ratios for NADP-MDH activation restricts export of reductant as malate, even in the light (138).

The mitochondria export reductant via a malate-OAA shuttle to supply electrons to peroxisomes for the reduction of β-hydroxy pyruvate in photorespiration (37–39, 109). Malate could be oxidized in the cytosol by the cytosolic NAD-MDH, but only if OAA is removed fast enough to pull the reaction in favor of NAD reduction (97). A translocator with a high affinity for OAA has been identified in the mitochondria of pea seedlings and roots (39, 109).

Plant mitochondrial activity during photosynthesis relates to its potential role in NO_3^- assimilation. The high rate of biosynthesis during photosynthesis requires that TCA carbon flow continue in the light (57). It has been argued that the mitETC is inhibited in the light by the high level of ATP produced by photophosphorylation; therefore, the TCA cycle–produced reductant is available to NR (37, 135). There is substantial evidence, however, that mitochondrial respiration is not light inhibited (38). Respiratory O_2 uptake can be measured in the light when the complicating photorespiration and the Mehler reactions are inhibited (177). Further, inhibition of the mitETC by oligomycin shows that mitochondria contribute significantly to cytosolic ATP levels in the light (89, 90).

The increased respiratory quotient during NO_3^- assimilation and the stimulation of dark NO_3^- reduction when use of mitETC electron is inhibited chemically or by anaerobiosis supports a role for a mitochondrial supply of energy to NO_3^- reduction (58, 179). A significant coupling of respiration to dark NO_3^- assimilation in barley plants has been demonstrated (12). In addition, calculation of the electron requirement necessary to support the rate of NO_3^- assimilation observed in N-limited *S. minutum* shows that photosynthetic electron supply is inadequate to meet the needs of NO_3^- reduction and that some reductant must be provided by respiration (179).

The redox equilibrium between plant cell compartments is maintained by tight regulation of chloroplastic and mitochondrial reductant translocation (37, 61). Factors affecting one source of cytosolic NADH should trigger other sources to compensate for the change. Nitrate reduction is one of many reactions in the total cytosolic redox equilibrium; thus, any of the cytosolic NADH producing pathways can potentially supply energy for NO_3^- reduction (e.g. 45, 97).

Energy Supply in the Plastid

THE CHLOROPLAST IN THE LIGHT AND DARK   Two of the three reductive steps in NO_3^- assimilation (NiR and GOGAT) take place in the chloroplast and require Fd. Light stimulation and DCMU inhibition of NO_3^- reduction in green tissue indicate a link between linear photosynthetic electron flow and nitrogen assimilation (58, 123, 184). Because both NO_2^- reduction and carbon assimilation occur in the chloroplast, the partitioning of electrons between these two pathways must be controlled. In vitro experiments have demonstrated that the addition of NiR to a reconstituted light system lowers the rate of NADPH formation by Fd-NADP oxidoreductase (FNR) (10). In saturating light, the supply of electron acceptors, rather than energy, limits photosynthetic electron transport (48). Enhancement of non-cyclic electron transport during NO_3^- assimilation (12, 34) indicates Fd can cycle faster because there is a new sink for photosynthetic electrons. Although electron flow is stimulated during photosynthetic NO_3^- assimilation, CO_2 fixation is suppressed in N-limited algae (43, 44). Ammonium causes less suppression of CO_2 fixation. These two observations demonstrate a direct effect of the energy demands of NO_3^- reduction on the carbon assimilation rate. Suppression of CO_2 fixation by nitrogen assimilation is less frequent in intact higher plants (12, 123), perhaps because CO_2 availability rather than energy often limits the rate of CO_2 fixation in higher plants.

Nitrate assimilation in a photosynthetic cell does not always require light energy (1). Rates of NO_3^- assimilation in darkened leaves as high as 76% of the rate in the light have been reported (126). Reduction of $^{15}NO_2^-$ has been demonstrated in the dark under anaerobic and aerobic conditions (187). Dark inhibition of NO_3^- assimilation can often be overcome in plants and algae if they have an alternative carbon source (7, 139, 163).

Cytoplasmic carbon oxidation could supply chloroplastic reductant if the redox shuttles that export excess photosynthetic reductant to the cytosol were reversed in the dark. However, two of the chloroplastic enzymes involved, NADP-MDH of the malate/OAA shuttle and the NADP-GAPDH of the DHAP/3-PGA shuttle, are light-regulated enzymes and should be inactive in the dark (24, 136). Reductant could be produced in situ by the chloroplastic OPP pathway, one of two routes available for the degradation of starch (152). Chloroplastic glucose 6-phosphate dehydrogenase (G6PDH), which controls the OPP pathway, is inhibited by both low NADP/NADPH ratios and reduction via the Fd-thioredoxin system (24, 136). In the dark, G6PDH is oxidized so the OPP pathway can operate (137; TJ Farr, manuscript in preparation).

In N-limited algae, changes in NADP/NADPH and 6-phosphogluconate/G6P ratios and the oxidative activation of G6PDH demonstrate activation of the OPP pathway upon resupply of NO_3^- in either the dark or light (75, 170; HC Huppe, manuscript in preparation) (Figure 1c). The need for energy can apparently stimulate carbon oxidation in the dark and even override inhibition of the OPP pathway in the light. How the NADPH produced reduces Fd is not clear. Catalysis by leaf FNR strongly favors reduction of NADPH by Fd; reversing the FNR reaction would require a very high turnover of Fd (45).

REDUCTANT SUPPLY IN THE ROOT PLASTID   The OPP pathway probably supplies energy for NO_3^- assimilation in the root (134). All enzymes of this pathway (G6PDH, 6-phosphogluconate dehydrogenase, transketolase, and transaldolase) are present in both root plastids and cytosol (46), and their activities increased when roots were incubated with NO_3^- (134). The ratio of $^{14}CO_2$

released from C_6- and C_1-labeled glucose fed to these roots during NO_3^- assimilation was also consistent with increased OPP pathway activity (134).

Reduction of NO_2^- to Glu in isolated plastids and the plastidic location of both NiR and GOGAT (46) indicate that the plastidic OPP pathway is involved. Isolated root plastids transport G6P and TP (14, 15, 19, 46). There is a stoichiometric relationship between the rate of NO_2^- reduction and G6P oxidation in isolated plastids from pea and barley (15, 19, 108). Lee (97) has estimated that the potential OPP pathway activity in the plastid is insufficient to support the potential for NO_2^- reduction (97); however, this is based on in vitro kinetics, which may not accurately predict the in vivo potentials of either pathway.

The OPP pathway produces NADPH, whereas NiR and GOGAT are Fd-dependent enzymes. Fd has been isolated from roots of radish, mung bean, and spinach (63, 104, 105, 175). A FNR-like protein has been detected in both corn (157) and pea roots (20) and isolated from spinach and radish root (104, 105). The identification of these two proteins provides a mechanism to reduce Fd in roots.

Root FNR activity was detected by NADPH oxidation rather than reduction (157). Fd- and FNR-like components in maize roots could substitute for leaf Fd or FNR, respectively, in cytochrome c reduction by NADPH, but not in the photoreduction of NADP (157), which is the kinetically favored direction of leaf FNR. Bean sprout FNR worked in both directions, but there was a kinetic advantage in NADP reduction when the spinach Fd was used, whereas the bean sprout Fd worked better to reduce cytochrome c (163). The primary structures of the spinach root FNR and Fd differ from their leaf counterparts (105). It will be interesting to learn how these observations relate to the apparent role of the root enzyme in Fd reduction rather than oxidation.

ATP Supply

Whether the mitETC, glycolysis, or photophosphorylation supplies ATP to nitrogen metabolism depends on when and where nitrogen is assimilated. Dark and heterotrophic tissues must rely on glycolysis and the mitETC, whereas photophosphorylation can contribute ATP to both the chloroplast and cytosol of photosynthesizing cells. Although chloroplastically produced ATP transported across the chloroplast membrane as TP is thought to be a major supplier of cytosolic ATP, it is now apparent that mitochondria contribute significant amounts of ATP in the cytosol in the light (90).

Primary nitrogen assimilation requires ATP at the GS step and probably involves the chloroplastic GS isozymes, but the source of ATP in darkened chloroplasts has not been identified. The onset of NH_4^+ assimilation causes a transient drop in ATP levels that is correlated with an activation of PK as evidenced by metabolite changes (166), but the compartmentalization of these changes is unknown. Chloroplasts transport adenylates directly at a slow rate. Starch degradation may be able to support ATP production because PK and most of the other glycolytic enzymes are present in the chloroplast (36). A supply of ATP has not been demonstrated in the root plastid either, but it may not be required because the cytosolic GS is probably the major isozyme in roots (107).

In the light, cells may modulate the capacity for ATP and NADPH production from photosynthesis by controlling the relative activities of cyclic and non-cyclic electron flow by state transitions (181). State transitions may alter ATP and NADPH production in response to changes in carbon (64–67) and nitrogen metabolism (167). When NH_4^+ is assimilated, the cellular ATP/NADPH requirements increase dramatically relative to those of CO_2 fixation. Following 20 min of NH_4^+ assimilation by S. *minutum,* the 77K fluorescence spectra showed a decrease in the ratio of PSII to PSI fluorescence typical of a state 2 transition possibly favoring ATP production by cyclic electron transport. Reduction and assimilation of NO_3^-, which demands more NADPH than NH_4^+ assimilation, did not invoke a similar state 2 transition (103, 167). These changes imply that the energy demands of nitrogen assimilation can modulate the quantum yield of PSII.

REGULATION IN DIFFERENT SYSTEMS

Understanding the regulation of rate-limiting enzymes during the integration of nitrogen and carbon assimilation requires examination of the metabolite fluxes that occur in relation to the physiological responses combined with characterization of the isolated enzymes. The physiological and metabolite changes that occur within seconds of the onset of nitrogen assimilation have been measured in unicellular algae. These observations can be applied to the longer time courses measurable in higher plants. Consideration of the differences in the metabolic requirements of the cells allows speculation of how the regulation might differ and provides clues for future work.

The Algae

In N-limited algae, the onset of NH_4^+ assimilation causes a rapid drop in Glu and ATP and an increase in ADP, AMP, and Gln (146, 166)). Levels of 2-OG decline, reflecting the metabolism of Gln by GOGAT (166). The first change in glycolytic metabolites is a drop in PEP and an increase in Pyr within seconds of NH_4^+ addition, consistent with rapid activation of PK. The ADP concentration and results of uncoupler experiments indicated that PK is adenylate restricted in these cells before NH_4^+ addition (98); therefore, PK activation may result initially from the transient rise in ADP levels (166). After ADP levels recover, the drop in Glu and increase in TP would sustain PK activation.

PEPcase is stimulated rapidly, perhaps responding to the decreased Glu/Gln ratio (171).

Nitrogen enrichment causes a decreased flow of fixed CO_2 to carbohydrates and can induce starch degradation and suppression of photosynthetic CO_2 fixation (8, 44, 146, 165). A drop in RuBP and F6P indicates slowing of the regeneration phase of the reductive pentose phosphate pathway, while increasing FBP and TP levels demonstrate rerouting of fixed carbon to the cytosol (42, 146, 166). The drop in PEP and 3-PGA mediated by activation of PK and PEPcase should relieve inhibition on the plastidic PFK (17) and aid the export of carbon from the chloroplast in exchange for Pi. Increased chloroplastic Pi favors starch breakdown.

Partitioning carbon in the cytosol of unicellular algae is simplified because there is no translocation of sucrose to other tissues. *S. minutum,* for example, has no cytosolic PFK (17). Any drain on TP by the cytosolic FBPase would be inhibited by F26P2, which increases initially at the onset of nitrogen assimilation in *S. minutum* (18, 166).

Nitrate assimilation induces similar, but slower changes in most key metabolites in *S. minutum* (170). There is also a rapid drop in G6P levels and the NADPH reduction state, and an increase in 6-phosphogluconate levels and G6PDH activity at the onset of NO_3^- assimilation, which does not occur during NH_4^+ assimilation (75, 170). These changes indicate that stimulation of the OPP pathway to provide reductant occurs in both the light and the dark before carbon flow increases to supply amino acid biosynthesis.

The Photosynthetic Higher Plant Cell

Similar regulatory signals operate in higher plants and algae to activate carbon flow to amino acid biosynthesis during nitrogen assimilation. In leaf discs and isolated cells, nitrogen stimulates increased ^{14}C labeling into amino acids, Pyr, and TCA intermediates (8, 185), which indicates increased flow of carbon from photosynthetic fixation and activation of anaplerotic CO_2 fixation, possibly involving phosphorylation of the PEPcase (27). Shorter time courses using protoplasts showed a transient drop in Glu labeling with a rise in Gln (113). In N-limited corn plants, an increased Gln/Glu ratio was found within 1 h of resupplying nitrogen (53).

The most consistent effect of nitrogen assimilation on leaf carbon metabolism is that sucrose synthesis decreases while starch or reductive pentose phosphate pathway components are often not affected (8, 91, 113). N-limitation in plants induces an overall increase in carbohydrate levels that is often greater in roots than in leaves (132). The major function of mature leaves is to supply sucrose for the rest of the plant. The cytosol is the primary site for the regulation of carbon partitioning in leaves; hence, photosynthesis can be unaffected even when there are large changes in the direction of carbon flow in the cytosol (151). Whether a cell exports sucrose or stores starch might help explain why leaves have been reported to stimulate, to suppress, and to not change O_2 evolution or photosynthetic CO_2 fixation upon nitrogen resupply.

The key regulators directing carbon to gluconeogenesis and respiration in the leaf cytosol are $F26P_2$ and Pi, which affect SPS and the enzymes regulating the balance of F6P and FBP (32, 70, 151). Few details of the regulatory fluxes in mature source leaves during nitrogen assimilation have been reported. In the developing cotyledon of *R. communis,* which is also a sucrose sink, analysis of metabolite changes during nitrogen assimilation demonstrated consumption of organic acids (2-OG and Pyr) in support of amino acid synthesis (56). A drop in PEP and PGA indicated activation of PK and PEPcase. Relieving the PEP inhibition of $F26P_2$kinase increased $F26P_2$, thereby encouraging glycolytic carbon flow via the PFP and inhibiting flow to gluconeogenesis via cytosolic FBPase.

During N assimilation in source leaves, high levels of $F26P_2$ and Pi would be expected to inhibit the flow of carbon to sucrose via FBPase and SPS. Decreasing SPS activity would lower carbon translocation from source tissues, whereas an inhibited SPS in sink tissues would favor the import and degradation of sucrose.

The regulation of SPS and NR by phosphorylation may have a role in directing carbon flow during NO_3^- assimilation (73). Detached leaves accumulate photosynthate. Excision causes more rapid phosphorylation (inhibition) of SPS during the day than in attached leaves whereas the NR activity does not decrease (72). Greater availability of carbon may favor NO_3^- assimilation. The effect of carbohydrate status on the light-related dephosphorylation of SPS and NR could be important in understanding the differential dark assimilation rates found in plants.

Heterotrophic Tissues

Roots must use different strategies than do photosynthetic tissues to support nitrogen assimilation because translocated photosynthate provides both carbon skeletons and all the energy to roots. Cytosolic energy for nitrogen assimilation is probably provided by glycolysis, the TCA cycle, and the mitETC, as it is in the leaf (45). The cytosolic OPP pathway also could be important in roots with NR that uses both NADPH and NADH (106).

Roots contain isozymes of all the regulatory enzymes that integrate carbon and nitrogen metabolism in leaves and algae; however, there has been less research done on the regulation of root isozymes (35, 36). In sucrose-storing roots, some regulatory mechanisms might vary considerably from those in source leaves. It will be interesting to investigate the regulation of the light-regulated enzymes, such as SPS and NR, to determine if carbohydrate status is the important regulatory signal in roots.

The plastidic OPP pathway is the most likely source of reductant for nitrite reduction (45) and a major route for degrading starch, which supplies carbon for amino acid synthesis in some plastids, yet the regulation of root G6PDH is not understood. The mechanism of Fd reduction via the root FNR also requires further characterization.

Although our knowledge of nitrogen assimilation in roots is improving, there is much work to be done on the metabolic regulation in this tissue.

CONCLUDING REMARKS

We are only beginning to realize the complexity of carbon and nitrogen interactions in photosynthetic eukaryotes. Despite different carbon sources (starch, sucrose, or recent photosynthate) for nitrogen assimilation in various tissues, some unifying features are emerging. The initial steps in nitrogen assimilation result in metabolite changes that activate glycolysis at the level of PK and PEPcase. This activation causes metabolite changes that signal the need to direct carbon to respiration from either starch, sucrose, or the reductive pentose phosphate pathway, depending on the tissue. During photosynthesis, the metabolic demands of nitrogen assimilation are transduced to the photosynthetic apparatus to balance ATP and NADPH production with the energy requirements. The extra reductant demanded in assimilating NO_3^- changes the initial priority of carbon flow in tissues where photosynthetic energy is inadequate to meet the electron demands. In such cases, the OPP pathway is activated to provide reducing power. The nitrogen and carbon metabolic pathways are interconnected through regulatory signals that include allosteric metabolite effectors, phosphorylation, and redox regulations. Carbon and nitrogen metabolism are interdependent processes, so neither can continue to operate to the detriment of the other. Nutrient status, either of carbohydrate levels or nitrogen compounds, are important factors in determining the priority of resource use between these two pathways. The ideas and concepts that are emerging from studies of carbon and nitrogen interactions in plants will continue to provide much insight into how each of these processes acts individually and how they interact in response to the environment of the organism.

ACKNOWLEDGMENTS

We wish to thank Drs. M-L Champingy, C Foyer, SC Huber, A Oaks, and WC Plaxton for providing unpublished manuscripts and preprints. We are indebted to Drs. R Guy and WC Plaxton for critical reading of the manuscript, to S Congdon for assistance in preparing the manuscript, and to TJ Farr for overall support and for doing so much, so well.

Literature Cited

1. Abrol YP, Sawhney SK, Naik MS. 1983. Light and dark assimilation of nitrate in plants. *Plant Cell Environ.* 6:595–99
2. Amory AM, Vanlerberghe GC, Turpin DH. 1991. Demonstration of both a photosynthetic and nonphotosynthetic CO_2 requirement for NH_4^+ assimilation in the green alga *Selenastrum minutum. Plant Physiol.* 95:192–96
3. Andrews M. 1986. The partitioning of nitrate assimilation between root and shoot of higher plants. *Plant Cell Environ.* 9:511–19
4. ap Rees T. 1990. Carbon metabolism in mitochondria. In *Plant Biochemistry, Physiology and Molecular Biology,* ed. DT Dennis, DH Turpin, pp. 106–23. Essex: Longman Sci. Tech.
5. ap Rees T, Morrell S, Edwards J, Wilson PM, Green JH. 1985. Pyrophosphate and glycolysis in higher plants. In *Regulation of Carbon Partitioning in Photosynthetic Tissue,* ed. RL Heath, J Preiss, pp. 76–92. Rockville, MD: Am. Soc. Plant Physiol.
6. Aslam M, Huffaker RC. 1982. *In vivo* nitrate reduction in roots and shoots of barley (*Hordeum vulgare* L.) seedlings in light and darkness. *Plant Physiol.* 70:1009–13
7. Aslam M, Huffaker RC, Rains DW, Rao KP. 1979. Influence of light and ambient carbon dioxide concentration on nitrate assimilation by intact barley seedlings. *Plant Physiol.* 63:1205–9
8. Bassham JA, Larsen PO, Lawyer AL, Cornwell KL. 1981. Relationships between nitrogen metabolism and photosynthesis. In *Nitrogen and Carbon Metabolism,* ed. JD Bewley, pp. 135–63. London: Junk
9. Baysdorfer C, Bassham JA. 1984. Spinach pyruvate kinase isoforms. Partial purification and regulatory properties. *Plant Physiol.* 74:374–79
10. Baysdorfer C, Robinson JM. 1985. Metabolic interactions between spinach leaf nitrite reductase and ferredoxin-NADP reductase. *Plant Physiol.* 77:318–20
11. Beevers L, Hageman RH. 1980. Nitrate and nitrite reduction. In *Amino Acids and Derivatives,* ed. BJ Miflin, pp. 115–68. New York: Academic
12. Bloom AJ, Caldwell RM, Finazzo J, Warner RL, Weissbart J. 1989. Oxygen and carbon dioxide fluxes from barley shoots depend on nitrate assimilation. *Plant Physiol.* 91:352–56
13. Bloom AJ, Sukrapanna SS, Warner RL. 1991. Root respiration associated with ammonium and nitrate absorption and assimilation by barley. *Plant Physiol.* 99:1294–301
14. Borchert S, Grosse H, Heldt HW. 1989. Specific transport of inorganic phosphate, glucose 6-phosphate, dihydroxyacetone phosphate and 3-phosphoglycerate into amyloplasts from pea roots. *FEBS Lett.* 253:183–86
15. Borchert S, Harborth J, Schünemann D, Hoferichter P, Heldt HW. 1993. Studies of the enzymic capacities and transport properties of pea root plastids. *Plant Physiol.* 101:303–12
16. Botha FC, Cawood MC, Small JGC. 1988. Kinetic properties of the ATP-dependent phosphofructokinase isoenzymes from cucumber seeds. *Plant Cell Physiol.* 29(3): 415–21
17. Botha FC, Turpin DH. 1990. Molecular, kinetic, and immunological properties of the 6-phosphofructokinase from the green alga *Selenastrum minutum.* Activation during biosynthetic carbon flow. *Plant Physiol.* 93:871–79
18. Botha FC, Turpin DH. 1990. Fructose 1,6-bisphosphatase in the green alga *Selenastrum minutum.* I. Evidence for the presence of isoenzymes. *Plant Physiol.* 93: 1460–65
19. Bowsher CG, Hucklesby DP, Emes MJ. 1989. Nitrite reduction and carbohydrate metabolism in plastids purified from roots of *Pisum sativum* L. *Planta* 177:359–66
20. Bowsher CG, Hucklesby DP, Emes MJ. 1993. Induction of ferredoxin-NADP oxidoreductase and ferredoxin synthesis in pea root plastids during nitrate assimilation. *Plant J.* 3:463–67
21. Brouquisse R, Nishimura M, Gaillard J, Douce R. 1987. Characterization of a cytosolic aconitase in higher plant cells. *Plant Physiol.* 84:1402–7
22. Bryan KK. 1990. Advances in the biochemistry of amino acid biosynthesis. In *Advances in Nitrogen Assimilation,* ed. BJ Miflin, PJ Lea, pp. 161–95. San Diego: Academic
23. Buchanan BB. 1980. Role of light in the

chloroplast enzymes. *Annu. Rev. Plant Physiol.* 31:341–74
24. Buchanan BB. 1991. Regulation of CO_2 assimilation in oxygenic photosynthesis: the ferredoxin/thioredoxin system. Perspective on its discovery, present status, and future development. *Arch. Biochem. Biophys.* 288:1–9
25. Budde RJA, Randall DD. 1987. Regulation of pea mitochondrial pyruvate dehydrogenase complex activity: inhibition of ATP-dependent inactivation. *Arch. Biochem. Biophys.* 258:600–6
26. Budde RJA, Randall DD. 1990. Light as a signal influencing the phosphorylation status of plant proteins. *Plant Physiol.* 94: 1501–4
27. Champigny M-L, Foyer C. 1992. Nitrate activation of cytosolic protein kinases diverts photosynthetic carbon from sucrose to amino acid biosynthesis. Basis for a new concept. *Plant Physiol.* 100:7–12
28. Champigny M-L, Talouizte A. 1986. Dependence of nitrate reduction on root soluble carbohydrates in wheat seedlings. In *Fundamental, Ecological and Agricultural Aspects of Nitrogen Metabolism in Higher Plants,* ed. H Lambers, JJ Neeteson, I Stulen, pp. 279–82. Netherlands: Martinus Nijhoff
29. Chen R-D, Bismuth E, Champigny M-L, Gadal P. 1989. Chromatographic and immunological evidence that chloroplastic and cytosolic pea (*Pisum sativum* L.) NADP-isocitrate dehydrogenases are distinct isoenzymes. *Planta* 178:157–63
30. Chen R-D, Gadal P. 1990. Do the mitochondria provide the 2-oxoglutarate needed for glutamate synthesis in higher plant chloroplasts? *Plant Physiol. Biochem.* 28:141–45
31. Chen R-D, Gadal P. 1990. Structure, functions and regulation of NAD and NADP dependent isocitrate dehydrogenases in higher plants and in other organisms. *Plant Physiol. Biochem.* 28:411–27
32. Cséke C, Buchanan BB. 1986. Regulation of the formation and utilization of photosynthate in leaves. *Biochim. Biophys. Acta* 853:43–63
33. Dailey FA, Kuo T, Warner RL. 1982. Pyridine nucleotide specificity of barley nitrate reductase. *Plant Physiol.* 69:1196–99
34. de la Torre A, Delgado B, Lara C. 1991. Nitrate-dependent O_2 evolution in intact leaves. *Plant Physiol.* 96:898–901
35. Dennis DT, Greyson MF. 1987. Fructose 6-phosphate metabolism in plants. *Physiol. Plant.* 69:395–404
36. Dennis DT, Miernyk JA. 1982. Compartmentation of nonphotosynthetic carbohydrate metabolism. *Annu. Rev. Plant Physiol.* 33:27–50
37. Douce R. 1985. *Mitochondria in Higher Plants. Structure, Function, Biogenesis,* pp. 77–153; 205–35. Orlando, FL: Academic
38. Dry IB, Dimitriadis E, Ward AD, Wiskich JT. 1987. The photorespiratory hydrogen shuttle. Synthesis of phthalonic acid and its use in the characterization of the malate/aspartate shuttle in pea (*Pisum sativum*) leaf mitochondria. *Biochem. J.* 245:669–75
38a. Duff SMG, Moorehead GBG, Lefebvre DD, Plaxton WC. 1989. Phosphate starvation inducible 'bypasses' of adenylate and phosphate dependent glycolytic enzymes in *Brassica nigra* suspension cells. *Plant Physiol.* 90:1275–78
39. Ebbighausen H, Jia C, Heldt HW. 1985. Oxaloacetate translocator in plant mitochondria. *Biochim. Biophys. Acta* 810: 184–99
40. Edwards JW, Walker EL, Coruzzi GM. 1990. Cell-specific expression in transgenic plants reveals nonoverlapping roles for chloroplast and cytosolic glutamine synthetase. *Proc. Natl. Acad. Sci. USA* 87: 3459–63
41. Elias BA, Givan CV. 1977. Alpha-ketoglutarate supply for amino acid synthesis in higher plant chloroplasts. Intrachloroplastic localization of NADP-specific isocitrate dehydrogenase. *Plant Physiol.* 59:738–40
42. Elrifi IR, Holmes JJ, Weger HG, Mayo WP, Turpin DH. 1988. RuBP limitation of photosynthetic carbon fixation during NH_3 assimilation. Interactions between photosynthesis, respiration, and ammonium assimilation in N-limited green algae. *Plant Physiol.* 87:395–401
43. Elrifi IR, Turpin DH. 1985. Transient photosynthetic responses of nitrogen limited microalgae to nitrogen addition. *Mar. Ecol. Prog. Ser.* 20:253–58
44. Elrifi IR, Turpin DH. 1986. Nitrate and ammonium induced photosynthetic suppression in N-limited *Selenastrum minutum. Plant Physiol.* 81:273–79
45. Emes MJ, Bowsher CG. 1991. Integration and compartmentation of carbon and nitrogen metabolism in roots. In *Compartmentation of Plant Metabolism in Non-Photosynthetic Tissues,* ed. MJ Emes, pp. 147–65. Cambridge: Cambridge Univ. Press
46. Emes MJ, Fowler MW. 1979. The intracellular location of the enzymes of nitrate assimilation in the apices of seedling pea roots. *Planta* 144:249–53
47. Entwistle G, ap Rees T. 1990. Lack of fructose-1,6-bisphosphatase in a range of higher plants that store starch. *Biochem. J.* 271:467–72
48. Farquhar GD, von Caemmerer S, Berry JA. 1980. A biochemical model of photosynthetic CO_2 fixation in leaves of C_3 species. *Planta* 149:78–90
49. Deleted in proof

50. Fischer P, Klein U. 1988. Localization of nitrogen-assimilating enzymes in the chloroplast of *Chlamydomonas reinhardtii*. *Plant Physiol.* 188:947–52

51. Foyer CH, Lefebvre C, Provot M, Vincentz M, Vaucheret H. 1993. Modulation of nitrogen and carbon metabolism in transformed *Nicotiana plumbaginifolia* mutant E23 lines expressing either increased or decreased nitrate reductase activity. In *Physiology of Varities,* ed. E White, PS Kettlewell, MA Parry, RP Ellis, pp. 137–45. Wellesbourne Warwick, UK: Assoc. Appl. Biol.

52. Foyer CH, Lescure J-C, Lefebvre C, Morot-Gaudry J-F, Vincentz M, Vaucheret H. 1994. Adaptations of photosynthetic electron transport, carbon assimilation and carbon partitioning in transgenic *Nicotiana plumbaginifolia* plants to changes in nitrate reductase activity. *Plant Physiol.* 104:171–78

53. Foyer CH, Noctor G, Lelandais M, Lescure J-C, Valadier MH, Boutin JP, Horton P. 1994. Short-term effects of nitrate, nitrite and ammonium assimilation on chlorophyll a fluorescence, thylakoid protein phosphorylation, net CO_2 assimilation and amino acid biosynthesis in maize. *Planta* 192:211–20

54. Gardemann A, Schimkat D, Heldt HW. 1986. Control of CO_2 fixation. Regulation of stromal fructose-1,6-bisphosphatase in spinach by pH and Mg^{2+} concentration. *Planta* 168:536–45

55. Garland WJ, Dennis DT. 1980. Plastid and cytosolic phosphofructokinases from the developing endosperm of *Ricinus communis*. II. Comparison of the kinetic and regulatory properties of the isoenzymes. *Arch. Biochem. Biophys.* 204:310–17

56. Geigenberger P, Stitt M. 1991. Regulation of carbon partitioning between sucrose and nitrogen assimilation in cotyledons of germinating *Ricinus communis* L. seedlings. *Planta* 185:563–68

57. Graham D. 1980. Effects of light on 'dark' respiration. In *Metabolism and Respiration,* ed. DD Davies, pp. 525–79. New York: Academic

58. Gray VM, Cresswell CF. 1984. The effect of inhibitors of photosynthetic and respiratory electron transport on nitrate reduction and nitrite accumulation in excised *Z. mays* L. leaves. *J. Exp. Bot.* 35:1166–76

59. Guy RD, Vanlerberghe GC, Turpin DH. 1989. Significance of phosphoenolpyruvate carboxylase during ammonium assimilation. Carbon isotope discrimination in photosynthesis and respiration by the N-limited green alga *Selenastrum minutum*. *Plant Physiol.* 89:1150–57

60. Hanning I, Heldt HW. 1992. Provision of carbon skeletons for nitrate assimilation by spinach leaf mitochondria. In *Molecular, Biochemical and Physiological Aspects of Plant Respiration,* ed. H Lambers, LHW van der Plas, pp. 249–53. The Hague: SPB Academic

61. Heineke D, Riens B, Grosse H, Hoferichter P, Peter U, et al. 1991. Redox transfer across the inner chloroplast envelope membrane. *Plant Physiol.* 95:1131–37

62. Heupel R, Heldt HW. 1992. Redox transfer between mitochondria and peroxisomes. In *Molecular, Biochemical and Physiological Aspects of Plant Respiration,* ed. H Lambers, LHW van der Plas, pp. 243–47. The Hague: SPB Academic

63. Hirasawa M, Chang K-T, Knaff DB. 1990. Characterization of a ferredoxin:$NADP^+$ oxidoreductase from a nonphotosynthetic plant tissue. *Arch. Biochem. Biophys.* 276:251–58

64. Horton P. 1987. Interplay between environmental and metabolic factors in the regulation of electron transport in higher plants. In *Progress in Photosynthesis,* ed. J Biggins, pp. 681–88. Dordrecht: Martinus Nijhoff

65. Horton P. 1989. Interaction between electron transport and carbon assimilation: regulation of light harvesting and photochemistry. In *Photosynthesis,* ed. WR Briggs, pp. 393–406. New York: Liss

66. Horton P, Lee P. 1986. Observation of enhancement and state transitions in isolated intact chloroplasts. *Photosynth. Res.* 10:297–302

67. Horton P, Lee P, Fernyhough P. 1990. Emerson enhancement, photosynthetic control and protein phosphorylation in isolated maize mesophyll chloroplasts: dependence upon carbon metabolism. *Biochim. Biophys. Acta* 1017:160–66

68. Huber JL, Huber SC, Campbell WH, Redinbaugh MG. 1992. Reversible light/dark modulation of spinach leaf nitrate reductase activity involves protein phosphorylation. *Arch. Biochem. Biophys.* 296:58–65

69. Huber JLA, Huber SC, Nielsen TH. 1989. Protein phosphorylation as a mechanism for regulation of spinach leaf sucrose-phosphate synthase activity. *Arch. Biochem. Biophys.* 270:681–90

70. Huber SC, Huber JL.1992. Role of sucrose-phosphate synthase in sucrose metabolism in leaves. *Plant Physiol.* 99:1275–78

71. Huber SC, Huber JL, Campbell WH, Redinbaugh MG. 1992. Apparent dependence of the light activation of nitrate reductase and sucrose-phosphate synthase activities in spinach leaves on protein synthesis. *Plant Cell Physiol.* 33:1–8

72. Huber SC, Huber JL, Campbell WH, Redinbaugh MG. 1992. Comparative studies of the light modulation of nitrate reduc-

tase and sucrose-phosphate synthase activities in spinach leaves. *Plant Physiol.* 100: 706–12

73. Huber SC, Huber JL, McMichael JRW, Campbell WH, Redinbaugh MG. 1992. Regulation of cytoplasmic C- and N- metabolism by protein phosphorylation. In *Research in Photosynthesis*, ed. N Murata, pp. 675–82. Dordrecht: Kluwer Academic

74. Deleted in proof

75. Huppe HC, Vanlerberghe GC, Turpin DH. 1992. Evidence for activation of the oxidative pentose phosphate pathway during photosynthetic assimilation of NO_3^- but not NH_4^+ by a green alga. *Plant Physiol.* 100:2096–99

76. Ireland R. 1990. Amino acid and ureide biosynthesis. In *Plant Biochemistry, Physiology and Molecular Biology,* ed. DT Dennis, DH Turpin, pp. 407–21. Essex: Longman Sci. Tech.

77. Jiao J, Vidal J, Echevarria C, Chollet R. 1991. In vivo regulatory phosphorylation site in C_4-leaf phosphoenolpyruvate carboxylase from maize and sorghum. *Plant Physiol.* 96:297–301

78. Kaiser WM, Brendle-Behnisch E. 1991. Rapid modulation of spinach leaf nitrate reductase activity by photosynthesis. I. Modulation in vivo by CO_2 availability. *Plant Physiol.* 96:363–67

79. Kaiser WM, Spill D. 1991. Rapid modulation of spinach leaf nitrate reductase by photosynthesis. II. *In vitro* modulation by ATP and AMP. *Plant Physiol.* 96:368–75

80. Kanazawa T, Distefano M, Bassham JA. 1983. Ammonia regulation of intermediary metabolism in photosynthesizing and respiring *Chlorella pyrenoidosa*: comparative effects of methylamine. *Plant Cell Physiol.* 24:979–86

81. Kanazawa T, Kanazawa K, Kirk MR, Bassham JA. 1972. Regulatory effects of ammonia on carbon metabolism in *Chlorella pyrenoidosa* during photosynthesis and respiration. *Biochim. Biophys. Acta* 256:656–69

82. Kelly GJ, Latzko E. 1977. Chloroplast phosphofructokinase II. Partial purification, kinetic and regulatory properties. *Plant Physiol.* 60:295–99

83. Klein U. 1986. Compartmentation of glycolysis and of the oxidative pentose-phosphate pathway in *Chlamydomonas reinhardtii*. *Planta* 167:81–86

84. Kleinhofs A, Warner RL. 1990. Advances in nitrogen assimilation. In *Intermediary Nitrogen Metabolism*, ed. BJ Miflin, PJ Lea, pp. 89–120. San Diego: Academic

85. Klepper L, Flesher D, Hageman RH. 1971. Generation of reduced nicotinamide adenine dinucleotide for nitrate reduction in green leaves. *Plant Physiol.* 48:580–90

86. Knowles VL, Greyson MF, Dennis DT.

1990. Characterization of ATP-dependent fructose 6-phosphate 1-phosphotransferase isozymes from leaf and endosperm tissues of *Ricinis communis*. *Plant Physiol.* 92: 155–59

87. Kombrink E, Wöber G. 1982. Chloroplast phosphofructokinase in the green alga, *Dunaliella marina*: partial purification and kinetic and regulatory proerties. *Arch. Biochem. Biophys.* 213:602–19

88. Kombrink E, Kruger NJ, Beevers H. 1984. Kinetic properties of pyrophosphate: fructose 6-phosphate phosphotransferase from germinating castor bean endosperm. *Plant Physiol.* 74:395–401

89. Krömer S, Heldt HW. 1991. On the role of mitochondria oxidative phosphorylation in photosynthesis metabolism as studied by the effect of oligomycin on photosynthesis in protoplasts and leaves of barley (*Hordeum vulgare*). *Plant Physiol.* 95:1270–76

90. Krömer S, Stitt M, Heldt HW. 1988. Mitochondrial oxidative phosphorylation participating in photosynthetic metabolism of a leaf cell. *FEBS Lett.* 226:352–56

91. Larsen PO, Cornwell KL, Gee SL, Bassham JA. 1981. Amino acid synthesis in photosynthesizing spinach cells. Effects of ammonia on pool sizes and rates of labeling from $^{14}CO_2$. *Plant Physiol.* 68:292–99

92. Larsson C-M, Larsson M. 1987. Regulation of nitrate utilization in green algae. In *Inorganic Nitrogen Metabolism*, ed. W Ullrich, PJ Aparicio, PJ Syrett, F Castillo, pp. 203–7. New York: Springer-Verlag

93. Larsson M, Ingemarsson B, Larsson C-M. 1982. Photosynthetic energy supply for NO_3^- assimilation in *Scenedesmus*. *Physiol. Plant.* 55:301–8

94. Larsson M, Olsson T, Larsson C-M. 1985. Distribution of reducing power between photosynthetic carbon and nitrogen assimilation in *Scenedesmus*. *Planta* 164:246–53

95. Le VQ, Foyer C, Champigny M-L. 1991. Effect of light and NO_3^- on wheat leaf phosphoenolpyruvate carboxylase activity. *Plant Physiol.* 97:1476–82

96. Lea PJ, Robinson SA, Stewart G. 1990. The enzymology and metabolism of glutamine, glutamate, and asparagine. In *Advances in Nitrogen Assimilation,* ed. BJ Miflin, PJ Lea, pp. 121–59. San Diego: Academic

97. Lee RB. 1980. Sources of reductant for nitrate assimilation in non-photosynthetic tissue: a review. *Plant Cell Environ.* 3:65–90

98. Lin M, Turpin DH, Plaxton WC. 1989. Pyruvate kinase isozymes from the green alga, *Selenastrum minutum*. II. Kinetic and regulatory properties. *Arch. Biochem. Biophys.* 269:228–38

99. Mann AF, Hucklesby DP, Hewitt EJ. 1978. Sources of reducing power for nitrate re-

duction in spinach leaves. *Planta* 140:261–63

100. Marczewski W. 1989. Kinetic properties of phosphoenolpyruvate carboxylase from lupin nodules and roots. *Physiol. Plant.* 76:539–43

101. Miflin BJ, Lea PJ. 1980. Ammonia assimilation. In *Amino Acids and Derivatives,* ed. BJ Miflin, pp. 169–202. New York: Academic

102. Miyachi S, Miyachi S. 1985. Ammonia induces starch degradation in *Chlorella* cells. *Plant Cell Physiol.* 26:245–52

103. Mohanty N, Bruce D, Turpin DH. 1991. Dark ammonium assimilation reduces the plastoquinone pool of photosystem II in the green alga *Selenastrum minutum. Plant Physiol.* 96:513–17

104. Morigasaki S, Takata K, Suzuki T, Wada K. 1990. Purification and characterization of a ferredoxin-NADP$^+$ oxidoreductase-like enzyme from radish root tissues. *Plant Physiol.* 93:896–901

105. Morigasaki S, Takata K, Sanada Y, Wada K, Yee BC, et al. 1990. Novel forms of ferredoxin and ferredoxin-NADP reductase from spinach roots. *Arch. Biochem. Biophys.* 283:75–80

106. Naik MS, Nicholas DJD. 1981. Relation between CO_2 evolution and *in situ* reduction of nitrate in wheat leaves. *Aust. J. Plant Physiol.* 8:515–24

107. Oak A. 1992. The function of roots in the synthesis of amino acids and amides. In *Biosynthesis and Molecular Regulation of Amino Acids in Plants,* ed. BK Singh, HE Flores, JC Shannon, pp. 111–20. Rockville, MD: Am. Soc. Plant Physiol.

108. Oji Y, Watanabe M, Wakiuchi N, Okamoto S. 1985. Nitrite reduction in barley-root plastids: dependence on NADPH coupled with glucose-6-phosphate and 6-phosphogluconate dehydrogenases, and possible involvement of an electron carrier and a diaphorase. *Planta* 165:85–90

109. Oliver DJ, Walker GH. 1984. Characterization of the transport of oxaloacetate by pea leaf mitochondria. *Plant Physiol.* 76:409–13

110. Pace GM, Volk RJ, Jackson WA. 1990. Nitrate reduction in response to CO_2-limited photosynthesis. Relationship to carbohydrate supply and nitrate reductase activity in maize seedlings. *Plant Physiol.* 92:286–92

111. Palmer JM, Ward JA. 1985. The oxidation of NADH by plant mitochondria. In *Higher Plant Respiration,* ed. R Douce, D Day, pp. 173–201. Berlin: Springer-Verlag

112. Pate JS. 1983. Patterns of nitrogen metabolism in higher plants and their ecological significance. In *Nitrogen as an Ecological Factor,* ed. JA Lee, S McNeill, IH Rorison, pp. 225–55. Oxford: Blackwell Scientific

113. Paul JS, Cornwell KL, Bassham JA. 1978. Effects of ammonia on carbon metabolism in photosynthesizing isolated mesophyll cells from *Papaver somniferum* L. *Planta* 142:49–54

114. Peak JG, Peak MJ. 1981. Heterotrophic carbon fixation by *Euglena.* Function of phosphoenolpyruvate carboxylase. *Biochim. Biophys. Acta* 677:390–96

115. Peltier G, Thibault P. 1985. O_2 uptake in the light in *Chlamydomonas.* Evidence for persistent mitochondrial respiration. *Plant Physiol.* 79:225–30

116. Peterson JB, Evans HJ. 1979. Phosphoenolpyruvate carboxylase from soybean nodule cytosol. Evidence for isoenzymes and kinetics of the most active component. *Biochim. Biophys. Acta* 567:445–52

117. Podestá FE, Plaxton WC. 1991. Kinetic and regulatory properties of cytosolic pyruvate kinase from germinating castor oil seeds. *Biochem. J.* 279:495–501

118. Podestá FE, Plaxton WC. 1992. Plant cytosolic pyruvate kinase: a kinetic study. *Biochim. Biophys. Acta* 1160:213–20

119. Deleted in proof

120. Preiss J, Robinson N, Spilatro S, McNamara K. 1985. Starch synthesis and its regulation. In *Regulation of Carbon Partitioning in Photosynthetic Tissue,* ed. RL Heath, J Preiss, pp. 1–26. Rockville, MD: Am. Soc. Plant Physiol.

121. Proudlove MO, Thurman DA, Salisbury J. 1984. Kinetic studies on the transport of 2-oxoglutarate and L-malate into isolated pea chloroplasts. *New Phytol.* 96:1–5

122. Randall DD, Givan CV. 1981. Subcellular location of NADP$^+$-isocitrate dehydrogenase in *Pisum sativum* leaves. *Plant Physiol.* 68:70–73

122a. Randall DD, Rubin PM, Fenko M. 1977. Plant pyruvate dehydrogenase complex, purification, characterization and regulation by metabolites and phosphorylation. *Biochim. Biophys. Acta* 485:336–49

123. Rathnam CKM. 1978. Malate and dihydroxyacetone phosphate-dependent nitrate reduction in spinach leaf protoplasts. *Plant Physiol.* 62:220–23

124. Raven JA, Farquhar GD. 1990. The influence of N metabolism and organic acid synthesis on the natural abundance of isotopes of carbon in plants. *New Phytol.* 110:505–29

125. Redinbaugh MG, Campbell WH. 1981. Purification and characterization of NAD(P)H:nitrate reductase and NADH:nitrate reductase from corn roots. *Plant Physiol.* 68:115–20

126. Reed AJ, Canvin DT, Sherrard JH, Hageman RH. 1983. Assimilation of [^{15}N]nitrate and [^{15}N]nitrite in leaves of five plant species under light and dark conditions. *Plant Physiol.* 71:291–94

127. Reeves RE, Serrano R, South DJ. 1976. 6-Phosphofructokinase (pyrophosphate). Properties of the enzyme from *Entamoeba histolytica* and its reaction mechanism. *J. Biol. Chem.* 251:2958–62

128. Riens B, Heldt HW. 1992. Decrease of nitrate reductase activity in spinach leaves during a light-dark transition. *Plant Physiol.* 98:573–77

129. Robinson JM, Baysdorfer C. 1985. Interrelationships between photosynthetic carbon and nitrogen metabolism in mature soybean leaves and isolated leaf mesophyll cells. In *Regulation of Carbon Partitioning in Photosynthetic Tissue*, ed. RL Heath, J Preiss, pp. 333–57. Rockville, MD: Am. Soc. Plant Physiol.

130. Robinson SA, Slade AP, Fox GG, Phillips R, Ratcliffe RG, Stewart GR. 1991. The role of glutamate dehydrogenase in plant metabolism. *Plant Physiol.* 95:509–16

131. Robinson SA, Stewart GR, Phillips R. 1992. Regulation of glutamate dehydrogenase activity in relation to carbon limitation and protein catabolism in carrot cell suspension cultures. *Plant Physiol.* 98:1190–95

132. Rufty TW, Huber SC, Volk RJ. 1988. Alterations in leaf carbohydrate metabolism in response to nitrogen stress. *Plant Physiol.* 88:725–30

133. Rufty TW, MacKown CT, Volk RJ. 1989. Effects of altered carbohydrate availability on whole-plant assimilation of $^{15}NO_3^-$. *Plant Physiol.* 89:457–63

134. Sarkissian GS, Fowler MW. 1974. Interrelationship between nitrate assimilation and carbohydrate metabolism in plant roots. *Planta* 119:335–49

135. Sawhney SK, Naik MS, Nicholas DJD. 1978. Regulation of NADH supply for nitrate reduction in green plants via photosynthesis and mitochondrial respiration. *Biochem. Biophys. Res. Commun.* 81:1209–16

136. Scheibe R. 1991. Redox-modulation of chloroplast enzymes. A common principle for individual control. *Plant Physiol.* 96:1–3

137. Scheibe R, Geissler A, Fickenscher K. 1989. Chloroplast glucose-6-phosphate dehydrogenase: K_m shift upon light modulation and reduction. *Arch. Biochem. Biophys.* 274:290–97

138. Scheibe R, Jacquot J-P. 1983. NADP regulates the light activation of NADP-dependent malate dehydrogenase. *Planta* 157:548–53

139. Schlee J, Cho B-H, Komor E. 1985. Regulation of nitrate uptake by glucose in *Chlorella. Plant Sci.* 39:25–30

140. Schuller KA, Plaxton WC, Turpin DH. 1990. Regulation of phosphoenolpyruvate carboxylase from the green alga *Selenastrum minutum.* Properties associated with replenishment of tricarboxylic acid cycle intermediates during ammonium assimilation. *Plant Physiol.* 93:1303–11

141. Schuller KA, Randall DD. 1990. Mechanism of pyruvate inhibition of plant pyruvate dehydrogenase kinase and synergism with ADP. *Arch. Biochem. Biophys.* 278:211–16

142. Schuller KA, Turpin DH, Plaxton WC. 1990. Metabolite regulation of partially purified soybean nodule phosphoenolpyruvate carboxylase. *Plant Physiol.* 94:1429–35

143. Schürmann P, Wolosuik RA. 1978. Studies of the regulatory properties of chloroplast fructose 1,6-bisphosphate. *Biochim. Biophys. Acta* 522:130–38

144. Sechley KA, Yamaya T, Oaks A. 1992. Compartmentation of nitrogen assimilation in higher plants. *Int. Rev. Cytol.* 134:85–163

145. Smirnoff N, Stewart GR. 1985. Nitrate assimilation and translocation by higher plants: comparative physiology and ecological consequences. *Physiol. Plant.* 64:133–40

146. Smith RG, Vanlerberghe GC, Stitt M, Turpin DH. 1989. Short-term metabolite changes during transient ammonium assimilation by the N-limited green alga *Selenastrum minutum. Plant Physiol.* 91:749–55

147. Stitt M. 1990. Fructose-2,6-bisphosphate as a regulatory molecule in plants. *Annu. Rev. Plant Physiol. Plant Mol. Biol.* 41:153–85

148. Stitt M. 1984. Degradation of starch in the chloroplast: a buffer to sucrose metabolism. In *Storage Carbohydrates in Vascular Plants*, ed. DH Lewis, pp. 205–29. Cambridge: Cambridge Univ. Press

149. Stitt M, Cséke C, Buchanan BB. 1984. Regulation of fructose 2,6-phosphate concentration in spinach leaves. *Eur. J. Biochem.* 143:89–93

150. Stitt M, Heldt HW. 1981. Simultaneous synthesis and degradation of starch in spinach chloroplasts in the light. *Biochim. Biophys. Acta* 638:1–11

151. Stitt M, Quick WP. 1989. Photosynthetic carbon partitioning: its regulation and possibilities for manipulation. *Physiol. Plant.* 77:633–41

152. Stitt M, Steup M. 1985. Starch and sucrose degradation. In *Higher Plant Respiration*, ed. R Douce, D Day, pp. 347–90. Berlin: Springer-Verlag

153. Stitt M, Wilke I, Feil R, Heldt HW. 1988. Course control of sucrose-phosphate synthase in leaves: alterations of kinetic properties in response to the rate of photosynthesis and the accumulation of sucrose. *Planta* 174:217–30

154. Subularse DC, Anderson RL. 1981. D-Fructose 2,6-bisphosphatase: a naturally occurring activator for inorganic pyrophosphate: D-fructose 6-phosphate 1-phosphotransferase in plants. *Biochem. Biophys. Res. Commun.* 103:848–55

155. Suzuki A, Carrayol E, Zehnacker C, Deroche M-E. 1988. Glutamate synthase in *Medicago sativa* L.. Occurrence and proerties of the Fd-dependent enzyme in plant cell fraction during root nodule development. *Biochem. Biophys. Res. Commun.* 156:1130–38

156. Suzuki A, Gadal P. 1984. Glutamate synthase: physiochemical and functional properties of different forms in higher plants and other organisms. *Physiol. Vég.* 22:461–71

157. Suzuki A, Oaks A, Jacquot J-P, Vidal J, Gadal P. 1985. An electron transport system in maize roots for reduction of glutamate synthase and nitrite reductase. *Plant Physiol.* 78:374–78

158. Syrett PJ. 1953. The assimilation of ammonia by nitrogen-starved cells of *Chlorella vulgaris*. Part 1. The correlation of assimilation with respiration. *Ann. Bot.* 65:1–19

159. Syrett PJ. 1956. The assimilation of ammonia and nitrate by nitrogen-starved cells of *Chlorella vulgaris*. IV. The dark fixation of carbon dioxide. *Plant Physiol.* 9:165–71

160. Syrett PJ. 1956. The assimilation of ammonia and nitrate by nitrogen-starved cells of *Chlorella vulgaris*. II. The assimilation of large quantities of nitrogen. *Physiol. Plant.* 9:19–27

161. Syrett PJ. 1981. Nitrogen metabolism of microalgae. *Can. Bull. Fish. Aquat. Sci.* 210:182–210

162. Syrett PJ. 1988. Uptake and utilization of nitrogen compounds. In *Biochemistry of the Algae and Cyanobacteria*, ed. LJ Rogers, JR Gallon, pp. 23–39. Oxford: Clarendon

163. Thacker A, Syrett PJ. 1972. The assimilation of nitrate and ammonium by *Chlamydomonas reinhardtii*. *New Phytol.* 71:423–33

164. Turner JF, Turner DH. 1980. The regulation of glycolysis and the pentose phosphate pathway. In *Metabolism and Respiration*, ed. DD Davies, pp. 279–316. New York: Academic

165. Turpin DH. 1991. Effects of inorganic N availability on algal photosynthesis and carbon metabolism. *J. Phycol.* 27:14–20

166. Turpin DH, Botha FC, Smith RG, Feil R, Horsey AK, Vanlerberghe GC. 1990. Regulation of carbon partitioning to respiration during dark ammonium assimilation by the green alga *Selenastrum minutum*. *Plant Physiol.* 93:166–75

167. Turpin DH, Bruce D. 1990. Regulation of photosynthetic light harvesting by nitrogen assimilation in the green alga *Selenastrum minutum*. *FEBS Lett.* 263:99–103

168. Turpin DH, Miller AG, Canvin DT. 1985. Chemostats in the study of inorganic metabolism in microalgae. In *Inorganic Carbon Uptake by Aquatic Photosynthetic Organisms*, ed. WJ Lucas, JA Berry, pp. 437–47. Rockville, MD: Am. Soc. Plant Physiol.

169. Ullrich WR, Lesch S, Jarczyk L, Harterich M, Trogisch GD. 1990. Transport of inorganic nitrogen compounds: physiological studies on uptake and assimilation. In *Inorganic Nitrogen Metabolism in Plants and Microorganisms, Uptake and Metabolism*, ed. WR Ullrich, C Rigano, A Fuggi, PJ Aparico, pp. 44–50. New York: Springer-Verlag

170. Vanlerberghe GC, Huppe HC, Vlossak KDM, Turpin DH. 1991. Activation of respiration to support dark NO_3^- and NH_4^+ assimilation in the green algae *Selenastrum minutum*. *Plant Physiol.* 99:495–500

171. Vanlerberghe GC, Schuller KA, Smith RG, Feil R, Plaxton WC, Turpin DH. 1990. Relationship between NH_4^+ assimilation rate and *in vivo* phosphoenolpyruvate carboxylase activity. *Plant Physiol.* 94:284–90

172. Vaughn KC, Campbell WH. 1988. Immunogold localization of nitrite reductase in maize leaves. *Plant Physiol.* 88:1354–57

173. Vézina L-P, Hope HJ, Joy KW. 1987. Isoenzymes of glutamine synthetase in roots of pea (*Pisum sativum* L. cv Little Marvel) and alfalfa (*Medicago media* Pers. cv Saranac). *Plant Physiol.* 83:58–62

174. Vlossak KDM. 1993. *Effect of different rates of assimilation on carbon metabolism in the green alga,* Selenastrum minutum. MSc thesis. Queen's Univ., Kingston, Ontario, Can. 122 pp.

175. Wada K, Onda M, Matsubara H. 1989. Amino acid sequences of ferredoxin isoproteins from radish roots. *J. Biochem.* 105:19–25

176. Weger HG, Chadderton AR, Lin M, Guy RD, Turpin DH. 1990. Cytochrome and alternative pathway respiration during transient ammonium assimilation by N-limited *Chlamydomonas reinhardtii*. *Plant Physiol.* 94:1131–36

177. Weger HG, Birch DG, Elrifi IR, Turpin DH. 1988. Ammonium assimilation requires mitochondrial respiration in the light. *Plant Physiol.* 86:688–92

178. Weger HG, Guy RD, Turpin DH. 1990. Cytochrome and alternative pathway respiration in green algae. Measurements using inhibitors and $^{18}O_2$ discrimination. *Plant Physiol.* 93:356–60

179. Weger HG, Turpin DH. 1989. Mitochondrial respiration can support NO_3^- and NO_2^- reduction during photosynthesis. *Plant Physiol.* 89:409–15

180. Weiner H, McMichael RW, Huber SC. 1992. Identification of factors regulating the phosphorylation status of sucrose-phosphate synthetase. *Plant Physiol.* 99:1435–42

181. Williams WP, Allen JF. 1987. State 1/State 2 changes in higher plants and algae. *Photosynth. Res.* 13:19–45

182. Wong JH, Yee BC, Buchanan BB. 1987. A novel type of phosphofructokinase from plant. *J. Biol. Chem.* 262:3185–91

183. Woo KC, Boyle FA, Flügge UI, Heldt HW. 1987. ^{15}N-Ammonia assimilation, 2-oxoglutarate transport, and glutamate export in spinach chloroplasts in the presence of dicarboxylates in the light. *Plant Physiol.* 85:621–25

184. Woo KC, Canvin DT. 1980. The role of malate in nitrate reduction in spinach leaves. *Can. J. Bot.* 58:517–21

185. Woo KC, Canvin DT. 1980. Effect of ammonia on photosynthetic carbon fixation in isolated spinach leaf cells. *Can. J. Bot.* 58:505–10

186. Wu HB, Turpin DH. 1992. Purification and characterization of pyruvate kinase from the green alga *Chlamydomonas reinhardtii. J. Phycol.* 28:472–81

187. Yoneyama T. 1981. ^{15}N Studies on the in vitro assay of nitrate reductase in leaves: occurrence of underestimation of the activity due to dark assimilation of nitrate and nitrite. *Plant Cell Physiol.* 22:1507–20

188. Zimmerman G, Kelly GJ, Latzko E. 1978. Purification and properties of spinach leaf cytoplasmic fructose-1,6-bisphosphatase. *J. Biol. Chem.* 253:5952–56

Annu. Rev. Plant Physiol. Plant Mol. Biol. 1994. 45:609–31

ENDOCYTOSIS IN PLANTS

Philip S. Low and Sreeganga Chandra

Department of Chemistry, Purdue University, West Lafayette, Indiana 47907

KEY WORDS: receptor-mediated endocytosis, clathrin coated pits, fluid phase endocytosis, membrane trafficking, vesicle transport

CONTENTS

INTRODUCTION 609
SUMMARY OF ENDOCYTOSIS IN ANIMAL CELLS 610
   Multiple Pathways Contribute to High Endocytic Capacity 610
   Ligands and Receptors May Experience Different Fates Following
    Endocytosis 611
   The Clathrin Coated Pit Pathway 611
HISTORIC PERSPECTIVE ON ENDOCYTOSIS IN PLANTS 613
   Theoretical Arguments For and Against Endocytosis in Plants 613
   Experimental Obstacles to Endocytosis Studies 614
   Summary of Observations Supporting Endocytosis in Plants 615
EVIDENCE FOR MAJOR ENDOCYTIC STRUCTURES IN PLANTS 616
   Plasma Membrane Receptors 616
   Coated Pits 617
   Coated Vesicles 619
   Partially Coated Reticulum 621
   Multivesicular Bodies 621
   Vacuoles 622
THE ENDOCYTIC PATHWAY IN PLANT CELLS AND PROTOPLASTS 622
FLUID PHASE ENDOCYTOSIS IN PLANT CELLS AND PROTOPLASTS 623
RECEPTOR-MEDIATED ENDOCYTOSIS IN PLANT CELLS 624
FUTURE DIRECTIONS AND CONCLUSIONS 625

INTRODUCTION

Endocytosis is the process whereby the plasma membrane invaginates and pinches off to form a closed vesicle within the cytoplasm. Its functions may include the remodeling or repair of the cell membrane (91, 107), the removal

0066-4294/94/0601-0609$05.00

of signaling ligands (e.g. hormones, elicitors) from the cell surface following signal transduction (13, 115), the uptake of needed nutrients (e.g. vitamins) into the cell (3), the transport of molecules across an epithelial cell (transcytosis) (74), the delivery of active signaling components from the plasma membrane to internal organelles (12), or the return of plasma membrane to the cell interior following secretory activity (109). Much is known about endocytosis in animal cells (for excellent reviews, see 43, 97, 106), but until recently endocytosis was considered impossible in plants because of constraints imposed by high turgor pressure (22, 47). This review summarizes briefly the major features of endocytosis in animals and then examines the evidence for and against their occurrence in plants. Most classical endocytic structures and components are well represented in plant cells and endocytosis does occur, but the molecular details of the endocytic process in plants remain largely unexplored.

SUMMARY OF ENDOCYTOSIS IN ANIMAL CELLS

Multiple Pathways Contribute to High Endocytic Capacity

The clathrin coated pit constitutes a major membrane site involved in receptor-mediated endocytosis, the predominant form of endocytosis in many animal cells. These coated pits are lined with an electron dense polypeptide called clathrin and may cover 1–2% of the total surface area of the cell (42). Because their residence time on the plasma membrane is only 1–2 min (49, 86), the majority of plasma membrane material must internalize every hour. In baby hamster kidney cells, ~1500–3000 coated vesicles endocytose per min, resulting in total plasma membrane turnover every ~30 min (72). Invagination of coated pits to form coated vesicles is a constitutive process, proceeding unabated even in the absence of receptor-targeted ligands.

Receptor-mediated endocytosis also may occur at caveolae, which are slightly smaller membrane depressions coated by caveolin, a less easily stained protein (98). In some cells, caveolae are thought to invaginate and constrict but never completely sever from the plasma membrane (3). In other cells, complete separation from the membrane occurs (117). Glycosylphosphatidylinositol-anchored receptors may endocytose predominantly at caveolae.

Endocytosis at uncoated membrane indentations is also well documented. Thus, cells devoid of clathrin heavy chain (83, 84) and normal cells treated to disassemble clathrin coated pits (101, 122 and references within) still conduct endocytosis at a rapid rate. Furthermore, cells stimulated with growth factors frequently initiate uptake of extracellular fluid at nonclathrin coated pits (50, 87, 110). Because these latter invaginations are larger than clathrin coated pits, the internalization process is sometimes called macropinocytosis (11, 16). Fluid phase uptake can occur in coated pits, noncoated pits, and caveolae, since extracellular medium should readily fill any invaginations formed during endocytosis. Fluid phase endocytosis in the absence of ligand uptake can return plasma membrane to the cell interior during periods of intense secretory activity.

Ligands and Receptors May Experience Different Fates Following Endocytosis

The ultimate destination of the internalized ligands varies significantly with the putative function of the endocytic process. Hormones that are cleared from the cell surface following signaling (e.g. insulin, epidermal growth factor) are commonly routed to lysosomes for destruction (13, 115). Proteins transporting essential nutrients (e.g. transferrin) may recycle to the plasma membrane following release of their cargo within the cell (40, 76). Ligands required on the opposite sides of epithelial cell barriers (e.g. IgA, IgM) are transported there by a process called transcytosis (74), and storage proteins (e.g. vitellogenin) are commonly targeted to vacuoles and storage granules (121). Finally, small molecular weight nutrients required in the cytoplasm (e.g. folic acid) are thought to be deposited there following initial invagination at the cell surface (3).

Ligand receptors generally have two possible fates: either they are returned to the cell surface or they are degraded in lysosomes. A small fraction of signaling receptors may migrate to the nucleus or other organelles to continue signal transduction (12). The majority of receptors recycle several times before they are routed to lysosomes (e.g. transferrin receptor, low density lipoprotein receptor, insulin receptor, α_2-macroglobulin receptor, asialoglycoprotein receptor). A few receptors (e.g. epidermal growth factor receptor) are used only once before they are degraded (43).

The Clathrin Coated Pit Pathway

Because coated pits and vesicles are common structures in plant cells, a brief synopsis of the clathrin coated pit pathway in animal cells will be useful for later comparison. Only the more widely accepted features from the most prominent pathways will be discussed, with the caveat that refinement of this model is inevitable. In general, receptors localizing in coated pits are membrane-spanning polypeptides or proteins associated with such polypeptides. Some receptors contain cysteine-rich extracellular domains that stabilize the ligand binding site via disulfide bond formation (25). Although the N-terminus can be either exoplasmic or cytoplasmic, usually there is a coated pit localization signal in the cytoplasmic domain within ~10 residues of the membrane surface (106). Although this signal cannot be characterized by a consensus sequence, it almost always involves a tyrosine residue in a tight turn flanked by charged (largely basic) amino acids (116). NPXY, where X is any amino acid, is the minimum sequence found in several coated pit receptors (116 and references within).

Rates of receptor internalization vary significantly among receptor types in the same cell (106), indicating variations in receptor entry or capture by coated pits. Some receptors are excluded specifically from coated pits (10, 70, 86), either by association with a cortical skeleton (e.g. CD4) or by interaction with a unique membrane phase (e.g. Thy-1). Other receptors (e.g. epidermal growth factor receptor) localize in coated pits only upon ligand binding (32, 103). Most receptors, however, diffuse constitutively to coated pits and internalize regardless of their state of occupancy, although some data do indicate that ligation can enhance entrapment in coated pits (43).

Although distinct receptor types can colocalize in the same coated pit (40), they all must anchor there by interaction with an adaptor complex at their localization sequences (85, 106). This complex, called HAII in the case of plasma membrane coated pits, consists of α and β adaptins (M_r ~105,000) plus polypeptides of M_r ~50,000 and ~16,000. The adaptor complex, in turn, associates with clathrin to complete the connection between the receptor and the clathrin cage. Soon after coated pit formation, the pit invaginates and pinches off to generate a clathrin coated vesicle.

Although the clathrin cage forms spontaneously at the membrane surface, disassembly of the coat immediately following endocytosis and release of the uncoated vesicle from the membrane requires energy and an uncoating ATPase (99). The adaptor complex also dissociates during uncoating. The uncoated vesicle is not long-lived, but is thought to fuse rapidly with an early endosome in a GTP-dependent step (46). Several protein complexes including low molecular weight G proteins are thought to participate in this fusion (4, 45, 46).

Most researchers describe early endosomes as consisting of small vesicles and tubular membranes, usually located near the plasma membrane (106). Because early endosomes represent the site of receptor and ligand dissociation, they are also called CURL (Compartment of Uncoupling of Receptor and Ligand) (97). The uncoupling or dissociation event results from proton pump–mediated acidification of the compartment, causing a reduction in affinity of the receptor for its ligand. The soluble ligand and entrapped medium are then shuttled nonspecifically via vesicles to late endosomes and lysosomes. The membrane bound receptor, on the other hand, can either follow the same itinerary or be returned to the plasma membrane. These trafficking events again seem to require G proteins and other protein complexes (4, 46).

Late endosomes consist of multivesicular bodies and related structures. Their lumen is more acidic than that of early endosomes and their enzymatic activities include some lysosomal enzymes (106). Although trafficking from this compartment to the plasma membrane does occur, most flow is toward the lysosome.

HISTORIC PERSPECTIVE ON ENDOCYTOSIS IN PLANTS

Theoretical Arguments For and Against Endocytosis in Plants

The major argument against endocytosis in plants is based on their high turgor pressure (e.g. hydrostatic pressures up to 1 MPa may have to be overcome to invaginate the plasma membrane). Based on calculations of energy available to the cell and the pressure-volume work needed to create a vesicle of volume (V) against a pressure difference δp across the plasma membrane, Cram (22) has argued that endocytosis is not energetically feasible in turgid plant cells. The energy required for endocytosis far exceeds the calculated energy available from respiration and photosynthesis. In addition, Cram (22) noted that uptake of extracellular solutes accompanying endocytosis would further increase the turgor pressure, because active mechanisms for dissipating the pressure are lacking.

Others have made similar calculations of the cost of endocytosis to the plant cell and have arrived at different conclusions. Discrepancies with Cram's results (22) have arisen either because of differences in values assigned to variables in the calculations or because of additional parameters Cram did not consider. For example, Raven (89) noted that the energy needed for vesicle internalization is still a fraction of what the cell expends for cell wall biosynthesis. Saxton & Breidenbach (102) argued that the energy needed to internalize the plasma membrane via coated vesicles of diameter 100 nm (the size normally observed) is well within the metabolic capacity of a cell. Saxton & Breidenbach further emphasized the importance of vesicle size in endocytosis (i.e. the power needed to invaginate the plasma membrane is proportional to the radius of the nascent endocytic vesicle) and that fusion of exocytic vesicles with the plasma membrane might compensate for internalization of endocytic vesicles, i.e. two factors not considered by Cram. Finally, they showed that increases in turgor are minimal and consequent energy requirements are low during endocytosis. In contrast, Gradmann & Robinson (47) focused on the thermodynamics of coated vesicle formation and argued that energy requirements for endocytosis of 100 nm vesicles far exceed the available free energy. Therefore, they concluded that endocytosis could not occur in a fully turgid cell but would be limited to low turgor conditions. Subsequently, these authors extended their conclusions, noting that a plant cell may normally experience fluctuations in turgor as high as 10 bar in 1 h (94a). Taken together, their calculations now allow for the possibility of endocytosis in plants.

Experimental Obstacles to Endocytosis Studies

Besides uncertainties about the physical feasability of endocytosis in plants, research in the field also has been retarded by several methodological difficulties. First, there has been an unusual absence of appropriate membrane impermeable ligands and fluid phase markers for microscopic analysis of endocytosis. Although plant hormones have long been well characterized and commercially available, there is still no evidence that they can initiate their signaling cascades from plasma membrane receptors. Furthermore, because of their small sizes and moderate membrane permeabilities, their potential uptake via endocytosis cannot be easily distinguished from influx via facilitated transport or passive diffusion. Until recently, macromolecular ligands for cell surface receptors (e.g. elicitors, nutrients) also were not readily available for assay of receptor-mediated endocytosis.

Analysis of fluid phase uptake by endocytosis has been similarly handicapped by a dearth of impermeant fluid phase markers. Although salts of heavy metals can be seen readily in endocytic compartments of plant cells (63, 100), there has been concern that heavy metal toxicity might alter the pathways under scrutiny. Lucifer yellow, a common fluorescent probe of fluid phase endocytosis in animal cells, suffers from the unexpected ability to enter plant cells and cross vacuolar membranes by classical anion transport (17, 77, 79). Furthermore, macromolecular fluid phase markers, such as colloidal gold-labeled proteins and horseradish peroxidase, simply penetrate the pores of the cell wall too slowly to serve as indicators of fluid phase endocytosis. Thus, although cell wall porosity may permit slow penetration of macromolecules larger than 60,000 Daltons (8, 14, 41, 62, 114), endocytosis of such molecules in sufficient quantities to be visualized by microscopic methods should only occur when specific receptors concentrate the macromolecules on the membrane surface. Indeed, a reliable and convenient probe of fluid phase endocytosis in intact plant cells is still lacking.

Because the existence of endocytosis in turgid cells was doubted and suitable markers for receptor-mediated endocytosis were lacking, most of the early progress in characterizing endocytosis in plants was achieved on protoplasts, where neither turgor pressure nor cell wall porosity could hinder uptake. Therefore, with the exception of a few studies of heavy metal endocytosis by intact cells, the structural analyses described below deal almost exclusively with protoplast preparations.

Summary of Observations Supporting Endocytosis in Plants

Three fundamental lines of evidence support the occurrence of endocytosis in higher plants. First, the insertion of exocytic vesicles into the plasma membrane during periods of intense secretory activity requires that a pathway exists for return of membrane to intracellular compartments. From morphometric measurements of exocytic vesicle formation in secretion-inhibited cells, Steer (108) has calculated that higher plants must endocytose the equivalent of their entire plasma membranes within a period of 10 min to 3 h, depending on the growth rate of the cell. Similarly, by measuring the increase in cell wall volume deposited during growth of root cells of *Lobelia erinus,* Samuels & Bisalputra (100) estimated that to deliver the observed quantities of extracellular wall material, secretory vesicles totaling eightfold more membrane area than the entire plasma membrane must fuse with the cell membrane. Because Samuels & Bisalputra observed a direct correlation between high secretory activity and elevated rates of endocytosis, they hypothesized that endocytosis balances exocytosis during cell wall secretion. From a different perspective, Emons & Traas (33) determined coated pit densities in various plant cells, and from measurements of pit diameters and rates of internalization (~30–60 s), they estimated that the entire plasma membrane must internalize every 20–115 min, depending on cellular growth rates. Finally, Phillips et al (88) calculated that even in elongating epidermal cells of oat coleoptiles, where part of secretory vesicle infusion can be used for cell membrane expansion, two to three times more membrane is still inserted into the plasma membrane than is needed for cell elongation. These observations demonstrate a requirement for endocytosis to offset exocytosis in establishing a stable or moderately increasing plasma membrane surface area.

The second major argument supporting endocytosis in higher plants comes from observations that the major components and structures that conduct endocytosis in animals are abundant in plants. Plant cells contain plasma membrane receptors, coated pits, coated vesicles, early and late endosomes, the Golgi apparatus, and a degradative compartment analogous to lysosomes (i.e. vacuoles). Some of these structures may perform other essential functions, but if plant cells cannot engage in endocytosis, why have they assembled plasma membrane coated pits and endosomes for which no other functions are known?

Finally, evidence is accumulating that macromolecular ligands (e.g. lectins, elicitors, and vitamin-conjugated proteins) and impermeable solutes (e.g. heavy metals, ferritin) can rapidly enter whole plant cells and structurally intact protoplasts (*vide infra*). Impermeant molecules of this sort are known to enter animal cells only by endocytosis. Although data on uptake of impermeant molecules in whole plant cells are still scarce, the composite evidence from multiple studies in various systems now overwhelmingly supports endocytic pathways in plants.

EVIDENCE FOR MAJOR ENDOCYTIC STRUCTURES IN PLANTS

Plasma Membrane Receptors

In receptor-mediated endocytosis, the first stable interaction following penetration of the ligand through the cell wall is the association of the ligand with its plasma membrane receptor. Unfortunately, no receptor has been isolated or characterized from the plasma membrane of higher plants, except possibly the auxin-binding protein from soybean and maize (6, 53, 55). Because this polypeptide is found predominantly in the endoplasmic reticulum and it contains no membrane-spanning sequence, its role in signaling and receptor-mediated endocytosis remains uncertain. Still, the existence of other plasma membrane receptors can be inferred from the presence of specific ligand binding sites on plasma membranes and from the abilities of impermeable ligands to initiate signaling in intact cells. According to these criteria, plasma membrane receptors must exist for elicitors such as β-glucans (15, 20), polygalacturonic acid (60, 67), and polypeptides from *Verticillium dahliae* (60) and *Phytophthera megasperma* (82); for fungal toxins such as fusicoccin (26, 73); for vitamins like biotin (61, 71); and for symbionts such as *Rhizobium* spp. (104, 105).

Among the putative signaling receptors mentioned, one of the best characterized is the plasma membrane protein that binds β-glucan elicitor. According to measurements by Cheong & Hahn (15), the elicitor associates saturably with 1.2 pmol/mg membrane protein with a dissociation constant of 7.5×10^{-10} M. Similar studies by Ebel and coworkers (21) show a Kd $\sim 3.7 \times 10^{-8}$ M with the same total binding capacity. Attempts to isolate the native β-glucan receptor have yielded a plasma membrane–derived protein with a pI near neutrality and a $M_r \sim 300,000$ (19). Affinity labeling of the receptor in soybean root membrane extracts tags a protein of subunit $M_r \sim 70,000$ (20). Similarly detailed studies document the existence of a plasma membrane fusicoccin receptor (26, 27, 73).

Aside from biochemical data, ligand receptors on plant plasma membranes have been inferred from fluorescence and epipolarization microscopy studies. Using the former method, both elicitor (60) and biotin (61) binding to the surface of cultured soybean cells have been visualized via localization of fluorescent derivatives of the ligands at the membrane surface. Application of the latter, higher resolving technique has allowed direct observation of punctate fluorescent intensities evenly distributed over the entire plasma membrane (W Diekmann, B Herkt, PS Low, D Scheel, DG Robinson, submitted for publication). Because the putative binding sites in all the above studies could be blocked by unlabeled ligands, the sites were assumed to represent specific receptors for their respective elicitors/biotin.

Coated Pits

As with most endocytic structures, coated pits of plant and animal cell plasma membranes have roughly the same dimensions and morphology. Both are composed of a lipid bilayer lined by a mixed hexagonal/pentagonal protein lattice (37). Coated pits in plants (Figures 1 and 2) are characterized by bristles or spikes extending 10–20 nm from the cytoplasmic surface of the membrane (96). The coated regions may be flat or invaginated, and their diameters may range from 60–170 nm for planar regions to 74–124 nm for curved regions (33). The average *Equisetum hyemale* root hair coated pit contains ~ 15 clathrin polygons, with values of 5–23. The center-to-center distance of the polygons is 20–27 nm and the length of each side is ~ 16 nm (18, 33). Both pentagons and hexagons are observed, but the latter are more abundant (18, 37). Coated membranes are also found on the Golgi apparatus, the partially coated reticulum, and less abundantly on other internal membranes. In animal cells, coated pits associated with the Golgi apparatus are biochemically and morphologically distinct (96). Importantly, electron dense endocytic markers associate rapidly with coated pits after their addition to protoplasts or plant cells (37, 63, 100). The basic structural component of the coated pit is clathrin, a complex protein made of heavy and light chains. The heavy chain of plant clathrin (M_r ~190,000) is about 10,000 Daltons larger than the heavy chain of animal clathrin (18). The light chains of plant clathrin have not been well characterized, but most studies report a higher molecular weight for these subunits also (5, 18, 28). The heavy chain of clathrin folds into a three-legged structure called a triskelion, each leg averaging 61.5 nm in length with a globular terminal domain of 9.6 nm and a kink ~ 20 nm from the central hub. Although animal clathrin also contains this kink, the triskelion leg length is 11–17 nm shorter. Curiously, despite these clathrin dimensional differences, the assembled clathrin cages in plants and animals have identical dimensions (54, 123). Further, both plant and animal clathrin cages assemble spontaneously from their component proteins when dissociating reagents are removed (18, 123). Thus, the significance, if any, of the dimensional differences is unknown, but it would be interesting to learn whether the two isoforms of clathrin could form mixed cages if allowed to reassemble in the same flask.

Coated pits are more abundant in actively growing cells than in quiescent cells (33, 37, 100). Their membrane densities range from 0.1–0.6 pits/μm^2 in full grown (mature) root hairs to 2.5–4.5 pits/μm^2 in proliferating hairs. Cultured cells may maintain 5000–10,000 coated pits/cell, while protoplasts can express 35,000/cell (118). Even within individual cells, the density of coated

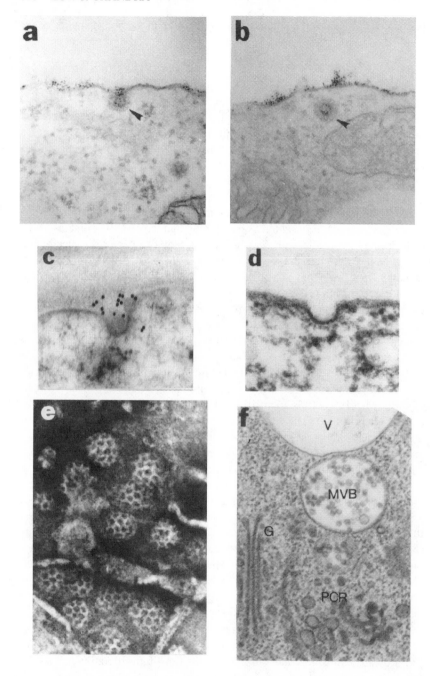

pits may vary, with higher values occurring at the growing or dividing regions of the cell, e.g. at the cell plate (33, 37, 75). These observations support a role for endocytosis in membrane recycling.

Coated Vesicles

Coated pits in plant cells have a lifetime of less than 60 s (112). In animal cells, coated pits endocytose to form coated vesicles. Consistent with this observation, coated vesicles in plants (Figures 1 and 2) are found most abundantly near the plasma membrane (18). Coated vesicles are enclosed in the polygonal lattice characteristic of the clathrin polymer. Their outer diameters are measured at 84–91 nm (18), ~ 100 nm (36, 100), or 72–96 nm (33), depending on the method of fixation and the plant species examined. Although a fraction of the vesicles can be seen in the process of severing from the plasma membrane, many vesicles show no connection to either the plasma membrane (38) or Golgi apparatus (113). Instead, they appear freely suspended in the cytoplasm. Soon after release from the plasma membrane, they are thought to undergo a rapid but stepwise uncoating. Vesicles at various stages of uncoating are commonly observed in the cortical cytoplasm (36, 113).

Isolated coated vesicles contain a vacuolar and perhaps E_1/E_2-type proton ATPase (29, 31). Curiously, they have not been shown to generate a proton gradient or membrane potential, despite considerable effort to examine this issue (31). Whether acidification is a diagnostic trait of an endosomal compartment in plants, as it is in animals, is an important matter, because it bears on the mechanism and compartmental location of receptor-ligand dissociation.

Coated vesicles are also seen near the Golgi apparatus and the partially coated reticulum (PCR) (18, 33, 96). In some cases they appear to be attached or in the process of budding off. These coated vesicles are characteristically smaller than those associated with the plasma membrane, and as in animal

---

Figure 1 Electron micrographs of organelles and structures involved in endocytosis. (*a*) Internalization of cationized ferritin by a coated pit (arrow head) in a soybean protoplast exposed to cationized ferritin (CF) for 30 s. (*b*) Cationized ferritin entrapped in a coated vesicle (arrow head) in a soybean protoplast exposed to CF for 30 s. Micrographs in (*a*) and (*b*) courtesy of LC Fowke, reproduced with permission from Cambridge University Press (37). (*c*) Coated pits decorated with gold conjugate antibodies against β(1–3) glucan at the site of penetration of *Uromyces viciae-fabe* basidiospores into leaves of *Vigna sinensis*. Micrograph kindly provided by K Mendgen and H Xu, *Universität Koblanz, Germany*. (*d*) Cross-section of coated pit in tobacco protoplast. Micrograph courtesy of LC Fowke, reproduced with permission from the National Research Council of Canada (118). (*e*) Negatively stained coated vesicles on the inner surface of a plasma membrane fragment from tobacco. Micrograph courtesy of LC Fowke, reproduced with permission from Plenum Publishing Corporation (34a). (*f*) Golgi (G), vacuole (V), PCR, and multivesicular body (MVB) are visible in freeze-fixed white spruce cells. Micrograph courtesy of LC Fowke, reproduced with permission from The Company of Biologists Limited (39).

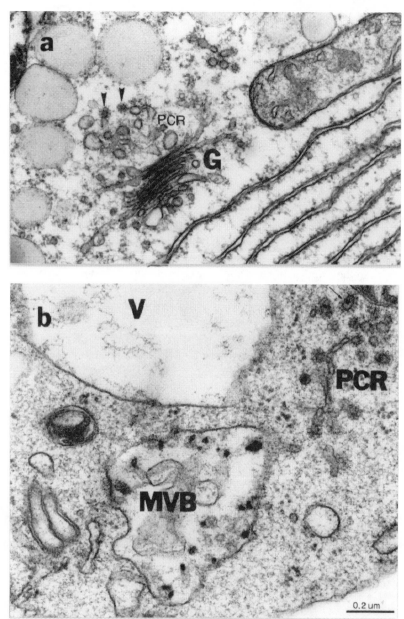

Figure 2   Electron micrographs of organelles and structures involved in endocytosis. (*a*) Barley aleurone protoplast sections showing a PCR and golgi (G). Arrows point to budding coated vesicles. Micrograph courtesy of DG Robinson, reproduced with permission from Springer-Verlag, Austria (124). (*b*) Lead nitrate stained maize root cap cell section showing multivesicular bodies (MVB), PCR, and vacuole (V). Micrograph courtesy of DG Robinson, reproduced with permission from Springer-Verlag, Berlin and Heidelberg (96).

cells, they may have different functions. The observation that Golgi coated vesicles may contain storage proteins or hydrolytic enzymes (52, 93) indicates that they may be enroute from the Golgi apparatus to a storage or lytic vacuole. Although some investigators have speculated that the larger coated vesicles near the plasma membrane may also be Golgi-derived and in the process of membrane fusion with the plasma membrane (35, 75, 92), we dismiss this possibility, because clathrin hinders rather than facilitates membrane fusion (2). Instead, clathrin probably catalyzes the invagination of cellular membranes before vesicle release, regardless of whether the vesicle stems from the Golgi apparatus or plasma membrane.

Partially Coated Reticulum

Soon after endocytic markers are seen in coated vesicles, the markers move to the partially coated reticulum (PCR; Figures 1 and 2), an interconnected system of tubular membranes surrounded by coated and partially uncoated vesicles (37). PCR are frequently found associated with Golgi, but whether a direct connection exists (57) or not (112, 113) is still debated. The 1 μm wide organelle has also been reported to communicate physically with small vacuoles (100), and is sometimes seen quite distant from the Golgi. Coated vesicles frequently are seen in the process of fusing or budding off from PCR (113), confirming that PCR participates in vesicle trafficking.

According to Robinson and colleagues (57), the animal cell counterpart of PCR is the trans Golgi network, a tubular membrane compartment proximal to the Golgi apparatus, which is involved in the sorting of Golgi-derived material destined for the plasma membrane or lysosomes. In contrast, Fowke and coworkers (39, 113) suggest the homology lies closer to CURL, the early endosomal compartment where receptor and ligand dissociate and are sorted to their next processing station (e.g. multivesicular bodies, trans Golgi network, plasma membrane). Because both the trans Golgi network and CURL can have similar morphologies, and both participate in sorting and dispatching endocytosed material, resolution of this controversy may require identification of better biochemical or histochemical markers. Based on electron microscopic analysis, however, Griffing (48) suggests that the PCR is distinct from the trans Golgi network.

Multivesicular Bodies

Virtually all workers agree that plant multivesicular bodies (MVB) occupy a prominent position in the endocytic pathway. These 250–700 nm organelles (Figures 1 and 2) are bounded by a limiting membrane that may have plaques or regions of clathrin-like coat and coated pits (37, 100, 111). Within the compartment are multiple 50–100 nm vesicles. Because endocytosed cationized ferritin is seen associated with both the internal surface of the limiting membrane and with the external surface of the entrapped vesicles, and because incoming vesicles frequently are observed either attached or in close proximity to multivesicular bodies, it is assumed that newly arrived vesicles are incorporated into MVB by a fusion-invagination sequence (37). There is also strong evidence that MVB contain some degradative enzymes (51, 90), providing evidence that they are not only morpologically but also biochemically similar to animal MVB (36).

Vacuoles

Endocytic markers are frequently but not invariably deposited in vacuoles (Figures 1 and 2) (for exceptions, see 63, 66, 112, 120). Although the tonoplast occasionally can be labeled, more frequently the markers are enclosed in smaller vacuoles (37, 39, 56). Functional specialization among isoforms of this organelle could exist, and because some vacuoles contain lytic enzymes, these forms may be the plant counterpart of the lysosome (48).

THE ENDOCYTIC PATHWAY IN PLANT CELLS AND PROTOPLASTS

Studies of the time course of occupancy by extracellular solutes in various endocytic compartments have used a variety of plant sources and electron-dense markers: cationized ferritin in soybean, white spruce, tobacco, and bean leaf protoplasts (39, 112, 113); lanthanum and lead salts in root cap cells of maize and *Lobelia erinus* (63, 100); colloidal gold-bovine serum albumin in soybean protoplasts (120); and lectin-gold in bean leaf protoplasts (56). Although these kinetic studies have not defined an endocytosis pathway, there is consensus that many internalized markers are initially captured at coated pits, engulfed quickly (< 6 min) in coated vesicles, and transferred gradually (< 20 min) to the partially coated reticulum (37, 39, 56, 90). However, subsequent destinations of the markers appear to diverge, indicating the possibility of multiple fates. There is clear evidence that some material moves from PCR to the Golgi. However, in the same cells, many of the markers move from PCR through MVB into vacuoles (39, 56, 63, 95, 113). The partitioning of endocytic markers between these two routes may also differ among systems. For example, in protoplasts, where secretory activity is high because of intense regeneration of the cell wall, endocytosis trafficking may be diverted toward recycling, i.e. coated pit →coated vesicle →PCR →Golgi apparatus or recycling compartment → plasma membrane (Figure 3). However, in more quiescent cells, where secretory activity is minimal, the itinerary may be unidirectional, i.e. coated pit → coated vesicle → PCR→MVB →vacuole.

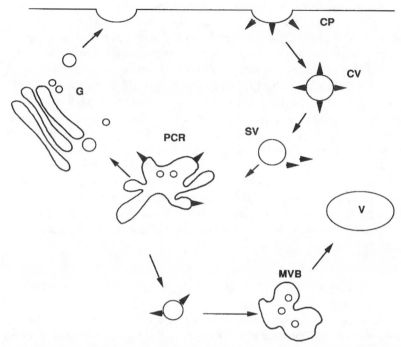

Figure 3   Diagram of two possible endocytic pathways in plants. Receptor-associated or fluid phase markers entering at coated pits (CP) are sequentially seen in coated vesicles (CV), smooth vesicles (SV), and partially coated reticulum (PCR). Then, depending on cell type and growth activity, markers may pass through multivesicular bodies (MVB) to the vacuole (V), or through the Golgi apparatus (G) or a recycling compartment to the plasma membrane.

FLUID PHASE ENDOCYTOSIS IN PLANT CELLS AND PROTOPLASTS

Internalization of heavy metal salts and impermeable fluorescent probes has been cited as proof of fluid phase endocytic pathways in plants. However, recent reports have contested these conclusions, noting that the toxicity of lanthanum, lead, and uranium may have altered cell behavior (1, 68), and that direct membrane permeation of the fluorescent probes may have led to false positive results (58, 59, 63, 80, 100). For example, lucifer yellow, a putatively impermeant fluorescent probe with ideal physicochemical properties (pH tolerance, strong negative charge, and high quantum yield), has been found to enter plant cells via an anion transporter (reviewed in 94, 78). Several groups have shown that probenecid, an organic acid transport inhibitor in animal cells, inhibits the accumulation of lucifer yellow in the vacuole (17, 77, 79). In addition, microinjection of lucifer yellow into the cytoplasm to examine the impermeability of the tonoplast membrane has shown that direct penetration of the tonoplast membrane is possible (34, 44). Taken with other findings implicating fluid phase endocytic uptake of the dye, these data point to the existence of at least two uptake pathways for lucifer yellow, one via endocytosis and the other via anion transport. The data also point to the unreliability of lucifer yellow as a probe for fluid phase endocytosis (FPE) in plants.

To more carefully define the route of entry of lucifer yellow into the plant cell, Owen et al (81) precipitated lucifer yellow with $BaCl_2$ and examined its intracellular location by electron microscopy. Precipitation products were readily seen in several endosomal compartments, including coated pits, coated and smooth vesicles, dictyosomes, and PCR, indicating that lucifer yellow can enter cells via an endocytic route. More recent results with other probes, such as silver enhanced BSA-gold, have also shown uptake into cytoplasmic vesicles (120). Curiously, neither investigation reported observing any labeling of the vacuole, raising questions about the terminal destination of solutes internalized by FPE.

RECEPTOR-MEDIATED ENDOCYTOSIS IN PLANT CELLS

Although there is ample evidence from nonspecific marker uptake for the occurrence of endocytosis in plants, evidence using biological ligands remains sparse. Initial indications that receptor-mediated endocytosis might occur in vivo came from studies of Rhizobium-legume interactions, where thin sections of nodule tissue showed symbiotic bacteria in the process of ingestion and encapsulation within host plant cytoplasmic vesicles (9, 64 and references within). Subsequent research has shown that rhicadhesin, a 14,000-Dalton calcium-binding protein, may facilitate recognition of the root hair cells by the bacteria (105). The plant receptor in rhicadhesin-mediated binding, however, has not been identified, although there are strong indications that a cell surface lectin may contribute significantly (65). These data obviously confirm that uptake via membrane invagination can occur in vivo, but the process of bacterial phagocytosis may differ from classical receptor-mediated endocytosis. Further confirmation that the latter event occurs in vivo is needed.

The only direct evidence for receptor-mediated endocytosis in plants comes from work on elicitor and vitamin uptake in suspension cultured plant cells (60–62, 71). Endocytosis of both a proteinaceous elicitor from *Verticillium dahliae* and a polygalacturonic acid elicitor from citrus pectin was demonstrated by labeling both elicitors with fluorescent or radioactive tags and observing their uptake by suspension cultured soybean cells (60, 71). Though the time course of internalization differed for the two macromolecules, they both appeared to be routed along the same pathway. Thus, binding was rapid at 23°C (< 20 min), followed by slow uptake into a vacuole. Only binding was seen at 4°C, but upon rewarming to 23°C, transport to the vacuole resumed and continued until little fluorescence could be detected on the plasma membrane. The uptake process was saturable, competitively inhibited by unlabeled elicitor, and not observed in cells treated with fluorescent nonspecific ligands. These results indicated a native receptor-mediated endocytic pathway for certain elicitors.

An endocytic pathway for biotin was also discovered when fluorescent, biotin-conjugated proteins were added to cultured soybean cells and observed to internalize (61, 62, 71). Although the kinetics, saturability, competitive blockade by free biotin, and temperature dependence of the uptake pathway were analogous to results obtained with elicitors, the final site of biotin conjugate deposition was unexpectedly different. Whereas fluorescent elicitors eventually accumulated in the vacuole, the biotin-linked proteins were delivered primarily to the cytoplasm (71). Legendre (67) confirmed G protein involvement in plant signal transduction by using biotin receptor-mediated endocytosis to deliver antibody fragments (Fab) to a G protein in the cytoplasm of soybean cells.

Biotin receptor-mediated endocytosis was also used to characterize parameters affecting endocytic uptake in vivo (62). Rates of endocytosis were found to be sensitive to macromolecular size, decreasing linearly with the log molecular weight of the conjugated protein. Uptake was also pH sensitive, displaying an optimum near pH 5 (i.e. the pH of the culture medium). Cell age (or perhaps proliferative activity) also had an impact on biotin endocytosis, with increasing culture age correlating with reduced uptake activity. Finally, and perhaps most surprisingly, internalization was strongly influenced by the osmolarity of the medium, with maximal endocytic rates at or slightly below physiological osmotic strength, rather than that at elevated osmotic pressures where cell turgor pressure is reduced. Clearly, the pH and osmotic strength of normal growth conditions support endocytosis optimally in soybean cells in culture.

FUTURE DIRECTIONS AND CONCLUSIONS

Three lines of evidence support the occurrence of endocytosis in higher plants. First, calculations of membrane flow during secretory vesicle fusion indicated that a pathway existed for return of membrane material to the cytoplasm. Second, examination of plant cells for major structures or components diagnostic of endocytosis in animal cells invariably revealed homologous structures with highly similar properties. Studies of electron-dense endocytic marker uptake in protoplasts and plant cells showed a chronological sequence of endocytic compartment occupancy that was analogous to the sequence found in animals. And finally, where endocytosis was monitored in intact plant cells with natural macromolecular ligands, specific saturable uptake was observed with final deposition in the vacuole or cytoplasm. Taken together, these arguments constitute compelling evidence that higher plants do conduct endocytosis.

Although the above observations represent significant progress along several research fronts, our current understanding of some areas of endocytosis is still deficient. For example, although the itineraries of cationized ferritin, colloidal gold-lectins, and heavy metal salts in plant cells or protoplasts have been scrutinized in detail, there is presently no information on the pathway followed by a natural ligand of a cell surface receptor in an intact plant cell. It would be instructive to learn whether ligands destined for destruction (e.g. pathogen-derived elicitors) follow distinct intracellular routes from ligands internalized for anabolic applications (e.g. biotin). As noted above, the final destinations of elicitors and biotin are different, but whether they follow totally unique pathways or diverge at some intracellular sorting compartment is unknown. It would also be useful to know how interconnected the PCR and Golgi apparatus are and what the final destination is for markers not targeted to vacuoles. Finally, it would be instructive to know more about the physical properties (e.g. pH, ion composition, electric potential) and biochemical characteristics (e.g. enzymatic activities, lipid composition, structural proteins) of each organelle along the endocytic pathway and whether distinctions in these properties arise via a process of gradual endosome maturation or vesicle transport (106).

A second major direction of future research might be to identify and characterize clathrin-independent pathways of endocytosis. As noted above, at least two additional endocytic pathways exist in animal cells, and clathrin-deficient yeast still conduct rapid endocytosis (84), but other questions remain. For example, do the observed smooth pits and vesicles constitute a distinct endocytic pathway in plants (39, 100, 112) or are they indications of a perturbed system seen only in protoplasts or cells treated with heavy metal salts? Do caveolae exist in plants and are they the collection sites for glycosylphosphatidyl inositol-anchored receptors (3)? How does plasma membrane return to the interior of actively secreting cells that engage in little or no fluid phase endocytosis (88)?

A third area that has received limited attention is the biochemistry and molecular and cell biology of the clathrin-dependent pathway in plants. Much has been accomplished at the morphological level, but many unanswered questions remain at the molecular level. For example, do plants require adaptor complexes to connect their clathrin to membrane receptors? Why does the uncoating enzyme dismantle only coated vesicles and not coated pits? What stimulus promotes dissociation of ligand-receptor complexes in the sorting compartment? If low pH performs this uncoupling function, as it does in animals, then how does the ligand initially bind in the acidic extracellular milieu? What is the molecular nature of the sorting process in the PCR, and what amino acid motifs target a receptor to different destinations? Do low molecular weight G proteins and other cytoplasmic components catalyze budding and fusing of vesicles during interorganelle transport? Do transcytosis pathways exist in plants? Are certain vacuoles specialized to receive late endosomal material, and if so, what are the recognition signals? How does the plant cytoskeleton participate in endocytosis? And how is endocytosis regulated at each step of the process? Clearly, the plant endocytosis field is fertile, and with novel tools and reagents that recently have become available, significant progress can be expected in the future.

Any *Annual Review* chapter, as well as any article cited in an *Annual Review* chapter, may be purchased from the Annual Reviews Preprints and Reprints service. 1-800-347-8007; 415-259-5017; email: arpr@class.org

Literature Cited

1. Abe S, Takeda J. 1988. Effects of La^{+3} on surface changes, dielectrophoresis, and electrofusion of barley protoplasts. *Plant Physiol.* 87:389–94
2. Alstiel L, Branton D. 1983. Fusion of coated vesicles with lysosomes: measurement with a fluorescence assay. *Cell* 32:921–29
3. Anderson RGW, Kamen BA, Rothberg KG, Lacey SW. 1992. Potocytosis: sequesteration and transport of small molecules by caveolae. *Science* 255:410–11
4. Balch WE. 1990. Small GTP-binding proteins in vesicular transport. *Trends Biochem. Sci.* 15:473–77
5. Balusek K, Depta H, Robinson DG. 1988. Two polypeptides (30 and 38 kDa) in plant coated vesicles with clathrin light chain properties. *Protoplasma* 146:174–76
6. Barbier-Brygoo H, Ephritikhine G, Klambt D, Ghislain M, Guern J. 1989. Functional evidence for an auxin receptor at the plasmalemma of tobacco mesophyll protoplasts. *Proc. Natl. Acad. Sci. USA* 86:891–95
7. Deleted in proof
8. Baron-Epel O, Gharyal PK, Schindler M. 1988. Pectins as mediators of wall porosity in soybean cells. *Planta* 175:389–95
9. Bassett B, Goodman RN, Novacky A. 1977. Ultrastructure of soybean modules. I: Release of rhizobia from the infection thread. *Can. J. Microbiol.* 23:573–82
10. Bretscher MS, Thomson JN, Pearse BMF. 1980. Coated pits act as molecular filters. *Proc. Natl. Acad. Sci. USA* 77:4156–59
11. Brunk U, Schellens J, Westermark B. 1976. Influence of epidermal growth factor (EGF) on ruffling activity, pinocytosis and proliferation of cultivated human glia cells. *Exp. Cell Res.* 103:295–302
12. Burwen SJ, Jones AL. 1987. The association of polypeptide hormones and growth factors with the nuclei of target cells. *Trends Biochem. Sci.* 12:159–63
13. Carpenter G, Cohen S. 1979. Epidermal growth factor. *Annu. Rev. Biochem.* 48:193–216
14. Carpita N, Sabularse D, Montezinos D, Delmer DP. 1979. Determination of the pore site of cell walls of living plant cells. *Science* 205:1144–47
15. Cheong J-J, Hahn MG. 1991. A specific, high-affinity binding site for the hepta-β-glucoside elicitor exists in soybean membranes. *Plant Cell* 3:137–47
16. Chinkers M, McKanna JA, Cohen S. 1979. Rapid induction of morphological changes in human carcinoma cells A-431 by epidermal growth factor. *J. Cell Biol.* 83:260–65
17. Cole L, Coleman J, Kearns A, Morgan G, Hawes C. 1991. The organic anion transport inhibitor, probenecid, inhibits the transport of lucifer yellow at the plasma membrane and the tonoplast in suspension-cultured plant cells. *J. Cell Sci.* 99:545–55
18. Coleman J, Evans D, Hawes C, Horsley D, Cole L. 1987. Structure and molecular organization of higher plant coated vesicles. *J. Cell Sci.* 88:35–45
19. Cosio EG, Frey T, Ebel J. 1990. Solubilization of soybean membrane binding sites for fungal β-glucans that elicit phytoalexin accumulation. *FEBS Lett.* 264:235–38

20. Cosio EG, Frey T, Ebel J. 1992. Identification of a high-affinity binding protein for a hepta-β-glucoside phytoalexin elicitor in soybean. *Eur. J. Biochem.* 204:1115–23
21. Cosio EG, Pöpper LH, Schmidt WE, Ebel J. 1988. High-affinity binding of fungal β-glucan fragments to soybean microsomal fractions and protoplasts (*Glycine max* L.). *Eur. J. Biochem.* 175:309–15
22. Cram WJ. 1980. Pinocytosis in plants. *New Phytol.* 84:1–17
23. Deleted in proof
24. Deleted in proof
25. Daniel TO, Schneider WJ, Goldstein JL, Brown MS. 1983. Visualization of lipoprotein receptors by ligand blotting. *J. Biol. Chem.* 258:4606–11
26. DeBoer AH, Watson BA, Cleland RE. 1989. Purification and identification of the fusicoccin binding protein from oat root plasma membrane. *Plant Physiol.* 89:250–59
27. deMichaelis MI, Pugliarello MC, Rasi-Caldogno R. 1989. Fusicoccin binding to its plasma membrane receptor and the activation of the plasma membrane H⁺ ATPase I. Characteristics and intracellular localization of the fusicoccin receptor in microsomes from radish seedlings. *Plant Physiol.* 90:133–39
28. Demmer A, Holstein SEH, Hinz G, Schauermann G, Robinson DG. 1993. Improved coated vesicle isolation allows better characterization of clathrin polypeptides. *J. Exp. Bot.* 44:23–33
29. Depta H, Holstein SEH, Robinson DG, Lützelschwab M, Michalke W. 1990. Membranes markers in highly purified clathrin-coated vesicles from *Cucurbita* hypocotyls. *Planta* 183:1–9
30. Deleted in proof
31. Drucker M, Hinz G, Robinson DG. 1993. ATPases in plant coated vesicles. *J. Exp. Bot.* 44:283–91
32. Dunn WA, Hubbard AL. 1984. Receptor mediated endocytosis of epidermal growth factor by hepatocyte in perfused rat liver: ligand and receptor dynamics. *J. Cell Biol.* 98:2148–59
33. Emons AMC, Traas JA. 1986. Coated pits and coated vesicles on the plasma membrane of plant cells. *Eur. J. Cell Biol.* 41:57–64
34. Fisher DG. 1988. Movement of lucifer yellow in leaves of *Coleus blumei* Benth. *Plant Cell Environ.* 11:639–44
34a. Fowke LC, Constabel F. 1982. Plant protoplasts: an experimental system for cell biologists. In *Plant Cell Culture in Crop Improvement*, ed. KL Giles, SK Sen, pp. 185–200. New York: Plenum
35. Fowke LC, Griffing LR, Mersey BG, Tanchak MA. 1985. Protoplasts for studies of cell organelles. In *Plant Protoplasts*, ed.

LC Fowke, F Constabel, pp. 39–52. Boca Raton, FL: CRC
36. Fowke LC, Tanchak MA. 1988. The structure and function of plant coated vesicles. In *Plant Cell Biotechnology, NATO ASI Ser. MS*, ed. MSS Pais, F Mavituna, JM Novais, H18:153–63. Berlin: Springer-Verlag
37. Fowke LC, Tanchak MA, Galway ME. 1991. Ultrastructural cytology of the endocytotic pathway in plants. In *Endocytosis, Exocytosis and Vesicle Traffic in Plants*, ed. CR Hawes, JOD Coleman, DE Evans. *Soc. Exp. Biol. Sem. Ser.* 45:15–40
38. Fowke LC, Tanchak MA, Rennie PJ. 1989. Serial section analysis of coated pits and coated vesicles in soybean protoplasts. *Cell Biol. Int. Rep.* 13:419–25
39. Galway ME, Rennie PJ, Fowke LC. 1993. Ultrastructure of the endocytotic pathway in glutaraldehyde-fixed and high pressure frozen/freeze substituted protoplasts of white spruce (*Picea glauca*). *J. Cell Sci.* 106:847–58
40. Geuze HJ, Slot JW, Strous GJAM, Peppard J, von Figura K. 1984. Intracellular receptor sorting during endocytosis: comparative immunoelectron microscopy of multiple receptors in rat liver. *Cell* 37:195–204
41. Gogarten JP. 1988. Physical properties of the cell wall of photoautorophic suspension cells from *Chenopodium rubrum* L. *Planta* 174:333–39
42. Goldstein JL, Anderson RGW, Brown MS. 1979. Coated pits, coated vesicles, and receptor-mediated endocytosis. *Nature* 279:679–85
43. Goldstein JL, Brown MS, Anderson RGW, Russell DW, Schneider WJ. 1985. Receptor-mediated endocytosis: concepts emerging from the LDL receptor system. *Annu. Rev. Cell Biol.* 1:1–39
44. Goodwin PB, Shepherd V, Erwee MG. 1990. Compartmentation of fluorescent tracers injected into the epidermal cells of *Egeria densa* leaves. *Planta* 181:129–36
45. Gorvel J, Chavrier P, Zerial M, Gruenberg J. 1991. rab 5 controls early endosome fusion *in vitro*. *Cell* 64:915–25
46. Goud B, McCaffrey M. 1991. Small GTP-binding proteins and their role in transport. *Curr. Opin. Cell Biol.* 3:626–33
47. Gradmann O, Robinson DG. 1989. Does turgor prevent endocytosis in plant cells? *Plant Cell Environ.* 12:151–54
48. Griffing LR. 1991. Comparisons of golgi structure and dynamics in plant and animal cells. *J. Electron Microsc. Tech.* 17:179–99
49. Griffiths G, Back R, Marsh M. 1989. A quantitative analysis of the endocytic pathway in baby hamster kidney cells. *J. Cell Biol.* 109:2703–20
50. Haigler HT, McKanna JA, Cohen S, 1979.

Rapid stimulation of pinocytosis in human carcinoma cells A-431 by EGF. *J. Cell Biol.* 83:82–90

51. Halperin W. 1969. Ultrastructural localization of acid phosphatase in cultured cells of *Daucus carota. Planta* 88:91–102

52. Harley SM, Beevers L. 1989. Coated vesicles are involved in the transport of storage proteins during seed development in *Pisum sativum* L. *Plant Physiol.* 91:674–78

53. Hesse T, Feldwisch J, Balshusemann D, Bauw G, Puype M, et al. 1989. Molecular cloning and structural analysis of a gene from *Zea mays* (L.) coding for a putative receptor for the plant hormone auxin. *EMBO J.* 8:2453–62

54. Heuser JE, Kirchhausen T. 1985. Deep-etch views of clathrin assemblies. *J. Ultrastruct. Res.* 92:1–27

55. Hicks GR, Rayle DL, Jones AM, Lomax TL. 1989. Specific photoaffinity labelling of two plasma membrane polypeptides with an azido-auxin. *Proc. Natl. Acad. Sci. USA* 86:4948–52

56. Hillmer S, Depta H, Robinson DG. 1986. Confirmation of endocytosis in higher plant protoplasts using lectin-gold conjugates. *Eur. J. Cell Biol.* 41:142–49

57. Hillmer S, Freundt H, Robinson DG. 1988. The partially coated reticulum and its relationship to the golgi apparatus in higher plants. *Eur. J. Cell Biol.* 47:206–12

58. Hillmer S, Hedrich R, Robert-Nicoud M, Robinson DG. 1990. Uptake of lucifer yellow CH in leaves of *Commelina communis* is mediated by endocytosis. *Protoplasma* 158:142–48

59. Hillmer S, Quader H, Robert-Nicoud M, Robinson DG. 1989. Lucifer yellow uptake in cells and protoplasts of *Daucus carota* visualized by laser scanning microscopy. *J. Exp. Bot.* 40:417–23

60. Horn MA, Heinstein PF, Low PS. 1989. Receptor-mediated endocytosis in plant cells. *Plant Cell* 1:1003–9

61. Horn MA, Heinstein PF, Low PS. 1990. Biotin-mediated delivery of exogenous macromolecules into soybean cells. *Plant Physiol.* 93:1492–96

62. Horn MA, Heinstein PF, Low PS. 1992. Characterization of parameters influencing receptor mediated endocytosis in cultured soybean cells. *Plant Physiol.* 98:673–79

63. Hübner R, Depta H, Robinson DG. 1985. Endocytosis in maize root cap cells: evidence obtained using heavy metal salt solutions. *Protoplasma* 129:214–22

64. Kijne JW. 1992. The rhizobium infection process. In *Biological Nitrogen Fixation,* ed. G Stacey, RH Burris, HJ Evans, pp. 349–99. New York: Chapman & Hall

65. Kijne JW, Smit G, Diaz CL, Lugtenberg BJJ. 1988. Lectin enhanced accumulation of manganese-limited *Rhizobium legu-*minosarum cells on pea root hair tips. *J. Bacteriol.* 170:2994–3000

66. Lazzaro MD, Thomson WW. 1992. Endocytosis of lanthanum nitrate in the organic acid-secreting trichomes of chickpea (*Cicer arietinum* L.). *Am. J. Bot.* 79:1113–18

67. Legendre L, Heinstein PF, Low PS. 1992. Evidence for the participation of GTP-binding proteins in elicitation of the rapid oxidative burst in cultured soybean cells. *J. Biol. Chem.* 267:20140–47

68. Leonard RT, Nagahashi G, Thomson WW. 1975. Effect of lanthanum on ion absorption in corn roots. *Plant Physiol.* 55:542–46

69. Deleted in proof

70. Low MG, Kincade PW. 1985. Phosphatidylinositol is the membrane-anchoring domain of the Thy-1 glycoprotein. *Nature* 318:62–64

71. Low PS, Legendre L, Heinstein PF, Horn MA. 1993. Comparison of elicitor and vitamin receptor-mediated endocytosis in cultured soybean cells. *J. Exp. Bot.* 44:269–74

72. Marsh M, Helenius A. 1980. Adsorptive endocytosis of semliki forest virus. *J. Mol. Biol.* 142:439–54

73. Meyer C, Feyerabend M, Weiler E. 1989. Fusicoccin binding proteins in *Arabidopsis thaliana* L. Heynh. *Plant Physiol.* 89:692–99

74. Mostov KE, Smister NE. 1985. Transcytosis. *Cell* 43:389–90

75. Nakamura S, Miki-Hirosige H. 1982. Coated vesicles and cell plate formation in microspore mother cell. *J. Ultrastruct. Res.* 80:302–11

76. Octave JN, Schneider YJ, Trouet A, Crichton RR. 1983. Iron uptake and utilization by mammalian cells. I: Cellular uptake of transferrin and iron. *Trends Biochem. Sci.* 8:217–20

77. O'Driscoll D, Wilson G, Steer MW. 1991. Lucifer yellow and fluorescein isothiocyanate uptake by cells of *Morinda citrifolia* in suspension cultures is not confined to the endocytotic pathway. *J. Cell Sci.* 100: 237–41

78. Oparka KJ. 1991. Uptake and compartmentation of fluorescent probes by plant cells. *J. Exp. Bot.* 42:565–79

79. Oparka KJ, Murant EA, Wright KM, Prior DAM. 1991. The drug probenecid inhibits the vacuolar accumulation of fluorescent anions in onion epidermal cells. *J. Cell Sci.* 99:557–63

80. Oparka KJ, Robinson D, Prior DAM, Derrick P, Wright KM. 1988. Uptake of lucifer yellow CH into intact barley roots: evidence for fluid-phase endocytosis. *Planta* 76:541–47

81. Owen TP Jr, Platt-Aloia KA, Thomson

WW. 1991. Ultrastructural localization of lucifer yellow and endocytosis in plant cells. *Protoplasma* 160:115–20

82. Parker JE, Schulte W, Hahlbrock K, Scheel D. 1991. An extracellular glycoprotein from *Phytophthora megasperma* f. sp. *glycinea* elicits phytoalexin synthesis in cultured parsley cells and protoplasts. *Mol. Plant Microbe Interact.* 4:19–27

83. Payne GS. 1990. Genetic analysis of clathrin function in yeast. *J. Membr. Biol.* 116:93–105

84. Payne GS, Baker D, Van Tuinen E, Schekman R. 1988. Protein transport to the vacuole and receptor mediated endocytosis by clathrin heavy chain-deficient yeast. *J. Cell Biol.* 106:1453–61

85. Pearse BMF, Robinson MS. 1990. Clathrin, adaptors, and sorting. *Annu. Rev. Cell Biol.* 6:151–71

86. Pelchen-Matthews A, Armes JE, Griffiths G, Marsh M. 1991. Differential endocytosis of CD4 in lymphocytic and non-lymphocytic cells. *J. Exp. Med.* 173:575–87

87. Phaire-Washington L, Silverstein SC, Wang E. 1980. Phorbol myristate acetate stimulates microtubule and 10 nm filament extension and lysosome redistribution in mouse macrophages. *J. Cell Biol.* 86:641–55

88. Phillips GD, Preshaw C, Steer MW. 1988. Dictyosome vesicle production and plasma membrane turnover in auxin-stimulated outer epidermal cells of coleoptile segments from *Avena sativa* (L.). *Protoplasma* 145:59–65

89. Raven JA. 1987. The role of vacuoles. *New Phytol.* 196:357–422

90. Record RD, Griffing LR. 1988. Convergence of endocytis and lysosomal pathways in soybean protoplasts. *Planta* 176:425–32

91. Regoeczi E, Chindemi PA, Debanne MT, Charlwood PA. 1982. Partial resialylation of human asialotransferrin type 3 in the rat. *Proc. Natl. Acad. Sci. USA* 79:2226–30

92. Robertson JG, Lyttleton P. 1982. Coated and smooth vesicles in the biogenesis of cell walls, plasma membranes, infection threads and peribacteroid membranes in root hairs and nodules of white clover. *J. Cell. Sci.* 58:63–78

93. Robinson DG, Balusek K, Freundt H. 1989. Legumin antibodies recognize polypeptides in coated vesicles isolated from developing pea cotyledons. *Protoplasma* 150:79–82

94. Robinson DG, Hedrich R. 1991. Vacuolar lucifer yellow uptake in plants: endocytosis or anion transport. A critical opinion. *Bot. Acta* 104:257–64

94a. Robinson DG, Hedrich R, Herkt B, Diekmann W, Robert-Nicoud M. 1992. Endocytosis in plants: problems and perspectives. In *Endocytosis, NATO ASI Ser.,* ed. PJ Courtoy, H62:459–66. Berlin: Springer-Verlag

95. Robinson DG, Hillmer S. 1990. Endocytosis in plants. *Physiol. Plant.* 79:96–104

96. Robinson DG, Hillmer S. 1990. Coated pits. In *The Plant Plasma Membrane*, ed. C Larsson, IM Moller, pp. 233–55. Berlin: Springer-Verlag

97. Rodman JS, Mercer RW, Stahl PD. 1990. Endocytosis and transcytosis. *Curr. Opin. Cell Biol.* 2:664–72

98. Rothberg KG, Heuser JE, Donell WC, Ying Y, Glenney JR, Anderson RGW. 1992. Caveolin, a protein component of caveolae membrane coats. *Cell* 68:673–82

99. Rothman JE, Schmid SL. 1986. Enzymatic recycling of clathrin from coated vesicles. *Cell* 46:5–9

100. Samuels AL, Bisalputra T. 1990. Endocytosis in elongating root cells of *Lobelia erinus. J. Cell. Sci.* 97:157–65

101. Sandvig K, van Deurs BV. 1991. Endocytosis without clathrin. *Cell Biol. Int. Rep.* 15:3–8

102. Saxton MJ, Breidenbach RW. 1988. Receptor-mediated endocytosis in plants is energetically possible. *Plant Physiol.* 86:993–95

103. Schlessinger J. 1980. The mechanism and role of hormone-induced clustering of membrane receptors. *Trends Biochem. Sci.* 5:210–14

104. Smit G, Kijne JW, Lugtenberg BJJ. 1986. Correlation between extracellular fibrils and attachment of *Rhizobium leguminosarum* to pea root hair tips. *J. Bacteriol.* 168:821–27

105. Smit G, Logman TJJ, Boerreigter METI, Kijne JW, Lugtenberg BJJ. 1989. Purification and partial characterization of *Rhizobium leguminosarum* biovar *viciae* Ca^{2+}-dependent adhesin, which mediates the first step in attachment of cells of the family Rhizobiceae to plant root hair tips. *J. Bacteriol.* 171:4054–62

106. Smythe E, Warren G. 1991. The mechanism of receptor-mediated endocytosis. *Eur. J. Biochem.* 202:689–99

107. Snider MD, Rogers OC. 1985. Intracellular movement of cell surface receptors after endocytosis: resialylation of asialo-transferrin receptor in human erythroleukemia cells. *J. Cell Biol.* 100:826–34

108. Steer MW. 1985. Vesicle dynamics. In *Biological Microscopy*, ed. AW Robards, pp. 129–55. Oxford: Oxford Univ. Press

109. Steinman RM, Mellman IS, Muller WA, Cohn ZA. 1983. Endocytosis and recycling of plasma-membrane. *J. Cell Biol.* 96:1–27

110. Swanson JA, Yirinec BD, Silverstein SC. 1985. Phorbol esters and horseradish peroxidase stimulate pinocytosis and redirect

the flow of pinocytosed fluid in macrophages. *J. Cell Biol.* 100:851–59

111. Tanchak MA, Fowke LC. 1987. The morphology of multivesicular bodies in soybean protoplasts and their role in endocytosis. *Protoplasma* 138:173–82

112. Tanchak MA, Griffing LR, Mersey BG, Fowke LC. 1984. Endocytosis of cationized ferritin by coated vesicles of soybean protoplasts. *Planta* 162:481–86

113. Tanchak MA, Rennie PJ, Fowke LC. 1988. Ultrastructure of the partially coated reticulum and dictyosomes during endocytosis by soybean protoplasts. *Planta* 175:433–41

114. Tepfer M, Taylor IEP. 1981. The permeability of plant cell walls as measured by gel filtration chromatography. *Science* 213: 761–63

115. Terris S, Hofmann C, Steiner DF. 1979. Mode of uptake and degradation of ^{125}I labelled insulin by isolated hepatocytes and H4 hepatoma cells. *Can. J. Biochem.* 57: 459–68

116. Trowbridge IS. 1991. Endocytosis of signals for internalization. *Curr. Opin. Cell Biol.* 3:634–41

117. Turek JJ, Leamon CP, Low PS. 1993. Endocytosis of folate-protein conjugates: ultrastructural localization in KB cells. *J. Cell Sci.* 106: 423–30

118. Vander Valk P, Fowke LC. 1981. Ultrastructural aspects of coated vesicles of coated vesicles in tobacco protoplasts. *Can. J. Bot.* 59:1307–13

119. Deleted in proof

120. Villanueva MA, Taylor J, Sui X, Griffing LR. 1993. Endocytosis in plant protoplasts: visualization and quantitation of fluid-phase endocytosis using silver-enhanced bovine serum albumin-gold. *J. Exp. Bot.* 44:275–81

121. Wall DA, Meleka J. 1985. An unusual lysosome compartment involved in vitellogenin endocytosis by Xenopus oocytes. *J. Cell Biol.* 101:1651–64

122. Watts C, Marsh M. 1992. Endocytosis: what goes in and how? *J. Cell. Sci.* 103:1–8

123. Wiedenhoeft RE, Schmidt GW, Palevitz BA. 1988. Dissociation and reassembly of soybean clathrin. *Plant Physiol.* 86:412–16

124. Zingen-Sell I, Hillmer S, Robinson DG, Jones RL. 1990. Localization of α-amylase isozymes within the endomembrane system of barley aleurone. *Protoplasma* 154: 16–24

Annu. Rev. Plant Physiol. Plant Mol. Biol. 1994. 45:633–62

PHOTOINHIBITION OF PHOTOSYNTHESIS IN NATURE[1]

S.P. Long and S. Humphries

Department of Biology, University of Essex, Colchester, CO4 3SQ, United Kingdom

P.G. Falkowski

Division of Oceanographic and Atmospheric Sciences, Building 318, Brookhaven National Laboratory, Upton, New York 11973

KEYWORDS: photosynthesis, photoinhibition, light stress, plant production, phytoplankton, energy dissipation, plant canopies

CONTENTS

INTRODUCTION 633
MECHANISMS... 636
   Photosystem II Inactivation, Damage and Recovery: The D1 Story 637
   Avoidance of PSII Damage: The Xanthophyll Cycle Story 640
   Damage and Avoidance: A Blurring of the Division 643
PHOTOINHIBITION IN THE FIELD AND OPEN OCEAN... 644
   Terrestrial Vegetation 644
   Phytoplankton 646
SIGNIFICANCE TO PRODUCTION: A MODELING APPROACH.................................... 648
OBSERVED CHANGES IN PRODUCTION 653
PHOTOINHIBITION AND PLANT DISTRIBUTIONS 654
CONCLUSIONS.. 655

INTRODUCTION

Light, as the energy source for photosynthesis, is the essential prerequisite for plant life. Excess light, however, can photoinhibit photosynthesis and may lead to photooxidative destruction of the photosynthetic apparatus (29, 66,

---

[1]
   The US Government has the right to retain a nonexcusive royalty-free license in and to any copyright covering this paper.

103, 112). Plant evolution has therefore had to tread a path between maximizing light interception for photosynthesis and minimizing the potential for damage arising from the over-excitation of the photosynthetic apparatus. In nature, light differs from all other major climate variables in the size and speed of its variation. Within a single day plants must cope with several orders of magnitude change in the quantity of light, fluctuating on a range of time scales (109). Superimposed on the daily and annual cycle (46) are stochastic fluctuations with frequencies ranging from seconds [e.g. the focusing and defocusing of light by waves (51)], to years [e.g. the life span of an overshadowing tree (90)]. Plants may respond to low frequency variation in light quantity and quality via development to adapt their photosynthetic capacity and ability to dissipate intercepted light energy (for review see 4, 5, 36, 39, 104, 109). Where high frequency variation or inability to adapt to prevailing light conditions produces excessive excitation of the photosynthetic apparatus, photoinhibition may result (103). In nature, plants show an array of responses to excess light, which may be partially categorized between damage and avoidance of damage, operating on a range of time-scales (Table 1).

Table 1 Mechanisms by which a higher plant leaf or unicellular alga avoids, tolerates, or is damaged by a step increase in the quantity of excess light absorbed.

Time-scale	Avoidance	Tolerance	Damage
< seconds		trans-thylakoid ΔpH (61)	
seconds–minutes	chloroplast movement (19)	state-transitions (22), increased activation of enzymes of photosynthetic C-metabolism	
minutes–hours	leaf movement (47, 114), cell movement (phytoplankton) (87)	PSII inactivation by zeaxanthin de-epoxidation (29), increased synthesis of D1 (8), inducible activity of active O_2 scavenging systems (62)	PSII inactivation by formation of nonfunctional PSII (8)
hours–days		developmental changes (phytoplankton)	decreased A_{sat} (112), bulk pigment loss with photo-oxidation (112), loss of stromal enzyme activity, including rubisco (128)
days–weeks	increased albedo (118)	developmental changes (higher plants)	cell death (112)

When is light in excess? If we assume that photosynthetic carbon metabolism, both via the Calvin cycle and the photorespiratory C_2 pathway, provides the major sink for NADPH and ATP generated on the photosynthetic membrane, we can predict utilization of quanta via well established biochemical models of steady-state photosynthesis (32, 44, 45). Utilization of absorbed quanta with respect to photon flux (*I*) describes the familiar non-rectangular hyperbolic response of photosynthesis to light, while the quanta absorbed describe a linear response (Figure 1). At low photon fluxes (> 100 μmol m^{-2} s^{-1}) more than 80% of the absorbed quanta would be utilized, in accordance with the maximum measured quantum yields of O_2 evolution (14). When *I* approaches half of full sunlight (~ 1000 μmol m^{-2} s^{-1}) as little as 25% of the absorbed quanta are utilized, and at full sunlight, utilization drops to approximately 10%. Under sub-optimal conditions (cf 10°C, Figure 1), the gap between absorption and utilization increases, illustrating the increased potential for over-excitation of the photosynthetic apparatus whenever environmental conditions suppress photosynthetic carbon metabolism. Chilling and freezing temperatures (80, 113), high temperatures (18), and nitrogen deficiency (60) (94) all strongly limit photosynthetic C-metabolism and have been shown to

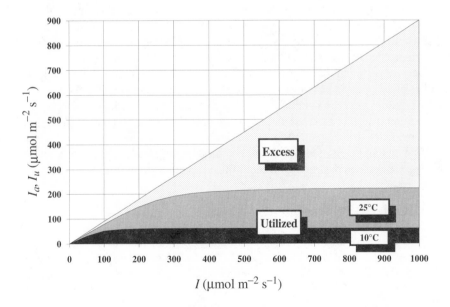

Figure 1   The predicted response of absorption (I_a), utilization (I_u) and excess of quanta ($I_a - I_u$) to incident photon flux (*I*). A leaf absorptance of 0.9 is assumed and utilization is calculated for an idealized and healthy C_3 leaf at 25°C and 10°C using a steady-state model of photosynthetic carbon metabolism (45, 78, 97).

intensify photoinhibition.

What is photoinhibition? We consider it the light dependent and slowly reversible retardation of photosynthesis, independent of any developmental change. Functionally, the consequences of photoinhibition of photosynthesis are a reduction in the maximum quantum yields for CO_2 uptake (ϕ) and O_2 evolution (ϕ_O) (112), a decrease in the convexity (θ) of the photosynthetic light response curve (75), and, with prolonged exposure to excessive light, a decreased rate of light saturated photosynthesis (A_{sat}) (80, 113, 121). Decrease in ϕ and θ has been observed to precede decrease in A_{sat}, and may often occur without decrease in A_{sat} (59, 82, 121, 140). Some reviews have used photoinhibition to mean damage to photosystem II (PSII) and photoprotection to mean changes assumed to protect PSII against damage (29). Although such a distinction has theoretical appeal, the distinction between damage and protection is not always clear (96). The term photoinhibition of photosynthesis has been "used almost synonymously with damage to PSII" (29), but it is important to recognize that light impairment of PSII is only photoinhibition, by definition, if a decrease in the overall photosynthetic rate results.

Photoinhibition has received intermittent attention for almost a century (33, 64, 66, 67, 91, 129), but wider recognition and interest has only developed in the last two decades. Powles (112) and Neale (92) reviewed the many, largely observational studies of photoinhibition on terrestrial plants through 1984 and aquatic plants through 1987, respectively. Several recent reviews have examined the mechanisms that may underlie photoinhibition (4, 21, 25, 70, 72, 103) and protect against damage to PSII in both visible (29, 61) and UV (26) light. Despite the plethora of laboratory studies on photoinhibition, little attention has been given to its relevance to biomass production and the distribution and survival of plants in nature. This review evaluates the significance of photoinhibition in nature by examining 1. mechanisms and their relevance in nature, 2. evidence for and patterns of occurrence in natural systems and crops, 3. the potential and measured significance to production, and 4. evidence that these losses of potential carbon gain can influence plant distributions.

MECHANISMS

Mechanisms of photoinhibition (8, 21, 25, 29, 70) and the closely related topic of environmental regulation of PSII (9, 61, 72) have both been reviewed recently. Here we outline these mechanisms and their potential relevance in nature.

As early as 1956, Kok (66) proposed that the primary site of damage in photoinhibition was at the photosynthetic reaction center. Since then, decreases in ϕ and ϕ_O frequently have been correlated with decreases in the variable fluorescence of PSII (usually parameterized as F_v/F_m) in vivo (10, 28, 30, 95), which implies decreased photochemical conversion efficiency of PSII (70). Observations that the action spectrum for photoinhibition follows that of photosynthesis and that the initial symptom of photoinhibition is decline in PSII photochemical efficiency support the assumption that photoinhibition results from energy absorbed by the photosynthetic pigments and channeled to PSII. The detailed knowledge of PSII structure, availability of molecular probes for components of the PSII complex, and wide recognition of the central role of PSII in photoinhibition have encouraged a major research effort on the mechanisms of both light impairment and protection of PSII (for review see 8, 29). Figure 2 is a simplified view of these mechanisms. A photoinhibitory decrease in ϕ through a decreased photochemical efficiency at PSII may be effected by two broad processes: (*a*) a cycle of PSII reaction center inactivation and repair; and (*b*) avoidance of over-excitation of the PSII reaction center by decreased absorption or internally by increased thermal dissipation of excitation energy, particularly in association with xanthophyll cycle activity.

Photosystem II Inactivation, Damage and Recovery: The D1 Story

Early research into mechanisms of photoinhibition noted an associated loss of the D1 32kDa-PSII reaction center polypeptide (74, 99), blockage of recovery by the protein synthesis inhibitor chloramphenicol (CAP) (52), and protection by herbicides that attach to the Q_B-binding site of the D1 polypeptide (73). Subsequent studies associated damage and loss of D1 with slowly reversible photoinhibition, and there is now consensus that when photoinhibition involves damage, D1 is the first site to be affected (105). Among the chloroplast-encoded polypeptides, D1 shows the most rapid turnover, even in low light (73), with an estimated $t_{1/2}$ of approximately 2 h (134). Photoinhibitory loss of D1 should only inhibit photosynthesis if the rate of repair fails to keep pace with the rate of damage. However, kinetic studies have shown a strong inhibition of electron transport during photoinhibition at low temperatures or under anaerobic conditions without loss of D1. If these thylakoids were transferred to room temperature in darkness, degradation of D1 occurred but with no further loss of electron transport capacity (7). These results imply that inactivation must precede any loss of D1; degradation may be arrested allowing an accumulation of inactivated centers; and initially D1 loss follows inactivation of electron transport and not vice versa (23). This may be explained by the cycle of inactivation, damage, and repair of PSII, as currently deduced from in vivo studies. Aro et al (8) propose a complex cycle involving more than 20 possible stages. Key stages are outlined in Figure 2: 1. Over-excitation may reversibly inactivate the center ($PSII_I$). 2. Non-functional centers ($PSII_{NF}$) with a damaged D1 may form from reversibly inactivated centers or directly from the pool of functional complexes ($PSII_F$). 3. Repair of $PSII_{NF}$

apparently requires migration of the complex into the stromal non-appressed thylakoids, where the damaged D1 is removed by a temperature sensitive serine protease (6). 4. The resulting $PSII_{-D1}$ is restored by insertion of a newly synthesized D1 polypeptide and then migrates back to the appressed thylakoids regaining photochemical competence. The extent of photoinhibition will not only depend on the rate of D1 damage, but also on the rates of migration of the damaged and repaired PSII within the membrane, degradation of the damage D1, and synthesis of new D1. All these factors can contribute to the extent and rate of recovery from photoinhibition and may explain part of the interaction between photoinhibition and other environmental variables, particularly with respect to recovery where replacement of D1 is necessary to restore the pool of $PSII_F$.

Why does damage appear to center on D1? A heterodimer of the D1 polypeptide and related D2 polypeptide are now considered to form the PSII reaction center (8). D1 has both donor and acceptor side roles in electron transport through PSII, with the binding sites for the pheophytin primary acceptor of electrons from the P_{680} reaction center molecule and for the Q_B quinone binding site, which accepts electrons for protonation of the plastoqui-

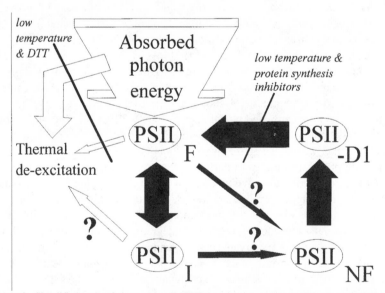

Figure 2   Simplified cycle of photosystem II (PSII) inactivation and repair. The forms of PSII are functional ($PSII_F$), reversibly inactivated ($PSII_I$), irreversibly inactivated or non-functional ($PSII_{NF}$), and D1 removed ($PSII_{-D1}$). Synthesis of D1 is blocked by inhibitors of chloroplast protein synthesis and by low temperature. Excitation of PSII is decreased by thermal de-excitation of both the PSII antenna and center. The component of de-excitation attributed to xanthophyll cycle activity may be inhibited by low temperatures and by the action of dithiothreitol (DTT).

none pool. The proximity of D1 to powerful oxidants with potential to form highly reactive radicals may explain the vulnerability suggested by the high turnover of the polypeptide.

Donor side and acceptor side photoinactivation have been demonstrated in vitro (13). Double reduction of the quinone acceptor Q_A leads to the formation of P_{680} in the triplet excited state ($^3Chl^*$), which can react with O_2 to form the highly toxic singlet oxygen radical ($^1O_2$), apparently resulting in acceptor-side inhibition of PSII electron transport. Combined fast fluorescence and EPR spectroscopy of acceptor-side photoinactivation under anaerobic conditions indicates that this oxidative damage may be preceded by the formation of a series of reversibly inactivated forms ($PSII_I$), leading to a final photoinactivated state, which is the first irreversible step and which may include or be immediately followed by D1 damage ($PSII_{NF}$ see Figure 2). On addition of O_2, thylakoids inactivated under anaerobic conditions show rapid consumption of $^3Chl^*$ with a concurrent loss of D1, indicating a critical role of O_2 in the conversion of $PSII_I$ to $PSII_{NF}$ (8). Inhibition of electron donation from the water splitting complex to the P_{680} reaction center prolongs the lifetime of the highly-oxidizing P_{680}^+ species, which is thought to cause donor-side inhibition, inactivating PSII electron transport and causing polypeptide damage (8, 13). Recent evidence now indicates that there may be a separate reversible donor side inhibition of PSII related to reversible Ca^{2+} exchange at low pH (142). It seems likely that both donor side and acceptor side inactivation are involved in photoinhibition in nature (8).

Although photoinhibition has, for nearly a decade, been almost synonymous with D1 loss, it is now apparent, both from inhibitor and protein studies in vivo, that inactivation and recovery of PSII can occur without the involvement of D1 degradation and replacement. In spinach (*Spinacia oleracea*) grown and studied at 25°C photoinhibition was exacerbated by the protein synthesis inhibitor streptomycin (SM), consistent with a requirement for D1 repair. SM had no effect on photoinhibition in plants grown and treated at 2–4°C, yet these plants were capable of full recovery. Based on in vitro studies, it was concluded that in vivo photoinhibition of PSII consists of two steps: 1. inactivation, in which D1 remains unaffected and from which fast reactivation is possible without D1 turnover; and 2. damage, requiring the removal of the altered D1 and replacement by newly synthesized D1 (71). Other studies of cold-hardened plants have noted substantial photoinhibition at low temperatures without loss or turnover of D1 (106, 107). In mangrove (*Avicennia marina*), photoinhibition of PSII at room temperature in vivo resulted in a marked inhibition of electron transport and O_2 evolution, with no detectable loss of D1 (21, 23).

Despite the large research effort in the biophysics, biochemistry, and molecular biology of D1 during photoinhibition, the relevance of D1 turnover to photoinhibition in nature remains uncertain. As Demmig-Adams & Adams (29) note, most of the studies demonstrating a concurrent decline in D1 with the development of photoinhibition have used treatments to which the plant material is not acclimated. For example, leaves, chloroplasts, or membranes were experimentally exposed to photon fluxes considerably greater than those in which they developed. Leaves of rape (*Brassica napus*) grown at a photon flux (*I*) of 750 µmol m^{-2} s^{-1} were capable of approximately double the rate of D1 polypeptide synthesis of leaves developed at 350 µmol m^{-2} s^{-1}. Further, D1 synthesis increased as the treatment photon flux increased, up to the maximum the leaves had experienced during their growth (134). One interpretation of these data is that capacity for D1 synthesis develops in response to light prehistory and that a significant decline in D1 would be avoided in plants acclimated to high light in the field. To date, we are unaware of any direct evidence of photoinhibitory loss of the D1 polypeptide of leaves in natural environments (9, 29). This may reflect the statistical difficulties of identifying such a change given the variability induced in the field by heterogeneous exposure to light.

Avoidance of PSII Damage: The Xanthophyll Cycle Story

Mechanisms that may avoid the over-excitation of PSII can be identified at several levels of organization (see Table 1) and may be divided between two broad categories: (*a*) external factors that physically decrease the absorption of light and (*b*) internal factors that increase the dissipation of excitation energy through processes other than photosynthetic C-metabolism.

In still waters unicellular algae may have the option of movement to avoid excess light by changing buoyancy or, in motile species, by vertical migration. Some higher plants, such as the succulent *Cotyledon orbicula,* can avoid excess photon flux by forming a highly reflective cuticle (118). Although this results in decreased light absorption at limiting photon fluxes, this effect may be inconsequential in habitats where light is rarely a limiting resource, for example, in hot semi-deserts.

Paraheliotropic movement maintains the leaf parallel to the direction of the sun, minimizing absorption of direct solar radiation. These same leaves often show a diaheliotropic response in the morning and evening (i.e. maintaining themselves perpendicular to the solar beam), and a paraheliotropic response around mid-day (65). This changing pattern of response to directional light maximizes interception when light is limiting and minimizes interception when light is excessive. This capacity apparently is limited to a few taxa of dicotyledonous plants (69). By artificially restraining leaf movement in the field, it has been shown that the paraheliotropic movement avoids photoinhibition in the sun species soybean (*Glycine max*) under conditions of water shortage (65) and in the shade species *Oxalis oregana* when exposed to sunflecks (114). These movements also minimize thermal radiation loads. An analysis of the thermal and photoinhibitory consequences of restraining leaf movement in *Vitis californica* indicated that avoidance of excessive leaf temperature rather than photoinhibition was the major benefit of paraheliotropic movement (47).

Chloroplast movements have long been recognized as a means of decreasing light interception by the photosynthetic apparatus and may reduce absorptance by as much as 20% (19). Nonphotosynthetic pigments in the epidermis may also prevent over-excitation. The chilling tolerant C_4 species *Spartina anglica* commonly shows an accumulation of anthocyanins in the leaf upper epidermis during the spring, when the combination of low early morning temperatures and clear skies would make it most vulnerable to chilling-dependent photoinhibition (77).

When an excess of light energy reaches the photosynthetic apparatus, over-excitation of PSII may be avoided by decreasing the probability of the energy of absorbed quanta reaching the reaction center. An increased probability of the absorbed quanta being dissipated as thermal energy or transferred to PSI is equivalent to a decrease in the effective absorption cross-section of PSII (σ_{PSII}). Functional decrease in σ_{PSII} may be effected by three inducible processes (see Table 1).

TRANSTHYLAKOID ΔpH The most rapid response to excess light is the increase in the rate constant for thermal energy dissipation from the PSII antenna and/or reaction center (k_{heat}) associated with the development of the transthylakoid ΔpH gradient (141). The mechanism of this affect of increasing ΔpH is uncertain, but may involve aggregation of LHCII particles, facilitating increased k_{heat} in the antenna (61). Alternatively, or in addition, the pH change may cause a reversible inactivation of the PSII reaction center involving loss of Ca^{2+} and increasing k_{heat} at the reaction center (142). Changes in the quantum efficiency of PSII have been associated with ΔpH, which afford some level of photoprotection (61). A 20% change in the quantum efficiency of fluorescence, ascribed to ΔpH, can occur at extremely low photon flux (40, 122); thus, the direct significance of ΔpH to the higher light levels in nature is unclear. It is assumed to act synergistically with the xanthophyll cycle, because the presence of zeaxanthin may act as a quenching amplifier, increasing k_{heat} and extending the significance of ΔpH to much higher photon fluxes (61, 93). The ΔpH-dependent increase in k_{heat} is related to the energy-dependent quenching of variable chlorophyll fluorescence (q_e), which forms and relaxes within seconds and is assumed to be the first response of the photosynthetic apparatus to excess light. Osmond (103) called this dynamic photoinhibition. Because it is present only when light is in excess, it is uncertain whether it affects potential carbon assimilation. It might only be of significance to C-uptake in rapidly fluctuating light (0.01–1 Hz), where any loss appears more than outweighed by the stimulation of photosynthesic efficiency, relative to constant light (51, 132; reviewed in 109).

STATE-TRANSITIONS   State-transitions in which the reversible phosphorylation of LHCII effectively increases transfer of excitation energy away from PSII may alter σ_{PSII} (17) by a maximum of about 25% using extreme conditions of phosphorylation in vitro. In the deep shade species *Monstera deliciosa*, state-transition provided some protection against photoinhibition in vitro (61) in the absence of other apparent mechanisms for decreasing σ_{PSII} (22). Although state transitions are interesting physiologically, significant relevance to photosynthetic C-uptake in nature has not been shown. In aquatic systems, red and far-red light (i.e. PSI-stimulating) is attenuated rapidly with depth (35). Cells near the surface could theoretically obtain sufficient stimulation of PSI to drive a state I transition, but such an effect would lead to an increase in σ_{PSII}, thereby further potentiating the photodamage. It is more likely that the state-transitions are driven by light intensity through the redox poise of the PQ pool.

XANTHOPHYLL CYCLE   A more profound change in σ_{PSII} can be inferred from changes in fluorescence yields and ϕ associated with the epoxidation state of xanthophyll cycle intermediates in both higher plants (28, 29) and algae (100, 101). The xanthophyll cycle in higher plants and green algae consists of conversion of the diepoxide violaxanthin to the epoxide-free zeaxanthin, via the monoepoxide antheroxanthin, when light absorption exceeds photochemical utilization. Epoxidation back to violaxanthin through antheroxanthin occurs in darkness or when light absorption is no longer in excess. In diatoms, dinoflagellates, and brown seaweeds, the xanthophyll cycle is limited to two intermediates, the monoepoxide diadinoxanthin (DD), which is converted to its de-epoxide, diatoxanthin (DT), when light is in excess and epoxidated back to DD in darkness or when light is no longer excessive. The epoxidation state (EPS) is expressed as the number of moles of epoxidated sites per mole of total xanthophyll cycle intermediates (29). Thus, EPS declines in excessive light and recovers in darkness and low light. The first order rate constant for DD/DT conversion is approximately 40 h^{-1}, which is about 10-fold greater than higher plant xanthophyll cycle rates (100, 101). Quantitative analysis of the relation between DT formation and a reduction in σ_{PSII} indicates that the cycle can produce approximately a 50% change in σ_{PSII}, which is significant, but not sufficient to reduce over-excitation of the reaction center under full sunlight (49, 100).

A strong correlation between decrease in EPS, σ_{PSII}, F_v/F_m, and ϕ, with an inferred increase in k_{heat}, has been demonstrated widely (29), but with some clear exceptions (e.g. 119). Dithiothreitol (DTT), a de-epoxidation inhibitor, promotes photoinhibitory damage to D1, providing further support of a role for the xanthophyll cycle in avoidance of over-excitation of PSII. This interpretation assumes, however, that DTT application has no other effects of significance. A theoretical mechanism of resonance coupling between xanthophylls and excited singlet chlorophylls ($^1Chl^*$), which would increase k_{heat}, has been proposed recently (108). Experimental evidence of the forbidden states necessary for this increased thermal deactivation in the pigment bed has been obtained (H Franks, manuscript in preparation).

Interaction between ΔpH and EPS provides a link between the rapid increase in k_{heat} on immediate exposure to photon flux (Figure 1) and the slower increase in k_{heat} associated with decrease in EPS on longer exposures to excess photon flux. The EPS-dependent increase in k_{heat} and resulting decrease in ϕ, which persists for some time after return to lower light levels (30), cause a loss of potential C-uptake and thus may be considered photoinhibitory.

An alternative mechanism for protecting PSII when absorption of quanta exceeds the requirements of photosynthetic carbon metabolism is an increase in the rates of electron transport processes. In supra-optimal irradiances there is the potential for cyclic electron flow around PSII, in which photochemically reduced Q_A^- can re-reduce the donor side of PSII. This pathway appears to be potentiated by the reduction of plastoquinol and can dissipate approximately 10–15% of the surplus electrons (38). The Mehler reaction of PSI may be coupled to super-oxide dismutases both in the thylakoid membrane and stroma, allowing the rapid conversion of O_2^- to H_2O_2, which can then react with ascorbate, via ascorbate peroxidase, to form the mono-dehydroascorbate radical (MDHA). In turn, MDHA may be either directly re-reduced with NADPH or metabolized further to dehydroascorbate and re-reduced by glutathione, followed by NADPH, completing a cycle for dissipation of both oxidizing radicals and energy (62, 115). Photosynthetic nitrogen metabolism might also consume up to about one tenth of the number of quanta utilized in photosynthetic C-metabolism (31, 58). Although the contribution of these reactions is small, relative to photosynthetic C-metabolism, it may be critical in allowing the establishment of the transthylakoid ΔpH, which is presumed to be a prerequisite for the thermal energy dissipation associated with the xanthophyll cycle (61).

Damage and Avoidance: A Blurring of the Division

To what extent can the functional changes in photosynthetic O_2 and CO_2 exchange be explained by current understanding of mechanisms? Two apparently ubiquitous mechanisms would both explain the decreases in ϕ and ϕ_O that are observed in photoinhibited plants in vivo.

Demmig-Adams & Adams (29) reported that for leaves that had developed in photon fluxes equivalent to those in the field, recovery from photoinhibition only became sensitive to the D1 synthesis inhibitor CAP if they were first treated with DTT. This indicates that the reaction center is normally protected from over-excitation by the increased thermal energy dissipation associated with a decreased epoxidation state (EPS) of the xanthophyll cycle intermediates. In the sun-species kiwifruit (*Actinidia deliciosa*), however, CAP exacerbated photoinhibition and blocked recovery, despite the presence of significant xanthophyll cycle activity (54).

In summary, there is widespread evidence for the occurrence of the xanthophyll cycle and a high capacity for xanthophyll cycle activity in terrestrial plants of exposed habitats (1–3, 30). This can now be coupled with a biophysical explanation of the increase in thermal energy dissipation associated with decreased EPS. It therefore seems likely that the xanthophyll cycle is a, or the, major initial mechanism of decrease in PSII efficiency and ϕ in sun leaves, and a first line of defense against PSII over-excitation during absorption of light in excess of photochemical utilization (29). Evidence of increased capacity for both D1 synthesis and xanthophyll cycle activity during acclimation to high light, indicates that both are probably critical to surviving excess light.

Shade plants appear to have a much lower capacity for xanthophyll cycle–related thermal energy dissipation (29) and for D1 repair (102, 138). This observation is paralleled by the situation at low temperatures, where enzyme activities within both the xanthophyll and D1-repair cycles will be limited (55, 106). In these situations, photoinhibited reaction centers are thought to accumulate in the thylakoids. Inactivated reaction centers (Figure 2) can be potent quenchers, converting absorbed excitation energy to heat (70). Based on this conclusion, Greer et al (55) proposed that in photoinhibition, the remaining fraction of active reaction centers are protected from damage by the accumulated fraction of inactive centers. This is supported by observations of a complete loss of photosynthetic activity without D1 damage in the evergreen leaves of Scots Pine (*Pinus sylvestris*) at sub-zero temperatures and their subsequent recovery on warming in the spring (107). When protective mechanisms in the antenna and repair processes in the PSII complex are impaired by extreme conditions, or in shade species where their capacity is low, accumulation of inactive PSII centers may provide an alternative route for increased thermal energy dissipation (Figure 2). Thus, what may in the past have been viewed as damage to the PSII reaction center might now be seen as avoidance through inactivation, blurring the distinction between avoidance/protection and damage in photoinhibition.

PHOTOINHIBITION IN THE FIELD AND OPEN OCEAN

Terrestrial Vegetation

Photon fluxes equivalent to about one half of full sunlight frequently produce a persistent photoinhibition when applied in controlled environments in combination with conditions that would limit photosynthetic C-metabolism (i.e. at low temperatures, in particular) (112). Does this occur in vegetation in the field? Controlled environment studies suggested that corn (*Zea mays*), in common with many crops of tropical origin, would be subject to photoinhibition during the cool conditions that frequently coincide with the crops' emergence in temperate regions (53, 80, 112). Weather records indicated that in western Europe, the crop would be at greatest risk shortly after emergence, when periods with low morning air temperatures and clear skies were frequent. To test this hypothesis, shades were positioned by randomized design each night so that individual plants within a crop in eastern England would be shaded from direct sunlight during the morning. At mid-morning the ϕ was determined using an integrating sphere chamber to separate any direct effect of photoinhibition on ϕ from any change in the light absorptance of the leaf. During the first 60 days following crop emergence, statistically significant reductions of up to 45% in ϕ were observed, whenever clear-sky conditions coincided with morning temperatures of approximately 10°C or lower. Recovery typically required 2–3 days of dull warm weather (41). The same approach was used to study a winter rape (*Brassica napus*) crop (42). Significant photoinhibitory reductions in ϕ were observed on a majority of winter days. From January to March, a 17% average reduction in ϕ could be attributed directly to photoinhibition. The largest photoinhibitory reductions in ϕ (~ 60%) occurred when sunny days coincided with freezing temperatures (42). Similarly, shades used in combination with water stress treatments showed that water stress significantly intensified photoinhibition of ϕ_O in stands of grain sorghum (*Sorghum bicolor*) (85). Artificial shades and natural shading have also been used to demonstrate a freezing dependent photoinhibition of F_v/F_m in *Pinus sylvestris* seedlings (133) and trees (107), respectively. Natural shading was also used to infer photoinhibition of ϕ_O and F_v/F_m during freezing in natural stands of snow gum (*Eucalyptus pauciflora*) seedlings (12) and holly (*Ilex aquifolium*) trees (57).

Because of the time required to complete measurements of ϕ (83) and ϕ_O (14), direct demonstration of a significant reduction in capacity for CO_2 or O_2 exchange is only practical if recovery is slow. Slowly recovering depressions of ϕ following photoinhibition at low temperatures may be detected readily by gas exchange techniques (41, 42, 81), but it is difficult to obtain statistical evidence of more rapidly reversible forms of photoinhibition acting in the field (30). Although potentially more ambiguous in interpretation than gas exchange measurements, the rapidity with which variable fluorescence may be measured with field-portable (15, 123) and submersible (68) fluorometers has been vital to the identification of more transient forms of photoinhibition. Application of fluorescence has shown that even in vegetation in near-optimal conditions, exposure to full sunlight can result in a mid-day or early afternoon slowly-reversible ($\geq$ 10 min) decrease in F_v/F_m of approximately 20–60%. Parallel reductions in ϕ_O confirm the significance of these findings. Recovery occurs in late afternoon and is often complete by dusk. This regular mid-day depression in PSII efficiency appears to occur whenever leaves are exposed to direct sunlight for most of the day. It has been demonstrated in vegetation ranging from desert CAM plants (1) and shrubs in the Mediterranean (30) to water replete stands of willow in northern Sweden (95) and a range of crops (16). A similar pattern of mid-day depression and recovery has even been observed in cold tolerant winter wheat cultivars at low temperatures in the field (56). This diurnal photoinhibition and recovery has been frequently correlated with increased zeaxanthin formation and increased thermal energy dissipation from PSII, as the assumed cause of depression of F_v/F_m and ϕ_O (29, 72). However, the correlation is not perfect, and in some species, recovery of F_v/F_m has required significantly more time than does recovery of EPS, indicating that xanthophyll cycle activity is only part of the cause of this decline (119).

Phytoplankton

In contrast to the situation in higher plants in nature, the potential for (63), and occurrence of, photoinhibition in phytoplankton has been known for some time. After the introduction of radiocarbon to measure phytoplankton photosynthetic rates in 1952 (130), there was an attempt to relate photoinhibitory responses in algal cultures to repression of primary production in the integrated water column of the ocean (120). The problem is confounded by vertical mixing and calculation of the exposure of cells to supraoptimal irradiance. Unlike terrestrial plants, phytoplankton are not fixed in space, and are moved continuously between the surface and the base of the upper mixed layer. Incubation of cells with radiocarbon requires enclosure of water samples in a bottle that will prevent natural mixing. This limitation has been partially overcome by simulating mixing in some measurements (87). Neither of these approaches allows accurate estimation of photoinhibition in the water column. Although it is possible to deduce the rate of mixing from the light history of the phytoplankton (35), photoinhibition does not obey simple rules of reciprocity, in which the effect can be calculated from the product of irradiance and duration of exposure (136). However, there are conditions that constrain the mixing process, and therefore allow experimental investigation of the susceptibility of natural phytoplankton communities to photoinhibition.

In the central subtropical oceans a diurnal thermocline of 2–5 m depth forms during the day and disperses at night. Cells trapped in the diurnal thermocline have no refuge from full sunlight for the day, and all measures of photosynthesis infer marked photoinhibition. For example, early workers found a mid-day suppression of phytoplankton carbon fixation in central ocean gyres (84, 86, 135). Involvement of photoinhibition was confirmed by suppression of DCMU-enhanced fluorescence (89, 139) and with pump-and-probe fluorescence (48, 68). Both showed large diurnal reductions in F_v/F_m in the upper 5 m, and pump-and-probe fluorescence measurements also indicated a marked decrease in PSII electron transport. Unlike DCMU-enhanced fluorescence measurements, the pump and probe technique is nondestructive and noninvasive. It follows the change in the quantum yield of fluorescence induced by a weak probe flash preceding and succeeding an actinic pump flash (40). If the time between the pump and second probe flash is sufficiently short (< 100 s) and the flash is saturating, the quantum efficiency of photochemistry in PSII can be determined with great precision in the dark or under ambient irradiance.

In temperate and in more turbulent waters, the effective exposure of cells to photoinhibitory radiation is dependent on the rate of vertical mixing. The rate of mixing is highly variable and depends on the rate at which turbulent energy, supplied primarily by wind shearing at the surface, is dissipated in the water column. On calm days, cells may only move the equivalent of 0.1 euphotic zone depths up or down. On windy days, the entire euphotic zone can mix completely in one hour. Phytoplankton have evolved strategies for coping with the rapid and variable changes in spectral irradiance. For example, to cope with day-to-day variations in irradiance, phytoplankton can show developmental photoacclimation (39).

Vertical profiles of phytoplankton photosynthesis based on radiocarbon incorporation into acid-stable organic material usually shows suppression of chlorophyll specific rates near the surface (111, 126). Studies of the diurnal cycle of photosynthetic rates have established that assimilation is higher from sunrise to mid-day than from mid-day to sunset. Thus, there is a hysteresis in the light response curve of photosynthesis in natural phytoplankton communities, especially in cells nearer the surface. Although this diurnal pattern may be partly the result of a circadian rhythm, damage to PSII also appears to contribute. Measures of changes in variable fluorescence with a pump-and-probe technique show marked reduction in Fv/Fm and a concomitant increase in the prompt fluorescence rise (F_o). Time course measurements of the relaxation of F_o and the maximum fluorescence emission (F_m) quenching in natural phytoplankton communities indicate two components. There is a short-term change (10–20 min) in fluorescence quenching, which appears to be correlated with xanthophyll cycle activity (36), and a much longer quenching component, which is consistent with impairment of reaction center function or the reduction in the efficiency of energy transfer to PSII. The more slowly recovering component of variable fluorescence in the dark shows a time constant on the order of 2–6 h, is temperature dependent, and can be inhibited by CAP, indicating that plastid-encoded protein synthesis is required for repair (36).

Thus, in contrast to field evidence for higher plants, both xanthophyll cycle and D1 loss may contribute significantly to photoinhibition in situ.

Algae may occur across a greater range of light environments, with respect to mean photon flux, than in higher plants. As a result, they can provide clues about the differences between readily photoinhibitable and photoinhibition-proof cells. Cells living under continuous low irradiance are much more susceptible to photoinhibition. For example, in polar seas, a rich and diverse community of phytoplankton is trapped in an ice matrix at the ice-water interface. The light response curves of photosynthesis in ice algae show a loss of photosynthetic capacity at irradiances above 5–15 μmol m^{-2} s^{-1} (24). Moreover, the cells are genetically adapted to low temperature and will not grow above approximately 2°C. Thus, although thermal stress may potentiate photoinhibition, it is not possible to relieve the stress, even experimentally, by elevated temperature. Zooxanthellate corals are at the other extreme. All hermatypic (reef-forming) corals are a symbiotic association between an oxygenic photoautotrophic dinoflagellate and a cnidarian invertebrate—the algae live inside the animal cells. On shallow coral reefs, many species are exposed to extremely high solar radiance fluxes on a daily basis, and the algae are physiologically acclimated to the high photon fluxes. Light response curves of photosynthesis for these high-light adapted corals show no sign of photoinhibition, even at irradiance levels of 150% of maximum sunlight (37).

Concern about an increase in incident solar UV flux has stimulated experimental field research on the photoinhibitory effects of UV (see 26). Gelbstoff, the dissolved yellow complexes derived from soil humic substances and carried by rivers into the sea, causes rapid attenuation of UV. Rising UV-B fluxes at the surface are unlikely to affect aquatic photosynthesis in coastal waters, because of high Gelbstoff concentrations. In open ocean waters, however, UV has been implicated in photoinhibitory damage (125). Most photosynthetic incubations of natural phytoplankton are performed in glass containers, which screen UV-B. Subsequent use of UV transparent containers has led several investigators to conclude that UV-B can significantly increase the photoinhibitory effects of visible light (27, 127).

In summary, photoinhibition detected in nature might be of two overlapping types: 1. A diurnal decline and complete recovery of ϕ, called diurnal photoinhibition (96), is now recognized to occur widely. This probably involves an increased rate constant for thermal energy dissipation (k_{heat}) associated with increased xanthophyll de-epoxidation, although it may also include significant D1 loss in algae. 2. Stress-dependent photoinhibition results in larger and more persistent decreases in ϕ, requiring days for complete recovery in higher plants. This can involve significant inactivation of PSII and accumulation of PSII$_I$, which may increase k_{heat}, thus providing additional protection for the remaining $PSII_F$. Recovery, with the possible exception of photoinhibition at low temperature, appears to require D1 synthesis.

SIGNIFICANCE TO PRODUCTION: A MODELLING APPROACH

In terms of photosynthetic C-uptake, the predominant effect at the cell and leaf level is a decrease in the light-limited rate characterized by decrease in the maximum quantum efficiency (ϕ) . Because photoinhibition is most likely to occur when light is saturating, a decrease in ϕ will have little or no immediate impact, unless the decrease is so great that it raises the point of light saturation beyond the photon flux incident at that instant in time. If the light level incident on the photoinhibited tissue becomes limiting to photosynthesis before recovery is complete, then potential C-uptake will decrease. In mixed water columns, where a proportion of the phytoplankton population will have been photoinhibited and subsequently transferred to light-limiting conditions, a decrease in potential C-uptake appears inevitable. In plant canopies, where leaves are fixed in three dimensional space, the significance of a decrease in ϕ developed under light saturating conditions is less obvious. Low temperatures allow the development of persistent photoinhibition over prolonged periods. Thus, ϕ may be decreased throughout a crop canopy for days or weeks (43, 81). What significance will this have to the net carbon gain? The photosynthetic rate of individual leaves will vary over the course of a day following the diurnal pattern of light. Empirical studies have shown that the response of CO_2 uptake (A) to photon flux (I) can be described effectively for both unicellular algae and higher plants by a nonrectangular hyperbola (34, 75, 88, 137). The constants of this relationship are the initial slope (maximum quantum efficiency, ϕ), asymptote (light saturated rate of CO_2 uptake, A_{sat}), and convexity (rate of bending toward the asymptote, θ):

$$A = \frac{\phi I + A_{sat} - \sqrt{(\phi I + A_{sat})^2 - 4\theta\,(\phi\,IA_{sat})}}{2\theta} \qquad 1.$$

At high values of θ ($\rightarrow 1$) there is an abrupt transition in the influence of ϕ and A_{sat}, with respect to I, while at low values of θ ($\rightarrow 0$) this transition may extend across the full range of I. Leverenz et al (75) have proposed that the heterogeneity in PSII complexes following photoinhibition will lead to a decrease in θ coupled to the decrease in ϕ. Such a coupled decrease θ and ϕ in photoinhibition occurs in leaves (76, 98) and in monolayers of algal cells, demonstrating that the effect is not solely the result of heterogeneity in cell layers (75). A decline in θ coupled to a decrease in ϕ in photoinhibition is significant because it will extend the influence of a decrease in ϕ to higher photon fluxes, as illustrated by the following example. By using Equation 1, with an A_{sat} of 25 μmol m^{-2} s^{-1}, it can be shown that a 50% decrease in φ (from 0.055 to 0.028) would cause a negligible 2% decrease in A at full sunlight (2000 μmol m^{-2} s^{-1}). If this decrease in φ is coupled with a 10% decrease in θ (from 0.095 to 0.0855), then A at full sunlight would decrease by 26%.

How significant are changes in φ and θ to crop canopies? Several widely validated models for predicting light within crop canopies have been developed. Forseth & Norman (46) provide a simple model to describe the heterogeneity of lighting in plant canopies. Given leaf area index (F), the angular distribution of leaves, and the amounts and angular distribution of light, this model predicts the dynamically changing proportions of the total leaf area index that are sunlit (F_{sun}) and shaded (F_{shade}), and the photon fluxes incident on each of these leaf categories (I_{sun}, I_{shade}). Using standard equations, based on solar angle and atmospheric transmissivity at a given latitude and time of day, the downward flux of light and its angular distribution can be computed (20). Combining this information with the canopy lighting model allows the calculation of photon flux on leaves within a canopy (I_{shade}, I_{sun}) throughout a 24 h period (46). In turn, by substituting the right-hand side of Equation 1 for $f(I)$ below (79):

$$A_C = \int_{t=0}^{t=24} [\, f(I_{sun,t}) \cdot F_{sun,t} + f(I_{shade,t}) \cdot F_{shade,t}\,] \qquad \qquad 2.$$

we can now assess the influence of photoinhibitory decreases in φ and θ at the leaf level on the daily integral of canopy photosynthesis (A_c). For these simulations, an A_{sat} of 25 μmol m^{-2} s^{-1}, φ of 0.055, and θ of 0.95 were assumed as typical values for healthy, nonphotoinhibited C$_3$ leaves (75, 78, 79). The simulations were for a clear sky day in June in southern England (Lat. 52°N). From Equation 2, a 50% decrease in φ decreases A_c by just 28% at a leaf area index (F) of 1, but the decrease becomes 42% for an F of 10 (Figure 3a). The effect of photoinhibition becomes much greater at low F if a simulated 50% decrease in φ is combined with a 25% decrease in θ. At an F of 1, the decrease in A_c is extended from 28% to 38% (Figure 3a).

To have significance for whole crop carbon assimilation, photoinhibitory decreases in φ must decrease the efficiency of utilization of the light absorbed by a canopy over the course of a day. This relationship is complex because φ will have the least influence over periods of the day and in portions of the canopy where light is saturating. The relationship between φ at the leaf level and light use efficiency by the canopy may be visualized by modeling. Based on the principles explained in deriving Equation 3, the integral of photon flux absorbed by a canopy over a day (I_c) may be determined (46) and an efficiency ($\varepsilon_c = A_c/I_c$) calculated. Simulating two canopies of $F = 3$ in January and

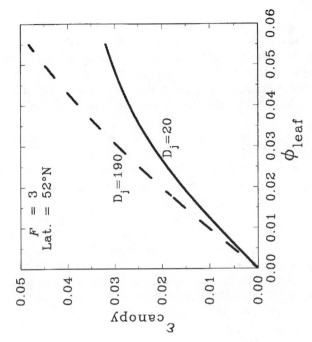

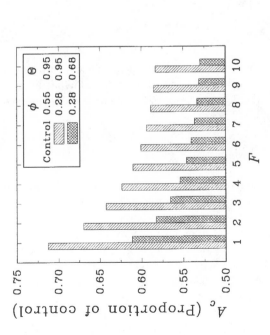

Figure 3a Simulated daily integral of photosynthetic carbon gain (A_c) for canopies of varying leaf area index (F) and suffering photoinhibition, as a proportion of carbon gain in control canopies. The effects of a 50% decrease in leaf quantum yield (ϕ) on A_c with and without a 25% decrease in convexity (θ) are illustrated. Simulations assume an A_{sat} of 25 μmol m^{-2} s^{-1}, a spherical distribution of leaf angles, and the diurnal course of direct and diffuse solar radiation for a clear sky day in June ($D_j = 190$) in eastern England (78).

Figure 3b Simulated response of the daily average quantum efficiency of canopy carbon uptake (ε_{canopy}) to variation in leaf quantum efficiency (ϕ). Parameters are as given in Figure 3a, except that leaf area index (ϕ_{leaf}) was set at 3 and A_{sat} was reduced to 10 μmol m^{-2} s^{-1} for the January day ($D_j = 20$).

in June in southern England, Figure 3b illustrates how variation in ϕ canopy would influence ε_c. Because of the strong dependence of A_{sat} on temperature, an A_{sat} of 10 μmol m^{-2} s^{-1} was assumed for the January simulation, compared to 25 μmol m^{-2} s^{-1} for June. On both days a partial dependence of ε_c on ϕ is clear, approaching a linear relationship at low ϕ, without simulating any coupled decrease in θ, which would increase the linearity of this relationship further.

So far we have assumed that decreases in ϕ effected by photoinhibition apply to the canopy as a whole. This may be applicable to crops at low temperatures, where recovery may be slow or absent. It is also relevant to some phytoplankton communities where mixing may continually transfer cells photoinhibited at the surface to lower levels in the water column. As we have seen, there is now much evidence for the widespread occurrence of moderate diurnal photoinhibition, which lowers ϕ for only a few hours, and in the leaves which are exposed to full sunlight. What significance can this transient decrease in the outermost leaves have on the potential carbon gain of the canopy? Such diurnal photoinhibition must represent a balance between inhibition and recovery, so that the influence of the absorbed photon flux at one point in the day will diminish with time. This may be simulated by calculating a weighted average light dosage (I_{int}):

$$I_{int} = \sum_{i=1}^{i=24} \frac{I_i}{i} \qquad\qquad 3.$$

where I, is the mean photon flux (μmol m^{-2} s^{-1}) incident on a leaf during hour i. The influence of I_{int} on ϕ may then be empirically simulated by assuming a linear relationship between the decrease in ϕ, as a proportion of its assumed maximum (0.055), and I_{int}:

$$\phi = \phi_{max} \cdot [1 - I_{int}/1.2 \times 10^4] \qquad\qquad 4.$$

For a leaf exposed to a natural pattern of diurnal variation in diffuse and direct sunlight, with a mid-day peak of 1800 μmol m^{-2} s^{-1}, this would simulate an approximately 50% decline in ϕ by mid-afternoon with a recovery to approximately 90% of ϕ_{max} by dusk, mimicking patterns of transient photoinhibition reported for vegetation in the field (3, 30, 95). Again using a model of canopy light distribution (46), the photon flux incident on the outermost unit of leaf area index and I_{int} may be calculated (Equation 3), and in turn the diurnal depression of ϕ simulated (equation 4). Figure 4 simulates the diurnal course of canopy carbon uptake for a hypothetical canopy with a leaf area index (F) of 3 on a clear sky day in southern England (latitude = 52°N). With respect to equation 1, parameters again simulate a healthy C3 leaf, A_{sat} = 25

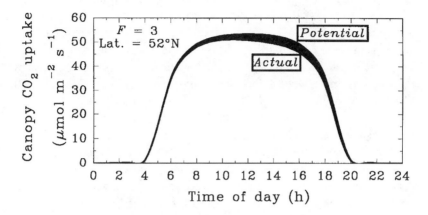

Figure 4   The simulated course of instantaneous canopy photosynthetic C-uptake over a day with clear sky conditions, assuming in the upper line no photoinhibition and in the lower a transient decrease in φ, simulating a diurnal photoinhibition of the outer canopy layer. Assumptions are the same as those used for the control in Figure 3a, except F equals 3.

$\mu mol\ m^{-2}\ s^{-1}$, and $\theta = 0.95$. The upper line simulates a constant φ (0.055) and the lower simulates the transient depression of φ in the outermost unit F of leaves. φ in the lower 2 units of F are assumed to be unaffected and maintain a maximum φ. This model predicts 9% loss of potential CO_2 uptake, predominantly in the early afternoon. An analysis based on photosynthetic and light microclimate measurements made on a willow (*Salix* sp.) stand in northern Sweden, indicated that transient photoinhibition decreased potential C gain by 10% (98). Although photoinhibition decreases capacity for light-limited photosynthesis at the leaf level, significant decreases in canopy photosynthetic C-uptake on sunny days will nevertheless result because of the continually changing pattern of lighting, which continually moves some leaves from high photoinhibiting light levels to low light levels, that limit photosynthesis, and vice versa.

OBSERVED CHANGES IN PRODUCTION

Given the frequent observations of significant photoinhibition of leaves in the field and the extension of these observations, through mathematical models, to decreased photosynthetic carbon gain, it follows that there should be a detectable decrease in dry matter production. Decrease in φ at the leaf level was predicted mathematically to decrease the efficiency of crop CO_2 uptake (ε_c) and, assuming that respiration was not increased under these conditions, dry matter production. The conversion efficiency of dry matter production from intercepted light (ε'_c) over a given interval is estimated from the ratio of the measured increments in biomass and solar radiation interception by the crop. Assuming a constant C-content of the dry matter and ratio of respiration to carbon gain, the predicted decreases in e_c should be apparent in measured ε'_c. The average leaf ϕ and crop ε'_c for a field of winter rape (*B. napus*) in eastern England was measured on 19 occasions from the late fall to early spring (42). The ϕ varied from 0.052 to 0.020 over this period. Parallel shading experiments showed that the majority of this decline was attributable to photoinhibition. ε'_c showed a strong positive correlation with ϕ ($r^2 = 0.90$) and a significant linear dependence on ϕ ($\varepsilon'_c = -3.31 + 134\phi$) (42). These data provide strong evidence for the expected decrease in efficiency of light use in dry matter production with photoinhibitory decrease in ϕ (Figure 3b). The leaf area index of the *B. napus* crop exceeded 3 and varied little throughout the period of measurement. Thus, change in the measured ε'_c has to reflect change in utilization of the intercepted radiation rather than any change that might be attributed to variation in light interception. A similar analysis in a corn (*Z. mays*) crop in eastern England also showed a strong linear dependence of ε'_c on ϕ during the period when chilling-dependent photoinhibition was frequent (81). This dependence was slight in a subsequent year with mild spring temperatures and little evidence of photoinhibition (131).

The quantitative influence of photoinhibition on production in open ocean phytoplankton communities is controversial and is complicated by the problem of assessing the rate of vertical mixing in situ (111). Although the most robust estimates of production are based on radiocarbon incorporation, such measurements require an incubation period (typically several hours) in an enclosure. During incubation, cells are artificially trapped and, although carbon fixation usually decreases in the upper water column as a result of photoinhibition, it is unclear whether the effect would occur in situ if the cells were allowed to mix vertically (87). In situ pump-and-probe fluorescence data indicate that the mid-day reduction in photosynthetic energy conversion efficiency occurs in the face of turbulent mixing, and reduces ϕ by as much as 25%. Integration of this loss over the water column indicates that photoinhibition can reduce carbon fixation from between 5 to 20%, depending on mixing rates (R Greene, manuscript in preparation). Although such a reduction may appear to represent a significant decrease in organic carbon flux, it is unclear whether phytoplankton biomass would be larger even if all cells were photoinhibition-proof. Because of the limitation of essential plant nutrients, especially nitrogen, and highly constrained carbon/nitrogen ratios in phytoplankton (6:1 by moles), any increases in carbon fixation would require a concomitant increase in the flux of inorganic nutrients.

PHOTOINHIBITION AND PLANT DISTRIBUTIONS

Although diurnal photoinhibition allows full recovery within a few hours, the recovery is apparently too slow to avoid significant losses of potential carbon gain. The expected losses (~ 10%) would clearly be disadvantageous in some competitive situations, but could these losses be avoided? Decreased C-gain resulting from light limitation often results in a disproportionate decrease in allocation toward reproduction (110). If the carbon gained by being photoinhibition-proof was allocated to reproduction, considerable selective pressure against photoinhibition might be expected. Results of a cost-benefit analysis indicate that the energetic costs of both repair and protective mechanisms are small relative to the potential losses through photoinhibition (116, 117). The widespread occurrence of the xanthophyll cycle in sun species, relative to shade, indicates strong selective pressure for mechanisms of avoiding damage to PSII at the cost of a decrease in potential carbon gain. Might there, however, be circumstances where not "down-regulating" and gaining this extra carbon could be advantageous? This might be possible in species with an increased capacity for repair. The oxygenic photoautotrophic dinoflagellates of corals, which are not photoinhibited even in a photon flux of 150% of full sunlight (37), show that photoinhibition-proof plants do exist, so this strategy is possible. Analyses of genetic variation within a single species or species complex show considerable variation in apparent susceptibility to photoinhibition (81, 82, 124). In a survey of *Zea* spp. genotypes, collections from high altitude sites, where the potential for chilling-dependent photoinhibition was probably greatest, had the highest resistance to photoinhibition, i.e. the smallest decrease in ϕ (81, 82). A collection from the northern edge of the range of the C_4 grass *Cyperus longus* in Europe was similarly found to be less susceptible to photoinhibition than were populations from further south. This lower susceptibility corresponded to a higher rate of recovery (50), indicating an increased capacity for repair.

Two studies of tree regeneration provide evidence that lost potential production resulting from photoinhibition is important for survival. Ball (11, 12) has shown an apparent link between early growth and low temperature–dependent photoinhibition in tree seedlings. Seedlings of both the snow gum (*Eucalyptus pauciflora*) and the mountain beech (*Nothofagus solandri*), in their native environments, show a distribution that reflects the compromise between (*a*) obtaining sufficient light to sustain growth and (*b*) protection from high photon fluxes during periods of low temperature. In sheltered positions, on the south shaded side of mature trees, over-wintering seedlings maintained a higher ϕ_O and a higher F_v/F_m than did seedlings on the exposed northern side, even though both gained equal protection from freezing by the frost shadow cast by the overlying tree canopy. In spring, the winter-shaded seedlings showed the most rapid growth, even though they received less light. Population size and mean seedling mass was negatively correlated with the potential for winter photoinhibition in the different microsites. These results provide the clearest evidence that photoinhibition, and the limitation it can have on production, also imposes a limitation on fitness and survival.

CONCLUSIONS

Despite a plethora of laboratory studies of the mechanism of photoinhibition, much uncertainty remains about which mechanisms are significant to plants in nature. Overcoming this uncertainty has and will continue to depend on the development of techniques that can deal with the heterogeneity of photoinhibition in nature. Photoinhibition has been demonstrated in situ in both crop and natural higher plant canopies, as well as in phytoplankton communities. Where stress factors limit photosynthetic carbon metabolism, persistent decreases in photosynthetic efficiency have been observed that may continue for days or even months in evergreen leaves. A milder, diurnal photoinhibition also occurs in most vegetation that is exposed to direct sunlight. Modeling studies indicate a clear cost of both forms of photoinhibition in terms of potential carbon acquisition. For stress-induced photoinhibition, the cost is decreased efficiency of conversion of intercepted light into crop dry matter. Although it remains to be proven whether plants could use any of the additional carbon that would be gained if they were resistant to photoinhibition, studies of natural regeneration of tree seedlings provide strong evidence of situations in nature where photoinhibition is detrimental to fitness and survival.

ACKNOWLEDGMENTS

This review is based to a significant extent on research supported by grants awarded to the authors from the Agricultural and Food Research Council (United Kingdom) and the Department of Energy (United States). The preprints of papers provided by colleagues were much appreciated. Finally we thank NR Baker and JR Bowyer for inviting us to participate in the 41st Harden Conference, "Photoinhibition of Photosynthesis—Molecular Mechanisms to the Field." This meeting had an invaluable influence on our final draft.

Literature Cited

1. Adams WW III. 1988. Photosynthetic acclimation and photoinhibition of terrestrial and epiphytic CAM tissues growing in full sunlight and deep shade. *Aust. J. Plant Physiol.* 15:123–34
2. Adams WW III, Demmig-Adams B. 1992. Operation of the xanthophyll cycle in higher plants in response to diurnal changes in incident sunlight. *Planta* 186:390–98
3. Adams WW III, Díaz M, Winter K. 1989. Diurnal changes in photochemical efficiency, the reduction state of Q, radiationless energy dissipation, and non-photochemical fluorescence quenching in cacti exposed to natural sunlight in northern Venezuela. *Oecologia* 80:553–61
4. Anderson JM, Chow WS, Goodchild DJ. 1988. Thylakoid membrane organisation in sun/shade acclimation. *Aust. J. Plant Physiol.* 15:11–26
5. Anderson JM, Osmond CB. 1987. Shade-sun responses: compromises between acclimation and photoinhibition. See Ref. 74a, pp. 1–38
6. Andersson B, Salter HA, Virgin I, Vass I, Styring S. 1992. Photodamage to photosystem II—primary and secondary events. *J. Photochem. Photobiol.* 15:15–31
7. Aro E-M, Hundal T, Carlberg I, Andersson B. 1990. In vitro studies of light-induced inhibition of photosystem II and D1-protein degradation at low temperature. *Biochim. Biophys. Acta* 1019:269–75
8. Aro E-M, Virgin I, Andersson B. 1993. Photoinhibition of photosystem II. Inactivation, protein damage and turnover. *Biochim. Biophys. Acta* 1143:113–34
9. Baker NR. 1991. A possible role for photosystem II in environmental perturbations of photosynthesis. *Physiol. Plant.* 81:563–70
9a. Baker NR, Bowyer JR, eds. 1994. Photoinhibition of Photosynthesis—Molecular Mechanisms to the Field. Oxford: Bios. Sci. In press
10. Baker NR, Bradbury M, Farage PK, Ireland CR, Long SP. 1989. Measurements of quantum yield of carbon assimilation and chlorophyll fluorescence for assessment of photosynthetic performance of crop plants in the field. *Philos. Trans. R. Soc. London Ser. B* 323:295–308
11. Ball MC. 1993. The role of photoinhibition during tree seedling establishment at low temperature. See Ref. 9a. In press
12. Ball MC, Hodges VS, Laughlin GP. 1991. Cold-induced photoinhibition limits regeneration of snow gum at tree-line. *Func. Ecol.* 5:663–68
13. Barber J, Andersson B. 1992. Too much of a good thing: light can be bad for photosyn-

thesis. *Trends Biochem. Sci.* 17:61–66
14. Björkman O, Demmig B. 1987. Photon yield of O_2 evolution and chlorophyll fluorescence characteristics at 77K among vascular plants of diverse origins. *Planta* 170:489–504
15. Bolhár-Nordenkampf HR, Baker NR, Long SP, Öquist G, Schreiber U, Lechner EG. 1989. Chlorophyll fluorescence as a probe of the photosynthetic competence of leaves in the field: a review of current instrumentation. *Func. Ecol.* 3:497–514
16. Bolhár-Nordenkampf HR, Hofer M, Lechner EG. 1991. Analysis of light-induced reduction of the photochemical capacity in field-grown plants. Evidence for photoinhibition? *Photosynth. Res.* 27:31–39
17. Bonaventura C, Myers J. 1969. Fluorescence and oxygen evolution from *Chlorella pyrenoidosa*. *Biochim. Biophys. Acta* 189: 366–83
18. Bongi G, Long SP. 1987. Light-dependent damage to photosynthesis in olive leaves during chilling and high temperature stress. *Plant Cell Environ.* 10:241–49
19. Brugnoli E, Björkman O. 1992. Chloroplast movements in leaves: influence on chlorophyll fluorescence and measurements of light-induced absorbance changes related to pH and zeaxanthin formation. *Photosynth. Res.* 32:23–35
20. Campbell GS. 1977. *An Introduction to Environmental Biophysics*. New York: Springer-Verlag. 159 pp.
21. Cleland RE. 1988. Molecular events of photoinhibitory inactivation in the reaction centre of photosystem II. *Aust. J. Plant Physiol.* 15:135–50
22. Cleland RE, Demmig-Adams B, Adams WW III, Winter K. 1990. Phosphorylation state of the light-harvesting chlorophyll-protein complex of photosystem II and chlorophyll fluorescence characteristics in *Monstera deliciosa* Liebm. and *Glycine max* (L.) Merrill in response to light. *Aust. J. Plant Physiol.* 17:589–99
23. Cleland RE, Ramage RT, Critchley C. 1990. Photoinhibition causes loss of photochemical activity without degradation of D1 protein. *Aust. J. Plant Physiol.* 17:641–51
24. Cota GF. 1985. Photoadaptation of high Arctic ice algae. *Nature* 315:219–22
25. Critchley C. 1988. The molecular mechanism of photoinhibition—facts and fiction. *Aust. J. Plant Physiol.* 15:27–41
26. Cullen JJ, Neale PJ. 1994. Ultaviolet radiation, ozone depletion and global biogeochemical cycles. *Photosynth. Res.* In press
27. Cullen JJ, Neale PJ, Lesser MP. 1992. Bio-

658   LONG ET AL

logical weighting function for the inhibition of phytoplankton photosynthesis by ultraviolet radiation. *Science* 258:646–50
28. Demmig B, Winter K, Kruger A, Czygan F-C. 1987. Photoinhibition and zeaxanthin formation in intact leaves. A possible role of the xanthophyll cycle in the dissipation of excess light energy. *Plant Physiol.* 84:218–24
29. Demmig-Adams B, Adams WW III. 1992. Photoprotection and other responses of plants to high light stress. *Annu. Rev. Plant Physiol. Plant Mol. Biol.* 43:599–626
30. Demmig-Adams B, Adams WW III, Winter K. 1989. Photochemical efficiency of photosystem II, photon yield of O_2 evolution, photosynthetic capacity, and carotenoid composition during the midday depression of net CO_2 uptake in *Arbutus unedo* growing in Portugal. *Planta* 177:377–87
31. Edwards GE, Walker DA. 1983. *C_3, C_4 Mechanisms, Cellular and Environmental Regulation of Photosynthesis.* Oxford: Blackwell. 552 pp.
32. Evans JR, Farquhar GD. 1991. Modeling canopy photosynthesis from the biochemistry of the C_3 chloroplast. In *Modeling Crop Photosynthesis—from Biochemistry to Canopy,* ed. KJ Boote, RS Loomis, pp. 1–16. Madison, WI: Crop Science Soc. Am.
33. Ewart AJ. 1895. On assimilatory inhibition in plants. *J. Linn. Soc. Bot.* 31:364–461
34. Falk S, Leverenz JW, Samuelsson G, Öquist G. 1992. Changes in photosystem II fluorescence in *Chlamydomonas reinhardtii* exposed to increasing levels of irradiance in relationship to the photosynthetic response to light. *Photosynth. Res.* 31:31–40
35. Falkowski PG. 1983. Light-shade adaptation and vertical mixing of marine phytoplankton: a comparative field study. *J. Mar. Res.* 41:215–37
36. Falkowski PG. 1992. Molecular ecology of phytoplankton photosynthesis. In *Primary Productivity and Biogeochemical Cycles in the Sea,* ed. PG Falkowski, A Woodhead, pp. 47–67. New York: Plenum
37. Falkowski PG, Dubinsky Z. 1980. Light-shade adaptation of *Stylophora pistillata,* a hematypic coral from the Gulf of Eilat. *Nature* 289:172–74
38. Falkowski PG, Fujita Y, Ley AC, Mauzerall DC. 1986. Evidence for cyclic electron flow around photosystem II in *Chlorella pyrenoidosa. Plant Physiol.* 81:310–12
39. Falkowski PG, LaRoche J. 1991. Adaptation to spectral irradiance in unicellular algae. *J. Phycol.* 27:8–14
40. Falkowski PG, Wyman K, Ley AC, Mauzerall D. 1986. Relationship of steady state photosynthesis to fluorescence in eucaryotic algae. *Biochim. Biophys. Acta* 849:183–92

41. Farage PK, Long SP. 1987. Damage to maize photosynthesis in the field during periods when chilling is combined with high photon fluxes. In *Progress in Photosynthesis Research,* ed. J Biggins, 4:139–42. Dordrecht: Nijhoff
42. Farage PK, Long SP. 1991. The occurrence of photoinhibition in an over-wintering crop of oil-seed rape (*Brassica napus* L.) and its correlation with changes in crop growth. *Planta* 185:279–86
43. Farage PK, Long SP, Lechner EG, Baker NR. 1991. The sequence of change within the photosynthetic apparatus of wheat following short-term exposure to ozone. *Plant Physiol.* 95:529–35
44. Farquhar GD, von Caemmerer S. 1982. Modelling of photosynthetic responses to environmental conditions. In *Physiological Plant Ecology. II. Encyclopedia of Plant Physiology (NS),* ed. OL Lange, PS Nobel, CB Osmond, H Ziegler, pp. 548–77. Berlin: Springer-Verlag
45. Farquhar GD, von Caemmerer S, Berry JA. 1980. A biochemical model of photosynthetic CO_2 assimilation in leaves of C_3 species. *Planta* 149:78–90
46. Forseth IN, Norman JM. 1993. Modelling of solar irradiance, leaf energy budget and canopy photosynthesis. In *Photosynthesis and Productivity in a Changing Environment,* ed. DO Hall, JMO Scurlock, HR Bolhár-Nordenkampf, RC Leegood, SP Long, pp. 207–19. London: Chapman & Hall
47. Gamon JA, Pearcy RW. 1989. Leaf movements, stress avoidance and photosynthesis in *Vitis californica. Oecologia* 79:475–81
48. Geider RJ, Greene RM, Kolber Z, MacIntyre HL, Falkowski PG. 1993. Fluorescence assessment of the maximum quantum efficiency of photosynthesis in the western North Atlantic. *Deep-Sea Res.* 40:1205–24
49. Gilmore AM, Yamamoto HY. 1993. Linear models relating xanthophylls and lumen acidity to non-photochemical fluorescence quenching. *Photosynth. Res.* 35:67–78
50. Gravett A, Long SP. 1990. Intraspecific variation in susceptibility to photoinhibition during chilling of *Cyperus longus* L. populations from Europe. In *Current Research in Photosynthesis,* ed. M Baltscheffsky, 2:475–78. Dordrecht: Kluwer
51. Greene RM, Gerard VA. 1990. Effects of high-frequency light fluctuations on growth and photoacclimation of the red alga *Chondrus crispus. Mar. Biol.* 105:337–44
52. Greer DH, Berry JA, Björkman O. 1986. Photoinhibition of photosynthesis in intact bean leaves: role of light and temperature, and requirement for chloroplast-protein synthesis during recovery. *Planta* 168:253–

60
53. Greer DH, Hardacre AK. 1989. Photoinhibition of photosynthesis and its recovery in two maize hybrids varying in low temperature tolerance. *Aust. J. Plant Physiol.* 16:189–98
54. Greer DH, Laing WA, Woolley DJ. 1993. The effect of chloramphenicol on photoinhibition of photosynthesis and its recovery in intact kiwifruit (*Actinidia deliciosa*) leaves. *Aust. J. Plant Physiol.* 20:33–43
55. Greer DH, Ottander C, Öquist G. 1991. Photoinhibition and recovery of photosynthesis in intact barley levels at 5 and 20°C. *Physiol. Plant.* 81:203–10
56. Groome QJ, Baker NR, Long S. 1990. Photoinhibition of photosynthesis in a winter wheat crop. In *Current Research in Photosynthesis,* ed. M Baltscheffsky, 2:463–66. Dordrecht: Kluwer
57. Groome QJ, Baker NR, Long S. 1991. Photoinhibition of holly (*Ilex aquifolium*) in the field during the winter. *Physiol. Plant.* 83: 585–90
58. Guerrero MG, Vega JM, Losada M. 1981. The assimilatory nitrate-reducing system and its regulation. *Annu. Rev. Plant Physiol.* 32:169–202
59. Henley WJ, Levavasseur G, Franklin LA, Osmond CB, Ramus J. 1991. Photoacclimation and photoinhibition in *Ulva rotundata* as influenced by nitrogen availability. *Planta* 184:235–43
60. Herzig R, Falkowski PG. 1989. Nitrogen limitation of *Isochrysis galbana.* I. Photosynthetic energy conversion and growth efficiencies. *J. Phycol.* 25:462–71
61. Horton P, Ruban AV. 1992. Regulation of photosystem II. *Photosynth. Res.* 34:375–85
62. Jahnke LS, Hull MR, Long SP. 1991. Chilling stress and oxygen metabolizing enzymes in *Zea mays* and *Zea diploperennis.* *Plant Cell Environ.* 14:97–104
63. Jenkin PM. 1937. Oxygen production by the diatom *Coscinodiscus excentrucus* Ehr. in relation to submarine illumination in the English Channel. *J. Mar. Biol. Assoc. UK* 22:301–43
64. Jones LW, Kok B. 1966. Photoinhibition of chloroplast reactions. I. Kinetics and action spectra. *Plant Physiol.* 41:1037–43
65. Kao W-Y, Forseth IN. 1992. Diurnal leaf movement, chlorophyll fluorescence and carbon assimilation in soybean grown under different nitrogen and water availabilities. *Plant Cell Environ.* 15:703–10
66. Kok B. 1956. On the inhibition of photosynthesis by intense light. *Biochim. Biophys. Acta* 21:234–44
67. Kok B, Spruit CJP. 1957. Report on some recent results at Wageningen. In *Research in Photosynthesis,* ed. H Gaffron, AH Brown, CS French, pp. 353–65. New York: Interscience
68. Kolber Z, Wyman KD, Falkowski PG. 1990. Natural variability in photosynthetic energy conversion efficiency: a field study in the Gulf of Maine. *Limnol. Oceanogr.* 35:72–79
69. Koller D. 1990. Light-driven leaf movements. *Plant Cell Environ.* 13:615–32
70. Krause GH. 1988. Photoinhibition of photosynthesis. An evaluation of damaging and protective mechanisms. *Physiol. Plant.* 74: 566–74
71. Krause GH. 1994. Photoinhibition at low temperatures in vivo. See Ref. 9a. In press
72. Krause GH, Weis E. 1991. Chlorophyll fluorescence and photosynthesis: the basics. *Annu. Rev. Plant Physiol. Plant Mol. Biol.* 42:313–49
73. Kyle DJ. 1987. The biochemical basis of photoinhibition of photosystem II. See Ref. 74a, pp. 197–226
74. Kyle DJ, Ohad I, Arntzen CJ. 1984. Membrane protein damage and repair: selective loss of a quinone-protein function in chloroplast membranes. *Proc. Natl. Acad. Sci. USA* 81:4070–74
74a. Kyle DJ, Osmond CB, Arntzen CJ, eds. 1987. *Photoinhibition.* Amsterdam: Elsevier
75. Leverenz JW, Falk S, Pilström C-M, Samuelsson G. 1990. The effects of photoinhibition on the photosynthetic light-response curve of green plant cells (*Chlamydomonas reinhardtii*). *Planta* 182:161–68
76. Leverenz JW, Öquist G, Wingsle G. 1992. Photosynthesis and photoinhibition in leaves of chlorophyll *b*-less barley in relation to absorbed light. *Physiol. Plant.* 85:494–502
77. Long SP. 1983. C_4 photosynthesis at low temperatures. *Plant Cell Environ.* 6:345–63
78. Long SP. 1991. Modification of the response of photosynthetic productivity to rising temperature by atmospheric CO_2 concentrations: Has its importance been underestimated? *Plant Cell Environ.* 14:729–39
79. Long SP. 1993. The significance of light-limiting photosynthesis to crop canopy carbon gain and productivity—a theoretical analysis. In *Photosynthesis Photoreactions to Plant Productivity,* ed. YP Abrol, P Mohanty, Govindjee, pp. 548–59. New Delhi: Oxford Univ. Press & IBH
80. Long SP, East TM, Baker NR. 1983. Chilling damage to photosynthesis in young *Zea mays.* *J. Exp. Bot.* 34:177–88
81. Long SP, Farage PK, Aguilera C, Macharia JMN. 1992. Damage to photosynthesis during chilling and freezing, and its significance to the photosynthetic productivity of field crops. In *Trends in Photosynthesis Research,* ed. J Barber, MG Guerrero, H Medrano, pp. 345–56. Andover: Intercept

82. Long SP, Nugawela A, Bongi G, Farage PK. 1987. Chilling-dependent photoinhibition of photosynthetic CO_2 uptake. In *Progress in Photosynthesis Research*, ed. J Biggins, 4:131–38. Dordrecht: Nijhoff

83. Long SP, Postl WF, Bolhár-Nordenkampf HR. 1993. Quantum yields for uptake of carbon dioxide in C_3 vascular plants of contrasting habitats and taxonomic groupings. *Planta* 189:226–34

84. Lorenzen CJ. 1963. Diurnal variation in photosynthetic activity of natural phytoplankton populations. *Limnol. Oceanogr.* 8:56–62

85. Ludlow MM, Powles SB. 1988. Effects of photoinhibition induced by water stress on growth in yield of grain sorghum. *Aust. J. Plant Physiol.* 15:178–94

86. Malone TC. 1971. Diurnal rhythms in netplankton and nannoplankton assimilation numbers. *Mar. Biol.* 10:285–89

87. Marra J. 1978. Phytoplankton photosynthetic response to vertical movement in a mixed layer. *Mar. Biol.* 46:203–8

88. Marshall B, Biscoe PV. 1980. A model for C_3 leaves describing the dependence of net photosynthesis on irradiance. *J. Exp. Bot.* 120:29–39

89. Matorin DN, Vasilev IR, Verdernikov VI. 1992. Photoinhibition of primary photosynthetic reactions in natural phytoplankton populations of the Black Sea. *Sov. Plant Physiol.* 39:285–90

90. Mulkey SS, Pearcy RW. 1992. Interactions between acclimation and photoinhibition of photosynthesis of a tropical forest understory herb, *Alocasia macrorrhiza*, during simulated canopy gap formation. *Func. Ecol.* 6:719–29

91. Myers J, Burr G. 1940. Studies on photosynthesis. Some effects of light of high intensity on *Chlorella. J. Gen. Physiol.* 24: 45–67

92. Neale J. 1987. Algal photoinhibition and photosynthesis in the aquatic environment. See Ref. 74a, pp. 39–65

93. Noctor G, Rees D, Young A, Horton P. 1991. The relationship between zeaxanthin, energy-dependent quenching of chlorophyll fluorescence, and trans-thylakoid pH gradient in isolated chloroplasts. *Biochim. Biophys. Acta* 1057:320–30

94. Nunes MA, Ramalho JDC, Dias MA. 1993. Effect of nitrogen supply on the photosynthetic performance of leaves from coffee plants exposed to bright light. *J. Exp. Bot.* 44:893–99

95. Ögren E. 1988. Photoinhibition of photosynthesis in willow leaves under field conditions. *Planta* 175:229–36

96. Ögren E, Evans JR. 1992. Photoinhibition of photosynthesis *in situ* in six species of *Eucalyptus. Aust. J. Plant Physiol.* 19:223–32

97. Ögren E, Rosenqvist E. 1992. On the significance of photoinhibition of photosynthesis in the field and its generality among species. *Photosynth. Res.* 33:63–71

98. Ögren E, Sjöstrom M. 1990. Estimation of the effect of photoinhibition on the carbon gain in leaves of a willow canopy. *Planta* 181:560–67

99. Ohad I, Kyle DJ, Arntzen CJ. 1984. Membrane protein damage and repair: removal and replacement of inactivated 32-kilodalton polypeptides in chloroplast membranes. *J. Cell Biol.* 99:481–85

100. Olaizola M. 1993. *Laboratory and field studies of diatoxanthin cycling and nonphotochemical quenching in marine phytoplankton*. PhD thesis. State Univ. NY, Stony Brook. 285 pp.

101. Olaizola M, Yamamoto HY. 1993. Shortterm response of the diadinoxanthin cycle and fluorescence yield to high PFD in *Chaetoceros gracilis* (Bacillariophyceae). *J. Phycol.* In press

102. Öquist G, Anderson JM, McCaffery S, Chow WS. 1992. Mechanistic differences in photoinhibition of sun and shade plants. *Planta* 188:422–31

103. Osmond CB. 1993. What is photoinhibition? Some insights from comparisons of shade and sun plants. See Ref. 9a. In press

104. Osmond CB, Chow WS. 1988. Ecology of photosynthesis in the sun and shade: summary and prognostications. *Aust. J. Plant Physiol.* 15:1–9

105. Osmond CB, Ramus J, Levavasseur G, Franklin LA, Henley WJ. 1993. Fluorescence quenching during photosynthesis and photoinhibition of *Ulva rotundata* Blid. *Planta* 106:97–106

106. Ottander C, Hundal T, Andersson B, Huner NPA, Öquist G. 1993. Photosystem II reaction centres stay intact during low temperature photoinhibition. *Photosynth. Res.* 35: 191–200

107. Ottander C, Öquist G. 1991. Recovery of photosynthesis in winter-stressed Scots pine. *Plant Cell Environ.* 14:345–49

108. Owens TG, Shreve AP, Albrecht AC. 1992. Dynamics and mechanism of singlet energy transfer between carotenoids and chlorophylls: light harvesting and non-photochemical fluorescence quenching. In *Research in Photosynthesis*, ed. N Murata, pp. 179–86. Dordrecht: Kluwer

109. Pearcy RW. 1990. Sunflecks and photosynthesis in plant canopies. *Annu. Rev. Plant Plant Physiol. Mol. Biol.* 41:421–53

110. Pfitsch WA, Pearcy RW. 1992. Growth and reproductive allocation of *Adenocaulon bicolor* following experimental removal of sunflecks. *Ecology* 73:2109–17

111. Platt T, Gallegos CL, Harrison WG. 1980. Photoinhibition of photosynthesis in natural assemblages of marine phytoplankton.

J. Mar. Res. 38:687–701

112. Powles SB. 1984. Photoinhibition of photosynthesis induced by visible light. *Annu. Rev. Plant Physiol.* 1984:15–44

113. Powles SB, Berry JA, Björkman O. 1983. Interaction between light and chilling temperature on the inhibition of photosynthesis in chilling-sensitive plants. *Plant Cell Environ.* 6:117–23

114. Powles SB, Björkman O. 1981. Leaf movement in the shade species *Oxalis oregana*. II. Role in protection against injury by intense light. *Carnegie Inst. Washington Yearb.* 80:63–66

115. Radmer RJ, Kok B. 1976. Photoreduction of O_2 primes and replaces CO_2 assimilation. *Plant Physiol.* 58:336–40

116. Raven JA. 1989. Fight or flight: the economics of repair and avoidance of photoinhibition. *Func. Ecol.* 3:5–19

117. Raven JA. 1993. The cost of photoinhibition to plant communities. See Ref. 9a. In press

118. Robinson SA, Lovelock CE, Osmond CB. 1993. Wax as a mechanism for protection against photoinhibition—a study of *Cotyledon orbiculata*. *Bot. Acta* 106:In press

119. Rosenqvist E, Wingsle G, Ögren E. 1991. Photoinhibition of photosynthesis in intact willow leaves in response to moderate changes in light and temperature. *Physiol. Plant.* 83:390–96

120. Ryther JH. 1956. Photosynthesis in the ocean as a function of light intensity. *Limnol. Oceanogr.* 1:61–70

121. Sassenrath GF, Ort DR. 1990. The relationship between inhibition of photosynthesis at low temperature and the inhibition of photosynthesis after rewarming in chill-sensitive tomato. *Plant Physiol. Biochem.* 28:457–65

122. Schreiber U, Bilger W, Schliwa U. 1986. Continuous recording of photochemical and non-photochemical quenching with a new type of modulation fluorometer. *Photosynth. Res.* 10:51–62

123. Schroeter B, Green TGA, Seppelt RD, Kappen L. 1992. Monitoring photosynthetic activity of crustose lichens using a PAM-2000 fluorescence system. *Oecologia* 92:457–62

124. Smillie RM, Hetherington SE, He J, Nott R. 1988. Photoinhibition at chilling temperatures. *Aust. J. Plant Physiol.* 15:207–22

125. Smith RC, Baker KS. 1979. Penetration of UV-B and biologically effective dose-rates in natural waters. *Photochem. Photobiol.* 29:311–23

126. Smith RC, Baker KS, Holm-Hansen O, Olson R. 1980. Photoinhibition of photosynthesis in natural waters. *Photochem. Photobiol.* 31:585–92

127. Smith RC, Prezelin BB, Baker KS. 1992. Ozone depletion: ultraviolet radiation and phytoplankton biology in Antarctic waters.

Science 255:952–59

128. Stamp P. 1987. The expression of photosynthetic traits during and following severe chilling stress of European and tropical maize genotypes. *Physiol. Plant.* 71:73–76

129. Steemann-Nielsen E. 1952. On detrimental effects of high light intensities on the photosynthetic mechanism. *Physiol. Plant.* 5:334–44

130. Steemann-Nielsen E. 1952. The use of radio-active carbon (C14) for measuring organic production in the sea. *J. Cons. Int. Explor. Mer.* 18:117–40

131. Stirling CM, Nie G-Y, Aguilera C, Nugawela A, Long SP, Baker NR. 1991. Photosynthetic productivity of an immature maize crop: changes in quantum yield of CO_2 assimilation, conversion efficiency and thylakoid proteins. *Plant Cell Environ.* 14:947–54

132. Stitt M. 1986. Limitation of photosynthesis by carbon metabolism. I. Evidence for excess electron transport capacity in leaves carrying out photosynthesis in saturating light and CO_2. *Plant Physiol.* 81:1115–22

133. Strand M, Lundmark T. 1987. Effects of low night temperature and light on chlorophyll fluorescence of field-grown seedlings of Scots pine. *Tree Physiol.* 3:211–24

134. Sundby C, McCaffery S, Anderson JM. 1993. Turnover of the photosystem II D1 protein in higher plants under photoinhibitory and nonphotoinhibitory irradiance. *J. Biol. Chem.* 268:In press

135. Taguchi S. 1976. Short-term variability of photosynthesis in natural marine phytoplankton populations. *Mar. Biol.* 37:197–207

136. Takahashi M, Shimura S, Yamaguchi Y, Fujita Y. 1971. Photo-inhibition of phytoplankton photosynthesis as a function of exposure time. *J. Oceanogr. Soc. Jpn.* 27:43–50

137. Terashima I, Saeki T. 1985. A new model for leaf photosynthesis incorporating the gradients of light environment and of photosynthetic properties of chloroplasts within a leaf. *Ann. Bot.* 56:489–99

138. Tyystjarvi E, Ali-Yrkkö K, Kettunen R, Aro E-M. 1992. Slow degradation of the D1 protein is related to the susceptibility of low-light grown pumpkin plants to photoinhibition. *Plant Physiol.* 100:1310–17

139. Vincent WF, Neale PJ, Ricerson PJ. 1984. Photoinhibition: algal responses to bright light during diel stratification and mixing in a tropical alpine lake. *J. Phycol.* 20:201–11

140. Walker DA. 1989. Automated measurement of leaf photosynthetic O_2 evolution as a function of photon flux density. *Philos. Trans. R. Soc. London Ser. B* 323:313–26

141. Walters RG, Horton P. 1993. Theoretical assessment of alternative mechanisms for

non-photochemical quenching of PS II fluorescence in barley leaves. *Photosynth. Res.* 36:119–39

142. Weis E, Kreiger A, Spiedel D, Drücke M, Jordan U. 1993. The role of Ca^{2+} in energy quenching and photoinhibition. In *Photoinhibition of Photosynthesis—from Molecular Mechanisms to the Field. 41st Harden Conf.,* ed. NR Baker, JR Bowyer. London: Biochem. Soc. (Abstr.)

Annu. Rev. Plant Physiol. Plant Mol. Biol. 1994. 45:663–74

TAXOL

P. F. Heinstein and C.-j. Chang

Department of Medicinal Chemistry and Pharmacognosy, Purdue University, West Lafayette, Indiana 47907

KEY WORDS: *Taxus brevifolia,* antitumor drug, diterpene alkaloid, tissue culture, biosynthesis

CONTENTS

INTRODUCTION 663
IMPORTANCE.. 665
    Antitumor Activity and Clinical Efficacy 665
    Mechanism of Action 665
    Structure-Activity Relationships 666
ANALYTICAL CHEMISTRY.. 667
HPLC.. 667
    Enzyme-Linked Immunosorbent Assay.. 667
    Tandem Mass Spectrometry.. 668
SUPPLY.. 669
    Collection and Commercial Propagation 669
    Chemical Synthesis.. 670
    Cell Culture 670
BIOSYNTHESIS.. 671
SUMMARY AND CONCLUSIONS 672

INTRODUCTION

The systematic separation and testing of plant extracts can yield specific compounds that have physiological activities in mammalian cells. The bioassay-directed analysis (6) of *Taxus brevifolia* (the western yew) and other *Taxus* species has resulted in the isolation of the novel diterpene, taxol, (51) which has potent antileukemic activity (Figure 1). Taxol's in vivo inhibitory activity against murine and human solid tumors was demonstrated in 1977 and, as a result, taxol was selected for further evaluation (49). Clinical evaluation of taxol was delayed, however, primarily because of the difficulties in securing the large quantities of yew bark needed to isolate sufficient quantities of taxol and because of taxol's insolubility in aqueous solutions. Taxol was later shown to have a unique pharmacological activity—the inhibition of microtubule depolymerization (44). This finding stimulated renewed interest in the development of taxol as an antitumor drug. Taxol now is considered one of the most promising anticancer drugs to come from a plant source. This review

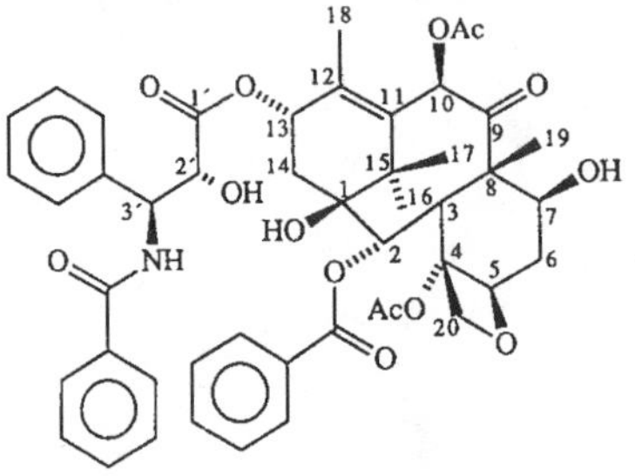

Taxol

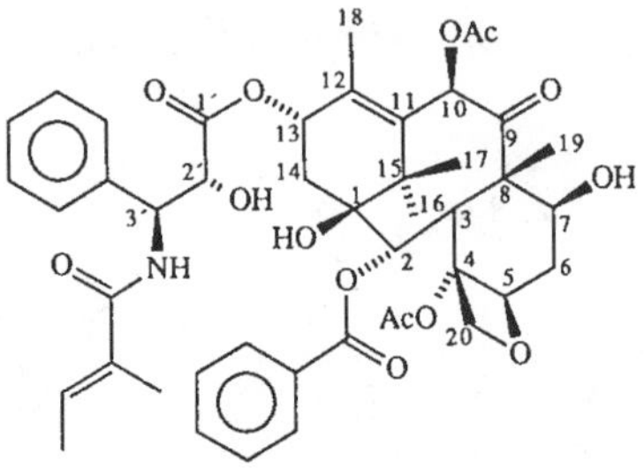

Cephalomannine

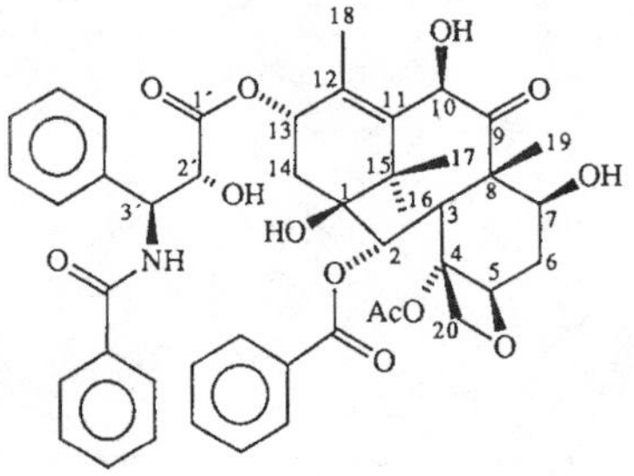

10-Deacetyltaxol

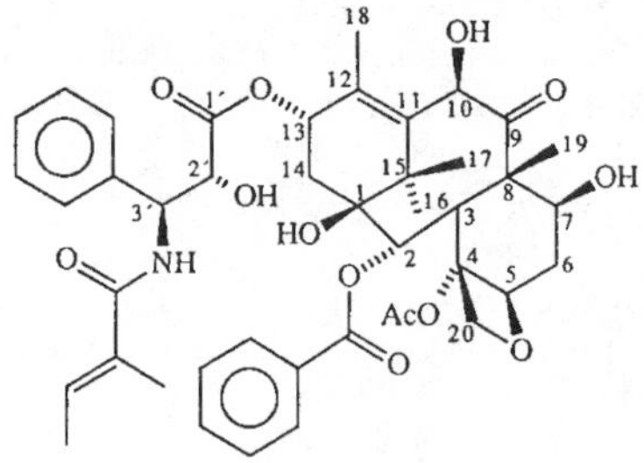

10-Deacetylcephalomannine

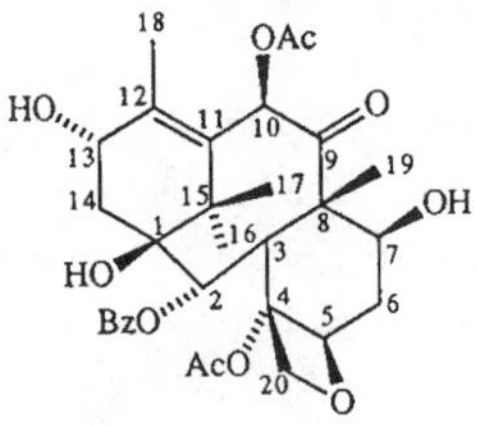

Baccatin III

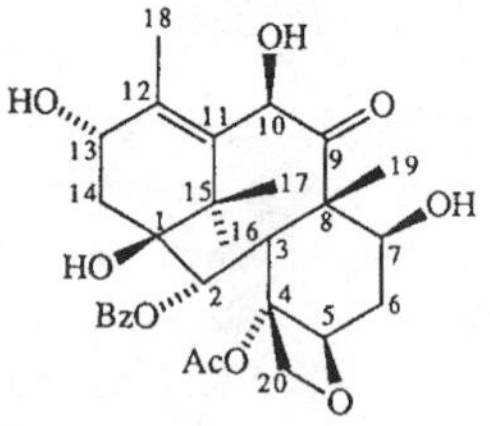

10-Deacetylbaccatin III

Figure 1   The structure of taxanes isolated from *Taxus* species.

discusses the supply of taxol, its formation in cell cultures, the biosynthesis of taxanes, and analytical procedures for the determination of taxol and derivatives. Its clinical activity, mechanism of action, structure-activity relationship, and its chemical synthesis are mentioned briefly. Reviews of the latter topics are available elsewhere (19, 28, 52).

IMPORTANCE

Antitumor Activity and Clinical Efficacy

The initial isolation of taxol was guided by bioassay in various murine leukemia systems and human carcinoma of the nasopharynx (9kB) (24). The activity against Lewis Lung tumor of partially pure taxol was also detected. In 1977, taxol was selected for further development as an antitumor agent because of its activity against B-16 murine melanoma and MX-1 mammary xenograft (49). The National Cancer Institute also found taxol to be active in the CX-1 colon xenograft, LX-1 lung xenograft, colon 26, L1210 leukemia, P388 leukemia, and P1534 leukemia systems (49). Jacrot et al showed the antitumor activity of taxol against human tumor xenograft (hepatic metastasis of breast cancer, cutaneous metastasis of bronchial cancer, and tumor of the base of the tongue) (24).

Clinical trials of taxol were begun in 1983. Phase I and II study results indicated that taxol was active against ovarian cancers (1, 32–40). McGuire et al reported that patients with platinum-refractory ovarian cancer showed a 30% response to taxol (34). Paclitaxel, a therapeutic formulation of taxol, recently has been approved by the United States Food and Drug Administration for treating ovarian cancer. Phase III studies of taxol and cisplatin versus cyclophosphamide and cisplatin in untreated patients with ovarian cancer are in progress. Holmes et al reported a 56% response rate in women with metastatic breast carcinoma who had received no more than one prior chemotherapy regimen (22). Subsequently, a 62% response rate was reported in women who had received no chemotherapy for metastatic breast cancer (45). A phase III trial in metastatic breast cancer is ongoing. Clinical data also suggest that taxol may be helpful in treating lung and head and neck cancers. Taxol does not appear effective for patients with colon, prostate, renal, or cervical cancers, or with melanoma.

Mechanism of Action

Taxol was demonstrated as an antimitotic agent by causing the G_2/M phase arrest of tumor cells. Unlike other naturally occurring antimitotic agents, such as colchicine, podophyllotoxin, and vinblastine, which disrupt microtubule assembly, taxol enhances the assembly of microtubules in vitro (44, 45). This unique mechanism of action promoted the further development of taxol as an antineoplastic agent.

The therapeutic efficacy of taxol is thought to be controlled primarily through its disruption of the tubulin-microtubule equilibrium. Other biochemical mediations may also play a role in modulating the bio-efficacy of taxol in tumor cells. Recent evidence indicates that taxol mimics the effect of lipopolysaccharide (LPS) by triggering tumor necrosis factor-α (TNF-2) and interleukin-1 release and by down-regulating the numbers of TNF-2 receptors (3, 12). These LPS-like effects might contribute to taxol's antitumor activity. In addition, taxol induces the tyrosine phosphorylation of a 41- and 42-Kda protein, which can also be induced by LPS (33).

Structure-Activity Relationship

Structure-activity relationships of taxol analogs have been discussed extensively (19, 27). The most widely used bioassay is microtubule assembly assay, which measures the microtubule assembly rate at 37°C and/or the microtubule disassembly rate at 0°C. All data indicate that (a) the C-13 acyl side chain and the oxetane ring are critical; (b) a free 2'-hydroxyl group and an acylated 3'-amino group are essential; (c) the absolute configurations at the C-2' and C-3' centers are important; and (d) the C-7, C-9, and C-10 functional groups may not be directly involved in the binding of taxol with tubulins.

Cytotoxicity against different mammalian cells [murine leukemic cells (p388); human carcinoma of nasopharynx (KB); and mouse macrophage-like cells (J779-2)] have also been used to measure the antitumor potential of taxol analogs. In vivo antitumor efficacy evaluations of a limited number of taxol analogs (Figure 2) were determined using murine leukemia (P388 and L1210), murine melanoma (B16), or human mammary carcinoma (M-1) cell lines. All the results have been described in Kingston's review (27).

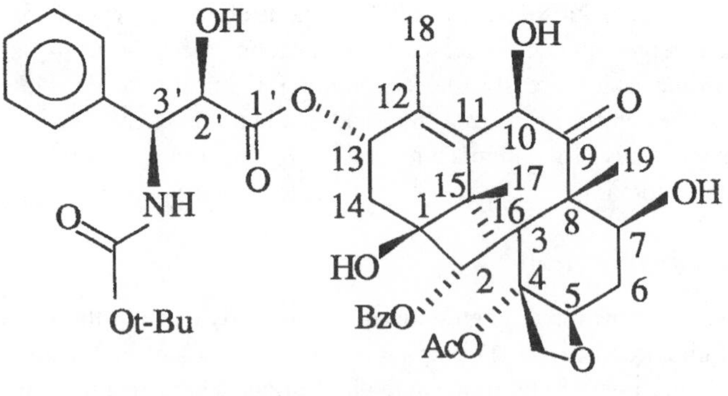

Figure 2   Taxotere, a semisynthetic taxol derivative.

ANALYTICAL CHEMISTRY

Accurate determination of taxol and its analogs in plant tissue samples requires highly sensitive methods. The usual chromatographic procedures, such as thin layer chromatography, are not applicable because taxol is present in yew bark at a concentration of about 0.01% of dry weight.

HPLC

A few HPLC procedures have been developed (7, 8, 55). In general, reverse phase separations with methanol-acetonitrile-methylene chloride-water solvent mixtures on phenyl-, cyano-, or C_{18}-bonded columns are used and give good separations of the six major compounds (taxol, cephalomannine, baccatin III, and the 10-deacetyl derivatives). Detection is by UV absorption at 227 nm because all the compounds listed above contain at least one aromatic or substituted olefinic moiety (Figure 1). These procedures do not detect saturated taxanes, they are analytical in scale, and they require modification for scale-up applications.

Enzyme-Linked Immunosorbent Assay

Specific polyclonal and monoclonal antibodies have been obtained against taxane-protein conjugates (18, 25, 30). Assays using these antibodies are highly specific and sensitive with a lower limit of detection of 1 nM or 0.085 ng of taxol per ml (30). These assays can be used to monitor serum level of taxol in cancer patients and to select taxol-producing *Taxus* spp. cells from callus or suspension cultures.

The general procedure for conjugate formation uses 7-succinyl- (25) or 2'-succinyl-taxol (30), prepared from succinic anhydride and taxol in the presence of pyridine (18), and bovine serum albumin, ovalbumin, or Keyhole Limpet hemocyanin.

Polyclonal anti-taxol antibodies are very specific for taxol and cephalomannine, but less specific for baccatin III, 7-epi, and 10-deacetyl derivatives (18, 30). Monoclonal anti-taxane antibodies, which are available commercially (25), have a broader specificity and have been used to detect taxanes with a 13-hydroxy function and the oxytane ring system. The commercial supplier of the monoclonal antibody suggests that the cross reactivity of this antibody may be broad and can be used to detect taxanes in general. However, some taxanes [e.g. derivatives of taxinine containing an exocyclic methylene group (at C-4) instead of the oxetane ring and derivatives of baccatin I with an oxirane (at C-4) instead of the oxitane ring (see Figure 3)] do not react with the monoclonal antibody (X Chen, C-j Chang, PF Heinstein, unpublished results).

Immunological determination of taxol, cephalomannine, and baccatin III in total is fast and sensitive because of crossreactivity, but analysis of the amounts of each individual compound in mixtures awaits development of compound-specific antibodies with little crossreactivity. In addition, the specificity of presently available antibodies for certain functionalities causes problems in the distinction and determination of potential biosynthetic intermediates with slightly different structures.

Tandem Mass Spectrometry

Tandem mass spectrometry (MS/MS) is a rapid procedure for analyzing specific compounds in crude plant extracts (21, 38). This method is highly specific and sensitive enough to allow independent determination of taxol, cephalomannine, and baccatin III in crude plant extracts at levels less than 500 pg (9). Furthermore, the presence and quantity of the above three taxanes in a single western yew needle can be determined in less than 1 hr (including extraction, sample preparation, and MS/MS analysis).

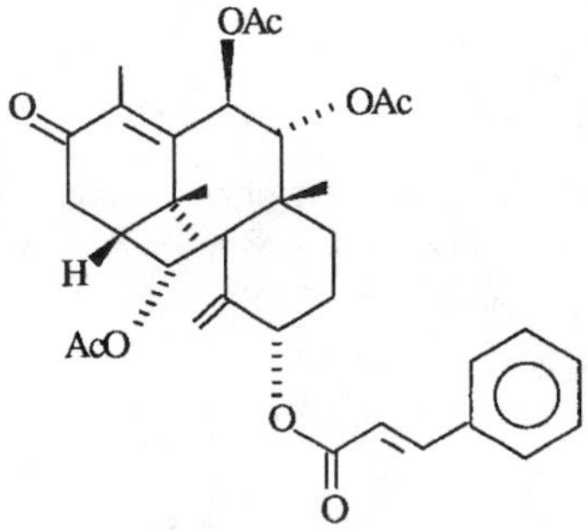

Taxinine

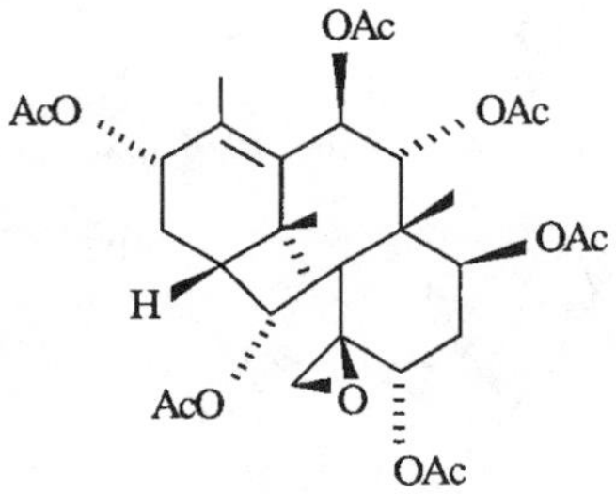

Baccatin I

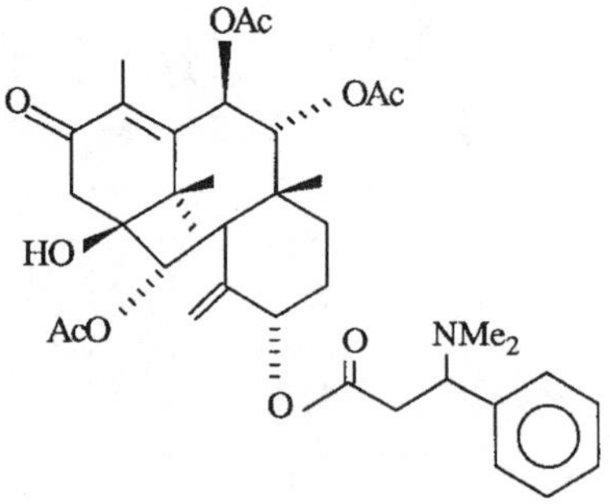

Taxine I

Figure 3   Structures of simple taxane derivatives.

SUPPLY

Collection and Commercial Propagation

The primary source of taxol is the bark of the western yew (*Taxus brevifolia*). Although the western yew is relatively plentiful and is not harvested for wood products, the isolation of sufficient taxol to treat 50,000–60,000 breast and ovarian cancer patients would eliminate the plant in a few years. Other sources have been proposed. For example, the Himalayan tree, *Taxus baccata*, contains taxol and baccatin III in the bark and needles. Similarly, extensive analysis showed a number of other *Taxus* species to contain taxol and derivatives in the needles (13, 26, 50). This would allow the propagation of *Taxus* spp. in fieldplots, and needles and small twigs could be harvested on a yearly basis without destroying the trees or bushes by removing the bark.

Although taxol concentrations in bark, needles, or twigs is dependent on genetic differences [e.g. *T. brevifolia* has more taxol than does *T. baccata, T. cuspidata*, or *T. media* (54)], much larger differences in taxol content appear to be the result of epigenetic and environmental factors (53). Taxol, cephalomannine, and baccatin III varied within a population of *T. brevifolia* trees with respect to type of tissue, age of tissue, location (i.e. sunlight or shade), and soil condition (53). For example, taxol content was 0.018% of dry weight in *T. brevifolia* trees growing on the western slopes of the Cascade Mountains in Oregon and Washington compared to 0.002% in eastern areas of the same states. These variations in taxane concentrations suggest a phytoalexin role for these compounds, which is supported further by the observation that crude preparations of phytopathogenic fungi can induce taxane formation in *T. brevifolia* cell suspension cultures (see below).

A promising development in securing sufficient quantities of taxol for large scale clinical application is the identification of a fungus, *Taxomyces andreanae*, which produces taxol (48). This fungus was isolated from the phloem-cambial tissue of a *T. brevifolia* tree located in Montana. A three-week mycelial culture in a semi-defined medium produced up to 50 ng of taxol per liter (47). Potera (37) has proposed that the fungus obtained its capacity to synthesize taxol from the host tree through a transfer of genetic material and that a critical component of the host is required for taxol formation.

An attractive alternative to obtaining taxol from the bark of the western yew is the collection and extraction of the trimmed plant material from ornamental *Taxus X media* 'Hicksii' and *Taxus* spp. hybrids. The needles and twigs of ornamental cultivars contain small amounts of taxol and larger quantities of its precursor, baccatin III (4). Baccatin III can be converted to taxol by an esterification reaction (11). It appears that the combined extraction of taxol and baccatin III from the yearly harvest of needles, shoots, and twigs and the semisynthesis of taxol from baccatin III will be the most economical and most environmentally feasible process for producing sufficient quantities of taxol for chemotherapy.

Chemical Synthesis

Taxol is a highly complex molecule with eleven asymmetric centers. It has 2048 potential steric isomers. Total synthesis of taxol is an extremely challenging problem and has yet to be accomplished. Taxol consists of two biosynthetically unrelated components: a geranylgeranyl pyrophosphate-derived diterpenoid component (baccatin II) and a phenylalanine-derived side-chain component (N-benzoylphenylisoserine). Several efficient syntheses of the phenylisoserine moiety have been reported (28). Brieva et al (5) and Gou et al (17) recently described an elegant chemoenzymatic synthesis of the optically active phenylisoserine moiety. Over 30 research groups have been actively involved in the total synthesis of various taxanes (19, 28). A detailed treatment of all synthetic approaches is beyond the scope of this review. Wender et al have reported a versatile strategy that may offer promise for the complete synthesis of taxol using optically active pinene as a precursor (52). It is still uncertain whether a commercially sound route to the total synthesis of taxol will be found in the near future. However, all these synthetic endeavors may lead to the discovery of other simpler, but similarly active taxanes.

Cell Culture

Because taxol is difficult to synthesize chemically and *T. brevifolia* and other *Taxus* species may be depleted by reliance on harvesting from whole plants, the formation of taxol in callus and cell suspension cultures has been of interest. Patents have been filed that promise an unlimited source of taxol and other taxanes. However, *Taxus* species cell cultures are very slow growing— they have a growth cycle of 15–20 days, and after 21 days, taxol production is only 0.02% of dry weight. *T. brevifolia* cell suspension cultures cannot compete with commercial field propagation of *Taxus* species.

T. brevifolia explants produced callus when plated on Gamborg's B5 medium (16) supplemented with Casamino acids (2 g/L) and 2 mg/L of 2,4-dichlorophenoxy-acetic acid (2,4D) (10). Taxol was produced after 2–4 weeks at a concentration of 1–3 mg/L (10). Other media produced similar results. Both Murashige & Skoog's (35) and Schenk & Hildebrandt's (42) basic media, with 1-naphthalene acetic acid (4 mg/L), 0.1–0.5 mg/L kinetin, and 10 g/L of sucrose, resulted in callus formation from stem explants (41). Taxol synthesis (0.05% of dry matter) was observed in suspension cultures derived from the above calli after 4 weeks of incubation at 20–25° and 100 rpm in the light (41). Kinetin was omitted from the suspension culture medium because in its presence, browning of the cell mass was observed.

The most extensive investigation of requirements for viable *T. cuspidata* and *T. canadensis* callus and cell suspension cultures established Gamborg's B5 salts and organics (16), supplemented with 1.5% (w/v) soluble poly-vinylpyrrolidone (for binding of phenolics and prevention of browning), 3% (w/v) sucrose, and 4 mg/L 2,4D as the optimum medium. Cultures were grown in the dark at 23°C on a rotary shaker at 110 rpm (46). Under these conditions the cell mass in one cell suspension line increased 2.8-fold in 25 days. Calli produced 0.02% of taxol on a dry weight basis and 0.043% baccatin III after 55 days (46). Immobilized cells produced 0.012% of dry weight of taxol after 6 months of culture (46). Taxanes are present in the culture medium and it has been proposed that this is a result of cell turnover rather than active secretion of these secondary metabolites.

Although it has been stated (10) that taxol formation can be induced by phytopathogenic fungi and additions of non-nutrients to the medium, experi-mental data is lacking. Because epigenetic and environmental factors can influence the taxane concentration in intact *Taxus* spp. plants (53), it should be possible to develop conditions and elicitors that induce taxane synthesis in cell suspension cultures.

Cell suspension cultures are limited by the quantity of taxol they produce and by the length of incubation required to yield taxol concentrations compa-rable to those produced by intact *Taxus* spp. tissue. As a result, cell culture systems probably cannot compete economically with field propagation of *Taxus* spp. plants and periodic harvesting of needles and twigs. If more rapid growth conditions are possible and taxol formation can be induced in these cultures, the commercial production of taxol in cell suspension cultures may be possible (14, 46).

BIOSYNTHESIS

The study of the enzymatic formation of taxol can be divided into three parts: the formation of the taxane ring system and substitutions, the synthesis of the sidechain, and the attachment of sidechain to the diterpene moiety.

The taxane ring system is a diterpenoid and, therefore, is derived from the mevalonate pathway. Incubation of ground *T. canadensis* needles in a nutrient solution with (3RS) (5RS) [5-^{3}H] mevalonate as a substrate resulted in 0.12% incorporation of the radioactivity into taxol purified to constant specific activ-ity (56). The ^{3}H/^{14}C ratio of taxol synthesized in similar incubations of ground needles with [ring-2,6-^{3}H] L-phenylalanine and (3R) [2-^{14}C] mevalonate (^{3}H/^{14}C = 3.6) was found to be 4.4. Upon hydrolysis of the taxol to baccatin III the ^{3}H/^{14}C ratio was reduced to 1.75 (56). These results are consistent with the hypothesis that the taxane ring system is derived from mevalonate via the isoprene pathway and the major side chains are derived from phenylalanine.

Cyclization reactions from geranylgeranyl pyrophosphate have been proposed, whereby the C-20 intermediate of the isoprene pathway and the generally accepted linear precursor of cyclic diterpenes is converted to the taxane ring system (20) via the bicyclic diterpene verticillene (23). However, biomimetic synthesis of a tricyclic taxane ring system from verticillene was unsuccessful (2). An enzyme preparation from yew bark was able to catalyze the cyclization of geranylgeranyl pyrophosphate to the cyclic hydrocarbon, taxa-4(20),11-diene (31). The yields of these reactions were very low and absolute stereochemical determination of the structure of the "cyclase" reaction product was not possible.

The synthesis of the phenylisoserine sidechain of taxol has been studied in detail (15, 29, 36). The 3-(dimethylamino)-3-phenylpropanoic acid moiety (Winterstein's acid) of taxine-I (Figure 2) is derived from phenylalanine. Similarly, phenylalanine gives rise to the side chain in taxol. From the retention of the *pro*-S hydrogen of the C-3 of the phenylalanine sidechain, it appears that cinnamic acid is not an intermediate and phenylalanine:ammonia-lyase is not involved as a catalyst in the pathway. Neither cinnamic acid nor its epoxide were incorporated into taxol by incubations with bark tissue (15). The incorporation of chemically synthesized potential precursors indicates that β-phenylalanine, phenylisoserine, and the intact sidechain N-benzoyl-phenylisoserine are incorporated into taxol (15). The enzymatic esterification of the diterpenoid moiety with the sidechain has not been studied.

SUMMARY AND CONCLUSIONS

All indications point to a major role for taxol as one of the most promising clinical agents for the treatment of ovarian, breast, and solid tumors of the head and neck (squamous cell carcinoma). Taxol can be secured through the cultivation of ornamental *Taxus* spp., which yield needles and twigs containing taxol and larger quantities of baccatin III. Through a chemical esterification reaction, the baccatin III can be converted to taxol. Formation of taxol in cell suspension cultures requires a long incubation period, which may be shortened through elicitation. The concentrations of taxol accumulating in these cultures are very low; however, these cultures may be used to elucidate the biosynthetic reactions leading to taxanes.

ACKNOWLEDGMENTS

The authors are thankful for support from the National Cancer Institute (CA55118) and appreciate Kathy Larsen's help in typing this manuscript.

Literature Cited

1. Arbuck S. 1992. *Current status of the clinical development of taxol.* Presented at 2nd Natl. Cancer Inst. Workshop on Taxol and *Taxus,* Alexandria, VA, Sept. 23–24, 1992

2. Begley MJ, Jackson CB, Pattenden G. 1985. Investigation of transannular cyclisations to the taxane ring system. *Tetrahedron Lett.* 26:3397–400

3. Bogdan C, Ding A. 1992. Taxol, microtubule-stabilizing antineoplastic agent, induces expression of tumor necrosis factor α and interleukin-1 in macrophages. *J. Leuk. Biol.* 52:119–21

4. Borman S. 1991. Scientists mobilize to increase supply of anticancer drug taxol. *Chem. Eng. News* 2:11–18

5. Brieva R, Crich J, Sih CJ. 1993. Chemoenzymatic synthesis of the C-13 side-chain of taxol: optically active 3-hydroxy-4-phenyl β-lactam derivatives. *J. Org. Chem.* 58:1068–75

6. Cancer Chemotherapy National Service Center. 1961. Protocols for screening chemical agents and natural products against animal tumors and other biological systems. *Cancer Chem. Rep.* 25:1–66

7. Cardellina JH II. 1991. HPLC separation of taxol and cephalomannine. *J. Liq. Chromatogr.* 14:659–65

8. Castor TP, Tyler TA. 1993. Determination of taxol in *Taxus media* needles in the presence of interfering components. *J. Liq. Chromatogr.* 16:723–31

9. Chang C-j, Hoke SH, Wood JM, Li X-H, Cooks RG. 1992. *Taxane analysis by tandem mass spectrometry.* Presented at 2nd Natl. Cancer Inst. Workshop on Taxol and *Taxus,* Alexandria, VA, Sept. 23–24

10. Christen AA, Gibson DM, Bland J. *US Patent No. 5019504*

11. Denis JN, Greene AE, Guénard D, Guéritte-Voegelein F, Mangatal L, Potier P. 1988. A highly efficient, practical approach to natural taxol. *J. Am. Chem. Soc.* 110: 5917–19

12. Ding AH, Porteu F, Sanchez E, Nathan CF. 1990. Shared actions of endotoxin and taxol on TNF receptor and TNF release. *Science* 248:370–72

13. Fett-Neto AG, DiCosmo F. 1992. Distribution and amounts of taxol in different shoot parts of *Taxus cuspidata. Planta Med.* 58: 464–66

14. Fett-Neto AG, DiCosmo F, Reynolds WF, Sakata K. 1992. Cell culture of *Taxus* as a source of the antineoplastic drug taxol and related taxanes. *Biotechnology* 10:1572–75

15. Fleming PE, Mocek U, Floss HG. 1993.

Biosynthesis of taxoids. Mode of formation of the side chain. *J. Am. Chem. Soc.* 115: 805–7

16. Gamborg OL, Miller RA, Ojima K. 1968. Nutrient requirements of suspension cultures of soybean root cells. *Exp. Cell Res.* 50:151–58

17. Gou DM, Liu YC, Chen CS. 1993. A practical chemotherapy synthesis of the taxol C-13 side-chain N-benzoyl-(2*R*, 3*s*)-3-phenylisoserine. *J. Org. Chem.* 58:1287–89

18. Grothaus PG, Bignami GS, Lazo CB, Byrnes JB, O'Mally S, et al. 1992. *Analysis and purification of taxol and taxanes using monoclonal antibodies.* Presented at 2nd Natl. Cancer Inst. Workshop on Taxol and *Taxus,* Alexandria, VA, Sept. 23–24

19. Guénard D, Guéritte-Voegelein F, Potier P. 1993. Taxol and taxotere: discovery, chemistry, and structure-activity relationship. *Acc. Chem. Res.* 26:160–67

20. Guéritte-Voegelein F, Guénard D, Potier P. 1987. Taxol and derivatives: a biogenetic hypothesis. *J. Nat. Prod.* 50:9–18

21. Hoke SH, Wood JM, Cooks RG, Li X-H, Chang C-j. 1992. Rapid screening of taxanes by tandem mass spectrometry. *Anal. Chem.* 64:2313–15

22. Holmes FA, Walters RS, Theriault RL. 1991. Phase II trial of taxol, an active drug in the treatment of metastatic breast cancer. *J. Natl. Cancer Inst.* 83:1795–805

23. Jackson CB, Pattenden G. 1985. Total synthesis of verticilline, the putative biogenic precursor of the taxane alkaloids. *Tetrahedron Lett.* 26:3393–96

24. Jacrot M, Riondel J, Picot F, Leroux D, Mouriquand C, et al. 1983. Action du taxol vis-a-vis de tumeurs humaines transplantées sur des Souris athymiques. *C. R. Acad. Sci. Paris Ser. III* 297:149–600

25. Jaziri M, Diallo BM, Vanhaelen MH, Vanhaelen-Fastre RJ, Zhiri A, et al. 1991. Enzyme-linked immunosorbent assay for the detection and semi-quantitative determination of taxane diterpenoids related to taxol in *Taxus* sp. and tissue cultures. *J. Pharm. Belg.* 46:93–99

26. Kelsey RG, Vance NC. 1992. Taxol and cephalomannine concentrations in the foliage and bark of shade-grown and sun-exposed *Taxus brevifolia* trees. *J. Nat. Prod.* 55:912–17

27. Kingston DGI, Samaranayake G, Ivey CA. 1990. The chemistry of taxol, a clinically useful anticancer agent. *J. Nat. Prod.* 53:1–12

28. Kingston DGI. 1991. The chemistry of taxol. *Pharmacol. Ther.* 52:1–34

29. Leete E, Bodem GB. 1966. Biosynthesis of

3-dimethylamino-3-phenyl propanoic acid. *Tetrahedron Lett.* 1966:3925–27

30. Leu J-G, Chen B-X, Schiff PB, Erlanger BF. 1993. Characterization of polyclonal and monoclonal anti-taxol antibodies and measurement of taxol in serum. *Cancer Res.* 53:1388–91

31. Lewis N, Croteau R. 1992. *Taxol biosynthesis.* Presented at 2nd Natl. Cancer Inst. Workshop on Taxol and *Taxus,* Alexandria, VA, Sept. 23–24

32. Lowe J. 1993. Pacific yew: draft environmental impact statement. USDA Forest Serv. Appendix K. 41–77

33. Manthey CA, Brandes ME, Perera PY, Vogel SN. 1992. Taxol increases steady-state levels of lipopolysaccharide inducible genes and protein-tyrosine phosphorylation in murine macrophages. *J. Immunol.* 149:2459–65

34. McGuire WP, Rowinsky EK, Rosenshein NB, Grumbine FC, Ettinger DS, Armstrong DK. 1989. Taxol: a unique antineoplastic agent with significant activity in advanced ovarian epithelial neoplasmas. *Ann. Intern. Med.* 111:273–79

35. Murashige T, Skoog F. 1962. A revised medium for rapid growth and bioassays with tobacco tissue cultures. *Physiol. Plant* 15:473–97

36. Platt RV, Opie CT, Haslam E. 1984. Biosynthesis of flavan-3-ols and other secondary plant products from (2S)-phenylalanine. *Phytochemistry* 23:2211–17

37. Potera C. 1993. Texas firm will make taxol from fungus. *Gen. Eng. News* 13:1

38. Roush RA, Cooks RG. 1984. Characterization of alkaloids and other secondary metabolites by multiple stage mass spectrometry. *J. Nat. Prod.* 47:197–214

39. Rowinsky EK, Cazenave LA, Donehower RC. 1990. Taxol: a novel investigational antineoplastic agent. *J. Natl. Cancer Inst.* 82:1247–59

40. Runowicz CD, Wiernik PH, Einzig AI, Goldberg GL, Horwitz SB. 1993. Taxol in ovarian cancer. *Cancer* 71:1591–96

41. Saito K, Ohashi H, Tahar M. 1992. *Int Patent No. WO92/13961*

42. Schenk RU, Hildebrandt AC. 1972. Medium and techniques for induction and growth of monocotyledonous and dicotyledonous plant cell cultures. *Can. J. Bot.* 50:199–204

43. Schiff PB, Fant J, Horwitz SB. 1979. Promotion of microtubule assembly in vitro by taxol. *Nature* 277:665–67

44. Schiff PB, Horwitz SB. 1981. Taxol assembles tubulin in the absence of exogenous guanosine 5′-triphosphate or microtubule-associated proteins. *Biochemistry* 20: 3242–52

45. Seidman A, Reichman B, Crown J. 1992. Activity of taxol with recombinant granulocyte colony stimulating factor as first chemotherapy of patients with metastatic breast cancer. *Proc. ASCO* 11:59 (Abstr. 64)

46. Shuler ML, Hirasuna TJ, Willard DM. 1992. *Kinetics of taxol production by tissue culture.* Presented at 2nd Natl. Cancer Center Inst. Workshop on Taxol and *Taxus,* Alexandria, VA, Sept. 23–24

47. Stierle A, Strobel G, Stierle D. 1993. Taxol and taxane production by *Taxomyces andreanae,* an endophytic fungus of pacific yew. *Science* 260:214–16

48. Strobel GA, Stierle A, Stierle D, Hess WM. 1993. *Taxomyces andreanae,* a proposed new taxon for a bulbilliferons hyphomycete associated with Pacific Yew (*Taxus brevifolia*). *Mycotaxon* 47:71–80

49. Suffness M, Cordell GA. 1985. Antitumor alkaloids. *Alkaloids* 25:6–18, 280–88

50. Vidensek N, Lim P, Campbell A, Carlson C. 1990. Taxol content in bark, wood, root, leaf, twig, and seedling from several *Taxus* species. *J. Nat. Prod.* 53:1609–10

51. Wani MC, Taylor HL, Wall ME, Coggon P, McPhail AT. 1971. Plant antitumor agents. VI. The isolation and structure of taxol, a novel antileukemic and antitumor agent from *Taxus brevifolia. J. Am. Chem. Soc.* 93:2325–27

52. Wender PA, Mucciaro TP. 1992. A new and practical approach to the synthesis of taxol and taxol analogues: the pinene path. *J. Am. Chem. Soc.* 114:5878–79

53. Wheeler NC, Jech K, Masters S, Brobst SW, Alvarado AB, Hoover AJ. 1992. Effects of genetic, epigenetic, and environmental factors on taxol content in *Taxus brevifolia* and related species. *J. Nat. Prod.* 55:432–40

54. Witherup KM, Look SA, Stasko MW, Ghiorzi TJ, Muschik GM. 1990. *Taxus* spp. needles contain amounts comparable to the bark of *Taxus brevifolia:* analysis and isolation. *J. Nat. Prod.* 53:1249–55

55. Witherup KM, Look SA, Stasko MW, McCloud TG, Issaq HJ, Muschik GM. 1989. High performance liquid chromatographic separation of taxol and related compounds from *Taxus brevifolia. J. Liq. Chromatogr.* 12:2117–32

56. Zamir LO, Nedea ME, Garneau FX. 1992. Biosynthetic building blocks of *Taxus canadensis* taxanes. *Tetrahedron Lett.* 33: 5235–36

AUTHOR INDEX

A

Abe A, 52, 53
Abe M, 117, 534
Abe S, 623
Abel S, 423, 424, 428, 433, 438
Abouzid AM, 84
Abrams SR, 115
Abrol YP, 591, 594
Accotto GP, 81–84, 86, 88, 94, 97
Acedo GN, 38
Acton GJ, 431, 434
Adams MJ, 495, 508
Adams WW, 238, 241, 246
Adams WW III, 288, 294, 296, 633, 634, 636, 637, 640, 642–46, 652
Adamse P, 158, 296
Adamska I, 157
Adejare GO, 92
Adler K, 239
Aebi R, 184
Aerts RJ, 272
AESCHBACHER RA, 25–45; 26, 27, 32, 35, 39
Agostino A, 458, 461
Agostoni-Carbone ML, 555
Aguan K, 117
Aguilar M, 126
Aguilera C, 645, 649, 654, 655
Ahl P, 80
Ahlquist P, 80, 84, 471
Ahmad M, 153
Ai Y, 425, 440
Aigle M, 555
Ainsworth C, 200
Aiso K, 559
Aizawa K, 370, 374, 376
Akashi T, 528, 530, 531, 533, 537, 539
Akiyama Y, 482, 484
Akiyoshi DE, 176
Albanese G, 480, 485
Alberghina L, 555
Alberti M, 291
Albrecht AC, 643
Alibert G, 453
Ali-Yrkkö K, 644
Allen JF, 596
Allen LN, 207
Allen RD, 122, 343, 347
Allen S, 448, 452
Alliotte T, 186, 559
Aloysius SKD, 207
Alstiel L, 621

Altabella T, 250
Altendorf K, 222, 228
Altschul MA, 304
Alvarado AB, 669, 671
Alwine JC, 51, 52
Amable RA, 204
Amann M, 264, 268, 269
Amano Y, 271
Amasino RM, 176, 186, 187
Amory AM, 579, 581, 583, 585, 592
Amrhein N, 246, 458
An G, 56, 313
An K, 56
Anderberg RJ, 117
Andersen MT, 83, 84
Anderson IC, 288, 293, 295
Anderson JA, 223, 224
Anderson JC, 343
Anderson JM, 308, 311, 315, 634, 636, 637, 640, 644
Anderson JW, 456
Anderson LE, 345
Anderson MA, 422, 423, 425, 428, 435, 437, 438
Anderson PM, 373
Anderson RGW, 610, 611, 626
Anderson RL, 357, 588
Anderson RS, 64
Anderson S, 62, 66
Andersson B, 634, 636–39, 644
Andersson I, 344
Andre B, 406
André C, 63
André CP, 63
Andreo CS, 345–47
Andrews DL, 314
Andrews M, 452, 578, 579
Andrews TJ, 324, 325, 344
Andrianov VM, 188
Ang D, 470–72, 479
Antoniw JF, 430
Aono M, 343
Apel GA, 434
Apel K, 157, 318
Appel K, 71
ap Rees T, 449, 586, 588
Arai M, 424, 425
Arakawa H, 272
Arata H, 458, 461, 494, 496, 511, 512, 514
Araya A, 70
Arbuck S, 665
Arkona C, 239
Arlotto MA, 267
Armes JE, 610, 612

Armstrong DJ, 177, 179
Armstrong DK, 665
Armstrong GA, 291
Arntzen CJ, 358, 637
Aro E-M, 634, 636, 637, 644
Arretz M, 555
Arrio-Dupont M, 449
Arron GP, 331, 349
Arts GJ, 69
Artus NN, 347
Asahi T, 559, 564
Asahi Y, 317
Asaine K, 273
Asard H, 149, 150
Ashbaugh M, 158, 160
Ashton NW, 181, 183
Askerlund P, 148
Aslam M, 578, 579, 583, 584, 594
Assmann SM, 163, 164
Aåström H, 537
Atencio DP, 550, 552
Atkinson A, 425, 428, 435
Atkinson AH, 438
Atkinson M, 81
Atkinson T, 476, 484, 486
Attardi G, 48
Attridge TH, 158
Auchincloss AH, 67
Auer CA, 177
Auf der Mauer A, 477, 480
Augur C, 177
Austin B, 200
Ausubel F, 158, 160, 161
Ausubel FM, 156, 158, 161, 312
Avadhani PN, 452
Aveni MT, 204
Avramova ST, 377
Avron M, 345
Axelos M, 186
Azzi A, 457

B

Baas PD, 93
Baas R, 452
Babiano MJ, 183
Back R, 610
Bacot KO, 290, 291
Badcoe IG, 478, 481, 482, 484
Baden CS, 117, 449
Bader SB, 125, 128
BADGER MR, 369–92; 247, 371, 373–77, 379, 385, 386
Baenziger PS, 436, 437
Bagnall DJ, 250

Baijot M, 227
Bailey HM, 408
Bak H, 424, 425
Bakalara N, 69
Baker A, 550, 551
Baker D, 610, 626
Baker JC, 348
Baker KP, 547, 550, 551, 554
Baker KS, 647, 648
Baker NR, 340, 342, 635, 636, 640, 645, 646, 649, 654
Bakhuizen R, 41
Bakker JA, 158
Bako L, 411, 412
Bakrim N, 449
Balch WE, 612
Baldi BG, 402
Baldwin WS, 411
Ball MC, 645, 655
Ballaré CL, 144
Ballo BL, 122, 123
Balsevich J, 271
Balshusemann D, 395, 398, 399, 401, 412, 616
Balusek K, 617, 621
Bamberger ES, 345
Bamforth CW, 207
Bancroft I, 229
Baneyx F, 478, 479, 482, 484, 485
Bankier AT, 62, 66
Bantroch DJ, 308, 311
Barber J, 519, 639
Barbier-Brygoo H, 400, 402, 404, 405, 412, 458, 459, 616
Barbieri L, 422
Bariola PA, 422, 424, 425, 428, 432, 433, 435
Barker GR, 434
Barker H, 81
Barker R, 83, 84, 102
Barker RDJ, 408
Barlow PW, 28, 37
Barna B, 433
Barnett L, 155
Baron-Epel O, 614
Baroncelli S, 293, 295
Barraclough R, 471
Barratt DHP, 117
Barrell BG, 62, 66
Barrera-Saldane H, 90
Barrett J, 149
Barrette T, 294
Barry GF, 176, 247
Barta A, 48
Bartel B, 221
Bartels D, 118, 127, 129
Bartholomew DM, 123
Bartkiewicz M, 439
Bartlett SG, 382, 384, 385
BARTLEY GE, 287–301; 123, 290–92, 294, 296, 297
Basile DV, 199–203
Basile MR, 199, 200, 203

Baskin TI, 26, 36, 37, 145, 146, 152
Basra AS, 134
Bassett B, 624
Bassett ME, 379
Bassford PJ, 483
Bassham JA, 236, 579, 583, 585, 587, 597
Bassi R, 295
Bassil NV, 177
Bathgate B, 559
Batschauer A, 156–58, 160
Battelli MG, 422
Battraw MJ, 122
Bauer W, 263
Baulcombe DC, 123
Baumann B, 199
Baumeister W, 471, 474, 478, 483
Baumgartner B, 428, 432
Bauw G, 395, 398, 399, 401, 412, 616
Bayley CC, 121
Baysdorfer C, 584, 587, 593
Bazlen I, 146
Beach LR, 123
Beachy RN, 83, 84, 87, 117, 122
Beadle BC, 495, 508
Beagley T, 65
Beale JM, 263
Beale MH, 394
Beams B, 314
Beasley EM, 548, 559, 562, 563
Beaudoin JD, 354
Beauregard M, 501, 512
Beavis AD, 458
Becerril JM, 358
Beck CF, 457
Beck E, 249, 343
Becker TW, 296
Bedford I, 89
Bedu S, 374
Beebe SJ, 148
Beer S, 379, 380
Beevers H, 459, 588
Beevers L, 591, 621
Begley MJ, 672
Begu D, 70
Begum D, 122
Behnke S, 157
Behrens M, 550, 555
Beinsberger S, 188
Beintema JJ, 424, 425, 428, 433
Bejarano ER, 98, 99, 101, 296
Belanger R, 507
Bell AW, 548
Bell E, 258, 312
Bell JC, 40
Belostotsky DA, 439
Beltzer JP, 556
Bendahmane M, 83, 84
Bendich AJ, 62–64

BENFEY PN, 25–45; 26, 27, 32, 35, 36, 39, 122
Benne R, 69, 73
Bennett AB, 216
Bennett JL, 429, 430
Bennetzen JL, 291, 293, 295
Benson AA, 204
Benveniste P, 400
Benz AJ, 216
Beopoulos N, 434
Berberián GE, 215, 217
Bergey DR, 270, 271
Bergfeld R, 528–30
Bergman A, 349
Berkelman T, 217
Berleth T, 26, 28–31
Berlin J, 279
Bernatzky R, 425, 428, 435
Berndt WA, 332, 333
Bernier D, 127
Bernier F, 122
Berrier AL, 377
Berry JA, 238, 240, 343, 375, 594, 635–37
Bertani A, 412
Bertsch U, 472, 473, 477
Bessoule J-J, 552
Best TR, 160
Bestman H, 238, 243, 246, 247
Bettelheim KA, 205
Bettey M, 458
Betzner AS, 26, 36, 37
Bevan MW, 188, 189
Bewley JD, 318, 434
Beyer P, 297
Beyreuther K, 267
Bhat GJ, 66
Bhatt AM, 229
Bhuvaneswari TV, 408
Bialek K, 402
Bibb MJ, 62, 66
Bicanic DD, 497
Bickel-Sandkötter S, 340, 341
Bignami GS, 667
Bijaisoradat N, 117
Bilang J, 410–12
Bilger W, 641
Binder BM, 226, 227
Binder S, 67–70, 546
Binns A, 184
BINNS AN, 173–96; 176, 185, 188, 190
Bintrim SB, 425
Birch DG, 580, 581, 593
Bird AS, 290, 291, 297
Bird CR, 290, 291, 297
Bird IF, 351, 352
Bird S, 66
Bisaccia F, 457
Bisalputra T, 614, 615, 617, 619, 621, 626
Bisaro DM, 81, 82, 86–99, 101, 103
Biscoe PV, 649
Bishop RW, 291

Bismuth E, 585
Bisseling T, 38, 408
Bisseret P, 205
Bjork M, 379, 380
Björkman O, 634–37, 640, 641, 645
Black CC, 346, 351, 357
Black DM, 90
Black M, 158
Black RC, 188
Blackburn P, 422
Blackman SA, 125
Blackwell D, 324
Blackwell RD, 350–53, 360
Blain I, 291
Blake TK, 423
Bland J, 670, 671
Blank A, 431, 432, 434
Blanz P, 63
Blatt MR, 178, 225
Bleecker AB, 180, 228
Blench IP, 87
Blevins DG, 180, 251, 350
Bligny R, 251
Blobel G, 550, 562
Bloch R, 37
Blom TJM, 270, 271
Blom-Zandstra M, 458
Blonstein AD, 183
Bloom AJ, 580, 593, 594
Bloom MV, 471, 478
Blum B, 69
Blum DE, 162
Blum H, 497
Blumberg H, 550, 552
Blume B, 428, 438
Boag S, 461
Bochardt A, 200
Bochkareva ES, 476, 478–80, 482, 483
Bock A, 269, 271, 278
Bock KR, 81
Bock RM, 179
Bodem GB, 672
Boehni P, 555
Boerjan W, 34
Boerreigter METI, 616, 624
Boersig M, 248
Bogaerts P, 216, 220
Bogdan C, 666
Böger P, 340–42
Boggio S, 472
Bogumil Z, 522
Bohnert HJ, 125, 130, 183, 186, 449, 451
Boivin P, 507
Bol JF, 80
Boland W, 394, 408
Bolhár-Nordenkampf HR, 645, 646
Boller T, 227, 428, 453
Bolliger L, 550, 551
Bolton MA, 395, 396, 412
Boman HG, 564
Bonaventura C, 642
Bonen L, 62, 66, 68, 70, 73,
546
Bongi G, 635, 636, 655
Bönig I, 435, 437
Boniwell JM, 291, 297
Bonnard G, 62, 67, 69, 70
Bonneaud N, 223
BONNER J, 1–23
Bonner WD, 457
Bonner WD Jr, 561
Boot CJM, 409
Borchert S, 355, 595
Borkovich KA, 128
Borman S, 669
Börner T, 68
Borstlap AC, 458
Bossert M, 84
Boteva R, 477, 478, 480, 484
Botha FC, 588, 595–97
Botterman J, 279
Bottino PJ, 182
Boudet AM, 163
Boulter D, 38, 304, 306
Boulton MI, 83, 87–89, 91–95, 97, 102
Bouma D, 384, 385
Bouman H, 408
Bourguignon J, 324–30, 332, 350, 559
Bousfield IJ, 200
Boutin JP, 581, 597
Boutry M, 54, 66, 216, 217, 219, 220, 546, 555, 559, 560, 564
Bouvier F, 290, 291, 297
Bouyssou H, 458
Bowers GR, 81
Bowes G, 343, 380
Bowlby NR, 294
Bowler C, 342, 343, 559
Bowsher CG, 593–95, 598, 599
Bowyer P, 88, 95
Boyes DC, 228
Boyle FA, 344, 585
Boyle SM, 258
Bozak KR, 266
Bracher D, 267, 278
Bradbury M, 636
Bradford KJ, 127, 406
Braig K, 471, 483
Brain RD, 149
Brambl R, 559
Bramley P, 291
Brand L, 86–88, 94, 95, 97–99, 176
Branden C-I, 344
Brandes ME, 665, 666
Brandt A, 546, 550, 562
Brandt U, 556
Branton D, 621
Braslavsky SE, 494–96, 504, 511, 512, 514, 520
Brauer M, 250
Brauer O, 293, 294
Braun AC, 188, 190
Braun H-P, 550, 555, 556,
559
Braus GH, 48
Bravdo BA, 522
Bray CM, 434
Bray EA, 113, 115, 117–19
Breekland AE, 114
Breidenbach RW, 613
Brendle-Behnisch E, 591
Brenner ML, 130
BRENNICKE A, 61–78; 62–73, 546
Bressan RA, 120
Bretscher MS, 612
Breu V, 126
Brewer PE, 358
Brewster JL, 229
Breyne P, 57
Brich M, 145–47
Briddon RW, 83, 84, 87, 89, 92, 94, 97
Brieva R, 670
BRIGGS WR, 143–71; 144–50, 152–56, 158, 159, 161, 163
Brinegar AC, 179
Brinkhorst-van der Swan DLC, 114, 130, 294
Briquet M, 546
Briskin DP, 212
Brisson N, 80, 265, 279
Britton G, 288, 295
Britz SJ, 149
Brobst SW, 669, 671
Brodmann P, 49, 56, 57
Broekaert WF, 120
Broger C, 457
Brooks A, 241–43
Brooks L, 49, 53, 54
Brosehart H, 438
Brough CL, 81, 82, 87, 88, 94, 98, 99, 103
Brouquisse R, 585
Brown BD, 439
Brown CR, 295
Brown GG, 62, 67
Brown HR, 346
Brown JW, 430
Brown MS, 610, 611
Brown PH, 423, 428, 434, 435
Brown SA, 157
Brown SC, 94
Browning CK, 98, 99
Browning KS, 81
Browse JA, 382, 384, 385, 552
Bruce D, 596
Brugnoli E, 634, 641
Brummell DA, 530
Brundish H, 83, 84
Bruneau J-M, 248
Brunk DG, 349
Brunk U, 611
Brunner J, 550, 551
Bruns B, 156
Brunschier R, 482, 484

Brutsch H, 550
Bryan KK, 585
Bryce JH, 382
Brzobohaty B, 401, 411, 412
Bubrick P, 80
Buchanan BB, 236, 239, 356, 357, 581, 588, 589, 594, 598
Buchholz WG, 102
Buchmann I, 176
Buchner J, 476, 481, 482, 484, 485
Buck KW, 81, 83, 84, 86, 87, 89, 91–95, 97–99, 101, 102, 105
Buckle KA, 297
Buckner B, 291, 293, 295
Budde RJA, 587
Buetelmann P, 181
Bults G, 494, 496, 497, 500, 502, 504, 505, 509, 512, 513, 518
Bunning E, 37
Burchartz N, 346
Burke JJ, 343, 347
Burnell JN, 347, 371, 381, 382, 384–86, 449
Burnett RJ, 270, 271
Burr G, 636
Burrell M, 98, 99, 101
Burris RH, 451
Burrows PR, 41
Burton JD, 354
Burton S, 343
Burwen SJ, 610, 611
Buschmann C, 495, 497, 501, 502, 508, 518, 521, 522
Buser C, 453
Buser-Sutter C, 458
Bush DR, 213, 225
Bush LP, 275
Büssis D, 250
Bustos MM, 122
Butcher G, 117
Butcher GW, 395, 396, 412
Butler WL, 149, 150
Buttner C, 311
Buttner G, 26, 29, 31
Buurmeijer WF, 497, 500, 507, 512, 518, 519
Byrd GT, 346
Byrne DH, 431
Byrnes JB, 667

C

Caboche M, 296
Cahen D, 5, 494–97, 500–6, 508, 509, 512, 513, 515, 516, 518, 519
Calderwood SK, 125, 128, 553
Caldwell MM, 144
Caldwell RM, 580, 593, 594
Caligari PDS, 207

Callaham DA, 539
Callis JB, 496, 511, 512
Callis R, 96–101
Callis RJ, 83, 84
Camara B, 288, 290–92, 297
Camm EL, 512, 513
Cammue BPA, 120
Campbell A, 669
Campbell D, 272
Campbell GS, 650
Campbell WH, 589, 591, 592, 598
Campbell WJ, 243, 344
Campell BR, 181
Campos N, 394, 398, 399, 411, 412
CANAANI O, 493–526; 494, 496, 500, 501, 504, 506–10, 512, 513, 516, 518, 519, 521, 522
Cannon F, 56
Cannon M, 56
Canotta H, 207
Canut H, 458
Canvin DT, 371, 373, 376, 377, 386, 579, 585, 592–94, 597
Cao W, 550, 551
Caplan AB, 176
Carbonero P, 48
Cardellina JH II, 667
Cardenas J, 378
Cardoso MIL, 267, 270
Carlberg I, 637
Carlson C, 669
Carlson JE, 70
Carlson PS, 207
Carlson SJ, 375
Carnal NW, 351, 357
Carpenter G, 610, 611
Carpenter JF, 125
Carpenter R, 275
Carpentier R, 501, 503, 504, 512, 513, 519
Carpita N, 614
Carrayol E, 591
Carrillo N, 472
Carroll KL, 439
Carson CB, 123, 131
Carswell S, 51, 52
Cashmore AR, 153, 156
Caspar T, 161, 247, 251, 552
Cassells AC, 198
Castor TP, 667
Caton J, 84
Cattaneo R, 73
Caubergs R, 149, 150
Caubergs RJ, 149, 150
Causton HC, 429
Cawood MC, 588
Cayley PJ, 430
Cazenave LA, 665
Ceccarelli EA, 472
Cegielski R, 504, 512
Cerdá-Olmedo E, 296
Cha Y, 501, 512, 515

Chadderton AR, 581
Chadha KC, 431
Chadwick AV, 533
Chakravorty AK, 433
Chaleff RS, 207
Chamberlin LCL, 83, 84
Chambon P, 90, 414
Chamovitz D, 290, 292
Champigny M-L, 248, 578, 584–86, 597
CHANDLER PM, 113–41; 115, 116, 118, 123, 127
CHANDRA S, 609–31
Chandrasekhar GN, 472, 475, 477, 479
Chang A, 227
Chang C, 228
CHANG C-j, 663–74; 668
Chang K, 451, 454
Chang K-T, 595
Chapdelaine Y, 68, 70, 546
Chaplin JF, 275
Charbonneau H, 226
Charland M, 504, 506, 512, 513, 515, 518, 520, 521
Charlebois D, 520
Charlebois DCN, 513, 515, 520
Charlwood PA, 610
Charpenteau M, 550, 552, 553, 559
Chatani M, 84, 88, 94
Chatfield JM, 177
Chatterton NJ, 249
Chaudhury AM, 181
Chaumont F, 555, 559, 560, 564
Chavrier P, 612
Chee PP, 200
Chen B-X, 667
Chen C, 176, 184, 186
Chen CS, 670
Chen H-H, 130
Chen J, 457
Chen L, 225
Chen R-D, 585, 586
Chen TH, 318
Chen WJ, 551, 560
Chen Z, 120, 394, 411
Cheng C-L, 38
Cheng MY, 472, 478, 547, 550, 553, 554, 562
Cheong J-J, 394, 616
Chesny RH, 200
Chetelat RT, 216
Cheyne VA, 207
Chindemi PA, 610
Chinkers M, 611
Chiou T-J, 225
Cho B-H, 579, 581, 584, 594
Cho C, 349
Cho H-Y, 354
Choe HT, 307
Choi HK, 56
Choi J, 313
Choi S-Y, 399, 401, 402

Chollet R, 344, 345, 449, 586
Chon HP, 152
Chory J, 38, 158, 160, 161, 333
Chow RL, 474, 481, 482, 484
Chow WS, 634, 636, 644
Chretien P, 127
Chriqui D, 178, 411
Chrispeels MJ, 39, 178, 271, 307, 314, 428
Christeller JT, 470, 481, 482, 485, 486
Christen AA, 670, 671
Christensen G, 215, 217, 219
Christiansen J, 184
Christoffersen RE, 266
Christopher DA, 159
Chu NM, 48
Chua N-H, 39, 48, 54, 55, 57, 117, 121, 122, 155, 158, 160, 333, 559, 564
Chuang DT, 472
Churio MS, 494, 497, 511, 514
Chvojka L, 529, 530, 533
Chye M-L, 559
Cianzio SR, 201, 202
Claeys D, 457
Clark AG, 425
Clark KR, 425, 435
Clark WG, 272
Clarke AE, 422, 425, 435, 437, 438
Clayton DA, 62, 66, 429, 430
Cleary AL, 536, 540
Cleland RE, 162, 394, 407, 616, 634, 636, 637, 639, 642
Clericus M, 552, 553, 559
Cline K, 565
Cloney LP, 471, 473
Close TJ, 116, 123, 125
Clugston C, 155
Cochran DS, 397, 401, 412
Cockburn W, 452
Codd GA, 373
Coen ES, 27, 275
Coenen C, 407
Coffin RS, 84
Coggon P, 663
Cohen A, 115, 118, 119
Cohen JD, 177, 397, 401, 402, 412
Cohen S, 610, 611
Cohen SJ, 92, 610, 611
Cohn ZA, 610
Colbert JT, 431
Cole GC, 200
Cole L, 614, 617, 619, 623
Coleman CE, 425
Coleman GD, 318
Coleman J, 614, 617, 619, 623
Coleman JR, 373, 374, 376–78, 382, 384–86
Collins GB, 275

Collins MKL, 434
Collins PD, 425, 435
Colman B, 378
Colwell G, 313
Comai L, 449
Combettes B, 67, 70, 71
Condron R, 472, 477, 485
Conkling MA, 38
Connelly JA, 354
Connolly JA, 539
Constabel F, 272, 619
Cook CM, 344
Cook SK, 117, 180
Cooke DL, 198, 200
Cooks RG, 665, 668
Cooper C, 39
Cooper G, 179
Copeman RJ,200
Corbin JD, 148
Cordell GA, 663, 665
Corell RA, 69
Cork A, 26, 36, 37
Cormier MJ, 226
Cornelissen B, 80
Cornelius MJ, 351, 352
Cornish EC, 423, 435
Cornwell KL, 579, 583, 585, 597
Corpe WA, 199–203
Coruzzi G, 155
Coruzzi GM, 591
Coscia CJ, 266, 267, 270, 272
Cosgrove D, 152, 160
Cosgrove DJ, 161, 162
Cosio EG, 616
Cosowsky L, 98, 99
Cosset A, 69
Costa MA, 56, 313
Costandy H, 73
Cota GF, 648
Coulson AR, 62, 66
Coutts RHA, 81, 83, 84, 87, 89, 91–95, 97, 99, 101, 102, 105
Cove DJ, 181, 183
Covello PS, 62, 64, 67, 69, 70, 72, 73
Cowan IR, 371, 382, 384, 385
Cowan NJ, 474, 481, 482, 484
Cox B, 83, 88
Cox BS, 88, 89, 91, 93, 97
Cox RP, 472
Crafts-Brandner SJ, 315
Craig EA, 127, 550, 552, 559, 560, 562, 563
Craig S, 181
Craig-Kennard AC, 354
Cram WJ, 610, 613
Crawford NM, 226
Creach E, 349
Creelman RA, 113, 124, 311, 312
Creighton TE, 470, 480–82
Creissen G, 81, 82, 99, 102
Crespi M, 176

Crespi S, 83, 84
Cresswell CF, 593
Cretin C, 449
Crich J, 670
Crichton RR, 611
Crispeels MJ, 250
Crispino JD, 122
Critchley C, 636, 637, 639
Cronan JE, 324
Cross JW, 394, 406
Croteau R, 266, 672
Crouch ML, 117, 121
Crowe JH, 125
Crowe LM, 125
Crowell DN, 186, 187
Crown J, 665
Croy RRD, 314
Crozier A, 411
Cséke C, 354, 356, 357, 581, 589, 598
Cudny H, 423
Culbertson MR, 227
Cullen JJ, 636, 648
Cummings DW, 66
Cumsky MG, 548, 556, 557, 562
Cunningham FX Jr, 292
Cunningham K, 546, 562
Cuozzo M, 54, 55
Curry GM, 152
Curry J, 115
Curson SJ, 87, 89
Curtin TF, 198
Cushman JC, 449, 451
Cutler AJ, 265
Cyr ER, 318
Cyr RJ, 536, 539
Czaja I, 190
Czosnek H, 83, 84
Czygan F-C, 636, 642

D

Dahlin C, 565
Dailey FA, 591
Dainese P, 295
Dairaghi DJ, 429, 430
D'Alessio G, 422
Dalling MJ, 306
Damsteegt VD, 87, 88, 94, 95, 97
Dance GSC, 429
Danger DP, 411
Daniel JM, 228
Daniel SG, 407
Daniel TO, 611
Danielson JDS, 496, 511, 512
Danner M, 482, 484
Dannhauer I, 155, 157
Darby LA, 294
Darr SC, 430
DAS OP, 79–112; 87, 94, 96, 104, 105
Das S, 99, 102
Dathe W, 311
Dau H, 505

Daum G, 560, 561
David C, 404
David NR, 553
Davidonis GH, 532
Davie JR, 472
Davies DR, 546
Davies JW, 81, 83, 84, 87, 88, 92–95, 97, 102
Davies KM, 291, 297
Davies PJ, 174
Davies WJ, 113, 115, 118
Davioud E, 404
Davis AW, 179
Davis DL, 275
Davis JM, 318
Davis KR, 312
Davis NW, 430
Davis PD, 148
Dawson AJ, 546
Dawson GW, 266
Dawson WO, 80
Day AG, 98, 99, 101
Day CD, 556, 559
Day DA, 331, 455–57, 459, 546
De A, 425
Deal L, 352
Dean BB, 295
Dean C, 48, 226, 229
Deardorff JA, 439
Debanne MT, 610
Debellé F, 185, 187
Debergh PC, 201
De Beuckeleer M, 440
De Block M, 440
deBoer AH, 394, 616
DeBoer SH, 200
de Bruijn FJ, 187
de Bruijn MHL, 62, 66
Debus G, 404, 406
de Carolis E, 268
Decottignies P, 449
DeFabo E, 152
de Gier J, 561
De Greef JA, 150, 188
De Greve H, 48
Dehio C, 187
de Kam RJ, 273
Dekeyser RA, 56
Dekker CA, 431
Dekker EL, 83, 88, 89, 91, 93, 97
de Klerk GJM, 182
de Koning P, 270
de Kruijff B, 561
Delanoy MR, 293, 294
de Larrinoa IF, 125
de la Serve BT, 186
de la Torre A, 581, 594
Delbarre D, 404, 405
delCardayré SB, 422, 424, 425, 428, 432, 433, 435
Delcher E, 69, 546
Deleens E, 449
Delgado B, 581, 594
Delgado E, 242

DeLisle AJ, 117, 121
DellaPenna D, 294
Dell'Orto P, 546
Delmer DP, 614
De Loose M, 186, 559
Delorme S, 62
del Rio LA, 342, 343
Delrot S, 355
de Luca V, 258, 265, 268, 271, 272, 279
deMichaelis MI, 616
Demmer A, 617
Demmers C, 318
Demmig B, 635, 636, 642, 645
Demmig-Adams B, 238, 241, 246, 288, 294, 296, 633, 634, 636, 637, 640, 642–46, 652
Dénarié J, 185, 187
den Boer B, 34
Dench JE, 327
Deng X-W, 161, 429
Denis JN, 669
Denis M-H, 355
Dennis DT, 471, 475, 478, 587, 588, 596, 598
Dennis ES, 181
Deom CM, 87
De Palma A, 457
Depaoli RAA, 227
de Pater S, 188
Depicker A, 57
Depicker AG, 56
Depta H, 614, 617, 619, 622, 623
Derbyshire E, 304, 306
Derksen J, 536
Deroche M-E, 591
DeRocher AE, 553
Derrick P, 623
De Santis A, 457
DeScenzo RA, 279
Deshaies RJ, 550, 552, 562
Desomer J, 176
de Souza AP, 546
Deurenberg-Vos HWJ, 199, 205, 206
Deus B, 246
Deutsch HF, 370
Deutscher MP, 439, 440
Devagupta R, 267, 278
De Valoir T, 229
de Waal A, 266, 267, 270, 271
DeWald DB, 311–13
Dewey RE, 546, 559
DeWit CT, 317
DeWitt ND, 39, 216, 219
DeZazzo JD, 51, 52
De Zoeten GA, 80
Dhaese P, 48
D'Harlingue A, 291
Diallo BM, 667
Dias MA, 635
Diaz CL, 624

Diaz E, 346
Díaz M, 644, 652
Dickinson CD, 250
DiCosmo F, 669, 671
Diekmann W, 614
Dietmeier K, 550, 551
Dietrich A, 62, 66, 69, 71, 175
Dietrich JT, 180
Dietrich MA, 148
Dietrich RA, 39, 449
Dietz K-J, 240, 242, 458
Dietze P, 279
Dillenschneider M, 550, 552, 553, 559
Dimitriadis E, 588, 592, 593
Ding A, 666
Ding AH, 666
Dionisio-Sese ML, 377, 378
Distefano M, 579
Ditto CL, 358
Dittrich H, 263, 269, 271, 272, 278
Dixon NE, 481, 485, 487
Dixon RA, 276
Dixon SC, 177
Dobres MS, 156
Doddema H, 225
Dodge AD, 342, 358
Dodge JD, 375
Doehlert DC, 248, 357
Dogbo O, 288, 291
Dohlman HG, 221
Dohrman U, 397, 400
Dolan L, 26, 31–34, 37, 564
Dolja VV, 87
Domdey H, 52, 53
Domec C, 70
Dominov JA, 186, 187, 410
Donaldson G, 478, 481, 486
Donaldson GK, 470–72, 477, 478, 481, 486
Donehower RC, 665
Donell WC, 610
Dong X, 312
Dong Z, 477, 480
Donnelly AE, 205
Donson J, 83, 84, 86–89, 92–95, 97, 102
Doohan ME, 539
Dorhout R, 41
Dosch DC, 222
Douce R, 62, 251, 324–32, 350, 448, 449, 457, 546, 565, 585, 592, 593, 559
Douglas MG, 550, 551, 560
Dräger B, 257, 260
Drechsler Z, 379
Drew MC, 121
Droog FNJ, 410
Dröscher L, 456
Drücke M, 639, 641
Drucker M, 619
Drumm-Herrel H, 158
Dry IB, 83, 84, 331, 349, 588, 592, 593

Drygas ME, 550, 562
Dube SK, 182
Dubinsky Z, 648, 655
Ducharme D, 500
Duck NB, 552, 553
Duckham SC, 130, 293, 294
Dudel A, 155, 157
Dudley MW, 185
Duell-Pfaff N, 156
Duff SMG, 588
Duffus JE, 92
Duine JA, 264
Duke SO, 358
Dumas R, 330
Dumont ME, 550, 551
Dunleavy JM, 200, 201
Dunn WA, 612
Dunsmuir P, 48
Dupont I, 504, 521
Durand B, 177
Durand R, 177
Dure L, 117, 125
Durnam DJ, 531
Dwyer JG, 270
Dwyer KG, 228
Dwyer MR, 370, 383, 386
Dwyer ND, 229
Dybing JK, 201
Dyer TA, 188, 189
Dymock D, 188
Dyson WH, 184

E

Earnshaw BA, 411
East TM, 635, 636, 645
Ebbighausen H, 331, 457, 592, 593
Ebel J, 616
Eberle J, 180
Ebert PR, 422, 423
Echevarria C, 586
Echols H, 473, 487
Ecker JR, 174, 180
Eckerskorn C, 474, 475, 481, 484
Eddy P, 88, 89
Edel H-G, 404, 406
Edelmann H, 530, 532
Edgerton MD, 400
Edmondson DL, 245
Edwards CG, 295
Edwards G, 382, 383, 448, 449
Edwards GE, 239, 242, 331, 344–47, 349–51, 355, 359, 360, 384, 385, 448–51, 643
Edwards J, 588
Edwards JW, 591
Edwards R, 276
Edwards YH, 73
Egli CM, 48
Egli DB, 305, 315
Egli MA, 354, 359
Ehmann B, 156, 158, 160
Ehrenschaft M, 559

Ehrlich M, 94, 98, 99, 103
Eichholz R, 184
Eilers M, 551, 560, 562
Eilert U, 272
Einset JW, 184
Einspanier R, 155, 157
Einzig AI, 665
Eisinger W, 533
Eisinger WR, 528, 533
Ekman P, 380
Ekstein I, 546
Elias BA, 585, 586
Elliott MC, 408
Ellis N, 332
Ellis RJ, 470, 471, 473, 478, 551
Elliston KO, 86, 93, 97
Ellwood S, 89, 96, 97, 99, 101
Elmer JS, 86–88, 93–95, 97–99, 105
Elmer S, 95, 97, 99
Elmlinger MW, 155, 158
Elrifi IR, 580, 581, 583, 593, 594, 597
El-Sawi Z, 188
El-Shora HM, 449
Elstner EF, 448
Elzenga JTM, 162, 379
Emes MJ, 591, 593–95, 598, 599
Emmermann M, 550, 552, 553, 555, 556, 559
Emons AMC, 615, 617, 619
Endara ME, 312
Endo T, 270, 562
Engel K, 484
Engelhardt K, 118
Englund PT, 63
Enomoto M, 537
Entwistle G, 588
Ephritikhine G, 402, 404, 405, 616
Epstein W, 222, 228
Erdei S, 432, 434
Erdjument-Bromage H, 474, 475, 481, 482, 484
Erdos GW, 471
Ericson I, 349
Ericson MC, 352
Eriksson A-C, 555, 556
Eriksson AE, 385
Erion JL, 179
Erlanger BF, 667
Ermak G, 98, 99, 103
Ernst JF, 550, 551
Ernst SG, 318
Ertl JR, 186
Erwee MG, 624
Esau K, 28, 92
Esnault R, 434
Espelund M, 116
Espie GS, 371, 373, 377
Estabrook RW, 267
Estelle MA, 26, 38, 40, 113, 174, 180, 183, 189

Estruch JJ, 178, 189, 190, 411
Etessami P, 86, 88, 96–98
Ettinger DS, 665
Eubanks S, 38
Evans D, 88, 95, 617, 619
Evans DM, 279
Evans HJ
Evans IM, 38
Evans JR, 384–86, 635, 636, 648
Evans LV, 380
Evans M, 397, 401, 412
Everett M, 156
Evstigneeva ZG, 471, 475
Ewart AJ, 636
Ewing NN, 216
Eyletters M, 521, 522

F

Fairbairn JW, 268
Fairley-Grenot KA, 164
Falco SC, 341, 352, 358
Falconet D, 62
Falk S, 636, 649, 650
Falkiner FR, 207
FALKOWSKI PG, 633–61; 634, 635, 641–43, 645–48, 655
Fall R, 204
Fambrini M, 293, 295
Fang HH, 224
Fantauzzo F, 113
Farage PK, 636, 645, 649, 653–55
Faria JC, 84
Farkas GL, 422, 423, 430, 432, 434, 437, 440
Farmer EE, 227, 312
Farquhar GD, 240, 244, 245, 250, 448, 460, 586, 594, 635
Farr GW, 474, 481, 482, 484
Faryar K, 230
Faucher C, 41
Fauron CM-R, 65
Favre D, 425
Favreau M, 48
Fawcett TW, 382, 384, 385
Fayet O, 470–73, 479
Feagin JE, 66
Featherstone DR, 181, 183
Fechheimer M, 226
Fecker LF, 279
Feenstra WJ, 225
Feierabend J, 348
Feil R, 244, 247, 357, 585, 589, 595–97
Feiler HS, 65
Feinbaum R, 161
Feinbaum RL, 156, 158
Feld A, 353, 354
Feldman KA, 180
Feldman LJ, 27, 28, 36
Feldmann KA, 226
Feldwisch J, 394, 395, 398,

399, 401, 408, 411, 412,
616
Felix G, 177, 183, 184, 186,
227, 410
Felle HH, 214, 229, 405,
458, 459
Feller U, 304, 305
Fellous A, 539
Fellows RJ, 248, 250
Felts JM, 354
Feng X-H, 182
Fenko M, 587
Fenoll C, 90
Fenton WA, 485, 486, 554
Ferenc N, 564
Ferguson C, 215, 217
Fernandes JA, 265, 272
Fernandez JM, 458, 462
Fernando M, 214
Fernbach E, 147, 148
Fernyhough P, 596
Ferraira GC, 559
Ferrant V, 216
Ferreyra RG, 472
Fersht AR, 476, 477
Fett-Neto AG, 645, 647
Feyerabend M, 394, 616
Fickenscher K, 594
Fieder B, 68
Fiedler F, 272
Figeys H, 522
Filipowicz W, 51, 91, 429,
430
Filippini F, 411
Filner P, 154
Finazzo J, 580, 593, 594
Findenegg G, 448, 453
Finer JJ, 206
Finkelstein R, 48, 117
Finkelstein RR, 117, 131
Finley D, 221
Firn RD, 131
Fischer P, 591
Fischer RL, 126
Fisher DG, 624
Fisher LM, 317
Fisher MT, 478, 481, 482,
484
Fisher RF, 41
Fitchen J, 268
Fladung M, 178, 411
Flanagan J, 473, 487
Fleming AJ, 258
Fleming PE, 672
Flesher D, 592
Fletcher JD, 291, 297
Fletcher LM, 205
Flieger K, 67
Flores S, 186
Floss HG, 263, 672
Flügge UI, 355, 448, 453,
455, 456, 458–60, 462, 585
Fluhr R, 54, 155, 158, 160
Flynn GC, 476, 479, 480, 483
Fock HP, 370, 374–77, 379,
386

Focke M, 353, 354, 358
Folk WR, 552
Follin A, 188
Fondy BR, 241, 247–49
Fong F, 117, 120, 383, 384
Fontes EPB, 87, 90, 93
Foote M, 307, 317
Ford CW, 451
Ford S, 38
Forde BG, 304
Fork DC, 497, 500, 512, 513,
515, 518–21
Forseth IN, 634, 640, 649,
650, 652
Förster J, 450
Fosket DE, 185
Foster R, 122, 184
Fougère F, 455
Fowke LC, 398, 399, 402,
539, 617, 619, 621, 622,
626
Fowler MW, 591, 594, 595
Fowler TJ, 39
Fox GG, 591
Fox JE, 179
Fox TD, 65
Foyer C, 238, 248, 251, 296,
356, 357, 585, 586, 597
Foyer CH, 248, 252, 581–83,
591, 597
Frackowiak D, 504, 512
Fragata M, 512, 513
Fraley RT, 54, 87, 96, 176
Franceschi VR, 307, 308,
312, 315, 317
Francis R, 99, 102
Franck M, 126
Frank MJ, 226
Franke-van Dijk MEI, 270
Franklin AE, 217
Franklin LA, 636, 637
Franklin RM, 227
Franssen H, 39
Fray RG, 293, 294
Freas S, 343
Fredrikson K, 148
Freeberg JA, 149
Freire MA, 116
French R, 80
Frenzel T, 263
Freundt H, 621
Frey S, 561
Frey T, 616
Fricaud AC, 565
Fried M, 438
Frischmuth S, 89–91
Frischmuth T, 81, 84, 87–89,
93, 95, 99, 101
Frommer WB, 225
Frydman J, 474, 475, 481,
482, 484
Frye RB, 179
Fuchigami L, 318
Fuchs JA, 373
Fuchs M, 481, 482, 485
Fujimori T, 534

Fujimura T, 559
Fujita T, 179, 258
Fujita Y, 643, 646
Fujiwara H, 268
Fujiwara K, 326, 328
Fujiwara S, 377, 378
Fujiwara T, 122
Fuks B, 522
Fukui H, 269, 277
Fukui T, 394, 398, 399, 401,
402, 412
Fukumoto H, 317
Fukuzawa H, 71, 373, 377,
378, 385, 386
Füllgraf H, 454
Funatsu G, 424, 425
Funck C, 257, 260
Furbank RT, 251, 346, 381,
458, 461
Furner IJ, 42
Furuya F, 471, 483
Furuya M, 38, 144, 159, 163,
530
Furze JM, 279
Fusakawa T, 52, 53
Futai M, 394, 395, 398, 399,
401, 402, 412

G

Gaba V, 150, 158
Gaber RF, 212, 223, 224
Gäbler L, 66
Gabrielli AC, 497, 508
Gadal P, 449, 585, 586, 591,
595
Gaestel M, 484
Gage DA, 113
Gagnon J, 327, 332
Gaillard J, 585
Gaiser JC, 407
Galasinski SC, 343
Galau GA, 116–18, 120
Gallagher S, 144, 145, 227
Galland P, 152, 154
Gallegos CL, 647, 654
Gallie DR, 178, 439
Galneder E, 261, 264, 268,
269, 277
Galston AW, 152
Galtier N, 248, 252
Galun M, 494, 508, 509, 512,
513, 516, 522
Galway M, 38
Galway ME, 535, 617, 619,
621, 622, 626
Gamble PE, 157
Gamborg OL, 670, 671
Gamon JA, 634, 641
Ganem D, 51, 52
Gantt E, 292
Gao Y, 474, 481, 482, 484
Garbers C, 398, 399, 401
Garcia-Olmeds F, 48
Garcia-Reina G, 379, 380
Gardemann A, 345, 589

Gardeström P, 251, 349
Gardiner WE, 86–89, 94–99, 103
Gardner G, 157
Gardner GM, 394
Gardner R, 88, 89
Garland WJ, 588
Garneau FX, 671
Garrett S, 227
Gartner F, 549, 550
Garty J, 494, 508, 509, 512, 513, 516, 522
Gass DA, 65, 546
Gasser S, 555
Gasser SM, 560, 561
Gast RT, 124
Gatehouse JA, 38
Gatehouse LN, 38
GATENBY AA, 469–91; 470–72, 475, 477–79, 481, 482, 484–86
Gatz C, 230
Gautier H, 163
Gaxiola R, 125
Gay MR, 81, 84, 87
Gay SL, 275
Gebhardt C, 410
Gee SL, 597
Gegenheimer P, 430
Gehrke P, 246
Geider RJ, 646
Geigenberger P, 598
GEIGER DR, 235–56; 236, 238, 239, 241–50, 252
Geissler A, 594
Gelvin SB, 55
Genchi G, 457
Genetello C, 186, 188
Gengenbach BG, 354, 359
George MW, 200
Georgopoulos C, 470–73, 475, 477–79, 481, 483
Geraghty D, 48
Gerard VA, 634, 642
Gerhardt R, 242, 244, 249, 448, 450, 453, 461
Gething M-J, 551
Geuze HJ, 611, 612
Gharyal PK, 614
Gheysen G, 186
Ghiorzi TJ, 669
Ghislain M, 404, 616
Giannattasio S, 556, 557
Giaquinta RT, 307, 308, 317
Gibbs MJ, 382, 384, 385
Gibbs SP, 373, 375, 376
Gibson DM, 670, 671
Giddings TH, 528, 539
Gidoni D, 90
Gielen J, 48, 186
Gierasch LM, 479, 482
Gietl C, 559, 564, 565
Gilbert JE, 207
Gilbertson RL, 84
Gilman GA, 151
Gilmore AM, 642

Gilmore-Heber M, 216
Gilmour SJ, 116, 121, 131
Giordano M, 379, 380
Giraudat J, 131
Giroud C, 267
Girshovich AS, 476, 478–80, 482, 483
GIULIANO G, 287–301; 290, 293–97
Givan CV, 344, 348–52, 354, 359, 360, 585, 586
Glaser E, 555, 556, 559, 564
Glaser SM, 548, 556, 557
Glass ADM, 214
Glaubitz JC, 70
Glenney JR, 610
Glick B, 546, 559, 560
Glick BS, 546, 548, 559, 562, 563
Gloudemans T, 38
Glover LA, 546
Glund K, 423–25, 428, 433, 438
Goday A, 116, 119
Goddijn OJM, 266, 267, 271, 273, 279
Goffeau A, 212, 216, 227, 546
Gogarten JP, 614
Goh CJ, 452
Gohlke JR, 424, 425
Goldberg GL, 665
Goldberg RB, 440
Goldmark PJ, 117
Goldschmidt EE, 238, 250, 251
Goldsmith MHM, 149, 150, 404
Goldstein AH, 433
Goldstein JL, 610, 611
Goldthwaite JJ, 186
Goller M, 454
Goloubinoff P, 470, 472, 475, 481, 482, 485, 486
Golz A, 353, 354
Gomez J, 115, 116, 119
Gommers FJ, 41
Gonzalez DH, 67, 345, 346
Goodall GJ, 51, 91
Goodbody A, 270
Goodchild DJ, 634, 636
Goodfellow M, 200
Goodlove PE, 87, 88, 95, 104
Goodman HM, 38, 131, 156, 161, 341, 342, 352, 358
Goodman RM, 81, 84, 92
Goodman RN, 624
Goodwin PB, 624
Goodwin TW, 288, 294
Gopalraj M, 124
Gorbalenya AE, 93
Gordon JF, 205
Gordon MP, 176, 318
Gorecki RJ, 204
Göringer U, 69
Gorton HL, 157

Gorvel J, 612
Gossele V, 440
Gottlieb LD, 244, 247, 357
Gou DM, 670
Goud B, 612
Gough S, 293, 295
Gougler JA, 247
Gouterman M, 496, 511, 512
Goyal A, 377, 378
Goyvaerts E, 54, 55
Gradmann O, 610, 613
Graham D, 370, 383, 386, 593
Graham IA, 449
Grant G, 314
Graves PR, 227
Graves PV, 70, 546
Gravett A, 655
Gray JE, 425, 435, 437
Gray MW, 62, 64, 65, 67–73
Gray TE, 476, 477
Gray VM, 593
Graybosch R, 49, 50, 52, 56
Greber RS, 85
Green JH, 588
Green PB, 35, 528, 533, 540
GREEN PJ, 421–45; 54, 55, 333, 422–25, 428, 431–33, 435, 439
Green PN, 200
Green TGA, 621, 645
Green TR, 119
Greenberg E, 350
Greenblatt IM, 293, 294
Greene AE, 669
Greene RM, 634, 642, 646
Greenwood JS, 308, 311, 314, 318
Greer DH, 637, 644, 645
Gregorini G, 179
Greppin H, 507, 521, 522
Gressel J, 150, 158
Gressel JB, 152
Gresshof PM, 459
Greyson MF, 587, 588, 598
Gribnau JH, 409
Griebenow S, 495, 520
Grienenberger J-M, 62, 67, 69, 70, 546
Grierson D, 126, 290, 291, 293, 294, 297
Grierson P, 297
Griffin DA, 177
Griffin JD, 201
Griffin WH, 326
Griffing LR, 619, 621, 622, 624, 626
Griffiths A, 293
Griffiths DJ, 380
Griffiths G, 549–51, 610, 612
Grignon C, 448, 451, 452
Grill E, 458
Grimes HD, 312, 315
Grimm B, 157
Grimm F, 199
Grimm R, 470, 474, 475,

478, 481, 484, 486
Grimsley N, 99, 102
Grimsley NH, 181
Grodzinski B, 385
Groen BW, 264
Grohmann L, 62, 66, 72, 73
Gronenborn B, 80, 83, 84,
   86–88, 90–94, 96–98, 101,
   102
Gronwald JW, 354, 359
Groome QJ, 645, 646
Gross GG, 448
Grosse H, 448, 453, 460,
   592, 593, 595
Grosskopf DG, 227
Grossman AR, 378
Grossmann K, 178, 411
Grothaus PG, 667
Grover SD, 347
Gruenberg J, 612
Gruissem W, 429
Grumbine FC, 665
Grumet R, 276
Grun P, 425
Grundler FMW, 41
Gruszecki WI, 504, 521, 522
Gualberto JM, 62, 67, 69, 70,
   546
Guarente L, 52, 126
Guénard D, 665, 666, 669,
   670, 672
Gubatz S, 288, 297
Guerbette F, 126
Guerineau F, 49, 53, 54, 83,
   88, 97
Guéritte-Voegelein F, 665,
   666, 669, 670, 672
Guern J, 400, 402, 404, 405,
   412, 448, 616
Guerrero MG, 643
Guiard B, 548, 557, 560, 567
Guilfoyle TJ, 148, 189
Guillemaut P, 66, 71
Guilloton MB, 373
Guiltinan MJ, 113, 121
Gundlach H, 272, 312
Gunn HV, 83, 84, 102
Gunning BES, 33, 36, 528,
   532, 537, 539
Günter G, 250
Gupta AS, 343, 347
Gupta RS, 471, 473, 475, 480
Gustin MC, 229
Guthrie EJ, 81
Gutteridge S, 343
Guy RD, 581, 585

H

Haas N, 38
Haber CL, 207
Haber S, 81
Habricot Y, 180, 182
Hack E, 472, 546, 550, 553
Hackert ML, 328
Hadley HH, 305

Hadziyev D, 428
Hageman RH, 246, 579, 591,
   592, 594
Hagen G, 148, 189
Hager A, 145, 146, 404, 406
Hagiwara S, 163
Haglund K, 379, 380
Hahlbrock K, 156, 261, 272,
   616
Hahn MG, 394, 616
Haigler HT, 610
Hainfeld JF, 471, 483
Hainline BE, 150
Hajek K, 412
Hajela RK, 121, 131
Hakim F, 268
Hakvoort TBM, 551
Haley A, 83, 84, 88–90
Haley BE, 397
Haley L, 101
Hall DO, 327
Hall JL, 274, 530
Hall MA, 394
Hall NP, 352, 353, 360
Hall RH, 184
Hall TC, 122
Hallahan BJ, 266
Hallahan DL, 266
Hallberg RL, 471, 472, 474,
   546, 550, 553, 562
Hallgren L, 118
Halliwell B, 327
Halperin W, 622
Halsall DJ, 476, 478, 481,
   482, 484, 486
Hamada H, 533, 536, 540
Hamaguchi N, 179
Hamblin MT, 122, 123
Hamill JD, 279
Hamilton AJ, 126
Hamilton WDO, 81, 83, 84
Hamm HE, 151, 164
Hammond SA, 425
Hampp N, 267
Hampp R, 448, 454
Hampsey DM, 52
Hanagata T, 409
Hangarter RP, 144, 152, 159,
   160
Hanhart CJ, 114, 134
Hanic-Joyce PJ, 62, 64, 67–
   69
Hankin L, 206
Haňkovská J, 529, 530, 533
Hanley-Bowdoin L, 86, 87,
   90, 93, 97
Hanning I, 585
Hannus R, 123
Hansen A, 118
Hansen CE, 184
Hansen U-P, 505
Hanson AD, 125, 276
Hanson MR, 65
Hanson RS, 200, 207
Hanson SF, 84
Hanway JJ, 305

Haouazine N, 546
Harada JJ, 39, 117, 449
Harbison S, 83, 84
Harborth J, 595
Harbron S, 356, 357
Hardacre AK, 645
Hardham AR, 528, 535–37,
   539, 540
Haring V, 422, 423, 425
Harland RM, 439
Harley SM, 621
Harmey MA, 552, 564, 566
Harmon AC, 226
Harper JF, 39, 184, 212, 213,
   216, 219, 226, 227
Harren FHM, 497
Harris RLN, 346
Harrison BD, 81, 85, 91, 92,
   101
Harrison JE, 266, 276
Harrison SC, 122
Harrison WG, 647, 654
Hart JW, 154
Harterich M, 581
Hartitz M, 81, 82
Hartitz MD, 88, 90, 91
Hartl F-U, 470–75, 477, 478,
   480–84, 487, 547, 550,
   551, 553–55, 559, 560,
   562, 563, 567
Hartman DJ, 472, 477, 481,
   485, 487
Hartmuth K, 48
Hartung W, 113, 123
Hasan M, 278
Haschke HP, 448
Hase T, 547, 549
Hasegawa K, 529, 535
Hasegawa PM, 120
Hasezawa H, 535
Hasezawa S, 538
HASHIMOTO T, 257–85;
   258, 260, 261, 271, 278,
   279
Hashimoto Y, 179
Haslam E, 665, 672
Hassel BA, 430
Hatano M, 258
Hatch MD, 346, 347, 371,
   381, 382, 384–86, 449,
   455, 456, 458, 461
Hattori T, 123, 131, 317
Hattum J, 458
Hatzopoulos P, 117, 120
Hauer CR, 179
Hauge BM, 131
Hauser M-T, 26, 32, 35, 36,
   39
Havaux M, 495, 496, 504,
   506, 507, 512, 515, 517,
   518, 521, 522
Havir EA, 348, 349, 359
Hawes C, 398, 399, 402,
   614, 617, 619, 623
Hawlitschek G, 550, 554, 555
Haworth P, 352

Hayashi A, 271
Hayashi H, 129, 190
Hayashi T, 411
Hayashida H, 86, 88, 95
Hayer MK, 473, 487
Hayes RJ, 86, 87, 92–95, 97, 102, 105
Hayward AC, 199
He J, 655
Head K, 88–91
Hearst JE, 291
Heath IB, 37, 539
Heath RL, 383, 384, 438
Heath TG, 113
Heber U, 239–42, 458, 461
Hecht SM, 179
Hedrich R, 448, 453, 458–60, 462, 614, 623
Heelis PF, 150
Heeringa GH, 158
Hegerl R, 474
Heidecker H, 48
Heidmann I, 94, 96
Heidmann S, 52, 53
Heihoff K, 496
Heineke D, 242, 244, 250, 345, 448, 453, 460, 592, 593
Heinen JL, 343, 347
Heino P, 116, 120, 131, 132
HEINSTEIN PF, 663–74, 614, 616, 624, 625
Heinz E, 129
Heisel R, 546
Heitefuss R, 433
Hejgaard J, 118
Heldt HW, 241, 242, 244, 249, 331, 345, 355, 448, 450, 451, 453–61, 585, 589, 592, 593, 595
Helenius A, 610
Hellen R, 332
Helmer GL, 222
Hemenway C, 101
Hemmerich P, 150
Hemmingsen SM, 471, 473, 475, 478
Hendrich W, 504, 512
Hendrick JP, 470, 471, 483, 551, 554, 556, 557
Hendrix R, 472, 475, 477, 479
Hendrix RW, 471, 474, 475
Henley WJ, 636, 637
Henrichs G, 355
Hepler PK, 528, 532, 536, 539
Herber B, 412
Herbert SK, 497, 500, 512, 513, 515, 518–21
Herbon LA, 63
Hergersberg C, 551
Herkt B, 614
Herman EB, 207, 402, 403
Herman LMF, 56
Herminghaus S, 279

Hernandez G, 56
Hernandez-Lucas C, 48
Hernould M, 70
Herold A, 238, 243, 244
Herrada G, 459
Herrmann RG, 556
Hershey HP, 230
Hertel R, 397, 400, 406
Herz U, 66
Herzig R, 635
Hess D, 423
Hess WM, 669
Hess WR, 68
Hesse F, 401, 408, 411, 412
Hesse JE, 228
Hesse T, 394, 395, 398, 399, 401, 412, 616
Hetherington AM, 113
Hetherington SE, 655
Heupel R, 448, 459, 592
Heuser JE, 610, 617
Hewitt EJ, 592
Heyraud F, 92, 93
Hibi N, 258
Hicks GR, 406–8, 412, 616
Hidayat SH, 84
Hidema J, 385
Hiebert E, 84
Hiesel R, 62–65, 67, 69–71, 73
Higashi S, 129
Higgins CF, 429
Higgins TJV, 123
Hildebrand DF, 315
Hildebrandt AC, 670
Hilf ME, 80
Hilhorst HWM, 114, 134
Hill CH, 148
Hill HM, 198
Hill RD, 115
Hillebrandt S, 279
Hillmer S, 617, 619–23
Hilson P, 439
Himelblau E, 39
Hind G, 501, 504, 512, 513
Hines ER, 535, 539
Hines V, 550
Hinton P, 293, 294
Hinz G, 617, 619
Hiraga K, 324, 326–28
Hiraoka Y, 52, 53
Hirasawa M, 595
Hirasuna TJ, 671
Hirata H, 448, 453
Hirsch AM, 398, 408
Hirschberg J, 290, 292, 546
Hirsch-Wyncott ME, 55
Hiser C, 546
Ho TD, 118
Ho T-HD, 395, 423, 428, 434, 435
Hobson GE, 297
Hoch B, 71
Hock B, 564, 565
Hockley DG, 370, 383, 386
Hodal L, 200

Hodge TP, 546
Hodges PE, 554, 556, 557
Hodges VS, 645, 655
Hofer JMI, 88, 89, 91, 93, 97
Hofer M, 646
Hoferichter P, 448, 453, 460, 592, 593, 595
Hoffland E, 448, 453
Hoffman NE, 217
Hoffmann A, 474
Hoffmann NL, 87, 96
Hoffmann S, 353, 354
Hoffmann-Benning S, 294
Hofhaus G, 471, 474
Hofmann C, 610, 611
Hofstra JJ, 225
Hofte H, 39
Hoge JHC, 190, 258, 266, 267, 270, 273, 279
Hogetsu T, 528–31, 533–35, 539
Hoggart R, 26, 36, 37, 423, 435
Höhler A, 257, 260
Hohn B, 93, 99, 102
Hohn T, 49, 55–57, 80, 99, 102
Hoj PB, 472, 477, 481, 485, 487
Hoke SH, 668
Holaday AS, 343, 347
Holbrook GP, 343
Holbrook JJ, 476, 478, 481, 482, 484, 486
Holder C, 382, 384, 385
HOLLAND MA, 197–210; 201, 202, 204, 206
Hollis JP, 200
Höll-Neugebauer B, 476, 481, 482
Holmes FA, 665
Holmes JJ, 581, 583, 597
Holm-Hansen O, 647
Holstein SEH, 617, 619
Holt JS, 358
Holtum JAM, 358, 451
Homble F, 522
Homma H, 411
Hondelmann W, 276
Hong B, 118
Hoogenraad NJ, 472, 477, 481, 485, 487
Hooley R, 123, 394
Hoover AJ, 669, 671
Hooykaas PJJ, 190, 410
Hope HJ, 591
Hopkins R, 407
Hoppe HH, 354
Horgan R, 113, 119, 183, 188, 293, 297
Horgan RH, 183
Horisberger MA, 80
Horiuchi H, 424, 425
Hormuzdi SG, 81, 82, 88, 89, 93, 94, 101

Horn MA, 614, 616, 624, 625
Horner HT Jr, 205
Horowitz PM, 476, 481, 483, 485
Horsch RB, 87, 96, 187, 198, 207
Horseele R, 184
Horsey AK, 595–97
Horsley D, 617, 619
Horsnell PR, 382, 384, 385
Horst M, 550, 551, 562
Horstmann H, 550, 551
Horton P, 581, 596, 597, 634, 636, 641
Horvath B, 38
Horvath DP, 121
Horvitz HR, 34
Horwich AL, 471–75, 477, 478, 480–84, 547, 548, 550, 554, 557, 562
Horwitz BA, 144, 152, 494, 496, 497, 500, 505, 509, 512, 513, 518
Horwitz SB, 664, 665
Houlne G, 217, 219
House CM, 456
Houwen PJW, 199, 205, 206
Howarth AJ, 84–86
Howell SH, 38, 83, 84, 90, 92, 93, 183, 312, 316, 410
Howitz KT, 356
Howlett B, 228
Hrazdina G, 239, 268, 276
Hsieh C-L, 80, 103, 104
Hsieh SC, 377
Hsu L-C, 225
Hsu-Ching C, 429
Hu CY, 200
Hu NT, 48
Huang J-F, 308, 310, 311, 315, 316, 559
Huang L, 217
Huang LK, 250
Huang S, 437
Hubbard AL, 612
Hubbard EJA, 227
Huber J, 248, 252
Huber JL, 248, 355, 357, 589, 591, 592, 598
Huber JLA, 589
Huber SC, 238, 247, 248, 250–52, 345, 355–57, 385, 448, 449, 582, 584, 589, 591, 592, 597, 598
Hübner R, 614, 617, 622, 623
Huchzermeyer B, 341
Hucklesby DP, 592, 595
Huffaker RC, 307, 578, 579, 583, 584, 594
Hughes DW, 116–18, 120
Hughes FL, 83–85, 92, 93
Hugosson M, 555, 564
Hugueney P, 288, 290–92, 297
Huiwen Z, 120
Hull MR, 634, 643

Hummel S, 225
HUMPHRIES S, 633–61
Hundal T, 637, 639, 644
Huner NPA, 639, 644
HUNT AG, 47–60; 48–50, 52, 54, 56
Hunt MD, 65
Hunter AS, 471, 479, 480, 483, 484, 486
HUPPE HC, 577–607; 590, 594, 597
Huprikar SS, 223
Hurt EC, 547, 557
Hüsemann W, 155, 157
Husic DW, 324, 325, 344, 348, 360
Husic HD, 324, 325, 344, 348, 360, 377, 378
Hutchinson EG, 471, 474
Hutchison RS, 242
Huttly AK, 123
Hwang S, 551
Hwang ST, 547, 550, 553, 559, 560, 562, 567
Hyeon S-B, 349
Hylton CM, 334, 564, 565
Hynes GM, 474

I

Iacobazzi V, 457
Iannace G, 556, 557
Ibenthal WD, 433
Ichikawa T, 182
Ide H, 424, 425
Igamberdiev AU, 348, 350
Iglesias AA, 345, 346, 355
Igloi GL, 71
Ihara M, 176
Iino M, 144, 146, 152, 155, 163
Ikegami M, 84
Il PC, 188
Ilag L, 399, 401, 402
Ilyina TV, 87
Imai Y, 267
Imaseki H, 409
Imbault N, 178
Imperiale MJ, 51, 52
Inada Y, 425
Inamdar NM, 94, 98, 99, 103
Indivieri C, 457
Ingelbrecht I, 57
Ingelbrecht ILW, 56
Ingemarsson B, 581
Inglis AS, 123
Innes CMJ, 207
Inohara N, 395, 398, 399, 401, 402, 412
Inokuchi N, 425
Inoue Y, 496
Inzé D, 188, 342, 343, 559
Ioerger TR, 424, 425
Ireland CR, 636
Ireland R, 326, 331, 332, 559, 578, 585

Irie M, 424, 425, 428
Irish VF, 42
Irniger S, 48
Isaac PG, 63, 546
Isaya G, 554, 555
Ishida K, 529, 534
Ishiguro Y, 326, 327
Ishii N, 479–81
Ishii R, 349
Ishikawa M, 80
Ishizaki-Nishizawa O, 129
Ishizuka K, 529, 535
Israel A, 379
Issakidis E, 449
Issaq HJ, 667
Ito A, 550, 554
Ito K, 482, 484
Iturriaga G, 129
Iversen T-H, 26, 40
Ivey CA, 666
Iwama M, 424
Iwamura J, 179
Iwanaga S, 548, 550
Iwanari S, 271
Iwasaki I, 458, 461
Iwata K, 529
Izumi K, 537

J

Jabben M, 497, 511, 512, 514
Jackson C, 327
Jackson CB, 672
Jackson GS, 476, 484, 486
Jackson JA, 144, 159, 160
Jackson WA, 583, 592
Jacob U, 484
Jacobi A, 408
Jacobs A, 57
Jacobs JD, 90
Jacobs M, 406
Jacobsen BS, 383, 384
Jacobsen H-J, 412
Jacobsen JV, 123, 125, 531
Jacobson BS, 384
Jacobson MK, 348
Jacquot J-P, 449, 592, 595
Jacrot M, 665
Jaenicke R, 470, 481, 482, 484–86
Jahnen W, 437
Jahnke LS, 634, 643
Jakob U, 484
Jakobsen KS, 116
Jamieson GC, 352
Jan LY, 223
Jan YN, 223
Janda M, 80
Janmaat K, 26, 31–34, 37
Jansen M, 506, 512, 518, 519
Jansz HS, 93
Jarchow E, 93
Jarczyk L, 581
Jaspers PAP, 158
Jaspers PAPM, 296
Jause C, 448

Jaworski EG, 304, 306
Jayasekera S, 266, 276
Jaziri M, 667
Jech K, 669, 671
Jenkin PM, 646
Jenkins CLD, 346, 347, 349,
   350, 381, 461
Jenkins G, 155
Jenkins GI, 144, 154, 155,
   159, 160
Jenkins JA, 293, 294
Jeno P, 550, 551
Jensen RA, 239, 246, 268,
   358
Jensen RE, 550, 554
Jensen RG, 130
Jeppson RG, 305
Jeske H, 81, 84, 88, 89, 95,
   103, 104
Jewess PJ, 348
Jia C, 331, 592, 593
Jiang CJ, 539
Jiang C-Z, 294
Jiao JA, 345, 449, 586
Johal S, 344
Johnson JL, 150
Johnson MA, 411
Johnson RA, 471, 473, 478
Johnson RR, 305
Johnson-Flanagan AM, 120
Joliot P, 517
Jones AL, 610, 611
JONES AM, 393–420; 158,
   394, 395, 397–403, 406–
   10, 412, 616
Jones DA, 334
Jones HG, 113
Jones JDG, 229
Jones LW, 636
Jones RL, 162, 531, 620
Jones TA, 385
Jones VP, 546
Jonniaux JL, 216, 220
Jordan BR, 334
Jordan R, 482
Jordan U, 639, 641
Jordana X, 546
Jorgensen KG, 118
Jorgensen RA, 156
Jorna ML, 130, 294
Joset F, 374
Joshi CP, 48, 53
Joshi HC, 540
Jost W, 424, 425, 428, 433
Jourdain A, 326, 327
Joy KW, 351–53, 591
Joyce CM, 38
Joyce PBM, 65, 71
Jubier M-F, 69, 546
Judd AK, 201
Julien B, 343
Juniper BE, 537, 539
Jurgens G, 26, 28–31
Just WW, 459

K

Kader J, 126
Kadlecek AT, 186, 187
Kadonaga JT, 90
Kadota A, 536, 538
Kagawa T, 538
Kaiser A, 230
Kaiser G, 453, 458
Kaiser WM, 450, 453, 454,
   591, 592
Kakimoto T, 528, 539
Kalkan FA, 122
Kallender H, 87
Kalousek F, 472, 485, 486,
   547, 550, 553–55, 562
Kalt W, 462
Kamen BA, 610, 611, 626
Kamen MD, 236
Kaminek M, 180
Kammann M, 87, 93, 94, 96–
   99, 101–3
Kamo T, 377
Kanazawa K, 394, 579
Kanazawa T, 579
Kaneta T, 528–32
Kanevski IF, 98, 99
Kang PJ, 550, 552, 562, 563
Kang SM, 304, 317
Kaniewski WK, 101
Kannangara CG, 126
Kanstad SO, 495, 506, 512,
   513, 515
Kao T-h, 422, 424, 425, 436–
   40
Kao W-Y, 640
Kaplan A, 371, 373, 375, 386
Kapoor HC, 412
Kapoor R, 57
Kappen L, 645
Kappes B, 227
Karavaiko NN, 394
Karlsson PE, 163
Karp F, 266
Karssen CM, 114, 130, 134,
   294
Kasamo K, 216
Kassenbrock CK, 550, 551
Kastner P, 414
Katagiri F, 39, 122
Katayama C, 48
Kato Y, 148
Katoh S, 512, 513
Katsumi M, 529–31, 534
Katsuta J, 538, 539
Katz D, 214
Katzman GL, 375
Kauffman JM, 54
Kaufman LS, 144, 151–57,
   164
Kaufmann H, 425
Kauppinen V, 123
Kaur H, 412
Kawachi E, 179
Kawalleck P, 261, 272
Kawasaki H, 327

Kawasaki S, 537, 539
Kawata T, 54
Kawata Y, 423, 482, 484
Kayser B, 398, 399, 401, 412
Kearns A, 614, 623
Kearns EV, 163
Kearsey S, 430
Keech E, 148
Keegstra K, 355, 471, 553
Keil P, 550
Keilani S, 507
Keist M, 304, 305
Keith B, 54, 57
Kellens JTC, 120
Keller B, 35
Keller H, 261
Keller W, 48
Kelly GJ, 588
Kelly MO, 406
Kelsey RG, 669
Kelson TL, 293, 295
Kendall A, 189
Kendall AC, 352, 353, 360
Kende H, 180, 394, 428,
   432, 530
Kendrick RE, 158, 296
Kenefick DG, 423
Kengen HMP, 536
Kerby NW, 380
Kern G, 484
Kerr IM, 430
Kerr MW, 348–50
Kerr PS, 248
Kettunen R, 644
Keys A, 189
Keys AJ, 242, 343, 344, 351,
   352
Khavari-Nejad RA, 348
Khavkin EE, 401
Kheir YE, 268
Kheyr-Pour A, 83, 84, 425
Khudairi AK, 297
Khurana JP, 146, 152, 153,
   159, 160
Kieber JJ, 180
Kiebler M, 550, 551
Kiefer H, 397
Kienzle BK, 291
Kijima H, 458, 461
Kijne JW, 616, 624
Kikichi M, 409
Kikuchi G, 324, 326, 327
Kikuchi M, 409, 411
Kikuno R, 86, 88, 95
Kim J-M, 349
Kim KS, 92
Kim M, 159
Kim S, 313
Kim Y, 324, 327, 332, 333
Kimura M, 424, 425
Kimura T, 559
Kincade PW, 612
King AA, 495, 508
King DI, 87, 88, 94, 95, 97
King J, 198, 207
King PJ, 180, 183, 410, 412

King RW, 114, 250
Kingston DGI, 665, 666, 670
Kingston-Smith AH, 343
Kinoshita T, 163
Kinzer SM, 294
Kipling D, 430
Kirby EA, 452
Kirby R, 83, 84, 92, 93
Kirch H-H, 422
Kirchhausen T, 617
Kirk MR, 579
Kirkbright GF, 495, 508
Kirkby EA, 448, 451, 452
Kishore G, 247
Kishore GM, 352, 355, 358
Kispal G, 561
Kiss T, 429, 430
Kitamura T, 277
Kitayama M, 375, 378
Kizawa S, 394
Klambt D, 394, 395, 397, 402, 404, 405, 412, 616
KLECZKOWSKI LA, 339–67; 344, 345, 347, 348, 350–52, 359, 360
Klee H, 113, 174, 176, 180, 183, 189
Klee HJ, 187, 188
Kleiber J, 550
Klein SM, 324
Klein U, 588, 591
Kleinhofs A, 591
Kleinig H, 297
Kleinjan DA, 480, 485
Klepper L, 592
Klessig DF, 394, 411
Klincare AA, 199, 203
Klinkenberg FA, 95, 97
Klock G, 375
Kloppstech K, 157
Knaff DB, 595
Knapp J, 297
Knapp JE, 291, 297
Kneifel W, 207
Knight AH, 448, 451, 452
Knight MR, 155
Knight S, 344
Knoop V, 62, 67, 68, 72, 73, 546
Knorpp C, 555, 559, 564
Knott TJ, 73
Knowles VL, 588
Knox JP, 39
Kobayashi M, 263
Kobayashi Y, 269
Kobek K, 353, 354, 358
Kobs G, 431
Kobza J, 239, 242, 343
Koch BD, 550, 552, 562
Koch GLE, 550, 553
Koch JL, 204
Kochi H, 326, 327
Kochi T, 71
Kochian LV, 223, 224
Kocsányi L, 518, 521
Koda Y, 534

Koetje DS, 315
Kohchi T, 63
Kohlhaw GB, 556
Kohno J, 261, 271
Kohorn BD, 228
Kok B, 633, 636, 643
Kolarov J, 227
Kolber Z, 645, 646
Kolbowski J, 496, 497, 501, 507, 510
Koll H, 548, 557
Koller D, 164, 640
Kolloffel C, 41
Kombrink E, 588
Komor E, 433, 579, 581, 584, 594
Komukai Y, 373, 385, 386
Konagai A, 277
Kondo N, 343
Konishi J, 479, 481
Konjevic R, 146, 152, 153, 159
Kooijman M, 500, 504, 506, 507, 517, 518
Koonin EV, 87, 93
Koornneef M, 114, 130, 134, 152, 158, 159, 293, 294, 296
Koot HTM, 458
Korte JJ, 373
Kortt AA, 116
Kosic Smithers J, 552
Kositratana W, 83, 84
Kosloff R, 371, 373, 386
Koslowsky DJ, 66, 69
Kössel H, 71
Kottmeier C, 453
Kovach DA, 127
Koyata H, 326, 327
Krake LR, 83, 84
Kramer J, 552
Krapp AR, 472
Kraus C, 327, 331
Krause GH, 238, 241, 246, 247, 513, 517, 634, 636, 637, 639, 644, 646
Kreiger A, 639, 641
Kreiling KM, 425
Kreimer G, 401
Kreis M, 304
Kremer DF, 248
Kreslina DJ, 199, 203
Kretovich WL, 471, 475
Kreutzberg K, 375
Kreuz K, 458
Kridl J, 48
Krieg UC, 547, 550, 553, 562
Krinsky NI, 288
Kristoffersen P, 412
Krömer S, 457, 458, 461, 592, 593, 595
Kronidou NG, 550, 551, 562
Kruckeberg AL, 244, 247, 357
Kruft V, 550, 555, 556, 559
Kruger A, 636, 642

Kruger NJ, 588
Kruse E, 157
Kubasek WL, 156, 161
Kubo A, 343
Kubota H, 474
Kuchitsu K, 375, 376
Kuchler K, 221
Kück U, 62
Kuhlemeier C, 54, 55, 258, 333
Kulaeva ON, 394
Kulpa J, 227
Kumar AM, 559
Kume A, 324, 326, 327
Kung S, 182
Kuntz M, 288, 290–92, 297
Kunzelmann P, 152
Kuo T, 591
Kure S, 324, 326, 327
Kurihara H, 425
Kurkdjian A, 448
Kurkela S, 126
Kurz WGW, 265, 271, 272, 279
Kusano T, 117
Kuss-Wymer CL, 536
Kutchan TM, 258, 263, 264, 267–69, 271, 272, 278, 312
Kutschera U, 534
Kwok SF, 228
Kwong PL, 550
Kyle DJ, 637

L

Laborie D, 458
Labriola J, 188
Lacey SW, 610, 611, 626
Lacroute F, 223
Laferriere A, 291
Laing WA, 246, 644
Lakhani S, 57
Laliberte J-F, 80
Lalonde BA, 228
Laloue M, 179
Lam E, 39, 54
Lamattina L, 62, 67, 69, 70, 546
Lamb CJ, 35, 154
Lambardi L, 412
Lambers H, 251
Lambert AM, 539
Lamblin AF, 373
Lamerson PL, 397, 398, 401, 402, 412
Laminet AA, 478, 481
Lammers PJ, 125
LaMotte CE, 205
Lan Y, 243
Lancaster J, 188
Lance C, 448
Lancelin D, 546
Lancelle SA, 536, 539
Lancellin D, 69
Landry J, 127
Landry SJ, 479, 482

Landsmann J, 279
Lane S, 228
Lang JM, 528, 533
Lang V, 131, 132
Langer T, 471, 473, 474,
477, 478, 480, 483, 484,
487
Langeveld SA, 93
Lannoye R, 521, 522
LaNoue JF, 456, 457
Lappi DA, 422
Lara C, 581, 594
Large PJ, 207
Larkins BA, 55
LaRoche J, 634, 647
LaRosa PC, 120
LaRossa RA, 472
Larsen F, 116
Larsen PO, 579, 583, 585,
597
Larsson C, 148, 149
Larsson C-M, 581
Larsson M, 581
LaRue B, 501, 503, 504, 512,
519
Laskowski MJ, 152, 163, 529
Lasser-Ross N, 496, 500,
502, 504, 512, 513
Laszlo A, 127
Latham JR, 89
Latzko E, 588
Lauer MJ, 251
Laufs J, 98, 99, 102, 103
Laughlin GP, 645, 655
Laurent B, 374
Lawlor D, 189, 242
Lawson EC, 101
Lawson VR, 529, 530
Lawton G, 148
Lawton JR, 537, 539
Lawton MA, 154
Lawyer AL, 579, 583, 585,
597
Lazar M, 123, 131
Lazarowitz SG, 81, 84, 87,
88, 93–95, 97–99, 104, 105
Lazarus CM, 398
Lazdins IB, 84
Lazo CB, 667
Lazzaro MD, 622
Le VQ, 585, 586
Lea PJ, 324, 350–53, 360,
591
Leach G, 565
Leamon CP, 610
Leaver CJ, 62, 63, 65, 66, 73,
449, 546, 555, 556, 559
Leblanc GA, 411
Leblanc RM, 5, 495, 500,
501, 503, 504, 506, 512,
513, 515, 519–22
Leblova S, 346
Lebrun M, 327, 332
Lechner EG, 645, 646, 649
Lecker S, 483
Leclerc RF, 471

Ledbetter M, 528
Lee FS, 440
Lee G-H, 474, 481, 482, 484
Lee H-S, 425, 436–39
Lee P, 596
Lee RB, 578, 591–93, 595
Lee S, 186, 187, 410
Lee TH, 297
Lee Y, 177
Leech RM, 564
Leegood RC, 451
Leegwater P, 69
Leemans J, 440
Leenhouts JM, 561
Leete E, 258, 672
Lefebvre C, 582, 583, 591
Lefebvre DD, 80, 588
Legatt IV, 198
Legendre L, 616, 624, 625
Legg PD, 275
Leggett JE, 305
Leggewie G, 49–54
Lehnen LP, 358
Lehner K, 455
Lehnerer M, 559
Leid M, 414
Leidreiter K, 250
Leifert C, 198–201, 207
Leisner S, 186, 187, 410
Lejeune B, 62, 69, 546
Leland TJ, 276
Lelandais M, 581, 597
Lemberg R, 149
Lennon AM, 550, 552, 564,
566
Leonard KR, 471, 474
Leonard NJ, 395
Leonard RT, 623
Leong T-Y, 149, 150, 154
Leonhardt W, 207
Leopold AC, 125
Lerbs-Mache S, 68
Lerner DR, 39
Lerner EI, 496, 500
Lernmark U, 376
Lerouge P, 41
Leroux B, 121, 122
Leroux D, 665
Lersten NR, 205
Le Rudulier D, 455
Lesch S, 581
Lescure J-C, 581, 582, 591,
597
Lessard PA, 122
Lesser MP, 648
Leterme S, 66
Letham DS, 174–77
Letham S, 181
Leu J-G, 667
Levavasseur G, 636, 637
Leverenz JW, 636, 649, 650
Levings CS III, 62, 546, 559
Levy A, 63
Levy AA, 63
Lewandowski DJ, 80
Lewis DH, 244

Lewis M, 398, 399, 402
Lewis N, 672
Lewis SD, 327
Lewis V, 474
Ley AC, 641, 643, 647
Li B, 241, 248, 249
Li GC, 127
Li K, 556, 559
Li PH, 130
Li QY, 199, 200, 203
Li W-Z, 52, 163
Li X-H, 668
Li Y, 189
Libbenga KR, 178, 394, 408–
11
Liberek K, 470–72, 479
Lichtenstein CP, 98, 99, 101
Lichtenthaler HK, 353, 354,
358, 522
Liddell A, 66
Lidstrom ME, 207
Lieber MR, 80, 103, 104
Lieman-Hurwitz J, 371, 373
Lignell Å, 380
Liljas A, 385
Lill R, 483, 561, 562
Lilley GG, 308, 312, 313
Lilley R McC, 241
Lim P, 669
Lin C, 126
Lin DX, 250
Lin M, 581, 586, 587, 596
Linden H, 290, 292
Lindquist S, 127, 128, 552
Lindquist SL, 128
Lindqvist Y, 344
Lindsay JG, 546
Linforth RST, 130, 293, 294
Ling M-L, 73
Linnemeyer PA, 162
Linstead PJ, 26, 31, 32, 34,
35, 37, 39
Lippok B, 65, 69, 70
Liscum E, 144, 152, 159, 160
Lissin NM, 476, 478–80, 483
Lister RM, 317
Listowsky I, 411
Litton CC, 275
Litvak S, 70, 546
Liu B, 540
Liu HY, 92
Liu J, 346
Liu W, 395, 398, 412
Liu YC, 670
Livrea MA, 414
Lloyd CW, 533, 535, 537,
539
Lloyd J, 384–86
Lloyd JC, 382, 384, 385
Lobler M, 394, 395, 397,
398, 404, 404, 412
Loboda T, 348
Loeffler M, 28
Loehr A, 341
Löffler A, 424, 425, 428, 433
Logman TJJ, 616, 624

Lomax TL, 406–8, 412, 616
Lonergan KM, 73
Long RD, 198
LONG SP, 633–61; 634–36,
641, 643, 645, 646, 649–
51, 653–55
Long SR, 41
Lönneborg A, 350
Lonsdale DM, 62, 71, 546,
555
Look LL, 354
Look SA, 667, 669
López-Juez E, 155
Lorenzen CJ, 646
Loretzl CJ, 63
Lorimer GH, 239, 324, 325,
343, 344, 470–72, 475–78,
480–83, 485, 486
Lorrain L, 495, 506, 512,
513, 515
Losada M, 643
Lo Schiavo F, 411
Lottspeich F, 267
Lou H, 97, 98, 100
Louarn J-M, 473
Louis J-P, 177
Lovelock CE, 634, 640
Low MG, 612
LOW PS, 609–31; 610, 614,
616, 624, 625
Lowe J, 641, 665
Lowe SB, 535, 539
Lu C, 473, 487
Lu J, 186
Lu LS, 241
Luan S, 163
Lubben TH, 470–72, 475,
477, 478, 481, 482, 485,
486
Lubenow H, 190
Lucas WJ, 87, 223, 224
Luckow VA, 87, 90, 93
Lucy A, 92–94
Ludevid D, 39
Ludlow MM, 645
Lugtenberg BJJ, 616, 624
Lundmark T, 645
Lundqvist T, 344
Lunness P, 83, 84
Lupold DS, 68
Lush WM, 437
Luster DG, 150
Lüthi E, 350, 351
Lüttge U, 448, 450, 451, 458,
459, 462
Lutz A, 448
Lutz J, 184
Lützelschwab M, 214, 619
Lycett GW, 126
Lydon J, 358
Lyman ML, 435
Lyndon RF, 431
Lynn DG, 185
Lyttleton P, 621

M

Maan A, 394
Maathuis FJM, 213, 224
Maberly SC, 379, 380
MacDonald H, 83, 92, 93,
97, 101, 102, 410, 412
MacDonald M, 49, 54
MacDonald MH, 49, 50, 52,
56
MacDonald RC, 204
MacDowell SW, 83, 93, 101
Macer DRJ, 550, 553
Macfarlane JL, 65
MacFarlane S, 88, 94
Macharia JMN, 645, 649,
654, 655
Macherel D, 327–29, 332,
559
MacIntyre HL, 646
Mack S, 68
MacKinney G, 293, 294
MacKown CT, 316, 579, 584
Madsen JP, 275
Madyastha KM, 266, 267,
270
Mae T, 385
Maekawa T, 539
Maeshima M, 317, 559, 564
Maghlaoui K, 480, 485
Magid E, 370
Mahalingam R, 66
Mahendran R, 73
Maher EP, 26, 38, 40
Mahkle K, 547, 559, 562
Mahlberg PG, 268, 270
Maier RM, 71
Maillet F, 41
Majeau N, 382, 384, 385
Major I, 343
Makaroff CA, 65, 68, 546
Makino A, 385
Maldiney R, 180, 182
MALKIN S, 493–526; 494–
97, 500, 501, 504–10, 512,
513, 515, 516, 518, 519,
521, 522
Malmberg RL, 258
Malone TC, 646
Maluszynski M, 423
Mandel T, 258
Manderscheid R, 352
Mangatal L, 669
Mangold U, 266
Mann AF, 592
Mann TJ, 274, 275
Mann V, 546
Manna E, 63, 69, 73
Mannaerts GP, 459
Manney L, 39, 216
Mannheim-Rodman LA,
125, 128
Manning-Krieg UC, 473,
553, 562
Mannteuffel R, 239
Manorama CJ, 186, 187

Mansfield MA, 120–23
Mansfield TA, 115, 163
Mantell SH, 200
Manthey CA, 665, 666
Mapelli S, 412
Marc J, 540
Marceau M, 327
Marceau N, 127
Marchant HJ, 535, 539
Marchfelder A, 69, 70, 546
Marcotte WR, 113, 121
Marcus A, 39
Marcus Y, 375
Marczewski W, 586
Maréchal-Drouard L, 62, 66,
71, 175
Margoliash E, 551
Margossian L, 126
Mariani C, 440
Marie D, 94
Marigo G, 458
Marineau C, 265
Markgraf T, 295, 459
Markham PG, 83–85, 87–89,
91, 92, 94, 95, 97, 102
Marks MS, 88, 95, 102
Marner F-J, 176, 266
Marquardt J, 295
Marquardt-Jarczyk G, 458
Marra E, 556, 557
Marra J, 634, 646, 654
Marrs KA, 156, 157
Marsden WJN, 373
Marsh M, 610, 612
Marshall B, 649
Marshall JS, 553
Marshall LC, 354
Marshallsay C, 429, 430
Martel R, 471, 473
Marten I, 448, 453, 460
Martin C, 275
Martin F, 352
Martin J, 471, 472, 474, 477,
478, 480, 483, 484, 547,
553, 559, 562
Martin RC, 177, 178
Martin RR, 177
Martindale SJB, 38
Martinez MC, 116
Martinez-Zapater JM, 48
MARTINOIA E, 447–67;
453, 458
Maruyama H, 411
Maslyar DJ, 449
Mason CB, 375
Mason HS, 307, 311–13
Massey V, 150
Masson K, 342
Masson PH, 439
Masters S, 669, 671
Masucci J, 38
Masuda Y, 528
Matile P, 428, 432, 438, 453,
458
Matorin DN, 646
Matsubara H, 595

Matsuda J, 261, 271, 278
Matsui C, 327
Matsuki T, 534
Matsumoto H, 358, 431
Matsumoto Y, 84
Matsuoka K, 271
Matsushiro A, 474
Matthijs HCP, 501, 504, 512, 513
Mattsson O, 200
Matzeit V, 83, 84, 86–88, 90–94, 96–99, 101–3
Matzinger DF, 274, 275, 559
Matzke AJM, 48
Matzke MA, 48
Mau S-L, 347, 423, 435
Mauk CS, 318
Maule AJ, 87
Maunders MJ, 290, 291, 297
Maurel C, 400, 404, 405, 412
Mauzerall D, 494, 496, 501, 507, 509, 510, 512, 515, 641, 647
Mauzerall DC, 641, 643, 647
Mayanil CSK, 475
Mayer U, 26, 28, 29, 31
Mayerbacher R, 412
Mayo WP, 581, 583, 597
Mazur BJ, 341, 352, 358
McCaffery S, 637, 640, 644
McCaffrey M, 612
McCarthy JEG, 279
McCarty DR, 123, 131
McCarty RE, 356
McCloud TG, 667
McClure BA, 422, 423, 425, 435, 437
McCullough AJ, 97, 98, 100
McCurdy DW, 226
McElwain EF, 183, 186
McFadden GI, 425, 428, 435
McFadden HG, 346
McFarlane JD, 352
McGaw BA, 174, 176, 177
McGuire WP, 665
McIntosh CA, 456
McIntosh L, 546, 559
McKanna JA, 610, 611
McKay RML, 373, 375, 376
McKellar JF, 150
McKeon TA, 431, 432, 434, 435
McKillop A, 156, 161
McKnight SL, 122
McLean BG, 38
McMacken R, 482
McMichael JRW, 598
McMichael RW, 589
McMullin TW, 471, 474
McNamara K, 589
McNight TD, 267, 270, 271
McNulty IB, 450
McPhail AT, 663
Meagher RB, 38, 439
Medford JI, 188
Medrano H, 242

Meehan LJ, 294
Meehan TD, 266
Meek J, 343
Mehdy MC, 267
Meier D, 353
Meijer AH, 258, 266, 267, 270
Meins F Jr, 177, 183, 184, 186, 205, 410
Melan MA, 312, 535, 536
Meleka J, 611
Melhado LL, 395
Melhorn RJ, 406
Melitz DK, 176, 184
Mellman IS, 610
Menassa R, 67
Mendel-Hartvig J, 66
Mendoza JA, 476, 481, 483, 485
Mendu N, 532
Mennes AM, 178, 394, 408, 409, 411
Menold MM, 227
Menzlaff E, 455, 462
Mercer RW, 610, 612
Meredith G, 81
Mersey BG, 619, 621, 622, 626
Mertens R, 119
Meshi T, 80
Messens E, 176
MESSING J, 79–112; 48, 49, 51, 56, 86, 87, 93, 94, 96, 97, 104, 105
Mettler IJ, 459
Meurs C, 134
Meyer-Bothling LE, 202
Meyer A, 311
Meyer C, 616
Meyer CR, 345, 394, 616
Meyer DJ, 216
Meyer G, 157, 449, 451
Meyer P, 94, 96
Meyerowitz EM, 27, 228
Michaelis G, 550, 555
Michalke W, 619
Michalowski CB, 449, 451
Michelet B, 216
Miernyk JA, 552, 553, 587, 588, 596, 598
Miersch O, 311
Miflin BJ, 304, 351, 352, 591
Miginiac-Maslow M, 449
Migliaccio F, 399, 401, 402
Mignery GA, 317
Mihaliak CA, 266
Mikami K, 54
Miki BL, 80
Miki-Hirosige H, 619, 621
Miklos D, 474, 481, 482, 484
Milani A, 184
Millard P, 305
Miller AD, 480, 485
Miller AG, 371, 373, 377, 579

Miller BR, 548, 556, 557, 562
Miller CO, 173, 176
Miller DL, 73
Miller IM, 205
Miller PD, 200
Miller RA, 670, 671
Miller RD, 275
Miller SG, 471
Milliman C, 123
Milos P, 471, 478
Mimeault M, 501, 504, 512, 513
Minami Y, 425
Minamikawa T, 57
Minet M, 223
Mino M, 121
Minocha SC, 279
Min-Wang L, 178
Miranda LCM, 497, 508
Mirza JI, 26, 40
Misawa M, 270
Misawa N, 292
Misera S, 28, 29
Mishke IV, 199, 203
Mita T, 530, 531, 533, 534, 536, 540
Mitani A, 260
Mitra A, 56
Mitsui Y, 425
Mittelsten-Scheid O, 98, 99, 103
Miyachi S, 370, 373–78, 382, 383, 385, 386, 581, 584
Miyata T, 86, 88, 95
Miziorko HM, 239
Mizobata T, 482, 484
Mizuno H, 425
Mizuno K, 532, 537, 538
Mizusaki S, 273
Mizuta H, 84
Mizzen LA, 127
Mocek U, 672
Moczko M, 549–51
Moenne A, 546
Mogen BD, 49, 50, 52, 56
Mohamed AK, 384, 385
Mohanty N, 596
Mohnen D, 177, 186, 410
Mohr H, 152, 155, 157, 158
Mok DWS, 174, 177–79
Mok MC, 177–79
Mol JNM, 274
Moll C, 531
Moller IM, 66
Moloshok TD, 227
Momotani E, 179
Monk BC, 227
Monnat RJ, 63
Monroe M, 471
Montezinos D, 614
Montoliu L, 48
Montoya J, 48
Mooney B, 552, 566

MOORE AL, 545–75; 327, 546, 550, 552, 553, 559, 561, 564–66
Moore Bd, 242, 343
Moore I, 394, 411, 412
Moore RC, 258
Moore S, 422
Moore TA, 497, 500, 502, 512
Moorehead GBG, 588
Morales FJ, 84
Moreland DE, 341
Morelli G, 54, 155
Morgan AJ, 87, 94
Morgan C, 331, 332, 559
Morgan G, 614, 623
Mori H, 553, 559
Mori M, 474, 548, 550
Mori S, 534
Moriau L, 216, 220
Morigasaki S, 595
Morikami A, 559
Moritz T, 178
Moromoto M, 148
Moroney JV, 375, 377, 378
Morot-Gaudry J-F, 582, 591
Morrell S, 588
Morris BAM, 83, 84, 88–90
Morris CF, 115, 117
Morris KL, 378
Morris RO, 176, 180
Morris SR, 556
Morris-Krsinich BAM, 83, 84, 87, 89, 92, 93
Moses MS, 115, 118
Moses P, 155
Moshkov IE, 394
Mostov KE, 610, 611
Mostowska A, 358
Motel A, 353, 354
Motita T, 474
Motokawa Y, 326–28
Mott KA, 243
Motzan Z, 507, 513, 516
Mouras A, 70
Moureau P, 290, 291, 297
Mouriquand C, 665
Mrachatz G, 257, 260
Mtolera M, 379
Mu J, 425
Mucciaro TP, 665, 670
Mudd A, 343
Mudd E, 429
Muemmler S, 268
Mulkey SS, 634
Muller B, 448, 451, 452
Muller CH, 204
Muller J-F, 404
Müller MJ, 261, 272, 312
Muller S, 546, 562
Muller U, 547, 549
Muller WA, 610
Muller-Hill B, 122
Mullet JE, 124, 157, 159, 307, 311–13
Mulligan RM, 344

Mullineaux CW, 495, 520
Mullineaux P, 49, 53, 54, 99, 102
Mullineaux PM, 81–84, 86–89, 91–95, 97, 102
Mummert E, 474, 475, 481, 484
Munck L, 118
Mundy J, 113, 115–18, 121–23, 126
Muneyuki E, 479
Muñoz V, 149
Munro D, 115
Murakami H, 550, 562
Murakami K, 548, 550
Murakoshi I, 267
Murant EA, 614, 623
Murashige T, 665, 670
Murata K, 474
Murata N, 129
Murata Y, 349
Murofushi H, 539
Murofushi N, 534
Murray AJS, 324, 352, 353, 360
Muschik GM, 667, 669
Musgrove JE, 471, 473, 478
Mushegian AR, 87
Muto S, 375, 377
Mutschler MA, 294
Myers AM, 471
Myers J, 636, 642
Myler PJ, 69

N

Näf U, 182
Nagahashi G, 623
Nagai R, 539
Nagakura N, 263, 264, 268
Nagano E, 537
Nagata R, 179
Nagata T, 410, 538
Nagel E, 522
Nagpal P, 38, 161
Nagy F, 48, 54
Naik MS, 591–94, 598
Naitoh A, 425
Nakagawa K, 277
Nakagawa S, 317
Nakagawa T, 559, 564
Nakajima K, 260
Nakajima M, 394
Nakajima N, 409
Nakamoto H, 449
Nakamoto RK, 212
Nakamura C, 408, 409, 411
Nakamura K, 271, 317, 559, 564
Nakamura KJ, 559
Nakamura KT, 425
Nakamura S, 619, 621
Nakamura Y, 62, 67, 71, 73
Nakatani H, 501, 504, 513
Nakayama T, 54
Nalborczyk E, 348

Nalcez MJ, 457
Nandi SK, 182
Napier RM, 229, 394–96, 398–400, 402, 405, 412
Naprstek J, 222, 228
Nargang FE, 550, 562
Nasrallah JB, 228
Nasrallah ME, 228
Nathan CF, 666
Nave J-F, 400
Navot N, 83, 84
Neale I, 636
Neale PJ, 636, 646, 648
Neckermann N, 556, 559
Nedea ME, 671
Neill SJ, 119
Neilsen JE, 200
Nelemans J, 448, 453
Nelson DE, 120
Nery JW, 497, 508
Nessler CL, 267, 268, 270, 271, 279
Nester EW, 102, 176
Neuburger M, 62, 324–32, 350, 448, 449, 457, 546, 565
Neuhaus HE, 244, 247, 355, 357
Neumann D, 552, 553, 559
Neupert W, 327, 472, 546, 547, 549–56, 559–63, 567
Nevins DL, 425
Nevo Y, 119
Newbigin E, 422, 437, 438
Newman TC, 439
Newton KJ, 62, 65, 546
Ngai PK, 425
Nguyen M, 548
Nguyen TT, 428
Nicholas DJD, 592, 593, 598
Nicholas J, 198
Nicholson DW, 550, 551
Nick P, 163, 529, 530
Nicole LM, 127
Nie G-Y, 654
Niedenhof I, 94, 96
Nielsen NC, 308, 312, 313
Nielsen OF, 293, 295
Nielsen TH, 589
Niemanis S, 239–42
Niitsu Y, 411
Nikaido H, 406, 412
Nilsson O, 178, 411
Nimmesgern E, 473–75, 481, 482, 484
Niogret MF, 121
Nishida I, 129
Nishimura M, 163, 458, 461, 553, 559
Nishimura S, 176
Nishizaki Y, 164
Nissen H, 546
Nitsch C, 497, 504, 511, 512, 514
Nixon JS, 148
Noctor G, 581, 597, 641

Noguchi M, 273
Noji M, 267
Nojiri H, 534
Nomoto H, 148
Nomura AM, 204
Nordin K, 116, 120, 131, 132
Norman JM, 634, 649, 650,
652
Norris BL, 206
Norris SR, 294
Northcote DH, 423, 435
Nott R, 655
Novacky A, 624
Novikova GV, 394
N'soukpoé-Kossi CN, 507,
522
Nugawela A, 636, 654, 655
Nugent JHA, 266
Nugent JM, 72, 546
Nunes MA, 635
Nürnberger T, 433
Nyren P, 559

O

Oaks A, 352, 579, 591, 595,
596
Obendorf RL, 125, 204
Obermaier B, 52, 53
O'Brien WT, 200
O'Connell KM, 101
Octave JN, 611
Oda K, 62–65, 67, 70, 71, 73
Odell JT, 48
O'Driscoll D, 614, 623
Oeda K, 122
Oehler O, 497
Oelck M, 352
Oeller PW, 178, 297
Oelmüller R, 158, 186, 295
Oetiker J, 410
Oettmeier W, 342
Ogawa T, 163
Ogg SC, 52, 53
Ogihara S, 539
Ögren E, 376, 635, 636, 642,
646, 648, 649, 652
Ogren WL, 239, 243, 246,
324, 343, 344, 353, 375,
382, 384, 385
Ohad I, 157, 512, 513, 519,
521, 522, 637
O'Hara EP, 497, 512
Ohashi H, 670
Ohgi K, 424, 425
Ohl S, 156
Ohme-Takagi M, 439
Ohmiya A, 411
Ohmori S, 267
Ohta E, 62, 64, 65, 67, 70,
71, 73
Ohta S, 550
Ohtsubo N, 54
Ohwaki Y, 448, 453
Ohyama K, 63–65, 70, 71
Ohyama T, 305

Ojala D, 48
Oji Y, 595
Ojima K, 385, 670, 671
Okabe K, 382
Okabe S, 261, 271
Okada H, 264
Okada K, 26, 27, 40
Okada NA, 65, 264
Okada Y, 80
Okamoto S, 595
Okamura-Ikeda K, 326, 328
Okazaki H, 550, 554
O'Keefe DP, 266, 472, 475,
481, 482, 485, 486
Okkels FT, 200
Okuley JJ, 425, 435
Olaizola M, 642
O'Laughlin JT, 346
O'Leary MH, 345–47
OLIVER DJ, 323–37; 324–
33, 349, 456, 457, 559,
592, 593
Oliver SG, 555
Olsen FL, 121, 123
Olsen GM, 26, 40
Olsen O, 559
Olsen O-A, 350, 351
Olson R, 647
Olsson O, 178, 411
Olsson T, 581
Olszewski NE, 124
Oltmann FL, 150
O'Mahony P, 564
O'Mally S, 667
Omura T, 550, 554
Onckelen HV, 184, 189, 190
Onda M, 595
Ongena G, 260
Ono H, 411, 548
Onur G, 341
Ooms G, 189
Oparka KJ, 614, 623
Opie CT, 665, 672
Oppenheim J, 471, 475, 480
Oppenheimer DG, 38
Oppliger W, 550, 551, 560
Öquist G, 639, 644, 645, 649
Orci L, 269
O'Riordan V, 555, 560, 564
Orpin T, 425, 428, 435
Orr JD, 185
Ort DR, 242, 244, 636
Osbourne BI, 52
Oshima Y, 529
Osman TAM, 93
Osmond B, 385
Osmond CB, 238, 241, 242,
250, 458, 462, 633, 634,
636, 637, 640, 641
Oster GF, 563
Ostermann J, 472, 548, 550,
552, 554, 557, 562, 563
Ottander C, 639, 644, 645
Otten H, 225
Ottersbach N, 160
Ou WJ, 550, 554

Ouyang LJ, 459
Ovadi J, 268
Ow DW, 90
Owen TP Jr, 624
Owens TG, 501, 504, 512,
513, 643
Ozeki Y, 182

P

Pace B, 430
Pace GM, 583, 592
Pace NR, 430
Packer L, 414
Pacold I, 345
Page T, 226
Pages M, 115, 116, 119
Pain D, 550, 562
Paiva E, 317
Paiva NL, 276
Palcic MM, 428
Palevitz BA, 226, 528, 539,
540, 617
Pallaghy CK, 88, 94
Palma JM, 342, 343
Palme K, 229, 394, 398, 399,
401, 404, 405, 408, 410–12
Palmer CE, 352
Palmer JD, 62, 63, 65, 68,
71, 72, 156, 546
Palmer JM, 145, 147, 148,
153, 227, 580
Palmgren MG, 148, 212,
215, 217, 219, 227
Palmieri F, 457
Palmqvist K, 376, 377
Palni LMS, 174, 182
Paluch U, 556, 559
Palva ET, 116, 120, 131, 132
Paneth P, 347
Panico M, 87
Papa C, 315
Parcy F, 131
Pardo JM, 212, 214, 216
Parets-Soler A, 212, 214
Park WD, 314, 317
Parker CW, 182
Parker JE, 616
Parker K, 152
Parker WB, 354
Parks B, 160
Parks BM, 158
Parr AJ, 279
Parry AD, 113, 119, 183,
293, 297
Parry MAJ, 242, 343
Parson WW, 494, 496, 511,
512, 514
Parsons BJ, 150
Parthier B, 113, 186, 311
Partridge BE, 344
Pascal E, 88, 95, 104
Pascal N, 330
Pasquali G, 267, 271
Paszkowski J, 80, 98, 99, 103
Paszkowski U, 98, 99, 103

Pate JS, 304, 579
Patel M, 564, 565
Patell VM, 69
Paton AM, 207
Patt TE, 200
Pattenden G, 672
Patterson BD, 370, 383, 386
Patterson DT, 385
Patterson SE, 228
Paul JS, 579, 597
Payne GS, 610, 626
Paz N, 351, 357
Paz-Ares J, 48
Peak JG, 586
Peak MJ, 586
Pearce CJ, 395
Pearcy RW, 242, 634, 641, 642, 654
Pearse BMF, 612
Peart J, 39
Pease RJ, 73
Péaud-Lenoë C, 186
Pecker I, 290, 292
Pedersén M, 379, 380
Pederson PL, 559
Peet MM, 385
Pelchen-Matthews A, 610, 612
Pelcher LE, 471, 473
Pelese F, 182
Pelham HRB, 398–400, 402
Peliska J, 346
Peltier G, 581
Pena-Cortes H, 119
Penarrubia L, 126
Pennings EJM, 264, 267
Penswick JR, 80
Peoples MB, 306
Peppard J, 611, 612
Percival MP, 340, 342
Pereira de Souza A, 69
Perera PY, 665, 666
Perez C, 216
Perez-Morga DL, 63
Perrollaz AL, 66
Perrot-Rechenmann C, 400, 405, 412
Perry KC, 181
Perryman RA, 552, 566
Peskin CS, 563
Pessoa O, 497, 508
Pessold-Hurt B, 547
Peter C, 539
Peter GF, 295
Peter U, 448, 453, 460, 592, 593
Peterman TK, 312
Peters JL, 158
Peters KS, 497
Peto C, 161
Peto CA, 158, 160
Petty ITD, 87, 89, 91, 97, 99, 105
Peumans WJ, 120, 314
Pey EBF, 199, 205, 206
Pfaller R, 549–51

Pfanner N, 327, 546, 547, 549, 550, 552, 553, 559, 560, 562, 563, 567
Pfeifer G, 471, 474, 478, 483
Pfitsch WA, 654
Pfitzner U, 267
Phaire-Washington L, 611
Phelps A, 551, 559
Phillips GD, 615, 626
Phillips GO, 150
Phillips IDJ, 173
Phillips R, 591
Phinney BO, 174, 180
Phipps BM, 474
Piatigorsky J, 125
Piatkowski D, 127
Pichersky E, 83, 84
Pickett FB, 26, 38, 40, 183
Picketts DJ, 475
Picot F, 665
Pierce J, 343, 344
Pierson ES, 536
Pietrzak M, 423
Pike DA, 178, 297
Pilartz M, 103, 104
Pilate G, 182
Pilström C-M, 636, 649, 650
Pinder AJ, 87, 88, 94, 95, 97
Pineau B, 295
Pinner MS, 83–85, 87, 89, 91, 92, 97, 102
Piruzian ES, 188
Pla M, 116, 119, 121, 122
Plant AL, 115, 118
Platt RV, 665, 672
Platt T, 48, 52, 647, 654
Platt-Aloia KA, 624
Plaxton WC, 251, 355, 585–88, 596, 597
Plesch G, 272
Plischke W, 423
Plückthun A, 478, 481, 483, 484, 486
Podestá FE, 355, 586, 587
Poethig S, 31, 32, 34, 37
Poff KL, 146, 147, 152–54, 159, 160, 227
Poincelot RP, 370, 383
POLACCO JC, 197–210; 201, 202, 204, 206
Polans NO, 156
Polarek JW, 228
Pollock RA, 472, 547, 550, 553, 554, 562
Polm MW, 497, 500, 507, 512, 518
Polston JE, 84
Polya GM, 179
Ponse G, 341
Ponz F, 48
Ponzone C, 457
Poole DS, 38
Popovic R, 501, 512, 513
Pöpper LH, 616
Porst M, 144, 147, 148, 154
Porter KR, 528

Porter MA, 385
Porteu F, 666
Portillo F, 212, 227
Portis AR, 371
Portis AR Jr, 239, 241–44, 343
Portsteffen A, 257, 260
Post LC, 425
Postl WF, 645
Poston JM, 203
Potera C, 665, 669
Potier P, 665, 666, 669, 670, 672
Potrykus I, 80, 187
Potten CS, 28
Potter R, 397
Poulet P, 494, 501, 505, 512, 516
Poulsen C, 54
Powell AJ, 535, 539
Powell LM, 73
Powell MJ, 552
Powell SJ, 559
Power JB, 200
Powles SB, 358, 633–36, 640, 645
Prasad PV, 394, 409, 412
Prasad TK, 472, 550, 553, 559
Prat S, 119
Pratje E, 550, 555
Pratt L, 158, 160, 161
Pratt LH, 145
Pratt RD, 559
Prehn H, 495, 497, 501, 502, 508
Preiss J, 247, 350, 351, 355, 589
Preisser J, 433
Prenger JP, 148
Prescott AG, 261
Preshaw C, 615, 626
Prestamo G, 435
Preston RD, 528
Prezelin BB, 648
Price BD, 125, 128
Price CD, 459
PRICE GD, 369–92; 371, 373, 374, 379, 384–86
Pring D, 62, 69
Pring DR, 62
Prins HBA, 162, 379
Prinsen E, 189, 190
Prior DAM, 614, 623
Prioul JL, 449
Pritchard AE, 66
Pritchard GG, 326
Prombona A, 68
Pronina NA, 378
Proteau H, 507
Proudfoot NJ, 48
Proudlove MO, 455, 585
Provart NJ, 385
Provot M, 582, 583
Pryer NK, 269
Psenak M, 272

Ptashne M, 121
Pugliarello MC, 616
Pugliesi G, 293, 295
Puigdomenech P, 48, 115
Pumfrey JE, 42
Punnett T, 520
Puppo A, 459
Purdue PE, 559
Pushkin AV, 471, 475
Pusztai A, 314
Putnam-Evans C, 226
Putnam-Evans CL, 226
Puype M, 395, 398, 399,
  401, 412, 616
Pyndji MM, 206

Q

Quader H, 528, 623
Quagliariello E, 556, 557
Quail PH, 144, 158, 161
Quarrie SA, 119
Quatrano RS, 113, 114, 117,
  121–23
Quetier F, 62
Quick WP, 239, 244, 247,
  248, 581, 598
Quint A, 409
Quy LV, 248

R

Racusen D, 307, 317
Racusen RH, 33
Radermacher E, 395, 402
Radin JW, 114
Radke SE, 39
Radmer RJ, 643
Radosevich SR, 144
Raffo AJ, 80
Raghavendra AS, 154, 163,
  461
Rahman FMM, 297
Rahnema S, 206
Raikhel NV, 39, 120, 121,
  271, 428
Raineri DM, 102
Raines CA, 382, 384, 385
Raines RT, 422, 424, 425,
  428, 432, 433, 435
Rains DW, 583, 584, 594
Rajagopalan KV, 150
Rakhaminova AB, 401
Ramage RT, 637, 639
Ramalho JDC, 635
Ramamonjisoa D, 69
Ramazanov Z, 378–80
Ramus J, 636, 637
Randall DD, 347, 350, 354,
  553, 585, 587
Randall PJ, 384, 385
Ranjeva R, 550, 552, 553,
  559
Ranty B, 550, 552, 553, 559
Rao KP, 583, 584, 594
Rao R, 212

Rapp WD, 67, 68, 308, 312,
  313
Raschke K, 241, 242, 244,
  448, 451, 460
Rasenick MM, 151, 164
Rasi-Caldogno R, 616
Rasmussen AG, 66
Rassow J, 546, 548, 550,
  553, 557, 560, 562, 567
Ratcliffe RG, 591
Rathnam CKM, 581, 583,
  592–94
Rau W, 296
Raudaskoski M, 537
Raven JA, 379, 380, 448,
  452, 586, 613, 655
Rawsthorne S, 326, 331,
  332, 334, 559, 564, 565
Ray JA, 290, 291, 297
Ray PM, 145, 397, 400, 402,
  404, 406
Rayle DL, 406, 407, 412, 616
Rea PA, 458, 461
Reading DS, 471
Reckmann U, 448
Record RD, 622
Redei GP, 436
Redinbaugh MG, 589, 591,
  592, 598
Reed AJ, 579, 594
Reed JW, 38, 472, 475, 481,
  482, 485, 486
Reed KE, 324
Reed LJ, 328
Reed ML, 370, 383, 386
Rees D, 641
Reeves RE, 588
Regenass M, 227
Regoeczi E, 610
Reichman B, 665
Reicosky DA, 305
Reid JB, 113
Reijnders W, 150
Reimann-Philipp U, 157
Reinbott TM, 180
Reiner J, 343
Reinhold L, 371, 373, 386
Reis FAM, 497, 508
Reising H, 496, 497, 501,
  507, 510, 517
Reiskind JB, 380
Remy R, 425
Ren L, 39
Ren Z, 160
Rendina AR, 354
Rennie PJ, 539, 619, 621,
  622, 626
Renoux JM, 205
RENTSCH D, 447–67; 458
Resemann A, 241
Reuling G, 130
Reumann S, 448, 459
Reuss J, 497
Reuter CC, 352
Revington GN, 81, 82, 89,
  92, 93

Reymond PR, 146–48, 153,
  159, 227
Reynolds H, 99, 102
Reynolds HV, 81, 82, 87–89,
  91, 93–95, 97
Reynolds WF, 671
Rezaian MA, 83, 84
Rhee Y, 310, 312, 313, 315,
  316
Rheem S, 201, 203
Rhoads DM, 546, 559
Rhodes D, 349, 352
Rhodes MJC, 279
Riabowol KT, 127
Ribeiro-Neto L, 224
Rice MS, 406, 407
Ricerson PJ, 646
Rich PR, 561
Rich TCG, 160
Richardson KA, 83, 84, 88–
  90
Richardson LI, 395, 396, 412
Richter G, 155, 157, 160
Ricigliano J, 394
Rick CM, 293, 294
Ridgway JE, 270
Ridley SM, 352
Rieger A, 448
Riens B, 448, 453, 460, 591–
  93
Riesmeier JW, 225
Riezman H, 547, 549
Rigau J, 48, 115
Rigaud J, 459
Rigden JE, 83, 84
Rijven AHGC, 431
Riondel J, 665
Risiott R, 188, 189
Ritchie DB, 294
Ritchie JY, 198, 199, 201,
  207
Rivas AL, 434
Roach PJ, 227
Robert-Nicoud M, 614, 623
Roberts DM, 226, 459
Roberts EJF, 84
Roberts IM, 92
Roberts IN, 533, 535, 537
Roberts JA, 40
Roberts JKM, 451, 454
Roberts K, 26, 32, 35, 39,
  533, 535, 537
Robertson DS, 288, 293, 295
Robertson JG, 621
ROBERTSON M, 113–41;
  115, 116, 118
Robins RJ, 279
Robinson DG, 454, 459, 528,
  610, 613, 614, 617, 619–23
Robinson DJ, 92
Robinson GW, 291
Robinson JM, 584, 593
Robinson LC, 227
Robinson MS, 612
Robinson N, 589
Robinson SA, 591, 634, 640

Robinson SP, 237, 240, 343
Roby C, 251
Roche P, 41
Rocher JP, 449
Rochester DE, 83, 84, 552
Rock CD, 113, 130, 293, 294
Rodermel SR, 239, 294
Rodionova LG, 348
Rodman JS, 610, 612
Rodrigues Pereira AS, 199, 205, 206
Rodriguez-Navarro A, 225
Roeder PE, 397
Roeske CA, 382, 384, 385
Roessner CA, 267, 278
Rogers E, 476, 481, 485
Rogers JC, 121, 123, 126
Rogers LJ, 198, 349, 350
Rogers OC, 610
Rogers SG, 54, 86–88, 93–99, 105, 176, 187
Rogers WJ, 334, 564, 565
Rohmer M, 205
Roise D, 561
Rolff E, 152, 159
Roman G, 180, 185
Romanov GA, 179
Romanowski M, 504, 512
Romeike A, 260
Römer S, 288, 290–92, 297
Roncarati R, 118
Ronen R, 494, 508, 509, 512, 513, 516, 522
Roseman AM, 471, 479, 480, 483, 484, 486
Rosenberg C, 185, 187
Rosenberg LE, 485, 486, 550, 554–57
Rosencwaig A, 494, 500, 508
Rosenqvist E, 635, 642, 646
Rosenshein NB, 665
Ross J, 431
Rotatore C, 378
Roth I, 258
Rothberg KG, 610, 611, 626
Rothenberg M, 174, 180
Rothman JE, 269, 476, 479, 480, 483, 612
Rotter M, 118
Roush RA, 665, 668
Rowinsky EK, 665
Rowse P, 37
Roy H, 471, 478
Royo J, 48
Ruban AV, 634, 636, 641
Rubenstein I, 48
Rubery PH, 394
Rubin PM, 587
Rubinstein B, 150
Rück A, 229, 405
Rüdelsheim P, 184
Rudich J, 522
Rudolph R, 476, 481, 482, 485
Rueffer M, 261, 263, 264, 268, 269

Rufty TW Jr, 248, 579, 582, 584, 597
Rugenhagen C, 279
Rühle W, 352
Rumpho ME, 355
Runowicz CD, 665
Runswick MJ, 457, 559
Ruoho AE, 397
Rushing AE, 96
Rusnak F, 163
Russell DW, 610–12
Russell SH, 121
Russnak R, 51, 52
Russo P, 52
Rustin P, 345, 448
Ryabov EV, 87
Ryan CA, 119, 227, 312, 313
Rybicki EP, 83–85, 92, 93
Ryther JH, 646

S

Sabbagh I, 180, 182
Sable CL, 66
Sabularse D, 614
Sabularse DC, 357, 614
Sachar RC, 57
Sachs AB, 439
Saeboe-Larssen S, 116
Saedler H, 94, 96
Saeki T, 649
Saganich R, 158, 160
Sage RF, 244, 245
Sagers RD, 324
Saibil H, 474, 477, 480
Saibil HR, 471, 479, 480, 483, 484, 486
Saier MH Jr, 406, 412
Saini HS, 120
Saito K, 267, 670
Saji H, 343
Sakai S, 394, 409, 411, 412
Sakakibara T, 326, 327
Sakamoto M, 559
Sakashita H, 385
Sakata K, 671
Sakiyama F, 423–25
Sakiyama M, 529, 533, 534, 537
Sakiyama-Sogo M, 529, 534
Sakoda M, 529, 535
Salamini F, 118, 127, 129, 425
Salinas J, 122
Salisbury J, 455, 585
Salmenkallio M, 123
Salmon JM, 223
Salter HA, 638
Salvacion FF, 222
Salvucci ME, 239, 242, 243, 315, 343, 380
Samaranayake G, 666
Sambrook J, 551
Samuels AL, 214, 614, 615, 617, 619, 621, 626
Samuelsson G, 636, 649, 650

Sanada Y, 595
Sanchez E, 666
Sánchez RA, 144
Sanchez Y, 128
Sanchez-Martinez D, 115
Sanchez-Serrano JJ, 119
Sandalio LM, 342, 343
Sandberg G, 178, 411
Sanders D, 213, 224, 458, 461
Sandman G, 131, 132
Sandmann G, 290, 292, 340
Sands DC, 206
Sandvig K, 610
Sanfaçon H, 48, 49, 55–57
San Miguel P, 291, 293, 295
Sano T, 71
Sarah CJ, 556, 559
Sarkissian GS, 594, 595
Sarojini G, 324, 326, 327, 329–31, 349
Sasaki T, 317
Sassenrath GF, 244, 636
Sathasivan K, 267
Sato F, 268
Sato H, 269
Sato M, 273
Sato N, 326
Sato T, 326, 327
Satoh K, 518
Sauer H, 352
Saunders JW, 275
Saunders K, 86, 88, 92
Sauter M, 530
Sawbridge TI, 155
Sawhney SK, 591–94
Sawhney VK, 531
Saxton MJ, 227, 613
Saxton RW, 206
Scandalios JG, 559, 564
Schachtman DP, 224
Schaefer S, 87, 92–94, 96–99, 101–3
Schäfer E, 152, 156–58, 160, 163, 475, 478, 481, 486, 529, 530
Schafer G, 341
Schaffner K, 497, 511, 512, 514
Schalk H-J, 83, 86–88, 90, 91, 93, 94, 96, 97, 101
Schaller GE, 214, 226–28
Schaniel A, 550, 551
Schatz G, 473, 546–51, 553–55, 557, 559–63, 567
Schatz GH, 497, 504, 511, 512, 514
Schauermann G, 617
Scheel D, 261, 616
Scheibe R, 239, 242, 245, 343, 355, 449, 460, 592, 594
Scheid OM, 98, 99, 103
Schekman R, 610, 626
Schekman RW, 269, 610, 626
Schell J, 48, 83, 86–88, 90–

94, 96–98, 101, 102, 178, 188–90, 394, 398, 399, 401, 408, 410–12
Schellenbaum P, 539
Schellens J, 611
Schenk RU, 670
Scherer GFE, 406
Scherer PE, 473, 547, 550, 553, 555, 562
Scheres B, 26, 33, 38, 39
Schiavone FM, 33
SCHIEFELBEIN JW, 25–45; 26, 27, 32, 35, 37–39
Schiff PB, 664, 665, 667
Schiller B, 83, 86, 88, 90, 91, 96
Schilperoort RA, 190, 267, 271, 273
Schiltz E, 474, 475, 481, 484
Schimkat D, 345, 589
Schindler M, 614
Schlee J, 579, 581, 584, 594
Schlesser A, 212
Schlessinger J, 612
Schleyer M, 560, 561
Schliwa U, 641
Schloss JV, 344
Schlossmann J, 550
Schmid SL, 612
Schmidhauser TJ, 292
Schmidt B, 98, 550, 554, 555, 560, 561
Schmidt C, 370, 374, 376, 377, 379, 386
Schmidt GW, 617
Schmidt M, 476, 481, 482, 484, 485
Schmidt R, 129, 250
Schmidt RJ, 436
Schmidt S, 158
Schmidt W, 150, 154
Schmidt WE, 616
Schmitt JM, 449, 451
Schmitt ME, 429, 430, 450
Schmitz RY, 179
Schmitz UK, 550, 554–56, 559
Schmülling T, 178, 188, 411
Schnabl H, 451, 453
Schneider A, 550, 555
Schneider G, 344
Schneider H, 550, 554, 555
Schneider K, 118, 127, 129
Schneider M, 90, 93
Schneider WJ, 610, 611
Schneider YJ, 611
Schnepf E, 37
Schobel W, 63, 64, 67, 546
Schobert CT, 551, 559
Schoolwerth AC, 456, 457
Schopfer P, 162, 434, 528–30, 532, 534
Schramel A, 477, 478, 480, 484
Schreiber SL, 163
Schreiber U, 496, 497, 501,

506, 507, 510, 512, 517–19, 641, 645
Schreier PH, 279
Schröder G, 176, 266
Schröder JI, 163, 176, 224, 226, 266, 448, 458
Schröder W, 66
Schroeder R, 422
Schroeter B, 645
Schröppel-Meier G, 453, 454
Schrott E, 149
Schubert B, 348
Schuch W, 290, 291, 297
Schuler MA, 97, 98, 100
Schuller KA, 585–87, 597
Schulte W, 616
Schultz A, 341, 342, 352, 358
Schultze M, 80
Schulze E-D, 239
Schumacher H-M, 272
Schünemann D, 595
Schürmann P, 588, 589
Schurr U, 239
Schuster AM, 546, 559
Schuster G, 429, 512, 513, 519, 521, 522
SCHUSTER W, 61–78; 62–64, 67–73, 546
Schwager SJ, 294
Schwarz JJ, 90, 186, 187, 410
Schwarz R, 371, 373
Schwob E, 399, 401, 402
Sciaky D, 190
Scofield S, 117, 229
Scofield SR, 188, 189
SCOLNIK PA, 287–301; 123, 290–94, 296, 297
Scopel AL, 144
Scott AI, 267, 278
Scott IM, 113, 180
Scott J, 73
Scrubb LA, 433
Sczepaniak A, 504, 512
Seagull RW, 530, 535, 536, 538, 539
Seamon PT, 380
Sechley KA, 579, 591
Seckler R, 482, 484
Secor J, 354
Seeley KA, 431
Seemann JR, 123, 242, 244, 245, 343
Seetharam R, 471–73, 475, 477, 480
Segui B, 550, 551
Segui-Real B, 561
Seidman A, 665
Seilhamer JJ, 66
Selivankina SY, 394
Selker JML, 528, 540
Sembdner G, 311
Semenenko VE, 378
Senger H, 144, 152, 377
Senn AP, 404
Sentenac H, 223
Seppelt RD, 645

Serrano E, 163
Serrano R, 125, 212, 214–17, 227, 404, 406, 588
SERVAITES JC, 235–56; 236, 238, 239, 241–50, 252, 343, 344, 359
Sesay A, 306
Seto H, 534
Seurinck J, 48
Shabanowitz J, 179
Shafer JA, 327
Shah DM, 39, 352, 355, 358, 552
Shah K, 327, 332, 333
Shalgi E, 506, 512, 518, 519
Sharkey TD, 123, 163, 238, 241–45, 247, 343
Sharrock RA, 158
Shatters RG, 552
Shaw DC, 125
Shaw G, 177
Shaw MJ, 144, 159, 160, 433
Sheahan JJ, 224
Sheets MD, 52, 53
Shen W-H, 99, 102, 404, 405
Shepherd V, 624
Sherman F, 52, 550, 551
Sherman TD, 358
Sherrard JH, 579, 594
Shet MS, 267
Shewry PR, 304
Shi L, 124
Shiba T, 80
SHIBAOKA H, 527–44; 528–31, 533–40
Shibata K, 496
Shibata K, 163
Shieh W-J, 236, 239, 241–47, 249, 250
Shields CR, 62
Shilling J, 550, 552, 562, 563
Shimabukuro RH, 341
Shimada H, 425, 559
Shimazaki K, 163
Shimizu-Takahama M, 239, 241, 242
Shimogawara K, 377
Shimokoriyama M, 530, 531, 533
Shimomura S, 394, 395, 398, 399, 401–3, 412
Shimura S, 646
Shimura Y, 26, 27, 40
Shinkle JR, 162
Shinmyo A, 264
Shinozaki K, 116
Shinshi H, 177, 186, 410
Shipley A, 37
Shira H, 71
Shiraiwa Y, 377, 378
Shirley BW, 156, 161
Shochat S, 494, 497, 511, 514
Shock TL, 81, 92
Shohet S, 207
Shore GC, 548

Short KC, 185
Short MN, 87, 89
SHORT TW, 143–71; 144–48, 153, 154, 159, 227
Shreve AP, 643
Shropshire W, 152
Shubin VV, 471, 475
Shudo K, 179
Shugar D, 439
Shuler ML, 671
Shumway JE, 117
Shuster R, 556, 559
Sicher RC, 248, 345
Siebers A, 222, 228
Siebke K, 461
Siedow JN, 462, 546, 565
Siefermann-Harms D, 295
Siegel A, 80
Siehl DL, 358
Siemester G, 398, 399, 401, 412
Sierakowska H, 439
Sierra M, 270
Sierzputowska-Gracz H, 251
Sievers A, 37
Sigler PB, 122
Sih CJ, 670
Sijmons PC, 41
Sikkema KD, 346
Silflow CD, 38, 532
Silva H, 394, 411
Silver PA, 550, 552
Silverman DN, 370, 385, 386
Silverman RH, 430
Silverstein SC, 611
Silverthorne J, 154, 333
Silvius JE, 249
Simmonds DH, 538
Simmons C, 399, 401, 402
Simoens C, 188
Simon A, 117
Simon K, 395
Simon M, 471, 483
Simon SM, 563
Simoncini L, 163
Simpson DJ, 297
Simpson L, 63
Simpson M, 69
Simpson RJ, 304, 306, 422, 423, 438
Sims TL, 422, 425, 435, 438
Sinclair JB, 206
Sinclair TR, 317
Singer SJ, 397
Singh A, 425, 436, 438–40
Singh B, 350
Singh M, 67
Singh NK, 120
Singh T, 206
Sinnot EW, 37
Sirevag R, 266
Sivak MN, 351, 355
Sjodin C, 229
Sjoling S, 555, 556
Sjöstrom M, 649, 652
Skoog F, 173, 179, 184, 665,

670
Skowyra D, 470–72, 479
Skriver K, 113, 115, 121, 123
Slabas AR, 345, 535, 539
Slack JMW, 413
Slade AP, 591
Slade LL, 202
Slater A, 291, 297
Slayman CL, 225
Slayman CW, 212, 227
Slife FW, 354
Sloof P, 69, 73
Slot JW, 611, 612
Slovik S, 113, 123
Sly WS, 374
Small ID, 63
Small JGC, 588
Smalle J, 131
Smart CM, 188, 189
Smigocki AC, 188
Smiley DW, 266
Smillie RM, 655
Smirnoff N, 579–81
Smister NE, 610, 611
Smit BA, 318
Smit G, 616, 624
Smith AM, 334
Smith BJ, 125, 128
Smith C, 99, 102, 480, 485
Smith CJ, 478, 481, 482, 484
Smith FA, 448
Smith H, 160, 414
Smith HH, 182
Smith JAC, 382, 458, 462
Smith RC, 647, 648
Smith RG, 581, 583, 585, 595–97
Smith S, 80
Smith SJ, 394
Smith SM, 449
Smith TA, 228
Smith-Monroy CA, 291
Smythe E, 610, 612, 613
Snel JFH, 497, 500, 504, 506, 507, 512, 517, 518
Snider MD, 610
Snustad DP, 38
Snyder GJ, 497
Solberg L, 117
Söll D, 401, 559
Soll J, 472, 473, 477
Sollner T, 549–51, 560, 567
Solovjeva NA, 471, 475
Somers DA, 354, 359
Somerville C, 26, 37, 38, 40, 180, 246, 247, 251, 552
Somerville CR, 48, 131, 324, 349, 352, 353, 360, 455
Somerville SC, 343, 455
Sommarin M, 148, 149
Sommergruber K, 48
Somssich IE, 261, 272
Song L, 373, 385
Song P-S, 152, 154
Song Y-R, 39
Songstad DD, 279

Sonnewald U, 250
Sonobe S, 535, 536, 538–40
Soria M, 422
Sossountzov L, 180, 182
Sotobayashi S, 394
Sotta B, 180, 182
Souer E, 266
South DJ, 588
Spaink H, 38
Spaink HP, 41
Spalding EP, 161, 162
Spalding MH, 371, 375
Spampinato CP, 347
Speer M, 450, 453, 454
Speijer D, 69
Spena A, 178, 189, 190, 404, 411
Speth V, 470, 474, 475, 481, 484, 528
Spiedel D, 639, 641
Spilatro SR, 308, 315, 589
Spill D, 591, 592
Spottswood MR, 73
Spray CR, 174, 180
Spreitzer RJ, 375
Sprinkle JR, 551
Sprinzl M, 556, 559
Spruit CJP, 152, 159, 636
Srere PA, 268
Srinivasan R, 327, 331–33, 559
Srivastava BIS, 431
Srivastava LM, 531
Staal M, 162
Stacey NJ, 87, 94, 95, 97
Stadler R, 263, 267
Staehelin LA, 528, 539
Stahl PD, 610, 612
Stamp P, 634
Staniforth RA, 476, 484, 486
Stanley J, 81, 83–89, 91–97, 99, 101
Stapleton AE, 144
Stark DM, 247
Stasko MW, 667, 669
STASWICK PE, 303–22; 124, 304, 307, 308, 310–13, 315, 316
Stebbins NE, 202
Steele JA, 206
Steemann-Nielsen E, 636, 646
Steen DA, 533
Steer MW, 614, 615, 623, 626
Steeves TA, 34–36
Steffens P, 263
Steger HF, 550, 551
Stein JC, 228
Stein VE, 81, 84, 87
Steinbiss H-H, 96, 102
Steiner DF, 610, 611
Steinitz B, 152, 160
Steinkellner H, 87, 88, 94, 95, 97
Steinman RM, 610

Steinrücken HC, 246
Stenger DC, 92, 93, 101
Stenzler L, 186, 187, 410
Steppuhn J, 556
Stern DB, 67, 68, 71, 429
Sternberg PW, 34
Sternlicht H, 474, 481, 482, 484
Sternlicht ML, 474, 481, 482, 484
Steup M, 589, 594
Stevens A, 179, 430, 439
Stevens LH, 267, 270, 271
Stevenson MA, 553
Stevenson MC, 92, 93, 101
Stewart CR, 349, 553, 559
Stewart GR, 352, 564, 565, 579–81, 591
Stewart JC, 314
Stiborova M, 346
Stidham MA, 355, 358
Stiefel V, 115
Stierle A, 200, 205, 669
Stierle D, 200, 205, 669
Stilling B, 121, 122
Stillman B, 80
Stinard PS, 293
Stinissen HM, 314
Stirling CM, 654
Stirling DI, 207
Stirpe F, 422
Stitt M, 241, 242, 244, 247–51, 345, 356, 357, 448, 450, 451, 453, 461, 581, 583, 587–90, 593–98, 642
Stoddart JL, 305
Stokes DM, 345
Stoner TD, 230
Storey HH, 91, 92
Storz G, 156, 158
Strand M, 645
Stransky H, 404, 406
Strasser RJ, 507, 521, 522
Streeter JG, 455
Strobel G, 200, 205, 669
Strong F, 173
Strotmann H, 340, 341
Strous GJAM, 611, 612
Struck F, 559
Stuart K, 66, 69
Stuart RA, 550, 551, 556, 562
Stumpf DK, 451
Sturm A, 410–12
Styring S, 638
Su W, 38, 183, 312, 316
Subramanian AR, 68
Subularse DC, 588
Suckow M, 122
Suda K, 549
Suffness M, 663, 665
Suffolk R, 63
Sugawara K, 117
Sugiharto B, 381
Sugimoto Y, 349
Sugiyama RH, 431
Sugiyama T, 381

Suharsono S, 70
Sui X, 622, 624
Suire C, 288, 290, 291, 297
Sukrapanna SS, 580
Sullivan ML, 439
Sültemeyer DF, 370, 374–77, 379, 386
Sumar N, 564, 565
Sumi M, 479, 480
Summerfelt K, 248
Summers MD, 314
Sun YJ, 276
Sundby C, 637, 640
Sundqvist C, 150
Sung ZR, 117, 120
Sünkel S, 67
Sunter G, 81, 82, 86–91, 93–99
Surif MB, 379, 380
Surin BP, 481, 485, 487
Surowy TK, 212, 214–16
Susek RE, 333
Süss K-H, 239
Sussex IM, 34, 42, 117
SUSSMAN MR, 211–34; 39, 212–16, 219, 223, 224, 226, 227
Sutton SH, 222
Suyama Y, 425
Suzuki A, 349, 591, 595
Suzuki E, 373, 385, 386
Suzuki H, 39
Suzuki I, 381
Suzuki M, 277
Suzuki T, 595
Svensson B, 126
Swader JA, 384
Swanson JA, 611
Swanson ST, 561
Swaroop M, 550
Swinhoe R, 38
Syono K, 182, 535
Syrett PJ, 579–81, 583, 585, 594, 595
Sze H, 458, 461
Szewczyk A, 457

T

Tabata M, 268, 269, 277
Tabata T, 54
Tabita FR, 475
Tachiki A, 377, 378
Tada K, 324
Taguchi H, 479–81
Taguchi S, 646
Tahar M, 670
Tak T, 41
Takagi M, 424, 425
Takahama U, 239, 241, 242
Takahashi I, 148
Takahashi K, 90
Takahashi M, 646
Takahashi N, 534
Takahashi S, 535, 539, 540
Takahashi Y, 410

Takamatsu N, 80
Takano M, 216
Takata K, 595
Takatsuji H, 39
Takeda J, 623
Takeda K, 528, 531
Takeda S, 559
Takemura M, 62, 64, 65, 67, 70, 71, 73
Takeshita N, 268
Tal M, 119
Talouizte A, 578, 584
Tamaki E, 273
Tamaki S, 48
Tamaoki H, 423–25
Tamaoki T, 148
Tamm LK, 561
Tamura S, 349
Tan C-T, 559
Tanabe Y, 273
Tanaka K, 343
Tanaka S, 269
Tanaka Y, 176
Tanchak MA, 617, 619, 621, 622, 626
Tang A, 307
Tanguay RM, 127
Tanksley SD, 294
Tappeser B, 395
Taran VY, 179
Tarczynski MC, 130
Tardieu F, 118
Tarlowski J, 348
Tarui Y, 57
Tasaka Y, 129
Tashian RE, 370, 385
Taulien J, 128
Taya Y, 176
Taylor CB, 422, 424, 425, 428, 432, 433, 435, 439
Taylor CG, 38
Taylor HL, 663
Taylor IB, 130, 293, 294
Taylor IEP, 614
Taylor J, 205, 622, 624
Taylor LP, 297
Taylor WC, 296, 333
Tazaki H, 534
Teeri TH, 123
Teller G, 177
Tempé J, 178, 404
Tempst P, 474, 475, 481, 482, 484
Tenbarge K, 117
Tenbarge KM, 117
Tepfer M, 614
Teramura AH, 144
Terashima I, 250, 649
Terlesky KC, 475
Terpstra P, 424, 425
Terris S, 610, 611
Terzi M, 411
Tessier A, 500
Teulieres C, 291
Teutonico RA, 185
Tevini M, 144

Thacker A, 579, 581, 594, 595
Thakur S, 98, 99
Thayer SS, 307
Theilmann J, 377
Thelen MP, 423, 435
Theodorou ME, 251
Theologis A, 178, 297
Theriault RL, 665
Thiagarajah MR, 120
Thibault P, 581
Thibodeau PS, 304, 306
Thiel G, 178
Thierfelder J-M, 552, 553, 559
Thimann KV, 152
Thomas CM, 546
Thomas EW, 402
Thomas H, 305
Thomas JC, 183, 186
Thomas JE, 83, 84
Thomas JO, 474, 481, 482, 484
Thomas TL, 121
Thomashow MF, 116, 121, 126, 131, 176
Thommes P, 93
Thompson AE, 293, 294
Thompson D, 48
Thompson RD, 422, 425
Thompson WF, 154–56, 296
Thomson JN, 612
Thomson MC, 65
Thomson WW, 622–24
Thornber JP, 295
Thorner J, 221
Thorpe JR, 334, 564, 565
Thurman DA, 455, 585
Tichelaar W, 471, 474
Tierney ML, 39, 312
Tillmann U, 398, 399, 401, 412
Tillson P, 188
Tilly K, 471, 472, 475, 477–79
Timmerman KP, 247, 341
TIMMERMANS MCP, 79–112; 86, 87, 93, 94, 96, 97, 104, 105
Timmis KN, 425
Timothy DH, 546, 559
Ting IP, 448–51
Titus JS, 304, 317
Tjian R, 90
Tobias CM, 228
Tobin AK, 334, 564, 565
Tobin EM, 154, 186, 333
Todd MJ, 470, 476, 477, 480, 485, 486
Togasaki RK, 375, 378
Toh H, 86, 88, 95
Tokunaga F, 548, 550
Tolbert NE, 324, 325, 344, 348, 360, 375, 377, 378
Tom RD, 497, 512
Tomita F, 148

Tommasini R, 458
Topfer R, 96, 102
Topper JN, 429
Topping JF, 73
Tor M, 200
Torisky RS, 201
Torrent M, 116
Torres Ruiz RA, 28, 29
Torrey JG, 408
Torti G, 412
Touraine B, 448, 451, 452
Town CD, 181
Townsend R, 87, 89, 93, 94, 96, 97
Traas JA, 615, 617, 619
Tranbarger TJ, 315
Travis AJ, 163
Trebal JP, 434
Trent JD, 473, 475, 481, 482
Tripepi RR, 200
Trogisch GD, 581
Tropschug M, 550, 554, 555
Tropsha A, 400
Trouet A, 611
Trowbridge IS, 612
Truchet G, 41
Truettner J, 440
Trumpower BL, 556
Tsai D-S, 425
Tsang A, 117, 119
Tsay YF, 226
Tsay YH, 291
Tsogai Y, 179
Tsugeki R, 553, 559
Tsuji H, 179
Tsuprun VL, 471, 475
Tsuzuki M, 370, 374–77, 382, 383, 450
Tuboi S, 548
Tucci MA, 241, 243, 246–48
Tukey HB Jr, 199
Turbeck BO, 370
Turek JJ, 610
Turgeon R, 304
Turner DH, 587
Turner JC, 26, 38, 40, 183
Turner JE, 177
Turner JF, 587
Turner SR, 326, 331, 332, 559
TURPIN DH, 577–607; 579–81, 583–88, 590, 592, 594–97
Tyler RT, 271
Tyler TA, 667
Typke D, 474
Tyystjärvi E, 644

U

Udvardi MK, 459
Ueda T, 49, 51, 56, 86, 93, 97
Ugaki M, 86, 93, 97
Uknes SJ, 118
Ulbrich B, 412
Ullrich WR, 581

Unseld M, 63, 64
Urwin N, 155
Urwin NAR, 144, 159, 160
Usuda H, 345, 349, 355, 450

V

Vacca RA, 556, 557
Vadell J, 242
Vainberg IE, 474, 481, 482, 484
Valadier MH, 581, 597
Valleau WD, 275
Vallee BL, 440
Vallone D, 411
Valon C, 131
Valsamakis A, 51, 52
Van Arkel GA, 93
Van Beusichem ML, 452
van Brussel AAN, 41
Vance NC, 669
van Cleve B, 318
Vancompernolle K, 57
Van de Boogard R, 448, 453
Van de Mark GJ, 85, 86
Vanden Boom TJ, 324
van den Burg J, 69
van der Art JM, 555
Vanderhoef LN, 395
van der Klei IJ, 546, 550, 553, 567
van der Linde PCG, 408, 409
Vanderschaeghe AM, 201
van der Valk P, 539, 617, 619
van der Vies SM, 470, 471, 475, 478, 481, 482, 486, 551
van der Vlugt R, 88, 95
van der Zaal EJ, 408–10
van Deurs BV, 610
van de Wiel C, 38, 39
Van Dyk TK, 472
Van Etten RA, 62, 66
Van Eycken F, 522
Vanhaelen MH, 667
Vanhaelen-Fastre RJ, 667
Vani T, 154, 163
van Lammeren A, 39
van Leperen W, 507, 518
Vanlerberghe GC, 579, 581, 583, 585, 590, 592, 594–97
van Lierop M-J, 39
Van Lijsbettens M, 188
van Loon LC, 134
Van Mansfeld ADM, 93
Van Montagu M, 34, 48, 57, 176, 342, 343, 559
Van Montagu MC, 56
Van Onckelen H, 188
Van Rensen JJS, 341, 342
Van Sanford DA, 316
Van Slogteren GMS, 190
van Spronsen PC, 41
Vantard M, 539
van Telgen H-J, 408, 409, 411

van Tuinen A, 158
Van Tuinen E, 610, 626
van Tunen AJ, 274
Van Veldhoven PP, 459
van Vliet TB, 270
Van Vloten-Doting L, 80
Van Volkenburgh E, 162
Varadi A, 216
Vargas H, 497, 508
Varner JE, 39
Varshavsky A, 221
Vasil V, 123, 131
Vasil'ev IR, 646
Vass I, 638
Vaucheret H, 582, 583, 591
Vaughn KC, 375, 591
Vavasseur AGL, 163
Veeranjaneyulu K, 513, 515, 520, 522
Vega JM, 643
Venis MA, 179, 229, 394–402, 405, 408, 412
Venken M, 150
Venuti SE, 66
Venyaminov SY, 478, 479
Vercesi AE, 458
Verdernikov VI, 646
Verjovski-Almeida S, 212
Verma N, 57
Vernieri P, 293, 295
Vernon DM, 125
Verpoorte R, 258, 264, 266, 267, 270, 271
Vestweber D, 547, 550, 551, 553, 562
Vetter H-P, 266
Vézina L-P, 591
Viale AM, 472
Vidal J, 586, 595
Vidal V, 550, 552, 553, 559
Vidensek N, 669
Vieira J, 86, 93, 97
Vierling E, 470, 551, 553
Vierstra RD, 149, 154
Viil J, 240, 242
VIITANEN PV, 469–91; 290–92, 470–73, 475–78, 480–82, 485, 486
Vilardell J, 116, 119, 121, 122
Villalba JM, 125, 212, 214, 215, 217
Villand P, 350, 351
Villanueva MA, 622, 624
Villemur R, 38
Vincent WF, 646
Vincentz M, 582, 583, 591
Vinha CA, 497, 508
Vinson CR, 122
Viola G, 395, 398, 399, 401, 412
Vioque A, 292
Virgin I, 634, 636, 638
Virtanen I, 537
Vivekananda J, 121, 457
Vlossak KDM, 584, 590,

594, 597
Voelker TA, 248, 252
Voelkner P, 228
Vogel K, 52, 53
Vogel SN, 665, 666
Vogt E, 458
Vojnikov V, 565
Volfová A, 529, 530, 533
Volk RJ, 579, 582–84, 592, 597
Volker S, 474
Volokita M, 382, 384, 385
von Ahsen U, 422
Von Arnim A, 87, 88, 95, 97
von Caemmerer S, 240, 244, 245, 247, 384–86, 594, 635
von Figura K, 611, 612
von Haeseler A, 70, 71, 73
von Heijne G, 556
von Mende N, 41
Von Saltza MH, 173
von Schaewen A, 250, 251
Von Wechmar MB, 85
von Wettstein-Knowles P, 126
von Wilcken-Bergmann B, 122
Voos W, 562, 563
Voskuilen JT, 267, 270
Voytas DF, 294
Vredenberg WJ, 497, 500, 504, 506, 507, 512, 517–19
Vukovic J, 270

W

Wach A, 212
Wachter C, 559, 560, 562, 563, 567
Wada K, 595
Wada M, 216, 536, 538
Wade RP, 556, 559
Wadsworth P, 532
Waffenschmidt S, 176
Wagner R, 341
Wahle E, 48
Waites B, 207
Waites WM, 198–201, 207
Wakabayashi E, 425
Wakiuchi N, 595
Walberg MW, 62, 66
Walbot V, 63
Walden R, 98, 99, 190
Walderhaug MO, 222, 228
Walesby NJ, 430
Walker DA, 237, 240, 242, 244, 251, 344, 345, 347, 351, 355–57, 359, 448, 449, 636, 643
Walker EB, 154
Walker EL, 591
Walker GH, 326, 331, 457, 592, 593
Walker JC, 228
Walker JE, 457, 559
Walker JL, 325, 326, 328

Walker KA, 352
Walker RP, 394
Walker-Simmons MK, 115, 117
Wall DA, 611
Wall JS, 473–75, 481, 482, 484
Wall ME, 663
Wallace DC, 556, 559
Wallis SC, 73
Wallsgrove RM, 351, 352
Walter TJ, 434
Walters AJ, 550, 552, 553, 559, 561, 564, 565
Walters RG, 641
Walters RS, 665
Walton JD, 397
Wang E, 611
Wang MJ, 430
Wang S, 73
Wang TL, 117, 180, 181
Wani MC, 663
Wanner G, 268, 269
Ward A, 98
Ward AD, 588, 592, 593
Ward BL, 64
Ward JA, 580
Ward K, 54, 55
Warner RL, 580, 591, 593, 594
Warpeha KMF, 148, 149, 151–54, 156, 164
Warren G, 610, 612, 613
Wasteneys GO, 532, 537
Watanabe A, 496
Watanabe H, 424, 425
Watanabe M, 162, 595
Watanabe S, 395, 398, 399, 403, 412
Watanabe Y, 80
Waters SP, 306
Watson BA, 394, 616
Watson GM, 471, 479, 480, 483, 484, 486
Watson JC, 154, 155
Watts C, 610
WATTS FZ, 545–75; 552, 553, 559, 561, 564, 565
Watts J, 86–89, 94, 96, 97
Wattts FZ, 550, 564
Weaver DC, 459
Weber CR, 305
Weber G, 150
Wedding RT, 345, 347
Weger HG, 580, 581, 583, 593, 597
Wehmeyer B, 156
Wei N, 161
Weidner M, 346
Weil J-H, 62, 66, 69–71, 175, 288, 290, 291, 297, 546
Weiler EW, 180, 394, 616
Weiner H, 451, 589
Weintraub RL, 529, 530
Weis E, 513, 517, 636, 639,

641, 646
Weisbeek P, 26, 33
Weisbeek PJ, 93
Weiss CV, 149
Weiss G, 297
Weiss H, 471, 474
Weissbart J, 580, 593, 594
Welch WJ, 127
Wellmann E, 156
Wellnitz D, 395
Wells B, 180
Wendenburg R, 230
Wender PA, 665, 670
Wengenmayer F, 341, 342, 352, 358
Wenzler HC, 317
Werck-Reichhart D, 266
Weretilnyk EA, 125
Werner D, 408
Werner-Washburne M, 550, 552, 562
Wernsman EA, 275
Wessel K, 155, 157
Wesselius JC, 158
Westermark B, 611
Wetzel CM, 294
Wetzel S, 318
Weybrew JA, 274, 275
Wheeler NC, 669, 671
Whelan J, 459, 546, 555, 559, 564
Whitaker DP, 348
White JA, 559, 564
White MJ, 154, 296
White PJ, 458
White RF, 430
Whitecross MI, 379
Whitehouse DG, 565
Whitelam GC, 160, 414
Whitford PN, 117
Whittingham CP, 326
Wickens MP, 52, 53, 439
Widell S, 148–50
Widholm JM, 354
Wiech H, 484
Wieczorek A, 55
Wiedenhoeft RE, 617
Wiedmann M, 550
Wiemken A, 453, 458
Wiermann R, 288, 297
Wiernik PH, 665
Wild A, 352
Wilke HL, 304
Wilke I, 589
Willard DM, 671
Willeford KO, 345
Willemsen V, 26, 31–34, 37
Williams B, 117, 119
Williams EG, 423, 435
Williams G, 228
Williams HJ, 278
Williams K, 66
Williams ME, 122
Williams T, 378
Williams W, 266, 276
Williams WP, 596

Williamson JD, 55, 117, 122, 123
Williamson RE, 26, 36, 37, 528, 537, 540
Williamson VM, 313
Willison K, 474
Willmitzer L, 119, 225, 250, 251
Wilson AK, 26, 38, 40, 183
Wilson CM, 422, 423, 430, 433, 434, 436
Wilson G, 614, 623
Wilson PM, 588
Wilson ZA, 200
Wimmers LE, 216
Winchcombe CF, 434
Winer JA, 552
Wingsle G, 642, 646, 649
Winning BM, 556, 559
Winspear MJ, 352
Winter E, 229
Winter H, 454
Winter K, 382, 448, 450, 451, 634, 636, 642–46, 652
Wintz H, 546
Wirsching P, 346
Wirtz U, 98, 99, 102, 103
Wiskich JT, 66, 331, 349, 456, 457, 546, 588, 592, 593
Wissinger B, 62–65, 68–70, 546
Wistow GJ, 125
Wistrom CA, 125
Witherup KM, 667, 669
Witt W, 318
Witte C, 550, 554
Wittenbach VA, 305, 307, 308, 315, 434
Wöber G, 588
Wohlmuth M, 98, 99, 103
Wohlrab H, 551, 559
Wojtczak L, 457
Wolf S, 87
Wolf YI, 93
Wolkenfelt H, 26, 33
Wolosuik RA, 588, 589
Wolstenholme DR, 65
Wolter FP, 129
Woltering EJ, 497
Wonders JHAM, 519
Wong JH, 588
Wong LM, 297
Wong SC, 250
Woo KC, 455, 585, 592, 593, 597
Wood C, 66
WOOD CK, 545–75
Wood HN, 188, 190
Wood JM, 668
Wood S, 477, 478, 480–82, 484
Woodrow IE, 238, 240, 243
Woolford C, 471, 472, 475, 477–79
Woolhouse H, 334

Woolhouse HW, 334
Woolley DJ, 644
Woolston CJ, 81–84, 87–89, 91, 93–95, 97, 102
Worrell AC, 248
Wright CT, 62, 66
Wright DJ, 304, 306
Wright KM, 614, 623
Wright SM, 207
Wróbel D, 504, 512
Wu CH, 552
Wu HB, 587
Wu L, 49, 51, 56
Wu LC, 87, 88, 93, 95, 98, 99, 104, 105
Wuestehube LJ, 269
Wyen NV, 432, 434
Wyman K, 641, 647
Wyman KD, 645, 646
Wyndaele R, 184
Wynn RM, 472
Wyse DL, 354, 359
Wyss U, 41

X

Xu B, 424, 425
Xue Y, 83, 88, 546

Y

Yaffe MB, 474, 481, 482, 484
Yaffe MP, 125, 128, 550, 552, 554
Yagawa S, 535, 539, 540
Yagawa Y, 375
Yaguzhinskii LS, 401
Yakir D, 522
Yamada H, 425
YAMADA Y, 257–85; 258, 260, 261, 264, 268, 271, 278, 279
Yamagishi A, 512, 513
Yamaguchi I, 394, 534
Yamaguchi Y, 646
Yamaguchi-Shinozaki K, 116, 121, 122
Yamamoto A, 474
Yamamoto H, 277
Yamamoto HY, 642
Yamamoto M, 327, 328
Yamamoto YT, 38, 377, 378
Yamane H, 534
Yamato K, 62, 67, 71, 73
Yamaya T, 579, 591
Yanagi H, 80
Yanai K, 425
Yang C-P, 295
Yang M, 554
Yang N, 73
Yang S, 382
Yano K, 425
Yanofsky C, 292
Yanofsky MF, 398, 399
Yasuhara H, 539
Ye Z-H, 39

Yee BC, 588, 595
Yelenosky G, 318
Yen Y, 423, 431, 436, 437
Yenofsky RA, 201
Yin ZH, 461
Ying Y, 610
Yirinec BD, 611
Yoneyama T, 594
Yoo MH, 39, 216
Yoshida M, 479–81
Young A, 641
Young AJ, 295, 617
Young EG, 65
Young JC, 152, 159, 160
Yousif AE, 355
Yu CA, 556
Yu H, 266
Yu J-W, 373, 377, 384–86
Yu L-X, 398, 556
Yukimune Y, 260
Yumoto N, 482, 484
Yun D-J, 279
Yun S, 349
Yusibov VM, 188

Z

Zacher H, 354
Zahler WL, 347
Zahn R, 483, 484, 486

Zaina S, 412
Zalensky A, 38
Zambryski PC, 176, 178
Zamir D, 83, 84
Zamir LO, 671
Zandomeni K, 162, 528
Zanetti A, 273
Zanlungo S, 546
Zaret KS, 52
Zeevaart JAD, 113, 130, 293, 294
Zehir C, 305
Zehnacker C, 591
Zeichner S, 51, 52
Zeidan M, 83, 84
Zeiger E, 163, 448, 460
Zeiher CA, 354
Zeilstra-Ryalls J, 470, 479
Zelitch I, 348–50
Zemlyanukhin AA, 348
Zenk MH, 258, 261, 263, 264, 267–69, 272, 277, 278, 312
Zerial M, 612
Zettl R, 394, 401, 408, 410, 411, 412
Zhan XC, 83, 84, 88–90
Zhang D, 532
Zhang N, 316
Zhang R, 228

Zhang X-Y, 94, 98, 99, 103, 250
Zhao S, 207
Zhao Y, 227
Zheng D, 471, 474, 479, 480, 483, 484, 486
Zheng X, 485, 486
Zhiri A, 667
Zhou A, 430
Zhou JH, 200
Zhu SL, 374
Ziegelhoffer T, 471, 472, 478, 481, 483
Ziegler H, 452
Zimmat G, 81, 84
Zimmerman BK, 152
Zimmerman G, 588
Zimmermann R, 484, 556, 559
Zingen-Sell I, 620
Zinnen TM, 206
Zipkin ID, 439
Zoglowek C, 457, 461
Zundel M, 205
zur Nieden U, 552, 553, 559
Zviman M, 371, 373, 386
Zwar JA, 123
Zylicz M, 470–72, 479

SUBJECT INDEX

A

Aba
loci, 131
mutants, 132–34
Abi mutants, 132–34
Ablution mosaic virus
viral vector, 81, 84, 88, 90, 95, 103
Abscisic acid, 180
carotenoid-ABA biosynthetic pathway, 289, 295
induces changes in microtubules, 534
Abscisic acid regulated gene expression, 113–41
gene products and stress tolerance, 124–34
cold tolerance, 130–33
desiccation tolerance, 133–34
function determination, 125–30
general considerations, 124–25
regulation of gene expression, 114–24
evidence, 115–16
gene nomenclature, 116–17
negative, 123–24
positive, 117–23
ACC oxidase, 126
Acetazolamide, 384
Acetyl-CoA carboxylase, 353–54
Acifluorfen, 150
Actin filaments, 536
Actinomycin D, 532
Adenine nucleotides
role protein folding, 475, 478–79, 484
Adenosine diphosphate
see ADP
Adenosine triphosphate
see ATP
Adenylate kinase, 347–48
Addicott, Frederick, 22
ADP/ATP translocator, 563
ADP-glucose pyrophosphorylase
regulates starch biosynthesis, 350–51
ADT
role protein folding, 476–78
Affinity labeling, 616
see also Photoaffinity labeling
African cassava mosaic virus
viral vector, 81, 86–92, 95–99, 101–2
Agrobacterium
cytokinin production, 203
AHA genes
plasma membrane proteins, 216, 220–22
Air pollution level
photoacoustic methods, 522
Aldose reductase, 118
Alfalfa isoflavone reductase, 276
Algae
Carbon and nitrogen metabolism in cells, 596–97
mitochondrial genome, 71

Alkaloid biogenesis, 257–85
biosynthetic enzymes and genes, 258–68
indole alkaloids, 264–68
isoquinoline alkaloids, 261–64
nicotine and tropane alkaloids, 258–61
biotechnological applications, 278–79
mutants and variants, 274–78
prospects, 279–80
regulation, 271–74
elicitors and growth regulators, 272–73
promoter analysis, 273–74
tissue and cell-specificity; light regulation, 271–72
subcellular compartmentation, 268–71
alkaloid vesicles, 268–71
other compartments, 270–71
Allicin, 353
Amberlite C6-50 chromatography, 21
American Journal of Botany, 13
Amino acid sequences, 153
berberine bridge enzyme, 263
chaperonin 60, 471
co-chaperonins, 472
dehydrin, 116
folded proteins, 469–70
gemini viruses, 85
H. Niger, 261
nitrate carrier, 225
oat protein, 215–19
P-type cation translocating ATPases, 218, 220, 222
RNases, 423–24
Amino acids
herbicide action, 358
vegetative storage proteins, 307
see also Carbon and nitrogen metabolism in cells
4-Amino-5-fluoropentanoic acid, 157
3-Aminomethylindole, 276
Amino-terminal sequence
auxin binding proteins, 401
Aminotransferases, 349
Ammonia toxicity, 353
Ammonium, 579
α-Amylase/protease inhibitor, 126
Anapleurotic carbon fixation, 585
Anderson, E.G., 7
Anthocyanin, 641
accumulation, 160
induction impeded, 158
Antibiotics
contaminant control, 206–7
Antibodies
anti-ABP1, 405
Anticytokinins, 179
Antisense
DNA technology, 359–60
transformants, 126
Antitumor drugs

see Taxol
Apoplast
   malate, 453
Aquatic macroalgae
   see Carbonic anhydrase
Arabidopsis
   ABA-regulated genes, 121, 126
   thaliana, 130, 132, 146, 152–53, 159
      mutants, 158, 181, 293, 312, 384
   see also Root development
Ardisia crispa
   cytokinin effects, 205
Arginase, 202
Arginine, 202
Arginine decarboxylase, 258–59
Aspartate, 586
Aspartyl residue
   histidine phosphate transferred to, 228
ATP, 19, 147
   binding site, 148
   cellular, 214
   role
      mitochondrial protein import, 545, 560,
         562–63, 565
      protein folding, 474–77, 480, 484–86
      role photosynthesis regulation, 238, 240–
         41, 243, 251
   supply, 595–96
   see also Carbon and nitrogen metabolism in
      cells
ATP synthase, 546
Atp genes
   mitochondrial genome, 66, 68, 70
ATPase
   activity, 214
   chaperonin-mediated protein folding, 475–
      77, 479
   hormone induced changes in microtubules,
      528–30
   hydrolysis, 476
   inhibitors, 162
   protein, 221
   regulation, 163
Atrazine, 341
Auxin-binding proteins, 393–420
   endomembrane system, 394–406
   nucleus, 408–10
   plasma membrane, 406–7
Auxins, 12
   role
      alkaloid biosynthesis, 272–73
      cytokinin accumulation, 173, 183, 187–88
      transport inhibitors, 40
   AXA clone, 217
   axis formation
      root development, 28–30
Azaserine, 352
5-Azidoindole-3-acetic acid, 395–96, 410–11

B

B vitamins
   role root growth, 13–14
Baccatin III, 669
Bacterial contaminants
   see Covert contaminants

Bark
   storage protein, 318
   yew
      see Taxol
Beadle, George Wells, 8
Beer-Lambert Law, 501
Beet curly top virus
   biology, 82–85, 89, 92, 101
Benedict, Chauncy R., 18
N-Benzoylphenylisoserine, 670
Benzyl adenosine, 177
Berberine
   biosynthetic pathways, 261–62, 277
   Berberine bridge enzyme, 263–64, 268–70,
      272, 278
Berra, Yogi, 198, 207
Biosynthetic pathways
   see Alkaloid biogenesis
Biotechnology
   see Alkaloid biogenesis
Biotin, 616, 625
Blu mutants, 153, 159 -60
Bisbenzylisoquinoline alkaloids, 272
Blue light photoreception, 227
Blue light signal transduction, 143–71
   blue light regulation of gene expression,
      154–61
      cis and *trans*-acting elements, 158–59
      transcription and transcript abundance
         measurements, 154–58
   GTP-binding proteins, 151
      with plasma membrane, 151
   phosphorylation of a plasma membrane pro-
      tein, 145–49
      biochemical properties of the reaction, 146
      correlation with phototropism, 145–46
      nature of the reaction, 146–47
      preliminary characterization, 145
      properties of the protein, 148–49
      properties of the reaction, 147–48
   photoperception, 151–54
      nature of B photoreceptors, 151–54
      reduction of a *b*-type cytochrome:the LIAC,
         149–50
Bonner, James: life, 1–23
Borsook, Henry, 5
Borthwick, Harry, 16
Botanical Gazette, 16, 18
Bridges, Calvin B., 5
Brome mosaic virus
   foreign gene expression, 80
Bromophosphoenolpyruvate, 346
Brownian ratchet mechanism, 563
Brutlag, Douglas, 22

C

C_2 cycle, 324
   photosynthetic inhibitors, 348–50
C_3 cycle, 324
   photosynthetic inhibitors, 343–45
C_3 plants
   photosynthesis
      see Diurnal regulation of photosynthesis
   see also Carbonic anhydrase
C_4 cycle, 345–48

C_4 plants
  photosynthetic inhibitors, 360
  see also Carbonic anhydrase
Ca^+-ATPase, 217, 220
Cab, 123
CAB, 155–56
CAH genes
  cDNA, 377
  mRNA transcript levels, 378
Calcium
  stomatal regulation, 163–64
Calcium-dependent protein kinasès, 212, 226–27
California Institute of Technology
  see Bonner, James: life
Calmodulin-dependent myosin light chain kinase like protein, 163
Calvin cycle
  see Diurnal regulation of photosynthesis
CAM plants
  (*S*)-canadine, 263, 264
  (*S*)-canadine oxidase, 264, 268–69
  see Carbonic anhydrase
Canadine synthase, 264
Canopies
  see Photoinhibition of photosynthesis
Capsanthin/capsorubin synthase, 292
Carbon
  concentrating mechanisms
    see Carbonic anhydrase
  in sinks, 304
  metabolism
    see Diurnal regulation of photosynthesis
  photorespiratory, 324–27
Carbon and nitrogen metabolism in cells in cells, 577–607
  approach, 579–80
  carbon requirements, 581–86
    anapleurotic carbon fixation, 585
    2-Oxoglutarate for glutamate synthesis, 585–66
    photosynthetic carbon fixation, 583–86
  energy considerations, 591–96
    ATP supply, 595
    chloroplast in light and dark, 593–94
    cytosolic reductant supply, 592–93
    energy supply in the plastid, 593–95
    enzymes of nitrate reduction, 591–92
    reductant supply in the root plastid, 594–95
    model for activation, 590
  physiological evidence, 580–81
  regulation in systems, 596–99
    algae, 596–97
    heterotrophic tissues, 598–99
    photosynthetic higher plant cell, 597–98
  regulation of carbon flow to the TCA cycle, 586–90
    fructose-1,6-bisphosphatase, 588–89
    phosphoenolpyruvate carboxylase, 586
    phosphofructokinase and pyrophosphate fructose-6-phosphate 1-phosphotransferase, 587–88
    pyruvate dehydrogenase complex, 587
    pyruvate kinase, 586–87
    starch metabolism, 589–90
    sucrose phosphate synthetase, 589

Carbon dioxide
  CO_2 permeable membrane, 500
  fixation, 22
Carbonic anhydrase, 374–79, 386
2-Carboxyarabinitol-1,5-bisphosphate, 544
2-Carboxyarabinitol-1-phosphate, 236, 242–43
  inhibits rubisco, 343–44, 359
Carboxysomal carbonic anhydrase, 373
Carotenoid biosynthesis, 287–301
  developmental expression, 331–34
    H-protein and the small subunit of rubisco, 332–33
    light-dependent, 331–32
    tissue-specific expression, 333–34
  gene cloning, 288–92
    geranylgeranyl pyrophosphate synthase, 288–91
    other genes, 292
    phytoene desaturase, 292
    phytoene synthase, 291
  genetic control, 293–95
    Arabidopsis thaliana, 293–94
    Lycopersicon esculentum, 294
    other species, 295
    Zea mays, 295
  regulation of expression, 295–97
    development, 296–97
    light, 295–96
    photooxidative stress, 296
Carotenoid pathway, 114
  biosynthesis, 120
Carriers
  see Photosynthetic inhibitors; Plasma membrane proteins
cat, 56
Catalase inhibitors, 348
Cauliflower mosaic virus, 135, 189
  Poly(A) signal
    see mRNA 3' end formation
  viral
    promoters, 39, 90
    vector, 80, 89
Caveolae, 610–11, 626
Cell
  biology
    see Bonner, James: life
  culture
    taxol production, 670–71
  growth
    herbicide action, 358
    role blue light transduction, 161–62
  localization
    role ABA, 113
  metabolism
    see Carbon and nitrogen metabolism in cells
  proteins
    see Chaperonin-mediated protein folding
Cell-cell interaction, 33
Cell cycle control
  role cytokinin, 185
Cell wall protein gene, 39
Channels
  membrane proteins, 405–6, 412
  mitochondrial genome, 62
CHI gene expression regulation, 156

Chaperonin-mediated protein folding, 469–91
 chaperonin molecules, 470–75
  chaperonin 10, 472–73
  chaperonin 60, 471–72
  cytosolic T-complex polypeptide-1 related
   chaperonins, 473–75
 mechanism, 475–87
  ATPase activity, 475–79
  ATP-dependent association of cpn10 with
   cpn60, 479–80
  binding of polypeptides to chaperonins,
   480–83
  release of polypeptides from chaperonins,
   483–87
Chemical disinfection
 contaminant control, 206
Chemotherapy
 see Taxol
Chitinase mRNA accumulation, 186
Chlorophyll
 fluorescence, 294, 641
 glyceraldehyde-3-phosphate dehydrogenase
  loss, 306
 NADP-dependent, 158
Chloroplasts, 471
 assimilate export
  inhibitors, 355–56
 biogenesis
  mutants, 293
 Carbon and nitrogen metabolism in cells,
  593–94, 596
 chaperonin 10, 472–73, 478–79
 CO2 concentration, 385
 differentiation, 157
 envelope, 384
 isolation, 18–19
 malate, 455–60
 movements, 641
 phosphate translocators, 355
 photosynthesis, 370–71, 374
 RNA polymerases, 68
 sequences
  mitochondrial genome, 71
 stroma
  C3, 344–45
Chlorsulfuron
 resistance to
Chromatin, 19, 21–22, 411
chs mRNA abundance, 160
CHS gene expression, 158
 regulation, 156
Ci transport system, 372, 374–80
Cigars
 nicotine content, 275
Circadian events
 photosynthesis regulation, 249
cis
 arrangements, 69
 elements,
  poly(A) sites, 54, 56
  role ABA, 121–22, 134
 sequences, 105
cis-trans zeatin isomerase, 177
cis-zeatin, 175
Citrate synthase, 481, 485

ckr mutants, 183
Clathrin coated pits
 see Endocytosis
Cloning
 blue light mediated gene expression, 158–
  59
 see also Carotenoid biosynthesis; cDNA
Coat proteins
 geminiviruses, 85, 87, 96–100, 105
Coated vesicles, 619
Cob genes, 64, 70
Cobra, 36–37
Colchicine, 528, 533
Cold
 disrupts microtubules, 536–37
 photosynthesis inhibition, 641, 645, 655
 tolerance
  role ABA, 114, 121, 126, 130–34
 see also Temperature
Columbamine, 277
β-Conglycinin, 117
Cor genes, 132–33
Cortical microtubules
 see Hormone-induced changes in microtubules
Cotyledons
 vegetative storage proteins, 309–10
Covert contaminants, 197–210
 life without contamination, 206–7
 plants and microbes, 198–200
 plants or microbes, 204–6
 PPFMs, 200–4
Cox genes
 mitochondrial genome, 64–67, 70, 72–73
 COXIV-DHFR, 557, 562, 566
Crop
 yield
  resistance to geminiviruses, 101
 see also Photoinhibition of photosynthesis
Crown gall tumors
 role cytokinins
  accumulation, 182, 184, 190
  biosynthetic, 175–76
Cruciferin, 117
Curtobacterium flaccumfaciens
 contaminant 200
Cyanobacteria
 see Carbonic anhydrase
Cycloheximide, 130, 592
Cyst nematodes
 role root development, 41
Cysteine, 21
Cytochrome, 73
 flavin complex, 149
Cytochrome oxidase pathway, 580
 see also Blue light signal transduction
Cytochrome c heme lyase, 551
Cytochrome c reductase, 55
 mitochondrial protein import, 547, 559, 561–
  62, 565
Cytokinin
 binding proteins, 179–80, 394
 effects, 205
 induces changes in microtubules, 533
 oxidase, 177
 receptors, 179–80

analysis, 179
Cytokinin accumulation and action, 173–96
  biochemical approaches, 174–80
  biosynthesis, 175–76
  metabolism, 176–78
  receptors, 178–80
  genetic approaches, 180–85
  accumulation mutants, 180–82
  response mutants, 182–85
  molecular approaches, 185–90
  control of gene expression, 186–87
  transgenic plants express the *ipt* gene, 187–90
Cytokinin glucosides, 411
Cytokinin-O-glucosidase, 411
Cytoplasmic binding proteins, 414
Cytoplasmic male sterility, 62, 64
Cytosol, 430
Cytosolic
  gluconeogenesis, 583
  HSP70s, 551–52
  presequence binding factors, 548
  reductant supply, 592–93
  T-complex
  polypeptide-1 related chaperonins, 473–75

D

Dark reactions, 507
Datura innoxia, 207
DCMU, 646
  photosynthetic inhibitors, 341
Dehydration stress
  role ABA, 118–19
Dehydrin, 116
Dehydrin-like proteins, 125, 127
Dehydrodiconiferyl alcohol glucosides, 185
De Jong, Bungenberg, 12
3-Deoxy-*D*-arabino-heptulosonate-7-phosphate, 236, 246–47
Depodding
  protein increase, 315
  slows chlorophyll loss, 308
Desacetoxyvindoline 4-hydroxylase, 268
Desiccation tolerance
  role ABA, 114–15, 118, 134
Development
  carotenoid levels, 296–97
  see also Blue light signal transduction; Root development
DFR genes
  expression regulation, 156
Dgt, 407–8
Dhn genes, 116
Diadinoxanthin, 642
Diarginine residues, 290
Diaspartic residues, 290
Diatoxanthin, 642
3,3-Dichloro-2(dihydroxyphosphinoyl-methyl)propenoate, 346
Diffusion, 507–8
Digitaria streak virus
  biology and applications, 83, 102
Dihydrofolate reductase, 480–83
Dihydrolipoamide, 326
Dihydrolipoamide acyltransferase, 328

3,4-Dihydroxyphenylalanine, 261
Dihydrozeatin, 175–76
Diamine oxidase, 260
Diiodebenzoic acid, 394–95
4,4′-diisothiocyanatostilbene-2,2′-disulfonic acid, 355
Dinoseb
  inhibitors, 341–42
Dio-9, 341
Diseases
  Rnase activity, 433–34
Diterpene alkaloids
  see Taxol
Diurnal regulation of photosynthesis
  acclimation during light regime, 241–42
  Calvin cycle, 236–41
  integration of assimilation and carbon allocation, 243–251
DNA, 63
  analysis
  see Plasma membrane proteins
  histone-free, 20
  photolyase activity, 153
  root cell expression, 38–39
cDNA, 555
  ABA regulated, 129
  auxin-binding proteins, 398–99
  CAH genes, 377
  cDNA expression libraries, 174–75
  probes, 155
  see also Carotenoid biosynthesis
mtDNA, 63–65, 71
ssDNA, 88
  see also Geminiviruses
T-DNA, 176, 190, 226
  geminivirus
  markers, 102
  vector delivery, 98–99
  hairy root disease, 178
  ipt gene, 186
DNases, 423
DNA-dependent RNA polymerase, 20
Dobzhansky, Theodosius, 5–7, 9
Dolk, Herman E., 8, 10
Drosophila, 5–6, 9
Drought stress
  ABA, 131–32
Ds, 229

E

Early light-induced proteins, 157
Edman degradation, 214, 226
Electron microscopy, 474, 479–80
  mitochondrial genome, 62–63
Electron transport, 500
  photosynthesis inhibitors, 535
Electroporation
  geminiviral vectors, 99
Elicitors
  alkaloid biogenesis, 272–73
ELISA
  auxin-binding proteins, 399–400
  taxol, 667–68
Elongation, 160
  blue light, 145

cell, 565
  role auxin, 404, 531, 533
rates, 124
stem
  role ABA, 534
Em expression
  regulated by ABA, 119
Em genes, 121
Endocytosis, 609–31
  endocytic pathway, 622–23
  evidence, 616–22
    coated pits, 617–19
    coated vesicles, 619–21
    multivesicular bodies, 621–22
    partially coated reticulum, 621
    plasma membrane receptors, 616–17
    vacuoles, 622
  fluid phase, 623–24
  future directions, 625–27
  historic perspective, 613–16
    experimental obstacles, 614
    summary of observations, 615–16
    theoretical arguments for and against, 613
  in animal cells, 610–13
    clathrin coated pit pathway, 611–13
    ligands and receptors, 611
    multiple pathways, 610–11
  receptor-mediated, 624–25
Endomembrane systems
  see Auxin-binding proteins
Endonucleases, 429
Endoplasmic reticulum, 225, 551
  alkaloid vesicles, 268, 270
  auxin-binding protein, 229
  enzyme markers, 394
  proteins, 398, 400, 402, 412–13
  RNases, 428
Endoribonuclease, 431
Energy
  storage, 496–97, 504
    quantification of light distribution, 519
    transients, 518
    see also Photoacoustic method and photo-
      synthesis
  see also Carbon and nitrogen metabolism in
    cells
Environmental stress
  see Photoinhibition of photosynthesis
Enzymes
  nitrate reduction, 591–92
  see also Carbon and nitrogen metabolism in
    cells; Photosynthetic inhibitors
Epoxidation state, 642–43, 646
Erwinia carotovora var. atroseptica 199
Escherichia coli
  carboxysomal CA, 373
Erxleben, Hanni, 11–12
  poly(A) signals, 52, 57
Ethoxyzolamide, 374, 380, 384
Ethyl *N*-phenylcarbamate, 528, 531
Ethylene, 126, 180, 261
  induces
    changes in microtubules, 533–34
    tomato gene expression, 297
    mRNA accumulation, 186
Ethylene response, 174

Etr, 213, 228

F

Far upstream elements
  see mRNA 3′ end formation
Fambrough, Douglas, 20, 22
Fate mapping
  root development, 42
Fatty acid biosynthesis
  photosynthetic inhibitors, 353–55
Ferredoxin-dependent glutamate synthase, 353
Ferredoxin-NADP⁺ reductase, 342
Ferredoxin-thioredoxin system, 589
Ferritin, 622, 626
Fick's law of diffusion, 219
Ferns
  mitochondrial genome, 71
Flavins, 149–50
  fluorescence, 149–50
  photoreceptors, 162
  role alkaloid biogenesis, 263–64
Floral
  development, 27
Fluridone, 120
Flowering
  role photoperiodism, 15–16
Fluorescence, 645–46
  quenching, 153–54, 517–18, 647
Folding
  protein
    see Chaperonin-mediated protein folding
Frey-Wyssling, A., 12
Fructokinase, 250
Fructose bisphosphatase
  Carbon and nitrogen metabolism in cells,
    588–89
  photosynthesis
    inhibition, 356–57
    regulation, 236–37, 242, 245, 248, 250
Fungi
  mitochondrial genome, 62–63
  phytopathogenic
    taxol production, 671
Fusicoccin, 616
Fusicoccin-binding proteins, 394

G

Geminiviruses, 79–112
  applications, 99–105
    DNA methylation, 103
    engineering resistance, 101–2
    expression: foreign genes, 100–1
    expression: viral genes, 100–1
    expression: whole plants, 99–100
    gene expression, 99–100
    markers for gene transfer, 102
    potential applications, 104
    transposition mechanisms, 102–3
  characteristics, 81–91
    atypical, 83–85
    bidirectional transcription, 89–90
    bipartite, 81–82
    coat protein, 87
    evolution factors, 86

gene expression regulation, 90–91
genomic organization, host range, insect
   vectors, 81–86
monopartite, 83
movement proteins, 87–88
other geminiviruses, 89–90
other ORFs, 88–89
replication associated protein, 86–87
sequence comparisons, 85–86
transcriptional activators, 88
viral gene transcription, 89–91
viral promoters, 90
viral protein, 86–89
life cycle, 91–96
   complimentary strand synthesis, 92–93
   host range and symptom development, 95–
      96
   local movement, 94–95
   long-distance movement, 95
   movement, symptom development, host
      range, 94–96
   replication, 92–94
   replication mechanism, 92
   replication regulation, 94
   replication site, 92
   transmission, 91–92
   viral strand synthesis, 93–94
vectors, 96–99
   advantages, 96
   bipartite, 96–97
   construction, 97–99
   delivery, 99
   design, 96–97
   disadvantages, 96
Gene
  expression
    see Geminiviruses; Mitochondrial glycine
      decarboxylase complex; mRNA 3′ end for-
      mation
  regulation
    see Abscisic acid regulated gene expres-
      sion; Alkaloid biogenesis; Blue light sig-
      nal transduction
Genetic
  analysis
    see Cytokinin accumulation and action
  determinants
    control RNase activity, 436–37
  engineering
    pathogen resistant plants, 440
Genetics
  see Root development
Geranyl phosphate, 266
Geranylgeranyl pyrophosphate, 670, 672
Geranylgeranyl pyrophosphate synthase, 288–
   92
Germination
  vegetative storage proteins, 309
Ghost mutant, 294, 296
Ghosts, 539–40
Gibberellic acid, 118, 123, 434
Gibberellin, 178, 180
  binding proteins, 394
  induces changes in microtubules, 529–33,
    537
  role ABA, 118, 124

synthesis, 174
Glackin, Carlotta, 22
β-1,3 glucanase, 184
Glucanase genes, 72
β-glucans, 616
Glucose 6-phosphate, 589
Glucose 6-phosphate dehydrogenase, 594
α-glucosidase, 476, 481–82
β-glucosidase activity, 178, 186
Glutamate, 587
Glutamate dehydrogenase, 591
Glutamate synthesis, 585–66
Glutamine synthase
  inhibitors, 352
Glutathione S-transferase, 410–11
Glyceraldehyde 3-phosphate
  photosynthetic regulation, 236, 239–42
Glycerol-3-phosphate acyltransferase, 129
Glycine decarboxylase, 349–50, 565
Glycine decarboxylase complex
  see Mitochondrial glycine decarboxylase
    complex
Glycolate oxidase
  inhibitors, 348, 360
S-Glycoproteins
  see Ribonucleases
Glycosylphosphatidyl inositol-anchored recep-
   tors, 626
Glyoxylate reductase, 350
Gnom gene, 29–31, 33
Golgi apparatus, 402
  endocytosis, 615–17, 619, 621
Gramine, 276
Gravity
  role root development, 39–40
GroE
  see Chaperonin-mediated protein folding
Growth
  hormones
  regulators
    alkaloid biogenesis, 272–73
    see also Auxin-binding proteins
    role cytokinin, 174
GTP binding proteins
  see Blue light signal transduction
Guard cell plasmalemma, 154
Guayule, 16–18
GUS
  activity
    role contaminants, 200
  genes, 100
GUS reporter genes, 38, 122, 156, 333
Gymnocolea inflata
  vitamin B stimulates growth, 203

H

Haagen-Smit, A.J., 11
Hairy root disease, 178
Haldane, J.B.S., 10
Hamner, Karl, 14–16
H⁺ATPase
  pumps, 404–5
  see also Plasma membrane proteins
Heat shock proteins, 125
  cytokinin accumulation, 188–89

role thermotolerance, 127–28
Heat shock proteins, 484
  see also Mitochondrial protein import
Heat stress, 507, 648
Hedin, Sven, 2
Helminthosporium maydis
  host-specific toxin, 434
Hendricks, Sterling, 16
Herbicides
  photoacoustic methods, 522
  photosynthetic inhibitors, 340–43,
    358–59
Heteroauxin, 12
Heyn, A.N.J., 11
Histidine residues, 399
Histone, 6, 21–22
Hl loci, 184
Hoagland, Dennis, 12
Homologies
  carbonic anhydrase proteins and genes, 385–
    86
  tRNAs, 147
Hormone-induced changes in microtubules,
  527–44
  changes in the orientation, 528–35
  abscisic acid, 534
  auxins, 528–30
  cytokinins, 533
  ethylene, 533–34
  gibberellins, 530–33
  other, 534–35
  interactions with the plasma membrane,
    535–40
  dissociation from the plasma membrane,
    535–36
  in vitro, 539–40
  mediating factors, 538–40
  microtubule-associated proteins, 538–39
  protein phosphorylation, 538
  role of cell wall, 537–38
  stabilization and destabilization, 537–38
Hormones
  endocytosis, 611
  mutants
    see Cytokinin accumulation and action
  production, 34
  receptors
    see Auxin-binding proteins
  with PPFMs, 203
  see also Abscisic acid regulated gene expres-
    sion; Bonner, James: life
Host
  range
    geminiviruses, 95–96
  resistance
    geminivirus engineered, 101–2
HPLC
  taxol, 667–68
Hsp
  cytokinin accumulation, 188–89
  mutants, 128
Huang, Ra-chic C., 19
HY genes, 159
Hy mutants, 153, 159–60
Hydrophobic table
  auxin ring system, 400–1

Hydroxy-pyruvate reductase, 350
Hyoscyamine, 261
Hyoscyamine 6-hydroxylase, 261, 273–74,
  278–79
Hyphomicrobium spp., 198
Hypocotyl growth
  inhibited, 164

I

IAA
  see Indoleacetic acid
Immunoglobulin binding protein, 552
Indole alkaloids
  see Alkaloid biogenesis
Indoleacetic acid, 9, 11, 395, 409, 411
  auxin binding proteins, 409, 411
  induces changes in microtubules, 528–30,
    533
Indole-3-carbinol, 411
Inhibitors
  manipulation of ABA levels, 119–20
  mRNA synthesis, 532
  see also Photosynthetic inhibitors; Protei-
    nase inhibitors
Initials
  role root development, 28, 32
Inner membrane peptidase, 555
Insect vectors
  see Geminiviruses
Intermembrane space
  HSP70s, 553
Invertase, 250
Ion channel blockers, 162
Ion channels
  see Plasma membrane proteins
Ion transport
  role ABA, 124
Ion uptake
  in root cells, 39
Ipt gene
  cytokinin accumulation, 176, 180, 184, 186–
    90
Isopentenyl adenine, 175–76
Isopentenyl transferase, 175–76
Isopropylmalate synthase, 556
Isoquinoline alkaloids
  see Alkaloid biogenesis

J

Jasmonate
  vegetative protein gene expression, 311–12
Jasmonic acid
  pathway, 14
  role microtubule orientation, 534
Journal of Molecular Biology, 21
Jungermannia leiantha
  vitamin B stimulates growth, 203

K

Kalckar, Herman, 4
Kanamycin selection, 98
KdpB gene, 222
Kerchoff Laboratories

see Bonner, James: life
Ketone tropinone, 260
Kinase function
  effect light, 147
Kinetin, 533, 670
Kraus, E.J., 14
Krebs cycle, 19

L

Lactate dehydrogenase, 483–84
Laser induced optoacoustic spectroscopy,
  496–97, 499, 511, 514
Lea, 116–17
Leaves
  lectin gene, 72
  malate, 445–53
  photoacoustic method and photosynthesis,
    508–10
  proteins, 19
  vegetative storage proteins, 305, 315, 318
  see also Diurnal regulation of photosynthe-
    sis; Photoinhibition of photosynthesis
Lectins
  root specific, 31
Leukemia
  see Taxol
Ligands
  endocytosis, 610–11, 614–16, 627
Light
  breaks 15–16
  reactions
    see Photosynthetic inhibitors
  regulation
    alkaloid biogenesis, 271–72
    role root development, 39–40
  signals
    see Blue light signal transduction; Carote-
      noid biosynthesis; Diurnal regulation of
      photosynthesis; Photoacoustic method and
      photosynthesis
  stress
    see Photoinhibition of photosynthesis
Light-dark cycles, 15–16
Light emitting diodes, 500
Light guide optics, 497
Light harvesting complex, 296, 642
  mRNA, 186
Lion's tail, 36
Lipid biosynthesis, 358
Lipoxygenase, 311, 315
Littleton, John, 19
Lucifer yellow, 614, 623
Lupin
  alkaloid mutants, 276
Lysophosphotidylcholine, 406

M

MacCallum, William, 17
Macropinocytosis, 611
Magnesium ions, 591–92
Maize streak virus
  biology and applications, 183–84, 188, 190–
    92, 195, 197, 103
Malate, 349, 585–86

Malate compartmentation, 447–66
  concentrations in tissues, 450–53
  apoplast and exudates, 453
  leaves, 45–53
  roots, 451–52
  xylem and phloem, 452
  final comments, 462–63
  intracellular compartmentation, 455–60
  regulation of distribution, 460–62
  chloroplasts, 460
  mitochondria, 461
  vacuoles, 461–62
  transport systems, 455–60
  chloroplasts, 455–60
  mitochondria, 456–58
  peroxisomes, 459
  plasma membrane anion channel, 460
  symbiosomes, 459
  vacuoles, 461–62
Malate dehydrogenase, 331, 480, 485
Malonyl-CoA, 354
Mannitol l-phosphate, 130
Mannose, 351
Mapping
  mitochondrial genome, 62
Marushige, Keije, 21
MAS genes, 554
McGinnies, William, 18
Meloidogyne incognita
  role in root development, 41
Membrane proteins, 550
Metabolism
  role ABA, 113
  see also Carbon and nitrogen metabolism in
    cells; Photosynthetic inhibitors
Methylation
  DNA, 103
Methanol, 203–4
Methionine sulphoximine, 352
Methylamine dihydrolipoamide, 316
N^5, N^{10}-methylene tetrahydrofolate, 324
Methyl jasmonate
  role microtubule orientation, 534
  vegetative storage protein gene expression,
    311–13
Methylobacterium organophilum
  production triterpenoids, 205
Methylobacterium
  see Covert contaminants
O-Methyltransferases, 263, 278
Mevalonate, 671
Microalgae
  see Carbonic anhydrase
Microbial contaminants
  see Covert contaminants
Microtubules
  assembly
    taxol enhances, 665–66
    blue light transduction, 162–63
    role cell expansion, 35
    see also Hormone-induced changes in micro-
      tubules
Migration
  see Mitochondrial genome
Millerd, Adele, 19
Millikan, R.A., 15

Mitochondria, 439–30, 471
  citrate export, 585
  cpn10, 477, 480
  electron transport chain, 580, 595
  malate, 456–58, 592–93
Mitochondrial genome, 61–78
  gene expression, 67–71
    biochemistry and specificity, 69–70
    cis- and trans-splicing, 68–69
    RNA editing, 69–71
    RNA editing: evolution, 70–71
    RNA editing: functional proteins, 70
    transcription initiation, 67–68
  information content, 63–67
    genes encodes, 65–67
    introns, 64–65
  interorganellar flux of nucleic acids, 71–73
    export to nucleus, 72–73
    import into mitochondria, 71–72
  physical structure, 63–63
Mitochondrial glycine decarboxylase complex, 323–27
  glycine decarboxylase in the photorespira-
    tory pathway, 324–25
  glycine decarboxylase reaction, 325–27
  metabolic control; substrate transporters,
    330–31
  protein chemistry, 327–29
    H-protein, 327–28
    L-protein, 328–29
    P-protein, 327
    T-protein, 328
  serine hydroxymethyltransferase, 331–34
  structure, 329–30
Mitochondrial intermediate protease, 554
Mitochondrial phosphate transporter, 551
Mitochondrial protein import, 545–575
  basic processes, 546–48
  developmental regulation, 564–67
    protein import, 565–67
    respiratory activity, 564–65
  energetic requirements, 560–61
    protein translocation, 563
    role of ATP, 562–63
    role of the membrane potential, 560–61
  presequences of nuclear-encoded plant mito-
    chondrial proteins, 556–57
  processing of imported proteins, 554–56
  proteins required, 548–53
    cytosolic HSP70s, 551–52
    cytosolic presequence binding factors, 548
    cytochrome c
      heme lyase, 551
    intermembrane space HSP70s, 553
    mitochondrial cpn60s, 553
    mitochondrial HSP70s, 552–53
    molecular chaperones, 551–53
    receptor and import channel proteins, 549–
      51
  respiratory chain, 546
  use of heterologous systems, 557–59
α-Mitochondrial processing peptidase, 554
β-Mitochondrial processing peptidase, 554–55
Modeling
  photoinhibition of photosynthesis, 648–53
Molecular chaperones, 548, 551–53

MOM receptor proteins, 549–51
Monoclonal antibodies
  ABA levels, 115
Monopteros gene, 29–31, 33
Morgan, Thomas H., 4–5, 9–10, 13
Morphine, 268
Morphogens
  auxin as morphogen, 413–14
Movement
  geminiviruses, 94–95
MRP
  RNA, 429–30
Multivesicular bodies
  endocytosis, 621–22
Murray, Kenneth, 21
Mutants
  ABA application, 118–19
  A. thaliana
    see Root development
  NADH:ubiquinone-oxidooreductase, 66
  see also Alkaloid biogenesis; Cytokinin ac-
    cumulation and action

N

NAD
Nad genes
  mitochondrial genome, 64–66, 68–69, 73
NAD⁺, 330
  glycine dependent reduction to NADH, 326,
    330
NADH, 326–27
NADH dehydroygenase, 546
NADP-malic enzyme, 346–47
NADPH, 592, 594, 595–96, 598–99
  photosynthesis regulation, 239, 241
NADPH-cytochrome P-450 reductase, 267,
  270
Napthalene-1-acetic acid, 395, 399
Naphthylphthalamic acid
  binding protein, 394
Near-upstream elements
  see mRNA 3′ end formation
Nematodes
  role root development, 41
Neomycin phosphotransferase II gene, 56
Niacin
  role root growth, 14
Neurospora
  see Mitochondrial protein import
NH₃
  photorespiration, 324, 327
Nicotine alkaloids
  see Alkaloid biogenesis
Nigericin, 560
Nitrate, 579–80
  assimilation in photosynthetic cells, 594
  carrier
    plasma membrane proteins, 212, 225
Nitrate reductase, 186, 581–82, 591
Nitrate reduction, 593
Nitrenes, 397
Nitrogen
  assimilation, 351–53
  metabolism
    role PPFMs, 202

see also Carbon and nitrogen metabolism in cells; Vegetative storage proteins
Nitrophen, 341
(S)-norcoclaurine, 261, 263
Norflurazon, 292, 333
Nornicotine, 274–75
Northern blots, 216–17
Noyes, 15
Nuclear magnetic resonance, 482
Nuclei, 431
Nucleic acids
information exchange, 71–73
Nucleotide sequences
geminiviruses, 85
tRNA genes
Nutrients
uptake, 610

O

Ocean environment
photoinhibition of photosynthesis, 644–48
Ocs genes
poly(A)signals, 49, 51, 53–54
Octopine synthase, 48
Oenothera berteriana
mitochondrial genome, 63, 67, 69
tRNA genes
Okadic acid, 592
Oligosaccharide, 187
Open reading frames
geminivirus biology, 81, 83, 85, 87–89, 94, 97, 100, 104
mitochondrial genome, 64, 67
Orf genes
mitochondrial genome, 67
ORFs
see Open reading frames
Organ development
see Root development
Organelles
source of cytosol reductant, 592
Ornithine decarboxylase, 258–59
Ornithine transcarbamylase, 485
Osmotic imbalance, 305
OVE mutants, 181
Oxidative pentose phosphate pathway, 589, 594–95, 584, 599
2-Oxoglutarate/ glutamate synthesis, 585–66
12-Oxo-phytodienoic acid, 272
Oxygen uptake, 580

P

Paclitaxel, 665
PAL genes
expression regulation, 156
Paraheliotropic movement, 640
Paraquat
photosynthetic inhibitors, 342–43
Partially coated reticulum
endocytosis, 617, 619, 621
Particle bombardment
geminiviral vectors, 99
Patatin, 314, 317
Patch clamp studies, 163, 223–24, 405

Pathogen attack
role root development, 39, 41
Pathogen resistance, 227
PCR
see Polymerase chain reaction
PEG, 118
geminiviral vectors, 99
Pectin
methylated, 204
Peroxisomes, 459
PH, 626–27, 641
optimum
auxin binding, 403
role
elongation
growth substances, 9–10
Phaseic acid, 115
Phenylactic acid, 150, 154
Phenylalanine, 671
L-Phenylglycolic acid, 356
Phloem
malate, 452
Phloridzin, 341
Phosphate
regulation starch synthesis, 351
starvation
see Ribonucleases
5-Phosphate-dependent amino acid decarboxylase, 325–27, 331
Phosphatidyl glycerol, 129
Phosphoamino acid analysis, 148
Phosphoenolpyruvate, 588, 590, 597–98
Phosphoenolpyruvate carboxylase, 186, 371, 380–82, 386
C_4 cycle, 345–46
Carbon and nitrogen metabolism in cells, 586
Phosphofructokinase, 357
Carbon and nitrogen metabolism in cells, 587–88, 590
Phosphoglucomutase, 247
Phosphoglycerate kinase
photosynthesis regulation, 236
Phosphoglyceric acid, 244, 247
Phospholipase A, 406
Phospholipids, 561
Phosphoribulokinase, 236, 239–40
photosynthesis regulation, 242, 244
Phosphorylation, 1–9, 212, 227, 553, 589, 593
see also Blue light signal transduction
Photoacoustic method and photosynthesis, 493–526
directly-coupled-detector configuration, 510–12
historical notes, 495–97
other applications, 497
photoacoustic and photothermal methods, 494–95
photosynthesis physiology studies, 516–22
light distribution between photosystems, 518–20
low light state, 520
PSI independent activities, 521
stress physiology and ecology, 521–22
stromal sinks, 520–21
transients, 516–18

results of photosynthetic activities, 512–16
  energy storage, 512–16
  light saturation curves, 516
  photobaric measurements, 516
signal generation and utilization, 497–510
  experimental, 497–501
  fluorescence, 504–5
  gas (air)-coupled microphone PA cell, 500–1
  locus in a leaf, 508–9
  mechanism, 501–2
  modulated photobaric signal, 505–9
  periodically modulated, 501
  photobaric signal, 506–7
  photocalorimetry, 502–3
  photothermal and photobaric signals in the time domain, 509–10
  reaction kinetics, 509
  referencing methods, 503–4
  response to saturation in leaves, 509–10
  response to weak flashes, 510
  time domain measurements, 501
Photoaffinity labeling
  auxin-binding proteins, 396–97, 401, 406, 408, 410
Photobaric signals
  see Photoacoustic method and photosynthesis
Photocalorimetry, 502–3
Photochemical loss, 503–4
Photoinhibition of photosynthesis, 633–61
  mechanisms, 636–44
    damage and avoidance, 643–44
    PSII damage avoidance: xanthophyll cycle, 640–43
    PSII inactivation: D1, 637–40
  observed changes, 653–54
    photoinhibition: field and ocean, 644–48
    phytoplankton, 646–48
    terrestrial vegetation, 644–46
    plant distributions, 654–55
    significance to production: modeling, 648–53
Photomorphogenesis
  see Blue light signal transduction
Photon flux, 640, 644, 652
Photooxidative stress
  carotinoid biosynthesis, 296
Photoperception
  see Blue light signal transduction
Photoperiodism, 14–16
Photorespiration
  see Mitochondrial glycine decarboxylase complex
Photosynthate, 312, 317
Photosynthetic carbon fixation
  see Carbon and nitrogen metabolism in cells
Photosynthetic inhibitors
  C_2, 348–50
  C_3, 343–45
  C_4, 360
  electron transport, 535
  fatty acid biosynthesis, 353-55
  herbicides, 340–43, 358–59
  PSI, 340
  PSII, 340–42

Photosynthesis
  ABA levels in water stress, 123
  see also Carbon and nitrogen metabolism in cells; Carbonic anhydrase; Diurnal regulation of photosynthesis; Mitochondrial glycine decarboxylase complex; Photoacoustic method and photosynthesis
Photosystems
  CAB
  gene expression, 155–56
  light distribution, 518–20
  PSI, 494, 497, 507, 512, 519, 521
    photosynthesis inhibitors, 340
  PSII, 494, 497, 519, 587, 596
    photosynthesis inhibitors, 340–42
  see also Photoinhibition of photosynthesis
Photothermal methods
  see Photoacoustic method and photosynthesis
Phototropic response
  role root development, 40
Phototropism, 146
Phylloplane
  bacteria, 199
Phytoalexins, 258, 669
Phytochrome, 296
  blue light signals 144, 152, 155, 157–59, 162, 475, 481, 483
  discovery, 16
Phytoene desaturase, 290, 292, 294, 296–97
Phytoene synthase
  role carotenoid biosynthesis, 290–91, 294, 296–97
Phytoplankton
  photoinhibition 646–49
Pink-pigmented facultative methyltrophs
  see Covert contaminants
Plant growth regulators
  production by bacteria, 205
Plant-microbe interactions, 198
Plasma membrane, 145, 148, 380
  anion channel, 460
  ATPase, 163
  H^+-ATPase, 39
  proteins, 402, 404, 407, 412
  redox activity, 150
  see also Endocytosis; Hormone-induced changes in microtubules
Plasma membrane proteins, 211–234
  carriers and channels, 222–26
  protein kinases, 226–29
  protein pump, 213–22
Plasmalemma, 154
Plasmodesmata, 381
Plastid
  energy supply, 593–95
  reductant supply in root, 594–95
  mRNA metabolism, 429
Pollen
  rRNA, 437–38
Polyacrylamide gel electrophoresis, 21
Polyadenylation signals
  see mRNA 3′ end formation
Polygalacturonic acid, 616, 624
Polyisoprene, 18
Polymerase chain reaction, 622

plasma membrane proteins, 216, 230
Polypeptides, 470
  with chaperonins, 480–87
Polyprenyl transferase, 18
Pom-pom mutant, 35–36
Potassium channel
  plasma membrane proteins, 212, 222–23
  pyrophosphate regulation, 163–64
PPFMs
  see Covert contaminants
Pre-β-lactamase, 481
Princeton University, 2
Proceedings of the National Academy of Sciences, 13
Proline residues, 399
Promoter proximity
  effect poly(A) signals, 57
Protease studies, 549
Proteins
  bodies, 307
  coding genes
    mitochondrial genome, 68
  folding
    see Chaperonin-mediated protein folding
  glycine decarboxylase reaction
  G-proteins
    role stomatal regulation, 163–64
  H-protein, 326–30, 332–33
  import
    see Mitochondrial protein import
  kinase
    activity inhibitors, 148
    role stomatal regulation, 163
    see also Plasma membrane proteins
  L-protein, 325, 328–29, 332, 334
  P-protein, 325–27, 329
  T-protein, 325, 328, 332
  targeting
    see Mitochondrial protein import
  see also Coat proteins; Auxin-binding proteins; Vegetative storage proteins
Proteinase inhibitors, 119, 312
  role RNA editing, 70
Proton pump
  see Plasma membrane proteins
Psb upstream sequences, 159
Pseudomonas syringae pv. syringae
  contaminant, 200
Pterins, 150–51, 154
Pulse field gel electrophoresis, 62
Putrescine, 258
Putrescine *N*-methyltransferase, 258–60, 271
Pyrenoid body
  photosynthesis, 380
  rubisco, 375
Pyrophosphate, 351
Pyrophosphate fructose-6-phosphate 1-phosphotransferase, 587–88
Pyruvate, 19
Pyruvate dehydrogenase complex, 328–30, 587
Pyruvate kinase
  Carbon and nitrogen metabolism in cells, 586–87

Q

Quiescent center, 28

R

Rab genes, 116, 118–19, 121
Rabbit liver
  expressed in tobacco, 267
Rain forest ecosystems, 199
Raphanusanin, 535
Rbc genes, 48–51, 54–55, 123
RBCS gene family, 155
  promoters, 158
RPK clone, 228
Receptor channel proteins, 549–51
Receptor protein kinases, 212, 227–28
Receptors
  endocytosis, 611
Red light, 401–14, 642
Regulation
  alkaloid biogenesis, 271–74
  Carbon and nitrogen metabolism in cells, 596–99
  geminivirus replication, 81, 86–87, 94
  gene expression
    see Abscisic acid regulated gene expression: Diurnal regulation of photosynthesis
Reporter genes
  expression
    role poly(A) signals, 50
    see also GUS reporter genes
Respiration
  activation
    see Carbon and nitrogen metabolism in cells
  role mitochondria, 19
  see also Mitochondrial glycine decarboxylase complex
Respiratory
  activity
    role mitochondrial protein import, 564–65
  chain
    role mitochondrial protein import, 546–47
Restriction site polymorphisms
  geminiviruses, 85
Reticuline, 268
(*S*)-reticuline, 263, 277
Retinoic acid, 414
Rhd mutants
  root development, 37–38
RHD 1, 37
Rhicadhesin, 624
Rhizobium
  effect root development, 41
  root cell specific gene expression, 38
Rhizopus suinus
  growth hormone source, 8
Rhodanese, 476, 480–82, 485
Ribonucleases, 421–45
  functional roles
    phosphate remobilization, 438
    plant defense, 438–39
    mRNA decay, 439
    self-incompatibility, 437–38
  localization, 428–31

in chloroplasts and mitochondria, 429–30
other intracellular, 430–31
secretory, 428–29
regulation, 431–37
genetic determinants, 436–37
light, 434–36
phosphate starvation, 433
plant disease, 433–34
senescence, 432
Ribosomes,431
Ribulose 1,5-bisphosphate, 507, 583, 597
Ribulose bisphosphate carboxylase, 19, 48, 324, 332–33
Ribulose-bisphosphate carboxylase-oxygenase, 476, 583
chloroplast cpn10, 472, 485
chloroplast cpn60, 471, 481, 485
folding, 486
leaf protein, 306–8, 315
photosynthesis, 370, 373, 375–76, 380, 382, 384–85
inhibitors, 343–44, 359
Ribulose-bisphosphate carboxylase-oxygenase bisphosphate, 324
Rieske FeS protein, 555–57
RNA
antisense, 104
degrading enzymes
see Ribonucleases
editing
see Mitochondrial genome
polymerase, 20, 47, 68, 93
replication
viral vectors, 80
splicing
geminiviruses, 90
synthesis, 20
mRNA, 68–69, 73, 216, 510
accumulation, 186–87
and ABA levels, 120–21, 123, 128, 132–34
CAB, 156–57
chs abundance, 160
CHSR, 161
decay
role RNases, 439–40
GUS, 156
H-protein, 333
levels in vegetative storage proteins, 310
light dependent accumulation, 155
mitochondrial genome
see Mitochondrial genome
PDS, 286–97
PSY, 296–97
probes, 220
thionin, 157
mRNA 3' end formation, 47–60
polyadenylation signals: structure, 48–54
cleavage/polyadenylation site, 53–54
far-upstream elements, 50–52
near-upstream elements, 52–53
polyadenylation signals: differential utilization, 54–56
rRNA, 120
tRNA-derived cytokinins, 175–77
RNA blots, 120

RNases, 431–39
see also Ribonucleases
Roberts, Paul, 18
Rockefeller Foundation, 5
Rol genes, 178, 182
Root development, 25–45
axis formation and meristem introduction, 28–31
lateral and adventitious, 34–35
meristem structure: seed, primary root, 31–33
meristematic program: cell division, expansion, 35–36
response to environmental stimuli, 39–41
root cell specific gene expression, 38–39
root hair cell differentiation, 36–38
Root-knot nematodes
effect root development, 41
Roots
exudates
growth inhibition
cytokinin, 183
hair cells, 36–38
initiation
role cytokinin, 173
malate, 451–52
nitrogen transport, 305
organ culture, 13–14
Rps gene
mitochondrial genome, 72–73
Rsw, 36
Rubber
production from guayule, 16–18
Rubisco
see Ribulose-bisphosphate carboxylase oxygenase
Rust
RNase activity, 433–34

S

Salicylate-binding protein, 394
Salicylhydroxamic acid, 150
Salicylic acid, 411
Salt
stress
cytokinin interactions, 186
tolerance, 125
Scanning transmission electron microscopy, 483
Scopolamine, 261, 278–79
(S)-Scoulerine, 263, 269, 272
Scoulerine-methyltransferase, 268–69
SDS-PAGE, 328, 308, 564
plasma membrane proteins, 214, 219
Seeds
development
role ABA, 114–15, 133–34
germination
role PPFMs, 203–4
mitochondrial genome, 72
root development, 31–33
see also Vegetative storage proteins
Self-incompatibility
reaction
Brassica pollen, 228

see also Ribonucleases
Senescence, 305–6
    role cytokinin, 186
L-Serine dehydratase, 198
Serine hydroxymethyltransferase, 349–50
    see also Mitochondrial glycine decarboxy-
        lase complex
Serine residues, 212
Serological cross-reactivity
    geminiviruses, 85
Shade plants
    xanthophyll cycle activity, 644
Shaker, 223–24
Shoot initiation
    role cytokinin, 173
Short-root mutant, 36
Signal transduction
    role cytokinin, 178, 185
Signals
    mitochondrial genome, 73
    plasma membrane, 610
    poly(A)
        see mRNA 3' end formation
    root development, 40
    see also Auxin-binding proteins; Blue light
        signal transduction
Silver ions
    antagonize ethylene action, 297
Sink tissues
    carbon metabolism regulation, 248–50
    removal elevates soybean protein, 314–15
Smith, Emil, 21
Sodium chloride treatment
    ABA levels, 118, 130
Sodium pump, 213–14
Solvent extraction, 398
Sounds
    see Photoacoustic method and photosynthe-
        sis
Southern blots, 217, 229
soybean
    Bradyrhizobium japonicum symbiosis, 198
    fungal contaminant control, 206
    PPFM interactions, 202
    see also Vegetative storage proteins
Spermidine synthase, 258, 260
Sporamin, 317–18
Squash leaf curl virus, 84, 92, 104
Stanford University, 20–21
Starch
    biosynthesis, 350–51
    metabolism, 589–90
    degradation, 596
    synthesis
        photosynthesis regulation, 249
Steroid hormone
    receptors, 408
Stomata
    closure, 123
    regulation
        role blue light transduction, 163–64
Storage proteins
    ABA regulation, 117, 124
    see also Vegetative storage proteins
Stress responses
    environmental

see Environmental stress
hormones
    see Abscisic acid regulated gene expres-
        sion; Carotenoid biosynthesis
    physiology
    photo-acoustic measurements, 521–22
Strictosidine, 266, 268, 271
Strictosidine synthase, 265, 267, 270–71, 273,
    278
Stromal sinks
    photosynthesis, 520–21
Sturtevant, Alfred H., 5, 10
Succinate, 19
Sucrose
    export
        photosynthesis regulation, 249
    inhibitors, 356–57
Sucrose phosphate synthase, 248, 356
Sucrose phosphate synthetase, 589
Sur mutant, 34
Swiss Institute of Technology, 12
Symbiosomes, 459
Symptom development
    geminiviruses, 95–96
Synechococcus PCC7942, 373
Systematic point mutation analysis, 52

T

Tamiya, Hiroshi, 12
Tandem mass spectrometry
    taxol, 668
Target proteins, 359, 481, 483–84
    chaperonin protein folding, 486
    see also Mitochondrial protein import; Pro-
        teins
Taxol, 200, 205, 535, 540, 663–74
    analytical chemistry, 667
    biosynthesis, 671–72
    HPLC, 667–68
    ELISA, 667–68
    tandem mass spectrometry, 668
    importance, 665–66
    antitumor activity, 665
    mechanism of action, 665–66
    structure-activity relationships, 666
    supply, 669–70
    cell culture, 670–71
    chemical synthesis, 670
    collection and commercial propagation,
        669–70
Taxomyces andreanae
    source taxol, 200
Taxus brevifolia
    see Taxol
TCP1
    see Chaperonin-mediated protein folding
Temperature
    role rubber production, 18
    see also Cold; Heat stress
Tetraethyl-ammonium, 162
Tetrahydrofolate transferase
    glycine decarboxylase reaction, 325–26,
        328, 331
(S)-tetrahydroprotoberberine oxidase, 264,
    268–69

Thal ethylene response system, 180
Thermal
  deactivation spectrum, 504
  diffusivity, 502
Thiamine, 14
Thimann, Kenneth V., 8
Thionins
  leaf down regulation, 157
Threonine residues, 212
Thylakoid membranes, 340
Time domain measurements, 501
TIP gene, 2
Tip mutants, 38
Tobacco
  nicotine content, 275–76
  transgenic, 385
    GUS expression, 332
Tobacco mosaic virus
  foreign gene expression, 80
Tobacco yellow dwarf virus
  biology and applications, 83–85
TOM5, 291, 294
Tomato acid phosphatase
  relationship to vegetative storage proteins,
    313–14
Tomato golden mosaic virus, 81–82, 84, 87,
  89, 90–91, 94–97, 100, 102
Tomato leaf curl virus, 83–86
Tomato yellow leaf curl virus
  biology and applications, 82–86, 94
Touch
  role root development, 39
Tran-acting elements
  ABA, 121–22
  B-mediated gene expression, 158
Trans-Golgi network, 269
Trans-zeatin, 175–77
Transcription factors
  poly(A) signals, 57
Transcription initiation
  mitochondrial genome, 67–68
Transcriptional activation
  geminiviruses, 88
Trans-splicing, 68–69
Transfer RNA
  see tRNA
Transgenic plants, 39, 56, 250, 385
  breeding for geminivirus resistance, 101–2
  see also Cytokinin accumulation and action;
    cDNA; Engineering, genetic
Translocation
  protein, 563, 567
Transmission
  geminiviruses, 91–92
Traumatic acid
  wound hormone, 14
Triazine herbicides, 341
Tricarboxylic acid cycle
  see Carbon and nitrogen metabolism in cells
Triose-P, 345
  isomerase
    photosynthesis regulation, 236, 238, 246–
      49
Triskelion, 617
Triterpenoids
  production by *M. organophilum*, 205

Triton solubilization, 149
Trk genes, 223–25
Tropane alkaloids
  see Alkaloid biogenesis
Tubers
  vegetative storage proteins, 317
Tubulin, 481, 532, 537, 539, 666
Turgid cells, 614
Trypanosomes
  RNA editing, 73
Tryptophan, 9
Tryptophan decarboxylase, 264–65
Tyramine, 279
Tyrosine, 261
Tyrosine decarboxylases, 261, 273, 279
Tyrosine residues, 212, 612
  α-tubulin genes, 38
  β-tubulin genes, 38

U

United States Department of Agriculture, 1–17
United States Forest Service, 17
University of California, Berkeley, 3, 12
University of Leiden
  Medical School, 12
University of Minnesota, 2
University of Toronto, 2, 5
University of Utah, 2–3, 7
Urease isozymes
  soybean, 201–2
Ureides, 305
UV-dependent photosensory systems
  see Blue light signal transduction
UV flux
  photosynthesis inhibition, 648

V

Vacuoles
  endocytosis, 622–23
  malate, 461–62
  RNases, 428, 438
Varner, Joe, 19
Vectors
  see Geminiviruses
Vegetable oil treatment
  contaminant control, 206
Vegetative storage proteins, 303–22
  nitrogen storage and redistribution, 304–6
  protein storage: variations, 317–19
  proteins related to soybean VSP in other spe-
    cies, 316
  sink removal elevates other proteins, 314–15
  soybean, 307–14
    jasmonate's role in regulation, 311–13
    nitrogen's role in regulation, 310–11
    regulation of VSP genes, 310
    relationship to tomato acid phosphatase,
      313–14
    stress responses, 313
  storage protein definition, 306–7
Vinblastine, 264
Vindoline biosynthesis, 268, 271
Violaxanthin, 294, 296
Viral vectors

see Geminiviruses
Vitamin B, 199, 203
Vitamins
  bacterial production, 204
Von Wettstein, 10
Vp mutants, 120, 295
Vspa, 313
VSPa, VSP
  see Vegetative storage proteins

W

Waste products
  fate, 204
Water shortage
  photosynthesis inhibition, 640
Water stress, 252
  ABA levels, 114–15, 123, 125
Went, Frits, 8–10, 17
Western blotting, 148, 221, 401, 403, 565
Wheat
  mitochondrial genome, 69
Wheat dwarf virus
  biology and applications, 82–84, 88, 90, 96–
    99, 100, 102
  shuttle vector, 98
White, Philip, 13
White fly
  viral vector, 81, 91
Wildman, Samuel G., 18–19
World War II
  rubber production, 16–18
Wound hormone, 14
Wound sites
  root development, 30
Wounding

experiments
  ABA deficient mutants, 119
  role
    VSP gene expression, 310, 312–13
    WIN gene expression, 318

X

Xanothophyll, 294
  accumulation, 295
  cycle, 640–44, 647
XRN, 430
Xylem
  malate, 452
Xylogenesis, 440

Y

Yeast
  growth enhancer, 13
  poly(A) signals, 52
  see also Mitochondrial protein import
Yew bark
  see Taxol
Yost, Don, 3–4

Z

Zeatin reductase, 177
Zeaxanthin, 641–42
Zein gene
  poly(A) signals, 49, 51, 53–55, 57
Zeta-carotene desaturase, 292
Zinc
  see Carbonic anhydrase

CUMULATIVE INDEXES

CONTRIBUTING AUTHORS, VOLUMES 36–45

A

Abeles FB, 37:49–72
Adams WW III, 43:599–626
Aeschbacher RA, 45:25–45
Akazawa T, 36:441–72
Aloni R, 38:179–204
Anderson JM, 37:93–136
Andréasson L, 39:379–411
Apel K, 42:227–40
Appels R, 43:117–43
Atkinson CJ, 41:55–75

B

Badger MR, 36:27–53; 45:369–92
Barber MJ, 41:225–53
Bartley GE, 45:287–301
Baskin TI, 41:277–315
Baum M, 43:117–43
Beard WA, 38:347–89
Beck E, 40:95–117
Beevers H, 44:1–12
Benfey PN, 45:25–45
Bennett AB, 42:675–703
Bennett J, 42:281–311
Benson DR, 37:209–32
Benveniste P, 37:275–308
Bernier G, 39:175–219
Berry JA, 39:533–94
Binns AN, 45:173–96
Bishop PE, 41:109–25
Blatt MR, 44:543–67
Bohlmann H, 42:227–40
Boller T, 37:137–64
Bonner J, 45:1–23
Boudet AM, 38:73–93
Bouton JH, 44:435–56
Bowes G, 44:309–32
Bowler C, 43:83–116
Boyer JS, 36:473–516
Brady CJ, 38:155–78
Brennicke A, 45:61–78
Briggs WR, 45:143–71
Broekaert WF, 44:591–615
Brown RH, 44:435–56
Browse J, 42:467–506
Buikema WJ, 44:33–52
Burnell JN, 36:255–86
Bush DR, 44:513–42

C

Cairns AJ, 42:77–101
Canaani O, 45:493–526

Cande WZ, 41:277–315
Cassab GI, 39:321–53
Chandler PM, 45:113–41
Chandra S, 45:609–31
Chang C-j, 45:663–74
Chang M, 41:497–526
Chrispeels MJ, 42:21–53
Chua N, 38:221–57
Clarkson DT, 36:77–115
Clegg MT, 38:391–418
Coen ES, 42:241–79
Cosgrove D, 37:377–405
Cote GG, 44:333–56
Covello PS, 43:145–75
Crain RC, 44:333–56
Creelman RA, 39:439–73
Cullis CA, 36:367–96

D

Dainty J, 41:1–20
Dale JE, 39:267–95
Das OP, 45:79–112
Davies WJ, 42:55–76
Dawson WO, 43:527–55
Dean C, 40:415–39
Delmer DP, 38:259–90
Demmig-Adams B, 43:599–626
Depta H, 39:53–99
Dietrich A, 44:13–32
Dilley RA, 38:347–89
Dixon RA, 41:339–67
Douce R, 40:371–414
Dring MJ, 39:157–74
Dunsmuir P, 40:415–39
Dutcher FR, 38:317–45

E

Edwards GE, 36:255–86
Ehleringer JR, 40:503–38
Erickson RO, 39:1–22
Estelle M, 42:529–51
Etzler ME, 36:209–34
Evans PT, 40:235–69
Evenari M, 36:1–25

F

Falco SC, 40:441–70
Falkowski PG, 45:633–61
Farmer EE, 42:651–74
Farquhar GD, 40:503–37
Fincher GB, 40:305–46
Fischer RL, 42:675–703

Flügge U, 42:129–44
Fork DC, 37:335–61
Fosket DE, 43:201–40
Fry SC, 37:165–86
Furuya M, 44:617–45

G

Gallie DR, 44:77–105
Gasser CS, 42:621–49
Gatenby AA, 45:469–91
Geiger DR, 45:235–56
Ghanotakis DF, 41:255–76
Gianinazzi-Pearson V, 39:221–44
Giuliano G, 45:287–301
Glazer AN, 38:11–45
Golbeck JH, 43:293–324
Good NE, 37:1–22
Graebe JE, 38:419–65
Gray MW, 43:145–75
Green PJ, 38:221–57; 45:421–45
Gresshoff PM, 39:297–319
Grignon C, 42:103–28
Guern J, 40:271–303
Guy CL, 41:187–223

H

Hahlbrock K, 40:347–69
Halstead TW, 38:317–45
Hanic-Joyce PJ, 43:145–75
Hanson AD, 44:357–84
Hara-Nishimura I, 36:441–72
Hardham AR, 43:491–526
Harmon A, 43:375–414
Harris N, 37:73–92
Harwood JL, 39:101–38
Haselkorn R, 44:33–52
Hashimoto T, 45:257–85
Hatch MD, 36:255–86
Hayashi T, 40:139–68
Hedden P, 44:107–29
Hedrich R, 40:539–69
Heichel GH, 42:373–92
Heidecker G, 37:439–66
Heinstein PF, 45:663–74
Heldt HW, 42:129–44
Hepler PK, 36:397–439
Herman EM, 39:139–55
Hetherington AM, 41:55–75
Hilf ME, 43:527–55
Hirel B, 36:345–65
Ho LC, 39:355–78
Ho T-H. D., 37:363–76
Holland MA, 45:197–210

Holt JS, 44:203–29
Holtum JAM, 44:231–51
Honegger R, 42:553–78
Horsch R, 38:467–86
Hrazdina G, 43:241–67
Huang A, 43:177–200
Huber SC, 37:233–46
Hubick KT, 40:503–37
Hull R, 38:291–315
Humphries S, 45:633–61
Hunt AG, 45:47–60
Hunt S, 44:483–511
Huppe HC, 45:577–607

I

Inzé D, 43:83–116

J

Jackson MB, 36:145–74
Jäger K, 43:325–49
Jensen RA, 43:241–67
Joerger RD, 41:109–25
Jones AM, 45:393–420

K

Kadota A, 40:169–91
Kamiya N, 40:1–18
Kauss H, 38:47–72
Keegstra K, 40:471–501
Kende H, 44:283–307
King RW, 36:517–68
Kirst GO, 41:21–53
Kleczkowski LA, 45:339–67
Klee H, 38:467–86; 42:529–51
Kleinig H, 40:39–59
Koide RT, 43:557–81
Krause GH, 42:313–49
Kuhlemeier C, 38:221–57
Kurkdjian A, 40:271–303

L

Lagudah E, 43:117–43
Lamb CJ, 41:339–67
Langdale JA, 43:25–47
Lara M, 42:507–28
Layzell DB, 44:483–511
Lee H, 44:591–615
Lee M, 39:413–37
Leong SA, 37:187–208
Lewis NG, 41:455–97
Lin W, 37:309–34
Lloyd CW, 38:119–39
Long SP, 45:633–61
Low PS, 45:609–31
Lucas WJ, 41:369–419
Lumsden PJ, 42:351–71
Luster DG, 44:131–55
Lynn DG, 41:497–526

M

Malkin S, 45:493–526

Malmberg RL, 40:235–69
Mandava NB, 39:23–52
Mansfield TA, 41:55–75
Maréchal-Drouard L, 44:13–32
Marrè E, 42:1–20
Martinoia E, 45:447–67
Mascarenhas JP, 41:317–38
Matzke AJM, 44:53–76
Matzke M, 44:53–76
Mazur BJ, 40:441–70
Meeks JC, 40:193–210
Melis A, 38:11–45
Messing J, 37:439–66; 45:79–112
Mimura T, 38:95–117
Møller IM, 37:309–34
Moore AL, 45:545–75
Morejohn LC, 43:201–40
Morris RO, 37:509–38
Mullet JE, 39:475–502

N

Nakamoto H, 36:255–86
Nasrallah JB, 42:393–422
Nasrallah ME, 42:393–422
Neilands JB, 37:187–208
Nelson T, 43:25–47
Neuburger M, 40:371–414
Newton KJ, 39:503–32
Nishio T, 42:393–422

O

Oaks A, 36:345–65
Oliver DJ, 45:323–37
Olsen LJ, 40:471–501
Ort DR, 43:269–91
Oxborough K, 43:269–91

P

Padilla JE, 42:507–38
Parthier B, 44:569–89
Passioura JB, 39:245–65
Payne PI, 38:141–53
Pearcy RW, 41:421–53
Pérez H, 42:507–28
Peters GA, 40:193–210
Pharis RP, 36:517–68
Phillips RL, 39:413–37
Pichersky E, 40:415–39
Pickard BG, 36:55–75
Polacco JC, 45:197–210
Pollock CJ, 42:77–101
Poole RJ, 44:157–80
Portis A Jr, 43:415–37
Potrykus I, 42:205–25
Powles SB, 44:203–29
Press MC, 41:127–51
Price GD, 45:369–92

R

Raikhel N, 44:591–615
Ranjeva R, 38:73–93
Raskin I, 43:439–63

Rea P, 44:157–80
Rentsch D, 45:447–67
Rhodes D, 44:357–84
Robards AW, 41:369–419
Roberts DM, 43:375–414
Robertson M, 45:113–41
Robertson RN, 43:1–24
Robinson D, 39:53–99
Rogers S, 38:467–86
Rolfe BG, 39:297–319
Rubinstein B, 44:131–55
Russell SD, 42:189–204
Ryan CA, 42:651–74

S

Sachs MM, 37:363–76
Sánchez F, 42:507–28
Sanders D, 41:77–107
Satoh K, 37:335–61
Scheel D, 40:347–69
Schiefelbein JW, 45:25–45
Schmidt A, 43:325–49
Schnepf E, 37:23–47
Schreiner RP, 43:557–81
Schroeder JI, 40:539–69
Schubert KR, 37:539–74
Schulze E-D., 37:247–74
Schuster W, 45:61–78
Schwintzer CR, 37:209–32
Scolnik PA, 45:287–301
Sembdner G, 44:569–89
Sentenac H, 42:103–28
Serrano R, 40:61–94
Servaites JC, 45:235–56
Shibaoka H, 45:527–44
Shimmen T, 38:95–117
Short TW, 45:143–71
Siedow JN, 42:145–88
Silverthorne J, 36:569–93
Smith SE, 39:221–44
Smith TA, 36:117–43
Snell WJ, 36:287–315
Solomonson LP, 41:225–53
Somerville C, 42:467–506
Somerville CR, 37:467–507
Sperry JS, 40:19–38
Spiker S, 36:235–53
Spreitzer RJ, 44:411–34
Staswick PE, 45:303–22
Steffens JC, 41:553–75
Stewart GR, 41:127–51
Stitt M, 41:153–85
Sussman MR, 45:211–34
Sweeney BM, 38:1–9
Sze H, 36:175–208

T

Taylor WC, 40:211–33
Tazawa M, 38:95–117
Theg SM, 38:347–89; 40:471–501
Theologis A, 37:407–38
Thiel G, 44:543–67
Thompson WF, 42:423–66

Thorne JH, 36:317–43
Timmermans MCP, 45:79–112
Ting IP, 36:595–622
Tjepkema JD, 37:209–32
Tobin EM, 36:569–93
Turgeon R, 40:119–38
Turpin DH, 45:577–607
Tyerman SD, 43:351–73
Tyree MT, 40:19–38

V

Van Bel AJE, 44:253–81
Vance CP, 42:373–92
van Huystee RB, 38:205–19
Van Montagu M, 43:83–116
Vänngaård T, 39:379–411
Varner J, 39:321–53
Verbeke JA, 43:583–98

Vermaas W, 44:457–81
Vierling E, 42:579–620
Vierstra RD, 44:385–410
Viitanen PV, 45:469–91
Vogelmann TC, 44:231–51

W

Wada M, 40:169–91
Walbot V, 36:367–96; 43:49–82
Watts FZ, 45:545–75
Wayne RO, 36:397–439
Weil CF, 41:527–52
Weil JH, 44:13–32
Weis E, 42:313–49
Wessler SR, 41:527–52
White MJ, 42:423–66
Wiemken A, 37:137–64
Williamson RE, 44:181–202

Wood CK, 45:545–75
Woodrow IE, 39:533–94

Y

Yamada Y, 45:257–85
Yamamoto E, 41:455–97
Yocum CF, 41:255–76

Z

Zaitlin M, 38:291–315
Zambryski PC, 43:465–90
Zeevaart JAD, 39:439–73
Zhang J, 42:55–76
Ziegler P, 40:95–117
Zurawski G, 38:391–418

CHAPTER TITLES, VOLUMES 36–45

PREFATORY CHAPTERS

A Cat Has Nine Lives	M. Evenari	36:1–25
Confessions of a Habitual Skeptic	N. E. Good	37:1–22
Living in the Golden Age of Biology	B. M. Sweeney	38:1–9
Growth and Development of a Botanist	R. O. Erickson	39:1–22
My Early Career and the Involvement of World War II	N. Kamiya	40:1–18
Prefatory Chapter	J. Dainty	41:1–20
Short Story of a Plant Physiologist and Variations on the Theme	E. Marrè	42:1–20
A Dilettante Australian Plant Physiologist	R. N. Robertson	43:1–24
Forty Years in the New World	H. Beevers	44:1–12
Chapters From My Life	J. Bonner	45:1–23

BIOCHEMISTRY & BIOPHYSICS

Photosynthesis

Photosynthetic Oxygen Exchange	M. R. Badger	36:27–53
The Control by State Transitions of the Distribution of Excitation Energy in Photosynthesis	D. C. Fork, K. Satoh	37:335–61
Analysis of Photosynthesis with Mutants of Higher Plants and Algae	C. R. Somerville	37:467–507
Photochemical Reaction Centers: Structure, Organization, and Function	A. N. Glazer, A. Melis	38:11–45
Membrane-Proton Interactions in Chloroplast Bioenergetics: Localized Proton Domains	R. A. Dilley, S. M. Theg, W. A. Beard	38:347–89
Photosynthetic Electron Transport in Higher Plants	T. Vänngård, L. Andréasson	39:379–411
Carbon Isotopes Discrimination and Photosynthesis	G. D. Farquhar, J. R. Ehleringer, K. T. Hubick	40:503–38
Photosystem II and the Oxygen-Evolving Complex	D. F. Ghanotakis, C. F. Yocum	41:255–76
Chlorophyll Fluorescence and Photosynthesis: The Basics	G. H. Krause, E. Weis	42:313–49
In situ Regulation of Chloroplast Coupling Factor Activity	D. R. Ort, K. Oxborough	43:269–91
Structure and Function of Photosystem I	J. H. Golbeck	43:293–324
Physiology and Genetics of Interspecific Hybrids Between Photosynthetic Types	R. H. Brown, J. H. Bouton	44:435–56
Molecular-Biological Approaches to Analyze Photosystem II Structure and Function	W. Vermaas	44:457–81
Diurnal Regulation of Photosynthetic Carbon Metabolism in C_3 Plants	D. R. Geiger, J. C. Servaites	45:235–56
The Role of Carbonic Anhydrase in Photosynthesis	M. R. Badger, G. D. Price	45:369–92
The Use and Characteristics of the Photoacoustic Method in the Study of Photosynthesis	S. Malkin, O. Canaani	45:493–526

Respiration

The Uniqueness of Plant Mitochondria	R. Douce, M. Neuburger	40:371–414

Metabolic Pathways/Secondary Metabolites

Plant Chemiluminescence	F. B. Abeles	37:49–72
Fructose 2,6-Bisphosphate as a Regulatory Metabolite in Plants	S. C. Huber	37:233–46
Sterol Biosynthesis	P. Benveniste	37:275–308
Cellulose Biosynthesis	D. P. Delmer	38:259–90
Fatty Acid Metabolism	J. L. Harwood	39:101–38
Biosynthesis and Degradation of Starch in Higher Plants	E. Beck, P. Ziegler	40:95–117
Physiology and Molecular Biology of Phenylpropenoid Metabolism	K. Hahlbrock, D. Scheel	40:347–69
Fructose-2,6-Bisphosphate as a Regulatory Molecule in Plants	M. Stitt	41:153–85
Lignin: Occurrence, Biogenesis, and Degradation	N. G. Lewis, E. Yamamoto	41:455–97
Fructan Metabolism in Grasses and Cereals	C. J. Pollock, A. J. Cairns	42:77–101
The Biochemistry and the Physiological and Molecular Actions of Jasmonates	G. Sembdner, B. Parthier	44:569–89
Alkaloid Biosynthesis: Molecular Aspects	T. Hashimoto, Y. Yamada	45:257–85

Nitrogen Metabolism and Fixation

Polyamines	T. A. Smith	36:117–43
Nitrogen Metabolism in Roots	A. Oaks, B. Hirel	36:345–65
Physiology of Actinorhizal Nodules	J. D. Tjepkema, C. R. Schwintzer, D. R. Benson	37:209–32
Genetic Analysis of Legume Nodule Initiation	B. G. Rolfe, P. M. Gresshoff	39:297–319
Genetics and Molecular Biology of Alternative Nitrogen Fixation Systems	P. E. Bishop, R. D. Joerger	41:109–25
Assimilatory Nitrate Reductase: Functional Properties and Regulation	M. J. Barber, L. P. Solomonson	41:225–53
Open Questions of Sulfur Metabolism in Plants	A. Schmidt, K. Jäger	43:325–49
Gas Exchange of Legume Nodules and the Regulation of Nitrogenase Act	S. Hunt, D. B. Layzell	44:483–511

Transport

The Role of Plastids in Isoprenoid Biosynthesis	H. Kleinig	40:39–59
Kinetic Modeling of Plant and Fungal Membrane Transport Systems	D. Sanders	41:77–107
The Heavy Metal Binding Peptides of Plants	J. C. Steffens	41:553–75
Carbon in N_2 Fixation: Limitation or Exquisite Adaptation?	C. P. Vance, G. H. Heichel	42:373–92
Glycerolipid Synthesis: Biochemistry and Regulation	J. Browse, C. Somerville	42:467–506
Anion Channels in Plants	S. D. Tyerman	43:351–73
Vacuolar H^+-Translocating Pyrophosphatase	P. A. Rea, R. J. Poole	44:157–80
Proton-Coupled Sugar and Amino Acid Transporters in Plants	D. R. Bush	44:513–42
Hormonal Control of Ion Channel Gating	M. R. Blatt, G. Thiel	44:543–67
Molecular Analysis of Proteins in the Plant Plasma Membrane	M. R. Sussman	45:211–34
Molecular Biology of Carotenoid Biosynthesis in Plants	G. E. Bartley, P. A. Scolnik, G. Giuliano	45:287–301
Malate Compartmentalization—Response to a Complex Metabolism	E. Martinoia, D. Rentsch	45:447–67

Protein Structure/Function/Regulation/Synthesis

H⁺-Translocating ATPases: Advances Using Membrane Vesicles	H. Sze	36:175–208
Plant Lectins: Molecular and Biological Aspects	M. E. Etzler	36:209–34
Pyruvate,Pᵢ Dikinase and NADP-Malate Dehydrogenase in C Photosynthesis: Properties and Mechanism of Light/Dark Regulation	G. E. Edwards, H. Nakamoto, J. N. Burnell, M. D. Slack	36:255–86
Membrane-Bound NAD(P)H Dehydrogenases in Higher Plant Cells	I. M. Møller, W. Lin	37:309–34
Some Molecular Aspects of Plant Peroxidase Biosynthetic Studies	R. B. van Huystee	38:205–19
Cell Wall Proteins	J. Varner, G. I. Cassab	39:321–53
Structure and Function of Plasma Membrane ATPase	R. Serrano	40:61–94
Plant Lipoxygenase: Structure and Function	J. N. Siedow	42:145–88
Thionins	H. Bohlmann, K. Apel	42:227–40
Protein Phosphorylation in Green Plant Chloroplasts	J. Bennett	42:281–311
The Roles of Heat Shock Proteins in Plants	E. Vierling	42:579–620
Superoxide Dismutase and Stress Tolerance	C. Bowler, D. Inzé, M. Van Montagu	43:83–116
Calcium-Modulated Proteins: Targets of Intracellular Calcium Signals in Higher Plants	D. M. Roberts, A. Harmon	43:375–414
Regulation of Ribulose 1,5-Bisphosphate Carboxylase/Oxygenase Activity	A. Portis, Jr.	43:415–37
Protein Degradation in Plants	R. D. Vierstra	44:385–410
Genetic Dissection of Rubisco Structure and Function	R. J. Spreitzer	44:411–34
Structure and Function of Chitin-Binding Proteins	N. V. Raikhel, H.-I. Lee, W. F. Broekaert	44:591–615
Phytochromes: Their Molecular Species, Gene Families, and Functions	M. Furuya	44:617–45
Storage Proteins of Vegetative Plant Tissues	P. E. Staswick	45:303–22
The Glycine Decarboxylase Complex from Plant Mitochondria	D. J. Oliver	45:323–37
Inhibitors of Photosynthetic Enzymes/Carriers and Metabolism	L. A. Kleczkowski	45:339–67
Auxin-Binding Proteins	A. M. Jones	45:393–420
The Ribonucleases of Higher Plants	P. J. Green	45:421–45
Structural and Functional Aspects of Chaperonin-Mediated Protein Folding	A. A. Gatenby, P. V. Viitanen	45:469–91

GENETICS & MOLECULAR BIOLOGY

Structure and Function of Nucleic Acids

Plant Chromatin Structure	S. Spiker	36:235–53
Structural Analysis of Plant Genes	G. Heidecker, J. Messing	37:439–66
Molecular Genetics of Cyanobacteria Development	W. J. Buikema, R. Haselkorn	44:33–52
Genomic Imprinting in Plants: Parental Effects and *Trans-* Inactivation Phenomena	M. Matzke, A. J. M. Matzke	44:53–76
The Genetic and Molecular Basis of Root Development	R. A. Aeschbacher, J. W. Schiefelbein, P. N. Benfey	45:25–45
Messenger RNA 3′ End Formation in Plants	A. G. Hunt	45:47–60
Geminiviruses and Their Uses as Extrachromosomal Replicons	M. C. P. Timmermans, O. P. Das, J. Messing	45:79–112

Role/Regulation/Organization of Nuclear Genes

Light Regulation of Gene Expression in Higher Plants | E. M. Tobin, J. Silverthorne | 36:569–93

Regulation of Gene Expression in Higher Plants | C. Kuhlemeier, P. J. Green, N. Chua | 38:221–57

Structure, Evolution, and Regulation of RbcS Genes in Higher Plants | C. Dean, E. Pichersky, P. Dunsmuir | 40:415–39

The Effects of Plant Transposable Element Insertion on Transcription Initiation and RNA Processing | C. F. Weil, S. R. Wessler | 41:527–52

Physiological and Molecular Studies of Light-Regulated Nuclear Genes in Higher Plants | W. F. Thompson, M. J. White | 42:423–66

Posttranscriptional Regulation of Gene Expression in Plants | D. R. Gallie | 44:77–105

Gene Expression Regulated by Abscisic Acid and its Relation to Stress Tolerance | P. M. Chandler, M. Robertson | 45:113–41

Role/Regulation/Organization of Organellar Genes

Chloroplast Development and Gene Expression | J. E. Mullet | 39:475–502

Plant Mitochondrial Genomes: Organization, Expression, and Variation | K. J. Newton | 39:503–32

Transcription, Processing, and Editing in Plant Mitochondria | M. W. Gray, P. J. Hanic-Joyce, P. S. Covello | 43:145–75

Transfer RNAs and Transfer RNA Genes in Plants | L. Maréchal-Drouard, J. H. Weil, A. Dietrich | 44:13–32

The Plant Mitochondrial Genome: Physical Structure, Information Content, RNA Editing, and Gene Migration to the Nucleus | W. Schuster, A. Brennicke | 45:61–78

CELL DIFFERENTIATION

Structure/Function/Development of Plastids and Mitochondria

Photoregulation of the Composition, Function, and Structure of Thyla Membranes | J. M. Anderson | 37:93–136

Metabolite Translocators of the Chloroplast Envelope | U. Flügge, H. W. Heldt | 42:129–44

Organelle Movements | R. E. Williamson | 44:181–202

Protein Import into Plant Mitochondria | A. L. Moore, C. K. Wood, F. Z. Watts | 45:545–75

Structure/Function/Development of Other Organelles

Organization of the Endomembrane System | N. Harris | 37:73–92

Dynamics of Vacuolar Compartmentation | T. Boller, A. Wiemken | 37:137–64

Cross-Linking of Matrix Polymers in the Growing Cell Walls of Angiosperms | S. C. Fry | 37:165–86

Biophysical Control of Plant Cell Growth | D. Cosgrove | 37:377–405

Membrane Control in the Characeae | M. Tazawa, T. Shimmen, T. Mimura | 38:95–117

The Plant Cytoskeleton: The Impact of Fluorescence Microscopy | C. W. Lloyd | 38:119–39

Coated Vesicles | D. Robinson, H. Depta | 39:53–99

Xyloglucans in the Primary Cell Wall | T. Hayashi | 40:139–68

The Physiology of Ion Channels and Electrogenic Pumps in Higher Plants | R. Hedrich, J. I. Schroeder | 40:539–69

The Structures and Function of the Mitotic Spindle in Flowering Plants | T. I. Baskin, W. Z. Cande | 41:277–315

Plasmodesmata   A. W. Robards, W. J. Lucas   41:369–419

Sorting of Proteins in the Secretory System   M. J. Chrispeels   42:21–53

pH and Ionic Conditions in the Apoplast   C. Grignon, H. Sentenac   42:103–28

Isolation and Characterization of Sperm Cells in Flowering Plants   S. D. Russell   42:189–204

Oil Bodies and Oleosins in Seeds   A. Huang   43:177–200

Structure and Function Organization of Tubulin   D. E. Fosket, L. C. Morejohn   43:201–40

Plasma Membrane Redox Activity: Components and Role in Plant Processes   B. Rubinstein, D. G. Luster   44:131–55

Integration of Metabolism

Crassulacean Acid Metabolism   I. P. Ting   36:595–622

Some Aspects of Calcium-Dependent Regulation in Plant Metabolism   H. Kauss   38:47–72

Enzymatic Regulation of Photosynthetic CO_2 Fixation in C_3 Plants   I. E. Woodrow, J. A. Berry   39:533–94

Spatial Organization of Enzymes in Plant Metabolic Pathways   G. Hrazdina, R. A. Jensen   43:241–67

Integration of Carbon and nitrogen metabolism in cells in Plant and Algal   H. C. Huppe, D. H. Turpin   45:577–607

Intracellular Communication

Topographic Aspects of Biosynthesis, Extracellular Secretion, and Intracellular Storage of Proteins in Plant Cells   T. Akazawa, I. Hara-Nishimura   36:441–72

Regulatory Interactions between Nuclear and Plastid Genomes   W. C. Taylor   40:211–33

Intracellular pH: Measurement and Importance in Cell Activity   A. Kurkdjian, J. Guern   40:271–303

Chloroplastic Precursors and Their Transport across the Envelope   K. Keegstra, L. J. Olsen, S. M. Theg   40:471–501

Role of Cell Wall Hydrolases in Fruit Ripening   R. L. Fischer, A. B. Bennett   42:675–703

Endocytosis in Plants   P. S. Low, S. Chandra   45:609–31

Cell Development

Plant Hormone-Induced Changes in the Orientation of Cortical Microtubules: Alterations in the Cross-Linking Between Microtubules and the Plasma Membrane, H. Shibaoka   45:527–44

TISSUE, ORGAN, AND WHOLE PLANT EVENTS

Signal Transduction in the Plant/Hormonal Regulation

Cell-Cell Interactions in *Chlamydomonas*   W. J. Snell   36:287–315

Gibberellins and Reproductive Development in Seed Plants   R. P. Pharis, R. W. King   36:517–68

Rapid Gene Regulation by Auxin   A. Theologis   37:407–38

Phosphorylation of Proteins in Plants: Regulatory Effects and Potential Involvement in Stimulus Response Coupling   R. Ranjeva, A. M. Boudet   38:73–93

Gibberellin Biosynthesis and Control   J. E. Graebe   38:419–65

Plant Growth-Promoting Brassinosteroids   N. B. Mandava   39:23–52

Metabolism and Physiology of Abscisic Acid   J. A. D. Zeevaart, R. A. Creelman   39:439–73

Do Polyamines Have Roles in Plant Development?   P. T. Evans, R. L. Malmberg   40:235–69

Molecular and Cellular Biology Associated with Endosperm Mobilization in Germinating Cereal Grains   G. B. Fincher   40:305–46

Root Signals and the Regulation of Growth and Development of Plants in Drying Soils — W. J. Davies, J. Zhang — 42:55–76

Oligosaccharide Signals in Plants: A Current Assessment — C. A. Ryan, E. E. Farmer — 42:651–74

Role of Salicylic Acid in Plants — I. Raskin — 43:439–63

Ethylene Biosynthesis — H. Kende — 44:283–307

Biochemistry of Phosphoinositides — G. G. Coté, R. C. Crain — 44:333–56

Cytokinin Accumulation and Action: Biochemical, Genetic, and Molecular Approaches — A. N. Binns — 45:173–96

Assimilation

Sunflecks and Photosynthesis in Plant Canopies — R. W. Pearcy — 41:421–53

Transport and Integration

Factors Affecting Mineral Nutrient Acquisition by Plants — D. T. Clarkson — 36:77–115

Phloem Unloading of C and N Assimilates in Developing Seeds — J. H. Thorne — 36:317–43

Water Transport — J. S. Boyer — 36:473–516

Products of Biological Nitrogen Fixation in Higher Plants: Synthesis, Transport, and Metabolism — K. R. Schubert — 37:539–74

Water Transport in and to Roots — J. B. Passioura — 39:245–65

Metabolism and Compartmentation of Imported Sugars in Sink Organs in Relation to Sink Strength — L. C. Ho — 39:355–78

Vulnerability of Xylem to Cavitation and Embolism — M. T. Tyree, J. S. Sperry — 40:19–38

The Sink-Source Transition in Leaves — R. Turgeon — 40:119–38

The Azolla-Anabaena Symbiosis: Basic Biology — G. A. Peters, J. C. Meeks — 40:193–210

Strategies of Phloem Loading — A. J. E. Van Bel — 44:253–81

Environmental Responses

Early Events in Geotropism of Seedling Shoots — B. G. Pickard — 36:55–75

Ethylene and Responses of Plants to Soil Waterlogging and Submergence — M. B. Jackson — 36:145–74

Alteration of Gene Expression During Environmental Stress in Plants — M. M. Sachs, T.-H. D. Ho — 37:363–76

Plants in Space — T. W. Halstead, F. R. Dutcher — 38:317–45

Photocontrol of Development in Algae — M. J. Dring — 39:157–74

Photomorphogenesis in Lower Green Plants — M. Wada, A. Kadota — 40:169–91

Some Current Aspects of Stomatal Physiology — T. A. Mansfield, A. M. Hetherington, C. J. Atkinson — 41:55–75

Circadian Rhythms and Phytochrome — P. J. Lumsden — 42:351–71

Facing the Inevitable: Plants and Increasing Atmospheric CO_2 — G. Bowes — 44:309–32

Quaternary Ammonium and Tertiary Sulfonium Compounds in Higher lants — D. Rhodes, A. D. Hanson — 44:357–84

The Transduction of Blue Light Signals in Higher Plants — T. W. Short, W. R. Briggs — 45:143–71

Plant Responses to Biotic Factors/Symbiosis/Toxins

Siderophores in Relation to Plant Growth and Disease — J. B. Neilands, S. A. Leong — 37:187–208

Genes Specifying Auxin and Cytokinin Biosynthesis in Phytopathogens — R. O. Morris — 37:509–38

Plant Virus-Host Interactions — M. Zaitlin, R. Hull — 38:291–315

Physiological Interactions Between Symbionts in Vesicular-Arbuscular — S. E. Smith, V. Gianinazzi-Pearson — 39:221–44

The Physiology and Biochemistry of Parasitic Angiosperms — G. R. Stewart, M. C. Press — 41:127–51

Molecular Communication in Interactions be-
tween Plants and Microbial Pathogens   C. J. Lamb, R. A. Dixon   41:339–67
Phenolic Signals in Cohabitation: Implications
for Plant Development   D. G. Lynn, M. Chang   41:497–526
Functional Aspects of the Lichen Symbiosis   R. Honegger   42:553–78
Cell Biology of Pathogenesis   A. R. Hardham   43:491–526
Chronicles From the Agrobacterium-Plant Cell
DNA Transfer Story   P. C. Zambryski   43:465–90
Host Range Determinants of Plant Viruses   W. O. Dawson, M. E. Hilf   43:527–55
Regulation of the Vesicular-Arbuscular Mycor-
rhizal Symbiosis   R. T. Koide, R. P. Schre-
iner   43:557–81

PPFMs and other Covert Contaminants: Is
There More to Plant Physiology than Just
Plant?   M. A. Holland, J. C. Po-
lacco   45:197–210

Morphogenesis

Calcium and Plant Development   P. K. Hepler, R. O.
Wayne   36:397–439
Cellular Polarity   E. Schnepf   37:23–47
Fruit Ripening   C. J. Brady   38:155–78
Differentiation of Vascular Tissues   R. Aloni   38:179–204
The Control of Floral Evocation and Morpho-
genesis   G. Bernier   39:175–219
The Control of Leaf Expansion   J. E. Dale   39:267–95
Gene Activity During Pollen Development   J. P. Mascarenhas   41:317–38
Molecular Studies on the Differentiation of Flo-
ral Organs   C. S. Gasser   42:621–49
Control of Nodulin Genes In Root-Nodule De-
velopment and Metabolism   F. Sanchez, J. E. Padilla,
H. Perez, M. Lara   42:507–28
Fusion Events during Floral Morphogenesis   J. A. Verbeke   43:583–98

ACCLIMATION AND ADAPTATION

Economic Botany

Taxol   P. F. Heinstein,
C.-j. Chang   45:663–74

Physiological Ecology

Carbon Dioxide and Water Vapor Exchange in
Response to Drought in the Atmosphere and
in the Soil   E.-D. Schulze   37:247–74
Salinity Tolerance of Eukaryotic Marine Algae   G. O. Kirst   41:21–53
Cold Acclimation and Freezing Stress Toler-
ance: Role of Protein Meta   C. L. Guy   41:187–223
Gene Transfer to Plants: Assessment of Publish-
ed Approaches and Results   I. Potrykus   42:205–25
Photoprotection and Other Responses of Plants
to High Light Stress   B. Demmig-Adams, W.
W. Adams III   43:599–626
Plant Tissue Optics   T. C. Vogelmann   44:231–51
Photoinhibition of Photosynthesis in Nature   S. P. Long, S. Hum-
phries, P. G. Falkowski   45:633–61

Plant Genetics/Evolution

Rapid Genomic Change in Higher Plants   V. Walbot, C. A. Cullis   36:367–96
Genetics of Wheat Storage Proteins and the Ef-
fect of Allelic Variation on Bread-Making
Quality   P. I. Payne   38:141–53
Evolution of Higher Plant Chloroplast DNA-En-
coded Genes: Implications for Structure-Func-
tion and Phylogenetic Studies   G. Zurawski, M. T. Clegg   38:391–418

The Chromosomal Basis of Somaclonal Variation   M. Lee, R. L. Phillips   39:413–37

The Role of Homeotic Genes in Flower Development and Evolution   E. S. Coen   42:241–79

The Self-Incompatibility Genes of Brassica: Expression and Use in Genetic Ablation of Floral Tissues   J. B. Nasrallah, T. Nishio, M. E. Nasrallah   42:393–422

Molecular Genetic Approaches to Plant Hormone Biology   H. Klee, M. Estelle   42:529–51

Developmental Genetics of C_4 Photosynthesis   T. Nelson, J. A. Langdale   43:25–47

Wide Crosses in Cereals   R. Appels, M. Baum, E. Lagudah   43:117–43

Plant Improvement

Agrobacterium-Mediated Plant Transformation and Its Further Applications to Plant Biology   H. Klee, R. Horsch, S. Rogers   38:467–86

The Development of Herbicide Resistant Crops   B. J. Mazur, S. C. Falco   40:441–70

Mechanisms and Agronomic Aspects of Herbicide Resistance   J. S. Holt, S. B. Powles, J. A. M. Holtum   44:203–29

METHODS

Immunocytochemical Localization of Macromolecules with the Electron Microscope   E. M. Herman   39:139–55

Strategies for Mutagenesis and Gene Cloning Using Transposon Tagging and T-DNA Insertional Mutagenesis   V. Walbot   43:49–82

Modern Methods for the Quantitative Analysis of Plant Hormones   P. Hedden   44:107–29

ANNUAL REVIEWS

a nonprofit scientific publisher

4139 El Camino Way
P.O. Box 10139
Palo Alto, CA 94303-0139 • USA

Annual Reviews publications may be ordered directly from our office; through booksellers and subscription agents, worldwide; and through participating professional societies. **Prices are subject to change without notice. We do not ship on approval.**

- **Individuals:** Prepayment required on new accounts. in US dollars, checks drawn on a US bank.

- **Institutional Buyers:** Include purchase order.   Calif. Corp. #161041 • ARI Fed. I.D. #94-1156476

- **Students / Recent Graduates:** $10.00 discount from retail price, per volume. *Requirements:* **1.** be a degree candidate at, or a graduate within the past three years from, an accredited institution;   **2.** present proof of status (photocopy of your student I.D. or proof of date of graduation);   **3.** Order direct from Annual Reviews;   **4.** prepay.   This discount **does not** apply to standing orders, *Index on Diskette*, Special Publications, ARPR, or institutional buyers.

- **Professional Society Members:** Many Societies offer *Annual Reviews* to members at reduced rates. Check with your society or contact our office for a list of participating societies.

- **California orders** add applicable sales tax. • **Canadian orders** add 7% GST. Registration #R 121 449-029.

- **Postage paid** by Annual Reviews (4th class bookrate/surface mail). UPS ground service is available at S2.00 extra per book within the contiguous 48 states only. UPS air service or US airmail is available to any location at actual cost. UPS requires a street address. P.O. Box, APO, FPO, not acceptable.

- **Standing Orders:** Set up a standing order and the new volume in series is sent automatically each year upon publication. Each year you can save 10% by prepayment of prerelease invoices sent 90 days prior to the publication date. Cancellation may be made at any time.

- **Prepublication Orders:** Advance orders may be placed for any volume and will be charged to your account upon receipt. Volumes not yet published will be shipped during month of publication indicated.

N O T E	For copies of individual articles from any *Annual Review*, or copies of any article cited in an *Annual Review*, call **Annual Reviews Preprints and Reprints (ARPR)** toll free 1-800-347-8007 (fax toll free 1-800-347-8008) from the USA or Canada. From elsewhere call 1-415-259-5017.

ANNUAL REVIEWS SERIES *Volumes not listed are no longer in print*	Prices, postpaid, per volume. USA/other countries	Regular Order Please send Volume(s):	Standing Order Begin with Volume:
☐ *Annual Review of* **ANTHROPOLOGY**			
Vols. 1-20	(1972-91).....................................$41 / $46		
Vols. 21-22	(1992-93).....................................$44 / $49		
Vol. 23	(avail. Oct. 1994)..... $47 / $52	Vol(s). _____	Vol. _____
☐ *Annual Review of* **ASTRONOMY AND ASTROPHYSICS**			
Vols. 1, 5-14, 16-29	(1963, 67-76, 78-91) ....$53 / $58		
Vols. 30-31	(1992-93)...............................$57 / $62		
Vol. 32	(avail. Sept. 1994).....................$60 / $65	Vol(s). _____	Vol. _____
☐ *Annual Review of* **BIOCHEMISTRY**			
Vols. 31-34, 36-60	(1962-65,67-91) ....$41 / $47		
Vols. 61-62	(1992-93) ....$46 / $52		
Vol. 63	(avail. July 1994).....................$49 / $55	Vol(s). _____	Vol. _____
☐ *Annual Review of* **BIOPHYSICS AND BIOMOLECULAR STRUCTURE**			
Vols. 1-20	(1972-91)...............................$55 / $60		
Vols. 21-22	(1992-93)...............................$59 / $64		
Vol. 23	(avail. June 1994)..................$62 / $67	Vol(s). _____	Vol. _____